Principles of Digital Audio

Ken C. Pohlmann

Sixth Edition

New York Chicago San Francisco Lisbon London Madrid
Mexico City Milan New Delhi San Juan Seoul

The McGraw-Hill Companies

Cataloging-in-Publication Data is on file with the Library of Congress

McGraw-Hill books are available at special quantity discounts to use as premiums and sales promotions, or for use in corporate training programs. To contact a representative, please e-mail us at bulksales@mcgraw-hill.com.

Principles of Digital Audio, Sixth Edition

1 2 3 4 5 6 7 8 9 0 DOC/DOC 1 6 5 4 3 2 1 0

ISBN 978-0-07-166346-5
MHID 0-07-166346-0

This book is printed on acid-free paper.

Sponsoring Editor
Judy Bass

Editorial Supervisor
Stephen M. Smith

Production Supervisor
Pamela A. Pelton

Acquisitions Coordinator
Michael Mulcahy

Project Manager
Tania Andrabi, Glyph International

Copy Editors
Carol Loomis, Sharon Green

Proofreader
Linda Leggio

Art Director, Cover
Jeff Weeks

Composition
Glyph International

for
Leslie Hope
and her
headwind tenacity
and tailwind exuberance

About the Author

Ken C. Pohlmann is a professor emeritus at the University of Miami in Coral Gables, Florida. He was the director of the Music Engineering program in the university's Frost School of Music from 1983 to 2007. He initiated new undergraduate and graduate courses in digital audio, advanced digital audio, Internet audio, acoustics and psychoacoustics, and studio production. In 1986, he founded the first master's degree program in Music Engineering in the United States. Mr. Pohlmann holds Bachelor of Science and Master of Science degrees in Electrical Engineering from the University of Illinois in Urbana–Champaign.

Mr. Pohlmann is the author of *Principles of Digital Audio* (McGraw-Hill); this book has appeared in six editions and has been translated into Dutch, Spanish, and Chinese. He is also author of *The Compact Disc Handbook* (A-R Editions); this book has appeared in two editions and has been translated into German. He is co-author of the most recent (fifth) edition of *The Master Handbook of Acoustics* (McGraw-Hill). He is also co-author of *Writing for New Media* (John Wiley & Sons), and editor and co-author of *Advanced Digital Audio* (Howard W. Sams). Since 1982, he has written over 2500 articles for publications including *Audio, Broadcast Engineering, dB, Car Stereo Review, CD Review, Edmunds.com, Electronics Australia, Guitar Player, Handbook for Sound Engineers, IEEE Spectrum, Journal of the Audio Engineering Society, Laserdisk Professional, McGraw-Hill Encyclopedia of Science and Technology, Mix, Mobile Entertainment, National Association of Broadcasters Handbook, NARAS Journal, PC, Road & Track Road Gear, Sound & Vision, Scientific American, Spektrum der Wissenschaft, Stereo Review, Video*, and *World Book Encyclopedia*. He is a contributing technical editor and columnist for *Sound & Vision*.

Mr. Pohlmann has worked extensively in the research, development, and testing of new audio technology. He serves as a consultant in the theory and application of digital audio systems and the development of sound systems for automobile manufacturers. Some of his consulting clients include: Alpine Electronics, Analog Devices, Apple Computer, Bertlesmann Music Group, Blockbuster Entertainment, BMW, Canadian Broadcasting Corporation, Cirrus Logic, DaimlerChrysler, Eclipse, Ford, Fujitsu Ten, Harman International, Hughes Electronics, Hyundai, IBM, Kia, Lexus, Lucent Technologies, Microsoft, Mitsubishi Electronics, Motorola, Nippon Columbia, Onkyo America, Philips, RealNetworks, Recording Industry Association of America, Samsung, Sensormatic, Sonopress, Sony, TDK, Time Warner, Toyota, and United Technologies.

Mr. Pohlmann has served as a consultant or expert witness for patent infringement and other issues with law firms such as Arnold & Porter; Baker & McKenzie; Barnes & Thornburg; Christie Parker & Hale; Cushman Darby & Cushman; Darby & Darby; Dewey Ballantine; Firmstone & Feil; Fish & Neave; Fish & Richardson; Greenberg, Glusker, Fields, Claman, Machtinger & Kinsella; Howrey; Hunton & Williams; Jenner & Blocker; Jones Day; Kenyon & Kenyon; Kirkland & Ellis; McDermott Will & Emery; McDonnell Boehnen Hulbert & Berghoff; Paul, Weiss, Rifkind, Wharton & Garrison; Young & Thompson; and the U.S. Justice Department Anti-Trust Division.

Mr. Pohlmann co-founded Microcomputer Arts, Inc. (1980), International Business Information Systems Inc. (1982), and U.S. Digital Disc Corporation (1985). He chaired the Audio Engineering Society's (AES) International Conference on Digital Audio in Toronto in 1989 and co-chaired the Society's International Conference on Internet Audio in Seattle in 1997. He was presented two AES Board of Governors Awards (1989 and 1998) and was named an AES Fellow in 1990 for his work as an educator and author. He was elected to the AES Board of Governors in 1991. He was presented the University of Miami Philip Frost Award for Excellence in Teaching and Scholarship in 1992. He served as AES convention papers chairman in 1984 and papers co-chairman in 1993. He was elected as the Vice President of the AES Eastern U.S. and Canada Region in 1993. He served as a Non-Board Member of the National Public Radio Distribution/Interconnection Committee (2000–2003). He served on the Board of Directors of the New World Symphony (2000–2005). Mr. Pohlmann joined the Advisory Board of SRS Labs in 2009 as the charter member.

Contents

Preface

It seems inconceivable, but the editions of this book have arrived at number six. Clearly, decades ago, I had no idea what I was getting myself into. In 1984, at an Audio Engineering Society convention in New York, I presented a lecture on digital audio theory, using material from a course I had recently inaugurated at the University of Miami. After the lecture, I was thronged by engineers seeking more information on this wondrous new topic. An astute acquisitions editor noted the response and, in true Mephistophelean style, persuaded me to author a book on digital audio. After all, he explained, it would be simple to put everything down on paper. I naively accepted his bargain, and *Principles of Digital Audio* was published in 1985. The book's publication fortuitously coincided with a tremendous surge in the development of digital audio technology.

Digital audio's commercial popularity was matched only by the demand for explanations of its workings. Audio engineers, and students wishing to join their ranks, assimilated every scrap of information they could find, further propelling development of the science. The first edition gave way to the second, which stepped aside for the third, which submitted to the fourth, which yielded to the fifth. Despite occasional half-hearted promises that I would never write another edition, I could not ignore the fact that digital audio technology continued to expand, and there were increasing numbers of people eager to learn about it. Clearly, new editions became inevitable. As I slavishly authored one edition after another, I had only the royalty checks to console me. With the arrival of this sixth edition, I suppose I should start planning for the seventh.

Readers familiar with earlier editions will see the same patterns in the renovations in this latest installment. The essential nature of topics such as discrete time sampling has not changed, but our appreciation of them has. Similarly, the relative importance of topics continually changes; advances in technology diminish the magnitude of some issues, while magnifying others. Moreover, and even more significantly, in the years since the last edition digital audio theory has evolved and entirely new applications have been developed.

Examination of the contents of this sixth edition will show both losses and gains. A number of topics have been eliminated to conserve the page count; most notably and not surprisingly, the magnetic tape, DAT, and MiniDisc chapters have been dropped. In their place, numerous fresh topics have been introduced. For example, readers will find two new chapters on Blu-ray disc and speech coding, and the perceptual audio chapter (renamed "Low Bit-Rate Coding") has been split into two chapters to allow more thorough discussions of this important topic. Also, the bibliography has been updated as needed.

As in past editions, a user-friendly approach has been retained, but the greater depth given to some material has somewhat increased the reading sophistication required. Of course, readers may pick and choose according to their level and need. Also, the wider scope of topics should satisfy a broader range of professional practitioners and students. One thing has not changed. This book is neither a compendium of every possible fact nor an advanced treatise. It is an introductory text that attempts to provide lucid explanations, and to strike that all-important balance between mere information and understanding. In other words, this is a learning tool, sharpened by years of refinement.

A final note: The material in this book stems from the work of the many pioneers and leaders in the field of digital audio technology. We owe a tremendous debt to those wonderfully creative and hard-working engineers who breathe life into digital audio. Clearly, their vision of the potential of this science is transforming both our industry and our society.

Ken C. Pohlmann
Durango, Colorado

CHAPTER 1

Sound and Numbers

Digital audio is a highly sophisticated technology. It pushes the envelope of many diverse engineering and manufacturing disciplines. Although the underlying concepts were well understood in the 1920s, commercialization of digital audio did not begin until the 1970s because theory had to wait 50 years for technology to catch up. The complexity of digital audio is all the more reason to start with the basics. This chapter begins our exploration of ways to numerically encode the information contained in an audio event.

Physics of Sound

It would be a mistake for a study of digital audio to ignore the acoustic phenomena for which the technology has been designed. Music is an acoustic event. Whether it radiates from musical instruments or is directly created by electrical signals, all music ultimately finds its way into the air, where it becomes a matter of sound and hearing. It is therefore appropriate to briefly review the nature of sound.

Acoustics is the study of sound and is concerned with the generation, transmission, and reception of sound waves. The circumstances for those three phenomena are created when energy causes a disturbance in a medium. For example, when a kettledrum is struck, its drumhead disturbs the surrounding air (the medium). The outcome of that disturbance is the sound of a kettledrum. The mechanism seems fairly simple: the drumhead is activated and it vibrates back and forth. When the drumhead pushes forward, air molecules in front of it are compressed. When it pulls back, that area is rarefied. The disturbance consists of regions of pressure above and below the equilibrium atmospheric pressure. Nodes define areas of minimum displacement, and antinodes are areas of maximum (positive or negative) displacement.

Sound is propagated by air molecules through successive displacements that correspond to the original disturbance. In other words, air molecules colliding one against the next propagate the energy disturbance away from the source. Sound transmission thus consists of local disturbances propagating from one region to another. The local displacement of air molecules occurs in the direction in which the disturbance is traveling; thus, sound undergoes a longitudinal form of transmission. A receptor (like a microphone diaphragm) placed in the sound field similarly moves according to the pressure acting on it, completing the chain of events.

We can access an acoustical system with transducers, devices able to change energy from one form to another. These serve as sound generators and receivers. For example, a kettledrum changes the mechanical energy contributed by a mallet to acoustical energy. A microphone responds to the acoustical energy by producing electrical energy. A loudspeaker reverses that process to again create acoustical energy from electrical energy.

The pressure changes of sound vibrations can be produced either periodically or aperiodically. A violin moves the air back and forth periodically at a fixed rate. (In practice, things like vibrato make it a quasi-periodic vibration.) However, a cymbal crash has no fixed period; it is aperiodic. One sequence of a periodic vibration, from pressure rarefaction to compression and back again, determines one cycle. The number of vibration cycles that pass a given point each second is the frequency of the sound wave, measured in Hertz (Hz). A violin playing a concert A pitch, for example, generates a waveform that repeats 440 times per second; its frequency is 440 Hz. Alternatively, the reciprocal of frequency, the time it takes for one cycle to occur, is called the period. Frequencies in nature can range from very low, such as changes in barometric pressure around 10^{-5} Hz, to very high, such as cosmic rays at 10^{22} Hz. Sound is loosely described to be that narrow, low-frequency band from 20 Hz to 20 kHz—roughly the range of human hearing. Audio devices are generally designed to respond to frequencies in that general range. However, digital audio devices can be designed to accommodate audio frequencies much higher than that.

Wavelength is the distance sound travels through one complete cycle of pressure change and is the physical measurement of the length of one cycle. Because the velocity of sound is relatively constant—about 1130 ft/s (feet per second)—we can calculate the wavelength of a sound wave by dividing the velocity of sound by its frequency. Quick calculations demonstrate the enormity of the differences in the wavelength of sounds. For example, a 20-kHz wavelength is about 0.7 inch long, and a 20-Hz wavelength is about 56 feet long. Most transducers (including our ears) cannot linearly receive or produce that range of wavelengths. Their frequency response is not flat, and the frequency range is limited. The range between the lowest and the highest frequencies a system can accommodate defines a system's bandwidth. If two waveforms are coincident in time with their positive and negative variations together, they are in phase. When the variations exactly oppose one another, the waveforms are out of phase. Any relative time difference between waveforms is called a phase shift. If two waveforms are relatively phase shifted and combined, a new waveform results from constructive and destructive interference.

Sound will undergo diffraction, in which it bends through openings or around obstacles. Diffraction is relative to wavelength; longer wavelengths diffract more apparently than shorter ones. Thus, high frequencies are considered to be more directional in nature. Try this experiment: hold a magazine in front of a loudspeaker—higher frequencies (short wavelengths) will be blocked by the barrier, while lower frequencies (longer wavelengths) will go around it.

Sound also can refract, in which it bends because its velocity changes. For example, sound can refract because of temperature changes, bending away from warmer temperatures and toward cooler ones. Specifically, velocity of sound in air increases by about 1.1 ft/s with each increase of 1°F. Another effect of temperature on the velocity of sound is well known to every woodwind player. Because of the change in the speed of sound, the instrument must be warmed up before it plays in tune (the difference is about half a semitone).

The speed of sound in air is relatively slow—about 740 mph (miles per hour). The time it takes for a sound to travel from a source to a receptor can be calculated by dividing the distance by the speed of sound. For example, it would take a sound about one-sixth of a second to travel 200 feet in air. The speed of sound is proportional to elasticity of the medium and inversely proportional to its density. For example, steel is 1,230,000 times more elastic than air thus the speed of sound in steel is 14 times greater than the speed in air, even though the density of steel is 6000 times greater than air. Sound is absorbed as it travels. The mere passage of sound through air acts to attenuate the sound energy. High frequencies are more prominently attenuated in air; a nearby lightning strike is heard as a sharp clap of sound, and one faraway is heard as a low rumble, because of high-frequency attenuation. Humidity affects air attenuation—specifically, wet air absorbs sound better than dry air. Interestingly, moist air is less dense than dry air (water molecules weigh less than the nitrogen and oxygen they replace) causing the speed of sound to increase.

Sound Pressure Level

Amplitude describes the sound pressure displacement above and below the equilibrium atmospheric level. In absolute terms, sound pressure is very small; if atmospheric pressure is 14.7 psi (pounds per square inch), a loud sound might cause a deviation from 14.699 to 14.701 psi. However, the range from the softest to the loudest sound, which determines the dynamic range, is quite large. In fact, human ears (and hence audio systems) have a dynamic range spanning a factor of millions. Because of the large range, a logarithmic ratio is used to measure a sound pressure level (SPL). The decibel (dB) uses base 10 logarithmic units to achieve this. A base 10 logarithm is the power to which 10 must be raised to equal the value. For example, an unwieldy number such as 100,000,000 yields a tidy logarithm of 8 because $10^8 = 100{,}000{,}000$. Specifically, the decibel is defined to be 10 times the logarithm of a power ratio:

$$\text{Intensity level} = 10\log\left(\frac{P_1}{P_2}\right)\text{dB}$$

where P_1 and P_2 are values of acoustical or electrical power.

If the denominator of the ratio is set to a reference value, standard measurements can be made. In acoustic measurements, an intensity level (IL) can be measured in decibels by setting the reference intensity to the threshold of hearing, which is 10^{-12} W/m^2 (watts per square meter). Thus the intensity level of a loud rock band producing sound power of 10 W/m^2 can be calculated as:

$$\begin{aligned}\text{Intensity level} &= 10\log\left(\frac{P_1}{P_2}\right)\text{dB}\\ &= 10\log\left(\frac{10^1}{10^{-12}}\right)\\ &= 130\ \text{dB SPL}\end{aligned}$$

When ratios of currents, voltages, or sound pressures are used (quantities whose square is proportional to power), the above decibel formula must be multiplied by 2.

The zero reference level for an acoustic sound pressure level measurement is a pressure of 0.0002 dyne/cm^2. This level corresponds to the threshold of hearing, the lowest SPL humans can perceive, which is nominally equal to 0 dB SPL. The threshold of feeling, the loudest level before discomfort begins, is 120 dB SPL. Sound pressure levels can be rated on a scale in terms of SPL. A quiet home might have an SPL of 35 dB, a busy street might be 70 dB SPL, and the sound of a jet engine in close proximity might exceed 150 dB SPL. An orchestra's pianissimo might be 30 dB SPL, but a fortissimo might be 110 dB SPL. Thus its dynamic range is 80 dB.

The logarithmic nature of these decibels should be considered. They are not commonly recognizable, because they are not linear measurements. Two motorcycle engines, each producing an intensity level of 80 dB, would not yield a combined IL of 160 dB. Rather, the logarithmic result would be a 3-dB increase, yielding a combined IL of 83 dB. In linear units, those two motorcycles each producing sound intensities of 0.0001 W/m^2 would combine to produce 0.0002 W/m^2.

Harmonics

The simplest form of periodic motion is the sine wave; it is manifested by the simplest oscillators, such as pendulums and tuning forks. The sine wave is unique because it exists only as a fundamental frequency. All other periodic waveforms are complex and comprise a fundamental frequency and a series of other frequencies at multiples of the fundamental frequency. Aperiodic complex waveforms, such as the sound of motorcycle engines, do not exhibit this relationship. Many musical instruments are examples of the special case in which the harmonics are related to the fundamental frequency through simple multiples. For example, a complex pitched waveform with a 150-Hz fundamental frequency will have overtones at 300, 450, 600, 750 Hz, and so on.

Overtones extend through the upper reaches of human hearing. The relative amplitudes and phase relationships of those overtones account for the timbre of the waveform. For example, a cello and trumpet can both play a note with the same fundamental pitch; however, their timbres are quite different because of their differing harmonic series. When a cellist plays the note D4 as a natural harmonic, the open D string is bowed, which normally produces a note of pitch D3, and the string is touched at its midpoint. The pitch is raised by an octave because the player has damped out all the odd-numbered harmonics, including the fundamental frequency. The pitch changes; because the harmonic structure changes, the timbre changes as well. Harmonic structure explains why the ear has limited ability to distinguish timbre of high-frequency sounds. The first overtone of a 10-kHz periodic tone is at 20 kHz; most people have trouble perceiving that overtone, let alone others even higher in frequency. Still, to record a complex waveform properly, both its fundamental and harmonic structure must be preserved, at least up to the limit of hearing.

The harmonic nature of periodic waveforms is summarized by the Fourier theorem. It states that all complex periodic waveforms are composed of a harmonic series of sine waves; complex waveforms can be synthesized by summing sine waves. Furthermore, a complex waveform can be decomposed into its sine-wave content to analyze the nature of the complex waveform. A mathematical transform can be applied to a waveform represented in time to convert it to a representation in frequency. For example, a square wave would be transformed into its fundamental sine wave and higher order odd harmonics. An inverse transform reverses the process. Likewise the information in any audio signal can thus be represented in either the time domain or the frequency domain. Some digital audio systems (such as the Compact Disc) code the audio signal as time-based samples. Other systems (such as MP3 players) code the audio signal as frequency coefficients.

Given the evident complexity of acoustical signals, it would be naive to believe that analog or digital audio technologies are sufficiently advanced to fully capture the complete listening experience. To complicate matters, the precise limits of human perception are not known. One thing is certain: at best, even with the most sophisticated technology, what we hear reproduced through an audio system is an approximation of the actual sound.

Digital Basics

Acoustics and analog audio technology are mainly concerned with continuous mathematical functions, but digital audio is a study of discrete values. Specifically, a waveform's amplitude can be represented as a series of numbers. That is an important first principle, because numbers allow us to manage audio information very efficiently. Using digital techniques, the capability to process information is greatly enhanced. The design nature of audio recording, signal processing, and reproducing hardware has followed the advancement of digital technology; the introduction of software programming into the practical audio environment has been revolutionary. Thus, digital audio is primarily a numerical technology. To understand it properly, let's begin with a review of number systems.

The basic problem confronting any digital audio system is the representation of audio information in numerical form. Although many possibilities present themselves, the logical choice is the binary number system. This base 2 representation is ideally suited for storing and processing numerical information. Fundamental arithmetic operations are facilitated, as are logic operations.

Number Systems

It all begins with numbers. With digital audio, we deal with information and numbers, as opposed to an analog representation. Numbers offer a fabulous way to code, process, and decode information. In digital audio, numbers entirely represent audio information. We usually think of numbers as symbols. The symbology is advantageous because the numerical symbols are highly versatile; their meaning can vary according to the way we use them.

For example, consider my classic 1962 BMW R50/2 motorcycle, 500 cubic centimeters, registered as 129907, and shown in Fig. 1.1. Several numbers describe this machine; not so obvious is the important context of each. R50/2 represents the motorcycle's model number, 1962 is the year of manufacture, and 500 represents the quantity of cubic centimeters of engine displacement. The license number 129907 represents coded information that allows my speeding tickets to be properly credited to my account. These various numbers are useful only by virtue of their arbitrarily assigned contexts. If that

Figure 1.1 The author's classic 1962 BMW R50/2 motorcycle.

context is confused, then information encoded by the numbers goes awry. I could end up with a motorcycle with license number 1962, manufactured in the year 500, with an engine displacement of 129907 cubic centimeters.

Similarly, the numerical operations performed on numbers are matters of interpretation. The tally of my moving violations determines when my license will be suspended, but the sum of my license plate numerals is less problematic. Numbers, if properly defined, provide a good method for storing and processing data. The negative implication is that numbers and their meanings have to be used carefully.

For most people, the most familiar numbers are those of the base 10 system, apparently devised in the ninth century by Hindu astronomers who conceived of the 0 numeral to represent nothing and appended it to the nine other numerals already in use. Earlier societies were stuck with the unitary system, which used one symbol in a series of marks to answer the essential question: how many? That is an unwieldy system for large numbers; thus, higher-base systems were devised. Babylonian mathematicians invented a number system that used 60 symbols. It was a little cumbersome, but even today, 3700 years later, the essence of their system is still used to divide an hour into 60 minutes, a minute into 60 seconds, and a circle into 360 degrees.

Selection of a number system is a question of preference, because any integer can be expressed using any base. Choosing a number system simply questions how many different symbols we think are most convenient. The base 10 system uses 10 numerals; the radix of the system is 10. In addition, the system uses positional notation; the position of the numerals shows the quantities of ones, tens, hundreds, thousands, and so on. In other words, the number in each successive position is multiplied by the next higher power of the base. A base 10 system is convenient for 10-fingered organisms such as humans, but other number bases might be more appropriate for other applications. In any system, we must know the radix; the numeral 10 in base 10 represents the total number of fingers you have, but 10 in base 8 is the number of fingers minus the thumbs. Similarly, would you rather have 10,000 dollars in base 6, or 100 dollars in base 60? Table 1.1 shows four of the most popular number systems.

Binary Number System

Gottfried Wilhelm von Leibnitz, philosopher and mathematician, devised the binary number system on March 15, 1679. That day marks the origin of today's digital systems. Although base 10 is handy for humans, a base 2, or binary, system is more efficient for digital computers and digital audio equipment. Only two numerals are required to satisfy the machine's principal electrical concern of voltage being on or off. Furthermore, these two conditions can be easily represented as 0 and 1; these binary digits are called bits (*bi*nary digi*ts*). From a machine standpoint, a binary system is ruthlessly efficient, and it is fast. Imagine how quickly we can turn a switch on and off; that speed represents the rate at which we can process information. Imagine a square wave; the wave could represent a machine operating the switch for us. Consider the advantages in storage. Instead of saving infinitely different analog values, we must only remember two values. Only through the efficiency of binary data can digital circuits process the tremendous amount of information contained in an audio signal.

Whatever information is being processed—in this case, an audio signal that has been converted to binary form—no matter how unrelated it might appear to be to numbers, a digital processor codes the information in the form of numbers, using the base 2 system. To better understand how audio data is handled inside a digital audio system,

Hexadecimal (Base 16)	Decimal (Base 10)	Octal (Base 8)	Binary (Base 2)
0	0	0	0000
1	1	1	0001
2	2	2	0010
3	3	3	0011
4	4	4	0100
5	5	5	0101
6	6	6	0110
7	7	7	0111
8	8	10	1000
9	9	11	1001
A	10	12	1010
B	11	13	1011
C	12	14	1100
D	13	15	1101
E	14	16	1110
F	15	17	1111

TABLE 1.1 Four common number systems.

a brief look at the arithmetic of base 2 will be useful. In fact, we will consistently see that the challenge of coding audio information in binary form is a central issue in the design and operation of digital audio systems.

In essence, all number systems perform the same function; thus, we can familiarize ourselves with the binary system by comparing it to the decimal system. A given number can be expressed in either system and converted from one base to another. Several methods can be used. One decimal-to-binary conversion algorithm for whole numbers divides the decimal number by 2 and collects the remainders to form the binary number. Similarly, binary-to-decimal conversion can be accomplished by expressing the binary number in a power of 2 notation, then expanding and collecting terms to form the decimal number.

The conversion points out the fact that the base 2 system also uses positional notation. In base 2, each successive position represents a doubling of value. The right-most column represents 1s, the next column is 2s, then 4s, 8s, 16s, and so on. It is important to designate the base being used; for example, in base 2 the symbol 10 could represent a person's total number of hands.

Just as a decimal point is used to delineate a whole number from a fractional number, a binary point does the same for binary numbers. The fractional part of a decimal number can be converted to a binary number by multiplying the decimal number by 2. Conversion often leads to an infinitely sustaining binary number, so the number of terms must be limited.

As in the base 10 system, the standard arithmetic operations of addition, subtraction, multiplication, and division are applicable to the base 2 system. As in any base, base 2 addition can be performed using the addition rules needed to form an addition table. The procedure is the same as in the decimal system; however, the addition table is simpler. There are only four possible combinations, compared to the more than 100 possible combinations resulting from the rules of decimal addition. The generation of the carry, as in the decimal system, is necessary when the result is larger than the largest digit in the system. The algorithms for subtraction, multiplication, and division in the binary system are identical to the corresponding algorithms in the decimal system.

Binary numbers are unwieldy for humans, so we often represent them as base 16 (hexadecimal) numbers. Hexadecimal numbers use numerals 0 through 9 and A, B, C, D, E, F (0 to 15 in decimal), and each hexadecimal number can represent four binary digits (see Table 1.1). Thus an 8-bit byte can be represented by 2 hex digits. For example, the value 0110 1110 would be represented as 6E.

A number is what we make it, and the various systems—differing only by base—operate in essentially the same way. A computer's use of the binary system is a question of expediency; it presents no real barrier to an understanding of digital techniques. It is simply the most logical approach. Ask yourself, would you rather deal with 10, 60, an infinite analog number, or 2 voltage levels? Fortunately, most digital systems automatically choose the best of both number-base worlds; you punch base-10 numbers into a calculator, it uses base 2 to perform its operations, then displays the result in base 10.

Binary Codes

Although the abstractions of binary mathematics form the basis of digital audio systems, the implementation of these primitives requires higher-level processing. Specifically, the next step up the evolutionary ladder is the coding of binary information. For example, individual binary bits or numbers can be ordered into words with specific connotations attached. In this way, both symbolic and numeric information are more easily processed by digital systems.

Just as the digits in a motorcycle license number carry a specially assigned meaning, groups of binary numbers can be encoded with special information. For example, a decimal number can be converted directly to its equivalent binary value; the binary number is encoded as the binary representation of the decimal number. Obviously, there is a restriction on the number of possible values that can be encoded. Specifically, an n-bit binary number can encode 2^n numbers. Three bits, for example, could encode eight states: 000, 001, 010, 011, 100, 101, 110, and 111. These could correspond to the decimal numbers 0, 1, 2, 3, 4, 5, 6, and 7.

Negative numbers present a problem because the sign must be encoded (with bits) as well. For example, a 1 in the left-most position could designate a negative number, a 0 could designate a positive number, and the remaining bits could represent the absolute value of the number. This kind of coding is called a signed-magnitude representation. The 3-bit words 000, 001, 010, 011, 100, 101, 110, and 111 might correspond to +0, +1, +2, +3, −0, −1, −2, and −3. An irregularity is the presence of both +0 and −0. Other methods can be used to better represent negative numbers.

Because we live in a decimal world, it is often useful to create binary words coded to decimal equivalents, preserving the same kind of decimal characteristics. Unfortunately, there is no binary grouping that directly represents the 10 decimal digits. Three bits handle the first seven decimal numbers, and four bits handle 16. For greater efficiency, a more sophisticated coding method is desirable. This method is easily accomplished with groups of four bits each, with each group representing a decimal digit:

First decimal group	Second decimal group	nth decimal group
$a_3a_2a_1a_0$	$b_3b_2b_1b_0 \ldots$	$n_3n_2n_1n_0$

Given this approach, there are many ways that the 10 decimal digits can be encoded as 4-bit binary words. It makes sense to find a method that provides as many benefits as possible. For example, a good code should facilitate arithmetic operations and error correction, and minimize storage space and logic circuitry. Similarly, whenever digital audio designers select a coding method, they examine the same criteria.

Weighted Binary Codes

In some applications, weighted codes are more efficient than other representations. In a weighted code, each binary bit is assigned a decimal value, called a weight. Each number represented by the weighted binary code is calculated from the sum of the weighted digits. For example, weights w_3, w_2, w_1, w_0 and bits a_3, a_2, a_1, a_0 would represent the decimal number $N = w_3 \times a_3 + w_2 \times a_2 + w_1 \times a_1 + w_0 \times a_0$.

The binary coded decimal (BCD) code is commonly used in digital applications. BCD is a positional weighted code in which each decimal digit is binary coded into 4-bit words. It is sometimes called the 8-4-2-1 code, named after the values of its weights. The BCD representation is shown in Table 1.2, along with several other binary codes. The 8-4-2-1 and 6-3-1-1 codes are weighted; the others are not. Any decimal number can be represented with the BCD code. For example, the number 5995 would be 0101 1001 1001 0101. Note that the resulting binary number is quite different from that obtained by direct decimal-to-binary conversion. The BCD code solves the binary problem of representing large decimal numbers; systems interfacing with decimal numbers often

Decimal Digit	8-4-2-1 Code	6-3-1-1 Code	Excess-3 Code	2-out-of-5 Code	Gray Code
0	0000	0000	0011	00011	0000
1	0001	0001	0100	00101	0001
2	0010	0011	0101	00110	0011
3	0011	0100	0110	01001	0010
4	0100	0101	0111	01010	0110
5	0101	0111	1000	01100	1110
6	0110	1000	1001	10001	1010
7	0111	1001	1010	10010	1011
8	1000	1011	1011	10100	1001
9	1001	1100	1100	11000	1000

Table 1.2 Examples of several binary codes. The 8-4-2-1 and 6-3-1-1 codes are weighted; the others are unweighted.

use BCD code for this reason. Thus, they incorporate binary-to-BCD, and BCD-to-binary conversion programs. The 6-3-1-1 code is another example of a weighted code. For example, using the weights assigned to the 6-3-1-1 code, the codeword 1001 represents the decimal number $N = 6 \times 1 + 3 \times 0 + 1 \times 0 + 1 + 1 = 7$.

Unweighted Binary Codes

In some applications, unweighted codes are preferred. Several examples of unweighted codes are excess-3, 2-out-of-5, and Gray code. The excess-3 code is derived from the 8-4-2-1 code by adding 3 (0011) to each codeword. In other words, the decimal digit d is represented by the 4-bit binary number $d + 3$. In this way, every codeword has at least one 1. The 2-out-of-5 code is defined so that exactly two out of the five bits are 1 for every valid word. This definition provides a simple way to check for errors; an error could result in more or less than two 1s.

Other unweighted codes can be defined, for example, so that no codeword has less than one 1 or more than two 1s. This minimizes transitions in logic states when changing words, a potential cause of errors or distortion in output circuitry. In a Gray code, sometimes called a reflected code system, only one digit can change value when counting from one state to the next. A disadvantage of an unweighted code is that generally the corresponding decimal value cannot be easily computed from the binary values.

Two's Complement

Although it is comforting to define binary operations in terms familiar to humans, clearly such an enterprise would not be expedient. It makes more sense to specify numbers and operations in ways most easily handled by machines. For example, when binary numbers are stored in complemented form, an adding operation can be used to perform both addition and subtraction. Furthermore, when binary numbers are stored and processed through registers, a modular number system is often preferable.

Simple binary arithmetic operations can present problems when the result is stored. For example, given a 3-bit system, suppose the numbers 110 and 100 are added. The answer is 1010. The left-most bit is a carry digit from the addition process. However, in a 3-bit system, the carry digit would be lost, and the remaining answer would be 010. This is problematic unless the system can identify that an overflow has occurred. With modular arithmetic, such a result is easily reconciled.

Overflow is inherent in any system with a finite number of digits. Overflow requires an adjustment in our thinking. When an infinite number of digits is available, the task of adding, for example, 6 and 4 could be visualized by joining two straight line segments of lengths 6 and 4 together to obtain a line segment of 10. However with a finite number of digits, it is better to visualize a circular system. For example, when we add 6_{10} (110_2) and 4_{10} (100_2), the appropriate circular segments are joined, leading to a result of 2_{10} (010_2). The resulting number 2 is the remainder obtained by subtracting 8, that is (2^3), from the sum of 10. Two numbers A and B are equivalent in modulo N if A divided by N yields a remainder that equals the remainder obtained when B is divided by N. For example, if $A = 12$ and $B = 20$, then $A = B \pmod 8$ because $12/8 = 1 +$ remainder 4 and $20/8 = 2 +$ remainder 4. The remainders are equal, thus the numbers are equivalent in modulo 8. Modulo 2 arithmetic is used in many binary applications; in general, when performing numerical operations with n-bit words, modulo 2^n arithmetic is used.

As noted, negative numbers can be expressed in binary form with a sign bit. Although convenient for humans, it is illogical for machines. Independent circuits would be required for addition and subtraction. Instead, using a special property of modulo number systems, an addition operation can be used to perform subtraction as well. Although a carry operation is still required, a borrow is not. Specifically, rather than store a negative number as a sign and magnitude, it can be coded in a base-complemented form. Two kinds of complements exist for any number system. A radix-minus-one complement is formed by subtracting each digit of the number from the radix-minus-one of the number system. This is known as the nine's complement in the decimal system and the one's complement in the binary system. A true complement is formed by subtracting each digit of the number from the radix-minus-one, then adding 1 to the least significant digit. The true complement is called the ten's complement in the decimal system, and the two's complement in the binary number system. With base complement, a number added to an existing number results in the full next power of the base. For example, the ten's complement of 35 is 65 because $35 + 65 = 10^2 = 100$, and the two's complement of 1011 is 0101 because $1011 + 0101 = 2^4 = 10000$. From this, note that a complement is the symbol complement plus 1, and the symbol complement is found by subtracting the original number from the highest symbol in the system.

Forming the one's and two's complements in the binary system is easily accomplished and expedites matters tremendously. Because the radix is 2, each bit of the binary number is subtracted from 1. Thus, a one's complement is formed by replacing 1s by 0s, and 0s by 1s, that is, complementing the digits. The two's complement is formed by simply adding 1 to that number and observing any carry operations. For example, given a number 0100, its one's complement is 1011, and its two's complement is 1100. It is the two's complement, formed from the one's complement, that is most widely used. For example, Table 1.3 shows

Decimal Number	Binary Number
7	0111
6	0110
5	0101
4	0100
3	0011
2	0010
1	0001
0	0000
–1	1111
–2	1110
–3	1101
–4	1100
–5	1011
–6	1010
–7	1001

TABLE 1.3 A table of values showing a two's complement representation of positive and negative numbers.

a representation of numbers from +7 to –7 with two's complement representation of negative numbers. A positive two's complement number added to its negative value will always equal zero.

Recalling the discussion of modulo number systems, we observe how subtraction can be replaced by addition. If $A + B = 0$ in a modulo system N, then B is the negative of A. There are in fact many Bs such that $B = kN - A$ where $k = 0, 1, 2, \ldots$. When $k = 1$ and A is less than N, then $B = N - A$ behaves as a positive number less than N. We can use $B = N - A$ in place of $-A$ in any calculation in modulo N. For example, $C = D - A$ is equivalent to $C = D + (N - A)$. In other words, if we can obtain $N - A$ without performing subtraction, then subtraction can be performed using addition. After some thought on the matter, we observe that, conveniently, $N - A$ is the two's complement of A.

Complement subtraction can be performed with addition. With binary, first consider standard subtraction:

$$\begin{array}{r} 10001 \\ -01011 \\ \hline 00110 \end{array}$$

However, the same operation can be performed by adding the two's complement of the subtrahend:

$$\begin{array}{rl} 10001 & \\ +01011 & \\ \hline 100110 & \text{(the carry is discarded)} \end{array}$$

When a larger number is subtracted from a smaller one, there is no carry. For example, $2 - 8$ becomes $2 + (10 - 8)$ or $2 + 2 = 4$. Notice that the answer 4 is the ten's complement of the negative result that is $-(10 - 4) = -6$. Similarly, when a larger number is subtracted from a smaller one in binary:

$$\begin{array}{r} 101 \\ -11011 \\ \hline -10110 \end{array}$$

In two's complement:

$$\begin{array}{r} 101 \\ +00101 \\ \hline 01010 \end{array}$$

The answer is negative, but in two's complement form. Taking the two's complement and assigning a negative sign results in the number –10110. When performing two's complement subtraction, the final carry provides the sign of the result. A final carry of 1 indicates a positive answer, and a carry of 0 indicates a negative answer, in its two's complement, positive form.

If base complementing seems tedious, it is redeemed by its advantages when handling positive and negative (bipolar) numbers, which might, for example, represent an audio waveform. The most significant bit (MSB) is the sign bit. When it is 0, the number is positive, and when it is 1, the number is negative. In true binary form, the number 5 can be represented by 00000101 and –5 by 10000101. By representing negative numbers in two's complement form, –5 becomes 11111011, and the sign is handled automatically. All additions and subtractions will result in positive numbers in true binary form and all negative numbers in two's complement form, with the MSB automatically in the proper sign form. Humans appreciate two's complement because the left digit always denotes the sign. Digital processors appreciate it because subtraction can be easily performed through addition.

Boolean Algebra

The binary number system presents tremendous opportunities for the design of electronic hardware and software, including, of course, digital audio applications. Boolean algebra is the method used to combine and manipulate binary signals. It is named in honor of its inventor, George Boole, who published his proposal for the system in 1854 in a book entitled *An Investigation of the Laws of Thought, on Which Are Founded the Mathematical Theories of Logic and Probabilities*. Incidentally, historians tell us that Boole's formal education ended in the third grade. Also, the lunar crater Boole is named after him.

Boolean logic is essential to digital applications because it provides the basis for decision making, condition testing, and performing logical operations. Using Boolean algebra, all logical decisions are performed with the binary digits 0 and 1, a set of operators, and a number of laws and theorems. The on/off nature of the system is ideally suited for realization in digital systems. The set of fundamental logic operators provides the tools necessary to manipulate bits and hence design the logic that comprises useful digital systems. Everything from vending machines to super computers can be designed with this powerful mathematics. It was Claude Shannon, introduced to Boolean logic in a philosophy class, who first used Boolean algebra to establish the theory of digital circuits.

The Boolean operators are shown in Fig. 1.2. The operators OR, AND, and EXCLUSIVE OR (XOR) combine two binary digits to produce a single-digit result. The Boolean operator NOT complements a binary digit. NAND and NOR are derived from the other operators. The operators can be used singly or in combinational logic to perform any possible logical operation.

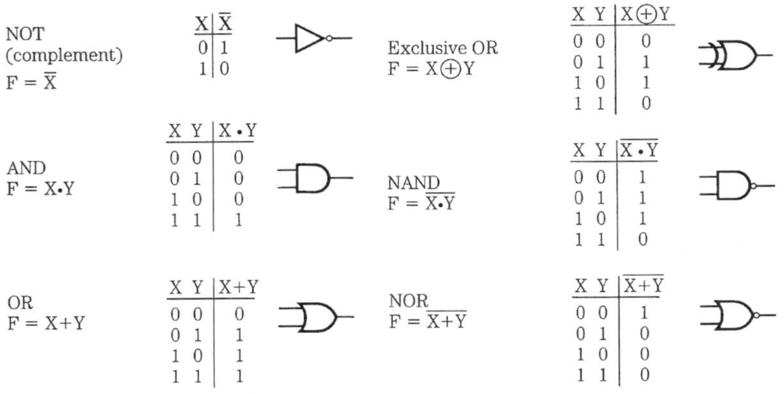

NOT (complement) $F = \overline{X}$

X	$\overline{X}$
0	1
1	0

Exclusive OR $F = X \oplus Y$

X	Y	$X \oplus Y$
0	0	0
0	1	1
1	0	1
1	1	0

AND $F = X \cdot Y$

X	Y	$X \cdot Y$
0	0	0
0	1	0
1	0	0
1	1	1

NAND $F = \overline{X \cdot Y}$

X	Y	$\overline{X \cdot Y}$
0	0	1
0	1	1
1	0	1
1	1	0

OR $F = X + Y$

X	Y	$X+Y$
0	0	0
0	1	1
1	0	1
1	1	1

NOR $F = \overline{X+Y}$

X	Y	$\overline{X+Y}$
0	0	1
0	1	0
1	0	0
1	1	0

Figure 1.2 The six basic Boolean operators can be used to manipulate logical conditions.

The NOT operation complements any set of digits. The complement of 0 is 1, and the complement of 1 is 0. A bar is placed over the digit to represent a complement.

The AND operation is defined by the statement: If X AND Y are both 1, then the result is 1; otherwise the result is 0. Either a dot symbol or no symbol is used to denote AND.

The OR operation is defined by the statement: If X OR Y, or both, are 1 then the result is 1; otherwise the result is 0. A plus sign is usually used to denote OR.

EXCLUSIVE OR differentiates whether binary states are the same or different. Its output is 1 when X differs from Y, and is 0 when X is the same as Y. Importantly, the XOR function thus performs modulo 2 binary addition. A circled plus sign is used to denote XOR.

Combining AND and NOT produces NAND, and combining OR and NOT produces NOR; their results are the NOT of AND and OR, respectively.

The Boolean operators can be combined into meaningful expressions, giving statement to the condition at hand. Moreover, such statements often lead to greater insight of the condition, or its simplification. For example, a digital system needs only the OR and NOT functions because any other function can be derived from those functions. This relationship can be shown using De Morgan's theorem, which states:

$$\overline{A \cdot B} = \overline{A} + \overline{B}$$

$$\overline{A + B} = \overline{A} \cdot \overline{B}$$

Using De Morgan's theorem, observe that the expression:

$$A \cdot B = \overline{\overline{A} + \overline{B}}$$

generates AND from OR and NOT, and the expression:

$$A \oplus B = \overline{(\overline{A} + B)} + \overline{(A + \overline{B})}$$

generates XOR from OR and NOT.

This example shows that Boolean operators can be combined into expressions. In this case, De Morgan's theorem is used to form the complement of expressions. This ability to form logical expressions allows us to use Boolean operators, along with one or more variables or constants, to solve applications problems. Parentheses are used to define the order in which operations are performed; operations are initiated within parentheses. When parentheses are omitted, complementation is performed first, followed by AND and then OR.

Logical expressions correspond directly to networks of logic gates, realizable in hardware or software. For example, Fig. 1.3A shows a logical expression and its equivalent network of logic gates. An expression can be evaluated by substituting a value of 0 or 1 for each variable, and carrying out the indicated operations. Each appearance of a variable or its complementation is called a literal.

A truth table, or table of combinations, can be used to illustrate all the possible combinations contained in an expression. In other words, the output can be expressed in terms of the input variables. For example, the truth table in Fig. 1.3B shows the results of the logic circuit in Fig. 1.3A.

Given a set of operators and ways to combine them into expressions, the next step is to develop a system of Boolean algebraic relations. That set forms the basis of digital processing in the same way that regular algebra governs the manipulation of the familiar

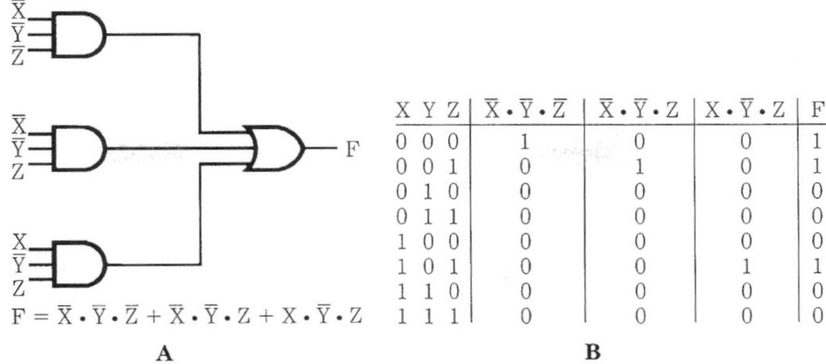

X Y Z	$\overline{X}\cdot\overline{Y}\cdot\overline{Z}$	$\overline{X}\cdot\overline{Y}\cdot Z$	$X\cdot\overline{Y}\cdot Z$	F
0 0 0	1	0	0	1
0 0 1	0	1	0	1
0 1 0	0	0	0	0
0 1 1	0	0	0	0
1 0 0	0	0	0	0
1 0 1	0	0	1	1
1 1 0	0	0	0	0
1 1 1	0	0	0	0

FIGURE 1.3 Boolean expressions can be realized through hardware or software, as circuits or logical statements. A. The logic realization of a Boolean expression. B. A truth table verifying the solution to the Boolean expression.

base 10 operations. In fact, the base 2 and base 10 algebraic systems are very similar, to the point of confusion. Relations such as complementation, commutation, association, and distribution form the mathematical logic needed to create logical systems. These laws hold true for Boolean algebra, but they are often uniquely defined.

Double complementation results in the original value. In other words, if 1 is complemented it becomes 0, and when complemented again, it becomes 1 again. Commutative laws state that the order in which terms are combined (with addition and multiplication operators) does not affect the result. Associative laws state that when several terms are added or multiplied, the order of their selection for the operation is immaterial. Distributive laws demonstrate that the product of one term multiplied by a sum term equals the sum of the products of the first term multiplied by each product term.

Using Boolean theorems and other reduction theorems, such as De Morgan's theorem, complex logical expressions can often be untangled to provide simple results. For example, consider these expressions:

$$F = X\cdot Y + X\cdot\overline{Y} + \overline{Y}(X+\overline{X})$$

$$F = X\cdot Y + X\cdot\overline{Y} + \overline{X}\cdot\overline{Y}$$

$$F = X\cdot Y + X\cdot\overline{Y} + \overline{Y}$$

$$F = \overline{Y} + Y\cdot X$$

$$F = X + \overline{X}\cdot\overline{Y}$$

$$F = X + \overline{Y}$$

All these functions call for different Boolean manipulations, but their outputs are identical. Whether the end result of an expression is a hardware circuit or software program, simplification is understandably critical. For example, Boolean algebra can be applied to the expression in Fig. 1.3, resulting in a much simpler expression, as shown in Fig. 1.4. In this case, the original circuit required three AND gates, one OR gate, and three NOT circuits. The simplified circuit requires one AND, one OR, and two NOTS. The important point is that both circuits perform the same function, with identical outputs for any possible inputs.

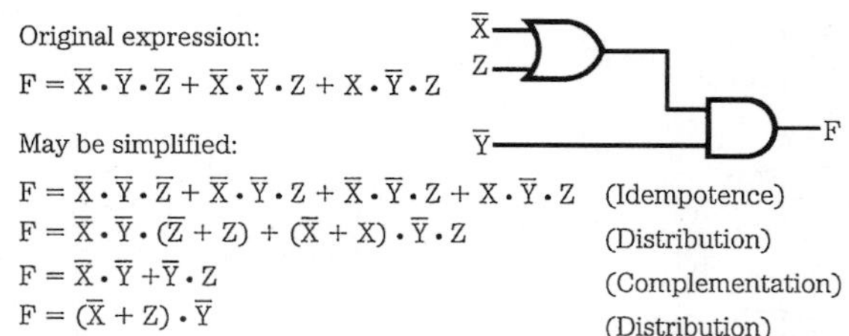

FIGURE 1.4 Complex logical expressions can be simplified with Boolean laws to permit simpler hardware or software realization.

Using these tools, useful digital hardware and software can be conceived to manage information or perform processing. More generally, Boolean algebra can be applied to many types of problems. Even word problems posed in English can be converted to Boolean form and solved. For example, consider the syllogism (by Lewis Carroll) with these three premises:

1. All babies are illogical.
2. Nobody is despised who can manage a crocodile.
3. Illogical persons are despised.

By applying the principles of symbolic logic, the absolutely logical conclusion is:

4. A person able to manage a crocodile is not a baby.

Whether the problem involves babies, crocodiles, motorcycle engines, or violin tones, binary logic can be applied with great finesse when the problem can be reduced to a binary state. From that elemental simplicity, tremendous complexity can be achieved. In this respect, all digital systems are alike. They differ, however, according to the kinds of information they process and what data manipulations they are called upon to accomplish. A digital audio system is thus a unique digital system, which is specially configured to process audio data. Nonintuitively, digital audio systems have more in common with syllogisms than they do with analog audio systems.

Analog versus Digital

Digital audio entails entirely new concepts and techniques, distinct from those employed in analog technology. From the very onset, we must think of information and its storage and processing in a new light. In that respect, a few comparisons between analog and digital technology might be helpful in showing the vast differences between them, as well as some of the advantages and disadvantages of each technology.

An analog signal can be compared to a digital signal; consider a bucket of water compared to a bucket of ball bearings. Both water and ball bearings fill their respective containers. The volume of each bucket can be characterized by measuring its contents, but the procedures for each bucket would be different. With water, we could weigh the bucket and water, pour out the water, weigh the bucket alone, subtract to find the weight of the water, then calculate the volume represented. Or we could dip a measuring cup into the bucket and withdraw measured amounts. In either case, we run the risk of spilling some water or leaving some at the bottom of the bucket; our measurement would be imprecise.

With a bucket of ball bearings, we simply count each ball bearing and calibrate the volume of the bucket in terms of the number of ball bearings it holds. The measurement is relatively precise, if perhaps a little tedious (we might want to use a computer to do the counting). The ball bearings represent the discrete values in a digital system, and demonstrate that with digital techniques we can quantify the values and gain more accurate information about the measurement. In general, precision is fundamental to any digital system. For example, a bucket that has been measured to hold 1.6 quarts of water may be less useful than a bucket that is known to hold 8263 ball bearings. In addition, the counted ball bearings provide repeatability. We might fill another bucket with water, but that bucket might be a different size and we could easily end up with 1.5 quarts. On the other hand, we could reliably count out 8263 ball bearings anywhere, anytime.

The utility of a digital system compared to an analog system is paradoxical. Conceptually, the digital system is much simpler, because counting numbers is easier than measuring a continuous flow. That is, it is easier to process a representation of a signal rather than process the signal itself. However, in practice, the equipment required to accomplish that simple task must be more technologically sophisticated than any analog equipment.

The comparison between analog and digital audio can be simply summarized. An analog signal chain, from recording studio to living room, must convey a continuous representation of amplitude change in time. This is problematic because every circuit and storage medium throughout the chain is itself an analog device, contributing its own analog distortion and noise. This unavoidably degrades the analog signal as it passes through. In short, when an analog signal is processed through an analog chain with inherent analog artifacts, deterioration occurs. In addition, it is worth noting that analog systems tend to be bulky, power hungry, prone to breakdown, and cannot be easily modified to perform a different task.

With digital audio, the original analog event is converted into a quantity of binary data, which is processed, stored, and distributed as a numerical representation. The reverse process, from data to analog event, occurs only at playback, thus eliminating many occasions for intervening degradation. The major concern lies in converting the analog signal into a digital representation, and back again to analog. Nevertheless, because the audio information is carried through a numerical signal chain, it avoids much spurious analog distortion and noise. In addition, digital circuits are less expensive to design and manufacture, they offer increased noise immunity, age immunity, temperature immunity, and greatly increased reliability. Even more important, the power of software algorithms can flexibly process numerically represented audio signals, yielding precise and sophisticated results.

In theory, digital audio systems are quite elegant, suggesting that they easily surpass more encumbered analog systems. In practice, although digital systems are free of many analog deficiencies, they can exhibit some substantial anomalies of their own. The remaining chapters only scratch the surface of the volume of knowledge needed to understand the complexities of digital audio systems. Moreover, as with any evolving science, every new insight only leads to more questions.

CHAPTER 2

Fundamentals of Digital Audio

The digital techniques used to record, reproduce, store, process, and transmit digital audio signals entail concepts foreign to analog audio methods. In fact, the inner workings of digital audio systems bear little resemblance to analog systems. Because audio itself is analog in nature, digital systems employ sampling and quantization, the twin pillars of audio digitization, to represent the audio information. Any sampling system is bound by the sampling theorem, which defines the relationship between the message and the sampling frequency. In particular, the theorem dictates that the message be bandlimited. Precaution must be taken to prevent a condition of erroneous sampling known as aliasing. Quantization error occurs when the amplitude of an analog waveform is represented by a binary word; effects of the error can be minimized by dithering the audio waveform prior to quantization.

Discrete Time Sampling

With analog recording, a tape is continuously modulated or a groove is continuously cut. With digital recording, discrete numbers must be used. To create these numbers, audio digitization systems use time sampling and amplitude quantization to encode the infinitely variable analog waveform as amplitude values in time. Both of these techniques are considered in this chapter. First, let's consider discrete time sampling, the essence of all digital audio systems.

Time seems to flow continuously. The hands of an analog clock sweep across the clock face covering all time as it passes by. A digital readout clock also tells time, but with a discretely valued display. In other words, it displays sampled time. Similarly, music varies continuously in time and can be recorded and reproduced either continuously or discretely. Discrete time sampling is the essential mechanism that defines a digital audio system, permits its analog-to-digital (A/D) conversion, and differentiates it from an analog system.

However, a nagging question immediately presents itself. If a digital system samples an audio signal discretely, defining the audio signal at distinct times, what happens between samples? Haven't we lost the information present between sample times? The answer, intuitively surprising, is no. Given correct conditions, no information is lost due to sampling between the input and output of a digitization system. The samples contain the same information as the conditioned unsampled signal. To illustrate this, let's try a conceptual experiment.

Suppose we attach a movie camera to the handlebars of a BMW motorcycle, go for a ride, and then return home and process the film. Auditioning this piece of avant-garde cinema, we discover that the discrete frames of film reproduce our ride. But when we traverse bumpy pavement, the picture is blurred. We determine that the quick movements were too fast for each frame to capture the change. We draw the following conclusion: if we increased the frame rate, using more frames per second, we could capture quicker changes. Or, if we complained to city hall and the bumpy pavement was smoothed, there would be no blur even at slower frame rates. We settle on a compromise—we make the roads reasonably smooth, and then we use a frame rate adjusted for a clean picture.

The analogy is somewhat clumsy. (For starters, cinema comprises a series of discontinuous still images—it is the brain itself that creates the illusion of a continuum. An audio waveform played back from a digital source really is continuous because of the interpolation function used to create it.) Nevertheless, the analogy shows that the discrete frames of a movie create a moving picture, and similarly the samples of a digital audio recording create a continuous signal. As noted, sampling is a lossless process if the input signal is properly conditioned. Thus, in a digital audio system, we must smooth out the bumps in the incoming signal. Specifically, the signal is lowpass filtered; that is, the frequencies that are too high to be properly sampled are removed. We observe that a signal with a finite frequency response can be sampled without loss of information; the samples contain all the information contained in the original signal. The original signal can be completely recovered from the samples. Generally, we observe that there exists a method for reconstructing a signal from its amplitude values taken at periodic points in time.

The Sampling Theorem

The idea of sampling occurs in many disciplines, and the origin of sampling theorems comes from many sources. Most audio engineers recognize American engineer Harry Nyquist as the author of the sampling theorem that founded the discipline of modern digital audio. The recognition is well-founded because it was Nyquist who expressed the theorem in terms that are familiar to communications engineers. Nyquist, who was born in Sweden in 1889, and died in Texas in 1976, worked for Bell Laboratories and authored 138 U.S. patents. However, the story of sampling theorems predates Nyquist.

When he was not busy designing military fortifications for Napoleon, French mathematician Augustin-Louis Cauchy contemplated statistical sampling. In 1841, he showed that functions could be nonuniformly sampled and averaged over a long period of time. At the turn of the century, it was thought (incorrectly) that a function could be successfully sampled at a frequency equal to the highest frequency. In 1915, Scottish mathematician E. T. Whittaker, working with interpolation series, devised perhaps the first mathematical proof of a general sampling theorem, showing that a bandlimited function can be completely reconstructed from samples. In 1920, Japanese mathematician K. Ogura similarly proved that if a function is sampled at a frequency at least twice the highest function frequency, the samples contain all the information in the function, and can reconstruct the function. Also in 1920, American engineer John Carson devised an unpublished proof that related the same result to communications applications.

It was Nyquist who first clarified the application of sampling to communications, and published his work. In 1925, in a paper titled "Certain Factors Affecting Telegraph Speed," he proved that the number of telegraph pulses that can be transmitted over a telegraph line per unit time is proportional to the bandwidth of the line. In 1928, in a paper titled "Certain Topics in Telegraph Transmission Theory," he proved that for complete signal reconstruction, the required frequency bandwidth is proportional to the signaling speed, and that the minimum bandwidth is equal to half the number of code elements per second. Subsequently, Russian engineer V. A. Kotelnikov published a proof of the sampling theorem in 1933.

American mathematician Claude Shannon unified and proved many aspects of sampling, and also founded the larger science of information theory in his 1948 book, *A Mathematical Theory of Communication*. Shannon's 1937 master's thesis, "A Symbolic Analysis of Relay and Switching Circuits," showed that circuits could use Boolean algebra to solve logical or numerical problems; his work was called "possibly the most important, and also the most famous, master's thesis of the century." Shannon, a distant relative of Thomas Edison, could also juggle three balls while riding a unicycle. Today, engineers usually attribute the sampling theorem to Shannon or Nyquist. The half-sampling frequency is usually known as the Nyquist frequency.

Whoever gets the credit, the sampling theorem states that a continuous bandlimited signal can be replaced by a discrete sequence of samples without loss of any information and describes how the original continuous signal can be reconstructed from the samples; furthermore, the theorem specifies that the sampling frequency must be at least twice the highest signal frequency. More specifically, audio signals containing frequencies between 0 and $S/2$ Hz can be exactly represented by S samples per second. Moreover, in general, the sampling frequency must be at least twice the bandwidth of a sampled signal. When the lowest frequency of the bandwidth of interest is zero, then the signal's bandwidth equals the highest frequency. The sampling theorem is applied widely and diversely throughout engineering, science, and mathematics.

Nyquist Frequency

When the sampling theorem is applied to audio signals, the input audio signal is lowpass filtered, so that it is bandlimited with a frequency response that does not exceed the Nyquist ($S/2$) frequency. Ideally, the lowpass filter is designed so that the only signals removed are those high frequencies that lie above the high-frequency limit of human hearing. The signal can now be sampled to define instantaneous amplitude values. The sampled bandlimited signal contains the same information as the unsampled bandlimited signal. At the system output, the signal is reconstructed, and there is no loss of information (due to sampling) between the output signal and the input filtered signal. From a sampling standpoint, the output signal is not an approximation; it is exact. The bandlimited signal is thus re-created, as shown in Fig. 2.1.

Consider a continuously changing analog function that has been sampled to create a series of pulses. The amplitude of each pulse, determined through quantization, yields a number that represents the signal amplitude at that instant. To quantify the situation, we define the sampling frequency as the number of samples per second. Its reciprocal, sampling rate, defines the time between each sample. For example, a sampling frequency of 48,000 samples per second corresponds to a rate of 1/48,000 seconds. A quickly changing waveform—that is, one with high frequencies—requires a higher sampling frequency. Thus, the digitization system's sampling frequency determines the high frequency limit of the system. The choice of sampling frequency is thus one of the most important design criteria of a digitization system, because it determines the audio bandwidth of the system.

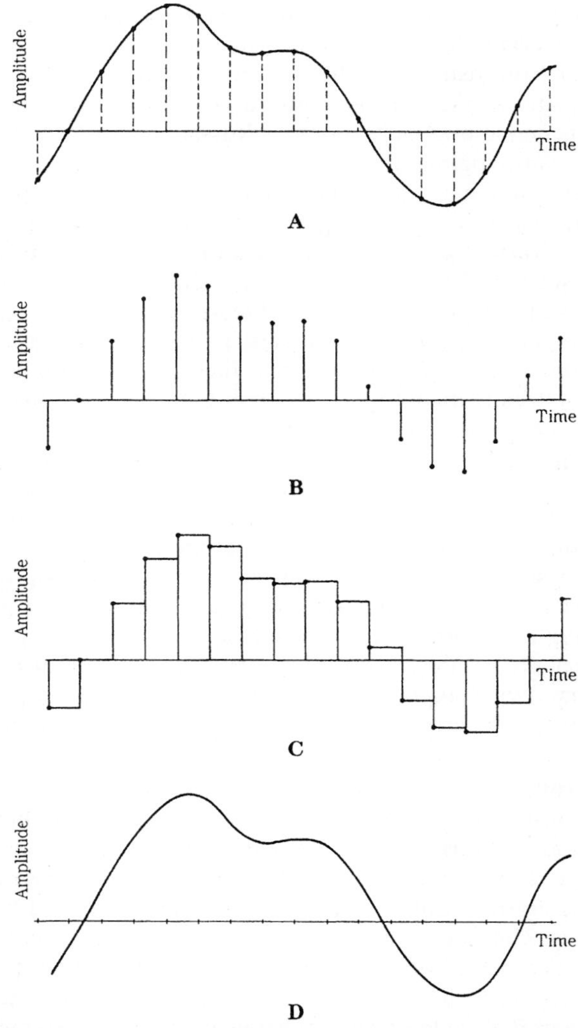

FIGURE 2.1 With discrete time sampling, a bandlimited signal can be sampled and reconstructed without loss because of sampling. A. The input analog signal is sampled. B. The numerical values of these samples are stored or transmitted (effect of quantization not shown). C. Samples are held to form a staircase representation of the signal. D. An output lowpass filter interpolates the staircase to reconstruct the input waveform.

The sampling theorem precisely dictates how often a waveform must be sampled to provide a given bandwidth. Specifically, as noted, a sampling frequency of S samples per second is needed to completely represent a signal with a bandwidth of $S/2$ Hz. In other words, the sampling frequency must be at least twice the highest audio frequency to achieve lossless sampling. For example, an audio signal with a frequency response of 0 to 24 kHz would theoretically require a sampling frequency of 48 kHz for proper sampling. Of course, a system could use any sampling frequency as needed. It is crucial to observe the sampling theorem's criteria for limiting the input signal to no more than half the sampling frequency (the Nyquist frequency). An audio frequency above this would cause aliasing distortion, as described later in this chapter. A lowpass filter must be used to remove frequencies above the half-sampling frequency limit. A lowpass filter is also placed at the output of a digital audio system to remove high frequencies that are created internally in the system. This output filter reconstructs the original waveform. Reconstruction is discussed in more detail in Chap. 4.

Another question presents itself with respect to the sampling theorem. We observe that when low audio frequencies are sampled, because of their long wavelengths, many samples are available to represent each period. But as the audio frequency increases, the periods are shorter and there are fewer samples per period. Finally, in the theoretical limiting case of critical sampling, at an audio frequency of half the sampling frequency, there are only two samples per period. However, even two samples can represent a waveform. For example, consider the case of a 48-kHz sampling frequency and an audio input of 24-kHz sine wave. The sampler produces two samples, which will yield a 24-kHz square wave. In itself, this waveform is quite unlike the original sine wave. However, a lowpass filter at the output of the digital audio system removes all frequencies higher than the half-sampling frequency. (The 24-kHz square wave consists of odd harmonics—sine waves starting at 24 kHz.) With all higher frequency content removed, the output of the system is a reconstructed 24-kHz sine wave, the same as the sampled waveform. We know that the sampled waveform was a sine wave because the input lowpass filter will not pass higher waveform frequencies to the sampler. Similarly, a digitization system can reproduce all information from 0 to $S/2$ Hz, including sine wave reproduction at $S/2$ Hz; even in the limiting case, the sampling theorem is valid. Conversely, all information above $S/2$ is removed from the signal. We can state that higher sampling frequencies permit recording and reproduction of higher audio frequencies. But given the design criteria of an audio frequency bandwidth, higher sampling frequencies will not improve the fidelity of those signals already within the bandlimited frequency range.

For critical sampling, there is no guarantee that the sample times will coincide with the maxima and minima of the waveform. Sample times could coincide with lower-amplitude parts of the waveform, or even coincide with the zero-axis crossings of the waveform. In practice, this does not pose a problem. Critical sampling is not attempted; a sampling margin is always present. As we have seen, to satisfy the sampling theorem, a lowpass filter must precede the sampler. Lowpass filters cannot attenuate the signal precisely at the Nyquist frequency so a guard band is employed. The filter's cutoff frequency characteristic starts at a lower frequency, for example, at 20 kHz, allowing several thousand Hertz for the filter to attenuate the signal sufficiently. This ensures that no frequency above the Nyquist frequency enters the sampler. The waveform is typically not critically sampled; there are always more than two samples per period. Furthermore, the phase relationship between samples and waveforms is never exact because acoustic waveforms do not synchronize with a sampler. Finally, when we examine the sampling theorem more rigorously in Chap. 4, we will see that parts of the waveform lying between samples can be captured and reproduced by sampling. We shall see that the output signal is not reconstructed sample by sample; rather, it is formed from the summation of the response of many samples. It is also worth noting that the bandwidth of any practical analog audio signal is also limited. No analog audio system has infinite bandwidth. The finite bandwidth of audio signals shows that the continuous waveform of an analog signal or the samples of a digital signal can represent the same information.

The need to bandlimit the audio signal is not as detrimental as it might first appear. The upper frequency limit of the audio signal can be extended as far as needed, so long as the appropriate sampling frequency is employed. For example, depending on the application, sampling frequencies from 8 kHz to 192 kHz may be used. The trade-off, of course, is the demand placed on the speed of digital circuitry and the capacity of the storage or transmission medium. Higher sampling frequencies require that circuitry operate faster and that larger amounts of data be conveyed. Both are ultimately questions of economics. Manufacturers selected a sampling frequency of 44.1 kHz for the Compact Disc, for example, because of its size, playing time, and cost of the medium. On the other hand, DVD-Audio and Blu-ray discs can employ sampling frequencies up to 192 kHz.

The entire sampling (and desampling) process is summarized in Fig. 2.2. The signals involved in sampling are shown at different points in the processing chain.

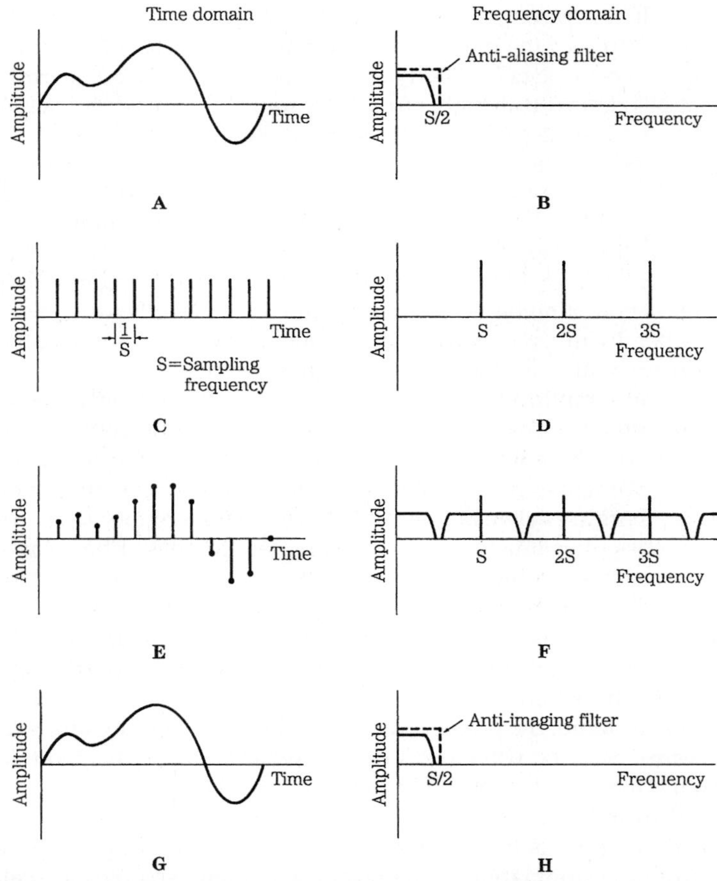

FIGURE 2.2 Time domain (left column) and frequency domain (right column) signals illustrate the process of bandlimited waveform sampling and reconstruction. A. Input signal after anti-aliasing filter. B. Spectrum of input signal. C. Sampling signal. D. Spectrum of the sampling signal. E. Sampled input signal. F. Spectrum of the sampled input signal. G. Output signal after anti-imaging filter. H. Spectrum of the output signal.

Moreover, the left half of the figure shows the signals in the time domain and the right half portrays the same signals in the frequency domain. In other words, we can observe a signal's amplitude over time, as well as its frequency response. We observe in Figs. 2.2A and B that the input audio signal must be bandlimited to the half-sampling frequency $S/2$, using a lowpass anti-aliasing filter. This filter removes all components above the Nyquist frequency of $S/2$. The sampling signal in Figs. 2.2C and D recurs at the sampling frequency S, and its spectrum consists of pulses at multiples of the sampling frequency: S, $2S$, $3S$, and so on. When the audio signal is sampled, as shown in Figs. 2.2E and F, the signal amplitude at sample times is preserved; however, this sampled signal contains images of the original spectrum centered at multiples of the sampling frequency. To reproduce the sampled signal, as in Figs. 2.2G and H, the samples are passed through a lowpass anti-imaging filter to remove all images above the $S/2$ frequency. This filter interpolates between the samples of the waveform, recreating the input, bandlimited audio signal. As described in Chap. 4, the output filter's impulse response uniquely reconstructs the sample pulses as a continuous waveform.

The sampling theorem is unequivocal: a bandlimited signal can be sampled; stored, transmitted, or processed as discrete values; desampled; and reconstructed. No bandlimited information is lost through sampling. The reconstructed waveform is identical to the bandlimited input waveform. Sampling theorems such as the Nyquist theorem prove this conclusively. Of course, after it has time-sampled the signal, a digital system also must determine the numerical values it will use to represent the waveform amplitude at each sample time. This question of quantization is explained subsequently in this chapter. For a more detailed discussion of discrete time sampling, and a concise mathematical demonstration of the sampling theorem, refer to the Appendix.

Aliasing

Aliasing is a kind of sampling confusion that can originate in the recording side of the signal chain. Just as people can take different names and thus confuse their identity, aliasing can create false signal components. These erroneous signals can appear within the audio bandwidth and are impossible to distinguish from legitimate signals. Obviously, it is the designer's obligation to prevent such distortion from ever occurring. In practice, aliasing is not a serious limitation. It merely underscores the importance of observing the criteria of the sampling theorem.

We have observed that sampling is a lossless process under certain conditions. Most important, the input signal must be bandlimited with a lowpass filter. If this is not done, the signal might be undersampled. Consider another conceptual experiment: use your motion picture camera to film me while I drive away on my motorcycle. In the film, as I accelerate, the spokes of the wheels rotate forward, appear to slow and stop, then begin to rotate backward, rotate faster, then slow and stop, and appear to rotate forward again. This action is an example of aliasing. The motion picture camera, with a frame rate of 24 frames per second, cannot capture the rapid movement of the wheel spokes.

Aliasing is a consequence of violating the sampling theorem. The highest audio frequency in a sampling system must be equal to or less than the Nyquist frequency. If the audio frequency is greater than the Nyquist frequency, aliasing will occur. As the audio frequency increases, the number of sample points per period decreases. When the Nyquist frequency is reached, there are two samples per period, the minimum needed to record the audio waveform. With higher audio frequencies, the sampler will continue to produce samples at its fixed rate, but the samples create false information in the form of alias frequencies. As the audio frequency increases, a descending alias frequency is created. Specifically, if S is the sampling frequency, F is a frequency higher than the half-sampling frequency, and N is an integer, then new frequencies F_f are created at $F_f = \pm NS \pm F$. In other words, alias frequencies appear back in the audio band (and the images of the audio band), folded over from the sampling frequency. In fact, aliasing is sometimes called foldover. Although disturbing, this is not totally surprising. Sampling is a kind of modulation; in fact, sampling is akin to the operation of a heterodyne demodulator in an amplitude modulation (AM) radio. A local oscillator multiplies the input signal to move its frequency down to the standard intermediate frequency (IF). Although the effect is desirable in radios, aliasing in digital audio systems is undesirable.

Consider a digitization system sampling at 48 kHz. Further, suppose that a signal with a frequency of 40 kHz has entered the sampler, as shown in Fig. 2.3. The primary alias component results from $S - F = F_f$ or $48 - 40 = 8$ kHz. The sampler produces the improper samples, faithfully recording a series of amplitude values at sample times. Given those samples, the device cannot determine which was the intended frequency: 40 kHz or 8 kHz. Furthermore, recall that a lowpass filter at the output of a digitization system smoothes the staircase function to reconstruct the original signal. The output filter removes content above the Nyquist frequency. In this case, following the output filter, the 40-kHz signal would be removed, but the 8-kHz alias signal would remain, containing samples as innocuous as a legitimate 8-kHz signal. That unwanted signal is a distortion in the audio signal.

There are other manifestations of aliasing. Although only the $S - F$ component appears as an interfering frequency in the audio band, an alias component will appear in the audio band, no matter how high in frequency F becomes. Consider a sampling

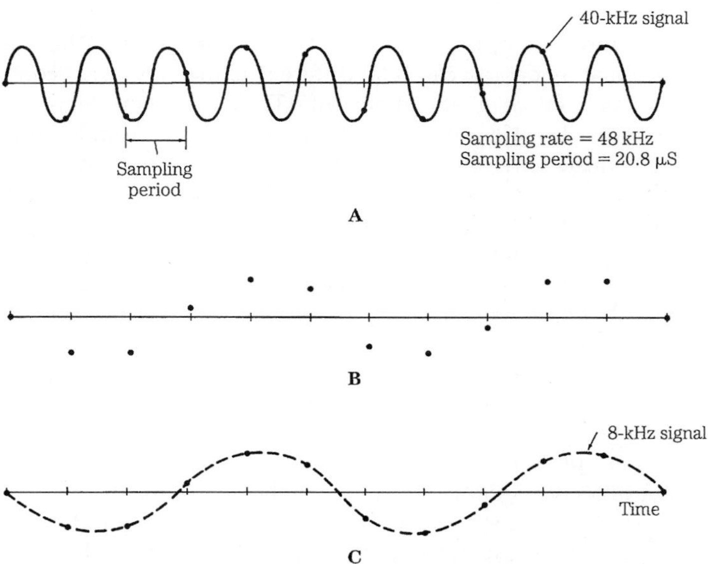

FIGURE 2.3 An input signal greater than the half-sampling frequency will generate an alias signal, at a lower frequency. A. A 40-kHz signal is sampled at 48 kHz. B. Samples are saved. C. Upon reconstruction, the 40-kHz signal is filtered out, leaving an aliased 8-kHz signal.

frequency of 48 kHz; a sweeping input frequency from 0 to 24 kHz would sound fine, but as the frequency sweeps from 24 kHz to 48 kHz, it returns as a frequency descending from 24 kHz to 0. If the input frequency sweeps from 48 kHz to 72 kHz, it appears again from 0 to 24 kHz, and so on.

Alias components occur not only around the sampling frequency, but also in the multiple images produced by sampling (see Fig. 2.2F). When the sampling theorem is obeyed, the audio band and image bands are separate, as shown in Figs. 2.4A and B. However, when the audio band extends past the Nyquist frequency, the image bands overlap, resulting in aliasing as shown in Figs. 2.4C and D. All these components would be produced in an aliasing scenario: $\pm S \pm F$, $\pm 2S \pm F$, $\pm 3S \pm F$, and so on. For example, given a 48-kHz sampler and a 29-kHz input signal, some of the resulting alias frequencies would be 19, 67, 77, 115, 125, 163, and 173 kHz, as shown in Fig. 2.4D. With a sine wave, aliasing is limited to the one and only partial of a sine wave. With complex tones, content is generated for all spectra above the Nyquist frequency.

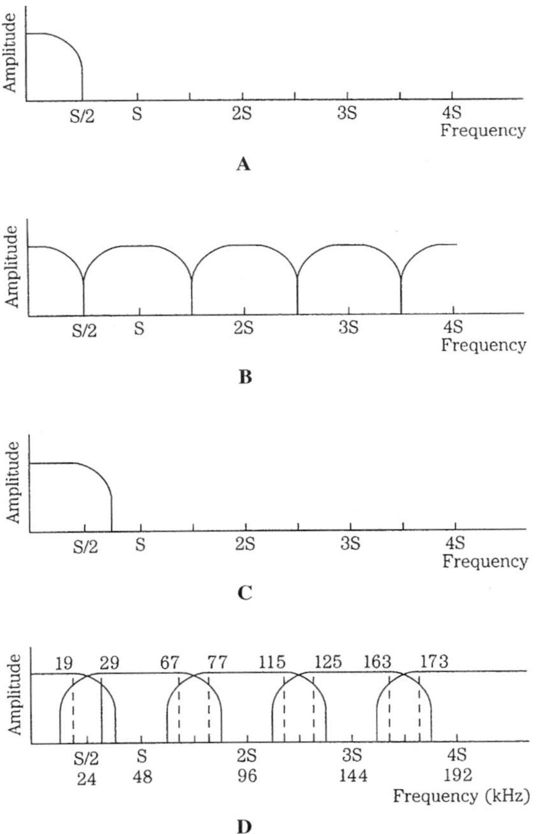

FIGURE 2.4 Spectral views of correct sampling and incorrect sampling causing aliasing. A. An input signal bandlimited to the Nyquist frequency. B. Upon reconstruction, images are contained within multiples of the Nyquist frequency. C. An input signal that is not bandlimited to the Nyquist frequency. D. Upon reconstruction, images are not contained within multiples of the Nyquist frequency; this spectral overlap is aliasing; for example, a 29-kHz signal will alias in a 48-kHz sampler.

In practice, aliasing can be overcome. In fact, in a properly designed digital recording system, aliasing does not occur. The solution is straightforward: the input signal is bandlimited with a lowpass (anti-aliasing filter) that provides significant attenuation at the Nyquist frequency to ensure that the spectral content of the sampled signal never exceeds the Nyquist frequency. An ideal anti-aliasing filter would have a "brick-wall" characteristic with instantaneous and infinite attenuation in the stopband. Practical anti-aliasing filters have a transition band above the Nyquist frequency, and attenuate stopband frequencies to below the resolution of the A/D converter. In practice, as described in Chap. 3, most systems use an oversampling A/D converter with a mild lowpass filter, high initial sampling frequency, and decimation processing to prevent aliasing at the downsampled output sampling frequency. This ensures that the system meets the demands of the sampling theorem; thus, aliasing cannot occur.

It is critical to observe the sampling theorem, and lowpass filter the input signal in a digitization system. If aliasing is allowed to occur, there is no technique that can remove the aliased frequencies from the original audio bandwidth.

Quantization

A measurement of a varying event is meaningful if both the time and the value of the measurement are stored. Sampling represents the time of the measurement, and quantization represents the value of the measurement, or in the case of audio, the amplitude of the waveform at sample time. Sampling and quantization are thus the fundamental components of audio digitization, and together can characterize an acoustic event. Sampling and quantization are variables that determine, respectively, the bandwidth and resolution of the characterization. An analog waveform can be represented by a series of sample pulses; the amplitude of each pulse yields a number that represents the analog value at that instant. With quantization, as with any analog measurement, accuracy is limited by the system's resolution. Because of finite word length, a quantizer's resolution is limited, and a measuring error is introduced. This error is akin to the noise floor in an analog audio system; however, perceptually, it can be more intrusive because its character can vary with signal amplitude.

With uniform quantization, an analog signal's amplitude at sample times is mapped across a finite number of quanta of equal size. The infinite number of amplitude points on the analog waveform must be quantized by the finite number of quanta levels; this introduces an error. A high-quality representation requires a large number of levels; a high-quality music signal might require, for example, 65,536 amplitude levels or more. However, a few pulse-code modulation (PCM) levels can still carry information content; for example, just two amplitude levels can (barely) convey intelligible speech.

Consider two voltmeters, one analog and one digital, each measuring the voltage corresponding to an input signal. Given a good meter face and a sharp eye, we might read the analog needle at 1.27 V (volts). A digital meter with only two digits might read 1.3 V. A three-digit meter might read 1.27 V, and a four-digit meter might read 1.274 V. Both the analog and digital measurements contain errors. The error in the analog meter is caused by the ballistics of the mechanism and the difficulty in reading the meter. Even under ideal conditions, the resolution of any analog measurement is limited by the measuring device's own noise.

With the digital meter, the nature of the error is different. Accuracy is limited by the resolution of the meter—that is, by the number of digits displayed. The more digits, the greater the accuracy, but the last digit will round off relative to the actual value; for example, 1.27 V would be rounded to 1.3 V. In the best case, the last digit would be completely accurate; for example, a voltage of exactly 1.3000 V would be shown as 1.3 V. In the worst case, the rounded off digit will be one-half interval away; for example, 1.250 V would be rounded to 1.2 V or 1.3 V. Similarly, if a binary system is used for the measurement, we say that the error resolution of the system is one-half of the least significant bit (LSB). For both analog and digital systems, the problem of measuring an analog phenomenon such as amplitude leads to error. As far as voltmeters are concerned, a digital readout is an inherently more robust measurement. We gain more concise information about an analog event when it is characterized in terms of digital data. Today, an analog voltmeter is about as common as a slide rule.

Quantization is thus the technique of measuring an analog audio event to form a numerical value. A digital system uses a binary number system. The number of possible values is determined by the length of the binary data word—that is, the number of bits available to form the representation. Just as the number of digits in a digital voltmeter determines resolution, the number of bits in a digital audio recorder also determines resolution. Clearly, the number of bits in the quantizing word is an arbitrary gauge of accuracy; other limitations may exist. In practice, resolution is primarily influenced by the quality of the A/D converter.

Sampling of a bandlimited signal is theoretically a lossless process, but choosing the amplitude value at the sample time certainly is not. No matter what the choice of scales or codes, digitization can never perfectly encode a continuous analog function. An analog waveform has an infinite number of amplitude values, but a quantizer has a finite number of intervals. The analog values between two intervals can only be represented by the single number assigned to that interval. Thus, the quantized value is only an approximation of the actual.

Signal-to-Error Ratio

With a binary number system, the word length determines the number of quantizing intervals available; this can be computed by raising the word length to the power of 2. In other words, an n-bit word would yield 2^n quantization levels. The number of levels determined by the first $n = 1$ to 24 bits are listed in Table 2.1. For example, an 8-bit word provides $2^8 = 256$ intervals and a 16-bit word provides $2^{16} = 65{,}536$ intervals. Note that each time a bit is added to the word length, the number of levels doubles. The more bits, the better the approximation; but as noted, there is always an error associated with quantization because the finite number of amplitude levels coded in the binary word can never completely accommodate an infinite number of analog amplitudes.

It is difficult to appreciate the accuracy achieved by a 16-bit measurement. An analogy might help: if sheets of typing paper were stacked to a height of 22 feet, a single sheet of paper would represent one quantization level in a 16-bit system. Longer word lengths are even more impressive. In a 20-bit system, the stack would reach 352 feet. In a 24-bit system, the stack would tower 5632 feet in height—over a mile high. The quantizer could measure that mile to an accuracy equaling the thickness of a piece of paper. If a single page were removed, the least significant bit would change from 1 to 0. Looked at in another way, if the driving distance between the Empire State Building and Disneyland was measured with 24-bit accuracy, the measurement would be accurate to within 11 inches. A high-quality digital audio system thus requires components with similar tolerances—not a trivial feat.

$2^1 = 2$	$2^{13} = 8192$
$2^2 = 4$	$2^{14} = 16{,}384$
$2^3 = 8$	$2^{15} = 32{,}768$
$2^4 = 16$	$2^{16} = 65{,}536$
$2^5 = 32$	$2^{17} = 131{,}072$
$2^6 = 64$	$2^{18} = 262{,}144$
$2^7 = 128$	$2^{19} = 524{,}288$
$2^8 = 256$	$2^{20} = 1{,}048{,}576$
$2^9 = 512$	$2^{21} = 2{,}097{,}152$
$2^{10} = 1024$	$2^{22} = 4{,}194{,}304$
$2^{11} = 2048$	$2^{23} = 8{,}388{,}608$
$2^{12} = 4096$	$2^{24} = 16{,}777{,}216$

TABLE 2.1 The number (N) of quantization intervals in a binary word is $N = 2^n$, where n is the number of bits in the word.

At some point, the quantizing error approaches inaudibility. Most manufacturers have agreed that 16 to 20 bits provide an adequate representation; however, that does not rule out longer data words or the use of other signal processing to optimize quantization and thus reduce quantization error level. For example, the DVD and Blu-ray formats can code 24-bit words and many audio recorders use noise shaping to reduce in-band quantization noise.

Word length determines the resolution of a digitization system and hence provides an important specification for evaluating system performance. Sometimes the quantized interval will be exactly at the analog value; usually it will not. At worst, the analog level will be one-half interval away—that is, the error is half the least significant bit of the quantization word. For example, consider Fig. 2.5. Suppose the binary word 101000 corresponds to the analog interval of 1.4 V, 101001 corresponds to 1.5 V, and the analog

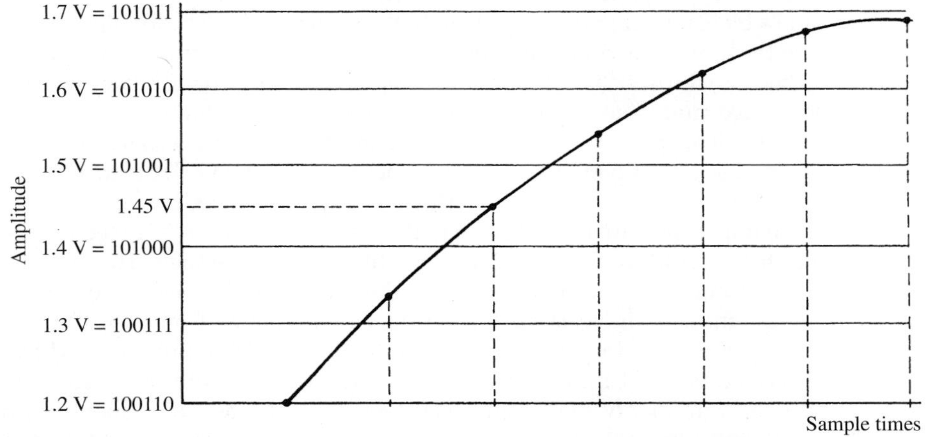

FIGURE 2.5 Quantization error is limited to one-half of the least significant bit.

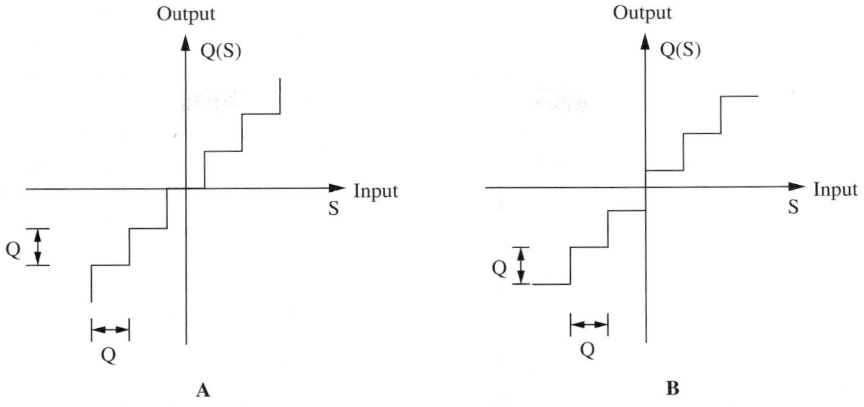

FIGURE 2.6 Signals can be quantized in one of two ways. A. A midtread quantizer. B. A midrise quantizer. Q (or 1 LSB) is the quantizer step size.

value at the sample time is unfortunately 1.45 V. Because 101000 1/2 is not available, the quantizer must round up to 101001 or down to 101000. Either way, there will be an error with a magnitude of one-half of an interval.

Generally, uniform step-size quantization is accomplished in one of two ways, as shown in the staircase functions in Fig. 2.6. Both methods provide equal numbers of positive and negative quantization levels. A midtread quantizer (Fig. 2.6A) places one quantization level at zero (yielding an odd number of steps or 2^n–1 where n is the number of bits); this architecture is generally preferred in many converters. A midrise quantizer (Fig. 2.6B), with an even number of steps $2n$, does not have a quantization level at zero. A/D converter architecture is described in Chap. 3.

Quantization error is the difference between the actual analog value at sample time and the selected quantization interval value. At sample time, the amplitude value is rounded to the nearest quantization interval, as shown in Fig. 2.7. At best (sample points 11 and 12 in

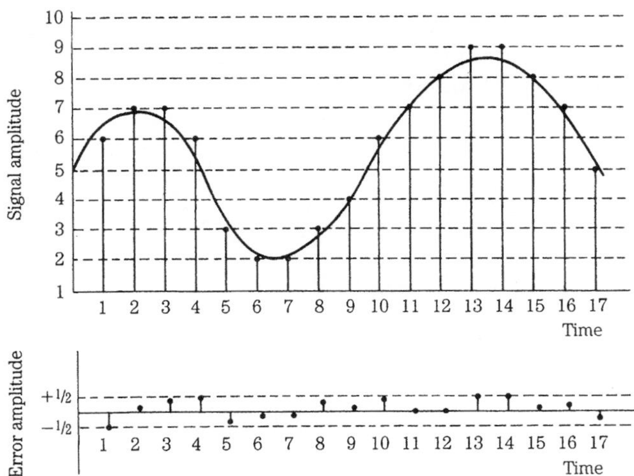

FIGURE 2.7 The amplitude value is rounded to the nearest quantization step. Quantization error at sample times is less than or equal to 1/2 LSB.

the figure), the waveform coincides with quantization intervals. At worst (sample point 1 in the figure), the waveform is exactly between two intervals. Quantization error is thus limited to a range between $+Q/2$ and $-Q/2$, where Q is one quantization interval (or 1 LSB). Note that this selection process, of one level or another, is the basic mechanism of quantization, and occurs for all samples in a digital system. Moreover, the magnitude of the error is always less than or equal to 1/2 LSB. This error results in distortion that is present for an audio signal of any amplitude. When the signal is large, the distortion is proportionally small and likely masked. However, when the signal is small, the distortion is proportionally large and might be audible.

In characterizing digital hardware performance, we can determine the ratio of the maximum expressible signal amplitude to the maximum quantization error; this determines the signal-to-error (S/E) ratio of the system. The S/E ratio of a digital system is similar, but not identical to the signal-to-noise (S/N) ratio of an analog system. The S/E relationship can be derived using a ratio of S/E voltage levels.

Consider a quantization system in which n is the number of bits, and N is the number of quantization steps. As noted:

$$N = 2^n$$

Half of these 2^n values are used to code each part of the bipolar waveform. If Q is the quantizing interval, the peak values of the maximum signal levels are $\pm Q2^{n-1}$. Assuming a sinusoidal input signal, the maximum root mean square (rms) signal S_{rms} is:

$$S_{rms} = \frac{Q2^{n-1}}{(2)^{1/2}}$$

The energy of the quantization error can also be determined. When the input signal has high amplitude and wide spectrum, the quantization error is statistically independent and uniformly distributed between the $+Q/2$ and $-Q/2$ limits, and zero elsewhere, where Q is one quantization interval. This dictates a uniform probability density function with amplitude of $1/Q$; the error is random from sample to sample, and the error spectrum is flat. Ignoring error outside the signal band, the rms quantization error E_{rms} can be found by summing (integrating) the product of the error and its probability:

$$\begin{aligned} E_{rms} &= \left[\int_{-\infty}^{+\infty} e^2 p(e)\, de\right]^{1/2} \\ &= \left[\frac{1}{Q}\int_{-Q/2}^{+Q/2} e^2\, de\right]^{1/2} \\ &= \left[\frac{Q^2}{12}\right]^{1/2} \\ &= \frac{Q}{(12)^{1/2}} \end{aligned}$$

The power ratio determining the signal to quantization error is:

$$S/E = \left[\frac{S_{\text{rms}}}{E_{\text{rms}}}\right]^2$$

$$= \frac{\left[\dfrac{Q2^{n-1}}{(2)^{1/2}}\right]^2}{\left[\dfrac{Q}{(12)^{1/2}}\right]^2}$$

$$= \frac{3}{2}(2^{2n})$$

Expressing this ratio in decibels:

$$S/E\ \text{dB} = 10\ \log\left[\frac{3}{2}(2^{2n})\right]$$

$$= 20\ \log\left[\left(\frac{3}{2}\right)^{1/2}(2^n)\right]$$

$$= 6.02n + 1.76\ \text{dB}$$

Using this approximation, we observe that each additional bit increases the *S/E* ratio (that is, reduces the quantization error) by about 6 dB, or a factor of two. For example, 16-bit quantization ideally yields an *S/E* ratio of about 98 dB, but 15-bit quantization is inferior at 92 dB. Looked at in another way, when the word length is increased by one bit, the number of quantization intervals is doubled. As a result, the distance between quantization intervals is halved, so the amplitude of the quantization error is also halved. Longer word lengths increase the data signal bandwidth required to convey the signal. However, the signal-to-quantization noise power ratio increases exponentially with data signal bandwidth. This is an efficient relationship that approaches the theoretical maximum, and it is a hallmark of coded systems such as pulse-code modulation (PCM) described in Chap. 3. The value of 1.76 is based on the statistics (peak-to-rms ratio) of a full-scale sine wave of peak amplitude; it will differ if the signal's peak-to-rms ratio is different from that of a sinusoid.

It also is important to note that this result assumes that the quantization error is uniformly distributed, and quantization is accurate enough to prevent signal correlation in the error waveform. This is generally true for high-amplitude complex audio signals where the complex distortion components are uncorrelated, spread across the audible range, and perceived as white noise. However, this is not the case for low-amplitude signals, where distortion products can appear.

Quantization Error

Analysis of the quantization error of low-amplitude signals reveals that the spectrum is a function of the input signal. The error is not noise-like (as with high-amplitude signals);

it is correlated. At the system output, when the quantized sample values reconstruct the analog waveform, the in-band components of the error are contained in the output signal. Because quantization error is a function of the original signal, it cannot be described as noise; rather, it must be classified as distortion.

As noted, when quantization error is random from sample to sample, the rms quantization error $E_{rms} = Q/(12)^{1/2}$. This equation demonstrates that the magnitude of the error is independent of the amplitude of the input signal, but depends on the size of the quantization interval; the greater the number of intervals, the lower the distortion. However, the relevant number of intervals is not only the number of intervals in the quantizer, but also the number of intervals used to quantize a particular signal. A maximum peak-to-peak signal (as used in the preceding analysis) presents the best case scenario because all the quantization intervals are exercised. However, as the signal level decreases, fewer and fewer levels are exercised, as shown in Fig. 2.8. For example, given a 16-bit quantizer, a half-amplitude signal would be mapped into half of the intervals. Instead of 65,536 levels, it would see 32,768 intervals. In other words, it would be quantized with 15-bit resolution.

The problem increases as the signal level decreases. A very low-level signal, for example, might receive only single-bit quantization or might not be quantized at all. In other words, as the signal level decreases, the percentage of distortion increases. Although the distortion percentage might be extremely small with a high level, 0 dBFS (dB Full Scale) signal, its percentage increases significantly at low-amplitude, for example, –90 dBFS levels. As described in a following section, dither must be used to alleviate the problem.

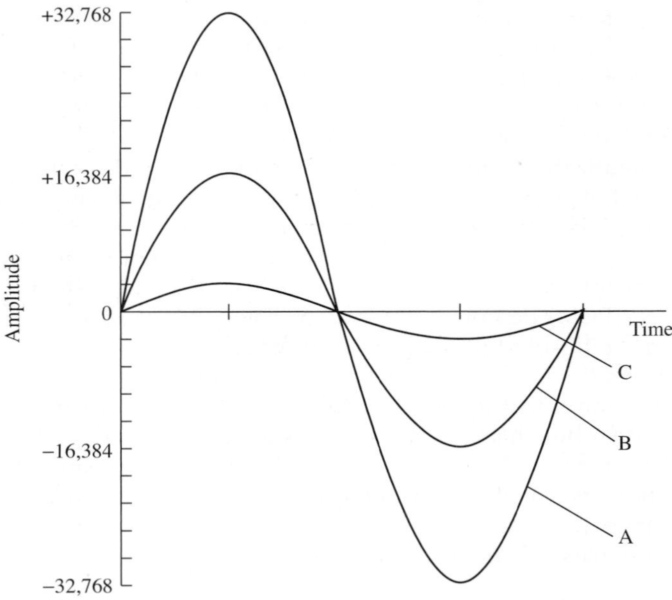

FIGURE 2.8 The percentage of quantization error increases as the signal level decreases. Full-scale waveform A has relatively low error (16-bit resolution). Half-scale waveform B has higher error (effectively 15-bit resolution). Low-amplitude waveform C has very high error.

The error floor of a digital audio system differs from the noise floor of an analog system because in a digital system the error is a function of the signal. The nature of quantization error varies with the amplitude and nature of the audio signal. For broadband, high-amplitude input signals (such as are typically found in music), the quantization error is perceived similarly to white noise. A high-level complex signal may show a pattern from sample to sample; however, its quantization error signal shows no pattern from sample to sample. The quantization error is thus independent of the signal (and thus assumes the characteristics of noise) when the signal is high level and complex. The only difference between this error noise and analog noise is that the range of values is more limited for a constant rms value. In other words, all values are equally likely to fall between positive and negative peaks. On the other hand, analog noise is Gaussian-distributed, so its peaks are higher than its rms value.

However, the perceptual qualities of the error are less benign for low-amplitude signals and high-level signals of very narrow bandwidth. This is based on the fact that white noise is perceptually benign because successive values of the signal are random, whereas predictable noise signals are more readily perceived. For broadband high-level signals, the statistical correlation between successive samples is very low; however, it increases for broadband low-level signals and narrow bandwidth, high-level signals. As the statistical correlation between samples increases, error initially perceived as benign white noise become more complex, yielding harmonic and intermodulation distortion as well as signal-dependent modulation of the noise floor.

Quantization distortion can take many guises. For example, the quantized signal might contain components above the Nyquist frequency; thus, aliasing might occur. The components appear after the sampler, but are effectively sampled. The effects of sampling the output of a limiter or limiting the output of a sampler are indistinguishable. If the signal is high level or complex, the alias components will add to the other complex, noise-like errors. If the input signal is low level and simple, the aliased components might be quite audible. Consider a system with sampling frequency of 48 kHz, bandlimited to 24 kHz. When a 5-kHz sine wave of amplitude of one quantizing step is applied, it is quantized as a sampled 5-kHz square wave. Harmonics of the square wave appear at 15, 25, and 35 kHz. The latter two alias back to 23 and 13 kHz, respectively. Other harmonics and aliases appear as well.

The aliasing caused by quantization can create an effect called granulation noise, so called because of its gritty sound quality. With high-level signals, the noise is masked by the signal itself. However, with low-level signals, the noise is audible. This blend of gritty, modulating noise and distortion has no analog counterpart and is audibly unpleasant. Furthermore, if the alias components are near a multiple of the sampling frequency, beat tones can be created, producing an odd sound called "bird singing" or "birdies." A decaying tone presents a waveform descending through quantization levels; the error is perceptually changed from white noise to discrete distortion components. The problem is aggravated because even complex musical tones become more sinusoidal as they decay. Moreover, the decaying tone will tend to amplitude-modulate the distortion components. Dither addresses these quantization problems.

Other Architectures

Quantization is more than just word length; it also is a question of hardware architecture. There are many techniques for assigning quantization levels to analog signals. For example, a quantizer can use either a linear or nonlinear distribution of quantization intervals along the amplitude scale. One alternative is delta modulation, in which a one-bit quantizer is used to encode amplitude, using the single bit as a sign bit. In other cases, oversampling and noise shaping can be used to shift quantization error out of the audio band. Those algorithm decisions influence the efficiency of the quantizing bits, as well as the relative audibility of the error. For example, as noted, a linear quantizer produces a relatively high error with low-level signals that span only a few intervals. A nonlinear system using a floating point converter can increase the amplitude of low-level signals to utilize the greatest possible interval span. Although this improves the overall S/E ratio, the noise modulation by-product might be undesirable. Historically, after examining the trade-offs of different quantization systems, manufacturers determined that a fixed, linear quantization scheme is highly suitable for music recording. However, newer low bit-rate coding systems challenge this assumption. Alternative digitization systems are examined in Chap. 4. Low bit-rate coding is examined in Chaps. 10 and 11.

Dither

With large-amplitude complex signals, there is little correlation between the signal and the quantization error; thus, the error is random and perceptually similar to analog white noise. With low-level signals, the character of the error changes as it becomes correlated to the signal, and potentially audible distortion results. A digitization system must suppress any audible qualities of its quantization error. Obviously, the number of bits in the quantizing word can be increased, resulting in a decrease in error amplitude of 6 dB per additional bit. This is uneconomical because relatively many bits are needed to satisfactorily reduce the audibility of quantization error. Moreover, the error will always be relatively significant with low-level signals.

Dither is a far more efficient technique. Dither is a small amount of noise that is uncorrelated with the audio signal. Dither is added to the audio signal prior to sampling. This linearizes the quantization process. With dither, the audio signal is made to shift with respect to quantization levels. Instead of periodically occurring quantization patterns in consecutive waveforms, each cycle is different. Quantization error is thus decorrelated from the signal and the effects of the quantization error are randomized to the point of elimination. However, although it greatly reduces distortion, dither adds some noise to the output audio signal. When properly dithered, the number of bits in a quantizer determines the signal's noise floor, but does not limit its low-level detail. For example, signals at –120 dBFS can be heard and measured in a dithered 16-bit recording.

Dither does not mask quantization error; rather, it allows the digital system to encode amplitudes smaller than the least significant bit, in much the same way that an analog system can retain signals below its noise floor. A properly dithered digital system far exceeds the signal to noise performance of an analog system. On the other hand, an undithered digital system can be inferior to an analog system, particularly with low-level signals. High-quality digitization demands dithering at the A/D converter. In addition, digital computations should be digitally dithered to alleviate requantization effects.

Consider the case of an audio signal with amplitude of two quantization intervals, as shown in Fig. 2.9A. Quantization yields a coarsely quantized waveform, as shown in Fig. 2.9B. This demonstrates that quantization ultimately acts as a hard limiter; in

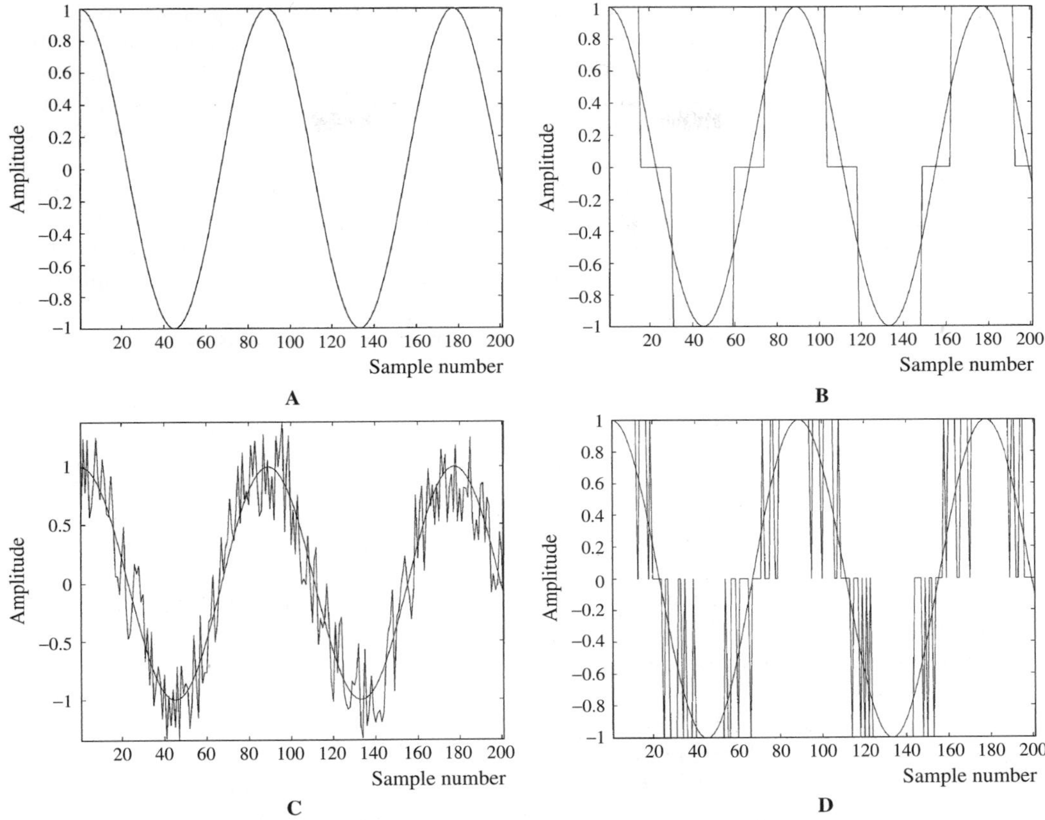

FIGURE 2.9 Dither is used to alleviate the effects of quantization error. A. An undithered input sine wave signal with amplitude of two LSBs. B. Quantization results in a coarse coding over three levels. C. Dither is added to the input sine wave signal. D. Quantization yields a PWM waveform that codes information below the LSB.

other words, severe distortion takes place. The effect is quite different when dither is added to the audio signal. Figure 2.9C shows a dither signal with amplitude of one quantization interval added to the input audio signal. Quantization yields a pulse signal that preserves the information of the audio signal, shown in Fig. 2.9D. The quantized signal switches up and down as the dithered input varies, tracking the average value of the input signal.

Low-level information is encoded in the varying width of the quantized signal pulses. This encoding is known as pulse-width modulation, and it accurately preserves the input signal waveform. The average value of the quantized signal moves continuously between two levels, alleviating the effects of quantization error. Similarly, analog noise would be coded as a binary noise signal; values of 0 and 1 would appear in the LSB in each sampling period, with the signal retaining its white spectrum. The perceptual result is the original signal with added noise—a more desirable result than a quantized square wave.

Mathematically, with dither, quantization error is no longer a deterministic function of the input signal, but rather becomes a zero-mean random variable. In other words, rather than quantizing only the input signal, the dither noise and signal are quantized together, and this randomizes the error, thus linearizing the quantization process. This particular technique is known as nonsubtractive dither because the dither signal is permanently added to the audio signal; the total error is not statistically independent of the audio signal, and errors are not independent sample to sample. However, nonsubtractive dither techniques do manipulate the statistical properties of the quantizer, statistically rendering conditional moments of the total error independent of the input, effectively decorrelating the quantization error of the samples from the signal, and from each other. The power spectrum of the total error signal can be made white. Subtractive dithering, in which the dither signal is removed after requantization, theoretically provides total error statistical independence, but is more difficult to implement.

John Vanderkooy and Stanley Lipshitz demonstrated the remarkable benefit of dither with a 1-kHz sine wave with a peak-to-peak amplitude of about 1 LSB, as shown in Fig. 2.10. Without dither, a square wave is output (Fig. 2.10A). When wideband Gaussian dither with an rms amplitude of about 1/3 LSB is added to the original signal before quantization, a pulse-width-modulated (PWM) waveform results (Fig. 2.10B). The encoded sine wave is revealed when the PWM waveform is averaged 32 times (Fig. 2.10C) and 960 times (Fig. 2.10D). The averaging illustrates how the ear responds

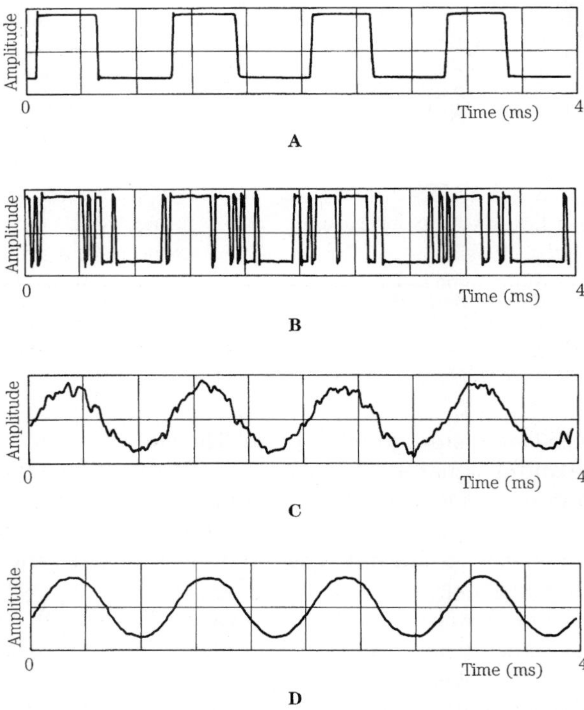

Figure 2.10 Dither permits encoding of information below the least significant bit. A. Quantizing a 1-kHz sine wave with peak-to-peak amplitude of 1 LSB without dither produces a square wave. B. Dither of 1/3 LSB rms amplitude is added to the sine wave before quantization, resulting in PWM modulation. C. Modulation conveys the encoded sine wave information, as can be seen after 32 averagings. D. The encoded sine wave information is more apparent after 960 averagings. (*Vanderkooy and Lipshitz, 1984*)

in its perception of acoustic signals; that is, the ear is a lowpass filter that averages signals. In this case, a noisy sine wave is heard, rather than a square wave.

The ear is quite good at resolving narrow-band signals below the noise floor because of the averaging properties of the basilar membrane. The ear behaves as a one-third octave filter with a narrow bandwidth; the quantization error, which is given a white noise character by dither, is averaged by the ear, and the original narrow-band sine-wave is heard without distortion. In other words, dither changes the digital nature of the quantization error into white noise, and the ear can then resolve signals with levels well below one quantization level.

This conclusion is an important one. With dither, the resolution of a digitization system is far below the least significant bit; theoretically, there is no limit to the low-level resolution. By encoding the audio signal with dither to modulate the quantized signal, that information can be recovered, even though it is smaller than the smallest quantization interval. Furthermore, dither can eliminate distortion caused by quantization by reducing those artifacts to white noise. Proof of this is shown in Fig. 2.11, illustrating a computer simulation performed by John Vanderkooy, Robert Wannamaker, and Stanley Lipshitz. The figure shows a 1-kHz sine wave of 4 LSB peak-to-peak amplitude. The first column shows the signal without dither. The second column shows the same signal with triangular probability density function dither (explained in the following paragraphs) of 2 LSB peak-to-peak amplitude. In both cases, the first row shows the input signal. The second row shows the output signal. The third row shows the total quantization error signal. The fourth row shows the power spectrum of the output signal (this is estimated from sixty 50% overlapping Hann-windowed 512-point records at 44.1 kHz). The undithered output signal (Fig. 2.11D) suffers from harmonic distortion, visible at multiples of the input frequency, as well as inharmonic distortion from aliasing. The error signal (Fig. 2.11G) of the dithered signal shows artifacts of the input signal; thus, it is not statistically independent. Although it clearly does not look like white noise, this error signal sounds like white noise and the output signal sounds like a sine wave with noise. This is supported by the power spectrum (Fig. 2.11H) showing that the signal is free of signal-dependent artifacts, with a white noise floor. The highly correlated truncation distortion of undithered quantization is eliminated. However, we can see that dither increases the noise floor of the output signal.

Types of Dither

There are several types of dither signals, generally differentiated by their probability density function (pdf). Given a random signal with a continuum of possible values, the integral of the probability density function describes the probability of the values over an interval. The probability of where the dither signal falls within an interval defines the area under the function. For example, the dither signal might have equal probability of falling anywhere over an interval, or it might be more likely that the dither signal is in the middle of the interval. An interval, for example, might be 1 or 2 LSBs wide. For audio applications, interest has focused on three dither signals: Gaussian pdf, rectangular (or uniform) pdf, and triangular pdf, as shown in Fig. 2.12. For example, we might speak of a statistically independent, white dither signal with a triangular pdf having a peak-to-peak level or width of 2 LSB. Figure 2.13 shows how triangular pdf dither of 2-LSB peak-to-peak level would be placed in a midrise quantizer.

Dither signals may have a white spectrum. However, for some applications, the spectrum can be shaped by correlating successive dither samples without modifying

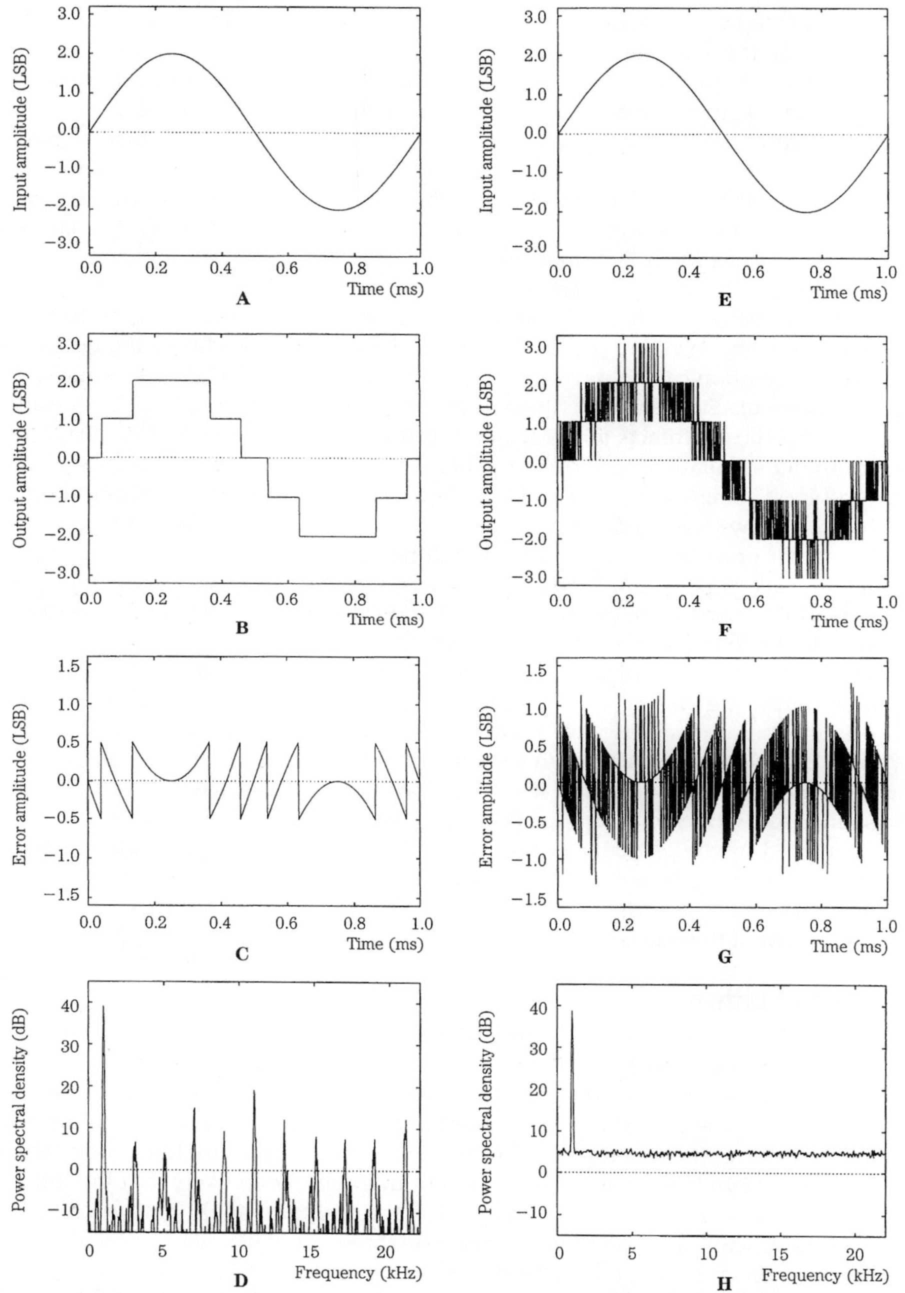

Figure 2.11 Computer-simulated quantization of a low-level 1-kHz sine wave without and with dither. A. Input signal. B. Output signal (no dither). C. Total error signal (no dither). D. Power spectrum of output signal (no dither). E. Input signal. F. Output signal (triangular dither). G. Total error signal (triangular dither). H. Power spectrum of output signal (triangular dither). (*Lipshitz, Wannamaker, and Vanderkooy, 1992*)

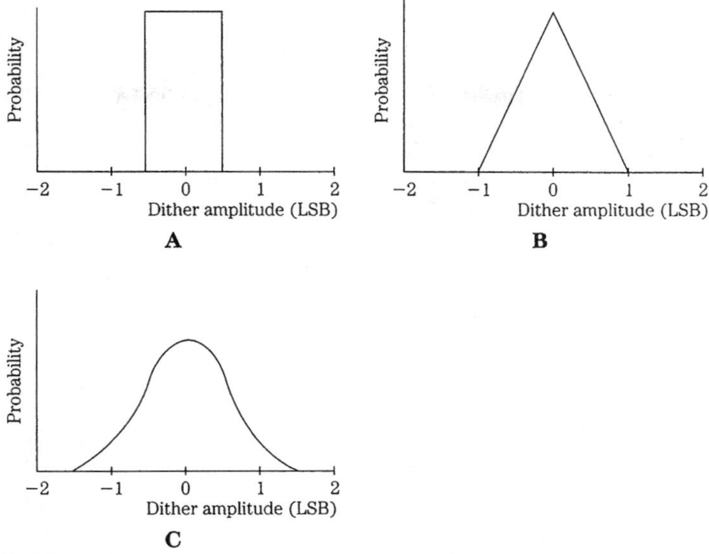

FIGURE 2.12 Probability density functions are used to describe dither signals. A. Rectangular pdf dither. B. Triangular pdf dither. C. Gaussian pdf dither.

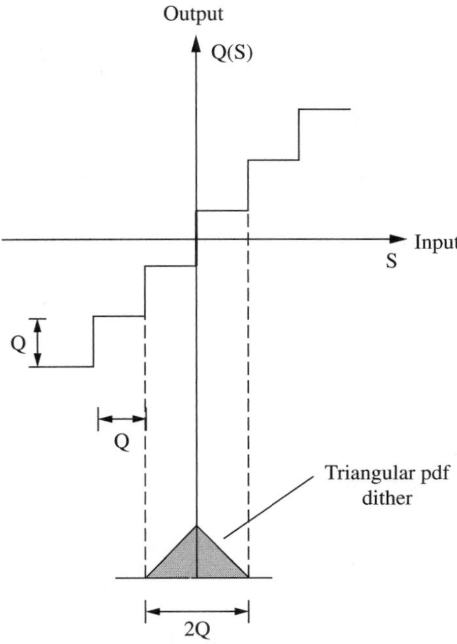

FIGURE 2.13 Triangular pdf dither of 2-LSB peak-to-peak level is placed at the origin of a midrise quantizer.

the pdf; for example, a highpass triangular pdf dither signal could easily be created. By weighting the dither to higher frequencies, the audibility of the noise floor can be reduced.

All three dither types are effective at linearizing the transfer characteristics of quantization, but differ in their results. In almost all applications, triangular pdf dither is preferred. Rectangular and triangular pdf dither signals add less overall noise to the signal, but Gaussian dither is easier to implement in the analog domain.

Gaussian dither is easy to generate with common analog techniques; for example, a diode can be used as a noise source. The dither noise must vary between positive and negative values in each sampling period; its bandwidth must be at least half the sampling frequency. Gaussian dither with an rms value of 1/2 LSB will essentially linearize quantization errors; however, some noise modulation is added to the audio signal. The undithered quantization noise power is $Q^2/12$ (or $Q/(12)^{1/2}$ rms), where Q is 1 LSB. Gaussian dither contributes noise power of $Q^2/4$ so that the combined noise power is $Q^2/3$ (or $Q/(3)^{1/2}$ rms). This increase in noise floor is significant.

Rectangular pdf dither is a uniformly distributed random voltage over an interval. Rectangular pdf dither lying between ±1/2 LSB (that is, a noise signal having a uniform probability density function with a peak-to-peak width that equals 1 LSB) will completely linearize the quantization staircase and eliminate distortion products caused by quantization. However, rectangular pdf does not eliminate noise floor modulation. With rectangular pdf dither, the noise level is more apt to be dependent on the signal, as well as width of the pdf. This noise modulation might be objectionable with very low frequencies or dynamically varied signals. If rectangular pdf dither is used, to be at all effective, its width must be an integer multiple of Q. Rectangular pdf dither of $\pm Q/2$ adds a noise power of $Q^2/12$ to the quantization noise of $Q^2/12$; this yields a combined noise power of $Q^2/6$ (or $Q/(6)^{1/2}$ rms).

It is believed that the optimal nonsubtractive dither signal is a triangular pdf dither of 2 LSB peak-to-peak width, formed by summing (convolving the density functions) of two independent rectangular pdf dither signals each 1 LSB peak-to-peak width. Triangular pdf dither eliminates both distortion and noise floor modulation. The noise floor is constant; however, the noise floor is higher than in rectangular pdf dither. Triangular pdf dither adds a noise power of $Q^2/6$ to the quantization noise power of $Q^2/12$; this yields a combined noise power of $Q^2/4$ (or $Q/2$ rms). The AES17 standard specifies that triangular pdf dither be used when evaluating audio systems. Because all analog signals already contain Gaussian noise that acts as dither, A/D converters do not necessarily use triangular pdf dither. In some converters, Gaussian pdf dither is applied.

Using optimal dither amplitudes, relative to a nondithered signal, rectangular pdf dither increases noise by 3 dB, triangular pdf dither increases noise by 4.77 dB, and Gaussian pdf dither increases noise by 6 dB. In general, rectangular pdf is sometimes used for testing purposes because of its expanded S/E ratio, but triangular pdf is far preferable for most applications including listening purposes, in spite of its slightly higher noise floor. Clearly, Gaussian dither has a noise penalty. Because rectangular and triangular pdf dither are easily generated in the digital domain, they are always preferable to Gaussian dither in requantization applications prior to D/A conversion. When measuring the low-level distortion of digital audio products, it is important to use dithered test signals; otherwise, the measurements might reflect distortion that is an artifact of the test signal and not of the hardware under test. However, a dithered test signal will limit measured noise level and distortion performance. In practical use, analog

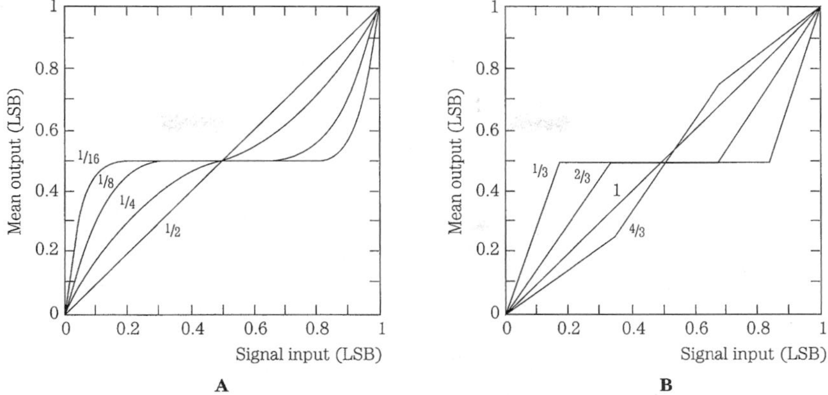

Figure 2.14 Input/output transfer characteristic showing the effects of dither of varying amplitudes. A. Gaussian pdf dither of 1/2 LSB rms linearizes the audio signal. B. Rectangular pdf dither of 1 LSB linearizes the audio signal. (*Vanderkooy and Lipshitz, 1984*)

audio signals contain thermal (Gaussian) noise; even when otherwise theoretically optimal dither is added, nonoptimal results are obtained.

The amplitude of any dither signal is an important concern. Figure 2.14 shows how a quantization step is linearized by adding different amplitudes (width of pdf) of Gaussian pdf and rectangular pdf dither. In both cases, quantization artifacts are decreased as relatively higher amplitudes of dither are added. As noted, a Gaussian pdf signal with an amplitude of 1/2 LSB rms provides a linear characteristic. With rectangular pdf dither, a level of 1 LSB peak-to-peak provides linearity. In either case, too much dither overly decreases the S/N ratio of the digital system with no additional benefit.

The increase in noise yielded by dither is usually negligible given the large S/E ratio inherent in a digital system, and its audibility can be minimized, for example, with a highpass dither signal. This can be easily accomplished with digitally generated dither. For example, the spectrum of a triangular pdf dither can be processed so that its amplitude rises at high audio frequencies. Because the ear is relatively insensitive to high frequencies, this dither signal will be less audible than broadband dither, yet noise modulation and signal distortion are removed. Such techniques can be used to audibly reduce quantization error, for example, when converting a 20-bit signal to a 16-bit signal. More generally, signal processing can be used to psychoacoustically shape the quantization noise floor to reduce its audibility. Noise-shaping applications are discussed in Chap. 18.

Designers have observed that the amplitude of a dither signal can be decreased if a sine wave with a frequency just below the Nyquist frequency, with an amplitude of 1 or 1/2 quantization interval, is added to the audio signal. The added signal must be above audibility yet below the Nyquist frequency to prevent aliasing. It alters the spectrum of quantization error to minimize its audibility and overall does not add as much noise to the signal as broadband dither. For example, discrete triangular pdf dither might yield a 2-dB penalty, as opposed to 4.77 dB. However, a discrete dither frequency might lead to intermodulation products with audio signals. Wideband dither signals alleviate this artifact.

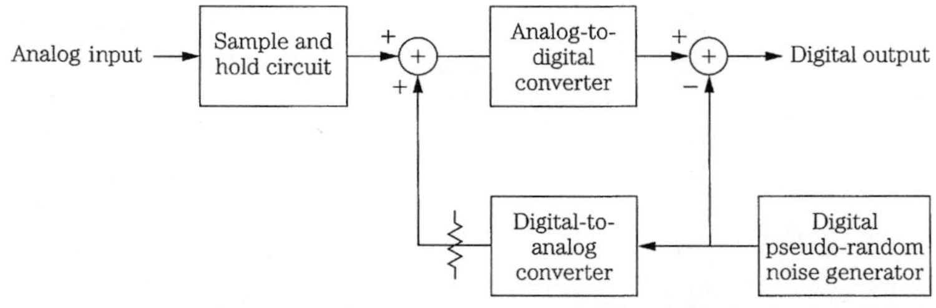

FIGURE 2.15 An example of a subtractive digital dither circuit using a pseudo-random number generator. (*Blesser, 1983*)

An additive dither signal necessarily decreases the S/E ratio of the digitization system. A subtractive dither signal proposed by Barry Blesser that would preserve the S/E ratio is shown in Fig. 2.15. Rectangular noise is a random-valued signal that can be simulated by generating a quickly changing pseudo-random sequence of digital data. This can be accomplished with a series of shift registers and a feedback network comprising EXCLUSIVE OR gates. This sequence is input to a D/A converter to produce analog noise which is added to the audio signal to achieve the benefit of dither. Then, following A/D conversion, the dither is digitally subtracted from the audio signal, preserving the dynamic range of the original signal. A further benefit is that inaccuracies in the A/D converter are similarly randomized. Other additive-subtractive methods call for two synchronized pseudo-random signal generators, one adding rectangular pdf dither at the A/D converter, and the other subtracting it at the D/A converter. Alternatively, in an auto-dither system, the audio signal itself could be randomized to create an added dither at the A/D converter, then re-created at the D/A converter and subtracted from the audio signal to restore the dynamic range.

Digital dither must be used to decrease distortion and artifacts created by round-off error when signal manipulation takes place in the digital domain. For example, the truncation associated with multiplication can cause objectionable error. Digital dither is described in Chap. 17.

For the sake of completeness, and although the account is difficult to verify, one of the early uses of dither came in World War II. Airplane bombers used mechanical computers to perform navigation and bomb trajectory calculations. Curiously, these computers (boxes filled with hundreds of gears and cogs) performed more accurately when flying on board the aircraft, and less well on terra firma. Engineers realized that the vibration from the aircraft reduced the error from sticky moving parts. Instead of moving in short jerks, they moved more continuously. Small vibrating motors were built into the computers, and their vibration was called dither from the Middle English verb "didderen," meaning "to tremble." Today, when you tap a mechanical meter to increase its accuracy, you are applying dither, and dictionaries define dither as "a highly nervous, confused, or agitated state." At any rate, in minute quantities, dither successfully makes a digitization system a little more analog in the good sense of the word.

Summary

Sampling and quantizing are the two fundamental elements of an audio digitization system. The sampling frequency determines signal bandlimiting and thus frequency response. Sampling is based on well-understood principles; the cornerstone of discrete-time sampling yields completely predictable results. Aliasing can occur when the sampling theorem is not observed. Quantization determines the dynamic range of the system, measured by the S/E ratio. Although bandlimited sampling is a lossless process, quantization is one of approximation. Quantization artifacts can severely affect the performance of a system. However, dither can eliminate quantization distortion, and maintain the fidelity of the digitized audio signal. In general, a sampling frequency of 44.1 kHz or 48 kHz and a dithered word length of 16 to 20 bits yields fidelity comparable to or better than the best analog systems, with advantages such as longevity and fidelity of duplication. Still higher sampling frequencies and longer word lengths can yield superlative performance. For example, a sampling frequency of 192 kHz and a word length of 24 bits is available in the Blu-ray disc format.

Postscript

A special note on high sampling frequencies: Before ending our discussion of discrete time sampling, consider a hypothesis concerning the nature of time. Time seems to be continuous. However, physicists have suggested that, just as this book consists of a finite number of atoms, time might come in discrete intervals. Specifically, the indivisible period of time might be 10^{-43} second, known as Planck time. One theory is that no time interval can be shorter than this because the energy required to make the division would be so great that a black hole would be created and the event would be swallowed up inside it. If any of you are experimenting in your basements with very high sampling frequencies, please be careful.

CHAPTER 3

Digital Audio Recording

The hardware design of a digital audio recorder embodies fundamental principles such as sampling and quantizing. The analog signal is sampled and quantized and converted to numerical form prior to storage, transmission, or processing. Subsystems such as dither generator, anti-aliasing filter, sample-and-hold, analog-to-digital converter, and channel code modulator constitute the hardware encoding chain. Although other architectures have been devised, the linear pulse-code modulation (PCM) system is the most illustrative of the nature of audio digitization and is the antecedent of other methods. This chapter and the next, focus on the PCM hardware architecture. Such a system accomplishes the essential pre- and postprocessing for either a digital audio recorder or a real-time digital processor.

The bandwidth of a recording or transmission medium measures the range of frequencies it is able to accommodate with an acceptable amplitude loss. An audio signal sampled at a frequency of 48 kHz and quantized with 16-bit words comprises 48 kHz × 16 bits, or 768 kbps (thousand bits per second). With overhead for data such as synchronization, error correction, and modulation, the channel bit rate might be 1 Mbps (million bits per second) for a monaural audio channel. Clearly, unless bit-rate reduction is applied, considerable throughput capacity is needed for digital audio recording and transmission. It is the task of the digital recording stage to encode the audio signal with sufficient fidelity, while maintaining an acceptable bit rate.

Pulse-Code Modulation

Modulation is a means of encoding information for purposes such as transmission or storage. In theory, many different modulation techniques could be used to digitally encode audio signals. These techniques are fundamentally identical in their task of representing analog signals as digital data, but in practice they differ widely in relative efficiency and performance. Techniques such as amplitude modulation (AM) and frequency modulation (FM) have long been used to modulate carrier frequencies with analog audio information for radio broadcast. Because these are continuous kinds of modulation, they are referred to as wave-parameter modulation.

When conveying sampled information, various types of pulse modulation present themselves. For example, a pulse width or pulse position in time might represent the signal amplitude at sample time; pulse-width modulation (PWM) is an example of the former, and pulse-position modulation (PPM) is an example of the latter. In both cases, the original signal amplitude is coded and conveyed through constant-amplitude pulses. A signal's amplitude can also be conveyed directly by pulses;

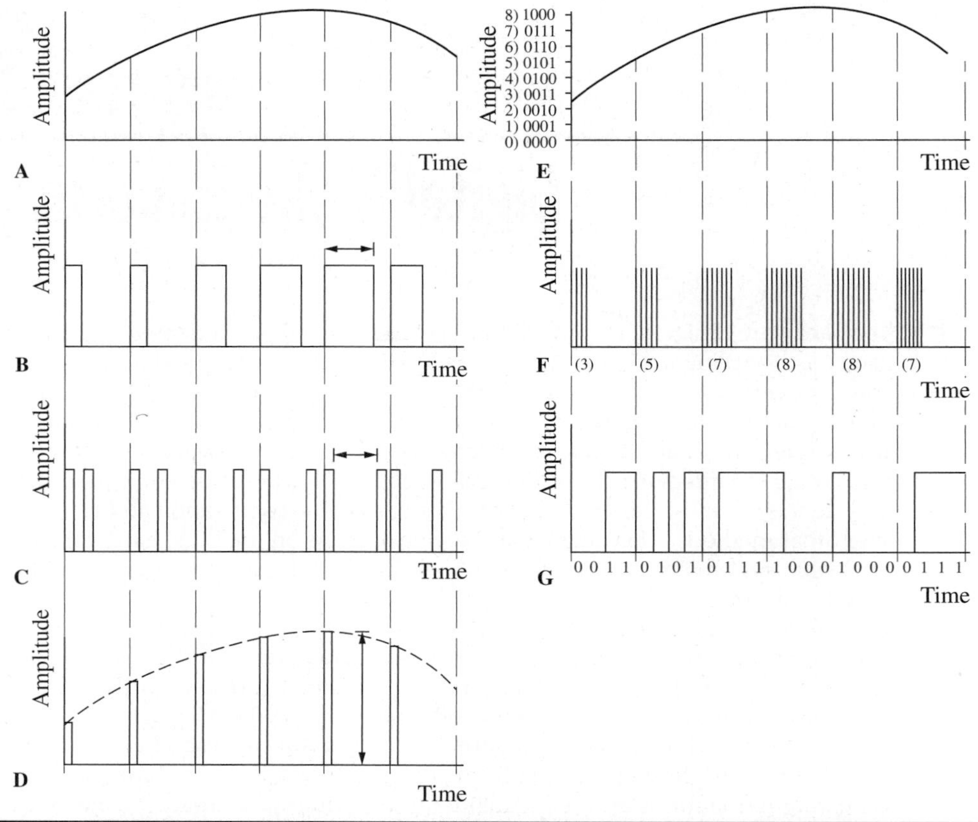

Figure 3.1 PWM, PPM, and PAM are examples of pulse-parameter modulation. PNM and PCM are examples of numerical pulse parameter modulation. A. Analog waveform. B. Pulse-width modulation. C. Pulse-position modulation. D. Pulse-amplitude modulation. E. Quantized analog waveform. F. Pulse-number modulation. G. Pulse-code modulation.

pulse-amplitude modulation (PAM) is an example of this approach. The amplitude of the pulses equals the amplitude of the signal at sample time. PWM, PPM, and PAM are shown in Figs. 3.1A through D. In other cases, sample amplitudes are conveyed through numerical methods. For example, in pulse-number modulation (PNM), the modulator generates a string of pulses; the pulse count represents the amplitude of the signal at sample time; this is shown in Fig. 3.1F. However, for high resolution, a large number of pulses are required. Although PWM, PPM, PAM, and PNM are often used in the context of conversion, they are not suitable for transmission or recording because of error or bandwidth limitations.

The most commonly used modulation method is pulse-code modulation (PCM). PCM was devised in 1937 by Alec Reeves while he was working as an engineer at the International Telephone and Telegraph Company laboratories in France. (Reeves also invented PWM.) In PCM, the input signal undergoes sampling, quantization, and coding. By representing the measured analog amplitude of samples with a pulse code, binary numbers can be used to represent amplitude. At the receiver, the pulse code is used to reconstruct an analog waveform. The binary words that represent sample amplitudes are directly coded into PCM waveforms as shown in Fig. 3.1G.

With methods such as PWM, PPM, and PAM, only one pulse is needed to represent the amplitude value, but in PCM several pulses per sample are required. As a result, PCM might require a channel with higher bandwidth. However, PCM forms a very robust signal in that only the presence or absence of a pulse is necessary to read the signal. In addition, a PCM signal can be regenerated without loss. Therefore the quality of a PCM transmission depends on the quality of the sampling and quantizing processes, not the quality of the channel itself. In addition, depending on the sampling frequency and capacity of the channel, several PCM signals can be combined and simultaneously conveyed with time-division multiplexing. This expedites the use of PCM; for example, stereo audio is easily conveyed. Although other techniques presently exist and newer ones will be devised, they will measure their success against that of pulse-code modulation digitization. In most cases, highly specialized channel codes are used to modulate the signal prior to storage. These channel modulation codes are also described in this chapter.

The architecture of a linear PCM (sometimes called LPCM) system closely follows a readily conceptualized means of designing a digitization system. The analog waveform is filtered and time sampled and its amplitude is quantized with an analog-to-digital (A/D) converter. Binary numbers are represented as a series of modulated code pulses representing waveform amplitudes at sample times. If two channels are sampled, the data can be multiplexed to form one data stream. Data can be manipulated to provide synchronization and error correction, and auxiliary data can be added as well. Upon playback, the data is demodulated, decoded, and error-corrected to recover the original amplitudes at sample times, and the analog waveform is reconstructed with a digital-to-analog (D/A) converter and lowpass filter.

The encoding section of a conventional stereo PCM recorder consists of input amplifiers, a dither generator, input lowpass filters, sample-and-hold circuits, analog-to-digital converters, a multiplexer, digital processing and modulation circuits, and a storage medium such as an optical disc or a hard-disk drive. An encoding section block diagram is shown in Fig. 3.2. This hardware design is a practical realization of the sampling theorem. In practice, other techniques such as oversampling may be employed.

An audio digitization system is really nothing more than a transducer, which processes the audio signal for digital storage or transmission, then processes it again for

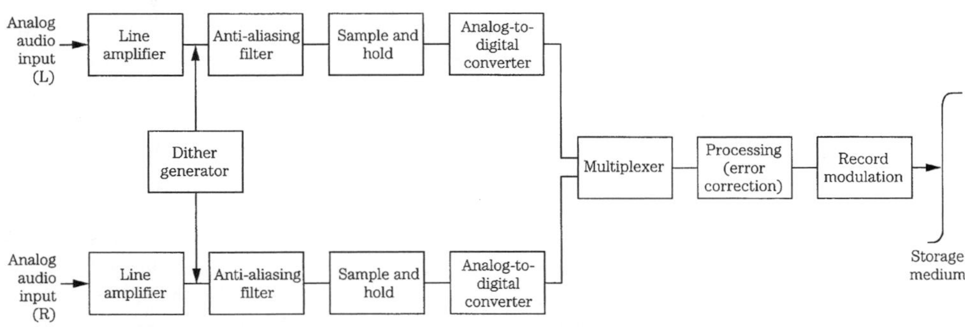

FIGURE 3.2 A linear PCM record section showing principal elements.

reproduction. Although that sounds simple, the hardware must be carefully engineered; the quality of the reproduced audio depends entirely on the system's design. Each subsystem must be carefully considered.

Dither Generator

Dither is a noise signal added to the input audio signal to remove quantization artifacts. As described in Chap. 2, dither causes the audio signal to vary between adjacent quantization levels. This action decorrelates the quantization error from the signal, removes the effects of the error, and encodes signal amplitudes below the amplitude of a quantization increment. However, although it reduces distortion, dither adds noise to the audio signal. Perceptually, dither is beneficial because noise is more readily tolerated by the ear than distortion.

Analog dither, applied prior to A/D conversion, causes the A/D converter to make additional level transitions that preserve low-level signals through duty cycle, or pulse-width modulation. This linearizes the quantization process. Harmonic distortion products, for example, are converted to wideband noise. Several types of dither signals, such as Gaussian, rectangular, and triangular probability density functions can be selected by the designer; in some systems, the user is free to choose a dither signal. The amplitude of the applied dither is also critical. In some cases, the input signal might have a high level of residual noise. For example, an analog preamplifier might have a noise floor sufficient to dither the quantizer. However, the digital system must provide a dynamic range that sufficiently captures all the analog information, including the signal within the analog noise floor, and must not introduce quantization distortion into it. The word length of the quantizer must be sufficient for the audio program, and its least significant bit (LSB) must be appropriately dithered. In addition, whenever the word length is reduced, for example, when a 20-bit master recording is transferred to the 16-bit format, dither must be applied, as well as noise shaping. Dither is discussed more fully in Chap. 2; psychoacoustically optimized noise shaping is described in Chap. 18.

Input Lowpass Filter

An input audio signal might have high-frequency content that is above the Nyquist (half-sampling) frequency. To ensure that the Nyquist theorem is observed, and thus prevent aliasing, digital audio systems must bandlimit the audio input signal, eliminating high-frequency content above the Nyquist frequency. This can be accomplished with an input lowpass filter, sometimes called an anti-aliasing filter. In a system with a sampling frequency of 48 kHz, the ideal filter cutoff frequency would be 24 kHz. The input lowpass filter must attenuate all signals above the half-sampling frequency, yet not affect the lower in-band signals. Thus, an ideal filter is one with a flat passband, an immediate or brick-wall filter characteristic, and an infinitely attenuated stopband, as shown in Fig. 3.3A. In addition to these frequency-response criteria, an ideal filter must not affect the phase linearity of the signal.

Although in practice an ideal filter can be approximated, its realization presents a number of engineering challenges. The filter's passband must have a flat frequency response; in practice some frequency irregularity (ripple) exists, but can be minimized. The stopband attenuation must equal or exceed the system's dynamic range, as determined by the word length. For example, a 16-bit system would require stopband attenuation of

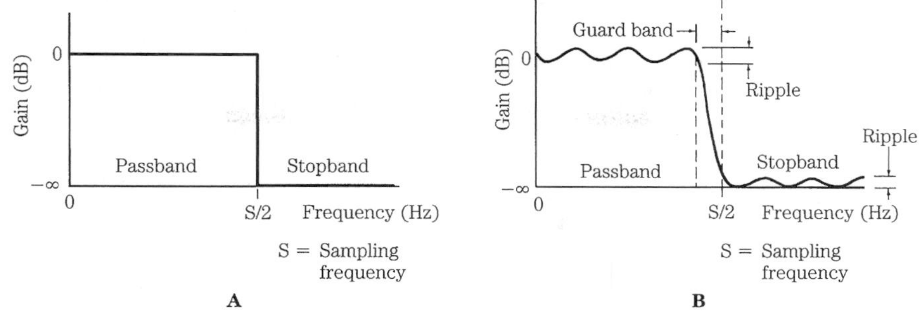

FIGURE 3.3 Lowpass filter characteristics. A. An ideal lowpass filter has a flat passband response and instantaneous cutoff. B. In practice, filters exhibit ripple in the stopband and passband, and sloping cutoff.

over –95 dB; a stopband attenuation of –80 dB would yield 0.01% alias distortion under worst-case conditions. Modern A/D converters use digital filtering and oversampling methods to perform anti-aliasing; this is summarized later in this chapter, and described in detail in Chap. 18. Early systems used only analog input lowpass filters. Because they clearly illustrate the function of anti-aliasing filtering, and because even modern converters still employ low-order analog filters, analog lowpass filters are described below.

In early systems (as opposed to modern systems that use digital filters) the input signal is lowpass filtered by an analog filter with a very sharp cutoff to bandlimit the signal to frequencies at or below the half-sampling frequency. A brick-wall cutoff demands compromise on specifications such as flat passband and low phase distortion. To alleviate the problems such as phase nonlinearities created by a brick-wall response, analog filters can use a more gradual cutoff. However, a low-order filter with a cutoff at the half-sampling frequency would roll off audio frequencies, hence its passband must be extended to a higher frequency. To avoid aliasing, the sampling frequency must be extended to ensure that the filter provides sufficient attenuation at the half-sampling frequency. A higher sampling frequency, perhaps three times higher than required for a brick-wall filter, might be needed; however, this would raise data bandwidth requirements. To limit the sampling frequency and make full use of the passband below the half-sampling point, a brick-wall filter is mandated. When a sampling frequency of 48 kHz is used, the analog input filters are designed for a flat response from dc to 22 kHz. This arrangement provides a guard band of 2 kHz to ensure that attenuation is sufficient at the half-sampling point. A practical lowpass filter characteristic is shown in Fig. 3.3B.

Several important analog filter criteria are overshoot, ringing, and phase linearity. Sharp cutoff filters exhibit resonance near the cutoff frequency and this ringing can cause irregularity in the frequency response. The sharper the cutoff, the greater the propensity to ringing. Certain filter types have inherently reduced ringing. Phase response is also a factor. Lowpass filters exhibit a frequency-dependent delay, called group delay, near the cutoff frequency, causing phase distortion. This can be corrected with an analog circuit preceding or following the filter, which introduces compensating delay to achieve overall phase linearity; this can yield a pure delay, which is inaudible. In the cases of ringing and group delay, there is debate on the threshold of audibility of such effects; it is unclear how perceptive the ear is to such high-frequency phenomena.

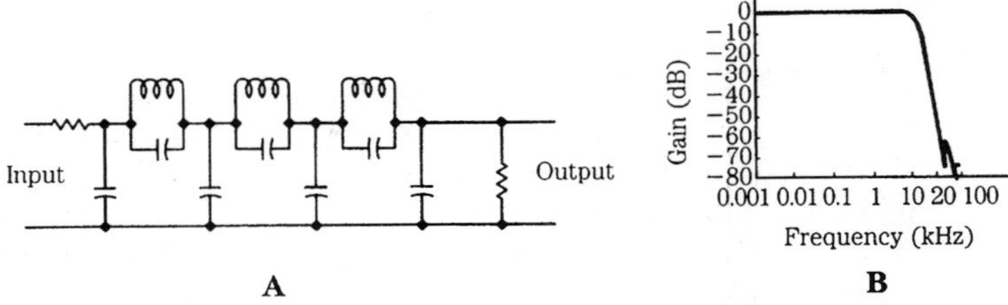

FIGURE 3.4 An example of a Chebyshev lowpass filter and its frequency response. A. A passive lowpass filter schematic. B. Lowpass filter frequency response showing a steep cutoff.

Analog filters can be classified according to the mathematical polynomials that describe their characteristics. There are many filter types; for example, Bessel, Butterworth, and Chebyshev filters are often used. For each of these filter types, a basic design stage can be repeated or cascaded to increase the filter's order and to sharpen the cutoff slope. Thus, higher-order filters more closely approximate a brick-wall frequency response. For example, a passive Chebyshev lowpass filter is shown in Fig. 3.4; its cutoff slope becomes steeper when the filter's order is increased through cascading. However, phase shift also increases as the filter order is increased. The simplest lowpass filter is a cascade of RC (resistor-capacitor) sections; each added section increases the roll-off slope by 6 dB/octave. Although the filter will not suffer from overshoot and ringing, the passband will exhibit frequency response anomalies.

Resonant peaks can be positioned just below the cutoff frequency to smooth the passband response of a filter but not affect the roll-off slope; a Butterworth design accomplishes this. However, a high-order filter is required to obtain a sharp cutoff and deep stopband. For example, a design with a transition band 40% of an octave wide and a stopband of –80 dB would require a 33rd-order Butterworth filter.

A filter with a narrow transition band can be designed at the expense of passband frequency response. This can be achieved by placing the resonant peaks somewhat higher than in a Butterworth design. This is the aim of a Chebyshev filter. A 9th-order Chebyshev filter can achieve a ±0.1-dB passband ripple to 20 kHz, and stopband attenuation of –70 dB at 25 kHz.

One characteristic of most filter types is that attenuation continues past the necessary depth for frequencies beyond the half-sampling frequency. If the attenuation curve is flattened, the transition band can be reduced. Anti-resonant notches in the stopband are often used to perform this function. In addition, reactive elements can be shared in the design, providing resonant peaks and anti-resonant notches. This reduces circuit complexity. The result is called an elliptical, or Cauer filter. An elliptical filter has the steepest cutoff for a given order of realization. For example, a 7th-order elliptical filter can provide a ±0.25-dB passband ripple, 40% octave transition band, and a –80-dB stopband. In practice, a 13-pole design might be required.

In general, for a given analog filter order, Chebyshev and elliptical lowpass filters give a closer approximation to the ideal than Bessel or Butterworth filters, but Chebyshev filters can yield ripple in the passband and elliptical filters can produce severe phase nonlinearities. Bessel filters can approximate a pure delay and provide excellent phase response; however, a higher-order filter is needed to provide a very high rate of attenuation. Butterworth filters are usually flat in the passband, but can exhibit slow transient response. No analog filter is ideal, and there is a trade-off between a high rate of attenuation and an acceptable time-domain response.

In practice, as noted, because of the degradation introduced by analog brick-wall filters, all-analog designs have been superseded by A/D converters that employ a low-order analog filter, oversampling, and digital filtering as discussed later in this chapter and described in detail in Chap. 18. Whatever method is used, an input filter is required to prevent aliasing of any frequency content higher than the Nyquist frequency.

Sample-and-Hold Circuit

As its name implies, a sample-and-hold (S/H) circuit performs two simple yet critical operations. It time-samples the analog waveform at a periodic rate, putting the sampling theorem into practice. It also holds the analog value of the sample while the A/D converter outputs the corresponding digital word. This is important because otherwise the analog value could change after the designated sample time, causing the A/D converter to output incorrect digital words. The input and output responses of an S/H circuit are shown in Fig. 3.5. The output signal is an intermediate signal, a discrete PAM staircase representing the original analog signal, but is not a digital word. The circuit is relatively simple to design; however, it must accomplish both of its tasks accurately. Samples must be captured at precisely the correct time and the held value must stay within tolerance. In practice, the S/H function is built into the A/D converter. The S/H circuit is also known as a track-hold circuit.

As we have seen, time and amplitude information can completely characterize an acoustic waveform. The S/H circuit is responsible for capturing both informational

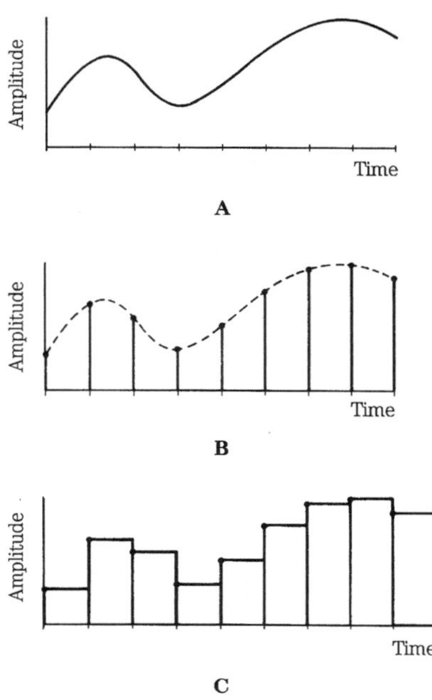

FIGURE 3.5 The sample-and-hold circuit captures an analog value and holds it while A/D conversion occurs. A. Analog input signal. B. Sampled input. C. Analog held output.

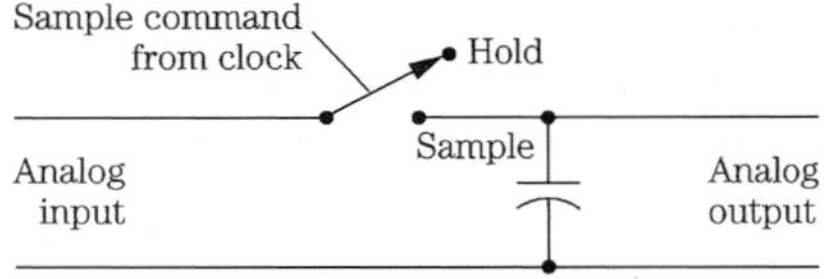

FIGURE 3.6 A conceptual sample-and-hold circuit contains a switch and storage element. The switch is closed to sample the signal.

aspects from the analog waveform. Samples are taken at a periodic rate and reproduced at the same periodic rate. The S/H circuit accomplishes this time sampling. A clock, an oscillator circuit that outputs timing pulses, is set to the desired sampling frequency, and this command signal controls the S/H circuit.

Conceptually, an S/H circuit is a capacitor and a switch. The circuit tracks the analog signal until the sample command causes the digital switch to isolate the capacitor from the signal; the capacitor holds this analog voltage during A/D conversion. A conceptual S/H circuit is shown in Fig. 3.6. The S/H circuit must have a fast acquisition time that approaches zero; otherwise, the value output from the A/D converter will be based on an averaged input over the acquisition time, instead of the correct sample value at an instant in time. In addition, varying sample times result in acquisition timing error; to prevent this, the S/H circuit must be carefully designed and employ a sample command that is accurately clocked.

Jitter is any variation in absolute timing; in this case, variation in the sampling signal is shown in Fig. 3.7. Jitter adds noise and distortion to the sampled signal, and must be limited in the clock used to switch the S/H circuit. Jitter is particularly significant in the case of a high-amplitude, high-frequency input signal. The timing precision required for accurate A/D conversion is considerable. Depending on the converter design, for example, jitter at the S/H circuit must be less than 200 ps (picosecond) to allow 16-bit accuracy from a full amplitude, 20-kHz sine wave, and less than 100 ps for 18-bit accuracy. Only then would the resulting noise components fall below the quantization noise floor. Clearly, S/H timing must be controlled by a clock designed with a highly accurate crystal quartz oscillator. Jitter is discussed in Chap. 4.

Acquisition time is the time between the initiation of the sample command and the taking of the sample. This time lag can result in a sampled value different from the one

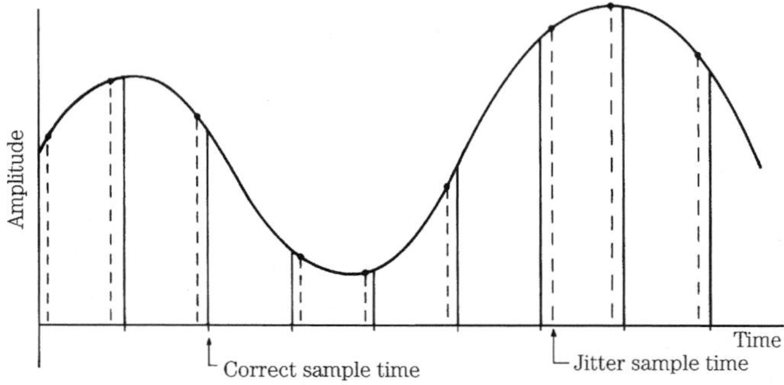

FIGURE 3.7 Jitter is a variation in sample times that can add noise and distortion to the output analog signal.

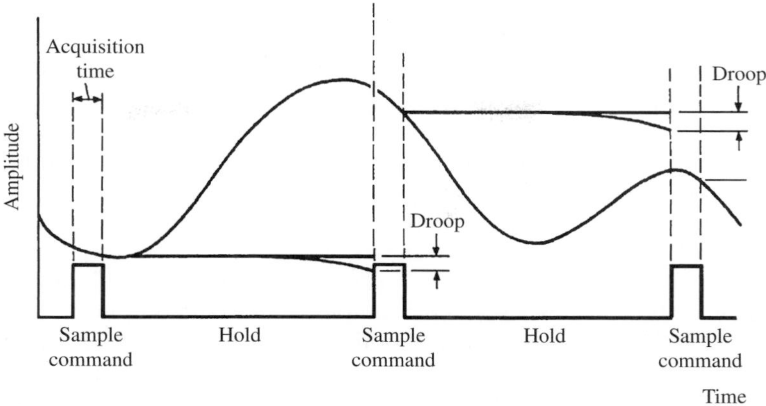

Figure 3.8 Acquisition time and droop are two error conditions in the sample-and-hold circuit.

present at the correct sample time. The effect of the delay is a function of the amplitude of the analog signal. It is therefore important to minimize acquisition time. The S/H circuit's other primary function is to hold the captured analog voltage while conversion takes place. This voltage must remain constant because any variation greater than a quantization increment can result in an error at the A/D output. The held voltage can be prone to droop because of current leakage. Droop is the decrease in hold voltage as the storage capacitor leaks between sample times. Care in circuit design and selection of components can limit droop to less than one-half a quantization increment over a 20-μs period. For example, a 16-bit, ±10-V range A/D converter must hold a constant value to within 1 mV during conversion. Acquisition time error and droop are illustrated in Fig. 3.8.

The demands of fast acquisition time and low droop are in conflict in the design of a practical S/H circuit. For fast acquisition time, a small capacitor value is better, permitting faster charging time. For droop, however, a large-valued capacitor is preferred, because it is better able to retain the sample voltage at a constant level for a longer time. However, capacitor values of approximately 1 nF can satisfy both requirements. In addition, high-quality capacitors made of polypropylene or Teflon dielectrics can be specified. These materials can respond quickly, hold charge, and minimize dielectric absorption and hysteresis—phenomena that cause voltage variations.

In practice, an S/H circuit must contain more than a switch and a capacitor. Active circuits such as operational amplifiers must buffer the circuit to condition the input and output signals, speed switching time, and help prevent leakage. Only a few specialized operational amplifiers meet the required specifications of large bandwidth and fast settling time. Junction field-effect transistor (JFET) operational amplifiers usually perform best. Thus, a complete S/H circuit might have a JFET input operational amplifier to prevent source loading, improve switching time, isolate the capacitor, and supply capacitor-charging current. The S/H switch itself may be a JFET device, selected to operate cleanly and accurately with minimal jitter, and the capacitor may exhibit low hysteresis. A JFET operational amplifier is usually placed at the output to help preserve the capacitor's charge.

Analog-to-Digital Converter

The analog-to-digital (A/D) converter lies at the heart of the encoding side of a digital audio system, and is perhaps the single most critical component in the entire signal chain. Its counterpart, the digital-to-analog (D/A) converter, can subsequently be improved for higher fidelity playback. However, errors introduced by the A/D converter will accompany the audio signal throughout digital processing and storage and, ultimately, back into its analog state. Thus the choice of the A/D converter irrevocably affects the fidelity of the resulting signal.

Essentially, the A/D converter must examine the sampled input signal, determine the quantization level nearest to the sample's value, and output the binary code that is assigned to that level—accomplishing those tasks in one sampling period (20 μs for a 48-kHz sampling frequency). The precision required is considerable: 15 parts per million for 16-bit resolution, 4 parts per million for 18-bit resolution, and 1 part per million for 20-bit resolution. In a traditional A/D design, the input analog voltage is compared to a variable reference voltage within a feedback loop to determine the output digital word; this is known as successive approximation. More common oversampling A/D converters are summarized in this chapter and discussed in detail in Chap. 18.

The A/D converter must perform a complete conversion on each audio sample. Furthermore, the digital word it provides must be an accurate representation of the input voltage. In a 16-bit successive approximation converter, each of the 65,536 intervals must be evenly spaced throughout the amplitude range, so that even the least significant bit in the resulting word is meaningful. Thus, speed and accuracy are key requirements for an A/D converter. Of course, any A/D converter will have an error of ±1/2 LSB, an inherent limitation of the quantization process itself. Furthermore, dither must be applied.

The conversion time is the time required for an A/D converter to output a digital word; it must be less than one sampling period. Achieving accurate conversion from sample to sample is sometimes difficult because of settling time or propagation time errors. The result of accomplishing one conversion might influence the next. If a converter's input moves from voltage *A* to *B* and then later from *C* to *B*, the resulting digital output for *B* might be different because of the device's inability to properly settle in preparation for the next measurement. Obviously, dynamic errors grow more severe with demand for higher conversion speed. In practice, speeds required for low noise and distortion can be achieved. Indeed, many A/D converters simultaneously process two waveforms, alternating between left and right channels. Other converters can process 5.1 channels of input audio signals.

Numerous specifications have been devised to evaluate the performance accuracy of A/D converters. Amplitude linearity compares output versus input linearity. Ideally, the output value should always correspond exactly with input level, regardless of level. To perform the test, a series of tones of decreasing amplitude, or a fade-to-zero tone, is input to the converter. The tone is dithered with rectangular pdf dither. A plot of device gain versus input level will reveal any deviations from a theoretically flat (linear) response.

Integral linearity measures the "straightness" of the A/D converter output. It describes the transition voltage points, the analog input voltages at which the digital output changes from one code to the next, and specifies how close they are to a straight line drawn through them. In other words, integral linearity determines the deviation of an actual bit transition from the ideal transition value, at any level over the range of the converter. Integral linearity is illustrated in Fig. 3.9A. Integral linearity is tested, and the

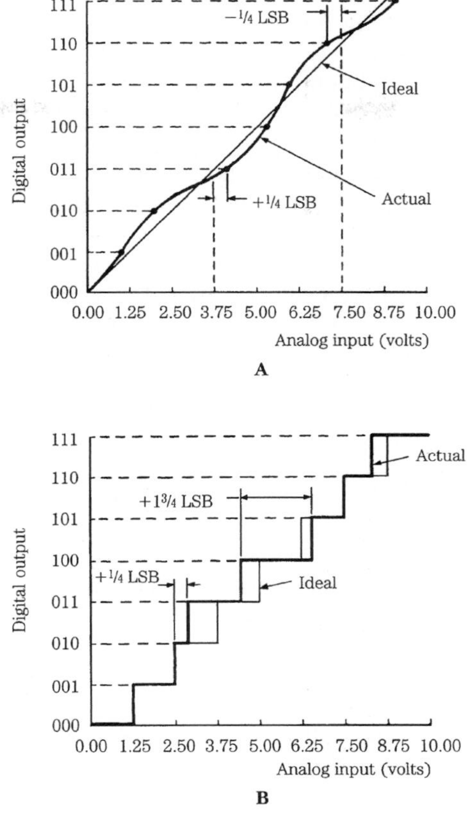

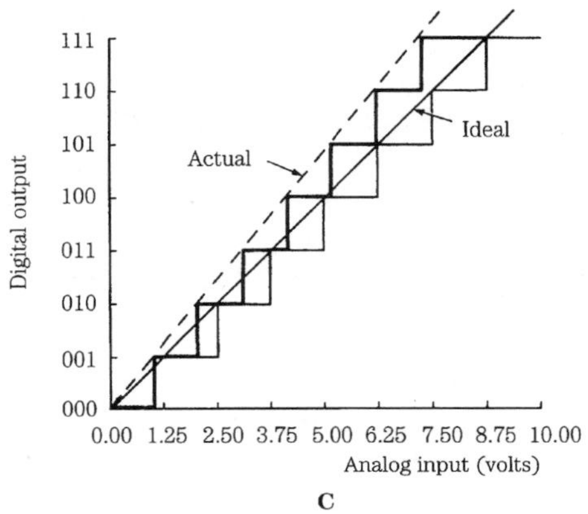

FIGURE 3.9 Performance of an A/D converter can be specified in a variety of ways. A. Integral linearity specification of an A/D converter. B. Differential linearity specification of an A/D converter. C. Absolute accuracy specification of an A/D converter.

reference line is drawn across the converter's full output range. Integral linearity is the most important A/D specification and is not adjustable. An n-bit converter is not a true n-bit converter unless it guarantees at least ±1/2 LSB integral linearity. The converter in Fig. 3.9A has a ±1/4 LSB integral linearity.

Differential linearity error is the difference between the actual step height and the ideal value of 1 LSB. It can be measured as the distance between transition voltages, that is, the widths of individual input voltage steps. Differential linearity is shown in Fig. 3.9B. Ideally, all the steps of an A/D transfer function should be 1 LSB wide. A maximum differential linearity error of ±1/2 LSB means that the input voltage might have to increase or decrease as little as 1/2 LSB or as much as 1 1/2 LSB before an output transition occurs. If this specification is exceeded, to perhaps ±1 LSB, some steps could be 2 LSBs wide and others could be 0 LSB wide; in other words, some output codes would not exist. High-quality A/D converters are assured of having no missing codes over a specified temperature range. The converter in Fig. 3.9B has an error of ±1/2 LSB; some levels are 1/4 LSB wide, others are 1 1/2 LSB wide. Conversion speed can affect both integral linearity and differential linearity errors. Quality A/D converters are guaranteed to be monotonic; that is, the output code either increases or remains the same for increasing analog input signals. If differential error is greater than 1 LSB, the converter will be nonmonotonic.

Absolute accuracy error, shown in Fig. 3.9C, is the difference between the ideal level at which a digital transition occurs and where it actually occurs. A good A/D converter should have an error of less than ±1/2 LSB. Offset error, gain error, or noise error can affect this specification. For the converter in Fig. 3.9C, each interval is 1/8 LSB in error. In practice, otherwise good successive approximation A/D converters can sometimes drift with temperature variations and thus introduce inaccuracies.

Code width, sometimes called quantum, is the range of analog input values for which a given output code will occur. The ideal code width is 1 LSB. A/D converters can exhibit an offset error as well as a gain error. An A/D converter connected for unipolar operation has an analog input range from 0 V to positive full scale. The first output code transition should occur at an analog input value of 1/2 LSB above 0 V. Unipolar offset error is the deviation of the actual transition value from the ideal value. When connected in a bipolar configuration, bipolar offset is set at the first transition value above the negative full-scale value. Bipolar offset error is the deviation of the actual transition value from the ideal transition value at 1/2 LSB above the negative full-scale value. Gain error is the deviation of the actual analog value at the last transition point from the ideal value, where the last output code transition occurs for an analog input value 1 1/2 LSB below the nominal positive full-scale value. In some converters, gain and offset errors can be trimmed at the factory, and might be further zeroed with the use of external potentiometers. Multiturn potentiometers are recommended for minimum drift over temperature and time.

Harmonic distortion is a familiar way to characterize audio linearity and can be used to evaluate A/D converter performance. A single pure sine tone is input to the device under test, and the output is examined for spurious content other than the sine tone. In particular, spectral analysis will show any harmonic multiples of the input frequency. Total harmonic distortion (THD) is the ratio of the summed root mean square (rms) voltage of the harmonics to that of the input signal. To further account for noise in the output, the measurement is often called THD+N. The figure is usually expressed as a decibel figure or a percentage; however, visual examination of the displayed spectral output is a valuable diagnostic. It is worth noting that in most analog systems, THD+N decreases as the signal level decreases. The opposite is true in digital systems. Therefore, THD+N should be specified at both high and low signal levels. THD+N should be evaluated versus amplitude and versus frequency, using FFT analysis.

Dynamic range can also be used to evaluate converter performance. Dynamic range is the amplitude range between a maximum-level signal and the noise floor. Using the EIAJ specification, dynamic range is typically measured by reading THD+N at an input amplitude of –60 dB; the negative value is inverted and added to 60 dB to obtain dynamic range. Also, signal-to-noise ratio (examining idle channel noise) can be measured by subtracting the idle noise from the full-scale signal. For consistency, a standard test sequence such as the ITU CCITT 0.33.00 (monaural) and CCITT 0.33.01 (stereo) can be used; these comprise a series of tones and are useful for measuring parameters such as frequency response, distortion, and signal to noise. Noise modulation is another useful measurement. This test measures changes in the noise floor relative to changes in signal amplitude; ideally, there should be no correlation. In practice, because of low-level nonlinearity in the converter, there may be audible shifts in the level or tonality of the background noise that correspond to changes in the music signal. Precisely because the shifts are correlated to the music, they are potentially much more perceptible than benign unchanging noise. In one method used to observe noise modulation, a low-frequency sine tone is input to the converter; the sine tone is removed at the output and the spectrum of the output signal is examined in 1/3-octave bands. The level of the input signal is decreased in 5-dB steps and the test is repeated. Deviation in the noise floor by more than a decibel in any band across the series of tested amplitudes may indicate potentially audible noise modulation.

As noted, an A/D converter is susceptible to jitter, a variation in the timebase of the clocking signal. Random-noise jitter can raise the noise floor and periodic jitter can create sidebands, thus raising distortion levels. Generally, the higher the specified dynamic range of the converter, the lower the jitter level. A simple way to test an A/D converter for jitter limitations is to input a 20-kHz, 0-dBFS (full-amplitude) sine tone, and observe an FFT of the output signal. Repeat with a 100-Hz sine tone. An elevated noise floor at 20 kHz compared to 100 Hz indicates a potential problem from random-noise jitter, and discrete frequencies at 20 kHz indicate periodic jitter. High-quality A/D converters contain internal clocks that are extremely stable, or when accepting external clocks, have clock recovery circuitry to reject jitter disturbance. It is incorrect to assume that one converter using a low-jitter clock will necessarily perform better than another converter using a high-jitter clock; actual performance depends very much on converter design. Even when jitter causes no data error, it can cause sonic degradation. Its effect must be carefully assessed in measurements and listening tests. Jitter is discussed in more detail in Chap. 4.

The maximum analog signal level input to an A/D converter should be scaled as close as possible to the maximum input conversion range, to utilize the converter's maximum signal resolution. Generally, a converter can be driven by a very low impedance source such as the output of a wideband, fast-settling operational amplifier. Transitions in a successive approximation A/D converter's input current might be caused by changes in the output current of the internal D/A converter as it tests bits. The output voltage of the driving source must remain constant while supplying these fast current changes.

Changes in the dc power supply can affect an A/D converter's accuracy. Power supply deviations can cause changes in the positive full-scale value, resulting in a proportional change in all code transition values, that is, a gain error. Normally, regulated power supplies with 1% or less ripple are recommended. Power supplies should be bypassed with a capacitor—for example, 1 to 10 μF tantalum—located near the converter, to obtain noise-free operation. Noise and spikes from a switching power supply must be carefully filtered. To minimize jitter effects, accurate crystal clocks must be used to clock all A/D and S/H circuits.

Sixteen-bit resolution was formerly the quality benchmark for most digital audio devices, and it can yield excellent audio fidelity. However, many digital audio devices now process or store more than 16 bits. A digital signal processing (DSP) chip might internally process 56-bit words; this resolution is needed so that repetitive calculations will not accumulate error that could degrade audio fidelity. In addition, for example, Blu-ray discs can store 20- or 24-bit words. Thus, many A/D and D/A converters claim conversion of up to 24 bits. However, it is difficult or impossible to achieve true 24-bit conversion resolution with current technology. A resolution of 24 bits ostensibly yields a quantization error floor of about –145 dBFS (dB Full Scale). If a dBFS of 2 V rms is used, then a level at –145 dBFS corresponds to about 0.1 μV rms. This is approximately the level of thermal noise in a 6-ohm resistor at room temperature. The ambient noise in any practical signal chain would preclude ideal 24-bit resolution. Internal processing does require longer word lengths, but it is unlikely that A/D or D/A converters will process signals at such high resolution.

Successive Approximation A/D Converter

There are many types of A/D converter designs appropriate for various applications. For audio digitization, the necessity for both speed and accuracy limits the choices to a few types. The successive approximation register (SAR) A/D converter (sometimes known as a residue converter) is a classical method for achieving good-quality audio digitization; a SAR converter is shown in Fig. 3.10. This converter uses a D/A converter in a feedback loop, a comparator, and a control section. In essence, the converter compares an analog voltage input with its interim digital word converted to a second analog voltage, adjusting its interim conversion until the two agree within the given resolution. The device follows an algorithm that, bit by bit, sets the output digital word to match the analog input.

For example, consider an analog input of 6.92 V and an 8-bit SAR A/D converter. The operational steps of SAR conversion are shown in Fig. 3.11. The most significant bit in the SAR is set to 1, with the other bits still at 0; thus the word 10000000 is applied to the internal D/A converter. This word places the D/A converter's output at its half value of 5 V. Because the input analog voltage is greater than the D/A converter's output, the comparator remains high. The first bit is stored at logical 1. The next most significant bit is set to 1 and the word 11000000 is applied to the D/A converter, with an interim output of 7.5 V. This voltage is too high, so the second bit is reset to 0 and stored. The third bit is set to 1, and the word 10100000 is applied to the D/A converter; this produces 6.25 V, so the third bit remains high. This process continues until the LSB is stored and the digital word 10110001, representing a converted 6.91 V, is output from the A/D converter.

This successive approximation method requires n D/A conversions for every one A/D conversion, where n is the number of bits in the output word. In spite of this recursion, SAR converters offer relatively high conversion speed. However, the converter must be precisely designed. For example, a 16-bit A/D converter ranging over ±10 V with 1/2-LSB error requires a conversion accuracy of 3 mV. A 10-V step change in the D/A converter must settle to within 0.001% during a period of 1 μs. This period corresponds to an analog time

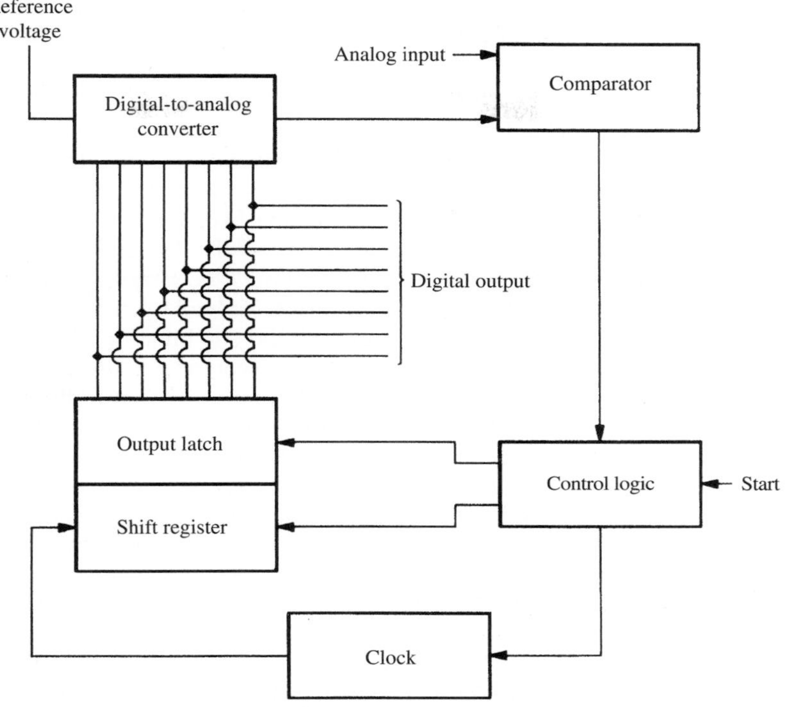

Figure 3.10 A successive approximation register A/D converter showing an internal D/A converter and comparator.

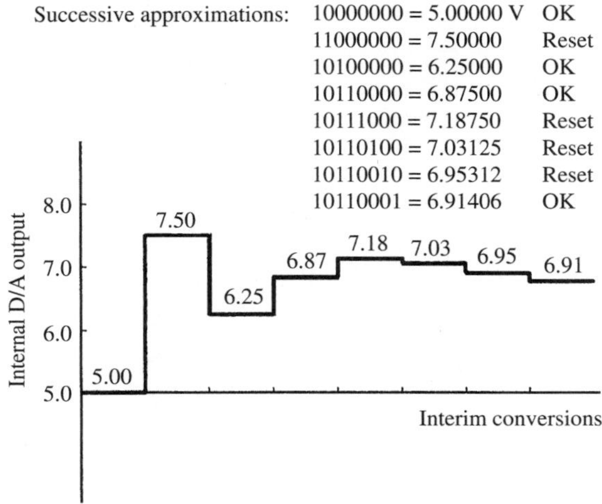

Figure 3.11 The intermediate steps in an SAR conversion showing results of interim D/A conversions.

constant of about 100 ns. The S/H circuit must be designed to minimize droop to ensure that the LSB, the last bit output to the SAR register, is accurate within this specification.

Oversampling A/D Converter

As noted, analog lowpass filters suffer from limitations such as noise, distortion, group delay, and passband ripple; unless great care is taken, it is difficult for downstream A/D converters to achieve resolution beyond 18 bits. In most applications, brick-wall analog anti-aliasing filters and SAR A/D converters have been replaced by oversampling A/D converters with digital filters. The implementation of a digital anti-aliasing filter is conceptually intriguing because the analog signal must be sampled and digitized prior to any digital filtering. This conundrum has been resolved by clever engineering; in particular, a digital decimation filter is employed and combined with the task of A/D conversion. The fundamentals of oversampling A/D conversion are presented here.

In oversampling A/D conversion, the input signal is first passed through a mild analog anti-aliasing filter which provides sufficient attenuation, but only at a high half-sampling frequency. To extend the Nyquist frequency, the filtered signal is sampled at a high frequency and then quantized. After quantization, a digital low-pass filter uses decimation to both reduce the sampling frequency to a nominal rate and prevent aliasing at the new, lower sampling frequency. Quantized data words are output at a lower frequency (for example, 48 or 96 kHz). The decimation low-pass filter removes frequency components beyond the Nyquist frequency of the output sampling frequency to prevent aliasing when the output of the digital filter is resampled (undersampled) at the system's sampling frequency.

Consider the oversampling A/D converter and D/A converter (both using two-times oversampling) shown in Fig. 3.12. An analog anti-aliasing filter restricts the bandwidth to

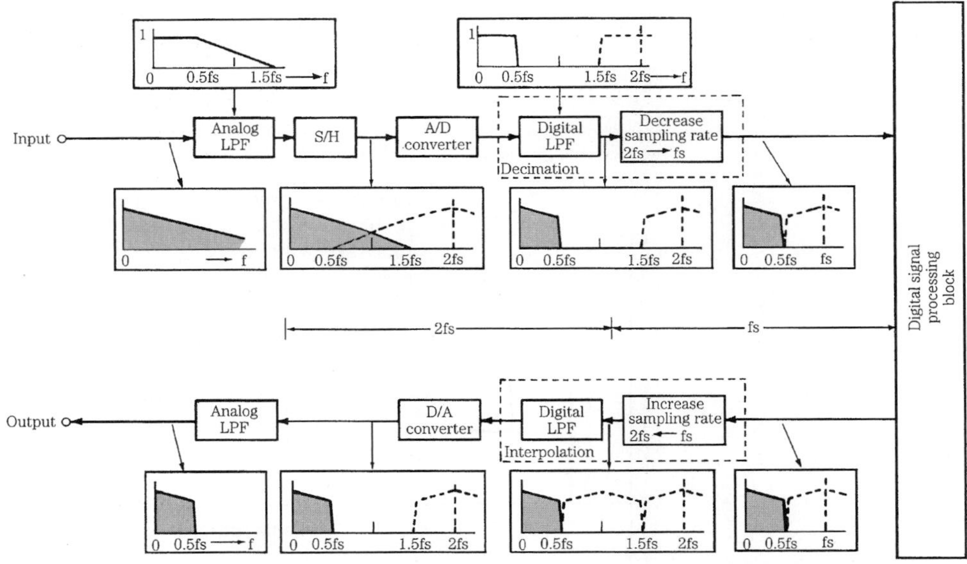

FIGURE 3.12 A two-times oversampling A/D and D/A conversion system. Decimation and interpolation digital filters increase and decrease the sampling frequency while removing alias and image signal components.

$1.5f_s$, where f_s is the sampling frequency. The relatively wide transition band, from 0.5 to $1.5f_s$, is acceptable and promotes good phase response. For example, a 7th-order Butterworth filter could be used. The signal is sampled and held at $2f_s$, and then converted. The digital filter limits the signal to $0.5f_s$. With decimation, the sampling frequency of the signal is undersampled and hence reduced from $2f_s$ to f_s. This action is accomplished with a linear-phase finite impulse response (FIR) digital filter with uniform group delay characteristics. Upon playback, an oversampling filter doubles the sampling frequency, samples are converted to yield an analog waveform, and high-frequency images are removed with a low-order lowpass filter.

Many oversampling A/D converters use a very high initial sampling frequency (perhaps 64- or 128-times 44.1 kHz), and take advantage of that high rate by using sigma-delta conversion of the audio signal. Because the sampling frequency is high, word lengths of one or a few bits can provide high resolution. A sigma-delta modulator can be used to perform noise shaping to lower audio band quantization noise. A decimation filter is used to convert the sigma-delta coding to 16-bit (or higher) coding, and a lower sampling frequency. Consider an example in which one-bit coding takes place at an oversampling rate R of 72; that is, 72×44.1 kHz = 3.1752 MHz, as shown in Fig. 3.13. The decimation filter provides a stopband from 20 kHz to the half-sampling frequency of 1.5876 MHz. One-bit A/D conversion greatly simplifies the digital filter design. An output sample is not required for every input bit; because the decimation factor is 72, an output sample is required for every 72 bits input to the decimation filter. A transversal filter can be used, with filter coefficients suited for the decimation factor. Following decimation, the result can be rounded to 16 bits, and output at a 44.1-kHz sampling frequency.

In addition to eliminating brick-wall analog filters, oversampling A/D converters offer other advantages over conventional A/D converters. Oversampling A/D converters can achieve increased resolution compared to SAR methods. For example, they extend the spectrum of the quantization error far outside the audio baseband. Thus the in-band noise can be made quite small. The same internal digital filter that prevents aliasing also removes out-of-band noise components. Increasingly, oversampling A/D converters are employed. This type of sigma-delta conversion is discussed in Chap. 18. Whichever A/D conversion method is used, the goal of digitizing the analog signal is accomplished, as data in two's complement or other form is output from the device.

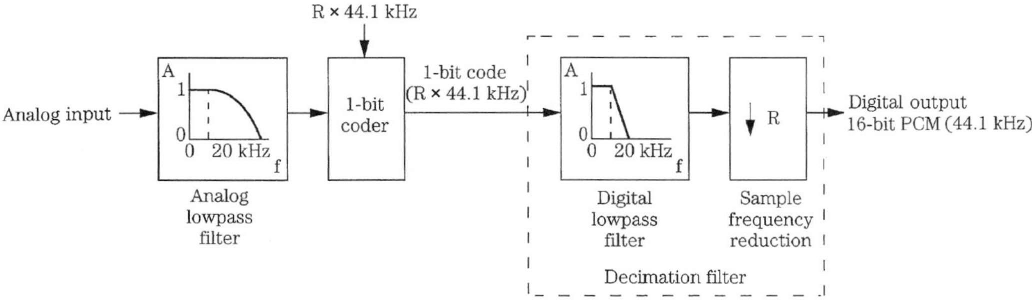

FIGURE 3.13 An oversampling A/D converter using one-bit coding at a high sampling frequency, and a decimation filter.

For digitization systems in which real-time processing such as delay and reverberation is the aim, the signal is ready for processing through software or dedicated hardware. In the case of a digital recording system, further processing is required to prepare the data for the storage medium.

Record Processing

After the analog signal is converted to binary numbers, several operations must occur prior to storage or transmission. Although specific processing needs vary according to the type of output channel, systems generally multiplex the data, perform interleaving, add redundancy for error correction and provide channel coding. Although there is an uninteresting element of bookkeeping in this processing, any drudgery is critical to prepare the data for the output channel and ensure that playback ultimately will be satisfactorily accomplished.

Some digital audio programs are stored or transmitted with emphasis, a simple means of reducing noise in the signal. Pre-emphasis equalization boosts high frequencies prior to storage or transmission. At the output, corresponding de-emphasis equalization attenuates high frequencies. The net result is a reduction in the noise floor. A common emphasis characteristic uses time constants of 50 and 15 μs, corresponding to frequency points at 3183 and 10,610 Hz, with a 6-dB/octave slope between these points, as shown in Fig. 3.14. Use of pre-emphasis must be identified in the program material so that de-emphasis equalization can be applied at the output.

In analog recording, an error occurring during storage or transmission results in degraded playback. In digital recording, error detection and correction minimize the effect of such defects. Without error correction, the quality of digital audio recording would be greatly diminished. Several steps are taken to combat the effects of errors. To prevent a single large defect from destroying large areas of consecutive data, interleaving is employed; this scatters data through the bitstream so the effect of an error is scattered when data is de-interleaved during playback. During encoding, coded parity data is added; this is redundant data created from the original data to help detect and correct errors. A discussion of parity, check codes, redundancy, interleaving, and error correction is presented in Chap. 5.

Multiplexing is used to form a serial bitstream. Most digital audio recording and transmission is a serial process; that is, the data is processed as a single stream of information. However, the output of the A/D converter can be parallel data; for example, two 16-bit words may be output simultaneously. A data multiplexer converts this parallel

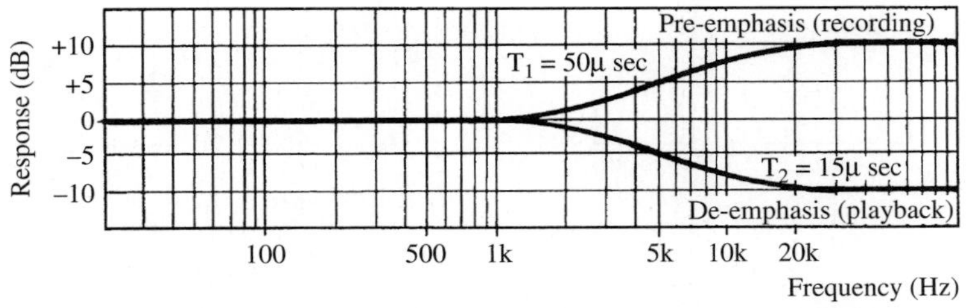

FIGURE 3.14 Pre-emphasis boosts high frequencies during recording, and de-emphasis reduces them during playback to lower the noise floor.

data to serial data; the multiplexing circuit accepts parallel data words and outputs the data one bit at a time, serially, to form a continuous bitstream.

Raw data must be properly formatted to facilitate its recording or transmission. Several kinds of processing are applied to the coded data. The time-multiplexed data code is usually grouped into frames. To prevent ambiguity, each frame is given a synchronization code to delineate frames as they occur in the stream. A synchronization code is a fixed pattern of bits that is distinct from any other coded data bit pattern in much the same way that a comma is distinct from the characters in a sentence. In many cases, data files are preceded by a data header with information defining the file contents.

Addressing or timing data can be added to frames to identify data locations in the recording. This code is usually sequentially ordered and is distributed through the recording to distinguish between different sections. As noted, error correction data is also placed in the frame. Identification codes might carry information pertinent to the playback processing. For example, specification of sampling frequency, use of pre-emphasis, table of contents, timing and track information, and copyright information can be entered into the data stream.

Channel Codes

Channel coding is an important example of a less visible, yet critical element in a digital audio system. Channel codes were aptly described by Thomas Stockham as the handwriting of digital audio. Channel code modulation occurs prior to storage or transmission. The digitized audio samples comprise 1s and 0s, but the binary code is usually not conveyed directly. Rather, a modulated channel code represents audio samples and other conveyed information. It is thus a modulation waveform that is interpreted upon playback to recover the original binary data and thus the audio waveform. Modulation facilitates data reading by further delineating the recorded logical states. Moreover, through modulation, a higher coding efficiency is achieved; although more bits might be conveyed, a greater data throughput can be achieved overall.

Storing binary code directly on a medium is inefficient. Much greater densities with high code fidelity can be achieved through methods in which modulation code fidelity is low. The efficiency of a coding method is the number of data bits transmitted divided by the number of transitions needed to convey them. Efficiencies vary from about 50% to nearly 150%. In light of these requirements, PCM, for example, is not suitable for transmission or recording to a medium such as optical disc; thus, other channel modulation techniques must be devised. Although binary recording is concerned with storing the 0s and 1s of the data stream, the signal actually recorded might be quite different. Typically, it is the transitions from one level to another, rather than the amplitude levels themselves, which represent the channel data. In that respect, the important events in a digitally encoded signal are the instants in time at which the state of the signal changes.

A channel code describes the way information is modulated into a channel signal, stored or transmitted, and demodulated. In particular, information bits are transformed into channel bits. The transfer functions of digital media create a number of specific difficulties that can be overcome through modulation techniques. A channel code should be self-clocking to permit synchronization at the receiver, minimize low-frequency content that could interfere with servo systems, permit high data rate transmission or high-density recording, exhibit a bounded energy spectrum, have immunity to channel noise, and reveal invalid signal conditions. Unfortunately, these requirements are largely mutually conflicting, thus only a few channel codes are suitable for digital audio applications.

The decoding clock in the receiver must be synchronized in frequency and phase with the clock (usually implicit in the channel bit patterns) in the transmitted signal. In most cases, the frames in a binary bitstream are marked with a synchronization word. Without some kind of synchronization, it might be impossible to directly distinguish between the individual channel bits.

Even then, a series of binary 1s or 0s form a static signal upon playback. If no other timing or decoding information is available, the timing information implicitly encoded in the channel bit periods is lost. Therefore, such data must often be recorded in such a way that pulse timing is delineated. Codes that provide a high transition rate, which are suitable for regenerating timing information at the receiver, are called self-clocking codes.

Thus, one goal of channel modulation is to combine a serial data stream with a clock pulse to produce a single encoded waveform that is self-clocking. Generally, code efficiency must be diminished to achieve self-clocking because clocking increases the number of transitions, which increases the overall channel bit rate. The high-frequency signal produced by robust clocking content will decrease a medium's storage capacity, and can be degraded over long cable runs. A minimum distance between transitions (T_{min}) determines the highest frequency in the code, and is often the highest frequency the medium can support. The ratio of T_{min} to the length of a single bit period of input information data is called the density ratio (DR). From a bandwidth standpoint, a long T_{min} is desirable in a code. T_{max} determines the maximum distance between transitions sufficient to support clocking. From a clocking standpoint, a shorter T_{max} is desirable.

Time-axis variations such as jitter are characterized by phase variations in a signal, observable as a frequency modulation of a stable waveform. The constraints of channel coding and data regeneration fundamentally limit the maximum number of incremental periods between transitions, that is, the number of transitions that can be detected between T_{min} and T_{max}. An important consideration in modulation code design is tolerance in locating a transition in the code. This is called the window margin, phase margin, or jitter margin and notated as T_w. It describes the minimum difference between code wavelengths: the larger the clock window, the better the jitter immunity. The efficiency of a code can be measured by its density ratio that is the ratio of the number of information bits to the number of channel transitions. The product of DR and T_w is known as the figure of merit (FoM); by combining density ratio and jitter margin, an overall estimate of performance is obtained: the higher the numerical value of FoM, the better the performance.

An efficient coding format must restrict dc content in the coded waveform, which could disrupt timing synchronization; dc content measures the time that the waveform is at logical 1 versus the time at logical 0; a dc content of 0 is ideal. Generally, digital systems are not responsive to direct current, so any dc component of the transmitted signal may be lost. In addition, dc components result in a baseline offset that reduces the signal-to-noise ratio. The dc content is the fraction of time that the signal is high during a string of 1s or 0s minus the fraction of time it is low. It results in a nonzero average amplitude value. For example, a nonreturn to zero (NRZ) signal (in which binary values are coded as high- or low-signal amplitudes) with all 0s or 1s would give a dc content of 100%.

The dc content can be monitored through the digital sum value (DSV). The DSV of a code can be thought of as the difference in accumulated charge if the code was passed through an ac coupling capacitor. In other words, it shows the dc bias that accumulates in a coded sequence. Figure 3.15 shows two different codes and their DSV; over the

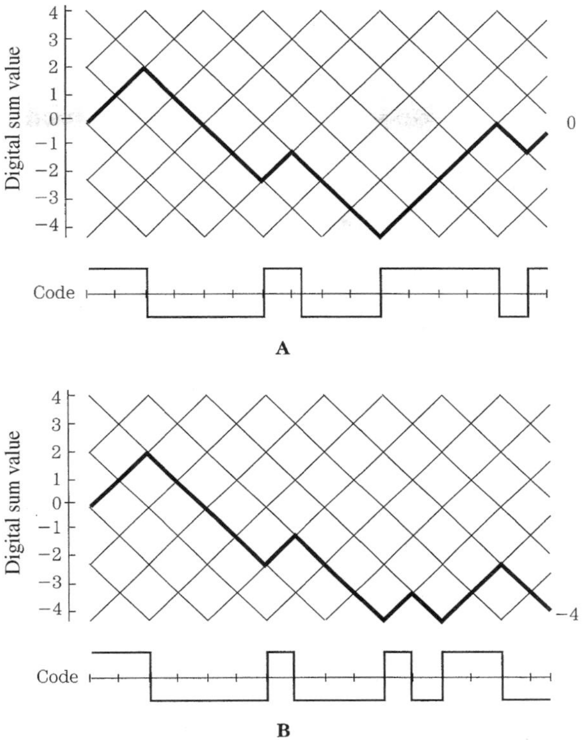

Figure 3.15 The digital sum value (DSV) monitors the dc content in a bitstream. A. A coded waveform that is free of dc content over the measured interval. B. A coded waveform that contains dc content.

measured interval, the first code does not show dc content, the second does. The dc content might cause problems in transformer-coupled magnetic recording heads; magnetic heads sense domains inductively and hence are inefficient in reading low-frequency signals. The dc content can present clock synchronization problems, and lead to errors in the servo systems used for radial tracking and focusing in an optical system. These systems generally operate in the low-frequency region. Low-frequency components in the readout signal cause interference in the servo systems, making them unstable. A dc-free code improves both the bandwidth and signal-to-noise ratio of the servo system. In the Compact Disc format, for example, the frequency range from 20 kHz to 1.5 MHz is used for information transmission; the servo systems operate on signals in the 0- to 20-kHz range.

A single sampling pulse is easy to analyze because of its periodic nature in the time domain; Fourier analysis clearly shows its spectrum. However, a data stream differs in that the data pulses occur aperiodically, and in fact can be considered to be random. The power spectrum density, or power spectrum, shows the response of the data stream. For example, Fig. 3.16 shows the spectral response of three types of channel coding with random data sequences: nonreturn to zero (NRZ), modified frequency modulation (MFM), and biphase. A transmission waveform ideally should have minimal energy at

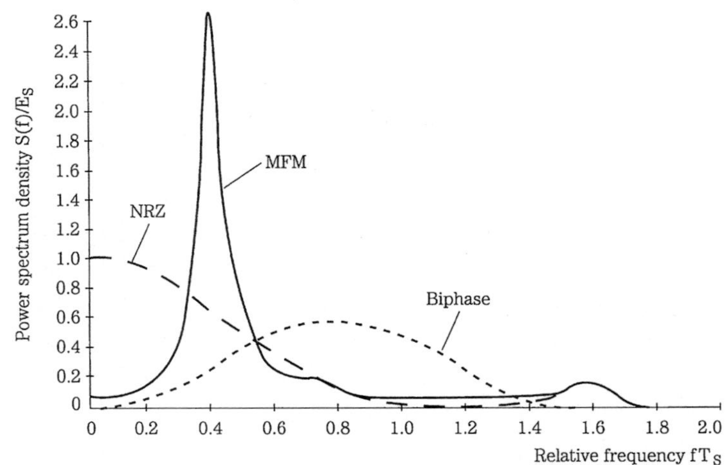

FIGURE 3.16 The power spectral density shows the response of a stream of random data sequences. NRZ code has severe dc content; MFM code exhibits a very narrow spectrum; biphase codes yield a broadband energy distribution.

low frequencies to avoid clocking and servo errors, and minimal energy at high frequencies to reduce bandwidth requirements. Biphase codes (there are many types, one being binary FM) yield a spectrum that has a broadband energy distribution. The MFM code exhibits a very narrow spectrum. The MFM and biphase codes are similar because they have no strong frequency components at low frequencies (lower than $0.2f$ where $f = 1/T$). If the value of f is 500 kHz, for example, and the servo signals do not extend beyond 15 kHz, these codes would be suitable. The NRZ code has a strong dc content and could pose problems for a servo system.

To minimize decoding errors, formats can be developed in which data is conveyed with data patterns that are as individually unique as possible. For example, in the eight-to-fourteen modulation (EFM) code devised for the Compact Disc format, 8-bit symbols are translated into 14-bit symbols, carefully selected for maximum difference between symbols. In this way, invalid data can be more easily recognized. Similarly, a data symbol could be created based on previous adjacent symbols and the receiver could recognize the symbol and its past history as a unique state. A state pattern diagram is used in which all transitions are defined, based on all possible adjacent symbols.

As noted, in many codes, the information is contained in the timing of transitions, not in the direction (low to high, or high to low) of the transitions. This is advantageous because the code is thus insensitive to polarity; the content will not be affected if the signal is inverted. The EFM code enjoys this property.

Simple Codes

The channel code defines the logical 1 and 0 of the input information. We might assume a direct relationship between a high amplitude and logical 1, and a low amplitude and logical 0. However, many other relationships are possible; for example, in one version of frequency-shift keying (FSK), a logical 1 corresponds to a sine burst of 100 kHz and a logical 0 corresponds to a sine burst of 150 kHz. Methods that use only two values take

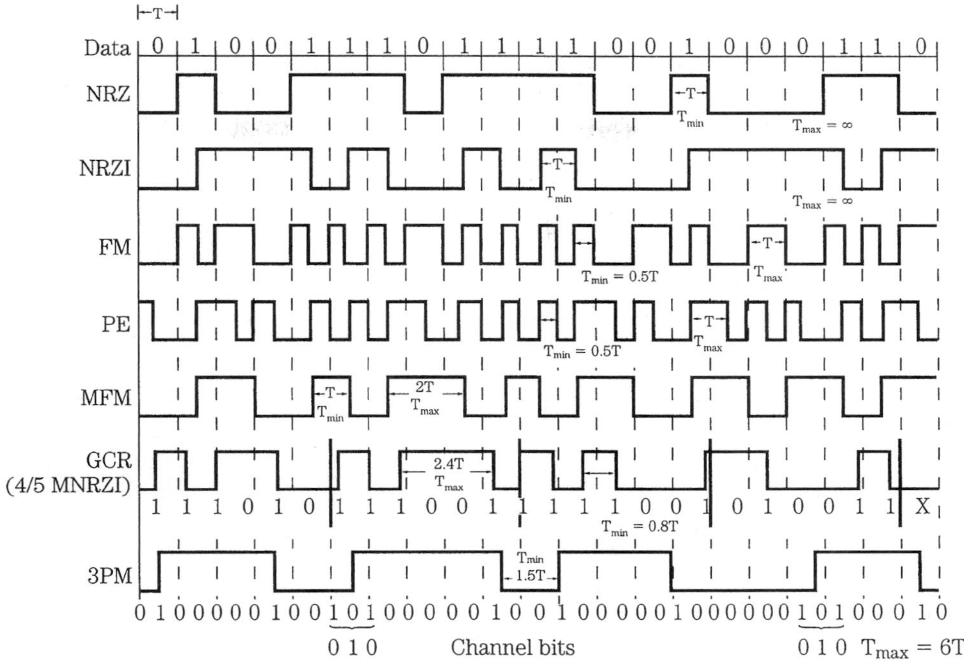

FIGURE 3.17 A comparison of simple and group-code waveforms for a common data input.

full advantage of digital storage; relatively large variations in the medium do not affect data recovery. Because digitally stored data is robust, high packing densities can be achieved. Various modulation codes have been devised to encode binary data according to the medium's properties. Of many, only a few are applicable to digital audio storage, on either magnetic or optical media. A number of channel codes are shown in Fig. 3.17.

Perhaps the most basic code sends a pulse for each 1 and does not send a pulse for a 0; this is called return to zero (RZ) code because the signal level always returns to zero at the end of each bit period.

The nonreturn to zero (NRZ) code is also a basic form of modulation: 1s and 0s are represented directly as high and low levels. The direction of the transition at the beginning or end of a bit period indicates a 1 or 0. The minimum interval is T, but the maximum interval is infinite (when data does not change) thus NRZ suffers from one of the problems that encourages use of modulation: strings of 1s or 0s do not produce transitions in the signal, thus a clock cannot be extracted from the signal. In addition, this creates dc content. The data density (number of bits per transition) for NRZ is 1.

The nonreturn to zero inverted (NRZI) code is similar to the NRZ code, except that only 1s are denoted with amplitude transitions (low to high, or high to low); no transitions occur for 0s. For example, any flux change in a magnetic medium indicates a 1, with transitions occurring in the middle of a bit period. With this method, the signal is immune to polarity reversal. The minimum interval is T, and the maximum is infinite; a clock cannot be extracted. A stream of 1s generates a transition at every clock interval; thus, the signal's frequency is half that of the clock. A stream of 0s generates no transitions. Data density is 1.

In binary frequency modulation (FM), also known as biphase mark code, there are two transitions for a 1 and one transition for a 0; this is essentially the minimum frequency implementation of FSK. The code is self-clocking. Biphase space code reverses the 1/0 rules. The minimum interval is $0.5T$ and the maximum is T. There is no dc content and the code is invertible. In the worst case, there are two transitions per bit, yielding a density ratio of 0.5, or an efficiency of 50%. FoM is 0.25. This code is used in the AES3 standard, described in Chap. 13.

In phase encoding (PE), also known as phase modulation (PM), biphase level modulation or Manchester code, a 1 is coded with a negative-going transition, and a 0 is coded with a positive-going transition. Consecutive 1s or 0s follow the same rule, thus requiring an extra transition. These codes follow phase-shift keying techniques. The minimum interval is $0.5T$ and the maximum is T. This code does not have dc content, and is self-clocking. Density ratio is 0.5. The code is not invertible.

In modified frequency modulation (MFM) code, sometimes known as delay modulation or Miller code, a 1 is coded with either a positive- or negative-going transition in the center of a bit period, for each 1. There is no transition for 0s; rather, a transition is performed at the end of a bit period only if a string of 0s occurs. Each information bit is coded as two channel bits. There is a maximum of three 0s and a minimum of one 0 between successive 1s. In other words, $d = 1$ and $k = 3$. The minimum interval is T and the maximum is $2T$. The code is self-clocking, and can have dc content. Density ratio is 1 and FoM is 0.5.

Group Codes

Simple codes such as NRZ and NRZI code one information bit into one channel bit. Group codes use more sophisticated methods for great coding efficiency and overall performance. Group codes use code tables to convert groups of input words (each with m bits) into patterns of output words (each with n bits); the output patterns are specially selected for their desirable coding characteristics, and uniqueness that helps detect errors. The code rate R for a group code is m/n. The value of the code rate equals the value of the jitter margin. In some group codes, the correspondence between the input information word and output codeword is not fixed; it might vary adaptively with the information sequence itself. These multiple modes of operation can improve code efficiency.

Group codes also can be considered as run-length limited (RLL) codes; the run length is the time between channel-bit transitions. This coding approach recognizes that transition spacings can be any multiple of the period, as shown in Fig. 3.18. This breaks the distinction between data and clock transitions and instead specifies a minimum number d and maximum number k of 0s between two successive 1s. These T_{min} and T_{max} values define the code's run length and specifically determine the code's spectral limits; clearly, data density, dc content, and clocking are all influenced by these values. The value of the density ratio equals the value of T_{min}, such that $DR = T_{min} = (d + 1)(m)/n$. Similarly, jitter margin $T_w = m/n$ and $FoM = (d + 1)(m^2)/n^2$.

As always, density must be balanced against other factors, such as clocking. Generally, d is selected to be as large as possible for high density, and k is selected to be as large as possible while maintaining stable clocking. Note that high k/d ratios can yield code sequences with high dc content; this is a shortcoming of RLL codes. The minimum and maximum lengths determine the minimum and maximum rates in the code waveform; by choosing specific lengths, the spectral response can be shaped.

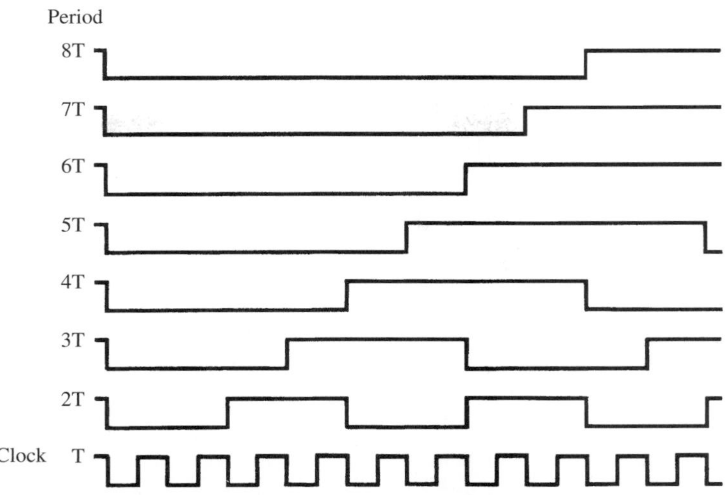

Figure 3.18 Run-length limited (RLL) codes regulate the number of transitions representing the channel code. In this way, transition spacings can be any multiple of the channel period, increasing data density.

RLL codes use a set of rules to convert information bits into a stream of channel bits by defining some relationship between them. A channel bit does not correspond to one information bit; the channel bits can be thought of as short timing windows, fractions of the clock period. Data density can be increased by increasing the number of possible transitions within a bit period. Channel bits are often converted into an output signal using NRZI modulation code; a transition defines a 1 in the channel bitstream. The channel bit rate is usually greater than the information bit rate, but if the run lengths between 1s can be distinguished, the overall information density can be increased. To ensure that consecutive codewords do not violate the run length and to ensure dc-free coding, alternate codewords or merging bits may be placed between codewords; however, this decreases density. Generally, RLL codes are viable only in channels with low noise; for example, the required S/N ratio increases as minimum length d increases. Fortunately, optical discs are suitable for RLL coding. Technically, NRZ and NRZI are RLL codes because $d = 0$ and $k = \infty$; MFM could be considered as a (1,3) RLL code.

In the group coded recording (GCR) code, data is parsed into four bits, coded into 5-bit words using a lookup table as shown in Table 3.1, and modulated as NRZI signals. This implementation is sometimes known as 4/5 MNRZI (modified NRZI) code. There is a transition every three bits, because of the 4/5 conversion, the minimum interval is $0.8T$, and the maximum interval is $2.4T$. Adjacent 1s are permitted ($d = 0$) and the maximum number of 0s is 2 ($k = 2$). The code is self-clocking with great immunity to jitter, but exhibits dc content. The density ratio is 0.8.

The three-position modulation (3PM) code is a (2,7) RLL adaptive code. Three input bits are converted into a 6-bit output word in which T_{min} is $1.5T$ and T_{max} is $6T$, as shown in Table 3.2. There must be at least two channel 0s between 1s ($d = 2$). When 3PM words are merged, a 101 pattern might occur; this violates the T_{min} rule. To prevent this, 101 is replaced by 010; the last channel bit in the coding table is reserved for this merging operation. The 3PM code is so called because of the three positions ($d + 1$) in the minimum

Data Bit	Channel Bit
0000	11001
0001	11011
0010	10010
0011	10011
0100	11101
0101	10101
0110	10110
0111	10111
1000	11010
1001	01001
1010	01010
1011	01011
1100	11110
1101	01101
1110	01110
1111	01111

Table 3.1 Conversion for the GCR (or 4/5 MNRZI) code. Groups of four information bits are coded as 5-bit patterns, and written in NRZI form.

Data Bit			Channel Bit					
			P1	P2	P3	P4	P5	P6
0	0	0	0	0	0	0	1	0
0	0	1	0	0	0	1	0	0
0	1	0	0	1	0	0	0	0
0	1	1	0	1	0	0	1	0
1	0	0	0	0	1	0	0	0
1	0	1	1	0	0	0	0	0
1	1	0	1	0	0	0	1	0
1	1	1	1	0	0	1	0	0
	|←T→|			|←T→|				

Table 3.2 Conversion for 3PM (2,7) code. Three input bits are coded into a 6-bit output word in which the minimum interval is maintained at 1.57 through pattern inversion.

distance. The code is self-clocking. For comparison, the duration between transitions for 3PM is 0.5T and for MFM is T. Thus, the packing density of 3PM is 50% higher than MFM. However, its maximum transition is 100% longer and its jitter margin 50% worse. 3PM exhibits dc content. Data density is 1.5.

In the 4/5 group code, groups of four input bits are mapped into five channel bits using a lookup table. The 16 channel codewords are selected (from the 32 possible) to yield a useful clocking signal while minimizing dc content. Adjacent 1s are permitted ($d = 0$) and the maximum number of 0s is 3 ($k = 3$). $T_{max} = 16/5 = 3.2T$. Density ratio is 0.8 and FoM is 0.64. The 4/5 code is used in the Fiber Distributed Data Interface (FDDI) transmission protocol, and the Multichannel Audio Digital Interface (MADI) protocol as described in Chap. 13.

As noted, RLL codes are efficient because the distance between transitions is changed in incremental steps. The effective minimum wavelength of the medium becomes the incremental run lengths. Part of this advantage is lost because it is necessary to avoid all data patterns that would put transitions closer together than the physical limit. Thus all data must be represented by defined patterns. The eight-to-fourteen modulation (EFM) code is an example of this kind of RLL pattern coding. The incremental length for EFM is one-third that of the minimum resolvable wavelength. Data density is not tripled, however, because 8 data bits must be expressed in a pattern requiring 14 incremental periods. To recover this data, a clock is run at the incremental period of 1T.

EFM code is used to store data on a Compact Disc; it is an efficient and highly structured (2,10) RLL code. Blocks of 8 data bits are translated into blocks of 14-bit channel symbols using a lookup table that assigns an arbitrary and unique word. The 1s in the output code are separated by at least two 0s ($d = 2$), but no more than ten 0s ($k = 10$). That is, T_{min} is 3 channel bits and T_{max} is 11 channel bits. A logical 1 causes a transition in the medium; this is physically represented as a pit edge on the CD surface. High recording density is achieved with EFM code. Three merging bits are used to concatenate 14-bit EFM words, so 17 incremental periods are required to store 8 data bits. This decreases overall information density, but creates other advantages in the code. T_{min} is 1.41T and T_{max} is 5.18T. The theoretical recording efficiency is thus calculated by multiplying the threefold density improvement by a factor of 8/17, giving 24/17, or a density ratio of 1.41. That is, 1.41 data bits can be recorded per shortest pit length. For practical reasons such as S/N ratio and timing jitter on clock regeneration, the ratio is closer to 1.25. In either case, there are more data bits recorded than are transitions on the medium. The merging bits completely eliminate dc content, but reduce efficiency by 6%. The conversion table was selected by a computer algorithm to optimize code performance. From a performance standpoint, EFM is very tolerant of imperfections, provides high density, and promotes stable clock recovery by a self-clocking decoder. EFM is used in the CD format and is discussed in Chap. 7. The EFMPlus code used in the DVD format is discussed in Chap. 8. The 1-7PP code used in the Blu-ray format is discussed in Chap. 9.

Zero modulation (ZM) coding is an RLL code with $d = 1$ and $k = 3$; it uses a convolutional scheme, rather than group coding. One information bit is mapped into two data bits, coded with rules that depend on the preceding and succeeding data patterns, and written in NRZI code. As with many RLL codes that followed it, ZM uses data patterns that are optimal for its application (magnetic recording) and were selected by computer search. Generally, the bitstream is considered as any number of 0s, two 0s separated by an odd number of 1s or no 1s, or two 0s separated by an even number of 1s. The first two types are coded as Miller code, and in the last, the 0s are coded as Miller code, but the 1s are coded as if they were 0s, but without alternate transitions. The density ratio is approximately 1. There is no dc content in ZM.

An eight-to-ten (8/10) modulation group code was selected for the DAT format in which 8-bit information words are converted to 10-bit channel words. The 8/10 code permits adjacent channel 1s, and there are no more than three channel 0s between 1s. The density ratio is 0.8, and FoM is 0.64. Bit synchronization and block synchronization are provided with a $3.2T + 3.2T$ synchronization signal, a prohibited 8/10 pattern. The ideal 8/10 codeword would have no net dc content, with equal durations at high and low amplitudes in its modulated waveform. However, there are an insufficient number of such 10-bit channel words to represent the 256 states needed to encode the 8-bit data input. Moreover, given the maximum run length limitation, only 153 channel codes satisfy both requirements. Thus 103 codewords must have dc content, or nonzero digital sum value (DSV). The DSV tallies the high-amplitude channel bit periods versus the low-amplitude channel bit periods as encoded with NRZI. Two patterns are defined for each of the 103 nonzero DSV codewords, one with a +2 DSV and one with a –2 DSV; to achieve this, the first channel bit is inverted. Either of these codewords can be selected based on the cumulative DSV. For example, if DSV ranges negatively, a +2 word is selected to tend toward a zero dc condition. Channel codewords are written to tape with NRZI modulation. The decoding process is relatively easy to implement because DSV need not be computed. Specifications for a number of simple and group codes are listed in Table 3.3.

Parameter	NRZ NRZI	PE FM	MFM	ZM	GCR	3PM	EFM	HDM-1	8/10
Window margin (T_w)	T	$0.5T$	$0.5T$	$\sim 0.5T$	$0.8T$	$0.5T$	$0.471T$	$0.5T$	$0.8T$
Minimum transition (T_{min})	T	$0.5T$	T	$\sim T$	$0.8T$	$1.5T$	$1.41T$	$1.5T$	$0.8T$
Maximum transition (T_{max})	∞	T	$2T$	$\sim 2T$	$2.4T$	$6T$	$5.18T$	$4.5T$	$3.2T$
DC content (dc)	yes	no	yes	no	yes	yes	no	yes	no
Clock rate (CLK)	no	$2/T$	$2/T$	$2/T$	$1.25T$	$2/T$	$17/8T$	$2/T$	$10/8T$
Density ratio (DR)	1	0.5	1	~1	0.8	1.5	1.41	1.5	0.8
Figure of Merit $T_w T_{min}$ (T^2)	1	0.25	0.5	0.5	0.64	0.75	0.664	0.75	0.64
Maximum/Minimum ratio (T_{max}/T_{min})	∞	2	2	2	3	4	3	3	4

Table 3.3 Specification for a number of simple and group codes.

Code Applications

Despite different requirements, there is similarity of design between codes used in magnetic or optical recording. Practical differences between magnetic and optical recording codes are usually limited to differences designed to optimize the code for the specific application. Some codes such as 3PM were developed for magnetic recording, but later applied to optical recording. Still, most practical applications use different codes for either magnetic or optical recording.

Optical recording requires a code with high density. Run lengths can be long in optical media because clock regeneration is easily accomplished. The clock content in the data signal provides synchronization of the data, as well as motor control. Because this clock must be regenerated from the readout signal (for example, by detecting pit edges) the signal must have a sufficient number of transitions to support regeneration, and the maximum distance between transitions ideally must be as small as possible. In an optical disc, dirt and scratches on the disc surface change the envelope of the readout signal, creating low-frequency noise. This decreases the average level of the readout signal. If the signal falls below the detection level, it can cause an error in readout. This low-frequency noise can be attenuated with a highpass filter, but only if the information data itself contains no low-frequency components. A code without dc content thus improves immunity to surface contamination by allowing insertion of a filter. Compared to simple codes, RLL codes generally yield a larger bit-error rate because of error propagation; a small physical error can affect proportionally more bits. Still, RLL codes offer good performance in these areas and are suitable for optical disc recording. Ultimately, a detailed analysis is needed to determine the suitability of a code for a given application.

In many receiving circuits, a phase-locked loop (PLL) circuit is used to reclock the channel code, for example, from a storage medium. The channel code acts as the input reference, the loop compares the phase difference between the reference and its own output, and drives an internal voltage-controlled oscillator to the reference frequency, decoupling jitter from the signal. The comparison occurs at every channel transition, and interim oscillator periods count the channel periods, thus recovering the code. A synchronization code is often inserted in the channel code to lock the PLL. In an RLL code, a pattern violating the run length can be used for synchronization; for example, in the CD, two $11T$ patterns precede an EFM frame; the player can lock to the channel data, and will not misinterpret the synchronization patterns as data.

Following channel coding, the data is ready for storage or transmission. For example, in a hard-disk recorder, the data is applied to a recording circuit that generates the current necessary for saturation recording. The flux reversals recorded on the disk thus represent the bit transitions of the modulated data. The recorded patterns might appear highly distorted; this does not affect the integrity of the data, and permits higher recording densities. In optical systems such as the Compact Disc, the modulation code results in pits. Each pit edge represents a binary 1 channel bit, and spaces between represent binary 0s. In any event, storage to media, transmission, or other real-time digital audio processing, marks the end of the digital recording chain.

CHAPTER 4

Digital Audio Reproduction

Digital audio recording and reproduction processors serve as input and output transducers to the digital domain. They convert an analog audio waveform into a signal suitable for digital processing, storage, or transmission, and then convert the signal back to analog form for playback. In a linear pulse-code modulation (PCM) system, the functions of the reproduction circuits are largely reversed from those in the record side. Reproduction functions include timebase correction, demodulation, demultiplexing, error correction, digital-to-analog (D/A) conversion, output sample-and-hold (S/H) circuit, and output lowpass filtering. Oversampling digital filters, preceding D/A conversion, are used universally.

Reproduction Processing

The reproduction signal chain accepts a coded binary signal and ultimately reconstructs an analog waveform, as shown in Fig. 4.1. The reproduction circuits must minimize any detrimental effects of data storage or transmission. For example, storage media suffer from limitations such as mechanical variations and potential for physical damage. With analog storage, problems generally must be corrected within the medium itself; for example, to minimize wow and flutter, a turntable's speed must be precise. With digital systems, the potential for degradation is even greater. However, digital processing offers the opportunity to correct many faults. For example, the reproduction circuits buffer the data to minimize the effects of timing variations in the data, and also perform error correction.

To achieve high data density, the physical fidelity of the stored or transmitted channel code waveform is allowed to deteriorate. Thus, the signal does not have the clean characteristics that the original data may have enjoyed. For example, a recorded digital signal as read from a magnetic-disk head is not sharply delineated, but instead is noisy and rounded from bandlimiting. A waveform shaper circuit identifies the transitions and reconstructs a valid data signal. In this way, data can be recovered without penalty for the waveform's deterioration.

Synchronization pulses and other clocking information in the bitstream are identified and used by timebase correction circuits to synchronize the playback signal, delineate individual frames, and thus determine the binary content of each pulse. In most cases, the signal contains timing errors such as jitter as it is received for playback. Phase-locked loops (PLLs) and data buffers can overcome this problem. A buffer can be thought of as a pail of water: water is poured into the pail carelessly, but a spigot at the bottom of the pail supplies a constant stream. Specifically, a buffer is a memory into which the data is fed irregularly, as it is received. However, data is clocked from the

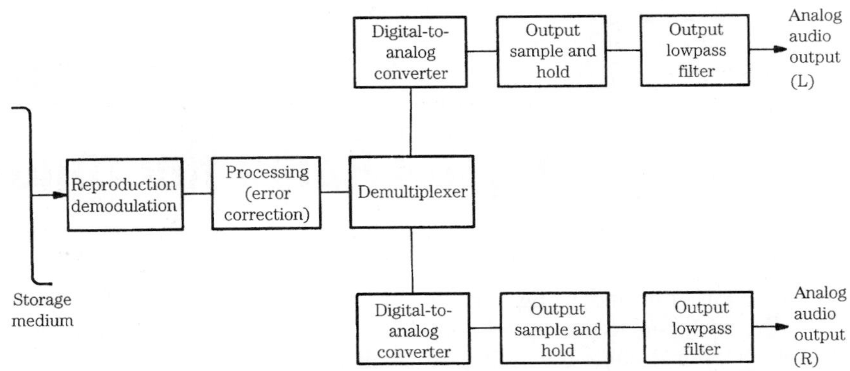

FIGURE 4.1 A conceptual linear PCM reproduction section. In practice, digital oversampling filters are used.

buffer at an accurately controlled rate, ensuring precise data timing. Samples can be reassembled at the same rate at which they were taken. Timebase correction is discussed in more detail later in this chapter.

The modulated audio data, whether it is eight-to-fourteen modulation (EFM), modified frequency modulation (MFM), or another code, is typically demodulated to non-return to zero (NRZ) code, that is, a simple code in which the amplitude level represents the binary information. The audio data thus regains a readily readable form and is ready for further reproduction processing. In addition, demultiplexing is performed to restore parallel structure to the audio data. The demultiplexer circuit accepts a serial bit input, counting as the bits are clocked in. When a full word has been received, it outputs all the bits of the word simultaneously, to form parallel data.

The audio data and coded error-correction data are identified and separated from other peripheral data. Using the error-correction data, the data signal is checked for errors that may have occurred following encoding. Because of the data density used in digital recording and transmission, many systems anticipate errors with certainty; only their frequency and severity vary. The error-correction circuits must de-interleave the data. Prior to recording, the data is scattered in the bitstream to ensure that a defect does not affect consecutive data. With de-interleaving, the data is properly reassembled in time, and errors caused by intervening defects are scattered through the bitstream, where they are more easily corrected. The data is checked for errors using redundancy techniques. When calculated parity does not agree with parity read from the bitstream, an error has likely occurred in the audio data. Error correction algorithms are used to calculate and restore the correct values. If within tolerance, errors can be detected and corrected with absolute fidelity to the original data, making digital recording and transmission a highly reliable technique. If the errors are too extensive for correction, error concealment techniques are used. Most simply, the last data value is held until valid data resumes. Alternatively, interpolation methods calculate new data to form a bridge over the error. A more complete discussion of error correction techniques can be found in Chap. 5. The serial bitstream consists of the original audio data, or at least data as close to the original as the error correction circuitry has yielded. On leaving the reproduction processing circuitry, the data has regained timing stability, been demultiplexed, been de-interleaved, and incurred errors have been corrected. The data is ready for digital-to-analog conversion. Digital filtering, which precedes D/A conversion, is discussed later in this chapter.

Digital-to-Analog Converter

The digital-to-analog (D/A) converter is one of the most critical elements in the reproduction system. Just as the analog-to-digital (A/D) converter largely determines the overall quality of the encoded signal, the D/A converter determines how accurately the digitized signal will be restored to the analog domain. The task of D/A conversion demands great precision; for example, with a ±10-V scale, a 16-bit converter will output voltage steps that measure 0.000305 V, and a 24-bit converter must output steps that measure only 0.00000119 V. Traditional D/A converters can exhibit nonlinearity similar to that of A/D converters; sigma-delta converters, operating on a time basis rather than an amplitude basis, can overcome some deficiencies but must use noise shaping to reduce the in-band noise floor. Fortunately, high-quality D/A converters are available at low cost.

Traditional D/A converters process parallel data words and are prone to many of the same errors as A/D converters, as described in Chap. 3. In practice, resolution is chiefly determined by absolute linearity error and differential linearity error. Absolute linearity error is a deviation from the ideal quantization staircase and can be measured at full signal level; it is smaller than ±1/2 LSB (least significant bit). Differential linearity error is a relative deviation from the ideal staircase by any individual step. The error is uncorrelated with high signal levels but correlated with low signal levels; as a result it is most apparent at low signal levels as distortion.

Differential nonlinearity appears as wide and narrow codes, and can cause entire sections of the transfer function to be missing. Differential nonlinearity is minor with high-amplitude signals but the errors can dominate low-level signals. For example, a signal at –80 dBFS (dB Full Scale) will pass through only six or seven codes in a 16-bit quantizer; if half of those codes are missing, the result will be 14-bit performance. Depending on bias, the differential linearity error in a 16-bit D/A converter for a –90 dBFS sine wave signal can result in generated levels ranging from –85.9 dB to –98.2 dB. Because the bits and their associated errors switch in and out throughout the transfer function, their effect is signal dependent. Thus, harmonic and intermodulation distortion and noise vary with signal conditions. Because this kind of error is correlated with the audio signal, it is more readily perceived. Nonmonotonicity is an extreme case of nonlinearity; an increase in a particular digital code does not result in an increase in the output analog voltage; most converters are guaranteed against this.

A linearity test measures the converter's ability to record or reproduce various signals at the proper amplitude. Specifically, linearity describes the converter's ability to output an analog signal that directly conforms to a digital word. For example, when a bit changes from 1 to 0 in a D/A converter, the analog output must decrease exactly by a proportional amount. Any nonlinearity results in distortion in the audio signal; the amount that the analog voltage changes will depend on which bit has changed. Every PCM digital word contains a series of binary digits, arranged in powers of two. The most significant bit (MSB) accounts for a change in fully half of the analog signal's amplitude, and the least significant bit accounts for the least change, for example, in an 18-bit word, an amplitude change of less than four parts per million. Physical realities conspire against this accuracy. Traditional ladder converters exhibit differential nonlinearity because of bit weighting errors; thermal or physical stress, aging, and temperature variations are also factors.

To help equipment manufacturers use D/A converters to their best advantage, some ladder D/A chips provide a means to calibrate the converter. Consider a 16-bit D/A converter offering calibration of the MSB. Because the MSB is so much larger than the other bits, an MSB error of only 0.01% (one part in 10,000) would completely negate the contributions of the two LSBs (which account for one part in 21,845 of the total signal amplitude). An error of 0.1% in the MSB would swamp the combined values of the five smallest bits. Overall, the tolerance of the MSB in a 16-bit converter is 1/65,536 of the LSB. Some converters offer calibration of the four most significant bits. It is interesting to note that fully 93% of the total analog output is represented through these four most significant bits. These MSBs largely steer the converter's output amplitude; when they are properly calibrated, the entire output signal of a well-designed D/A converter will be more accurate. This accuracy is most significant at low levels, usually below –60 dBFS. As noted, any nonlinearity in D/A conversion will be apparent as a deviation from nominal amplitude. Moreover, such nonlinearity can be heard; for example, by using a test disc containing a dithered fade to silence tone, poor D/A linearity is clearly audible.

The low-level performance of D/A converters can be evaluated using tests for D/A linearity. For example, Fig. 4.2 shows a D/A converter's low-level linearity. Reproduced signals lower than –100 dBFS in amplitude show nonlinearity. For example, a –100-dBFS signal is reproduced at an amplitude of –99 dBFS, and a –110-dBFS signal is reproduced at approximately –105 dBFS. Depending on signal conditions, this kind of error in low-level dynamics might audibly alter low-amplitude information.

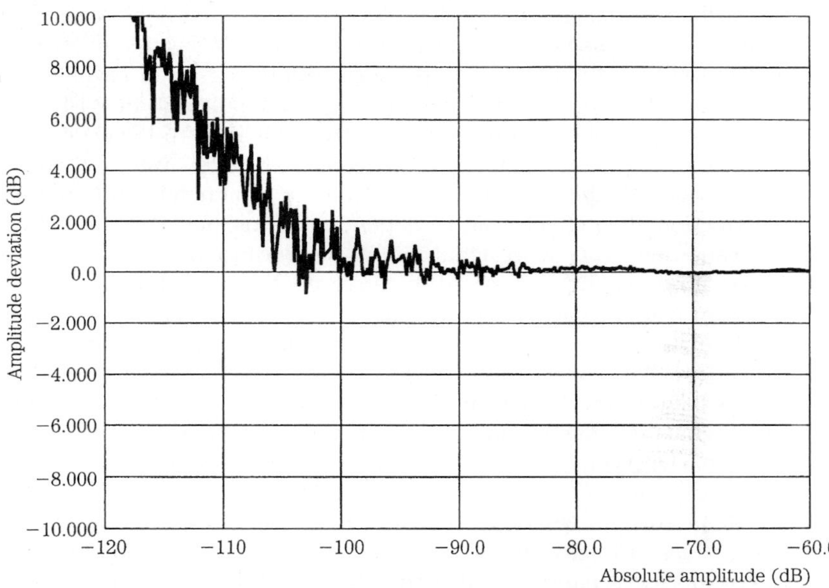

FIGURE 4.2 An example of a low-level linearity measurement of a D/A converter showing increasing nonlinearity with decreasing amplitude.

In practice, linear 16-bit conversion is insufficient for 16-bit data. Converters must have a greater dynamic range than the audio signal itself. No matter how many bits are converted, the accuracy of the conversion, and hence the fidelity of the audio signal, hinges on the linearity of the converter. Linearity errors in converters generally result in a stochastic deviation; it varies randomly from sample to sample. However, relative error increases as level decreases. The linearity of a D/A converter, and not the number of bits it uses, measures its accuracy.

A D/A converter must have a fast settling time. Settling time for a D/A converter is the elapsed time between a new input code and the time when the analog output falls within a specified tolerance (perhaps ±1/2 LSB). The settling time can vary with the magnitude of change in the input word.

Most D/A converters operate with a two's complement input. For example, an 8-bit D/A converter would have a most positive value of 01111111 and a most negative value of 10000000. In this format the MSB is complemented to serve as a sign bit. To accomplish this, the MSB can be inverted before the word is input to the D/A converter, or the D/A converter might have a separate, complementing input for the MSB.

Digitally generated test tones are often used to measure D/A converters; it is important to choose test frequencies that are not correlated with the sampling frequency. Otherwise, a small sequence of codes might be reproduced over and over, without fully exercising the converter. Depending on the converter's linearity at those particular codes, the output distortion might measure better, or worse, than typical performance. For example, when replaying a 1-second, 1-kHz, 0-dBFS sine wave sampled at 44.1 kHz, only 441 different codes would be used over the 44,100 points. A 0-dBFS sine wave at 997 Hz would use 20,542 codes, giving a much better representation of converter performance. Standard test tones have been selected to avoid this anomaly. For example, some standard test frequencies are: 17, 31, 61, 127, 251, 499, 997, 1999, 4001, 7993, 10,007, 12,503, 16,001, 17,989, and 19,997 Hz.

Weighted-Resistor Digital-to-Analog Converter

Various types of D/A converters are used for audio digitization. Operation of traditional converters is quite different from sigma-delta converters. The former are more easily understood; we begin with an illustration of the operation of a traditional ladder D/A converter. A D/A converter accepts an input digital word and converts it to an output analog voltage or current. The simplest kind of D/A converter, known as a weighted-resistor D/A converter, contains a series of resistors and switches, as shown in Fig. 4.3.

A weighted-resistor converter contains a switch for each input bit; the corresponding resistor represents the binary value associated with that bit. A reference voltage generates current through the resistors. A digital 1 closes a switch and contributes a current, and a digital 0 opens a switch and prevents current flow. An operational amplifier sums the currents and converts them to an output voltage. A low-value binary word with many 0s closes few switches thus a small voltage results. A high-value word with many 1s closes many switches; thus, a high voltage results. Consider this example of an 8-bit converter:

$$V_{\text{out}} = -V_{\text{ref}}\left(\frac{b1}{2}+\frac{b2}{4}+\frac{b3}{8}+\frac{b4}{16}+\frac{b5}{32}+\frac{b6}{64}+\frac{b7}{128}+\frac{b8}{256}\right)$$

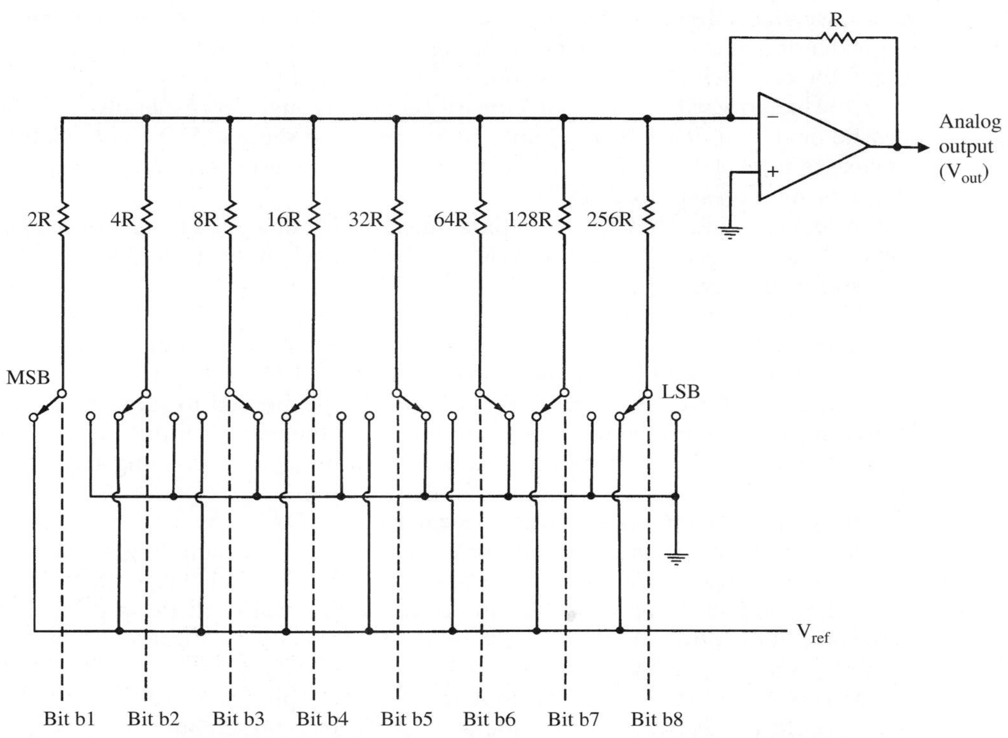

FIGURE 4.3 A weighted-resistor D/A converter uses resistors related by powers of 2, which limits resolution.

where *b*1 through *b*8 represent the input binary bits. For example, suppose the reference voltage $V_{ref} = 10$ V and the input word is 11010011:

$$V_{out} = -10\left(\frac{1}{2}+\frac{1}{4}+\frac{0}{8}+\frac{1}{16}+\frac{0}{32}+\frac{0}{64}+\frac{1}{128}+\frac{1}{256}\right)$$

$$= -10\left(\frac{1}{2}+\frac{1}{4}+\frac{1}{16}+\frac{1}{128}+\frac{1}{256}\right)$$

$$= -8.24\ V$$

Although this weighted-resistor design looks good on paper, it is rarely used in practice because of the complexity in manufacturing resistors with sufficient accuracy. It is extremely difficult to manufacture precise resistor values so that each next resistor value is exactly a power of 2 greater than the previous one. For example, in a 16-bit D/A converter, the largest-to-smallest ohm (Ω) resistor ratio is 65,536:1. If the smallest resistor value is 1 kΩ, the largest is over 65 MΩ. Similarly, the smallest current might be 30 nA (nanoampere) and the largest 2 mA (milliampere). In short, this design demands manufacturing conditions that cannot be efficiently met.

R-2R Ladder Digital-to-Analog Converter

A more suitable design approach for a D/A converter is the R-2R resistor ladder, as shown in Fig. 4.4. This circuit contains resistors and switches; in particular, there are two resistors per bit. Each switch contributes its appropriately weighted component to the output. The current splits at each node of the ladder, resulting in current flows through the switch resistors that are weighted by binary powers of 2. If current I flows from the reference voltage, $I/2$ flows through the first switch, $I/4$ through the second switch, $I/8$ through the third switch, and so on. Digital input bits are used to control ladder switches to produce an analog output. For example:

$$V_{out} = -V_{ref}\left(\frac{b1}{2}+\frac{b2}{4}+\frac{b3}{8}+\frac{b4}{16}+\frac{b5}{32}+\frac{b6}{64}+\frac{b7}{128}+\frac{b8}{256}\right)$$

With $V_{ref} = 10\ V$ and an input word of 01010110:

$$V_{out} = -10\left(\frac{1}{4}+\frac{1}{16}+\frac{1}{64}+\frac{1}{128}\right) = -3.36\ V$$

The R-2R network can be efficiently manufactured; only two values of resistors are needed. As noted, in many converters, one or several MSBs can be calibrated to improve linearity. In some designs, stability with respect to temperature is achieved with a compensation feedback loop. A high-precision signal is generated and compared to the signal generated by the D/A converter. The difference between the two

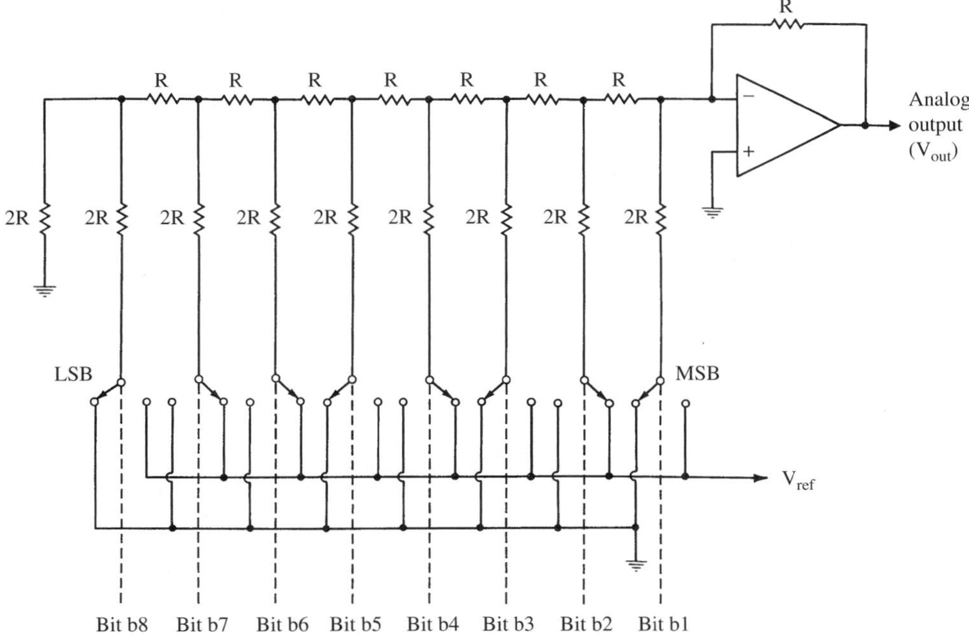

FIGURE 4.4 An R-2R D/A converter uses only two resistor values, improving resolution.

is applied to a memory, which in turn outputs a correction word to the input of the D/A converter. Errors caused by variations in the components are thus self-corrected and distortion is minimized.

Zero-Cross Distortion

A quality D/A converter must be highly accurate. As noted, a 16-bit D/A converter with an output range of ±10 V has a difference between quantization levels of 20/65,536 = 0.000305 V. For example, the output from the input word 1000 0000 0000 0000 should be 0.3 mV larger than that from the input word 0111 1111 1111 1111. In other words, the sum of the lower 15 bits must be accurate to that precision, compared to the value of the highest bit. The lower 15 bits should have a relative error of one-half quantization level, as should the MSB. However, differential linearity error is greatest for the MSB, which produces the largest output voltage. Moreover, the MSB changes each time the output waveform passes through zero. Difficulty in achieving accuracy at the center of a ladder D/A converter's range leads to zero-cross distortion.

Zero-cross distortion occurs at the zero-cross point between the positive and negative polarity portions of an analog waveform. When a resistor ladder D/A converter is switched around its MSB, that is, from 1000 0000 0000 0000 to 0111 1111 1111 1111 to reflect a polarity change, the internal network of resistors must be switched. Current fluctuations and variations in bit switching speeds can conspire to create differential nonlinearity and glitches, as shown in Fig. 4.5. Collectively, these defects are known as zero-cross distortion. Because musical waveforms continually change (twice each cycle) between positive and negative polarity, the zero axis is crossed repeatedly as the MSB is turned on and off. The error is particularly troublesome when reproducing low-level signals because the fixed-amplitude glitch may be proportionally large with respect to the signal. Furthermore, the audibility of zero-cross distortion can be aggravated by dithering because of the increase in the number of transitions around the MSB.

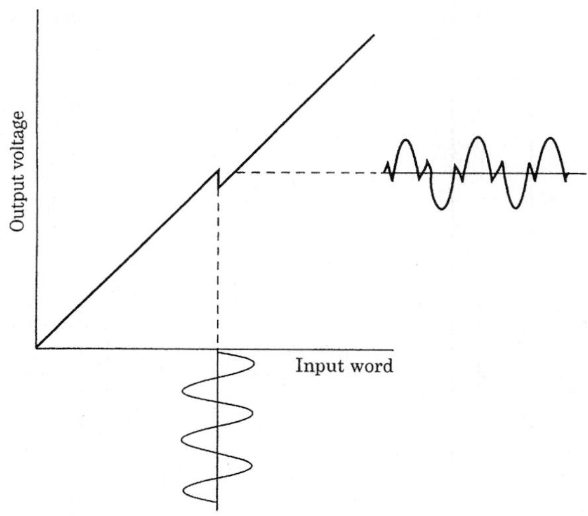

FIGURE 4.5 Crossover distortion occurs at the zero-cross point when the input word of a D/A converter changes polarity, resulting in a glitch in the output waveform.

Ideally, when there is no error difference between the smallest and largest resistors in the ladder, there is no output error. However, the MSB generally has large error compared to the LSB; the error can be larger than the value of the LSB itself. This is because the MSB in a 16-bit converter must be represented by a current value with an accuracy greater than 1/65,536 of the LSB. (This demonstrates why adjustment of the MSB is paramount in achieving converter linearity.) Similarly, error can occur when the 2nd, 3rd, and 4th bits are switched; however, the error proportionally decreases as the signal level increases. By the same token, error is relatively greatest for low-level audio signals. When longer word-length converters are used, it is proportionally difficult to match resistor values.

Zero-cross distortion can be alleviated by careful calibration of the converter's most significant bits. Alternatively, a sign-magnitude configuration can be used; the output amplifier switches between each positive and negative input code excursion. As a result, the LSB toggles around bipolar-zero, minimizing zero-cross distortion. Zero-cross distortion can also be reduced by providing a complementary D/A converter for each waveform polarity. Sign-magnitude code values are supplied to both parallel converters. In this way, total switching of digits across bipolar-zero never occurs. Zero-cross distortion is not present in sigma-delta D/A converters, as described in Chap. 18.

High-Bit D/A Conversion

Many manufacturers use D/A converters with 18-, 20- or 24-bit resolution in reproduction systems to provide greater playback fidelity for 16-bit recordings. The rationale for this lies in flaws that are inherent in D/A converters. Except in theory, 16-bit converters cannot fully decode a 16-bit signal without a degree of error. When, for example, 18 bits are derived from 16-bit data and converted with 18-bit resolution, errors can be reduced and reproduction specifications improved. To realize the full potential of audio fidelity, the signal digitization and processing steps must have a greater dynamic range than the final recording.

The choice of 16-bit words for the CD and other formats was determined primarily by the availability of 16-bit D/A converters and the fact that longer word lengths diminish playing time. However, 18-bit converters, for example, can provide better conversion of the stored 16 bits. When done correctly, 18-bit conversion improves amplitude resolution by ensuring a fully linear conversion of the 16-bit signal. An 18-bit D/A converter has 262,144 levels, exactly four times as many output levels as a 16-bit converter. Any nonlinearity is correspondingly smaller, and increasing the quantization word length at the conversion stage results in an increase in signal-to-noise (S/N) ratio. Simultaneously, any quantization artifacts are diminished. In other words, in this example, an 18-bit D/A converter gives better 16-bit conversion. In fact, the two extra bits of a linear 18-bit converter do not have to be connected to yield improved 16-bit performance.

The intent of high-bit D/A converters can be compared to that of oversampling: while the sampling frequency is increased, the method does not create new information; it makes better use of existing information. Oversampling provides the opportunity for high-bit conversion of 16-bit data. When a 44.1-kHz, 16-bit signal is oversampled, both the sampling frequency and number of bits are increased—the former because of oversampling, and the latter because of filter coefficient multiplication. The digital filter must be appropriately designed so the output word contains useful information in the bits below the 16-bit level.

It is not meaningful to gauge the performance of a D/A or A/D converter by its word length. More accurately, the S/N ratio measures the ratio between the maximum signal and the noise in the absence of signal. Because systems mute the output when there is a null signal, low-level error is removed. A dynamic range measurement is more useful when measured as the ratio of the maximum signal to the broadband noise (0 Hz to 20 kHz) using a –60-dBFS signal. This provides a measure of low-level distortion. Using the dynamic range measurement, a converter's ENOB (effective number of bits) can be calculated:

$$\text{ENOB} = \frac{\text{dynamic range} - 1.72}{6.02}$$

For example, a 16-bit converter with a dynamic range of 90 dB provides 14.7 bits of resolution; 1.3 bits are mired in distortion and noise.

In many designs, sigma-delta D/A converters are used. They minimize many problems inherent in traditional converter design. Sigma-delta systems use very high oversampling rates, noise shaping, and one- or multi-bit conversion. A true one-bit system outputs a binary waveform, at a very high rate, to represent the audio waveform. Other multi-bit systems output a multi-value step signal that forms a pulse-width modulation (PWM) representation of the audio signal. Because of inherently high noise levels, sigma-delta systems must use noise-shaping algorithms. Sigma-delta conversion is discussed in Chap. 18.

Although sigma-delta converters are widely used, traditional converters can offer some advantages. Because traditional converters do not employ noise shaping, they do not yield high out-of-band noise. In addition, at very low signal levels, the binary-weighted current sources are not heavily switched, so noise and glitching artifacts are very low.

Output Sample-and-Hold Circuit

Many digital audio systems contain two sample-and-hold (S/H) circuits. One S/H circuit at the input samples the analog value and maintains it while A/D conversion occurs. Another S/H circuit on the output samples and holds the signal output from the D/A converter, primarily to remove irregular signals called switching glitches. Because it can also compensate for a frequency response anomaly called aperture error, the output S/H circuit is sometimes called the aperture circuit.

Many D/A converters can generate erroneous signals, or glitches, which are superimposed on the analog output voltage. Digital data input to a D/A converter may require time to stabilize to the correct binary levels. In particular, in some converters, input bits might not switch states simultaneously. For example, during an input switch from 01111111 to 10000000, the MSB might switch to 1 before the other bits; this yields a momentary value of 11111111, creating an output voltage spike of one-half full scale. Even D/A converters with very fast settling times can exhibit momentary glitches. If these glitches are allowed to proceed to the digitization system's output, they are manifested as distortion.

An output S/H circuit can be used to deglitch a D/A converter's output signal. The output S/H circuit acquires voltage from the D/A converter only when that circuit has reached a stable output condition. That correct voltage is held by the S/H circuit during the intervals when the D/A converter switches between samples.

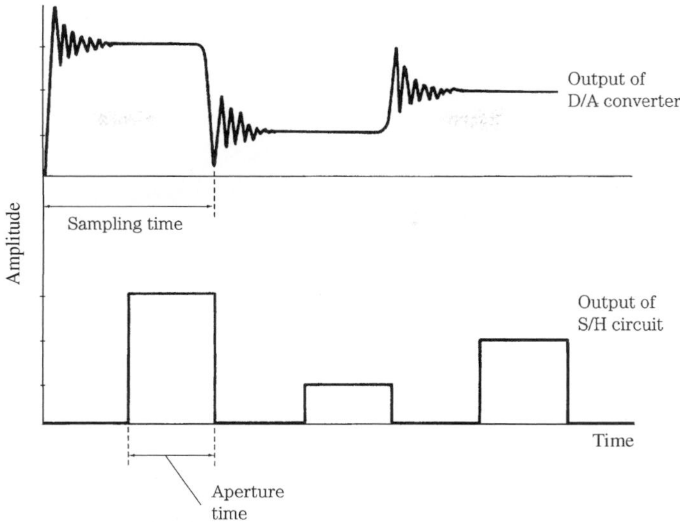

Figure 4.6 An output S/H circuit can be used to remove glitches in the signal output from some D/A converters.

This ensures a glitch-free output pulse-amplitude modulation (PAM) signal. The operation of an output S/H circuit is shown in Fig. 4.6.

From a general hardware standpoint, the output S/H circuit is designed similarly to the input S/H circuit. In some specifications, such as droop, the output S/H circuit might be less precise. Any droop results in only a dc shift at the digitization system's output, and this can be easily removed. In other respects, the output S/H circuit must be carefully designed and implemented. Because of its differing utility, the output S/H circuit requires attention to specifications such as hold time and transition speed from sample to sample.

The S/H circuit is occasionally used to correct for aperture error, an attenuation of high frequencies. Different approaches can also be used. Aperture error stems from the width of the output samples. In this case, the narrower the pulse width, the less the aperture error. Given ideal (instantaneous) A/D sampling, the output of an ideal D/A converter would be an impulse train corresponding to the original sample points; there would be no high-frequency attenuation. However, an ideal D/A converter is an impossibility. The PAM staircase waveform comprises pulses each with a width of one sample period. (Mathematically, the function is a convolution of the original samples by a square pulse with width of one sample period.) The spectrum of a series of pulses of finite width naturally attenuates at high frequencies. This differs from the original flat response of infinitesimal pulse widths; thus, a frequency response error called aperture error results, in which high audio frequencies are attenuated. Specifically, the frequency response follows a lowpass $|\sin(x)/x|$ function. When the output pulse width is equal to the sample period, the frequency response is zero at multiples of the sampling frequency, as shown in Fig. 4.7A. The attenuation of the in-band high-frequency response and that of high-frequency images can be observed.

At the Nyquist frequency, the function's value is 0.64, yielding an attenuation of 3.9 dB. This can be addressed with S/H circuits by approximating the impulse train output from an ideal D/A converter; the duration of the hold time is decreased in the S/H circuit.

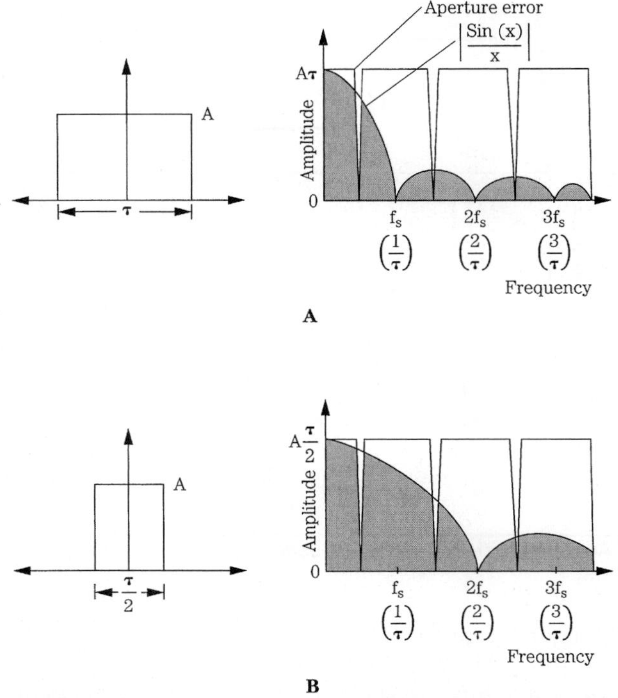

FIGURE 4.7 Aperture error can be minimized by decreasing the output pulse width. A. A pulse width equal to the sample period yields an in-band high-frequency attenuation. B. In-band high-frequency response improves when pulse width is one-half the sample period.

Sampling theory demonstrates that the bandwidth of the response is determined by the pulse width. The shorter the duration of the pulse, the greater the bandwidth. Specifically, if the output pulse width is narrowed with a shorter hold time, the attenuation at the half-sampling frequency can be decreased, as shown in Fig. 4.7B, where the pulse width is halved. If hold time is set to one-quarter of a sample period, the amplitude at the Nyquist frequency is 0.97, yielding an attenuation of 0.2 dB. This is considered optimal because a shorter hold time degrades the S/N ratio of the system.

Another remedy for aperture error is frequency compensation of the data prior to D/A conversion; this can be built into the digital lowpass filter that precedes the S/H circuit. This high-frequency boost, offset by aperture error, produces a net flat response, and the S/N ratio is not degraded. A boost could also be designed into the succeeding analog lowpass filter. Alternatively, a pre-emphasis high-frequency boost could be applied at the system input; the naturally occurring de-emphasis at the output S/H circuit would result in a flat response.

When an output S/H circuit is used to eliminate switching errors caused by the D/A converter during transition, the S/H circuit must avoid introducing transition errors of its own. An S/H circuit outputs a steady value while in the hold mode. When switching to the sample (or acquisition) mode, a slow transition introduces incorrect intermediate values into the staircase voltage. This problem is extraneous in the input

S/H circuit because the A/D converter accomplishes its digitization during the hold mode and ignores the transition mode. However, the output S/H circuit is always connected to the system output and any transition error appears at the output. In other words, not only are the levels themselves part of the output signal, but the way in which the S/H circuit moves from sample to sample is included as well.

Distortion is greatest for high frequencies because they have a large difference between values. For example, with a 48-kHz sampling frequency, a 20-Hz signal does not change appreciably in one sampling interval; however, a high-level 20-kHz signal will traverse almost the full amplitude range. Although the distortion products themselves can be removed by the output filter, these products can internally beat with the sampling frequency to generate in-band distortion as well. To overcome this problem, the output S/H circuit must switch as quickly as possible from hold to sample mode. A square-wave response would be ideal.

In theory, this eliminates the possibility of distortion caused by transition; however, in practice, it is impossible to achieve the necessary high slew rate, calculated to be as high as 5 V/ns (volts per nanosecond). Thus, an additional modification to the basic S/H circuit can be applied. An exponential change in amplitude from one quantization interval to the next does not create nonlinearity in the signal. Following output filtering, this exponential acquisition results in a linear response. It can be shown that an exponential transition from sample to sample causes only a slight high-frequency de-emphasis at the output, but no distortion or nonlinearity. An S/H circuit that integrates the difference between its present and next value yields such an exponential transition. The attenuation of high frequencies in an integrate-and-hold circuit is less than that produced by the sample-and-hold process itself, and also can be equalized.

The output S/H circuit thus removes switching glitches from the D/A converter's output voltage. Hold time can be set to less than a sample period to minimize aperture error. Many D/A converters are stringently designed to avoid switching glitches, and thus operate without an S/H circuit; aperture error is corrected in the digital filter. In some cases, the S/H function is included in the D/A converter chip. Whichever method is used, the PAM staircase analog signal is ready for output filtering, and final reconstruction.

Output Lowpass Filter

The first and last circuits in an audio digitization system are lowpass filters, known as anti-aliasing and anti-imaging filters, respectively. Although their analog designs can be almost identical, their functions are very different. In lieu of classic analog anti-imaging filters, digital filters using oversampling techniques have replaced brick-wall analog filters. However, even digital filters employ a low-order analog lowpass filter on the final output of the system.

Given the criteria of the Nyquist sampling theorem, the function of the input lowpass filter is clear: it removes all frequency content above the half-sampling frequency to prevent aliasing. Similarly, a lowpass filter at the output of the digitization system removes frequency content above the half-sampling frequency. However, its function is different. This filter converts the D/A converter's output pulse-amplitude modulation (PAM) staircase to a smoothly continuous waveform, thus reconstructing the bandlimited input signal. The PAM staircase is an analog waveform, but it contains modulation artifacts of the sampling process that create high-frequency components not present in the original signal. An output lowpass filter converts the staircase into a smoothly continuous waveform by removing the high-frequency components, leaving the original waveform. The staircase signal is smoothed; the output filter is sometimes called a smoothing filter.

The conceptual design criteria for an analog output lowpass filter are similar to those of the input filter. The passband should be flat and the stopband highly attenuated. The cutoff slope should be steep. Audibility of any phase shift must be considered. One criteria unique to output lowpass filters is transient response. Unlike the input filter, it must process the amplitude changes in the staircase waveform. Just as a slow output S/H circuit can introduce distortion, the output filter can create unwanted byproducts if its transient response is inadequate. One consideration not commonly addressed is the possible presence of extreme high-frequency components of several megahertz that might be contained in the output signal. Because of its high-speed operation, digital processing equipment can create this noise, and the filtering characteristics of some audio lowpass filters do not extend to those frequencies.

Viewing the output filtering process from a more mathematical standpoint, we can observe how sampling creates the need for filtering. Sampling multiplies the time domain audio signal with the time domain sampling (pulse) signal. In terms of the spectra of these two sampled signals, this convolution produces a new sampled spectrum identical to the original unsampled spectrum. However, additional spectra are infinitely repeated across the frequency domain at multiples of the sample frequency. For example, an original 1-kHz sine wave sampled at 48 kHz also creates components at 47, 49, 95, 97 kHz, and so on. Although the sample-and-hold process substantially reduces the amplitude of the extraneous frequency bands, significant components still remain after the S/H circuit, particularly in the region near the audio band (see Fig. 4.7). To convert the sampled information back into correct analog information, the image spectra must be removed, leaving only the original baseband spectrum. This is accomplished by output lowpass filtering.

Some might question the need to filter out frequencies above the Nyquist frequency because they lie above the presumed limit of human audibility. The original waveform is reproduced without filtering, but filtering is needed because the accompanying spectra could cause modulation in other downstream equipment through which the signal passes. This in turn could negatively affect the audio signal. Other digital systems might be immune because their input filters remove the high frequencies, but oscillators in analog recorders or transmitters could conceivably create difference frequencies in the audible band. Clearly, in systems with lower sampling frequencies (e.g., 8 kHz) any images above the Nyquist frequency would be audible, and thus must be removed.

Impulse Response

Logically, when an impulse is input to a device such as a lowpass filter, the output is the filter's impulse response. As explained in Chap. 17, the impulse response can completely characterize a system: a filter can be described by its time-domain impulse response or its frequency response; these are related by the Fourier transform. Furthermore, multiplying an input spectrum by a desired filter transfer function in the frequency domain is equivalent to convolving the input time-domain function with the desired filter's impulse response in the time domain. An ideal lowpass filter with a brick-wall response in the frequency domain, shown in Fig. 4.8A, displays an impulse response in the time

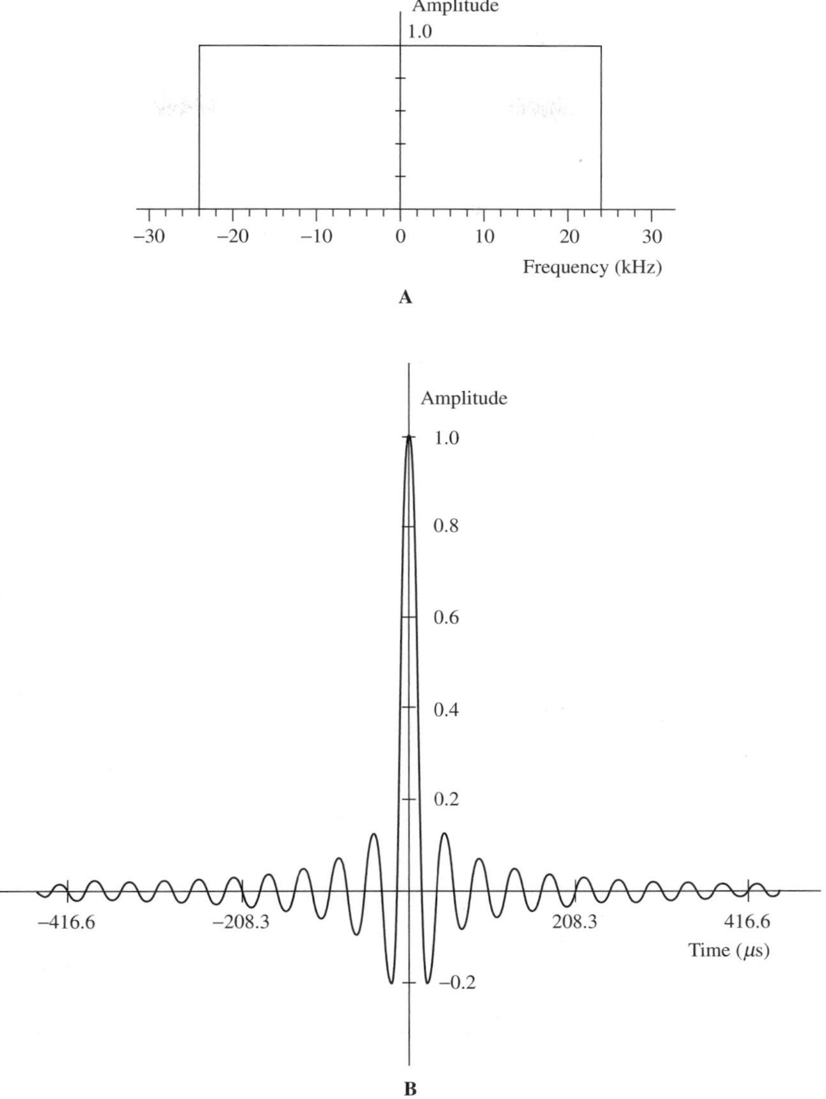

FIGURE 4.8 An impulse response completely characterizes a system. A. The frequency response of an ideal brick-wall lowpass filter with a 24-kHz cutoff frequency. B. The time-domain impulse response of the ideal brick-wall filter shown above. The $\sin(x)/x$ function passes through 0 at multiples of the sampling period of 1/48,000 Hz. The ringing thus occurs at 24 kHz.

domain that takes the form of $\sin(x)/x$, as shown in Fig. 4.8B. The $\sin(x)/x$ function is also known as the sinc function. As we will see, this is an important key to understanding how a digital audio system can reconstruct the output waveform.

The action of lowpass filtering, and the impulse response, is summarized in Fig. 4.9. In this example, the input signal comprises two sine waves of different frequencies

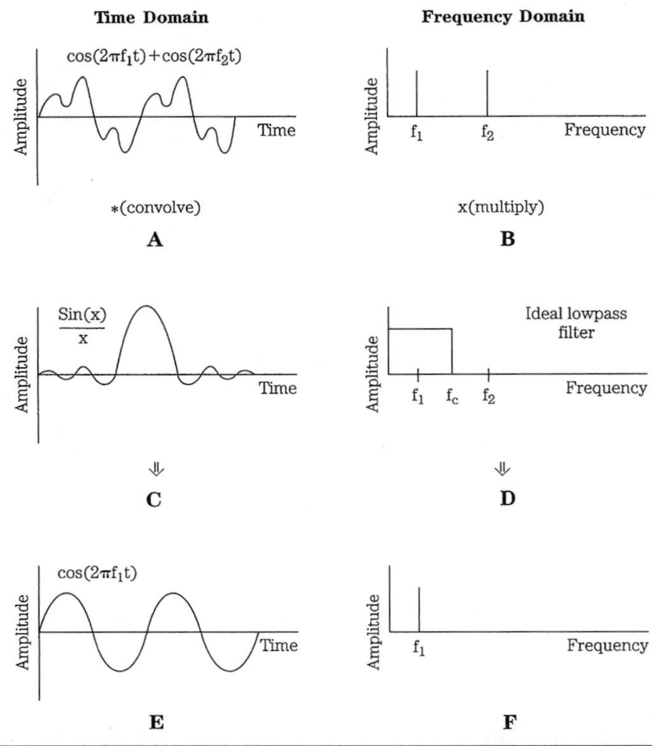

Figure 4.9 An example of lowpass filtering shown in the time domain (left column) and the frequency domain (right column). A. The input signal comprises two sine waves. B. Spectrum of the input signal. C. Impulse response of the desired lowpass filter is a $\sin(x)/x$ function. D. Desired lowpass filter transfer function. E. Output filtered signal is the convolution of the input with the impulse response. F. Spectrum of the output filtered signal is the multiplication of the input by the transfer function.

(Fig. 4.9A); their spectra consist of two lines at the sine wave frequencies (Fig. 4.9B). Suppose that we wish to lowpass filter the signal to remove the higher frequency. This is accomplished by multiplying the input spectra with an ideal lowpass filter characteristic (Fig. 4.9D), producing the single spectral line (Fig. 4.9F). The time-domain impulse response of this ideal lowpass filter follows a $\sin(x)/x$ function (Fig. 4.9C); if the input time-domain signal is convolved with the $\sin(x)/x$ function, the result is the filtered output time-domain signal (Fig. 4.9E). In other words, a time-domain, sampled signal can be filtered by applying the time-domain impulse response that describes the filter's characteristic. In digital systems, both the signal and the impulse response are represented by discrete values; the impulse response of an ideal reconstruction filter is zero at all sample times, except at one central point.

An ideal brick-wall output filter reconstructs the audio waveform. Although the idea of smoothing the output samples to remove high frequencies is essentially correct, a more analytical analysis shows exactly how waveform reconstruction is accomplished by lowpass filtering. An ideal brick-wall lowpass filter is needed to exactly reconstruct the output waveform from samples; the sampling theorem dictates this. As noted, an

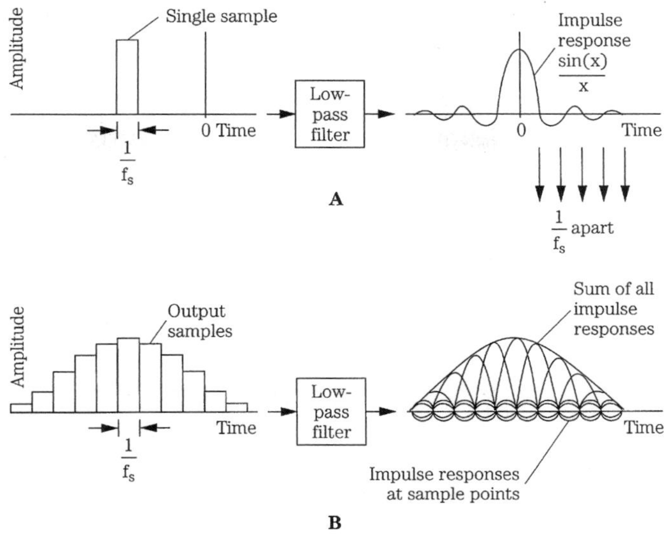

FIGURE 4.10 The impulse response of an ideal lowpass filter reconstructs the output analog waveform. A. The impulse response of an ideal lowpass filter is a sin(*x*)/*x* function; it is truncated in this drawing. B. When a series of impulses (samples) are lowpass-filtered, the individual impulse responses sum to form the output waveform.

ideal brick-wall lowpass filter has a $\sin(x)/x$ impulse response. When samples of a bandlimited input signal are processed (convolved) with a $\sin(x)/x$ function, the bandlimited input signal represented by the samples will be exactly reproduced; the sampling theorem guarantees this.

Specifically, as shown in Fig. 4.10A, when a single rectangular audio sample passes through an ideal lowpass filter, after a filter delay it emerges with a $\sin(x)/x$ response. If the lowpass filter has a cutoff frequency at the half-sampling point ($f_s/2$) then the $\sin(x)/x$ curve passes through zero at multiples of $1/f_s$. When a series of audio samples pass through the filter, the resulting waveform is the delayed summation of all the individual $\sin(x)/x$ contributions of the individual samples, as shown in Fig. 4.10B. Each sample's impulse response is zero at the $1/f_s$ position of any other sample's maximum response. The output waveform has the value of each sample at sample time, and the summed response of the impulse responses of all samples between sample times. This superposition, or summation, of individual impulse responses forms all the intermediate parts of the continuous reconstructed waveform. In other words, when the high-frequency staircase components are removed, the remaining fundamental passes exactly through the same points as the original filtered waveform.

Output filtering, with an ideal lowpass filter, yields an output audio waveform that is theoretically identical to the input bandlimited audio waveform. In other words, it is the filter's impulse response to the audio samples that reconstructs the original audio waveform. To summarize mathematically, we see that an ideal brick-wall filter outputs a $\sin(x)/x$ function at each sample time; moreover, the amplitude of each of these functions corresponds to the amplitude of the sample; when all the $\sin(x)/x$ functions are summed, the audio waveform is reconstructed.

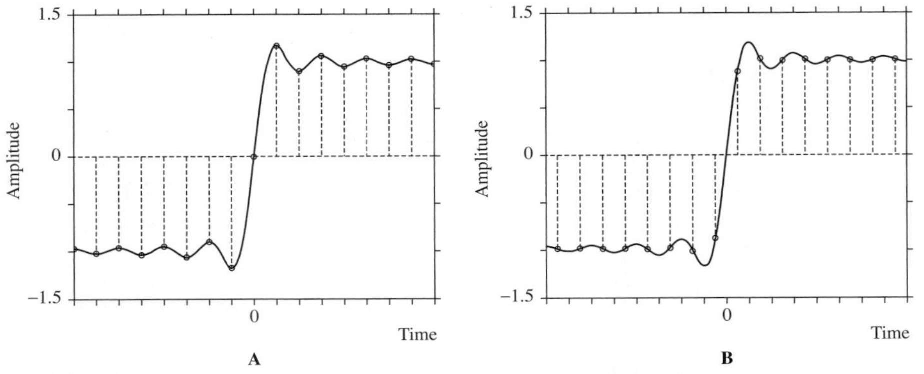

FIGURE 4.11 An audio signal is correctly reconstructed regardless of the sample location relative to the waveform timing. A. A bandlimited step is sampled with a sample at a zero-crossing time. B. A bandlimited step is sampled with samples symmetrical about the zero-crossing time. In these cases, and in other cases where samples occur at other times, the same waveform is reconstructed. (*Lipshitz and Vanderkooy, 2004*)

To return to a question posed in Chap. 2, we can again ask whether samples can capture waveform information that occurs between samples. The impulse response shows that they do. Moreover, the capture is successful no matter where the samples occur in relation to the waveform's timing. Stanley Lipshitz and John Vanderkooy have concisely illustrated this. In Fig. 4.11A, a sample happens to fall exactly at the zero-crossing of a step waveform. The bandlimited step waveform (the heavy line) is reproduced (by the summed $\sin(x)/x$ functions). In Fig. 4.11B, samples fall at either side of the zero-crossing. Again, the waveform (heavy line) is exactly reconstructed. Similarly, the same waveform would be reconstructed no matter where the samples lie, and the zero-crossing time is the same. Also, nonintuitively, a higher sampling frequency does not somehow improve the resolution of the successful capture and reconstruction of this bandlimited waveform.

Digital Filters

Because of the phase shift and distortion they introduce, analog brick-wall output filters have been abandoned by manufacturers, in favor of digital filters. A digital filter is a circuit (or algorithm) that accepts audio samples and outputs audio samples; the values of the throughput audio samples are altered to produce filtering. When used on the output of a digital audio system, the digital filter simulates the process of ideal lowpass filtering and thus provides waveform reconstruction. Rather than suppress high-frequency images after the signal has been converted to analog form, digital filters perform the same function in the digital domain, prior to D/A conversion. Following D/A conversion, a gentle, low-order analog filter removes remaining images that are present at very high frequencies. In most cases, finite impulse response (FIR) digital filters are used; oversampling techniques allow a less complex digital filter design. With oversampling, additional sample values are computed by interpolating between original samples. Because additional samples are generated (perhaps two, four, or eight times as many), the sampling frequency of the output signal is greater than the input signal. A transversal filter architecture is typically used; it comprises a series of delays, multipliers, and adders.

The task of an oversampling filter is twofold: to resample, and to filter through interpolation. The signal input to the filter is sampled at f_s, and has images centered around multiples of f_s, as shown in Figs. 4.12A and B. Resampling begins by increasing the sampling frequency. That is accomplished by injecting zero-valued samples between original samples at some interpolation ratio. For example, for four-times oversampling, three zero-valued samples are inserted for every original sample as shown in Fig. 4.12C. The oversampling sampling frequency equals the interpolation ratio times the input sampling frequency, but the spectrum of the oversampled signal is the same as the original signal spectrum, as shown in Fig. 4.12D. In this case, for example, with an input sampling frequency of 48 kHz, the oversampling frequency is 192 kHz. This data enters the lowpass digital filter with a cutoff frequency of $f_s/2$, operating at an effective frequency of 192 kHz. Although the original data was sampled at 48 kHz, with oversampling it is conceptually indistinguishable from data originally sampled at 192 kHz. The zero-packed signal is passed through a digital lowpass filter with the time-domain impulse response shown in Fig. 4.12E. The lowpass filter's frequency-domain transfer function is shown in Fig. 4.12F. The fixed sample values in Fig. 4.12E, occurring at the oversampling frequency, comprise the coefficients of the digital transversal lowpass filter. The input (oversampled) samples convolved with the impulse response coefficients yield the filter's output. In particular, the filter's output is an interpolated digital signal, as shown in Fig. 4.12G, with images centered around multiples of the higher oversampling frequency f_a, as shown in Fig. 4.12H.

To recapitulate, interpolation is used to create the intermediate sample points. In a four-times oversampling filter, the filter outputs four samples for every original sample. However, to be useful, these samples must be computed according to a certain algorithm. Specifically, each intermediate sample is multiplied by the appropriate $\sin(x)/x$ coefficient corresponding to its contribution to the overall impulse response of the lowpass filter in the time domain (see Fig. 4.10B). The $\sin(x)/x$ function in the time domain has zeros exactly aligned with sample times, except at the sample currently being interpolated. Thus, every interpolated sample is a linear combination of all other input samples, weighted by the $\sin(x)/x$ function. The multiplication products are summed together to produce the output-filtered sample. The conceptual operation of a digital filter thus corresponds exactly to the summed impulse response operation of an ideal analog brick-wall filter. Spectral images appear at multiples of the oversampled sampling frequency. Because the distance between the baseband and sidebands is larger, a gentle low-order analog filter can remove the images without causing a phase shift or other artifacts.

The oversampling ratio can be defined as

$$R = \frac{f_a}{f_s}$$

where f_a = the oversampling frequency and f_s = the input sampling frequency.

Oversampling initially requires insertion of $(R - 1)$ zero samples per input sample. The samples created by oversampling must be placed symmetrically between the input samples. A lowpass filter is used to bandlimit the input data to $f_s/2$, with spectral images

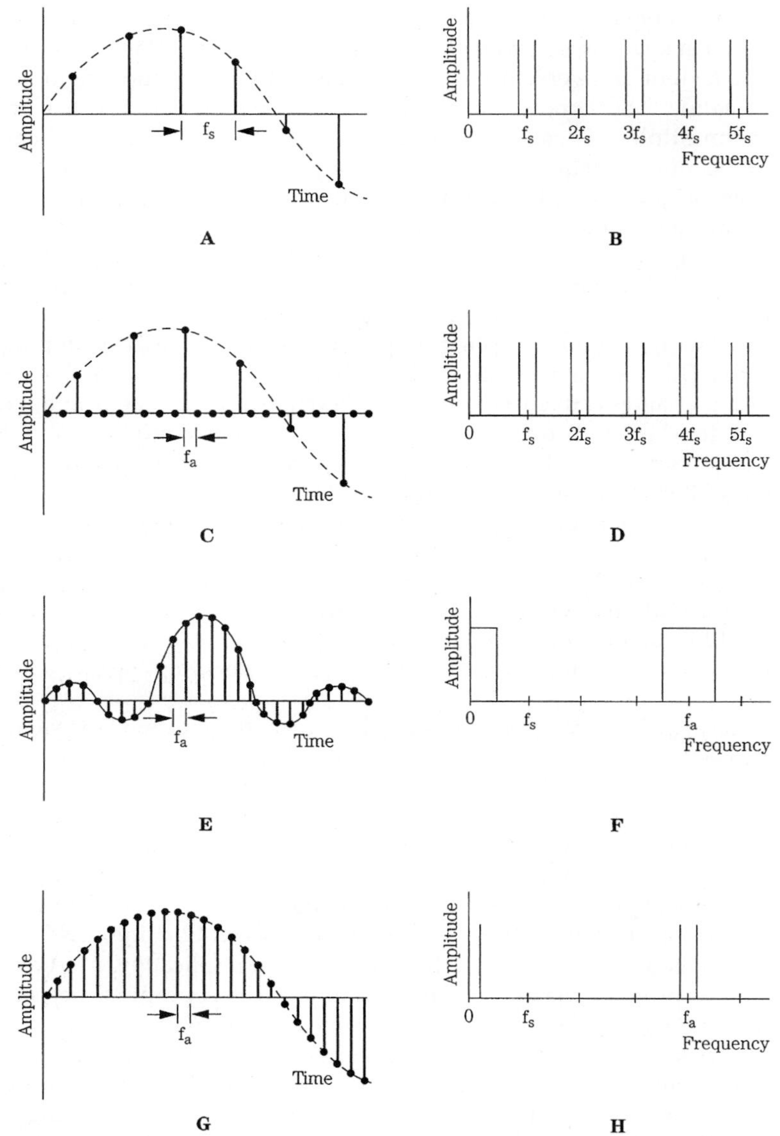

FIGURE 4.12 An oversampling filter resamples and interpolates the signal, using the impulse response; this is shown in the time domain (left column) and frequency domain (right column). A. The input signal is sampled at f_s. B. The signal spectrum has images centered around multiples of f_s. C. With resampling, zero-valued samples are placed between original samples at some interpolation ratio. D. The spectrum of the oversampled signal is the same as the original signal spectrum. E. The values of a sampled impulse response correspond to the coefficients of the digital filter. F. The transfer function of the filter shows passbands in the audio band and oversampling band. G. The digital filter performs interpolation to form new output sample values. H. The output-filtered signal has images centered around multiples of the oversampling frequency, f_a.

at integer multiples of ($R \times f_s$). Moreover, lowpass filtering creates the intermediate sample values through interpolation. Rather than perform multiplication on zero samples, redundancy can be used advantageously to design a more efficient filter.

FIR Oversampling Filter

As we have seen, a digital filter can use an impulse response as the basis of the computation to filter the audio signal. When the impulse response uses a finite number of points, the filter is called a finite impulse response (FIR) filter. In addition, in the case of an output lowpass filter, oversampling is used to extend the Nyquist frequency.

Figure 4.13 shows a four-times oversampling FIR digital filter, generating three intermediate samples between each input sample. The filter consists of a shift register of 24 delay elements, each delaying a 16-bit sample for one input sampling period. Thus each sample remains in each delay element for a sample period before it shifts to the next delay element. During this time each 16-bit sample is tapped off the delay line and multiplied four times by a 12-bit coefficient stored in ROM, a different coefficient for each multiplication, with different coefficients for each tap. In total, the four sets of coefficients are applied to the samples in turn, thus producing four output values. The 24 multiplication products are summed four times during each period, and are output from the filter. The filter characteristic ($\sin(x)/x$ in this case) determines the values of the interpolated samples. Each 16-bit data word is passed to the next delay (traversing the entire delay line), where the process is repeated. Many samples are simultaneously present in the delay line, and the computed impulse responses of these many samples are summed. The product of each multiplication is a 28-bit word ($16 + 12 = 28$). When these products are summed, a weighted average of a large number of samples is obtained. Four times as many samples are present after oversampling, with interpolation

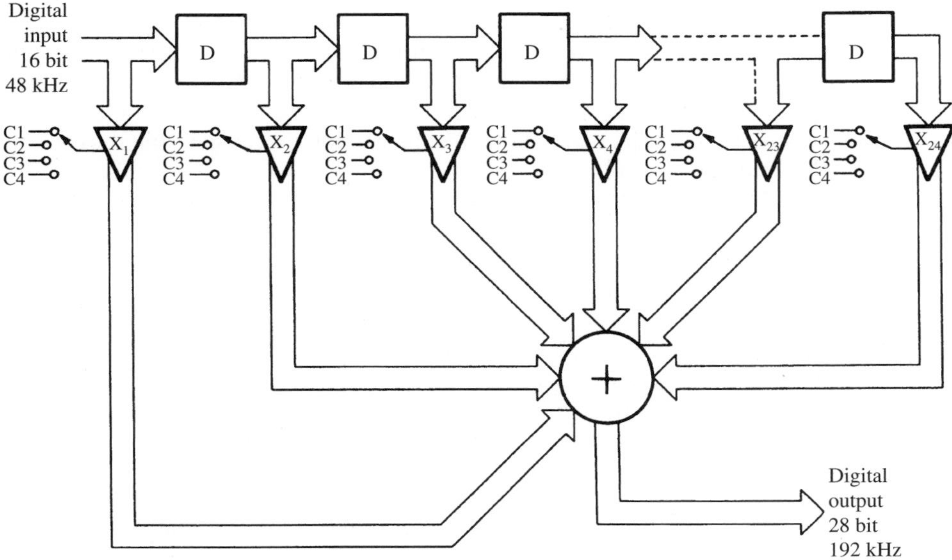

FIGURE 4.13 Twenty-four element digital transversal filter showing a tapped delay line, coefficient multipliers, and an adder.

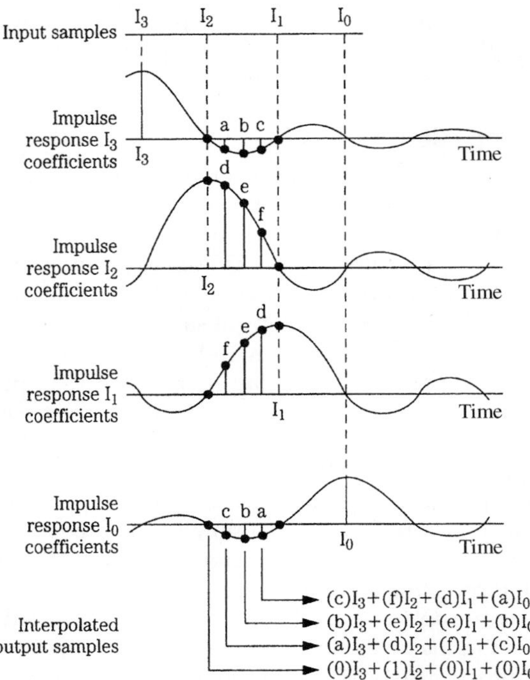

Figure 4.14 A four-times oversampling filter treats input samples as points on a sin(x)/x curve; this reconstructs the output waveform, as in an ideal lowpass filter. In practice, many more samples are needed.

values calculated by the filter. In this example, the sampling frequency is increased four times, to 192 kHz. The result is the multiplication of the sampling frequency, and a cutoff filter characteristic. Through proper selection of filter coefficients, length of delay line, and time interval between taps, the desired lowpass sin(x)/x response is obtained—yielding correct waveform reconstruction. Because of the movement of the data across the shift register, this design is often called a transversal filter.

Figure 4.14 illustrates how an oversampling digital filter simulates the effect of an analog filter in waveform reconstruction; specifically, it shows how interpolated samples are computed in a four-times oversampling filter. The input samples I_3, I_2, I_1, and I_0 are treated as sin(x)/x impulses placed relative to the center of the filter. Their maximum sin(x)/x impulse response amplitudes are equal to the original sample amplitudes, and the width of the impulse responses are determined by the response of the filter, in this case a filter with a cutoff frequency at the Nyquist frequency of the input samples. The summation of their unique contributions forms the interpolated samples (in practice, as noted, many more than four samples would be present in the filter). Each of the three interpolated samples is formed by adding together the four products, as shown in the figure. Original input samples pass through the filter unchanged by using one set of filter coefficients that contains three zero coefficients and one unity coefficient. In this way, the output sample that coincides with the input sample is unchanged. By multiplying each group of samples (four in this case) with this coefficient set, and three others, interpolated samples are output at a four-times rate.

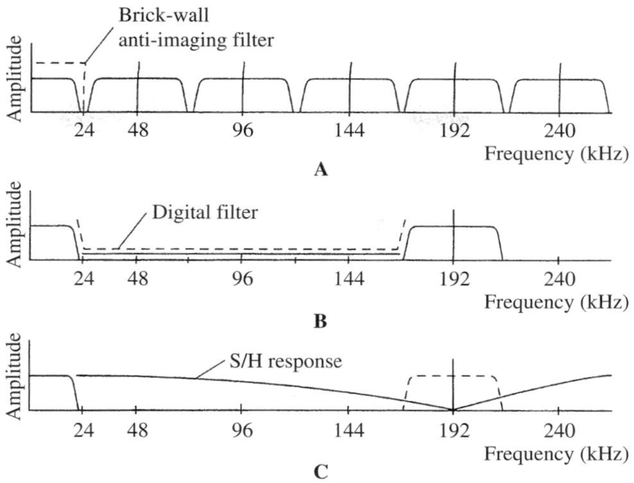

FIGURE 4.15 Image spectra in nonoversampled and oversampled reconstructions. A. A brick-wall filter must sharply bandlimit the output spectra. B. With four-times oversampling, images appear only at the oversampling frequency. C. The output S/H circuit can be used to further suppress the oversampling spectra.

The overall effect of four-times oversampling filtering is shown in Fig. 4.15. A brick-wall filter must sharply bandlimit the output spectra; with oversampling filtering, the images between 24 kHz and 168 kHz (centered at 48, 96, and 144 kHz) are suppressed, leaving the oversampling images. The output S/H circuit can be used to further suppress the oversampling spectra, as shown. In practice, a number of considerations determine the design of oversampling filters. The $\sin(x)/x$ waveform extends to infinity in both the positive and negative directions, so theoretically all the values of that infinite waveform would be required to reconstruct the analog signal. Although an analog filter can theoretically access an infinite number of samples for each reconstruction value, a finite impulse response filter, as its name implies, cannot. Thus the oversampling filter is designed to accommodate the number of samples required to maintain an error less than the system's overall resolution. In the example above, only four coefficients were used, but as many as 300 28-bit coefficients might be needed; in practice, perhaps 100 coefficients would suffice. However, because the $\sin(x)/x$ response is symmetrical, only half the coefficients need to be stored in memory; the table can be read bidirectionally. As noted, the multipliers in a digital filter increase the output word length. The output word cannot simply be truncated; this would increase distortion. The word should be redithered.

These examples have used a four-times oversampling filter. However, two-times, four-times, or (most commonly) eight-times digital filters can be used, in which a 48-kHz sampling frequency is oversampled to 96, 192, or 384 kHz, respectively. For example, in an eight-times oversampling filter, seven new audio samples are computed for each input sample, raising the output sampling frequency to 384 kHz. Images are shifted to a band centered at 384 kHz where they are easily removed with a simple analog lowpass filter. Whatever the rate, most oversampling digital filters are similar in operation. However, the characteristic of the analog lowpass filter varies. The lower the oversampling frequency, the closer to the audio band the image spectrum will be, hence the steeper the analog filter response required.

When using traditional D/A converters, a practical limit to the oversampling frequency is reached (around eight times) because most D/A converters cannot operate at faster speeds. On the other hand, D/A converters can more easily convert an oversampling waveform, because the successive changes in amplitude are smaller in an oversampled signal. More precisely, the slew rate, the rate of variation in the output waveform, is lower. This, along with less ringing and less overshoot, reduces intermodulation distortion. Some designers feel that in oversampling designs of eight times or more, the remaining image spectrum is at such a high frequency that analog filtering can be accomplished with a simple second-order lowpass filter.

In terms of performance, digital filters represent a great improvement over analog filters because a digital filter can be designed with considerable and stable precision. Because digital filtering is a purely numerical process, the characteristics of digital filtering, as opposed to analog filtering, cannot change with temperature, age, and so on. A digital filter might provide an in-band ripple of ±0.00001 dB, with a stopband attenuation of –120 dB. A digital filter can have a linear phase response. High-frequency attenuation due to aperture error can be compensated for in the transversal filter by choosing coefficients to create a slightly rising characteristic prior to cutoff. Digital filter theory is discussed in Chap. 17.

Noise Shaping

Another important benefit of oversampling is a decrease in audio band quantization noise. This is because the total noise power is spread over a larger (oversampled) frequency range. In particular, each doubling of the oversampling ratio R lowers the quantization noise floor by 3 dB ($10 \log(R) = 10 \log(2) = 3.01$ dB). For example, in a four-times oversampling filter, data leaves the filter at a four-times frequency, and the quantization noise power is spread over a band that is four times larger, reducing its power density in the audio band to one-fourth, as shown in Fig. 4.16. In this example, this yields a 6-dB reduction in noise (3 dB each time the sampling rate is doubled), which is equivalent to

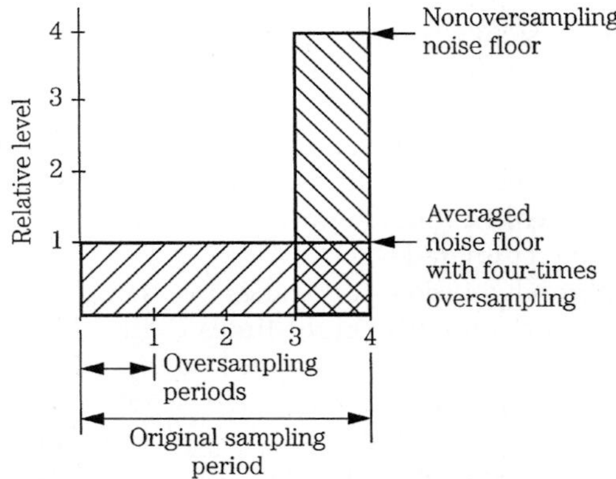

FIGURE 4.16 Oversampling extends quantization error over a larger band, correspondingly reducing in-band error.

adding one bit of word length. Higher oversampling ratios yield corresponding lower noise (for example, eight-times yields an additional 6 dB).

Noise shaping, also called spectral shaping, can significantly reduce the in-band noise floor by changing the frequency response of quantization error. The technique is widely used in both A/D and D/A conversion; for example, noise shaping can be accomplished with sigma-delta modulation. When quantization errors are independent, the noise spectrum is white; by selecting the nature of the dependence of the errors, the noise spectrum can be shaped to any desired frequency response, while keeping the frequency response of the audio signal unchanged. Noise shaping is performed by an error-feedback algorithm that produces the desired statistically dependent errors.

In a simple noise shaper, for example, 28-bit data words from the filter are rounded and dithered to create the most significant 16-bit words. The 12 least significant bits (the quantization error that is created) are delayed by one sampling period and subtracted from the next data word, as shown in Fig. 4.17A. The result is a shaped noise floor. The delayed quantization error, added to the next sample, reduces quantization error in the output signal; the quantization noise floor is decreased by about 7 dB in the audio band, as shown in Fig. 4.17B. As the input audio signal changes more rapidly, the effect of the error feedback decreases; thus quantization error increases with increasing audio frequency. For example, approaching 96 kHz, the error feedback comes back in phase with the input, and noise is maximized, as also shown in Fig. 4.17B. However, the out-of-band noise is high in frequency and thus less audible, and can be attenuated by the output filter. This trade-off of low in-band noise, at the expense of higher out-of-band noise, is inherent in noise shaping. Noise shaping is used, and must be used, in low-bit converters to yield a satisfactorily low in-band noise floor.

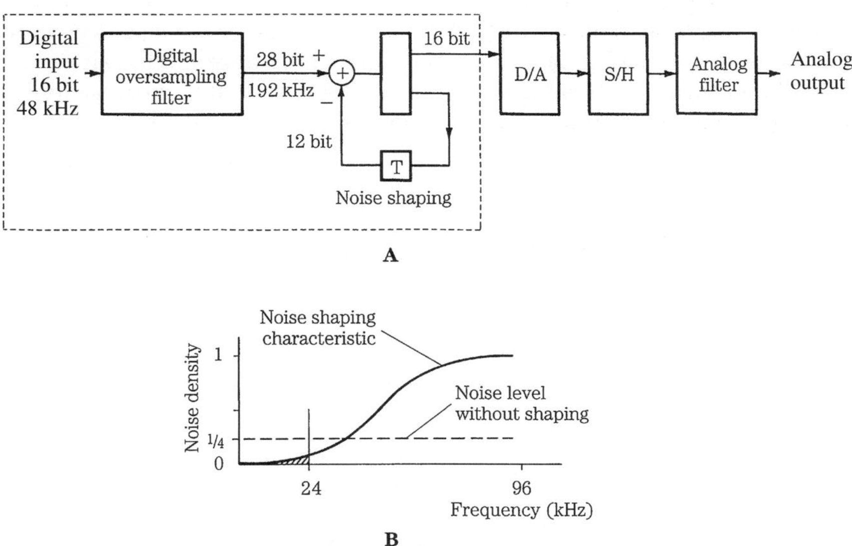

FIGURE 4.17 Noise shaping following oversampling decreases in-band quantization error. A. Simple noise-shaping loop. B. Noise shaping suppresses noise in the audio band; boosted noise outside the audio band is filtered out.

Noise shaping is often used in conjunction with oversampling. When the audio signal is oversampled, the bandwidth is extended, creating more spectral space for the elevated high-frequency noise curve.

In practice, oversampling sigma-delta A/D and D/A converters are widely used. For example, an oversampling A/D converter may capture an analog signal using an initial high sampling frequency and one-bit signal, downconvert it to 48 kHz and 16 bits for storage or processing, then the signal may be upconverted to a 64- or 128-times sampling frequency for D/A conversion. Sigma-delta modulation and noise shaping is discussed in Chap. 18.

Noise shaping can also be used without oversampling; the quantization error remains in the audio band and its average power spectral density is unchanged, but its spectrum is psychoacoustically shaped according to the ear's minimum audible threshold. In other words, the level of the noise floor is higher at some in-band frequencies, but lower at others; this can render it significantly less audible. Psychoacoustically optimized noise shaping is discussed in Chap. 18.

Output Processing

Following digital filtering, the data is converted back to analog form with a D/A converter. In the case of four-times oversampling, with a sampling frequency of 48 kHz, the aperture effect of an output S/H circuit creates a null at 192 kHz, further suppressing that oversampling image. As noted, designing a slight high-frequency boost in the digital filter can compensate for the slight attenuation of high audio frequencies. The remaining band around 192 kHz can be completely suppressed by an analog filter. This low-order anti-imaging filter follows the D/A converter. Because the oversampling image is high in frequency, a filter with a gentle, 12-dB/octave response and a –3 dB point between 30 and 40 kHz is suitable; for example, a Bessel filter can be used. It is a noncritical design, and its low order guarantees good phase linearity; phase distortion can be reduced to ±0.1° across the audio band. Oversampling can decrease in-band noise by 6 dB, and noise shaping can further decrease in-band noise by 7 dB or more. Thus, with an oversampling filter and noise shaping, even a D/A converter with 16-bit resolution can deliver good performance.

Alternate Coding Architectures

Linear PCM is considered to be the classic audio digitization architecture and is capable of providing high-quality performance. Other digitization methods offer both advantages and disadvantages compared to PCM. A linear PCM system presents a fixed scale of equal quantization intervals that map the analog waveform, while specialized systems offer modified or wholly new mapping techniques. One advantage of specialized techniques is often data reduction, in which fewer bits are needed to encode the audio signal. Specialized systems are thus more efficient, but a penalty might be incurred in reduced audio fidelity.

In a fixed linear PCM system, the quantization intervals are fixed over the signal's amplitude range, and they are linearly spaced. The quantizer word length determines the number of quantization intervals available to encode a sample's amplitude. The intervals are all of equal amplitude and are assigned codes in monotonic order. However, both of these parameters can be varied to yield new digitization architectures.

Longer word lengths reduce quantization error; however, this requires a corresponding increase in data bandwidth. Uniform PCM quantization is optimal for a uniformly distributed signal, but the amplitude distribution of most audio signals is not uniform. In some alternative PCM systems, quantization error is minimized by using nonuniform quantization step sizes. Such systems attempt to tailor step sizes to suit the statistical properties of the signal. For example, speech signals are best served by an exponential-type quantization distribution; this assumes that small amplitude signals are more prevalent than large signals. Many quantization levels at low amplitudes, and fewer at high amplitudes, should result in decreased error. Companding, with dynamic compression prior to uniform quantization, and expansion following quantization, can be used to achieve this result. Floating point systems use range-changing to vary the signal's amplitude to the converter, and thus expand the system's dynamic range. A greatly modified form of PCM is a differential system called delta modulation (DM). It uses only one bit for quantizing; however, a very high sampling frequency is required. Other forms of delta modulation include adaptive, companded, and predictive delta modulation. Each offers unique strengths and weaknesses. Low bit-rate, based on psychoacoustics, is discussed in Chaps. 10 and 11.

Floating-Point Systems

Floating-point systems use a PCM architecture modified to accept a scaling value. It is an adaptive approach, with nonuniform quantization. In true floating-point systems, the scaling factor is instantaneously applied from sample to sample. In other cases, as in block floating-point systems, the scale factor is applied to a relatively large block of data.

Instead of forming a linear data word, a floating-point system uses a nonuniform quantizer to create a word divided into two parts: the mantissa (data value) and exponent (scale factor). The mantissa represents the waveform's uniform value and its scaled amplitude; the quantization step size is represented by the exponent. In particular, the exponent acts as a scalar that varies the gain of the signal in the PCM A/D converter. By adjusting the gain of the signal, the A/D converter is used more efficiently. Low-level signals are boosted and high-level signals are attenuated; specifically, a signal's level is set to the highest possible level that does not exceed the converter's range. This effectively varies the quantization step size according to the signal amplitude and improves accuracy of low-level signal coding, the condition where quantization error is relatively more problematic. Following D/A conversion, the gain is again adjusted to correspond to its original value.

For example, consider a floating-point system with a 10-bit mantissa (A/D converter), and 3-bit exponent (gain select), as shown in Fig. 4.18. The 3-bit exponent provides eight different ranges for a 10-bit mantissa. This is the equivalent multiplicative range of 1 to 128. The maximum signals are –65,536 to 65,408. In this way, 13 bits cover the equivalent of a 17-bit dynamic range, but only a 10-bit A/D converter is required. However, large range and small resolution are not simultaneously available in a floating-point system because of its nonuniform quantization. For example, although 65,408 can be represented, the next smallest number is 65,280. In a linear PCM system, the next smallest number is, of course, 65,407. In general, as the signal level increases, the number of possible quantization intervals decreases; thus, quantization error increases and the S/N ratio decreases. In particular, the S/N ratio is signal-dependent, and less than the dynamic range. In various forms, an exponent/mantissa representation is used in many types of systems.

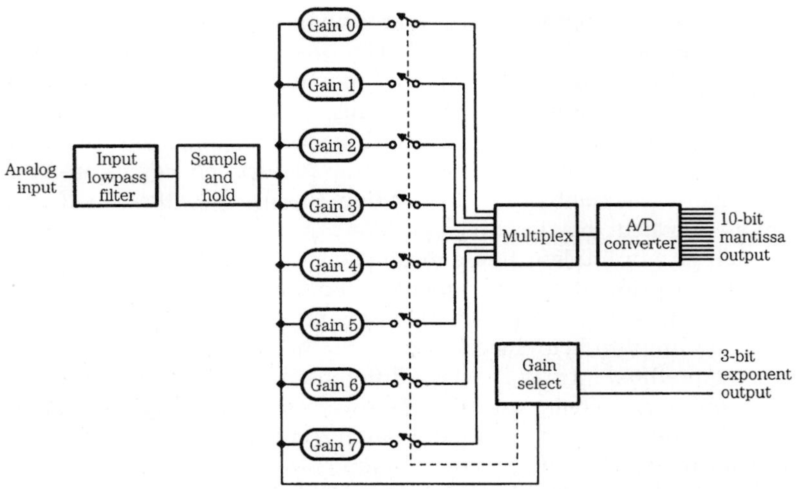

FIGURE 4.18 A floating-point converter uses multiple gain stages to manipulate the signal's amplitude to optimize fixed A/D conversion.

A floating-point system uses a short-word A/D converter to achieve a moderate dynamic range, or a longer-word converter for a larger dynamic range. For example, a floating-point system using a 16-bit A/D converter and a 3-bit exponent adjusted over a 42-dB range in 6-dB steps would yield a 138-dB dynamic range (96 dB + 42 dB). This type of system would be useful for encoding particularly extreme signal conditions. In addition, this floating-point system only requires a split 19-bit word, but the equivalent fixed linear PCM system would require a linear 23-bit word. In addition, when the gain stages are placed at 6-dB intervals, the coded words can be easily converted to a uniform code for processing or storage without computation. The mantissa undergoes a shifting operation according to the value of the exponent.

Although a floating-point system's dynamic range is large, the nature of its dynamic range differs from that of a fixed linear system; it is inherently less than its S/N ratio. This is because dynamic range measures the ratio between the maximum signal and the noise when no signal is present. With the S/N ratio, on the other hand, noise is measured when there is a signal present. In a fixed linear system, the dynamic range is approximately equal to the S/N ratio when a signal is present. However, in a floating-point system, the S/N ratio is approximately determined by the resolution of the fixed A/D converter (approximately $6n$), which is independent of the larger dynamic range. Changes in the signal dictate changes in the gain structure, which affect the relative amplitude of quantization error.

The S/N ratio thus continually changes with exponent switching. For example, consider a system with a 10-bit mantissa and 3-bit exponent with 6-dB gain intervals. The maximum S/N ratio is 60 dB. As the input signal level falls, so does the S/N ratio, falling to a minimum of 54 dB until the exponent is switched, and the S/N again rises to 60 dB. For longer-word converters, a complex signal will mask the quantization error. However, in the case of simple tones, the error might be audible. For example, modulation noise from low-frequency signals and quantization noise from nearly inaudible signals might result. Another problem can occur with gain switching;

inaccuracies in calibration might present discontinuities as the different amplifiers are switched.

The changes in the gain structure can affect the audibility of the error. Instantaneous switching from sample to sample tends to accentuate the problem. Instead, gain switching should be performed with algorithms that follow trends in signal amplitude, based on the type of program to be encoded. For example, syllabic algorithms are adapted to the rate at which syllables vary in speech. Gain decreases are instantaneous, but gain increases are delayed. This approximates a block floating-point system, as described below. In any event, the gain must be switched to prevent any overload of the A/D converter.

Block Floating-Point Systems

The block floating-point architecture is derived from the floating-point architecture. Its principal advantage is data reduction making it useful for bandlimited transmission or storage. In addition, a block floating-point architecture facilitates syllabic or other companding algorithms.

In a block floating-point system, a fixed linear PCM A/D converter precedes the scalar. A short duration of the analog waveform (1 ms, for example) is converted to digital data. A scale factor is calculated to represent the largest value in the block, and then the data is scaled upward so the largest value is just below full scale. This reduces the number of bits needed to represent the signal. The data block is transmitted, along with the single scale factor exponent.

During decoding, the block is properly rescaled. In the example in Fig. 4.19, 16-bit words are scaled to produce blocks of 10-bit words, each with one 3-bit exponent. Because only one exponent is required for the entire data block, data rate efficiency is increased over conventional floating-point systems. The technique of sending one scale factor per block of audio data is used in many types of systems.

Block floating-point systems avoid many of the audible artifacts introduced by instantaneous scaling. The noise amplitude lags the signal amplitude by the duration of the buffer memory (for example, 1 ms); because this delay is short compared to human perception time, it is not perceived. Thus, a block floating-point system can minimize any perceptual gain error.

Floating-point systems work best when the audio signal has a low peak-to-average ratio for short durations. Because most acoustical music behaves in this manner, system

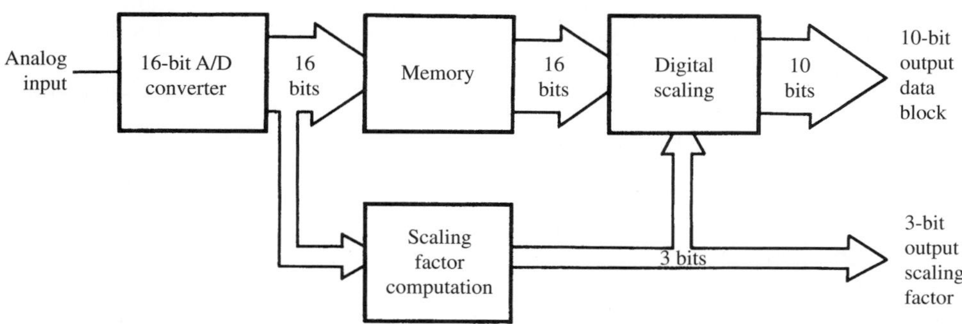

FIGURE 4.19 A block floating-point system uses a scaling factor over individual blocks of data.

performance is relatively good. An instantaneous floating-point system excels when the program varies rapidly from sample to sample (as with narrow, high-amplitude peaks) yet is inferior to a fixed linear system. Performance dependence on program behavior is a drawback of many alternative (non-PCM) digitization systems.

Nonuniform Companding Systems

With linear PCM, quantization intervals are spaced evenly throughout the amplitude range. As we have observed, the range changing in floating-point systems provides a nonuniform quantization. Nonuniform companding systems also provide quantization steps of different sizes, but with a different approach, called companding. Although companding is not an optimal way to achieve nonuniform quantization, its ease of implementation is a benefit.

In nonuniform companding systems, quantization levels are spaced far apart for high-amplitude signals, and closer together for low-amplitude signals. This follows the observation that in some types of signals, such as speech, small amplitudes occur more often than high amplitudes. In this way, quantization is relatively improved for specific signal conditions. This nonuniform distribution is accomplished by compressing and expanding the signal, hence the term, companding. When the signal is compressed prior to quantization, small values are enhanced and large values are diminished. As a result, perceived quantization noise is decreased.

A logarithmic function is used to accomplish companding. Within the compander, a linear PCM quantizer is used. Because the compressed signal sees quantization intervals that are uniform, the conversion is equivalent to one of nonuniform step sizes. On the output, an expander is used to inversely compensate for the nonlinearity in the reconstructed signal. In this way, quantization levels are more effectively distributed over the audio dynamic range.

A companding system is shown in Fig. 4.20. The encoded signal must be decoded before any subsequent processing. Higher amplitude signals are more easily encoded, and lower amplitude signals have reduced quantization noise. This results in a higher S/N ratio for small signals, and it can increase the overall dynamic range compared to fixed linear PCM systems. Noise increases with large amplitude audio signals and is correlated; however, the signal amplitude tends to mask this noise. Noise modulation audibility can be problematic for low-frequency signals with quickly changing amplitudes.

μ-Law and *A*-Law Companding

The μ-law and *A*-law systems are examples of quasi-logarithmic companding, and are used extensively in telecommunications to improve the quality of 8-bit quantization.

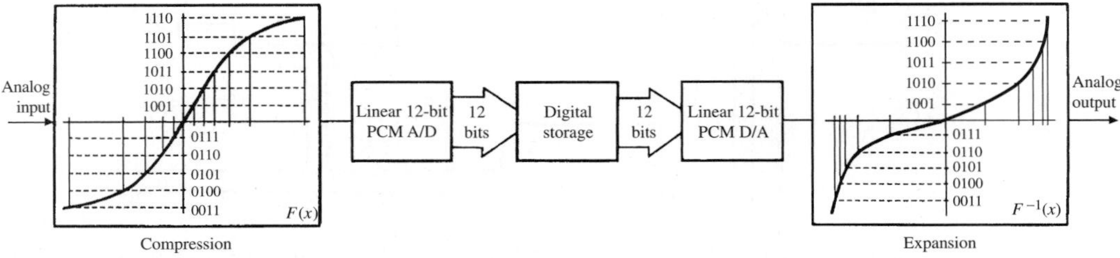

FIGURE 4.20 A nonlinear conversion system uses companding elements before and after signal conversion.

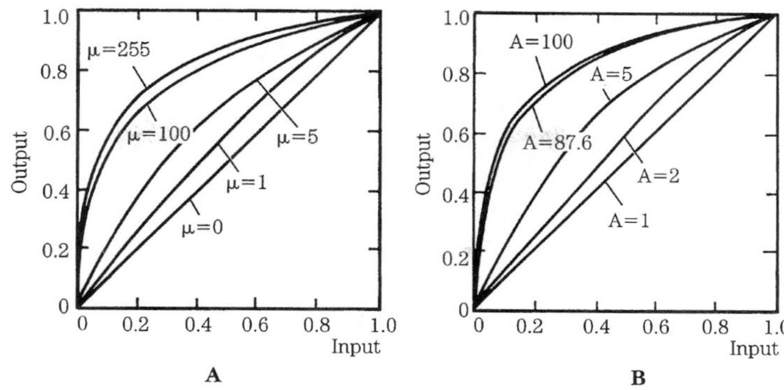

FIGURE 4.21 Companding characteristics determine how the quantization step size varies with signal level. A. μ-law characteristic. B. *A*-law characteristic.

Generally, the quantization distribution is linear for low amplitudes, and logarithmic for higher amplitudes. The μ-law and *A*-law standards are specified in the International Telecommunications Union (ITU) recommendation G.711.

The μ-law encoding method was developed to code speech for telecommunications applications. It was developed for use in North America and Japan. The audio signal is compressed prior to quantization and the inverse function is used for expansion. The value of 0 corresponds to linear amplification, that is, no compression or uniform quantization. Larger values result is greater companding. A $\mu = 255$ system is often used in commercial telecommunications. An 8-bit implementation can achieve a small-signal S/N ratio and dynamic range that are equivalent to that of a 12-bit uniform PCM system. The *A*-law is a quantization characteristic that also varies quasi-logarithmically. It was developed for use in Europe and elsewhere. An $A = 87.56$ system is often employed using a midrise quantizer. Figure 4.21 shows μ-law and *A*-law companding functions for several values of μ and *A*. The μ-law and *A*-law transfer functions are very similar, but differ slightly for low-level input signals.

Generally, logarithmic PCM methods such as these require about four fewer bits per sample for speech quality equivalent to linear PCM; for example, 8 bits might be sufficient, instead of 12. Because speech signals are typically sampled at 8 kHz, the standard data rate for μ-law or *A*-law PCM data is therefore 64,000 bits/second (or 64 kbps). The device used to convert analog signals to/from compressed signals is often called a codec (coder/decoder).

Differential PCM Systems

Differential PCM (DPCM) systems are unlike linear PCM methods. They are more efficient because they code differences between samples in the audio waveform. Intuitively, it is not necessary to store the absolute measure of a waveform, only how it changes from one sample to the next. The differences between samples are often smaller than the amplitudes of the samples themselves, so fewer bits should be required to encode the possible range of signals. Furthermore, a waveform's average value should change only slightly from one sample to the next if the sampling frequency is fast enough; most sampled signals show significant correlation between successive samples. Differential systems thus exploit the redundancy from sample to sample, using a few PCM bits to code the difference in amplitude between successive samples. The quantization error can be smaller than with traditional waveform PCM coding. Depending on the design, differential systems can use uniform or nonuniform quantization.

The rate at which signal voltage can change is inherently limited in DPCM systems; the coded signal amplitude decreases at 6 dB/octave; thus the S/N ratio decreases at 6 dB/octave. The frequency response of the coded signal can be filtered to improve the S/N ratio. For example, a signal with little high-frequency content can be filtered to increase high frequencies; this is reversed during decoding, and noise is relatively masked by the low-frequency content.

Predictive Differential Coding

Differential systems use a form of predictive coding. The technique predicts the value of the current sample based on values of previous samples, and then codes the difference between the predicted value and the unquantized input sample value. In particular, a predicted signal is subtracted from the actual input and the difference signal (the error) is quantized. The decoder produces the prediction from previous samples; using the prediction and the difference value, the waveform is reconstructed sample by sample. When the prediction is accurate and the error signal is small, predictive coding requires fewer bits to quantize an audio signal, but performance depends on the type of function used to derive the prediction signal and its ability to anticipate the changing signal. Compared to PCM systems, differential systems are cost-effective in hardware, provide data rate efficiency, and are perceptually less affected by bit errors.

Delta Modulation

As noted, differential systems encode the difference between the input signal and a prediction. As the sampling frequency increases, the possible amount of change between samples decreases and encoding resolution increases. Delta modulation (DM) is the simplest form of differential PCM. It uses a very high sampling frequency so that only a one-bit quantization of the difference signal is used to encode the audio waveform. Conceptual operation of a delta modulation system is shown in Fig. 4.22. Positive or

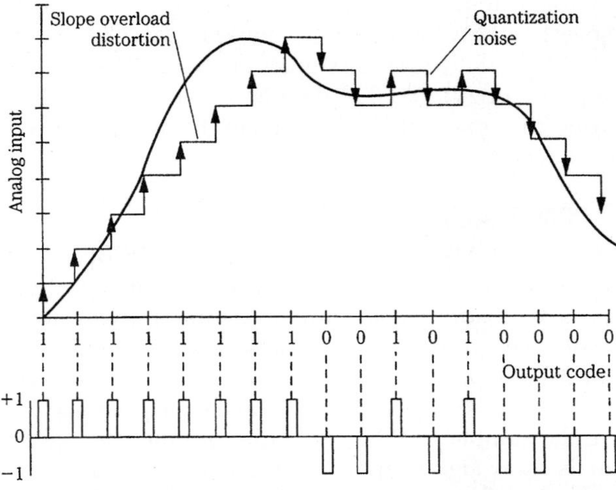

FIGURE 4.22 In a delta-modulation coder, one differential bit is used to encode the audio signal.

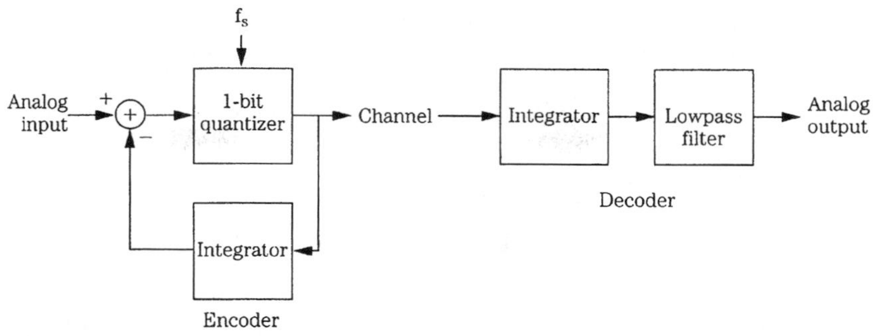

FIGURE 4.23 Delta-modulation encoder and decoder.

negative transitions in the quantized waveform are used to encode the audio signal. Because the staircase can move from sample to sample by only one fixed quantization interval, a fast sampling frequency is required to track the signal's transients. In the example in Fig. 4.22, the sampling frequency is insufficient to track the signal's rise time. In this slope-overload condition, the differential itself is encoded, rather than the original signal.

Delta modulation is efficient from a hardware standpoint, as shown in Fig. 4.23. Integrators are used as first-order predictors. The difference between the input signal and its predicted value is quantized as a one-bit correcting word and generated at sample time. The system determines if the sign of its error is positive or negative, and applies the sign (a positive or negative pulse) to the integrator that correspondingly moves its next value up or down one increment, always closer to the present value. The accuracy of the encoding rests on the size of the increment, or step. Also, the signal must change at each sample; it cannot be constant. At the output, the correction signal is decoded with an integrator to estimate the input signal. As with any DPCM system, the coded signal amplitude decreases with frequency, so the S/N ratio decreases by 6 dB/octave. Only one correction can occur per sample interval, but a very fast rate could theoretically allow tracking of even a fast-transient audio waveform. Delta modulation offers excellent error performance. In a linear PCM system, an uncorrected MSB error results in a large discontinuity in the signal. With delta modulation, there is no MSB. Each bit merely tracks the difference between samples, thus inherently limiting the amount of error to that difference. The possibility of degradation, however, necessitates error correction. Parity bits and interleaving are commonly used for this purpose.

From a practical standpoint, DM fails to perform well in high-fidelity applications because of its inherent trade-off between sampling frequency and step size. Encoding resolution is directly dependent on step size. The smaller the steps, the better the approximation to the audio signal, and the lower the quantization error. However, when the encoding bit cannot track a complex audio waveform that has low sample-to-sample correlation, slew rate limitations yield transient distortion. The sampling frequency can be increased with oversampling to compensate, but the rates required for successful encoding of wide bandwidth signals are extreme. Perceptually, when slope overload is caused by high-frequency energy, the signal tends to mask the distortion. Quantization error is always present, and is most audible for low-amplitude signals. To make the best of the circumstances, step size can be selected to minimize the sum of the mean squares of the two distortion values from slope overload and quantization.

Timothy Darling and Malcolm Hawksford have shown that the signal to quantization error ratio of a delta modulation system can be expressed as:

$$S/E = \left[-16 + 10\ \log\left(\frac{f_s^3}{f_b f_i^2}\right)\right] \text{dB}$$

where f_s = sampling frequency
f_b = noise bandwidth
f_i = audio (sine-wave) frequency

To achieve a 16-bit PCM S/N ratio of 96 dB over a 20 kHz bandwidth, a DM system would require clocking at 200 MHz. Although a doubling of bit rate in DM results in an increase in S/N of 9 dB, a doubling of word length in PCM produces an exponential S/N increase. From an informational standpoint, we can see that the nature of DM hampers its ability to encode audio information. A sampling frequency of 500 kHz, for example, would theoretically permit encoding of frequencies up to 250 kHz, but that bandwidth is largely wasted because of the relatively low frequency of audio signals. In other words, the informational encoding distribution of delta modulation is inefficient for audio applications.

On the other hand, the high sampling frequency required for delta modulation offers one benefit. As observed with noise shaping, for each doubling of sampling frequency, the noise in a fixed band decreases by 3 dB. The total noise remains constant, but it is spread over a larger spectrum thus in-band noise is reduced. This somewhat lowers the noise floor in a DM system. In addition, because of the high sampling frequency, brick-wall filters are not required. Low-order filters can provide adequate attenuation well before the half-sampling frequency without affecting audio response. Of course, conventional A/D and D/A converters are not required. Ultimately, because of its limitations, delta modulation is not often used for high-fidelity applications. However, a variation known as sigma-delta modulation offers excellent results and is widely used in oversampling A/D and D/A converters, as discussed in Chap. 18. The Super Audio CD (SACD) format uses Direct Stream Digital (DSD) coding, a one-bit pulse density method using sigma-delta modulation. SACD is discussed in Chap. 7.

Adaptive Delta Modulation

Adaptive delta modulation (ADM) systems vary quantization step sizes to overcome the transient response limitations of delta modulation. At the same time, quantization error is held to a reasonable value. A block diagram of an ADM encoder is shown in Fig. 4.24A. The encoder examines input data to determine how to best adjust step size. For example, with a simple adaptive algorithm, a series of all-positive or all-negative difference bits would indicate a rapid change from the approximation. The step size would increase to follow the change either positively or negatively. The greater the overload condition, the larger the step size selected. Alternating positive and negative difference bits indicate good tracking, and step size is reduced for even greater accuracy, as shown in Fig. 4.24B. This yields an increase in S/N, with no increase in sampling frequency or bit rate. As more bits in the stream are devoted to diagnosing signal behavior, step size selection can be improved.

ADM design is complicated because the decoder must be synchronized to the step size strategy to recognize the variation. Also, it can be difficult to change step size quickly and radically enough to accommodate sharp audio transients. As high-frequency and

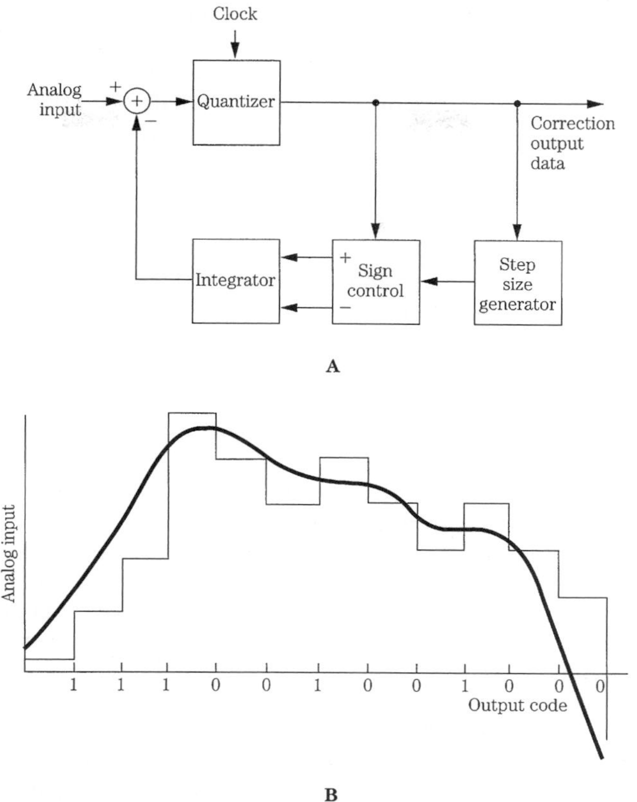

FIGURE 4.24 Operation of an adaptive delta modulation encoder. A. Block diagram of an adaptive delta modulation encoder with variable step size. B. Changes in step size are triggered by continuous 1s or 0s.

high-amplitude signals demand large increments, quantization noise is increased, producing noise modulation with a varying noise floor. In addition, it is difficult to inject a dither signal in an ADM system; since the step size changes, a fixed amount of dither is ineffective. Error feedback can reduce in-band noise. A pre-emphasis filter characteristic can reduce subjective noise in small-amplitude signals, mask the change in noise with changing step size, and reduce low-frequency noise in high-amplitude, high-frequency signals. As the audio slope increases, a control signal from the delta modulator, the same signal used to control step size, raises the frequency of the highpass filter and attenuates low frequencies. Another variation of ADM is continuously variable slope delta modulation (CVSDM), in which step size is continuously variable, rather than incremental.

Companded Predictive Delta Modulation

Companded predictive delta modulation (CPDM) rejects ADM in favor of a compander delta modulation scheme. Instead of varying the step size in relation to the signal, the signal's amplitude is varied prior to the constant step size delta modulator to protect against modulator overload. To reduce the quantization noise floor level, a linear predictive filter is used, in which an algorithm uses many past samples to better predict the next sample.

The companding subsystem consists of a digitally controlled amplifier in both the encoder and decoder, for controlling broadband signal gain. The bitstream itself controls both amplifiers to minimize tracking error. The digitally controlled amplifiers continually adjust the signal over a large range to best fit the fixed step size of the delta modulator. A transient "speed-up" circuit in the level sensing path allows faster gain reduction during audio transients. Strings of 1s or 0s indicate the onset of an overload, and trigger compression of broadband gain to ensure that the transients are not clipped at the modulator. The speed of the gain change can be either fast or slow, depending on the audio signal dynamics. Spectral compression can be used to reduce variations in spectral content; the circuit could reduce high frequencies when the input spectrum contains predominantly high frequencies, and boost high frequencies when the spectrum is weighted with low frequencies. The spectrum at the A/D converter is thus more nearly constant.

Adaptive Differential Pulse-Code Modulation

Adaptive differential pulse-code modulation (ADPCM) combines the predictive and adaptive difference signal of ADM with the binary code of PCM to achieve data reduction. Although designs vary, in many cases the difference signal to be coded is first scaled by an adaptive scale factor, and then quantized according to a fixed quantization curve. The scale factor is selected according to the signal's properties; for example, the quantizer step size can be varied in proportion to the signal's average amplitude. Signals with large variations cause rapid changes in the scale factor, whereas more stable signals cause slow adaptation. Step size can be effectively varied by either directly varying the step size, or by scaling the signal with a gain factor.

A linear predictor, optimized according to the type of signal to be coded, is used to output a signal estimate of each sample. This signal is subtracted from the actual input signal to yield a difference signal; this difference signal is quantized with a short PCM word (perhaps 4 or 8 bits) and output from the encoder. In this way, the signal can be adaptively equalized, and the quantization noise adaptively shaped, to help mask the noise floor for a particular application. For example, the noise can be shaped so that its spectrum is white after decoding. Noise shaping is described in Chap. 18.

The decoder performs the same operations as the encoder; by reading the incoming data stream, the correct step size is selected, and the difference signal is used to generate the output samples. Lowpass filtering is applied to the output signal. The benefit of ADPCM is bit-rate reduction based on the tendency for amplitude and spectrum distribution of audio signals to be concentrated in a specific region. The scale factors and other design elements in an ADPCM algorithm take advantage of these statistical properties of the audio signal. In speech transmission applications, a bit rate of 32 kbps (kilobits per second) is easily achieved.

ADPCM's performance is competitive, or relatively superior to that of fixed linear PCM. When the audio signal remains near its maximum frequency, ADPCM's performance is similar to PCM's. However, this is rarely the case with audio signals. Instead, the instantaneous audio frequency is relatively low, thus the signal changes more slowly, and amplitude changes are smaller. As a result, ADPCM's quantization error is less than that of PCM. In theory, given the same number of quantization bits, ADPCM can achieve better signal resolution. In other words, relatively fewer bits are needed to achieve good performance. In practice, a 4-bit ADPCM signal might provide fidelity subjectively similar to an 8-bit PCM signal.

Variations of the ADPCM algorithm are used in many telecommunications standards. The ITU-T Recommendation G.721 uses ADPCM operating at 32 kbps. G.722 contains another example of an ADPCM coder. G.723 uses ADPCM at bit rates of 24 to 40 kbps. G.726 specifies a method to convert 64 kbps μ-law or *A*-law PCM data to 16-, 24-, 32-, or 40-kbps ADPCM data. The bit rate of 32 kbps is used for speech applications. In the G.726 standard, both the quantizer and predictor adapt to input signal conditions. The G.727 standard uses ADPCM at bit rates of 16, 24, 32 and 40 kbps. The G.727 standard can also be used within the G.764 standard for packet network applications. Speech coding is described in more detail in Chap. 12.

The CD-ROM/XA format uses several ADPCM coding levels to deliver fidelity according to need. An 8-bit ADPCM quality level can yield an S/N ratio of 90 dB, and bandwidth of 17 kHz. The two 4-bit levels yield an S/N of 60 dB, and bandwidths of 17 kHz and 8.5 kHz. During encoding, the original audio data frequency (44.1 kHz) is reduced by a sampling rate converter to a lower frequency (37.8 kHz or 18.9 kHz) depending on the quality level selected. The original word length (16 bits) is reduced (4 or 8 bits) per sample with ADPCM encoding. Four different prediction filters can be selected to optimize the instantaneous S/N ratio; a filter is selected for a block of 28 samples depending on the spectral content of the signal. Companding and noise shaping are also used to increase dynamic range. The filter type is described in each data block. During ADPCM decoding, the audio data is block-decoded and expanded to a linear 16-bit form. Depending on audio quality level, this ADPCM encoder can output from 80 to 309 kbps/channel, yielding a reduced data rate. ADPCM with a 4:1 compression ratio is sometimes used in QuickTime software, and Windows software may use ADPCM coding in .aiff and .wav files. ADPCM also appears in video-game platforms.

Unlike perceptual coding algorithms, ADPCM encoding and decoding may be executed with little processor power. For example, the Interactive Multimedia Association (IMA) version of ADPCM is quite concise. Four bits can be used to store each sample. The encoder finds the difference between two samples, divides it by the current step size and outputs that value. To create the next sample, the decoder multiplies this value by the current step size and adds it to the result of the previous sample. The step size is not stored directly. Instead, a table of possible step sizes (perhaps 88 entries) is used; the entries follow a quasi-exponential progression. The decoder references the table, using previous values to correctly update the step size. When the scaled difference is small, a smaller step size is selected; when the difference is large, a larger step size is selected. In this way, ADPCM coding may be efficiently performed in software.

ADPCM, as well as other specialized designs, offer alternatives to classic linear PCM design. They adhere to the same principles of sampling and quantizing; however, their implementations are quite different. Perceptual coding systems are commonly used to code audio signals when bit rate is the primary concern; these systems are discussed in Chaps. 10 and 11.

Timebase Correction

As we observed in Chap. 3, modulation coding is used in storage and transmission channels to improve coding efficiency and, for example, make the data self-clocking. However, successful recovery of data is limited by the timebase accuracy of the received clock. For example, speed variations in the transport of an optical disc player, instability imposed on a data stream's embedded clock, and timing inaccuracies in the oscillator used to clock an A/D or D/A converter can all lead to degraded performance in the form of data errors, or noise and modulation artifacts in the converted waveform. For successful recovery of data, receiver circuits must minimize timebase errors that occur in the storage medium, during transmission, or within regeneration and conversion circuitry itself. For example, phase-locked loops (PLLs) are often used to resynchronize a receiver with the transmitted channel code's clock.

Timing accuracy is challenging in the digital environment, because of the noise and interference that is present. Moreover, tolerances needed for timebase control increase with word length; for example, 20-bit conversion requires much greater timebase accuracy than 16-bit conversion. Above all, timing stability can be problematic because of the absolute tolerances it demands. A clock might require timing accuracy of 20 ps (picoseconds). Note that a picosecond is the reciprocal of 1 THz (terahertz), which is 1000 GHz (gigahertz), which is 1,000,000 MHz.

Jitter

Any deviation in the zero-crossing times of a data waveform from the zero-crossing times of a perfectly stable waveform can be characterized as jitter. In particular, it is the time-axis instability of a waveform measured against an ideal reference with no jitter. Timing variations in an analog signal may be directly audible as pitch instability. However, jitter in a digital signal may cause bit errors in the bitstream or be indirectly audible as increased noise and distortion in the output analog waveform or, if the digital signal is correctly de-jittered, there may be no bit errors or audible effects at all. Jitter is always present; its effect and the required tolerance depend on where in the signal processing chain the jitter occurs. Relatively high jitter levels will not prevent error-free transfer of data from one device interfaced to another, but some interface devices are less tolerant of jitter than others. During A/D or D/A conversion, even low jitter levels can induce artifacts in the analog output waveform; some converters are also less tolerant of jitter than others.

Jitter manifests itself as variations in the transition times of the signal, as shown in Fig. 4.25. Around each ideal transition is a period of variation or uncertainty in arrival time; this range is called the peak-to-peak jitter. Jitter occurs in data in a storage medium, transmission channel, or processing or regeneration circuits such as A/D and D/A converters. Jitter can occur as random variations in clock edges (white phase jitter), it can be related to the width of a clock pulse (white FM jitter), or it can be related to other events (correlated jitter), sometimes periodically.

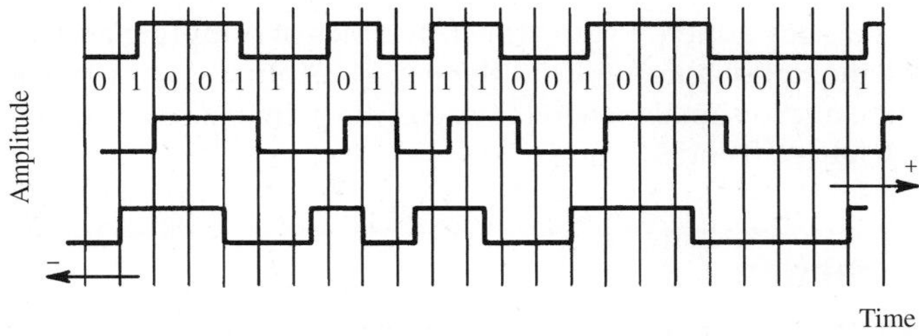

FIGURE 4.25 A time variation in the medium, or regeneration or processing circuits, results in a timebase jitter error.

Jitter is best described by its spectral characteristics; this shows the amplitude and frequency of the jitter signal. Random jitter will show a broadband spectrum; when the data is reconstructed as an analog waveform it will have an increased noise floor. Periodic jitter will appear as a single spectral line; FM sidebands or modulated noise could appear in the reconstructed signal, spaced on either side of the signal frequency. Jitter at frequencies less than the sampling frequency causes a timing error to accumulate; the error depends on the amplitude and frequency of the modulation waveform. Generally, peak-to-peak jitter is a valid measure; however, when the jitter is random, a root mean square (rms) jitter measure is valid. Care must be taken when specifying jitter; for example, if the deviation in a clock signal's period width is averaged over time, the average can converge to zero, resulting in no measured jitter.

Eye Pattern

An oscilloscope triggered from a stable reference clock, and timebase set at one unit interval (UI), will display a superimposed successive transition known as the eye pattern. (A reference clock can be approximated by a phase-locked low-jitter clock.) The eye pattern can be used to interpret the quality of the received signal. It will reveal noise as the waveform's amplitude variations become indistinct, and jitter as the transitions shift about time intervals of the code period. Peak shift, dc offset, and other faults can be observed as well. The success in regeneration can similarly be evaluated by examining the eye pattern after processing. Noise in the channel will tend to close the pattern vertically (noise margin), and jitter closes it horizontally (sampling time margin), as shown in Fig. 4.26, possibly degrading performance to the point where pulse shaping can no longer accurately retrieve the signal. The amount of deterioration can be gauged by measuring the extent of amplitude variations, and forming an eye opening ratio:

$$E = \frac{a_2}{a_1} = \frac{(a_1 - 2\Delta a)}{a_1}$$

where a_1 = outside amplitude
a_2 = inside amplitude
Δa = amplitude variation

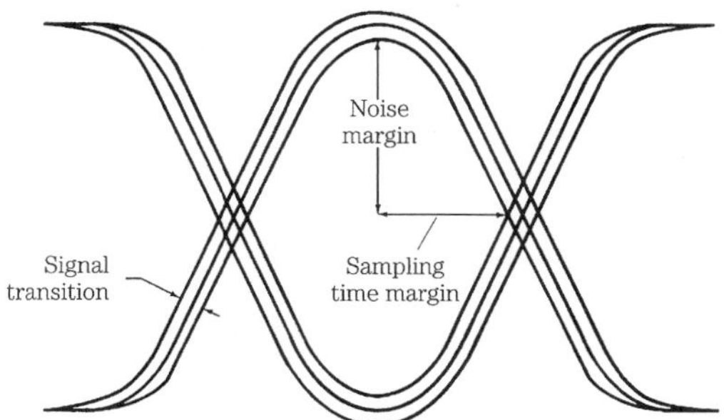

FIGURE 4.26 The eye pattern can be used to interpret the quality of an RF data signal. A minimum inside eye opening size determines the limit of a signal's usability. If the eye closes, errors will result.

Both variations are measured from the center of the opening in the eye pattern. The eye height is specified in volts, and eye width in bit time or percent of cell period. The width of the eye gives the percentage of the data period available to ascertain its logical value, and the height shows the maximum difference between these levels during the available time. The receiver is generally able to read the signal if the eye opening has at least a minimum interior size and is maintained at coded intervals; the center of the eye shows the amplitude level decision threshold and the sampling time. However, jitter introduced into the receiver's clock recovery circuits will cause the receiver to sample the data further from the center of the data cell.

Jitter observed on an oscilloscope shows the dynamic variations in the signal, but a more careful method applies the data signal to an FM demodulator, which is connected to a spectrum analyzer. Random jitter exhibits a broadband spectrum and raises the noise floor in the analog signal reconstructed from the data. Periodic jitter will appear as a single spectral line, at a low frequency for a slow clock variation or at a high frequency for a fast clock variation; the reconstructed signal may contain FM sidebands or modulated noise.

Interface Jitter and Sampling Jitter

The significance of jitter depends on where in the signal chain it is considered. For example, it is important to distinguish between interface jitter (which occurs in digital-to-digital data transfer) and sampling jitter (which occurs when converting data into and out of the digital domain). Interface timing errors are distinct from those caused by sampling jitter. Interface jitter occurs in transmitted data clocks, when conveying data from one device to another.

Interface jitter is a concern if it causes uncorrected errors in the recovered signal; the quality of transmitted data can be monitored by error detection circuits at the receiver. Many data streams are self-clocking; a receiver must recover the clock and hence the data by synchronizing to the transmitted clock. A receiving circuit with a fixed clock would not be able to lock onto a signal with unregulated timing variations in the received clock, even if its rate was nominally equal to the received clock rate. For this reason, receiving circuits commonly use a phase-locked loop (PLL) circuit to align their clocks with the data rate of the incoming signal. An interface PLL, as shown in Fig. 4.27, acts like a mechanical flywheel, using its lowpass filter characteristic to attenuate short-term fluctuations in the signal timing. It tracks the slowly changing data signal, but strips off its quickly changing jitter. A PLL accepts the input signal as a timing reference,

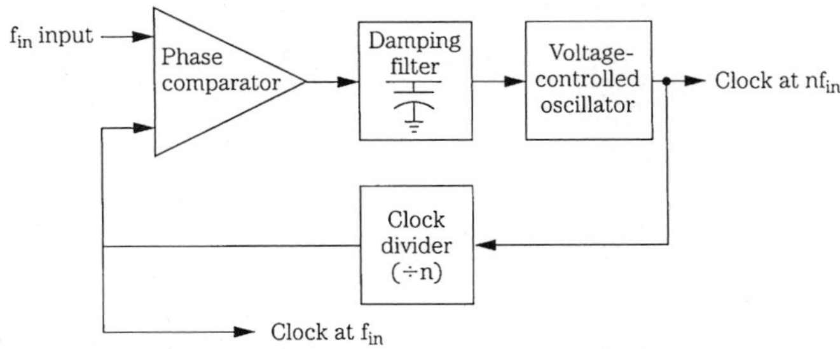

FIGURE 4.27 An example of an interface phase-locked loop. The PLL's limited bandwidth prevents jitter from passing through the PLL on the output data or output clock.

measures the phase error between the input signal and its own output through a feedback loop and uses the error to control a voltage-controlled oscillator (VCO) within the loop. In response, the VCO reaches equilibrium by minimizing this loop phase error. Once the VCO is locked to the phase of the input signal, the oscillator runs at the reference frequency or a multiple of it. The oscillator is decoupled from the reference, attenuating high-frequency jitter on the PLL's output data. A PLL can thus provide jitter reduction by reclocking signals to an accurate and stable timebase. Adversely, at lower jitter frequencies relative to the filter's corner frequency, a PLL's jitter attenuation will be reduced. At low jitter frequencies, the PLL will track the jitter and pass it on (a second PLL stage with a lower corner frequency might be needed). Moreover, any gain near the cutoff frequency of the PLL's lowpass filter function would increase jitter.

If the recovered data is free of errors, interface jitter has not affected it. However, if subsequent reclocking circuits do not remove jitter from the recovered data, potentially audible artifacts can result from sampling jitter at the D/A converter. Sampling jitter can affect quality of an audio signal as it is sampled or resampled with this timing error. Sampling jitter can be induced by sampling clocks, and affects the quality of the reproduced signal, adding noise and distortion. Timing tolerances must be very tight so that timing errors in a recovered clock are minimized. The clock generating circuits used in a sampling device may derive a timing reference from the interface signal. In this case, a PLL might be needed to remove interface jitter to yield a sample clock that is pure enough to avoid potentially audible jitter modulation products.

Jitter can occur throughout the signal chain so that precautions must be taken at each stage to decouple the bitstream from the jitter, so data can be passed along without error. This is particularly important in a chain of connected devices because jitter will accumulate in the throughput signal. Each device might contribute a small amount of jitter, as will the interface connections, eventually leading to data errors or conversion artifacts. Even a badly jittered signal can be cleaned by retiming it through jitter attenuation. With proper means, the output data is correct in frequency and phase, with propagated jitter within tolerance.

Jitter in Mechanical Storage Media

Jitter must be controlled throughout the audio digitization chain, beginning at the mechanical storage media. Storage media such as optical disc can impose timebase errors on the output data signal because of speed variations in the mechanical drives they use. Accurate clocks and servo systems must be designed to limit mechanical speed variations, and input and output data must be buffered to absorb the effects of data irregularities. Speed variations in the transport caused by eccentricities in the rotation of spindle motors will cause the data rate to vary; the transport's speed can slowly drift about the proper speed, or fluctuate rapidly. If the amount of variation is within tolerance, that is, if the proper value of the recorded data can be recovered, then no error in the signal results.

Servo control circuits are used to read timing information from the output data and generate a transport correction signal. In many cases, a PLL circuit is used to control the servo, as shown in Fig. 4.28. Speed control can be achieved with a PLL by comparing the synchronization words in the output bitstream (coded at a known rate) to a reference, and directing a speed control servo voltage to the spindle motor to dynamically minimize the difference. Fine speed control can use a second PLL, for example, to achieve constant linear velocity in an optical disc drive.

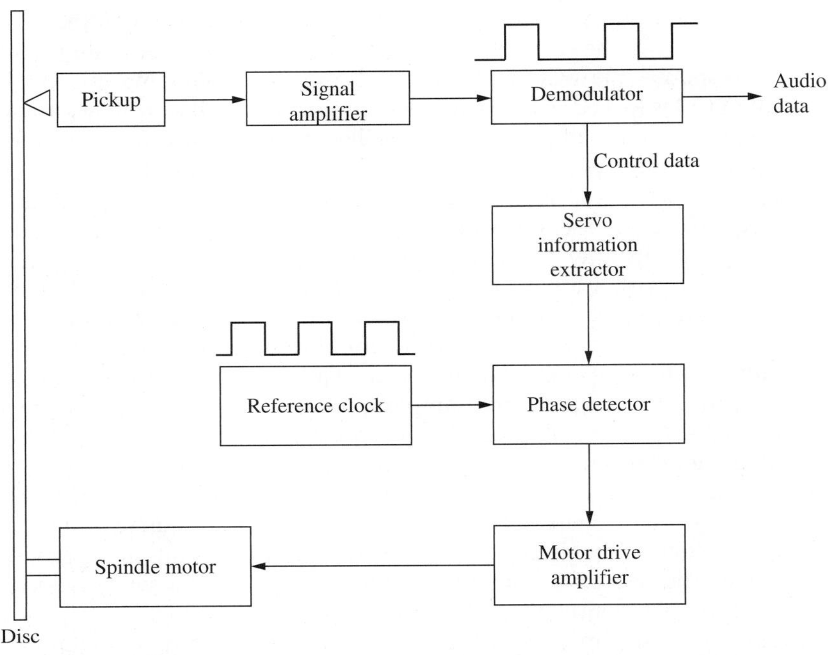

FIGURE 4.28 An example of a servo-pulse extractor/control loop used to control disc speed.

Although phase-locked servo systems act to maintain accurate and constant speed of the mechanical storage medium, timebase errors can still exist in the data. Likewise, jitter might be encountered whenever data is transmitted through an electrical or optical interconnection. To minimize the effect of timebase variations, data is often reclocked through a buffer. A buffer memory effectively increases the capture range for a data signal; for example, sampling frequency variations can be absorbed by the buffer. However, the longer the buffer, the greater the absolute time delay experienced by the throughput signal; delays greater than a few milliseconds can be problematic in real-time applications such as radio broadcasting where off-the-air monitoring is needed.

A buffer can be designed using RAM so that its address counter periodically overflows, resulting in a virtual ring structure. Because data output from the memory is independent from data being written, an inconsistent data input from the medium does not affect a precise data output. Clearly, the clock controlling the data readout from memory must be decoupled from the input clock. The amount of data in the buffer at any time can be used to control a transport's speed. For example, the difference between the input and output address, relative to buffer capacity, can be converted to an analog servo signal. If the buffer's level differs from the optimal half-full condition, the servo is instructed to either speed up or slow down the transport's speed. In this way, the audio data in the buffer neither overflows nor underflows.

Alternatively a buffer can be constructed using a first-in first-out (FIFO) memory. Input data is clocked into the top of the memory as it is received from the medium, and output data is clocked from the bottom of the memory. In addition to their application in reducing jitter, such timebase correction circuits are required whenever a discontinuous storage medium such as a hard disk is used. The buffer must begin filling when a sector is read from the disk, and continue to output data between sector read cycles. Likewise, when writing to memory, a continuous data stream can be supplied to the disk drive for each sector. If jitter is considerably less than a single bit period, no buffer is needed.

Jitter in Data Transmission

Interface jitter must be minimized during data transmission. No matter what jitter errors are introduced by the transmitting circuit and cable, the receiver has two tasks: data recovery and clock recovery. Jitter in the signal can affect the recovery of both, but the effect of jitter depends on the application. When data is transferred but will not be regenerated (converted to analog) at the receiver, only data recovery is necessary. The interface jitter is only a factor if it causes data errors at the receiver. The jitter tolerance is relatively low; for example, data with 5 to 10 ns of jitter could be recovered without error. However, when the data is to be regenerated or requantized, data recovery and particularly clock recovery are needed. High jitter levels may compromise a receiver's ability to derive a stable clock reference needed for conversion. Depending on the D/A converter design, clock jitter levels as low as 20 ps might be required.

For example, when digitally transferring data from a CD player, to a flash recorder, to a workstation, to a hard-disk recorder, only interface jitter is relevant to the data recovery. Jitter attenuation might be required at points in the signal path so that data errors do not occur. However, when the data is converted to an analog signal at a D/A converter, jitter attenuation is essential. Clock jitter is detrimental to the clock recovery process because it might compromise the receiver's ability to derive a stable clock reference needed for conversion.

As noted, a receiving PLL circuit separates the received clock from the received data, uses the received clock to recover the data, and then regenerates the clock (attenuating jitter), using it as the internal timebase to reclock the data (see Fig. 4.27). In some designs, a receiver might read the synchronizing signal from input data, place the data in a buffer, and then regenerate the clock and output the data with the frequency and phase appropriate to the destination. The buffer must be large enough to prevent underflow or overflow; in the former, samples must be repeated at the output, and in the latter, samples must be dropped (in both cases, ideally, during silences). The method of synchronizing from the embedded transmission clock is sometimes called "genlock." It works well in most point-to-point transmission chains. When the data sampling frequency is different from the local system sampling frequency, a sample rate converter is needed to convert the incoming timebase; this is described in Chap. 13. Sample rate converters can also be used as receivers for jitter attenuation.

Some receivers used in regeneration devices use a two-stage clock recovery process, as shown in Fig. 4.29. The first step is clock extraction; the received embedded clock is synchronized so the data can be decoded error-free in the presence of jitter. An initial PLL uses data transitions as its reference; the PLL is designed to track jitter well, but not attenuate it. At this stage, the sample clock might have jitter. The recovered data is placed in a FIFO buffer. A buffer is not needed if the jitter at this stage is considerably less than one bit period. The second step is jitter attenuation; a PLL with low-jitter clock characteristics locks to the sample clock and retimes it with an accurate reference. The new, accurate clock is used to read data from the FIFO. In other words, the second PLL is not required to track incoming jitter, but is designed to attenuate it. Overall, a receiver must decouple the digital interface from the conversion circuitry before regeneration.

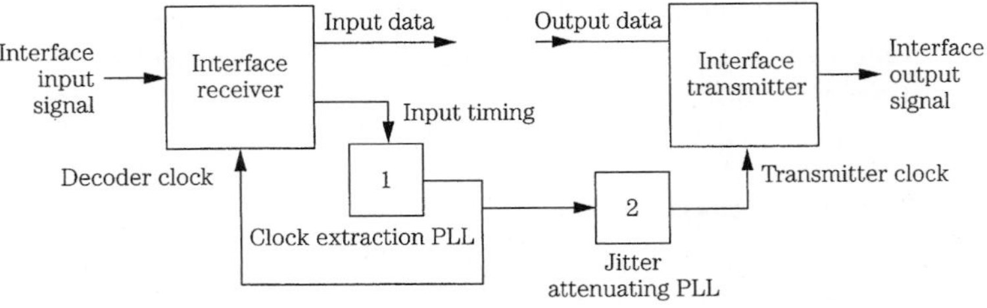

FIGURE 4.29 An example of dual-clock transceiver architecture, providing clock extraction and jitter attenuation.

Looked at in another way, degraded sound from a converter due to clock jitter might not be the converter's fault—it might be the receiver's.

Levels of transmission jitter often depend on cable characteristics. One type of cable-induced jitter, called pattern-dependent jitter, is a modulation that depends on the data values themselves. For example, patterns of 0s might produce more delay in transitions than patterns of 1s. The amount of modulation is a function of the high-frequency loss (bandwidth) of the cable. For example, jitter might be less than 1 ns if the cable has a 3-dB bandwidth of more than 4 MHz, but jitter increases to 9 ns if bandwidth is halved. For this reason, many transmission protocols use a fixed synchronization pattern or preamble; because the pattern is static, this cause of pattern-dependent jitter is removed. However, pattern-dependent jitter can be caused by other factors. For example, pattern-dependent jitter can be correlated to the polarity and frequency of the audio signal. When the coded audio signal level is low and at a single frequency, the more significant bits change together to reflect signal polarity, and pattern-dependent jitter can occur at that frequency. Any serial format that does not scramble its data word is prone to this phenomenon; with scrambling, this pattern-dependent jitter would be decorrelated and thus would be benign.

In more complex installations, with many devices, jitter protection is more sophisticated. If devices are cascaded without a master clock signal, clocking is derived from each previous device, which extracts it from its input data. Some devices can pass jitter, or even amplify some jitter components. For example, even if a PLL attenuates high-frequency jitter, it might pass low-frequency jitter. Jitter can thus accumulate, to the point where downstream devices lose lock. For more accurate transmission through a signal chain, each piece of equipment must be frequency-locked to a star-configuration distributed master clock. Devices must have both data inputs, as well as external clock inputs. The use of an external clock is sometimes called master clock synchronization. Each device ignores jitter on its input data and instead accepts a clock from the master reference; jitter cannot accumulate. Interconnection synchronization issues are discussed in Chap. 13.

Jitter in Converters

Jitter must be controlled throughout the digital audio chain, but it is more critical at conversion points. It must be carefully minimized in the clocks used for both A/D and D/A converters; faults can result in degradation of the output analog waveform. Audio samples must be acquired particularly carefully at the A/D converter. Simply put, clock jitter at the A/D converter results in the wrong samples (incorrect amplitude values) at the wrong time. Moreover, even if these samples are presented to a D/A converter with a jitter-free clock, the result will be the wrong samples at the right time. The magnitude of the error is proportional to the slope of the audio signal; the amplitude of the error increases at higher audio frequencies. Jitter is most critical in the A/D converter clock. Crystal oscillators typically have jitter of less than 10 ps rms; they must be used as reference for the entire digital system. Good circuit design and board layout are mandated. Clocks that use PLLs to divide down a high-frequency master clock to a usable average frequency are often prone to jitter.

Analyzed in the frequency domain, the effects of jitter on the sampling clock of an A/D converter are quite similar to FM modulation; the input frequency acts as the carrier, and clock jitter acts as the modulation frequency. Low-frequency periodic jitter reduces the amplitude of the input signal, and adds sideband components equally spaced at either side of the input frequency at a distance equal to multiples of the jitter frequency. As jitter increases, the amplitude of the sidebands increases. The effect of jitter increases as the input signal frequency increases; specifically, jitter amplitude error increases with input signal slew rate. In A/D converters, jitter must not interfere with correct sampling of the LSB. A 2-ns white-noise clock jitter applied to a successive approximation 16-bit A/D converter will degrade its theoretical dynamic range of 98 dB to 91 dB, as shown in Fig. 4.30.

The timing accuracy required for A/D conversion is considerable: the maximum rate of change of a sinusoidal waveform occurs at the zero crossing and can be calculated as $2\pi Af$, where A is the peak signal amplitude and f is the frequency in Hertz. By one estimation, a jitter specification of 250 ps would allow 16-bit accuracy from a full amplitude, 20-kHz sine wave. Only then would the jitter components fall below the quantization noise floor. A peak jitter of less than 400 ps would result in artifacts that decrease the dynamic range by less than 0.5 dB. Steven Harris has shown that oversampling sigma-delta A/D converters are equally susceptible to sinusoidal clock jitter as Nyquist nonoversampling successive approximation A/D converters. Oversampling sigma-delta A/D converters are less susceptible to random clock jitter than Nyquist sampling A/D converters because the jitter is extended over the oversampling range, and lowpass filtered.

D/A converters are also susceptible to jitter. The quality of samples taken from a perfectly clocked A/D converter will be degraded if the D/A converter's clock is nonuniform, creating the scenario of the right samples at the wrong time. Even though the data values are numerically accurate, the time deviations introduced by jitter will result in increased noise and distortion in the output analog signal. Fortunately, the distortion in the output waveform is a playback-only problem; the data itself might be correct, and only awaits a more accurate conversion clock. The samples are not wrong, they are only being converted at the wrong times. Not all data receivers (such as some S/PDIF receivers discussed in Chap. 13) provide sufficiently low jitter. As noted, in improved designs, data from a digital interconnection or hardware source is resynchronized to a new and accurate clock to remove jitter from the data signal prior to D/A conversion; phase-locked loops or other circuits are used to achieve this.

The effect of jitter on the output of a resistor ladder D/A converter can be observed by subtracting an output staircase affected with jitter from an ideal, jitter-free output. The difference signal contains spikes corresponding to the timing differences between

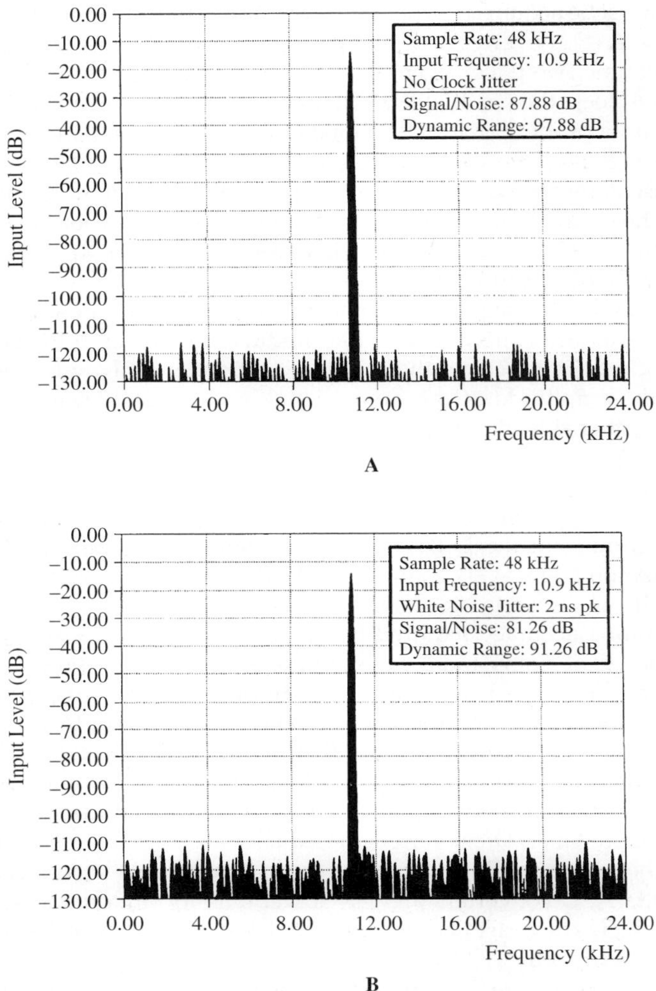

Figure 4.30 Simulations showing the spectrum output of nonoversampling A/D conversion. A. No clock jitter. B. White noise clock jitter with 2-ns peak value. (*Harris, 1990*)

samples; the different widths correspond to the differences in arrival time between the ideal and jittered clocks. The amplitudes of the error pulses correspond to the differences in amplitude from the previous sample to the present; large step sizes yield large errors. The signal modulates the amplitudes, yielding smaller values at signal peaks. The error noise spectrum is white because there is no statistical relationship between error values. The noise amplitude is a function of step size, specifically, the absolute value of the average slope of the signal. Thus, the worst case for white phase jitter on a resistor ladder D/A converter occurs with a full-amplitude signal at the Nyquist frequency. Depending on converter design (and other factors) a jitter level of at least 1 ns is necessary to obtain 16-bit performance from a resistor ladder converter. A tolerance of half that level, 500 ps, is not unreasonable. Unfortunately, consumer devices might contain clocks with poor stability; jitter error can cause artifact peaks to appear 70 dB or 80 dB below the maximum level.

The error caused by jitter is present in the jitter spectrum. Further, jitter bandwidth extends to the Nyquist frequency. Oversampling thus spreads the error over a larger spectrum. When an oversampling filter is used prior to resistor ladder conversion, the converter's sensitivity to random (white phase) jitter is reduced in proportion to the oversampling rate; for example, an eight-times oversampling D/A converter is four times less sensitive to jitter as a two-times converter in the audio band of interest. However, low-frequency correlated jitter is not reduced by oversampling.

Sigma-delta D/A converters, discussed in more detail in Chap. 18, can be very sensitive to clock jitter, or not particularly, depending on their architecture. When a true one-bit rail-to-rail signal is output, jitter pulses have constant amplitude. In a one-bit converter in which the output is applied to a continuous-time filter, random jitter is signal-independent, and in fact jitter pulses will be output even when no signal is present. A peak jitter level below 20 ps might be required to achieve 16-bit noise performance in a one-bit converter with a continuous-time filter. As Robert Adams notes, this is because phase modulation causes the ultrasonic, out-of-band shaped noise to fold down into the audio band, increasing the in-band noise level. This is avoided in most converters. Some one-bit converters use a switched-capacitor (discrete-time) output filter; because a switched capacitor will settle to an output value regardless of when a clock edge occurs, it is inherently less sensitive to jitter. The jitter performance of this type of converter is similar to that of a resistor ladder converter operating at the same oversampling rate. However, to achieve this, the switched-capacitor circuit must remove all out-of-band noise. Multi-bit converters are generally less susceptible to jitter than true one-bit converters. Because multi-bit converters use multiple quantization levels, ultrasonic quantization noise is less, and phase error on the clock used to generate timing for the sigma-delta modulator will have less effect.

Audiophiles have sometimes reported hearing differences between different kinds of digital cables. That could be attributed to a D/A converter with a design that is inadequate to recover a uniformly stable clock from the input bitstream. But, a well-designed D/A converter with a stable clock will be immune to variations in the upstream digital signal path, as long as data values themselves are not altered.

Jitter must be controlled at every stage of the audio digitization chain. Although designers must specially measure clock jitter in their circuits, traditional analog measurements such as total harmonic distortion plus noise (THD+N) and spectrum analysis can be used to evaluate the quality of the output signal, and will include effects caused by jitter. For example, if THD+N measured at 0 dB, 20 kHz is not greater than THD+N measured at 1 kHz, then jitter is not significant in the converter. Indeed, measurements such as THD+N may more properly evaluate jitter effects than jitter measurements themselves. As noted, if a receiving circuit can recover a data signal without error, interface jitter is not a factor. This is why data can be easily copied without degradation from jitter. However, because of sampling jitter, an A/D converter must be accurately clocked, and clock recovery is important prior to D/A conversion.

Jitter does not present serious shortcomings in well-designed, high-quality audio equipment. With 1-ns rms of jitter, distortion will be at –81 dBFS for a full-scale 20-kHz sine wave, and at –95 dBFS for a 4-kHz sine wave. For lower frequencies, the distortion is even lower, and the threshold of audibility is higher, thus at normal listening levels the effects of jitter will be inaudible, even in the absence of a signal. When masking is considered, higher jitter levels and higher distortion levels are acceptable. For example, with 10-ns rms of jitter, distortion will be at –75 dBFS for a 4-kHz sine wave; however, the distortion is still about 18 dB below the masking threshold created by the sine wave. One test found that the threshold of audibility for jitter effects with pure tones was 10-ns rms at 20 kHz, and higher at lower frequencies. Using the most critical program material, the threshold of audibility ranged from 30- to 300-ns rms for sinusoidal jitter. However, as components with longer word lengths and higher sampling frequencies come to market, jitter tolerances will become more critical. Ultimately, jitter is one of several causes of distortion in an output audio signal that must be minimized through good design practices.

CHAPTER 5

Error Correction

The advent of digital audio changed audio engineering forever and introduced entirely new techniques. Error correction was perhaps the most revolutionary new technique of them all. With analog audio, there is no opportunity for error correction. If the conveyed waveform is disrupted or distorted, then the waveform is usually irrevocably damaged. It is impossible to exactly reconstruct an infinitely variable waveform. With digital audio, the nature of binary data lends itself to recovery in the event of damage. When a bit is diagnosed as incorrect, it is easy to invert it. To permit this, when audio data is stored or transmitted, it can be specially coded and accompanied by redundancy. This enables the reproduced data to be checked for errors. When an error is detected, further processing can be performed to absolutely correct the error. In the worst case, errors can be concealed by synthesizing new data. In either case, error correction makes digital data transmission and storage much more robust and indeed makes the systems more efficient. Strong error-correction techniques allow a channel's signal-to-noise ratio to be reduced; thus, for example, permitting lower power levels for digital broadcasting. Similarly, error correction relaxes the manufacturing tolerances for mass media such as CD, DVD, and Blu-ray.

However, error correction is more than a fortuitous opportunity; it is also an obligation. Because of high data densities in audio storage media, a petty defect in the manufactured media or a particle of dust can obliterate hundreds or thousands of bits. Compared to absolute numerical data stored digitally, where a bad bit might mean the difference between adding or subtracting figures from a bank account, digital audio data is relatively forgiving of errors; enjoyment of music is not necessarily ruined because of occasional flaws in the output waveform. However, even so, error correction is mandatory for a digital audio system because uncorrected audible errors can easily occur because of the relatively harsh environment to which most audio media are subjected.

Error correction for digital audio is thus an opportunity to preserve data integrity, an opportunity not available with analog audio, and it is absolutely necessary to ensure the success of digital audio storage, because errors surely occur. With proper design, digital audio systems such as CD, DVD, and Blu-ray can approach the computer industry standard, which specifies an error rate of 10^{-12}, that is, less than one uncorrectable error in 10^{12} (one trillion) bits. However, less stringent error performance is adequate for most audio applications. Without such protection, digital audio recording would not be viable. Indeed, the evolution of digital audio technology can be measured by the prerequisite advances in error correction.

Sources of Errors

Degradation can occur at every stage of the digital audio recording chain. Quantization error, converter nonlinearity, and jitter can all limit system performance. High-quality components, careful circuit design, and strict manufacturing procedures minimize these degradations. For example, a high-quality analog-to-digital (A/D) converter can exhibit satisfactory low-level linearity, and phase-locked loops can overcome jitter.

The science of error correction mainly attends to stored and transmitted data, because it is the errors presented by those environments that are most severe and least subject to control. For example, optical media can be affected by pit asymmetry, bubbles or defects in substrate, and coating defects. Likewise, for example, transmitted data is subject to errors caused by interfering signals and atmospheric conditions.

Barring design defects or a malfunctioning device, the most severe types of errors occur in the recording or transmission signal chains. Such errors result in corrupted data, and hence a defective audio signal. The most significant cause of errors in digital media is dropouts, essentially a defect in the medium that causes a momentary drop in signal strength. Dropouts can occur in any optical disc and can be traced to two causes: a manufactured defect in the medium or a defect introduced during use. Transmitted signals are vulnerable to disturbances from many types of human-made or natural interference. The effects of noise, multipath interference, crosstalk, weak signal strength, and fading are familiar to all cell phone users.

Defects in storage media or disruptions during transmission can cause signal transitions that are misrepresented as erroneous data. The severity of the error depends on the nature of the error. A single bad bit can be easily corrected, but a string of bad bits may be uncorrectable. An error in the least significant bit of a PCM word may pass unnoticed, but an error in the most significant bit may create a drastic change in amplitude. Usually, a severe error results in a momentary increase in noise or distortion or a muted signal. In the worst case, a loss of data or invalid data can provoke a click or pop, as the digital-to-analog converter's output jumps to a new amplitude while accepting invalid data.

Errors occur in several modes; thus, classifications have been developed to better identify them. Errors that have no relation to each other are called random-bit errors. These random-bit errors occur singly and are generally easily corrected. A burst error is a large error, disrupting perhaps thousands of bits, and requires more sophisticated correction. An important characteristic of any error-correction system is the burst length, that is, the maximum number of adjacent erroneous bits that can be corrected. Both random and burst errors can occur within a single medium; therefore, an error-correction system must be designed to correct both error types simultaneously.

In addition, because the nature of errors depends on the medium, error-correction techniques must be optimized for the application at hand. Thus, storage media errors, wired and wireless transmission errors must be considered separately. Finally, error-correction design is influenced by the kind of modulation code used to convey the data.

Optical discs can suffer from dropouts resulting from various manufacturing and handling problems. When a master disc is manufactured, an incorrect intensity of laser light, incorrect developing time, or a defect in the photoresist can create badly formed pits. Dust or scratches incurred during the plating or pressing processes, or pinholes or other defects in the discs' reflective coating can all create dropouts. Stringent manufacturing conditions and quality control can prevent many of these dropouts from leaving the factory. Optical discs become dirty or damaged with use. Dust, dirt, and oil can be wiped

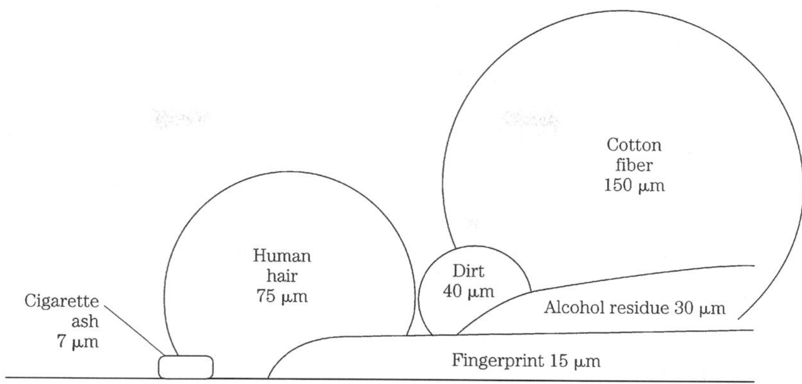

FIGURE 5.1 Small physical objects are large in the context of a 1-μm recorded bit size.

clean; however, scratches might interfere with the pickup's ability to read data. Although physically small to bulky humans, foreign objects such as those shown in Fig. 5.1 are large when it is considered that the recorded bit size can be much smaller than 1 μm (micrometer). When cleaning optical media such as a Compact Disc, a clean, soft cloth should be used to wipe the disc radially, that is, across the width of the disc and not around its circumference. Any scratches that result will be perpendicular to tracks and thus easier to correct, but a single scratch along one track could be impossible to correct because of the sustained consecutive loss of data.

Hard-disk magnetic media are enclosed in a sealed, protective canister and are mainly immune to dust and dirt contamination. Moreover, bad sectors are marked and not used for recording; only minimal error correction is needed.

A number of factors conspire to degrade the quality of wired data channels including limited bandwidth and intersymbol interference, attenuation, noise, ringing, reflection, and substandard cables and connectors. For example, as data travels down a length of cable, the cable acts as a low pass filter, reducing bandwidth and adding group delay. A square-wave signal will show a gradual rise and fall time, and if the data rate is too high, the signal might not reach its full logical 1 and 0 levels. In other words, the level actually representing one bit can partly depend on the level of the previous bit; this is intersymbol interference. Similarly, interference can be caused by signal reflection, echoes from the terminating circuits that can reinforce, or cancel signal levels. Reflections can be caused by impedance mismatches between cables and terminating circuits. Other error sources include external low-frequency noise that can shift data transitions, causing jitter. Radio-frequency (RF) interference must be prevented with proper shielding and grounding. Fiber-optic cables are largely immune to radio-frequency interference and other problems that plague copper cables. The quality of a digital signal can be evaluated from its eye pattern, as discussed in Chap. 4.

Cellular telephones, digital radio, and other audio technologies using wireless transmission and reception are subject to numerous error conditions. These systems use radio waves traveling through air for communication, and these signals may be attenuated or otherwise unreliable, resulting in poor reception, dropped calls, and dead zones. Many types of electronic devices emit unwanted electromagnetic signals which can interfere with wireless communications. Adjacent-channel interference is impairment at one frequency caused by a signal on a nearby channel frequency. Co-channel interference is created by other transmitters using the same channel frequency, for example, in nearby cells. Radio signals can be disrupted by multipath interference created when signals are reflected by a large surface such as a nearby building. Signal strength is affected by location distance or placement of antennas in cellular network architectures, and low network density. Weather conditions as well as solar flares and sunspots can affect signal strength and quality. Signal strength can also be reduced by many kinds of obstructions; this commonly occurs in some buildings, subways, and tunnels, as well as on hilly terrain. Some kinds of obstructions are not intuitively obvious. Excessive foliage can block signals; for example, mobile systems operating in the 800-MHz band are not used in forested areas because pine needles can absorb and obstruct signals at that wavelength.

Quantifying Errors

A number of parameters have been devised to quantify the integrity of data. The bit-error rate (BER) is the number of bits received in error divided by the total number of bits received. For example, a BER of 10^{-9} specifies one bit error for 10^9 (1 billion) bits transmitted. In other words, BER is an error count. An optical disc system, for example, may contain error-correction algorithms able to handle a BER of 10^{-5} to 10^{-4}. The BER specifies the number of errors, but not their distribution. For example, a BER value might be the same for a single large error burst, as for several smaller bursts. Thus the BER is not a good indicator for an otherwise reliable channel subject to burst errors. For example, over 80% of the errors in an optical disc might be burst errors.

Taking a different approach, the block-error rate (BLER) measures the number of blocks or frames of data that have at least one occurrence of uncorrected data. In many cases, BLER is measured as an error count per second, averaged over a 10-second interval. For example, the CD standard specifies that a CD can exhibit a maximum of 220 BLER errors per second, averaged over a 10-second interval. These values do not reflect the literal number of errors, but gauge their relative severity. BLER is discussed in more detail below, in the context of the Reed–Solomon error-correction code. The burst-error length (BERL) counts the number of consecutive blocks in error; it can be specified as a count per second.

Because burst errors can occur intermittently, they are difficult to quantify with a bit-error rate. In some cases, the errored second is measured; this tallies each second of data that contains an error. Errors can be described as the number of errored seconds over a period of time, or the time since the last errored second. As with BLER and BERL, this does not indicate how many bits are in error, but rather counts the occasions of error.

When designing error correction for an audio system, benchmarks such as raw BER and BLER must be accurately estimated because they are critical in determining the extent of error correction needed. If the estimates are too low, the system may suffer from uncorrected errors. On the other hand, if the estimates are too high, too much channel capacity will be given to error-correction redundancy, leaving relatively less capacity for audio data, and thus overall audio fidelity may suffer. No matter how errors are addressed, it is the job of any system to detect and correct errors, keeping uncorrected errors to a minimum.

Objectives of Error Correction

Many types of error correction exist for different applications. Designers must judge the correctability of random and burst errors, redundancy overhead, probability of misdetection, maximum burst error lengths to be corrected or concealed, and the cost of an encoder and decoder. For digital audio, errors that cannot be corrected are concealed. However, sometimes a misdetected error cannot be concealed, and this can result in an audible click in the audio output. Other design objectives are set by the particular application; for example, optical discs are relatively tolerant of fingerprints because the transparent substrate places them out of focus to the optical pickup. Because of the varied nature of errors—some predictable and some unpredictable—error-correction systems ultimately must use various techniques to guard against them.

Many storage or transmission channels will introduce errors in the stored or transmitted data; for accurate recovery, it is thus necessary for the data to include redundancy that overcomes the error. However, redundancy alone will not ensure accuracy of the recovered information; appropriate error-detection and correction coding must be used.

Although a perfect error-correction system is theoretically possible, in which every error is detected and corrected, such a system would create an unreasonably high data overhead because of the amount of redundant data required to accomplish it. Thus, an efficient audio error-correction system aims for a low audible error rate after correction and concealment, while minimizing the amount of redundant data and data processing required for successful operation. An error-correction system comprises three operations:

- Error detection uses redundancy to permit data to be checked for validity.
- Error correction uses redundancy to replace erroneous data with newly calculated valid data.
- In the event of large errors or insufficient data for correction, error concealment techniques substitute approximately correct data for invalid data.

When not even error concealment is possible, digital audio systems can mute the output signal rather than let the output circuitry attempt to decode severely incorrect data, and produce severely incorrect sounds.

Error Detection

All error-detection and correction techniques are based on data redundancy. Data is said to be redundant when it is entirely derived from existing data, and thus conveys no additional information. In general, the greater the likelihood of errors, the greater the redundancy required. All information systems rely heavily on redundancy to achieve reliable communication; for example, spoken and written language contains redundancy. If the garbled lunar message "ONC GIAVT LE?P FOR MHNKIND" is received, the information could be recovered. In fact, Claude Shannon estimated that written English is about 50% redundant. Many messages incorporate both the original message as well as redundancy to help ensure that the message is properly understood.

Redundancy is required for reliable data communication. If a data value alone is generated, transmitted once, and received, there is no absolute way to check its validity at the receiving end. At best, a word that differs radically from its neighbors might be suspect. With digital audio, in which there is usually some correlation from one 48,000th of a second to the next, such an algorithm might be reasonable. However, we could not absolutely detect errors or begin to correct them. Clearly, additional information is required to reliably detect errors in received data. Since the additional information usually originates from the same source as the original data, it is subject to the same error-creating conditions as the data itself. The task of error detection is to properly code transmitted or stored information, so that when data is lost or made invalid, the presence of the error can be detected.

In an effort to detect errors, the original message could simply be repeated. For example, each data word could be transmitted twice. A conflict between repeated words would reveal that one is in error, but it would be impossible to identify the correct word. If each word was repeated three times, probability would suggest that the two in agreement were correct while the differing third was in error. Yet the opposite could be true, or all three words could agree and all be in error. Given enough repetition, the probability of correctly detecting an error would be high; however, the data overhead would be enormous. Also, the increased data load can itself introduce additional errors. More efficient methods are required.

Single-Bit Parity

Practical error detection uses techniques in which redundant data is coded so it can be used to efficiently check for errors. Parity is one such method. One early error-detection method was devised by Islamic mathematicians in the ninth century; it is known as "casting out 9s." In this technique, numbers are divided by 9, leaving a remainder or residue. Calculations can be checked for errors by comparing residues. For example, the residue of the sum (or product) of two numbers equals the sum (or product) of the residues. It is important to compare residues, and sometimes the residue of a sum or product residue must be taken. If the residues are not equal, it indicates an error has occurred in the calculation, as shown in Fig. 5.2. An insider's trick makes the method even more efficient; the sum of digits in a number always has the same 9s residue as the number itself. The technique of casting out 9s can be used to cast out any number, and forms the basis for a binary error-detection method called parity.

Casting out 9s:

$240 + 578 \stackrel{?}{=} 818 \rightarrow$ $(2 + 4 + 0 = 6)$
$(5 + 7 + 8 = 20, 2 + 0 = 2)$
$(8 + 1 + 8 = 17, 1 + 7 = 8)$

$6 + 2 \stackrel{\checkmark}{=} 8$ Casting out 9s sum agrees, thus no error

$227 \times 67 \stackrel{?}{=} 15209$
$2 \times 4 \stackrel{\checkmark}{=} 8$ Casting out 9s product agrees, thus no error

$154 \times 95 \stackrel{?}{=} 14613$
$1 \times 5 \neq 6$ Casting out 9s product does not agree calculation is in error

Casting out 2s:

Sum of 11001011 is 5, which is odd.
Cast out 2s to get 1 and append to word:110010111. In this way, the number of 1s is always even.

FIGURE 5.2 Casting out of 9s and 2s provides simple error detection.

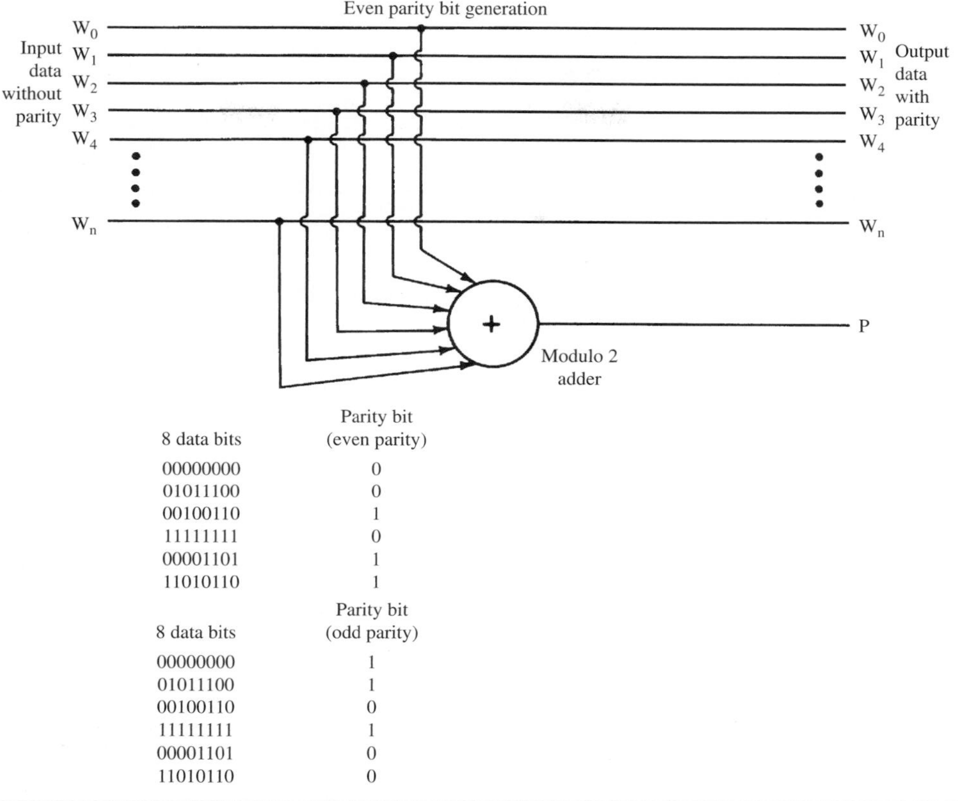

8 data bits	Parity bit (even parity)
00000000	0
01011100	0
00100110	1
11111111	0
00001101	1
11010110	1

8 data bits	Parity bit (odd parity)
00000000	1
01011100	1
00100110	0
11111111	1
00001101	0
11010110	0

FIGURE 5.3 Parity can be formed through the modulo 2 addition of data bits.

Given a binary number, a residue bit can be formed by casting out 2s. This extra bit is formed when the word is transmitted or stored, and is carried along with the data word. This extra bit, known as a parity bit, permits error detection, but not correction. Rather than cast out 2s, a more efficient algorithm can be used. An even parity bit is formed with two simple rules: if the number of 1s in the data word is even (or zero), the parity bit is a 0; if the number of 1s in the word is odd, the parity bit is a 1. In other words, with even parity, there is always an even number of 1s. To accomplish this, the data bits are added together with modulo 2 addition, as shown in Fig. 5.3. Thus, an 8-bit data word, made into a 9-bit word with an even parity bit, will always have an even number of 1s (or none). This method results in even parity. Alternatively, by forcing an odd number of 1s, odd parity results. Both methods are functionally identical.

At playback, the validity of the received data word is tested using the parity bit; that is, the received data bits are added together to calculate the parity of the received data. If the received parity bit and the calculated parity bit are in conflict, then an error has occurred. This is a single-bit detector with no correction ability. With even parity, for example, the function can determine when an odd number of errors (1, 3, 5, and so on) are present in the received data; an even number of errors will not be

Transmitted word		Received word		Parity calculated from received data word	
Data	Parity	Data	Parity		
00011001	1	00001001	1	0	Error detected
10101011	1	11001011	1	1	Errors not detected
01110100	0	01110100	1	0	Parity error detected
01101011	1	00000011	0	0	Errors not detected

FIGURE 5.4 Examples of single-bit parity error detection.

detected, as shown in Fig. 5.4. Probability dictates that the error is in the data word, rather than the parity bit itself. However, the reverse could be true.

In many cases, errors tend to occur as burst errors. Thus, many errors could occur within each word and single-bit parity would not provide reliable detection. By itself, a single-bit parity check code is not suitable for error detection in many digital audio storage or transmission systems.

ISBN

For many applications, simple single-bit parity is not sufficiently robust. More sophisticated error detection codes have been devised to make more efficient use of redundancy. One example of coded information is the International Standard Book Number (ISBN) code found on virtually every book published. No two books and no two editions of the same book have the same ISBN. Even soft- and hard-cover editions of a book have different ISBNs.

An ISBN number is more than just a series of numbers. For example, consider the ISBN number, 0-14-044118-2 (the hyphens are extraneous). The first digit (0) is a country code; for example, 0 is for the United States and some other English-speaking countries. The next two digits (14) is a publisher code. The next six digits (044118) is the book title code. The last digit (2) is particularly interesting; it is a check digit. It can be used to verify that the other digits are correct. The check digit is a modulo 11 weighted checksum of the previous digits. In other words, when the digits are added together in modulo 11, the weighted sum must equal the number's checksum. (To maintain uniform length of 10 digits per ISBN, the Roman numeral X is used to represent the check digit 10.) Given this code, with its checksum, we can check the validity of any ISBN by adding together (modulo 11) the series of weighted digits, and comparing them to the last checksum digit.

Consider an example of the verification of an ISBN code. To form the weighted checksum of a 10-digit number *abcdefghij*, compute the weighted sum of the numbers by multiplying each digit by its digit position, starting with the leftmost digit:

$$10a + 9b + 8c + 7d + 6e + 5f + 4g + 3h + 2i + 1j$$

For the ISBN number 0-14-044118-2, the weighted sum is:

$$10\times 0 + 9\times 1 + 8\times 4 + 7\times 0 + 6\times 4 + 5\times 4 + 4\times 1 + 3\times 1 + 2\times 8 + 1\times 2 = 110$$

The weighted checksum modulo 11 is found by taking the remainder after dividing by 11:

$$\frac{110}{11} = 10, \text{ with a 0 remainder}$$

The 0 remainder suggests that the ISBN is correct. In this way, ISBNs can be accurately checked for errors. Incidentally, this 10-digit ISBN system (which has the ability to assign 1 billion numbers) is being replaced by a 13-digit ISBN to increase the numbering capacity of the system; it adds a 978 prefix to existing ISBNs, a 979 prefix to new numbers, and retains a checksum. In any case, the use of a weighted checksum compared to the calculated remainder with modulo arithmetic is a powerful way to detect errors. In fact, error-detection codes in general are based on this principle.

Cyclic Redundancy Check Code

The cyclic redundancy check code (CRCC) is an error-detection method used in some audio applications because of its ability to detect burst errors in the recording medium or transmission channel. The CRCC is a cyclic block code that generates a parity check word. The bits of a data word can be added together to form a sum of the bits; this forms a parity check word. For example, in 1011011010, the six binary 1s are added together to form binary 0110 (6 in base 10), and this check word is appended to the data word to form the codeword for transmission or storage. As with single-bit parity, any disagreement between the received checksum and that formed from the received data would indicate with high probability that an error has occurred.

The CRCC works similarly, but with a more sophisticated calculation. Simply stated, each data block is divided by an arbitrary and constant number. The remainder of the division is appended to the stored or transmitted data block. Upon reproduction, division is performed on the received word; a zero remainder is taken to indicate an error-free signal, as shown in Fig. 5.5. A more detailed examination of the encoding and decoding steps in the CRCC algorithm is shown in Fig. 5.6. A message m in a k-bit data block is operated upon to form an $n-k$ bit CRCC detection block, where n is the length of the complete block.

The original k-bit data block is multiplied by X^{n-k} to shift the data in preparation to appending the check bits. It is then divided by the generation polynomial g to form the quotient q and remainder r. The transmission polynomial v is formed from the

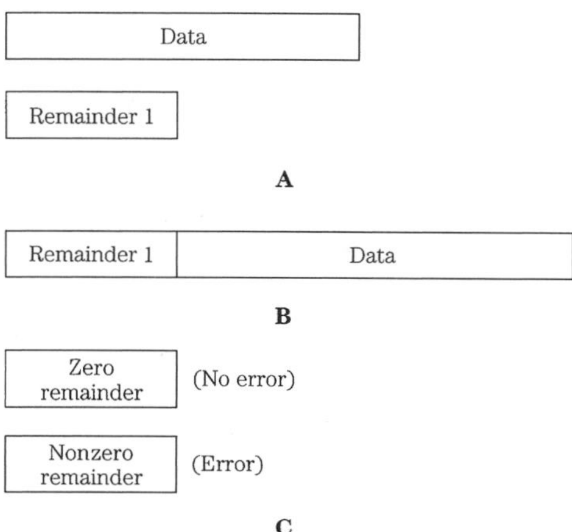

FIGURE 5.5 CRCC in simplified form showing the generation of the remainder. A. Original block of data is divided to produce a remainder. The quotient is discarded. B. The remainder is appended to the data word and both words are transmitted or stored. C. The received data is again divided to produce a remainder, used to check for errors.

← k bits →

Message m

$\mathbf{m} = (m_0, m_1, m_2, \cdots m_{k-1})$
In polynomial form,
$\mathbf{m}(X) = (m_0 + m_1X + m_2X^2 + \cdots + m_{k-1}X^{k-1}$
Multiplying $\mathbf{m}(X)$ by X^{n-k}
$X^{n-k}\,\mathbf{m}(X) = m_0\,X^{n-k} + m_1\,X^{n-k+1} + \cdots + m_{k-1}X^{n-1}$
Dividing $X^{n-k}\,\mathbf{m}(X)$ by $\mathbf{g}(X)$, the generation polynomial,
$X^{n-k}\,\mathbf{m}(X) = \mathbf{q}(X)\,\mathbf{g}(X) + \mathbf{r}(X)$
where $\mathbf{q}(X)$ and $\mathbf{r}(X)$ are quotient and remainder respectively,
where $\mathbf{r}(X) = r_0 + r_1X + r_2X^2 + \cdots + r_{n-k-1}X^{n-k-1}$
Arranging previous equation and adding $\mathbf{r}(X)$,
$\mathbf{r}(X) + Xr^{n-k}\,\mathbf{m}(X) = \mathbf{q}(X)\,\mathbf{g}(X) + \mathbf{r}(X) + \mathbf{r}(X)$
However $\mathbf{r}(X) + \mathbf{r}(X) = 0$ thus,
$\mathbf{r}(X) + X^{n-k}\,\mathbf{m}(X) = \mathbf{q}(X)\,\mathbf{g}(X)$
Thus $\mathbf{r}(X) + X^{n-k}\,\mathbf{m}(X)$ is a multiple of $\mathbf{g}(X)$.
$\mathbf{r}(X) + X^{n-k}\,\mathbf{m}(X)$ is the transmitted code polynomial $\mathbf{v}(X)$:

$$\mathbf{v}(X) = \mathbf{r}(X) + X^{n-k}\,\mathbf{m}(X) = r_0 + r_1X + r_2X^2 + \cdots + r_{n-k-1}X^{n-k-1} + m_0X^{n-k} + m_1X^{n-k+1} + \cdots m_{k-1}X^{n-1}$$

This corresponds to the transmitted codeword:
$(r_0, r_1, r_2, \cdots r_{n-k-1}, m_0, m_1, m_2, \cdots m_{k-1})$

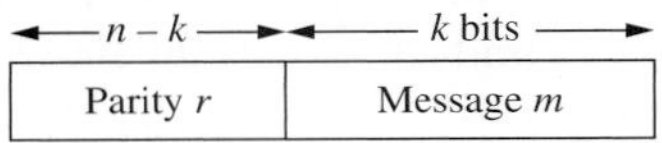

A

← n bits →

Received data

$\mathbf{u} = (u_1, u_2, u_3, u_4, \cdots u_{n-1})$
In polynomial form:
$\mathbf{u}(X) = u_0 + u_1X + u_2X^2 + \cdots + u_{n-1}X^{n-1}$
where $u_0, u_1, u_2, \cdots u_{n-k-1}$ are parity check bits and $u_{n-k}, \cdots u_{n-1}$ are information bits. The syndrome $\mathbf{s}$ is calculated by taking the mod 2 sum of the received parity bits and the parity bits formed from the received information. Thus, syndrome $\mathbf{s}(X)$ is equal to the remainder of $\mathbf{u}(X)$ divided by $\mathbf{g}(X)$:
$\mathbf{u}(X) = \mathbf{p}(X)\mathbf{g}(X) + \mathbf{s}(X)$
A nonzero value for $\mathbf{s}$ detects an error. The difference between received ($\mathbf{u}$) and transmitted ($\mathbf{v}$) information is an error pattern $\mathbf{e}$. From $\mathbf{e}$, we can recover $\mathbf{v}$ by using the syndrome for error correction:

$$\mathbf{u}(X) = \mathbf{v}(X) + \mathbf{e}(X)$$

Since $\mathbf{v}(X) = \mathbf{m}(X)\mathbf{g}(X)$,

$$\mathbf{u}(X) = \mathbf{m}(X)\mathbf{g}(X) + \mathbf{e}(X) = \mathbf{p}(X)\mathbf{g}(X) + \mathbf{s}(X)$$

Thus $\mathbf{e}(X) = [\mathbf{p}(X) + \mathbf{m}(X)]\mathbf{g}(X) + \mathbf{s}(X)$

Thus when the error pattern is divided by the generation polynomial, the remainder is the syndrome, which can be used to correct errors. Note that the generation polynomial was chosen so that the error polynomial consists of an error pattern not divisible by $\mathbf{g}$. The above derivations utilize the properties of modulo 2 arithmetic.

B

Figure 5.6 CRCC encoding and decoding algorithms. A. CRCC encoding. B. CRCC decoding and syndrome calculation.

original message m and the remainder r; it is thus a multiple of the generation polynomial g. The transmission polynomial v is then transmitted or stored. The received data u undergoes error detection by calculating a syndrome, in the sense that an error is sign of malfunction or disease. Specifically, the operation creates a syndrome c with modulo 2 addition of received parity bits and parity that is newly calculated from the received message. A zero syndrome shows an error-free condition. A nonzero syndrome denotes an error.

Error correction can be accomplished by forming an error pattern that is the difference between the received data and the original data to be recovered. This is mathematically possible because the error polynomial e divided by the original generation polynomial produces the syndrome as a remainder. Thus, the syndrome can be used to form the error pattern and hence recover the original data. It is important to select the generation polynomial g so that error patterns in the error polynomial e are not exactly divisible by g. CRCC will fail to detect an error if the error is exactly divisible by the generation polynomial.

In practice, CRCC and other detection and correction codewords are described in mathematical terms, where the data bits are treated as the coefficients of a binary polynomial. As we observed in Chap. 1, the binary number system is one of positional notation such that each place represents a power of 2. Thus, we can write binary numbers in a power of 2 notation. For example, the number 1001011, with MSB (most significant bit) leading, can be expressed as:

$$1\times 2^6+0\times 2^5+0\times 2^4+1\times 2^3+0\times 2^2+1\times 2^1+1\times 2^0$$

or:

$$2^6+2^3+2^1+2^0$$

In general, the number can be expressed as:

$$x^6+x^3+x+1$$

With LSB (least significant bit) leading, the notation would be reversed. This polynomial notation is the standard terminology in the field of error correction. In the same way that use of modulo arithmetic ensures that all possible numbers will fall in the modulus, polynomials are used to ensure that all values fall within a given range. In modulo arithmetic, all numbers are divided by the modulus and the remainder is used as the number. Similarly, all polynomials are divided by a modulus polynomial of degree n. The remainder has n coefficients, and is recognizable as a code symbol. Further, just as we desire a modulus that is a prime number (so that when a product is 0, at least one factor is 0), we desire a prime polynomial—one that cannot be represented as the product of two polynomials. The benefit is added structure to devised codes, resulting in more efficient implementation.

Using polynomial notation, we can see how CRCC works. An example of cyclic code encoding is shown in Fig. 5.7. The message 1001 is written as the polynomial $x^3 + 1$. So that three parity check bits can be appended, the message is multiplied by x^3 to shift the data to the left. The message is then divided by the generation polynomial $x^3 + x^2 + 1$. The remainder $x + 1$ is appended to create the polynomial $x^6 + x^3 + x + 1$.

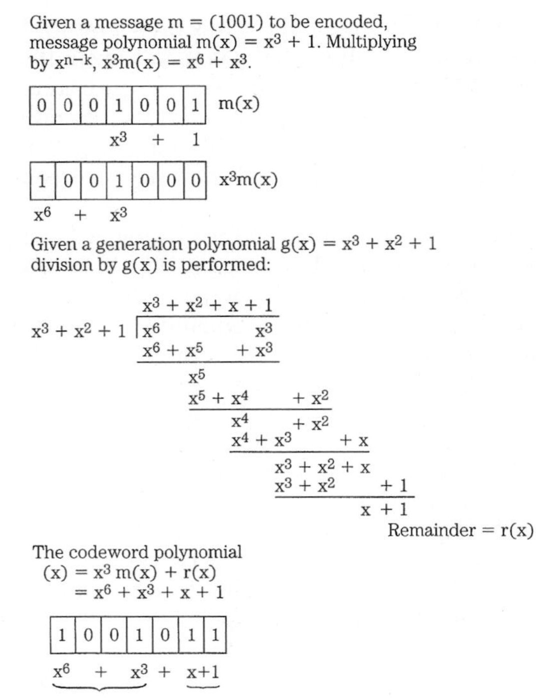

Given a message m = (1001) to be encoded,
message polynomial m(x) = x3 + 1. Multiplying
by xn−k, x3m(x) = x6 + x3.
0 0 0 1 0 0 1
m(x)
x3 + 1
1 0 0 1 0 0 0
x3m(x)
x6 + x3
Given a generation polynomial g(x) = x3 + x2 + 1
division by g(x) is performed:
x3 + x2 + x + 1
x3 + x2 + 1
x6 x3
x6 + x5 + x3
x5
x5 + x4 + x2
x4 + x2
x4 + x3 + x
x3 + x2 + x
x3 + x2 + 1
x + 1
Remainder = r(x)
The codeword polynomial
(x) = x3 m(x) + r(x)
= x6 + x3 + x + 1
1 0 0 1 0 1 1
x6 + x3 + x+1
Message
Parity

Figure 5.7 An example of cyclic code encoding, with message 1001 written as the polynomial $x^3 + 1$. The encoder outputs the original message and a parity word.

A shift register implementation of this encoder is shown in Fig. 5.8. The register is initially loaded with 0s, and the four message bits are sequentially shifted into the register, and appear at the output. When the last message bit has been output, the encoder's switches are switched, 0s enter the register, and three parity bits are output. Modulo 2 adders are used. This example points up an advantage of this type of error-detection encoding: the implementation can be quite simple.

The larger the data block, the less redundancy results, yet mathematical analysis shows that the same error-detection capability remains. However, if random or short duration burst errors tend to occur frequently, then the integrity of the detection is decreased and shorter blocks might be necessitated. The extent of error detectability of a CRCC code can be summarized. Given a k-bit data word with m (where $m = n - k$) bits of CRCC, a codeword of n bits is formed, and the following are true:

- Burst errors less than or equal to m bits are always detectable.
- Detection probability of burst errors of $m + 1$ bits is $1 - 2^{-m+1}$.
- Detection probability of burst errors longer than $m + 1$ bits is $1 - 2^{-m}$. (These first three items are not affected by the length n of the codeword.)
- Random errors up to three consecutive bits long can be detected.

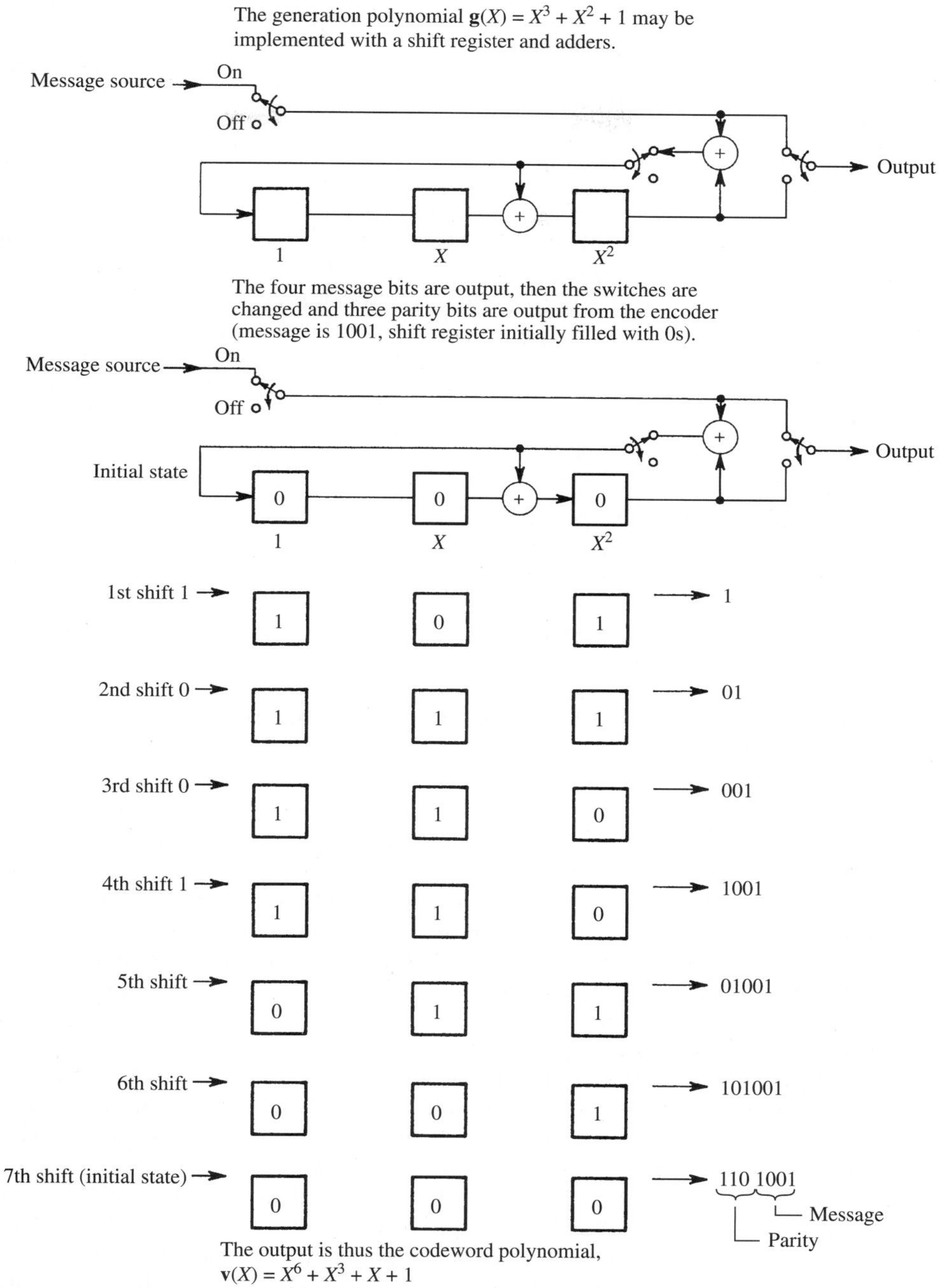

The generation polynomial $\mathbf{g}(X) = X^3 + X^2 + 1$ may be implemented with a shift register and adders.

Message source → On / Off → Output

1 X X^2

The four message bits are output, then the switches are changed and three parity bits are output from the encoder (message is 1001, shift register initially filled with 0s).

Message source → On / Off → Output

	1	X	X^2	Output
Initial state	0	0	0	
1st shift 1 →	1	0	1	→ 1
2nd shift 0 →	1	1	1	→ 01
3rd shift 0 →	1	1	0	→ 001
4th shift 1 →	1	1	0	→ 1001
5th shift →	0	1	1	→ 01001
6th shift →	0	0	1	→ 101001
7th shift (initial state) →	0	0	0	→ 110 1001

110 = Parity; 1001 = Message

The output is thus the codeword polynomial, $\mathbf{v}(X) = X^6 + X^3 + X + 1$

FIGURE 5.8 An implementation of a cyclic encoder using shift registers. Modulo 2 (XOR) adders are used.

CRCC is quite reliable. For example, if 16 parity check bits are generated, the probability of error detection is $1 - 2^{-16}$ or about 99.99%. Ultimately, the medium determines the design of the CRCC and the rest of the error-correction system. The power of the error-correction processing following the CRCC also influences how accurate the CRCC detection must be. The CRCC is often used as an error pointer to identify the number and extent of errors prior to other error-correction processing. In some cases, to reduce codeword size, a shortened or punctured code is used. In this technique, some of the leading bits of the codeword are omitted.

The techniques used in the CRCC algorithm apply generally to cyclic error-detection and error-correction algorithms. They use codewords that are polynomials that normally result in a zero remainder when divided by a generation polynomial. To perform error checking on received data, division is performed; a zero remainder syndrome indicates that there is no error. A nonzero remainder syndrome indicates an error. The nonzero syndrome is the error polynomial divided by the check polynomial. The syndrome can be used to correct the error by multiplying the syndrome by the check polynomial to obtain the error polynomial. The error in the received data appears as the error polynomial added to the original codeword. The correct original codeword can be obtained by subtracting (adding in modulo 2) the error polynomial from the received polynomial.

Error-Correction Codes

With the help of redundant data, it is possible to correct errors that occur during the storage or transmission of digital audio data. In the simplest case, data is merely duplicated. For example, instead of writing only one data track, two tracks of identical data could be written. The first track would normally be used for playback, but if an error was detected through parity or other means, data could be taken from the second track. To alleviate the problem of simultaneously erroneous data, redundant samples could be displaced with respect to each other in time.

In addition, channel coding can be used beneficially. For example, 3-bit words can be coded as 7-bit words, selected from 2^7 possible combinations to be as mutually different as possible. The receiver examines the 7-bit words and compares them to the eight allowed codewords. Errors could be detected, and the words changed to the nearest allowed codeword, before the codeword is decoded to the original 3-bit sequence. Four correction bits are required for every three data bits; the method can correct a single error in a 7-bit block. This minimum length concept is important in more sophisticated error-correction codes.

Although such simple methods are workable, they are inefficient because of the data overhead that they require. A more enlightened approach is that of error-correction codes, which can achieve more reliable results with less redundancy. In the same way that redundant data in the form of parity check bits is used for error detection, redundant data is used to form codes for error correction. Digital audio data is encoded with related detection and correction algorithms. On playback, errors are identified and corrected by the detection and correction decoder. Coded redundant data is the essence of all correction codes; however, there are many types of codes, different in their designs and functions.

The field of error-correction codes is a highly mathematical one. Many types of codes have been developed for different applications. In general, two approaches are used: block codes using algebraic methods, and convolutional codes using probabilistic methods. Block codes form a coded message based solely on the message parsed into a data block. In a convolutional code, the coded message is formed from the message present in the encoder at that time as well as previous message data. In some cases, algorithms use a block code in a convolutional structure known as a Cross-Interleave Code. For example, such codes are used in the CD format.

Block Codes

Block error-correction encoders assemble a number of data words to form a block and, operating over that block, generate one or more parity words and append them to the block. During decoding, an algorithm forms a syndrome word that detects errors and, given sufficient redundancy, corrects them. Such algorithms are effective against errors encountered in digital audio applications. Error correction is enhanced by interleaving consecutive words. Block codes base their parity calculations on an entire block of information to form parity words. In addition, parity can be formed from individual words in the block, using single-bit parity or a cyclic code. In this way, greater redundancy is achieved and correction is improved. For example, CRCC could be used to detect an error, then block parity used to correct the error.

A block code can be conceived as a binary message consolidated into a block, with row and column parity. Any single-word error will cause one row and one column to be in error; thus, the erroneous data can be corrected. For example, a message might be grouped into four 8-bit words (called symbols). A parity bit is added to each row and a parity word added to each column, as shown in Fig. 5.9. At the decoder, the data is

Transmitted data block	Transmitted single-bit parity	
00010111	0	
01101010	0	
10010111	1	
11010110	1	
00111100		Transmitted parity word

Received data block	Received parity bit	
00010111	0	
01101010	0	
11100100	1	
11010110	1	
00111100		Received parity word
01110011		Parity word calculated from received data and parity word

Parity calculated on received data block	
0	
0	
0	Indicates error in word 3
1	

01110011	Calculated parity word
+11100100	Incorrect word 3
10010111	Corrected word 3

FIGURE 5.9 An example of block parity with row parity bits and a column parity word.

checked for correct parity, and any single-symbol error is corrected. In this example, bit parity shows that word three is in error, and word parity is used to correct the symbol. A double-word error can be detected, but not corrected. Larger numbers of errors might result in misdetection or miscorrection.

Block correction codes use many methods to generate the transmitted codeword and its parity; however, they are fundamentally identical in that only information from the block itself is used to generate the code. The extent of the correction capabilities of block correction codes can be simply illustrated with decimal examples. Given a block of six data words, a 7th parity word can be calculated by adding the six data words. To check for an error, a syndrome is created by comparing (subtracting in the example) the parity (sum) of the received data with the received parity value. If the result is zero, then most probably no error has occurred, as shown in Fig. 5.10A. If one data word is detected and

Original data words and parity

W_1 10
W_2 30
W_3 20
W_4 25
W_5 30
W_6 15
P $130 = W_1 + W_2 + W_3 + W_4 + W_5 + W_6$

Received data words and parity

W_1 10
W_2 30
W_3 20
W_4 25
W_5 30
W_6 15
P 130

Syndrome $S = W_1 + W_2 + W_3 + W_4 + W_5 + W_6 - P$
$= 10 + 30 + 20 + 25 + 30 + 15 - 130 = 0$

Thus no error is indicated

A

Received data and parity word

W_1 10
W_2 30
W_3 20
W_4 15 — CRCC error pointer
W_5 30
W_6 15
P 130

Syndrome $S = 10 + 30 + 20 + 15 + 30 + 15 - 130 = -10$

Error correction: $W_4 = W_4' - S$
$= 15 - (-10)$
$= 25$

B

Received data and parity word

W_1 10
W_2 30
W_3 20
W_4 25
W_5 30 — False error pointer
W_6 15
P 130

Syndrome $S = 10 + 30 + 20 + 25 + 30 + 15 - 130 = 0$

Error correction: $W_5 = W_5' - S$
$= 30 - 0$
$= 30$

C

FIGURE 5.10 Examples of single-parity block coding. A. Block correction code showing no error condition. B. Block correction code showing a correction with a pointer. C. Block correction code showing a false pointer.

the word is set to zero, a condition called a single erasure, a nonzero syndrome indicates that; furthermore, the erasure value can be obtained from the syndrome. If CRCC or single-bit parity is used, it points out the erroneous word, and the correct value can be calculated using the syndrome, as shown in Fig. 5.10B. Even if detection itself is in error and falsely creates an error pointer, the syndrome yields the correct result, as shown in Fig. 5.10C. Such a block correction code is capable of detecting a one-word error, or making one erasure correction, or correcting one error with a pointer. The correction ability depends on the detection ability of pointers. In this case, unless the error is identified with a pointer, erasure, or CRCC detection, the error cannot be corrected.

For enhanced performance, two parity words can be formed to protect the data block. For example, one parity word might be the sum of the data and the second parity word the weighted sum as shown in Fig. 5.11A. If any two words are erroneous and marked with pointers, the code provides correction. Similarly, if any two words are marked with erasure, the code can use the two syndromes to correct the data. Unlike

Received data and two parity words

W_1	10
W_2	30
W_3	20
W_4	25
W_5	30
W_6	15
P	$130 = W_1 + W_2 + W_3 + W_4 + W_5 + W_6$
Q	$440 = 6W_1 + 5W_2 + 4W_3 + 3W_4 + 2W_5 + W_6$

Syndrome $S_1 = W_1 + W_2 + W_3 + W_4 + W_5 + W_6 - P = 10 + 30 + 20 + 25 + 30 + 15 - 130 = 0$

$S_2 = 6W_1 + 5W_2 + 4W_3 + 3W_4 + 2W_5 + W_6 - Q = 60 + 150 + 80 + 75 + 60 + 15 - 440 = 0$

A

Received data and two parity words

W_1	10	
W_2	30	
W_3	20	$S_1 = -20$
W_4	25	$S_2 = -40$
W_5	10	
W_6	15	
P	130	
Q	440	

Algebraically we see that

If $6S_1 = S_2$ then W_1 is erroneous
If $5S_1 = S_2$ then W_2 is erroneous
If $4S_1 = S_2$ then W_3 is erroneous
If $3S_1 = S_2$ then W_4 is erroneous
If $2S_1 = S_2$ then W_5 is erroneous
If $S_1 = S_2$ then W_6 is erroneous
If $S_1 \neq 0$ and $S_2 = 0$ then P is erroneous
If $S_1 = 0$ and $S_2 \neq 0$ then Q is erroneous

In this case $2S_1 = S_2$, W5 is erroneous, thus (as in single erasure case):

$S_1 = 10 + 30 + 20 + 25 + 0 + 15 - 130 = -30$

$W_5 = W_5' - S$
$= 0 - (-30)$
$= 30$ Corrected

B

FIGURE 5.11 Examples of double-parity block coding. A. Block correction code with double parity showing no error condition. B. Block correction code with double parity showing a single-error correction without a pointer.

the single parity example, this double parity code can also correct any one-word error, even if it is not identified with a pointer, as shown in Fig. 5.11B. This type of error correction is well-suited for audio applications.

Hamming Codes

Cyclic codes such as CRCC are a subclass of linear block codes, which can be used for error correction. Special block codes, known as Hamming codes, create syndromes that point to the location of the error. Multiple parity bits are formed for each data word, with unique encoding. For example, three parity check bits (4, 5, and 6) might be added to a 4-bit data word (0, 1, 2, and 3); seven bits are then transmitted. For example, suppose that the three parity bits are uniquely defined as follows: parity bit 4 is formed from modulo 2 addition of data bits 1, 2, and 3; parity bit 5 is formed from data bits 0, 2, and 3; and parity bit 6 is formed from data bits 0, 1, and 3. Thus, the data word 1100, appended with parity bits 110, is transmitted as the 7-bit codeword 1100110. A table of data and parity bits is shown in Fig. 5.12A.

This algorithm for calculating parity bits is summarized in Fig. 5.12B. An error in a received data word can be located by examining which of the parity bits detects an error. The received data must be correctly decoded; therefore, parity check decoding equations must be written. These equations are computationally represented as a parity check matrix H, as shown in Fig. 5.12C. Each row of H represents one of the original encoding equations. By testing the received data against the values in H, the location of the error can be identified. Specifically, a syndrome is calculated from the modulo 2 addition of the parity calculated from the received data and the received parity. An error generates a 1; otherwise a 0 is generated. The resulting error pattern is matched in the H matrix to locate the erroneous bit. For example, if the codeword 1100110 is transmitted, but 1000110 is received, the syndromes will detect the error and generate a 101 error pattern. Matching this against the H matrix, we see that it corresponds to the second column; thus, bit 1 is in error, as shown in Fig. 5.12D. This algorithm is a single-error correcting code; therefore, it can correctly identify and correct any single-bit error.

Returning to the design of this particular code, we can observe another of its interesting properties. Referring again to Fig. 5.12A, recall that the 7-bit data words each comprise four data bits and three parity bits. These seven bits provide 128 different encoding possibilities, but only 16 of them are used; thus, 112 patterns are clearly illegal, and their presence would denote an error. In this way, the patterns of bits themselves are useful. We also may ask the question: How many bits must change value for one word in the table to become another word in the table? For example, we can see that three bits in the word 0101010 must change for it to become 0110011. Similarly, we observe that any word in the table would require at least three bit changes to become any other word. This is important because this dissimilarity in data corresponds to the code's error-correction ability. The more words that are disallowed, the more robust the detection and correction. For example, if we receive a word 1110101, which differs from a legal word 1010101 by one bit, the correct word can be determined. In other words, any single-bit error leaves the received word closer to the correct word than any other.

The number of bits that one legal word must change to become another legal word is known as the Hamming distance, or minimum distance. In this example, the Hamming distance is 3. Hamming distance thus defines the potential error correctability of a code. A distance of 1 determines simple uniqueness of a code. A distance of 2 provides single-error detectability. A distance of 3 provides single-error correctability or double-error detection. A distance of 4 provides both single-error correction and double-error

A

X_0	X_1	X_2	X_3	X_4	X_5	X_6
0	0	0	0	0	0	0
0	0	0	1	1	1	1
0	0	1	0	1	1	0
0	0	1	1	0	0	1
0	1	0	0	1	0	1
0	1	0	1	0	1	0
0	1	1	0	0	1	1
0	1	1	1	1	0	0
1	0	0	0	0	1	1
1	0	0	1	1	0	0
1	0	1	0	1	0	1
1	0	1	1	0	1	0
1	1	0	0	1	1	0
1	1	0	1	0	0	1
1	1	1	0	0	0	0
1	1	1	1	1	1	1

B

X_0, X_1, X_2, X_3	Data bits
$X_4 = X_1 + X_2 + X_3$	(Mod 2) parity check bits
$X_5 = X_0 + X_2 + X_3$	(Mod 2)
$X_6 = X_0 + X_1 + X_3$	(Mod 2)
$X_0, X_1, X_2, X_3, X_4, X_5, X_6$	Transmitted codeword

C

$$\begin{aligned} X_1 + X_2 + X_3 + X_4 \quad &= 0 \\ X_0 \quad + X_2 + X_3 \quad + X_5 \quad &= 0 \\ X_0 + X_1 \quad + X_3 \quad + X_6 &= 0 \end{aligned}$$

Decoding algorithm

$$\downarrow \quad \downarrow \quad \downarrow \quad \downarrow \quad \downarrow \quad \downarrow \quad \downarrow$$

$$\begin{bmatrix} 0 & 1 & 1 & 1 & 1 & 0 & 0 \\ 1 & 0 & 1 & 1 & 0 & 1 & 0 \\ 1 & 1 & 0 & 1 & 0 & 0 & 1 \end{bmatrix} = H$$

Parity-check matrix

D

Example: 1100110 Transmitted word
1000110 Received word

$$\begin{aligned} P_4 &= X_1 + X_2 + X_3 = 0 + 0 + 0 = 0 \\ P_5 &= X_0 + X_2 + X_3 = 1 + 0 + 0 = 1 \\ P_6 &= X_0 + X_1 + X_3 = 1 + 0 + 0 = 1 \end{aligned}$$

Parity of received data

Syndromes are calculated with mod 2 addition of parity of received data and received parity bits:

$P_4 = 0,\ X_4 = 1,\ 0 + 1 = 1$ (Error)
$P_5 = 1,\ X_5 = 1,\ 1 + 1 = 0$ (Correct)
$P_6 = 1,\ X_6 = 0,\ 1 + 0 = 1$ (Error)

The resulting syndromes form the error pattern $\begin{bmatrix} 1 \\ 0 \\ 1 \end{bmatrix}$ which corresponds to the second column of H

$$H = \begin{bmatrix} 0 & 1 & 1 & 1 & 1 & 0 & 0 \\ 1 & 0 & 1 & 1 & 0 & 1 & 0 \\ 1 & 1 & 0 & 1 & 0 & 0 & 1 \end{bmatrix}$$

Thus bit X_1 is in error

FIGURE 5.12 With Hamming codes, the syndrome points to the error location. A. In this example with four data bits and three parity bits, the code has a Hamming distance of 3. B. Parity bits are formed in the encoder. C. Parity-check matrix in the decoder. D. Single-error correction using a syndrome to point to the error location.

detection, or triple-error detection alone. A distance of 5 provides double-error correction. As the distance increases, so does the correctability of the code.

The greater the correctability required, the greater the distance the code must possess. In general, to detect t_d number of errors, the minimum distance required is greater than or equal to $t_d + 1$. In the case of an erasure e (when an error location is known and set to zero), the minimum distance is greater than or equal to $e + 1$. To correct all combinations of t_c or fewer errors, the minimum distance required is greater than or equal to $2t_c + 1$. For a combination of error correction, the minimum distance is greater than or equal to $t_d + e + 2t_c + 1$. These hold true for both bit-oriented and word-oriented codes. If m parity blocks are included in a block code, the minimum distance is less than or equal to $m + 1$. For the maximum-distance separable (MDS) codes such as B-adjacent and Reed–Solomon codes, the minimum distance is equal to $m + 1$.

Block codes are notated in terms of the input relative to the output data. Data is grouped in symbols; the smallest symbol is one-bit long. A message of k symbols is used to generate a larger n-bit symbol. Such a code is notated as (n, k). For example, if 12 symbols are input to an encoder and 20 symbols are output, the code would be (20,12). In other words $n - k$, or eight parity symbols are generated. The source rate R is defined as k/n; in this example, $R = 12/20$.

Convolutional Codes

Convolutional codes, sometimes called recurrent codes, differ from block codes in the way data is grouped for coding. Instead of dividing the message data into blocks of k digits and generating a block of n code digits, convolutional codes do not partition data into blocks. Instead, message digits k are taken a few at a time and used to generate coded digits n, formed not only from those k message digits, but from many previous k digits as well, saved in delay memories. In this way, the coded output contains a history of the previous input data. Such a code is called an (n, k) convolutional code. It uses $(N - 1)$ message blocks with k digits. It has constraint length N blocks (or nN digits) equal to $n(m + 1)$, where m is the number of delays. Its rate R is k/n.

As with linear block codes, encoding is performed and codewords are transmitted or stored. Upon retrieval, the correction decoder uses syndromes to check codewords for errors. Shift registers can be used to implement the delay memories required in the encoder and decoder. The amount of delay determines the code's constraint length, which is analogous to the block length of a block code. An example of a convolutional encoder is shown in Fig. 5.13. There are six delays, thus the constraint length is 14. The other parameters are $q = 2$, $R = 1/2$, $k = 1$, $n = 2$, and the polynomial is $x^6 + x^5 + x^2 + 1$. As shown in the diagram, message data passes through the encoder, so that previous bits affect the current coded output.

Another example of a convolutional encoder is shown in Fig. 5.14A. The upper code is formed from input data with the polynomial $x^2 + x + 1$, and the lower with $x^2 + 1$. The data sequence enters the circuit from the left and is shifted to the right one bit at a

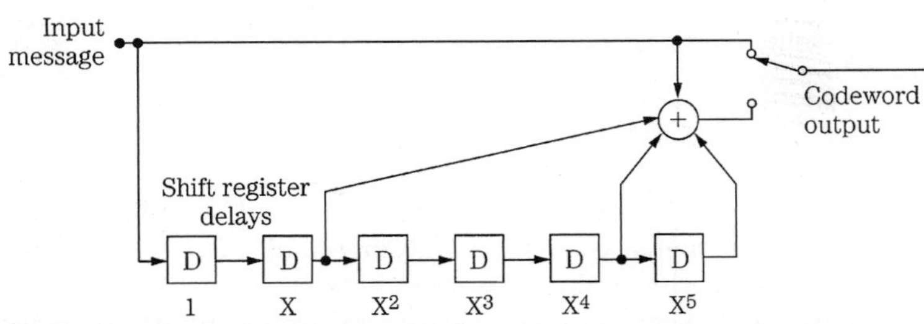

FIGURE 5.13 A convolutional code encoder with six delay blocks.

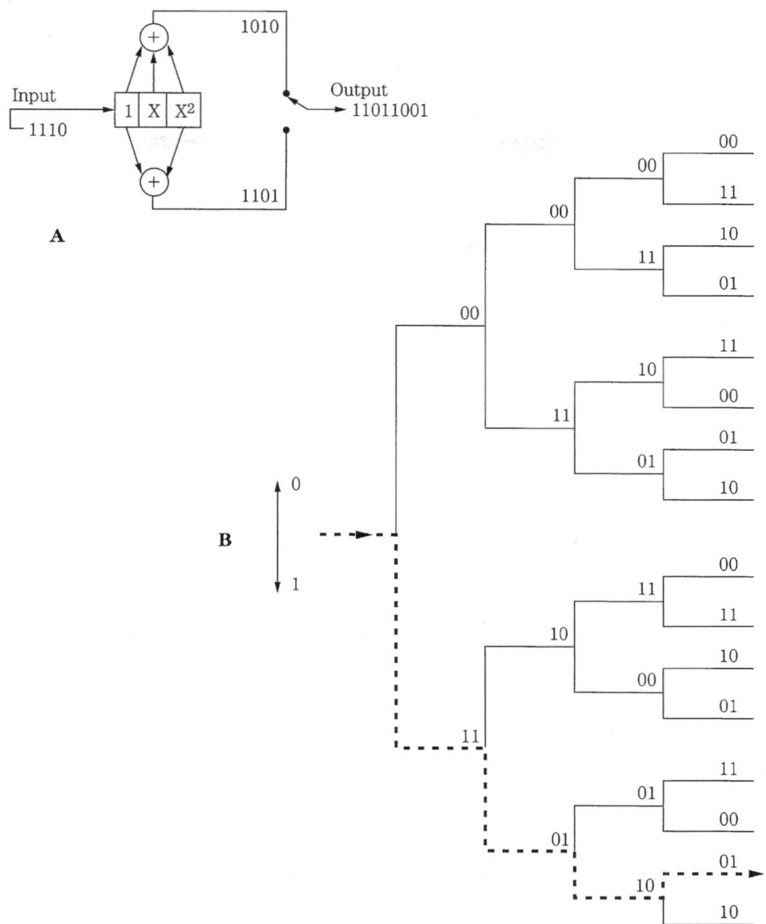

FIGURE 5.14 An example of convolutional encoding. A. Convolutional encoder with $k = 3$ and $R = 1/2$. B. Convolutional code tree diagram. (*Viterbi, 1983*)

time. The two sequences generated from the original sequence with modulo 2 addition are multiplexed to again form a single coded data stream. The resultant code has a memory of two because, in addition to the current input bit, it also acts on the preceding two bits. For every input bit, there are two output bits; hence the code's rate is 1/2. The constraint length of this code is $k = 3$.

A convolutional code can be analyzed with a tree diagram as shown in Fig. 5.14B. It represents the first four sequences of an infinite tree, with nodes spaced n digits apart and with $2k$ branches leaving each node. Each branch is an n-digit code block that corresponds to a specific k-digit message block. Any codeword sequence is represented as a path through the tree. For example, the encoded sequence for the previous encoder example can be traced through the tree. If the input bit is a 0, the code symbol is obtained by going up to the next tree branch; if the input is a 1, the code symbol is obtained by going down. The input message thus dictates the path through the tree, each input digit

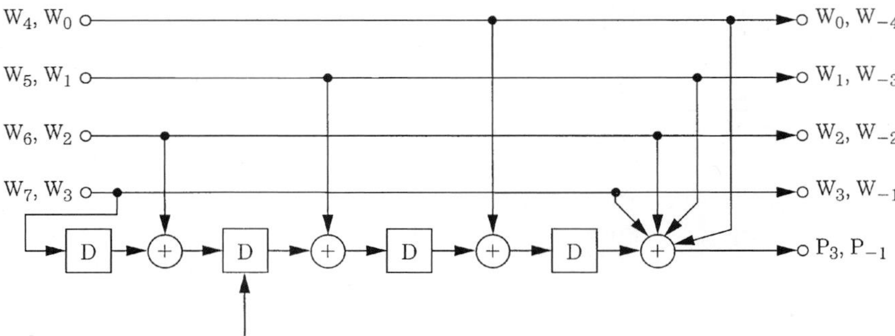

$$P_0 = W_0 + W_1 + W_2 + W_3 + W_{-4} + W_{-7} + W_{-10} + W_{-13}$$
$$P_4 = W_4 + W_5 + W_6 + W_7 + W_0 + W_{-3} + W_{-6} + W_{-9}$$
$$P_8 = W_8 + W_9 + W_{10} + W_{11} + W_4 + W_1 + W_{-2} + W_{-5}$$
$$P_{12} = W_{12} + W_{13} + W_{14} + W_{15} + W_8 + W_5 + W_2 + W_{-1}$$
$$P_{18} = W_{16} + W_{17} + W_{18} + W_{19} + W_{12} + W_9 + W_6 + W_3$$

Figure 5.15 An example of a convolutional code encoder, generating one check word for every four data words. (*Doi, 1983*)

giving one instruction. The sequence of selections at the nodes forms the output codeword. From the previous example, the input 1110 generates the output path 11011001. Upon playback, the data is sequentially decoded and errors can be detected and recovered by comparing all possible transmitted sequences to those actually received. The received sequence is compared to transmitted sequences, branch by branch. The decoding path through the tree is guided by the algorithm, to find the transmitted sequence that most likely gave rise to the received sequence.

Another convolutional code is shown in Fig. 5.15. Here, the encoder uses four delays, each with a duration of one word. Parity words are generated after every four data words, and each parity word has encoded information derived from the previous eight data words. The constraint length of the code is 14. Convolutional codes are often inexpensive to implement, and perform well under high error conditions. One disadvantage of convolutional codes is error propagation; an error that cannot be fully corrected generates syndromes reflecting this error, and this can introduce errors in subsequent decoding. Convolutional codes are not typically used with recorded data, particularly when editing is required; the discrete nature of block codes makes them more suitable for this. Convolutional codes are more often used in broadcast and wireless applications.

Interleaving

Error correction depends on an algorithm's ability to overcome erroneous data. When the error is sustained, as in the case of a burst error, large amounts of continuous data as well as redundant data may be lost, and correction may be difficult or impossible. To overcome this, data is often interleaved or dispersed through the data stream prior to storage or transmission. If a burst error occurs after the interleaving stage, it damages a continuous section of data. However, on playback, the bitstream is de-interleaved; thus, the data is returned to its original sequence and conversely the errors are distributed through the bitstream. With valid data and valid redundancy now surrounding the

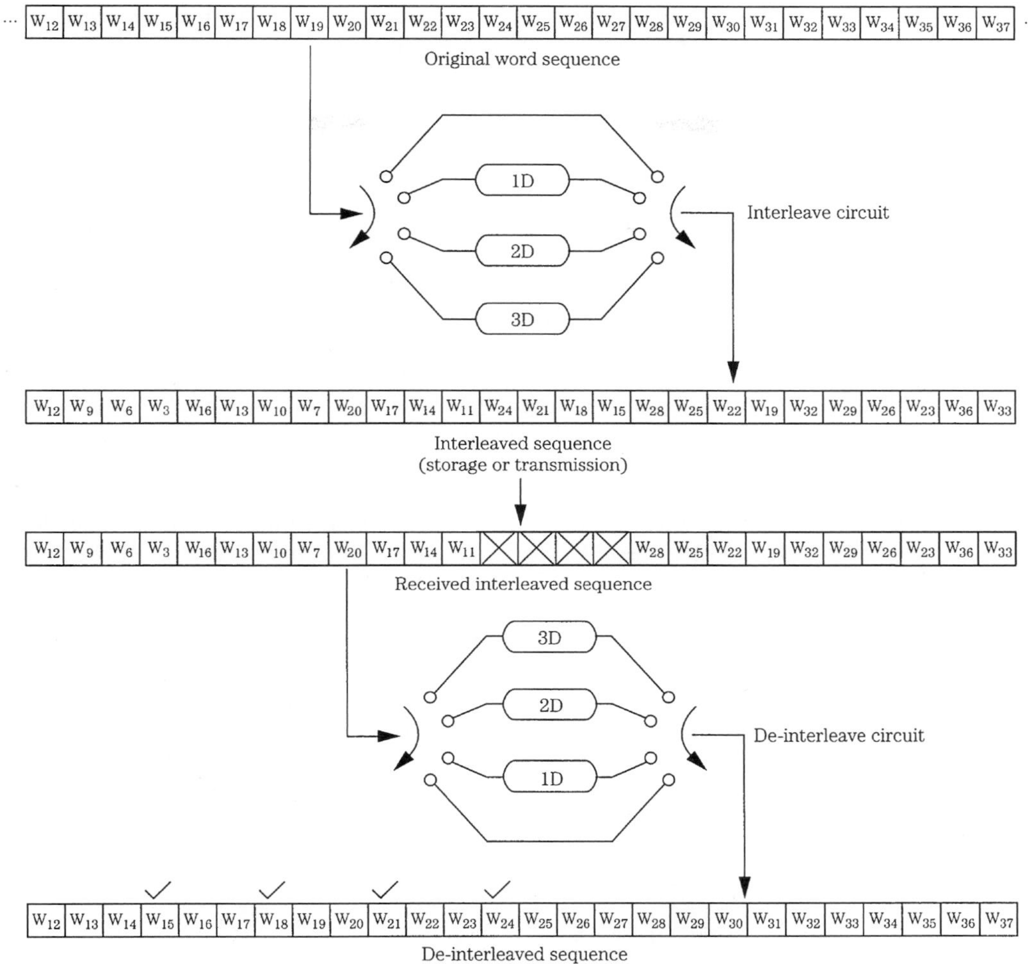

Figure 5.16 Zero-, one-, two-, and three-word delays perform interleaving and de-interleaving for error dispersion prior to correction. (*Doi, 1983*)

damaged data, the algorithm is better able to reconstruct the damaged data. Figure 5.16 shows an example of interleaving and de-interleaving, with a burst error occurring during storage or transmission. Following de-interleaving, the errors have been dispersed, facilitating error correction.

Interleaving provides an important advantage. Without interleaving, the amount of redundancy would be dictated by the size of the largest correctable burst error. With interleaving, the largest error that can occur in any block is limited to the size of the interleaved sections. Thus, the amount of redundancy is determined not by burst size, but by the size of the interleaved section. Simple delay interleaving effectively disperses data. Many block checksums work properly if there is only one word error per block. A burst error exceeds this limitation; however, interleaved and de-interleaved data may yield only one erroneous word in a given block. Thus, interleaving greatly increases burst-error correctability of block codes. Bit interleaving accomplishes much the same purpose as block interleaving: it permits burst errors to be handled as shorter burst errors or random errors. Any interleaving process requires a buffer long enough to hold the distributed data during both interleaving and de-interleaving.

Cross-Interleaving

Interleaving may not be adequate when burst errors are accompanied by random errors. Although the burst is scattered, the random errors add additional errors in a given word, perhaps overloading the correction algorithm. One solution is to generate two correction codes, separated by an interleave and delay. When block codes are arranged in rows and columns two-dimensionally, the code is called a product code (or cross word code). The minimum distance is the product of the distances of each code. When two block codes are separated by both interleaving and delay, cross-interleaving results. In other words, a Cross-Interleave Code (CIC) comprises two (or more) block codes assembled with a convolutional structure, as shown in Fig. 5.17. The method is efficient because the syndromes from one code can be used to point to errors, which are corrected by the other code. Because error location is known, correctability is enhanced. For example, a random error is corrected by the interleaved code, and a burst error is corrected after de-interleaving. When both codes are single-erasure correcting codes, the resulting code is known as a Cross-Interleave Code. In the Compact Disc format, Reed–Solomon codes are used and the algorithm is known as the Cross-Interleave Reed–Solomon code (CIRC). Product codes are used in the DVD and Blu-ray formats.

An example of a simple CIC encoder is shown in Fig. 5.18. The delay units produce interleaving, and the modulo 2 adders generate single-erasure correcting codes. Two parity words (P and Q) are generated, and with two single-erasure codes, errors are efficiently corrected. Triple-word errors can be corrected; however, four-word errors produce double-word errors in all four of the generated sequences, and correction is impossible. The CIC enjoys the high performance of a convolutional code but without

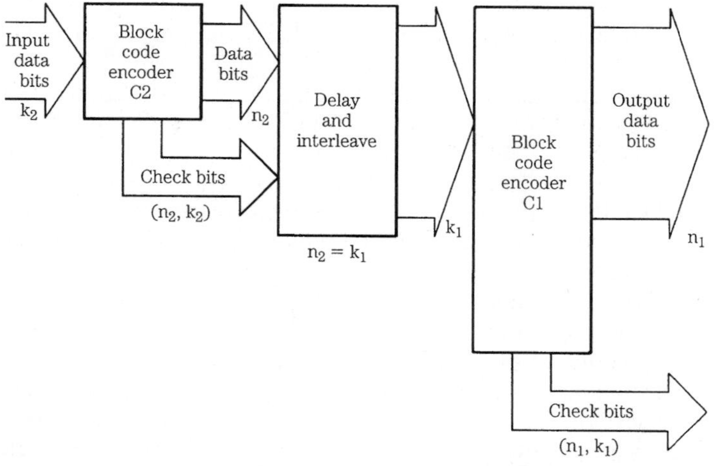

FIGURE 5.17 A cross-interleave code encoder. Syndromes from the first block are used as error pointers in the second block. In the CD format, $k_2 = 24$, $n_2 = 28$, $k_1 = 28$, and $n_1 = 32$; the C1 and C2 codes are Reed–Solomon codes.

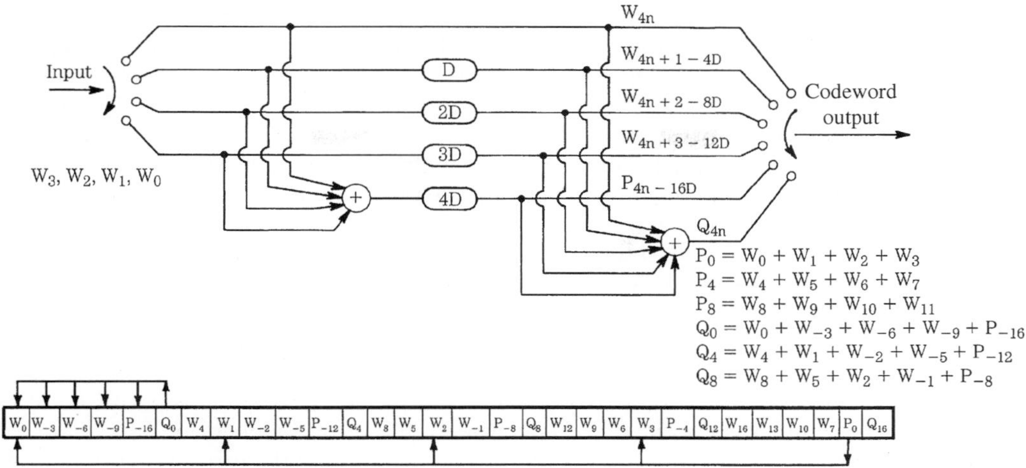

FIGURE 5.18 An example of a CIC encoder and its output sequence. (*Doi, 1983*)

error propagation, because any uncorrectable error in one sequence always becomes a one-word error in the next sequence and thus can be easily corrected.

Reed–Solomon Codes

Reed–Solomon codes were devised by Irving Reed and Gustave Solomon in 1960, at the Massachusetts Institute of Technology, Lincoln Laboratories. They are an example of an important subclass of codes known as q-ary BCH codes, devised by Raj Chandra Bose, Dwijendra Kumar Ray-Chaudhuri, and Alexis Hocquenghem in 1959 and 1960. BCH codes are a subclass of Hamming codes. Reed–Solomon codes are cyclic codes that provide multiple-error correction. They provide maximum correcting efficiency. They define code symbols from n-bit bytes, with codewords consisting of $2^n - 1$ of the n-bit bytes. If the error pattern affects s bytes in a codeword, $2s$ bytes are required for error correction. Thus $2^n - 1 - 2s$ bytes are available for data. When combined with cross-interleaving, Reed–Solomon codes are effective for audio applications because the architecture can correct both random and burst errors.

Reed–Solomon codes exclusively use polynomials derived using finite field mathematics known as Galois Fields (GF) to encode and decode block data. Galois Fields, named in honor of the extraordinary mathematical genius Evariste Galois (who devised them before his death in a duel at age 20), comprise a finite number of elements with special properties. Either multiplication or addition can be used to combine elements, and the result of adding or multiplying two elements is always a third element contained in the field. For example, when an element is raised to higher powers, the result is always another element in the field. Such fields generally only exist when the number of elements is a prime number or a power of a prime number. In addition, there exists at least one element called a primitive such that every other element can be expressed as a power of this element.

For error correction, Galois Fields yield a highly structured code, which ultimately simplifies implementation of the code. In Reed–Solomon codes, data is formed into symbols that are members of the Galois Field used by the code; Reed–Solomon codes are thus nonbinary BCH codes. They achieve the greatest possible minimum distance for the specified input and output block lengths. The minimum distance, the number of nonbinary symbols in which the sequences differ, is given by: $d = n - k + 1$. The size of the Galois Field, which determines the number of symbols in the code, is based on the number of bits comprising a symbol; 8-bit symbols are commonly used. The code thus contains $2^8 - 1$ or 255 8-bit symbols. A primitive polynomial often used in GF(2^8) systems is $x^8 + x^4 + x^3 + x^2 + 1$.

Reed–Solomon codes use multiple polynomials with roots that locate multiple errors and provide syndromes to correct errors. For example, the code can use the input word to generate two parity polynomials, P and Q. The P parity can be a modulo 2 sum of the symbols. The Q parity multiplies each input word by a different power of the GF primitive element. If one symbol is erroneous, the P parity gives a nonzero syndrome S_1. The Q parity yields a syndrome S_2; its value is S_1, raised to a power—the value depending on the position of the error. By checking this relationship between S_1 and S_2, the Reed–Solomon code can locate the error. When a placed symbol is in error, S_2 would equal S_1 multiplied by the element raised to the placed power. Correction is performed by adding S_1 to the designated location. This correction is shown in the example below. Alternatively, if the position of two errors is already known through detection pointers, then two errors in the input can be corrected. For example, if the second and third symbols are flagged, then S_1 is the modulo 2 sum of both errors, and S_2 is the sum of the errors multiplied by the second and third powers.

To illustrate the operation of a Reed–Solomon code, consider a GF(2^3) code comprising 3-bit symbols. In this code, α is the primitive element and is the solution to the equation:

$$F(x) = x^3 + x + 1 = 0$$

such that an irreducible polynomial can be written:

$$\alpha^3 + \alpha + 1 = 0$$

where + indicates modulo 2 addition. The elements can be represented as ordinary polynomials:

$$
\begin{aligned}
000 &= \qquad\qquad = 0 \\
001 &= \qquad\quad +1 = 1 \\
010 &= \quad +x \quad = x \\
011 &= \quad +x +1 = x+1 \\
100 &= x^2 \qquad\quad = x^2 \\
101 &= x^2 \qquad +1 = x^2+1 \\
110 &= x^2 + x \quad = x^2 + x \\
110 &= x^2 + x + 1 = x^2 + x + 1
\end{aligned}
$$

Because $\alpha = x$, using the properties of the Galois Field and modulo 2 (where $1 + 1 = \alpha + \alpha = \alpha^2 + \alpha^2 = 0$) we can create a logarithmic representation of the irreducible polynomial elements in the field where the bit positions indicate polynomial positions:

$$
\begin{aligned}
&0 && = 000 \\
&1 && = 001 \\
&\alpha && = 010 \\
&\alpha^2 && = 100 \\
&\alpha^3 = \alpha + 1 && = 011 \\
&\alpha^4 = \alpha \cdot \alpha^3 = \alpha(\alpha + 1) = \alpha^2 + \alpha && = 110 \\
&\alpha^5 = \alpha^2 + \alpha + 1 && = 111 \\
&\alpha^6 = \alpha \cdot \alpha^5 = \alpha(\alpha^2 + \alpha + 1) = \alpha^3 + \alpha^2 + \alpha && \\
&\quad = \alpha + 1 + \alpha^2 + \alpha = \alpha^2 + 1 && = 101 \\
&\alpha^7 = \alpha(\alpha^2 + 1) = \alpha^3 + \alpha = \alpha + 1 + \alpha = 1 && = 001 = 1
\end{aligned}
$$

In this way, all possible 3-bit symbols can be expressed as elements of the field ($0, 1 = \alpha^7, \alpha, \alpha^2, \alpha^3, \alpha^4, \alpha^5$, and α^6) where α is the primitive element (010). Elements can be multiplied by simply adding exponents, always resulting in another element in the Galois Field. For example:

$$\alpha \cdot \alpha = \alpha^2 = (010)(010) = 100$$

$$1 \cdot \alpha^2 = \alpha^2 = (001)(100) = 100$$

$$\alpha^2 \cdot \alpha^3 = \alpha^5 = (100)(011) = 111$$

The complete product table for this example GF(2^3) code is shown in Fig. 5.19; the modulo α^7 results can be seen. For example, $\alpha^4 \cdot \alpha^6 = \alpha^{10}$, or α^3. Using the irreducible polynomials and the product table, the correction code can be constructed. Suppose that A, B, C, and D are data symbols and P and Q are parity symbols. The Reed–Solomon code will satisfy the following equations:

$$A + B + C + D + P + Q = 0$$

$$\alpha^6 A + \alpha^5 B + \alpha^4 C + \alpha^3 D + \alpha^2 P + \alpha^1 Q = 0$$

Using the devised product laws, we can solve these equations to yield:

$$P = \alpha^1 A + \alpha^2 B + \alpha^5 C + \alpha^3 D$$

$$Q = \alpha^3 A + \alpha^6 B + \alpha^4 C + \alpha^1 D$$

Bits		000	001	010	011	100	101	110	111
	Elements	0	1	α	α^3	α^2	α^6	α^4	α^5
000	0	0	0	0	0	0	0	0	0
001	$\alpha^7 = 1$	0	1	α	α^3	α^2	α^6	α^4	α^5
010	α	0	α	α^2	α^6	α^3	1	α^5	α^6
011	α^3	0	α^3	α^4	α^6	α^5	α^2	1	α
100	α^2	0	α^2	α^3	α^5	α^4	α	α^6	1
101	α^6	0	α^6	1	α^2	α	α^5	α^3	α^4
110	α^4	0	α^4	α^5	1	α^6	α^3	α	α^2
111	α^5	0	α^5	α^6	α	1	α^4	α^2	α^3

FIGURE 5.19 The product table for a GF(2^3) code with $F(x) = x^3 + x + 1$ and primitive element 010.

For example, given the irreducible polynomial table, if:

$$A = 001 = 1$$

$$B = 101 = \alpha^6$$

$$C = 011 = \alpha^3$$

$$D = 100 = \alpha^2$$

we can solve for P and Q using the product table:

$$P = \alpha^1 \cdot 1 + \alpha^2 \cdot \alpha^6 + \alpha^5 \cdot \alpha^3 + \alpha^3 \cdot \alpha^2 = \alpha + \alpha + \alpha + \alpha^5$$

$$= \alpha + \alpha + \alpha + (\alpha^2 + \alpha + 1) = \alpha^2 + 1 = 101$$

$$Q = \alpha^3 \cdot 1 + \alpha^6 \cdot \alpha^6 + \alpha^4 \cdot \alpha^3 + \alpha^1 \cdot \alpha^2 = \alpha^3 + \alpha^5 + 1 + \alpha^3$$

$$= (\alpha + 1) + (\alpha^2 + \alpha + 1) + 1 + (\alpha + 1) = \alpha^2 + \alpha = 110$$

Thus:

$$P = 101 = \alpha^6$$

$$Q = 110 = \alpha^4$$

Errors in received data can be corrected using syndromes, where a prime (′) indicates received data:

$$S_1 = A' + B' + C' + D' + P' + Q'$$

$$S_2 = \alpha^6 A' + \alpha^5 B' + \alpha^4 C' + \alpha^3 D' + \alpha^2 P' + \alpha^1 Q'$$

If each possible error pattern is expressed by E_i, we write the following:

$$S_1 = E_A + E_B + E_C + E_D + E_P + E_Q$$

$$S_2 = \alpha^6 E_A + \alpha^5 E_B + \alpha^4 E_C + \alpha^3 E_D + \alpha^2 E_P + \alpha^1 E_Q$$

If there is no error, then $S_1 = S_2 = 0$.

If symbol A' is erroneous, $S_1 = E_A$ and $S_2 = \alpha^6 S_1$.
If symbol B' is erroneous, $S_1 = E_B$ and $S_2 = \alpha^5 S_1$.
If symbol C' is erroneous, $S_1 = E_C$ and $S_2 = \alpha^4 S_1$.
If symbol D' is erroneous, $S_1 = E_D$ and $S_2 = \alpha^3 S_1$.
If symbol P' is erroneous, $S_1 = E_P$ and $S_2 = \alpha^2 S_1$.
If symbol Q' is erroneous, $S_1 = E_Q$ and $S_2 = \alpha^1 S_1$.

In other words, an error results in nonzero syndromes; the value of the erroneous symbols can be determined by the difference of the weighting between S_1 and S_2. The ratio of weighting for each word is different thus single-word error correction is possible. Double erasures can be corrected because there are two equations with two unknowns. For example, if this data is received:

$$A' = 001 = 1$$

$$B' = 101 = \alpha^6$$

$$C' = 001 = 1 \text{ (erroneous)}$$

$$D' = 100 = \alpha^2$$

$$P' = 101 = \alpha^6$$

$$Q' = 110 = \alpha^4$$

We can calculate the syndromes (recalling that $1 + 1 = \alpha + \alpha = \alpha^2 + \alpha^2 = 0$):

$$\begin{aligned} S_1 &= 1 + \alpha^6 + 1 + \alpha^2 + \alpha^6 + \alpha^4 \\ &= 1 + (\alpha^2 + 1) + 1 + \alpha^2 + (\alpha^2 + 1) + (\alpha^2 + \alpha) \\ &= \alpha \\ &= 010 \end{aligned}$$

$$\begin{aligned} S_2 &= \alpha^6 \cdot 1 + \alpha^5 \cdot \alpha^6 + \alpha^4 \cdot 1 + \alpha^3 \cdot \alpha^2 + \alpha^2 \cdot \alpha^6 + \alpha^1 \cdot \alpha^4 \\ &= \alpha^6 + \alpha^4 + \alpha^4 + \alpha^5 + \alpha + \alpha^5 \\ &= (\alpha^2 + 1) + (\alpha^2 + \alpha) + (\alpha^2 + \alpha) + (\alpha^2 + \alpha + 1) + \alpha + (\alpha^2 + \alpha + 1) \\ &= \alpha^2 + \alpha + 1 \\ &= \alpha^5 \\ &= 111 \end{aligned}$$

Because $S_2 = \alpha^4 S_1$ (that is, $\alpha^5 = \alpha^4 \cdot \alpha$), symbol C' must be erroneous and because $S_1 = E_C = 010$, $C = C' + E_C = 001 + 010 = 011$, thus correcting the error.

In practice, the polynomials used in CIRC are:

$$P = \alpha^6 A + \alpha^1 B + \alpha^2 C + \alpha^5 D + \alpha^3 E$$

$$Q = \alpha^2 A + \alpha^3 B + \alpha^6 C + \alpha^4 D + \alpha^1 E$$

and the syndromes are:

$$S_1 = A' + B' + C' + D' + E' + P' + Q'$$

$$S_2 = \alpha^7 A' + \alpha^6 B' + \alpha^5 C' + \alpha^4 D' + \alpha^3 E' + \alpha^2 P' + \alpha^1 Q'$$

Because Reed–Solomon codes are particularly effective in correcting burst errors, they are highly successful in digital audio applications when coupled with error-detection pointers such as CRCC and interleaving. Reed–Solomon codes are used for error correction in applications such as CD, DVD, Blu-ray, direct broadcast satellite, digital radio, and digital television.

Cross-Interleave Reed–Solomon Code (CIRC)

The Cross-Interleave Reed–Solomon Code (CIRC) is a quadruple-erasure (double-error) correction code that is used in the Compact Disc format. CIRC applies Reed–Solomon codes sequentially with an interleaving process between C2 and C1 encoding. Encoding carries data through the C2 encoder, then the C1 encoder. Decoding reverses the process.

In CIRC, C2 is a (28,24) code, that is, the encoder inputs 24 symbols, 28 symbols (including four parity symbols) are output. C1 is a (32,28) code; the encoder inputs 28 symbols, and 32 symbols (with four additional parity symbols) are output. In both cases, because 8-bit bytes are used, the size of the Galois Field is GF(2^8); the calculation is based on the primitive polynomial: $x^8 + x^4 + x^3 + x^2 + 1$. Minimum distance is 5. Up to four symbols can be corrected if the error location is known, and two symbols can be corrected if the location is not known.

Using a combination of interleaving and parity to make the data more robust against errors encountered during storage on a CD, the data is encoded before being placed on the disc, and then decoded upon playback. The CIRC encoding algorithm is shown in Fig. 5.20; the similarity to the general structure shown in Fig. 5.17 is evident. Using this encoding algorithm, symbols from the audio signal are cross-interleaved, and Reed–Solomon encoding stages generate *P* and *Q* parity symbols.

The CIRC encoder accepts twenty-four 8-bit symbols, that is, 12 symbols (six 16-bit samples) from the left channel and 12 from the right channel. Interestingly, the value of 12 was selected by the designers because it has common multiples of 2, 3, and 4; this would have allowed the CD to more easily also offer 3- or 4-channel implementations—a potential that never came to fruition. An interleaving stage assists interpolation. A two-symbol delay is placed between even and odd samples. Because even samples are delayed by two blocks, interpolation is possible where two uncorrectable blocks occur. The symbols are scrambled to separate even- and odd-numbered symbols; this process assists concealment. The 24-byte parallel word is input to the C2 encoder that produces four symbols of *Q* parity. *Q* parity is designed to correct one erroneous symbol, or up to four erasures in one word.

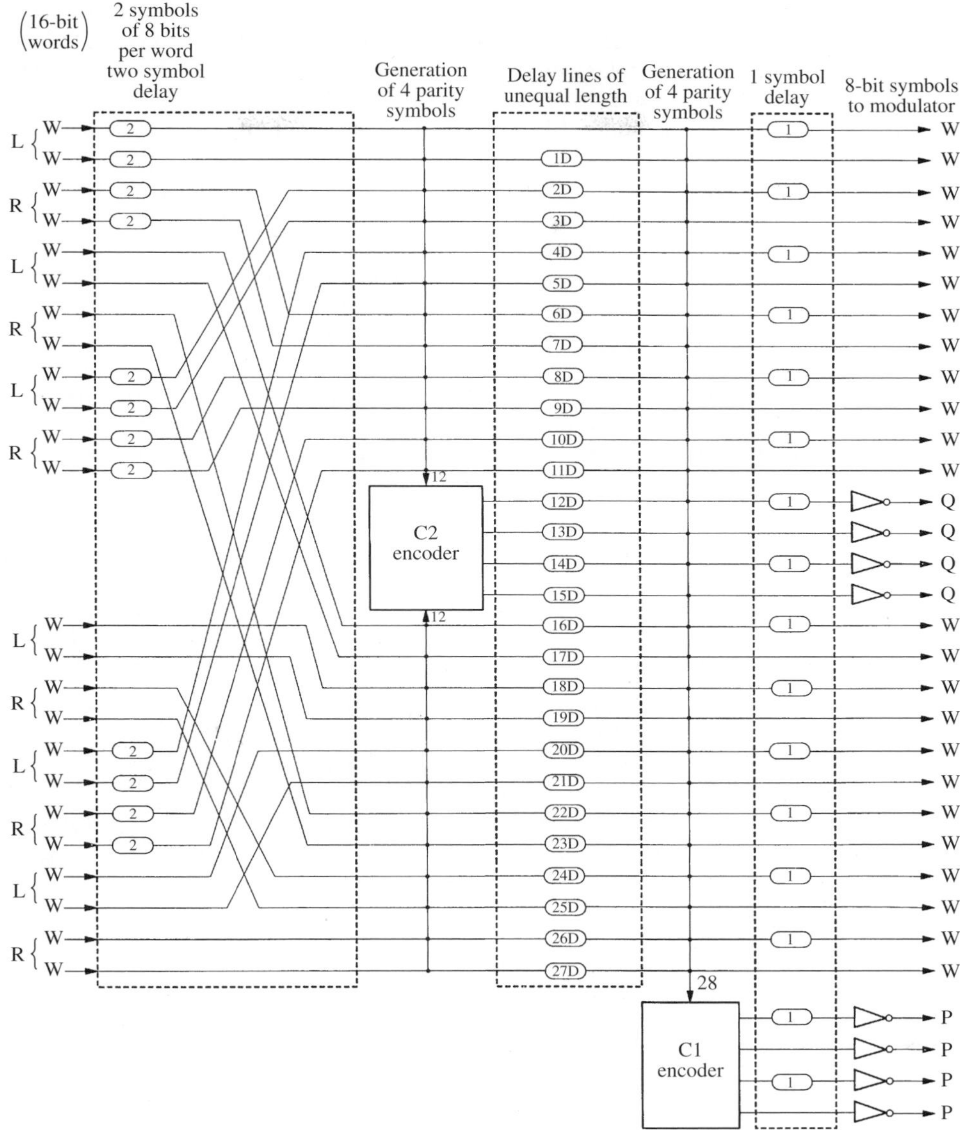

FIGURE 5.20 The CIRC encoding algorithm.

By placing the parity symbols in the center of the block, the odd/even distance is increased. This permits interpolation over the largest possible burst error.

In the cross-interleaving stage, 28 symbols are delayed by differing periods. These periods are integer multiples of four blocks. This convolutional interleave stores one C2 word in 28 different blocks, stored over a distance of 109 blocks. In this way, the data array is crossed in two directions. Because the delays are long and of unequal duration, correction of burst errors is facilitated. Twenty-eight symbols (from 28 different C2 words) are input to the C1 encoder, producing four *P* parity symbols. *P* parity is designed to correct single-symbol errors and detect and flag double and triple errors for *Q* correction.

An interleave stage delays alternate symbols by one symbol. This odd/even delay spreads the output symbols over two data blocks. In this way, random errors cannot corrupt more than one symbol in one word even if there are two erroneous adjacent symbols in one block. The *P* and *Q* parity symbols are inverted to provide nonzero *P* and *Q* symbols with zero data. Thirty-two 8-bit symbols leave the CIRC encoder.

The error processing must be decoded each time the disc is played to de-interleave the data, and perform error detection and correction. When a Reed–Solomon decoder receives a data block (consisting of the original data symbols plus the parity symbols) it uses the received data to recalculate the parity symbols. If the recalculated parity symbols match the received parity symbols, the block is assumed to be error-free. If they differ, the difference syndromes are used to locate the error. Erroneous words are flagged, for example, as being correctable, uncorrectable, or possibly correctable. Analysis of the flags determines whether the errors are to be corrected by the correction code, or passed on for interpolation. The CIRC decoding algorithm is shown in Fig. 5.21.

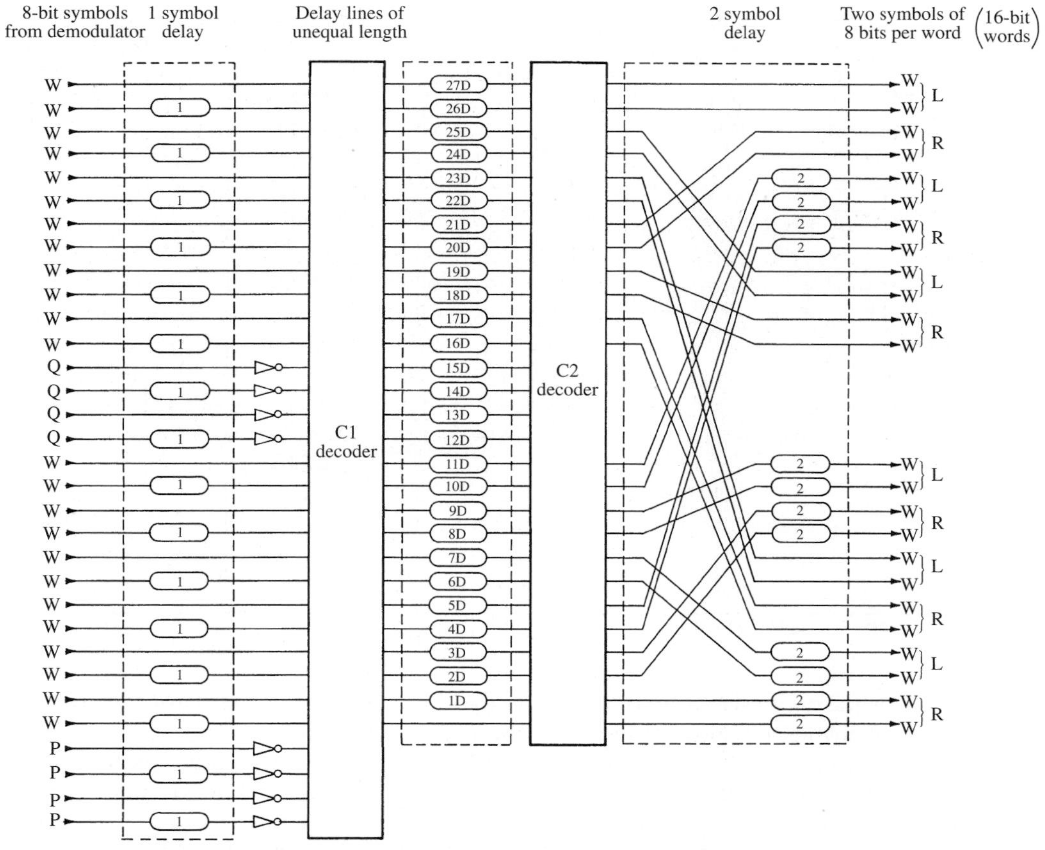

Figure 5.21 The CIRC decoding algorithm.

A frame of thirty-two 8-bit symbols is input to the CIRC decoder: twenty-four are audio symbols and eight are parity symbols. Odd-numbered symbols are passed through a one-symbol delay. In this way, even-numbered symbols in a frame are de-interleaved with the odd-numbered symbols in the next frame. Audio symbols are restored to their original order and disc errors are scattered. This benefits C1 correction, especially for small errors in adjoining symbols. Following this de-interleaving, parity symbols are inverted.

Using four *P* parity symbols, the C1 decoder corrects random errors and detects burst errors. The C1 decoder can correct one erroneous symbol in each frame. If there is more than one erroneous symbol, the 28 data symbols are marked with an erasure flag and passed to the C2 decoder. Valid symbols are passed along unprocessed. Convolutional de-interleaving between the decoders enables the C2 decoder to correct long burst errors. The frame input to C2 contains symbols from C1 decoded at different times; thus, symbols marked with an erasure flag are scattered. This assists the C2 decoder in correcting burst errors. Symbols without a flag are assumed to be error-free, and are passed through unprocessed.

Given precorrected data, and help from de-interleaving, C2 can correct burst errors as well as random errors that C1 was unable to correct. Using four *Q* parity symbols, C2 can detect and correct single-symbol errors and correct up to four symbols, as well as any symbols miscorrected by C1 decoding. C2 also can correct errors that might occur in the encoding process itself, rather than on the disc. When C2 cannot accomplish correction, for example, when more than four symbols are flagged, the 24 data symbols are flagged and passed on for interpolation. Final descrambling and delay completes CIRC decoding.

CIRC Performance Criteria

The integrity of data stored on a CD and passing through a CIRC algorithm can be assessed through a number of error counts. Using a two-digit nomenclature, the first digit specifies the number of erroneous symbols (bytes), and the second digit specifies at which decoder (C1 or C2) they occurred. Three error counts (E11, E21, and E31) are measured at the output of the C1 decoder. The E11 count specifies the frequency of occurrence of single-symbol (correctable) errors per second in the C1 decoder. E21 indicates the frequency of occurrence of double-symbol (correctable) errors in the C1 decoder. E31 indicates the frequency of triple-symbol (uncorrectable at C1) errors in the C1 decoder. E11 and E21 errors are corrected in the C1 stage. E31 errors are uncorrected in the C1 stage, and are passed along to the second C2 stage of correction.

There are three error counts (E12, E22, and E32) at the C2 decoder. The E12 count indicates the frequency of occurrence of a single-symbol (correctable) error in the C2 decoder measured in counts per second. A high E12 count is not problematic because one E31 error can generate up to 30 E12 errors due to interleaving. The E22 count indicates the frequency of two-symbol (correctable) errors in the C2 decoder. E22 errors are the worst correctable errors; the E22 count indicates that the system is close to producing an uncorrectable error; a CD-ROM with 15 E22 errors would be unacceptable even though the errors are correctable. A high E22 count can indicate localized damage to the disc, from manufacturing defect or usage damage. The E32 count indicates three or more symbol (uncorrectable) errors in the C2 decoder, or unreadable data in general; ideally, an E32 count should never occur on a disc. E32 errors are sometimes classified as noise interpolation (NI) errors; when an E32 error occurs, interpolation must be performed. If a disc has no E32 errors, all data is output accurately.

The block-error rate (BLER) measures the number of frames of data that have at least one occurrence of uncorrected symbols (bytes) at the C1 error-correction decoder input (E11 + E21 + E31); it is thus a measure of both correctable and uncorrectable errors at that decoding stage. Errors that are uncorrectable at this stage are passed along to the next error-correction stage. BLER is a good general measure of disc data quality. BLER is specified as rate of errors per second. The CD standard sets a maximum of 220 BLER errors per second averaged over 10 seconds for audio discs; however, a well-manufactured disc will have an average BLER of less than 10. A high BLER count often indicates poor pit geometry that disrupts the pickup's ability to read data, resulting in many random-bit errors. A high BLER might limit a disc's lifetime as other scratches accumulate to increase total error count. Likewise a high BLER may cause playback to fail in some players. The CD block rate is 7350 blocks per second; hence, the maximum BLER value of 220 counts per second shows that 3% of frames contain a defect. This defines the acceptable (correctable) error limit; greater frequency might lead to audible faults. The BLER does not provide information on individual defects between 100 and 300 μm, because the BLER responds to defects that are the size of one pit. BLER is often quoted as a 1-second actual value or as a sliding 10-second average across the disc, as well as the maximum BLER encountered during a single 10-second interval during the test.

The E21 and E22 signals can be combined to form a burst-error (BST) count. It counts the number of consecutive C2 block errors exceeding a specified threshold number. It generally indicates a large physical defect on the disc such as a scratch, affecting more than one block of data. For example, if there are 14 consecutive block errors and the threshold is set at seven blocks, two BST errors would be indicated. This count is often tabulated as a total number over an entire disc. A good quality control specification would not allow any burst errors (seven frames) to occur. In practice, a factory-fresh disc might yield BLER = 5, E11 = 5, E21 = 0, and E31 = 0. E32 uncorrectable errors should never occur in a new disc. Figure 5.22 shows an example of the error counts measured from a Compact Disc with good-quality data readout. Error counts are measured vertically, across the duration of this 50-minute disc. As noted above, the high E12 error count is not problematic. Following CIRC error correction, this disc does not have any errors in its output data.

The success of CIRC error correction ultimately depends on the implementation of the algorithm. Generally, CIRC might provide correction of up to 3874 bits, corresponding to a track-length defect of 2.5 mm. Good concealment can extend to 13,282 bits, corresponding to an 8.7-mm defect, and marginal concealment can extend to approximately 15,500 bits.

Product Codes

The CIRC is an example of a cross-interleaved code in which two codes are separated by convolutional interleaving. In two-dimensional product codes, two codes are serially separated by block interleaving. The code that is the first to encode and the last to decode is called the outer C2 code. The code that is the second to encode and the first to decode is called the inner C1 code. Product codes thus use the method of crossing two error-correction codes. If C2 is a (n_2, k_2) block code and C1 is a (n_1, k_1) block code, then

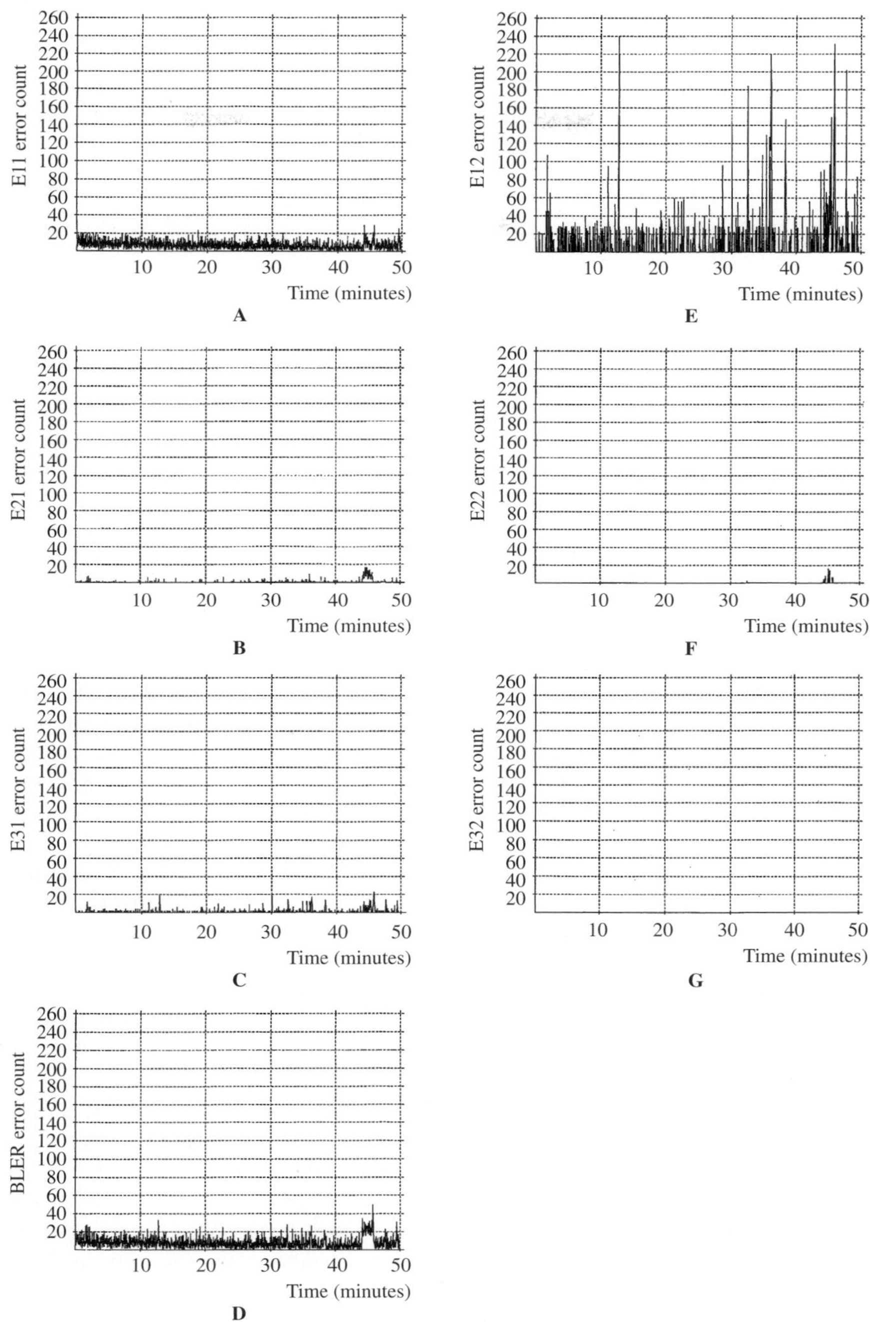

Figure 5.22 Example of error counts across a 50-minute Compact Disc. A. E11 error count. B. E21 error count. C. E31 error count. D. BLER error count. E. E12 error count. F. E22 error count. G. E32 error count (no errors).

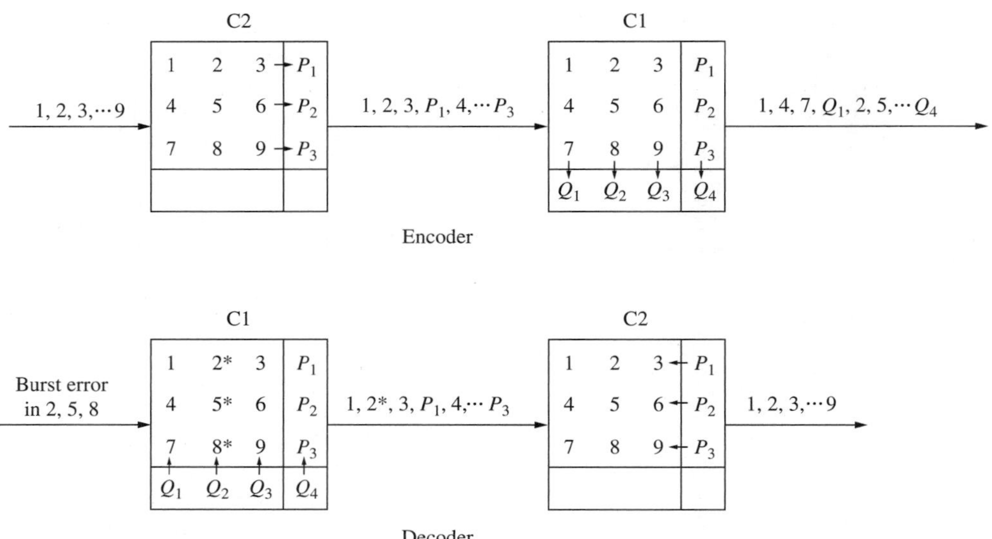

FIGURE 5.23 Block diagram of a product code showing outer-row (*P*) redundancy and inner-column (*Q*) redundancy. A burst error is flagged in the C1 decoder and corrected in the C2 decoder.

they yield a (n_1n_2, k_1k_2) product code with k_1k_2 symbols in an array. The C2 code adds *P* parity to the block rows, and the C1 code adds *Q* parity to the block columns as shown in Fig. 5.23. Every column is a codeword of C1. When linear block codes are used, the bottom row words are also codewords of C2. The check data in the lower right-hand corner (intersection of the *P* column and the *Q* row) is a check on the check data and can be derived from either column or row check data.

During decoding, *Q* parity is used by the C1 decoder to correct random-bit errors, and burst errors are flagged and passed through de-interleaving to the C2 decoder. Symbols are passed through de-interleaving so that burst errors in inner codewords will appear as single-symbol errors in different codewords. The *P* parity and flags are used by the C2 decoder to correct errors in the inner codewords.

Product codes are used in many audio applications. For example, the DAT format provides an example of a product code in which data is encoded with Reed–Solomon error-correction code over a Galois Field GF(2^8) using the polynomial $x^8 + x^4 + x^3 + x^2 + 1$. The inner code C1 is a (32,28) Reed–Solomon code with a minimum distance of 5. Four bytes of redundancy are added to the 28 data bytes. The outer code C2 is a (32,26) Reed–Solomon code with a minimum distance of 7. Six bytes of redundancy are added to the 26 data bytes. Both C1 and C2 codes are composed of 32 symbols and are orthogonal with each other. C1 is interleaved over two blocks, and C1 is interleaved across an entire PCM data track, every four blocks. As with the Compact Disc, the error-correction code used in DAT endeavors to detect and correct random errors (using the inner code) and eliminate them prior to de-interleaving; burst errors are detected and flagged. Following de-interleaving, burst errors are scattered and more easily corrected. Using the error flags, the outer code can correct remaining errors. Uncorrected errors must be concealed.

Two blocks of data are assembled in a frame, one from each head, and placed in a memory arranged as 128 columns by 32 bytes. Samples are split into two bytes to form 8-bit symbols. Symbols are placed in memory, reserving a 24-byte wide area in the middle columns. Rows of data are applied to the first (outer C2 code) Reed–Solomon encoder, selecting every fourth column, finishing at column 124 and so yielding 26 bytes. The Reed–Solomon encoder generates six bytes of parity yielding 32 bytes. These are placed in the middle columns, at every fourth location (52, 56, 60, and so on). The encoder repeats its operation with the second column, taking every fourth byte, finishing at column 125. Six parity bytes are generated and placed in every fourth column (53, 57, 61, and so on). Similarly, the memory is filled with 112 outer codewords. (The final eight rows require only two passes because odd-numbered columns have bytes only to row 23.)

Memory is next read by columns. Sixteen even-numbered bytes from the first column, and the first 12 even-numbered bytes from the second column are applied to the second (inner C1 code) encoder, yielding four parity bytes for a total codeword of 32 bytes. This word forms one recorded synchronization block. A second pass reads the odd-numbered row samples from the first two columns of memory, again yielding four parity bytes and another synchronization block. Similarly, the process repeats until 128 blocks have been recorded.

The decoding procedure utilizes first the C1 code, then the C2 code. In C1 decoding, a syndrome is calculated to identify data errors as erroneous symbols. The number of errors in the C1 code is determined using the syndrome; in addition, the position of the errors can be determined. Depending on the number of errors, C1 either corrects the errors or flags the erroneous symbols. In C2 decoding, a syndrome is again calculated, the number and position of errors determined, and the correction is carried out. During C1 decoding, error correction is performed on one or two erroneous bytes due to random errors. For more than two errors, erasure correction is performed using C1 flags attached to all bytes in the block prior to de-interleaving. After de-interleaving, these errors are distributed and appear as single-byte errors with flags. The probability of uncorrectable or misdetected errors increases along with the number of errors in C1 decoding. It is negligible for a single error because all four syndromes will agree on the error. A double error increases the probability.

The C2 decoding procedure is selected to reduce the probability of misdetection. For example, the optimal combination of error correction and erasure correction can be selected on the basis of error conditions. In addition, C2 carries out syndrome computation even when no error flags are received from C1. Because C1 has corrected random errors, C2's burst-error correction is not compromised. C2 independently corrects one- or two-byte errors in the outer codeword. For two to six erroneous bytes in the outer codeword, C2 uses flags supplied by C1 for erasure correcting. Because the outer code undergoes four-way interleaving, four erroneous synchronization blocks result in only one erroneous byte in an outer codeword. Because C2 can correct up to six erroneous bytes, burst errors up to 24 synchronization blocks in duration can thus be corrected.

Product codes can outperform other codes in modern audio applications; they demand more memory for implementation, but in many applications this is not a serious detriment. Product codes provide good error-correction performance in the presence of simultaneous burst and random errors and are widely used in high-density optical recording media such as DVD and Blu-ray discs. DVD is discussed in Chap. 8 and Blu-ray is discussed in Chap. 9.

Error Concealment

As noted, a theoretically perfect error-detection and error-correction method could be devised in which all errors are completely supplanted with redundant data or calculated with complete accuracy. However, such a scheme would be impractical because of the data overhead. A practical error-correction method balances those limitations against the probability of uncorrected errors and allows severe errors to remain uncorrected. However, subsequent processing—an error-concealment system—compensates for those errors and ensures that they are not audible. Several error-concealment techniques, such as interpolation and muting, have been devised to accomplish this.

Generally, there are two kinds of uncorrectable errors output from correction algorithms. Some errors can be properly detected; however, the algorithm is unable to correct them. Other errors are not detected at all or are miscorrected. The first type of error, detected but not corrected, can usually be concealed with properly designed concealment methods. However, undetected and miscorrected errors often cannot be concealed and may result in an audible click in the audio output. These types of errors, often caused by simultaneous random and burst errors, must be minimized. Thus, an error-correction system aims to reduce undetected errors in the error-correction algorithm, and then relies on the error-concealment methods to resolve detected but not corrected errors.

Interpolation

Following de-interleaving, most errors, even burst errors, are interspersed with valid data words. It is thus reasonable to use techniques in which surrounding valid data is used to calculate new data to replace the missing or uncorrected data. This technique works well, provided that errors are sufficiently dispersed and there is some correlation between data values. Fortunately, digital data comprising a musical selection can often undergo interpolation without adverse audibility. Although it is nonintuitive, studies have shown that within limits, the time duration of an error does not overly affect perception of the error.

In its simplest form, interpolation holds the previous sample value and repeats it to cover the missing or incorrect sample. This is called zero-order or previous-value interpolation. In first-order interpolation, sometimes called linear-order interpolation, the erroneous sample is replaced with a new sample derived from the mean value of the previous and subsequent samples. In many digital audio systems, a combination of zero- and first-order interpolation is used. If consecutive sample errors occur in spite of interleaving, then previous-value interpolation is used to replace consecutive errors, but the final held sample's value is calculated from the mean value of the held and subsequent sample. If the errors are random, that is, valid samples surround the errors, then mean value calculations are used. One interpolation strategy is shown in Fig. 5.24. Other higher-order interpolation is sometimes used; *n*th-order interpolation uses a higher-order polynomial to calculate substituted data. In practice, third- and fifth-order interpolations are sometimes used. Clearly, any interpolation calculations must be accomplished quickly enough to maintain the data rate. The results of one experiment in relative values of interpolation noise are shown in Fig. 5.25.

Muting

Muting is the simple process of setting the value of missing or uncorrected words to zero. This silence is preferable to the unpredictable sounds that can result from decoding

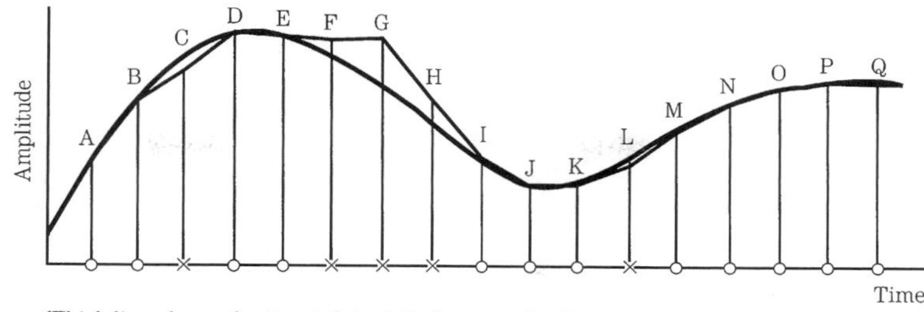

(Thick lines for analog input data. × is for error data)

Interpolation by maintenance of previous value

G = F(=E)

Interpolation by maintenance of mean value

$$C = \frac{1}{2}(B + D)$$

$$H = \frac{1}{2}(G + I) = \frac{1}{2}(E + I)$$

$$L = \frac{1}{2}(K + M)$$

FIGURE 5.24 Interpolation is used to conceal errors. For example, a previous value can be held, followed by a calculation of the mean value.

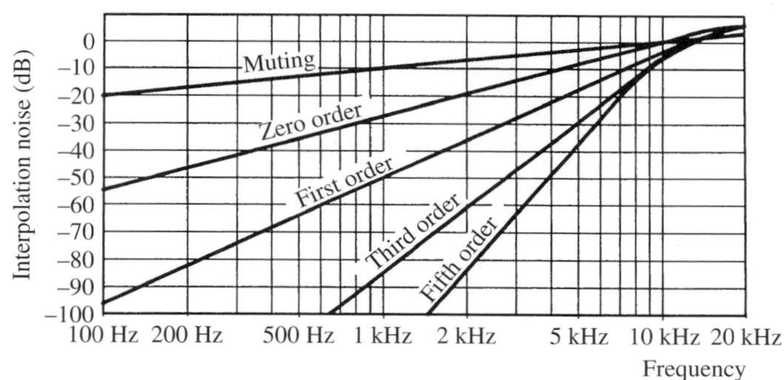

FIGURE 5.25 Noise characteristics for different interpolation methods (for 100-bit burst length, pure tone signals, and uncorrectable errors). (*Doi, 1983*)

incorrect data. Muting might be used in the case of uncorrected errors, which would otherwise cause an audible click at the output. The momentary interruption from a brief mute might be imperceptible, but a click would typically be audible. Also, in the case of severe data damage or player malfunction, it is preferable to mute the data output. To minimize audibility of a mute, muting algorithms gradually attenuate the output signal's amplitude prior to a mute, and then gradually restore the amplitude afterward. For example, gain can be adjusted by multiplying successive samples by attenuation coefficients over a few milliseconds. Of course, gain reduction must begin prior to the error itself, to allow time for a smooth fade. This is easily accomplished by delaying all audio samples briefly and feeding the mute flag forward. Very short-duration muting cannot be perceived by the human ear. Concealment strategies are assisted when audio channels are processed independently, for example, by muting only one channel rather than both.

Duplication

One of the benefits of digital audio recording is the opportunity to copy recordings without the inevitable degradation of analog duplication. Although digital audio duplication can be lossless, its success depends on the success of error correction. Although error-correction methods provide completely correct data, error concealment does not. Under marginal circumstances, error-concealment techniques do introduce audible errors into copied data. Thus, subsequent generations of digital copies could contain an accumulation of concealed errors not present in the original. As a result, errors must be corrected at their origin. Routine precautions of clean media, clean machine mechanisms, and proper low-jitter interfacing are important for digital audio duplication, particularly when many generations of copying are anticipated. In practice, digital audio data can be copied with high reliability. When a file is copied and contains exactly the same values as the original file, it is a clone of the original, with an implicit sampling frequency. Neither the original nor the cloned values have control over the storage media or playback conditions. The copy will sound the same as the original, but only if its playback conditions equal those of the original file's playback conditions. A simple way to test bit-for-bit accuracy is to load both files into a workstation and exactly synchronize the files to sample accuracy, invert one file and add them. A null result shows bit-for-bit accuracy. Likewise, a residue signal shows the difference between the files.

CHAPTER 6

Optical Disc Media

Optical disc storage technology was pioneered in the 1960s by inventors who devised numerous ways to store analog and digital signals on reflective discs. Today, optical discs are widely used for computer, audio, and video applications. They provide high storage capacity; for example, with the development of blue lasers, a multilayer Blu-ray disc can hold 50 Gbytes or more. Optical media are ideal for mass distribution of data; the manufacturing cost of optical media is a fraction of the cost per byte of magnetic-disk or solid-state media. The life expectancy of optical discs is much longer than that of magnetic media. With proper storage, a Compact Disc should last for 100 years or more. Optical storage of digital audio data is also universally employed for motion picture film soundtracks on optical film. The design of optical disc systems such as CD, DVD, and Blu-ray, as well as fiber-optic systems, rests on the fundamental principles of optics.

Optical Phenomena

Light is an electromagnetic vibration that can be characterized by wavelength, frequency, propagation velocity, propagation direction, vibration direction, and intensity. The optical spectrum, within the context of the electromagnetic spectrum, is shown in Fig. 6.1. Light ranges from 7.5×10^{10} to 6.0×10^{16} Hz. Wavelength is the distance between identical points on a waveform; the wavelengths of visible light extend from about 400 to 800 nm. Wavelength equals velocity of propagation divided by frequency. Likewise, velocity of propagation is the product of wavelength and frequency. The velocity of light in a vacuum is 3×10^8 m/s; light travels slower in other materials. Frequency measures the number of vibrations per second; the frequency of an electromagnetic wave is constant, and does not change in matter, but its velocity and wavelength are reduced. Light at different wavelengths travels at different velocities in the same medium. Light can exist at a single wavelength (monochromatic light) or a mixture of various wavelengths (for example, natural light). The intensity of a light wave is the amount of energy that flows per second across a unit area perpendicular to the direction of propagation. It is proportional to the square of the amplitude and to the square of the frequency.

Refraction occurs when light passes into a medium with a different index of refraction; light changes speed, which causes a deflection in its path. For example, when light in air strikes a glass prism, the velocity of propagation is decreased, the wavelength in the denser medium is shorter, thus the light travels through the receiving medium at a different angle, as shown in Fig. 6.2. When light leaves the prism, refraction occurs again. When white light (comprising many wavelengths) strikes a prism,

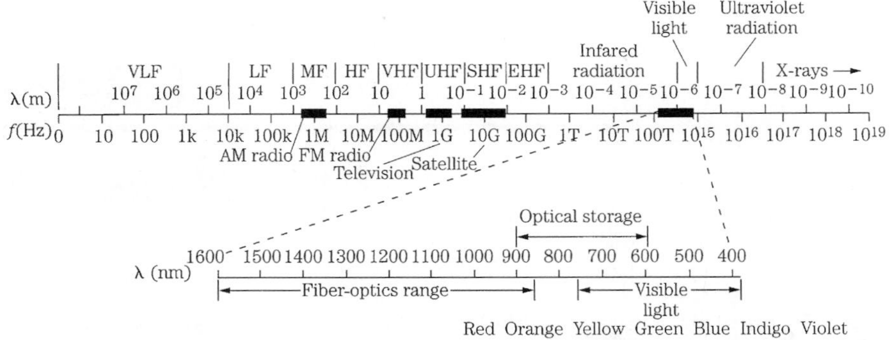

FIGURE 6.1 The electromagnetic spectrum, showing the wavelengths used in optical storage and fiber-optic systems.

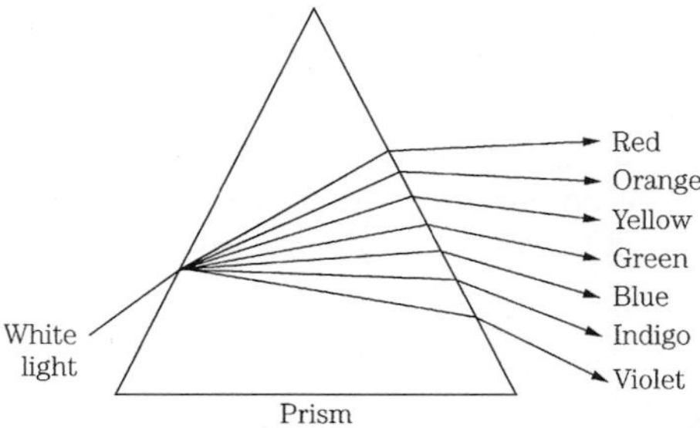

FIGURE 6.2 When white light passes through a prism, refraction occurs at both surfaces. Refraction disperses the beam into its component wavelengths. Similarly, water droplets in the air act as prisms, creating rainbows.

each wavelength changes speed differently, and is refracted at a different angle. Light emerges from the prism dispersed into colors of the visible spectrum. The index of refraction is smallest for red light and increasingly larger for smaller wavelengths.

A medium's index of refraction (n) is the ratio of the light's velocity (c) in a vacuum, to its velocity (v) in a medium; in other words, $n = c/v$. The index of refraction of light in a vacuum is 1.00; in air is 1.0003 (rounded to 1.0); in water is 1.33; in glass is 1.5. The indexes of refraction of the incident and receiving mediums determine the angle of the refracted beam relative to the incident beam. The angle of incidence is the angle between the incident ray and the normal, an imaginary line perpendicular to the materials' interface; the angle of refraction is the angle between the refracted ray and the normal.

When light passes into a medium with higher index of refraction, it is refracted toward the normal, as shown in Fig. 6.3A. Conversely, when light passes into a material with a lower index of refraction, it refracts away from the normal, as shown in Fig. 6.3B. As the angle of incidence increases, the angle of refraction approaches 90°; this is the critical angle, as shown in Fig. 6.3C.

At any incident angle greater than this critical angle, no refraction takes place, and all light is reflected back into the first medium at an angle identical to that of the incident

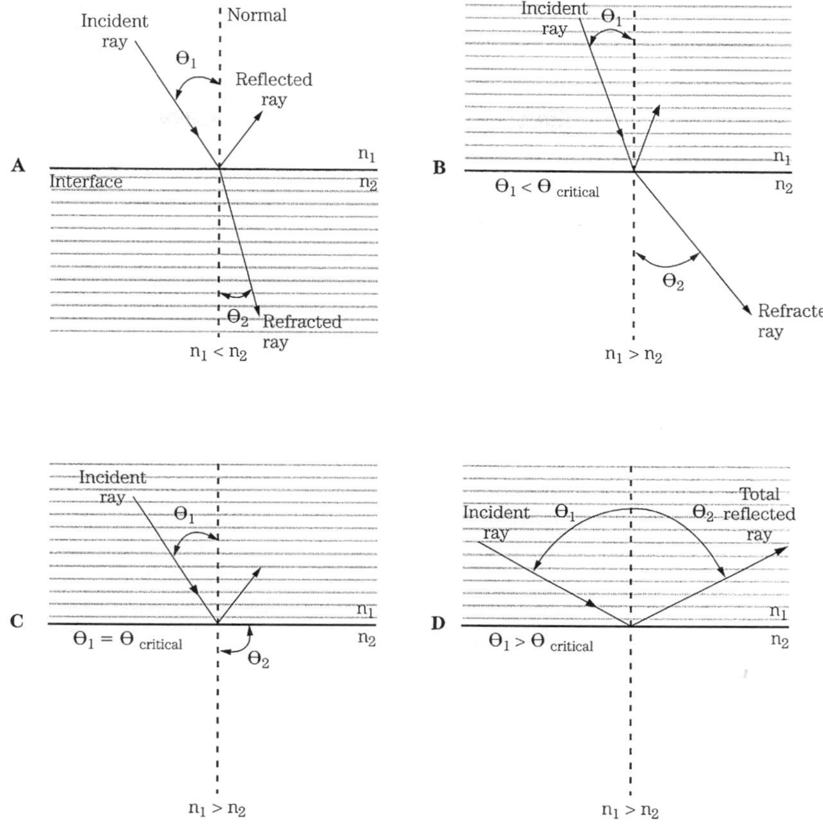

FIGURE 6.3 Light refracts when it passes from one medium to another with different indexes of refraction. A. When n_1 is less than n_2, the refracted ray is bent toward the normal. B. When n_1 is greater than n_2, the refracted ray is bent away from the normal. C. When n_1 is greater than n_2 and the incident angle equals the critical angle, light does not enter n_2. D. When n_1 is greater than n_2 and the incident angle is greater than the critical angle, all light is reflected back into n_1; this is total internal reflection.

angle, as shown in Fig. 6.3D. Specifically, when the angle of incidence is greater than the critical angle, the result is total internal reflection. This phenomenon is essential for light propagation in fiber optics, as described in Chap. 13.

More specifically, Snell's law states:

$$\frac{n_1}{n_2} = \frac{\sin\theta_2}{\sin\theta_1}$$

where n_1 = index of refraction of incident medium (such as a fiber-optic core)
θ_1 = angle of incidence
n_2 = index of refraction of receiving medium (such as a fiber-optic cladding)
θ_2 = angle of refraction

The critical angle of incidence is $\theta_{critical}$ where $\theta_2 = 90°$:

$$\theta_{critical} = \sin^{-1}\left(\frac{n_2}{n_1}\right)$$

Light is totally reflected at angles greater than $\theta_{critical}$ and the angle of total reflection equals the angle of incidence.

Diffraction

Light can be modeled to propagate via rays (which exist only in theory) in the direction of propagation. Surfaces perpendicular to the rays are called wavefronts (which exist in actuality). Wavefront normals, similar to rays, indicate the direction of propagation. The advance of light can be viewed as a summed collection of an infinite number of spherical waves emitted by point sources; this is described in Huygens' principle. These spherical waves are in phase, thus creating a wavefront. However, destructive interference occurs between the spherical waves at all other angles.

When a wavefront passes through an aperture that is small relative to the wavelength (an order of a half wavelength), diffraction occurs, and the wavefront emerges as a point source. For example, diffraction occurs when a single slit is placed in a barrier. A diffraction grating, invented by Joseph von Fraunhofer in 1821, contains a series of identical equidistant slits. Because of interference, a wavefront will only leave the grating in directions where light from all the slits is in phase. This happens straight ahead, and at certain other angles. The angle of the first oblique wave is a function of the wavelength of the light, and the spacing of the slits. If the spacing of the slits is reduced (the spatial frequency is increased), the oblique ray will leave at a greater angle to the centerline. Similarly, the smaller a physical object, the larger the angle over which light must be collected to view the object; fine detail can be resolved only if the diffraction wavefront is collected by the lens.

As described in the following sections, this angle is specified as the numerical aperture of the lens. A diffraction pattern shows the maxima and minima intensities corresponding to the phase differences resulting from different path lengths. The central maximum is the zero-order maxima. Other light is diffracted into a series of higher-order maxima. The diffraction pattern formed behind a circular aperture is known as the Airy pattern, proposed by British astronomer George Airy in 1835.

Specifically, even if a lens is perfectly free of aberration, a laser beam, for example, cannot be focused to a point with a circular lens of finite aperture. The focused laser spot is actually an area where the intensity of light varies as an Airy pattern function, as shown in Fig. 6.4. This is a circular diffraction pattern of maximum and minimum light intensities corresponding to phase differences. The central spot is the zero-order maximum, and the surrounding rings are higher-order maxima. About 83% of the total light falls in the central spot, and the brightest intensity in the first ring is only 1/60 that of the central spot.

The size of the Airy pattern is determined by the light wavelength and the numerical aperture of the lens. Laser pickup optics are said to be diffraction-limited as opposed to tolerance-limited. The size of the spot is dictated by diffraction; a higher-quality lens would still result in the same spot size. As a result, for example, an optical microscope cannot image the diffraction-limited pit spiral on an optical disc (it would show dark spots, not the three-dimensional contour of the pits); instead, a scanning electron

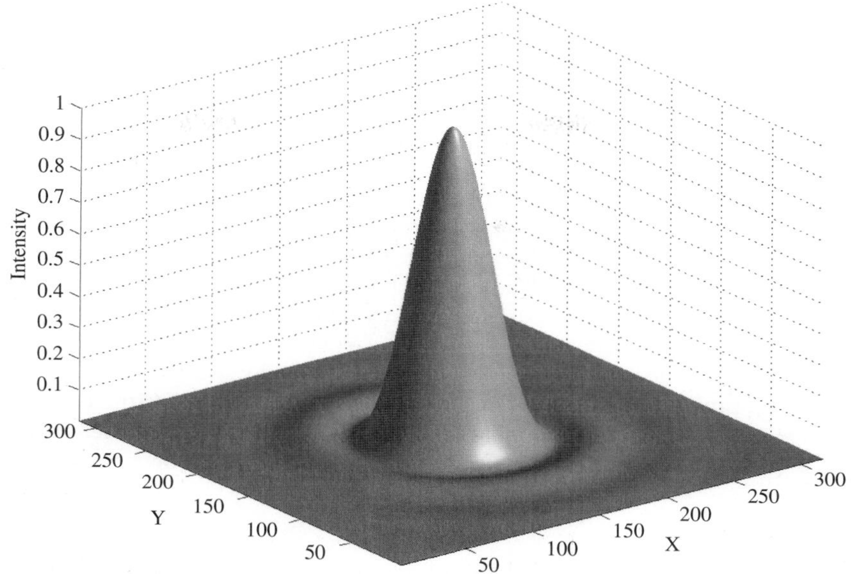

Figure 6.4 An Airy pattern is a diffraction pattern for a point source emerging from an aperture. This figure shows the pattern from a circular aperture.

microscope is needed. Similarly, a laser pickup operating at a short wavelength is needed to read the data pits on an optical disc.

A simple hands-on experiment can be used to demonstrate diffraction. In particular, we can show that the CD data surface causes diffraction, and we can use the result to calculate a CD's track pitch. Fetch a CD and a ruler, and seat yourself with a light bulb—60 W or so—about a meter behind you. Hold the CD about 1/3 meter from your face (reflective side facing you) and angle it so the light bulb's reflection disappears in the disc center hole. Now slowly move the disc toward your face, keeping the bulb's reflection centered in the hole. The brightly colored display is a result of the diffraction created by the pit track.

Now move the disc away from you until the violet ring is located on the edge of the CD. If you measure this, you might find the disc to be, for example, 20 cm from your eye. Now you can calculate the track pitch using the equation:

$$d = \lambda\left(1+\frac{a^2}{r^2}\right)^{1/2}$$

where d = track pitch

λ = wavelength of the violet light (450 nm)

α = distance between eye and disc (e.g., 20 cm)

r = radius of the violet ring (5.5 cm)

Calculating, in this example you will obtain a track-pitch value of 1.7 μm—only 1/10,000,000 of a meter away from the actual pitch (on most CDs) of 1.6 μm.

Resolution of Optical Systems

As described in previous sections, whenever parallel light passes through a circular aperture, it cannot be focused to a point image, but instead yields a diffraction pattern in which the central maximum has a finite width—an Airy pattern. The aperture of an objective lens system is circular. Consider the images of two equally bright point sources of light. Now, consider each point as an Airy pattern. If the points are close, the diffraction patterns will overlap. When the separation is such that it is just possible to determine that there are two points and not one, the points are said to be resolved, as shown in Fig. 6.5. This is the case when the center of one diffraction pattern coincides with the first minimum of the other. The distance between the two diffraction pattern maxima equals the radius of the first dark ring. This condition is called the Rayleigh criterion.

Similarly, the smaller the object that is viewed, the greater the angle over which light must be collected. In other words, the diffracted wavefront must be collected. The resolving power of the lens is determined by its numerical aperture (NA). The numerical aperture of a lens is the diameter of the lens in relation to its focal length and describes the angle over which it collects light. NA is calculated as follows:

$$\mathrm{NA} = n \sin \theta_{max}$$

where θ_{max} is the angle of incidence of a light ray focused through the margin of a lens and n = index of refraction of the medium.

A lens is a spatial lowpass filter and the cutoff of the spatial frequency is determined by the numerical aperture; the modulation transfer function (MTF) describes the response of the lens. This defines the minimum size of a resolved object for a given wavelength. Because the finite NA results in a cutoff of spatial frequency, the Airy pattern is analogous to the impulse response of a lowpass filter. An impulse (Dirac delta function) input to an electrical circuit yields its impulse response. In an optical system, the spatial equivalent of the impulse is approximated by a pinhole aperture. The impulse response can be viewed as the cross section of the intensity of the spot—an Airy pattern.

The spot size (d) is often defined as the half-intensity diameter of the Airy pattern such that $d = 0.61\ \lambda/\mathrm{NA}$, where λ is the wavelength of the laser light in a vacuum. For high data density on an optical disc, d must be small. For example, in the CD format, λ is fixed at 780 nm; thus NA must be as large as possible. However, as NA is increased, tolerances become severe: the depth-of-focus tolerance is proportional to NA^{-2}, skew (disc tilt)-tolerance is proportional to NA^{-3}, and disc-thickness tolerance is proportional to NA^{-4}. Clearly, for system stability, the NA should be as low as possible. Balancing

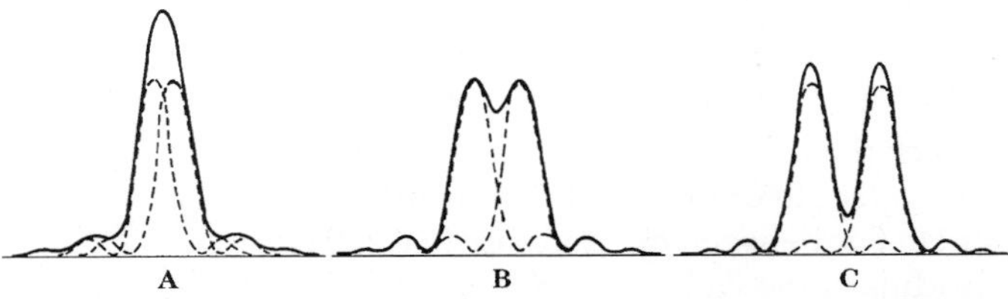

FIGURE 6.5 The Rayleigh criterion determines the resolution of an optical system. A. Overlapping diffraction patterns cannot be resolved. B. Partially overlapping patterns are resolved (Rayleigh criterion). C. Nonoverlapping patterns are resolved.

these factors, for example, the CD designers selected NA = 0.45 (NA is a dimensionless quantity). Thus, CD spot size is approximately 1.0 μm. The DVD format uses λ of 635 or 650 nm, and NA of 0.60, whereas the Blu-ray format uses λ of 405 nm and NA of 0.85. When NA exceeds about 0.5, the spot intensity pattern begins to deviate from the Airy pattern, but in relatively minor ways.

Similarly, λ and NA determine other specifications such as track pitch, cutoff frequency, and track velocity; ultimately, from these parameters, disc storage capacity can be deduced. For example, the spatial cutoff frequency in a CD system can be determined:

$$f_{sc} = \frac{2\text{NA}}{\lambda} = \frac{2(0.45)}{780 \times 10^{-9}} = 1.15 \times 10^{6}$$

Thus formations with a higher spatial frequency (for example, lines smaller than 1.15 lines per micrometer) cannot be resolved. Optical systems must be designed to operate within this constraint. For example, in the CD system, the shortest pit/land length is 0.833 μm. A series of short pit/lands would thus yield a spatial frequency of $1/(2)(0.833 \times 10^{-6}) = 0.600 \times 10^{6}$, or about half the cutoff frequency and hence would be easily readable. Furthermore, given a CD with a track velocity of 1.2 m/s, the temporal cutoff frequency is:

$$f_{tc} = \frac{2\text{NA}\upsilon}{\lambda} = \frac{2(0.45)(1.2)}{780 \times 10^{-9}} = 1.38 \text{ MHz}$$

In terms of an optical channel, the amplitude of the modulated signal is maximum at 0 Hz and linearly decreases to 0 at 1.38 MHz.

Polarization

Electromagnetic waves such as light waves (and radio waves) consist of electric and magnetic fields that are perpendicular to each other, and both oscillating transversely to the direction of propagation. (The reader will recall that sound propagates longitudinally.) Either field can be used to characterize light, but the electric field is generally used. The transverse electric field can oscillate with vertical, horizontal, diagonal, or other more complex geometry, but is always perpendicular to the direction of travel. Light from a light bulb is said to be unpolarized because an infinite number of electric fields exist and are randomly perpendicular to the direction of travel. Any one of these electric (E) fields, at any angle, can be considered as a vector represented by two orthogonal components E_x and E_y. Examination shows that the component waves are phase-incoherent.

However, when only one electric field is allowed to oscillate, in a single direction that is perpendicular to the direction of travel, the light is said to be linearly polarized or plane polarized. A polarizer accomplishes this, as shown in Fig. 6.6. The electric field lies in one plane, and does not change direction; moreover, one plane contains the electric vector and the direction of propagation. When the orthogonal components representing the E field are examined, we observe that each of them represent linearly polarized light, and are in phase with each other. Combined, they produce a single linearly polarized wave. In addition, the linearly polarized E_x and E_y components can be combined with a relative phase shift. In particular, when the phase difference is 90° (phase quadrature),

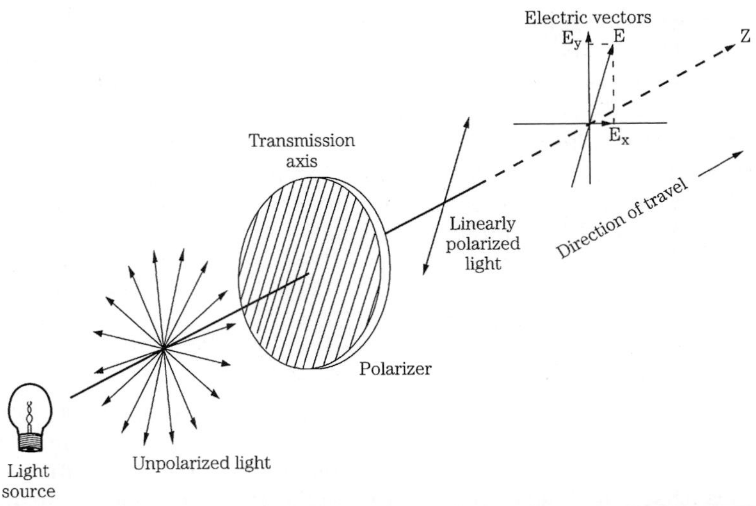

Figure 6.6 A polarizer is used to create linearly polarized light. Only light waves with an electric (E) field vector parallel to the transmission axis can pass through. The E vector can be represented by two components E_x and E_y.

the light is circularly polarized. The magnitude of the resulting electric vector is constant, but instead of remaining in one plane, it varies from point to point and from time to time. Specifically, it rotates through a helix once per wavelength; viewed from the direction of propagation, it traces a circle. At other component phase shifts, the light is elliptically polarized; the electric vector varies as it rotates, and traces out an ellipse. In any case, the wave can be resolved into its linearly polarized components.

Isotropic materials have one velocity of transmission independent of the plane of propagation. A material is anisotropic when the light entering it does not have the same velocity in all planes; in other words, the index of refraction depends on the direction of travel through it. Anisotropic materials such as calcite can be used to create polarized light. When unpolarized light enters an anisotropic medium, light rays split into two part rays. One ray passes through the object as in an isotropic medium, diffracting normally; the second ray is refracted more strongly and is thus displaced from the first as it emerges. The first ray is called the ordinary ray, and the second is called the extraordinary ray. The two rays are linearly polarized in mutually perpendicular planes. This is known as double refraction or birefringence. The direction along which no birefringence occurs, where the material behaves exactly like an isotropic medium, is called the optic axis.

The plastic used in optical-disc substrates can exhibit birefringence after it is subjected to the stress of melting and injection molding during disc manufacture. This birefringence is an unwelcome effect of anisotropy. A wavefront (perhaps emitted from a laser pickup) traveling through the substrate is distorted because the two orthogonal components travel at different velocities, creating a birefringent image. The velocity difference depends on the direction of the light ray passing through the birefringence material. In practice, birefringence is minimized through careful manufacturing techniques. With shorter laser wavelength and higher NA, birefringence increases and its effects become more severe. On the other hand, thinner substrates (such as in DVD) reduce birefringence and optical aberrations.

A rotation in the plane of polarization, resulting from different velocities in different planes, can be usefully employed. A retardation plate is a slice cut from a crystal in such a manner that the slice contains the optic axis. A beam of unpolarized light normally incident on the plate will create an ordinary and extraordinary beam. The phase difference between these beams is proportional to the distance traveled within the plate. When the beams emerge, they are not separated, but they are out of phase. If the thickness of the plate is such that the phase difference at emergence between the superimposed ordinary and extraordinary beam is $\lambda/4$, the plate is called a quarter-wave plate (QWP). Similarly, if the phase difference is $\lambda/2$, the plate is a half-wave plate.

By passing linearly polarized light through a QWP, it can be converted to circularly or elliptically polarized light, depending on the angle between the incident vibration plane and the optic axis. If the angle between the plane of linear polarization and the optic axis is exactly 45°, light is transformed from linear to circular polarization (or vice versa), as shown in Fig. 6.7. For other angles, the transformation is from linear to elliptical (or vice versa).

This rotation in the plane of polarization can be used to distinguish between light beams in an optical disc pickup. For example, a linearly polarized light beam might pass through the QWP to become circularly polarized, strike an optical disc, and return through the QWP again becoming linearly polarized. Because the resulting plane of polarization is perpendicular to the incident linearly polarized beam, it can be separated from the incident light by a polarizing prism, acting transparently to the incident light, but as a mirror to the reflected light. Many optical pickups use this technique, as described in Chap. 7. A half-wave plate operates similarly, but the plane rotates at twice the angle of the plane of polarization of the incident beam.

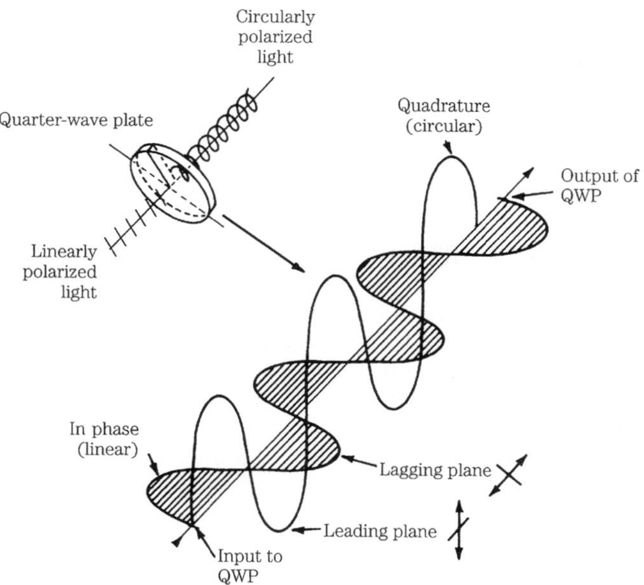

FIGURE 6.7 Linearly polarized light passing through a quarter-wave plate is converted to circularly polarized light. The angle between the plane of linear polarization and the optic axis must be 45°.

Design of Optical Media

Most optical storage systems store data across the surface of a flat disc. This allows random access of data, as well as ease of manufacturing replication. In most designs, because data is written and read via optical means, there is no physical contact between the pickup and the media. This ensures long pickup and media life and minimizes damage from head crashes or other catastrophes. In addition, because there is no need for physical contact between the pickup and the data surface, data can be embedded within a transparent protective layer to minimize the effect of surface contamination and damage on data readout. Also, multiple data layers can be placed within one substrate. However, stored data must undergo both modulation and error-correction encoding to maximize data density and guard against errors.

Data can be stored along either a spiral track or concentric tracks. Most optical disc pickups shine a laser on the media, and the reflected light is detected by a sensor and decoded to recover the carried data. To accomplish this, the media must present two states so the change between them varies the reflected light, and thus the data can be recognized in much the same way that the black characters on this page stand in contrast to the white paper. Data can be represented as a phase change, polarization change, or change in the intensity of reflected light. For example, pits in a reflective surface produce diffraction in the reflected light beam, decreasing its intensity. The resulting variation in intensity, from high intensity in reflecting areas, to low intensity in pit areas, can be converted to a varying electrical signal for data recovery.

To expedite data writing and recovery, a laser beam is used. Laser light, unlike incoherent light, yields a high signal-to-noise (S/N) ratio from the photodetector. The short wavelength permits a high information density to be achieved; for example, a data pit length might be 0.2 μm. As noted, track pitch, minimum spot size, and other dimensions are defined by the wavelength of the reading laser. Discs are thus diffraction-limited. Shorter wavelength laser light would yield greater storage density. A laser light source is also required to provide a sufficient S/N ratio for a high bit rate, given an illumination of an area on the order of 1 μm^2, for a period of 1 ms or less. Either analog or digital signals can be encoded and stored on a disc, within the bandwidth constraints of the media.

Any optical media must be supported by a sophisticated servo system to provide positioning, tracking, and focusing of the pickup, as well as accurate disc rotation. Focus tolerance, for example, can be on the order of 1 μm, and should be maintained in spite of mechanical shock and vibration. The pickup must generate a set of correction signals derived from the signal from the optical media itself and use a set of actuators to maintain proper conditions. Radial tracking correction signals can be derived using methods such as the twin spot, which uses a diffraction grating to create additional scanning spots, or wobble, in which the scanning spot is given a sinusoidal movement. Similarly, focus correction signals can be generated through methods such as the Foucault knife edge, using unequal distribution of illumination in a split beam; astigmatism, using a cylindrical lens to create spot asymmetry; or critical angle, using angle of incidence and a split beam. Laser pickup design is discussed in more detail in Chap. 7.

The shelf life of optical discs is longer than that of magnetic media, optical discs are less susceptible to damage from heat, and they are impervious to magnetic fields. The raw (uncorrected) error rate of an optical disc can be 10^{-6}, perhaps 10 to 30% of the disc capacity may be needed for error-correction coding to bring the corrected rate to 10^{-13}.

Any optical recording material must exhibit long-term stability, high absorptivity at the recording wavelength, low writing energy, high S/N ratio, good forming characteristics, low thermal conductivity, and low manufacturing cost. It is advantageous to design optical storage with specific applications in mind. Specifically, three separate systems, read-only, write-once, and erasable media have been developed for various applications.

Nonerasable Optical Media

In nonerasable optical disc systems, a laser light shines on the data surface, and the reflected light is detected by a photodiode and decoded to recover the carried data. To accomplish this, the surface presents two states, for example, to vary the intensity of the reflected light. In this way, data can be recognized by the pickup. For example, a reflective disc might have pits embossed into its surface to vary reflected intensity. The actual technology used depends on whether the media is read-only or write-once. Several mechanisms are available to achieve both of these results, but questions of durability, density, and feasibility of mass production ultimately define the most appropriate method.

Read-Only Optical Storage

CD-Audio, CD-ROM, DVD-Video, DVD-Audio, DVD-ROM, and Blu-ray BD-ROM discs are examples of read-only optical media. Whether a disc holds audio or software data, the data is permanently formed on the media during manufacture; it is a playback-only format. A plastic disc is impressed with a spiral track of pits set to a depth calculated to decrease the intensity of the laser light of the reading pickup. To provide reflectivity, the data surface is metallized. The reflective surface of the disc and the data pits are embedded between the transparent plastic substrate and a protective layer. The effects of scratches and dust particles on the reading surface are minimized because they are separated from the data surface and thus made out of focus with respect to the laser beam focused on the inner reflective data surface.

In a read-only optical media, the pit surface and laser readout forms a sophisticated optical system. The numerical aperture of the lens, wavelength of the laser light, thickness and refractive index of the disc substrate, and size and height of the pits all interact. As noted, when a laser beam is focused on the data surface, an Airy pattern results, as shown in Fig. 6.8. The spot diameter can be specified at half-intensity (for example, 1.0 μm) or at the first dark ring at $d = 1.22\ \lambda/\mathrm{NA}$ (for example, 2.1 μm). Allowable crosstalk between tracks determines track pitch (for example, 1.6 μm). Crosstalk must be acceptable even in the worst case of a slightly defocused beam and slight tracking error. In any case, the beam is focused so that approximately half its center area falls on a pit, and half falls on the surrounding reflective land.

Many theories are employed to model the readout of optical discs. Perhaps the most direct approach models the disc as a diffraction grating. An optical storage surface (such as a read-only CD, DVD, or Blu-ray) that uses a phase structure is a reflective phase grating, and acts similarly to a diffraction grating. Specifically, the pits cause diffraction. The smaller the pit in relation to the light wavelength, the greater is the angle at which light leaves. The area of light striking a pit is about equal to that striking the surrounding land. Using a simple model, light diffracted by the grating consists of a single zero-order

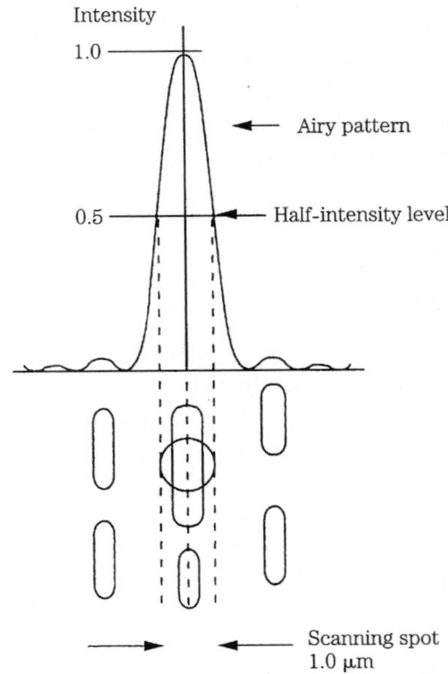

Figure 6.8 In write-once optical disc systems, the reading laser forms an Airy pattern on the data surface. The half-intensity spot diameter covers the pit track.

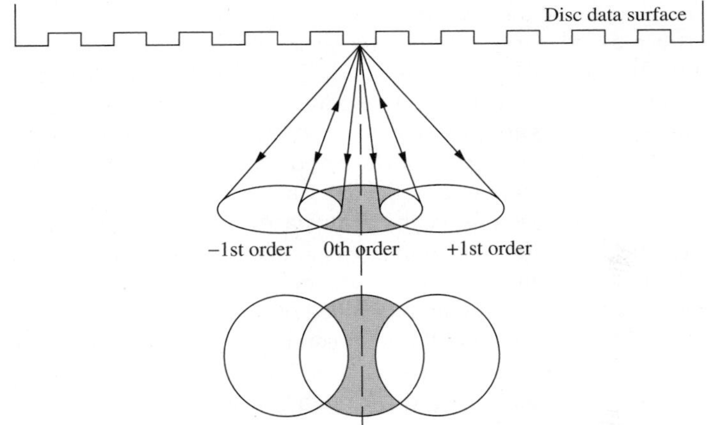

Figure 6.9 In write-once optical disc systems, the data surface forms a phase structure that acts like a reflective phase grating. Diffraction causes cancellation through interference in the reflected beams.

and multiple first-order beams, as shown in Fig. 6.9. These beams partly overlap each other; the resulting destructive interference between the zero- and first-order beams in the light ray returning to the lens yields cancellation due to interference. Effectively, a pit thus reduces the intensity of light returning to the lens. The image of the interference pattern from the multiple reflection orders is sometimes called the "baseball" pattern. For good contrast, the power in the two beams with different phases should be equal. Pit depth does not require great accuracy. Equality of pit/land areas is more important. A plane-wave model (with a large magnification system) predicts that pit height should be one-quarter of the apparent wavelength of the light, so that light striking a pit is out of phase by a half wavelength with that striking the land. A more complex spherical-wave model that accounts for effects of the converging focused beam (low magnification), devised by Charles Mecca, Yajun Li, and Emil Wolf, shows that the optimum pit depth should be one-half wavelength.

The CD, DVD, and Blu-ray systems are diffraction-limited because they operate at dimensions that are as small as permitted by the wave nature of light at their respective wavelengths. For example, with a CD spot diameter of 1.0 μm, details this size would give diffracted rays that only just fall within the lens aperture. Finer details yield greater convergence thus rays diffracted by the pits must fall outside the lens aperture. This is in fact the case with the pits on the data surface because they are narrower than the diameter of the spot. Rays diffracted by the pits must consequently fall outside the lens aperture, decreasing intensity reflected back into the lens, promoting a robust playback signal.

This type of diffraction media, using physical pits to store data, is attractive because it can be economically mass-produced. For example, on 12-cm-diameter discs, a CD can hold 680 Mbytes of formatted data, a DVD can hold 4.7 Gbytes on a data layer (multiple layers are possible), and a dual-layer Blu-ray disc can hold 50 Gbytes. The media is open-ended with respect to the type of data to be encoded. For example, the CD-Audio standard was joined by CD-ROM, Video-CD, SACD, and other formats. Manufacturers have developed higher-density DVD and Blu-ray discs with many times the storage capacity of a CD; this is accomplished with a shorter wavelength laser, higher NA, and thinner substrate, which permits smaller pit and track dimensions. The CD format is discussed in Chap. 7, the DVD format in Chap. 8, and the Blu-ray format in Chap. 9.

Write-Once Optical Recording

Recordability is essential for many applications; the simplest recordable optical systems are write-once (WO). The user records data permanently, until the disc capacity is filled. When only a few discs are needed for distribution, a write-once system is cost-efficient and write-once discs are widely used throughout the computer, and audio and video industries. A write-once optical disc can be implemented in a variety of ways, as summarized in Fig. 6.10. Dye-polymer recording uses a recording layer containing a heat-absorptive organic dye; the dye is absorptive at the wavelength of a writing laser. When the layer is heated, it melts and forms a depression. Simultaneously, a reflective layer is deformed. Data is read by shining a laser light on the surface; the physical formations decrease the intensity of the reflected beam. Several types of disc recorders use a dye-polymer method. The CD-WO format (better known as CD-R) is discussed in Chap. 7. The DVD-R format is discussed in Chap. 8. Some Blu-ray BD-R discs use dye-polymer recording; Blu-ray is discussed in Chap. 9. Instead of the "pit and land" nomenclature used to describe prerecorded optical media, some texts refer to the "marks and spaces" of recordable optical media.

In some systems, an irreversible phase change is used to alter the reflectivity of the media at the point where a writing laser is focused. In this way, a reading laser can differentiate between data. Some systems use a thin metallic recording layer that varies its physical property from crystalline to amorphous when it is heated by a writing laser.

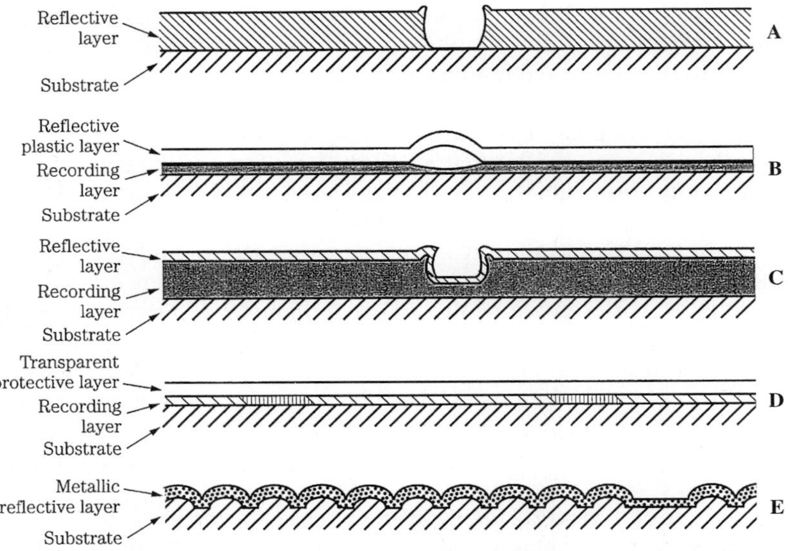

Figure 6.10 A variety of methods can be used in write-once storage. A. Pit formation. B. Bubble formation. C. Dye polymer. D. Phase change. E. Texture change.

The crystalline state is more translucent, thus more light reflects from the metal layer above, yielding high reflectivity. The amorphous state absorbs the reading laser light, yielding low reflectivity. The phase transition can triple the reflectivity of the recording layer at written spots, thus allowing laser reading of the data. The recording layer can use an antimony selenium (Sb-Se) metallic film and the heat-absorbing layer can be a bismuth-tellurium (Bi-Te) metallic film.

Other writing methods include pit formation, bubble formation, and texture change. With pit formation, a mechanism called ablation uses a laser writer of approximately 10 mW to burn holes in a reflective layer; the material is melted or vaporized by the heated spot. Melting is preferred because no residue is created around the pit. In either case, data can be read by monitoring the change in intensity. Similarly, bubble formation uses a laser to vaporize a recording layer, causing a bubble to form in an adjacent reflective layer. The bubble can be read with a laser, by monitoring reflecting light levels. The texture change method uses a reflective surface with small aberrations with dimensions and spacing designed to diffract a reading laser. When the layer is heated and melted, it forms a smooth face that increases reflectivity at that point.

Erasable Optical Media

Erasable optical disc systems provide versatile data storage. Data can be written, read, erased, and written again. In most cases, the number of erasures is essentially unlimited. Several recordable/erasable optical media technologies have been introduced in a variety of formats, varying broadly in cost. These erasable technologies include magneto-optical, phase-change, and dye-polymer recording. Magneto-optical recording uses a laser beam and magnetic bias field to record and erase data, and a laser beam alone to read data. Phase-change media use a reversible change in the index of reflectivity of materials. Dye-polymer media use reversible changes in physical formations induced by heating the recording layer.

Magneto-Optical Recording

Magneto-optical (MO) recording (sometimes known as optically assisted magnetic recording) technology combines magnetic recording and laser optics, utilizing the record/erase benefits of magnetic materials with the high density of optical technology. Magneto-optical recording uses vertical magnetic media in which magnetic particles are placed perpendicularly to the surface of a pregrooved disc. Vertical recording provides greater particle density and shorter recorded wavelengths; however, the high recording density is not fully utilized by conventional magnetic heads because their flux fields cannot be narrowed sufficiently. The recorded area is thus larger than necessary. Optical assistance increases the recording density.

With magneto-optics, a magnetic field is used to record data, but the applied magnetic field is much weaker than conventional recording fields. It is not strong enough to orient the magnetic fields of the particles, thus a unique property of magnetic materials is used. As the oxide particles are heated to their Curie temperature, their coercivity (the minimum magnetic field strength required to reverse an induced magnetization) decreases radically. In other words, when heated, a magnetic material loses its resistance to change in its magnetic orientation; its orientation can be affected by a small applied magnetic field. (Similarly, variations in the earth's magnetic field over time can be traced by studying the magnetic fields imprinted in ancient volcanic rocks.)

In the case of MO recording, this allows data to be written with a weak field. For example, coercivity falls almost to zero as the temperature rises to 150°C. A laser beam focused through an objective lens heats a spot of magnetic material to its Curie temperature. At that temperature, only the particles in that spot are affected by the magnetic field from the recording coil, as shown in Fig. 6.11A. When the beam is turned off, or the area moves away from the beam, it cools below the Curie temperature as the absorbed energy is dissipated to the substrate by thermal conduction. The applied magnetic field is withdrawn, and the magnetic orientation of the field is "frozen" and retained as data. In this way, the laser beam creates a recorded spot much smaller than otherwise possible thus increasing recording density. Moreover, at room temperature, the recording layer's high coercivity makes it highly resistant to the effect of stray magnetic fields.

The Kerr effect is used to read data; it describes the slight rotation of the plane of polarization of polarized light as it reflects from a magnetized material. (The Faraday effect describes the same phenomenon as light passes through a material.) The rotation of the plane of polarization of light reflected from the reverse-oriented regions differs slightly (by perhaps ±0.5°) from that reflected from unreversed regions. To read the disc, a laser is focused on the data surface, and the angle of rotation of reflected light is monitored, as shown in Fig. 6.11B. An analyzer distinguishes between rotated and unrotated light, and converts that information into a beam with varying light intensity. For example, by passing the polarized light through a polarizing prism, light with a polarization plane parallel to that of the prism will pass through, while relatively rotated light will be blocked, resulting in an intensity difference. Data is then recovered from that modulated signal. The power of the reading laser is much lower than the recording laser (perhaps by a factor of 10) so the recorded magnetic information is not affected.

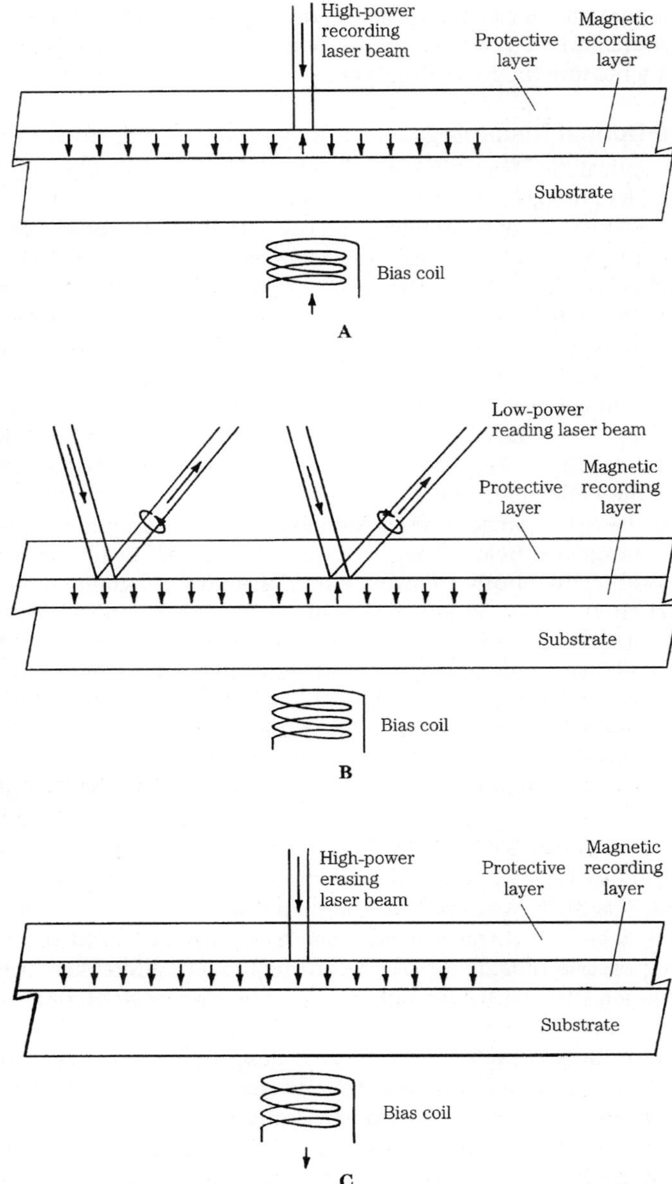

FIGURE 6.11 Techniques used in magneto-optical recording, reading, and erasing. A. In magneto-optical recording, a high-power laser beam heats the data surface, and magnetic recording is induced by a bias coil. B. In magneto-optical reading, data is read by reflecting a low-power laser beam from the data surface and observing the rotation in the plane of polarization of the reflected light. C. In magneto-optical erasing, data is erased by heating the data surface and reapplying the bias field.

To erase data, a magnetic field is applied to the disc along with the laser heating spot, and new data is written, as shown in Fig. 6.11C. New data can be rerecorded.

The magneto-optical recording layer can be placed between a transparent substrate and a protective layer. The laser light can shine through either the substrate or the protective layer, placing surface dust and scratches out of focus with respect to the data surface. A coil can be wrapped about the laser lens structure to produce the magnetic field. In some cases, its perpendicular alignment is assisted by a metal plate located on the opposite side of the disc, or the coil can be placed on the opposite side of the disc. A variety of magnetic materials can be used, selected on the basis of S/N ratio, orientation properties, and long-term stability. In general, amorphous, thin-film magnetic materials are used. Some applications use a material such as terbium ferrite cobalt. At room temperatures, the coercivity of the recording layer can be more than 10,000 oersteds, effectively eliminating the possibility of accidental erasure at room temperature. Magneto-optics is thus potentially more stable and reliable than other magnetic media. Tests indicate that MO data can be erased and rewritten 10 million times or more, equivalent to conventional magnetic media. Accelerated life measurements suggest MO discs will last at least 10 years. In one test, when exposing discs to 95°C at 95% humidity for 1000 hours, MO discs survived better than conventional hard disks.

To achieve mechanical compatibility from one recorder to another within the high tolerances of magneto-optical media, blank discs can be manufactured with prerecorded and nonerasable addressing. One method, called hardware address sectoring, uses a disc with spiral or concentric grooves, in which address information is physically formed in the groove and detected by light beam reflection. Using this system, a magneto-optical player can automatically track both address and data information contained on an MO disc. Storage capacity is not reduced because the recorded data signal is superimposed on the hardware addressing information.

The MiniDisc format used a small MO disc to provide recordability and portability. To provide 74 minutes of recording time on the 2.5-inch disc, the Adaptive TRansform Acoustic Coding (ATRAC) data-reduction algorithm was used.

Phase-Change Optical Recording

Erasable phase-change systems use technology similar to that used in write-once systems. They use materials that exhibit a reversible crystalline/amorphous phase change when recorded at one temperature and erased at another. For erasable media, a crystalline (translucent and yielding high reflectivity from the metal layer) to amorphous (absorptive and yielding low reflectivity) phase change is typically used to record data and the reverse to erase. Information is recorded by heating an area of the crystalline layer to a temperature slightly above its melting point. When the area rapidly cools below the crystallizing temperature and solidifies, it is amorphous, and the change in reflectivity can be detected. Because the crystalline form is more stable, the material will tend to change back to this form. Thus, when the area is heated to a point above its glass transition temperature but below its melting temperature and cooled gradually, it will return to a crystalline state, erasing the data. The temperature during melting might reach 800°C. The dielectric layers mediate the thermal stress, and the aluminum layer acts as a heat sink. The phase-change recording mechanism is shown in Fig. 6.12.

A number of materials have been devised for the recording layer. For example, layers comprising gallium antimonide and indium antimonide have been developed.

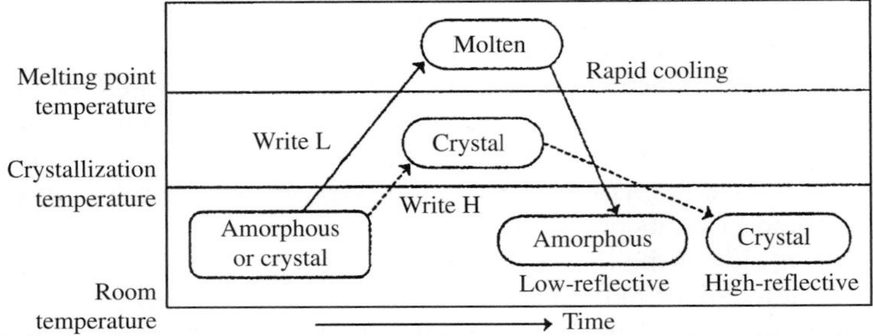

FIGURE 6.12 In phase-change recording, the recording media may be placed in an amorphous or crystalline state, depending on the heating temperature applied.

Some systems use tellurium alloyed with elements such as germanium and indium. Unlike the organic dyes used in write-once discs, phase-change recording materials are not wavelength-specific. Permanent recording can be achieved by simply increasing the power of the writing laser; this burns holes in the recording layer rather than changing its phase. Phase-change media have a long shelf life and are not affected by ambient temperatures and humidity. A large number of erasures can be achieved. Phase-change technology is used in the CD-RW format as described in Chap. 7. The DVD-RW and DVD-RAM formats use phase-change recording and are discussed in Chap. 8. Most recordable Blu-ray BD-R and BD-RE discs use phase-change technology; the Blu-ray format is discussed in Chap. 9.

Dye-Polymer Erasable Optical Recording

In dye-polymer recording, light-absorbing dyes are placed in a bi-layer disc structure. The two layers, an expansion layer and retention layer, are sensitive to different wavelengths. The physical deformation that the polymer layers undergo results in bumps; in this way, data can be written and read. For example, the top expansion layer can be composed of an elastomer containing a dye sensitive to light with an 840-nm wavelength. The inner retention layer can be sensitive to 780-nm wavelength light. During recording, infrared laser light of 840-nm wavelength is absorbed in the top recording layer, causing its temperature to rise; the bottom layer is transparent to this light. The top layer's heat causes it to expand, pushing against the bottom layer to form a bump. When the temperature cools, the bump is retained. This formation can be read by a low-power laser beam through diffraction. To overwrite new data, light of a 780-nm wavelength is absorbed by the bottom layer, but not the top layer. The retention layer is softened, reducing its modulus of elasticity. The expansion layer pulls itself back to its original condition, restoring the surface to a flat condition, ready for new data.

The system is efficient because the recording surface changes state each time laser light with the correct wavelength and power strikes the surface. One wavelength creates a bump; another wavelength flattens it. In addition, the wavelength that flattens a bump does not affect an already smooth surface. Data can thus be overwritten on a single revolution of the media, with data passing under the smoothing laser just before it passes under the forming laser.

Digital Audio for Theatrical Film

Several theatrical film formats have been developed to provide conventional optical storage of frame images, as well as multichannel digital audio storage. In double systems, external audio playback devices are synchronized to the picture using a timecode stripe added to conventional motion picture film. Single systems use optically encoded digital audio data on the film itself to produce a modulated signal. The latter approach, although technically more difficult, is preferred. Motion pictures generally use a stereo optical soundtrack (often matrixed for four channels), called the stereo variable area (SVA) printed along the frame edge; some digital systems retain this track for compatibility with legacy motion picture projection systems. In addition, in the event of catastrophic damage to the digital soundtrack, the system can automatically and momentarily switch to the analog soundtracks. New digital audio cinema systems have supplanted the older 70-mm six-track analog audio format that used magnetic striping.

Any theatrical film audio system must provide multichannel playback, minimally with left, right, and center channels, two surround channels for ambience, as well as a subwoofer channel. Other channels can encode foreign language dialogue or nonaudio data for timecode, control of theater curtains, and so on. Of the six (or more) audio channels, five must provide wide audio bandwidth, with the subwoofer channel reproducing signals below 100 Hz; all must provide a high dynamic range. Optical tracks must be robust, able to withstand hundreds or thousands of trips through the projector; raw error rates of 10^{-3} are typical for a worn film. Reliable error correction is mandated. In addition, the method must support high-speed copying for mass replication of films. Several imaging dye methods have been developed in which binary data is recorded as a series of transparent and opaque dots, using conventional film dyes and layers. The mosaic data pattern can be placed along the film edges or between the sprocket holes, as shown in Fig. 6.13. During playback, light from a source is focused on the tracks, and read with a sensor array on the opposite side. To encode sufficient data on the film, data reduction methods must be used, as described in Chaps. 10 and 11. Film using this method can be copied at high speed using conventional methods.

The Dolby Digital system retains analog soundtracks for compatibility, and adds an optical data track between sprocket holes, on the same side as the analog tracks. Six audio channels (left/center/right in front, left/right stereo surround in rear, and subwoofer) are sampled at 48 kHz, encoded through Dolby Digital (AC-3) data reduction, multiplexed together and with a 9.6 kbps (thousand bits per second) data channel, and written to film as 96 data blocks. Data blocks are formed from a 76×76-pixel array matrix; each pixel is 32 μm^2. The composite bit rate is 320 kbps, excluding error-correction data. The data is placed in the green layer of the track negative. A CCD scanner reads the optical information, and then it is demultiplexed, decoded, and output to the theater sound system. Audio data is recorded 2.5 seconds prior to picture to allow time for processing, and so the timebase can be adjusted to achieve subjectively correct synchronization of sound to picture for a given theater size. The subwoofer channel covers a range of 3 Hz to 125 Hz. Printing laboratories are able to perform high-speed duplication using special heads. Dolby Digital (AC-3) is discussed in Chap. 11. The Dolby Digital system was developed by Dolby Laboratories.

The DTS (Digital Theater Systems) format is a double system using external storage. A timecode track is placed between the picture and the standard analog SVA soundtrack; this code is used to synchronize external CD-ROM drives which hold audio data.

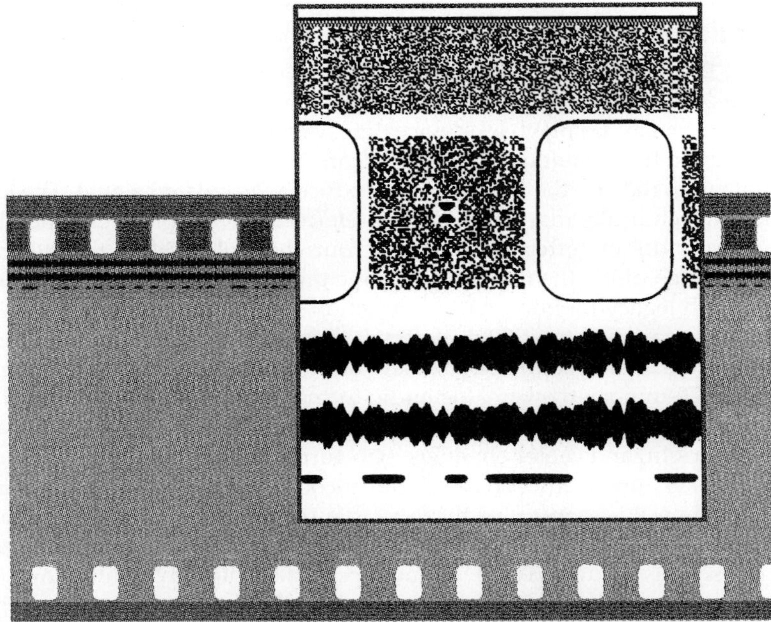

Figure 6.13 Detail showing one edge of a 35-mm film stock holding three digital audio tracks and one stereo analog audio track: Sony SDDS data on the top edge (and bottom edge), Dolby Digital data between sprocket holes, analog stereo SVA track, and a DTS timecode track synchronizing to external CD-ROM drives.

The timecode also contains a film serial number and a reel number that corresponds to a matching code on the CD-ROM disc. As with other double systems, timecode is read in advance of the picture so that any edits in the picture (such as missing footage or reel changes) can be anticipated and compensated for by the sound source. The timecode track can be placed in either the green or red layer of the track negative. DTS discs contain five data-compressed audio channels; the subwoofer track and the 20-Hz to 80-Hz information from the stereo surround channels is summed and recorded in that region of both surround tracks. The surround tracks are thus full-range while the surround speakers reproduce only 80 Hz and above. The subwoofer channel covers a range of 20 Hz to 80 Hz. Audio data is output on an SCSI bus to external compression decoders that drive movie house sound systems. Discs are placed in shipping containers that fit within the standard cases used to ship 2000-foot projection pancakes to theaters. Data compression is performed with the apt-X100 algorithm. The DTS system was developed by Digital Theater Systems.

In the SDDS (Sony Dynamic Digital Sound) system, eight audio channels are encoded and placed in two data tracks running outside the perforation holes, one stripe on each side of the film. Existing optical analog tracks are retained. Each stripe comprises 64 rows of 24 × 22-μm pixels; the "*P*" stripe (adjacent to the picture) carries the left and left-center channels as well as center and subwoofer. The "*S*" stripe (adjacent to the optical soundtrack) carries data for the right and right-center channels, as well as center and subwoofer. All audio data is offset by 17.8 frames; the *P* data leads while the

S data is in edit sync with the picture. If a data track is damaged, a concealment mode accesses additional data tracks. The data is placed in the red layer of the track negative. This system adds full-range left-center and right-center tracks to the standard 5.1-channel format. Five loudspeakers are placed behind the screen, along with one subwoofer loudspeaker, also usually placed behind the screen, and two surround sound loudspeakers; smaller playback configurations are supported as well. A professional version of the ATRAC data reduction algorithm, originally devised for the MiniDisc format, is used to code the audio channels. The subwoofer channel covers a range of 4 Hz to 500 Hz. The ATRAC algorithm is discussed in Chap. 11. The decoder also provides digital-domain equalization, level trims, and surround delay trims. The SDDS system was developed by Sony Corporation.

CHAPTER 7

Compact Disc

The introduction of the Compact Disc (CD) system was perhaps the most remarkable development in audio technology since the birth of audio recording technology in 1877 with Edison's invention of the tinfoil recorder. The Compact Disc system contains numerous technologies original to the audio field; when combined, these technologies formed a storage means that was unprecedented at its invention.

A Compact Disc contains digitally encoded data that is read by a laser beam. Because the reflective data layer is embedded within the disc, dust and fingerprints on the reading surface do not normally affect reproduction. The effect of most errors can be minimized by error-correction algorithms. Because no stylus touches the disc surface, there is no disc wear, no matter how often the disc is played. Thus, digital storage, error correction, and disc longevity result in a robust digital storage medium. In addition, the CD offers high-density data storage providing long playing time with a small disc size. Whereas the (analog) Edison cylinder stored the equivalent of 100 bits/mm^2, the CD stores about 1 million bits/mm^2. Above all, the CD established a new fidelity standard that was unprecedented for the consumer with flat frequency response and low distortion. In addition, the CD is highly effective for storing other types of data beyond digital music. But as impressive as the CD is, it is surpassed by its optical disc successors, the Super Audio CD, DVD, and Blu-ray formats.

Development

The chronology of events in the development of the Compact Disc spans almost a decade from inception to introduction. Even then, the development of optical disc storage predates the CD by several more decades. The CD incorporates many technologies pioneered by many individuals and corporations; however, Philips Corporation of The Netherlands and Sony Corporation of Japan must be credited with its primary development. Optical disc technology developed by Philips and error-correction techniques developed by Sony, when merged, resulted in the successful CD format. The original standard established by these two companies guarantees that discs and players made by different manufacturers are compatible. The CD-Audio or CD-DA (Compact Disc Digital Audio) format is sometimes called the Red Book standard (after the color of the notebook used to hold the original specification); it was formalized in 1980. In 1987, it was subsequently also specified in the IEC 908 standard (International Electrotechnical Commission) available from the American National Standards Institute (ANSI).

Philips began working on optical disc storage of images in 1969. It first announced the technique of storing audio material optically in 1972. Analog modulation methods used for video storage were deemed unsuitable, and the possibility of digital signal encoding was examined. Furthermore, Philips established laser readout and small disc diameter as a design prerequisite. Sony, similarly had explored the possibility of an optical, large-diameter audio disc, and had extensively researched the error-processing and channel-coding requirements for a practical realization of the system. Other manufacturers such as Mitsubishi, Hitachi, Matsushita, JVC, Sanyo, Toshiba, and Pioneer advanced proposals for a digital audio disc. By 1977, numerous manufacturers had shown prototype optical disc audio players. In 1978, Philips and Sony designated disc characteristics, signal format, and error-correction methods; and in 1979 they reached an agreement in principle to collaborate (with design meetings from August 1979 through May 1980) with decisions on signal format and disc material. In June 1980, they jointly proposed the Compact Disc Digital Audio system, which was subsequently adopted by the Digital Audio Disc Committee, a group representing more than 25 manufacturers.

Following the development of a semiconductor laser pickup and LSI (large-scale integration) circuits for signal processing and digital-to-analog conversion, the Compact Disc system was introduced in October of 1982 in Japan and Europe. In March 1983, the Compact Disc was made available in the United States. Over 350,000 players and 5.5 million discs were sold worldwide in 1983, and 900,000 players and 17 million discs in 1984, making the CD one of the most successful electronic product launches ever. Starting with the original CD-Audio format, the CD family was expanded to include CD-ROM (1984), CD-i (1986), CD-WO (1988), Video-CD (1994), and CD-RW (1996) with a host of applications in data, audio, and video. The SACD, introduced in 1999, incorporates aspects of the CD.

Overview

The Compact Disc is an efficient information storage system. An audio disc stores a stereo audio signal comprising two 16-bit data words sampled at 44.1 kHz; thus, 1.41 million bits per second (Mbps) of audio data are output from the player along with other nonaudio data. Altogether, the channel bit rate, the rate at which data is read from the disc, is 4.3218 Mbps. A disc containing an hour of music thus holds about 15.5 billion channel bits—a respectable capacity for a disc that costs a few cents to manufacture. Apart from overhead (33% for error correction, 7% for synchronization, and 4% for control and display), a CD-Audio disc holds a maximum of 6.3 billion bits, or 783 million bytes of user information.

A standard Compact Disc measures 12 cm in diameter and has a maximum playing time of 74 minutes, 33 seconds. By varying the CD standards slightly, longer playing times can be achieved. For example, a track pitch of 1.5 m and a linear velocity of 1.2 m/s would yield a playing time of about 82 minutes.

Information is contained in a pit track impressed into one side of the disc's plastic substrate. The substrate is made of polycarbonate plastic (also used for eyeglass lenses). The data surface is metallized to reflect the laser beam used to read the data from underneath the disc. A pit is about 0.6 μm wide (it is worth remembering that a micrometer [1 micron] equals 1-millionth of a meter, or about 40 millionths of an inch) and a disc might hold about 2 billion pits. If a disc were enlarged so that its pits were the size of grains of rice, the disc would be half a mile in diameter. Along the track, each pit edge represents a binary 1; flat areas between pits or areas within pits are decoded as binary 0s.

Data is read from the disc as a change in intensity of reflected laser light. Reading a CD causes no more wear to the recording than your reading causes to the words printed on this page (also conveyed to your eyes via reflected light).

The pits are aligned in a spiral track running from the inside radius of the disc to the outside. CDs with maximum playing times contain data to within 3 mm of the outer disc edge. CDs with shorter playing times have an unused area at the outer edge. This allows a greater manufacturing yield because errors tend to increase at the outer radius, and the disc is oblivious to fingerprints on the empty outer radius. If unwound, a CD track would run for about 3.5 miles. The pitch (distance between adjacent track revolutions) of the CD spiral is nominally 1.6 μm. There are 22,188 track revolutions across the disc's signal surface of 35.5 mm. The period at the end of this sentence would cover more than 200 tracks.

Data is retrieved with an optical pickup. A laser beam is emitted and is guided through optics to the disc data surface. The reflected light is detected by the pickup, and the data from the disc conveyed on the beam is converted to an electrical signal. Because nothing touches the disc except light, light itself and electrical servo circuits are used to keep the laser beam properly focused on the disc surface and properly aligned with the spiral track. The pits are encoded with eight-to-fourteen modulation (EFM) for greater storage density and Cross-Interleave Reed–Solomon code (CIRC) for error correction; algorithms in players provide demodulation and error correction. When the audio data has been properly recovered from the disc and converted into a binary signal, it is input to digital oversampling filters and digital-to-analog converters to reconstruct the analog signal.

Music CDs deliver high-fidelity sound with excellent performance specifications. With 16-bit quantization sampled at 44.1 kHz, players typically exhibit a frequency response of 5 Hz to 20 kHz with a deviation of 0.2 dB. Dynamic range exceeds 100 dB, signal-to-noise ratio exceeds 100 dB, and channel separation exceeds 100 dB at 1 kHz. Harmonic distortion at 1 kHz is typically less than 0.002%. Rotational speed deviation is limited to the tolerances of quartz accuracy, which is essentially unmeasurable. With digital filtering, phase shift is less than 0.5°. D/A converters provide linearity to within 0.5 dB at –90 dB. Excluding unreasonable abuse, a disc will remain in satisfactory playing condition indefinitely, as the medium does not significantly age. Electrical measurements of CD players may be carried out with a variety of techniques, such as those described in the AES17 specification.

One might reasonably ask why 44.1 kHz was selected as the sampling frequency for the Compact Disc. Professional video recorders were originally used to prepare CD master tapes because they were the only recorders capable of handling the high bandwidth requirements of digital audio signals. Because 16-bit digital audio signals (and error correction) were encoded as a video signal, the sampling frequency had to relate to television standards' line and field rate, storing a few samples per scan line. The NTSC (National Television Systems Committee) format used 525 lines in 30 frames per second; only 490 are available for storage. With two samples per line, $490 \times 30 \times 2 = 29.4$ kHz, a too-low sampling frequency. With four samples per line, $490 \times 30 \times 4 = 58.8$ kHz, was considered too high. With three samples per line, $490 \times 30 \times 3 = 44.1$ kHz—it is just right. Moreover, the PAL/SECAM (phase-alteration line/sequential-and-memory) format used 625 lines (588 active lines) in 25 frames per second, and $588 \times 25 \times 3 = 44.1$ kHz as well. Therefore, 44.1 kHz became the universal sampling frequency for CD master tapes. Because sampling-frequency conversion was difficult, and 44.1 kHz was appropriate, the same sampling frequency was used for finished discs.

Disc Design

The CD provides reasonable data density using a combination of the optical design of the disc and the method of coding the data impressed on it. For example, the wavelength of the reading laser and numerical aperture of the objective lens are selected to achieve a small spot size. This allows small pit/land dimensions. In addition, the pit/land track uses a constant linear velocity, and that velocity is set low, to increase the track's linear data density. Also, EFM is used to encode the stored data. Although it creates more channel bits to be stored, the net result is a 25% increase in audio data capacity.

Disc Optical Specification

The Red Book specifies both the physical and logical characteristics of a Compact Disc. The physical characteristics of a CD are shown in Fig. 7.1. Disc diameter is 120 mm, center hole diameter is 15 mm, and disc thickness is 1.2 mm. The innermost part of the disc does not hold data; it provides a clamping area for the player to hold the disc firmly to the spindle motor shaft. Data is recorded on an area that is 35.5 mm wide. A lead-in area rings the innermost data area, and a lead-out area rings the outermost area. The lead-in and lead-out areas contain nonaudio data used to control the player.

A transparent plastic substrate forms most of a disc's 1.2-mm thickness. Data is physically contained in pits that are impressed along its top surface and are covered with a very thin (50 nm to 100 nm) metal (typically aluminum) layer. Another thin (10 μm to

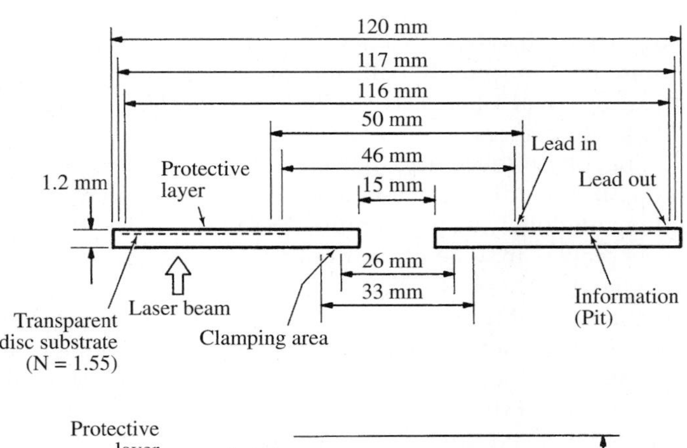

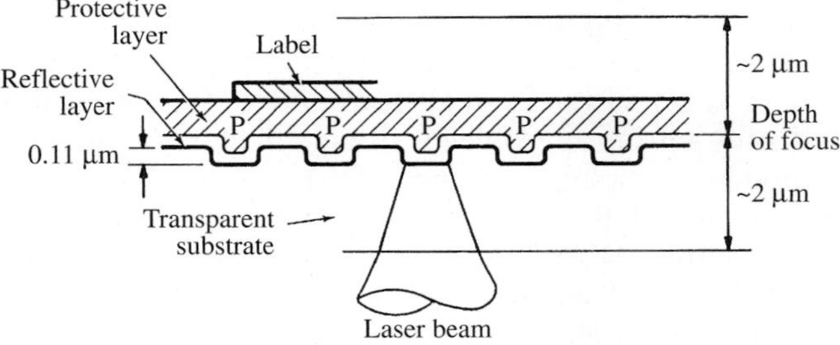

FIGURE 7.1 Physical specification of the Compact Disc showing disc dimensions and the relief structure of data pits.

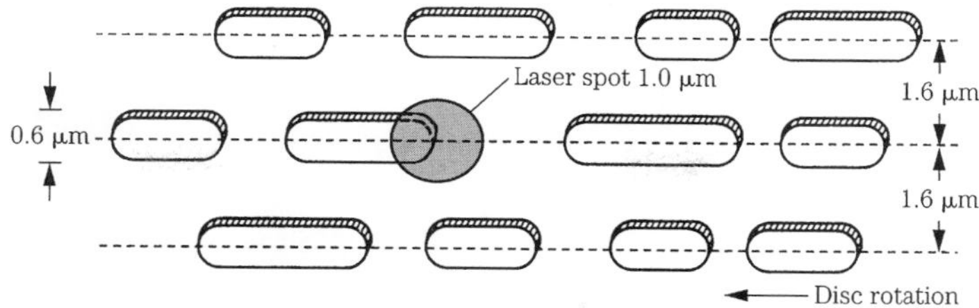

FIGURE 7.2 Data pits are aligned along a spiral track. The laser spot on the data surface has a diameter of approximately 1.0 μm (half-intensity level of the Airy pattern), covering the 0.6-μm pit width.

30 μm) plastic layer protects the metallized pit surface, on top of which the identifying label (5 μm) is printed.

The laser beam used to read data operates at a wavelength of 780 nm. The beam is applied from below the disc and passes through the transparent substrate and back again. The velocity of light decreases when it passes from air to the substrate. The substrate has a refractive index of 1.55 (as opposed to 1.0 for air); the velocity of light slows from 3×10^5 km/s to 1.9×10^5 km/s. When the velocity of light slows, the beam refracts, and focusing occurs. Because of the wavelength of the laser light, refractive index, thickness of the disc, and numerical aperture of the laser lens, the approximately 800-μm diameter of the laser beam on the disc surface is focused to a spot measuring approximately 1.0 μm in diameter (Airy pattern half-intensity level) at the pit surface. The CD is diffraction-limited; that is, the choices of the wavelength of the laser light and numerical aperture of the lens will not permit a smaller spot size.

The laser beam is thus focused to a spot that is slightly larger than the 0.6-micron pit width, as shown in Fig. 7.2. The effects of dust or scratches on the substrate's outer surface are minimized because their size at the data surface is effectively reduced along with the laser beam. Specifically, any obstruction less than 0.5 mm is insignificant and causes no error in the readout. On the other hand, because the disc substrate is part of the playback optics, its optical quality, in terms of birefringence and thickness, must be specified. In addition, because of the relatively large distance between the objective lens and the data surface, disc tilt can cause an error in refraction angle.

Data is physically stored as a phase structure, a metallized surface comprising pits and land. In theory, when the beam strikes the land between pits, virtually all of its light is reflected, and when it strikes a pit, virtually all of its light is canceled, so that virtually none is reflected. As noted in Chap. 6, complete destructive interference in the reflected beam results when the pit depth is such that the intensity of light reflected from a pit equals the intensity of light reflected from the surrounding land, as shown in Fig. 7.3. Specifically, the phase difference forms a diffraction pattern in the reflected light; this causes destructive interference in the main reflected beam. A pit thus reduces the intensity of the reflected light returning to the objective lens. A plane wave model suggests that pit height should be $\lambda/4$ where λ is the apparent wavelength of light. The model predicts that a pit height equal to $\lambda/4$ creates a phase difference of $\lambda/2$ ($1/4 + 1/4$ wavelength path differences) between the part of the beam reflected from the pit and the part reflected from the surrounding land. However, a more complex spherical wave model that accounts for effects of the converging focused beam predicts that the optimum pit depth should be $\lambda/2$.

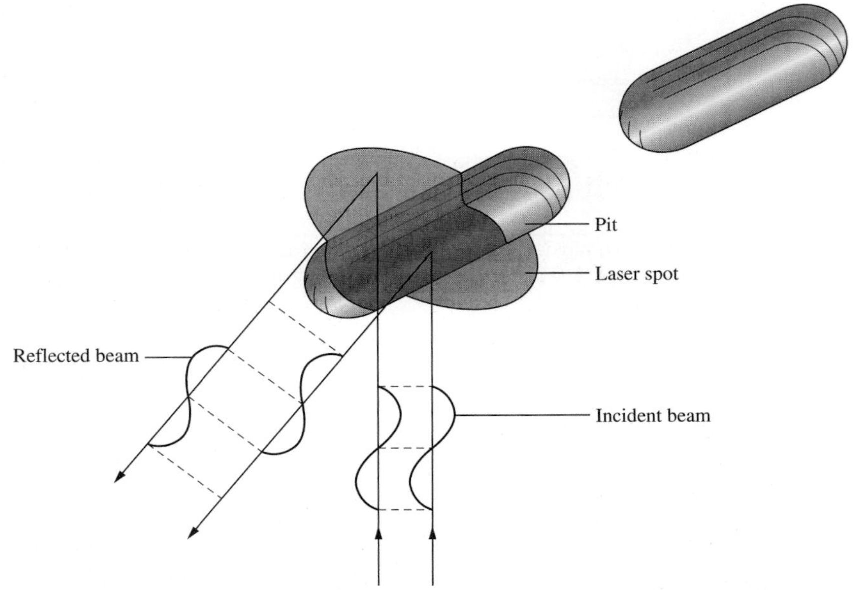

FIGURE 7.3 The laser spot reads data as an intensity modulation of its reflected beam. The phase structure of the data surface places the pit height above the land surface; this creates destructive interference in the reflected beam.

In either case, destructive interference causes an absence of reflected light when there is a pit, distinguishing it from the almost total reflection when the spot strikes the land between pits. In practice, a balance must be made between the data readout advantages of zero reflected pit light, and the reflected intensity that is conducive for signal tracking which requires a $\lambda/8$ pit depth for most pickups. In fact, the specifications for both pit depth and width are a compromise among several factors including optimal high-frequency readout signal, optimal radial tracking signal, and allowance for mass replication. For example, the readout signal should provide good contrast between pit and land areas, but for tracking, the reflected light should not be completely extinguished during a long pit. Moreover, pit geometry must allow the disc to be released from the mold. In practice, pits are made shallower than the theoretically optimal figure and the laser spot is larger than is required for complete cancellation between pit and land reflections. Most CD pressing plants use a pit depth that is approximately one-quarter of the laser's wavelength in the substrate. The laser beam's wavelength in air is 780 nm. Inside the polycarbonate substrate, with a refractive index of 1.55, the laser's wavelength is about 500 nm. Generally, the pit depth may be between 0.11 and 0.13 μm. A long pit causes about 25% of the power of the incident light to be reflected. The reflective flat land typically causes 90% of the laser light to be reflected. When viewed from the laser's perspective (underneath), the pits appear as bumps. In any case, the presence of pits and land is thus read by the laser beam; specifically, the disc surface modulates the intensity of the light beam. Thus, the data that is physically encoded on the disc can be recovered by the laser and then converted to an electrical signal.

Examination of a pit track reveals that the linear dimensions of the track are the same at the beginning of its spiral as at the end. Specifically, a CD rotates with constant linear velocity (CLV), a condition in which a uniform relative velocity is maintained between the disc and the pickup. CLV allows high data density, but necessitates more complex mechanics and also dictates slower access times. The player must adjust the disc's rotational speed to maintain a constant velocity as the spiral radius changes. Because the disc plays from the inner radius to the outer, and each outer track revolution contains more pits than each inner track revolution, the disc rotation must slow down as it plays. When the pickup is reading the inner circumference, the disc rotates at a speed of about 539 rpm (revolutions per minute), and as the pickup moves outward, the rotational speed gradually decreases to about 210 rpm. Thus a constant linear velocity is maintained along the pit track. Moreover, with CLV, the spindle motor must be able to change speed quickly, for example, when a user skips from track 1 at the inner radius to track 12 at the outer.

In other words, all the pits are read at the same speed, regardless of the circumference of that part of the spiral. This is accomplished by a CLV servo system; the player reads frame synchronization words from the data and adjusts the disc speed to maintain a constant data rate.

Although the CLV of any particular CD is fixed, the CLVs used on different discs can range from 1.2 to 1.4 m/s. In general, discs with playing times of less than 60 minutes are recorded at 1.4 m/s, and discs with longer playing times use a slower velocity, to a minimum of 1.2 m/s. The CD player is indifferent to the actual CLV; it automatically regulates the disc rotational speed to maintain a constant channel bit rate of 4.3218 MHz.

Data Encoding

The channel bits, the data physically encoded on the disc, are the end product of a coding process accomplished prior to disc mastering, and then decoded as a disc is played. Whether the original is an analog or digital recording, the audio program is represented as 16-bit pulse-code modulation (PCM) data. The data stream must undergo CIRC error correction encoding and eight-to-fourteen modulation (EFM), and subcode and synchronization words must be incorporated as well.

All data on a CD is formatted with frames. By definition, a frame is the smallest complete section of recognizable data on a disc. The frame provides a means to distinguish between audio data and its parity, the synchronization word and the subcode. Frame construction prior to EFM coding is shown in Fig. 7.4. All the required data is

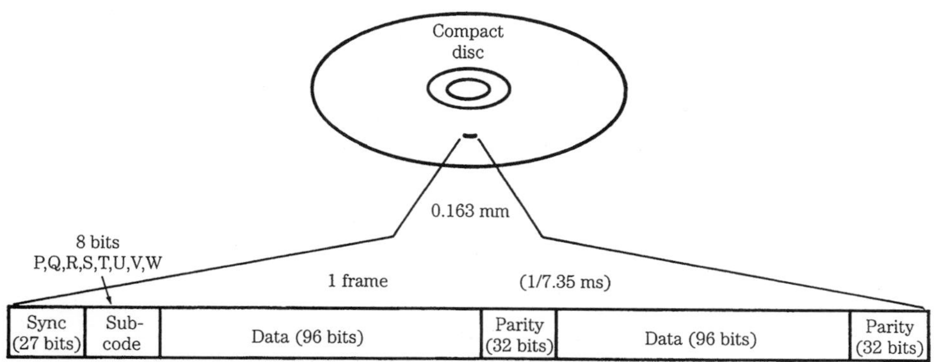

FIGURE 7.4 Elements of a CD frame shown without EFM modulation and interleaving. All data except the synchronization word undergo EFM modulation to create a total of 588 channel bits.

placed into the frame format during encoding. The end result of encoding and modulation is a series of frames, each frame consisting of 588 channel bits.

To begin assembly of a frame, six 32-bit PCM audio sampling periods (alternating between left and right channels) are grouped in a frame. This places 192 audio bits in the frame. The 32-bit sampling periods are divided to yield four 8-bit audio symbols. To scatter possible errors, the symbols from different frames are interleaved so that the audio signals in one frame originate from different frames. In addition, eight 8-bit parity symbols are generated per frame, four in the middle of the frame and four at the end. The interleaving and generation of parity bits constitute the error correction encoding based on the Cross-Interleave Reed–Solomon Code (CIRC). CIRC is discussed in Chap. 5.

One subcode symbol is added per frame; two of these subcode bits (*P* and *Q*) contain information detailing the total number of selections on the disc, their beginning and ending points, index points within a selection, and other information. Six of these subcode bits (*R*, *S*, *T*, *U*, *V*, and *W*) are available for other applications, such as encoding text or graphics information on audio CDs. After the audio, parity, and subcode data is assembled, this data is modulated using EFM. This gives the bitstream specific patterns of 1s and 0s, thus defining the lengths of pits and lands to facilitate optical reading of the disc. EFM permits a high number of channel bit transitions for arbitrary pit and land lengths. This increases data density and helps facilitate control of the spindle motor speed. To accomplish EFM, blocks of 8 data bits are translated into blocks of 14 channel bits using a dictionary that assigns an arbitrary and unambiguous word of 14 channel bits to each 8-bit word. The 8-bit symbols require $2^8 = 256$ unique patterns, and of the possible $2^{14} = 16,384$ patterns in the 14-bit system, 267 meet the pattern requirements; therefore, 256 are used and 11 discarded. A portion of the conversion table is shown in Table 7.1. EFM is discussed in Chap. 3.

8-Bit Data	14-Bit EFM
01100100	01000100100010
01100101	00000000100010
01100110	01000000100100
01100111	00100100100010
01101000	01001001000010
01101001	10000001000010
01101010	10010001000010
01101011	10001001000010
01101100	01000001000010
01101101	00000001000010
01101110	00010001000010
01101111	00100001000010
01110000	10000000100010

Table 7.1 Excerpt from the EFM conversion table. Data bits are translated into channel bits.

Blocks of 14 channel bits are linked by three merging bits. With the addition of merging bits, the ratio of bits before and after modulation is 8:17. The merging bits maintain the proper run length between words, suppress dc content, and aid clock synchronization. Successive EFM words cannot simply be concatenated; this might violate the run length of the code by placing binary 1s closer than 3 periods, or further than 11 periods. To prevent the former, a 0-merging bit is used, and the latter is prevented with a 1-merging bit. Two merging bits are sufficient to maintain proper run length. A third merging bit is used to more effectively control low-frequency content of the output signal. A 1 can be used to invert the signal and minimize accumulating dc offset in the signal's polarity. This is monitored by the digital sum value (DSV); it tallies the number of 1s by adding a +1 to its count, and the number of 0s by adding a –1. The Red Book uses a simple one-symbol look-ahead strategy when choosing a DSV merging bit. An example of a merging bit determination, observing run length and DSV criteria, is shown in Fig. 7.5. Low-frequency content must be avoided because it can interfere with the operation of tracking and focusing servos that operate at low frequencies (below 20 kHz); in addition, low-frequency signals such as from fingerprints on the disc can be filtered out without affecting the data signal itself.

The channel stream produces pits and lands that are at least 2 but no more than 10 successive 0s long. The EFM pit/land family portrait is shown in Fig. 7.6. This collection of pit/land lengths encodes all user data contained on a CD. These pit/land lengths are described as 3T, 4T, 5T, . . . , 11T with T referring to the period of one channel bit. The signal is sometimes called the 3T-11T signal. Physically, pit and land lengths vary incrementally from 0.833 μm to 3.054 μm at a track velocity of 1.2 m/s, and from 0.972 μm to 3.56 μm at a velocity of 1.4 m/s.

The 3T-11T signal represents EFM channel bits on the CD surface. This is accomplished by coding the channel bits as nonreturn to zero (NRZ), and then as nonreturn to zero inverted (NRZI) data. Each logical transition in the NRZI stream represents a pit

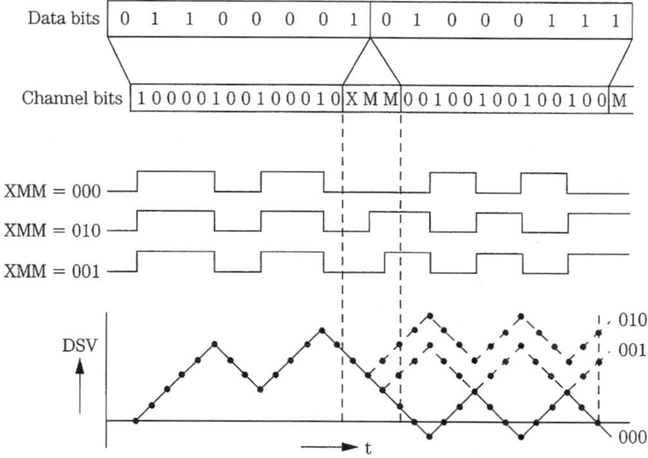

Figure 7.5 An example of merging-bit determination. With this data sequence, the first merging bit is set to 0 to satisfy EFM run-length rules; the two remaining bits are set to 00 to minimize DSV. (*Heemskerk and Schouhamer Immink, 1982*)

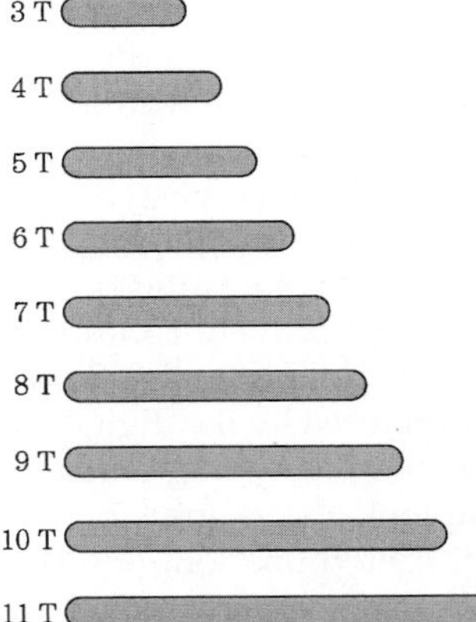

FIGURE 7.6 The complete collection of pit (and land) lengths created by EFM modulation ranges from 3T to 11T. Minimum pit length is 0.833 μm to 0.972 μm; maximum pit length is 3.054 μm to 3.56 μm, depending on velocity (1.2 m/s to 1.4 m/s).

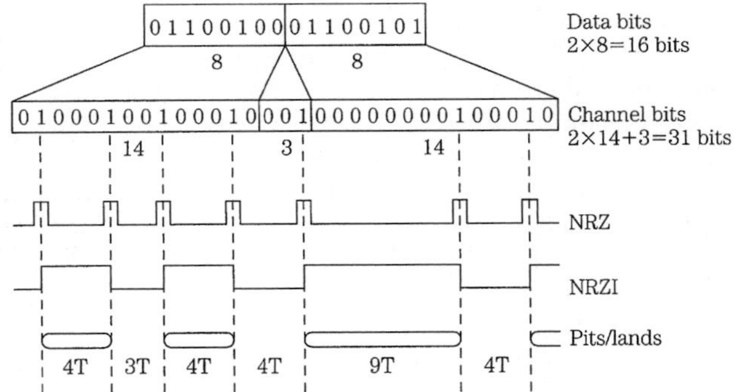

FIGURE 7.7 Each 8-bit half-sample undergoes EFM modulation, three merging bits concatenate 14-bit words, the NRZ representation is converted to NRZI, and transitions are represented as pit edges on the disc.

edge, as shown in Fig. 7.7. The code is invertible; pits and lands represent channel bits equally; inversions caused by merging bits do not affect the data content. When the signal is decoded, the merging bits are discarded. After EFM, there are more channel bits to accommodate, but acceptable pit and land patterns become available. With this modulation, the highest frequency in the signal is decreased; therefore, a lower track velocity can be utilized. One important benefit is conservation of disc real estate.

The resulting EFM data must be delineated, so a synchronization word is placed at the beginning of each frame. The synchronization word is uniquely identifiable from any other data configuration. Specifically, the 24 channel bit synchronization word is 100000000001000000000010 plus three merging bits. With the synchronization word, the player can identify the start of data frames. A complete frame contains one 24-bit synchronization word, 14 channel bits of subcode, 24 words of 14 channel bit audio data, eight words of 14 channel bit parity, and 102 merging bits, for a total of 588 channel bits

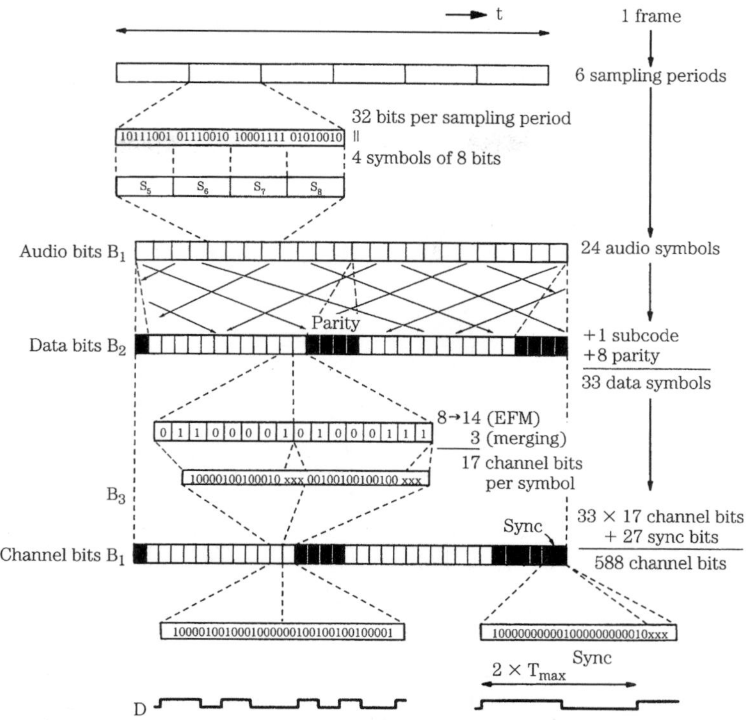

Figure 7.8 The algorithm used in the CD encoding process. Subcode and parity are added to the audio data, the data undergoes interleaving and modulation, and a synchronization word is added. *(Heemskerk and Schouhamer Immink, 1982)*

per frame. Because each 588-bit frame contains twelve 16-bit audio samples, the result is 49 channel bits per audio sample. Thus when the data manipulation is completed, the original audio bit rate of 1.41 million bits per second is augmented to 4.3218 million channel bits per second. This resulting channel bitstream is physically stored on the disc. The entire encoding process is summarized in Fig. 7.8.

A finished CD must contain a lead-in area, program area, and a 90-second lead-out area of silence. The program area holds from 1 to 99 tracks. In addition, each track can contain up to 100 time markers called index points.

Player Optical Design

The function of a Compact Disc player is to recover the data encoded on discs. That task begins at the laser pickup used to read data. In addition, automatic optical tracking and focusing systems must be used. Players generally use either three-beam or one-beam pickup designs. We will consider the more common three-beam design first.

Optical Pickup

The data is recovered from a Compact Disc with an optical pickup, which moves across the surface of the rotating disc. A disc might contain 2 billion pits precisely arranged on

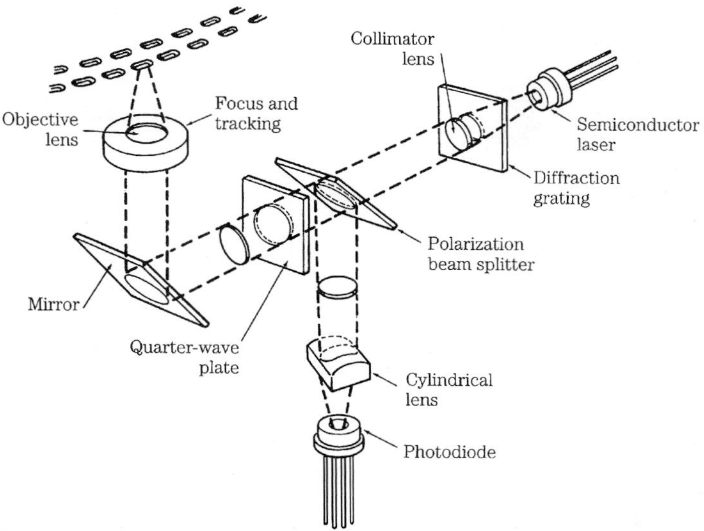

FIGURE 7.9 The optical elements in a three-beam laser pickup.

a spiral track; the optical pickup must focus, track, and read that data track with submicron precision. The entire lens structure, laser source, and reader must be small enough to move laterally underneath the disc surface, moving in response to linear tracking information or user random access track programming. Although particulars vary among manufacturers, pickups are similar in design and operation. A three-beam optical pickup contains a laser diode, diffraction grating, polarization beam splitter, quarter-wave plate, and several lens systems, as shown in Fig. 7.9.

A semiconductor laser is used as the light source. Laser light is monochromatic; the optical system is designed for one wavelength and this minimizes chromatic aberrations. Laser light is coherent and can be focused to a small spot. It also yields a concise interference pattern, and can be manipulated via polarization. The laser beam originates from a laser diode. A CD pickup uses a semiconductor laser with approximately a 5-mW (milliwatt) optical output irradiating a coherent AlGaAs beam with a 780-nm wavelength to yield a spot power on the disc of about 0.5 mW. The light emitting properties of semiconductors have been utilized for many years. By adding forward bias to a PN junction, the injected part of the carrier is recombined to emit light; light-emitting diodes (LEDs) use this phenomenon. However, laser light is significantly different from ordinary light in that it comprises a single wavelength and is coherent with respect to phase. Thus, a modified device is required.

The injection laser diode used in CD players uses a double heterojunction structure. It contains a thin (perhaps 0.1 μm) active layer of GaAs semiconductor, sandwiched between heavily doped P- and N-type AlGaAs materials, sometimes called cladding layers. Forward bias creates a high concentration of electrons (from the N layer) and holes (from the P layer) in the active layer. An inverted population condition is created with many electrons in a high-energy state band and many holes in a low-energy band. Electrons fall to a lower energy band, releasing a photon; this reaches equilibrium with the input energy pumping rate. Stimulated light emission is thus induced. However, the light must be amplified, so several steps are taken. Both sides of the activating layer are sandwiched within materials with a large band gap to enclose the carrier, and the refraction ratio at both boundaries of the activating layer is different to provide enclosure. Also, for amplification within the layer, the crystal surface in the direction of the light emission is made reflective, and acts as a light resonator for continuous wave emission. A monitor photodiode is placed next to the laser diode to control power to the laser, compensating for temperature changes. The monitor diode conducts current proportionally to the laser's light output. If the monitor diode's current output is low with respect to a reference, current to the laser's drive transistors is increased to increase the laser's light output. Similarly, if the monitor current is too high, supply current to the laser is decreased to compensate. The laser diodes used in CD players have a very long life expectancy, from hundreds of thousands, to millions of operating hours.

In a three-beam pickup, the light from the laser point source passes through a diffraction grating. This is a screen with slits spaced only a few laser wavelengths apart. As the beam passes through the grating, it diffracts at different angles. When the resulting collection is again focused, it appears as a bright center beam with successively less intense beams on either side. In a three-beam pickup design, the center beam is used for reading data and focusing, and two secondary beams, the first-order beams, are used for tracking.

A polarization beam splitter (PBS) directs laser light to the disc surface, then angles the reflecting light to the photodiode. For incident light approaching the polarization beam splitter, it acts as a transparent window, but for reflected light with a rotated plane of polarization, it acts as a prism redirecting the beam. The PBS comprises two orthogonal prisms with a common face with a dielectric membrane between them. A collimator lens follows the PBS (in some designs it precedes it). Its purpose is to take the divergent light rays and make them parallel. The light then passes through a quarter-wave plate (QWP), a crystal material with anisotropic properties of double refraction. It rotates the plane of polarization of the incident and reflected laser light; plane of polarization is rotated 45° as light passes through the plate, and then rotates another 45° as reflected light returns through it. The reflected light is thus polarized in a plane at a right angle relative to that of the incident light, thus allowing the PBS to properly deflect the reflected light.

The final piece of optics in the light path to the disc is the objective lens with a numerical aperture of 0.45. It is used to focus the beam to about 1.0 μm (half-intensity level) at the reflective surface, somewhat wider than the pit width of 0.6 μm. The objective lens is attached to a two-axis actuator and servo system for up/down focusing motion and lateral tracking motion.

As noted, when the spot strikes a land interval between two pits, the light is almost totally reflected. When it strikes a pit (a bump from the reading side), a lower-intensity light is returned. Ultimately, a change in intensity is deciphered as a 1 and unchanged intensity as 0. The varying intensity light returns through the objective lens, the QWP (to further rotate plane of polarization), and the collimator lens, and strikes the angled surface of the PBS. The light is deflected and passes through a collective lens and cylindrical lens. These optics are used to direct the operation of the focusing servo system to keep the objective lens at the proper depth of focus. The beam's main function, however, is to carry the data via reflected light to a four-quadrant photodiode. The electrical signals derived from that device are ultimately decoded into an audio waveform.

Autofocus Design

Nothing except laser light touches the data surface. That poses the engineering challenge of focusing on the pit surface and tracking the spiral pit sequence with nothing tangible to guide the pickup. To properly distinguish between pits and land, the laser beam must rely on interference in the reflected beam created by the height of the bumps, a 110-nm difference. The focus of the beam on the data surface is therefore critical; an unfocused condition might result in inaccurate or lost data. Specifically, the laser must stay focused within ±0.5 μm. A disc can contain deviations approaching ±0.4 mm. Thus, the objective lens must be able to refocus as the disc surface deviates. This is accomplished with a servo-driven autofocus system, which utilizes the center laser beam, a four-quadrant photodiode, control electronics, and a servo motor to move the objective lens. An operational diagram of the autofocus system is shown in Fig. 7.10.

Many methods have been devised to maintain focus on the pit track. In many pickups, the optical property of astigmatism is used to achieve autofocus. An astigmatic cylindrical lens has two different focal lengths and this performs the essential trick needed to detect an out-of-focus condition. As the distance between the objective lens

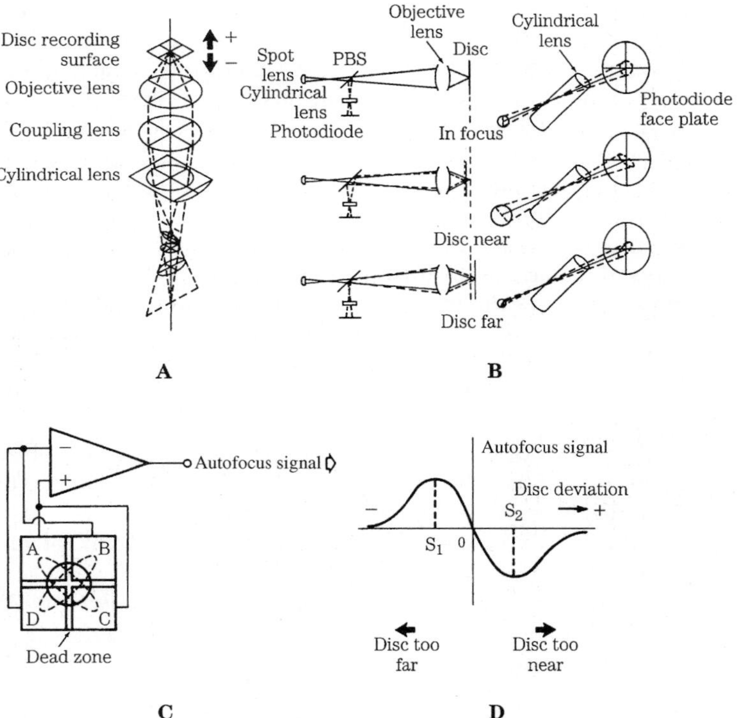

FIGURE 7.10 Astigmatism produced by a cylindrical lens is used to create a correction signal in an autofocus pickup. A. The main beam passes through a cylindrical lens; the image distorts and rotates relative to path length. B. Astigmatism creates an asymmetrical optical pattern because of path-length errors. C. A four-quadrant photodiode converts the optical pattern into an autofocus correction signal. D. The autofocus signal represents disc position and controls a servo to dynamically maintain focus.

and the reflective disc surface varies, the focal point of the system changes, and the image projected by the cylindrical lens changes shape. The change in the image on the photodiode is used to generate the focus correction signal.

When the disc surface lies at the focal point of the objective lens, the reflected image through the intermediate convex lens and the cylindrical lens is unaffected by the astigmatism of the cylindrical lens, and a circular spot strikes the center of the photodiode. When the distance between the disc and the objective lens decreases, the focal points of the objective lens, convex lens, and cylindrical lens move farther from the cylindrical lens, and the pattern becomes elliptical. Similarly, when the distance between the disc and the objective lens increases, the focal points are closer to the lens, and an elliptical pattern again results, but rotated 90° from the first elliptical pattern.

A four-quadrant photodiode reads an intensity level from each of the quadrants to generate four voltages. The value $(A + B + C + D)$ creates an audio data signal. If a focus correction signal is mathematically created to be $(A + C) - (B + D)$, the output error voltage is a bipolar *S* curve, centered around zero. Its value is zero when the beam is precisely focused on the disc; a positive-going focus correction signal is generated as the disc moves away, and a negative-going signal is generated as the disc moves closer. Using a closed-loop system, the difference signal continually corrects the mechanism to achieve a zero-difference signal, and hence a properly focused laser beam.

A servo system moves the objective lens up and down, to maintain a depth of focus within tolerance. A circuit deciphers the focus correction signal and generates a servo control voltage, which in turn controls the actuator to move the objective lens. The objective lens is displaced in the direction of its optical axis by a coil and a permanent magnet structure; it is similar to that used in a loudspeaker except that the objective lens takes the place of the speaker cone. A two-axis actuator incorporates these elements. The top assembly of the pickup is mounted on a base with a circular magnet ringing it. A circular yoke supports a bobbin with both the focus and tracking coils inside. Control voltages from the focus drive circuit are applied to the bobbin focus coil; this moves up and down with respect to the magnet. The objective lens thus maintains its proper depth of focus. The other axis of movement, from side to side, is used to maintain tracking.

Autotracking Design

An autotracking system is used to track the spiral pit sequence. The spiral pit track has a 1.6-μm pitch. An off-center disc might exhibit track eccentricity of over 100 μm. Vibration can further challenge the pickup's ability to track within a ±0.1-μm tolerance. A laser beam system is appropriately used for tracking; any purely mechanical tracking system would be inordinately costly. Many different autotracking methods have been devised. In a three-beam pickup, a design that is widely used, the center beam is split by a diffraction grating to create a series of secondary beams of diminishing intensity. The first-order beams are conveyed to the disc surface along with the central beam. The central beam spot covers the pit track, while the two tracking beams are aligned above and below, and offset to either side of the center beam. During proper tracking, part of each tracking beam illuminates a pit, while the other part illuminates the land between pit tracks. The three beams are reflected back through the QWP and PBS; the main beam strikes the four-quadrant photodiode and the two tracking beams strike two separate photodiodes mounted to either side of the main photodiode. The complete photodiode assembly for data reading, tracking, and focusing is shown in Fig. 7.11.

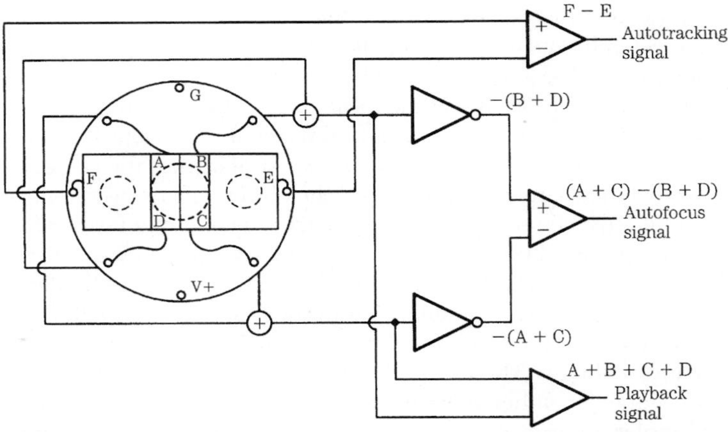

Figure 7.11 The four-quadrant photodiode (*A, B, C, D*) is used for autofocus and data playback. Photodiodes *E* and *F* are used for autotracking.

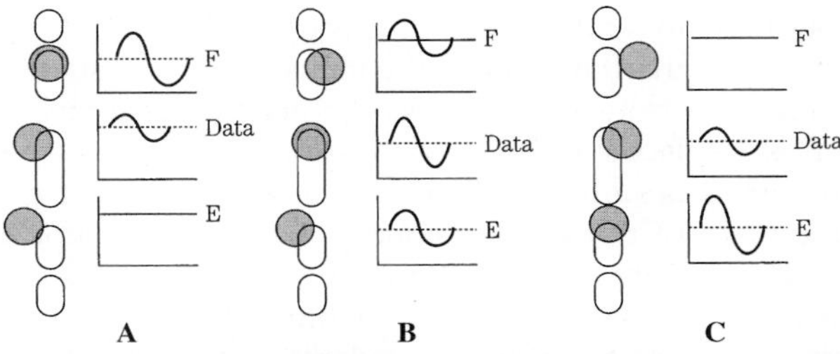

Figure 7.12 A tracking-correction signal is generated from an intensity imbalance in the two secondary beams. A servo system dynamically maintains tracking. A. Left mistracking (F < E). B. Correct tracking (F = E). C. Right mistracking (F > E).

If the three spots drift to either side of the pit track, the amount of light reflected from the tracking beams varies as one of the beams encounters more pit area; this results in less average light intensity. Meanwhile, the other beam encounters less pit area, returning greater reflected intensity. The relative voltage outputs from the two tracking photodiodes thus form a correction signal, as shown in Fig. 7.12. If tracking is precisely aligned, the difference between the tracking signals is zero. If the beams drift, a difference signal is generated, for example, varying positively for a left drift and negatively for a right drift, to create a tracking correction signal. That signal is applied to the two-axis actuator assembly containing the permanent magnet and focus/tracking coil. To correct for a tracking error, the correction voltage is applied to the coil; the bobbin swings around a shaft to laterally move the objective lens so that the main laser spot is again centered, and the tracking correction signal is again zeroed.

One-Beam Pickup

The optical components of a one-beam pickup are shown in Fig. 7.13A, along with the photodiode array used to generate tracking and focusing signals, and read the data

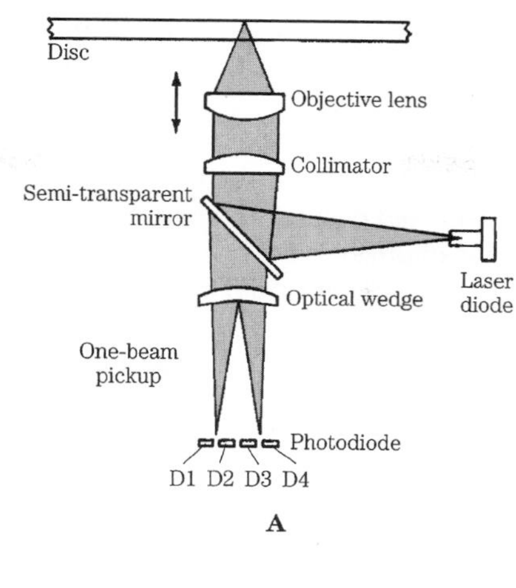

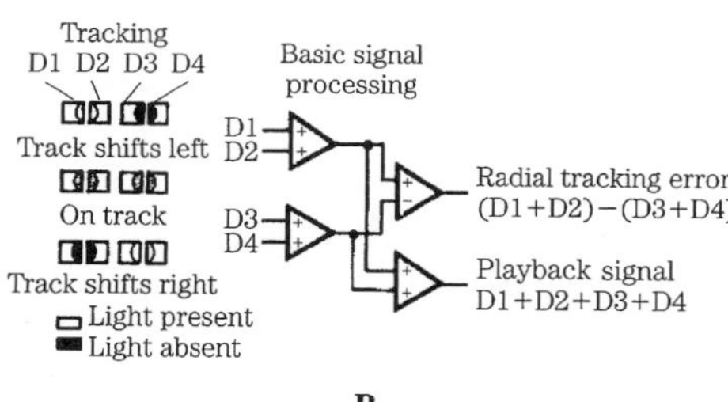

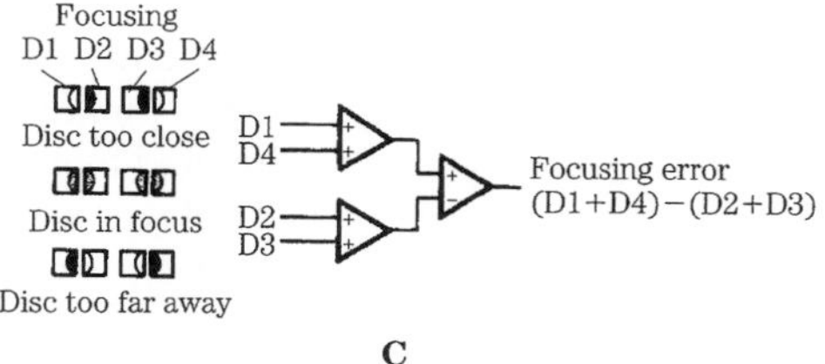

FIGURE 7.13 Design and operation of a one-beam optical pickup. A. The reflected beam is split by a wedge lens and directed to four photodiodes. B. Tracking is accomplished using intensity asymmetry in the beam. C. Focusing is maintained using the angle of deflection between the split beams.

signal. A semi-transparent mirror is used to direct light from the laser diode to the disc surface. Light reflected from the disc passes through the mirror and is directed through a wedge lens. The wedge lens splits the beam into two beams, adjusted to strike an array of four horizontally arranged photodiodes. The outputs of all the photodiodes are summed to provide the data signal ($D_1 + D_2 + D_3 + D_4$), which is demodulated to yield both audio data and control signals for the laser servo system.

Autotracking uses a push-pull technique. A symmetrical beam is reflected when the laser spot is centered on the pit track. If the laser beam deviates from the pit track, interference creates intensity asymmetry in the beam. This results in an intensity difference between the split beams. If the beam is off track, one side of the beam encounters more pit area; hence, greater interference occurs on that side of the beam, and reflected light is less intense there, as shown in Fig. 7.13B. As a result, the split beam derived from that side of the beam is less intense, and the photodiode's output is decreased. The difference between the pairs $(D_1 + D_2) - (D_3 + D_4)$ is used to generate an error signal to correct the pickup's tracking.

The intensity of the reflected beam could become asymmetrical from dirt in the optical system. This would create an offset in the tracking-correction signal, causing the pickup to remain slightly off track. To prevent this, a second tracking-error signal is generated. A low-frequency (for example, 600 Hz) signal is applied to the tracking servo. This signal modulates the output signal from the four photodiodes. If the pickup mistracks, a deviation occurs in the modulated signal. This signal is rectified and used to correct the primary tracking signal with a direct voltage. In this way, the effect of an offset is negated.

Autofocusing uses a Foucault technique. As shown in Fig. 7.13C, when correct focus is achieved, two images are centered between photodiode pairs. When focus varies, the focal point of the system is shifted. When the disc is too far, the split beams draw together; when the disc is too near, the beams move apart. The difference in intensity between diode pairs D_1/D_4 and D_2/D_3 forms a focus error signal $(D_1 + D_4) - (D_2 + D_3)$ that maintains focus of the servo-driven objective lens.

Pickup Control

A motor must precisely move the pickup across the disc surface to track the entire pit spiral. The pickup must also be able to jump from one location on the disc to another, find the desired location on the spiral, and resume tracking. These functions are handled by separate circuits using control signals. Three-beam pickups are mounted on a sled that moves radially across the disc surface. Linear motors are used to position the pickup according to user commands, and bring the pickup within capture range of the autotracking circuit. Most one-beam pickups are mounted on a pivoting arm, which describes an arc across the disc surface. A coil and a magnet are placed around the pivot point of the arm. When the coil is energized, the pickup can be positioned anywhere across the pit track and its precise position corrected by the autotracking circuit. In both three- and one-beam designs, tracking in a CD player is similar to that of an analog LP record player. In the same way that a record groove pulls the stylus across an LP record, the autotracking system pulls the pickup across a CD, keeping the pickup on track.

For fast forward or reverse, a microprocessor assumes control of the tracking servo to provide faster motion than is possible during normal tracking. When the correct location is reached, the *S* curve generated by the tracking correction signal is referenced to a microprocessor-generated control signal, and a signal signifies that proper tracking alignment is imminent. Just prior to alignment, a brake pulse is generated to compensate for the inertia of the pickup. The actuator comes to rest on the correct track, and normal autotracking is resumed.

The reflectivity of discs can vary because of manufacturing process differences, soiling of the player optics, and so on. It is important to maintain a constant voltage level for proper data recovery; thus, the gain of the output control amplifier is variable, depending on the intensity of the reflected laser beam. This gain adjustment is automatically accomplished during the initial reading of the disc table of contents and is maintained while the disc is played. This occurs under control of a microprocessor. For example, the amplifier's gain might be varied by ±10 dB. A control signal from the detection circuit can alert the focus servo system to defective or damaged discs. In severe cases, the objective lens is pulled away from the disc to prevent damage to the pickup.

Player Electrical Design

A CD player's task of reproducing the audio signal requires demodulation and error-correction processing, as well as digital filtering and D/A conversion. Only then is the data recovered from the disc suitable for playback. In addition, controls and displays are required to interface the player with the human user. To simplify operation, and control the many subsystems, players incorporate one or more microprocessors in their design. A block diagram of a CD player is shown in Fig. 7.14.

EFM Demodulation

The voltage from the central photodiode array is output as an electrical data signal. This data signal resembles a sinusoid and is a radio-frequency (RF) signal. The RF signal represents the EFM code and thus contains the data stored on the disc. A collection of EFM waveforms is called the eye pattern, and is shown in Fig. 7.15. The eye pattern is always present whenever a player is tracking data, and the quality of the signal can be observed from the pattern. The RF signal is also used to maintain proper CLV-rotation velocity of the disc. The RF signal is first amplified, and applied to a phase-locked loop to establish the correct timebase and read a valid data signal. The data signal is encoded with EFM, which specifies that the signal be composed of not less than 2 or more than 10 successive 0s between 0/1 or 1/0 transitions. This results in nine different incremental pit lengths from 3 channel bits long to 11 channel bits long. The shortest pit/land length of 3T describes a 720-kHz signal and the longest length of 11T describes a 196-kHz signal (at 1.2 m/s). The large range of pit/land lengths, a range of nearly 400% of the smallest length, allows a substantial tolerance for jitter error (50 ns) during data playback.

The information contained in the eye pattern is shown in Fig. 7.16. Although this signal is comprised of sinusoids, it contains digital information. It undergoes processing

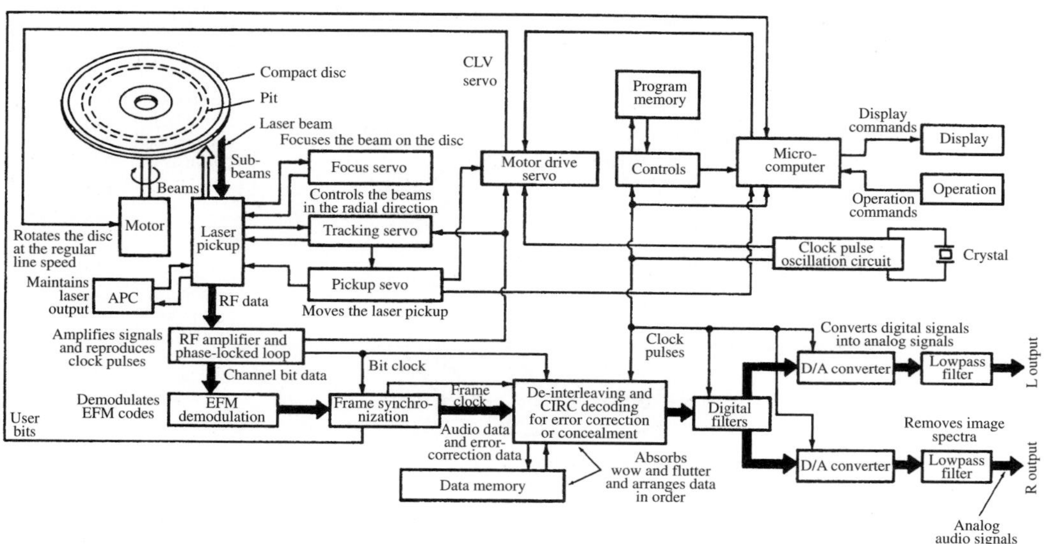

FIGURE 7.14 A CD player's architecture, showing optical processing and output signal processing.

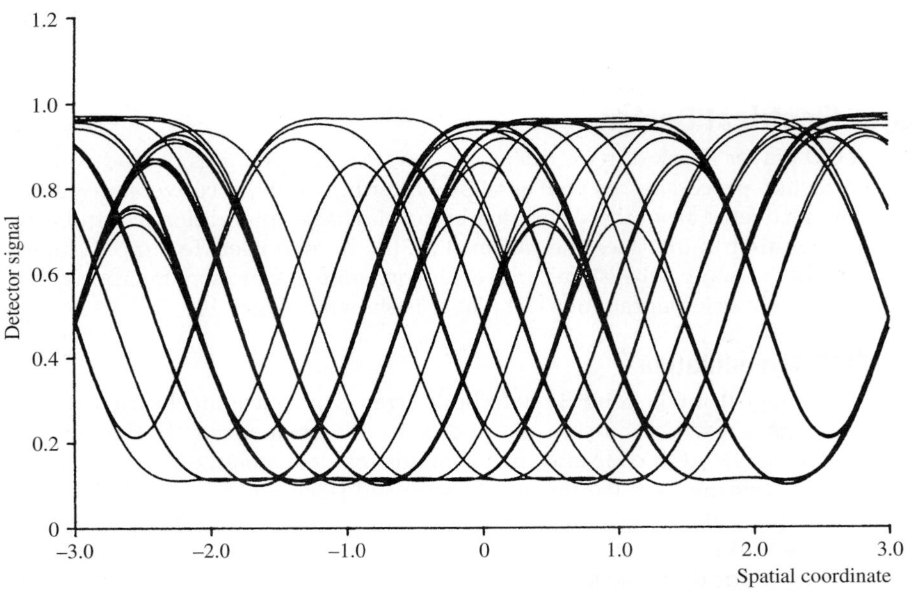

FIGURE 7.15 The modulated EFM data is read from the disc as an RF signal. The RF signal can be monitored through an eye pattern by simultaneously displaying successive waveform transitions. (*Bouwhuis et al., 1985*)

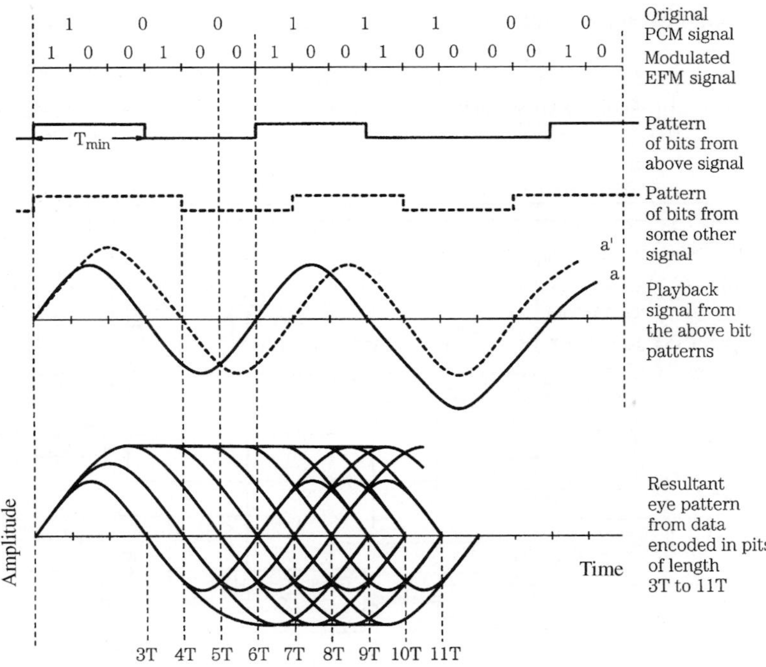

FIGURE 7.16 The RF signal contains all the audio and nonaudio data placed on the disc. The data is recovered by reading the EFM words in the eye pattern.

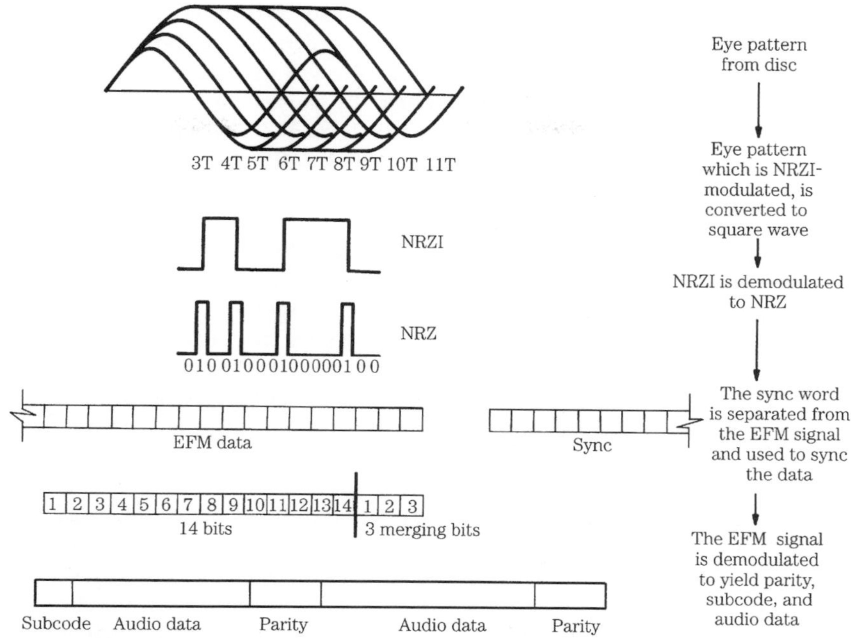

FIGURE 7.17 Demodulation of the EFM eye pattern signal permits recovery of synchronization, subcode, audio, and error-correction data.

to convert it into an NRZI signal, in which the preceding polarity is reversed whenever there is a binary 1. This does not affect the encoded data because the width of the EFM periods holds the pertinent values. The NRZI signal is further converted to NRZ.

Frame synchronization words that were added to each frame during encoding are extracted from the NRZ signal. They are used to synchronize the 33 symbols of channel data in each frame. Merging bits are discarded, and the individual channel bits are used to generate a synchronization pulse. The EFM code is demodulated so that every 14-bit EFM word is converted to 8 bits. Demodulation is accomplished by logic circuitry or a lookup table, using the recorded data to reference back to the original patterns of eight bits. The process from eye pattern to demodulated data is summarized in Fig. 7.17.

During decoding, data is applied to a buffer memory. Disc rotational irregularities might make data input irregular, but clocking ensures that the buffer output is precise. In addition, the buffer can also be used for data de-interleaving. To guarantee that the buffer neither overflows nor underflows, a synchronization control signal controls the disc rotation velocity. By varying the rate of data from the disc, the buffer level is properly maintained. Timebase correction is discussed in Chap. 4.

Error Detection and Correction

Following demodulation, data is sent to a Cross-Interleave Reed–Solomon Code (CIRC) algorithm for error detection and correction. Any error on a disc, for example, a 6T pit misinterpreted as a 7T pit, requires correction. The CIRC error correction decoding strategy uses a combination of two Reed–Solomon code decoders, C1 and C2. The CIRC is based on the use of parity bits and interleaving of the digital audio samples. Depending on implementation, CIRC can enable complete correction of burst errors up to 3874 bits (a 2.5-mm section of pit track). In practice, physical disc damage that would exceed the power of the error-correction algorithm usually causes laser mistracking anyway.

Theoretically, the raw bit-error rate (BER) on a CD is between 10^{-5} and 10^{-6}; that is, there is one incorrect bit for every 10^5 (100,000) to 10^6 (1 million) bits on a disc. Following CIRC error correction, the BER is reduced to 10^{-10} or 10^{-11}, or less than one bad bit in 10 billion to 100 billion bits. In practice, because of the high data density, even a mildly defective disc can exhibit a much higher BER. As discussed in Chap. 5, data is corrected through two CIRC decoders, C1 and C2. The C1 decoder corrects minor errors and flags uncorrectable errors. The C2 decoder corrects larger errors, aided by the error flags. Uncorrected errors leaving C2 are flagged as well. Error-correction flags generated from the CIRC algorithm during CD playback can represent the error rate (from sources such as poor pit geometry and uneven reflectivity) present on a disc.

If the CIRC decoder cannot correct all errors, it outputs the data symbols uncorrected (the parity symbols have been dropped), but marked with an erasure flag. Most of these symbols can be reconstructed with linear interpolation, using the combination of error flags to aid interpolation. The function of these error concealment circuits is to reduce such errors to inaudibility. Only uncorrected symbols, marked with erasure flags, are processed. All valid audio data passes through the concealment circuitry unaffected, except in the case of data surrounding a mute point, which is attenuated to minimize audibility of the mute. Concealment methods vary according to the degree of error encountered, and from player to player. In its simplest form, when a single sample is flagged between two correct samples, mean value interpolation is used to replace the erroneous sample. For longer consecutive errors, the last valid sample value is held, then the mean value is taken between the final held value and the next sample value. The system might permit recovery through adjacent sample interpolation of losses of up to 13,282 bits (8.7-mm track length).

If large numbers of adjacent samples are flagged, the concealment circuitry performs muting on one or more CD frames (1/75 second each). A number of previous valid samples (perhaps 30) are gradually attenuated with a cosine function to avoid the introduction of high-frequency components. Gain is kept at zero for the duration of the error, and then gain is gradually restored. Errors that escape the CIRC decoder without being flagged are not detected by the concealment circuitry, and therefore do not undergo concealment and may produce an audible click in the audio reproduction. Not all CD players are alike in error correction. Any CD player's error correction ability is determined by the success of the strategy devised to decode the CIRC, as well as the concealment algorithm.

The AES28 standard describes a method to estimate the life expectancy of CDs (excluding recordable media) based on the effects of temperature and humidity. In AES28, block-error rate (BLER) is the measured response and the end-of-life criterion is a 10-second average of maximum BLER of 220. The ISO/IEC 10149 and ANSI/NAPM IT9.21-1996 standards also specify this error count.

Output Processing

Following error correction, the digital data is processed to recover subcode information. During encoding, eight bits of subcode information per frame are placed in the bitstream. During decoding, subcode data from 98 frames is read and grouped together to form one block, then assigned eight different channels to provide control and (optionally) text or other information.

Output anti-imaging filtering is accomplished in the digital domain with oversampling filters. In oversampling, data is demultiplexed into left and right channels, and applied to an FIR transversal filter. Through interpolation, additional samples are inserted between disc samples, thus raising the sampling rate. An eight-times rate is common. As a result of oversampling, the output image spectra are raised to the corresponding multiple of the sampling frequency. When shifted to this higher frequency range, they can be easily removed by a low-order analog filter, free of phase distortion. Oversampling filters are discussed in Chap. 4. Following this processing, the data is converted into a format appropriate for the type of D/A converter used in the player. In most CD players, sigma-delta D/A converters are used, as described in Chap. 18.

Subcode

Each demodulated CD frame contains eight subcode bits, containing information describing where tracks begin and end, track numbers, disc timing, index points, and other parameters. The player uses the subcode bits to interpret the information on the disc, and facilitate user control of the player in accessing disc contents.

The eight subcode bits in every frame are designated as *P*, *Q*, *R*, *S*, *T*, *U*, *V*, and *W* as shown in Fig. 7.18A. Only the *P* and *Q* subcode bits are defined in the CD-Audio format. (There is no relation to the *P* and *Q* codes in CIRC.) A subcode block is constructed sequentially from 98 successive frames. Thus the eight subcode bits (*P* through *W*) are used as eight different channels, with each frame containing 1 *P* bit, 1 *Q* bit, and so on. This interleaving minimizes the effect of disc errors on subcode data. The subcode block rate can be determined: a CD codes 44,100 left and right 16-bit audio samples per second, so the 8-bit byte rate is 44,100 × 4, or 176.4 kbytes per second. With 24 audio symbols in every frame, the frame rate is 176.4/24 or 7350 Hz. Because 98 frames form one subcode block, the subcode block rate is 7350/98 or 75 Hz; that is, 75 subcode blocks per second. Parenthetically, 7350 frames per second multiplied by the number of channel bits, 588, results in 4.3218 MHz, the overall channel bit rate.

A subcode block is complete with its own synchronization word, instruction, data, commands, and parity. The start of each subcode block is denoted by the presence of S_0 and S_1 synchronization bits in the first symbol positions of two successive blocks. On most audio discs, only the *P* and *Q* subcode channels contain information; the others are recorded with 0s.

The *P* channel contains a flag bit. It designates the start of a track, as well as the lead-in and lead-out areas on a disc, as shown in Fig. 7.19. The music data is denoted by 0, and the start flag as 1. The length of a start flag is a minimum of 2 seconds, but equals the pause length between two tracks if this length exceeds 2 seconds. Lead-in and lead-out signals tell the player where the music program on the disc begins and ends. A lead-in signal consists of all 0s appearing just prior to the beginning of the music data. At the end of the lead-in, a start flag that is 2- to 3-seconds long appears just prior to the start of music. During the last music track, preceding the lead-out, a start flag of 2 to 3 seconds appears. The end of that flag designates the start of the lead-out and the flag remains at 0 for 2 to 3 seconds. Following that time, a signal consisting of alternating 1s and 0s (at a 2-Hz rate) appears. These signals could be used by players of basic design to control the optical pickup. For example, a player could count start flags placed

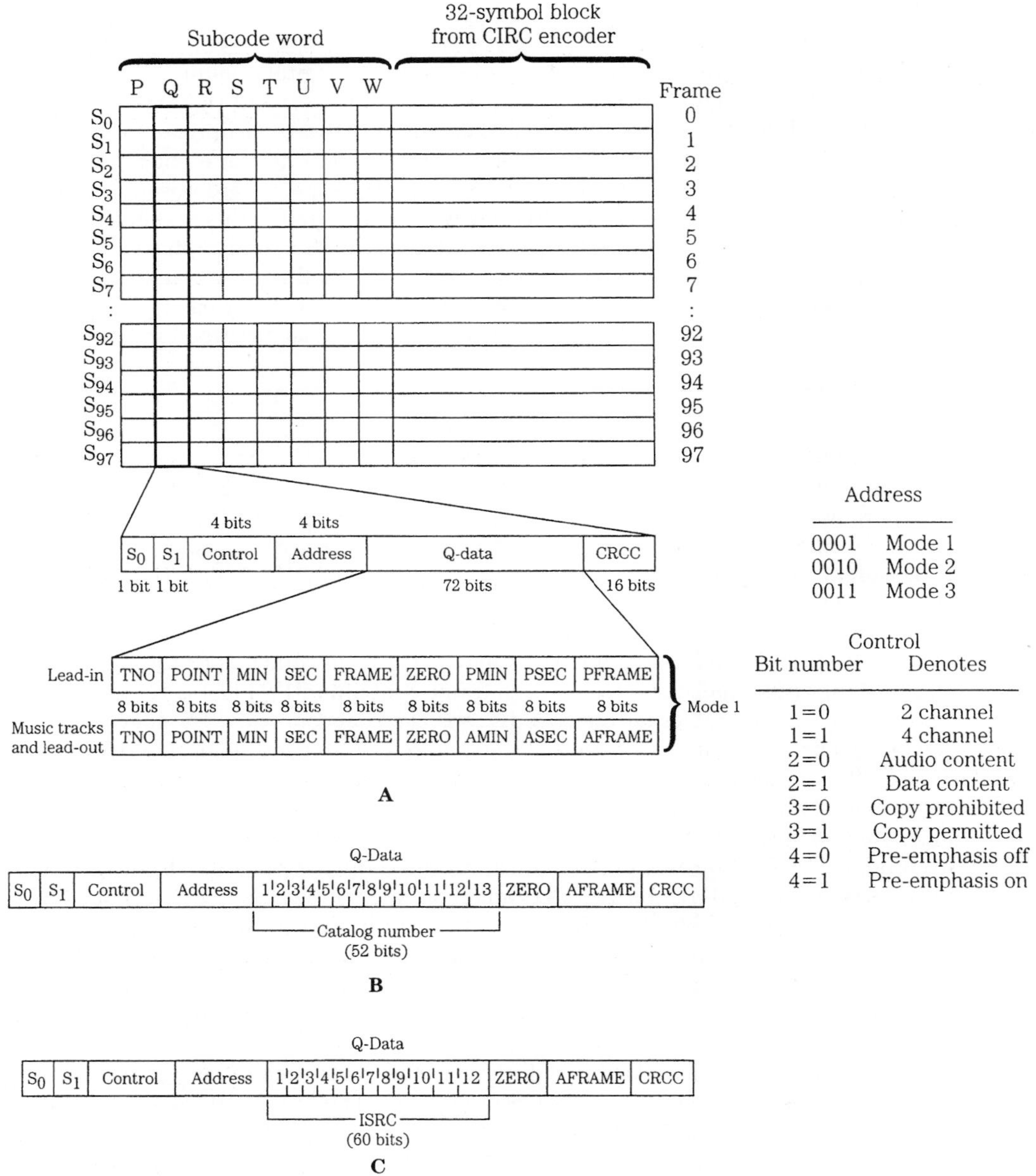

FIGURE 7.18 The data structure of the CD subcode block, showing detail of the *Q* subcode channel. A. Mode 1 has provisions for a lead-in and program format. B. Mode 2 format stores UPC codes. C. Mode 3 format stores ISRC code.

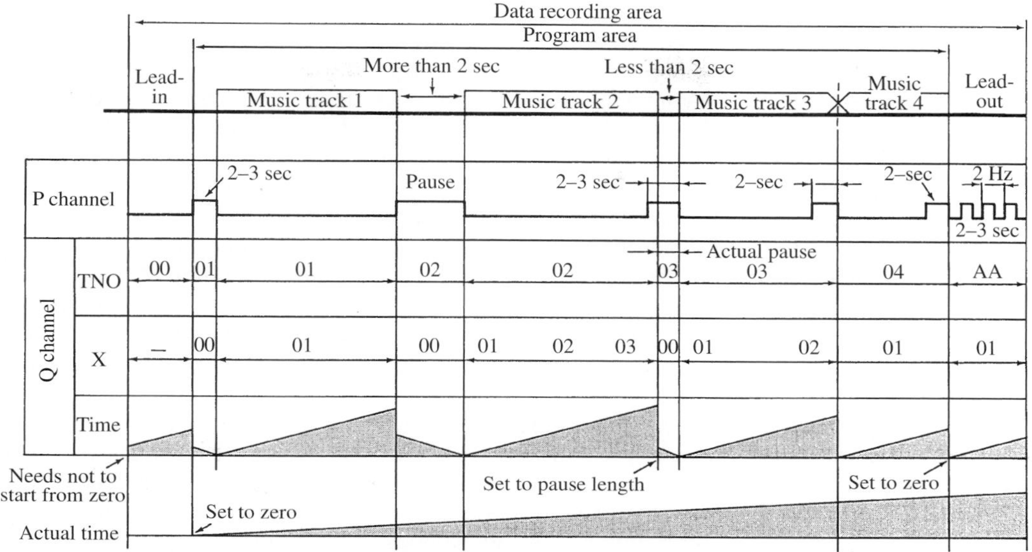

FIGURE 7.19 An example of the program information contained in the *P* and *Q* subcode channels across a disc surface.

in the blank interval between tracks to locate any particular track on a disc. In practice, players use only the more sophisticated *Q* code.

The *Q* channel (see Fig. 7.18) contains four basic kinds of information: control, address, *Q* data, and an error detection code. The control information (four bits) handles several player functions. The number of audio channels (2 or 4) is indicated; this distinguishes between two- and four-channel CD recordings (the latter was never implemented). The digital copy (permit/deny) bit regulates the ability of other digital recorders to record the CD's data digitally. Pre-emphasis (on/off) is also coded. When indicated, the player reads the code and switches to the de-emphasis circuit.

The address information consists of four bits designating the three modes for the *Q* data bits. Primarily, Mode 1 contains number and start times of tracks, Mode 2 contains a catalog number, and Mode 3 contains the International Standard Recording Code (ISRC). Each subcode block contains 72 bits of *Q* data, as described below, 16 bits for the cyclic redundancy check code (CRCC) generation polynomial $x^{16} + x^{12} + x^5 + 1$, used for error detection on the control, address, and *Q* data information in each block.

As noted, there are three modes of *Q* data. Mode 1 stores information in the disc lead-in area, program area, and lead-out area. The data content in the lead-in area (see Fig. 7.18A) differs from that in the other areas. Mode 1 lead-in information is contained in the table of contents (TOC). The TOC stores data indicating the number of music selections (up to 99) as a track number (TNO) and the starting times (P times) of the tracks. The TOC is read during disc initialization, before audio playback begins, so that the player can respond to any programming or program searching that is requested by the user. In addition, most players display this information.

In the lead-in area, the TNO is set to 00, indicating that the data is part of a TOC. The TOC is assembled from the point field; it designates a track number and the absolute starting time of that point in minutes, seconds, and frames (75 frames per second). The times of a multiple disc set can also be designated in the point field. When the point field is set to A0 (instead of a track number) the minute field shows the number of the first track on the disc. When the point field is set to A1, the minute field shows the number of the last track on the disc. When set to A2, the absolute running time of the start of the lead-out track is designated. During lead in, running time is counted in minutes, seconds, and frames. The TOC is repeated continuously in the lead-in area, and the point data is repeated in three successive subcode blocks.

In the program and lead-out area (see Fig. 7.18A) Mode 1 contains track numbers, index numbers (X) within a track, time within a track, and absolute time (A-time). TNO designates individual tracks and is set to AA during lead-out. Running time is set to zero at the beginning of each track (including lead-in and lead-out areas) and increases to the end of the track. Starting at the beginning of a pause, time counts down, ending with zero at the end of the pause. The absolute time is set to zero at the beginning of the program area (the start of the first music track) and increases to the start of the lead-out area. Program time and absolute time are expressed in minutes, seconds, and frames. Index numbers both separate and subdivide tracks. When set to 00, X designates a pause between tracks, and countdown occurs. Nonzero X values set index points inside tracks. A 01 value designates a lead-out area. Using indexing, up to 100 locations within tracks can be indexed. Index 0 marks the onset of the pre-gap (pause) that precedes the audio portion of the track and index 1 marks the beginning of the audio portion. The pre-gap is nominally 2 seconds long. Mode 1 information occupies at least 9 out of 10 successive subcode blocks. (Fig. 7.19 summarizes the timing relationships contained in Mode 1 *Q* channel information.)

In *Q* data Modes 2 and 3 (see Figs. 9.18B and C) the program and time information is replaced by other kinds of data. Mode 2 contains a catalog number of the disc, such as the UPC/EAN (Universal Product Code/European Article Number) codes. The UPC/EAN code is unchanged for an entire disc. Mode 2 also continues absolute time count from adjacent blocks. Mode 3 provides an ISRC number for each track. The ISRC number includes the country code, owner code, year of recording, and serial number. The ISRC code can change for each track. Mode 3 also continues absolute time.

Modes 2 and 3 can be omitted from the subcode if they are not required. If they are used, Mode 2 and Mode 3 must occupy at least one out of 100 successive subcode blocks, with identical contents in each block. In addition, Mode 2 and 3 data can be present only in the program area. The remaining six subcode bits (*R*, *S*, *T*, *U*, *V*, and *W*) are packed with zeros on most CDs. However, they are available for CD+G/M or CD Text data as described below.

Unlike newer formats such as DVD and Blu-ray, the CD was not originally designed to hold extensive text or menu information. Thus, the CD Text feature was appended to the original Red Book specification in June 1996. CD Text allows the album title, song titles, artist, composer, producer, and other text information to be added to a disc at the time of manufacture. Compatible players can use CD Text to display this textual information and also to search for particular album titles. CD Text data is placed in the subcode *R-W* subchannels; it supports a color display of 21 lines of 40 characters each and the option of displaying bitmaps and JPEG pictures. It also permits levels of menus, as well as scrolling lyrics. CD Text was envisioned for numerous applications. For example, catalog number, song title, and artist name can be automatically broadcast via an FM subcode data service. Also, record companies could mark highlighted disc areas for playback at record store listening kiosks. In practice, CD Text is used to display basic album text information.

Other unique approaches can be used to access text information. For example, when using a compatible software CD player, the database at www.cddb.com can be accessed to create metadata files with title, artist, and timing information. When a new disc is loaded, the specific information is accessed over the Internet and then stored locally for subsequent use each time the disc is played. The system creates a unique identifier for every title, based on its running times and number of tracks.

Disc Manufacturing

The Compact Disc manufacturing process enjoys the advantages of the disc medium, in which the information is placed on the disc simultaneously with its creation. However, the CD requires sophisticated manufacturing processes and stringent quality control to guarantee a satisfactory yield. Although manufacturers use different techniques to produce CDs, the manufacturing process always involves three general steps: premastering, disc mastering, and disc replication. Premastering can be accomplished in a recording studio or even on a personal computer. However, disc mastering and replication require specialized equipment found only in disc manufacturing plants. DVD and Blu-ray disc manufacturing is similar to CD manufacturing. The principal difference is the dual-substrate construction of DVD discs, and the presence of additional data layers in some DVD and Blu-ray discs. These differences are discussed in Chaps. 8 and 9. SACD discs are manufactured similarly to DVD discs.

Premastering

Premastering is the culmination of the recording process and the prelude to disc mastering and replication. In premastering, an audio media is prepared prior to creating a glass master disc. This media contains the final, edited version of the content to be replicated. This version should be recorded at the highest resolution possible, on a media suitable for robust storage. A variety of media are used to hold these recordings. Originally, most audio CDs were manufactured from data on 3/4-in U-Matic videotape cassettes; as noted, this accounts for the selection of 44.1 kHz as the CD sampling frequency. Data was formatted using a digital audio processor such as the PCM-1630 recording to a videocassette recorder. The videocassette contained the following information: video format tracks with digital audio data; analog audio channel 1 with *PQ* subcode; and analog audio channel 2 with continuous SMPTE (nondrop frame) timecode.

In many cases, Exabyte data tapes are used to hold the audio recording. Exabyte tapes use specially formulated 8-mm Hi-8 videotape and are also used to archive computer data. Exabyte is attractive because glass masters may be created at faster than real time speeds. In other cases, audio data is held on a hard-disk drive or is delivered by the Internet or other network protocols. For audio content, the Disk Description Protocol (DDP) file format is employed to hold an image file of Red Book data and *PQ* subcode information. Both DDP 1.0 and DDP 2.0 are used; the 2.0 specification writes the table of contents to the end of the tape. Generally, it is recommended to supply a replication plant with an Exabyte 8-mm tape with DDP files (including *PQ* and ISRC data) that has been verified by the artist and producer. In some cases, audio data is written to CD-ROM disc as 24-bit WAV or AIFF files. On request, a test disc can be sent to the artist and producer.

DAT tapes and CD-R discs can be used to deliver audio content to a replication facility, but their relatively higher error rates and susceptibility to damage make them less than ideal. If they are used, finished media must be checked to ascertain the error count. An analog tape can be used, but it must be converted to an interim digital format.

In some cases, the digital source media may not be compatible with the mastering equipment; for example, the sampling frequencies may differ. Although the digital recording could be converted to analog and then to a compatible format, degradation would result; hence, a sampling frequency converter should be used for a digital-to-digital transfer without significant deterioration in signal quality.

In many cases, no matter what media resource is used, data is converted to a DDP file with the necessary *PQ* and ISRC codes and stored on Exabyte 8-mm tape or a hard-disk drive. Care must be taken to ensure that equipment uses a stable central clock, and that the signal path is free of defects. From there, data passes to the laser beam recorder (LBR) for disc mastering.

Disc Mastering

Compact Disc mastering is the first process in disc manufacturing; the entire process is shown in Fig. 7.20. In many cases, a photoresist process is used to create a master disc, employing techniques similar to the microlithography used to manufacture integrated

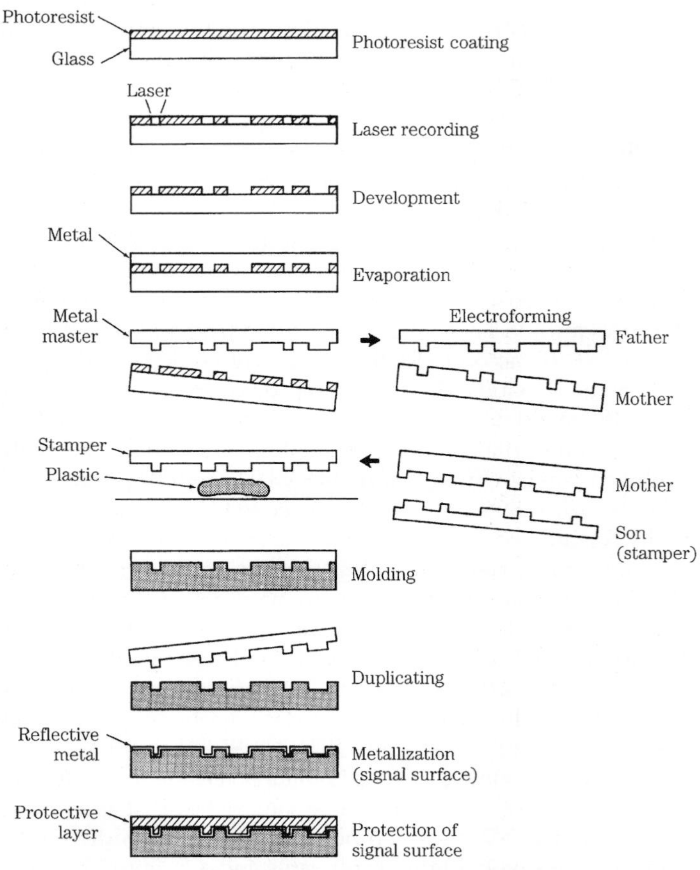

FIGURE 7.20 A summary of the principal steps in the CD disc manufacturing process.

circuits. A glass plate about 240 mm in diameter and 6 mm thick, composed of simple float glass, is washed in alkali and freon, lapped, and polished with a CeO_2 optical polisher. The plate is prepared in a clean room with extremely stringent dust filtering. After inspection and cleaning, the plate is tested for optical dropouts with a laser; any burst dropouts in reflected intensity are cause for rejection of the plate. To prepare the plate for photoresist mastering, an adhesive is applied, followed by a coat of photoresist applied by spin coating. The depth of the photoresist coating is critical because it determines the ultimate pit depth. The plate is cured in an oven and stored, with a shelf life of several weeks. The plate is ready for mastering.

The laser beam recorder is a device that photographically exposes the data spiral into the photoresist on the master glass disc. The LBR may have a control rack consisting of a computer, tape transport, hard-disk drive, CD and subcode encoders, and diagnostic equipment. The recorder may use an HeCd or argon ion gas laser, at 442 nm and 458 nm, or 488 nm, respectively, with a numerical aperture (NA) of 0.9. The laser is intensity-modulated by an acousto-optical modulator to create the exposing signal corresponding to the encoded data. Another laser, which does not affect the master disc photoresist, is used for focusing and tracking. The master glass plate coated with photoresist is placed on the LBR and exposed with the laser to create the spiral track, creating the disc contents as the audio signal is played and CD-encoded. During disc mastering, the *PQ* subcode is uniquely created for the glass master using a subcode editor, and is modulated into the CD bitstream.

Quality of the production discs depends directly on the LBR's signal characteristics such as eye pattern symmetry, signal modulation amplitude, and track following. The length, width, and edge angle of the pits on the replicated discs are subject to the exposure intensity and developing time of the photoresist. To guard against disc contamination, stringent air filtering is used inside the LBR. Although the optics are similar to those found inside consumer CD players (laser source, polarization beam splitter, objective lens), the mechanisms are built on a grander scale, especially for isolation from vibration. For example, the stylus block may be supported and moved by an air-float slider. The entire mastering process is accomplished automatically, under computer control.

After exposure in the laser beam recorder, the glass master is developed by an automatic developing machine. Developing fluid washes the rotating disc surface, removing the exposed areas of photoresist, leaving pits in the photoresist. During development, a laser monitors photoresist depth and stops development when the proper engraving depth has been reached; the developing pits form a diffraction grating, reflecting multiple beams. The relative intensity of the beams is monitored to indicate pit geometry. As noted, compromises must be made to determine the optimum practical pit depth. A production pit depth of 0.11 μm to 0.13 μm is typical. Following etching, a metal coating, usually of silver, is evaporated onto the photoresist layer. The master glass plate is ready for electroforming.

In some cases, a nonphotoresist (or "direct-effect") mastering technique is used. Using a glass plate coated with a dye-polymer recording layer, the LBR directs the input signal to an acousto-optical modulator that controls the recording laser. Pits are physically cut directly into the recording layer on the master disc using a blue laser. The system provides a direct-read-after-write (DRAW) function so that the recorded signal can be continuously monitored during mastering. A trailing red laser is focused on the disc just behind the recording laser, but does not affect the recording. Instead, it reads data so that analysis equipment can dynamically control cutting laser power and other critical parameters to ensure optimum pit geometry and decode the EFM signal to evaluate data error rates; this decreases production time.

Electroforming

The metallized master disc is transferred to an electroplating room where the plating process produces metal stampers. First, evaporation is used to coat the master disc with a silver or electroless nickel layer to make it electrically conductive. The master electroplating process imparts a nickel coating on the metallized glass master. The metal part is separated from the glass master, and any photoresist is removed. This metal disc is called the metal master, or father. Using the same electroplating process, the resulting metal father is used to generate a number of positive nickel impressions, called mothers. The process is repeated to produce a number of negative impression stampers, or sons. Disc substrates are replicated from these stampers. A center hole is precisely punched into the stamper. Stampers are about 300 μm thick—a thick metal foil. Perhaps 30,000 discs may be made from a stamper before wear limits its use.

Disc Replication

Mass production of discs can be accomplished with injection molding to produce disc substrates. A polycarbonate material is used. It is rugged, is one of the most stable of polymeric plastics, and in particular has a low vapor absorption coefficient that is about 70% less than that of polymethyl methacrylate (PMMA) plastic, also known as Plexiglass. Polycarbonate material has an inferior birefringence specification, especially when produced by injection molding; however, injection molding is a more efficient production method. After experimentation with different kinds of mold shapes and molding conditions, techniques for producing a single piece polycarbonate substrate were achieved. CD birefringence is specified to be less than 100 nm (measured double pass through the substrate). Polycarbonate pellets are heated to about 300°C; molten polycarbonate is injected into the mold cavity, faced on one side by the metal stamper, and the disc substrate (with pits) is produced; water channels in the molds help cool the substrate in less than 5 seconds. The substrate center hole is formed simultaneously.

After molding, a metal layer (about 50 nm to 100 nm thick) is placed over the pit surface to provide reflectivity. In most cases, aluminum is used; however, silver or gold or another metal can be used. The reflection coefficient of this layer, including the polycarbonate substrate (note that the CD player laser must pass through the substrate to the metal layer), is specified to be at least 70%. Aluminum evaporation can be accomplished with vapor deposition in a vacuum chamber. Alternatively, high-voltage magnetron sputtering can be used to deposit the reflective layer. A cold solid target is bombarded with ions, releasing metal molecules that coat the disc. Using high voltages, a discharge is formed between a cathode target and an anode. Permanent magnets behind the cathode form a concentrated plasma discharge above the target area. Argon ions are extracted from the plasma; these ions bombard the target surface, thus sputtering it; the disc is placed opposite the target and outside the plasma region. Metallization may take 3 seconds. The metal layer is covered by an acrylic plastic layer with a spin-coating machine, and cured with an ultraviolet light. This layer protects the metal layer from scratches and oxidation. The label is printed directly upon this layer.

The final step in CD manufacturing is inspection and packaging. Finished discs are inspected for continuous and random defects, using automated checking. Discs can be scanned to check for physical defects such as inclusions or bubbles in the substrate, missing metallization and staining, to evaluate the replication process. Scanners can use laser light, regular light, and cameras to quickly check discs. Because no data is read, scanning is fast and thus is often incorporated in the production line to check every disc. Disc readers check error rates, tracking, jitter, and data signal levels. This is more time-consuming and is typically done off-line on selected discs. Other off-line tests can check thickness variation, dynamic balance, and other parameters.

A number of optical, mechanical, and electrical criteria have been established. Molded discs are checked for correct dimensions, lack of flash and burrs, birefringence, reflectivity, flatness (skew angle), and general appearance. The pit surface is checked for correct pit depth, correct pit volume, and pit form and dimensions. The metallized coating is checked for pinholes and uneven thickness, and uneven or incorrect reflectivity. Birefringence can be checked with a circularly polarized light used to convert the phase change to an intensity variation measured with a photodiode.

Angle deviation measures the angle formed by the normal to the disc in the radial direction. This angle is critical because any deviation causes the reflected laser beam to deviate from its return path through the objective lens. This angle deviation could result from an improper manufacturing method; specifications call for a maximum angle of ±0.6°. Disc eccentricity measures the deviation from circularity of the pit track and the positioning of the center hole. Eccentricity may result if the stamper is not punched correctly or if the stamper is not precisely centered in the injection mold. Also, the electroforming or molding processes introduce some eccentricity in the shape of the pit track. In addition, the player's positioning of the disc in the drive might introduce eccentricity. If it is excessive, it could exceed the ability of the radial tracking servo of the player. Tolerances for deviation from circularity call for maximum eccentricity of ±70 μm. Disc eccentricity must also account for alignment of the center hole. Specifications call for a center hole tolerance of 0.1 mm. A hole that is off-center can lead to disc imbalance, and noise and resonance errors. Push-pull tracking evaluates the intensity of light returning from the left and right sides of the pit track; it can be used to monitor pit geometry, which affects the overall gain of the tracking servo.

Disc quality can be evaluated by examining the analog RF 3T-11T signal output from a pickup. The I11 signal is derived from reading an 11T pit/land and an I3 signal is derived from a 3T pit/land. ITOP measures the distance from the signal's baseline to the top of the amplitude of an I11 signal. The I3/ITOP ratio must be between 0.3 and 0.7, and the I11/ITOP ratio must be greater than 0.6. The higher the value, the better the signal's condition. Radial noise measures how much a pickup moves side to side to maintain tracking, and thus evaluates the straightness of a pit track. The maximum value is 30, and lower figures are better. Push-pull magnitude measures the magnitude of the tracking signal, which is determined by pit depth. Shallow pits yield a high push-pull magnitude, and deep pits yield a low push-pull. The minimum value for push-pull is 0.04 and the maximum value is 0.09. Crosstalk measures the interference from adjacent pit tracks; it increases as the track pitch is reduced. The maximum value for crosstalk is specified at 50%. Jitter measurements can monitor pit accuracy; maximum peak-to-peak jitter should be less than 50 ns (for modulation frequency of the channel bit clock frequency greater than 4 kHz).

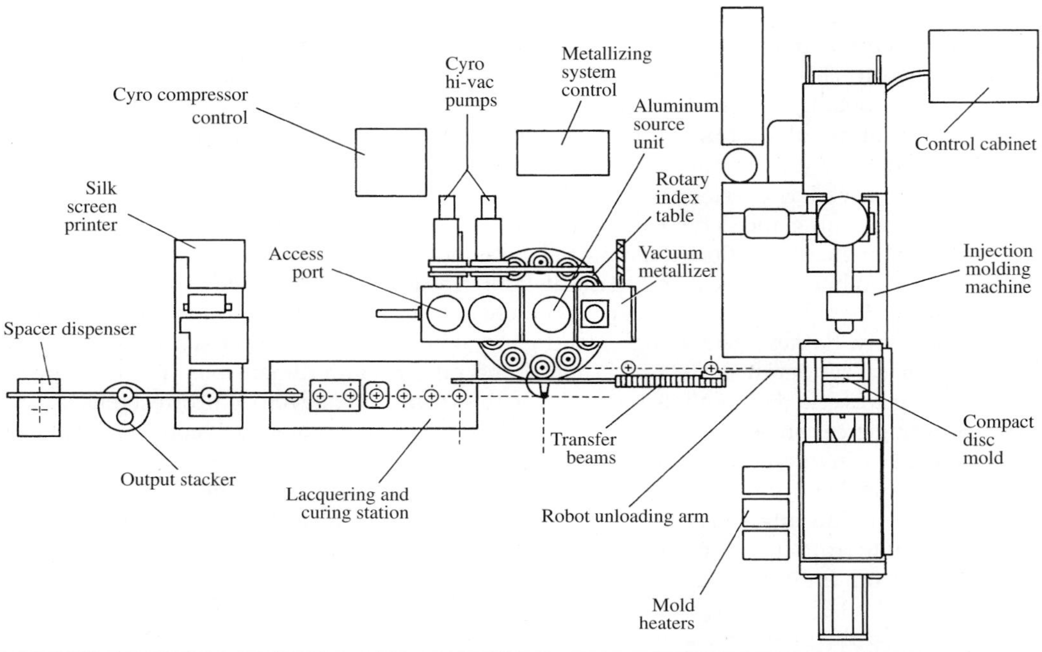

Figure 7.21 An example of the in-line hardware used to manufacture Compact Discs. Critical processes are enclosed in small clean enclosures.

Following packaging and wrapping, discs are ready for distribution. In most cases, the injection machine, sputtering machine, spin coater, and label printer are consolidated into one production unit; it might take a disc 2 minutes to travel from the injection machine to labeling; one unit can produce 2 million discs per year. Equipment used for disc replication is shown in Fig. 7.21.

Alternative CD Formats

The Compact Disc is an efficient storage medium allowing user information to be reliably stored on a low-cost disc using CIRC error correction and EFM coding techniques. Fortuitously, that medium is available for other storage applications beyond the CD-Audio format (also called CD-DA or CD-Digital Audio). Computer software, published material, or audio and video files can be stored in the CD-ROM file format. The CD-R standard specifies a write-once disc format, and the CD-RW standard describes a fully erasable disc format. As with many families, the interrelationships between members of the CD family are somewhat complicated, as shown in Fig. 7.22. Complications and incompatibilities arose because the CD was originally conceived only as a music carrier; subsequent evolutions occurred in a piecemeal fashion. In contrast, newer formats such as DVD and Blu-ray disc were initially designed for multiple uses. Despite their drawbacks, these alternative CD formats greatly expand the range of applications open to the CD.

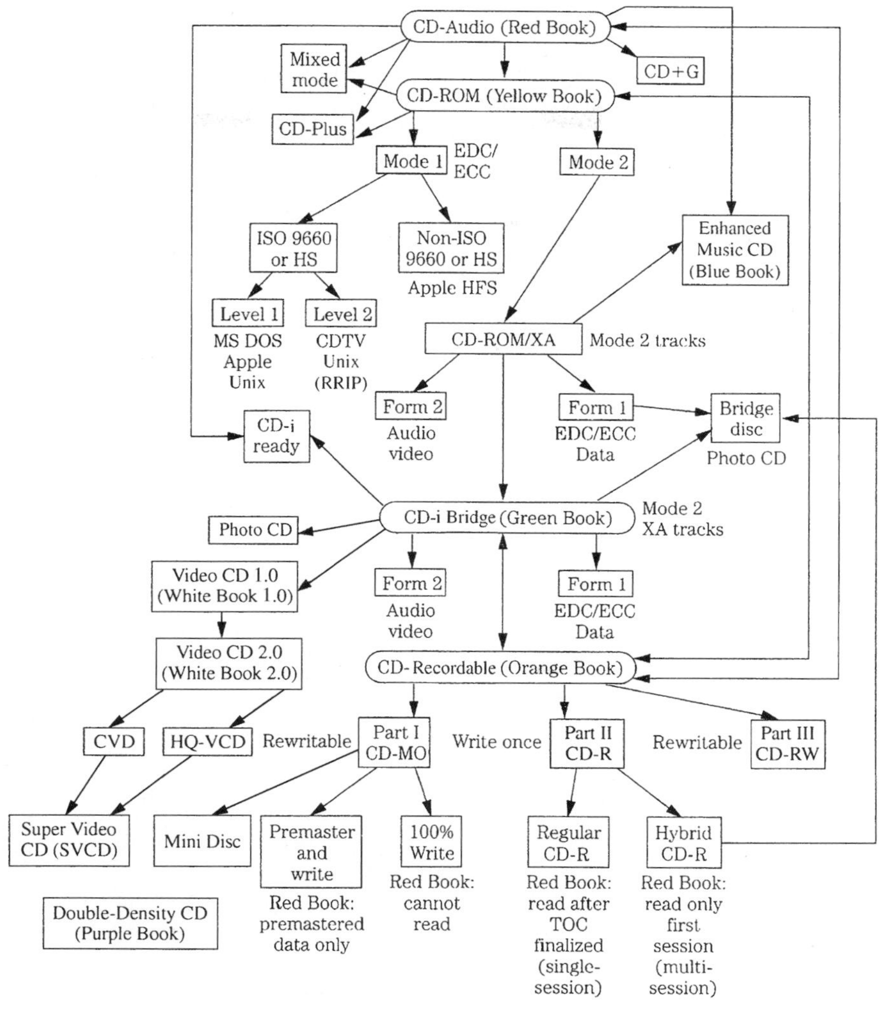

FIGURE 7.22 A simplified road map showing the complex interrelationships between CD formats.

CD-ROM

The Compact Disc Read-Only Memory (CD-ROM) format extends the digital audio CD format to the broader application of information storage in general. Rather than store only music, the CD-ROM format is used for diverse data. The CD-ROM format forms the basis for a read-only electronic publishing medium applicable to computer applications and for information distribution such as book publishing, dictionaries, technical manuals, business catalogs, and so on. Its advent represented an entirely new technology of information dissemination. CD-ROM discs use the same disc construction as audio discs, and can be mass produced with the same replication equipment; however, more stringent quality control may be required.

The CD-ROM standard is derived from the CD-Audio standard, but specifically defines a file format for general data storage. Unlike CD-Audio, CD-ROM is not tied to any specific application. Both standards use the 120-mm-diameter disc, but with different data formats. The CD-ROM standard, sometimes called the Yellow Book, was issued in 1983. In 1989, it was also specified in the ISO/IEC 10149 (ECMA-130) standard (International Organization for Standardization/ International Electrotechnical Commission).

A Mode 1 CD-ROM disc nominally holds 682 million bytes of user information (333,000 blocks × 2048 bytes). This storage area is roughly equivalent to 275,000 pages of alphanumerics. The CD-ROM format can store information such as computer applications software, audio files such as WAV or MP3, video files, operating systems, online databases, published reference materials, directories, encyclopedias, libraries of still pictures, parts catalogs, or other types of information. Read-only CD-ROM discs form a publishing medium that is much more efficient than paper. For example, the U.S. Navy investigated the use of CD-ROM to reduce the paperwork on naval ships. They found that a cruiser carries about 5.32 million pages of documentation, weighing almost 36 tons. That mass of paperwork could be reduced to about 20 CD-ROM discs, weighing 280 grams. On the other hand, the CD-ROM is not ideal for computer applications. The file sizes are not an exact power of 2 as computers prefer, interleaving dictates that large amounts of data must be read to recover any useful information, CLV rotation requires motor speed changes as data is accessed across the disc radius, and CD-ROM is not erasable. However, the CD-ROM data format is widely used to write data to CD-R and CD-RW discs.

The CD-ROM standard uses a data format modified from the CD-Audio standard. Ninety-eight CD frames are summed (as in CD subcode) to form a data block of 2352 bytes (24 byte × 98 frames) in length. A disc is divided into a maximum of 330,000 blocks; a 60-minute disc holds 283,500 blocks. The first 12 bytes of a block form a synchronization pattern, and the next 4 bytes comprise a header field for time and address flags. The remaining 2336 bytes can store user data, or data plus extended error correction, depending on the mode selected. The header contains three address bytes and a mode byte. Addresses are stored as a disc playing time. One address byte stores minutes, the second byte stores seconds, and the third stores block numbers within the second. For example, an address of 62-13-08 identifies the 8th block in the 13th second of the 62nd minute on the disc.

The mode byte identifies three modes, used for two different types of data. There are two data modes, as shown in Fig. 7.23. The Mode 1 format assigns 2048 bytes of each block to user data. Each block contains 2 kbytes (2 × 1024) of user data; 280 bytes are given to extended error detection and correction (EDC/ECC), which is an extra layer of coding in addition to the basic Red Book CIRC code. The Mode 2 format allows for the full 2336 bytes to be used for user data (14% more data), but in practice is rarely used except when coded in CD-ROM/XA mode (described below). There is also a null mode, Mode 0. In all cases, after sector data is created, the CD-ROM bitstream is applied to conventional CD encoding such that CIRC and EFM, and other processing is applied just as in an audio CD. For example, Mode 1 data thus has two independent layers of error correction (EDC + ECC and CIRC) whereas Mode 2 uses only CIRC coding.

Because of its extended error correction, Mode 1 has the greatest applications. The EDC/ECC field is essential for high-density numerical data storage, which is more demanding than audio data. A GF(2^8) Reed–Solomon Product Code (RS-PC) is used to encode each block. It produces *P* and *Q* parity bytes with (26,24) and (45,43) codewords, respectively. Because the EDC/ECC field is independent and supplements the CIRC error correction code applied to the frame structure, the error rate is improved over that

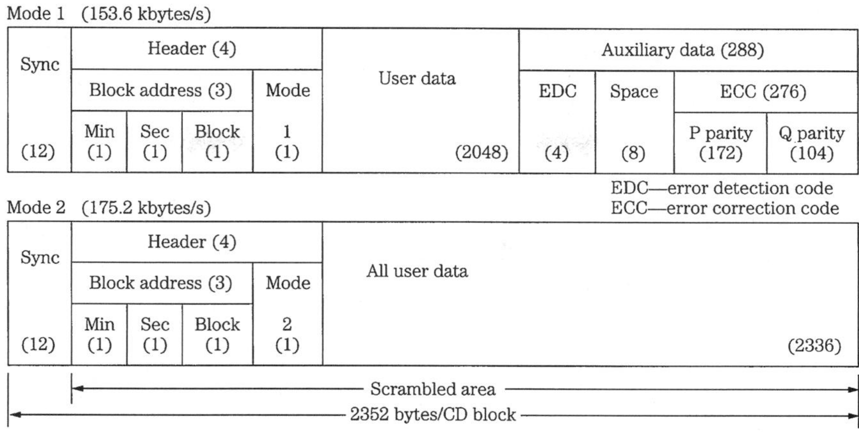

This area is equivalent to 1/75 s in CD audio, i.e., 16 bits × 2 channel × 44.1 kHz × 1/75 • 18,816 bits • 2352 bytes

1 mode per track
1—99 tracks per disc

Figure 7.23 The CD-ROM specification contains two modes of data block structures. Mode 1 allows for extended error detection and correction, and Mode 2 provides capacity for additional user data.

of CD-Audio. In Mode 1, the typical CD-ROM bit-error rate is approximately 10^{-15}, one uncorrectable bit in every 10^{15} bits.

CD-ROM/XA (eXtended Architecture) is an extension to the Yellow Book Mode 2 standard and defines a new type of data track; computer data, compressed audio data, and video and picture data can all be contained on one XA track. CD-ROM/XA Mode 2 differs from CD-ROM Mode 2 because it provides a subheader that defines the block type, as shown in Fig. 7.24. In this way, the XA track can interleave Form 1 and Form 2 blocks; this is useful in some applications. Specifically, XA defines two types of blocks: Form 1 for computer data and Form 2 for compressed audio/video data. The former provides a 2048-byte user area, and the latter provides 2324 bytes. CD-Audio data cannot be placed on an XA track. The XA data rate is 1.4 Mbps. Clearly, special processing is needed to decode the various data types found on an XA disc. Some players are dedicated to specific types of CD-ROM/XA discs; the Video CD and Photo CD are types of CD-ROM/XA. The CD-ROM/XA format is defined in the White Book. Not all CD-ROM drives support CD-ROM/XA; in some cases a special interface board must be used.

Hybrid audio/data CD formats, sometimes called CD Extra (formerly CD Plus), Enhanced CD, and Stamped Multisession or Mixed Mode, combine several different format types (such as CD-Audio and CD-ROM/XA) on a single disc. For example, a CD Extra disc has Red Book audio data in the first session, with Yellow Book ROM/XA Mode 2, Form 1 format in the second session. Each individual session must use the same data type. A CD-Audio player plays the first session, but will not play the second. A CD-ROM drive reads both the audio and nonaudio sessions, the latter containing, for example, programming relating to the audio session. For PC and Macintosh compatibilities, a hybrid disc would contain both ISO 9660 and HFS directories with common files such as video that can be shared between platforms.

An Enhanced CD disc is essentially a replicated multisession Orange Book disc, in which each session has lead-in and lead-out areas. CD Extra discs must contain the

Mode 2 Form 1

Sync	Header (4)				Sub-header 1	User data	Auxiliary data (280)		
	Block address (3)			Mode 2			EDC	ECC (276)	
	Min	Sec	Block					P parity	Q parity
(12)	(1)	(1)	(1)	(1)	(8)	(2048)	(4)	(172)	(104)

Mode 2 Form 2

Sync	Header (4)				Sub-header 2	User data	EDC (optional)
	Block address (3)			Mode 2			
	Min	Sec	Block				
(12)	(1)	(1)	(1)	(1)	(8)	(2324)	(4)

← 2352 bytes/CD block →

Subheader

File	Channel	Submode	Data type
(2)	(2)	(2)	(2)

Figure 7.24 The CD-ROM/XA data format is based on the CD-ROM Mode 2 format. It provides two forms: extended error detection and correction, and increased user data capacity.

AUTORUN.INF file to start the multimedia application, as well as CDPLUS and PICTURES folders. The former contains album title, artist, record company, catalog number, track titles with pointers to lyrics, and MIDI files. The latter contains a JPEG file of the album cover, and other files. CD Extra is described in the Blue Book (issued in 1995).

Alternatively, in Mixed Mode CDs, ROM data is placed in Track 1, while CD-Audio data is placed in subsequent tracks. However, with this design, an audio player may access the ROM track and erroneously output noise rather than muting. To avoid this, a "pre-gap" technique may be used such that ROM data is "hidden" by placing it after the disc TOC, but before the Red Book first track (containing music). ROM data (up to 40 minutes of equivalent playing time) is placed between Index 0 and Index 1 of Track 1, while the music starts at Track 1, Index 1. An audio player thus skips over the data, starting playback at the first music track. However, the pre-gap area is not accessible to all PC software. The track layout of the Red Book CD and several alternative CD types are summarized in Fig. 7.25.

As noted, the CD-ROM data format is similar to that of music CDs; music discs can be played on ROM players, but ROM discs are not playable on audio players. A CD-ROM drive typically includes D/A conversion and audio output stages, but requires an interface and an external computer for nonaudio data output. In most designs, the consolidation of both functions into one player is ideally cost effective. To permit this, a CD-ROM disc automatically identifies itself as differing from an audio CD (through the *Q* subcode channel).

Unlike the CD-Audio standard, the CD-ROM standard does not stipulate how data is to be defined. In an effort to provide compatibility, the ad hoc High Sierra group (meeting at Del Webb's High Sierra Hotel and Casino) developed a standard logical file structure; the High Sierra standard was issued in 1985. It was adopted with minor revisions by the ISO as standard ISO 9660 (ECMA-119), "Volume and File Structure of CD-ROM for Information Exchange." It universally specifies how computer data is placed on a CD-ROM disc; to read the data, the computer operating system must be capable of reading

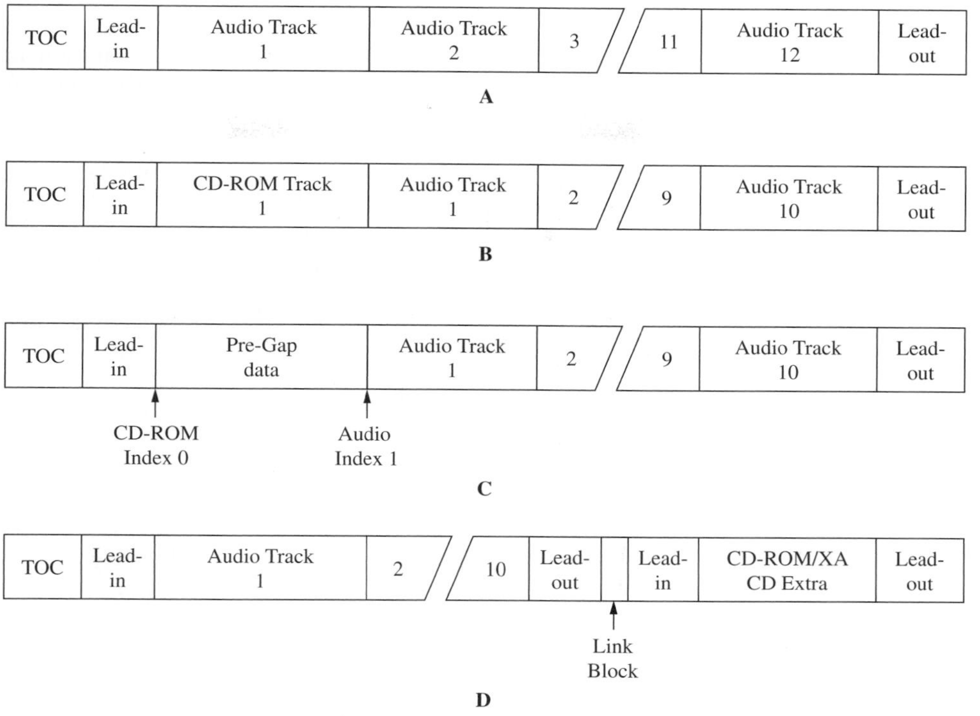

FIGURE 7.25 Several alternate CD types have been developed to allow both audio and nonaudio data to be placed in CD tracks. A. Red Book. B. Mixed Mode. C. Pre-gap. D. CD Extra.

the ISO 9660 file structure. Level One 9660 requires that files be written as a continuous stream with file name restrictions similar to the MS-DOS file system. Level Two allows longer file names, and is not usable in MS-DOS systems. Level Three is open-ended. MSCDEX.EXE is an MS-DOS Extension available from Microsoft; it contains extension programming and drivers so that an MS-DOS program can access a CD-ROM. The computer requires both Microsoft Extensions and the device driver for the MPC ROM drive. MSCDEX.EXE is often placed in the AUTOEXEC.BAT file, and the device driver is loaded from CONFIG.SYS. The computer can then read 9660 file directories and files from the disc.

Using extensions to ISO 9660, directories and files can be accessed from diverse platforms. The extension of ISO 9660 for the Unix platform is sometimes called the Rock Ridge extension; this is incorporated in the IEEE P1281 and P1282 standards. The Joliet extension is supported by 32-bit Windows 95/98/NT/2000/XP as well as Macintosh and Linux systems. The El Torito extension can create bootable discs for systems with the proper BIOS. HFS is Macintosh's native Hierarchical Filing System; most CD-ROMs authored for the Macintosh adhere to this format. CD-ROM discs can be authored for multiple platforms; however, executable files can only run on the appropriate platform. Additional incompatibility, such as file formats, file headers, bit resolutions, and sampling frequencies, exists within each platform, with competing CD-ROM systems. In some cases, cross-platform compatibility can be achieved; for example, hybrid CD-ROM titles can be played on PC and Apple platforms. The different data types are physically partitioned on the disc surface.

CD-R

The Compact Disc Recordable (CD-R) format allows users to record their own audio or other digital data to a CD. The format is officially named CD-WO (Write-Once) and it is defined in the Orange Book Part II, issued in 1988. It is a write-once format; the recording is permanent, and can be read indefinitely, but can never be erased or overwritten with new data. Text, audio, video, multimedia, and other executable data can be recorded and applications for CD-R are diverse. For example, the monthly phone bill for a large corporation might run 50,000 pages or more, but can be recorded on one CD-R disc. CD-R is ideal for distributing data to a few users or for archiving data. CD-R discs that are used to carry audio and nonaudio data prior to CD replication are written with the PMCD (premastered CD) format; the disc contains an index and other information normally found on a CD master tape. CD-R discs with up to 80 minutes (or about 700 Mbytes) of playing time are available. A complete subcode table is written in the disc TOC, and appropriate flags are placed across the playing surface.

CD-R discs that are used to record CD-Audio data can be played in Red Book players. However, they differ from prerecorded CD-Audio discs. All user data is recorded as a reflectivity change in a pregrooved track. Two areas are written to the inner portion (22.35 mm to 23 mm radius) of the disc before the Red Book lead-in radius, as shown in Fig. 7.26. Because these areas are inside the normal lead-in radius, conventional CD players do not read them. The PMA (program memory area), starting at –13 seconds (–00:13:25) relative to the start of the lead-in at 0 seconds, contains data describing the recorded tracks, a temporary TOC, as well as track skip information. When the disc is finalized, this data is transferred to the TOC. Disc-at-once (DAO) recording, described below, does not use the PMA area.

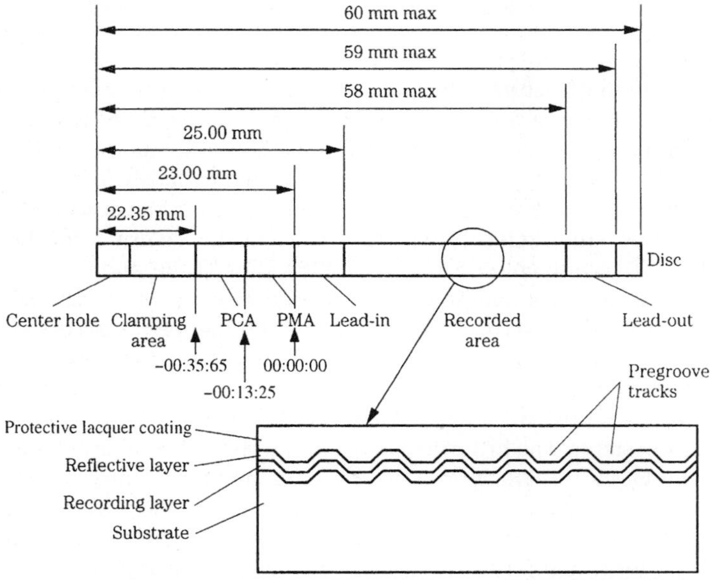

FIGURE 7.26 A CD-R disc holds data in pregrooved tracks. Data is permanently written into an organic dye-recording layer. The PCA area is used to calibrate the writing laser, and the PMA holds a temporary table of contents.

In addition, the PCA (power calibration area), starting at –35 seconds (–00:35:65), allows the laser to automatically make an optimal power calibration (OPC) test recording to determine proper laser power for data recording. The PCA contains a test area and a count area. The count area keeps track of available space in the test area; it contains 100 numbered partitions, each being one ATIP (absolute time in pregroove) frame long. Calibration test data at different (perhaps 15) power levels is written to one partition. This data is read back and an analog signal (not an error rate) from each test recording is compared to an optimal value and used to determine writing power. This is usually performed once each time a disc is loaded, and a count is incremented (up to 100) by filling a count area partition with random data. After this count is reached, no additional data can be written to the disc, even though there may be an open data area. Thus, only 100 recording operations are available; in some recorders, the count is filled after 100 insertions of a given disc. Several methods to more effectively use the count area, and increase recording sessions, have been devised. In some drives, the laser power is continually monitored and adjusted using a method known as Running OPC. The pregrooved program area holds user-recorded information such as track numbers, and start and stop times. A recording is complete when a lead-in area (with TOC), user data, and lead-out area have been written. A maximum of 99 tracks can be recorded on a disc.

As with prerecorded CDs, CD-R discs are built on a polycarbonate substrate, and contain a reflective layer and a protective top layer. However, they are otherwise substantially different. A recording layer comprising an organic dye is sandwiched between the substrate and reflective layer (see Fig. 7.26). During manufacture, it is applied by spin coating and cured. Together with the reflective layer, it allows a typical in-groove reflectivity of 73% and a carrier-to-noise ratio (CNR) of 47 dB. To achieve the minimum 70% reflectivity standard of CD, as the beam passes through the recording layer and substrate twice, a gold or less-costly silver halide reflective layer is typically used. The dyes employed would corrode an aluminum layer as normally used in prerecorded CDs. The thickness of the metal layer is typically 50 nm to 100 nm. A CD-R disc may look like a regular CD, but is usually distinguished by its recording layer that appears green, yellow-green, or blue. (A gold metal layer is often distinguishing as well; in some cases, when a silver layer is used, it is topped by gold paint.)

Unlike prerecorded CDs, CD-R discs are manufactured with a pregrooved 1.6-μm pitch spiral track, used to guide the recording laser along the track; this greatly simplifies recorder hardware design and helps ensure disc compatibility. Drives maintain radial position by detecting an 8% reduction in reflected intensity that occurs because of diffraction when a beam is correctly tracking the pregroove. The 0.6-μm wide track is physically modulated with a ±0.03-μm sinusoidal wobble with a frequency of 22.05 kHz as shown in Fig. 7.27. The wobble allows the recorder to control disc CLV rotation speed (a task accomplished with Red Book discs from the prerecorded data). Furthermore, the 22.05-kHz groove wobble excursion is frequency modulated with a ±1-kHz signal; this is used to create an ATIP absolute time clocking signal. A writing drive reads the ATIP in the lead-in area to determine recommended write power and other information such as write strategies and allowable speeds needed to optimize recording quality. ATIP also specifies the maximum start of the lead-out, which sets recording capacity. Track velocity is set according to disc capacity; for example, 63-minute discs use a 1.4 m/s track velocity and 74-minute discs use 1.2 m/s.

The recording mechanism itself can be described as heat-mode memory. Laser light is used to create heat to affect the change in the recording media. For example, an 8-mW

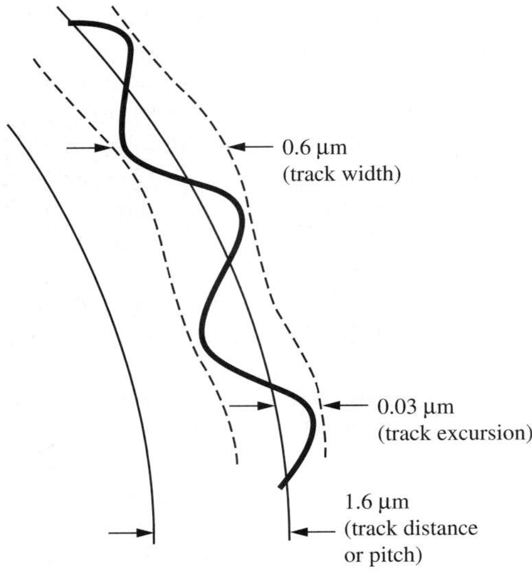

Figure 7.27 The CD-R pregroove track is modulated with a ±0.03-μm sinusoidal wobble with a frequency of 22.05 kHz.

laser spot power focused to a diameter of 1 μm yields a power density of 1×10^{10} W/m^2; the temperature can rise hundreds of degrees in a microsecond. The recording layer is a photo absorption surface which absorbs this heat energy from the recording laser. A 1× writing laser beam nominally with 4 mW to 8 mW of spot power (higher power is used for faster writing speeds, for example, 40 mW at 50×) and a wavelength of 775 nm to 795 nm passes through the polycarbonate substrate, and heats the organic dye recording layer to approximately 250°C, causing it to melt and/or chemically decompose to form a depression or mark in the recording layer. These depressions or marks create the decreased change in reflectivity (for example, 75% to 25% for an 11T pit) required by standard CD player pickups. During readout, the same laser, reduced to 0.5 mW of spot power, is reflected from the data surface and its changing intensity is monitored. The result is an eye pattern and modulation amplitude essentially identical to that of prerecorded CDs.

Generally, three types of organic dye polymers are used to form the recording layer: cyanine, phthalocyanine, or metal azo. These dye polymers are all chemically tuned to absorb light at 780 nm. Metal-stabilized cyanine-based media are usually recognized by an emerald green or blue-green color when a gold metal layer is used; cyanine dye is actually intense blue in color. The green appearance is a combination of the blue dye and gold metal layer; when a silver metal layer is used, with its wavelength independent reflectivity, the cobalt blue color is apparent. When heated by the writing laser, the dye degrades to create a mark with decreased optical reflectivity. Very generally, because the CD-R standard was originally devised using cyanine dyes, discs using cyanine dye are reliable in a wide range of recorders and laser powers, and at a wide range of writing speeds. In addition, cyanine dye has a relatively broad range of sensitivity to light resulting in a broader spot power margin for the writing laser (6.0 mW ± 1.0 mW). This makes cyanine more suitable for a range of recording speeds and laser powers, and also offers greater compatibility. Generally, when writing to cyanine, recorders can use longer laser pulses to create 3T–11T marks.

Phthalocyanine-based media have a yellow-green or gold color appearance (it is colloquially called "gold") when using a gold metal layer; the dye itself has a semi-transparent, nearly colorless yellow-green color. An advanced phthalocyanine dye is also used; it has an aqua color. During recording, the heated dye layer spot melts and shrinks, and the polycarbonate substrate expands to create a pit or mark. Very generally, phthalocyanine media is said to have greater longevity because it is stable and less sensitive to ordinary light. However, this lower sensitivity results in a small spot power margin for the writing laser (5.0 mW ± 0.5 mW), thus the writing speed and laser power must be more carefully controlled. Generally, when writing to phthalocyanine, recorders can use shorter pulses to create 3T–11T marks.

In some cases, metallized azo dye is used as the recording layer in CD-R media; its deep blue color is preserved when backed with a silver layer, and it appears green when backed by a gold layer. Even discs that use the same type of recording layer material can perform differently. Variations in recording layer thickness, reflective layer thickness, and different protective layers can affect disc-recording characteristics.

Organic dye layers are affected by aging. The dye layer will deteriorate over time because of oxidation and material impurities. In addition, the organic dye is sensitive to ultraviolet light and will degrade. Cyanine dye is more prone to degradation than phthalocyanine, which is inherently stable. To evaluate life expectancy of CD-R discs, discs are subjected to a variety of conditions that accelerate aging. For example, unrecorded and recorded media can be subjected to 65°C and 85% RH (relative humidity) for 2 months. (This equates to a 45-year duration at 22°C and 55% RH.) Discs can also be subjected to bending and scratch tests. Criteria such as BLER, E22, E32, and burst errors, described in Chap. 5, can be measured to determine end-of-life. In one test, errors were higher on media recorded after age testing than on media recorded prior to testing. Age testing degraded the recordability of the media more than its storage capability. Both unrecorded and recorded media should be stored in clean jewel cases in a stable environment of 10 to 15°C and 20 to 50% RH, and protected from sunlight and other radiation courses.

Shelf life of cyanine media is said to be from 10 to 100 years with nominal storage conditions. The typical BLER error rate is less than 20 per second, well below the Red Book CIRC tolerance of 220. In an accelerated aging test, the life expectancy of phthalocyanine discs was calculated to be 240 years. Ultimately, human carelessness, resulting in scratches, is probably the single biggest threat to CD-R longevity. The U.S. Mail is selectively irradiated by high-energy electronic beams. Irradiation can tint polycarbonate substrates, decrease reflectance, and increase error rates. However, radiation-induced errors are small and are corrected by error-correction codes; during testing, no uncorrectable errors resulted.

The Orange Book Part II defines both single session (regular) and multisession (hybrid) recording; a session is defined as a recording with lead-in, data, and lead-out areas. With single session recording, sometimes called disc-at-once (DAO) recording, a disc is recorded in its entirety, without interruption. The recorder records a TOC in the lead-in portion of the disc, data tracks, and a lead-out area, so any standard player can read the disc. A PMCD (premastered CD) recording used as a master is an example of a DAO application. Tracks can be recorded back to back without a gap; this allows for crossfades between tracks. When recording DAO, it is recommended to first create a disc image, a file that includes all the data to be recorded.

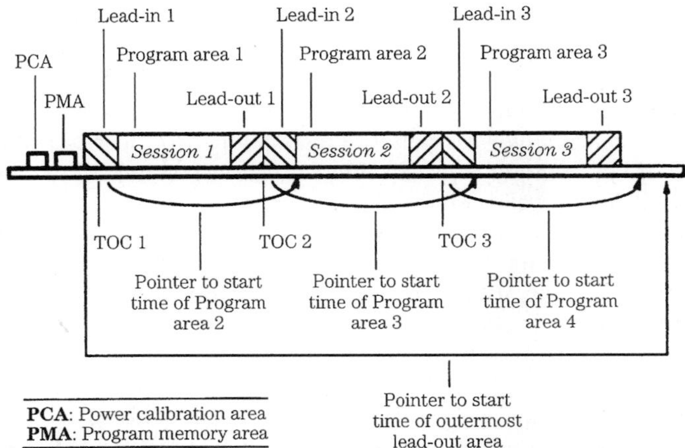

Figure 7.28 CD-R multisession recordings form a series of sessions across the disc surface, each with its own lead-in, program, and lead-out areas.

Alternatively, track-at-once (TAO) recording allows single or multiple tracks to be written in a session; this is the most widely supported recording method. After the program tracks are written, the recorder writes the TOC and lead-in areas, and lead-out areas. The writing laser is turned off after each track; this creates a gap between tracks. A partially recorded disc can be played on the CD-R recorder, but cannot be played on a CD-Audio player until the session ends when the final TOC and lead-out areas are recorded. Most recorders permit an unwanted track (such as a false start) to be marked and deleted from the TOC so the CD-R player (and CD-Audio players recognizing skip-ID flags) will skip over it. Recorders using TAO can also write a single-session CD-R.

The Orange Book Part II also specifies multisession (hybrid) recording in which sessions can be recorded one or a few at a time. Tracks can be written one at a time and recording can be stopped after each track. Separate recording sessions are permitted, each with its own lead-in TOC, data, and lead-out areas, as shown in Fig. 7.28. This session structure is required so Red Book players will recognize the beginning and end of segments. Each time a session is created, about 13.5 Mbytes (6750 blocks) of capacity is lost to lead-in and lead-out areas (22 Mbytes is used for the first session). The lead-in for a session occupies about 8.8 Mbytes, and the lead-out for a session occupies about 4.4 Mbytes (the lead-out for the first session occupies about 13.2 Mbytes). Clearly, this is inefficient for adding small amounts of data. TAO recorders allow multisession recording, in addition to single-session recording. With TAO recording, multiple tracks can be written to a session, adding data one track at a time; no lead-in or lead-out is written until the session is closed. This saves disc space, but the session cannot be read by most players until the session is closed. Older CD-ROM drives and all CD-Audio players can read only the first session on a multisession disc. Thus, multisession recording typically is not used for CD-R audio discs. Photo CD and some CD-ROM titles, are examples of multisession discs.

In multitrack recording, data is appended to a disc in tracks that are at least 2 seconds long (about 300 sectors or 700 kbytes). Tracks can contain one or more files and the session is left open between write operations. Individual tracks are separated by a 150-sector gap. After a track is written, track numbers and timings are written in the PMA; a link block is written where the laser turns off before it is temporarily moved to the PMA. When the session is closed, the TOC is written, the link blocks are hidden, and the disc can be removed from the drive.

By using the CD portion of the Universal Disk Format (CD-UDF), CD-R discs can perform packet writing so that small amounts of data can be efficiently written. For example, whereas multisession recording requires large data overhead, packet-writing overhead might consume less than 4 Mbytes per disc. Packet writing can be performed on CD-R media, making them functionally similar to small hard-disk drives. Data in a file can be appended and updated without rewriting the entire file. Because the data structures are so small, buffer underrun is alleviated. Packets of data (variable or fixed length) are written to a disc without closing either a track or a session, and without updating the TOC or PMA. Instead, system information about the partial track is placed in the Track Descriptor Block in the pre-gap before the track. A packet contains user data along with associated link blocks. Special blocks called run-in and run-out allow a recorder to synchronize data, and they also contain interleaved data from other blocks. Written data comprises a link block, four run-in blocks, user data, and two run-out blocks. Fixed-length packets allow data to be randomly erased and rewritten without accounting for different packet size; however, disc capacity is decreased to about 500 Mbytes. Variable-length packets allow greater disc capacity because mapping is fixed when data is written. Not all CD recorders support packet writing. Packet writing, also called block append, is defined in the Orange Book Part II for CD-WO. Packet writing is also defined in the ISO 13490 specification. The UDF Bridge file format is used in DVD, as described in Chap. 8.

Both stand-alone and peripheral CD-R recorders have been developed. Stand-alone recorders allow users to record discs and perform simple editing of tracks and subcode. These are intended for audio use and apply Serial Copy Management System (SCMS) data to the recorded program. Peripheral CD-R recorders interface to a host computer via a SCSI or other interface; many recorders are packaged as half-height drives. The recorders operate at speeds much faster than real time, generating all synchronization, header, CIRC, and EFM processing required by the CD standard. Depending on the software package, various degrees of data manipulation are possible; for example, a software application can consolidate fragmented files and specify the physical location of CD-ROM data on a disc so that retrieval time is shorter. When a CD-R disc is authored according to the ISO 9660 file format, the disc can be read on multiple platforms.

Real-time recorders operate at a 150-kbps rate; however, higher-speed recorders are widely used (higher laser power is needed at high speeds). Higher-speed recorders may provide lower error rates than single-speed recorders; this may be because in single-speed writing, the laser remains focused for a longer time and an unwanted annealing process may be caused by the added heat. Recorders with OPC can avoid this effect. Discs suitable for high-speed recording are specially approved for reliability.

At 1× speed, a disc spins at about 539 rpm when the head is placed on the inner radius and it slows to 210 rpm at the outer radius. At 16×, for example, the speeds are about 8000 and 3200 rpm respectively; a 50× drive might reach speeds in excess of 12,000 rpm. CLV is efficient when reading an audio disc at 1× speed; it is relatively easy to change speeds when accessing different tracks at different radii. However, at high speeds, it is difficult to quickly change high disc speeds over a range of radii. To accommodate high disc velocities, some disc drives use partial constant angular velocity (PCAV) to spin the disc at a lower fixed speed near the inner radius, and then shift to CLV near the outer radius. For example, a 20× drive might start reading the inner

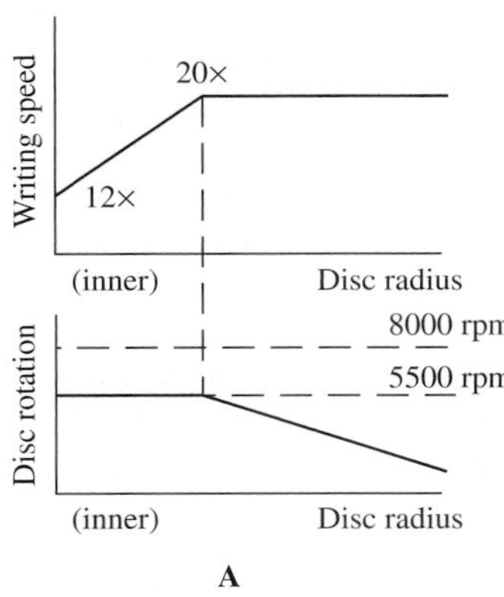

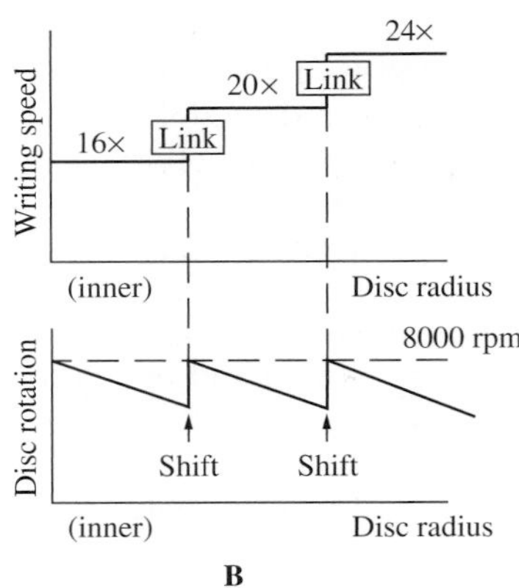

FIGURE 7.29 To optimize data transfer speed, disc drives can employ several techniques. A. In a PCAV drive, inner radii are read using CAV and outer radii are read using CLV. B. In a ZCLV drive, different CLV speeds are used across the disc radius.

radius at 12× using a CAV method, as shown in Fig. 7.29A. As reading progresses and the pickup moves outward, the transfer rate increases. When a 20× speed is reached, the drive switches to CLV and the data rate is constant at 20× (3000 kbps). Some drives use zoned constant linear velocity (ZCLV), in which different fixed writing speeds (operating in a CLV mode) are used in discrete regions of the disc as shown in Fig. 7.29B. For example, a 24× drive might use 16×, 20×, and 24× in different radii of the disc. Writing is suspended as the drive shifts speeds; to provide writing continuity, data is held in a buffer. In any high-speed drive, high-speed writing is not possible across the entire disc radius.

In some high-speed drives, the laser beam is diffracted to create multiple laser spots, one spot per groove. Multiple pickups receive simultaneous data, thus increasing the effective data-reading rate, particularly when reading sequential data.

Discs designed for high-speed writing are designed to accept the thermal effects of higher-power lasers, and the substrate and track must be mechanically sound and accurate. High-speed disc writers must use modified write strategies to allow for thermal effects; for example, the drive can read parameters encoded on the blank media to optimize the level and duration of the power region that initiates a recorded mark. The drive must also be mechanically able to withstand high velocities. For example, vibration caused by an eccentric or unbalanced disc must be minimized. Drives are relatively unaffected by vibration when reading because autotracking can use a robust error signal based on large (60%) changes in intensity of reflected light. However, drives are more sensitive to vibration when writing because tracking is performed with push-pull methods that must rely on small (5 to 10%) variations in reflected light intensity from the unrecorded pregroove. Also, contamination on a disc surface is relatively unimportant when reading data, but any obstruction to a writing laser will result in a permanent error. The Orange Book specifies that jitter for CD-R discs should be less than 35 ns.

Two types of CD-R discs are sold: for computer use, or for music use. The discs are otherwise physically identical, but during mastering (of the blank) a mandatory Disc Application Code is embedded in the ATIP information contained in the pregroove lead-in area. Three types of discs are defined: Types 1a and 1b for restricted use and Type 2 for nonrestricted use. Type 1a is used for CD-ROM or professional audio recording.

Type 1b is used for special purpose applications such as a Photo CD and can only be written to by those specialized recorders. Type 2 discs are for consumer audio recording. Stand-alone consumer CD-R audio recorders will not record unless that code is present. Thus, only music-use discs can be used; their higher cost is used to compensate artists. Computer-use CD-R discs are used in CD-R drives connected to a host computer; these discs can also be used to record music to a disc, using computer-use discs. Clearly, copyright laws should be observed. Music-only CD-R discs are sometimes referred to as CD-R-DA (Digital Audio).

Although CD writing on a PC is relatively simple, some care must be taken; the computer must provide a steady stream of data, while simultaneously interleaving, error-correcting, and formatting the data. Many systems can transfer data from either tape or hard disks to the recorder, and can produce CD-Audio, CD-ROM (including CD-ROM/XA), and Photo CD discs. Generally, a hard-disk drive with fast access (10-ms seek time) is required, connected to the PC via a fast (10 to 20 Mbytes/s) interface to the PC bus. During recording, any interruption in the data stream at the recording laser will render a disc unusable. Most CD-R recorders contain a cache memory (1.2 to 4 Mbytes); this helps prevent data stream problems from buffer underrun. Some users recommend several measures to help ensure successful writing: partition the staging drive to hold the disc image, create a real ISO image file (as opposed to a virtual image comprising lookup addresses of data to be written) of the data, defragment the hard drive, and test before writing. Some users recommend hard-disk drives with embedded servo tracks so that automatic thermal recalibration will not interrupt the data stream; alternatively, hard drives with unobtrusive recalibration procedures can be used. In addition, network services, auto-answer fax software, sound utilities, screen savers, virus checkers in resident memory, and TSR (terminate and stay resident) programs should be turned off during recording to prevent glitches in the bitstream.

CD-RW

The Compact Disc ReWritable (CD-RW) format allows data to be written, read, and erased and rewritten. The format is officially named CD-E and it is described in the Orange Book Part III standard, issued in 1996. A CD-RW drive can read, write, and erase CD-RW media, read and write CD-R media, and read CD-ROM and CD-Audio media. The data can comprise computer programs, text, pictures, video, audio, or other files. CD-RW disc capacity is about 700 Mbytes. A CD-RW disc has an embedded aluminum layer and a recording layer that appears gray. Altogether, there are five layers built on the polycarbonate substrate: a dielectric layer, a recording layer, another dielectric layer, a reflective aluminum layer, and a top acrylic protective layer, as shown in Fig. 7.30. This phase-change recording technology allows thousands (on the order of 105) of rewrite cycles.

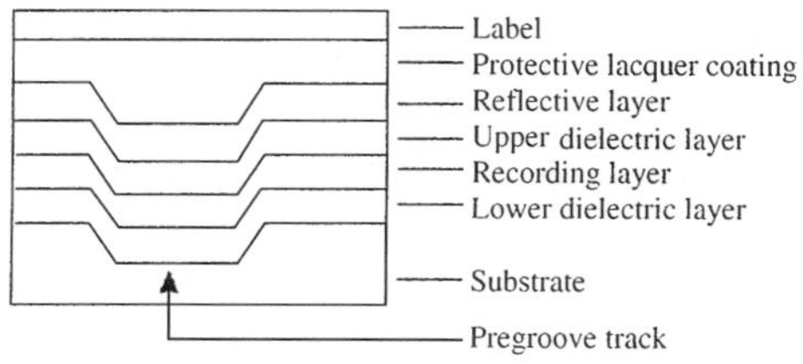

FIGURE 7.30 The CD-RW recording layer is sandwiched between two dielectric layers.

As in CD-R, the writing and reading laser follows a pregroove across the disc radius. However, whereas CD-R uses a dye-recording layer, the CD-RW format employs a phase-change recording layer comprising an alloy of silver, indium, antimony, and tellurium. This metal exhibits a reversible crystalline/amorphous phase change when recorded at one temperature and erased at another, as described in Chap. 6. The recording layer on the blank disc is in crystalline form; it is translucent thus light is reflected from the metal layer above it. Data is recorded by directing a laser (8-mW to 14-mW spot power) to heat an area of the crystalline layer to a temperature above its melting point (500 to 700°C). When the area revitrifies rapidly, it becomes amorphous and absorbs light, and the decreased change in reflectivity can be detected. A low-power laser (perhaps 0.5-mW spot power) is used to read data.

Because the crystalline form is more stable, the material will tend to change back to this form; thus, data can be erased using an annealing process. When the recording surface is heated by a lower laser spot power of perhaps 4 mW to 8 mW to its transition temperature (200°C) and cooled gradually, it returns to its original crystalline state. Unlike dye-polymer technologies, phase-change recording is not wavelength-specific.

Rewriting is accomplished through direct overwriting. Rather than completely erase a disc side, this "on the fly" erase feature allows the last recorded audio track to be erased simply by erasing the subcode reference to that track while leaving the recorded data in the recording layer. With this method, recorded tracks can be erased individually, working sequentially backward from the last recorded track, to provide editing control without requiring total erasure.

A technique called Running Optical Power Calibration determines the correct laser power levels when individual discs are loaded, and monitors and adjusts the power level to compensate for surface contamination such as fingerprints. Unlike CD-R recording in which the laser is turned on for the duration of the pit formation, in CD-RW, the laser is repeatedly switched between its write or erase power, and a low bias power (less than 1 mW) that is equal to the power used to read the disc. This switching is performed so that the recording alloy layer will not accumulate excess heat, thus creating overly large marks. The dielectic layers comprise silicon, oxygen, zinc, and sulfur; they control the optical response of the media and increase the efficiency of the laser by containing the heat that is used to record data on the recording layer. They also thermally insulate and protect the pregroove, substrate and reflective layers, and mechanically restrain the recording layer material.

As with CD-R, two types of CD-RW discs are sold: for computer use, or for music use. The discs are physically identical, but a permanently recorded flag is placed on music-use discs; stand-alone consumer CD-RW audio recorders will not record unless that flag is read. Music-only CD-RW discs are sometimes referred to as CD-RW-DA (Digital Audio).

The reflectivity of CD-RW discs is only about 15 and 25% (amorphous and crystalline states, respectively). They generally cannot be played in conventional CD players (many DVD players do play CD-RW discs) or CD-ROM drives. A CD-RW drive is required, or a MultiRead drive capable of reading lower reflectivity discs. Such drives contain an automatic gain control (AGC) circuit to compensate for the lower reflectivity and signal modulation. The AGC boosts the gain of the signal output from the photodiodes. To facilitate this, CD-RW discs carry a code that identifies them as CD-RW discs to the player. CD-RW drives are commonly found as PC peripherals. Software supports TAO, DAO, and multisession recording. When CD-RW discs are appropriately formatted, the CD-UDF specification permits easy file-by-file rewriting. In particular, users can write to the CD-RW drive by simply dragging and dropping.

CD-MO

The Orange Book Part I defines a Compact Disc Magneto-Optical (CD-MO) standard; data can be written, erased, and rewritten. Two types of discs are defined: a disc with a premastered area (recorded with pits) containing CD-ROM data plus a writable area, and a disc that is completely writable. Because writable data is read via changes in light polarization rather than intensity, CD-MO discs are not playable in CD-Audio or CD-R drives (however, CD-MO drives can play CD-Audio and CD-R discs). A CD-Audio player can read the premastered area on a CD-MO disc. In some ways, the MiniDisc was an evolution of the CD-MO specification.

CD-i

The Compact Disc Interactive (CD-i) standard was devised as a product-specific application of the CD-ROM format. CD-i permits storage of a simultaneous combination of audio, video, graphics, and text, and defines specific data formats for these. In addition, titles can function with real-time interactivity. For example, a CD-i dictionary might contain a word and its definitions, as well as spoken pronunciation, pictures, and translations into foreign languages. The CD-i standard, codified in the Green Book (issued in 1986), defines how each type of information is encoded as well as logical layout of files on the disc. It also specifies how hardware reads discs and decodes information.

The CD-i data format is derived from the CD-ROM Mode 2 format. CD-i data is arranged in 2352-byte blocks, as in the CD-ROM/XA format. The CD-i format accepts either PCM or ADPCM (adaptive differential pulse-code modulation) data. The full-motion video (FMV) extension allows storage of 74 minutes of full-motion digital video and stereo audio. The MPEG-1 coding standard is used to reduce the video bit rate to 1.15 Mbps and the audio rate to 0.22 Mbps; lower rates can also be used. CD-i players can also play Video CDs coded with MPEG-1. MPEG-1 audio is described in Chap. 11 and MPEG-1 video in Chap. 16. To ensure universal compatibility, dedicated hardware and interfaces are defined. The CD-Bridge format adds information to a CD-ROM/XA disc so it can be played on a CD-i player. Bridge tracks use Mode 2 data, tracks are listed in the TOC as a CD-ROM/XA track, and block layout is identical to CD-i and CD-ROM/XA. The Photo CD is an example of a Bridge disc. The CD-i format did not enjoy success among its targeted consumers.

Photo CD

The Photo CD is used to professionally store, manipulate, and display photographic images. Photographs can be viewed or reproduced as high-quality prints of images using a color printer. The 35-mm version of the Photo CD provides three to four times the resolution required in any high-definition television (HDTV) standard. Conventional photographic images can be scanned to the Photo CD, with 2048 scan lines across the short dimension of a 35-mm frame, with 3072 pixels on each line to yield a 3:2 aspect ratio. Data compression and decomposition are used to increase storage efficiency. During authoring, high-resolution image files are subjected to a 4:1 data reduction. In addition, file sizes can be reduced without significant visual loss by using chroma sub-sampling to take advantage of limitations in human visual perception. The Photo CD was developed by Kodak and is defined in the Beige Book.

Photo CD discs conform to the Orange Book Part II standard and are physically identical to CD-R audio discs; however, different data headers make them incompatible. Data blocks are written according to the CD-ROM/XA, Mode 2, Form 1 standard. Because discs use the CD-Bridge format, they are playable on CD-ROM/XA players. Because the Orange Book Part II permits additional multisession recording to a disc, images can be added over time. Pacs initially recorded on a disc are structured as a file using the ISO 9660 structure. Subsequently recorded Pacs use a CD-R Volume and File Structures format, using the multisession method. All Pacs are addressed through the block-addressing method used by CD-ROM discs and defined by the ISO/IEC 10149 standard. Because the Photo CD adheres to the CD-ROM/XA format, audio and video data can be interleaved; in this way, a soundtrack can accompany visuals. The Picture CD consumer format similarly stores photographic files on a CD-R disc; it provides 1024×1536 resolution using JPEG compression. The disc also contains software used to view and edit the photographs.

CD+G and CD+MIDI

The CD+G and CD+MIDI formats were devised to encode graphics or MIDI software on CDs, in addition to regular audio data. Special hardware or software is required to access this data. Eight subcode channels are accumulated over 98 frames; thus, each 98-bit subcode word is output at a 75-Hz rate. Subcode synchronization occupies the first two frames, thus a subcode block contains eight channels with 96 data bits. This data block is called a packet, and each quarter of a packet is called a pack. A pack is generated every 3.3 ms. Only *P* and *Q* are reserved for audio control information. Over the length of a CD, the remaining channels, *R* to *W*, provide about 25 Mbytes of 8-bit data. Utilization of that capacity has been promoted as CD+G or CD+Graphics, and CD+MIDI, sometimes known as CD+G/M. The player decodes the graphics or MIDI data separately from the audio data. In CD+G discs, data is collected over thousands of CD frames to form video images or other data fields. For example, a CD+G audio disc can contain video images, liner notes, librettos, or other information. Because video images require a large amount of data for storage, CD+G images provide limited resolution.

In the CD+MIDI application, MIDI (Musical Instrument Digital Interface) information is stored in the subcode field, and output synchronously with the audio playback. External MIDI instruments can synchronize to the melody or other musical parameters of an encoded disc. The subcode capacity is sufficient to store up to 16 channels of MIDI information. MIDI information can be supplemented with graphics information; for example, music notation could be supplied. Another variation can encode music notation in the subcode area to allow print out of sheet music. CD+G/M discs are compatible with any CD player, but only players equipped with CD+G/M output ports can retrieve the information from the disc. Alternatively, an external decoder can be connected to any CD player with a digital output port, provided that the full subcode data is available from the port. CD+G is sometimes used for karaoke applications.

CD-3

In addition to regular 120-mm-diameter CD discs, the CD-3 format describes 80-mm-diameter discs. The name derives from the approximately 3-in diameter. This small size promotes greater portability and the format is useful for short audio programs. A CD-3 disc holds a maximum of 20 minutes of music. Because a CD data track begins at the innermost radius, CD-3 discs are compatible with regular discs and players. Some players have concentric rings in their disc drawers to center both diameter discs over the spindle. The CD-3 format is also used to hold over 200 Mbytes of CD-ROM data. The CD-3 format is also used for CD-R and CD-RW discs.

Video CD

The Video CD format is an outgrowth of the CD-i standard; full-motion video was added to the original CD-i standard and that feature was subsequently revised in 1992 to form the Video CD standard. The Video CD uses the MPEG-1 coding standard for audio and video. The audio signal is coded with the Layer II standard at 44.1 kHz. A disc stores about 74 minutes of full-motion digital video and audio; a feature film is placed on two discs. The video decoder chip permits full-motion video (FMV) to be shown at either 29.97 (NTSC) or 25 (PAL/SECAM) frames per second at 352 pixels by 240 lines and 352 pixels by 288 lines, respectively, one-fourth the resolution of DVD's normal mode. The Video CD may be shown as a quarter-screen image. The video bit rate is 1.15 Mbps and the audio bit rate is 0.22 Mbps. The Video CD format is a CD-ROM/XA Bridge disc, Mode 2, Form 2; this allows a Video CD to play on a CD-ROM drive. A Video CD disc will not play on a CD-Audio player, but will play in many DVD players. Video CD is described in the White Book; version 1.0 of this specification was originally developed in 1992 for karaoke discs and in 1995 it was extended to version 2.0, which supported interactive video. The Video CD is different from the CD-Video format, now abandoned. The MPEG-1 video algorithm is discussed in Chap. 16. MPEG-1 audio is discussed in Chap. 11.

The Super Video CD (SVCD) is an enhanced version of the Video CD designed primarily for higher-quality movie playback. SVCD uses MPEG-2 coding for video compression to store about 70 minutes on a disc. The NTSC resolution is 480 × 480, and PAL resolution is 470 × 576—about three-fourths that of DVD's normal mode. Dual mono, stereo or 5.1-channel soundtracks can be used at bit rates ranging from 32 kbps to 384 kbps using MPEG-1 Layer II or MPEG-2 multichannel codecs. Uncompressed audio cannot be stored. The maximum data rate is 2.2 Mbps by virtue of a 2× drive. However, at the higher data rate, playback time is halved to about 35 minutes; a movie might occupy three discs. Copy Generation Management System (CGMS) copy protection can be enabled. SVCD's development was sponsored by the Chinese government as a low-cost alternative to DVD. Other technical aspects were derived from the Video CD format and the China Video CD (CVD). The SVCD specification was ratified by the China National Committee of Recording Standards in September 1998. SVCD is also standardized in the IEC-62107 document. A similar specification, the Chao-Ji ("Super") VCD standard, was developed to support both China Video CD and SVCD; many SVCD players and changers support the Chao-Ji standard and most discs use the SVCD format. The DSVCD (Double SVCD) format uses a smaller track pitch to permit longer high-quality playing times of about 60 minutes.

Super Audio CD

When the Compact Disc was launched in 1982, it was rightly heralded as a data carrier of immense storage capacity. However, over time the CD seemed increasingly small. Moreover, some audiophiles argued that its specifications constrained audio fidelity. In particular, the CD was insufficient for the large file sizes and high bit rates required by surround sound and high sampling frequency audio. In 1999, Philips and Sony introduced the high-density Super Audio CD standard, known as SACD. The SACD format supports discrete-channel (two-channel and multichannel) audio recordings, using the proprietary one-bit Direct Stream Digital (DSD) coding method. DSD uses a high sampling frequency and achieves a flat frequency response to 100 kHz and a dynamic range of 120 dB in the 0- to 20-kHz band.

SACD players can play both SACD and CD discs. SACD is not compatible with the DVD or Blu-ray formats. The mechanical and optical properties of an SACD disc are similar to those of a DVD-5 disc; however, the logical layout of content, the data format, and the copy protection measures are different. DSD data is not playable in standard DVD or Blu-ray drives, but a CD layer, if present on an SACD disc, is playable. Some players may include decoders to accommodate multiple disc formats. Other data such as text and graphics (but not video) can be included on an SACD disc; this content follows the Blue Book "Enhanced CD" standard. The SACD standard is sometimes known as the Scarlet Book, published in March 1999.

Disc Design

SACD discs use the same dimensions as a CD: 12-cm diameter and 1.2-mm thickness. The laser wavelength is 650 nm, the lens NA is 0.60, the minimum pit/land length is 0.40 μm, and the track pitch is 0.74 μm. (The pertinent CD figures are 780 nm, 0.45, 0.83 μm, and 1.6 μm.) Software providers may choose from three disc types specified in the SACD format: single-layer, dual-layer, and hybrid disc construction. The single-layer disc contains one layer of high-density DSD content (4.7 Gbytes); for two-channel stereo, this provides about 110 minutes of playing time. The dual-layer disc contains two layers of high-density content (8.5 Gbytes total). The hybrid disc is a dual-layer disc that contains one layer of high-density DSD content (4.7 Gbytes) and one layer of Red Book compatible stereo content (680 Mbytes), as shown in Fig. 7.31. The semi-reflective high-density layer must be reflective (readable) at the 650-nm wavelength of SACD, and transparent at the 780-nm wavelength used by conventional CD players; in other words, it acts as a color filter. The high-density layer is 0.6 mm from the readout surface and the CD layer is 1.2 mm from the surface. An SACD player can read both layers, and a CD player can read the CD layer.

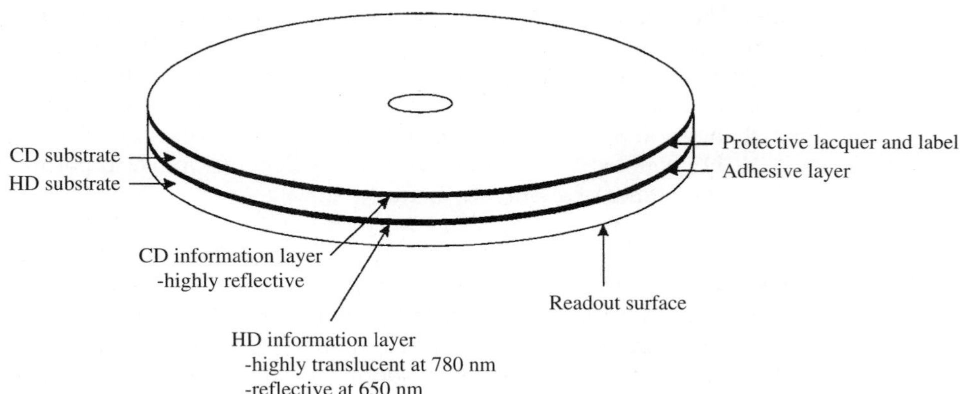

FIGURE 7.31 A hybrid SACD disc contains two data layers (high-density and CD). The two layers are bonded together to form a disc with a thickness of 1.2 mm. (*Verbakel et al., 1998*)

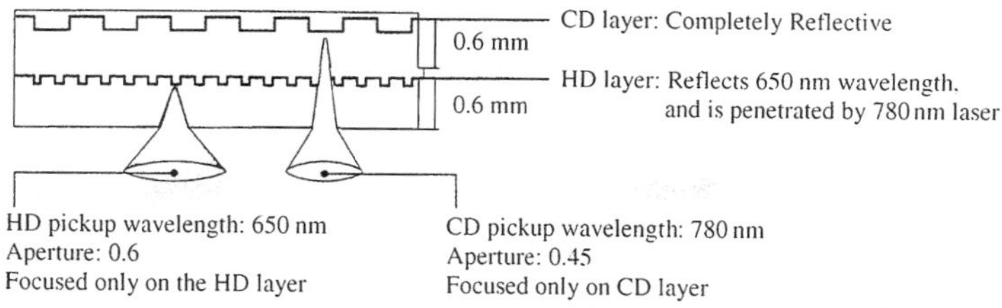

FIGURE 7.32 Both the high-density layer and CD layer in a hybrid SACD are read from one side by a laser. The high-density layer is semi-reflective, while the CD layer is fully reflective.

In dual-layer discs, two 0.6-mm substrates are bonded together. In all implementations, there is only one data side. A semi-reflective layer (20 to 40% reflective and approximately 0.05 μm in thickness) is used on the embedded inner data layer; in some cases, a silicon-based dielectric film is used. A fully reflective top metal layer (at least 70% reflective and approximately 0.05 μm in thickness) is used on the outer data surface. This surface is protected by an acrylic layer (approximately 10 μm in thickness) and a printed label. Care must be taken to seal a hybrid disc to limit water absorption and evaporation from the substrate; unequal absorption between the two disc sides could cause disc warpage. The back side is inherently protected by a metal layer and a lacquer layer while the front side is nominally unprotected, thus a front-side transparent silicon-based coating (10 nm to 15 nm) is needed. A hybrid disc in which a dual pickup (650 nm and 780 nm) is used to read both SACD and CD data is shown in Fig. 7.32.

The data on an SACD disc is grouped into sectors of 2064 bytes. This comprises: Identification Data (ID) of 4 bytes, ID Error Detection (IED) of 2 bytes, Reserved of 6 bytes, Main Data of 2048 bytes, and Error Detection Code (EDC) of 4 bytes. During encoding, following scrambling, 16 sectors form an error-correction code block, which is processed with a scheme using a Reed–Solomon Product Code. Rows of ECC blocks are interleaved and grouped into recording frames. Frames undergo EFMPlus modulation. Data is then placed in Physical Sectors and recorded to disc.

The radius of the high-density layer is segmented for different kinds of data, as shown in Fig. 7.33. The innermost radius contains the disc lead-in area, followed by the data area. It is divided into several areas including a Master Table of Contents (Master TOC) containing information on tracks and timing, as well as text data on the title and artist. The Master TOC is stored in three places (sectors 510, 520, and 530) to ensure readability. The next two radial areas are given to two-channel and multichannel recordings (up to six channels). The two-channel and multichannel areas use the same basic structure. The Area TOC for each audio area is placed at the beginning and end of each area. They contain track, sampling frequency, timing, and text information about the tracks included in that section. The SACD standard permits up to 255 tracks. Audio tracks contain two types of streams: audio elementary stream and supplementary data elementary stream; they are multiplexed. In addition, there are sequences of audio frames each with a timecode, and supplementary data frames for pictures, text, and graphics; each frame represents 1/75 second. Following the audio tracks, there is an area for optional data such as text, graphics, and video. This data can only be accessed by a file system; its format is not specified in the SACD specification. The outermost radius holds the disc lead-out. SACD discs can be read using a hierarchical TOC, or by optionally using a UDF or ISO 9660 file system.

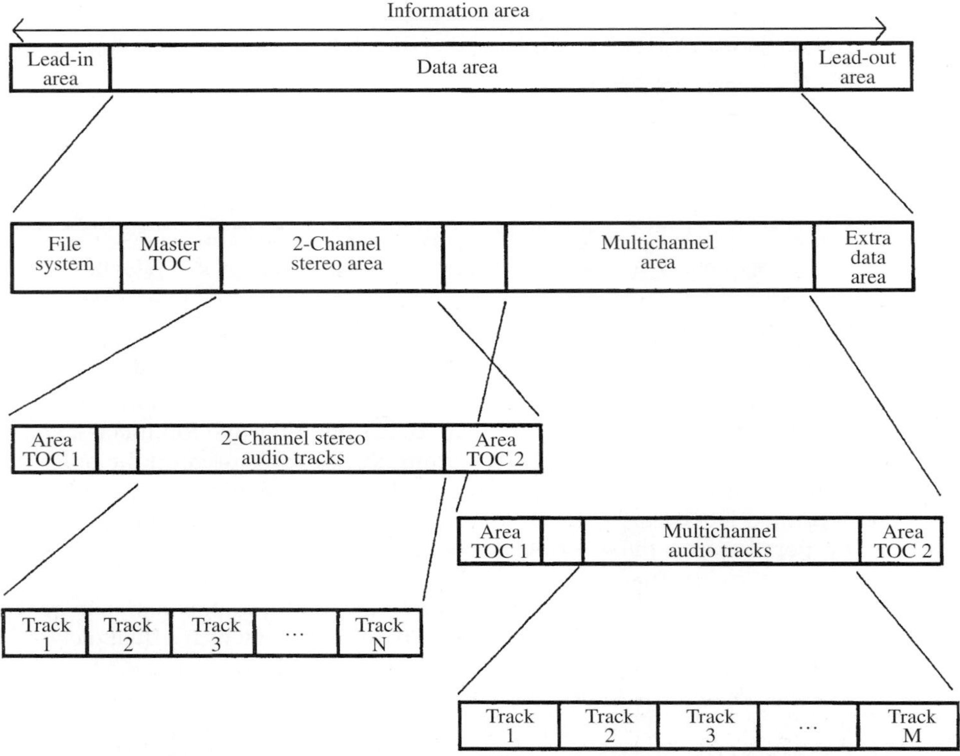

Figure 7.33 The SACD high-density data layer is designed to carry both two-channel and multichannel audio data.

All SACD discs incorporate an invisible watermark that is physically embedded in the substrate of the disc. Virtually impossible to copy, the watermark is used to conduct mutual authentication of the player and the disc. SACD players read the watermark and will reject any discs that do not bear an authentic watermark. Visible watermarks on the signal side of the disc in the form of faint images or letters may also be employed. A process called Pit Signal Processing (PSP) uses a controlled array of pit widths to create both invisible and visible watermarks; user data stored as pit/land lengths is unaffected by this watermarking.

DSD Modulation

Whereas all CD discs carry PCM data, all SACD discs carry Direct Stream Digital (DSD) data, in which audio signals are coded in one-bit pulse density form using sigma-delta modulation. Most conventional analog-to-digital (A/D) converters use sigma-delta techniques in which the input signal is upsampled to a high sampling frequency. The signal is passed through a decimation filter and also quantized for output as a PCM signal at a nominal sampling frequency of 44.1 kHz (for CD) and up to 192 kHz (for DVD-Audio or Blu-ray). Likewise, many D/A converters use oversampling to increase the sampling frequency of the output signal, to move the image spectra from the audio band. As in PCM systems, DSD begins with a high sampling frequency, but unlike

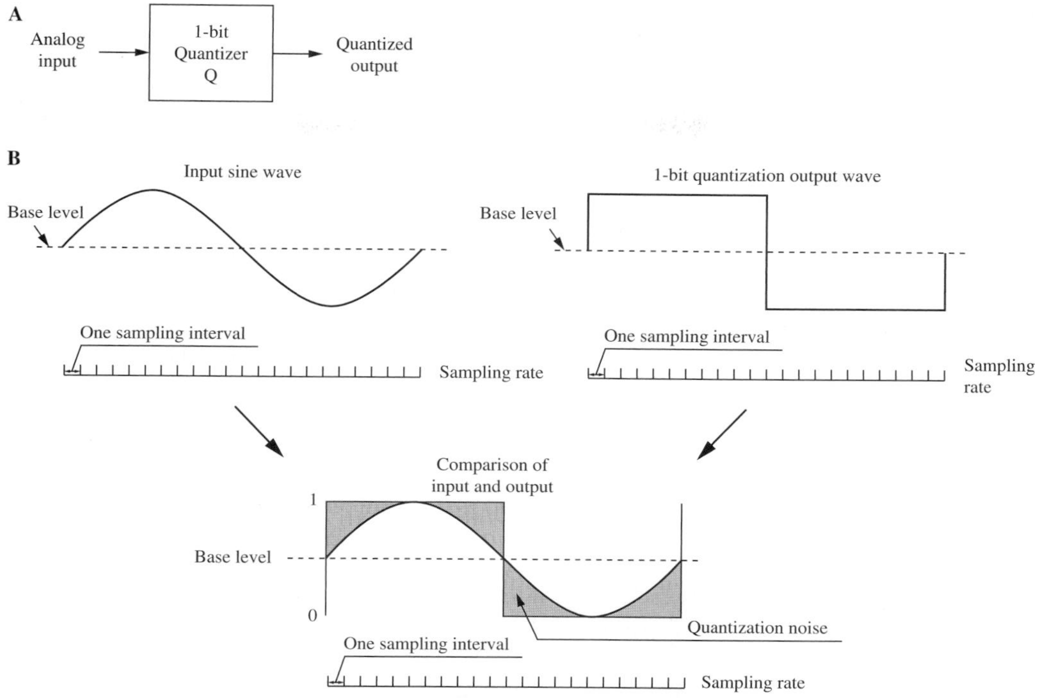

FIGURE 7.34 In principle, DSD coding is based on a one-bit quantization method. A. A one-bit quantizer produces a square wave output. B. The output square wave from a one-bit quantizer yields a large difference signal.

PCM systems, DSD does not require decimation filtering and PCM quantization in the recording process; instead, the original sampling frequency of 2.8224 MHz is retained. One-bit data is recorded directly on the disc. Unlike PCM, DSD does not employ interpolation (oversampling) filtering in the playback process. In other words, the basic DSD specification is based on the direct output of a typical sigma-delta A/D converter at 64 × 44.1 kHz.

DSD uses sigma-delta modulation and noise shaping. A simple one-bit quantizer is shown in Fig. 7.34A, and the output waveform resulting from a sine-wave input is shown in Fig. 7.34B. The shaded portion shows the difference error between the input waveform and the quantized output waveform. An example of a simple sigma-delta encoder is shown in Fig. 7.35A. The one-bit output signal is also used as an error signal and delayed by one sample and subtracted from the input analog signal. If the input waveform, accumulated over one sampling period, rises above the value accumulated in the negative feedback loop during previous samples, the converter outputs a 1 value. Similarly, if the waveform falls relative to the accumulated value, a 0 value is output. Fully positive waveforms will generate all 1 values and fully negative waveforms will generate all 0 values. This method of returning output error data to the input signal to be subtracted as compensation data is called negative feedback.

Figure 7.35B shows an input sine wave applied to a sigma-delta encoder and the resulting output signal. The pulses of the output signal reflect the magnitude of the input signal; this is a pulse density modulation representation in which a 0 value has no

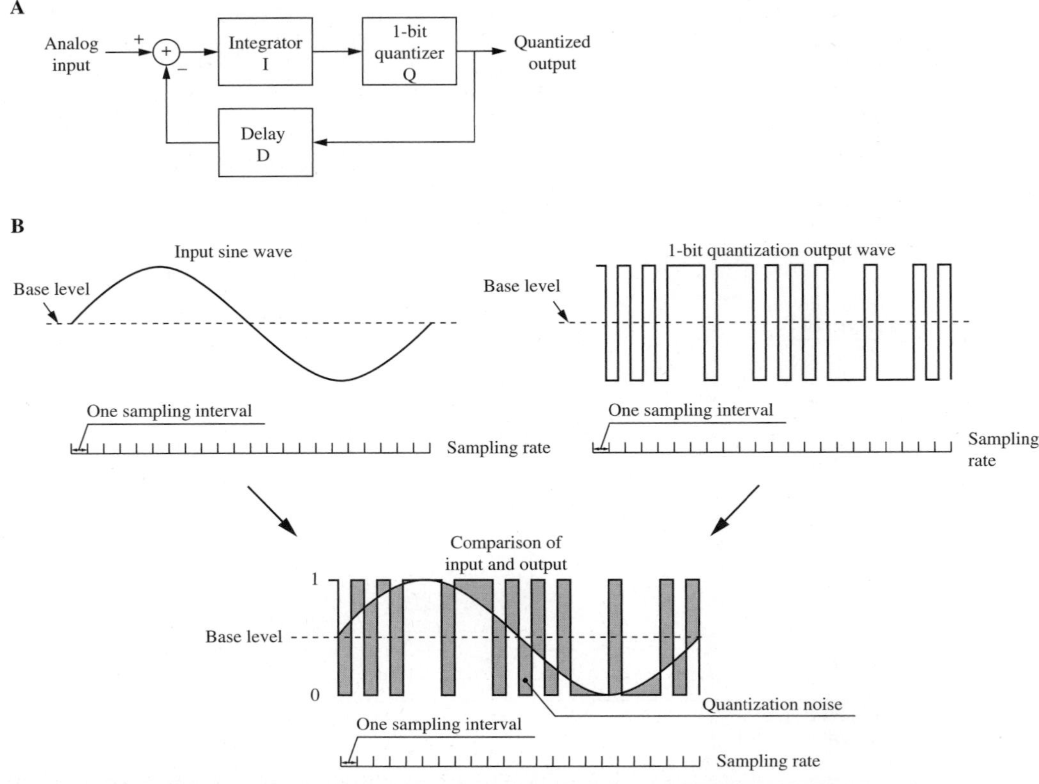

FIGURE 7.35 DSD coding uses a sigma-delta coding technique. A. A sigma-delta modulator uses negative feedback to subtract a compensation signal from the input. B. The output signal from a sigma-delta modulator is a pulse-density waveform.

pulse output while a 1 value does. The shaded portion shows the difference error; analysis shows that the volume of error is the same as in a simple quantizer; however, because the integrator (sigma) in the sigma-delta encoder acts as a lowpass filter, the amount of low-frequency error is reduced while the amount of high-frequency error is increased, as shown in Fig. 7.36. The system's designers note that the ear is sensitive to very high-frequency signals only if they are correlated to lower in-band signals. At frequencies higher than 20 kHz, they state that signal-to-noise ratios become less important. Thus, they argue that the uncorrelated high-frequency shaped noise is perceptually unimportant. This noise shaping property can be developed with higher-order (perhaps 5th order) noise shaping feedback filters to further decrease error in the audible range of frequencies. In principle, a lowpass filter can decode sigma-delta signals. Such a lowpass filter would also remove high-frequency noise resulting from noise shaping. The principles of sigma-delta modulation and noise shaping are discussed more fully in Chap. 18.

The DSD modulation used in the SACD format uses a sampling frequency that is 2.8224 MHz. In other words, the analog signal is sampled at a 2.8224 MHz rate and each

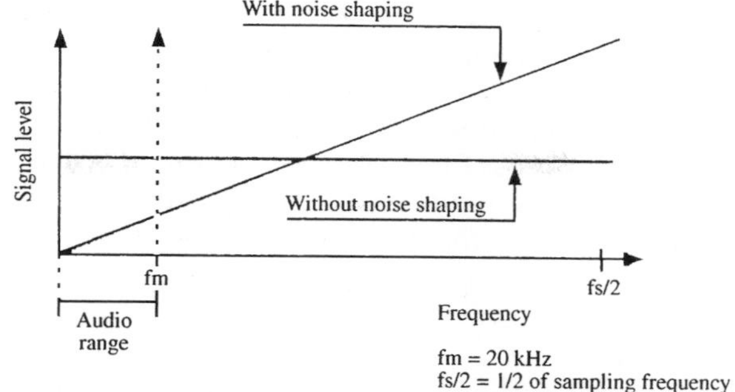

FIGURE 7.36 Noise-shaping algorithms are designed to reduce the low-frequency (in-band) quantization error, but also increase high-frequency (out-of-band) content.

sample is quantized as a one-bit word. Overall, the bit rate is thus four times higher than on a CD. In principle, the Nyquist frequency is thus 1.4112 MHz. However, in practice, to remove high-frequency noise introduced by high-order noise shaping, the high frequency response is limited to 100 kHz or less by analog filters. As shown in Fig. 7.37, a significant noise-shaping component is present in the 100-kHz band, as anticipated by the SACD standard.

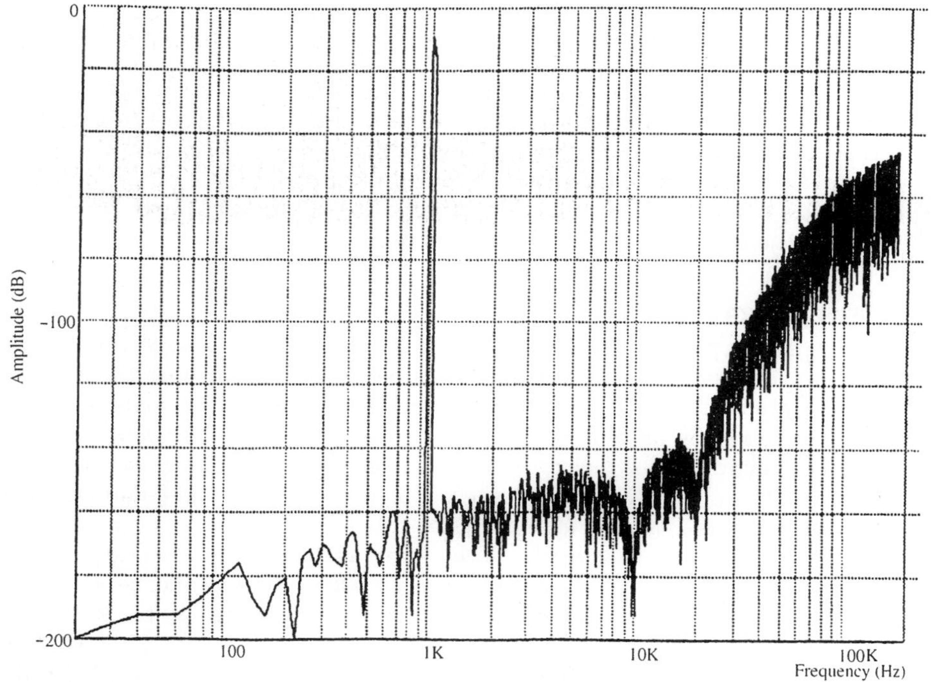

FIGURE 7.37 DSD coding used in the SACD format requires significant noise shaping to reduce low-frequency noise. However, this significantly increases high-frequency noise above 20 kHz.

The SACD standard specifies that noise power in the 100-kHz band should be 20 dB below the standard reference level. When a 100-kHz lowpass filter is used, at a volume level that achieves a 100-watt output, this noise component is thus 1 watt or less. However, at higher volume levels, the SACD standard recommends that SACD players incorporate a lowpass filter with a corner frequency of 50 kHz and a minimum 30-dB/octave slope for use with most conventional power amplifiers and speakers. When making audio measurements of the SACD, a 20-kHz lowpass filter (such as the 3344A filter by NF Electronic Instruments with 60 dB of attenuation above 24.1 kHz) is recommended to avoid the effects of the shaped components in the higher frequency range.

The 2.8224 MHz (64 × 44.1 kHz) sampling frequency of the one-bit DSD signal can be converted to a variety of standard PCM sampling frequencies with integer computation. Division by 64 and 32 yields 44.1 and 88.2 kHz. Following multiplication by 5, division by 441, 294, and 147 yields 32, 48, and 96 kHz, respectively. Also, an extended sampling frequency of 128 × 44.1 kHz is possible.

DST Lossless Coding

A lossless coding algorithm known as Direct Stream Transfer (DST) is employed in the SACD format to more than double effective disc capacity. Eight DSD channels (six multichannel plus a stereo mix) on a 4.7-Gbyte data layer are allowed a playing time of 27 minutes, 45 seconds. With DST, a 74-minute playing time is accommodated, effectively increasing storage capacity to about 12 Gbytes. As with other lossless compression methods, the compression achieved by DCT depends on the audio signal itself. In one survey, DCT yielded a coding gain of 2.4 to 2.5 for pop music, and 2.6 to 2.7 for classical music.

The DST encoder and decoder are shown in Fig. 7.38. DST uses data framing, an adaptive prediction filter and entropy coding. The use of lossless coding can be decided on a frame-by-frame basis; the flag information for the decoder is contained in each frame header. An area without any DST frames can be marked accordingly in the area TOC. DST coding yields variably sized frames; a buffer model is used to output a fixed bit rate. The theory of lossless coding is discussed in Chap. 10.

Player Design

SACD players play back both SACD and CD discs. Their design is similar to that of CD players. Dual laser pickups are required to operate at both the SACD 650-nm wavelength and the CD 780-nm wavelength. In some player designs, a single processor accepts the amplified RF signal from the dual pickup and performs clock signal extraction and synchronization, as well as demodulation and error correction for both CD and SACD signals. A servo chip controls the pickup and motor systems. CD data is passed along to the digital filter. SACD data is applied to the DSD decoder; this circuit first reads the invisible watermark, then intermittent data is rearranged and ordered in a buffer memory according to a master clock. This chip also reads subcode data, including TOC information such as track number, time, and text data.

DSD data is output as a one-bit signal at a frequency of 2.8224 MHz and applied to a pulse-density modulation processor in which the data signal is converted to a

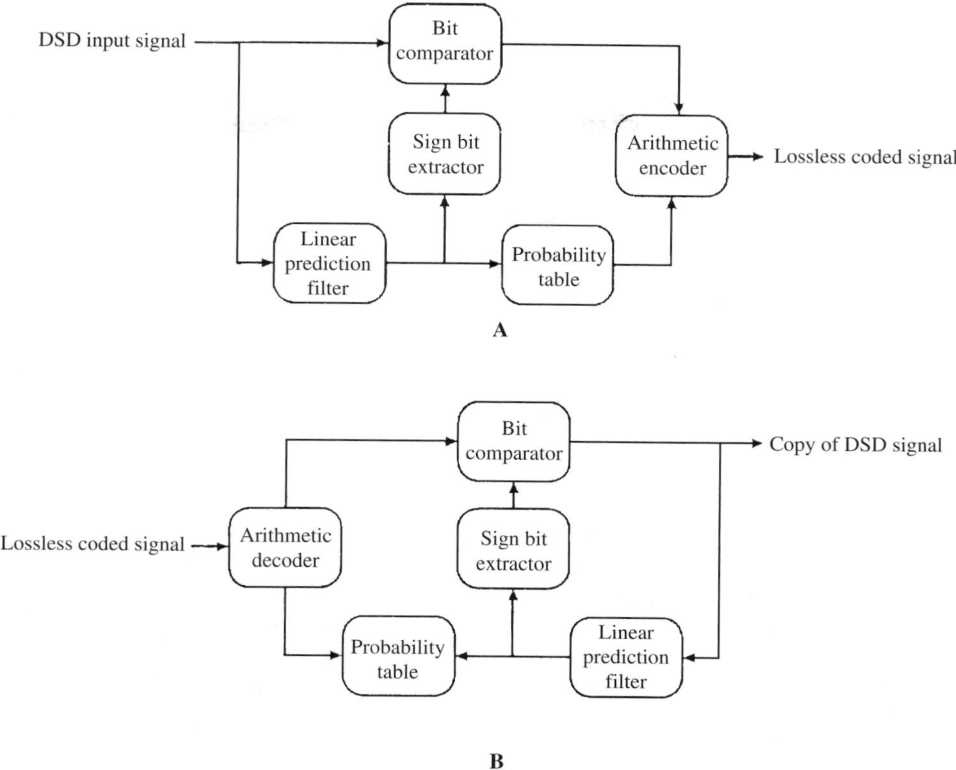

FIGURE 7.38 Direct Stream Transfer (DST) can be used for lossless coding of DSD data using an adaptive prediction filter and entropy (arithmetic) coding. A. DST encoder. B. DST decoder.

complementary signal in which each 1 value creates a wide pulse and each 0 value creates a narrow pulse. A current pulse D/A converter converts the voltage pulse output into a current pulse. This current pulse signal is passed through an analog lowpass filter to create the analog audio waveform. In some designs, this filter's response measures –3 dB at 50 kHz.

CHAPTER 8
DVD

The DVD format is a successor to the CD, and also creates new market opportunities for digital optical disc technology. DVD improves on the CD in all respects and in particular provides larger storage capacity and faster reading and writing. The technological improvement provided by the DVD specification allows filmmakers, musicians, and programmers to expand their creative horizons beyond old data carriers, and lets users enjoy the result. For everyone, DVD provides greatly upgraded capacity and flexibility of data storage . In particular, compared to previous consumer video formats, the DVD provides a significant improvement in the video and sound quality of the playback of motion pictures.

Development and Overview

Beginning with the CD-Audio format, and the subsequent development of CD-ROM, CD-R, and CD-RW, the CD revolutionized data storage. However, the CD's limited capacity and slow throughput bit rate made it unsuitable for high bandwidth or large volume applications such as high-quality digital video. Thus, the motion picture industry sought to develop a small optical disc holding a feature film coded with high-quality digital video. In December 1994, Sony and Philips proposed the MultiMedia Compact Disc (MMCD). In January 1995, Toshiba and Time Warner proposed the Super Density disc (SD). The DVD Technical Working Group subsequently outlined these criteria for the new video disc: A single standard that was fully interchangeable for TV and PC applications; fully backward-compatible with existing CD media, forward-compatible with write-once and rewritable media; low-cost and high-quality disc replication; a single universal file structure; support for both linear and nonlinear applications; and high capacity that was expandable for high definition media.

The SD and MMCD formats were similar but incompatible, and computer industry representatives urged the two sides to produce a unified standard. Subsequently, the SD and MMCD formats were merged. Moreover, the scope of the video-centric format was expanded to include digital audio and both playback-only and recordable media for general computer applications. The DVD family of formats was thus developed by a consortium of manufacturers known as the DVD Forum. Several working groups were charged with the development of different formats and aspects within the family. The preliminary DVD format was announced in December 1995. The DVD family includes different formats for video, audio, and computer applications. Because the scope of the applications far exceeded digital video, the original name of Digital Video Disc was changed to Digital Versatile Disc, but that name was never fully accepted. Instead, the format is simply called DVD.

Whatever the jargon, DVD supersedes the CD in the music and computer software markets and supersedes the LaserDisk and VHS tape in the video market. The DVD-Video and DVD-ROM formats were the first of the family to be introduced, early in 1997. Approximately 5 million DVD-ROM drives and 1.2 million DVD-Video players were sold in 1998. At the end of 1998, about 2000 DVD-Video titles were available. By mid-2004, over 34,000 DVD titles were in release, and since launch, more than 103 million DVD players had been sold and over 3 billion discs had been shipped. These early growth rates surpassed those of any other existing entertainment medium. DVD is also credited as a key factor behind the migration to digital television.

The DVD family portrait is shown in Fig. 8.1. There are six DVD books: Book A is DVD-ROM (Read-Only Memory); Book B is DVD-Video; Book C is DVD-Audio; Book D is DVD-R (Recordable); Book E is DVD-RAM (Random-Access Memory); and Book F is DVD-RW (ReWritable). The specification is further classified in several parts. Part 1 defines the physical specifications, Part 2 defines the file system specifications, and subsequent parts define specific applications and extensions. For example, Part 3 defines the Video application, Part 4 defines the Audio application, and Part 5 defines the Video Audio Navigation (VAN) extension. DVD-ROM, DVD-Video, and DVD-Audio discs (all read-only in nature) share the same disc specifications and physical format. Likewise they all share the same file system. The DVD-R, DVD-RAM, and DVD-RW formats

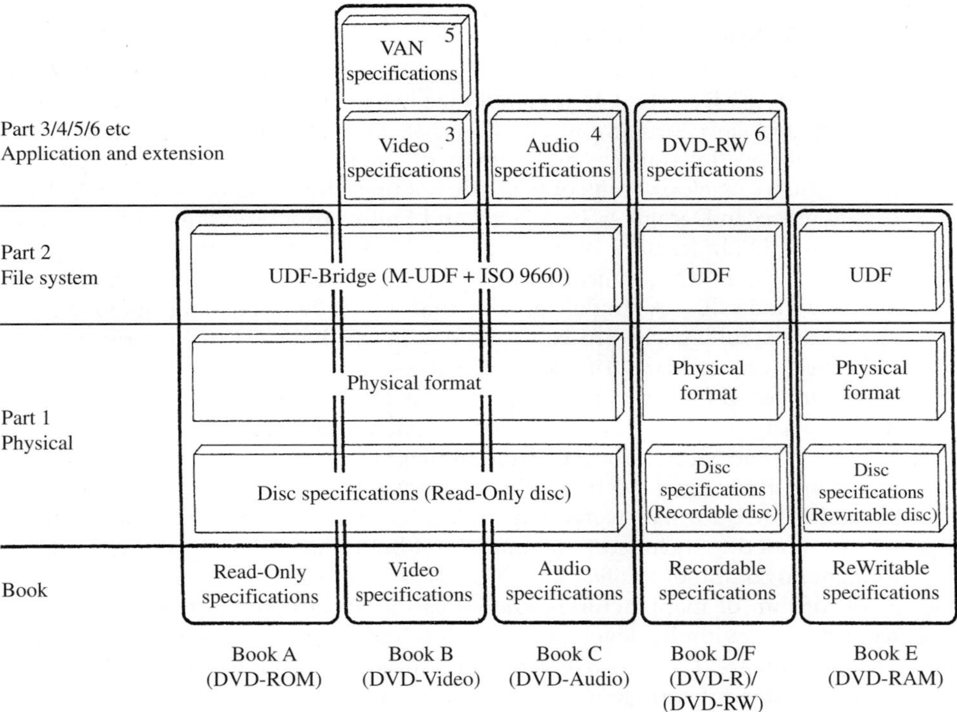

FIGURE 8.1 The DVD family of specifications includes six books for read-only and recordable discs. Some physical and file system attributes are shared, but specific application details are distinct to each specification book.

are more unique. The DVD specification borrows from other existing specifications. For example, the DVD file system uses elements of the UDF, ISO 9660, and ISO 13346 specifications. DVD-Video generally uses MPEG video coding and Dolby Digital (AC-3) or DTS audio coding, and DVD-Audio generally uses PCM and MLP (Meridian Lossless Packing) coding, as well as Dolby Digital.

Although based on CD technology, DVD employs new physical specifications and a new file format. Philosophically, DVD differs considerably from the CD. Whereas the CD was originally designed exclusively as an audio storage format, and incrementally adapted to other applications, DVD was wholly designed as a universal storage platform. The CD is also a simple format, designed to work with or without microprocessors in the player. In contrast, DVD is based on sophisticated microprocessor control to read its file structure and interact with the disc and its contents. Perhaps most importantly, Red Book CD-Audio was designed to play back a continuous stream of data thus addressing was not needed. Yellow Book CD-ROM only subsequently added addressing capability. In contrast, DVD is founded on the premise that all data will be addressable and randomly accessible. In short, all DVD contents are essentially viewed as software data. In that respect, DVD is more akin to CD-ROM than CD-Audio.

In all respects, DVD surpasses the CD format. Perhaps most strikingly, although its outer physical dimensions are identical, one DVD data layer provides seven times the storage capacity of CD. This increase is due to the shorter wavelength laser, higher numerical aperture, smaller track pitch, and other aspects, as illustrated in Fig. 8.2. In addition, overall disc and player tolerances are more stringent compared to those stipulated for the CD; this takes into account the improvement in manufacturing precision gained over the intervening 15 years. In all, the recording density of CD is 1 bit/μm^2 whereas the DVD recording density is 6 to 7 bits/μm^2. Even so, time marches on. The specifications of the DVD format have been eclipsed by the Blu-ray format.

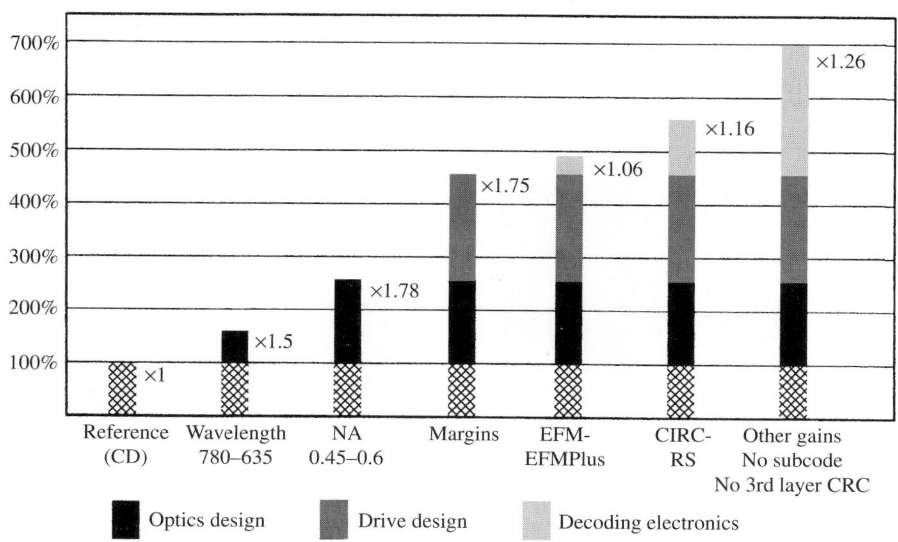

FIGURE 8.2 A DVD disc layer holds seven times the data capacity of CD. This is accomplished through improvements in optics design, improved drive design and precision, and more sophisticated decoding electronics. (*Schouhamer Immink, 1996*)

Disc Design

Part 1 of the DVD specification defines the physical specifications of DVD discs. The specifications for the DVD-ROM, DVD-Video, and DVD-Audio discs are identical; thus, Part 1 applies to all three formats. These read-only formats thus share disc construction, modulation code, error correction, and so on. The physical parameters call for 120-mm- and 80-mm-diameter discs, and single and dual layers per substrate. As with CD, DVD discs use a pit/land structure to store data. The DVD track pitch is 0.74 μm. The constant linear velocity (CLV) track velocity is 3.49 m/s on a single layer and 3.84 m/s on a dual layer. The pits and land that store binary data are as short as 0.4 μm. Minimum/maximum pit length is 0.40/1.87 μm (single layer) and 0.44/2.05 μm (dual layer).

These small dimensions are possible because the laser beam used to read DVD discs uses a visible red wavelength of 635 nm or 650 nm (both wavelengths are supported) compared to 780 nm in a CD. The standard specifies a lens with a numerical aperture (NA) of 0.6, compared to a CD's NA of 0.45. Together, the shorter laser wavelength and higher NA allow smaller pit dimensions which in turn allow a data density increase of 467% over that of CD. A DVD data layer holds about four times as many pits as a CD layer as shown in Fig. 8.3. If unwound, a DVD track would run for almost 7.5 miles. Combined with other coding efficiencies, a DVD layer can store 4.7 Gbytes of data and multiple data layers provide greater capacity. It is worth noting that the quoted DVD capacity of 4.7 Gbytes is more precisely 4.7 billion bytes (measured in multiples of 1000); in computer terms (measured in multiples of 1024) the capacity is 4.38 Gbytes.

Disc Optical Specification

A DVD disc appears similar to a CD and is the same diameter (120 mm) and thickness (1.2 mm). Whereas a CD uses a single polycarbonate substrate, a DVD disc employs two 0.6-mm substrates bonded together with the data layers placed near the internal interface. All data is held in a spiral data track that is read with counterclockwise rotation. The layer closest to the readout surface is Layer 0, and the layer further from the

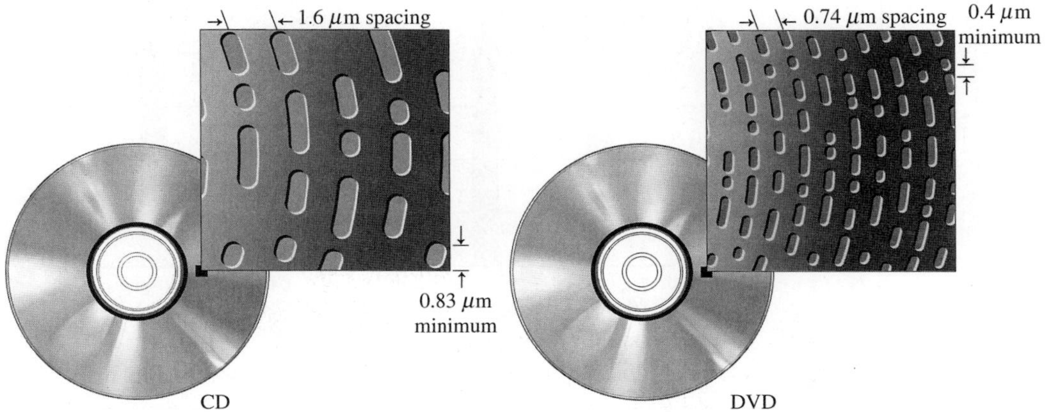

Figure 8.3 A comparison of CD and DVD data surfaces shows that a DVD holds about four times as many pits as a CD. With other improved efficiencies, overall capacity of a DVD is increased seven times.

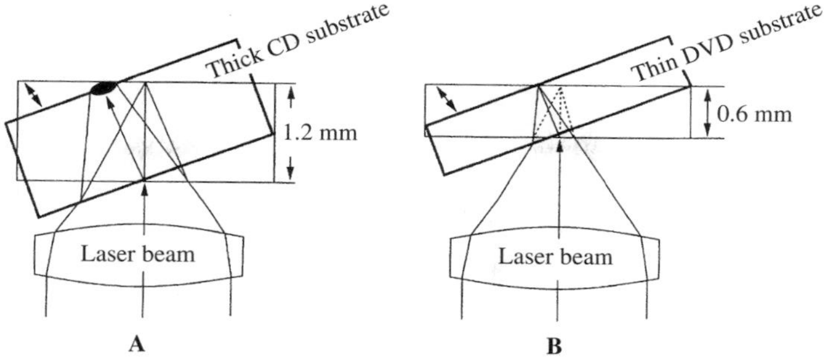

FIGURE 8.4 The thin (0.6-mm) DVD substrate is less sensitive to tracking and detection errors due to disc tilt. A. Thick CD substrate allows greater deviation. B. Thin DVD substrate has less deviation.

readout surface is Layer 1. Layer 0 is read first. A reading laser starts at the innermost radius of Layer 0 and reads to the disc outer edge, where it reaches a middle area, stops and refocuses, and reads Layer 1 from the outer edge to the inner radius; this is called opposite track path (OTP) organization. Alternatively, Layer 1 can be independently written from an inner to outer radius; this is called parallel track path (PTP). In PTP, each layer requires its own lead-in and lead-out areas.

Whereas a CD data layer is near the disc's top surface and thus somewhat vulnerable to damage, DVD data layers are embedded deeply within the disc and thus are more protected. Thinner substrates are advantageous because they are inherently more resistant to tracking errors that result when a disc is slightly tilted relative to the laser pickup, as shown in Fig. 8.4. However, because the thin substrate places the data layer closer to the outer disc surface, surface contamination is not placed as far out of focus as with a CD; this is compensated for with more powerful error correction. The DVD finished disc specification for radial tilt is ±0.8°, and for refractive index is 1.55 ± 0.10. A summary of specifications for CD and DVD is given in Table 8.1.

The dual substrate construction allows manufacturing variants, yielding five different capacities of playback-only discs. They are known as DVD-5, DVD-9, DVD-10, DVD-14, and DVD-18. As the nomenclature loosely suggests, the disc capacities are: 4.7, 8.54, 9.4, 13.24, and 17.08 billions of bytes, respectively. Expressed in terms of 8-bit bytes, they hold 4.37, 7.95, 8.75, 12.33, and 15.91 Gbytes. This is roughly a 7.4% difference (1 billion is 1,000,000,000 or 10^9, and 1 Gbyte is 1,073,741,824 or 2^{30}). When the average data output bit rate is 4.8 Mbps, the approximate playing times are: 133, 241, 266, 375, and 482 minutes, respectively. Of course, the bit rate can vary widely, and thus playing times vary too. For example, a DVD-5 disc can hold from 1 to 9 hours of video. Generally, a DVD-Video disc that holds a 2-hour movie and modest bonus features is a DVD-9 disc.

A single-layer, single-sided DVD-5 disc uses one substrate with a data surface and one blank substrate. The substrates are bonded together with either a hot-melt glue or an ultraviolet-cured photopolymer (2P); the former is generally preferred. Two substrates with data surfaces can be bonded together to form a single-layer, double-sided DVD-10 disc; the disc is turned over to access the opposite layer. The DVD standard also allows data to be placed on two layers in a substrate, one embedded beneath the other to create a dual-layer disc that is read from one side; this comprises a DVD-9 disc. Two dual-layer substrates comprise a double-sided DVD-18 disc. The layers are separated

	CD	DVD
Disc thickness (mm)	1.2	1.2 (2 × 0.6)
Disc diameter (mm)	80 or 120	80 or 120
Disc mass (g)	14–33	13–20
Disc spindle hole diameter (mm)	15	15
Lead-in diameter (mm)	46–50	45.2–48
Data diameter (mm)	50–116 (120 mm)	48–116 (120 mm)
Lead-out diameter (mm)	76–117	70–117
Outer guard-band diameter (mm)	117–120 (120 mm)	117–120 (120 mm)
Burst cutting area diameter (mm)	—	44.6–47.0
Track pitch (μm)	1.6	0.74
Pit length (μm)	0.833–3.054 (1.2 m/s)	0.400–1.866 (single layer)
	0.972–3.560 (1.4 m/s)	0.440–2.054 (dual layer)
Pit width (μm)	0.6	0.3
Pit depth (μm)	0.11	0.16
Average data bit length (μm)	0.6 (1.2 m/s)	0.2667 (single layer)
	0.7 (1.4 m/s)	0.2934 (dual layer)
Average channel bit length (μm)	0.3	0.1333 (single layer)
		0.1467 (dual layer)
Channel data rate (Mbps)	4.3218	26.15625
User data rate (Mbps)	1.41 (CD-DA)	10.08
	1.23 (CD-ROM Mode 1)	
Rotational speed (rpm)	200–500	570–1630
Scanning speed (m/s)	1.2–1.4	3.49 (single layer)
		3.84 (dual layer)
Reflectivity (%)	>70	>45–85 (single layer)
		>18–30 (dual layer)
Laser wavelength (nm)	780	635 or 650
Numerical aperture	0.45	0.60
Focus depth (μm)	1.0	0.47
Modulation code	EFM 8/14 +3 merging bits	EFMPlus 8/16
Error correction code	CIRC	RS-PC
Error correction overhead	23% (CD-DA)	13%
	34% (CD-ROM Mode 1)	
Bit error rate	10^{-14}	10^{-15}
Correctable error (mm)	2.5	6.0 (single layer)
		6.5 (dual layer)
Capacity (Gbytes)	0.783 (CD-DA)	1.4–8.0 per side
	0.635 (CD-ROM Mode 1)	

Table 8.1 Comparison of CD and DVD disc and player specifications.

by a clear resin and a very thin semi-reflective (from 25 to 40%) layer of gold or silicon; this layer is sometimes referred to as a semi-transparent layer. Gold is sputtered with conventional metallization techniques. Silicon is similarly sputtered, using argon gas. Clearly, gold is more expensive than silicon. The environmental stability and playability performance of the silicon semi-reflective layer meets or exceeds that of gold-based discs. When a layer is read through a bonding layer, the bonding material must be transparent; a 2P bonding agent can be used.

When using a semi-reflective and fully reflective layer, both layers can be read from one disc side by simply focusing the reading laser on either layer. The beam either reflects from the lower semi-reflective layer or passes through it and reflects from the top fully reflective layer. The laser light can be switched to either data layer in a few milliseconds (they are about 40 μm to 70 μm apart) by simply moving the objective lens; a buffer memory makes the transition indiscernible. Because reflectivity of the embedded layer is reduced, as is the signal-to-noise ratio (because of out-of-focus imaging from the inner layer), for reliable playback the embedded layer is formed with a faster linear velocity and thus holds somewhat less data than the top data layer. The interior data surface uses a faster scanning velocity of 3.84 m/s versus 3.49 m/s, thus the pit length is longer; for example, the minimum pit length is 0.44 μm versus 0.4 μm.

Conventional DVD discs cannot be played in CD players. However, the DVD-Forum devised specifications for a hybrid CD/DVD-Audio disc that plays in CD and DVD-Audio capable players. The DualDisc format comprises a double-sided disc that allows DVD-Audio data (multichannel audio or video) to be read from one side, and CD data (stereo music) from the other. For example, a movie could be accompanied by its (CD) soundtrack, or DVD bonus features could accompany a CD album. DualDiscs are slightly thicker (perhaps 1.4 mm to 1.5 mm) than the 1.2-mm nominal thickness of CD and DVD discs, but are playable in most DVD-Audio and CD players.

Disc Manufacturing and Playback

DVD manufacturing is very similar to CD manufacturing, as described in Chap. 7, but tighter tolerances and some new manufacturing steps are needed. Following authoring, disc content is typically imaged on a hard-disk drive, and then transferred to another delivery medium. In many cases, the content image is delivered to the disc mastering facility on Digital Linear Tape (DLT) using ANSI format. A DLT Type III tape cartridge can hold up to 10 Gbytes of uncompressed data; DLT Type IV tapes are also widely used and can hold up to 80 Gbytes of uncompressed data. With a transfer rate of 1.25 Mbytes/sec, a 135-minute program can be transferred in about an hour. A separate DLT is used for each physical disc layer and copy protection such as Content Scrambling System (CSS) and Content Protection for Prerecorded Media (CPPM) can be enabled. Alternatively, other authoring media such as DVD-R or DVD+R (single or dual layer) using the Cutting Master Format or Exabyte tape are sometimes used for simpler projects. In some cases, electronic file delivery via private secure networks is used to move files from authoring studios to replication plants. Content that will be copy protected cannot be submitted on CD-R.

Disc Description Protocol (DDP) files may accompany the master; this data provides the laser beam recorder (LBR) with disc identification information. Other data files (such as DVDID/DDPID and CONTROL.DAT) also accompany the content image file, supplying disc type, copy protection, and other information. DVD-Video and DVD-Audio authoring systems often use the Joliet extensions (supporting longer file names) in the ISO 9660 file format (within UDF Bridge); this assists compatibility in legacy operating systems that are unable to read UDF Bridge. Authoring systems must also read Macintosh file formats to allow conversion to the UDF Bridge format.

In DVD mastering, shorter wavelengths must be used in the laser beam recorder to create the smaller formations; blue, ultraviolet, or violet krypton lasers may be used. In some cases, solid-state lasers (instead of gas lasers) with frequency-doubling crystals are used. Either photoresist mastering or direct dye-polymer mastering may be used. All DVD discs use two substrates. Even in single-sided discs, two substrates, one holding data and the other a dummy substrate (with a cosmetic metal layer) must be manufactured. The thin DVD substrates require greater care in the molding process. For example, it is more difficult to uniformly flow molten polycarbonate into a thinner mold with minimal stress. The finer pit structure and the geometry of the pits (shorter and narrower than CD pits resulting in a steeper height-to-width ratio) may require injection molding machines with higher tonnage. Finally, it is more difficult to separate the disc from the stamper mold without strain.

In the case of double-sided discs, two substrates are independently formed, and then bonded together using a hot-melt adhesive or UV-curable bonding agent (the latter is preferred). As noted, dual-layer discs can be manufactured by independently molding two 0.6-mm polycarbonate substrates with one layer receiving full metallization and the other receiving semi-reflective metallization. The two substrates are then bonded together with a layer of UV-cured optically clear photopolymer as shown in Fig. 8.5. The reading laser can focus through the semi-reflective layer (and the bonding

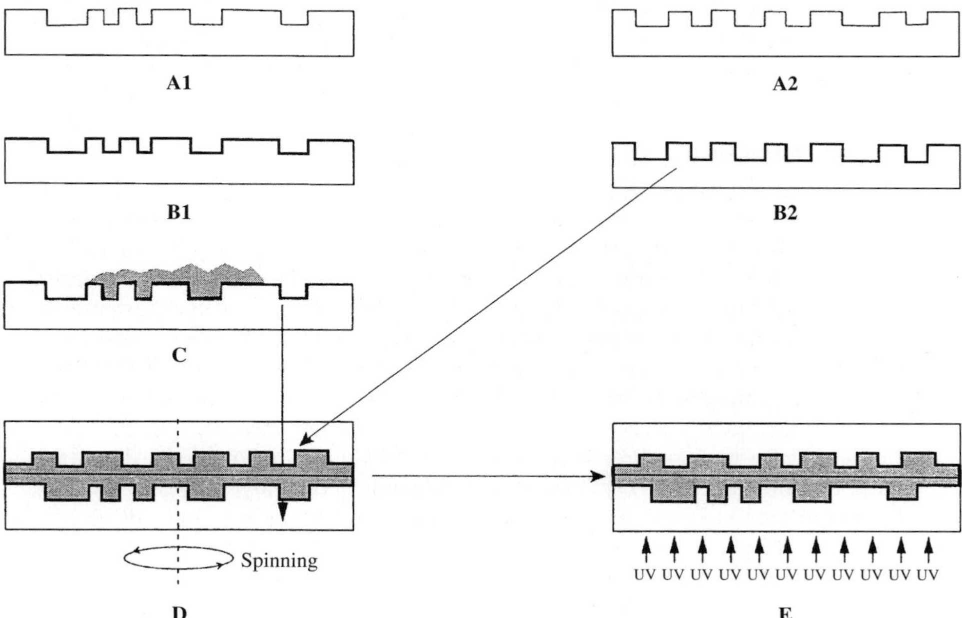

FIGURE 8.5 Single-sided, dual-layer DVD-9 discs can be manufactured with data layers on two substrates, one with a semi-reflective surface and another with a fully reflective surface. The replication steps are shown. A1. Replicate first-layer substrate. A2. Replicate second-layer substrate. B1. Deposit semi-reflective layer. B2. Deposit fully reflective layer. C. Put UV-hardened bonding resin on substrate. D. Bond substrates. E. Harden with UV light.

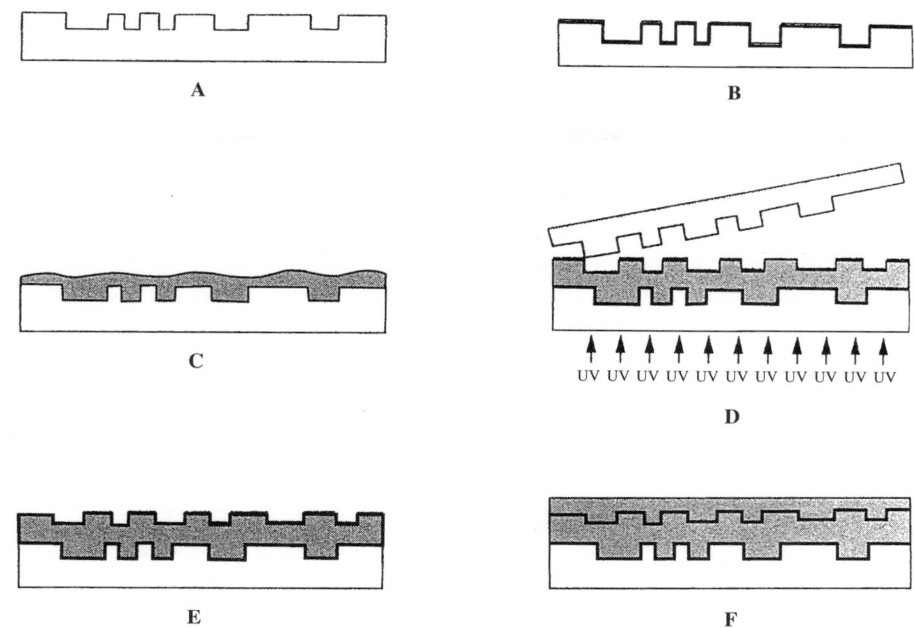

Figure 8.6 Dual-layer substrates can be manufactured by pressing a second data layer into an intermediate resin layer. This technique can be used to produce substrates for single-sided, dual-layer (DVD-9) discs and double-sided, dual-layer (DVD-18) discs. The replication steps are shown. A. Replicate first layer substrate. B. Deposit semi-reflective layer. C. Put UV-hardened resin on substrate. D. Replicate pits of second layer on resin by stamper. E. Deposit fully reflective layer. F. Apply UV-hardened resin to form protective layer.

material with a thickness of about 55 μm) to read the upper substrate; this design can be used to manufacture single-sided discs (such as some DVD-9 discs).

Alternatively, a molded single-layer substrate can be coated with a semi-reflective layer followed by a layer of liquid photopolymer that is molded by a second stamper and hardened by exposure to ultraviolet light; after the layer is hardened, the stamper is removed and a fully reflective metal layer is applied and the substrate is ready for bonding to a second substrate, as shown in Fig. 8.6. This technique is used for some DVD-9 and DVD-18 discs. In another approach, the first information layer is molded on an interim 0.6-mm polymethyl methacrylate (PMMA) substrate, and it is sputtered with aluminum. The PMMA substrate (which shares no strong molecular bonds with aluminum) is peeled away and recycled, leaving the resultant aluminum information layer that is transferred onto a basic gold metal/polycarbonate DVD substrate with an adhesive. The six disc types are shown in Fig. 8.7.

Optical Playback

Although backward compatibility with CD is not required (but is recommended) by the DVD specification, in practice all DVD players can read CD discs with molded plastic pits. To accomplish this, the laser must be able to focus on CD data layers at about 1.2 mm from the readout surface, as well as DVD layers at about 0.6 mm from the surface. Some

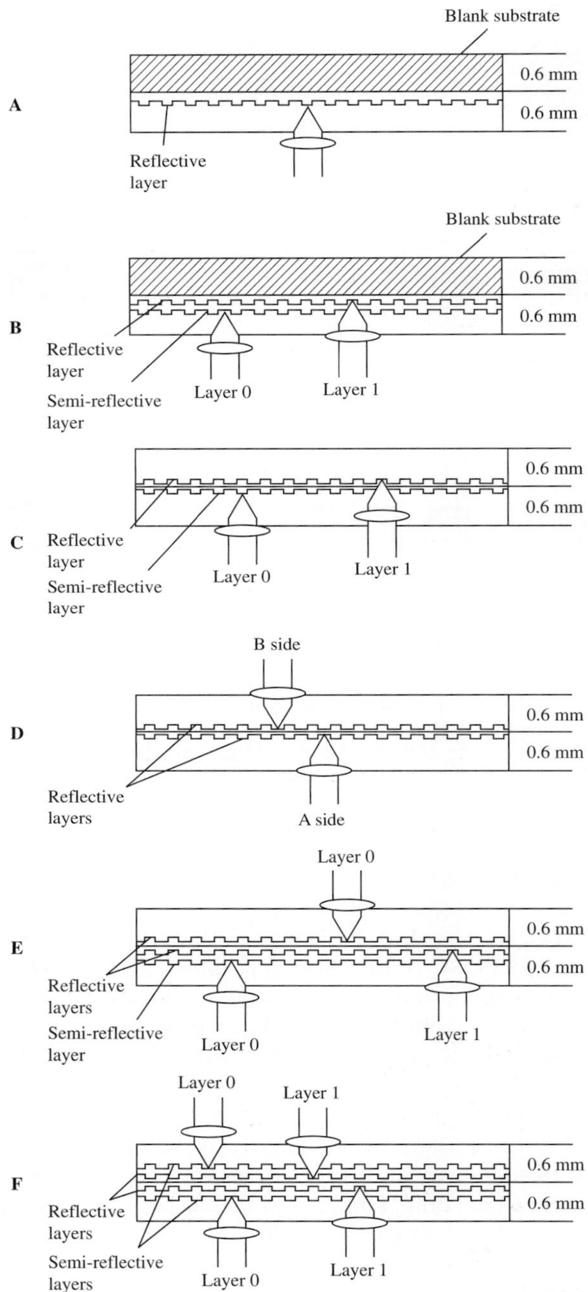

FIGURE 8.7 The DVD specification allows multiple disc types. A. Single-sided, single-layer DVD-5. B. Single-sided, dual-layer DVD-9. C. Single-sided, dual-layer (alternate version) DVD-9. D. Double-sided, single-layer DVD-10. E. Double-sided, single-/dual-layer DVD-14. F. Double-sided, dual-layer DVD-18.

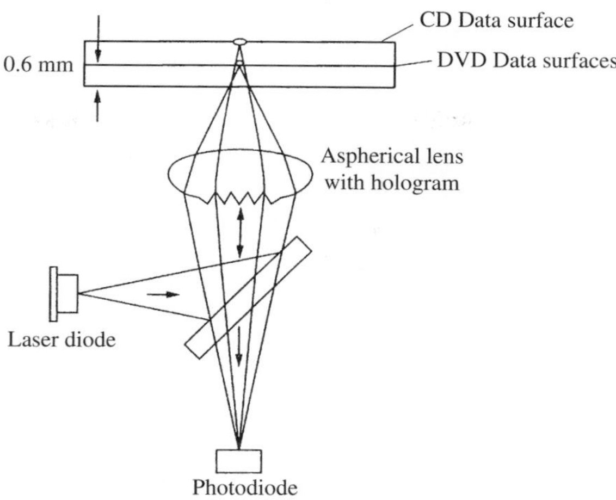

FIGURE 8.8 A DVD pickup can be designed to focus on either CD or DVD data layers. Different focal lengths can be achieved with a variety of techniques including a holographic lens.

pickups use a single shared objective lens for the CD and DVD wavelengths. The pickup has both 780-nm and 635/650-nm laser diodes, and a largely shared light path. In some designs, as shown in Fig. 8.8, holographic pickups use an aspherical lens; annular rings are cut into the center of the lens so that light passing through it is diffracted, yielding a longer focal length. Light passing through the outer smooth part of the lens has a shorter focal length. The optical spots appear to the photodiode array as concentric circles; light reflecting from 1.2 mm forms an inner circle and light reflecting from 0.6 mm forms an outer circle. Alternatively, a dual pickup can be designed with independent sections for each wavelength source including two separate objective lenses—one for CD and another for DVD.

Playback of CD-R discs is problematic; the optical response of the organic dye recording layer is extremely wavelength-dependent, with high absorption below a narrow range of around 780 nm. For example, a CD-R disc may have 65% reflectivity at 780 nm and only 10% reflectivity at 650 nm. As a result, contrast is low and the disc is difficult to read at 635 nm or 650 nm. CD-R-compatible DVD pickups are designed with two discrete optical paths at two wavelengths, or may employ one objective lens with two lasers. In one design, the numerical aperture is adjusted by coating the lens' outer circumference with a material that is opaque at 780 nm but transparent at 635 nm and 650 nm. When a 780-nm laser is used, the coating restricts the NA to 0.45, but when a 635-nm or 650-nm laser is used, since the coating is transparent, the NA is increased to 0.6 for reading DVD discs. Alternatively, for example, a dual-laser pickup could mount two objective lenses on a rotating head that places the appropriate lens in the optical path.

Many DVD drives use a differential phase-detection (DPD) method for autotracking. The pickup monitors asymmetry in the intensity pattern of the diffraction from the edges of pits. In particular, as an off-center spot moves from the leading edge to the trailing edge of a pit, the intensity of the pattern rotates. The pattern illuminates a four-quadrant photodiode, as shown in Fig. 8.9. In this example, at the leading edge of the pit, diode pairs B + D receive less light than pairs A + C; when the spot is in the middle

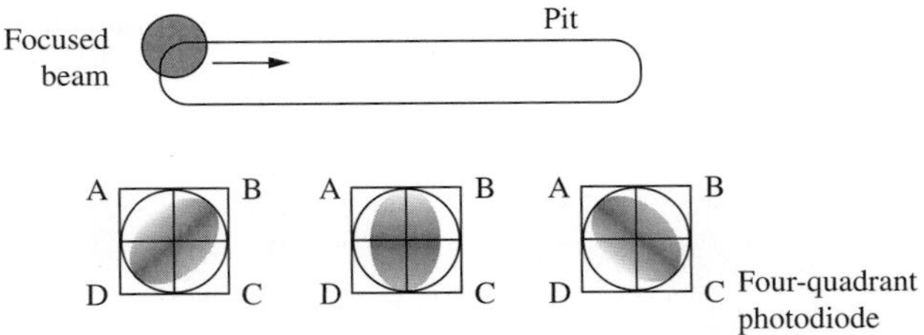

Figure 8.9 In the DPD autotracking method, an off-center beam creates a rotation in the intensity of the reflected beam as a pit is scanned. This is sensed by a four-quadrant photodiode. In this figure, darkened areas in the photodiode represent lower light intensity. (*Carriere et al., 2000*)

of the pit, the diode pairs receive equal intensity; at the trailing edge, B + D receive more light than A + C. This is used to generate a tracking error signal.

Data Coding

The disc lead-in area is the innermost area of the Information Area. It consists of the Initial zone, Reference code zone, Buffer zone 1, Control data zone, and Buffer zone 2. A Control data block comprises 16 sectors; information includes disc size, minimum readout rate, single/dual layer, track path, disc manufacturing information, and copyright. A Burst Cutting Area (BCA) is located inside the lead-in area, between 44.6 mm and 47 mm from center. BCA can create a unique serial number or ID code comprising a series of low-reflectance stripes, similar to a bar code, extending along the radial direction. It holds up to 188 bytes, in 16-byte increments. The code can be recorded by a high-power YAG laser that melts the aluminum sputtering layer to create the lines; codes can be written in both single and dual layers. The code is read by a drive's optical pickup. Recordable DVD media use the BCA to uniquely identify each disc; this can be used to encrypt recorded data for copy-protection purposes.

The read-only DVD data structure is similar to that of a CD-ROM in which data is stored in files within directories; this increases data density. DVD data is placed on a disc in physical sectors that run continuously without a gap from the lead-in to the lead-out areas. The lead-in area ends at address 02FFFF and data begins at address 030000. Two types of dual-layer discs are defined: parallel track and opposite track. In a parallel track path disc, addresses of both layers ascend from the inner radius, and there are two lead-in and lead-out areas. In an opposite track path disc, Layer 1 addresses ascend as the laser moves toward the inner radius, and there is one lead-in and lead-out area, and two middle areas. In a DVD disc, the lead-in area starts 2 mm closer to the center than on a CD.

A data sector comprises 2064 bytes, called Data Unit 1. It consists of 2048 bytes of main user data and 16 bytes of header; the latter comprises four bytes of sector identification (ID), two bytes of ID error detection (IED), six bytes reserved for copyright management, and four bytes of error detection code (EDC) data. A data sector can be viewed as a block of 2064 bytes with 172 bytes in 12 lines, as shown in Fig. 8.10. The four bytes of ID contain one byte of sector information and three bytes of sector number. A synchronization code is added to the head of every 91 bytes in the recording sector; this forms a physical sector; in all, 52 bytes of synchronization code are added. The initial 2048 bytes of user data is thus increased to 2418 bytes (2048 user + 16 header + 302 error correction + 52 synchronization).

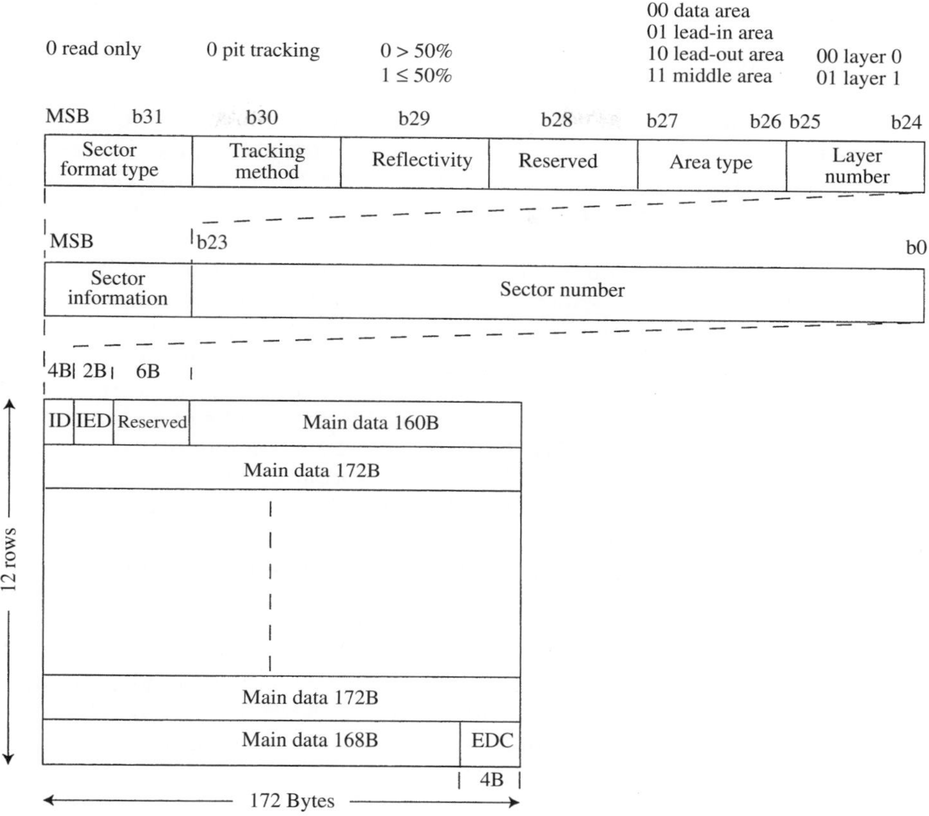

Figure 8.10 DVD data is placed in sectors, each with 2064 bytes of data. Four bytes of ID data contain sector information.

Reed–Solomon Product Code

The DVD format employs a Reed–Solomon Product Code (RS-PC) for error correction. This code uses a combination of two Reed–Solomon codes (denoted by C1 and C2) as a product code. It differs from the CD's CIRC code and is more similar to the code used in the DAT format. In the CD format, CD-ROM data can be coded with additional error correction; in the DVD format, all disc types use the same level of error correction. Moreover, in DVD, error concealment is not used. Instead, the data reliability of all DVD data must approach computer standards. The CIRC code uses a convolutional structure that is suited to long streams of data. In contrast, the matrix structure of the RS-PC code is suited to small blocks of data. It is a product code in which rows of outer parity are crossed with columns of inner parity. A small disadvantage of an RS-PC code is its larger memory requirement.

Error correction is more challenging on a DVD disc than on a CD because the pit size is smaller. In addition, because of the thin substrates, surface defects can more readily obscure the data surface (although the outer surface is out of focus to the reading laser in relation to the interior data layer). However, the superior RS-PC error correction code provides improved overall error protection compared to CIRC and is also more powerful than the double error correction used in the CD-ROM format. Because RS-PC is more efficient than CIRC in terms of overhead, its use increases data density by 16%.

In RS-PC, the two C1 and C2 product codes are (208,192) and (182,172) in length. The rate of the code is thus $(172 \times 192)/(182 \times 208)$, or 0.872. RS-PC is applied to the 2048 bytes of main data, each error-correction code (ECC) block providing error correction encoding over 16 data sectors. A total of 302 bytes of error correction code are added to each sector. Each ECC block thus contains 32,768 bytes of user data, 4832 bytes of ECC, 96 bytes of sector EDC, and 160 bytes of ID and copy protection, totaling 37,856 bytes. A 16-byte outer code parity (PO) and 10-byte inner code parity (PI) are added to form recording sectors. PO is formed from 172 columns and yields 16 new rows. PI is formed from 208 rows (192 + 16). The data block is parsed into groups of 12 data rows plus one parity row to create recording sectors of 182 bytes. Overall, error detection and correction data requires an overhead of about 13% of the recorded sectors. Single-byte synchronization codes are placed in the middle of each recording sector. This Data Unit 3 thus holds 2418 bytes (2366 + 52); this unmodulated physical sector is used to record data in all the recordable DVD formats.

The principal error criterion for CDs is the BLER measurement. In DVD discs, PI and PO error rates are used. PI errors use an 8 ECC block running sum. It counts the number of PI rows with any bad symbols. PI failures are the number of uncorrectable PI rows per ECC block. PO failures are the number of uncorrectable PO columns per ECC block. C1 and C2 can be decoded multiple times to improve performance. The maximum correctable burst length for CIRC is about 500 bytes (2.4 mm), while it is about 2200 bytes (4.6 mm) for RS-PC. The RS-PC can reduce a random input error rate of 2×10^{-2} to a data error rate of 10^{-15}. This is better than a CD by a factor of 10. For example, on a DVD-ROM, a burst error of about 2800 bytes can be corrected; this corresponds to an obstruction that is 6 mm in length.

EFMPlus Modulation

Data on DVD discs is recorded with EFMPlus modulation. It is an 8/16 RLL code that writes each 8-bit data byte as a 16-bit modulated channel byte. The modulated physical sector is thus 4836 bytes. EFMPlus is similar to the EFM code used in CDs; for example, it uses the same minimum (2) and maximum (10) run length and represents binary one channel bits as pit/land or land/pit transitions, and binary channel zero bits as no transition. EFM uses 8/14 coding with three merging bits to yield an 8/17 ratio. EFMPlus provides a 6% increase in user storage capacity compared to EFM because its coding is more efficient than EFM. As with many other codes, EFMPlus promotes timing recovery and suppression of low-frequency components. However, whereas EFM uses merging bits, a single lookup table, and simple concatenation rules to suppress low-frequency content, EFMPlus does not require merging bits and uses a more sophisticated lookup method. The EFMPlus encoder defines four lookup tables each with 351 possible source words. In practice, the source codebook size is 344; seven possible words are discarded to allow for a unique 26-bit synchronization word. Of these, 256 words are used to code input data. The remaining excess 88 words (344 – 256 = 88) are used to control low-frequency content.

DC suppression is accomplished by monitoring the digital sum value (DSV); the surplus words are used as alternative channel representations for codewords 0 through 87 to create alternative tables. Either main or alternative codewords are actively selected to minimize the running DSV. The decoder uses an array to examine the current 16-bit codeword and two positions of the upcoming codeword to translate 16 + 2 channel bits into 8-bit data words. Data-to-clock jitter is measured from the EFMPlus signal to the PLL clock for one disc revolution. Data-to-data jitter (effect length) is measured as timing between pits and lands. This differs somewhat from the data-to-data jitter measurement often used in CDs. The DVD-ROM specification calls for jitter of less than 8% of the channel bit clock period; if this period is 38 ns, then jitter must be less than 3 ns. DVD discs do not have a separate subcode area as in CDs; the subcode functions are intrinsically contained in the data format.

When a DVD disc is read, data passes through a buffer and then is evaluated by a navigator/splitter that separates the bitstream into video, sub-picture, audio, and navigational information. If necessary, the video, sub-picture, and audio data is descrambled and decoding takes place; for example, MPEG-2 video data is decoded as is Dolby Digital (AC-3) audio data. This can occur in a dedicated hardware chip or with software via a computer CPU. Video data is routed to the display monitor, audio is sent to other outputs, and navigational information is used by a controller for the user interface.

Universal Disc Format (UDF) Bridge

Part 2 of the DVD specification specifies a file format called UDF Bridge. It is fundamentally based on a simplified version of UDF called Micro UDF and includes ISO 9660. Read-only DVD discs must use the Universal Disc Format (UDF) Bridge for volume structure and file format; it is designed specifically for optical disc storage. It is common to the DVD-ROM, DVD-Video, and DVD-Audio formats (and applies to the write-once and rerecordable disc formats). However, application-specific parameters are unique to each of the B (Video) and C (Audio) books. The DVD format is unlike CD-Audio, the DVD being fundamentally computer-based with a file format defined for all its applications. And, whereas CD-ROM was designed without designating a specific file format, a DVD disc must use UDF Bridge. UDF Bridge is a simplified version based on Part 4 of ISO/IEC 13346 and conforms to both UDF and ISO 9660 (the file format used in CD-ROM). In other words, DVD-Video and DVD-Audio are much closer in concept to CD-ROM, than to CD-Audio.

UDF Bridge defines data structures such as volumes, files, blocks, sectors, CRCCs, paths, records, allocation tables, partitions, and character sets, as well as methods for reading, writing, and other operations. It is a flexible, multi-platform, multi-application, multi-language, multi-user oriented format that has been adapted to DVD. It is backward-compatible to existing ISO 9660 operating system software; however, a DVD-Video or DVD-Audio player supports only Micro UDF and not ISO 9660. UDF Bridge standardizes many file and directory names to simplify operation. For example, certain subdirectories are always read first. UDF Bridge permits use of DVD in Windows, Macintosh, Unix, OS/2, and DOS operating systems as well as dedicated players. Because the UDF Bridge file system is unified with the DVD format, host computers can be programmed to use DVDs. Conversely, a dedicated player can ignore files within UDF Bridge that it does not need for operation.

UDF Bridge defines the following: a Sector is the smallest addressable data field (2048 bytes); a Volume is a sector address space; a Volume Set is a collection of one or more volumes; a Volume Group within a volume consists of one or more consecutively numbered volumes; a File is a set of sectors with sector numbers in a continuously ascending sequence; an Application is a program that processes the contents of a file; a Descriptor contains information about a volume or file. UDF Bridge supports multiple extent files in which parts may be located noncontiguously on a disc. It supports file names with mixed cases, up to 256 characters long. It also supports Unicode, which supports all character sets. UDF Bridge also supports the resource fork and data fork in Macintosh files. UDF Bridge specifies a time stamp: year (1 to 9999), month, day, hour, minute, second, centiseconds, hundreds of microseconds, and microseconds. The UDF specification was developed by the Optical Storage Technology Association.

Because of its diverse applications, the UDF Bridge file format specification is quite detailed. However, its basic directory and file structure is quite explicit. For example, Fig. 8.11A shows how data is organized in a DVD-Video "Video" disc. Under a root directory, a DVD-Video zone and DVD-Other zone are defined. Within the DVD-Video zone, the VIDEO_TS directory contains both menu and program data. In particular, a Video Manager defines file types and organization of both video and audio data, and Video Title Set (VTS) subdirectories contain video and audio data files (such as MPEG-2 video and Dolby Digital audio). One Video Manager can contain up to 99 VTS subdirectories. Other computer data may be contained in the DVD-Other zone; this data may be used by DVD-ROM drives, and is ignored by DVD-Video players. This information may allow Internet connectivity, enhanced music, and other features.

Figure 8.11B shows how data is organized in a DVD-Audio "Audio-Only" disc. Information is contained in the DVD-Audio zone. Audio data, such as PCM, is contained in an Audio Title Set (ATS). An Audio Manager defines file types and organizes both audio and video data. Both menu and program data is included. Figure 8.11C shows how data is organized in a DVD-Audio AV "Audio with Video" disc. Audio data is contained in an Audio Title Set and video data in a Video Title Set. The Audio Manager and Video Manager define file types and organize both audio and video data. Both menu and program data is included. The Audio Manager can control a subset of the DVD-Video data. Link Info shows that a DVD-Audio player can play audio components of video contents. Figure 8.11D shows how data is organized in a DVD-Video VAN "Video Audio Navigation" disc. Audio data is contained in an Audio Title Set, and video data in a Video Title Set. The Audio Manager and Video Manager define file types and organize both audio and video data; both menu and program data is included. "Link Info" shows that a DVD-Audio player can play audio components of video contents.

DVD-Video

DVD-Video was the first DVD format to be launched and is the most widely used DVD format. In a typical application, a DVD-Video disc stores a feature film with 5.1-channel and stereo soundtracks. The film industry participated in the development of DVD-Video, and it was designed according to recommendations from the Studio Advisory Committee: approximately 135 minutes of digital video on one disc side, approaching CCIR-601 broadcast picture quality, stereo or multichannel digital audio, multiple aspect ratios, up to 8 language soundtracks, up to 32 subtitle streams, parental control options, and copy protection.

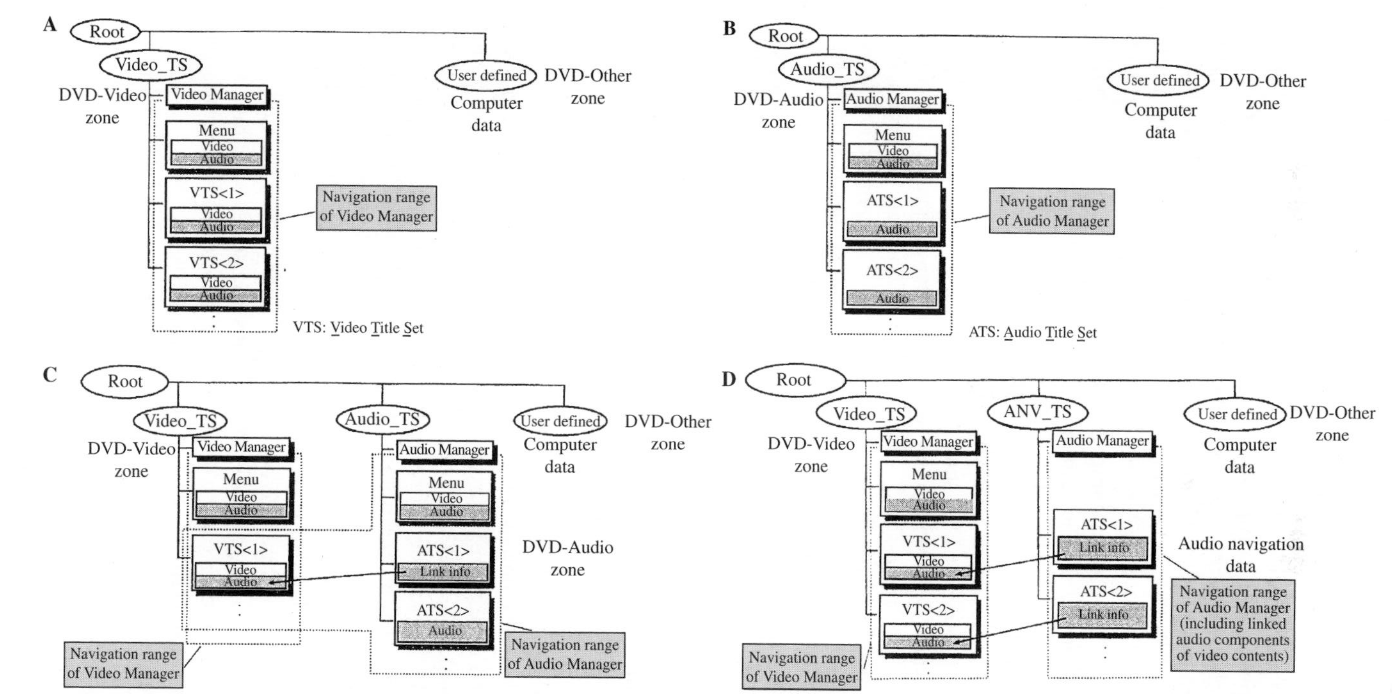

Figure 8.11 The same directory and file structure is used for different types of DVD discs. A. In a "Video" DVD-Video disc, the DVD-Video zone contains the Video Manager and VTS subdirectories that contain data files. B. In an "Audio-Only" DVD-Audio disc, the DVD-Audio zone contains the Audio Manager and ATS subdirectories. C. In an "AV" DVD-Audio disc, the Audio Manager can control a subset of the DVD-Video data. D. In a "VAN" DVD-Video disc, Link Info allows a DVD-Audio player to play audio components of video contents.

DVD-Video Video Coding

The task of storing a feature film on an optical disc is far from trivial. A CD disc is woefully inadequate for high-quality video storage. For example, if a movie was coded at a video bit rate of 166 Mbps, a CD would hold only about 40 seconds of this uncompressed video and would have to spin at a rate of 58,850 rpm, or 118 times faster than its normal speed. To accomplish its task, a DVD-Video disc takes advantage of the inherently higher capacity and output bit rate of DVD, and more importantly employs data reduction techniques. Although a DVD-Video data layer can hold seven times the data of a CD (4.7 Gbytes) it is insufficient to store a feature film; at a bit rate of 166 Mbps, a layer would hold less than 5 minutes of video. To overcome this, the DVD-Video standard uses the MPEG-2 data compression algorithm to encode its video program. The algorithm used for DVD-Video is based on the MPEG-2 Main Profile at Main Level protocol, also known as MP@ML.

MP@ML is an intermediate level, and below the High Level sometimes used for digital television (DTV). However, MP@ML yields a high-quality picture that equals that of the professional CCIR-601 (or D-1) standard operating at a rate of 270 Mbps. Even if it could be recorded to a DVD-Video disc without compression (it can't), this CCIR-601 video bitstream would fill a single-sided, single-layer DVD disc in 140 seconds. To instead store over 2 hours of an audio/video program would imply an overall reduction of about 60:1. However, several prefiltering operations reduce the burden of algorithmic compression.

An NTSC CCIR-601 signal assumes a sampling rate of 858 samples per scan line, 525 lines per frame (yielding a 858-by-525 display) and 30 frames per second; however, many of these samples are offscreen in blanking intervals. Thus, DVD-Video reduces the number of pixels to a 720-by-480-pixel display. The video bit rate is further decreased by decreasing the word length (for example, from 10 bits per sample to 8 bits per sample). Furthermore, rather than code RGB components, a YCrCb representation can be used more efficiently. For example, both the vertical and horizontal resolution of the chrominance information can be halved; the video program is stored as 4:2:0 component video instead of 4:4:2. These steps reduce the bit rate by 54%.

Additional efficiency can be realized when coding movies. Movies are filmed at 24 frames per second, whereas DVD-Video operates at 30 frames per second; this means that after conversion, 6 out of every 30 video frames are repeated, and this need not be separately coded. Overall, this type of prefiltering on the input signal may decrease the bit rate by 63% (for movie sources). Although the bit rate may be "only" 100 Mbps, it still requires algorithmic compression. For example, to place a 133-minute movie on a single-sided, single-layer disc, an average compression ratio of 21:1 is still needed.

The MPEG-2 video compression algorithm uses psychovisual models to analyze the video signal to determine how a human viewer will perceive it. Image data that is deemed redundant, not perceived, or marginally perceived, is not coded. This analysis is carried out for both individual video frames and series of frames. Over time, as much as 95% of the video data can be omitted without significant degradation of the picture. Video compression is discussed further in Chap. 16.

An important aspect of MPEG-2 coding is its variable bit rate (MPEG-1 uses a fixed bit rate). Because some pictures are more difficult to code than others, MPEG-2 allows for a variable bit rate. The bit rate needed to code motion pictures can vary greatly from scene to scene. For example, a scene with a "talking head" surrounded by a static background would require a relatively low bit rate. A fast action scene with a changing complex picture would require a higher bit rate. MPEG-2 encoders output a changing bit rate that reflects the changing degree of picture complexity and coding difficulty; in this way, bits are not wasted on low complexity frames, and artifacts are avoided in high complexity frames. In some encoders, video content may be coded in three passes, to optimize coding efficiency.

The MPEG-2 algorithm is specifically engineered so that improvements can be made in the encoding algorithm while retaining complete compatibility with existing decoders. Thus, the look of video software titles can improve, and the improvement will be seen on current and future players. The picture quality of a particular DVD-Video title is also influenced by the expertise used in the picture encoding. An overview of the DVD-Video properties is shown in Table 8.2.

Navigation Structure		
Playback Control	**Information/Control File**	
Navigation player model (command and user operation)	VMGI (Video Manager)	
	VTSI (Video Title Set)	
	PGC (Program Chain)	
	PCI (Presentation Control Information)	
Presentation Structure		
Multiplex System: MPEG-2 Program Stream		
Video	**Audio**	**Sub-Picture**
1 Stream MPEG-1, 2 MP@ML Bit rate MPEG-2 9.8 Mbps MPEG-1 1.856 Mbps	Max 8 streams 525 system AC-3, PCM, (MPEG) 625 system MPEG, PCM, (AC-3) AC-3 f_s = 48 kHz Max 448 kbps Max 5.1 ch surround MPEG-1, 2 f_s = 48 kHz Max 384/912 kbps Max 7.1 ch surround PCM f_s = 48,96 kHz 16/20/24 bit Max 8 ch	Max 32 streams Run-length coded bitmap 2 bit/pixel

Table 8.2 Summary of the principal characteristics of the DVD-Video format.

DVD-Video Audio Coding

The audio portion of the DVD-Video standard accommodates both multichannel and stereo soundtracks. DVD-Video titles can accommodate up to eight independent audio bitstreams. These audio bitstreams can be 1 to 8 channels of PCM, 1 to 6 channels of 5.1-channel (five main channels plus a low-frequency effects channel) of Dolby Digital, or 1 to 8 channels (5.1 or 7.1) of MPEG-2 audio. An NTSC title must include at least one Dolby Digital or PCM audio track. A PAL title must contain at least one Dolby Digital, PCM, or MPEG-2 audio track. A disc can also optionally employ DTS, SDDS, or other audio coding. Neither MP3 or AAC coding is allowed in the DVD-Video or DVD-Audio zone; however, these codings may be used in areas outside the zones. Dolby Digital is the standard coding used for multichannel soundtracks in the United States and Canada (Region 1) and other regions as well. Typically, Dolby Digital soundtracks are either stereo or 5.1-channel. For strictly commercial reasons, NTSC players generally play only NTSC discs, but many PAL players can play both NTSC and PAL discs. In either case, the corresponding display (NTSC or PAL) must be connected. Most computers can play both NTSC and PAL discs.

As used with DVD-Video, Dolby Digital (AC-3) codes 1 to 5 main discrete channels plus a discrete low-frequency effects (LFE) channel. A rear center channel can be added using Dolby Digital Surround EX which uses phase matrix encoding. The Dolby Digital sampling frequency is 48 kHz (the norm in digital video applications), the nominal output bit rate is 384 kbps, and the maximum bit rate is 448 kbps. Dolby Digital accommodates resolution of up to 24 bits. Dolby Digital decoders must also be able to downmix 5.1-channel soundtracks to stereo PCM. The center and surround channels are matrixed with the main stereo channels in the Dolby Surround format (that can be decoded by Dolby Pro Logic decoders). The LFE channel is not included in the downmix (for that reason, surround mixes should have bass content that takes full advantage of the low-frequency response of the main channels). An alternative to downmixing in the player is to simply create a new stereo mix and place that on the disc.

In the DVD-Video format, DTS can be optionally used to code 1 to 6 discrete main channels of audio data plus a discrete LFE channel. A rear center channel can be added discretely or by matrixing. The DTS sampling frequency is 48 kHz, and resolution is up to 24 bits. Typical bit rates for a 5.1-channel program are 768 kbps or 1536 kbps (the maximum rate). MPEG-1 stereo audio (Layer II) is sampled at 48 kHz with a maximum resolution of 20 bits and a maximum bit rate of 384 kbps. MPEG-2 multichannel audio (BC matrix mode) codes up to 7.1 channels. It is also coded at 48 kHz with a maximum resolution of 20 bits and a maximum bit rate of 912 kbps. The AAC codec is not supported. Audio codecs such as Dolby Digital, DTS, MPEG-1, and MPEG-2 are discussed in more detail in Chap. 11.

For compatibility, DVD-Video movies also carry a redundant PCM digital stereo soundtrack. These PCM audio tracks can employ sampling rates of either 48 kHz or 96 kHz, and word lengths of 16, 20, or 24 bits. The CD sampling frequency of 44.1 kHz is not supported on DVD-Video; files must undergo sampling-rate conversion. These PCM audio configurations are supported: 16/48 (up to eight channels), 20/48 (up to six channels), 24/48 (up to five channels), 16/96 (up to four channels), 20/96 (up to three channels), and 24/96 (up to two channels). Because up to eight independent PCM channels are permitted, for example, movies can be released in eight different languages. PCM coding can also employ a dynamic range control (the same provision as in the DVD-Audio specification). This is a disc option and player requirement. The maximum

PCM bit rate allowed on a DVD-Video disc is 6.144 Mbps. Of course, an increase in the audio bit rate decreases the bit rate available to the digital video signal.

To summarize, on a DVD-Video disc, the maximum channel bit rate for DVD is 26.16 Mbps. Following demodulation, this rate is halved to 13.08 Mbps. Following error correction, the maximum bit transfer rate is 11.08 Mbps. The maximum user data bit rate for video, audio, and sub-picture is 10.08 Mbps. The maximum bit rate into the respective buffers is 9.8 Mbps for MPEG-2 video, 1.856 Mbps for MPEG-1 video, 6.144 Mbps for PCM audio (DVD-Video), 9.6 Mbps for MLP/PCM audio (DVD-Audio), and 3.360 Mbps for sub-picture. In addition, these limits apply: 448 kbps for Dolby Digital, 384 kbps for MPEG-1 audio, 912 kbps for MPEG-2 audio, 1536 kbps for DTS, and 1280 kbps for SDDS. From these bit rates, playing times can be calculated; for example, a DVD-9 disc would hold 42 hours and 22 minutes of 5.1-channel Dolby Digital data. Similarly, disc authors can calculate bit budgets, to ensure that contents can fit on a particular disc format.

Figure 8.12 shows an example of how the bit rates of various contents are accommodated on a DVD-Video disc. In this example, the average video bit rate is 3.5 Mbps, there are three audio soundtracks each at 0.384 Mbps, and four subtitle streams each at 0.01 Mbps. This yields a total bit rate of 4.692 Mbps, which is within the DVD-Video specification. At this average bit rate, one 4.7-Gbyte layer would hold a 133-minute program comprising 4.68 Gbytes. This capacity can accommodate over 90% of all feature films; longer titles can use dual-layer discs. A 4.7-Gbyte disc might hold 133 minutes of program, an 8.5-Gbyte disc might hold 241 minutes, a 9.4-Gbyte disc might hold 266 minutes, and a 17-Gbyte disc might hold 482 minutes. These timings are only representative examples; different combinations of content streams and their bit rates

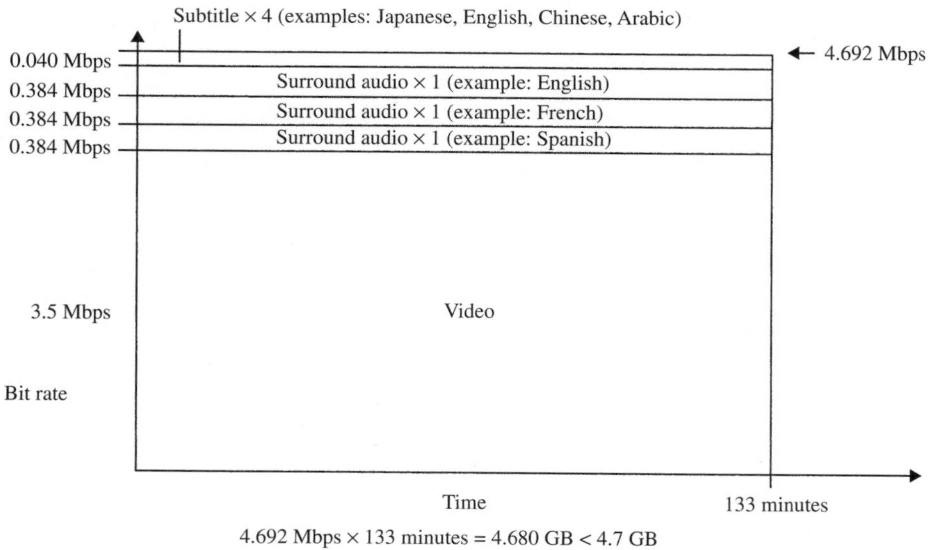

FIGURE 8.12 An example of how a video program, three audio programs, and four subtitle streams can be placed on a DVD-Video disc, while maintaining an overall capacity requirement of less than 4.7 Gbytes.

would yield different durations. During authoring, data reduction is applied so that content fits available disc capacity, within the constraints of sound and picture quality.

DVD-Video Playback Features

A DVD-Video player is connected to a home theater system and is used to play motion pictures and other video programs from DVD-Video and VAN discs (see below), as well as the video contents on a DVD-Audio AV "Audio with Video" disc and other compatible audio portions such as Dolby Digital tracks. In addition, a DVD-Video player can play audio CDs (not all players can play CD-R discs). Universal DVD players can play both DVD-Audio and DVD-Video discs. DVD-Video discs cannot be played in CD players. Most DVD-Video players provide a component video output (a professional video standard that avoids the carrier frequencies used in composite video signals) as well as a HDMI output. The 8-bit video signal is generally reproduced with D/A converters with 10-bit resolution. The DVD-Video standard provides a parental lockout; movies can be coded to play different versions, skipping potentially offensive scenes or using alternate scenes and dialogue tracks. A block diagram of the principal elements and signal flow in a DVD-Video player is shown in Fig. 8.13.

Hybrid DVD-Video discs may contain a movie that is playable in a dedicated DVD-Video player and a DVD-ROM-enabled PC; in addition, the ROM drive may be used to access disc contents such as the movie's screenplay, interactive games, screen savers,

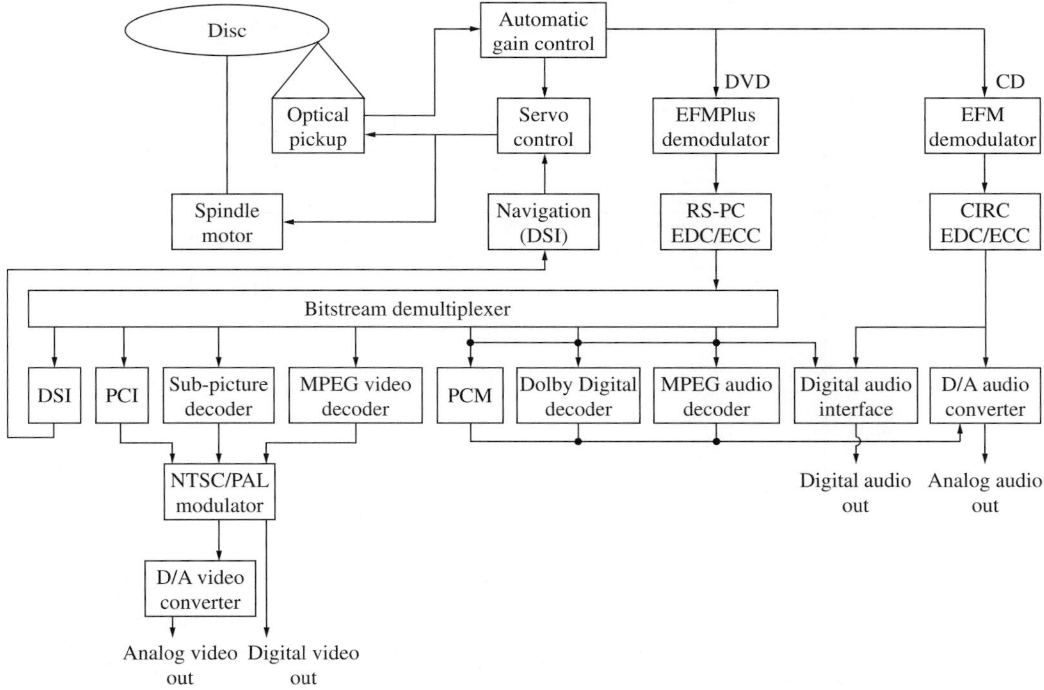

FIGURE 8.13 A DVD-Video player reads data from a disc, distinguishes between CD and DVD data, and directs data types to specific decoding circuits for processing and signal output.

and hyperlinks to Web sites. The DVD-Video specification (Part 5) also describes a hybrid video-audio disc. Because it contains "video audio navigation" information, it is sometimes known as a VAN disc. VAN discs are video discs but they contain audio information that can be played on DVD-Audio players. The content provider selects which portions of the audio tracks can be played on DVD-Audio players.

DVD-Video supports both normal (4:3) and widescreen (16:9) aspect ratios and an automatic pan-and-scan feature. To perform pan-and-scan, the player must use data specially inserted in the bitstream to display portions of a 16:9 picture on a 4:3 screen. In 16:9 mode, the player ignores the pan-and-scan data and instead produces a widescreen image. Alternatively, a 16:9 image can be letterboxed on a 4:3 screen with black stripes at the top and bottom of the screen. Other features include chapter division, forward and reverse scanning, up to nine camera angles and interactive story lines. These features are disc options, and implementation is left to the content provider. Digital audio data stored at a 96 kHz sampling rate is not output through a player's conventional digital audio outputs; it is downsampled to 48 kHz.

From an audio standpoint, the most significant feature of DVD-Video (and DVD-Audio) is multichannel playback of surround sound. Although discs can be coded with different audio channel outputs, the de facto standard is 5.1-channel playback. Figure 8.14 shows a loudspeaker configuration recommended for reproducing Dolby Digital surround sound. The front left/right speakers are placed at ±30° and the surround speakers at ±110° from the center. Optimal placement of the LFE channel speaker depends on room acoustics. One or two additional rear channel speakers are sometimes added.

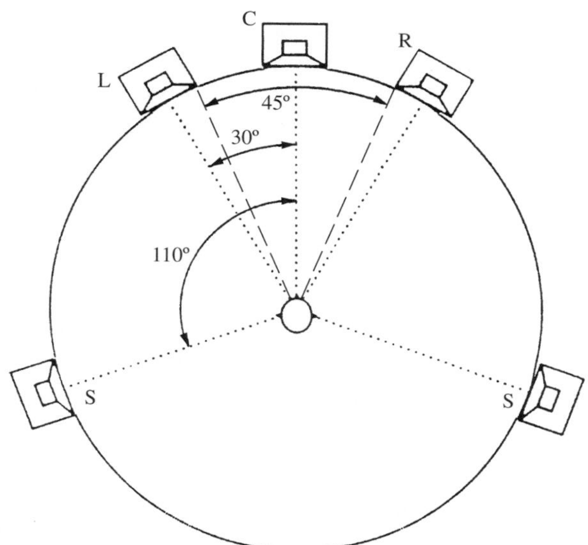

FIGURE 8.14 This loudspeaker configuration, taken from the ITU-R BS.775-2 recommendation, is often used for playback of multichannel audio from DVD, HDTV, and other surround sources. Other configurations are also used. Optimal placement of the LFE subwoofer depends on room acoustics. In some cases, additional rear speakers are added.

The DVD-Video format supports up to 32 channels of sub-picture information. Sub-pictures are graphic files that are bitmapped or overlayed onto the picture. Sub-pictures are generally used for captions, subtitles, or other text. Sub-pictures can be scrolled and faded and can change in every video field. The color palette comprises 16 colors and contrast values; four colors and four contrast values can be displayed in one channel at a time. Sub-picture data is a run-length coded bitmap with 2 bits/pixel, each at a bit rate of 10 kbps. Sub-picture information can be accessed by timecode or user button depressions to produce graphics and simple animation.

Discs and players contain regional coding flags so players will only play properly coded discs. For example, a Region 2 (Europe and Japan) player will not play discs intended for the North American (Region 1) market. This allows movie studios to control release of titles to different global markets. Regional codes on discs are optional; support is mandatory on players. Discs can be made with multiple codes, or no codes; discs with no codes will play on all players, regardless of country. There are six geographic regions: (1) Canada and the United States and its territories; (2) Japan, Western Europe, South Africa, Turkey, and the Middle East; (3) South Korea, Southeast and East Asia including Hong Kong; (4) Australia, New Zealand, Pacific Islands, Central America, Mexico, South America, and the Caribbean; (5) Russia, Indian Subcontinent, Africa, North Korea, and Mongolia; (6) People's Republic of China and Tibet. Region 7 is reserved. Region 8 coding is used for nontheatrical venues such as airplanes, cruise ships, hotels, and so on.

DVD-Video Authoring

A DVD-Video disc can hold video content, multiple soundtracks, subtitles, bonus features, and Web links, and allow elaborate navigation features. A DVD-Audio disc can hold stereo and multichannel music as well as videos, still photographs, graphics, animation, lyrics, and other features. In addition, most DVD-Audio discs hold a Dolby Digital mix for playback in DVD-Video players. A simple DVD title can be authored on a personal computer. A complex title can entail use of diverse professional authoring tools; in many cases, specialists in audio, video, graphic design, and interactivity handle different aspects of the workflow. Whether simple or complex, all title production is computer-based. First, the contents and functionality of the title are determined, along with navigational use of the contents. Also, a bit budget is determined to ensure that all contents will fit on a disc's layers, and peak output bit rates are checked. In most cases, content is already prepared before authoring begins. However, authoring steps must ensure that content adheres to the DVD standards. Audio and video content is supplied and coding is applied when necessary. Artwork, transition sequences, and still or animated menu graphics are created. Web-based multimedia programs are created. Following postproduction, all video and audio content is processed with appropriate coding such as Dolby Digital, DTS, or MPEG-2.

In DVD-Video authoring, program chains for parental lockout, camera angles, alternate endings, regional coding information, and supplemental information is prepared. In some cases, movie soundtracks need to be reconformed to the video being used, or remixed to a surround format. Also in DVD-Video authoring, it is essential that time synchronization of video and audio content is assured on the final disc. In DVD-Audio authoring, MLP coding and downmixing data is applied as necessary. Menu buttons are linked to content or other menus. Elementary streams are created and checked. The simulation is checked for any violations to the DVD specification and thoroughly tested and debugged. The streams are input to an authoring system to create a DVD disc image. The image is complied to a virtual disc format; data is multiplexed into a single composite stream, navigation files are generated, and data is formatted according to Micro UDF Bridge/ISO 9660 format. The image is placed on a hard-disk drive and playback is emulated with a DVD player application. The image is burned to DVD-R or another disc type and tested again. All audio and video elements are reviewed, sound-to-picture synchronization is checked, along with disc navigation and functionality. In some cases, the image is played from the DVD-R into the authoring workstation and the emulation is further checked and debugged there. A finished, encrypted version is written to DLT tape along with DDP data. A manufacturing plant produces test discs that are checked for functionality and player compatibility. After approval, finished discs can be replicated. A list of the principal steps in DVD-Video authoring (which is similar to DVD-Audio authoring) is given in Table 8.3.

Authoring Step	Description
Asset management and subsystem control	Multimedia database for asset management and project control
	Control over remote video and audio encoding subsystems for media capture
Media layout and assembly	Audio and video streams are assembled and synchronized
	Multiple language and subtitle tracks are synced
	Interactive branch points are determined
Title and menu page creation	Still images are compressed into MPEG-2 format and placed into the scenario editor
	User playback control is determined for menu items
Layout of interactive elements	Branch points for multi-angle and multi-story lines are determined and links are created
DVD simulation	Playback of DVD scenario to ensure that the content plays correctly
DVD stream multiplexing	All audio, video, and sub-picture data is multiplexed into a DVD-compliant bitstream
Data verification	Check multiplexed bitstream against the DVD specification for compliance
DVD player emulation	Emulation and playback of final DVD bitstreams for quality assurance
Disc image creation	Convert multiplexed stream into Micro-UDF/ISO 9660 format disc image
Dump to DLT premaster tape	Disc image is dumped to a "premaster" DLT tape to be sent to the DVD manufacturing plant for replication

Table 8.3 Principal steps in the authoring of a DVD-Video disc.

DVD-Video Developer's Summary

Programmers and others developing DVD-Video products should know the specific definitions of terms, file structures, and the interrelationships of file types. This section provides such overview information. As noted, the Part 3 DVD-Video format adheres to Parts 1 and 2 of the DVD specification. It employs the UDF file format; Part 3 specifically defines how the user can access disc contents (Navigation) and how the video data itself is structured (Video Objects). Discs can contain multiple titles; for example, there might be a movie and a trailer, or perhaps several short films. The Title is a disc's highest level of navigation, and users select which title they wish to view. Internally, the title manager contains one or more Video Title Sets (VTSs) and Program Chains (PGCs) that also contain audio and video elements. Within each title, a main menu shows particular contents, possibly leading to submenus. Typically, four directional buttons are used for onscreen selection, and an enter button activates selections within a menu. Other dedicated buttons access specific features such as audio tracks and subtitles. This and other navigation structure is defined in Part 3.

Part 3 defines a video disc for moving pictures. The Presentation data structure complies with the MPEG-1 and MPEG-2 specifications. A Pack is a pack header followed by one or more packets. A Pack is a layer in the system coding that is described in ISO/IEC 13818-1 (the MPEG-2 stream layer specification). MPEG 13818-1 defines disc program stream and broadcast transmission stream; 13818-2 defines video compression; 13818-3 defines audio compression for surround sound. (MPEG-1 is defined in ISO 11172-1, -2, and -3.)

A Packet is the elementary data stream following the header; there are five kinds of Packets. A Stream-ID defines the type of packet and is defined in ISO/IEC 13818-1. An ISO/IEC 13818-1 stream contains five packetized elementary streams of video, audio, sub-picture, presentation control information (PCI), and data search information (DSI). A Cell is a group of MPEG frames of indeterminate length starting and ending with an I-frame. An I-frame is an intra-coded picture without temporal prediction, as opposed to P (forward-predicted) frames and B (bidirectional) frames. Cells are used by PGCs, which are part of the Navigation system. (There are also audio cells.)

A Program Chain (PGC) contains navigation pointers to cells and cell groups. These PGCs can be selected by the viewer. For example, one PGC pointing to cells 1 to 20 and 35 to 50 would show the G-rated version of a movie, and another PGC pointing to cells 1 to 50 would show the complete R-rated version.

The Volume Space of a DVD-Video disc consists of the Volume and File structure, a single DVD-Video zone, and DVD-Other zone (see Fig. 8.11A). A DVD-Video zone consists of one Video Manager (VMG) and one or more Video Title Sets (VTSs). The VMG is the table of contents for all Video Title Sets; each VTS is a collection of titles. The VMG contains a main menu for disc title, text data, and so on. A Video Title Set is a collection of titles. A VTS contains a menu for title chapter, language for audio/sub-picture, playback control information (PGCI), and audio-video VOBS data. A Video Object Set (VOBS) is a collection of Video Objects (VOBs) that hold presentation data such as video, audio, or sub-picture data. For example, a VOB might contain all of an MPEG-2 video program. A Title consists of one or more PGCs, each containing Program Chain Information and VOBs. Titles with multiple PGCs permit branching, multiple story lines, etc. The DVD-Video data structure is shown in Fig. 8.15.

A DVD-Video zone also contains Navigation data (playback control) and Presentation data (the video program to be played back). The Navigation Manager handles

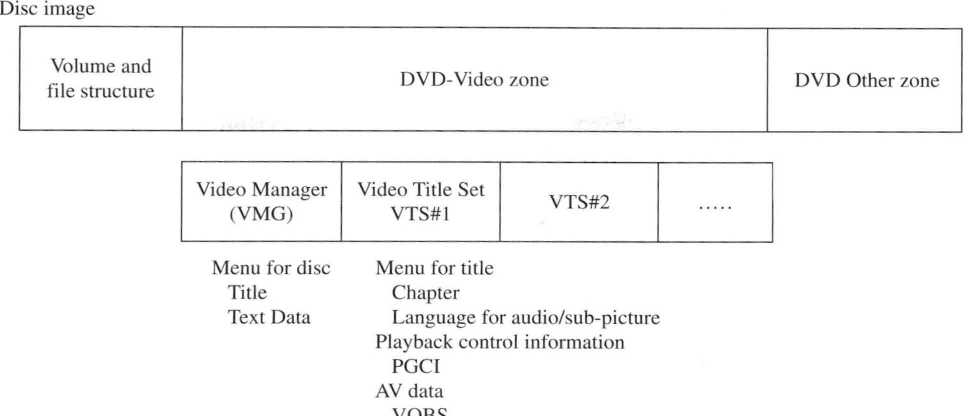

FIGURE 8.15 The DVD-Video data structure can be viewed as a disc image with the DVD-Video zone holding the Video Manager and Video Title Sets. DVD-Audio follows the same structure.

navigation data (VMGI, VTSI, PCI, and DSI) to control the user interface, control playback, interpret user actions, and determine how the Presentation Engine should play back Presentation data. A Button is an onscreen user control; Buttons are defined in PCI. A Menu is an onscreen display that includes Buttons.

The Presentation Engine follows instructions issued by the Navigation Manager to play Presentation data from the disc and control the displayed output. Presentation data is divided into cells. Presentation data consists of VOBs. For example, a chapter may be a VOB. Different VOBs may be used for different scenes, cuts, and so on (for director's cut, angles, parental lockout, and the like). A VOBS consists of one or more Video Object blocks.

Navigation data consists of attributes and playback control for the Presentation data. Navigation data allows the user to access disc contents. Content providers can use this data to code branching and interactivity. There are four types: Video Manager Information (VMGI), Video Title Set Information (VTSI), Presentation Control Information (PCI), and Data Search Information (DSI). VMGI is described in the Video Manager (VMG). It describes information in the VIDEO_TS directory. Data includes a video copy flag, an audio copy flag, number of volumes, disc side identifier, NTSC/PAL, aspect ratio, picture resolution, number of audio streams, audio-coding method, quantization, sampling rate, number of channels, sub-picture coding, menu language, and parental management.

Video Title Set Information (VTSI) is described in the Video Title Set (VTS). It describes information for one or more Video titles and the Video Title Set Menu. Its data is similar to VMGI. PGCI pointers are the Navigation data used to control presentation of the PGC and order of cell playback. PGCI pointers are usually played sequentially, but can be played in random or shuffled sequence. PGC is composed of PGCI and VOBs. For example, PGC may be used to create interactive programs. PGC has data such as presentation time in hours, minutes, seconds, frames, and cell information. PCI is dispersed in the VOBS along with Presentation data. PCI is the Navigation data used to control the presentation of a VOB Unit (VOBU). PCI is used by the playback engine to control what is seen and heard. PCI has data such as angle information, highlight information, relation between sub-picture and highlight, and buttons.

Data Search Information (DSI) is dispersed in the VOBS along with Presentation data. DSI is the Navigation information used to search and seamlessly play back the VOBU. DSI is also used for navigation and search control (such as branching). DSI has data such as interleaving, start address, and synchronization. Navigation Commands are used by content providers to allow changes in player operation including branching and interactivity (as opposed to linear playback). Navigation commands appear in PCIs and PGCIs. DVD-Video uses Navigation commands to provide a high degree of interactivity. Linking, looping, jumping, searching, and decision making are built into the specification as navigation commands. Software developers use this standardized command set. There are twenty-four 16-bit System Parameters (SPRMs) registers (such as angle number, video capability, audio capability, parental level, and language code) for player settings and sixteen 16-bit General Parameters (GPRMs) registers (such as go to, jump, link, and compare) to memorize the user's operational history and modify the player's operation.

A hardware splitter/navigator controls DVD playback. It uses PCI information that describes the stream contents, and then splits the data to the appropriate decoders. The navigation engine uses DSI information and user input to control playback. Omitting PCI and DSI, the three remaining stream types have a maximum bit rate of 9.8 Mbps (variable). In practice, the average rate might be 4.7 Mbps.

A VOB contains Presentation data (video data, audio data, sub-picture data, and VBI data) and part of the Navigation data (PCI and DSI). Video Objects are defined as pack types and have restrictions on data transfer rate. The video stream (maximum of 1) has a maximum transfer rate of 9.80 Mbps; the PCM audio stream (maximum of 8) has a maximum transfer rate of 6.144 Mbps; the sub-picture stream (maximum of 32) has a maximum transfer rate of 3.36 Mbps.

A Video Object Set is a collection of Video Objects, as shown in Fig. 8.16. Each VOB can be divided into cells (or scenes). Each cell contains Video Object Units that are groups of audio or video blocks; a cell is the smallest addressable data chunk, but may last for a moment or for a movie's entire duration. Each VOBU contains VOB pack types such as V_PCK (video packs), A_PCK (audio packs), and NV_PCK (navigation packs). VOBUs may also contain analog copy-protection data for Macrovision. A pack comprises packets and both comply with ISO/IEC 13818-1 (the MPEG-2 bitstream format, not the MPEG-2 audio or video coding formats). A pack is 2048 bytes total with up to 2034 bytes of user information. Figure 8.17 shows the structure of a pack. Navigation Packs (NV_PCK) contain PCI (979 bytes) and DSI (1017 bytes) data. Sub-picture packs (SP_PCK) contain sub-picture data (2024 bytes). Video Blanking Information (data placed in the video blanking period) packs (BVI_PCK) contain VBI (640 bytes). Video packs (V_PCK) contain video data (2025 bytes). Audio packs (A_PCK) contain audio data such as PCM (2013 bytes), AC-3 (2016 bytes), and MPEG (2020 bytes). An audio pack contains data such as audio emphasis, audio mute, audio frame number, quantization word length, audio sampling frequency, number of audio channels, dynamic range control, copyright, and so on.

To summarize, audio and video data as well as presentation and control information are held in packets. Usually, one MPEG Group of Pictures (GOP) occupies one VOBU. VOBUs are collected into cells. Sequences of cells comprise a program that is stored in a VOB. Sequences of programs comprise a Presentation Control Block (PCB). This,

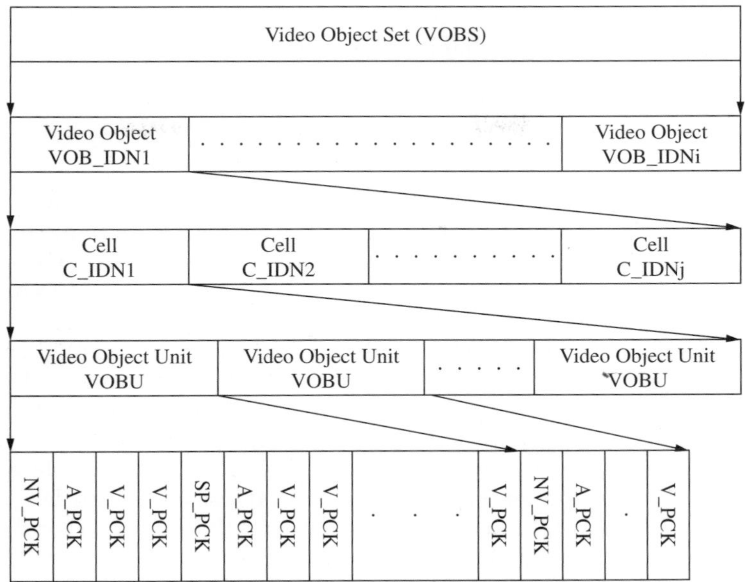

FIGURE 8.16 A Video Object Set (VOBS) is a collection of Video Objects, which in turn contain cells, and Video Object Units, which contain pack data.

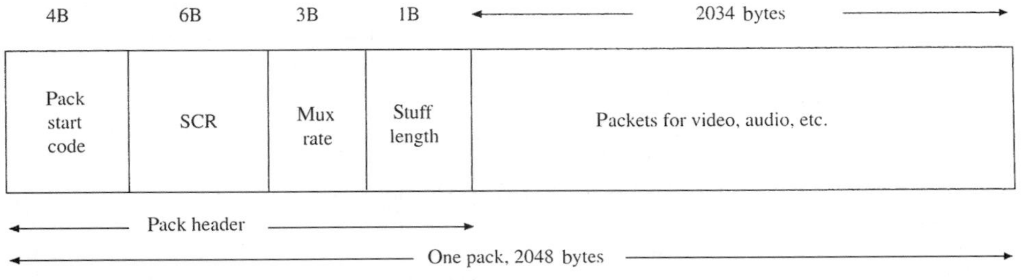

FIGURE 8.17 The structure of packs and packets used in DVD-Video adheres to the MPEG-2 standard. Presentation data is contained in packets.

along with command information, creates a Program Chain. Its audio and video content is stored in a VOBS. Titles are grouped to form a VTS.

In the DVD-Video player, packs in the program stream are received from the disc and transferred to the appropriate decoder. A buffer is used to ensure a continuous supply of data to the decoders. DSI data is treated separately. Video presentation data complies with ISO/IEC 13818-2 (MPEG-2 video standard) or ISO/IEC 11172-2 (MPEG-1 video standard). MPEG places constraints on the picture coding. Video data is split into VOBUs. Video data in a VOBU consists of one or more GOP. Audio presentation data comprises PCM or compression coded data such as AC-3, DTS, or MPEG. The audio stream is divided into packs and recorded on the disc. Sub-picture presentation data comprises the sub-picture header unit, pixel data, and display control sequence table. Video Blanking Information Unit presentation data is placed in the blanking period.

DVD-Audio

The DVD-Audio portion of the DVD specification describes a high-quality audio storage format that provides a wide variety of channels, sampling frequencies, word lengths, and other features. Although primarily an audio specification, it also provides for incorporation of video and other elements. In many ways, DVD-Audio is based on the DVD-Video specification. Development of DVD-Audio began in December 1995, with the formation of the DVD Forum's Working Group 4 (WG-4). Its first meeting was held in January 1996, and Version 0.9 of the specification was released in June 1998. The DVD-Audio Version 1.0 specification was finalized in February 1999, and was the last of the original DVD formats to be ratified by the DVD Forum. DVD-Audio products were introduced in early 2000. WG-4 received input from the International Steering Committee (ISC) representing the interests of the major record labels through trade associations (RIAA, IFPI, and RIAJ). The ISC established 15 criteria for DVD-Audio such as high-quality sound, multichannel audio, scalable parameters, CD compatibility, long playing time, optional video content, simple or menu-based disc navigation, and copy protection.

Although the DVD-Video format can provide high-quality audio (such as six channels of 48-kHz/20-bit audio) its maximum audio bit rate of 6.144 Mbps cannot support the highest quality levels. DVD-Audio's maximum bit rate of 9.6 Mbps increases its abilities. For example, with PCM coding, six channels of 96-kHz/16-bit audio is allowed. However, six channels of 96 kHz/24-bit PCM audio exceeds the maximum bit rate. In any case, high bit-rate PCM streams reduce playing time. Thus lossless and lossy compression algorithms can be optionally employed to reduce bit rate demands, and increase playing time.

The primary intent of the developers was to create an audio format that would retain compatibility with other DVD disc formats, some backward compatibility with the CD format, and introduce improved sound quality and multichannel playback. In addition, DVD-Audio would protect its content with stringent anti-piracy measures. To augment the already large capacity of DVD, DVD-Audio also provides for lossless data compression of audio. This option allows storage of over 74 minutes of high-quality multichannel music on a single data layer. DVD-Audio discs must contain an uncompressed or MLP-compressed PCM version of the DVD-Audio portion of the program. For further flexibility and added compatibility with existing DVD-Video players, DVD-Audio discs may also include video programs with Dolby Digital, DTS and/or PCM tracks. In most cases, in addition to high-resolution PCM tracks, DVD-Audio discs also contain Dolby Digital tracks, so that discs are playable in both DVD-Audio and DVD-Video players. Dolby Digital tracks are mandatory on discs that contain associated video tracks.

Two types of DVD-Audio discs are defined. An Audio-Only disc (see Fig. 8.11B) contains only music information. An Audio-Only disc can optionally include still pictures (one per track), text information, and a visual menu. In addition, an Audio with Video (AV) disc is defined (see Fig. 8.11C); it can contain motion video information formatted as a subset of the DVD-Video format.

DVD-Audio Coding and Channel Options

The DVD-Audio format supports a variety of coding methods and recording parameters, as shown in Table 8.4. PCM tracks are mandatory on all discs. Optional disc coding methods include MLP, Dolby Digital, MPEG-1, or MPEG-2 without extension bitstream, MPEG-2 with extension bitstream, DTS, DSD, and SDDS. DVD-Audio is said to be extensible; it is open-ended and can be adapted to future coding technologies. All DVD-Audio players must support MLP decoding. DVD-Audio is a "scalable" format; that is, its specification provides considerable flexibility for content providers. When PCM coding is used, the number of channels (1 to 6), the word length (16, 20, 24 bit), and the sampling frequency (44.1, 48, 88.2, 96, 176.4, or 192 kHz) can be interchanged. At the highest sampling frequencies of 176.4 kHz and 192 kHz, only two-channel playback is possible. To limit the output bit rate to the 9.6-Mbps maximum, other restrictions may apply to lower sampling frequencies. Audio attributes such as sampling frequency and word length can be set differently for each track.

The coding options, range of sampling frequencies and word lengths, and number of titles (stereo and/or multichannel) create a range of playback times. In addition, the number of disc layers determines playing times. For example, depending on its recording parameters, a stereo PCM program on a data layer might play for 422 or 65 minutes. Similarly, different configurations of multichannel recordings will yield a range of playing times, as shown in Table 8.5. MLP lossless compression can effectively almost double the disc capacity, thus increasing playing times. The compression achieved by MLP

Audio Coding	Sampling Frequency (kHz)	Word Length (bits)	Number of Channels
PCM	192	16, 20, 24	2
	176.4	16, 20, 24	2
	96	16, 20, 24	1 to 6[1]
	88.2	16, 20, 24	1 to 6[1]
	48	16, 20, 24	1 to 6
	44.1	16, 20, 24	1 to 6
MLP	192	16, 20, 24	2
	176.4	16, 20, 24	2
	96	16, 20, 24	1 to 6
	88.2	16, 20, 24	1 to 6
	48	16, 20, 24	1 to 6
	44.1	16, 20, 24	1 to 6
Dolby Digital	48	16, 20, 24	1 to 6
DTS	96	16, 20, 24	1 to 6
	48	16, 20, 24	1 to 6

[1]Number of channels less than 6 at 20- and 24-bit resolution.

TABLE 8.4 The DVD-Audio specification supports a variety of coding methods, each with many possible recording parameters. Some examples are shown here; this table is not inclusive of all possibilities.

Audio Coding	Sampling Frequency (kHz)	Word Length (bits)	Number of Channels	Playing Time (minutes)
PCM	192	24	2	65
	192	20	2	78
	96	24	2	129
	48	24	6	86
	44.1	16	2	422
MLP	192	24	2	117
	192	20	2	141
	96	24	2	234
	96	20	2	282
	48	24	2	468
	44.1	16	2	764
Dolby Digital	48	24	6	1550
DTS	48	24	6	425

TABLE 8.5 Examples of coding methods and recording parameters and resulting playing times per disc layer (DVD-5).

depends on the music being coded. Very approximately, it gives about a 1.85:1 compression ratio; thus, it can almost halve bit rate, and double playing time with no loss of audio quality. Similarly, lossy compression increases playing time. Dolby Digital, DTS, MPEG, MLP, and other audio codecs are discussed in more detail in Chap. 11.

The use of high sampling frequencies such as 96 kHz and 192 kHz may seem unnecessary. In rare cases, a person may be able to hear frequencies of 24 kHz or 26 kHz, far below the cutoff frequencies of 48 kHz and 96 kHz. In most cases, high-frequency hearing response is below 20 kHz. Thus, for steady-state tones, the higher frequency response may not be useful. However, it can be argued that high sampling frequencies improve the binaural time response, leading to improved imaging. For example, if short pulses are applied to each ear, a 15-μs difference between the pulses can be heard, and that time difference is shorter than the time between two samples at 48 kHz. Some people can hear a 5-μs difference, and that corresponds to the time difference between two samples at 192 kHz. In theory, this high sampling frequency may improve spatial imaging. Thus, it may take two ears to distinguish between a recording at 48 kHz, and one at 192 kHz. Its designers hoped that the DVD-Audio specification would offer improvements in fidelity and in any case its specifications would not be a limiting factor.

Various channel assignments can be made by placing the channels into two Channel Groups (CGs); examples of channel assignments are shown in Table 8.6. This prioritizes mixes that use the front L and R channels; front L, R, and C channels; and the corner L, R, Ls, and Rs channels. The sampling frequency and word length of CG1 is always greater than or equal to those of CG2, as shown in Table 8.7. Generally, CG1 assignments are for the front channels, and CG2 assignments are for the rear channels. There are numerous ways to assign channels, ranging from monaural to six channels,

		Channel Number					
		0	**1**	**2**	**3**	**4**	**5**
Mono/stereo playback	1	C					
	2	L	R				
Lf, Rf weighted	3	Lf	Rf	S			
	4	Lf	Rf	Ls	Rs		
	5	Lf	Rf	LFE			
	6	Lf	Rf	LFE	S		
	7	Lf	Rf	LFE	Ls	Rs	
	8	Lf	Rf	C			
	9	Lf	Rf	C	S		
	10	Lf	Rf	C	Ls	Rs	
	11	Lf	Rf	C	LFE		
	12	Lf	Rf	C	LFE	S	
	13	Lf	Rf	C	LFE	Ls	Rs
Front weighted	14	Lf	Rf	C	S		
	15	Lf	Rf	C	Ls	Rs	
	16	Lf	Rf	C	LFE		
	17	Lf	Rf	C	LFE	S	
	18	Lf	Rf	C	LFE	Ls	Rs
Corner weighted	19	Lf	Rf	Ls	Rs	LFE	
	20	Lf	Rf	Ls	Rs	C	
	21	Lf	Rf	Ls	Rs	C	LFE
				Channel Group 1		Channel Group 2	

Table 8.6 DVD-Audio channel assignments are made with two Channel Groups (CG1 and CG2). Assignments enable front mixes, front mixes with center channel, and four-channel corner mixes. There are other possible assignments beyond the 21 examples shown here.

and different word lengths and sampling frequencies can be employed on the front and rear channels. For example, front channels could be coded at 24/96 with the rear channels coded at 16/48. Coding the rear channels at a lower bit rate, for example, would allow longer playing times, or would allow disc capacity to be budgeted to other content such as videos or stereo mixes.

Sampling frequencies in channel groups must be in the same family, that is, related by a simple integer such as 48/96/192 kHz or 44.1/88.2/176.4 kHz. Table 8.8 shows examples of different channel configurations for 5.0- and 5.1-channel playback using PCM coding. Unlike some 5.1-channel systems (Dolby Digital, DTS, MPEG), the PCM coding used in DVD-Audio does not bandlimit the LFE channel, it is a full-bandwidth channel. The choice of sampling frequency family is probably best determined by the

	Channel Group 1	Channel Group 2
Sampling frequency (kHz)	48 96 192[1] 44.1 88.2 176.4[1]	48 96 or 48 192, 96, or 48 44.1 88.2 or 44.1 176.4, 88.2, or 44.1
Word length (bits)	16 20 24	16 20 or 16 24, 20, or 16

[1]More than two channels coded with MLP.

Table 8.7 The Channel Groups are scalable. The sampling frequency and word lengths of CG1 must be greater than or equal to those of CG2.

A. 48 kHz/96 kHz, 5.1 Channels			B. 48 kHz/96 kHz, 5.0 Channels		
Configuration (sampling frequency/ word length/number of channels)	**Bit Rate (Mbps)**	**Playback Time (minutes)[1]**	**Configuration (sampling frequency/ word length/number of channels)**	**Bit Rate (Mbps)**	**Playback Time (minutes)[1]**
48/24/6	6.912	86, 156	48/24/5	5.760	103, 187
96/16/6	9.216	64, 117	96/16/5	7.680	77, 140
96/20/2 and 48/20/4	7.680	77, 140	96/24/2 and 48/20/3	7.488	79, 144
96/24/2 and 48/24/4	9.216	64, 117	96/24/2 and 48/24/3	8.064	73, 134
96/24/2 and 48/20/4	8.448	70, 128	96/24/2 and 96/16/3	9.216	64, 117
96/16/3 and 48/16/3	6.912	86, 156	96/20/3 and 48/20/2	7.680	77, 140
96/20/3 and 48/16/3	8.064	73, 134	96/20/3 and 96/16/2	8.832	67, 122
96/20/3 and 48/20/3	8.640	68, 125	96/24/3 and 48/20/2	8.832	67, 122
96/24/3 and 48/16/3	9.216	64, 117	96/24/3 and 48/24/2	9.216	64, 117
96/20/4 and 48/16/2	9.216	64, 117	96/20/4 and 48/20/1	8.640	68, 125

[1]Playback time (minutes): single-layer/single-side, dual-layer/single-side discs.

Table 8.8 Examples of multichannel PCM channel configurations with multiple sampling rates, showing bit rate and playing time (on single-layer/dual-layer discs). A. 5.1-channels coded at 48 kHz/ 96 kHz. B. 5.0-channels coded at 48 kHz/96 kHz.

sampling frequency of the original recording; noninteger sample rate conversion might introduce audible artifacts. The frame rate is defined to be 1/600 second at sampling frequencies of 48, 96, and 192 kHz, and 1/551.25 second at 44.1, 88.2, and 176.4 kHz.

PCM coding also provides for an emphasis characteristic (zero at 50 μs and pole at 15 μs, for sampling frequencies of 48 kHz and 44.1 kHz); this boosts high frequencies during encoding and correspondingly cuts high frequencies during decoding to reduce the noise floor. This can be applied when all channels use the same sampling frequency. Use of pre-emphasis is optional on discs, but provision for de-emphasis is mandatory in players. PCM coding can also employ a dynamic range control (the same provision as in the DVD-Video specification). This is a disc option and player requirement.

A DVD-Audio disc can contain one or several selections, as provided by a content provider. For example, a disc might contain one selection coded as PCM; every player could play back this selection. Another disc might contain two selections, one coded as PCM multichannel and the other coded as PCM stereo; the provider can choose the order of the selections; only players with multichannel capability could play the multichannel selection. Another disc might contain two selections, one coded as PCM stereo and the other coded in an optional format such as Dolby Digital; the optional selection could be played by players equipped with that circuitry. It is advantageous to place Dolby Digital tracks on a DVD-Audio disc so they can be played in a DVD-Video player. A single-inventory disc may include: DVD-Audio stream of up to six channels of MLP at 96/24, stereo PCM stream, Dolby Digital 5.1-channel stream on the DVD-Video portion, and possibly even a Red Book layer at 44.1/16.

DVD-Audio discs can employ the SMART (System Managed Audio Resource Technique) feature with PCM tracks. SMART provides automatic downmixing so that a multichannel audio program can be mixed down to two channels by the player during playback and thus replayed over a stereo playback system. The content provider can program how the downmixing will occur, by selecting one of 16 coefficient tables, stored along with the audio data on the disc. Each coefficient table defines level, pan position, and phase. The level mixing ratio can vary from 0 to –60 dB. Coefficient tables can be varied on a track-by-track basis in each Audio Title Set. The SMART feature eliminates the need to include a separate stereo mix on a multichannel disc, thus wasting disc space. However, SMART may not allow the creative flexibility demanded by some content providers. When a separate stereo mix is coded along with a multichannel mix, the separate mix is automatically selected instead of a folded-down mix. Use of SMART downmixing is optional on discs, but its support is mandatory in players.

MLP supports the downmix feature. The player first decodes the multichannel signal and then accesses the coefficients to provide a two-channel playback. Optionally, the downmix can be created by the MLP encoder rather than the player so both two-channel and multichannel mixes can be conveyed separately by MLP substreams. The MLP decoder in the player reads the two-channel substream and outputs two channels. This reduces the computation required in stereo-only players. However, the two-channel mix increases the bit rate by about one bit per sample. For multichannel playback, the player extracts both substreams and the MLP decoder decodes all channels prior to possible downmixing.

Other "value-added" content on DVD-Audio discs may include artist names, song titles, liner notes, artist commentary, biographies, discographies, music videos, and Internet URLs. Nonreal-time information (such as content) is recorded in Information areas while real-time information (such as lyrics) is recorded in Data areas. Two character sets are supported: ISO 8859-1 for European languages and Music Shift JIS for Japanese. Multiple languages may be supported. Still images may be tagged to individual tracks; they may be displayed like a slide show (manual or automatic) while the music plays. Likewise, text and sound effects may be played in real time. This extra information is a disc option (track names are mandatory if there is any text information), but decoding must be supported by Universal players (described below). In some cases, the player uses the text information to construct a text menu.

Full motion video can be added to a DVD-Audio AV disc as an independent video portion; it is defined as a subset of the DVD-Video specification. Several restrictions apply: there is a maximum of two audio streams, at least one of which must be PCM and the PCM stream is limited to six channels with restricted channel assignments. In addition, there is no multi-story, multi-angle, parental control, or region control features. There is no mandatory PCM audio in the DVD-Video portion of disc (there is already a PCM version in the DVD-Audio part). Dolby Digital is mandatory in the DVD-Video portion (PCM is optional). DVD-Audio also defines a DVD-ROM zone that can contain compressed audio files. These files can be moved, for example, to portable music players. MPEG-4 High-Efficiency AAC (HE AAC), also known as aacPlus, files can be placed in this compressed audio zone.

DVD-Audio Disc Contents

DVD-Audio disc contents are arranged hierarchically as shown in Fig. 8.18. One album (or volume) describes the entire contents of one disc side. An album can contain up to nine groups. A group can contain up to 99 tracks. A track may contain up to 99 indices (in an Audio-Only disc). Figure 8.19 shows an example of the contents of a DVD-Audio Audio-Only disc. This disc contains two groups; in this case, Group 1 holds five main tracks and Group 2 holds two alternate remixed tracks. In addition, each group has two selections (labeled #1 and #2); for example, tracks 1 and 2 in Group 1 have two selections. In track 1, selection #1 is a multichannel mix and selection #2 is a stereo mix (downmixing is not used). Tracks 3, 4, and 5 may use downmixing. In this example, Group 2 selection #2 tracks use optional coding. Each group has one or more Audio Title Sets, and the tracks are objects within the Audio Title Sets.

An example of the contents of a DVD-Audio AV disc is shown in Fig. 8.20. There is one group in this album. It has five audio-only tracks and two AV tracks. Tracks 1 and

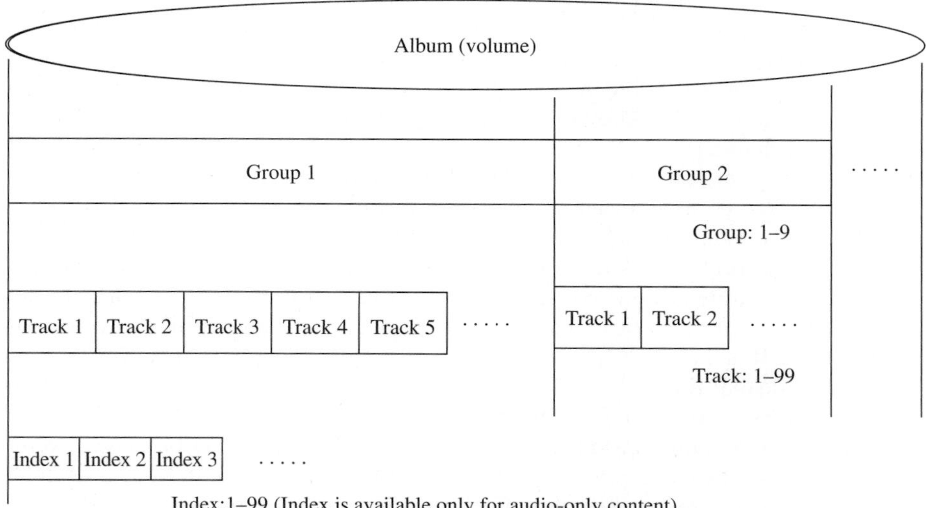

FIGURE 8.18 Contents of DVD-Audio discs are arranged hierarchically. Any track in an album is accessible using a group number and track number.

Group 1	#1	#2	Time
Track 1	48/20/5	48/20/2	4:00
Track 2	48/20/5	48/20/2	4:30
Track 3	96/20/3 and 48/20/3		5:10
Track 4	96/20/3 and 48/20/3		4:00
Track 5	96/24/2		3:50
		Group 1 total time	21:30
Group 2	#1	#2	time
Track 1	48/24/2	MPEG 5.1	4:20
Track 2	96/24/2	MPEG 5.1	5:00
		Group 2 total time	9:20
Visual Menu is available for U-Player.		Album total time	30:50

Figure 8.19 An example of the contents of a DVD-Audio Audio-Only disc, showing two groups with a total of seven tracks.

Group 1		#1	#2	Time
Track 1		48/20/2	48/20/5	4:00
Track 2		48/20/2	48/20/5	4:30
Track 3		96/20/3 and 48/20/3		5:10
Track 4		96/20/3 and 48/20/3		4:00
Track 5		96/24/2		3:50
Track 6	*<with Video>*	48/16/2	AC-3 5.1	4:40
Track 7	*<with Video>*	48/16/2	AC-3 5.1	3:10
			Album total time	29:20

Video component of Track 6,7 is not presented by A-Player.

Visual Menu is available for U-Player.

Figure 8.20 An example of the contents of a DVD-Audio AV disc, showing one group with seven tracks.

Disc	Contents		Audio-Only Player	Video Capable Audio Player	Universal Player	Video Player
DVD-Video	Title menu		NP	NP	M	M
	Video title		NP	NP	M	M
DVD-Audio	Visual menu		NP	M	M	NP
	Audio track	Audio	M	M	M	NP
		Still pictures	NP	M	M	NP
		Real-time text	O	O	O	NP
	Video track	Video	NP	M	M	M
		Audio	M[1]	M	M	M
	Text manager		O	O	O	NP

Note: M = mandatory, O = optional, NP = not playable.
[1]Only portions allowed are playable.

Table 8.9 Compatibility between DVD-Video and DVD-Audio discs, and four types of players: Audio-Only player, Video Capable Audio player, Universal player, and Video player.

2 have two selections; for example, in track 1, selection #1 is a stereo mix and selection #2 is a multichannel mix. Tracks 6 and 7 have video components (not playable on an Audio-Only player). The group has three Titles (two Audio Title Sets and one Video Title Set). The tracks are objects within the ATS and VTS.

DVD-Audio players (without video capability) can play back the audio contents and audio components of video contents of DVD-Audio AV discs. They can play selected audio components on DVD-Video VAN discs. Disc and player compatibility is illustrated in Table 8.9. DVD-Video players cannot play the high-resolution PCM tracks in the DVD-Audio zone. However, the video zone in DVD-Audio discs adheres to the DVD-Video format. Thus, for partial compatibility with DVD-Video players, many DVD-Audio discs contain a stereo PCM or Dolby Digital version of the album in their video zone. DVD-Audio players can also play hybrid DVD-Audio discs that contain both a DVD-Audio data layer and a Red Book CD layer. The DVD-Audio format also supports a SACD-like disc (Super Audio CD) as an optional format, thus some players can play both DVD and SACD discs. Universal DVD players can play the spectrum of DVD-Audio and DVD-Video discs. Mandatory player functions include user transport controls, selection of groups and tracks, and track searches. Optional features include group search, index search, visual menu, random play, and highlight selection. The visual menu is a subset of the DVD-Video menu specification; it is used to select groups and tracks, view multiple languages, and view still information such as liner notes and images. The visual menu is optional for Audio players but mandatory for Universal players.

The Simple Audio Play Pointer (SAPP) facilitates user navigation of the disc contents. SAPP information is contained in a table located in the disc lead-in area (it is similar to the TOC in the Red Book specification). SAPP is a subset of the more sophisticated and general Audio Navigation table; SAPP provides basic information for monaural and stereo PCM playback only, usually in simple players.

As with other DVD discs, DVD-Audio uses a robust RS-PC error correction system that includes a wide interleave. Local disc damage can cause unreadable data; the player's response depends on how the navigator is implemented. If the navigator attempts to reread the damaged sector, its success partly depends on the data rate. If the rate is low, the player may have time to reread without an interruption in the output data; if the rate is high and especially near the maximum rate of 9.6 Mbps, there may not be time for a reread, and the data output may be interrupted.

DVD-Audio Developer's Summary

Programmers and others developing DVD-Audio products should know the specific definitions of terms, file structures, and the interrelationships of file types. This section provides such overview information. Part 4 of the DVD specification describes the DVD-Audio format. It uses all the specifications from Part 1 and Part 2, and many of the specifications from Part 3. In particular, there are strong parallels between DVD-Video and DVD-Audio in terms of file management for navigation and presentation data. For example, instead of a Video Manager and Video Title Sets, an Audio Manager and Audio Title Sets are used. Likewise, many of the same navigation features are used. As noted, the complexity in the DVD standards is not in the coding itself (UDF, MPEG-2, and AC-3 are complex unto themselves). The DVD standards focus on how data is organized, and how the player should function. However, in some ways, DVD-Audio is somewhat more complex than DVD-Video. Whereas DVD-Video uses one kind of video coding (MPEG-2), DVD-Audio allows for many kinds of coding, with one to six channels. Parenthetically, it is worth noting that the terminology used in DVD-Audio differs from that used in DVD-Video. For example, in DVD-Video we refer to Title, Program, and Cell; whereas, in DVD-Audio we refer to Group, Track, and Index, respectively.

DVD-Audio provides stereo and multichannel playback. A Volume space includes a DVD-Volume zone and DVD-Audio zones as well as DVD-Video zones and DVD-Other zones (see Fig. 8.11B). The DVD-Volume zone complies with the UDF Bridge structure defined in Part 2. The DVD-Video zone and DVD-Other zone are defined in Part 3. A DVD-Audio zone is an area to record the audio contents in a volume of a DVD-Audio disc. It contains one Audio Manager (AMG) and one or more (maximum of 99) Audio Title Sets (ATSs). The AMG is the table for all contents in the DVD-Audio zone (and DVD-Video zone if present) and Navigation data. The AMG is composed of Audio Manager Information (AMGI), optional Video Object Set for AMG Menu (AMGM VOBS) and a backup of AMGI. ATT is a general name given to Audio Only Title (AOTT) and Audio with Video Title (AVTT). An AOTT title has no video data except for still pictures and is defined in the PGCI in the ATS. An AVTT title (otherwise known as AV) has video data and is defined in the PGCI in the VTS. AOTTs are playable by Audio players and Universal players, and AVTTs are playable by Universal players. The DVD-Audio file structure is the same as that used in DVD-Video (see Fig. 8.15).

The Audio Title Set defines the Audio Only Titles. The ATS contains Audio Objects (AOBs) and can contain visual menus. Generally, one AOB comprises one track on the disc. Further, an AOB can have two streams, such as stereo and multichannel. (AVTTs are defined in the Video Title Set with links from the ATS). An AOB track can also optionally contain still images, stored in Audio Still Video (ASV) files. There are two kinds of ATSs. One kind is composed of Audio Title Set Information (ATSI), Audio

Object Set for Audio Only Title (AOTT AOBS), and a backup of ATSI. The other is composed of ATSI and its backup.

The Presentation hierarchical structure is an Album (one disc side), a Group, an Audio Title (ATT), a Track, and Index. ATT is not accessible by the user. A Group contains one or more Audio Titles. There are several types of ATTs. AOTT is playable by all Audio players. AVTT is playable by a Video Capable Audio player.

Both an AOTT and AVTT contain a Program Chain (PGC). A Track is a Program (PG) defined in the PGC of ATS. The attribute is the definition of sampling frequency, quantization word length, and so on. An Index is a cell defined for audio contents in the PGC. An index may consist of two or more cells.

Presentation of contents starts from the track (or index) selected by the user. This is the same as the playback of PGC as defined in the ATS. Video data including sub-picture in video contents and still picture in audio contents are played by a Video Capable Audio player. Some types of Real-Time information can be recorded within the audio contents, if desired. Real-Time Text Data (RTXTDT) contains text data such as lyrics, and explanations of contents. One Page consists of 4 lines with 30 characters per line or 2 lines with 15 characters per line. Text is presented onscreen one page at a time. Eight languages are available. Two kinds of audio data (such as multichannel or not, or PCM or another coding) may be defined in an ATT; this is also known as a selection. (For AOTTs, they are defined as the PGC block, and for AVTTs, they are defined as two streams of audio in a Video Object.)

The Visual Menu is the menu for the AMG. Presentation of the Visual Menu is the same as playback of one or more PGCs that are defined for the AMG Menu. It is played back by Video Capable Audio players but ignored by Audio-Only players which use a simpler SAPP (Simple Audio Play Pointer) feature.

The AMG is the table of contents for all ATSs that exist in the DVD-Audio zone, and all VTSs for audio titles that exist in the DVD-Video zone. The AMG contains Audio Manager Information that contains Navigation data for every audio title and its backup. It may also contain Video Objects used for the Visual Menu. The AMGI is composed of the Audio Manager Information Management Table, Audio Title Search Pointer Table, Audio Only Title Search Pointer Table, Audio Manager Menu PGCI Unit Table, and optional Audio Text Data Manager. The AMGI describes information in the AUDIO_TS directory. The Audio Manager Information Management Table (AMGI_MAT) is a table that describes the size of the AMG and AMGI, starting addresses of information in the AMG and other attribute information such as number of volumes, disc side where a volume is recorded, video display mode, aspect ratio, audio coding mode, and sampling frequency. The Audio Title Search Pointer Table is a table with search information (starting and ending addresses) for Audio Titles and is used by a Video Capable Audio player.

The Audio Only Title Search Pointer Table is a table that contains search information (addresses) of AOTTs and is used by Audio-Only players; this is part of the SAPP feature. The Audio Manager Menu PGCI Unit Table is a table that describes the audio menu. The optional Audio Text Data Manager contains information such as album, group, and track names.

The Video Object for Audio Manager Menu (AMGM_VOB) contains Presentation data (video, audio, and sub-picture data) and some of the Navigation data (PCI and DSI). The AMGM_VOB is the same as the Video Object, with the same contents, same pack structure, and same data transfer rate. The Presentation data is essentially the same as in Part 3. Its PCI and DSI data are essentially the same, with some added restrictions. The AMGM_VOB contains the ISRC code.

The Audio Title Set defines the Audio Only Titles that are defined by the Navigation data and the Audio Objects in the ATS, or by the Navigation data in the ATS and the audio part of Video Objects in the VTS. The ATSs are recorded in the DVD-Audio zone along with the Audio Manager, and the VTSs to be used for the audio title are recorded in the DVD-Video zone along with the Video Manager. The ATSs contain Audio Title Set Information (ATSI), an Audio Object Set for Audio Only Title (AOTT_AOBS), and a backup. The ATSI contains the Navigation data needed to play back every ATT in the ATS and provides information to support User Operation. ATSI contains the Audio Title Set Information Management Table (ATSI_MAT), and Audio Title Set PCI Table (ATS_PGCIT). The AOTT_AOBS is a collection of Audio Objects for Audio Only Title that contains Presentation data such as audio data, optional still picture data, and some kinds of optional Real-Time Information (RTI).

The Audio Title Set Information Management Table (ATSI_MAT) describes the size and starting addresses of ATS and ATSI, as well as attributes. It describes the audio coding method, downmix mode, quantization word length, and sampling frequency of two channel groups. The ATSI_MAT also describes the coefficients to mix down the audio data from multichannel to two-channel. An area containing 16 coefficient tables is provided. The ATS also contains Audio Title Set PCI (ATS_PGCIT), which is the Navigation data to control the presentation of the Audio Title Set Program Chain (ATS_PGC). This information describes the addresses of data, as well as the presentation order of programs and cells.

The Audio Object for Audio Only Title (AOTT_AOB) contains the Presentation data that are audio data, Real-Time Information (RTI) data and still picture data. The AOTT_AOB is an elementary program stream described by the ISO/IEC 13818-1 standard. The AOTT_AOB uses three types of packs: Audio pack, Real-Time Information pack, and Still Picture pack. The maximum length of a pack is 2048 bytes. The maximum transfer rate of the audio stream is 9.6 Mbps. The maximum video transfer rate for still pictures is 9.8 Mbps.

The AOTT_AOBS Structure is a collection of AOTT_AOB files whose attributes are the same or different up to eight. It is composed of one or more cells that are made up of packs. The pack structure of the AOTT_AOB follows the general 2048-byte DVD pack layout (see Fig. 8.10) with 14 bytes of pack header and 2034 bytes of packets. In many ways, the AOTT_AOB is the core of a DVD-Audio disc. It contains the Presentation data (mainly audio). It adheres to the ISO/IEC 13818-1 standard (the MPEG-2 stream layer specification). As noted, 13818-1 compliance does not mean that audio is coded as MPEG-2. Rather, the format of the bitstream itself adheres to MPEG-2. Linear PCM data is held in A_PKT packets. As in Part 3, a Pack is a header, followed by packets. A Pack adheres to ISO/IEC 13818-1; a Packet is the elementary data stream following the header. An AOTT_AOB Audio pack (A_PCK) has up to 2013 bytes of user data. A LPCM packet (A_PKT) comprises a packet header, the private header, and the audio data. The private header has information such as ISRC, audio emphasis, downmix code, quantization word length, sampling frequency, multichannel type, and dynamic range control.

An AOTT_AOB Real-Time Information pack (RTI_PCK) contains up to 2015 user bytes. An RTI packet (RTI_PKT) comprises a packet header, the private header and RTI data. RTI data (such as real-time text, and ISRC) is used synchronously with audio data.

An AOTT_AOB Still Picture pack (SPCT_PCK) contains up to 2025 user bytes. A Still Picture pack (SPCT_PKT) comprises a packet header and video data for the still picture. The still picture is one GOP (with one I-frame), which complies with ISO/IEC 13818-2 (MEG-2 video). The following audio data is mandatory in an AOTT: PCM of one or two audio channels or PCM for three to six audio channels with downmix coefficients, or PCM data of three to six channels without downmix coefficients. (In some cases, an Audio player may not play back this PCM data.) Other types of audio data (such as compressed and lossless compressed) may be contained in an AOTT_AOB as an option. The audio stream is divided into packs of 2048 bytes.

The general DVD-Audio specification for linear PCM describes the number of channels, sampling frequency, quantization, and emphasis. Total maximum bit rate is 9.6 Mbps. Coding is two's complement. When there are three or more audio channels, they are classified into two groups: Channel Group 1 (CG1) and Channel Group 2 (CG2). The data in each group may use different sampling frequencies and word lengths. The word length is identical in every channel of the same CG, but each CG may have different word lengths (16, 20, 24). Sampling frequencies of 176.4 kHz and 192 kHz are only used when there are two channels (L and R) or less. CG1 generally defines stereo and front channels, and CG2 defines rear channels.

When an ATS has AOTT_AOBS, the structure of CG1 and CG2 and the relation between the audio channel and audio signal is described in the ATS Multi Channel Type area (it must be Type 1) according to an Assignment for Audio Object table. When an ATS has no AOTT_AOBS, the relation between the audio channel and audio signal and the number of audio channels is described in the ATS Multi Channel Type area (it must be Type 1) according to an Assignment for Video Object table. The channels supported are: Left Front, Right Front, Center, Low Frequency Effects, Surround, Left Surround, and Right Surround.

The downmix procedure outputs channel signals from input signals that have more channels than outputs. Mixing phase and mixing coefficients are used to produce a two-channel output from six-channel input signals. In this way, it is not necessary to place a separate two-channel mix on a multichannel disc. Each gain controller is programmed with a coefficient value. It also manages the phase (polarity) of the signal. The downmix coefficient (DM_COEFT) is defined in the disc ATS and SAPP lead-in area. It is used by the player to control the downmix procedure. The decimal coefficient value is 0 to 255. It is calculated according to a formula for a specific gain control curve.

The video contents of an Audio with Video Title (AVTT) (also known as AV) are recorded in the DVD-Video zone (one with one VMG and one or more VTSs). VMG is not used by Audio players but may be used by DVD-Video players, which cannot recognize AMG. The inclusion of a VMG allows DVD-Video players to play video tracks on a DVD-Audio disc that may include PCM or Dolby Digital content. VTS is used to define the AVTT of a DVD-Audio disc. Titles defined by VTS are pointed to by the Audio Manager. Various restrictions apply to the VMG of DVD-Audio; for example, a DVD-Audio disc has no region management or parental management. Restrictions also apply to the VTS of DVD-Audio.

DVD-Audio provides a degree of navigation interactivity, such as branching. However, these features are operational only for Video Capable Audio players. Audio-Only players may ignore these features. The navigation features for DVD-Audio are a subset of those specified for DVD-Video. Some special navigation features are added to the reserved areas of the Part 3 specification to support specific audio needs. Part 4 navigation parameters are classified as General Parameters (GPRM) and System Parameters (SPRM). There are sixteen 16-bit GPRMs (such as go to, jump, link, and compare) to memorize the user's operational history and modify operation of Video Capable Audio players. In addition, there are twenty-four 16-bit SPRMs (such as audio selection number, sub-picture stream number, highlighted button number, audio player configuration, and track number) for player settings.

The Simple Audio Play Pointer (SAPP) provides TOC-like data for simple Audio-Only players that may use a simple alphanumeric readout instead of a video display. The SAPPT is a table (a subset of navigation) recorded in the control data in the disc lead-in area and consists of one or more SAPPs. Each SAPP is information for the track presented by a simple Audio player that plays back only PCM not using the Program Chain defined in the Audio Title Set. Every audio program defined in the ATS_PGC that satisfies certain conditions is a SAPP. The conditions are: audio coding mode of PCM, and stereo or monaural output. A SAPPT specifically describes the number of SAPPs and the end address of the SAPPT. The size of an SAPPT must be less than 16,384 bytes. A SAPP describes address information and playback information for a track such as track number, start time of the first audio cell, playback time, track attributes such as word length and sampling frequency of CG, downmix coefficient, and end address.

At the hardware player, packs in the program stream are received from the disc and transferred to the appropriate decoder. A buffer is used to ensure continuous supply of data to the decoders. In a Video Capable Audio player, DSI data (navigation used to search and seamlessly play back branching) in AVTT is treated separately. The audio stream may include the Audio Gap that is the discontinuous period of the audio stream during the presentation of a Still Picture. During the Audio Gap, the player's audio output is muted.

Alternative DVD Formats

In addition to the DVD-Video format defined in Book B and the DVD-Audio format defined in Book C, the DVD family includes DVD-ROM (Read-Only Memory) defined in Book A, DVD-R (Recordable) defined in Book D, DVD-RAM (Random-Access Memory) defined in Book E, and DVD-RW (ReWritable) defined in Book F. DVD Books A, B, and C use a UDF Bridge file format (M-UDF+ISO 9660) and Books D, E, and F use the UDF format. The DVD-ROM, DVD-R, DVD-RAM, DVD-RW, and DVD+RW formats are employed as computer peripherals, in professional authoring environments, or in consumer applications. Single-sided and double-sided recordable discs are available. DVD-ROM is a read-only format, DVD-R is a write-once format, while the others are rewritable. The specifications for DVD-R, DVD-RW, and DVD-RAM are supported by the DVD Forum (www.dvdforum.org). The DVD+R and DVD+RW specifications are supported by the DVD+RW Alliance (www.dvdrw.com). The family of recordable DVD formats is summarized in Table 8.10.

The recordable DVD formats can store diverse types of data; however, several specifications for specific data types have been defined: DVD Video Recording (DVD-VR), DVD Audio Recording (DVD-AR), and DVD Stream Recording (DVD-SR). The DVD-VR recording format is borrowed from DVD-Video. DVD-VR recorders allow real-time recording of video, stereo PCM audio, as well as still pictures, and users can create custom play lists. A new VOBU map, located in the VOBI area, stores

	DVD-RAM	DVD-RW	DVD+RW	DVD-R	DVD+R
Book version	Version 2.1	Version 1.1	Version 1.1	Version 2.0/ for general	Version 1.0
Wavelength	650 nm	650 nm	655 nm	650 nm	655 nm
Numerical aperture	0.6	0.6	0.65	0.6	0.65
Channel bit length	0.14 ~ 0.146 μm	0.133 μm	0.133 μm	0.133 μm	0.133 μm
Track pitch	0.615 μm	0.74 μm	0.74 μm	0.74 μm	0.74 μm
Depth	~ 70 nm	~ 30 nm	~ 30 nm	~ 180 nm	~ 180 nm
Media type	Phase change	Phase change	Phase change	Dye	Dye
Active layer material	GeSbTe	AgInSbTe	AgInSbTe	Cyanine/Azo	Cyanine/Azo
Reflectivity	15 ~ 25%	18 ~ 30%	18 ~ 30%	45 ~ 85%	45 ~ 85%
Recording speed (KB/s)	2700	1350	1350 ~ 3240	1350/2700	1350 ~ 3240
Tracking method	Land/ groove	Groove	Groove	Groove	Groove
Addressing	CAPA	LPP	Wobble	LPP	Wobble
Format	ZCLV	CLV	CLV/CAV	CLV	CLV
Tracking	P-P	DPD	DPD	DPD	DPD
Recording	Random	Random sequential	Random sequential	Sequential	Sequential
Finalization	Without	With	With	With	With
Defect management	Yes	No	No	No	No
Copy protection	CPRM	CPRM	CPRM	—	—
Cartridge	Selectable	No	No	No	No
Cyclability	100,000 times	1000 times	1000 times	Once	Once

TABLE 8.10 Specifications for recordable DVD disc formats (4.7-Gbyte capacity).

time stamps; when a recording is completed, users can create menus to easily access programs. DVD-AR is derived from the DVD-Video and DVD-Audio formats; it supports real-time audio recording, as well as still pictures and text. The various DVD-Audio sampling frequencies are supported as is PCM, Dolby Digital, and MPEG formats. Users can create custom play lists to access disc contents. The DVD-SR format is derived from the DVD-Video format. It acts as a bit bucket to allow recording of streaming data from digital sources such as camcorders, cable boxes, and satellite receivers. An IEEE 1394 interface can be used.

DVD-ROM

At their base level, all DVD discs are DVD-ROM (Read-Only Memory) discs. That is, all DVD discs use the UDF format. Different DVD applications, such as DVD-Video, place specialized material in a specific place, such as the DVD-Video zone. Content contained in the DVD-Other zone may be quite varied, and DVD-ROM uses that opportunity for open-ended storage. In that respect, DVD-ROM is a large capacity bit bucket formatted as UDF. DVD-ROM discs are playback-only media used for high-capacity storage of data, software, games, and so on. DVD-ROM drives are connected to personal computers and function much like CD-ROM drives. With appropriate software, DVD-ROM drives can play DVD-Video and DVD-Audio discs. As with other DVD players, to play back CD-R discs at a 780-nm wavelength and DVD discs at a 635-nm or 650-nm wavelength, DVD-ROM drives must use pickups with dual lasers and other appropriate optical design. DVD-ROM drives support DVD-Video regional coding as well as CSS copy protection. The various recordable formats are not mutually compatible, and there is variability in disc-to-drive compatibility.

DVD-R and DVD+R

DVD-R (Recordable) discs, like CD-R, offer write-once capability to permanently record data. The DVD-R(A) Authoring format is often used for professional authoring and testing of DVD titles. The DVD-R(G) General format is used for business and consumer applications. Because DVD-R(A) uses a 635-nm laser for writing and DVD-R(G) uses a 650-nm laser, the two media are not write-incompatible. However, discs are playable in both types of drives. DVD-R(A) has a Cutting Master Format (CMF) functionality that allows a Disc Description Protocol (DDP) file to be written in the lead-in area for mastering applications. Replication plants can use these discs directly. DVD-R(G) discs include measures to limit piracy; for example, some decryption keys are blanked out. It is thus impossible to copy CSS-encrypted data to a disc. Also, DDP data cannot be written to DVD-R(G) discs.

DVD-R discs comprise two substrates bonded together. A single-sided disc uses one pregrooved substrate bonded to one pregrooved dummy substrate. The recording side of a single-sided disc comprises a polycarbonate substrate, organic dye recording layer, reflective layer, and protective lacquer overcoat, as shown in Fig. 8.21. The dummy side

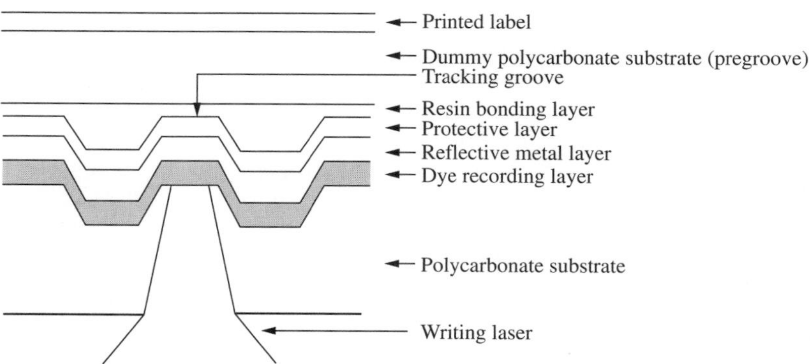

Figure 8.21 DVD-R discs contain a dye recording layer backed by a reflective metal and protective layer, all sandwiched between two substrates.

comprises a substrate, cosmetic reflective layer, and protective lacquer overcoat. The CLV wobbled pregroove generates a carrier signal used for motor control, tracking, and focus. However, whereas CD-R discs use a physical frequency modulation of the pregroove carrier signal to encode the Absolute Time In Pregroove address and prerecorded signal, DVD-R discs use pits and land (known as land pre-pits) molded into land areas between grooves. Placed at the beginning of each sector, the pre-pits contain addressing, laser writing power, and synchronization information. The reading laser tracks the pregroove, but the light shines on the pre-pits peripherally to create a secondary signal that can be extracted from the main signal. As with CD-R, DVD-R uses a pulsed laser to create marks in the organic dye, controlling the duration and intensity of the laser bursts. However, whereas the CD-R write strategy typically simply turns the laser on and off, during DVD-R writing the laser is modulated between a recording and reading bias power to create a multi-pulse train to write one mark. This efficiently controls heat, and creates smaller and more accurate marks. Disc manufacturers can optionally place a write strategy code in the lead-in pre-pits to modify the player's write strategy. Reflectivity for DVD-R and DVD+R discs is about 45 to 85%.

DVD-R discs contain a power calibration area (PCA) for testing laser power. A recording management area (RMA) stores calibration information, disc contents and recording locations, remaining capacity information, and recorder and disc identifiers for copy protection. Recorders perform an optimum power calibration (OPC) procedure to determine the correct laser writing power for particular discs. The PCA can hold 7088 different calibrations, and the RMA can hold OPC information for as many as four different recorders. The remainder of the disc comprises the Information Area. It contains the lead-in, data recordable area, and lead-out. The lead-in contains information on disc format, specification version, physical size and structure, minimum readout rate, recording density, and pointers to the location of the data recordable area where user data is recorded. The lead-out marks the end of the recording area.

DVD-R discs can use the same reference velocity and track pitch as molded discs to achieve the same unformatted storage capacity; user capacity of a "4.7 Gbyte" Version 2.0 disc holds 4.7 billion bytes, or 4.35 Gbytes of user data per side. A cyanine, phthalocyanine or azo dye recording layer may be used, with a 635-nm or 650-nm laser. Both sequential (disc-at-once) and incremental writing can be performed. Once recorded, DVD-R discs are highly compatible and can be played in many DVD-ROM, DVD-Video, and DVD-Audio players. Longevity of a recorded disc is similar to that of a CD-R disc; estimates range from 50 to 300 years. On the other hand, as with most recordable media, the shelf life of unrecorded discs might be only 10 years. Single-sided, dual-layer discs (using the same physical parameters as DVD-ROM discs) hold 8.5 billion bytes. Most drives can read both layers; however, a dual-layer (DL) recorder is needed to write to the second layer.

The DVD+R format is another write-once format. It is not officially a part of the DVD specification written by the DVD Forum. It uses a dye recording layer and CLV rotation. Discs are available in a 4.7- and 8.5-Gbyte (DL) capacity. DVD+R discs are highly compatible and can be played in many DVD-ROM, DVD-Video, and DVD-Audio players. However, DL recorders are needed to record to the added layer.

DVD-RW and DVD+RW

DVD-RW (ReWritable) allows rewriting of data; the specification is essentially an extension to the DVD-R format. It is similar to the CD-RW format. DVD-RW is used for both

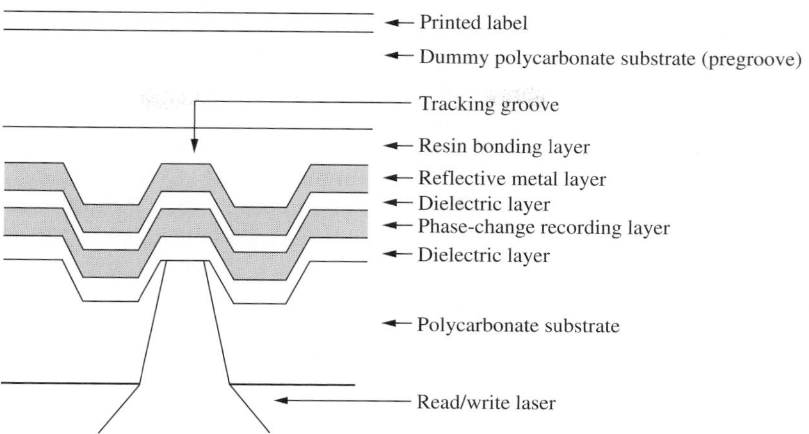

FIGURE 8.22 DVD-RW discs contain a phase-change recording layer backed by a reflective metal layer. The recording layer is sandwiched between two dielectric layers to control thermal properties. Two substrates are bonded together. DVD+RW discs use the same physical construction.

professional authoring and consumer applications. Discs use a phase-change recording mechanism and a multilayer disc structure shown in Fig. 8.22. The recording layer may use a silver, indium, antimony, and tellurium compounded layer, and perhaps 1000 read/erase cycles are possible. Unlike dye-polymer technologies, phase-change recording is not wavelength-specific. Reflectivity for DVD-RW discs (and other phase-change discs) is about 18 to 30%. The disc uses a wobbled pregroove, and pre-pits with addressing and synchronization information. Data is recorded inside the pregroove, and in relatively large blocks. As with DVD-R, there are PMA and RMA zones. A DVD-RW disc may hold 4.37 Gbytes per side. DVD-RW uses CLV rotation; thus, it is particularly used for sequential writing, as in mastering applications. Because it has less robust error protection and a relatively small number of rewrite cycles, DVD-RW is not intended for general purpose data storage and distribution. Although not required, some players use a protective disc caddy. DVD-RW was previously known as DVD-R/W. DVD-RW discs are highly compatible and can be played in many DVD drives. As with other recordable DVD media, the longevity of a DVD-RW disc might be as long as 100 years.

DVD+RW is another rewritable format. It is not officially a part of the DVD specification written by the DVD Forum. It uses phase-change media and a wobbled pregroove; the frequency modulation in the wobble provides address in pregroove (ADIP) addressing information. Data is written inside the groove and there is no pre-embossed addressing data. Disc layer construction is the same as in the DVD-RW format (see Fig. 8.22). Data is written and read in relatively large blocks compared to DVD-RAM. CLV or CAV rotation is allowed for recording, for either sequential data transfer (as in audio/video recording) or faster random access (as in computer data), respectively. Optional defect management features, similar to those found on DVD-RAM, are available. One thousand rewrite cycles are possible. Nominal capacity of a CLV disc is 4.7 Gbytes, and a double-layer (DL) disc holds 8.5 Gbytes. DVD+RW discs recorded with CLV can be played in some DVD drives.

DVD-RAM

DVD-RAM (Random–Access Memory) is a rewritable format. It uses a phase-change recording mechanism and a wobbled land and groove disc design. Using this structure, data may be recorded on both planar surfaces of the groove and land, as shown in the upper part of Fig. 8.23. This technique doubles disc capacity, but deep grooves with steep walls are needed to avoid crosstalk interference between adjacent data. In addition, servos must be employed to switch the pickup's focus between the groove and land area on each revolution. In addition, the tracking signal is inverted when the switch occurs. However, designers contend that the wider groove pitch provided by the groove/land recording technique allows easier tracking and faster recovery from physical shock. Discs also contain pre-embossed pit areas (for every 2k sector) to provide addressing header information, as shown in the lower part of Fig. 8.23. A zoned constant linear velocity (ZCLV) rotational control is used. This technique divides the disc surface into a number of zones, each with a different CLV, but with the same CAV within each zone. There are a total of 35 recording zones across a 120-mm Version 2.0 disc. Successive zones contain more sectors, with 39,200 sectors in the first zone and 105,728 sectors in the last zone. The ZCLV feature enables DVD-RAM to be used as a true random-access, nonsequential medium. Thus, DVD-RAM is well-suited for writing and reading chores done from computer drives.

DVD-RAM provides advanced error correction and defect management features. In the latter feature, defective sectors are identified during manufacture or formatting (or reformatting) and preallocated spare sectors can substitute for them. To reduce wear and tear on specific portions of the disc that are repeatedly written to, the system automatically shifts data placement on the disc surface.

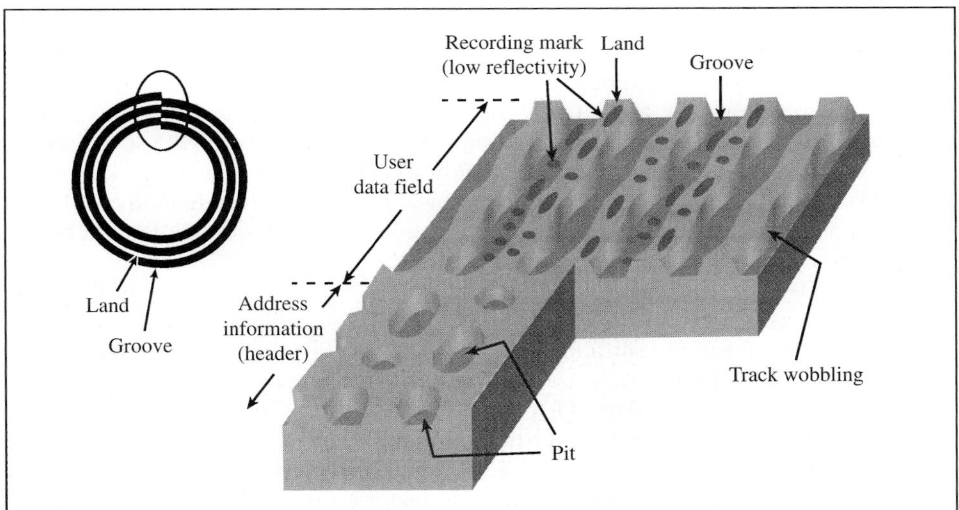

FIGURE 8.23 DVD-RAM discs use a land and groove recording technique. As seen in the upper portion of the figure, phase change data may be placed on planar surfaces of both the wobbled pregroove and land. As shown in the lower part of the figure, rewritable areas are separated by pre-embossed sector headers holding addressing information. (*Parker, 1998*)

DVD-RAM Version 2.0 discs marketed with capacities of 4.7 and 9.4 billion bytes hold 4.37 Gbytes (single-sided) and 8.74 Gbytes (double-sided), respectively. A recording rate of 22.16 Mbps is possible. A disc allows perhaps 100,000 rewrite cycles, and offers a high degree of stability for archiving integrity. DVD-RAM is designed primarily for professional DVD authoring and other post-production work but some consumers use DVD-RAM discs. DVD-RAM discs may be played in many DVD-ROM drives and in some DVD-Video and DVD-Audio players. Higher-capacity discs are usually held in protective cartridges that require slot-loading drives. Some discs can be removed from their cartridges for playback in tray-loading drives. Because of its relatively unique design features it is not as compatible as other recordable DVD formats.

DVD Multi is not a disc format; rather, it is a specification that promotes compatibility within the DVD family. A read-only drive denoted as DVD Multi is capable of reading DVD-ROM, DVD-R, DVD-RW, and DVD-RAM media. Likewise, a writeable drive denoted as a DVD Multi can read all of these media, and can also write on DVD-R, DVD-RW, and DVD-RAM media.

DVD Content Protection

The intellectual property potentially stored on DVD discs has a monetary value that is almost incalculable. Prerecorded formats such as DVD-Video and DVD-Audio provide content owners with the option of securing and monitoring their data in a variety of ways including encryption and watermarking. Likewise, recordable media such as DVD-R, DVD-RW, DVD-RAM, and DVD+RW can employ content protection to prohibit or limit copying. A delicate balance is required so that while content is secured, the user is not unnecessarily inconvenienced; these requirements are mutually contradictory in any content protection system. A number of copy-protection mechanisms, summarized in Table 8.11, are optionally available to content owners.

DVD-Video Copy Protection

A group known as the 4C entity, comprising Intel, IBM, Matsushita, and Toshiba, developed the Content Protection System Architecture (CPSA) that encompasses security issues for DVD formats. In all, eight different security features are used in DVD formats.

In cooperation with 4C, the Copy Protection Technical Working Group (CPTWG) representing 60 companies and interest groups developed the Content Scrambling

Hardware	Copy Protection
DVD-Video	CGMS, CSS, CPPM, Macrovision
DVD-Audio	CPPM, watermark
DVD recorders	CPRM
Analog video outputs	Macrovision
Digital video outputs	HDCP for DVI
IEEE 1394	DTCP

Table 8.11 Summary of copy-protection systems.

System (CSS) copy-protection system that is standard in DVD-Video discs. The CPTWG also established an independent, nonprofit group to oversee nominal cost-based licensing of the CSS technology. Data encryption is used so that content is self-protecting, but use of the technology is voluntary; discs can be distributed with or without copy protection. Likewise, manufacturers could offer a player without decryption hardware; however, it could only play nonencrypted discs. To obtain the algorithms and keys needed to decrypt data in their players, manufacturers must first obtain a DVD license. The data stream is flagged so computer programs can properly interpret the encryption. A Matsushita proposal is the basis for the CSS system. With CSS, content is self-protecting; that is, content cannot be digitally copied because software keys needed to decrypt the data are missing in any copy. Although it is a different technology, regional coding must be implemented in any CSS device.

When encrypted data is decoded in software (as opposed to a dedicated hardware chip) care must be taken so that excessive demands are not placed on the microprocessor. CSS is designed to minimize the burden, allowing efficient decryption (descrambling) without compromising integrity. Moreover, CSS limits the processor overhead required to perform decryption. In the variable-rate MPEG-2 video coding algorithm, video frames are stored as data sectors. There are 2048 bytes/sector. At a fast data rate of 10 Mbps, there might be 600 sectors/second or 20 sectors/frame; at a slow data rate of 2 Mbps, there might be 120 sectors/second or 4 sectors/frame. Instead of encrypting all video sectors, the CPTWG sets an upper limit (and lower limit) on the rate of sectors encrypted by CSS. For example, 10 or 15% of sectors might be encrypted. Even with a low rate, the picture will be unviewable, but the limit minimizes microprocessor overhead.

The proprietary CSS system encrypts data during encoding and then uses authentication to verify that the player's decoder is authorized to decrypt the data. Moreover, communication within the system is encrypted to maintain security over keys. Most DVD-Video players have dedicated authentication hardware. CSS decoding can be performed in hardware or software and every decoder has a 40-bit player key used to decrypt a disc key, and uses the result with the title key to decrypt the movie contents.

CSS features two copy-protection methods. The first, the "Content Scrambled DVD" method, is designed for DVD-Video players. Sectors containing audio and video signals are encrypted; navigation data in sector headers is not encrypted. Content providers must select two encryption keys—one disc key and one title key—jointly used to encrypt the data prior to storage on a DVD-Video disc. The title key is placed in a disc sector header, and the disc key is concealed in a control area of a disc that cannot be read by a DVD-ROM drive unless instructed by authentication commands. Each licensed manufacturer is assigned one of 400 unique player keys; all 400 keys are stored in every disc using CSS encryption. If a license lapses, that manufacturer's key can be omitted from future disc pressings.

The DVD-Video player's hardware decrypting chip is placed in the bitstream between the source data, and the internal Dolby Digital and MPEG-2 decoders. The person viewing the program via an analog output will not know that decryption is taking place. However, if the player contains a digital output, that output will be tapped off prior to decrypting. Copies made from the output digital stream cannot be decrypted because any subsequent decoders will not be able to retrieve the encryption keys and use them to decrypt the data.

The second, the "Bus Authentication and Encryption" method, is designed for use in the computer environment, where encrypted 128-bit keys must be transmitted from a DVD-Video disc across a computer bus to decryption software or hardware. An authentication key is used in addition to the disc and title keys, and each key is checked by elaborately sending data between the disc and the decrypter. This method is more sophisticated because during playback it performs additional encryption on the keys themselves.

In addition, CSS requires that an analog protection system (APS) be employed. Macrovision copy protection, similar to that used in set-top boxes and video networks (which in turn is similar to that protection used in analog videocassettes) is typically used. The Macrovision system can prevent digital-to-analog copying, for example, attempting to use the analog output from a DVD-Video player to make a VHS tape copy. This system uses automatic gain control (AGC) and Colorstripe methods. The AGC portion is virtually identical to that used in prerecorded videocassettes; bipolar pulse signals are added to the video vertical blanking signal causing a VCR to record a weak, noisy, and unstable signal. Because the AGC of a television works quite differently from the AGC of a VCR, VCR playback is disrupted, whereas television display is not. The Colorstripe method is similar to that used in digital set-top boxes; it modulates the phase of the colorburst signal in a rapid, controlled manner, creating horizontal stripes in a copy. A recording VCR recognizes the colorburst phase changes as timebase errors and acts to correct them, thus inducing color errors in the picture. An unauthorized copy shows stripes of color, distortion, rolling, a black and white picture, and dark/light cycling. Colorburst is not present in a component video signal. Use of Macrovision is optional and per-disc licensing fees are paid. The disc identifies its Macrovision protection to the player. A player or drive that does not contain APS would not play DVD-Video discs encrypted with CSS. Likewise, video cards may use APS.

CSS technology is used primarily by the motion picture industry (but its use on a disc is optional). Many computer software providers do not use CSS, even for their audio/video content. Importantly, CSS does not protect other types of data such as software programs. Manufacturers who want to accommodate playback of CSS-coded titles may apply for a license and place CSS decoders in their products. Products without CSS decoding would not play back CSS-coded titles. For example, a DVD-ROM drive might not contain a CSS descrambler; the drive could be used to play back non-scrambled data, but could not be used to watch scrambled movies on a computer. Some professional pirates use DVD replication lines to produce bit-for-bit accurate DVD-Video discs—complete with CSS encryption. Another piracy method rips a DVD-Video disc into its component video and audio contents and re-codes them to Video CD. Both methods violate copyright law. The Data Hiding Sub-Group (DHSG) of the CPTWG is charged with the development and evaluation of watermarks. Encryption and watermarking are also discussed in Chap. 15.

DVD-Audio Copy Protection

To protect against unauthorized copying, the DVD-Audio format uses an optional CPPM content protection framework employing encryption and embedded watermark technology. Copy-protected DVD-Audio discs can only be played on licensed players. The Content Protection for Prerecorded Media (CPPM) was devised in March 1999, by IBM, Intel, Matsushita, and Toshiba in conjunction with music industry companies such as BMG, EMI, Sony Music, Universal Music Group, and Warner Music Group. WG-9 is charged with copy-protection issues. CPPM is similar in intent to the CSS system used in the DVD-Video format and CPPM uses the same authentication measures as CSS. However, CPPM's protection is more sophisticated.

The CPPM encryption code is stronger than that used in the DVD-Video format. A secret album identifier is placed in a control area of the disc that cannot be read by recordable drives, and so cannot be copied to a blank media. Each player or drive has 16 device keys. A media key block is placed on every disc, and the player's device keys interact with the media key block to generate a media key. It is used with the album identifier to decrypt encrypted portions of the disc contents. In the event of hacking, there is capability to revoke, expire, or recover encryption keys.

The CPPM content protection system provides a number of options to content providers of prerecorded media; for example, consumers can make one CD-quality digital copy, per recorder, of the original content. Related content such as supporting text and images is not copied. Content providers can also allow additional copies at various quality levels, up to and including the full quality of the DVD-Audio multichannel original. The encryption used in DVD-Audio can allow two-channel CD-quality, real-time copying along the IEC-958 interface. It also allows both two-channel and multichannel, CD-quality and higher quality, high speed copying along the IEEE 1394 interface. The recorder receives ISRC data that identifies the original recording along with copy permission information describing, for example, how many copies are permitted.

The CPPM watermark is designed to identify content through unencrypted digital (and analog) links. It is not used in high-speed encrypted links and instead verifies copy status of unencrypted signals. The watermark is embedded in the audio signal and is robust over analog and data-compressed transmission links. The watermark operates similarly to SCMS in the digital domain, but it operates in the analog domain or unencrypted digital domain. A copy-permit is the default status; when a copy is made, the embedded watermark signal is updated to mark the copy as a second-generation source. Watermark-compliant recorders will check this mark prior to recording. The watermark can also identify the manufacturer, artist, copyright holder, and other characteristics. The encryption and watermarking technologies are independent; for example, watermarking is optional in encrypted discs.

Content Protection for Recordable Media

Recordable media are protected by the Content Protection for Recordable Media (CPRM) protocol. CPRM links content to the media it is recorded to, so that the recording is playable but copies of the recording are not. CPRM is similar to the CPPM system used specifically for the DVD-Audio format. All blank DVD media have a 64-bit media identifier placed in the burst cutting area at the time of manufacture that uniquely identifies each disc. With CPRM, when protected content is recorded to the disc, the media identifier is used to encrypt a title key, which in turn encrypts content. When the disk is played, the media identifier is again used along with other keys to decrypt a title key, which in turn is used to decode the contents. If the content is moved to another disc, its media identifier will not correctly decode the content. Only audio/video sectors are encrypted; navigation and other data is not encrypted.

The Copy Generation Management System (CGMS) controls the copying of digital and analog video signals. Discs can specify whether any copying is permitted; this is conveyed in data in the output analog and digital signal and interpreted by recorders.

Copy instructions in analog signals are placed in the XDS section of the NTSC signal, and in digital signals it is conveyed via DTCP and HDCP protocols. Copy control information (CCI) includes no copies, one copy, and unlimited copies. When one copy is permitted, the second-generation copy then contains a no-copy instruction; however, multiple copies may be made from the original copy. When a disc carries CSS, CPPM, or CPRM, then a "no copy" condition is assumed. Furthermore, when copying to unprotected media such as CD-R, DVD-Audio limits authorized copies to no more than two channels, 48 kHz and 16 bits.

Secure Digital Transmission

Many applications require a secure link between two devices, such as a computer video card and a display. Two principal systems have been developed. DVD data can be conveyed along these paths, but the disc itself does not participate; the player and display perform the necessary operations independent of the disc contents. The High-Bandwidth Digital Content Protection (HDCP) system defines a secure digital interface for players and displays designed according to the Digital Visual Interface (DVI) specification. DVI can support transmission at 4.95 Gbps; this provides 1600 × 1200 resolution that encompasses HDTV formats. Twin links can support even higher resolution. HDCP for DVI makes DVI a secure interface. Connected devices, such as a video display card and a monitor, exchange keys to authenticate the devices; the system uses forty 56-bit device keys and 40-bit key selection. Data is encrypted at the transmitting device and decrypted at the receiving device. If the receiving device is not HDCP equipped, the transmitting device may send a lower resolution version of the content. HDCP was proposed by Intel and ratified by the Digital Display Working Group in 1998.

The Digital Transmission Content Protection (DTCP) system provides secure transmission over bidirectional digital lines such as the IEEE 1394 bus. For example, a DVD player could be digitally and securely connected to an LCD display, and DTCP would resist unauthorized copying by another connected device. DTCP is described in Chap. 14 in the context of IEEE 1394.

DVD Watermarking

Watermarking can be used to intertwine data into DVD contents so that the watermark can later be retrieved to identify the contents on the disc. Furthermore, most watermarks resist tampering or removal. Watermarking does not prevent copying; it merely identifies the content. In some cases, a fragile watermark is used; analog copying degrades the watermark and thus identifies the content as a copy. A watermark is only useful if downstream equipment recognizes it. In the case of DVD, a license agreement needed to play encrypted contents may also legally bind the manufacturer to detect watermarks. DVD-Audio uses a watermark system developed by Verance. The license that enables the drive to play CPPM or CPRM discs obligates the manufacturer to detect the watermark. DVD-Audio recorders recognize CCI copy-generation watermarks.

HD DVD

The HD DVD format was envisioned as the successor to the DVD-Video format. This high-density disc format was designed primarily to deliver high-definition playback of motion pictures. Players and discs were introduced in March and April 2006, but the HD DVD format ultimately did not find commercial success against the competing

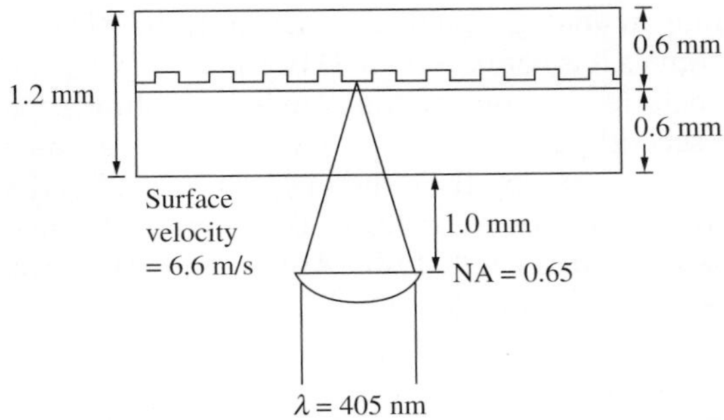

FIGURE 8.24 The abandoned HD DVD format provided greater storage capacity and higher output bit rate than DVD. HD DVD discs use two 0.6-mm substrates. A single-sided disc is shown, but the specification also supports double-sided discs.

Blu-ray system. In February 2008, its principal backer, Toshiba, announced that it would no longer develop or manufacture HD DVD players or drives. Soon thereafter, the format was abandoned in the marketplace.

The HD DVD format (High Density) uses a 405-nm blue-light laser and NA of 0.65 to achieve high storage capacity. An HD DVD-ROM disc holds 15 Gbytes on a single-layer disc and 30 Gbytes on a dual-layer disc. The structure of the HD disc is shown in Fig. 8.24. The VC-9 video codec used in Microsoft Windows Media 9 (WM9), MPEG-4 H.264 Advanced Video Codec (AVC), and MPEG-2 are mandatory video codecs for all licensed HD DVD players. Dolby Digital Plus and DTS are mandatory audio-coding formats. Lossless MLP 2-channel coding is mandatory and lossless DTS coding is optional. AES encryption is used to copy-protect contents. In addition, rewritable HD DVD discs have been developed. A single-sided, single-layer HD DVD-RW disc holds 20 Gbytes and a double-sided, single-layer disc holds 40 Gbytes. A single-sided, single-layer HD DVD-R disc holds 15 Gbytes. HD DVD was supported by the DVD Forum.

CHAPTER 9
Blu-ray

The Blu-ray optical disc system was developed as a successor to the DVD-Video format and as a competitor to the failed HD DVD format. Blu-ray is a consumer disc system that is widely used to play motion pictures with high-definition resolution. Its picture quality is greater than that of standard-definition DVD and meets broadcast high-definition DTV standards. In addition, a stereoscopic (3D) video Blu-ray specification has been introduced. Blu-ray is also used to distribute video games, and recordable and rewritable media are available. Blu-ray accommodates a variety of lossy and lossless audio formats. It can reproduce very high-quality multichannel sound and also provides long playing times. Blu-ray employs content protection that is more robust than that used in DVD. Over its product lifetime, the Blu-ray format is expected to outsell the DVD format.

Development and Overview

The group of companies developing the Blu-ray disc system (also called BD system) was headed by Sony Corporation. The Blu-ray Disc Founders group was established in early 2002. The group was renamed the Blu-ray Disc Association in June 2004. The first Blu-ray disc specification was announced in February 2002. The first Blu-ray disc recorders were introduced in Japan in April 2003, and the first standard Blu-ray players and discs were introduced worldwide in June 2006. Blu-ray overcame early competition from the HD DVD format and now enjoys wide commercial success. In naming the format, the *e* was omitted from *Blue* to allow trademarking; common words cannot be registered as trademarks.

As with the CD and DVD, Blu-ray disc media are available as prerecorded, recordable, and rewritable formats. The specifications are known respectively as BD-ROM, BD-R, and BD-RE. The three disc types have the same data capacity. All three types can hold a single data layer or dual data layers. The layers are independent, and both layers are read from the same side of the disc. The specifications that define the Blu-ray family are shown in Fig. 9.1; the specifications are authored across three parts: Part 1, Physical Standard, that defines the disc construction; Part 2, File System Standard, that defines the data format and file management system; and Part 3, Application Standard, that defines the data structure and mechanism for stream management and user presentation.

A key improvement in the design of the Blu-ray format is the use of a 405-nm wavelength laser for reading and recording. This wavelength is shorter than those used in the CD or DVD and allows greater data density. Because blue-violet lies at the "end" of the

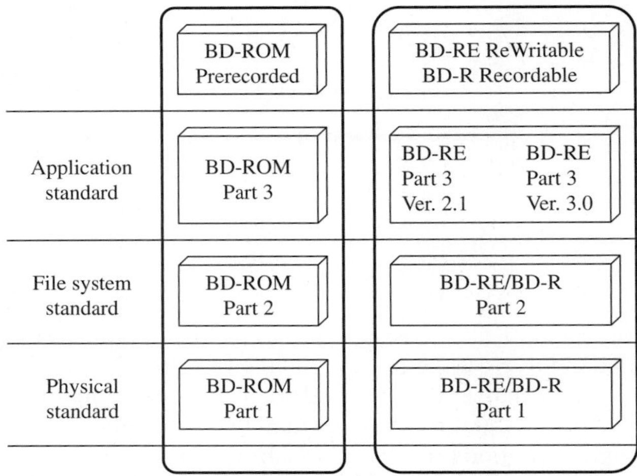

FIGURE 9.1 Three layers of standards comprise the Blu-ray specification: Application, File System, and Physical. They define the family of Blu-ray formats: BD-ROM (prerecorded), BD-RE (rewritable), and BD-R (recordable).

visible-light spectrum, it can be confidently predicted that Blu-ray will be the last optical disc format to use a visible-light source. Interestingly, BD designers considered even shorter laser wavelengths, but encountered "sunburn" durability problems in the plastic substrate. Moreover, the plastics typically used in substrates rapidly lose transparency at wavelengths shorter than about 400 nm.

A single-layer Blu-ray disc provides a storage capacity that is about 35 times greater than the capacity of a CD disc, and about five times greater than the capacity of a DVD disc. Blu-ray disc diameter is either 12 cm or 8 cm; these disc dimensions are identical to those of CD and DVD discs. The greater storage capacity is due to several improvements, most noticeably the shorter wavelength laser and an objective lens with a higher numerical aperture. These permit a narrower track pitch and smaller pit sizes. The three disc formats are compared in Table 9.1. The Blu-ray Disc Association recommends that Blu-ray players be able to play CD and DVD discs, and this is typically the case; however, this is not a technical requirement of the Blu-ray specification.

	CD	DVD (single layer)	Blu-ray (single layer)
Storage capacity	0.7 Gbytes	4.7 Gbytes	25 Gbytes
Track pitch	1.6 μm	0.74 μm	0.32 μm
Minimum pit length	0.8 μm	0.4 μm	0.15 μm
Storage density	0.41 Gbits/in^2	2.77 Gbits/in^2	14.73 Gbits/in^2

TABLE 9.1 Principal specifications of the CD, DVD, and Blu-ray formats.

Disc Capacity

A single-layer Blu-ray BD-ROM disc known as BD-25 can store about 25 Gbytes; it is capable of holding at least 2 hours of high-definition video. A dual-layer disc known as BD-50 holds about 50 Gbytes of data. Similarly, BD-27 and BD-54 disc formats are specified. A Mini Blu-ray BD-8 disc with 8-cm diameter and a single layer holds about 7.8 Gbytes, and a Mini Blu-ray BD-16 disc with dual layers holds about 15.6 Gbytes. These capacities provide compatibility with the earlier BD-RE specification. Subsequently, for development expediency, BD-ROM Version 1.3 excluded proposed capacities of BD-23 and BD-46; these formats are not implemented. In January 2010, it was announced that the storage capacity of 25-Gbyte discs could be increased to 35.4 Gbytes; the higher capacity discs are readable on current players with a firmware update.

Table 9.2 compares different Blu-ray disc types in terms of storage capacity and typical playing time. Disc capacities can be expressed as the decimal size of billions of bytes, and as the actual binary value of gigabytes. The difference of 7.4% accounts for the notational fact that a billion bytes in common usage is 10^9 and a gigabyte in computer terms is 2^{30}. Thus, the difference is 1,000,000,000 versus 1,073,741,824.

The Blu-ray format accommodates high-definition video with a maximum of 1920 × 1080 pixel resolution. Frame rate is 60 frames per second (fps) with interlaced scan, or 24 fps with progressive scan. The Blu-ray format has a maximum transport stream bit rate of 48 Mbps and maximum video bit rate of 40 Mbps. Variable bit-rate coding can be used to optimize data transfer rates. Recordable Blu-ray media support a bit rate of 36 Mbps. The maximum playing times depend on which codecs are used and their relative efficiency, video and audio bit rates, number of tracks, the content itself, and

Disc Type	Diameter (cm)	Disc Structure	Capacity (billions of bytes) (10^9)[1]	Capacity (gigabytes) (2^{30})[1]	Typical Video Playback Time (hours)[2]	2-Channel Audio-Only Playback Time (hours)[3]	5.1-Channel Audio-Only Playback Time (hours)[4]
BD-8	8	1 layer	7.791	7.256	0.7	3.8	27.1
BD-16	8	2 layers	15.582	14.512	1.4	7.5	54.1
BD-25	12	1 layer	25.025	23.306	2.3	12.1	86.9
BD-27	12	1 layer	27.020	25.164	2.5	13.0	93.8
BD-50	12	2 layers	50.050	46.613	4.6	24.1	173.8
BD-54	12	2 layers	54.040	50.329	5.0	26.1	187.6

[1] Capacities can be increased slightly if the track pitch is decreased.
[2] High-definition MPEG-2 video with audio tracks, bit rate of 24 Mbps.
[3] PCM, bit rate of 4.608 Mbps (coded at 24-bit/96-kHz).
[4] Dolby Digital, bit rate of 0.640 Mbps.

Table 9.2 Disc storage capacities and typical playing times.

other factors. A single-layer disc using MPEG-2 compression (less efficient than other allowed codecs) can hold, for example, 135 minutes of high-definition video with additional capacity for audio tracks and 2 hours of bonus material in standard definition. When coded with MPEG-4 AVC (High Profile), the same disc could hold about 4 hours of video along with audio and 100 minutes of bonus material. A dual-layer disc doubles those capacities. In terms of audio storage, a 50-Gbyte disc can hold over 10 hours of 192-kHz/24-bit PCM stereo audio, or over 200 hours of 5.1-channel Dolby Digital audio.

BD-ROM Player Profiles

The Blu-ray BD-ROM specification defines player profiles that differentiate types of players, and also describe a planned evolution of the player profiles following the format's launch. The three initial BD-Video profiles are described as Profile 1.0 Grace Period (Initial Standard), Profile 1.1 Bonus View (Final Standard), and Profile 2.0 BD-Live. The Grace Period profile was superseded by the Bonus View profile in November 2007, as the minimum profile for BD-Video players.

As shown in Table 9.3, the levels of hardware support vary with each profile; however, all BD-Video players must support the BD-J Java specification. BD-Live players are differentiated in that they have a user-accessible Internet connection for accessing additional features such as games, chat, and downloadable content. Also, BD-Live players have additional local memory to store this content. Discs with Bonus View or BD-Live interactive features may not play or may have limited capability when played on Grace Period players.

BD-Audio (Profile 3.0) players play audio-only content on Blu-ray discs; this profile uses the HD Movie (HDMV) format, but without video. BD-Audio players do not require support of BD-J or video decoding. Profile 4.0 (not shown in the table) describes the Blu-ray Disc Movie (BDMV) Recording format. Profile 5.0 (not shown in the table) defines the Blu-ray 3D stereoscopic format, as described later in this chapter.

	Profile 1.0 Grace Period	**Profile 1.1 Bonus View**	**Profile 2.0 BD-Live**	**Profile 3.0 Audio-Only**
Primary video decoder	Mandatory	Mandatory	Mandatory	No
Secondary video decoder	Optional	Mandatory	Mandatory	No
Primary audio decoder	Mandatory	Mandatory	Mandatory	Mandatory
Secondary audio decoder	Optional	Mandatory	Mandatory	No
Audio mixing	Optional	Mandatory	Mandatory	No
192-kHz PCM	Optional	Optional	Optional	Optional
Platform	BD-J, HDMV	BD-J, HDMV	BD-J, HDMV	HDMV
Internet access	No	No	Mandatory	No
Virtual file system	Optional	Mandatory	Mandatory	No

Table 9.3 A series of player profiles defines the evolution of player design and capabilities.

Disc Design

As with CD-ROM and DVD-ROM discs, BD-ROM discs store binary data using pits embossed on a substrate. During playback, the pits yield a change in reflected intensity in the reading laser and those changes, occurring at about 980,000 per second, are decoded to produce the stored data content.

For comparison, CD, DVD, and Blu-ray discs are shown in Fig. 9.2. All three disc formats are read from underneath the disc (a DVD disc can also be double-sided). In the CD, the data layer is on the far side of the substrate away from the reading laser. In the DVD, the data layer(s) are placed in the disc interior between two substrates. In Blu-ray, the data layer(s) are placed on the near side of the substrate. The latter provides advantages, but also presents challenges as described later.

The structures of single-layer and dual-layer Blu-ray BD-ROM discs are shown in Fig. 9.3. Both discs use a substrate with a nominal thickness of 1.1 mm. In a single-layer disc, the data layer of the substrate is covered by a reflective layer which is topped by a cover layer of 0.1-mm thickness. In a dual-layer disc, the substrate is covered by two data layers each separated by a transparent separation layer of about 0.025 mm. The inner data layer (L0) is covered by a reflective layer and the outer data layer (L1) is covered by a semi-reflective (or semi-transparent) layer, which in turn is covered by a transparent cover layer of 0.075 mm. The disc is read through the transmission stack; this is through the cover layer (or through the cover layer, outer data layer, and separation layer). The reading laser can be focused on either data layer to permit this.

Unlike CD and DVD discs, in Blu-ray discs, the optical path is through the cover layer, not the substrate. Thus the substrate's optical characteristics are not critical; for example, transparency and birefringence are not a concern. The substrate can be opaque. Because the objective lens is close to the data layer, optical aberration caused by disc tilt is limited. The cover layer provides scratch resistance and reduces the need for a disc cartridge. Unlike the CD or DVD specifications, the Blu-ray disc specification requires that the cover layer be scratch resistant. A separate protective layer on top of the cover layer is optional.

CD, DVD, and Blu-ray discs all use a spiral pit track. Also, all three formats begin reading from the inner radius and read outward. If a DVD or Blu-ray disc has two layers, when the end of the outer layer is reached, the laser can refocus on the inner layer and begin reading inward. This type of tracking is known as opposite track path, or reverse spiral dual layer. In DVDs, the inner layer can optionally be read from the inner radius in a technique known as parallel track path; however, Blu-ray does not support this. A buffer memory is used to allow continuous data output when switching layers. In addition, to effectively extend the buffer size, the data at the switch point can be written with a lower bit rate.

Disc Optical Specification

As described in earlier chapters, optical disc capacity is determined by specifications such as laser wavelength and numerical aperture (NA) of the objective lens. In particular, disc capacity can be increased by decreasing laser wavelength and increasing NA. Blu-ray uses a 405-nm wavelength blue-violet laser. For example, an indium gallium nitride (InGaN) laser can be used. Numerical aperture is 0.85. Shorter light wavelength, higher NA, and a thin cover layer over the data layer allow a (diffraction limited) spot size of 580 nm. Along with efficient data modulation, this permits high data density.

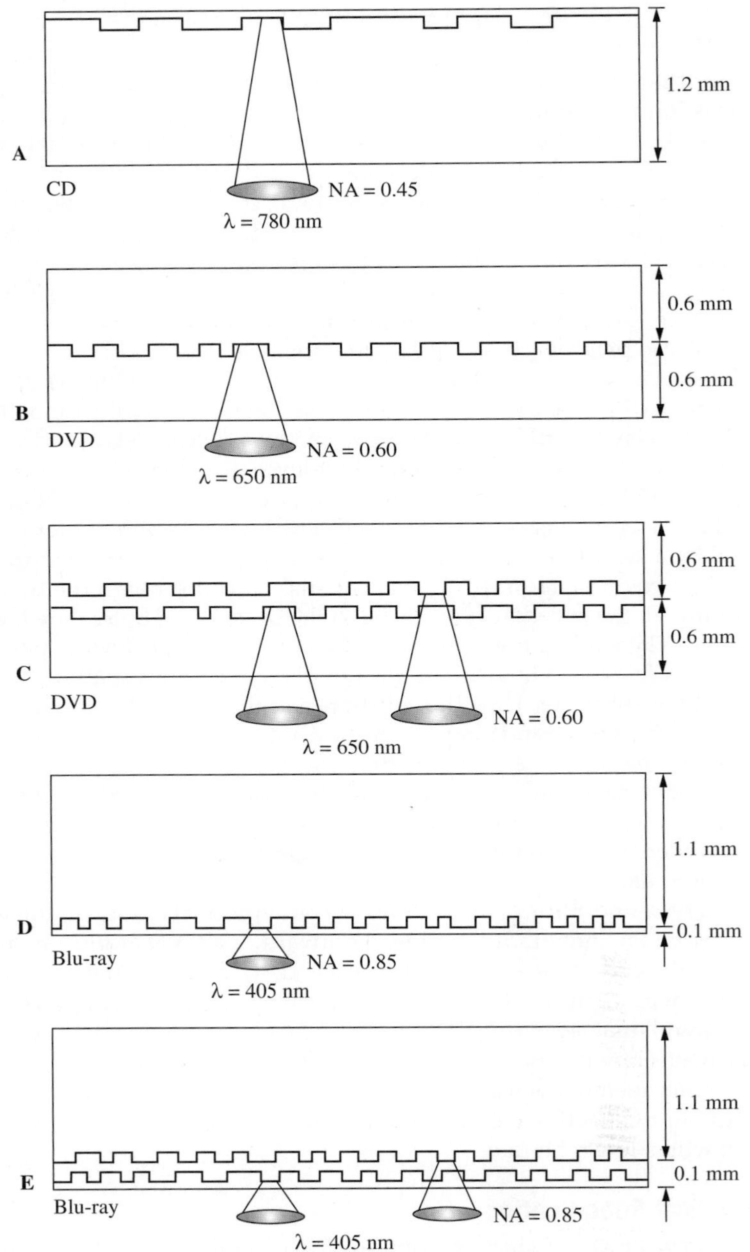

FIGURE 9.2 The structure of CD, DVD, and Blu-ray discs is markedly different particularly in the locations of the data layers. A. CD disc. B. Single-layer DVD disc. C. Dual-layer DVD disc. D. Single-layer Blu-ray disc. E. Dual-layer Blu-ray disc.

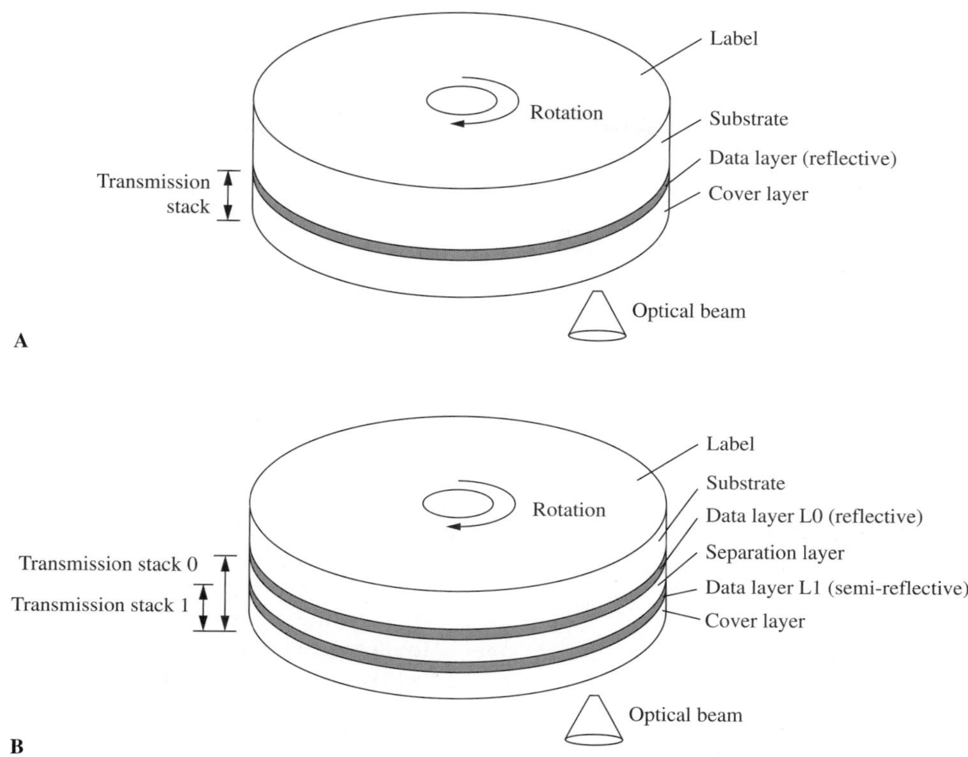

FIGURE 9.3 In the structure of Blu-ray discs, the data layer(s) are placed near the side of the reading laser, within a cover layer. A. Single-layer Blu-ray disc. B. Dual-layer Blu-ray disc.

Track pitch is 0.32 μm. Constant linear velocity (CLV) disc rotation is used. Minimum mark length (2T) is 0.149 μm for 25- and 50-Gbyte discs. Linear recording density varies for other disc capacities.

Compared to a DVD, the higher NA of the Blu-ray objective lens increases areal density two times and the shorter wavelength contributes an additional factor of 2.6. Thus, the Blu-ray laser spot size is about one-fifth that of the DVD so Blu-ray's capacity is about five times greater. Moreover, the output bit rate of Blu-ray is about five times greater than a DVD, but the rotation rate is only doubled.

As noted, a large value for objective-lens NA is desirable because laser spot diameter is inversely proportional to NA, and a small spot allows smaller pit formations as well as lower laser output power. However, higher NA increases the effect of optical aberrations and decreases allowable tolerances. For example, when disc tilt or other factors cause the optical axis of the objective lens to deviate from perpendicular to the disc surface, convergence deteriorates because of coma aberration. Decreasing the cover thickness compensates for increased coma aberration resulting from shorter wavelength and increased NA. In addition, spherical aberration occurs because of deviations in the relative thickness of the cover layer over the data layer; the tolerance for cover-layer thickness error of Blu-ray is about one-tenth that of a DVD. Thus the cover layer must be applied precisely.

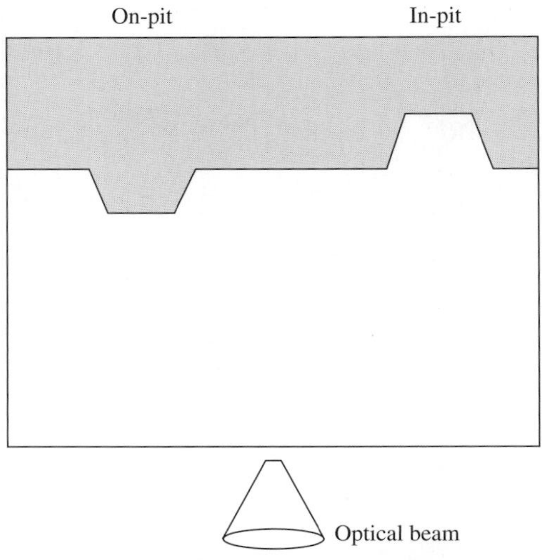

FIGURE 9.4 Two pit configurations are permitted for Blu-ray discs. As seen from the reading side of the disc, they are convex on-pit and concave in-pit.

A thin cover layer over the data layer is advantageous because NA can be increased; however, one benefit of a thick layer, the fact that dust and dirt on the outer surface is defocused relative to the data layer, is decreased. In DVD discs, the cover layer (substrate) is 0.6-mm thick and the laser wavelength is 650 nm. However, with the same thickness, even with a 405-nm wavelength laser, Blu-ray capacity would have been only 12 Gbytes. Moreover, the shorter wavelength and thick layer would have necessitated lowering the NA from 0.60 to about 0.55, yielding a BD capacity of only about 10 Gbytes. To overcome this and other limitations, BD designers determined that a cover-layer thickness of 0.1 mm and NA of 0.85 were needed to yield a 25-Gbyte capacity and tilt margin equal to that of a DVD.

Two pit configurations are permitted for Blu-ray discs as shown in Fig. 9.4. As seen from the optical beam, the concave pit is defined as the in-pit, and the convex pit is defined as the on-pit. Generally, replicated BD-ROM pits use an in-pit configuration. However, for the inner layer of a dual disc, one manufacturing method creates the pits by replicating them in the space layer that yields in-pits. Another method replicates on-pits in the outer layer. Both manufacturing methods yield discs within the Blu-ray specification. In Blu-ray, a pit depth of about $\lambda/4$ yields a low jitter figure.

To maintain compatibility, the reflectivity of BD-ROM discs was designed to be similar to that of the previously released BD-RE specification. As with the CD and DVD, aluminum was selected as the preferred metallization material. Two ranges of reflectivity are specified for the BD-ROM data layer: 35 to 70% for a single-layer disc, and 12 to 28% for a dual-layer disc. These figures are somewhat higher than those used in the BD-RE format.

Optical Pickup Design

As noted, for Blu-ray discs, the cover layer over the data layer(s) is thin compared to that of a CD or DVD, and the wavelength of laser light is much shorter. For these and other reasons, it is not possible to use the same optical pickup elements to read these three kinds of discs. However, by using multiple elements, some pickups are able to read all three disc types. These pickups must control the NA differently for each format and must compensate for any spherical aberration caused by an uneven cover-layer thickness. In addition, coma aberration caused by differences in wavefront curvature and lens movement must be accounted for; for example, different photodetectors must be used.

Various pickup designs can be used. In one design, an integrated three-format pickup uses a single aspherical objective lens and a polarized Holographic Optical Element (HOE) device. The HOE uses a birefringent material sandwiched between two substrates. The birefringent material has the same index of refraction as the bonding material for a certain polarization direction, but a different index of refraction for a different perpendicular polarizing direction. The HOE does not affect the 405-nm (Blu-ray) wavefront, but at 650 nm (DVD) and 780 nm (CD) the polarization direction is perpendicular to that at 405 nm, resulting in phase distributions. This yields a non-diffracted beam for Blu-ray and diffracted beams for the DVD and CD. The HOE is designed to control this phase distribution and also compensates for spherical aberration caused by thickness differences in the CD and DVD substrates. The numerical aperture needed to read CD and DVD discs can also be controlled with an aperture filter.

An alternative prototype recording/playback pickup is shown in Fig. 9.5. This design is capable of recording and playing back CD, DVD, and Blu-ray discs.

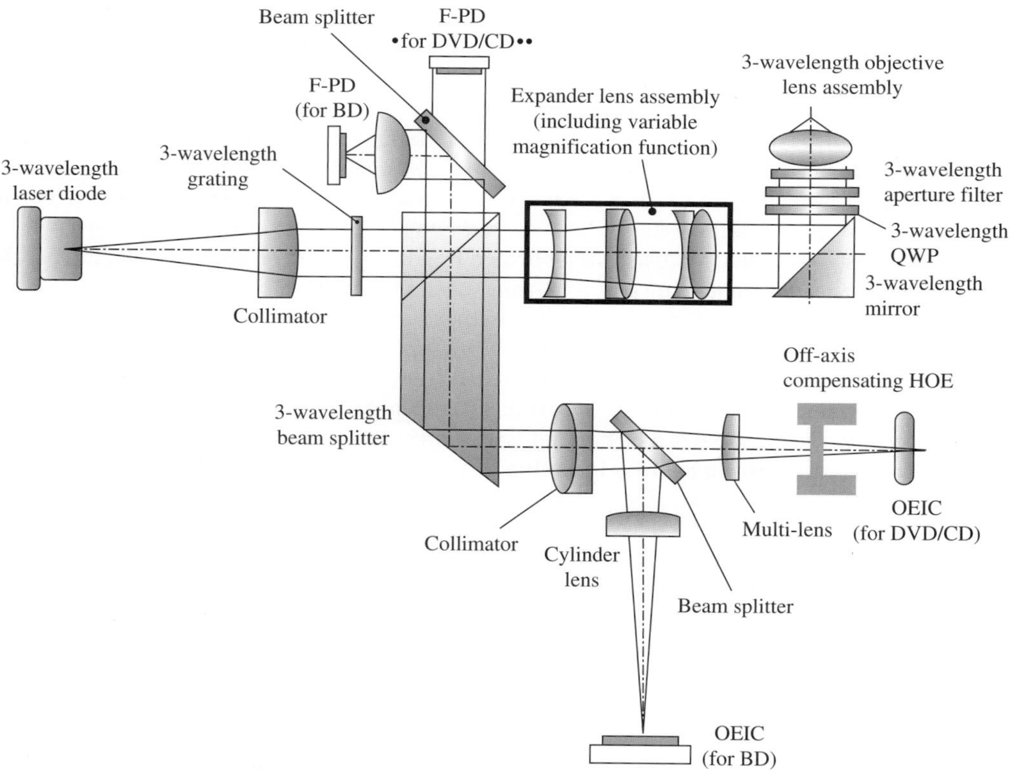

Figure 9.5 Diagram of an optical laser pickup able to record and play back CD, DVD, and Blu-ray discs. (*Blu-ray Disc Founders, 2004*)

The pickup outputs three laser wavelengths; a red laser diode, infrared laser diode, and blue laser diode are co-mounted and share a common optical path; the lasers are not powered simultaneously. A variable magnification function uses an expander lens with different magnifications for the corresponding objective lens apertures for each of the three disc formats. This improves laser efficiency for each of the disc types. The objective lens assembly uses an aspherical design and a holographic lens to yield spherical aberration correction for each laser wavelength.

Disc Manufacturing

The manufacture of BD-ROM discs uses techniques and workflow that are similar to those used to manufacture CD and DVD discs. However, many of the tolerances for Blu-ray discs are more critical, and several new manufacturing techniques are used. The process begins with mastering, in which a master recording is used to produce a master disc which in turn is used to produce stampers. These are used to replicate substrates with injection molding which are then metallized and covered to produce finished discs.

Three types of disc mastering methods were developed for Blu-ray. The Phase Transition Metal (PTM) system uses a blue-laser diode light source. A phase transition (amorphous to crystalline) inorganic recording material is used. It acts as a photoresist; the crystalline regions are soluble. Since the phase change is through a heat-chemical reaction, and not photochemical, the written spot area is only that area heated above the threshold temperature; thus, the resulting pit is smaller than the overall laser spot and smaller than when using a photoresist method. A silicon wafer with a layer of sputtered inorganic material is used as the mastering substrate. By monitoring changes in reflectivity during the phase change, the mastering process can be optimized in real time.

A deep-ultraviolet (UV) liquid immersion method known in microscopy, can also be used for mastering. The method uses water injected between a high NA lens and the photoresist. An electron beam recorder (EBR) can also be used for mastering in which a photoresist layer is exposed.

In most cases, Blu-ray discs are replicated by forming a substrate with pits, then adding finishing layers to it. Typically, a 1.1-mm substrate is formed by injection molding. The reflective layer is deposited by sputtering. In the case of a single-layer disc, the outer cover layer can be applied over the fully reflective layer with a UV-curable resin using spin coating. Alternatively, cover sheets can be punched from a roll, resin applied to the substrate, and the cover sheet bonded to the substrate by curing the resin with UV irradiation. Or for the latter, the cover sheet can be bonded with a pressure-sensitive adhesive (PSA).

The manufacturing process for a dual-layer disc adds several steps. The inner data layer on the substrate is fully metallized by sputtering and a separation layer (a UV-curable adhesive called HPSA) is formed on the substrate by pressure bonding. A stamper is pressed into the HPSA to replicate a second pit surface. UV light irradiates the underside. The stamper is removed and a semi-reflective layer is sputtered on the HPSA. The cover layer is applied using any of the same methods as for single-layer discs. The manufacturing process for a dual-layer disc is shown in Fig. 9.6.

The thinness of the cover layer and its thickness error tolerance of ± 3 μm necessitate precision. For example, unless care is taken, a UV-resin coating can have resin upheaval on the outer peripheral radius due to surface tension. To maintain constant thickness, the outer perimeter of the resin layer can be heated so that its viscosity is reduced. In another replication method, data is placed on a 0.1-mm polycarbonate layer by injecting the

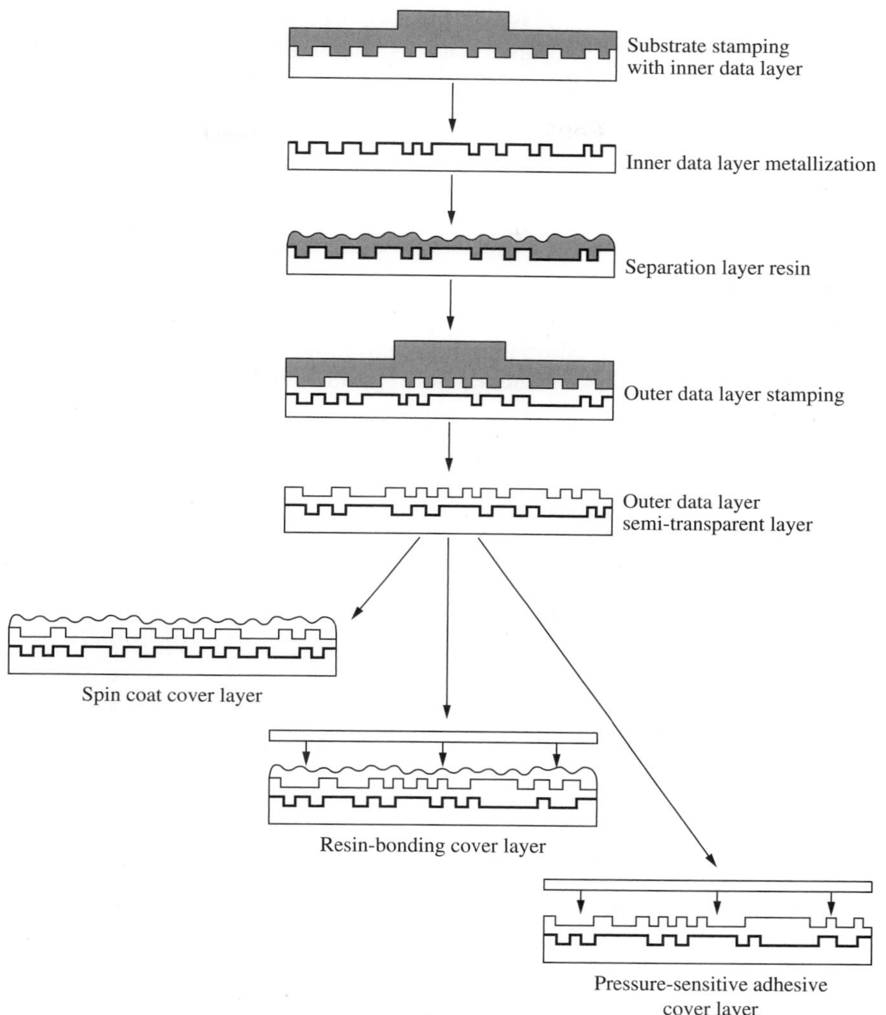

Figure 9.6 Flowchart for the manufacturing process for a dual-layer disc showing how both data layers are formed, and the three ways in which a cover layer can be applied. A single-layer disc uses the same techniques, but omits the separation layer and outer data layer.

polycarbonate into a polymethyl methacrylate (PMMA) mold; the layer is then bonded to a 1.1-mm substrate. In some cases, single-layer discs can be replicated in a single machine.

The Source Identification (SID) code is placed on the inner perimeter of a disc. It comprises visible characters that identify the disc manufacturer. There are two kinds of SID codes: the mastering code and the mold code. The mastering code is created by the master recorder and is present on the stamper. The mold code is etched on the mirror block side of the mold. On the replicated substrate, the master code appears on the same side of the substrate as the embossed pits, and the mold code appears on the opposite side of the substrate.

The Burst Cutting Area (BCA) is a reserved zone between a disc radii of 21.0 and 22.2 mm. The BCA is used to add unique information such as a serial number to individual discs after replication. The BCA code can be used as part of the Advanced Access Content System (AACS) copy-protection system to hold an encrypted key. The BCA code can be applied by a high-power diode laser. The code comprises up to 64 bytes divided into four 16-byte words. The BCA is recorded in CAV mode and appears as low-reflectance stripes in a circumferential orientation.

BD-ROM Specifications

The representative specifications for the BD-ROM format are summarized in Table 9.4 through Table 9.9. These tables show disc structure, capacity related items, measuring conditions, mechanical parameters, optical parameters, and operational signals.

Recording Method	In-Pit or On-Pit
Diameter (12-cm or 8-cm disc) (mm)	120 ± 0.3 or 80 ± 0.3
Center hole diameter (mm)	15 + 0.10/0.00
Clamping area diameter (mm)	23.0 to 33.0
Burst cutting area diameter (mm)	42.0 (+0.0/−0.6) to 44.4 (+0.4/−0.0)
Lead-in diameter (mm)	44.0 to 44.4 (+0.4/−0.0)
Data diameter (12-cm or 8-cm disc) (mm)	48.0 (+0.0/−0.2) to 116.2 or 48.0 (+0.0/−0.2) to 76.2
Lead-out diameter (12-cm or 8-cm disc) (mm)	117 or 77
Disc mass (12-cm or 8-cm disc) (g)	12 to 17 or 5 to 8
Rotation	Counterclockwise to readout surface
Track spiral (outer layer)	Clockwise (hub to edge)
Track spiral (inner layer)	Counterclockwise (edge to hub)
Rotational scanning velocity (CLV) (m/s)	5.280 (23.3 Gbytes), 4.917 (25.0 Gbytes), 4.544 (27.0 Gbytes)

TABLE 9.4 Disc structure for single-layer (SL) and dual-layer (DL) discs.

Total capacity SL/DL (Gbytes)	25.0 / 50.0
Video bit rate (Mbps)	40.0
Channel clock period (ns)	15
Pit length (2T to 8T) (μm)	0.149 to 0.695
Data bit length (average) (μm)	0.11175
Channel bit length (average) (μm)	0.0745
Linear velocity at 1× (m/s)	4.917
Track pitch (μm)	0.32 ± 0.003

TABLE 9.5 Capacity-related items.

Readout wavelength (nm)	405 ± 5
Read power (mW)	0.35 ± 0.1 (SL) and 0.70 ± 0.1 (DL)
NA of objective lens	0.85 ± 0.01
Polarization	Circular
Rim intensity	
Radial (%)	60 ± 5
Tangential (%)	65 ± 5
Wavefront aberration of ideal substrate (λ rms)	≤ 0.033
Relative intensity noise of laser (dB/Hz)	≤ −125
Normalized detector size (μm²)	$S/M^2 \leq 25$

TABLE 9.6 Measuring conditions.

Radial tracking	
Run-out SL and DL (mm)	≤ 0.050 and 0.075 p-p
LF residual error SL (nm)	≤ 9
HF residual error SL (nm)	≤ 6.4 rms
Axial tracking	
Run-out (12-cm and 8-cm disc) (mm)	≤ 0.3 and ≤ 0.2
LF residual error (nm)	45 maximum
HF residual error (nm)	32 rms maximum
Disc thickness (mm)	0.9 to 1.4
Disc unbalance (gmm)	≤ 4.0
Disc radial tilt margin (α) (°)	< 1.60 p-p
Disc tangential tilt margin (α) (°)	< 0.60 p-p
System radial tilt margin (α angular deviation) (°)	± 0.60
System tangential tilt margin (α angular deviation) (°)	± 0.30

TABLE 9.7 Mechanical parameters.

Cover layer	
Thickness SL (μm)	100
Thickness DL (μm)	75
Thickness variation (μm)	≤ 3.0
Spacer layer	
Thickness DL (μm)	25 ± 5
Thickness variation including cover layer (μm)	≤ 3.0
Hard coat (μm)	2
Substrate	No requirement
Reflectivity (%)	35 to 70 (SL) and 12 to 28 (DL)
Birefringence (μm)	≤ 0.030
Refractive index	1.45 to 1.70
Beam diameter (μm)	0.58
Optical spot diameter (μm)	0.11

TABLE 9.8 Optical parameters.

Limit equalizer jitter (channel clock period)	
SL disc (%)	≤ 6.5 (0.98 ns)
DL Layer 0 (%)	≤ 6.5 (0.98 ns)
DL Layer 1 (%)	≤ 8.5 (1.28 ns)
Symbol error rate without defects	$< 2 \times 10^{-4}$
Defect size (μm)	< 100 (air bubble), 150 (black spot)
Correctable burst error (mm)	7
Asymmetry	–0.10 to +0.15
DPD	0.28 to 0.62
Push-pull	0.10 to 0.35
Track cross	≥ 0.10
Storage temperature (°C)	–10 to +55, change of ≤ 15°C/hour
Storage humidity (% RH)	5 to 90
Operating temperature (°C)	–5 to + 55, change of ≤ 15°C/hour
Operating humidity (% RH)	3 to 90, change of ≤ 10%/hour

TABLE 9.9 Operational signals.

Audio Codecs

The Blu-ray specification specifies several audio and video codecs to allow flexibility in authoring, and to provide wide compatibility between disc content and player decoders. The BD-ROM specification supports, as either mandatory or optional, seven different audio formats. In particular, players must fully support PCM, Dolby Digital, and DTS (Coherent Acoustics) codecs. Optional codecs are Dolby Digital Plus, Dolby TrueHD lossless, DTS-HD High Resolution Audio, and DTS-HD Master Audio lossless.

However, in terms of mandatory or optional support, several qualifications must be noted. In part, this is because some Dolby and DTS codec formats are extensions to legacy formats and whether they are considered to be mandatory or optional depends on the bitstream configurations. For example, players support Dolby Digital Plus only for more than 5.1 channels. The base 5.1 channels are coded with core Dolby Digital; only higher channels are coded with Dolby Digital. For Dolby TrueHD, players must mandatorily support the core Dolby Digital bitstream up to 640 kbps; support of the lossless bitstream is optional. The core portions of DTS-HD extensions (High Resolution and Master Audio) are mandatorily supported up to 1.509 Mbps; support of the HD portion is optional. It is also important to note that the Blu-ray specification imposes various technical constraints on audio codecs, for example, in terms of maximum bit rate and number of channels. Specifications for the family of audio codecs are shown in Table 9.10.

Audio Codec	Mandatory	Compression	Sampling Frequency (kHz)	Word Length (bits)	Maximum Bit Rate (Mbps)	Maximum Number of Channels
PCM	Yes	Lossless	48, 96, 192[1]	16,20,24	27.648	8 (48 kHz and 96 kHz) 6 (192 kHz)
Dolby Digital	Yes	Lossy	48	16–24	0.640	5.1
Dolby Digital Plus	No	Lossy	48	16–24	1.7	7.1[2]
Dolby TrueHD	No[3]	Lossless	48, 96, 192[1]	16–24	18.64	8 (48 kHz and 96 kHz) 6 (192 kHz)
DTS	Yes	Lossy	48	16,20,24	1.509	5.1
DTS-HD High Resolution	No[4]	Lossy	48, 96	16–24	6	7.1
DTS-HD Master Audio	No[4]	Lossless	48, 96, 192[1]	16–24	24.5	8 (48 kHz and 96 kHz) 6 (192 kHz)

[1] A 192-kHz sampling frequency only allowed for 2, 4, and 6 channels.

[2] Codec is only supported for more than 5.1 channels. Core 5.1 channels are coded with Dolby Digital portion of stream.

[3] Codec is mandatory for a Dolby Digital core up to 640 kbps. Support of lossless portion of stream is optional.

[4] Codec is mandatory for a DTS core of up to 1.509 Mbps. Support of HD portion of stream is optional.

Table 9.10 Specifications for audio codecs supported by the Blu-ray system.

The Blu-ray format allows up to 32 primary audio bitstreams (such as main soundtracks) and up to 32 secondary audio bitstreams (such as commentary). It supports up to eight channels coded with PCM, Dolby, or DTS formats and may be coded in a variety of channel configurations such as monaural, stereo, 5.1, 7.1, and so on. Sampling frequencies up to 96- and 192-kHz are allowed. Sampling frequencies of 44.1 kHz and 88.2 kHz are not supported.

PCM soundtracks are fully supported. However, PCM soundtracks incur certain limitations. For example, downmix features and dialnorm (dialogue normalization) are only available when the PCM data is placed within a Dolby or DTS bitstream. Otherwise, for example, the player's default downmix mode is used. Also, at 192 kHz, as with other codecs, only a maximum of six channels is permitted.

The Dolby Digital (AC-3) and DTS (Coherent Acoustics) codecs are legacy formats, and are discussed in Chap. 11. The Blu-ray format accommodates a higher bit rate of 640 kbps for the Dolby Digital legacy format; Dolby Digital coding on DVD was constrained to 448 kbps. The DTS legacy format supports 768 kbps and 1.509 Mbps on both DVD and Blu-ray. Both legacy formats support up to 5.1 channels.

Dolby Digital Plus, also known as Enhanced AC-3 (or E-AC-3), is an extension format. It primarily adds spectral coding that is used at low bit rates, and it also supports two additional channels (up to 7.1) and higher Blu-ray bit rates. A core bitstream is coded with Dolby Digital, and the two-channel extension is coded as Dolby Digital Plus. Legacy devices that decode only Dolby Digital will not decode Dolby Digital Plus. The Dolby Digital Plus bitstream can be downconverted at a bit rate of 640 kbps to yield a Dolby Digital bitstream. However, if the original bitstream contains more than 5.1 channels, the additional channel content appears as a downmix to 5.1 channels.

The Dolby TrueHD lossless format in the Blu-ray specification uses an improved version of the Meridian Lossless Packing (MLP) format used in the DVD-Audio standard; for example, a higher bit rate of 18.64 Mbps is allowed. MLP is described in Chap. 11. For Blu-ray compatibility, Dolby TrueHD uses legacy Dolby Digital encoding as its core. Dolby TrueHD can accommodate up to 24-bit word lengths and a 192-kHz sampling frequency. Advantageously, multichannel MLP bitstreams include lossless downmixes; for example, a MLP bitstream holding eight channels also contains 2- and 6-channel lossless downmixes for compatibility with playback systems with fewer channels.

The DTS-HD bitstreams used in Blu-ray contain a core of DTS legacy 5.1-channel, 48-kHz data. The DTS-HD High Resolution Audio codec is a lossy extension format to DTS. It takes advantage of the extension option in the DTS (Coherent Acoustics) bitstream structure. The core component is typically coded at 768 kbps or 1.509 Mbps and contains 5.1-channel, 48-kHz content. DTS-HD High Resolution Audio adds an extension substream that may code data supporting additional (up to 7.1) channels and higher (96 kHz or 192 kHz) sampling frequencies. A legacy decoder will operate on the core bitstream whereas an HD-compatible decoder will operate on both the core and extension.

DTS-HD Master Audio is a lossless codec extension format to DTS. As with DTS-HD High Resolution Audio, DTS-HD Master Audio uses the extension substream to accommodate lossless audio compression to yield additional channels. DTS Master Audio lossless coding can accommodate up to 24-bit word lengths and a 192-kHz sampling frequency. As with MLP, multichannel bitstreams include lossless downmixes.

The legacy DTS codec supports XCH and X96 extensions. XCH (Channel Extension) is also known as DTS-ES; it adds a discrete monaural channel as a rear output. X96 (Sampling Frequency Extension) is also known as Core+96k or DTS-96/24; it extends the sampling frequency from 48 kHz to 96 kHz by secondarily encoding a residual signal following the baseband encoding; the residual is formed by decoding the encoded baseband and subtracting it from the original signal. The extension outputs 24-bit data.

The DTS-HD codec supports XXCH, XBR, and XLL extensions. XXCH (Channel Extension) adds additional discrete channels; for Blu-ray, two channels are added (7.1). XBR (High Bit-Rate Extension) allows an increase in bit rate. It is implemented as a residual signal as with X96. XLL (Lossless Extension) is also known as DTS-HD Master Audio. It allows lossless compression of up to eight channels at 192 kHz and 24 bits using a substream. Specifications for the bit rates of audio codecs and their extensions are shown in Table 9.11.

Codec	Minimum (Mbps)	Typical (Mbps)	Maximum (Mbps)
Primary Audio			
Dolby Digital 2.0	0.064	0.192	0.64
Dolby Digital 5.1	0.384	0.448	0.64
Dolby Digital Plus 7.1	0.640	1.024	1.7
Dolby TrueHD 5.1	0.80[1]	3.90[4]	18.64
DTS 5.1 (CBR, core)[2]	0.192	1.509	1.509
DTS 6.1 (CBR, core+XC, DTS ES)[2]	0.640	1.509	1.509
DTS-HD High Resolution Audio 7.1 (CBR, core+XXCH)[2]	0.768	1.509	6.0
DTS-HD High Resolution Audio 5.1 96 kHz (CBR, core+X96)[2]	0.30	1.509	6.0
DTS-HD Master Audio 5.1 (VBR, core+XLL)[2]	0.80[1]	4.0[4]	24.5
DTS-HD Master Audio 7.1 (VBR, core+XLL)[2]	0.80[1]	4.0[4]	24.5
Secondary Audio (streaming)[3]			
Dolby Digital Plus 1.0 (VBR)	0.032	0.064	0.256
Dolby Digital Plus 2.0 (VBR)	0.032	0.128	0.256
Dolby Digital Plus 5.1 (VBR)	0.032	0.128	0.256
DTS-HD 2.0 (LBR)	0.048	0.064	0.256
DTS-HD 5.1 (LBR)	0.192	0.256	0.256

[1] Practical lower limit for acceptable sound quality.
[2] DTS core bit rates are 768, 960, 1152, 1344, and 1509 kbps.
[3] Bit rates for secondary streaming audio are shown. Bit rates of secondary audio from disc or local storage are much higher, for example, a maximum of 2 Mbps.
[4] Lossless compression bit rates depend on content; these figures are for movie content, which is usually more compressed than music.

Table 9.11 Bit rates of audio codecs and their extensions. (*after Taylor, et al., 2009*)

As noted, the Blu-ray specification supports both primary and secondary audio streams. In this way, two streams can be played simultaneously and mixed; for example, a primary stream may contain a movie soundtrack while a secondary stream contains a commentary narration. Thus, players contain two audio decoders and a panner/mixer is used to mix the two audio streams at the output. Primary audio streams on a BD-ROM disc must use a mandatory codec. Secondary audio streams may use Dolby Digital Plus or DTS-HD LBR (low bit rate) (also known as DTS Express) codecs; the sampling frequency is 48 kHz, up to 5.1 channels are available, and the bit rate must be 256 kbps or less. The secondary stream also includes metadata for panning and mixing. It is possible to access a secondary audio stream from another source other than the disc; for example, a stream could be downloaded and stored locally and played in synchronization, or not, with the primary stream.

Audio data can be output in various downmixing modes and directed to different physical outputs; for example, audio could be output to the analog or HDMI output and content can be downsampled to particular outputs. This is controlled through metadata flags such as Lt/Rt (Left Total/Right Total) and Lo/Ro (Left Only/Right Only). AAF metadata can be placed in a secondary audio bitstream. AAF allows fader movements in the primary bitstream over a range of +12 dB to –50 dB; this can be used to vary main soundtrack levels relative to commentary dialogue.

Video Codecs

BD-ROM players must support multiple video codecs. In particular, players mandatorily support the MPEG-2 Part 2 (ISO/IEC 13818-2), MPEG-4 Advanced Video Codec (AVC) also known as ITU H.264 Part 10 (ISO/IEC 14496-10), and SMPTE VC-1 codecs. The latter is an "open standard" codec based on the Microsoft Windows Media Video 9 format. Both the AVC and VC-1 codecs provide more efficient compression and relatively higher performance than MPEG-2. The specifications for video streams are shown in Table 9.12.

A BD-ROM disc with video content must use one of the mandatory codecs. However, a disc may have files coded with multiple codecs; for example, the main title might use VC-1 coding and bonus content might use MPEG-2 coding. However, most commercial discs use one codec, generally MPEG-4 AVC or VC-1. Because of varying compression performance, the choice of video codec affects playing time for a given picture quality; MPEG-2 limits playing time to about 2 hours of high-definition video on a single-layer disc, and MPEG-4 AVC and VC-1 can provide about 4 hours. Because MPEG-2 is the video codec used in the DVD standard, Blu-ray players are thus backward-compatible with DVDs. The bit rates of the three video codecs, at high and standard definition, are shown in Table 9.13.

BD-ROM players must support two types of graphics streams. The Presentation Graphics stream is used for subtitles and animated graphics, and the Interactive Graphics stream is used for menu graphics. BD-ROM also supports text subtitle streams. A Text subtitle is defined by a series of character codes plus font and style information. Bitmap-based subtitles are provided by Presentation Graphics streams.

Video codecs	MPEG-4 AVC: HP@4.1/4.0 and MP@4.1/4.0/3.2/3.1/3.0, SMPTE VC-1: AP@L3 and AP@L2, MPEG-2: MP@ML and MP@HL/H1440L
Picture (high definition)	1920 × 1080 × 59.94i, 50i (16:9) 1920 × 1080 × 24p, 23.976p (16:9) 1440 × 1080 × 59.94i, 50i (16:9) MPEG-4 AVC/SMPTE VC-1 only 1440 × 1080 × 24p, 23.976p (16:9) MPEG-4 AVC/SMPTE VC-1 only 1280 × 720 × 59.94i, 50i (16:9) 1280 × 720 × 24p, 23.976p (16:9)
Picture (standard definition)	720 × 576 × 50i (4:3/16:9) 720 × 480 × 59.94i (4:3/16:9)
Display aspect ratio	4:3 and 16:9
Frame size	24/23.976 fps, 29.97/59.94 fps, 25/50 fps (PAL regions)
Audio codecs	PCM, Dolby Digital, Dolby Digital Plus,[1] Dolby TrueHD,[1] DTS, DTS-HD High Resolution,[1] DTS-HD Master Audio[1]
Audio channel configuration	1.0 to 7.1 channels
Audio sampling frequency	48, 96, 192 kHz
Subtitles and graphics	8-bit Interactive Graphics stream for menus, 8-bit Presentation Graphics stream for subtitles, HDMV Text Subtitle stream, 32-bit RGBA graphics for BD-J menus
Stream structure	MPEG-2 System Transport Stream
Maximum transport stream bit rate	48 Mbps
Maximum video stream bit rate	40 Mbps
Maximum number of video streams	9 primary/32 secondary
Maximum number of audio streams	32 primary/32 secondary
Maximum number of graphics streams	32
Maximum number of text streams	255

[1] Optional support in players.

Table 9.12 Specifications for video codecs supported by the Blu-ray system.

Codec	Minimum (Mbps)	Typical (Mbps)	Maximum (Mbps)
Primary Video (high definition)			
MPEG-4 AVC	4.0[1]	16.0	40.0
SMPTE VC-1	4.0[1]	18.0	40.0
MPEG-2	4.0[1]	24.0	40.0
Secondary Video (standard definition)			
MPEG-4 AVC	1.5[1]	2.0	8.0
SMPTE VC-1	1.0[1]	2.0	8.0
MPEG-2	1.0[1]	3.5	8.0

[1] Practical lower limit for acceptable sound quality.

TABLE 9.13 Bit rates of video codecs.

Modulation and Error Correction

Data on a Blu-ray disc, except for frame sync bits, is recorded using 1-7PP (Parity preserve/Prohibit repeated minimum transition run length) modulation. This is an RLL (1,7) code with minimum-pit $d = 1$ and k-constant of $k = 7$ yielding 2T to 8T run lengths. This differs from the $d = 2$ codes used in the CD and DVD formats. Data bits are converted to modulation bits with a modulation conversion table, and the modulation bits are converted to nonreturn to zero inverted (NRZI) channel bits for recording to disc. With the Parity preserve feature, as shown in Fig. 9.7, dc-control bits are inserted into the source

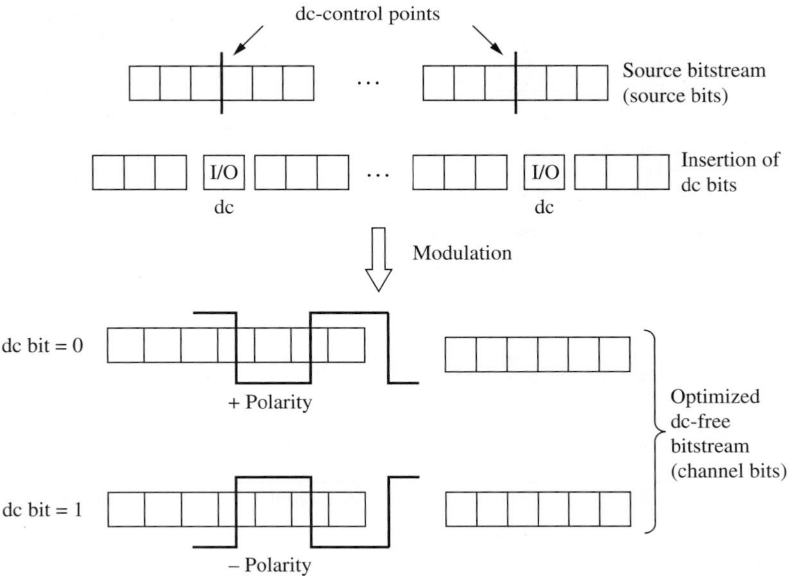

FIGURE 9.7 During modulation, with the Parity preserve feature, dc-control bits are inserted into the source bitstream at dc-control points to change the parity of the channel bitstream.

bitstream at dc-control points to change the parity of the source bitstream and hence the nonreturn to zero (NRZ) bitstream. This allows selection of the polarity of the NRZI channel bitstream so that the parity of the data stream is the same as the parity of the modulated stream. This controls dc content. In addition, the Prohibit repeated minimum transition run length feature prevents long consecutive runs (seven or more) of the minimum (2T) length; this restricts the occurrence of low signal levels.

The error-correction algorithm used in Blu-ray discs is quite efficient; for example, the error-correction data consumes less than 17% of the data signal (before modulation) yet its error-correcting abilities are considerable. As noted, on a Blu-ray disc, the cover layer over the data layer is relatively thin. This places dirt and dust particles closer to the data layer and thus they are less out of focus, so they have a relatively greater impact on data reading. This necessitates robust error correction. Blu-ray combines a deep interleave with a Long-Distance Code (LDC), a Reed–Solomon code with a size of 64 kbytes with a Burst Indicator Subcode (BIS) burst indicator.

It is instructive to compare the magnitude of a defect on both DVD and Blu-ray discs, as shown in Fig. 9.8. The figure shows the same 30-μm defect and a 138-μm beam spot size, which corresponds to the Blu-ray spot size on the cover layer, over error-correction code (ECC) blocks for both DVD and Blu-ray. The size of the Blu-ray ECC is twice that of a DVD. Also, Blu-ray ECC has 32 bytes of parity, which is twice that of the DVD, and data is de-interleaved twice. As a result, Blu-ray provides improved error correction over the DVD.

As described in previous chapters, error-correction performance is greatly improved if the decoder is given information showing the location of errors. A burst-error indicator can provide this prior to correction. The error-correction algorithm used in the Blu-ray system employs a picket code to indicate burst errors. The structure is shown in Fig. 9.9. Pickets are columns that are regularly inserted between columns of main data, which

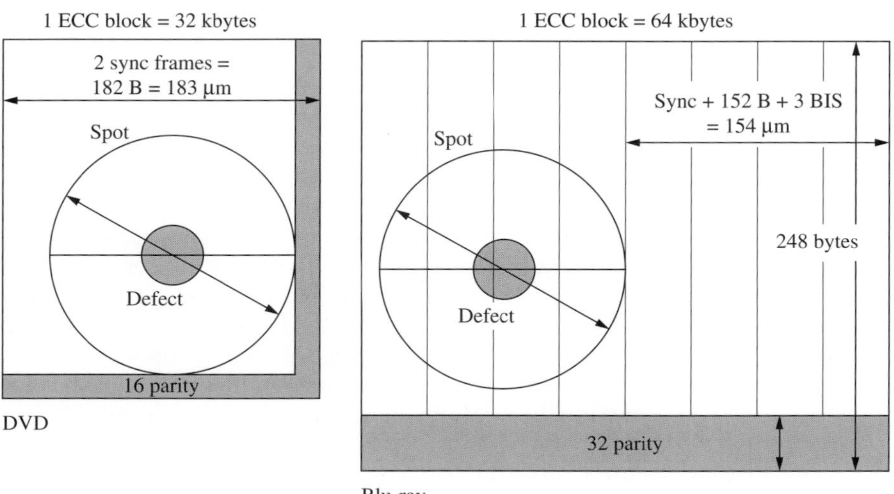

FIGURE 9.8 Comparison of a Blu-ray error-correction block and DVD error-correction block, applied to a 0.1-mm cover layer. In both cases, the reference spot size on the disc surface is 138 μm, and the defect size on the disc surface is 30 μm.

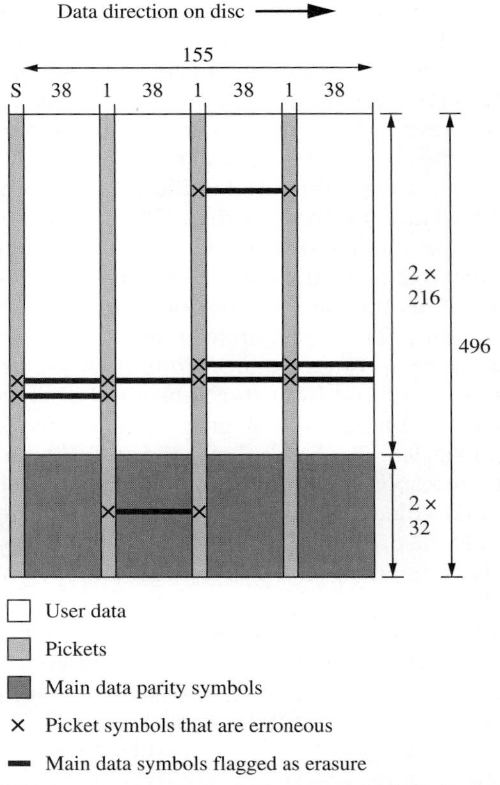

FIGURE 9.9 The error-correction algorithm used in the Blu-ray format employs a picket code to indicate burst errors. Pickets are regularly inserted between columns of main data that are protected by a Reed–Solomon code.

are protected by a first Reed–Solomon code. The pickets in turn are protected by a second Reed–Solomon code. During decoding, the picket columns are first corrected. Correction information is used to estimate the locations of possible burst errors in the main data. These symbols can be flagged as erasures when main-data correction is performed.

In the Blu-ray system, data is stored in blocks called clusters, each with 64 kbytes of data. Each cluster contains 32 frames, each with 2048 bytes of user data and 4 bytes of error-detection code (EDC). Clusters are divided into 16 address units. Clusters are protected by an error-correction algorithm. Following scrambling, clusters are processed by a LDC; this uses a Reed–Solomon code with 304 codewords with 216 information symbols and 32 parity symbols yielding a codeword length of 248. These codewords are interleaved two by two in a vertical direction; blocks of 152 columns by 496 rows are formed.

The LDC is divided into four groups of 38 columns. The leftmost picket is formed by frame sync patterns starting each row. The remaining three groups are protected by the BIS; each of three BIS columns is inserted between them. The BIS is a Reed–Solomon code with 30 information symbols and 32 parity symbols yielding a codeword length of 62. The BIS codewords are interleaved into three columns of 496 bytes each. With the addition of dc-control bits, a recording frame comprises 496 rows by 155 columns. Since the LDC and BIS codewords have the same number of parity symbols per word, one Reed–Solomon decoder can decode both codes. The information symbols of the BIS code form an additional data channel aside from the main data channel; it contains addressing information. The addressing information is protected against errors by an independent Reed–Solomon code with codewords with five information symbols and four parity symbols.

Audio-Video Stream Format and Directory

The Blu-ray system stores multiplexed audio, video, and other data in a container format known as the Blu-ray Disc Audio/Visual (BDAV) MPEG-2 Transport Stream. This format is based on the MPEG-2 System Transport Stream (ISO/IEC 13818-1); the filename extension is .m2ts. Video discs with movie content and menu use the Blu-ray Disc Movie (BDMV) directory and format with various data types stored in the BDAV container. The BDAV format is used for audio/video recording on BD-RE and BR-R discs. Because BDAV is compatible with broadcast DTV, these transport streams can be recorded directly and edited easily. The PES packet payload of the BDAV MPEG-2 Transport Stream contains the video, audio, graphics, and text subtitle elementary streams.

All BD-ROM application files are stored in a Blu-ray Disc Movie (BDMV) directory as shown in Fig. 9.10. The BDMV format is used in prerecorded Blu-ray discs. It was based on, and is compatible with, the BDAV bitstream format; for example, audio-video data is stored as an MPEG-2 transport stream. BDMV allows a wide range of presentation data; for example, a variety of video and audio codecs are permitted.

The BDMV directory contains the PlayList, Clipinf, Stream, Auxdata, and Backup directories. The PlayList directory contains the Database files for Movie PlayLists. The Clipinf directory contains the Database files for Clips. The Stream directory contains AV stream files. The Auxdata directory contains Sound data files and Font files. The Backup directory contains copies of the Index.bdmv file, the MovieObject.bdmv file, all files in the PlayList directory, and all files in the Clipinf directory. The Index.bdmv file stores information describing the contents of the BDMV directory. There is only one Index.bdmv file in the BDMV directory, and its file name is fixed. The MovieObject.bdmv file stores information for one or more Movie Objects. There is only one MovieObject.bdmv file in the BDMV directory, and its file name is fixed.

The xxxxx.mpls files ("xxxxx" is a five-digit number corresponding to the Movie PlayList) store information corresponding to the Movie PlayLists; one file is created for each Movie PlayList. The zzzzz.clpi files ("zzzzz" is a five-digit number corresponding to the Clip) store Clip information associated with a Clip AV stream file. The zzzzz.m2ts files (the same five-digit number is used for an AV stream and its associated Clip information file) contain a BDMV MPEG-2 transport stream. The sound.bdmv file stores data relating to one or more sounds associated with HDMV Interactive Graphic streams applications; this file may or may not exist in the Auxdata directory. If it exists, there is only one sound.bdmv file, and its file name is fixed. The aaaaa.otf file ("aaaaa" is a five-digit number corresponding to the Font) stores font information associated with Text subtitle applications.

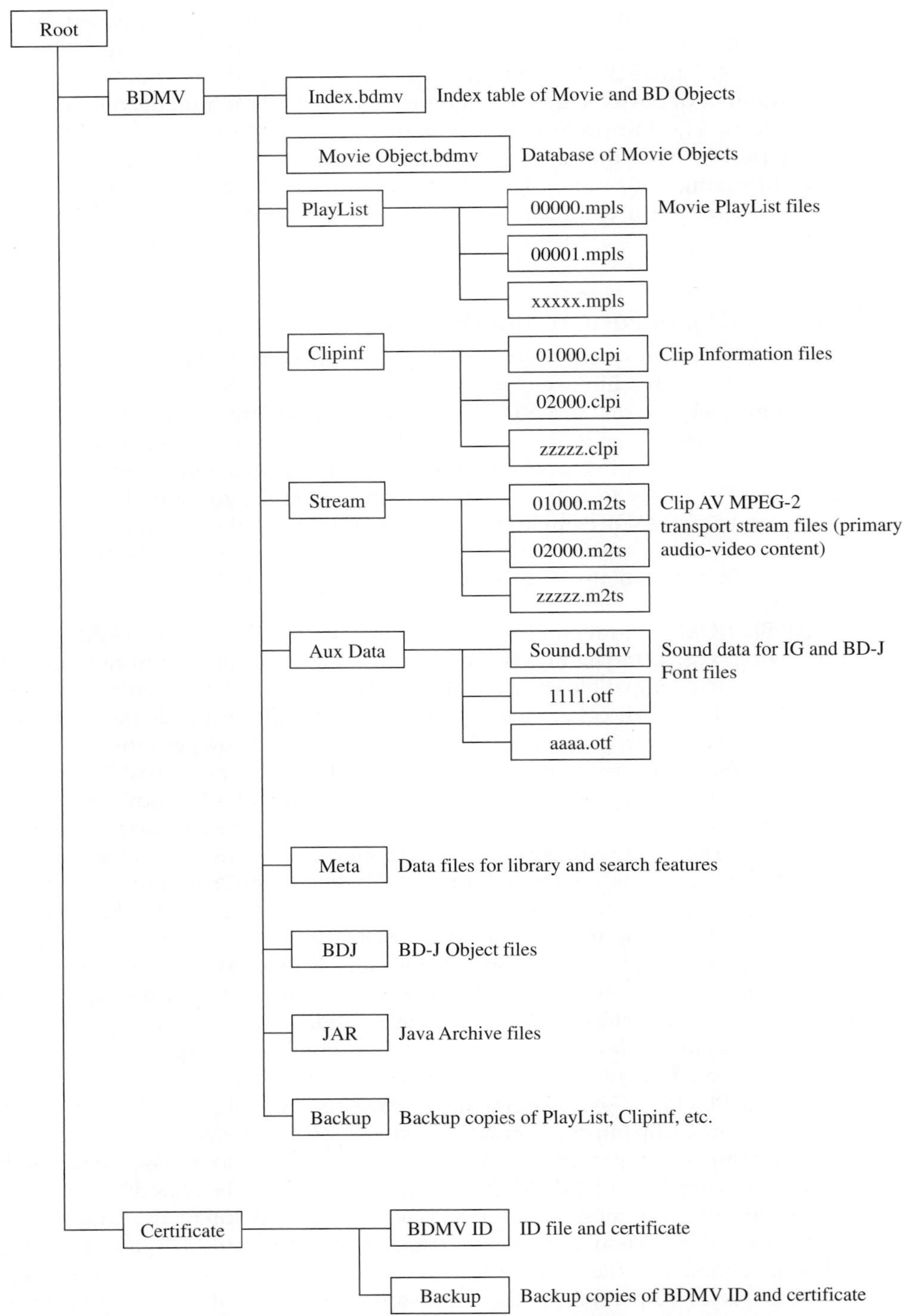

FIGURE 9.10 Directory structure of a BDMV disc.

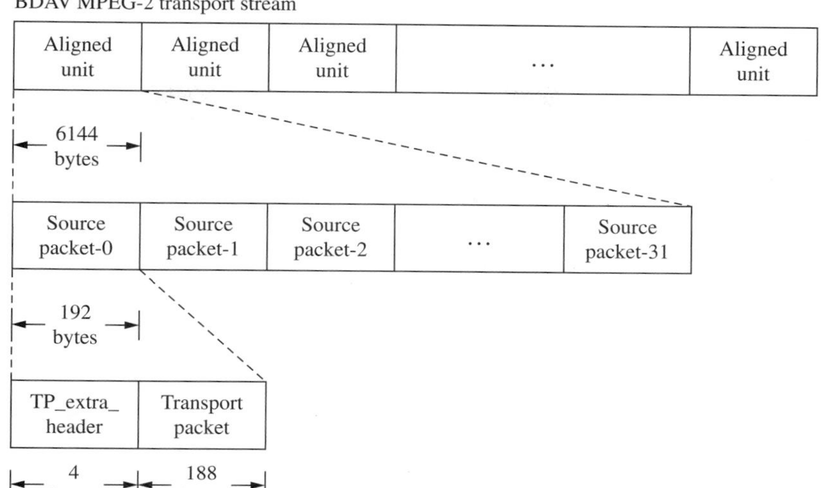

FIGURE 9.11 Structure of the BDMV MPEG-2 transport stream.

A BDMV stream is shown in Fig. 9.11. It is constructed from one or more Aligned units, each of which contains 6144 bytes (2048 × 3 bytes); an Aligned unit starts with the first byte of the source packets. A source packet comprises 192 bytes; one source packet consists of one TP_extra_header structure and one MPEG-2 transport packet structure. A TP_extra_header structure comprises 4 bytes, and the transport packet structure comprises 188 bytes. An Aligned unit comprises 32 source packets. Aligned units are placed in three consecutive logical sectors on a BD-ROM disc; each logical sector is 2048 bytes. The maximum multiplex rate of the BDAV MPEG-2 Transport Stream is 48 Mbps.

The Blu-ray BD-ROM specification describes four layers for managing AV stream files, as shown in Fig. 9.12. The layers are: Index Table, Movie Object/BD-J Object, PlayList, and Clip. The Index Table is a top-level table that defines and contains entry points for all the Titles and the Top Menu of a BD-ROM disc. The player references the table whenever a Title Search or Menu Call operation is called to determine the corresponding Movie Object/BJ Object to be executed. The Index Table also has an entry to a Movie Object/BD-J Object designated for First Playback; this can be used to provide automatic playback. A Movie Object consists of an executable navigation command program that enables a dynamic scenario description. Movie Objects exist in the layer above PlayLists. Navigation commands in a Movie Object can launch PlayList playback or another Movie Object. This can be used to define a set of Movie Objects for managing playback of PlayLists in accordance with interactive preferences.

When a Title associated with a BD-J Object is selected, the corresponding Java Xlet application is launched. The Xlet is controlled by the player's Application Manager through its Xlet interface. A Movie PlayList is a collection of playing intervals in the Clips. One playing interval is the PlayItem that consists of an In-point and Out-point; each refers to time positions (start and stop points) in the Clip. A Clip is an object comprising a Clip AV stream file and its corresponding Clip information file. An AV stream file together with its associated database attributes is considered to be one object. The AV Stream File is the Clip AV stream file, and the associated database attribute file is the

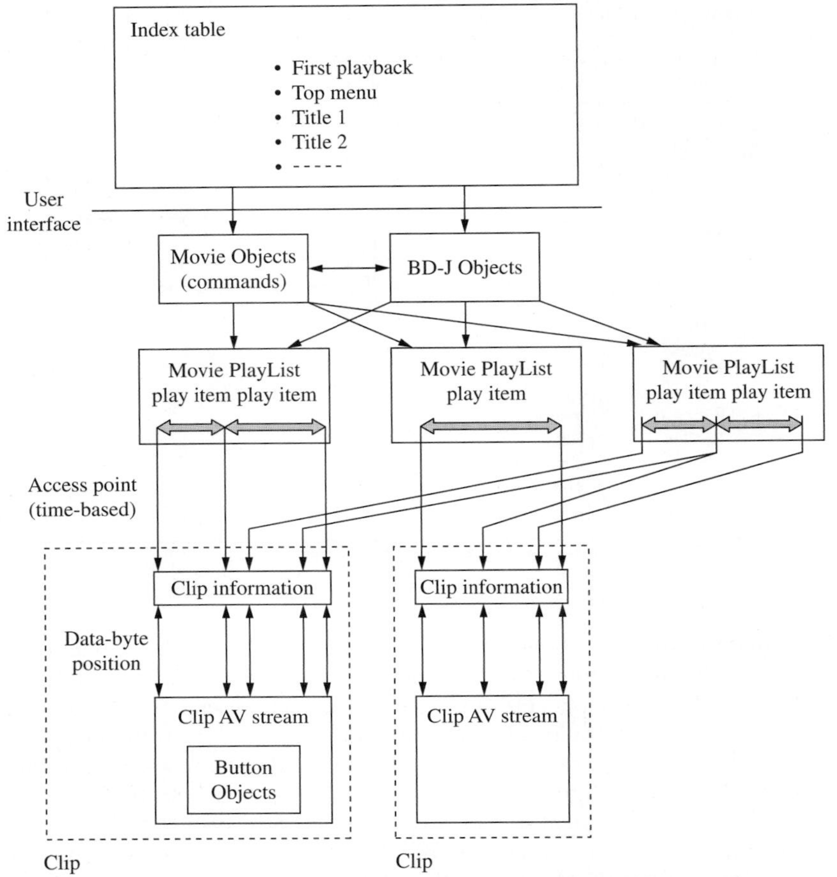

Figure 9.12 The BD-ROM specification describes four layers for managing AV stream files: Index Table, Movie Object/BD-J Object, PlayList, and Clip.

Clip Information File. A Clip AV Stream File is an MPEG-2 transport stream segment called by the BDAV specification. The Clip Information File stores the time stamps of the access point into the corresponding AV stream file. The player reads the information file to determine the start position in the AV stream file.

Blu-ray UDF File System

The Blu-ray specification specifies the Universal Disc Format (UDF) file system developed by the Optical Storage Technology Association (OSTA). UDF is widely used for optical discs and is independent of media type, hardware platform, or operating system. It also allows interchangeability between computer systems. In particular, Blu-ray employs UDF 2.5; this revision allows large capacity media and adds the Metadata File feature as well as the optional Metadata Mirror File feature. Metadata File allows clustering of the file system file management information such as file entries and directories and allows accessing of multiple directories. This allows faster disc start-up and scanning and better disc utilities support. The Metadata Mirror File increases robustness. Requirements for UDF volume and file structure are individually defined in the BD-ROM, BD-R, and BD-RE specifications. All Blu-ray discs have a logical sector size of 2 kbytes and an ECC block size of 64 kbytes.

HDMV and BD-J Application Programming Modes

The Blu-ray specification supports two types of application programming for features such as interactive menus: HD Movie (HDMV) and Blu-ray Disc Java (BD-J). HDMV offers relatively limited programming options that allow simple declarative program execution. HDMV is similar in concept to the type of programming used in the DVD-Video standard but it makes it easier to create content and to ensure compliance across many players.

The Blu-ray specification also mandates that players must support Java cross-platform software. Java is used for procedural applications programming. BD-J applications include AV playback control, movie trailer downloads, subtitle updates, online gaming, online shopping, chat, and so on. The BD-J system model is shown in Fig. 9.13. The BD-J standard is a part of the Globally Executable MHP (GEM) standard, which in turn is part of the Multimedia Home Platform (MHP) standard. BD-J is based on the Java 2 Micro-Edition (J2ME) Personal Basis Profile (PBP), a Java profile developed for consumer electronics. A BD-J Object is a Java Xlet that is registered in the Application Management Table (AMT). Each title can have an associated AMT. At least one application in the AMT must be designated as "autostart." It is started when the corresponding title is selected, and then the BD-J application uses the BD-J platform. BD-J includes a media framework for playback of media content on a BD-ROM disc and other sources.

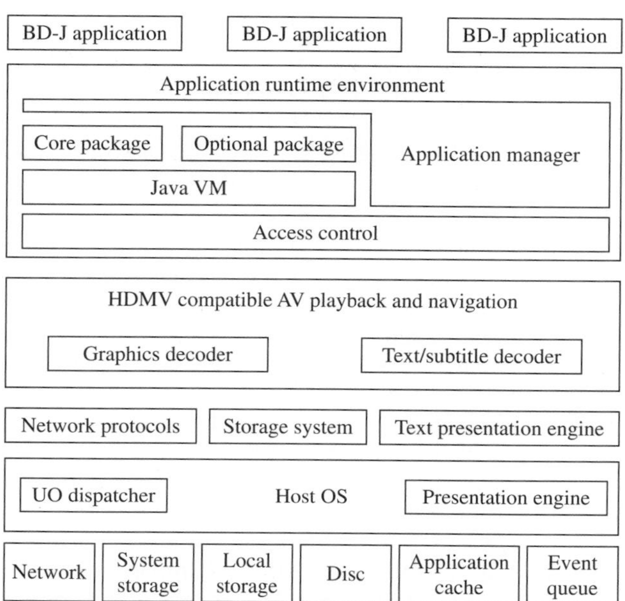

FIGURE 9.13 Summary of BD-J system model. BD-J is based on the Java 2 Micro-Edition (J2ME) Personal Basis Profile (PBP), a Java profile that was developed for consumer electronics devices.

As with HDMV, BD-J uses a PlayList as its unit of playback; all features of HDMV except Interactive Graphics can be used by a BD-J application. (HDMV graphics are replaced by BD-J graphics.) Features such as video, audio, Presentation Graphics, Text Subtitle component selection, media time, and playback-rate control are supported. BD-J includes standard Java libraries for decoding and displaying various image formats. Similarly, text can be rendered using standard Java.

BD-J uses a Java 2 security model to authenticate signed applications and to grant permissions beyond core functions. BD-J also contains the Java network package; applications can connect to Internet servers. TCP/IP is supported and the HTTP protocol may be used; visited Web sites are controlled by the content provider. BD-J also provides mandatory system storage and optional local storage; they are accessed with the Java IO package. BD-J includes a GUI framework that allows remote control navigation and easy customization. It is similar to the HAVi user interface, and is based on the core of AWT, but it is not a desktop GUI.

BD-J also allows flexible control of audio streams. The primary and secondary audio streams can be manipulated, as well as interactive menu sound effects. Only one audio stream can be output from a player thus players mix different streams internally. BD-J applications can supply mixing parameters to the software mixer.

The Blu-ray specification provides an option for interactive sounds, for example, button sounds on menus. Player support of interactive audio is mandatory in BD-J mode and optional in HDMV mode. When a user action is taken, an audio clip is played. This interactive audio feature uses PCM coding; the sampling frequency is 48 kHz, word length is 16 bits, and two channels are supported. Sound clips are stored in a sound.bdmv file; it can contain up to 128 clips, with a maximum overall file capacity of 2 Mbytes.

Blu-ray 3D

The Blu-ray 3D specification allows storage of stereoscopic (3D) video content on Blu-ray discs for playback in Blu-ray 3D players. The specification provides backward compatibility for both discs and players. When 3D discs include a 2D version of the content, the 2D version can be viewed on existing 2D players. 3D players will play 3D content, and allow users to play existing libraries of 2D content. The 3D specification is agnostic with respect to display; Blu-ray 3D will deliver 3D images to any compatible 3D display, regardless of whether the display uses LCD, plasma, or other display technology, and regardless of what 3D technology the display uses to deliver the image to the viewer's eyes. The Blu-ray 3D specification was released in December 2009.

The Blu-ray 3D system uses a Multiview Video Coding (MVC) codec that is an extension to the ITU-T H.264/MPEG-4 Advanced Video Coding (AVC) codec that is supported by all Blu-ray players. In particular, an MVC bitstream is backward-compatible with AVC codecs. The MVC codec compresses right- and left-eye views with approximately a 50% overhead compared to 2D content. It can provide full 1080p resolution to each eye. The specification also contains provisions for enhanced 3D graphics features for menus and subtitles. A 50-Gbyte disc can hold approximately 2 hours of full HD 3D content.

In the 3D format, video, audio, graphics, and text elementary streams are coded in the PES packet payload of the BDAV MPEG-2 transport stream. The video stream is composed of two parts. One part is the MVC Base view video stream and the other is the MVC Dependent view video stream. The maximum bit rate of the Base stream is 48 Mbps and the maximum bit rate of the Dependent stream is 48 Mbps; the maximum bit rate of the combined bit stream is 64 Mbps. The Base stream is compatible with the AVC stream format so that 2D players can decode MVC Base video streams. The content creator can choose that either the left or right eye stream is the Base view stream. Both streams are decoded simultaneously for 3D playback. The audio stream is coded the same as with 2D content. The graphics stream can be coded as a one plane plus offset Presentation graphics steam, a stereoscopic Presentation graphics stream, a one plane plus offset Interactive graphics stream, and a stereoscopic Interactive graphics stream.

Region Playback Code

The Blu-ray system supports global regional codes; the system is known as Region Playback Control (RPC). Players sold in a geographical region are permanently coded and will only play discs authorized for that region or discs that are not regionally coded. All Blu-ray players must support regional coding, but regional coding of discs is optional for content providers. Players and discs carry a logo that designates the region setting. The system designates three global regions:

- Region A: North, Central, and South America; Japan; Korea; Taiwan; Hong Kong; and Southeast Asia
- Region B: Europe, Middle East, Africa, Australia, and New Zealand
- Region C: China, India, Russia, and rest of the world

A BD-ROM disc may be coded for one, two, or three regions. The regional code is verified by the player software or firmware. Region control playback is managed by the disc using BD-J or HDMV software to check the player's authorization. For software players, the operating system or player application manages region control, allowing the user to change the region up to five times. Recordable discs are not region-coded.

Content Protection

The Blu-ray system supports three types of digital rights management (DRM) systems that restrict unauthorized distribution and copying: Advanced Access Content System (AACS), BR-ROM Mark, and BD+. The BD+ system operates independently of AACS and BD-ROM Mark and provides unique kinds of protection, but in a complementary way.

The Advanced Access Content System is the primary copy-protection system used to protect Blu-ray disc contents from unauthorized copying; its use is mandatory. It is a cryptographic system with 128-bit keys using the Advanced Encryption Standard (AES). It derives encryption keys, performs decryption of disc contents, and allows authentication of discs and players. It can also restrict how content is output from the player. AACS was first published in April 2005, and the final AACS agreements were ratified in June 2009.

In AACS, a Media Key Block (MKB) is used for cryptographic key management. The MKB contains keys for each model of AACS-licensed Blu-ray player. A tree structure is used so a compromised player cannot decrypt content. If a player is compromised, new MKBs can be written so that the keys associated with the compromised player cannot be used to derive valid keys to play new titles. Each new movie title contains an updated MKB that details compromised players. Principally, the AACS licensing administrator provides a MKB to replicators to be placed on each disc, and provides Device keys to player manufacturers. Each protected disc contains a Media Key Block, Volume ID, and encrypted title keys. The Volume ID is a unique serial number or identifier (128-bit key) embedded in a disc in the BD-ROM Mark. AACS uses Volume ID in the BD-ROM Mark to derive keys needed for decryption.

The BD-ROM Mark is physically stored on the disc. The mark must be present on every BD-ROM disc, and BD recorders cannot output the mark; the mark is never present outside the player. The Volume ID is encoded, and the mark is physically written during replication by licensed BD-ROM Mark Inserter hardware. The mark is placed on the disc apart from other Blu-ray content. An otherwise bit-exact copy of the original content that lacks the mark will not be decodable. A cryptographic certificate called the Private Host Key is needed to read the Volume ID. The BD-ROM Mark can also be used to trace the manufacturing source of discs.

An elaborate series of key authentications and decryptions are performed to play a protected disc. These steps help protect the Title key, which is ultimately used to decrypt the disc contents. Summarizing an intricate series of steps during playback, a hardware player uses its set of Device keys and the most recent MKB from the disc or the player to calculate a Media key. The player uses the Media key with the correctly read Volume ID on the disc to calculate a Key Variant Unit using a one-way AES-G encryption scheme. The Key Variant Unit is used with Unit Key Files on the disc to decrypt a Title key. Finally, the Title key is used to decrypt the disc contents. A summary of principal steps needed to play content protected with AACS is shown in Fig. 9.14.

Additional security is employed when using a computer drive and a software player. The software player on a host computer and the drive mutually verify validity and create a Bus key, which in turn is used for Bus encryption. Also, the drive creates a Read Data key that is encrypted by the Bus key and used by the software player for decryption. Data is read from the disc using the Read Data key so that only the specific player can decrypt it. In addition, software players, servers, and other devices must

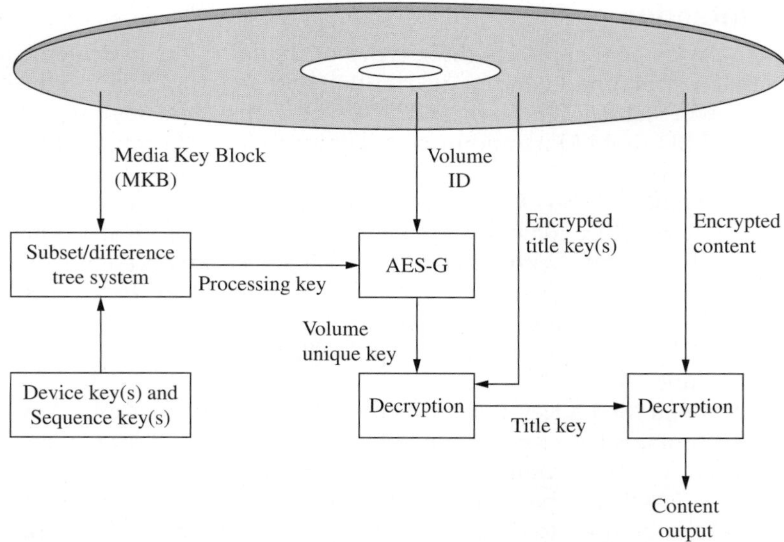

FIGURE 9.14 A summary of principal steps needed to play content protected with AACS.

proactively and regularly authenticate themselves during playback to verify that they have not been compromised.

The Media Key Block on each disc contains a host revocation list for software players and a drive revocation list for hardware players. Each player stores the most current revocation list in memory. Playback is allowed only after checking that both player and disc have uncompromised keys. If a player's device keys are compromised and distributed, the AACS licensing administrator can revoke keys by distributing updated MKB revocation lists on future content that prevent playback of new content. To accomplish this, the MKB is encrypted with a difference tree such that any key can find every other key except its parent keys. Thus, a device key can be revoked by encrypting the MKB with that device key's parent key.

As noted, revocation uses a player's device keys. However, a set of device keys could be used by a manufacturer in many different player models; thus it may be difficult to identify particular compromised players. This is addressed with a feature called Sequence Keys. Content such as a movie can be divided into six or fewer Sequence Key Blocks, and each of these can contain up to 256 PlayLists. Each of these lists contains a set of PlayItems with Key Sequence segments and non-Key Sequence segments. The former segments contain different versions of the same segment and the parts are encrypted with different keys; thus a player can only decrypt one version of each part. Moreover, different forensic marks are placed in each part. Each player thus plays the movie with a unique sequence path, and different forensic marks are read. In this way, the marks from unauthorized copies can be analyzed to determine which player was used to make the copies. Armed with this information, the device keys of that player can be revoked.

A mandatory feature known as AACS Managed Copy can be used to allow legal copies. As permitted by the content owner, an original BD-ROM disc can be copied identically to a recordable disc or other media, or an original disc can be copied in some modified form, for example, with lower picture resolution. In some cases, a Blu-ray disc may contain a low-resolution file that can be copied, for example, to a computer for viewing. Also, in some cases, the original disc can be used to authorize downloading of additional content. Copies of original content are protected, so copies cannot themselves be copied unless allowed. To make an authorized copy using AACS Managed Copy, the player uses its Managed Copy Machine and a link to obtain authorization from a Managed Copy Server via the Internet. Discs are authenticated using an embedded unique Prerecorded Media Serial Number placed in the burst cutting area. Following authentication, the user may copy disc contents for free or for a fee.

In addition to AACS, Blu-ray systems can also employ BD+ for additional content protection. With BD+, small portions of the audio/video bitstream are scrambled during authoring in a manner that is unique to that title. Title-specific code can be extracted and used for additional security review before the bitstream is unscrambled and output. In addition, BD+ can renew protection measures on compromised players. BD+ is mandatory in hardware and software Blu-ray players and optional for Blu-ray discs. BD+ uses software known as Security Virtual Machine which is embedded in all players. It loads and runs executable content security code from the disc. BD+ follows a technique of "self-protecting digital content" with a variety of possible countermeasures. For example, if a player has been compromised, BD+ code can be placed in new disc releases that will detect and repair the fault. Also, for example, BD+ programs can check a player for tampering by analyzing the memory footprint in the player, check that a player's keys have not been altered, and alter the output signals so that the program must unscramble them for correct playback. In addition, forensics can be used to identify the serial number of individual compromised players.

To provide further security in addition to AACS and BD+, an audio watermarking system from Verance can be used; this system was originally used in the DVD-Audio format. All Blu-ray players must implement the audio watermark, but its use on discs is optional. The watermark is created by varying audio waveforms in a regular-coded pattern that is not noticed by listeners. Movie studios can place a watermark in the soundtracks of theatrical films. If a Blu-ray player detects this watermark, it is assumed that the title is a copy of the theatrical film (perhaps made with a camcorder in a theater) and the player will not play the title. Also, different audio watermarks can be used in nontheatrical titles to prevent unauthorized disc copying. Watermarks can be placed in different versions and by examining what parts are illegally distributed, administrators can identify compromised keys and revoke them. In addition, traitor-tracing allows administrators to discover and track compromised keys.

The audio/video output bitstream on digital buses and networks is protected with encryption via DTCP (Digital Transmission Content Protection). It is used on the IEEE 1394 interface, USB bus, the Internet, and other transmission paths. An authentication protocol with key exchanges and certificates establishes a secure channel for connected devices. Keys can be updated or revoked using new content or devices. Playback-only devices are differentiated from recording devices; the latter can only receive data that is not protected or is marked as copy-permitted. DTCP is discussed in more detail in Chap. 14 in the context of IEEE 1394.

Audio/video output via DVI (Digital Visual Interface) or HDMI (High-Definition Multimedia Interface) interfaces can be protected with HDCP (High-bandwidth Digital Content Protection). HDCP is used for point-to-point connections. It uses encryption, key exchanges, and revocation to establish a secure path between sources and receivers. Each device has a set of forty 56-bit keys and a 40-bit key selection vector supplied by the licensing administrator. Authentication occurs, and then recurs every few seconds. HDCP is not mandatory and some DVI devices may not support it; when an HDCP-enabled device is connected to a non-HDCP device, the output image quality is reduced.

Some display devices do not have digital inputs. Thus, an analog signal must be conveyed to them. AACS can control content via two flags: Image Constraint Token (ICT) and Digital Only Token (DOT). ICT limits the resolution of video content of analog outputs to 960 × 540 pixels (in a 16:9 aspect ratio). If implemented, DOT prevents AACS-decrypted content from appearing at analog outputs; digital interfaces must be protected by DTCP, HDCP, or other means. In addition, AACS can define which devices can pass analog content prior to the December 2010 date. Analog output can be protected via CGMS-A; it establishes copy-prohibited or copy-permitted guidelines. As part of CGMS-A, the Redistribution Control (RC) option can be implemented. The Macrovision Analog Protection System (APS) as used in DVD-Audio can be employed. Analog connectivity will be phased out. Products manufactured and sold after December 31, 2010, can only convey standard-definition video through analog outputs; high-definition digital video outputs must be encrypted. Products manufactured and sold after December 31, 2013, cannot pass AACS-protected content through analog outputs. This timetable is sometimes referred to as the "analog sunset."

Blu-ray Recordable Formats

Two recordable Blu-ray disc formats are in common use. The recordable (BD-R) format allows discs to be written to once, and the rewritable (BD-RE) format allows discs to be written to, erased, and written to multiple times. Most recordable Blu-ray discs use

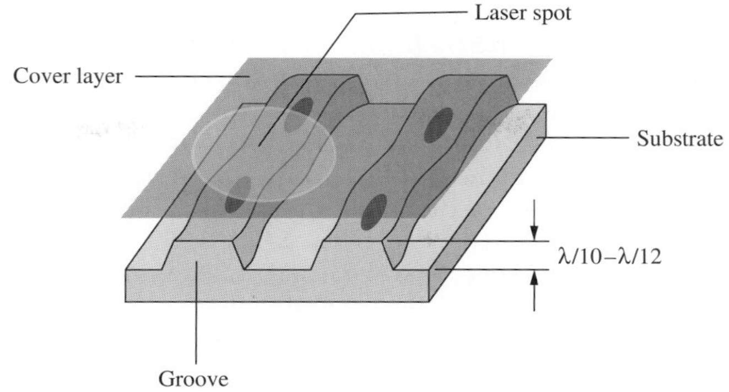

Figure 9.15 Recordable Blu-ray discs use a pregroove structure. Both on-groove recording (as shown here) and in-groove recording are permitted in the BD-R specification.

	Data Rate		Write Time (minutes)	
Drive Speed	Mbit/s	Mbyte/s	Single Layer	Dual Layer
1×	36	4.5	90	180
2×	72	9	45	90
4×	144	18	23	45
6×	216	27	15	30
8×	288	36	12	23
12×	432	54	8	15

Table 9.14 Data rates and writing times of recordable Blu-ray discs.

phase-change technology; either GST (GeSbTe stoichiometrical composition) or eutectic phase-change media can be used. BD-R discs may alternately use organic-dye or inorganic-alloy recording. BD-RE discs use an in-groove method whereas BD-R discs can use in-groove or on-groove recording; the latter is shown in Fig. 9.15. High-speed recording of about 12× is possible; this compares to maximum speeds of 20× and 52×, respectively, of standard DVD and CD discs. Table 9.14 shows recording data rates and writing times for different disc speeds.

Recordable disc capacities are generally 25 Gbytes for single-layer discs and 50 Gbytes for dual-layer discs. The BD-RE Version 1.0 specification describes disc capacities of 23.3, 25, and 27 Gbytes/layer; the latter capacity was reserved for future use. The BD5 format stores up to 4.5 Gbytes on a single-layer disc, and the BD9 format stores up to 9 Gbytes on a dual-layer disc. Both use inexpensive DVD-type discs. Some Blu-ray players will not play these discs. Prototype 400-Gbyte discs have been shown, and a 1-Tbyte disc is contemplated.

There are three versions of BD-RE discs. Version 1.0 has a unique BD file system and is not computer-compatible. Version 2.0 has a UDF 2.5 file system, is used for computers, and can use AACS. Version 3.0 adds an 8-cm camcorder disc diameter and is backward-compatible with Version 2.0. There are also three versions of BD-R discs. Version 1.0 has a UDF 2.5 file system, is used for computers, and can use AACS (the same as BD-RE Version 2.0). Version 1.2 adds a Low to High (BD-R LTH) standard. Version 2.0

adds an 8-cm camcorder disc diameter and is backward-compatible with Version 1.0 (the same as BD-RE Version 2.0). BD-LTH discs may use different recording media for write-once operation. However, some Blu-ray players cannot play these discs; in some cases, a firmware upgrade suffices to make such discs playable.

In the BD-RE and BD-R formats, a wobbled pregroove is used for addressing, similar to the method employed in the DVD+RW format. In particular, wobbling addressing is in the radial direction based on minimum shift keying (MSK) modulation and formatted in blocks of 64 kbytes. The wobble frequency (1×) is 956.522 kHz. An ADIP (address in pregroove) method is used over 56 wobble periods. Nominal wobble length is 5.1405 μm with 69 channel bits per wobble. Binary ADIP information is expressed as the position where the sinusoidal wobble is deviated by minimum shift-keying modulation. The shift keying can be influenced by reading defects. To overcome that, a sawtooth wobble (STW) signal is used to add secondary harmonics to the sinusoidal wobble and binary 0 and 1 correspond to the polarity of the added harmonics. Unlike the MSK energy, the energy of the STW signal is distributed on the disc and is detected by integration, so it is more robust against individual defects. Thus the addressing is more robust.

In the case of the CD and DVD formats, read-only discs were developed first. This led to complications when recordable and rewritable formats were subsequently developed; for example, it was difficult to link discontinuous recorded segments to the continuously recorded read-only standards. In Blu-ray, the rewritable BD-RE specifications were created before the BD-ROM specifications. This allowed efficient design of the data-linking methods. In particular, the data block unit recorded on a BD-RE disc is an LDC block of 64 kbytes. Between the adjacent two recording unit blocks, a two-sync frame length (run-in and run-out) area is prepared to accommodate variations in linear velocity. The signal from the wobbled groove is used for clocking during a link. Thus for BD-RE, the linking sequence is a physical cluster (498 frames), run-out (0.57 frame), run-in (1.43 frame), and a physical cluster (498 frames). BD-ROM does not use a wobble groove so two linking frames are used with the same length as the run-in and run-out areas of BD-RE. To maintain clocking in BD-ROM, the sync frame in the linking frames uses the same interval as the sync frames in the data area. The sync frame in the linking frame has a unique pattern for identification of the linking frame area. Thus for BD-ROM, the linking sequence is a physical cluster (498 frames), link frame 1 (1 frame), link frame 2 (1 frame), and a physical cluster (498 frames).

The AACS content protection system used in the BD-ROM format can also be employed in recordable discs; this application is technically known as AACS Protection for Prepared Video. This system uses AACS decryption techniques similar to those used for prerecorded media. Recordable media do not contain a BD-ROM Mark so the Volume ID is supplied differently. The system for prepared video adds features such as electronic sell-through; purchase, download, and store content; and manufacturing on demand. It is also used, for example, to create discs in a kiosk, and for permitting managed copy to an optical disc, in which content owners allow users to make legal but secure copies.

AVCHD Format

The Advanced Video Codec High Definition (AVCHD) format is used to record AVC-coded video onto DVD discs and other media such as SD/SDHC memory cards and hard-disk drives. Many Blu-ray players will play DVD discs and other format recorded in the AVCHD format. The AVCREC rewritable format enables recording and playback of BDAV content on DVD media with playback compatibility between the DVD and

Blu-ray formats. AVCREC Part 3 Version 1.0 requires that all players and recorders must play a MPEG-4 AVC video stream when the stream is recorded in Transcode mode. As stated in BD-RE Part 3 Version 2.1, playback of such streams is optional. Transcoding allows longer recording times in the smaller capacity DVD format. AVCREC uses the BDAV container to record ISDB video on DVD discs.

The Blu-ray disc format builds on the technology and market success of the CD and DVD to further extend the opportunities of optical disc storage for professional and consumer applications. Information is available from the Blu-ray Disc Association at www.blu-raydisc.info and www.blu-raydisc.com.

CHAPTER 10

Low Bit-Rate Coding: Theory and Evaluation

In a world of limitless storage capacity and infinite bandwidth, digital signals could be coded without regard to file size, or the number of bits needed for transmission. While such a world may some day be approximated, today there is a cost for storage and bandwidth. Thus, for many applications, it is either advantageous or mandated that audio signals be coded as bit-efficient as possible. Accomplishing this task while preserving audio fidelity is the domain of low bit-rate coding. Two approaches are available. One approach uses perceptual coding to reduce file size while avoiding significant audible loss and degradation. The art and science of lossy coding combines the perceptual qualities of the human ear with the engineering realities of signal processing. A second approach uses lossless coding in which file size is compressed, but upon playback the original uncompressed file is restored. Because the restored file is bit-for-bit identical to the original, there is no change in audio fidelity. However, lossless compression cannot achieve the same amount of data reduction as lossy methods. Both approaches can offer distinct advantages over traditional PCM coding.

This chapter examines the theory of perceptual (lossy) coding, as well as the theory of lossless data compression. In addition, ways to evaluate the audible quality of perceptual codecs are presented. Chapter 11 more fully explores the details of particular codecs, both lossy and lossless. The more specialized nature of speech coding is described in Chap. 12.

Perceptual Coding

Edison cylinders, like all analog formats, store acoustical waveforms with a mimicking pattern—an analog—of the original sonic waveform. Some digital media, such as the Compact Disc, do essentially the same thing, but replace the continuous mechanical pattern with a discrete series of numbers that represents the waveform's sampled amplitude. In both cases, the goal is to reconstruct a waveform that is physically identical to the original within the audio band. With perceptual (lossy) coding, physical identity is waived in favor of perceptual identity. Using a psychoacoustic model of the human auditory system, the codec (encoder-decoder) identifies imperceptible signal content (to remove irrelevancy) as bits are allocated. The signal is then coded efficiently (to avoid redundancy) in the final bitstream. These steps reduce the quantity of data needed to represent an audio signal but also increase quantization noise. However, much of the quantization noise can be shaped and hidden below signal-dependent thresholds of hearing. The method of lossy coding asks the conceptual question—how much noise can be introduced to the signal without becoming audible?

Through psychoacoustics, we can understand how the ear perceives auditory information. A perceptual coding system strives to deliver all of perceived information, but no more. A perceptual coding system recognizes that sounds that are reproduced have the human ear as the intended receiver. A perceptual codec thus strives to match the sound to the receiver. Logically, the first step in designing such a codec is to understand how the human ear works.

Psychoacoustics

When you hear a plucked string, can you distinguish the fifth harmonic from the fundamental? How about the seventh harmonic? Can you tell the difference between a 1000-Hz and a 1002-Hz tone? You are probably adept at detecting this 0.2% difference. Have you ever heard "low pitch" in which complex tones seem to have a slightly lower subjective pitch than pure tones of the same frequency? All this and more is the realm of psychoacoustics, the study of human auditory perception, ranging from the biological design of the ear to the psychological interpretation of aural information. Sound is only an academic concept without our perception of it. Psychoacoustics explains the subjective response to everything we hear. It is the ultimate arbitrator in acoustic concerns because it is only our response to sound that fundamentally matters. Psychoacoustics seeks to reconcile acoustic stimuli and all the scientific, objective, and physical properties that surround them, with the physiological and psychological responses evoked by them.

The ear and its associated nervous system is an enormously complex, interactive system with incredible powers of perception. At the same time, even given its complexity, it has real limitations. The ear is astonishingly acute in its ability to detect a nuance or defect in a signal, but it is also surprisingly casual with some aspects of the signal. Thus the accuracy of many aspects of a coded signal can be very low, but the allowed degree of diminished accuracy is very frequency- and time-dependent.

Arguably, our hearing is our most highly developed sense; in contrast, for example, the eye can only perceive frequencies over one octave. As with every sense, the ear is useful only when coupled to the interpretative powers of the brain. Those mental judgments form the basis for everything we experience from sound and music. The left and right ears do not differ physiologically in their capacity for detecting sound, but their respective right- and left-brain halves do. The two halves loosely divide the brain's functions. There is some overlap, but the primary connections from the ears to the brain halves are crossed; the right ear is wired to the left-brain half and the left ear to the right-brain half. The left cerebral hemisphere processes most speech (verbal) information. Thus, theoretically the right ear is perceptually superior for spoken words. On the other hand, it is mainly the right temporal lobe that processes melodic (nonverbal) information. Therefore, we may be better at perceiving melodies heard by the left ear.

Engineers are familiar with the physical measurements of an audio event, but psychoacoustics must also consider the perceptual measurements. Intensity is an objective physical measurement of magnitude. Loudness, first introduced by physicist Georg Heinrich Barkhausen, is the perceptual description of magnitude that depends on both intensity and frequency. Loudness cannot be empirically measured and instead is determined by listeners' judgments. Loudness can be expressed in loudness levels called phons. A phon is the intensity of an equally loud 1-kHz tone, expressed in dB SPL. Loudness can also be expressed in sones, which describe loudness ratios. One sone corresponds to the loudness of a 40 dB SPL sine tone at 1 kHz. A loudness of 2 sones corresponds to 50 dB SPL. Similarly, any doubling of loudness in sones results in a 10-dB increase in SPL. For example, a loudness ratio of 64 sones corresponds to 100 dB SPL.

The ear can accommodate a very wide dynamic range. The threshold of feeling at 120 dB SPL has a sound intensity that is 1,000,000,000,000 times greater than that of the threshold of hearing at 0 dB SPL. The ear's sensitivity is remarkable; at 3 kHz, a threshold sound displaces the eardrum by a distance that is about one-tenth the diameter of a hydrogen atom. For convenience of expression, it is clear why the logarithmic decibel is used when dealing with the ear's extreme dynamic range. The ear is also fast; within 500 ms of hearing a maximum-loudness sound, the ear is sensitive to a threshold sound. Thus, whereas the eye only slowly adjusts its gain for different lighting levels and operates over a limited range at any time, the ear operates almost instantaneously over its full range. Moreover, whereas the eye can perceive an interruption to light that is 1/60 second, the ear may detect an interruption of 1/500 second.

Although the ear's dynamic range is vast, its sensitivity is frequency-dependent. Maximum sensitivity occurs at 1 kHz to 5 kHz, with relative insensitivity at low and high frequencies. This is because of the pressure transfer function that is an intrinsic part of the design of the middle ear. Through testing, equal-loudness contours such as the Robinson–Dadson curves have been derived, as shown in Fig. 10.1. Each contour describes a range of frequencies that are perceived to be equally loud. The lowest contour describes the minimum audible field, the minimum sound pressure level across the audible frequency band that a person with normal hearing can perceive. For example, a barely

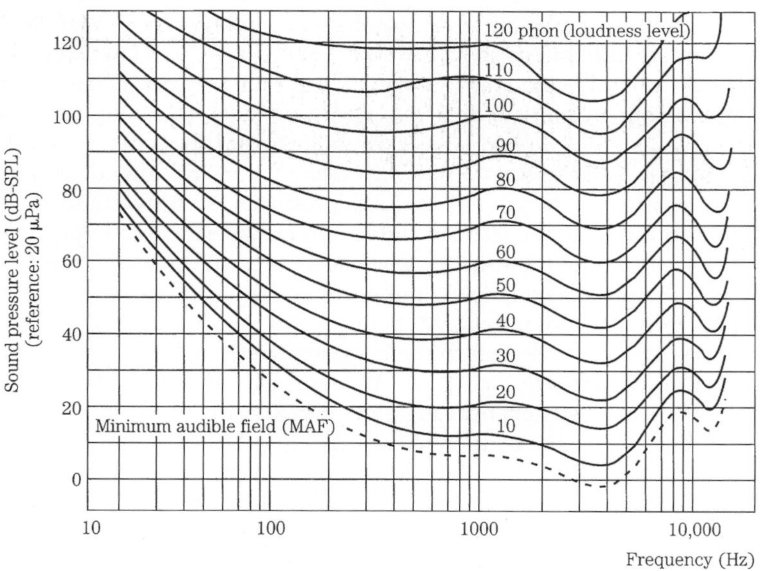

FIGURE 10.1 The Robinson–Dadson equal-loudness contours show that the ear is nonlinear with respect to frequency and level. These contours are based on psychoacoustic studies, using sine tones. (*Robinson and Dadson, 1956*)

audible 30-Hz tone would be 60 dB louder than a barely audible 4-kHz tone. The response varies with respect to level; the louder the sounds, the flatter our loudness response. The contours are rated in phons, measuring the SPL of a contour at 1 kHz.

Frequency is a literal measurement. Pitch is a subjective, perceptual measure. Pitch is a complex characteristic based on frequency, as well as other physical quantities such as waveform and intensity. For example, if a 200-Hz sine wave is sounded at a soft then louder level, most listeners will agree that the louder sound has a lower pitch. In fact, a 10% increase in frequency might be necessary to maintain a listener's subjective evaluation of a constant pitch at low frequencies. On the other hand, in the ear's most sensitive region, 1 kHz to 5 kHz, there is almost no change in pitch with loudness. Also, with musical tones, the effect is much less. Looked at in another way, pitch, quite unlike frequency, is purely a musical characteristic that places sounds on a musical scale.

The ear's response to frequency is logarithmic; this can be demonstrated through its perception of musical intervals. For example, the interval between 100 Hz and 200 Hz is perceived as an octave, as is the interval between 1000 Hz and 2000 Hz. In linear terms, the second octave is much larger, yet the ear hears it as the same interval. For this reason, musical notation uses a logarithmic measuring scale. Each four and one-half spaces or lines on the musical staff represent an octave, which might be only a few tens of Hertz apart, or a few thousands, depending on the clef and ledger lines used.

Beat frequencies occur when two nearly equal frequencies are sounded together. The beat frequency is not present in the audio signal, but is an artifact of the ear's limited frequency resolution. When the difference in frequency between tones is itself an audible frequency, a difference tone can be heard. The effect is especially audible when the frequencies are high, the tones fairly loud, and separated by not much more than a fifth. Although debatable, some listeners claim to hear sum tones. An inter-tone can also occur, especially below 200 Hz where the ear's ability to discriminate between simultaneous tones diminishes. For example, simultaneous tones of 65 Hz and 98 Hz will be heard not as a perfect fifth, but as an 82-Hz tone. On the other hand, when tones below 500 Hz are heard one after the other, the ear can differentiate between pitches only 2 Hz apart.

The ear-brain is adept at determining the spatial location of sound sources, using a variety of techniques. When sound originates from the side, the ear-brain uses cues such as intensity differences, waveform complexity, and time delays to determine the direction of origin. When equal sound is produced from two loudspeakers, instead of localizing sound from the left and right sources, the ear-brain interprets sound coming from a space between the sources. Because each ear receives the same information, the sound is stubbornly decoded as coming from straight ahead. Similarly, stereo is nothing more than two different monaural channels. The rest is simply illusion.

There is probably no limit to the complexity of psychoacoustics. For example, consider the musical tones in Fig. 10.2A. A scale is played through headphones to the right and left ears. Most listeners hear the pattern in Fig. 10.2B, where the sequence of pitches is correct, but heard as two different melodies in contrary motion. The high tones appear to come from the right ear, and the lower tones from the left. When the headphones are reversed, the headphone formerly playing low tones now appears to play high tones, and vice versa. Other listeners might hear low tones to the right and high tones to the left, no matter which way the headphones are placed. Curiously, right-handed listeners tend to hear high tones on the right and lows on the left; not so with lefties. Still other listeners might perceive only high tones and little or nothing of the low tones. In this case, most right-handed listeners perceive all the tones, but only half of the lefties do so.

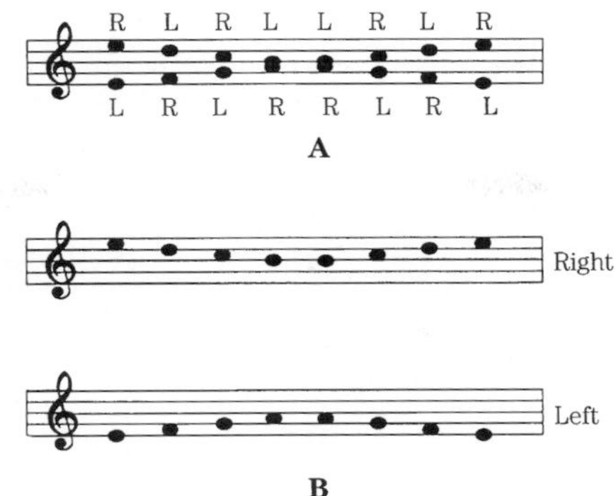

FIGURE 10.2 When a sequence of two-channel tones is presented to a listener, perception might depend on handedness. A. Tones presented to listener. B. Illusion most commonly perceived. (*Deutsch, 1983*)

The ear perceives only a portion of the information in an audio signal; that perceived portion is the perceptual entropy—estimated to be as low as 1.5 bits/sample. Small entropy signals can be efficiently reduced; large entropy signals cannot. For this reason, a codec might output a variable bit rate that is low when information is poor, and high when information is rich. The output is variable because although the sampling rate of the signal is constant, the entropy in its waveform is not. Using psychoacoustics, irrelevant portions of a signal can be removed; this is known as data reduction. The original signal cannot be reconstructed exactly. A data reduction system reduces entropy; by modeling the perceptual entropy, only irrelevant information is removed, hence the reduction can be inaudible. A perceptual music codec does not attempt to model the music source (a difficult or impossible task for music coding); instead, the music signal is tailored according to the receiver, the human ear, using a psychoacoustic model to identify irrelevant and redundant content in the audio signal. In contrast, some speech codecs use a model of the source, the vocal tract, to estimate speech characteristics, as described in Chap. 12.

Traditionally, audio system designers have used objective parameters as their design goals—flat frequency response, minimal measured noise, and so on. Designers of perceptual codecs recognize that the final receiver is the human auditory system. Following the lead of psychoacoustics, they use the ear's own performance as the design criterion. After all, any musical experience—whether created, conveyed, and reproduced via analog or digital means—is purely subjective.

Physiology of the Human Ear and Critical Bands

The ear uses a complex combination of mechanical and neurological processes to accomplish its task. In particular, the ear performs the transformation from acoustical energy to mechanical energy and ultimately to the electrical impulses sent to the brain, where information contained in sound is perceived. A simplified look at the human ear's physiological design is shown in Fig. 10.3. The outer ear collects sound, and its intricate folds help us to assess directionality. The ear canal resonates at around 3 kHz to 4 kHz, providing extra sensitivity in the frequency range that is critical for speech intelligibility.

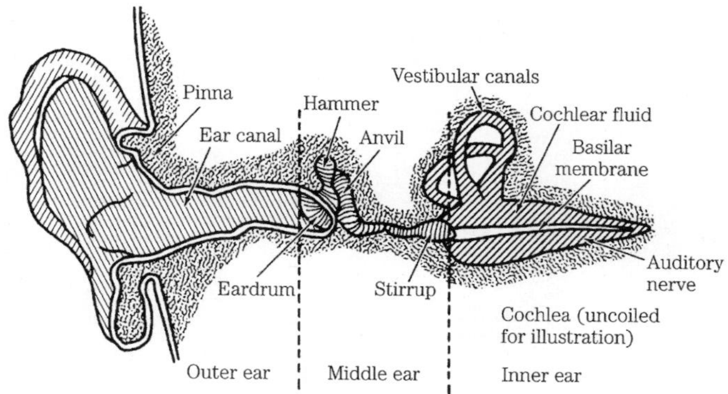

FIGURE 10.3 A simplified look at the physiology of the human ear. The coiled cochlea and basilar membrane are straightened for clarity of illustration.

The eardrum transduces acoustical energy into mechanical energy; it reaches maximum excursion at about 120-dB SPL, above which it begins to distort the waveform. The three bones in the middle ear, colloquially known as the hammer, anvil, and stirrup (the three smallest bones in the body) provide impedance matching to efficiently convey sounds in air to the fluid-filled inner ear. The vestibular canals do not affect hearing, but instead are part of a motion detection system providing a sense of balance. The coiled basilar membrane detects the amplitude and frequency of sound; those vibrations are converted to electrical impulses and sent to the brain as neural information along a bundle of nerve fibers. The brain decodes the period of the stimulus and point of maximum stimulation along the basilar membrane to determine frequency; activity in local regions surrounding the stimulus is ignored.

Examination of the basilar membrane shows that the ear contains roughly 30,000 hair cells arranged in multiple rows along the basilar membrane, roughly 32 mm long; this is the Organ of Corti. The cells detect local vibrations of the basilar membrane and convey audio information to the brain via electrical impulses. The decomposition of complex sounds into constituent components is analogous to Fourier analysis and is known as tonotopicity. Frequency discrimination dictates that at low frequencies, tones a few Hertz apart can be distinguished; however, at high frequencies, tones must differ by hundreds of Hertz. In any case, hair cells respond to the strongest stimulation in their local region; this region is called a critical band, a concept introduced by Harvey Fletcher in 1940.

Fletcher's experiments showed that, for example, when noise masks a pure tone, only frequency components of the noise that are near the frequency of the tone are relevant in masking the tone. Energy outside the band is inconsequential. This frequency range of relevancy is the critical band. Critical bands are much narrower at low frequencies than at high frequencies; three-fourths of the critical bands are below 5 kHz; in terms of masking, the ear receives more information from low frequencies and less from high frequencies. When critical bandwidths are plotted with respect to critical-band center frequency, critical bandwidths are approximately constant from 0 Hz to 500 Hz, and then approximately proportional to frequency from about 500 Hz upward, as shown in Fig. 10.4. In other words, at higher frequencies, critical bandwidth increases approximately linearly as the center frequency increases logarithmically.

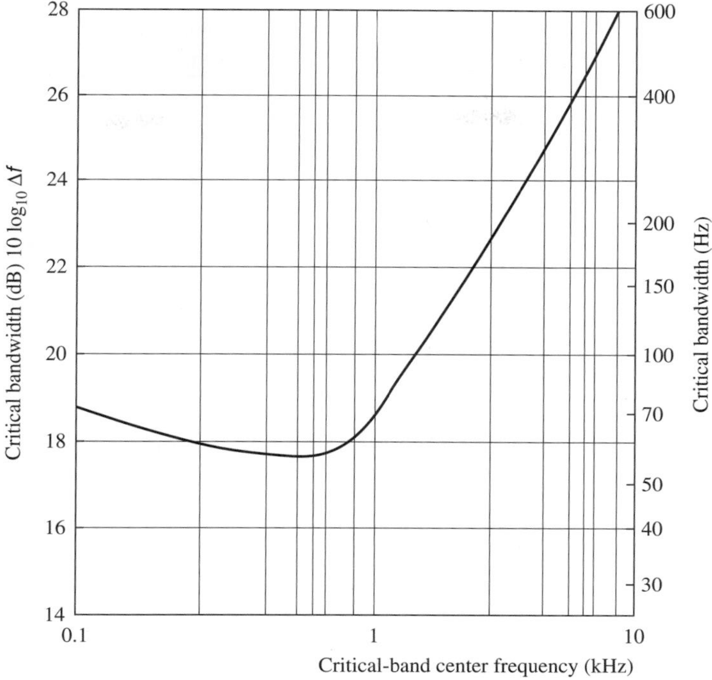

FIGURE 10.4 A plot showing critical bandwidths for monaural listening. (*Goldberg and Riek, 2000*)

Critical bands are approximately 100 Hz wide for frequencies from 20 Hz to 500 Hz and approximately 1.5 octaves in width for frequencies from 1 kHz to 7 kHz. Alternatively, bands can be assumed to be 1.3 octaves wide for frequencies from 300 Hz to 20 kHz; an error of less than 1.5 dB will occur. Other research shows that critical bandwidth can be approximated with the equation:

$$\text{Critical bandwidth} = 25 + 75[1 + 1.4(f/1000)^2]^{0.69} \text{ Hz}$$

where f = center frequency in Hz.

The ear was modeled by Eberhard Zwicker with 24 arbitrary critical bands for frequencies below 15 kHz; a 25th band occupies the region from 15 kHz to 20 kHz. An example of critical band placement and width is listed in Table 10.1. Physiologically, each critical band occupies a length of about 1.3 mm, with 1300 primary hair cells. The critical band for a 1-kHz sine tone is about 160 Hz in width. Thus, a noise or error signal that is 160 Hz wide and centered at 1 kHz is audible only if it is greater than the same level of a 1-kHz sine tone. Critical bands describe a filtering process in the ear; they describe a system that is analogous to a spectrum analyzer showing the response patterns of overlapping bandpass filters with variable center frequencies. Importantly, critical bands are not fixed; they are continuously variable in frequency, and any audible tone will create a critical band centered on it. The critical band concept is an empirical phenomenon. Looked at in another way, a critical band is the bandwidth at which subjective responses

Critical Band Number (Bark)	Center Frequency (Hz)	Critical Band (Hz)	Lower Cutoff Frequency (Hz)	Upper Cutoff Frequency (Hz)
1	50	—	—	100
2	150	100	100	200
3	250	100	200	300
4	350	100	300	400
5	450	110	400	510
6	570	120	510	630
7	700	140	630	770
8	840	150	770	920
9	1000	160	920	1080
10	1170	190	1080	1270
11	1370	210	1270	1480
12	1600	240	1480	1720
13	1850	280	1720	2000
14	2150	320	2000	2320
15	2500	380	2320	2700
16	2900	450	2700	3150
17	3400	550	3150	3700
18	4000	700	3700	4400
19	4800	900	4400	5300
20	5800	1100	5300	6400
21	7000	1300	6400	7700
22	8500	1800	7700	9500
23	10,500	2500	9500	12,000
24	13,500	3500	12,000	15,500
25	19,500	6550	15,500	22,050

TABLE 10.1 An example of critical bands in the human hearing range showing an increase in bandwidth with absolute frequency. A critical band will arise at an audible sound at any frequency. (*after Tobias, 1970*)

change. For example, if a band of noise is played at a constant sound-pressure level, its loudness will be constant as its bandwidth is increased. But as its bandwidth exceeds that of a critical band, the loudness increases.

Most perceptual codecs rely on amplitude masking within critical bands to reduce quantized word lengths. Masking is the essential trick used to perceptually hide coding noise. Indeed, in the same way that Nyquist is honored for his famous sampling frequency relationship, modern perceptual codecs could be called "Fletcher codecs."

Interestingly, critical bands have also been used to explain consonance and dissonance. Tone intervals with a frequency difference greater than a critical band are generally more consonant; intervals less than a critical band tend to be dissonant with intervals of about 0.2 critical bandwidth being most dissonant. Dissonance tends to increase at low frequencies; for example, musicians tend to avoid thirds at low frequencies. Psychoacousticians also note that critical bands play a role in the perception of pitch, loudness, phase, speech intelligibility, and other perceptual matters.

The Bark (named after Barkhausen) is a unit of perceptual frequency. Specifically, a Bark measures the critical-band rate. A critical band has a width of 1 Bark; 1/100 of a Bark equals 1 mel. The Bark scale relates absolute frequency (in Hertz) to perceptually measured frequencies such as pitch or critical bands (in Bark). Conversion from frequency to Bark can be accomplished with:

$$z(f) = 13\arctan(0.00076f) + 3.5\arctan[(f/7500)^2] \text{ Bark}$$

where f = frequency in Hz.

Using a Bark scale, the physical spectrum can be converted to a psychological spectrum along the basilar membrane. In this way, a pure tone (a single spectral line) can be represented as a psychological-masking curve. When critical bands are plotted using a Bark scale, they are relatively consistent with frequency, verifying that the Bark is a "natural" unit that presents the ear's response more accurately than linear or logarithmic plots. However, the shape of masking curves still varies with respect to level, showing more asymmetric slopes at louder levels.

Some researchers prefer to characterize auditory filter shapes in terms of an equivalent rectangular bandwidth (ERB) scale. The ERB represents the bandwidth of a rectangular function that conveys the same power as a critical band. The ERB scale portrays auditory filters somewhat differently than the critical bandwidth representation. For example, ERB argues that auditory filter bandwidths do not remain constant below 500 Hz, but instead decrease at lower frequencies; this would require greater low-frequency resolution in a codec. In one experiment, the ERB was modeled as:

$$\text{ERB} = 24.7[4.37(f/1000) + 1] \text{ Hz}$$

where f = center frequency in Hz.

The pitch place theory further explains the action of the basilar membrane in terms of a frequency-to-place transformation. Carried by the surrounding fluid, a sound wave travels the length of the membrane and creates peak vibration at particular places along the length of the membrane. The collective stimulation of the membrane is analyzed by the brain, and frequency content is perceived. High frequencies cause peak response at the membrane near the middle ear, while low frequencies cause peak response at the far end. For example, a 500-Hz tone would create a peak response at about three-fourths of the distance along the membrane. Because hair cells tend to vibrate at the frequency of the strongest stimulation, they will convey that frequency in a critical band, ignoring lesser stimulation. This excitation curve is described by the cochlear spreading function, an asymmetrical contour. This explains, for example, why broadband measurements cannot describe threshold phenomena, which are based on local frequency conditions. There are about 620 degrees of differentiable frequencies equally distributed along the basilar membrane; thus, a resolution of 1.25 Bark is reasonable. Summarizing, critical bands are important in perceptual coding because they show that the ear discriminates between energy in the band, and energy outside the band. In particular, this promotes masking.

Threshold of Hearing and Masking

Two fundamental phenomena that govern human hearing are the minimum-hearing threshold and amplitude masking, as shown in Fig. 10.5. The threshold of hearing curve describes the minimum level (0 sone) at which the ear can detect a tone at a given frequency. The threshold is referenced to 0 dB at 1 kHz. The ear is most sensitive in the 1-kHz to 5-kHz range, where we can hear signals several decibels below the 0-dB reference. Generally, two tones of equal power and different frequency will not sound equally loud. Similarly, the audibility of noise and distortion varies according to frequency. Sensitivity decreases at high and low frequencies. For example, a 20-Hz tone would have to be approximately 70 dB louder than a 1-kHz tone to be barely audible. A perceptual codec compares the input signal to the minimum threshold, and discards signals that fall below the threshold; the signals are irrelevant because the ear cannot hear them. Likewise, a codec can safely place quantization noise under the threshold because it will not be heard. The absolute threshold of hearing is determined by human testing, and describes the energy in a pure tone needed for audibility in a noiseless environment. The contour can be approximated by the equation:

$$T(f) = 3.64(f/1000)^{-0.8} - 6.5e^{-0.6[(f/1000)-3.3]^2} + 10^{-3}(f/1000)^4 \text{ dB SPL}$$

where f = frequency in Hz.

This threshold is absolute, but a music recording can be played at loud or soft levels—a variable not known at the time of encoding. To account for this variation, many codecs conservatively equate the decoder's lowest output level to a 0-dB level or alternatively to the –4-dB minimum point of the threshold curve, near 4 kHz. In other words, the ideal (lowest) quantization error level is calibrated to the lowest audible level.

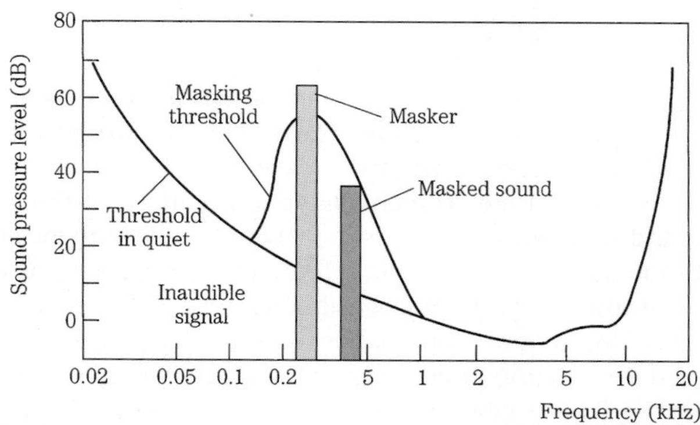

FIGURE 10.5 The threshold of hearing describes the softest sounds audible across the human hearing range. A masker tone or noise will raise the threshold of hearing in a local region, creating a masking curve. Masked tones or noise, perhaps otherwise audible, that fall below the masking curve during that time will not be audible.

Conversely, this corresponds to a maximum value of about 96 dB SPL for a 16-bit PCM signal. Some standards refer to the curve as the threshold of quiet.

When tones are sounded simultaneously, amplitude masking occurs in which louder tones can completely obscure softer tones. For example, it is difficult to carry on a conversation in a nightclub; the loud music masks the sound of speech. More analytically, for example, a loud 800-Hz tone can mask softer tones of 700 Hz and 900 Hz. Amplitude masking shifts the threshold curve upward in a frequency region surrounding the tone. The masking threshold describes the level where a tone is barely audible. In other words, the physical presence of sound certainly does not ensure audibility and conversely can ensure inaudibility of other sound. The strong sound is called the masker and the softer sound is called the maskee. Masking theory argues that the softer tone is just detectable when its energy equals the energy of the part of the louder masking signal in the critical band; this is a linear relationship with respect to amplitude. Generally, depending on relative amplitude, soft (but otherwise audible) audio tones are masked by louder tones at a similar frequency (within 100 Hz at low frequencies). A perceptual codec can take advantage of masking; the music signal to be coded can mask a relatively high level of quantization noise, provided that the noise falls within the same critical band as the masking music signal and occurs at the same time.

The mechanics of the basilar membrane explain the phenomenon of amplitude masking. A loud response at one place on the membrane will mask softer responses in the critical band around it. Unless the activity from another tone rises above the masking threshold, it will be swamped by the masker. Figure 10.6A shows four masking curves (tones masked by narrow-band noise) at 60 dB SPL, on a logarithmic scale in

A

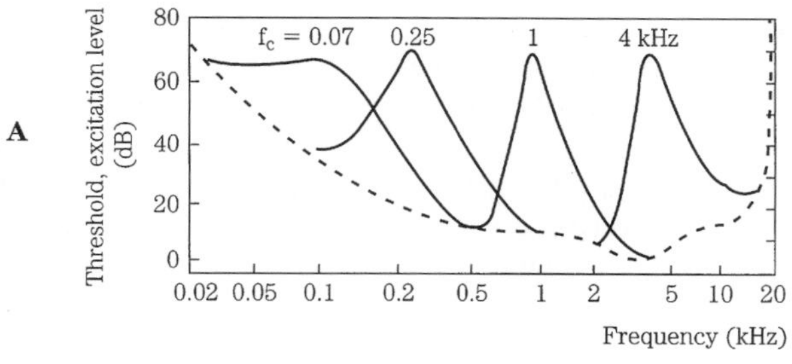

B

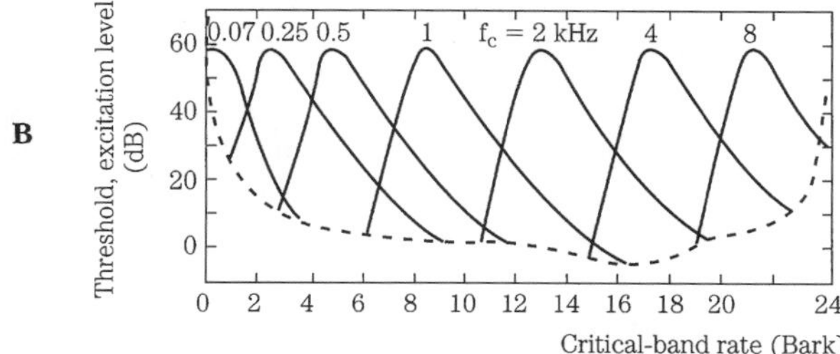

FIGURE 10.6 Masking curves describe the threshold where a tone or noise is just audible in the presence of a masker. Threshold width varies with frequency when plotted logarithmically. When plotted on a Bark scale, the widths and slopes are similar, reflecting response along the basilar membrane. A. Masking thresholds plotted with logarithmic frequency. B. Masking thresholds plotted with critical-band rate. (*Zwicker and Zwicker, 1991*)

Hertz. Figure 10.6B shows seven masking curves on a Bark scale; using this natural scale, the consistency of the critical-band rate is apparent. Moreover, this plot illustrates the position of critical bands along the basilar membrane.

Masking thresholds are sometimes expressed as an excitation level; this is obtained by adding a 2-dB to 6-dB masking index to the sound pressure level of the just-audible tone. Low frequencies can interfere with the perception of higher frequencies. Masking can overlap adjacent critical bands when a signal is loud, or contains harmonics; for example, a complex 1-kHz signal can mask a simple 2-kHz signal. Low amplitude signals provide little masking. Narrow-band tones such as sine tones also provide relatively little masking. Likewise, louder, more complex tones provide greater masking with masking curves that are broadened, and with a greater high-frequency extension.

Amplitude-masking curves are asymmetrical. The slope of the threshold curve is less steep on the high-frequency side. Thus it is relatively easy for a low tone to mask a higher tone, but the reverse is more difficult. Specifically, in a simple approximation, the lower slope is about 27 dB/Bark; the upper slope varies from –20 dB/Bark to –5 dB/Bark depending on the amplitude of the masker. More detailed approximations use spreading functions as described in the discussion of psychoacoustic models below. Low-level maskers influence a relatively narrow band of masked frequencies. However, as the sound level of the masker increases, the threshold curve broadens, and in particular its upper slope decreases; its lower slope remains relatively unaffected. Figure 10.7 shows a series of masking curves produced by a narrow band of noise centered at 1 kHz, sounded at different amplitudes. Clearly, the ear is most discriminating with low-amplitude signals.

Many masking curves have been derived from studies in which either single tones or narrow bands of noise are used as the masker stimulus. Generally, single-tone maskers produce dips in the masking curve near the tone due to beat interference between the masker and maskee tones. Narrow noise bands do not show this effect. In addition, tone maskers seem to extend high-frequency masking thresholds more readily than noise maskers. It is generally agreed that these differences are artifacts of the test itself. Tests with wideband noise show that only the frequency components of the masker that lie in the critical band of the maskee are effective at masking.

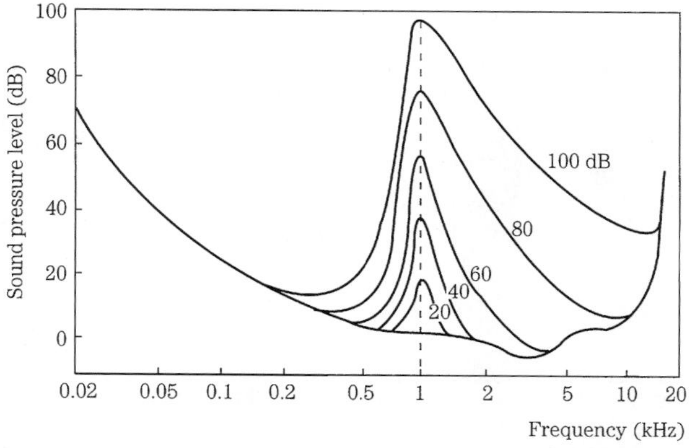

FIGURE 10.7 Masking thresholds vary with respect to sound pressure level. This test uses a narrow-band masker noise centered at 1 kHz. The lower slope remains essentially unchanged.

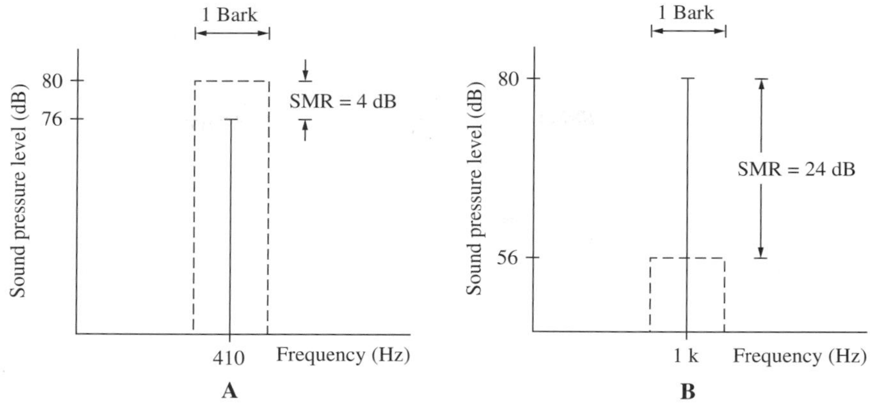

FIGURE 10.8 Noise maskers have more masking power than tonal maskers. A. In NMT, a noise masker can mask a centered tone with an SMR of only 4 dB. B. In TMN, a tonal masker can mask centered noise with an SMR of 24 dB. The SMR increases as the tone or noise moves off the center frequency. In each case, the width of the noise band is one critical bandwidth.

As noted, many masking studies use noise to mask a tone to study the condition called noise-masking-tone (NMT). In perceptual coding, we are often more concerned with quantization noise that must be masked by either a tonal or nontonal (noise-like) audio signal. The conditions of tone-masking-noise (TMN) and noise-masking-noise (NMN) are thus more pertinent. Generally, in noise-masking-tone studies, when the masker and maskee are centered, a tone is inaudible when it is about 4 dB below a 1/3-octave masking noise in a critical band. Conversely, in tone-masking-noise studies, when a 1/3-octave band of noise is masked by a pure tone, the noise must be 21 dB to 28 dB below the tone. This suggests that it is 17 dB to 24 dB harder to mask noise. The two cases are illustrated in Fig. 10.8. NMN generally follows TMN conditions; in one NMN study, the maskee was found to be about 26 dB below the masker. The difference between the level of the masking signal and the level of the masked signal is called the signal-to-mask ratio (SMR); for example, in NMT studies, the SMR is about 4 dB. Higher values for SMR denote less masking. SMR is discussed in more detail below.

Relatively little scientific study has been done with music as the masking stimulus. However, it is generally agreed that music can be considered as relatively tonal or nontonal (noise-like) and these characterizations are used in psychoacoustic models for music coding. The determination of tonal and nontonal components is important because, as noted above, the masking abilities are quite different and this greatly affects coding. In addition, sine-tone masking data is generally used in masking models because it provides the least (worst case) masking of noise; complex tones provide greater masking. Clearly, one musical sound can mask another, but future work in the mechanics of music masking will result in better masking algorithms.

Temporal Masking

Amplitude masking assumes that tones are sounded simultaneously. Temporal masking occurs when tones are sounded close in time, but not simultaneously. A signal

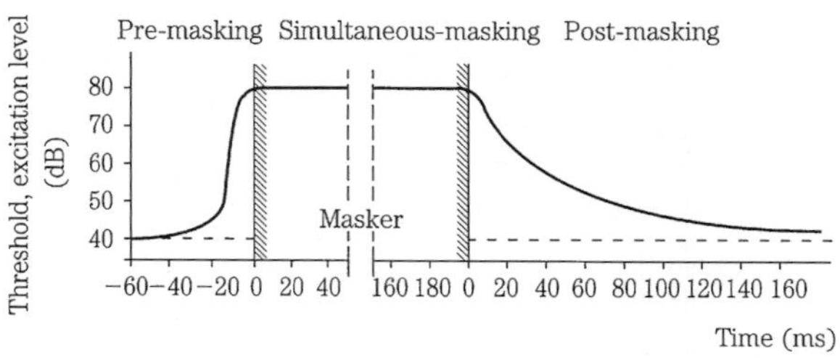

FIGURE 10.9 Temporal masking occurs before and, in particular, after a masker sounds; the threshold decreases with time. The dashed line indicates the threshold for a test-tone impulse without a masker signal. (*Zwicker and Zwicker, 1991*)

can be masked by a noise (or another signal) that occurs later. This is pre-masking (sometimes called backward masking). In addition, a signal can be masked by a noise (or another signal) that ends before the signal begins. This is post-masking (sometimes called forward masking). In other words, a louder masker tone appearing just after (pre-masking), or before (post-masking) a softer tone overcomes the softer tone. Just as simultaneous amplitude masking increases as frequency differences are reduced, temporal masking increases as time differences are reduced. Given an 80-dB tone, there may be 40 dB of post-masking within 20 ms and 0 dB of masking at 200 ms. Pre-masking can provide 60 dB of masking for 1 ms and 0 dB at 25 ms. This is shown in Fig. 10.9. The duration of pre-masking has not been shown to be affected by the duration of the masker. The envelope of post-masking decays more quickly as the duration of the masker decreases or as its intensity decreases. In addition, a tone is better post-masked by an earlier tone when they are close in frequency or when the earlier tone is lower in frequency; post-masking is slight when the masker has a higher frequency. Logically, simultaneous amplitude masking is stronger than either temporal pre- or post-masking because the sounds occur at the same time.

Temporal masking suggests that the brain integrates the perception of sound over a period of time (perhaps 200 ms) and processes the information in bursts at the auditory cortex. Alternatively, perhaps the brain prioritizes loud sounds over soft sounds, or perhaps loud sounds require longer integration times. Whatever the mechanism, temporal masking is important in frequency domain coding. These codecs have limited time resolution because they operate on blocks of samples, thus spreading quantization error over time. Temporal masking can help overcome audibility of the artifact (called pre-echo) caused by a transient signal that lasts a short time while the quantization noise may occupy an entire coding block. Ideally, filter banks should provide a time resolution of 2 ms to 4 ms. Acting together, amplitude and temporal masking form a contour that can be mapped in the time-frequency domain, as shown in Fig. 10.10. Sounds falling under that contour will be masked. It is the obligation of perceptual codecs to identify this contour for changing signal conditions and code the signal appropriately.

Although a maskee signal exists acoustically, it does not exist perceptually. It might seem quite radical, but aural masking is as real as visual masking. Lay your hand over this page. Can you see the page through your hand? Aural masking is just as effective.

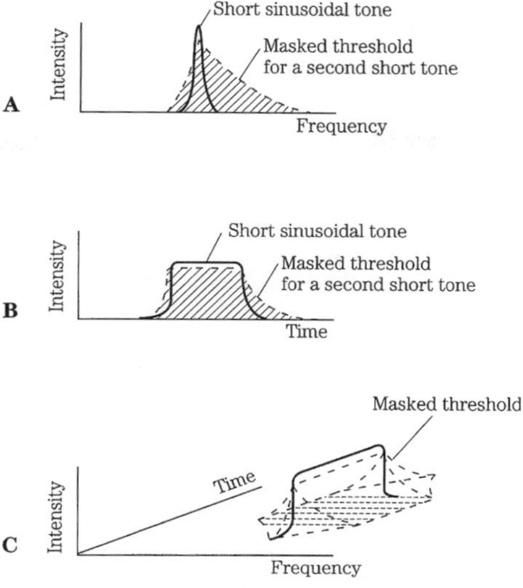

Figure 10.10 When simultaneous and temporal masking are combined, a time-frequency contour results. A perceptual codec must place quantization noise and other artifacts within this contour to ensure inaudibility. A. Simultaneous masking. B. Temporal masking. C. Combined masking effect in time and frequency. (*Beerends and Stemerdink, 1992*)

Psychoacoustic Models

Psychoacoustic models emulate the human hearing system and analyze spectral data to determine how the audio signal can be coded to render quantization noise as inaudible as possible. Most models calculate the masking thresholds for critical bands to determine this just-noticeable noise level. In other words, the model determines how much coding noise is allowed in every critical band, performing one such analysis on each frame of data. The difference between the maximum signal level and the minimum masking threshold (the signal-to-mask ratio) thus determines bit allocation for each band. An important element in modeling masking curves is determining the relative tonality of signals, because this affects the character of the masking curve they project. Any model must be time-aligned so that its results coincide with the correct frame of audio data. This accounts for the filter delay and need to center the analysis output in the current data block.

In most codecs, the goal of bit allocation is to minimize the total noise-to-mask ratio over the entire frame. The number of bits allocated cannot exceed the number of bits available for the frame at a given bit rate. The noise-to-mask ratio for each subband is calculated as:

$$\text{NMR} = \text{SMR} - \text{SNR dB}$$

The SNR is the difference between the masker and the noise floor established by a quantization level; the more bits used for quantization, the larger the value of SNR. The SMR is the difference between the masker and the minimum value of the masking threshold within a critical band. More specifically, the pertinent masking threshold is

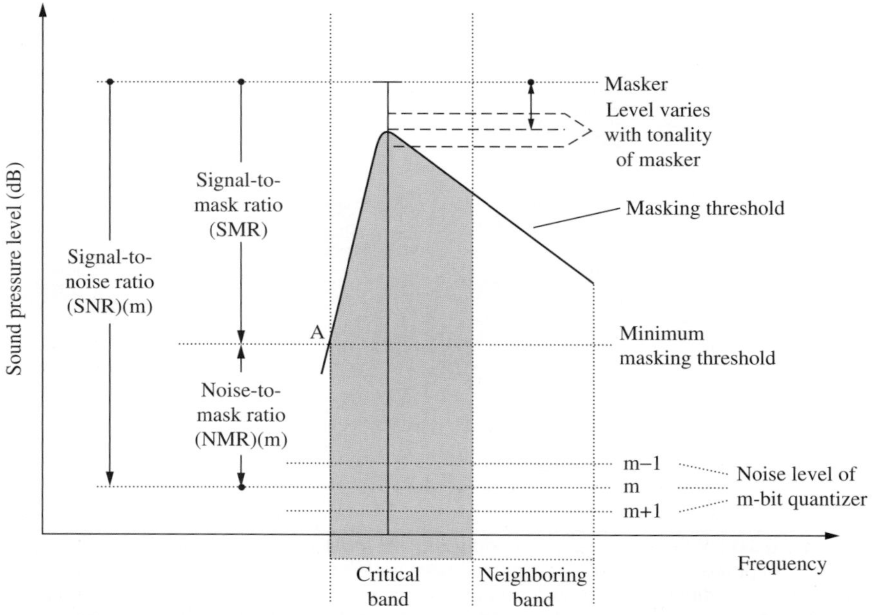

FIGURE 10.11 The NMR is the difference between the SMR and SNR, expressed in dB. The masking threshold varies according to the tonality of the signal. (*Noll, 1997*)

the global masking threshold (also known as the just-noticeable distortion or JND) within a critical band. The SMR determines the number of bits needed for quantization. If a signal is below the threshold, then the signal is not coded. The NMR is the difference between the quantization noise level, and the level where noise reaches audibility. The relationship is shown in Fig. 10.11

Within a critical band, the larger the SNR is compared to the SMR, the less audible the quantization noise. If the SNR is less than the SMR, then the noise is audible. A codec thus strives to minimize the value of the NMR in subbands by increasing the accuracy of the quantization. This figure gauges the perceptual quality of the coding. For example, NMR values of less than 0 may indicate transparent coding, while values above 0 may indicate audible degradation.

Referring again to Fig. 10.11, we can also note that the masking threshold is shifted downward from the masking peak by some amount that depends most significantly on whether the masker is tonal or nontonal. Generally, these expressions can be applied:

$$\Delta_{\text{TMN}} = 14.5 + z \text{ dB}$$

$$\Delta_{\text{NMT}} = S \text{ dB}$$

where z is the frequency in Bark and S can be assumed to lie between 3 and 6 but can be frequency-dependent.

Alternatively, James Johnston has suggested these expressions for the tonal-nontonal shift:

$$\Delta_{TMN} = 19.5 + z(18.0/26.0) \text{ dB}$$

$$\Delta_{NMT} = 6.56 - z(3.06/26.0) \text{ dB}$$

where z is the frequency in Bark.

The codec must place noise below the JND, or more specifically, taking into account the absolute threshold of hearing, the codec must place noise below the higher of JND or the threshold of hearing. For example, the SNR may be estimated from table data specified according to the number of quantizing levels, and the SMR is output by the psychoacoustic model. In an iterative process, the bit allocator determines the NMR for all subbands. The subband with the highest NMR value is allocated bits, and a new NMR is calculated based on the SNR value. The process is repeated until all the available bits are allocated.

The validity of the psychoacoustic model is crucial to the success of any perceptual codec, but it is the utilization of the model's output in the bit allocation and quantization process that ultimately determines the audibility of noise. In that respect, the interrelationship of the model and the quantizer is the most proprietary part of any codec. Many companies have developed proprietary psychoacoustic models and bit-allocation methods that are held in secret; however, their coding is compatible with standards-compliant decoders.

Spreading Function

Many psychoacoustic models use a spreading function to compute an auditory spectrum. It is straightforward to estimate masking levels within a critical band by using a component in the critical band. However, masking is usually not limited to a single critical band; its effect spreads to other bands. The spreading function represents the masking response of the entire basilar membrane and describes masking across several critical bands, that is, how masking can occur several Bark away from a masking signal. In crude models (and the most conservative) the spreading function is an asymmetrical triangle. As noted, the lower slope is about 27 dB/Bark; the upper slope may vary from –20 to –5 dB/Bark. The masking contour of a pure tone can be approximated as two slopes where S_1 is the lower slope and S_2 is the upper slope, plotted as SPL per critical-band rate. The contour is independent of masker frequency:

$$S_1 = 27 \text{ dB/Bark}$$

$$S_2 = [24 + 0.23(f_v/1000)^{-1} - 0.2L_v/\text{dB}] \text{ dB/Bark}$$

where f_v is the frequency of the masking tone in Hz, and L_v is the level of the masking tone in dB.

The slope of S_2 becomes steeper at low frequencies by the $0.23(f_v/1000)^{-1}$ term because of the threshold of hearing, while at masking frequencies above 100 Hz, the slope is almost independent of frequency. S_2 also depends on SPL.

A more sophisticated spreading function, but one that does not account for the masker level, is given by the expression:

$$10 \log_{10} SF(dz) = 15.81 + 7.5(dz + 0.474) - 17.5[1 + (dz + 0.474)^2]^{1/2} \text{ dB}$$

where dz is the distance in Bark between the maskee and masker frequency.

To use a spreading function, the audio spectrum is divided into critical bands and the energy in each band is computed. These values are convolved with the spreading function to yield the auditory spectrum. When offsets and the absolute threshold of hearing are considered, the final masking thresholds are produced. When calculating a global masking threshold, the effects of multiple maskers must be considered. For example, a model could use the higher of two thresholds, or add together the masking threshold intensities of different components. Alternatively, a value averaged between the values of the two methods could be used, or another nonlinear approach could be taken. For example, in the MPEG-1 psychoacoustic model 1, intensities are summed. However, in MPEG-1 model 2, the higher value of the global masking threshold and the absolute threshold is selected. These models are discussed in Chap. 11.

Tonality

Distinguishing between tonal and nontonal components is an important feature of most psychoacoustic models because tonal and nontonal components demand different masking emulation. For example, as noted, noise is a better masker than a tone. Many methods have been devised to detect and characterize tonality in audio signals. For example, in MPEG-1 model 1, tonality is determined by detecting local maxima in the audio spectrum. All nontonal components in a critical band are represented with one value at one frequency. In MPEG-1 model 2, a spectral flatness measure is used to measure the average or global tonality. These models are discussed in Chap. 11.

In some tonality models, when a signal has strong local maxima tonal components, they are detected and withdrawn and coded separately. This flattens the overall spectrum and increases the efficiency of the subsequent Huffman coding because the average number of bits needed in a codebook increases according to the magnitude of the maximum value. The increase in efficiency depends on the nature of the audio signal. Some models further distinguish the harmonic structure of multitonal maskers. With two multitonal maskers of the same power, the one with a strong harmonic structure yields a lower masking threshold.

Identification of tonal and nontonal components can also be important in the decoder when data is conveyed across an error-prone transmission channel and error concealment is applied before the output synthesis filter bank. Missing tonal components can be replaced by predicted values. For example, predictions can be made using an FIR filter for all-pole modeling of the signal, and using an autocorrelation function, coefficients can be generated with the Levinson–Durbin algorithm. Studies indicate that concealment in the lower subbands is more important than in the upper subbands. Noise properly shaped by a spectral envelope can be successfully substituted for missing nontonal sections.

Rationale for Perceptual Coding

The purpose of any low bit-rate coding system is to decrease the data rate, the product of the sampling frequency, and the word length. This can be accomplished by decreasing the sampling frequency; however, the Nyquist theorem dictates a corresponding decrease in high-frequency audio bandwidth. Another approach uniformly decreases the word length; however, this reduces the dynamic range of the audio signal by 6 dB per bit, thus increasing broadband quantization noise. As we have seen, a more enlightened approach uses psychoacoustics. Perceptual codecs maintain sampling frequency, but selectively decrease word length. The word-length reduction is done dynamically based on signal conditions. Specifically, masking and other factors are considered so that the resulting increase in quantization noise is rendered as inaudible as possible. The level of quantization error, and its associated distortion from truncating the word length, can be allowed to rise, so long as it is masked by the audio signal. For example, a codec might convey an audio signal with an average bit rate of 2 bits/sample; with PCM encoding, this would correspond to a signal-to-noise ratio of 12 dB—a very poor result. But by exploiting psychoacoustics, the codec can render the noise floor nearly inaudible.

Perceptual codecs analyze the frequency and amplitude content of the input signal. The encoder removes the irrelevancy and statistical redundancy of the audio signal. In theory, although the method is lossy, the human perceiver will not hear degradation in the decoded signal. Considerable data reduction is possible. For example, a perceptual codec might reduce a channel's bit rate from 768 kbps to 128 kbps; a word length of 16 bits/sample is reduced to an average of 2.67 bits/sample, and data quantity is reduced by about 83%. Table 10.2 lists various reduction ratios and resulting bit rates for 48-kHz and 44.1-kHz monaural signals. A perceptually coded recording, with a conservative level of reduction, can rival the sound quality of a conventional recording because the data is coded in a much more intelligent fashion, and quite simply, because we do not hear all of what is recorded anyway. In other words, perceptual codecs are efficient because they can convey much of the perceived information in an audio signal, while requiring only a fraction of the data needed by a conventional system.

Part of this efficiency stems from the adaptive quantization used by most perceptual codecs. With PCM, all signals are given equal word lengths. Perceptual codecs

Bits/Sample	Reduction Ratio	Bit Rate (kbps) at 48 kHz	Bit Rate (kbps) at 44.1 kHz
16	1:1	768	705.6
8	2:1	384	352.8
4	4:1	192	176
2.67	6:1	128	117.7
2	8:1	96	88.2
1.45	11:1	69.6	64

Table 10.2 Bit-rate reduction for 48-kHz and 44.1-kHz sampling frequencies.

assign bits according to audibility. A prominent tone is given a large number of bits to ensure audible integrity. Conversely, fewer bits are used to code soft tones. Inaudible tones are not coded at all. Together, bit rate reduction is achieved. A codec's reduction ratio (or coding gain) is the ratio of input bit rate to output bit rate. Reduction ratios of 4:1, 6:1, or 12:1 are common. Perceptual codecs have achieved remarkable transparency, so that in many applications reduced data is audibly indistinguishable from linearly represented data. Tests show that reduction ratios of 4:1 or 6:1 can be transparent.

The heart of a perceptual codec is the bit-allocation algorithm; this is where the bit rate is reduced. For example, a 16-bit monaural signal sampled at 48 kHz that is coded at a bit rate of 96 kbps must be requantized with an average of 2 bits/sample. Moreover, at that bit rate, the bit budget might be 1024 bits per block of analyzed data. The bit-allocation algorithm must determine how best to distribute the bits across the signal's spectrum and requantize samples to minimize audibility of quantization noise while meeting its overall bit budget for that block.

Generally, two kinds of bit-allocation strategies can be used in perceptual codecs. In forward adaptive allocation, all allocation is performed in the encoder and this encoding information is contained in the bitstream. Very accurate allocation is permitted, provided the encoder is sufficiently sophisticated. An important advantage of forward adaptive coding is that the psychoacoustic model is located in the encoder; the decoder does not need a psychoacoustic model because it uses the encoded data to completely reconstruct the signal. Thus as psychoacoustic models in encoders are improved, the increased sonic quality can be conveyed through existing decoders. A disadvantage is that a portion of the available bit rate is needed to convey the allocation information to the decoder. In backward adaptive allocation, bit-allocation information is derived from the coded audio data itself without explicit information from the encoder. The bit rate is not partly consumed by allocation information. However, because bit allocation in the decoder is calculated from limited information, accuracy may be reduced. In addition, the decoder is more complex, and the psychoacoustic model cannot be easily improved following the introduction of new codecs.

Perceptual coding is generally tolerant of errors. With PCM, an error introduces a broadband noise. However, with most perceptual codecs, the error is limited to a narrow band corresponding to the bandwidth of the coded critical band, thus limiting its loudness. Instead of a click, an error might be perceived as a burst of low-level noise. Perceptual coding systems also permit targeted error correction. For example, particularly vulnerable sounds (such as pianissimo passages) may be given greater protection than less vulnerable sounds (such as forte passages). As with any coded data, perceptually coded data requires error correction appropriate to the storage or transmission medium.

Because perceptual codecs tailor the coding to the ear's acuity, they may similarly decrease the required response of the playback system itself. Live acoustic music does not pass through amplifiers and loudspeakers—it goes directly to the ear. But recorded music must pass through the playback signal chain. Arguably, some of the original signal present in a recording could degrade the playback system's ability to reproduce the audible signal. Because a perceptual codec removes inaudible signal content, the playback system's ability to convey audible music may improve. In short, a perceptual codec may more properly code an audio signal for passage through an audio system.

Perceptual Coding in Time and Frequency

Low bit-rate lossy codecs, whether designed for music or speech coding, attempt to represent the audio signal at a reduced bit rate while minimizing the associated increase in quantization error. Time-domain coding methods such as delta modulation can be considered to be data-reduction codecs (other time-domain methods such as PCM do not provide reduction). They use prediction methods on samples representing the full bandwidth of the audio signal and yield a quantization error spectrum that spans the audio band. Although the audibility of the error depends on the amplitude and spectrum of the signal, the quantization error generally is not masked by the signal. However, time-domain codecs operating across the full bandwidth of the time-domain signal can achieve reduction ratios of up to 2.5. For example, Near Instantaneously Companded Audio Multiplex (NICAM) codecs reduce blocks of 32 samples from 14 bits to 10 bits using a sliding window to determine which 10 of the 14 bits can be transmitted with minimal audible degradation. With this method, coding is lossless with low-level signals, with increasing loss at high levels. Although data reduction is achieved, the bit rate is too high for many applications; primarily, reduction is limited because masking is not fully exploited.

Frequency-domain codecs take a different approach. The signal is analyzed in the frequency domain, and only the perceptually significant parts of the signal are quantized, on the basis of psychoacoustic characteristics of the ear. Other parts of the signal that are below the minimum threshold, or masked by more significant signals, may be judged to be inaudible and are not coded. In addition, quantization resolution is dynamically adapted so that error is allowed to rise near significant parts of the signal with the expectation that when the signal is reconstructed, the error will be masked by the signal. This approach can yield significant data reduction. However, codec complexity is greatly increased.

Conceptually, there are two types of frequency-domain codecs: subband and transform codecs. Generally, subband codecs use a low number of subbands and process samples adjacent in time, and transform codecs use a high number of subbands and process samples adjacent in frequency. Generally, subband codecs provide good time resolution and poor frequency resolution, and transform codecs provide good frequency resolution and poor time resolution.

However, the distinction between subband and transform codecs is primarily based on their separate historical development. Mathematically, all transforms used in codecs can be viewed as filter banks. Perhaps the most practical difference between subband and transform codecs is the number of bands they process. Thus, both subband and transform codecs follow the architecture shown in Fig. 10.12; either time-domain samples or frequency-domain coefficients are quantized according to a psychoacoustic model contained in the encoder.

In subband coding, a hybrid of time- and frequency-domain techniques is used. A short block of time-based broadband input samples is divided into a number of frequency subbands using a filter bank of bandpass filters; this allows determination of the energy in each subband. Using a side-chain transform frequency analysis, the samples in each subband are analyzed for energy content and coded according to a psychoacoustic model.

In transform coding, a block of input samples is directly applied to a transform to obtain the block's spectrum in the frequency domain. These transform coefficients are

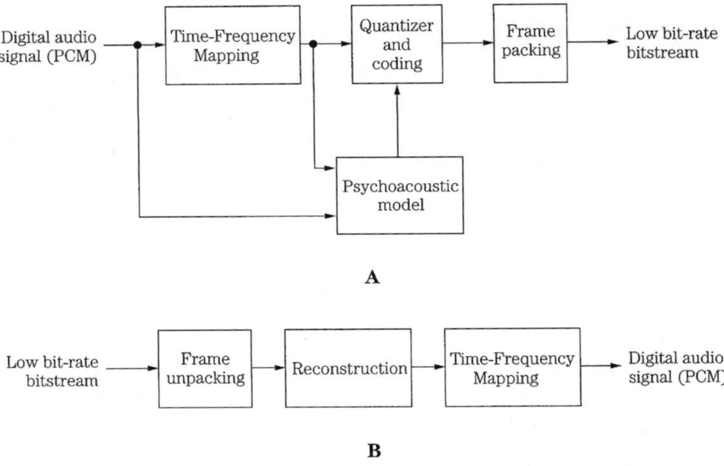

FIGURE 10.12 The basic structure of a time-frequency-domain encoder and decoder (A and B, respectively). Subband (time) codecs quantize time-based samples, and transform (frequency) codecs quantize frequency-based coefficients.

then quantized and coded according to a psychoacoustic model. Problematically, a relatively long block of data is required to obtain a high-resolution spectral representation. Transform codecs achieve greater reduction than subband codecs; ratios of 4:1 to 12:1 are typical. Transform codecs incur a longer processing delay than subband codecs.

As noted, most low bit-rate lossy codecs use psychoacoustic models to analyze the input signal in the frequency domain. To accomplish this, the time-domain input signal is often applied to a transform prior to analysis in the model. Any periodic signal can be represented as amplitude variations in time, or as a set of frequency coefficients describing amplitude and phase. Jean Baptiste Joseph Fourier first established this relationship between time and frequency. Changes in a time-domain signal also appear as changes in its frequency-domain spectrum. For example, a slowly changing signal would be represented by a low-frequency spectral content. If a sequence of time-based samples are thus transformed, the signal's spectral content can be determined over that period of time. Likewise, the time-based samples can be recovered by inverse transforming the spectral representation back into the time domain. A variety of mathematical transforms can be used to transform a time-domain signal into the frequency domain and back again. For example, the fast Fourier transform (FFT) gives a spectrum with half as many frequency points as there are time samples. For example, assume that 480 samples are taken at a 48-kHz sampling frequency. In this 10-ms interval, 240 frequency points are obtained over a spectrum from the highest frequency of 24 kHz to the lowest of 100 Hz, which is the period of 10 ms, with frequency points placed 100 Hz apart. In addition, a dc point is generated.

Subband Coding

Subband coding was first developed at Bell Labs in the early 1980s, and much subsequent work was done in Europe later in the decade. Blocks of consecutive time-domain samples representing the broadband signal are collected over a short period and applied to a digital filter bank. This analysis filter bank divides the signal into multiple (perhaps up to 32) bandlimited channels to approximate the critical band response of the human ear.

The filter bank must provide a very sharp cutoff (perhaps 100 dB/octave) to emulate critical band response and limit quantization noise within that bandwidth. Only digital filters can accomplish this result. In addition, the processing block length (ideally less than 2 ms to 4 ms) must be small so that quantization error does not exceed the temporal masking limits of the ear. The samples in each subband are analyzed and compared to a psychoacoustic model. The codec adaptively quantizes the samples in each subband based on the masking threshold in that subband. Ideally, the filter bank should yield subbands with a width that corresponds to the width of the narrowest critical band. This would allow precise psychoacoustic modeling. However, most filter banks producing uniformly spaced subbands cannot meet this goal; this points out the difficulties posed by the great difference in bandwidth between the narrowest critical band and the widest.

Each subband is coded independently with greater or fewer bits allocated to the samples in the subband. Quantization noise may be increased in a subband. However, when the signal is reconstructed, the quantization noise in a subband will be limited to that subband, where it is ideally masked by the audio signal in that subband, as shown in Fig. 10.13.

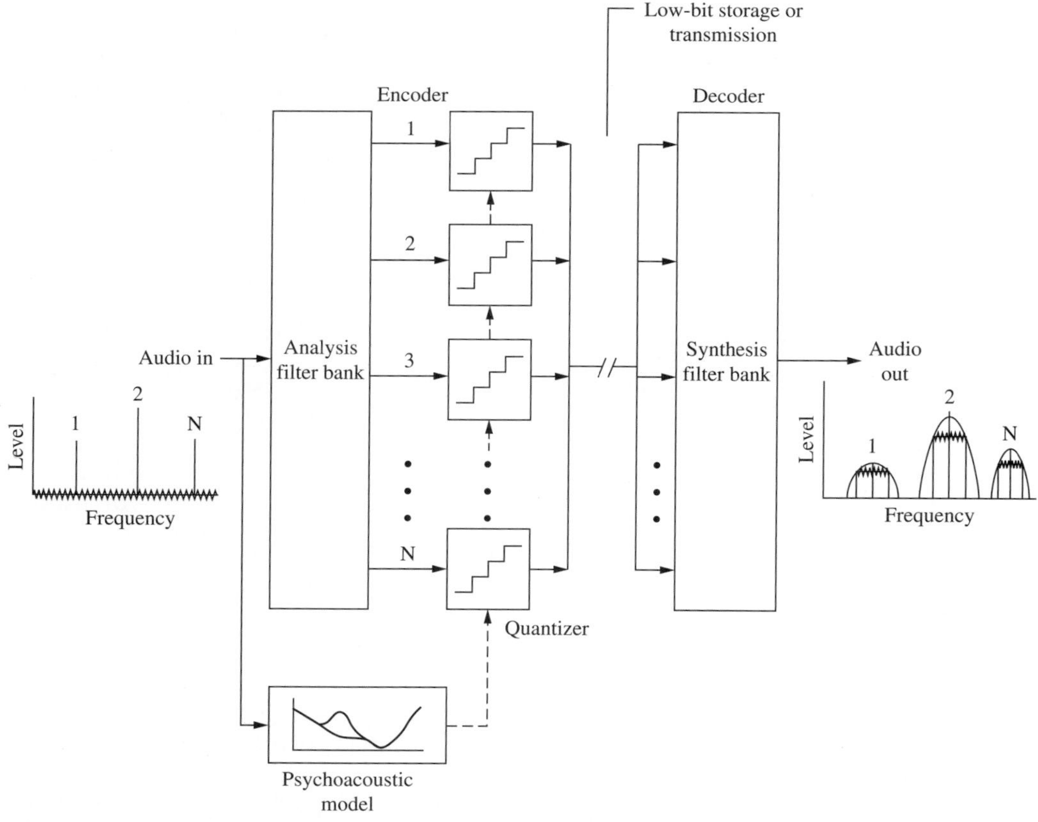

Figure 10.13 A subband encoder analyzes the broadband audio signal in narrow subbands. Using masking information from a psychoacoustic model, samples in subbands are coarsely quantized, raising the noise floor. When the samples are reconstructed in the decoder, the synthesis filter constrains the quantization noise floor within each subband, where it is masked by the audio signal.

Quantization noise levels that are otherwise intrusive can be tolerated in a subband with a signal contained in it because noise will be masked by the signal. Subbands that do not contain an audible signal are quantized to zero. Bit allocation is determined by a psychoacoustic model and analysis of the signal itself; these operations are recalculated for every subband in every new block of data. Samples are dynamically quantized according to audibility of signals and noise. There is great flexibility in the design of psychoacoustic models and bit-allocation algorithms used in codecs that are otherwise compatible. The decoder uses the quantized data to re-form the samples in each block; a synthesis filter bank sums the subband signals to reconstruct the output broadband signal.

A subband perceptual codec uses a filter bank to split a short duration of the audio signal into multiple bands, as depicted in Fig. 10.14. In some designs, a side-chain processor applies the signal to a transform such as an FFT to analyze the energy in each subband. These values are applied to a psychoacoustic model to determine the combined masking curve that applies to the signals in that block. This permits more optimal coding of the time-domain samples. Specifically, the encoder analyzes the energy in each subband to determine which subbands contain audible information. A calculation is made to determine the average power level of each subband over the block. This average level is used to calculate the masking level due to masking of signals in each subband, as well as masking from signals in adjacent subbands. Finally, minimum hearing threshold values are applied to each subband to derive its final masking level. Peak power levels present in each subband are calculated, and compared to the masking level. Subbands that do not contain audible information are not coded. Similarly, tones in a subband that are masked by louder nearby tones are not coded, and in some cases entire subbands can mask nearby subbands, which thus need not be coded.

Calculations determine the ratio of peak power to masking level in each subband. Quantization bits are assigned to audible program material with a priority schedule that allocates bits to each subband according to signal strength above the audibility curve. For example, Fig. 10.15 shows vertical lines representing peak power levels, and minimum and masking thresholds.

The signals below the minimum or masking curves are not coded, and the quantization noise floor is allowed to rise to those levels. For example, in the figure, signal A is below the minimum curve and would not be coded in any event. Signal C is also irrelevant in this frame because signal B has dynamically shifted the hearing threshold upward.

Signal B must be coded; however, its presence has created a masking curve, decreasing the relative amplitude above the minimum threshold curve. The portion of signal B between the minimum curve and the masking curve represents the fewer bits that are needed to code the signal when the masking effect is taken into account. In other words, rather than using a signal-to-noise ratio, a signal-to-mask ratio (SMR) is used. The SMR is the difference between the maximum signal and the masking threshold and is used to determine the number of bits assigned to a subband. The SMR is calculated for each subband.

The number of bits allocated to any subband must be sufficient to yield a requantizing noise level that is below the masking level. The number of bits depends on the SMR value, with the goal of maintaining the quantization noise level below the calculated masking level for each subband. In fixed-rate codecs, a bit-pool approach can be taken. A large number of subbands requiring coding and signals with large SMR values might empty the pool, resulting in less than optimal coding. On the other hand, if the pool is not empty after initial allocation, the process is repeated until all bits in the codec's data

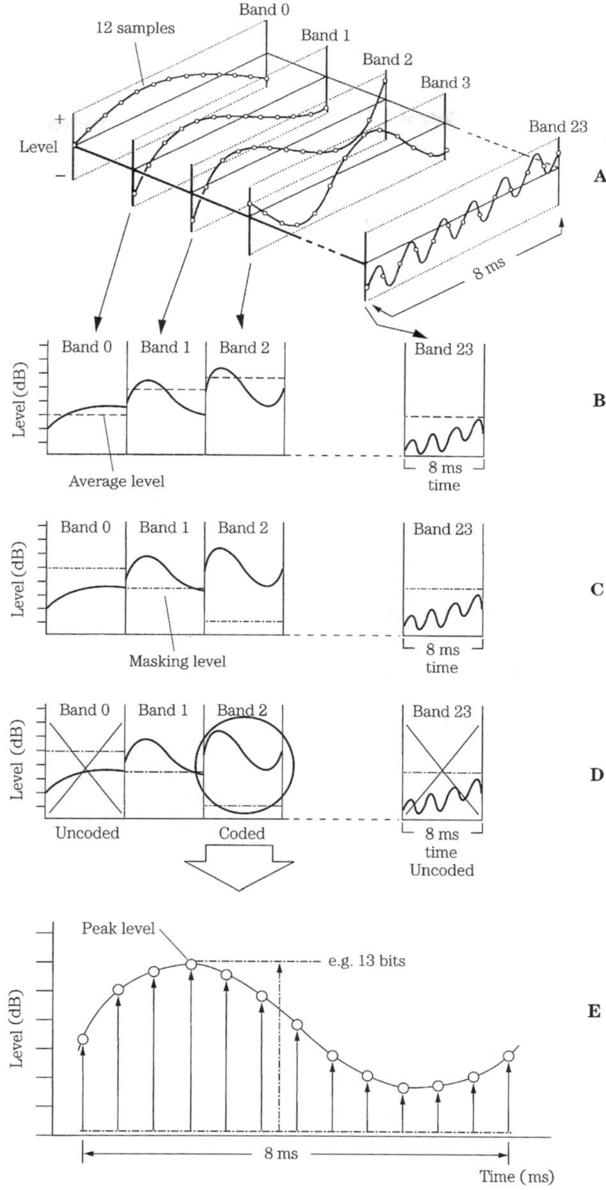

Figure 10.14 A subband codec divides the signal into narrow subbands, calculates average signal level, and masking level; and then quantizes the samples in each subband accordingly. A. Output of 24-band subband filter. B. Calculation of average level in each subband. C. Calculation of masking level in each subband. D. Subbands below audibility are not coded; bands above audibility are coded. E. Bits are allocated according to peak level above the masking threshold. Subbands with peak levels above the masking level contain audible signals that must be coded.

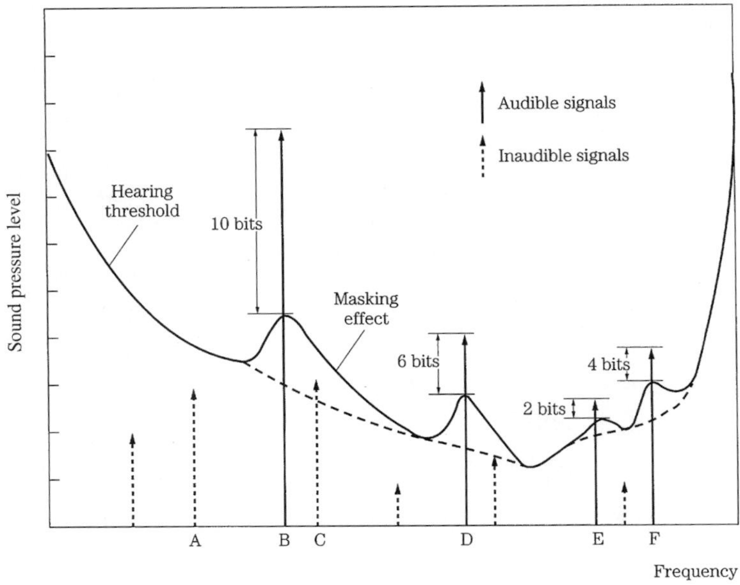

FIGURE 10.15 The bit-allocation algorithm assigns bits according to audibility of subband signals. Bits may not be assigned to masked or inaudible tones.

capacity have been used. Typically, the iterative process continues, allocating more bits where required, with signals with the highest SMR requirements always receiving the most bits; this increases the coding margin. In some cases, subbands previously classified as inaudible might receive coding from these extra bits. Thus, signals below the masking threshold can in practice be coded, but only on a secondary priority basis. Summarizing the concept of subband coding, Fig. 10.16 shows how a 24-subband codec might code three tones at 250 Hz, 1 kHz, and 4 kHz; note that in each case the quantization noise level is below the combined masking and threshold curve.

Transform Coding

In transform coding, the audio signal is viewed as a quasi-stationary signal that changes relatively little over short time intervals. For efficient coding, blocks of time-domain audio samples are transformed to the frequency domain. Frequency coefficients, rather than amplitude samples, are quantized to achieve data reduction. For playback, the coefficients are inverse-transformed back to the time domain.

The operation of the transform approximates how the basilar membrane analyzes the frequency content of vibrations along its length. The spectral coefficients output by the transform are quantized according to a psychoacoustic model; masked components are eliminated, and quantization decisions are made based on audibility. In contrast to a subband codec, which uses frequency analysis to code time-based samples, a transform codec codes frequency coefficients. From an information theory standpoint, the transform reduces the entropy of the signal, permitting efficient coding. Longer transform blocks provide greater spectral resolution, but lose temporal resolution;

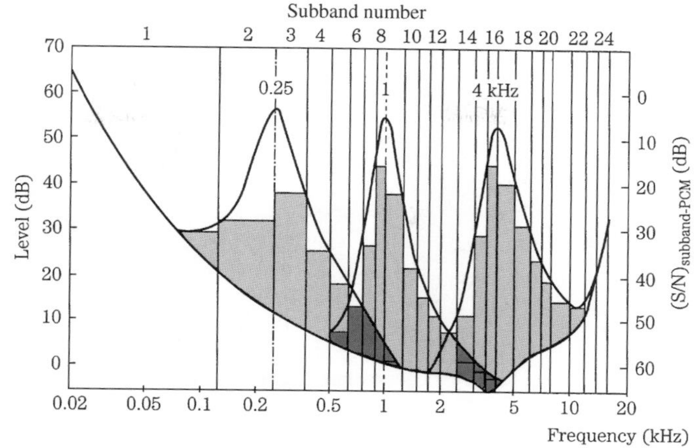

FIGURE 10.16 In this 24-band subband codec, three tones are coded so that the quantization noise in each subband falls below the calculated composite masking curves. (*Thiele, Link, and Stoll, 1987*)

for example, a long block might result in a pre-echo before a transient. In many codecs, block length is adapted according to audio signal conditions. Short blocks are used for transient signals, while long blocks are used for continuous signals.

Time-domain samples are transformed to the frequency domain, yielding spectral coefficients. The coefficient numbers are sometimes called frequency bin numbers; for example, a 512-point transform can produce 256 frequency coefficients or frequency bins. The coefficients, which might number 512, 1024, or more, are grouped into about 32 bands that emulate critical-band analysis. This spectrum represents the block of time-based input samples. The frequency coefficients in each band are quantized according to the codec's psychoacoustic model; quantization can be uniform, nonuniform, fixed, or adaptive in each band.

Transform codecs may use a discrete cosine transform (DCT) or modified discrete cosine transform (MDCT) for transform coding because of low computational complexity, and because they can critically sample (sample at twice the bandwidth of the bandpass filter) the signal to yield an appropriate number of coefficients. Most codecs overlap successive blocks in time by about 50%, so that each sample appears in two different transform blocks. For example, the samples in the first half of a current block are repeated from the second half of the previous block. This reduces changes in spectra from block to block and improves temporal resolution. The DCT and MDCT can yield the same number of coefficients as with non-overlapping blocks. As noted, an FFT may be used in the codec's side chain to yield coefficients for perceptual modeling.

All low bit-rate codecs operate over a block of samples. This block must be kept short to stay within the temporal masking limits of the ear. During decoding, quantization noise will be spread over the frequency of the band, and over the duration of the block. If the block is longer than temporal backward masking allows, the noise will be heard prior to the onset of the sound, in a phenomenon known as pre-echo. (The term pre-echo is misleading.) Pre-echo is particularly problematic in the case of a silence followed by a time-domain transient within the analysis window. The energy in the transient portion causes the encoder to allocate relatively few bits, thus raising

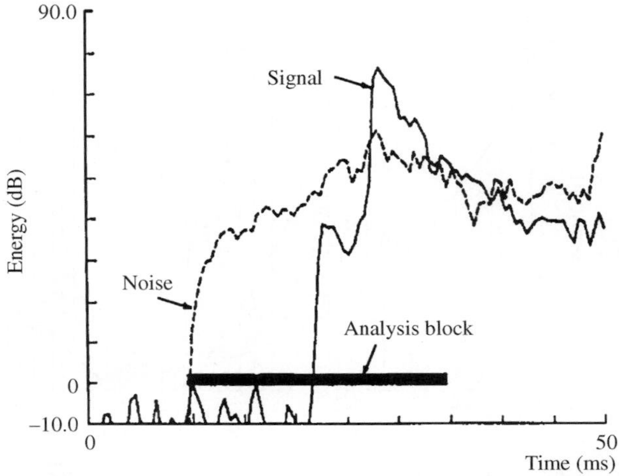

Figure 10.17 An example of a pre-echo. On reconstruction, quantization noise falls within the analysis block, where the leading edge is not masked by the signal. (*Herre and Johnston, 1996*)

the eventual quantization noise level. Pre-echoes are created in the decoder when frequency coefficients are inverse-transformed prior to the reconstruction of subband samples in the synthesis filter bank. The duration of the quantization noise equals that of the synthesis window, so the elevated noise extends over the duration of the window, while the transient only occurs briefly. In other words, encoding dictates that a transient in the audio signal will be accompanied by an increase in quantization noise but a brief transient may not fully mask the quantization noise surrounding it, as shown in Fig. 10.17. In this example, the attack of a triangle occurs as a transient signal. The analysis window of a transform codec operates over a relatively long time period. Quantization noise is spread over the time of the window and precedes the music signal; thus it may be audible as a pre-echo.

Transform codecs are particularly affected by the problem of pre-echo because they require long blocks for greater frequency accuracy. Short block length limits frequency resolution (and also relatively increases the amount of overhead side information). In essence, transform codecs sacrifice temporal resolution for spectral resolution. Long blocks are suitable for slowly changing or tonal signals; the frequency resolution allows the codec to identify spectral peaks and use their masking properties in bit allocation. For example, a clarinet note and its harmonics would require fine frequency resolution but only coarse time resolution. However, transient signals require a short block length; the signals have a flatter spectrum. For example, the fast transient of a castanet click would require fine time resolution but only coarse frequency resolution.

In most transform codecs, to provide the resolution demanded by particular signal conditions, and to avoid pre-echo, block length dynamically adapts to signal conditions. Referring again to Fig. 10.17, a shorter analysis block would constrain the quantization noise to a shorter duration, where it will be masked by the signal. A short block is also advantageous because it limits the duration of high bit rates demanded by transient encoding. Alternatively, a variable bit rate encoder can minimize pre-echo by briefly increasing the bit rate to decrease the noise level. Some codecs use temporal noise shaping (TNS) to minimize pre-echo by manipulating the nature of the quantization noise within a filter bank window. When a transient signal is detected, TNS uses a predictive coding method to shape the quantization noise to follow the transient's envelope. In this way, the quantization error is more effectively concealed by the transient. However, no matter what approach is taken, difficulty arises because most music simultaneously places contradictory demands on the codec.

In adaptive transform codecs, a model is applied to uniformly and adaptively quantize each individual band, but coefficient values within a band are quantized with the same number of bits. The bit-allocation algorithm calculates the optimal quantization noise in each subband to achieve a desired signal-to-noise ratio that will promote masking. Iterative allocation is used to supply additional bits as available to increase the coding margin, yet maintain limited bit rate. In some cases, the output bit rate can be fixed or variable for each block. Before transmission, the reduced data is often compressed with entropy coding such as Huffman coding and run-length coding to perform lossless compression. The decoder inversely quantizes the coefficients and performs an inverse transform to reconstruct the signal in the time domain.

An example of an adaptive transform codec proposed by Karlheinz Brandenburg is shown in Fig. 10.18. An MDCT transforms the signal to the frequency domain. Signal energy in each critical band is calculated using the spectral coefficients. This is used to determine the masking threshold for each critical band. Two iterative loops perform quantization and coding using an analysis-by-synthesis technique. Coefficients are initially assigned a quantizer step size and the algorithm calculates the resulting number of bits needed to code the signal in the block. If the count exceeds the bit rate allowed for the block, the loop reassigns a larger quantizer step size and the count is recalculated until the target bit rate is achieved. An outer loop calculates the quantization error as it will appear in the reconstructed signal. If the error in a band exceeds the error allowed by the masking model, the quantizer step size in the band is decreased. Iterations continue in both loops until optimal coding is achieved. Codecs such as this can operate at low bit rates (for example, 2.5 bits/sample).

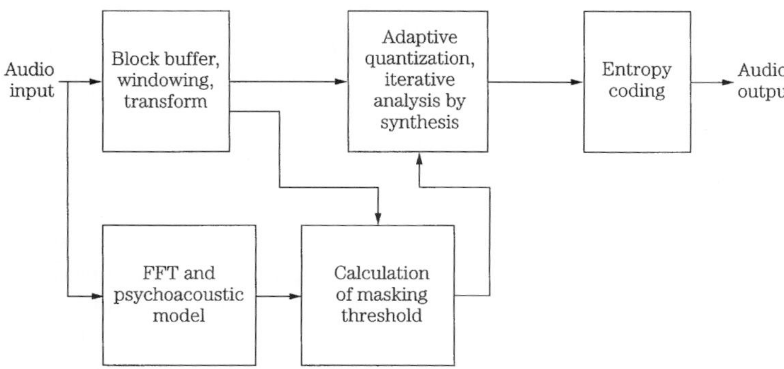

FIGURE 10.18 Adaptive transform codec using an FFT side-chain and iterative quantization to achieve optimal reduction. Entropy coding is additionally used for data compression.

Filter Banks

Low bit-rate codecs often use an analysis filter bank to partition the wide audio band into smaller subbands; the decoder uses a synthesis filter bank to restore the subbands to a wide audio band. Uniformly spaced filter banks downconvert an audio band into a baseband, then lowpass filter and subsample the data. Transform-based filter banks multiply overlapping blocks of audio samples with a window function to reduce edge effects, and then perform a discrete transform such as DCT. Mathematically, the transforms used in codecs can be seen as filter banks, and subband filter banks can be seen as transforms.

Time-domain filter banks should provide ideal lowpass and highpass characteristics with a cutoff of $f_s/2$, where f_s is the sampling frequency; however, real filters have overlapping bands. For example, in a two-band system, the subband sampling rates must be decreased (2:1) to maintain the overall bit rate. This decimation introduces aliasing in the subbands. In the lower band, signals above $f_s/4$ will alias to 0 to $f_s/4$. In the upper band, signals below $f_s/4$ alias up to $f_s/4$ to $f_s/2$. In the decoder, the sampling rate is restored (1:2) by adding zeros. Because of interpolation, in the lower band, signals from 0 to $f_s/4$ will image around $f_s/4$ into the upper band. Similarly, in the upper band, signals from $f_s/4$ to $f_s/2$ will image to the lower band.

Quadrature Mirror Filters

Generally, when N subbands are created, each subband is sub-sampled at $1/N$ to maintain an overall sampling limit. We recall from Chap. 2, that the sampling frequency must be at least twice the bandwidth of a sampled signal. As noted, most filter banks do not provide ideal performance because of the finite width of their transition bands; the bands overlap, and the $1/N$ sub-sampling causes aliasing. Clearly, bands that are spaced apart can avoid this, but will leave gaps in the signal's spectrum. Quadrature mirror filter (QMF) banks have the property of reconstructing the original signal from N overlapping subbands without aliasing, regardless of the order of the bandpass filters. The aliasing components are exactly canceled, in the frequency domain, during reconstruction and the subbands are output in their proper place in frequency. The attenuation slopes of adjacent subband filters are mirror images of each other. Ideally, alias cancellation is perfect only if there is no requantization of the subband signals. A QMF is shown in Fig. 10.19. Intermediate samples can be critically sub-sampled without loss; if the input signal is split into N equal subbands, each subband can be sampled at $1/N$; the sampling frequency for each subband filter is exactly twice the bandwidth of the filter. Generally, cascades of QMF banks may be used to create 4 to 24 subbands. By cascading some subbands unequally, relative bandwidths can be manipulated; delays are introduced to maintain time parity between subbands.

QMF banks can be implemented as symmetrical finite impulse-response filters with an even number of taps; the use of a reconstruction highpass filter with a z-transform of $-H(-z)$ instead of $H(-z)$ eliminates alias terms (when there is uniform quantizing in each subband). However, perfect reconstruction is generally limited to the case when $N = 2$, creating two equal-width subbands from one. These $f_s/2$ subbands can be further divided by repeating the QMF process, and splitting each subband into two more subbands, each with a $f_s/4$ sampling frequency. This can be accomplished with a tree structure; however, this adds delay to the processing. Other QMF architectures can be used to create multiple subbands with less delay. However, cascaded QMF banks

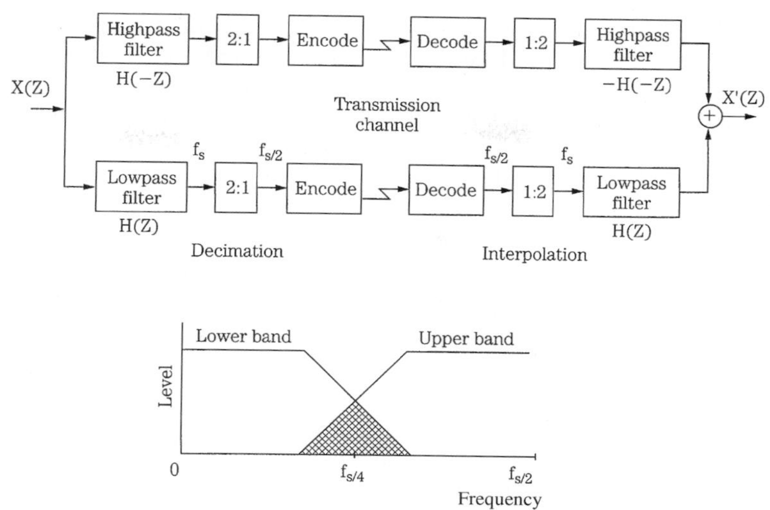

Figure 10.19 A quadrature mirror filter forms two equal subbands. Alias components introduced during decimation are exactly canceled during reconstruction. Multiple QMF stages can be cascaded to form additional subbands.

suitable for multi-band computation needed for codec design suffer from long delay and high complexity. For that reason, many codecs use a pseudo-QMF, also known as a polyphase filter, which offers a faster parallel approach that approximates the QMF. The QMF method is similar in concept to wavelet techniques.

Hybrid Filters

As noted, although filter banks with equally spaced bands are often used, they do not correlate well with the spacing of the ear's critical bands. Tree-structured filter banks can overcome this drawback. An example of a hybrid filter bank proposed by Karlheinz Brandenburg and James Johnston is shown in Fig. 10.20. It provides frequency analysis

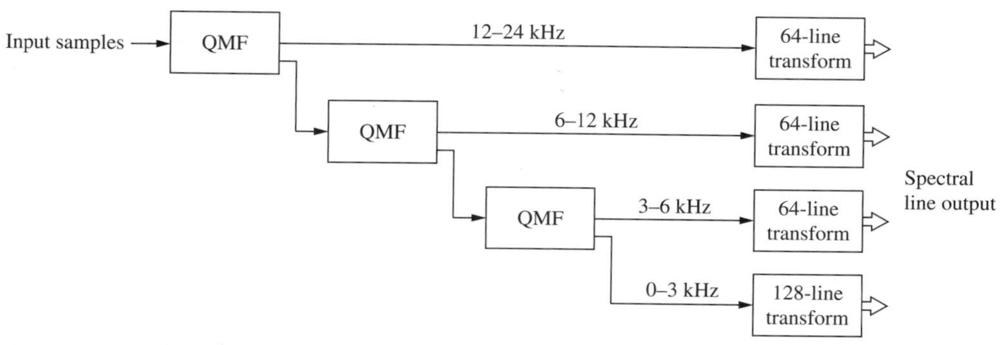

Figure 10.20 A hybrid filter using a QMF filter bank and transforms to yield unequally spaced spectral components. *(Brandenburg and Johnston, 1990)*

that corresponds more closely to critical-band spacing. Time-domain samples are applied to an 80-tap QMF filter bank to yield four bands with bandwidths of 3 kHz, 6 kHz, and 12 kHz. The bands are applied to a 64-line transform that is sine-windowed with a 50% overlap. The output of 320 spectral components has a frequency resolution of 23.4 Hz at low frequencies and 187.5 Hz at high frequencies. A corresponding synthesis filter is used in the decoder.

Polyphase Filters

A polyphase filter (also known as a pseudo-QMF or PQMF) yields a set of equally spaced bandwidth filters with phase interrelationships that permit very efficient implementation. A polyphase filter bank can be implemented as an FIR filter that consolidates interpolation filtering and decimation. The different filters are generated from a single lowpass prototype FIR filter by modulating the prototype filter over different phases in time. For example, using a 512-tap FIR filter, each block of input samples uses 32 sets of coefficients to create 32 subbands, as shown in Fig. 10.21. More generally, in an N-phase filter bank, each of the N subbands is decimated to a sampling frequency of $1/N$. For every N input, the filter outputs one value in each subband. In the first phase, the samples in an FIR transversal register and first-phase coefficients are used to compute the value in the first subband. In the second phase, the same samples in the register and second-phase coefficients are used to compute the value in the second subband, and so on. The impulse response coefficients in the first phase represent a prototype

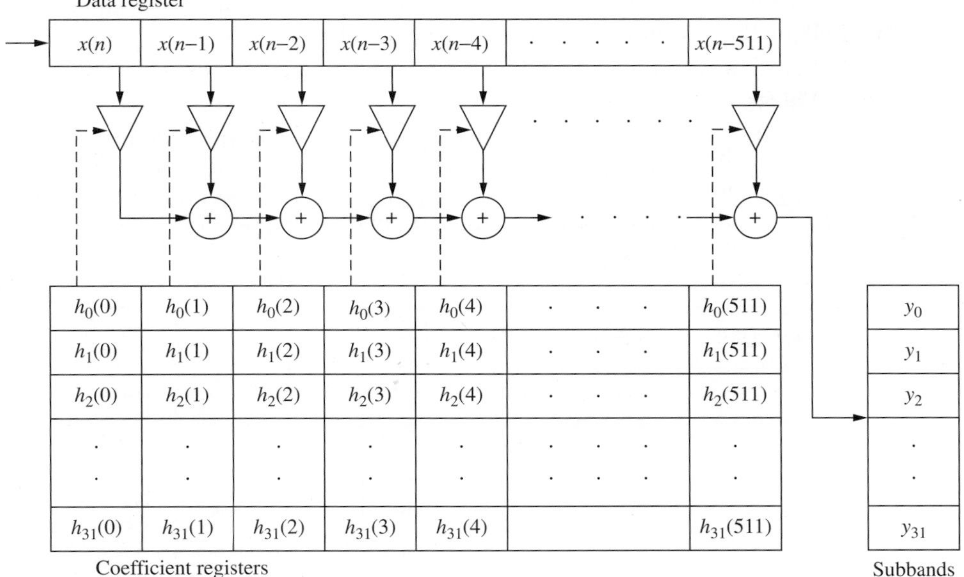

FIGURE 10.21 A polyphase filter uses a single lowpass prototype FIR filter to create subbands by modulating the prototype filter. This example shows a 512-tap filter with 32 sets of 512 coefficients, yielding 32 subbands.

lowpass filter, and the coefficients in the other subband phases are derived from the prototype by multiplication with a cosine modulating function that shifts the lowpass response to each next bandpass center frequency. After N samples are output, the samples in the register are shifted by N, and the process begins again. The center frequencies f_c are given by:

$$f_c = \pm\ f_s(k + 1/2)/2\text{N Hz}$$

where f_s is the sampling frequency and N is the number of subbands and $k = 0, 1, 2, \ldots, N-1$.

Each subband has a bandwidth of $f_s/64$. Polyphase filter banks offer good time-domain resolution, and good frequency resolution that can yield high stopband attenuation of 96 dB to minimize intraband aliasing. There is significant overlap in adjacent bands, but phase shifts in the cosine terms yield frequency-domain alias cancellation at the synthesis filter. The quantization process slightly degrades alias cancellation. The MPEG-1 and -2 standards specify a polyphase filter with QMF characteristics, operating as a block transform with $N = 32$ subbands and a 512-tap register. The flow chart of the MPEG-1 analysis filter and other details are given in Chap. 11. The particular analysis filter presented in the MPEG-1 standard requires approximately 80 real multiplies and 80 real additions for every output sample. Similarly, a polyphase synthesis filter is used in the decoder. The PQMF is a good choice to create a limited number of frequency bands. When higher-resolution analysis is needed, approaches such as an MDCT transform are preferred.

MDCT

The discrete Fourier transform (DFT) and discrete cosine transform (DCT) could be used in codecs to provide good frequency resolution. Also, the number of output points in the frequency domain equals the number of input samples in the time domain. However, to reduce blocking artifacts, an overlap-and-add of 50% doubles the number of frequency components relative to the number of time samples, and the DFT and DCT do not provide critical sampling. The increase in data rate is clearly counterproductive. The modified discrete cosine transform (MDCT) provides high-resolution frequency analysis and can allow a 50% overlap and provide critical sampling. As a result, it is used in many codecs.

The MDCT is an example of a time-domain aliasing cancellation (TDAC) transform in which only half of the frequency points are needed to reconstruct the time-domain audio signal. Frequency-domain sub-sampling is performed; thus, a 50% overlap results in the same number of output points as input samples. Specifically, the length of the overlapping windows is twice that of the block time (shift length of the transform) resulting in a 50% overlap between blocks. Moreover, the overlap-and-add process is designed to cancel time-domain aliasing and allow perfect reconstruction. The MDCT also allows adaptability of filter resolution by changing the window length. Windows such as a sine taper and Kaiser–Bessel can be used. The MDCT also lends itself to adaptive window switching approaches with different window functions for the first and second half of the window; the time-domain aliasing property must be independently valid for each window half. Many bands are possible with the MDCT with good efficiency, on the order of an FFT computation. The MDCT is also known as the modulated lapped transform (MLT).

As noted, a window function is applied to blocks of samples prior to transformation. If a block is input to a filter bank without a window, it can be considered as a time-limited signal with a rectangular window in which the samples in the block are multiplied by 1 and all other samples are multiplied by 0. The Fourier transform of this signal reveals that the sharp cutoff of the window's edges yields high-frequency content that would cause aliasing. To minimize this, a window can be applied that tapers the time-domain edge response down to zero. A window is thus a time function that is multiplied by an audio block to provide a windowed audio block. The window shape is selected to balance high-frequency resolution of the filter bank while minimizing spurious spectral components. The effect of the window must be compensated for to recover the original input signal after inverse-transformation. The overlap-and-add procedure accomplishes this by overlapping windowed blocks of the input signal and adding the result. The windows are designed specifically to allow this. Digital filters and windows are discussed in Chap. 17.

As noted, hybrid filter banks use a cascade of different filter types (such as polyphase and MDCT) to provide different frequency resolutions at different frequencies with moderate complexity. For example, MPEG-1 Layer III encoders use a hybrid filter with a polyphase filter bank and MDCT. The ATRAC algorithm is a hybrid codec that uses QMF to divide the signal into three subbands, and each subband is transformed into the frequency domain using the MDCT. Table 10.3 compares the properties of filter banks used in several low bit-rate codecs.

Feature	MPEG-1 Layer I	MPEG-1 Layer II	MPEG-1 Layer III	AC-2	AC-3	ATRAC[1]	PAC/ MPAC
Filter bank type	PQMF	PQMF	Hybrid PQMF/ MDCT	MDCT/ MDST	MDCT	Hybrid QMF/ MDCT	MDCT
Frequency resolution at 48 kHz	750 Hz	750 Hz	41.66 Hz	93.75 Hz	93.75 Hz	46.87 Hz	23.44 Hz
Time resolution at 48 kHz	0.66 ms	0.66 ms	4 ms	1.3 ms	2.66 ms	1.3 ms	2.66 ms
Impulse response (LW)	512	512	1664	512	512	1024	2048
Impulse response (SW)	—	—	896	128	256	128	256
Frame length at 48 kHz	8 ms	24 ms	24 ms	32 ms	32 ms	10.66 ms	23 ms

[1] ATRAC often operates at a sampling frequency at 44.1 kHz. For comparison, the frame length and impulse response figures are given for an ATRAC codec operating at 48 kHz.

Table 10.3 Comparison of filter-bank properties (48-kHz sampling frequency). (*Brandenburg and Bosi, 1997*)

Multichannel Coding

The problem of masking quantization noise is straightforward with a monaural audio signal; the masker and maskee are co-located. However, stereo and surround audio content also contain spatial localization cues. As a consequence, multichannel playback presents additional opportunities and obligations for low bit-rate algorithms. On one hand, for example, interchannel redundancies can be exploited by the coding algorithm to additionally and significantly reduce the necessary bit rate. On the other hand, great care must be taken because masking is spatially dependent. For example, quantization noise in one channel might be unmasked because its spatial placement differs from that of a masking signal in another channel. The human ear is fairly acute at spatial localization; for example, with the "cocktail-party effect" we are able to listen to one voice even when surrounded by a room filled with voices. The unmasking problem is particularly challenging because most masking studies assume monaural channels where the masker and maskee are spatially equal. In addition, perceptual effects are very different for loudspeaker versus headphone playback.

At frequencies above 2 kHz, within critical bands, we tend to localize sounds based on the temporal envelope of the signal rather than specific temporal details. Coding artifacts must be concealed in time and frequency, as well as in space. If stereo channels are improperly coded, coding characteristics of one channel might interfere with the other channel, creating stereo unmasking effects. For example, if a masking sum or difference signal is incorrectly processed in a matrixing operation and is output in the wrong channel, the noise in its original channel that it was supposed to mask might become audible.

Dual-mono coding uses two codecs operating independently. Joint-mono coding uses two monophonic codecs but they operate under the constraint of a single bit rate. Independent coding of multiple channels can create coding artifacts. For example, quantization noise might be audibly unmasked because it does not match the spatial placement of a stereo masking signal. Multichannel coding must consider subtle changes in the underlying psychoacoustics. For example, with the binaural masking level difference (BMLD) phenomenon, the masking threshold can be lower when listening with two ears, instead of one. At low frequencies (below 500 Hz), differences in phase between the masker and maskee at two ears can be readily audible.

Joint-stereo coding techniques use interchannel properties to take advantage of interchannel redundancy and irrelevance between stereo (or multiple) channels to increase efficiency. The data rate of a stereo signal is double that of a monaural signal, but most stereo programs are not dual-mono. For example, the channels usually share some level and phase information to create phantom images. This interchannel correlation is not apparent in the time domain, but it is readily apparent when analyzing magnitude values in the frequency domain. Joint-stereo coding codes common information only once, instead of twice as in left/right independent coding. A 256-kbps joint-stereo coding channel will perform better than two 128-kbps channels.

Many multichannel codecs use M/S (middle/side) coding of stereo signals to eliminate redundant monaural information. Coding efficiency can be high particularly with near-monaural signals and the technique performs well when there is a BMLD. With the M/S technique, rather than code discrete left and right channels, it is more efficient to code the middle (sum) and side (difference) signals using a matrix in either the time or frequency domains. The decoder uses reverse processing. In a multichannel codec, M/S coding can be applied to channel pairs placed left/right symmetrically to the listener. This configuration helps to avoid spatial unmasking.

As an alternative to M/S coding, intensity stereo coding (also known as dynamic crosstalk or channel coupling) can be used. High frequencies are primarily perceived as energy-time envelopes. With the intensity stereo technique, the energy-time envelope is coded rather than the waveform itself. One set of values can be efficiently coded and shared among multiple channels. Envelopes of individual channels can be reconstructed in the decoder by individually applying proper amplitude scaling. The technique is particularly effective at coding spatial information. In some codecs, different joint-stereo coding techniques are used in different spectral areas.

Using diverse and dynamically changing psychoacoustic cues and signal analysis, inaudible components can be removed with acceptable degradation. For example, a loud sound in one loudspeaker channel can mask other softer sounds in other channels. Above 2 kHz, localization is achieved primarily by amplitude; because the ear cannot follow fast individual waveform cycles, it tracks the envelope of the signal, not its phase. Thus the waveform itself becomes less critical; this is intensity localization. In addition, the ear is limited in its ability to localize sounds close in frequency. To convey a multichannel surround field, the high frequencies in each channel can be divided into bands and combined band by band into a composite channel. The bands of the common channel are reproduced from each loudspeaker, or panned between loudspeakers, with the original signal band envelopes. Use of a composite channel achieves data reduction. In addition, other masking principles can be applied prior to forming the composite channel. Many multichannel systems use a 5.1 format with three front channels, two independent surround channels, and a subwoofer channel. Very generally, the number of bits required to code a multichannel signal is proportional to the square root of the number of channels. A 5.1-channel codec, for example, would theoretically require only 2.26 times the number of bits needed to code one channel.

Tandem Codecs

In many applications, perceptual codecs will be used in tandem (cascaded). For example, a radio station may receive a coded signal, decode it to PCM for mixing, crossfading, and other operations, and then code it again for broadcast, where it is decoded by the consumer, who may make a coded recording of it. In all, many different coding processes may occur. As the signal passes through this chain, coding artifacts will accumulate and can become audible. The nature of the signal degrades the ability of subsequent codecs to suitably model the signal. Each codec will quantize the audio signal, adding to the quantization noise already permitted by previous encoders. Because many psychoacoustic models monitor the audio masking levels and not the underlying noise, noise can be allowed to rise to the point of audibility. Furthermore, when noise reaches audibility, the codec may allocate bits to code it, thus robbing bits needed elsewhere; this can increase noise in other areas.

In addition, tandem codecs are not necessarily synchronized, and will delineate audio data frames differently. This can yield audible noise-like artifacts and pre-echoes in the signal. When codecs are cascaded, it is important to begin with the highest-quality coding possible, then step down. A low-fidelity link will limit all subsequent processing. Highest-quality codecs have a high coding margin between the masking threshold, and coding noise; they tolerate more coding generations.

To reduce degradation caused by cascading, researchers are developing "inverse decoders" that analyze a decoded low bit-rate bitstream and extract the encoding parameters used to encode it. If the subsequent encoder is given the relevant coding decisions used by the previous encoder, and uses the same quantization parameters, the signal can be re-encoded very accurately with minimal loss compared to the first-generation coded signal. Conceivably, aspects such as type of codec, filter bank type, framing, spectral quantization, and stereo coding parameters can be extracted. Further, such systems could use the extracted parameters to ideally re-code the signal to its original bitstream representation, thus avoiding the distortion accumulated by repeatedly encoding/decoding content through the cascades in a signal chain. Some systems would accomplish this analysis and reconstruction task using only the decoded audio bitstream.

In some codec designs, metadata, sometimes called "mole" data because it is buried data that can burrow through the signal chain, is embedded in the audio data using steganographic means. Mole data such as header information, frame alignment, bit allocation, and scale factors may be conveyed. The mole data helps ensure that cascading does not introduce additional distortion. Ideally, the auxiliary information allows subsequent codecs to derive all the encoding parameters originally used in the first-generation encoding. This lets downstream tandem codecs apply the same framing boundaries, psychoacoustic modeling, and quantization processing as upstream stages. It is particularly important for the inverse decoder to use the same frame alignment as the original encoder so that the same sets of samples are available for processing.

With correct parameters, the cascaded output coded bitstream may be nearly identical to the input coded bitstream, within the tolerances of the filter banks. In another effort to reduce the effects of cascading, some systems allow operations to take place on the signal in the codec domain, without need for intermediate decoding to PCM and re-encoding. Such systems may allow gain-changing, fade-in and fade-outs, cross-fading, equalization, transcoding to a higher or lower bit rate, and transcoding to a different codec, as the signal passes through a signal chain.

Spectral Band Replication

Any perceptual codec must balance audio bandwidth, bit rate, and audible artifacts. By reducing coded audio bandwidth, relatively more bits are available in the remaining bandwidth. Spectral band replication (SBR) allows the underlying codec to reduce bit rate through bandwidth reduction, while providing bandwidth extension at the decoder. Spectral band replication is primarily a postprocess that occurs at the receiver. It extends the high-frequency range of the audio signal at the receiver. The lower part of the spectrum is transmitted. Higher frequencies are reconstructed by the SBR decoder based on the lower transmitted frequencies and control information. The replicated high-frequency signal is not analytically coherent compared to the baseband signal, but is coherent in a psychoacoustic sense; this is assisted by the fact that the ear is relatively less sensitive to variations at high frequencies.

The codec operates at half the nominal sampling frequency while the SBR algorithm operates at the full sampling frequency. The SBR encoder precedes the waveform encoder and uses QMF analysis and energy calculations to extract information that describes the spectral envelope of the signal by measuring energy in different bands. This process must be uniquely adaptive to its time and frequency analysis. For example, a transient signal might exhibit significant energy in the high band, but much less in the low band that will be conveyed by the codec and used for replication. The encoder also compares the original signal to the signal that the decoder will replicate. For example, tonal and nontonal aspects and the correlation between the low and high bands are analyzed. The encoder also considers what frequency ranges the underlying codec is coding. To assist the processing, control information is transmitted by the encoded bitstream at a very low bit rate. SBR can be employed for monaural, stereo, or multichannel encoding.

The SBR decoder follows the waveform decoder and uses the time-domain signal decoded by the underlying codec. This lowpass data is upsampled and applied to a QMF analysis filter bank and its subband signals (perhaps 32) are used for high-band replication based on the control information. In addition, the replicated high-band data is adaptively filtered and envelopment adjustment is applied to achieve suitable perceptual characteristics. The delayed low-band and high-band subbands are applied to a synthesis filter bank (perhaps 64 bands) operating at the SBR sampling frequency. In some cases, the encoder measures stereo correlation in the original signal and the decoder generates a corresponding pseudo-stereo signal. Replicated spectral components must be harmonically related to the baseband components to avoid dissonance.

At low and medium bit rates, SBR can improve the efficiency of a perceptual codec by as much as 30%. The improvement depends on the type of codec used. For example, when SBR is used with MP3 (as in MP3PRO), a 64-kbps stereo stream can achieve quality similar to that of a conventional MP3 96-kbps stereo stream. Generally, SBR is most effective when the codec bit rate is set to provide an acceptable level of artifacts in the restricted bandwidth. MP3PRO uses SBR in a backward- and forward-compatible way. Conventional MP3 players can decode an MP3PRO bitstream (without SBR) and MP3PRO decoders can decode MP3 and MP3PRO streams. SBR techniques can also be applied to Layer II codecs and MPEG-4 AAC codecs; the latter application is called High-Efficiency AAC (HE AAC) and aacPlus.

Low-complexity SBR can be accomplished with "blind" processing in which the decoder is not given control information. A nonlinear device such as a full-wave rectifier is used to generate harmonics, a filter selects the needed part of the signal, and gain is adjusted. This method assumes correlation between low- and high-frequency bands, thus its reconstruction is relatively inexact. Other SBR methods may extend low-frequency audio content by generating subharmonics, for example, in a voice signal passing through a telephone system that is bandlimited at low frequencies (as well as high frequencies). Other SBR methods are specifically designed to enhance sound quality through small loudspeakers. For example, a system might shift unreproducible low frequencies to higher frequencies above the speaker's cutoff, and rely on psychoacoustic effects such as residue pitch and virtual pitch to create the impression of low-frequency content.

Perceptual Coding Performance Evaluation

Whereas traditional audio coding is often a question of specifications and measurements, perceptual coding is one of physiology and perception. With the advent of digital signal processing, audio engineers can design hardware and software that "hears" sound the same way that humans hear sound.

The question of how to measure the sonic performance of perceptual codecs raises many issues that were never faced by traditional audio systems. Linear measurements might reveal some limitations, but cannot fully penetrate the question of the algorithm's perceptual accuracy. For example, a narrow band of noise introduced around a tone might not be audible, broadband white noise at the same energy would be plainly audible, but both would provide the same signal-to-noise measurement. Demonstrations of the so-called "13-dB miracle" by James Johnston and Karlheinz Brandenburg showed how noise that is shaped to hide under a narrow-band audio signal can be just barely noticeable or inaudible, yet the measured signal-to-noise ratio is only 13 dB. However, a wideband noise level with an S/N ratio of 60 dB in the presence of the same narrow-band audio signal would be plainly audible. As another example, a series of sine tones might provide a flat frequency response in a perceptual codec because the tones are easily coded. However, a broadband complex tone might be coded with a signal-dependent response.

Traditional audio devices are measured according to their small deviations from linearity. Perceptual codecs are highly nonlinear, as is their model, the human ear. The problem of how to determine the audibility of quantization noise levels and coding artifacts is not trivial. Indeed, the entire field of psychoacoustics must wrestle with the question of whether any objective or subjective measures can wholly quantify how an audible event affects a complex biological system—that is, a listener.

It is possible to nonquantitatively evaluate reduction artifacts using simple test equipment. A sine-wave oscillator can output test tones at a variety of frequencies; a dual-trace oscilloscope can display both the uncompressed and compressed waveforms. The waveforms should be time-aligned to compensate for processing delay in the codec. With a 16-bit system, viewed at 2 V peak-to-peak, one bit represents 30 μV. Any errors, including wideband noise and harmonic distortion at the one-bit level, can be observed. The idle channel signal performance can be observed with the coded output of a zero input signal; noise, error patterns, or glitches might appear. It also might be instructive to examine a low-level signal (0.1 V). In addition, a maximum level signal can be used to evaluate headroom at a variety of frequencies. More sophisticated evaluation can be performed with distortion and spectrum analyzers, but analysis is difficult. For example, traditional systems can be evaluated by measuring total harmonic distortion and noise (THD+N or SINAD) but such measurements are not meaningful for perceptual codecs. Figure 10.22 shows the spectral analysis of a 16-bit linear recorder and a 384-kbps perceptual recorder. When encoding a simple sine wave, although the perceptual codec adds noise within its masking curve around the tone, it easily codes the signal with low distortion. When using more complex stimuli, a perceptual codec will generate high distortion and noise measurements as anticipated by its inherent function. But, such measurements have little correlation to subjective performance. Clearly, traditional test tones and measurements are of limited use in evaluating perceptual codecs.

Richard Cabot devised a test that perceptually compares the codec output with a known steady-state test signal. A multi-tone signal is applied to the codec; the output is transformed into the frequency domain with FFT, and applied to an auditory model to

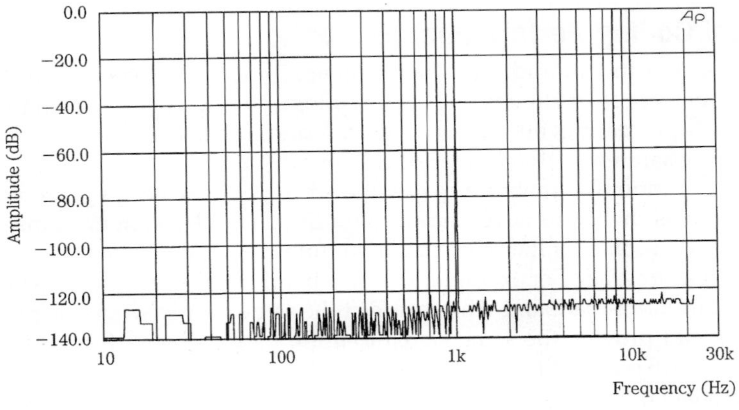

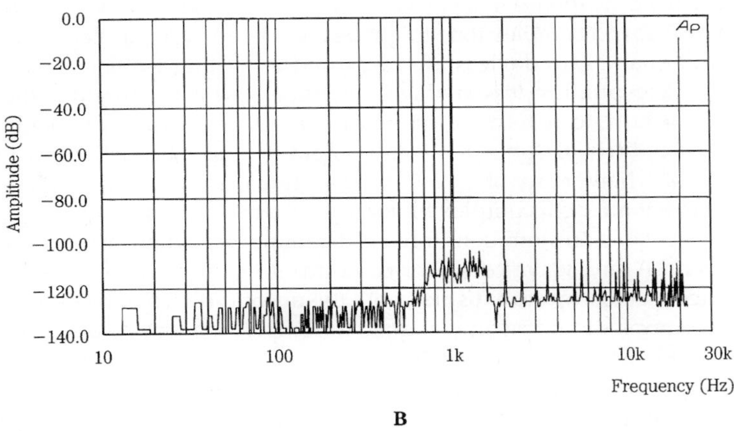

Figure 10.22 Spectral analysis of a single 1-kHz test tone reveals little about the performance of a perceptual codec. A. Analysis of a 16-bit linear PCM signal shows a noise floor that is 120 dB below signal. B. Analysis of a 384-kbps perceptually coded signal shows a noise floor that is 110 dB below signal with a slightly increased noise within the masking curve.

estimate masking effects of the signal. Because the spectrum of the test signal is known, any error products can be identified and measured. In particular, the error signal can be compared to internally modeled masking curves to estimate audibility. The NMR of the error signal level and masking threshold can be displayed as a function of frequency.

For example, Fig. 10.23A shows a steady-state test of a codec. The multi-tone signal consists of 26 sine waves distributed logarithmically across the audio spectrum, approximately one per critical band with a gap to allow for analysis of residual noise and distortion. It is designed to maximally load the codec's subbands, and consume its available bit rate. The figure shows the modeled masking curve based on the multi-tone, and distortion produced by the codec, consisting of quantization error, intermodulation products, and

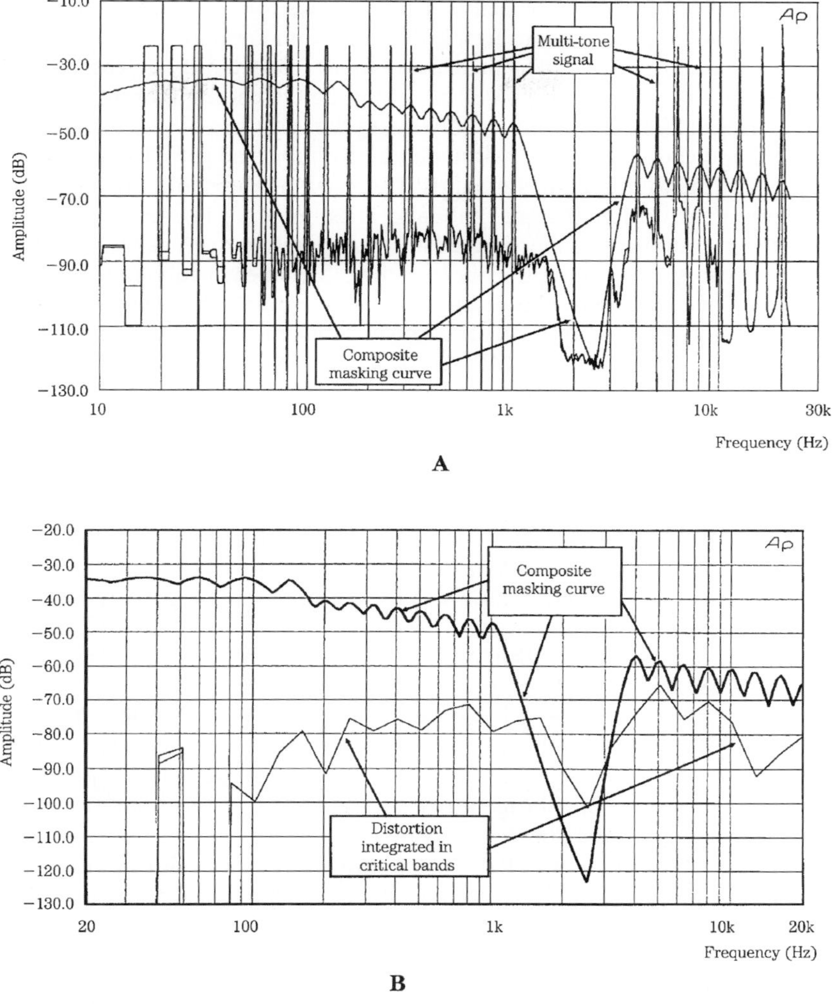

Figure 10.23 A multi-tone test signal is designed to deplete the bit capacity of a perceptual codec, emulating the action of a complex musical tone. A. Analysis shows the output test tone and the composite masking curve calculated by the testing device. B. The testing device computes the composite distortion in critical bands. Distortion above the calculated masking threshold might be audible.

noise sidebands. In Fig. 10.23B, the system has also integrated the distortion and noise products across the critical bandwidth of the human ear to simulate a perceived distortion level. Distortion products above the model's masking curve might be audible, depending on ambient listening conditions. In this case, some distortion might be audible at low levels. Clearly, evaluation of results depends on the sophistication of the analyzer's masking model. Multichannel test tones can be used to analyze interchannel performance, for example, to determine how high frequency information is combined between channels. This particular test does not evaluate temporal artifacts.

Theoretically, the best objective testing means for a perceptual codec is an artificial ear. To measure perceived accuracy, the algorithm contains a model that emulates the human hearing response. The measuring model can identify defects in the codec under test. A noise-to-mask ratio (NMR) can estimate the coding margin (see Fig. 10.11). The original signal (appropriately delayed) and the error signal (the difference between the original and coded signals) are independently subjected to FFT analysis, and the resulting spectra are divided into subbands. The masking threshold (maximum masked error energy) in each original signal subband is estimated. The actual error energy in each coded signal subband is determined and compared to the masking threshold. The ratio of error energy to masking threshold is the NMR in each subband.

A positive NMR would indicate an audible artifact. The NMR values can be linearly averaged and expressed in dB. This mean NMR measures remaining audibility headroom in the coded signal. NMR can be plotted over time to identify areas of coding difficulty. A masking flag, generated when the NMR exceeds 0 dB in a subband (artifact assumed to be audible) can be used to measure the number of impairments in a coded signal. A relative masking flag counts the number of impairments normalized over a number of blocks and subbands.

The ITU-R (International Telecommunication Union–Radiocommunication Bureau) Task Group 10/4 developed standardized objective perceptual measurement techniques that are described in a system called Perceptual Evaluation of Audio Quality (PEAQ). This system uses a multimode technique to compare an original signal to a processed signal and assess the perceived quality of the processed signal. Both signals are represented within an artificial ear and perceptually relevant differences are extracted and used to compute a quality measure. The standard describes both an FFT-based, and a filter bank-based ear algorithm; a simple implementation uses only the FFT-based model, while an advanced implementation uses elements of both. The system can evaluate criteria such as loudness of linear and nonlinear distortion, harmonic structure, masking ratios, and changes in modulation. For example, an NMR can specify the distance between a coded signal's error energy and the masked threshold. Mapping variables by a neural network yields quality measures that accurately correlate to the results of human testing. PEAQ is described in ITU-R Recommendation BS.1387-1, "Method for Objective Measurements of Perceived Audio Quality."

Use of an artificial ear to evaluate codec performance is limited by the quality of the reference model used in the ear. The codec's performance can only be as good as the artificial ear itself. Thus, testing with an artificial ear is inherently paradoxical. A reference ear that is superior to that in the psychoacoustic model of the codec under test would ideally be used to replace that in the codec. When the codec is supplied with the reference model, the codec could achieve "perfect" performance. Moreover, the criteria used in any artificial ear's reference model are arbitrary and inherently subjective because they are based on the goals of its designers.

Critical Listening

Ultimately, the best way to evaluate a perceptual codec is to exhaustively listen to it, using a large number of listeners. This kind of critical listening, when properly analyzed by objective means, is the gold standard for codec evaluation. In particular, the listening must be blind and use expert listeners, and appropriate statistical analysis must be performed to provide statistically confident results.

When a codec is not transparent, artifacts such as changes in timbre, bursts of noise, granular ambient sound, shifting in stereo imaging, and spatially unmasked noise can be used to identify the "signature" of the codec. Bandwidth reduction is also readily apparent, but a constant bandwidth reduction is less noticeable than a continually changing bandwidth. Changes in high-frequency content, such as from coefficients that come and go in successive transform blocks, create artifacts that are sometimes called "birdies." Speech is often a difficult test signal because its coding requires high resolution in both time and frequency. With low bit rates or long transform windows, coded speech can assume a strangely reverberant quality. A stereo or surround recording of audience applause sometimes reveals spatial coding errors. Subband codecs can have unmasked quantization noise that appears as a burst of noise in a processing block. In transform codecs, errors are reconstructed as basis functions (for example, a windowed cosine) of the codec's transform. A codec with a long block length can exhibit a pre-echo burst of noise just before a transient, or there might be a tinkling sound or a softened attack. Transform codec artifacts tend to be more audible at high frequencies. Changes in high-frequency bit allocation can result in a swirling sound due to changes in high-frequency timbre. In many cases, artifacts are discerned only after repeated listening trials, for example, after a codec has reached the marketplace.

For evaluation purposes, audio fidelity can be considered in four categories:

- *Large Impairments.* These sound quality differences are readily audible to even untrained listeners. For example, two identical speaker systems, one with normal tweeters and the other with tweeters disabled, would constitute a large impairment.
- *Medium Impairments.* These sound quality differences are audible to untrained listeners but may require more than casual listening. The ability to readily switch back and forth and directly compare two sources makes these impairments apparent. For example, stereo speakers with a midrange driver wired out of phase would constitute a medium impairment.
- *Small Impairments.* These sound quality differences are audible to many listeners, however, some training and practice may be necessary. For example, the fidelity difference between a music file coded at 128 kbps and 256 kbps, would reveal small impairments in the 128-kbps file. Impairments may be unique and not familiar to the listener, so they are more difficult to detect and take longer to detect.
- *Micro Impairments.* These sound quality differences are subtle and require patient listening by trained listeners over time. In many cases, the differences are not audible under normal listening conditions, with music played at normal levels. It may be necessary to amplify the music, or use test signals such as low-level sine tones and dithered silence. For example, slightly audible distortion on a −90 dBFS 1 kHz dithered sine wave would constitute a micro impairment.

When listening to large impairments such as from loudspeakers, audio quality evaluations can rely on familiar objective measurements and subjective terms to describe differences and find causes for defects. However, when comparing higher-fidelity devices such as codecs, smaller impairments are considerably more difficult to quantify. It is desirable to have methodologies to identify, categorize, and describe these subtle differences. Developing these will require training of critical listeners, ongoing listening evaluations, and discussions among listeners and codec designers to "close the loop." Further, to truly address the task, it will be necessary to systematically find thresholds of audibility of various defects. This search would introduce known defects and then determine the subjective audibility of the defects and thresholds of audibility.

It is desirable to correlate subjective impressions of listeners with objective design parameters. This would allow designers to know where audio fidelity limitations exist and thus know where improvements can be made. Likewise, this knowledge would allow bit rates to be lowered while knowing the effects on fidelity. There must be agreement on definitions of subjective terminology. This would provide language for listeners to use in their evaluations, it would bring uniformity to the language used by a broad audience of listeners, and it would provide a starting point in the objective qualification of their subjective comments.

The Holy Grail of subjective listening is the correlation between the listener's impressions, and the objective means to measure the phenomenon. This is a difficult problem. The reality is that correlations are not always known. The only way to correlate subjective impressions with objective data is with research—in particular, through critical listening. Over time, it is possible that patterns will emerge that will provide correlation. While correlation is desirable, critical listening continues to play an important role without it.

It is worth noting that within a codec type, for example, with MP3 codecs, the MPEG standard dictates that all compliant decoders should perform identically and sound the same. It is the encoders that may likely introduce sonic differences. However, some decoders may not properly implement the MPEG standard and thus are not compliant. For example, they may not support intensity stereo coding or variable rate bitstream decoding. MP3 encoders, for example, can differ significantly in audio performance depending on the psychoacoustic model, tuning of the nested iteration loops, and strategy for switching between long and short windows. Another factor is the joint-stereo coding method and how it is optimized for a particular number of channels, audio bandwidth and bit rate. Many codecs have a range of optimal bit rates; quality does not improve significantly above those rates, and quality can decrease dramatically below them. For example, MPEG Layer III is generally optimized for a bit rate of 128 kbps for a stereo signal at 48 kHz (1.33 bit/sample) whereas AAC is targeted at a bit rate of 96 kbps (1 bit/sample).

When differences between devices are small, one approach is to study the residue (difference signal) between them. The analog outputs could be applied to a high-quality A/D converter; one signal is inverted; the signals are precisely time-aligned with bit accuracy; the signals are added (subtracted because of inversion); and then the residue signal may be studied.

Expert listeners are preferred over average listeners because experts are more familiar with peculiar and subtle artifacts. An expert listener more reliably detects details and impairments that are not noticed by casual listeners. Listeners in any test should be trained on the testing procedure and in particular should listen to artifacts that the codec under test might exhibit. For example, listeners might start their training with very low bit-rate examples or left-minus-right signals, or residue signals with exposed artifacts so they become familiar with the codec's signature. It is generally felt that a 16-bit recording is not an adequate reference when testing high-quality perceptual codecs because many codecs can outperform the reference. The reference must be of the highest quality possible.

Many listening tests are conducted using high-quality headphones; this allows critical evaluation of subtle audible details. When closed-ear headphones are used, external conditions such as room acoustics and ambience noise can be eliminated. However, listening tests are better suited for loudspeaker playback. For example, this is necessary for multichannel evaluations. When loudspeaker playback is used, room acoustics play an important role in the evaluation. A proper listening room must provide suitable acoustics and also provide a low-noise environment including isolation from external noise. In some cases, rooms are designed and constructed according to standard reference criteria. Listening room standards are described below.

To yield useful conclusions, the results of any listening test must be subjected to accepted statistical analysis. For example, the ANOVA variance model is often used. Care must be taken to generate a valid analysis that has appropriate statistical significance. The number of listeners, the number of trials, the confidence interval, and other variables can all dramatically affect the validity of the conclusions. In many cases, several listening tests, designed from different perspectives, must be employed and analyzed to fully determine the quality of a codec. Statistical analysis is described below.

Listening Test Methodologies and Standards

A number of listening-test methodologies and standards have been developed. They can be followed rigorously, or used as practical guidelines for other testing. In addition, standards for listening-room acoustics have been developed.

Some listening tests can only ascertain whether a codec is perceptually transparent; that is, whether expert listeners can tell a difference between the original and the coded file, using test signals and a variety of music. In an ABX test, the listener is presented with the known A and B sources, and an unknown X source that can be either A or B; the assignment is pseudo-randomly made for each trial. The listener must identify whether X has been assigned to A or B. The test answers the question of whether the listener can hear a difference between A and B. ABX testing cannot be used to conclude that there is no difference; rather, it can show that a difference is heard. Short music examples (perhaps 15 to 20 seconds) can be auditioned repeatedly to identify artifacts. It is useful to analyze ABX test subjects individually, and report the number of subjects who heard a difference.

Other listening tests may be used to estimate the coding margin, or how much the bit rate can be reduced before transparency is lost. Other tests are designed to gauge relative transparency. This is clearly a more difficult task. If two low lossy codecs both exhibit audible noise and artifacts, only human subjectivity can determine which codec is preferable. Moreover, different listeners may have different preferences in this choice of the lesser of two evils. For example, one listener might be more troubled by bandwidth reduction while another is more annoyed by quantization noise.

Subjective listening tests can be conducted using the ITU-R Recommendation BS.1116-1. This methodology addresses selection of audio materials, performance of playback system, listening environment, assessment of listener expertise, grading scale, and methods of data analysis. For example, to reveal artifacts it is important to use audio materials that stress the algorithm under test. Moreover, because different algorithms respond differently, a variety of materials is needed, including materials that specifically stress each codec. Selected music must test known weaknesses in a codec to reveal flaws. Generally, music with transient, complex tones, rich in content around the ear's most sensitive region, 1 kHz to 5 kHz, is useful. Particularly challenging examples such as glockenspiel, castanets, triangle, harpsichord, tambourine, speech, trumpet, and bass guitar are often used.

Critical listening tests must use double-blind methods in which neither the tester nor the listener knows the identities of the selections. For example, in an "A-B-C triple-stimulus, hidden-reference, double-blind" test the listener is presented with a known

A uncoded reference signal, and two unknown B and C signals. Each stimulus is a recording of perhaps 10 to 15 seconds in duration. One of the unknown signals is identical to the known reference and the other is the coded signal under test. The assignment is made randomly and changes for each trial. The listener must assign a score to both unknown signals, rating them against the known reference. The listener can listen to any of the stimuli, with repeated hearings. Trials are repeated, and different stimuli are used. Headphones or loudspeakers can be used; sometimes one is more revealing than the other. The playback volume level should be fixed in a particular test for more consistent results. The scale shown in Fig. 10.24 can be used for scoring. This 5-point impairment scale was devised by the International Radio Consultative Committee (CCIR) and is often used for subjective evaluation of perceptual-coding algorithms. Panels of expert listeners rate the impairments they hear in codec algorithms on a 41-point continuous scale in categories from 5.0 (transparent) to 1.0 (very annoying impairments).

The signal selected by the listener as the hidden reference is given a default score of 5.0. Subtracting the score given to the actual hidden reference from the

Absolute grade		Difference grade
5.0	Imperceptible	0.0
4.9 4.8 4.7 4.6 4.5 4.4 4.3 4.2 4.1 4.0	Perceptible but NOT annoying	–0.1 –0.2 –0.3 –0.4 –0.5 –0.6 –0.7 –0.8 –0.9 –1.0
3.9 3.8 3.7 3.6 3.5 3.4 3.3 3.2 3.1 3.0	Slightly annoying	–1.1 –1.2 –1.3 –1.4 –1.5 –1.6 –1.7 –1.8 –1.9 –2.0
2.9 2.8 2.7 2.6 2.5 2.4 2.3 2.2 2.1 2.0	Annoying	–2.1 –2.2 –2.3 –2.4 –2.5 –2.6 –2.7 –2.8 –2.9 –3.0
1.9 1.8 1.7 1.6 1.5 1.4 1.3 1.2 1.1 1.0	Very annoying	–3.1 –3.2 –3.3 –3.4 –3.5 –3.6 –3.7 –3.8 –3.9 –4.0

Figure 10.24 The subjective quality scale specified by the ITU-R Rec. BS.1116 recommendation. This scale measures small impairments for absolute and differential grading.

score given to the impaired coded signal yields the subjective difference grade (SDG). For example, original, uncompressed material may receive an averaged score of 4.8 on the scale. If a codec obtains an average score of 4.8, the SDG is 0 and the codec is said to be transparent (subject to statistical analysis). If a codec is transparent, the bit rate may be reduced to determine the coding margin. A lower SDG score (for example, –2.6) assesses how far from transparency a codec is. Numerous statistical analysis techniques can be used. Perhaps 50 listeners are needed for good statistical results. Higher reduction ratios generally score less well. For example, Fig. 10.25 shows the results of a listening test evaluating an MPEG-2 AAC main profile codec at 256 kbps, with five full-bandwidth channels.

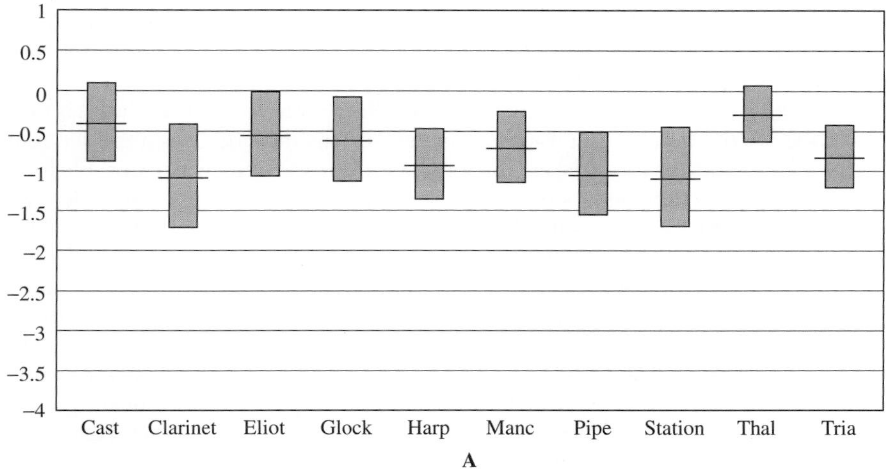

A

Name	Description
Cast	Castanets panned across front, noise in surround
Clarinet	Clarinet in center front, theater foyer ambience, rain on windows in surround
Eliot	Female and male speech in a restaurant, chamber music
Glock	Glockenspiel and timpani
Harp	Harpsichord
Manc	Orchestra—strings, cymbals, drums, horns
Pipe	Pitch pipe
Station	Male voice with steam-locomotive effects
Thal	Piano front left, sax front right, female voice center
Tria	Triangle

B

Figure 10.25 Results of listening tests for an AAC main profile codec at 256 kbps, five-channel mode showing mean scores and 95% confidence intervals. A. The vertical axis shows the AAC grades minus the reference signal grades. B. This table describes the audio tracks used in this test. (*ISO/IEC JTC1/SC29/WG-11 N1420, 1996*)

In another double-blind test conducted by Gilbert Soulodre using ITU-R guidelines, the worst-case tracks included a bass clarinet arpeggio, bowed double bass and harpsichord arpeggio (from an EBU SQAM CD), pitch pipe (Dolby recording), Dire Straits (Warner Brothers CD 7599-25264-2 Track 6), and a muted trumpet (University of Miami recording). In this test, when compared against a CD-quality reference, the AAC codec was judged best, followed by PAC, Layer III, AC-3, Layer II, and IT IS codecs, respectively. The highest audio quality was obtained by the AAC codec at 128 kbps and the AC-3 codec at 192 kbps per stereo pair. As expected, each codec performed relatively better at higher bit rates. In comparison to AAC, an increase in bit rate of 32, 64, and 96 kbps per stereo pair was required for PAC, AC-3, and Layer II codecs, respectively, to provide the same audio quality. Other factors such as computational complexity, sensitivity to bit errors, and compatibility to existing systems were not considered in this subjective listening test.

MUSHRA (MUltiple Stimulus with Hidden Reference and Anchors) is an evaluation method used when known impairments exist. This method uses a hidden reference and one or more hidden anchors; an anchor is a stimulus with a known audible limitation. For example, one of the anchors is a lowpass-coded signal. A continuous scale with five divisions is used to grade the stimuli: excellent, good, fair, poor, and bad. MUSHRA is specified in ITU-R BS.1534. Other issues in sound evaluation are described in ITU-T P.800, P.810, and P.830; ITU-R BS.562-3, BS.644-1, BS.1284, BS.1285, and BS.1286, among others.

In addition to listening-test methodology, the ITU-R Recommendation BS.1116-1 standard also describes a reference listening room. The BS.1116-1 specification recommends a floor area of 20 m^2 to 60 m^2 for monaural and stereo playback and an area of 30 m^2 to 70 m^2 for multichannel playback. For distribution of low-frequency standing waves, the standard recommends that room-dimension ratios meet these three criterion: $1.1(w/h) \leq (l/h) \leq 4.5(w/h) - 4$; $(w/h) < 3$; and $(l/h) < 3$ where l, w, h are the room's length, width, and height. The 1/3-octave sound pressure level, over a range of 50 Hz to 16,000 Hz, measured at the listening position with pink noise is defined by a standard-response contour. Average room reverberation time is specified to be: $0.25(V/V_0)^{1/3}$ where V is the listening room volume and V_0 is a reference volume of 100 m^3. This reverberation time is further specified to be relatively constant in the frequency range of 200 Hz to 4000 Hz, and to follow allowed variations between 63 Hz and 8000 Hz. Early boundary reflections in the range of 1000 Hz to 8000 Hz that arrive at the listening position within 15 ms must be attenuated by at least 10 dB relative to direct sound from the loudspeakers. It is recommended that the background noise level does not exceed ISO noise rating of NR10, with NR15 as a maximum limit.

The IEC 60268-13 specification (originally IEC 268-13) describes a residential-type listening room for loudspeaker evaluation. The specification is similar to the room described in the BS.1116-1 specification. The 60268-13 specification recommends a floor area of 25 m^2 to 40 m^2 for monaural and stereo playback and an area of 30 m^2 to 45 m^2 for multichannel playback. To spatially distribute low-frequency standing waves in the room, the specification recommends three criterion for room-dimension ratios: $(w/h) \leq (l/h) \leq 4.5(w/h) - 4$; $(w/h) < 3$; and $(l/h) < 3$ where l, w, h are the room's length, width, and height. The reverberation time (measured according to the ISO 3382 standard in 1/3-octave bands with the room unoccupied) is specified to fall within a range of 0.3 to 0.6 seconds in the frequency range of 200 Hz to 4000 Hz. Alternatively, average reverberation time should be 0.4 second and fall within a frequency contour given in the standard. The ambient noise level should not exceed NR15 (20 dBa to 25 dBA).

The EBU 3276 standard specifies a listening room with a floor area greater than 40 m^2 and a volume less than 300 m^3. Room-dimension ratios and reverberation time follow the BS.1116-1 specification. In addition, dimension ratios should differ by more than ±5%. Room response measured as a 1/3-octave response with pink noise follows a standard contour.

Listening Test Statistical Evaluation

As Mark Twain and others have said, "There are three kinds of lies: lies, damned lies, and statistics." To be meaningful, and not misleading, interpretation of listening test results must be carefully considered. For example, in an ABX test, if a listener correctly identifies the reference in 12 out of 16 trials, has an audible difference been noted? Statistic analysis provides the answer, or at least an interpretation of it. In this case, because the test is a sampling, we define our results in terms of probability. Thus, the larger the sampling, the more reliable the result. A central concern is the significance of the results. If the results are significant, they are due to audible differences. Otherwise they are due to chance. In an ABX test, a correct score 8 of 16 times indicates that the listener has not heard differences; the score could be arrived at by guessing. A score of 12/16 might indicate an audible difference, but could also be due to chance. To fathom this, we can define a null hypothesis H_0 that holds that the result is due to chance, and an alternate hypothesis H_1 that holds it is due to an audible difference. The significance level α is the probability that the score is due to chance. The criterion of significance α' is the chosen threshold of α that will be accepted. If α is less than or equal to α' then we accept that the probability is high enough to accept the hypothesis that the score is due to an audible difference. The selection of α' is arbitrary but a value of 0.05 is often used. Using this formula:

$$z = (c - 0.5 - np_1)/[np_1(1 - p_1)]^{1/2}$$

where z = standard normal deviate
c = number of correct responses
n = sample size
p_1 = proportion of correct responses in a population due to chance alone ($p_1 = 0.5$ in an ABX test)

We see that with a score of 12/16, $z = 1.75$. Binomial distribution thus yields a significance level of 0.038. The probability of getting a score as high as 12/16 from chance alone (and not from audible differences) is 3.8%. In other words, there is a 3.8% chance that the listener did not hear a difference. However, since α is less than α' ($0.038 < 0.05$) we conclude that the result is significant and there is an audible difference, at least according to how we have selected our criterion of significance. If α' is selected to be 0.01, then the same score of 12/16 is not significant and we would conclude that the score is due to chance.

We can also define parameters that characterize the risk that we are wrong in accepting a hypothesis. A Type 1 error risk (also often noted as α') is the risk of rejecting the null hypothesis when it is actually true. Its value is determined by the criterion of significance; if $\alpha' = 0.05$ then we will be wrong 5% of the time in assuming significant results. Type 2 error risk β defines the risk of accepting the null hypothesis when it is false. Type 2 risk is based on the sample size, value of α', the value of a chance score, and effect size or the smallest score that is meaningful. These values can be used to calculate sample size using the formula:

$$n = \{[z_1[p_1(1-p_1)]^{1/2} + z_2[p_2(1-p_2)]^{1/2}]/(p_2 - p_1)\}^2$$

where n = sample size

p_1 = proportion of correct responses in a population due to chance alone ($p_1 = 0.5$ in an ABX test)

p_2 = effect size: hypothesized proportion of correct responses in a population due to audible differences

z_1 = binomial distribution value corresponding to Type 1 error risk

z_2 = binomial distribution value corresponding to Type 2 error risk

For example, in an ABX test, if Type 1 risk is 0.05, Type 2 risk is 0.10, and effect size is 0.70, then the sample size should be 50 trials. The smaller the sample size, that is, the number of trials, the greater the error risks. For example, if 32 trials are conducted, $\alpha' = 0.05$, and the effect size is 0.70. To achieve a statistically significance result, a score of 22/32 is needed.

Binomial distribution analysis provides good results when a large number of samples are available. Other types of statistical analyses such as signal detection theory can also be applied to ABX testing. Finally, it is worth noting that statistical analysis can appear impressive, but its results cannot validate a test that is inherently flawed. In other words, we should never be blinded by science.

Lossless Data Compression

The principles of lossless data compression are quite different from those of perceptual lossy coding. Whereas perceptual coding operates mainly on data irrelevancy in the signal, data compression operates strictly on redundancy. Lossless compression yields a smaller coded file that can be used to recover the original signal with bit-for-bit accuracy. In other words, although the intermediate stored or transmitted file is smaller, the output file is identical to the input file. There is no change in the bit content, so there is no change in sound quality from coding. This differs from lossy coding where the output file is irrevocably changed in ways that may or may not be audible.

Some lossless codecs such as MLP (Meridian Lossless Packing) are used for standalone audio coding. Some lossy codecs such as MP3 use lossless compression methods such as Huffman coding in the encoder's output stage to further reduce the bit rate after perceptual coding. In either case, instead of using perceptual analysis, lossless compression examines a signal's entropy.

A newspaper with the headline "Dog Bites Man" might not elicit much attention. However, the headline "Man Bites Dog" might provoke considerable response. The former is commonplace, but the latter rarely happens. From an information standpoint, "Dog Bites Man" contains little information, but "Man Bites Dog" contains a large quantity of information. Generally, the lower the probability of occurrence of an event, the greater the information it contains. Looked at in another way, large amounts of information rarely occur.

The average amount of information occurring over time is called entropy, denoted as H. Looked at in another way, entropy measures an event's randomness and thus measures how much information is needed to describe it. When each event has the same probability of occurrence, entropy is maximum, and notated as H_{max}. Usually, entropy is less than this maximum value. When some events occur more often, entropy is lower. Most functions can be viewed in terms of their entropy. For example, the commodities market has high entropy, whereas the municipal bonds market has much lower entropy. Redundancy in a signal is obtained by subtracting from 1 the ratio of actual entropy to maximum entropy: $1 - (H/H_{max})$. Adding redundancy increases the data rate; decreasing redundancy decreases the rate: this is data compression, or lossless coding. An ideal compression system removes redundancy, leaving entropy unaffected; entropy determines the average number of bits needed to convey a digital signal. Further, a data set can be compressed by no more than its entropy value multiplied by the number of elements in the data set.

Entropy Coding

Entropy coding (also known as Huffman coding, variable-length coding, or optimum coding) is a form of lossless coding that is widely used in both audio and video applications. Entropy coding uses probability of occurrence to code a message. For example, a signal can be analyzed and samples that occur most often are assigned the shortest codewords. Samples that occur less frequently are assigned longer codewords. The decoder contains these assignments and reverses the process. The compression is lossless because no information is lost; the process is completely reversible.

The Morse telegraph code is a simple entropy code. The most commonly used character in the English language (e) is assigned the shortest code (.), and less frequently used characters (such as z) are assigned longer codes (_ _ ..). In practice, telegraph operators further improved transmission efficiency by dropping characters during coding and then replacing them during decoding. The information content remains unchanged. U CN RD THS SNTNCE, thanks to the fact that written English has low entropy; thus its data is readily compressed. Many text and data storage systems use data compression techniques prior to storage on digital media. Similarly, the abbreviations used in text messaging employ the same principles.

Generally, a Huffman code is a noiseless coding method that uses statistical techniques to represent a message with the shortest possible code length. A Huffman code provides coding gain if the symbols to be encoded occur with varying probability. It is an entropy code based on prefixes. To code the most frequent characters with the shortest codewords, the code uses a nonduplicating prefix system so that shorter codewords cannot form the beginning of a longer word. For example, 110 and 11011 cannot both be codewords. The code can thus be uniquely decoded, without loss.

Suppose we wish to transmit information about the arrival status of trains. Given four conditions, on time, late, early, and train wreck, we could use a fixed 2-bit codeword, assigning 00, 01, 10, and 11, respectively. However, a Huffman code considers the frequency of occurrence of source words. We observe that the probability is 0.5 that the train is on time, 0.35 that it is late, 0.125 that it is early, and 0.025 that it has wrecked. These probabilities are used to create a tree structure, with each node being the sum of its inputs, as shown in Fig. 10.26. Moreover, each branch is assigned a 0 or 1 value; the choice is arbitrary but must be consistent. A unique Huffman code is derived by following the tree from the 1.0 probability branch, back to each source word. For example, the code for early arrival is 110. In this way, a Huffman code is created so that the most probable status is coded with the shortest codeword and the less probable are coded

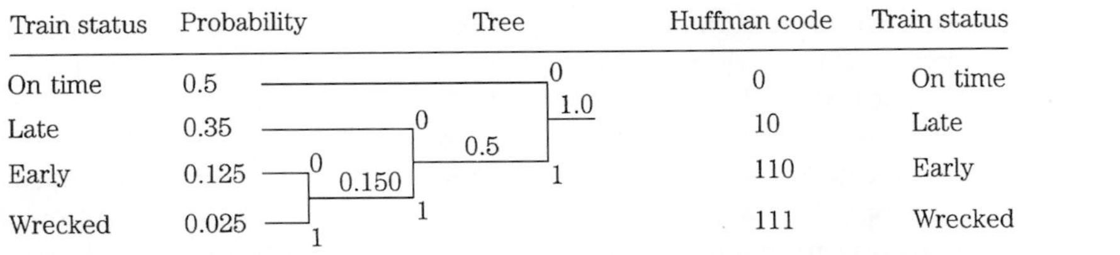

Train status	Probability	Tree	Huffman code	Train status
On time	0.5	0 1.0	0	On time
Late	0.35	0 0.5	10	Late
Early	0.125	0 0.150 1	110	Early
Wrecked	0.025	1 1	111	Wrecked

FIGURE 10.26 A Huffman code is based on a nonduplicating prefix, assigning the shorter codewords to the more frequently occurring events. If trains were usually on time, the code in this example would be particularly efficient.

with longer codewords. There is a reduction in the number of bits needed to indicate on-time arrival, even though there is an increase in the number of bits needed for two other statuses. Also note that prefixes are not repeated in the codewords.

The success of the code is gauged by calculating its average code length; it is the summation of each codeword length multiplied by its frequency of occurrence. In this example, the 1-bit word has a probability of 0.5, the 2-bit words have a probability of 0.35, and the 3-bit words have a combined probability of 0.15; thus the average code length is 1(0.5) + 2(0.35) + 3(0.15) = 1.65 bits. This compares favorably with the 2-bit fixed code, and approaches the entropy of the message. A Huffman code is suited for some messages, but only when the frequency of occurrence is known beforehand. If the relative frequency of occurrence of the source words is approximately equal, the code is not efficient. If an infrequent source word's probability approaches 1 (becomes frequent), the code will generate coded messages longer than the original. To overcome this, some coding systems use adaptive measures that modify the compression algorithm for more optimal operation. The Huffman code is optimal when all symbols have a probability that is an integral power of one-half.

Run-length coding also provides data compression, and is optimal for highly frequent samples. When a data value is repeated over time, it can be coded with a special code that indicates the start and stop of the string. For example, the message 6666 6666 might be coded as 86. This coding is efficient; run-length coding is used in fax machines, for example, and explains why blank sections of a page are transmitted more quickly than densely written sections. Although Huffman and run-length codes are not directly efficient for music coding by themselves, they are used for compression within some lossless and lossy algorithms.

Audio Data Compression

Perceptual lossy coding can provide a considerable reduction in bit rates. However, whether audible or not, the signal is degraded. With lossless data compression, the signal is delivered with bit-for-bit accuracy. However, the decrease in bit rate is more modest. Generally, compression ratios of 1.5:1 to 3.5:1 are possible, depending on the complexity of the data itself. Also, lossless compression algorithms may require greater processing complexity with the attendant coding delay.

Every audio signal contains information. A rare audio sample contains considerable information; a frequently occurring sample has much less. The former is hard to predict while the latter is readily predictable. Similarly, a tonal (sinusoidal) sound has considerable redundancy whereas a nontonal (noise-like) signal has little redundancy. For example, a quasi-periodic violin tone would differ from an aperiodic cymbal crash.

Further, the probability of a certain sample occurring depends on its neighboring samples. Generally, a sample is likely to be close in value to the previous sample. For example, this is true of a low-frequency signal. A predictive coder uses previous sample values to predict the current value. The error in the prediction (difference between the actual and predicted values) is transmitted. The decoder forms the same predicted value and adds the error value to form the correct value.

To achieve its goal, data compression inputs a PCM signal and applies processing to more efficiently pack the data content prior to storage or transmission. The efficiency of the packing depends greatly on the content of the signal itself. Specifically, signals with greater redundancy in their PCM coding will allow a higher level of compression. For that reason, a system allowing a variable output bit rate will yield greater efficiency than one with a fixed bit rate. On the other hand, any compression method must observe a system's maximum bit rate and ensure that the threshold is never exceeded even during low-redundancy (hard to compress) passages.

PCM coding at a 20-bit resolution, for example, always results in words that are 20 bits long. A lossless compression algorithm scrutinizes the words for redundancy and then reformats the words to shorter lengths. On decompression, a reverse process restores the original words. Peter Craven and Michael Gerzon suggest the example of a 20-bit word length file representing an undithered 4-kHz sine wave at –50 dB below peak level, sampled at 48 kHz. Moreover, a block of 12 samples is considered, as shown in Table 10.4. The file size is 240 bits. Observation shows that in each sample the four LSBs (least significant bits) are zero; an encoder could document that only the 16 MSBs (most significant bits) will be transmitted or stored. This is easily accomplished by right-justifying the data and then coding the shift count. Furthermore, the 9 MSBs in

Sample Number	Binary Value
1	00000000010000110000
2	00000000011000010000
3	00000000011001100000
4	00000000010011110000
5	00000000001000110000
6	11111111111011100000
7	11111111101111010000
8	11111111100111110000
9	11111111100110100000
10	11111111101100010000
11	11111111110111010000
12	00000000000100100000

TABLE 10.4 Twelve samples taken from a 20-bit audio file, showing limited dynamic range and resolution. In this case, simple data compression techniques can be applied to achieve a 60% decrease in file size. (*Craven and Gerzon, 1996*)

each sample of this low-level signal are all 1s or 0s; the encoder can simply code 1 of the 9 bits and use it to convey the other missing bits. With these measures, because of the signal's limited dynamic range and resolution, the 20-bit words are conveyed as 8-bit words, resulting in a 60% decrease in data. Note that if the signal were dithered, the dither bit(s) would be conveyed with bit-accuracy by a lossless coder, reducing data efficiency.

In practice, a block size of about 500 samples (or 10 ms) may be used, with descriptive information placed in a header file for each block. The block length may vary depending on signal conditions. Generally, because transients will stimulate higher MSBs, longer blocks cannot compress short periods of silence in the block. Shorter blocks will have relatively greater overhead in their headers. Such simple scrutiny may be successful for music with soft passages, but not successful for loud, highly compressed music. Moreover, the peak data rate will not be compressed in either case. In some cases, a data block might contain a few audio peaks. Relatively few high-amplitude samples would require long word lengths, while all the other samples would have short word lengths. Huffman coding (perhaps using a lookup table) can be used to overcome this. The common low-amplitude samples would be coded with short codewords, while the less common high-amplitude samples would be coded with longer codewords. To further improve performance, multiple codeword lookup tables can be established and selected based on the distribution of values in the current block. Audio waveforms tend to follow amplitude statistics that are Laplacian, and appropriate Huffman tables can reduce the bit rate by about 1.5-bit/sample/channel compared to a simple word-length reduction scheme.

A predictive strategy can yield greater coding efficiency. In the previous example, the 16-bit numbers have decimal values of +67, +97, +102, +79, +35, –18, –67, –97, –102, –79, –35, and +18. The differences between successive samples are +30, +5, –23, –44, –53, –49, –30, –5, +23, +44, and +53. A coder could transmit the first value of +67 and then the subsequent differences between samples; because the differences are smaller than the sample values themselves, shorter word lengths (7 bits instead of 8) are needed. This coding can be achieved with a simple predictive encode-decode strategy as shown in Fig. 10.27 where the symbol z^{-1} denotes a one-sample delay. If the value +67 has been previously entered, and the next input value is +97, the previous sample value of +67 is used as the predicted value of the current sample, the prediction error becomes +30,

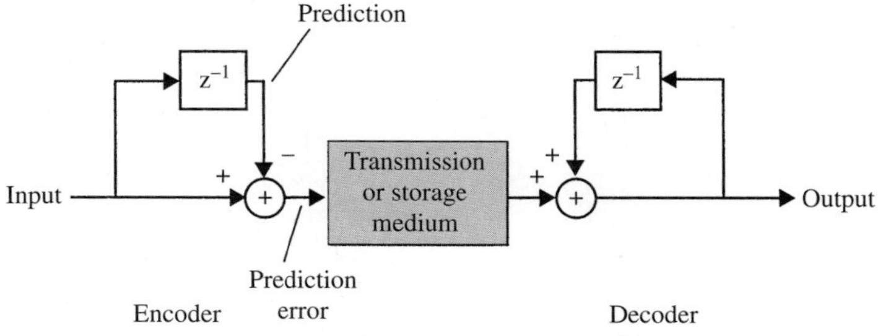

FIGURE 10.27 A first-order predictive encode/decode process conveys differences between successive samples. This improves coding efficiency because the differences are smaller than the values themselves. (*Craven and Gerzon, 1996*)

which is transmitted. The decoder accepts the value of +30 and adds it to the previous value of +67 to reproduce the current value of +97.

The goal of a prediction coder is to predict the next sample as accurately as possible, and thus minimize the number of bits needed to transmit the prediction error. To achieve this, the frequency response of the encoder should be the inverse of the spectrum of the input signal, yielding a difference signal with a flat or white spectrum. To provide greater efficiency, the one-sample delay element in the predictor coder can be replaced by more advanced general prediction filters. The coder with a one-sample delay is a digital differentiator with a transfer function of $(1 - z^{-1})$. An nth order predictor yields a transfer function of $(1 - z^{-1})^n$, where $n = 0$ transmits the original value, $n = 1$ transmits the difference between successive samples, $n = 2$ transmits the difference of the difference, and so on. Each higher-order integer coefficient produces an upward filter slope of 6, 12, and 18 dB/octave. Analysis shows that $n = 4$ may be optimal, yielding a maximum difference of 10. However, the high-frequency content in audio signals limits the order of the predictor. The high-frequency component of the quantization noise is increased by higher-order predictors; thus a value of $n = 3$ is probably the limit for audio signals. But if the signal had mainly high-frequency content (such as from noise shaping), even an $n = 1$ value could increase the coded data rate. Thus, a coder must dynamically monitor the signal content and select a predictive strategy and filter order that is most suitable, including the option of bypassing its own coding, to minimize the output bit rate. For example, an autocorrelation method using the Levinson–Durbin algorithm could be used to adapt the predictor's order, to yield the lowest total bit rate.

A coder must also consider the effect of data errors. Because of the recirculation, an error in a transmitted sample would propagate through a block and possibly increase, even causing the decoder to lose synchronization with the encoder. To prevent artifacts, audible or otherwise, an encoder must sense uncorrected errors and mute its output. In many applications, while overall reduction in bit rate is important, limitation of peak bit rate may be even more vital. An audio signal such as a cymbal crash, with high energy at high frequencies, may allow only slight reduction (perhaps 1 or 2 bits/sample/channel). Higher sampling frequencies will allow greater overall reduction and peak reduction because of the relatively little energy at the higher portion of the band. To further ensure peak limits, a buffer could be used. Still, the peak limit could be exceeded with some kinds of music, necessitating the shortening of word length or other processing.

The simple integer coefficient predictors described above provide upward slopes that are not always a good (inverse) match for the spectra of real audio signals. The spectrum of the difference signal is thus nonflat, requiring more bits for coding. Every 6-dB reduction in the level of the transmitted signal reduces its bit rate by 1 bit/sample. More successful coding can be achieved with more sophisticated prediction filters using, for example, noninteger-coefficient filters in the prediction loop. The transmitted signal must be quantized to an integer number of LSB steps to achieve a fixed bit rate. However, with noninteger coefficients, the output has a fractional value of LSBs. To quantize the prediction signal, the architecture shown in Fig. 10.28 may be employed. The decoder restores the original signal values by simply quantizing the output.

Different filters can be used to create a variety of equalization curves for the prediction error signal, to match different signal spectral characteristics. Different 3rd-order IIR filters, when applied to different signal conditions, may provide bit-rate reduction ranging from 2 to 4 bits, even in cases where the bit rate would be increased with simple

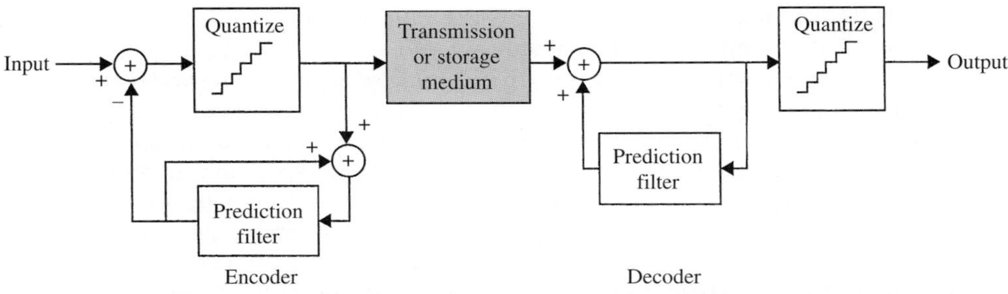

Figure 10.28 Noninteger-coefficient filters can be used in a prediction encoder/decoder. The prediction signal is quantized to an integer number of LSB steps. (*Craven and Gerzon, 1996*)

integer predictors. Higher-order filters increase the amount of overhead data such as state variables that must be transmitted with each block to the decoder; this argues for lower-order filters. It can also be argued that IIR filters are more appropriate than FIR filters because they can more easily achieve the variations found in music spectra. On the other hand, to preserve bit accuracy, it is vital that the filter computations in any decoder match those in any encoder. Any rounding errors, for example, could affect bit accuracy. In that respect, because IIR computation is more sensitive to rounding errors, the use of IIR predictor filters demands greater care.

Because most music spectral content continually varies, filter selection must be re-evaluated for each new block. Using some means, the signal's spectral content must be analyzed, and the most appropriate filter employed, by either creating a new filter characteristic or selecting one from a library of existing possibilities. Information identifying the encoding filter must be conveyed to the decoder, increasing the overhead bit rate. Clearly, processing complexity and data overhead must be weighed against coding efficiency.

As noted, lossless compression is effective at very high sampling frequencies in which the audio content at the upper frequency ranges of the audio band is low. Bit accuracy across the wide audio band is ensured, but very high-frequency information comprising only dither and quantization noise can be more efficiently coded. Craven and Gerzon estimate that whereas increasing the sampling rate of an unpacked file from 64 kHz to 96 kHz would increase the bit rate by 50%, a packed file would increase the bit rate by only 15%. Moreover, low-frequency effects channels do not require special handling; the packing will ensure a low bit rate for its low-frequency content. Very generally, at a given sampling frequency, the bit-rate reduction achieved is proportional to the input word length and is greater for low-precision signals. For example, if the average bit reduction is 9 bits/sample/channel, then a 16-bit PCM signal is coded as 7 bits (56% reduction), a 20-bit signal as 11 bits (45% reduction), and a 24-bit signal as 15 bits (37.5% reduction). Very generally, each additional bit of precision in the input signal adds a bit to the word length of the packed signal.

At the encoder's output, the difference signal data can be Huffman-coded and transmitted as main data along with overhead information. While it would be possible to hardwire filter coefficients into the encoder and decoder, it may be more expedient to explicitly transmit filter coefficients along with the data. In this way, improvements can be made in filter selection in the encoder, while retaining compatibility with existing decoders.

As with lossy codecs, lossless codecs can take advantage of interchannel correlations in stereo and multichannel recordings. For example, a codec might code the left channel, and frame-adaptively code either the right channel or the difference between the right and left channels, depending on which yields the highest coding gain. More efficiently, stereo prediction methods use previous samples from both channels to optimize the prediction.

Because no psychoacoustic principles such as masking are used in lossless coding, practical development of transparent codecs is relatively much simpler. For example, subjective testing is not needed. Transparency is inherent in the lossless codec. However, as with any digital-processing system, other aspects such as timing and jitter must be carefully engineered.

CHAPTER 11

Low Bit-Rate Coding: Codec Design

In the view of many observers, compared to newer coding methods, linear pulse-code modulation (PCM) is a powerful but inefficient dinosaur. Because of its gargantuan appetite for bits, PCM coding is not suitable for many audio applications. There is an intense desire to achieve lower bit rates because low bit-rate coding opens so many new applications for digital audio (and video). Responding to the need, audio engineers have devised many lossy and lossless codecs. Some codecs use proprietary designs that are kept secret, some are described in standards that can be licensed, while others are open source. In any case, it would be difficult to overstate the importance of low bit-rate codecs. Codecs can be found in countless products used in everyday life, and their development is largely responsible for the rapidly expanding use of digital audio techniques in storage and transmission applications.

Early Codecs

Although the history of perceptual codecs is relatively brief, several important coding methods have been developed, which in turn inspired the development of more advanced methods. Because of the rapid development of the field, most early codecs are no longer widely used, but they established methods and benchmarks on which modern codecs are based.

MUSICAM (Masking pattern adapted Universal Subband Integrated Coding And Multiplexing) was an early perceptual coding algorithm that achieved data reduction based on subband analysis and psychoacoustic principles. Derived from MASCAM (Masking pattern Adapted Subband Coding And Multiplexing), MUSICAM divides the input audio signal into 32 subbands with a polyphase filter bank. With a sampling frequency of 48 kHz, the subbands are each 750 Hz wide. A fast Fourier transform (FFT) analysis supplies spectral data to a perceptual coding model; it uses the absolute hearing threshold and masking to calculate the minimum signal-to-mask ratio (SMR) value in each subband. Each subband is given a 6-bit scale factor according to the peak value in the subband's 12 samples and quantized with a variable word ranging from 0 to 15 bits. Scale factors are calculated over a 24-ms interval, corresponding to 36 samples. A subband is quantized only if it contains audible signals above the masking threshold. Subbands with signals well above the threshold are coded with more bits. In other words, within a given bit rate, bits are assigned where they are most needed. The data rate is reduced to perhaps 128 kbps per monaural channel (256 kbps for stereo). Extensive tests of 128 kbps

MUSICAM showed that the codec achieves fidelity that is indistinguishable from a CD source, that it is monophonically compatible, that at least two cascaded codec stages produce no audible degradation, and that it is preferred to very high-quality FM signals. In addition, a bit-error rate of up to 10^{-3} was nearly imperceptible. MUSICAM was developed by CCETT, IRT, Matsushita, and Philips.

OCF (Optimal Coding in the Frequency domain) and PXFM (Perceptual Transform Coding) are similar perceptual transform codecs. A later version of OCF uses a modified discrete cosine transform (MDCT) with a block length of 512 samples and a 1024-sample window. PXFM uses an FFT with a block length of 2048 samples and an overlap of 1/16. PXFM uses critical-band analysis of the signal's power spectrum, tonality estimation, and a spreading function to calculate the masking threshold. PXFM uses a rate loop to optimize quantization. A stereo version of PXFM further takes advantage of correlation in the frequency domain between left and right channels. OCF uses an analysis-by-synthesis method with two iteration loops. An outer (distortion) loop adjusts quantization step size to ensure that quantization noise is below the masking threshold in each critical band. An inner (rate) loop uses a nonuniform quantizer and Huffman coding to optimize the word length needed to quantize spectral values. OCF was devised by Karlheinz Brandenburg in 1987, and PXFM was devised by James Johnston in 1988.

The ASPEC (Audio Spectral Perceptual Entropy Coding) standard described a MDCT transform codec with relatively high complexity and the ability to code audio for low bit-rate applications such as ISDN. ASPEC was developed jointly using work by AT&T Bell Laboratories, Fraunhofer Institute, Thomson, and CNET.

MPEG-1 Audio Standard

The International Organization for Standardization (ISO) and the International Electrotechnical Commission (IEC) formed the Moving Picture Experts Group (MPEG) in 1988 to devise data reduction techniques for audio and video. MPEG is a working group of the ISO/IEC and is formally known as ISO/IEC JTC 1/SC 29/WG 11; MPEG documents are published under this nomenclature. The MPEG group has developed several codec standards. It first devised the ISO/IEC International Standard 11172 "Coding of Moving Pictures and Associated Audio for Digital Storage Media at up to about 1.5 Mbit/s" for reduced data rate coding of digital video and audio signals; the standard was finalized in November 1992. It is commonly known as MPEG-1 (the acronym is pronounced "m-peg") and was the first international standard for the perceptual coding of high-quality audio.

The MPEG-1 standard has three major parts: system (multiplexed video and audio), video, and audio; a fourth part defines conformance testing. The maximum audio bit rate is set at 1.856 Mbps. The audio portion of the standard (11172-3) has found many applications. It supports coding of 32-, 44.1-, and 48-kHz PCM input data and output bit rates ranging from approximately 32 kbps to 224 kbps/channel (64 kbps to 448 kbps for stereo). Because data networks use data rates of 64 kbps (8 bits sampled at 8 kHz), most codecs output a data channel rate that is a multiple of 64.

The MPEG-1 standard was originally developed to support audio and video coding for CD playback within the CD's bandwidth of 1.41 Mbps. However, the audio standard supports a range of bit rates as well as monaural coding, dual-channel monaural coding, and stereo coding. In addition, in the joint-stereo mode, stereophonic irrelevance and redundancy can be optionally exploited to reduce the bit rate. Stereo audio bit rates below 256 kbps are useful for applications requiring more than two audio channels while maintaining full-screen motion video. Rates above 256 kbps are useful for applications requiring higher audio quality, and partial screen video images. In either case, the bit allocation is dynamically adaptable according to need. The MPEG-1 audio standard is based on data reduction algorithms such as MUSICAM and ASPEC.

Development of the audio portion of the MPEG-1 audio standard was greatly influenced by tests conducted by Swedish Radio in July 1990. MUSICAM coding was judged superior in complexity and coding delay. However, the ASPEC transform codec provided superior sound quality at low data rates. The architectures of the MUSICAM and ASPEC coding methods formed the basis for the ISO/MPEG-1 audio standard with MUSICAM describing Layers I and II and ASPEC describing Layer III. The 11172-3 standard describes three layers of audio coding, each with different applications. Specifically, Layer I describes the least sophisticated method that requires relatively high data rates (approximately 192 kbps/channel). Layer II is based on Layer I but is more complex and operates at somewhat lower data rates (approximately 96 kbps to 128 kbps/channel). Layer IIA is a joint-stereo version operating at 128 kbps and 192 kbps per stereo pair. Layer III is somewhat conceptually different from I and II, is the most sophisticated, and operates at the lowest data rate (approximately 64 kbps/channel). The increased complexity from Layer I to III is reflected in the fact that at low data rates, Layer III will perform best for audio fidelity. Generally, Layers II, IIA, and III have been judged to be acceptable for some broadcast applications; in other words, operation at 128 kbps/channel does not impair the quality of the original audio signal. The three layers (I, II, and III) all refer to audio coding and should not be confused with different MPEG standards such as MPEG-1 and MPEG-2.

In very general terms, all three layer codecs operate similarly. The audio signal passes through a filter bank and is analyzed in the frequency domain. The sub-sampled components are regarded as subband values, or spectral coefficients. The output of a side-chain transform, or the filter bank itself, is used to estimate masking thresholds. The subband values or spectral coefficients are quantized according to a psychoacoustic model. Coded mapped samples and bit allocation information are packed into frames prior to transmission. In each case, the encoders are not defined by the MPEG-1 standard, only the decoders are specified. This forward-adaptive bit allocation permits improvements in encoding methods, particularly in the psychoacoustic modeling, provided the data output from the encoder can be decoded according to the standard. In other words, existing codecs will play data from improved encoders.

The MPEG-1 layers support joint-stereo coding using intensity coding. Left/right high-frequency subband samples are summed into one channel but scale factors remain left/right independent. The decoder forms the envelopes of the original left and right channels using the scale factors. The spectral shape of the left and right channels is the same in these upper subbands, but their amplitudes differ. The bound for joint coding is selectable at four frequencies: 3, 6, 9, and 12 kHz at a 48-kHz sampling frequency; the bound can be changed from one frame to another. Care must be taken to avoid aliasing between subbands and negative correlation between channels when joint coding. Layer III also supports M/S sum and difference coding between channels, as described below. Joint stereo coding increases codec complexity only slightly.

Listening tests demonstrated that either Layer II or III at 2 × 128 kbps or 192 kbps joint stereo can convey a stereo audio program with no audible degradation compared to a 16-bit PCM coding. If a higher data rate of 384 kbps is allowed, Layer I also achieves transparency compared to 16-bit PCM. At rates as low as 128 kbps, Layers II and III can convey stereo material that is subjectively very close to 16-bit fidelity. Tests also have studied the effects of cascading MPEG codecs. For example, in one experiment, critical audio material was passed through four Layer II codec stages at 192 kbps and two stages at 128 kbps, and they were found to be transparent. On the other hand, a cascade of five codec stages at 128 kbps was not transparent for all music programs. More specifically, a source reduced to 384 kbps with MPEG-1 Layer II sustained about 15 code/decodes before noise became significant; however, at 192 kbps, only two codings were possible. These particular tests did not enjoy the benefit of joint-stereo coding, and as with other perceptual codecs, performance can be improved by substituting new psychoacoustic models in the encoder.

The similarity between the MPEG-1 layers promotes tandem operation. For example, Layer III data can be transcoded to Layer II without returning to the analog domain (other digital processing is required, however). A full MPEG-1 decoder must be able to decode its layer, and all layers below it. There are also Layer X codecs that only code one layer. Layer I preserves highest fidelity for acquisition and production work at high bit rates where six or more codings can take place. Layer II distributes programs efficiently where two codings can occur. Layer III is most efficient, with lowest rates, with somewhat lower fidelity, and a single coding.

MPEG-2 incorporates the three audio layers of MPEG-1 and adds additional features, principally surround sound. However, MPEG-2 decoders can play MPEG-1 audio files, and MPEG-1 two-channel decoders can decode stereo information from surround-sound MPEG-2 files.

MPEG Bitstream Format

In the MPEG elementary bitstream, data is transmitted in frames, as shown in Fig. 11.1. Each frame is individually decodable. The length of a frame depends on the particular layer and MPEG algorithm used. In MPEG-1, Layers II and III have the same frame length representing 1152 audio samples. Unlike the other layers, in Layer III the number of bits per frame can vary; this allocation provides flexibility according to the coding demands of the audio signal.

A frame begins with a 32-bit ISO header with a 12-bit synchronizing pattern and 20 bits of general data on layer, bit-rate index, sampling frequency, type of emphasis, and so on. This is followed by an optional 16-bit CRCC check word with generation polynomial $x^{16} + x^{15} + x^2 + 1$. Subsequent fields describe bit allocation data (number of bits used to code subband samples), scale factor selection data, and scale factors themselves. This varies from layer to layer. For example, Layer I sends a fixed 6-bit scale factor for each coded subband. Layer II examines scale factors and uses dynamic scale factor selection information (SCFSI) to avoid redundancy; this reduces the scale factor bit rate by a factor of two.

The largest part of the frame is occupied by subband samples. Again, this content varies among layers. In Layer II, for example, samples are grouped in granules. The length of the field is determined by a bit-rate index, but the bit allocation determines the actual number of bits used to code the signal. If the frame length exceeds the number of bits allocated, the remainder of the frame can be occupied by ancillary data (this feature is used by MPEG-2, for example). Ancillary data is coded similarly to primary frame data. Frames contain 384 samples in Layer I and

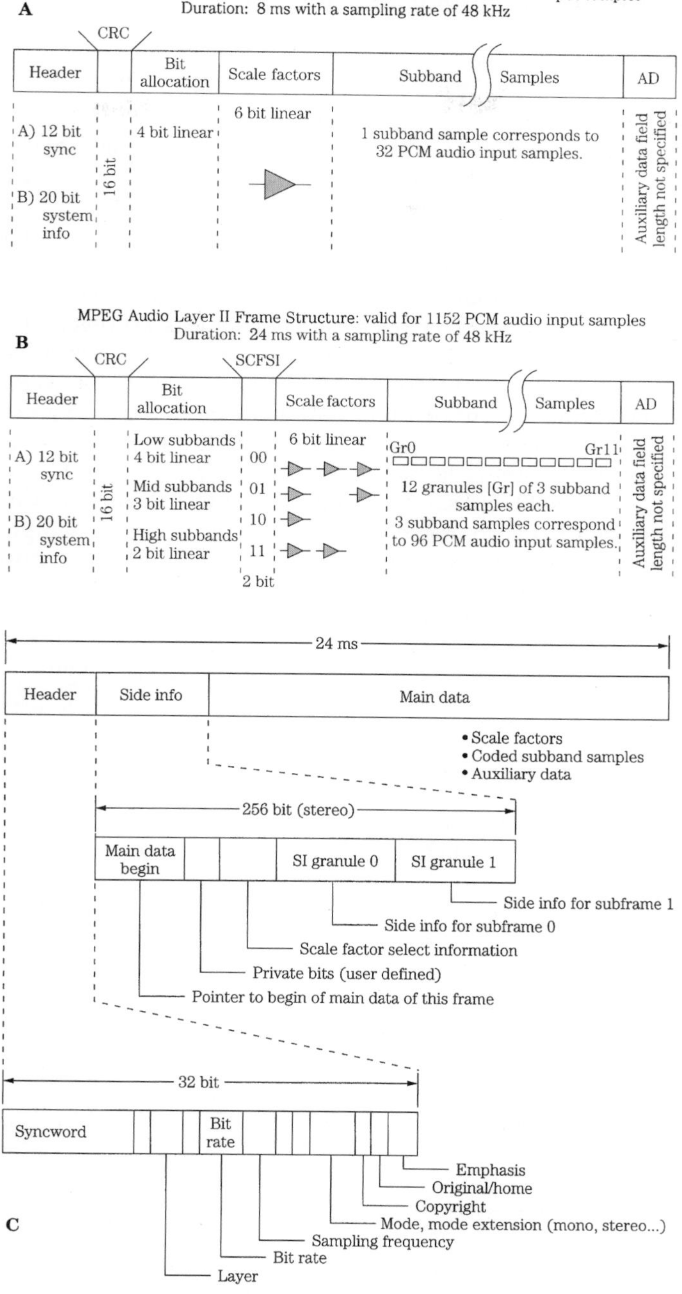

FIGURE 11.1 Structure of the MPEG-1 audio Layer I, II, and III bitstreams. The header and some other fields are common, but other fields differ. Higher-level codecs can transcode lower-level bitstreams. A. Layer I bitstream format. B. Layer II bitstream format. C. Layer III bitstream format.

1152 samples in II and III (or 8 ms and 24 ms, respectively, at a 48-kHz sampling frequency).

MPEG-1 Layer I

The MPEG-1 Layer I codec is a simplified version of the MUSICAM codec. It is a subband codec, designed to provide high fidelity with low complexity, but at a high bit rate. Block diagrams of a Layer I encoder and decoder (which also applies to Layer II) are shown in Fig. 11.2. A polyphase filter splits the wideband signal into 32 subbands of equal width. The filter is critically sampled; there is the same number of samples in the analyzed domain as in the time domain. Adjacent subbands overlap; a single frequency can affect two subbands. The filter and its inverse are not lossless; however, the error is small. The filter bank bands are all equal width, but the ear's critical bands are not; this is compensated for in the bit allocation algorithm. For example, lower bands are usually assigned more bits, increasing their resolution over higher bands. This polyphase filter bank with 32 subbands is used in all three layers; Layer III adds additional hybrid processing.

The filter bank outputs 32 samples, one sample per band, for every 32 input samples. In Layer I, 12 subband samples from each of the 32 subbands are grouped to form a frame; this represents 384 wideband samples. At a 48-kHz sampling frequency, this comprises a block of 8 ms. Each subband group of 12 samples is given a bit allocation;

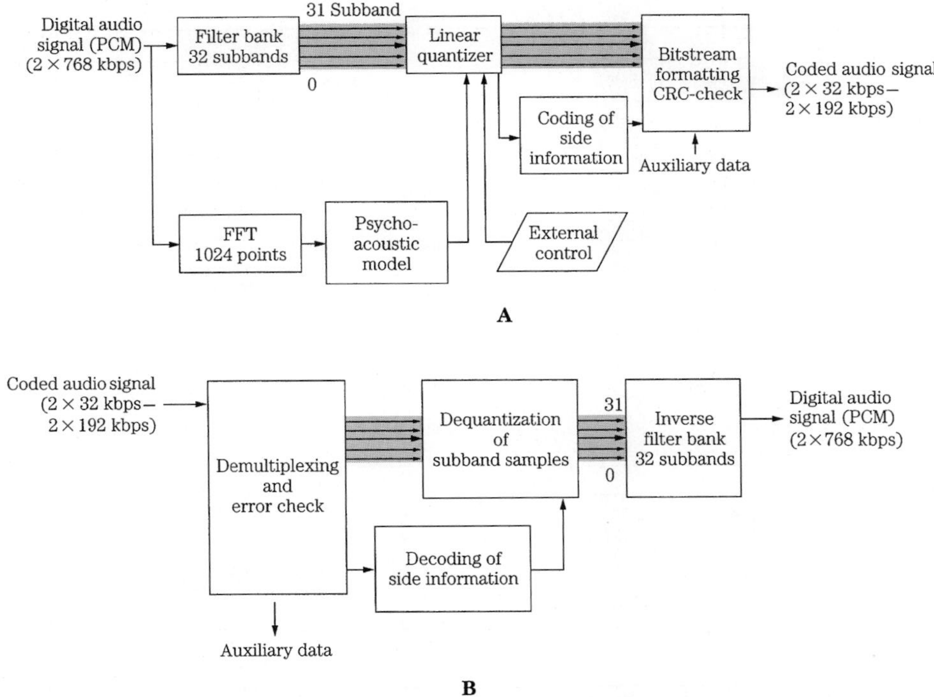

FIGURE 11.2 MPEG-1 Layer I or II audio encoder and decoder. The 32-subband filter bank is common to all three layers. A. Layer I or II encoder (single-channel mode). B. Layer I or II two-channel decoder.

subbands judged inaudible are given a zero allocation. Based on the calculated masking threshold (just audible noise), the bit allocation determines the number of bits used to quantize those samples. A floating-point notation is used to code samples; the mantissa determines resolution and the exponent determines dynamic range. A fixed scale factor exponent is computed for each subband with a nonzero allocation; it is based on the largest sample value in the subband. Each of the 12 subband samples in a block is normalized by dividing it by the same scale factor; this optimizes quantizer resolution.

A 512-sample FFT wideband transform located in a side chain performs spectral analysis on the audio signal. A psychoacoustic model, described in more detail later, uses a spreading function to emulate a basilar membrane response to establish masking contours and compute signal-to-mask ratios. Tonal and nontonal (noise-like) signals are distinguished. The psychoacoustic model compares the data to the minimum threshold curve. Using scale factor information, normalized samples are quantized by the bit allocator to achieve data reduction. The subband data is coded, not the FFT spectra.

Dynamic bit allocation assigns mantissa bits to the samples in each coded subband, or omits coding for inaudible subbands. Each sample is coded with one PCM codeword; the quantizer provides 2^n-1 steps where $2 \leq n \leq 15$. Subbands with a large signal-to-mask ratio are iteratively given more bits; subbands with a small SMR value are given fewer bits. In other words, the SMR determines the minimum signal-to-noise ratio that has to be met by the quantization of the subband samples. Quantization is performed iteratively. When available, additional bits are added to codewords to increase the signal-to-noise ratio (SNR) value above the minimum. Because the long block size might expose quantization noise in a transient signal, coarse quantization is avoided in blocks of low-level audio that are adjacent to blocks of high-level (transient) audio. The block scale factor exponent and sample mantissas are output. Error correction and other information are added to the signal at the output of the codec.

Playback is accomplished by decoding the bit allocation information, and decoding the scale factors. Samples are requantized by multiplying them with the correct scale factor. The scale factors provide all the information needed to recalculate the masking thresholds. In other words, the decoder does not need a psychoacoustic model. Samples are applied to an inverse synthesis filter such that subbands are placed at the proper frequency and added, and the resulting broadband audio waveform is output in consecutive blocks of thirty-two 16-bit PCM samples.

Example of MPEG-1 Layer I Implementation

As with other perceptual coding methods, MPEG-1 Layer I uses the ear's audiology performance as its guide for audio encoding, relying on principles such as amplitude masking to encode a signal that is perceptually identical. Generally, Layer I operating at 384 kbps achieves the same quality as a Layer II codec operating at 256 kbps. Also, Layer I can be transcoded to Layer II. The following describes a simple Layer I implementation without a psychoacoustic model; its design is basic compared to other modern codecs.

PCM data with 32-, 44.1-, or 48-kHz sampling frequencies can be input to an encoder. At these three sampling frequencies, the subband width is 500, 689, and 750 Hz, and the frame period is 12, 8.7, and 8 ms, respectively. The following description assumes a 48-kHz sampling frequency. The stereo audio signal is passed to the first stage in a Layer I encoder, as shown in Fig. 11.3. A 24-bit finite impulse response (FIR) filter with the equivalent of 512 taps divides the audio band into 32 subbands of equal 750-Hz width. The filter window is shifted by 32 samples each time (12 shifts) so all the 384 samples in

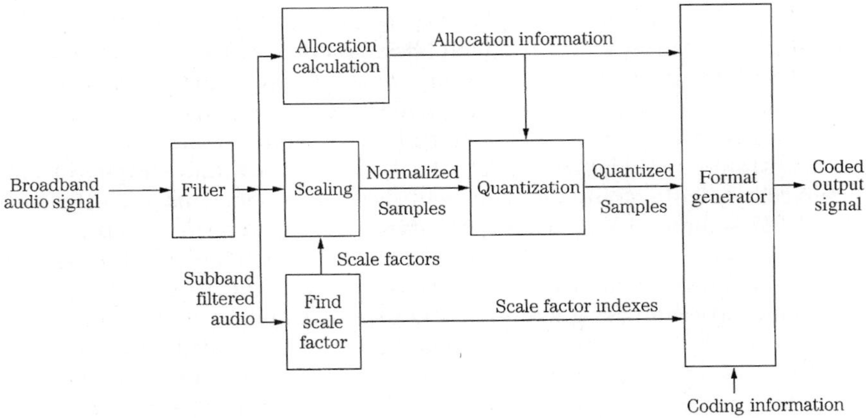

FIGURE 11.3 Example of an MPEG-1 Layer I encoder. The FFT side chain is omitted.

the 8-ms frame are analyzed. The filter bank outputs 32 subbands. With this filter, the effective sampling frequency of a subband is reduced by 32 to 1, for example, from a frequency of 48 kHz to 1.5 kHz. Although the channels are bandlimited, they are still in PCM representation at this point in the algorithm. The subbands are equal width, whereas the ear's critical bands are not. This can be compensated for by unequally allocating bits to the subbands; more bits are typically allocated to code signals in lower-frequency subbands.

The encoder analyzes the energy in each subband to determine which subbands contain audible information. This example of a Layer I encoder does not use an FFT side chain or psychoacoustic model. The algorithm calculates average power levels in each subband over the 8-ms (12-sample) period. Masking levels in subbands and adjacent subbands are estimated. Minimum threshold levels are applied. Peak power levels in each subband are calculated and compared to masking levels. The SMR value (difference between the maximum signal and the masking threshold) is calculated for each subband and is used to determine the number of bits N assigned to a subband (i) such that $N_i \geq (SMR_i - 1.76)/6.02$. A bit pool approach is taken to optimally code signals within the given bit rate. Quantized values form a mantissa, with a possible range of 2 to 15 bits. Thus, a maximum resolution of 92 dB is available from this part of the coding word. In practice, in addition to signal strength, mantissa values also are affected by rate of change of the waveform pattern and available data capacity. In any event, new mantissa values are calculated for every sample period.

Audio samples are normalized (scaled) to optimally use the dynamic range of the quantizer. Specifically, six exponent bits form a scale factor, which is determined by the signal's largest absolute amplitude in a block. The scale factor acts as a multiplier to optimally adjust the gain of the samples for quantization. This scale factor covers the range from –118 dB to +6 dB in 2-dB steps. Because the audio signal varies slowly in relation to the sampling frequency, the masking threshold and scale factors are calculated only once for every group of 12 samples, forming a frame (12 samples/subband × 32 subbands = 384 samples). For every subband, the absolute peak value of the 12 samples is compared to a table of scale factors, and the closest (next highest) constant is applied. The other sample values are normalized to that factor, and during decoding will be used as multipliers to compute the correct subband signal level.

A floating-point representation is used. One field contains a fixed-length 6-bit exponent, and another field contains a variable length 2- to 15-bit mantissa. Every block of 12 subband samples may have different mantissa lengths and values, but would share the same exponent. Allocation information detailing the length of a mantissa is placed in a 4-bit field in each frame. Because the total number of bits representing each sample within a subband is constant, this allocation information (like the exponent) needs to be transmitted only once every 12 samples. A null allocation value is conveyed when a subband is not encoded; in this case neither exponent nor mantissa values within that subband are transmitted. The 15-bit mantissa yields a maximum signal-to-noise ratio of 92 dB. The 6-bit exponent can convey 64 values. However, a pattern of all 1's is not used, and another value is used as a reference. There are thus 62 values, each representing 2-dB steps for an ideal total of 124 dB. The reference is used to divide this into two ranges, one from 0 to –118 dB, and the other from 0 to +6 dB. The 6 dB of headroom is needed because a component in a single subband might have a peak amplitude 6 dB higher than the broadband composite audio signal. In this example, the broadband dynamic range is thus equivalent to 19 bits of linear coding.

A complete frame contains synchronization information, sample bits, scale factors, bit allocation information, and control bits for sampling frequency information, emphasis, and so on. The total number of bits in a frame (with two channels, with 384 samples, over 8 ms, sampled at 48 kHz) is 3072. This in turn yields a 384-kbps bit rate. With the addition of error detection and correction code, and modulation, the transmission bit rate might be 768 kbps. The first set of subband samples in a frame is calculated from 512 samples by the 512-tap filter and the filter window is shifted by 32 samples each time into 11 more positions during a frame period. Thus, each frame incorporates information from 864 broadband audio samples per channel.

Sampling frequencies of 32 kHz and 44.1 kHz also are supported, and because the number of bands remains fixed at 32, the subband width becomes 689.06 Hz with a 44.1-kHz sampling frequency. In some applications, because the output bit rate is fixed at 384 kbps, and 384 samples/channel per frame is fixed, there is a reduction in frame rate at sampling frequencies of 32 kHz and 44.1 kHz, and thus an increase in the number of bits per frame. These additional bits per frame are used by the algorithm to further increase audio quality.

Layer I decoding proceeds frame by frame, using the processing shown in Fig. 11.4. Data is reformatted to PCM by a subband decoder, using allocation information and

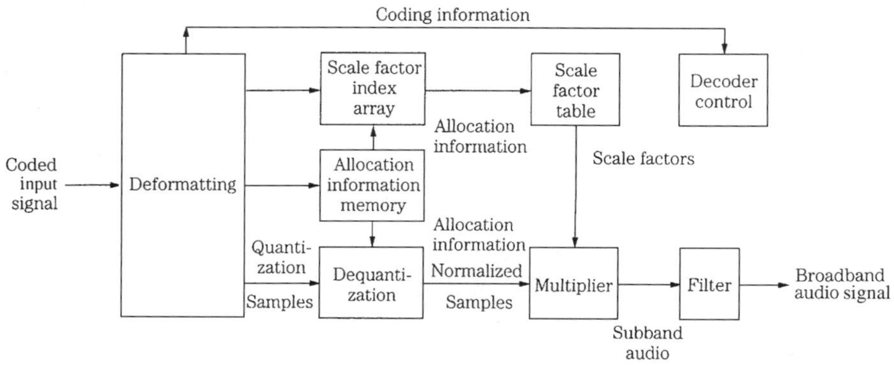

Figure 11.4 Example of an MPEG-1 Layer I decoder.

scale factors. Received scale factors are placed in an array with two columns of 32 rows, each six bits wide. Each column represents an output channel, and each row represents one subband. The decoded subband samples are multiplied by their scale factors to restore them to their quantized values; empty subbands are automatically assigned a zero value. A synthesis reconstruction filter recombines the 32 subbands into one broadband audio signal. This subband filter operates identically (but inversely) to the input filter. As in the encoder, 384 samples/channel represent 8 ms of audio signal (at a sampling frequency of 48 kHz). Following this subband filtering, the signal is ready for reproduction through D/A converters.

Because psychoacoustic processing, bit allocation, and other operations are not used in the decoder, its cost is quite low. Also, the decoder is transparent to improvements in encoder technology. If encoders are improved, the resulting fidelity would improve as well. Because the encoding algorithm is a function of digital signal processing, more sophisticated coding is possible. For example, because the number of bits per frame varies according to sampling rate, it might be expedient to create different allocation tables for different sampling frequencies.

An FFT side chain would permit analysis of the spectral content of subbands and psychoacoustic modeling. For example, knowledge of where signals are placed within bands can be useful in more precisely assigning masking curves to adjacent bands. The encoding algorithm might assume signals are at band edges, the most conservative approach. Such an encoder might claim 18-bit performance. Subjectively, at a 384-kbps bit rate, most listeners are unable to differentiate between a simple Layer I recording and an original CD recording.

MPEG-1 Layer II

The MPEG-1 Layer II codec is essentially identical to the original MUSICAM codec (the frame headers differ). It is thus similar to Layer I, but is more sophisticated in design. It provides high fidelity with somewhat higher complexity, at moderate bit rates. It is a subband codec. Figure 11.5 gives a more detailed look at a Layer II encoder (which also applies to Layer I). The filter bank creates 32 equal-width subbands, but the frame size is tripled to $3 \times 12 \times 32$, corresponding to 1152 wideband samples per channel. In other words, data is coded in three groups of 12 samples for each subband (Layer I uses one group). At a sampling frequency of 48 kHz, this comprises a 24-ms period. Figure 11.6 shows details of the subband filter bank calculation. The FFT analysis block size is increased to 1024 points. In Layer II (and Layer III) the psychoacoustic model performs two 1024-sample calculations for each 1152-sample frame, centering the first half and the second half of the frame, respectively. The results are compared and the values with the lower masking thresholds (higher SMR) in each band are used. Tonal (sinusoidal) and nontonal (noise-like) components are distinguished to determine their effect on the masking threshold.

A single bit allocation is given to each group of 12 subband samples. Up to three scale factors are calculated for each subband, each corresponding to a group of 12 subband samples and each representing a 2-dB step-size difference. However, to reduce the scale factor bit rate, the codec analyzes scale factors in three successive blocks in each subband. When differences are small or when temporal masking will occur, one scale factor can be shared between groups. When transient audio content is coded, two or three scale factors can be conveyed. Bit allocation is used to maximize both the subband and frame signal-to-mask ratios. Quantization covers a range from 3 to 65,535 (or none),

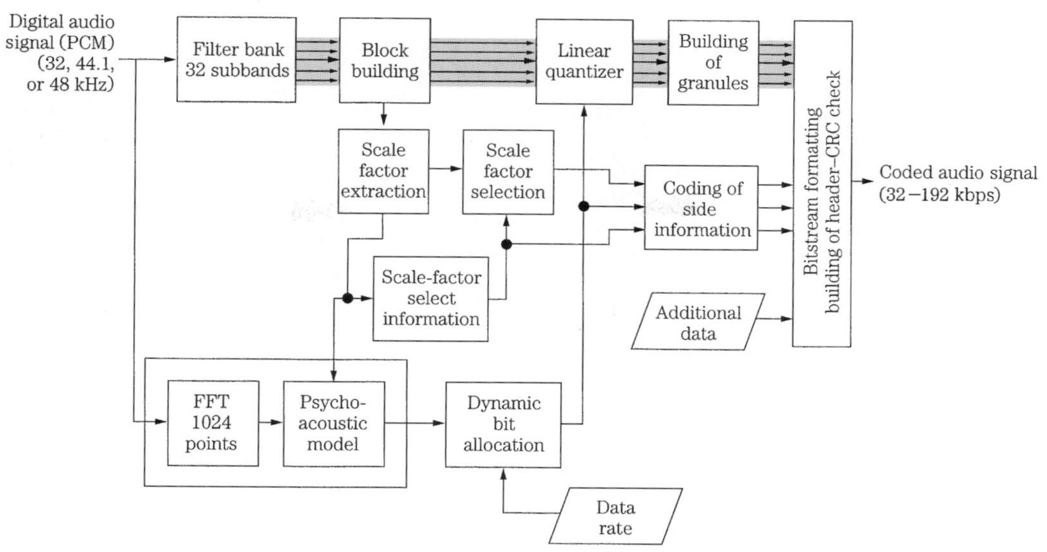

FIGURE 11.5 MPEG-1 Layer II audio encoder (single-channel mode) showing scale factor selection and coding of side information.

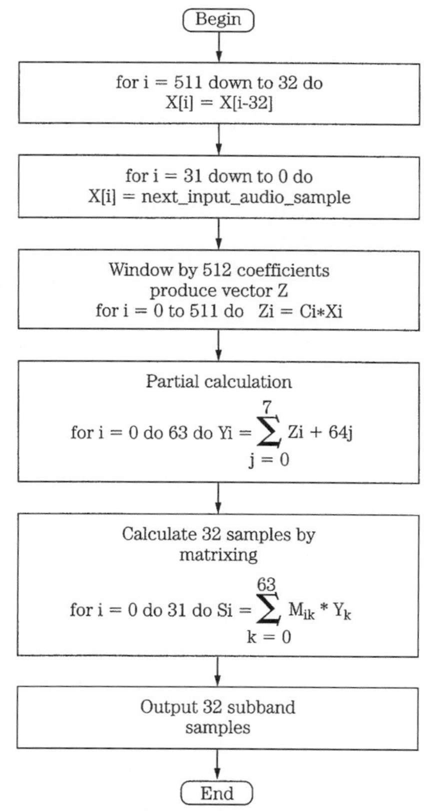

FIGURE 11.6 Flow chart of the analysis filter bank used in the MPEG-1 audio standard.

Parameter	MPEG Layer I	MPEG Layer II
Frame length (samples)	384	1152
Subbands	32	32
Subband samples	12	36
FFT (samples)	512	1024
Bit allocation (bits)	4 per	2 to 4 depending on subband
Scale factor select information (bits)	None	2 per subband
Scale factors (bits)	6 per subband	6 to 18 per subband (selectable)
Sample grouping	None	3 per subband (granule)

TABLE 11.1 Comparison of parameters in MPEG-1 Layer I and Layer II.

but the number of available levels depends on the subband. Low-frequency subbands can receive as many as 15 bits, middle-frequency subbands can receive seven bits, and high-frequency subbands are limited to three bits. In each band, prominent signals are given longer codewords. It is recognized that quantization varies with subband number; higher subbands usually receive fewer bits, with larger step sizes. Thus for greater efficiency, three successive samples (for all 32 subbands) are grouped to form a granule and quantized together.

As in Layer I, decoding is relatively simple. The decoder unpacks the data frames and applies appropriate data to the reconstruction filter. Layer II coding can use stereo intensity coding. Layer II coding provides for a dynamic range control to adapt to different listening conditions, and uses a fixed-length data word. Minimum encoding and decoding delays are about 30 ms and 10 ms, respectively. Layer II is used in some digital audio broadcasting (DAB) and digital video broadcasting (DVB) applications. Layer I and II are compared in Table 11.1. Figure 11.7 shows a flow chart summarizing the complete MPEG-1 Layer I and II encoding algorithm.

MPEG-1 Layer III (MP3)

The MPEG-1 Layer III codec is based on the ASPEC codec and contains elements of MUSICAM, such as a subband filter bank, to provide compatibility with Layers I and II. Unlike the Layer I and II codecs, the Layer III codec is a transform codec. Its design is more complex than the other layer codecs. Its strength is moderate fidelity even at low data rates. Layer III files are popularly known as MP3 files. Block diagrams of a Layer III encoder and decoder are shown in Fig. 11.8.

As in Layers I and II, a wideband block of 1152 samples is first split into 32 subbands with a polyphase filter; this provides backward compatibility with Layers I and II. Each subband's contents are transformed into spectral coefficients by either a 6- or 18-point modified discrete cosine transform (MDCT) with 50% overlap (using a sine window) so that windows contain either 12 or 36 subband samples. The MDCT outputs a maximum of $32 \times 18 = 576$ spectral lines. The spectral lines are grouped into scale factor bands that emulate critical bands. At lower sampling frequencies optionally provided by MPEG-2, the frequency resolution is increased by a factor of two; at a 24-kHz

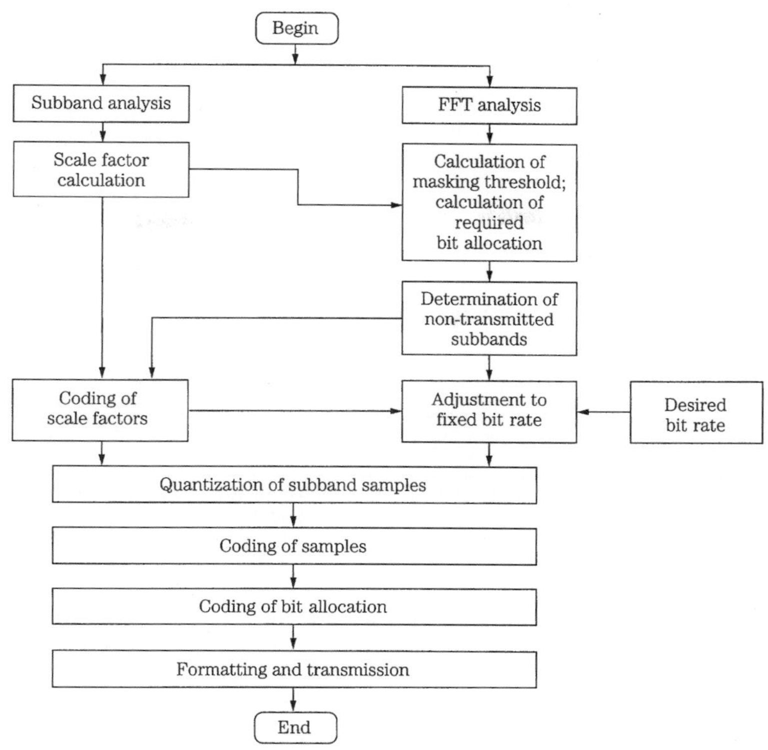

FIGURE 11.7 Flow chart of the entire MPEG-1 Layer I and II audio encoding algorithm.

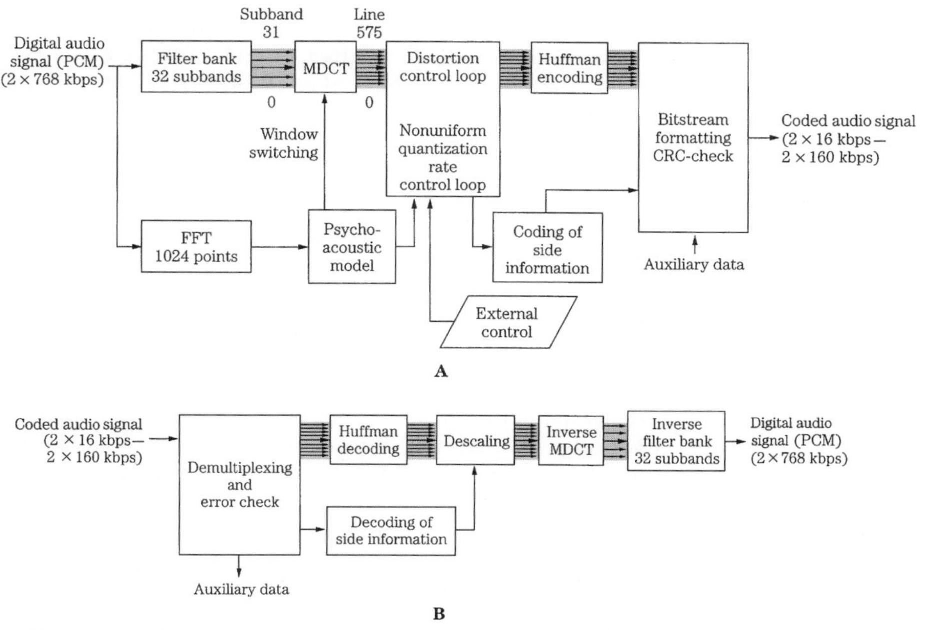

FIGURE 11.8 MPEG-1 Layer III audio encoder and decoder. A. Layer III encoder (single-channel mode). B. Layer III two-channel decoder.

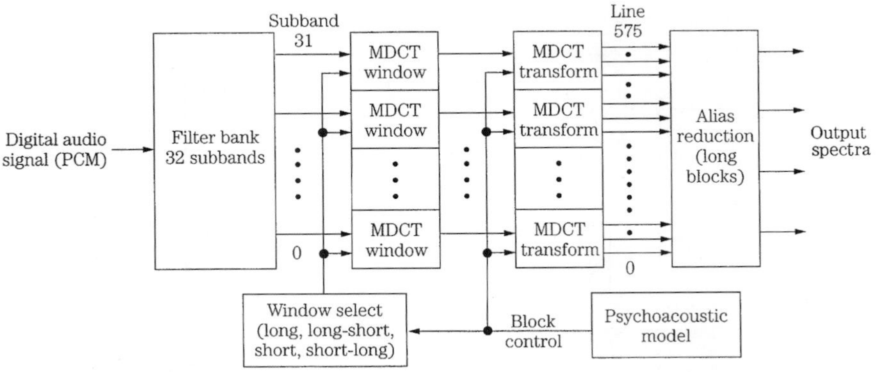

FIGURE 11.9 Long and short blocks can be selected for the MDCT transform used in the MPEG-1 Layer III encoder. Both long and short windows, and two transitional windows, are used.

sampling rate the resolution per spectral line is about 21 Hz. This allows better adaptation of scale factor bands to critical bands. This helps achieve good audio quality at lower bit rates.

Layer III has high frequency resolution, but this dictates low time resolution. Quantization error spread over a window length can produce pre-echo artifacts. Thus, under direction of the psychoacoustic model, the MDCT window sizes can be switched between frequency or time resolution, using a threshold calculation; the architecture is shown in Fig. 11.9. A long symmetrical window is used for steady-state signals; a length of 1152 samples corresponds 24 ms at a 48-kHz sampling frequency. Each transform of 36 samples yields 18 spectral coefficients for each of 32 subbands, for a total of 576 coefficients. This provides good spectral resolution of 41.66 Hz (24000/576) that is needed for steady state-signals, at the expense of temporal resolution that is needed for transient signals.

Alternatively, when transient signals occur, a short symmetrical window is used with one-third the length of the long window, followed by an MDCT that is one-third length. Time resolution is 4 ms at a 48-kHz sampling frequency. Three short windows replace one long window, maintaining the same number of samples in a frame. This mode yields six coefficients per subband, or a total of $32 \times 6 = 192$ coefficients. Window length can be independently switched for each subband. Because the switchover is not instantaneous, an asymmetrical start window is used to switch from long to short windows, and an asymmetrical stop window switches back. This ensures alias cancellation. The four window types, along with a typical window sequence, are shown in Fig. 11.10. There are three block modes. In two modes, the outputs of all 32 subbands are processed through the MDCT with equal block lengths. A mixed mode provides frequency resolution at lower frequencies and time resolution at higher frequencies. During transients, the two lower subbands use long blocks and the upper 30 subbands use short blocks. Huffman coding is applied at the encoder output to additionally lower the bit rate.

A Layer III decoder performs Huffman decoding, as well as decoding of bit allocation information. Coefficients are applied to an inverse transform, and 32 subbands are combined in a synthesis filter to output a broadband signal. The inverse modified discrete cosine transform (IMDCT) is executed 32 times for 18 spectral values each

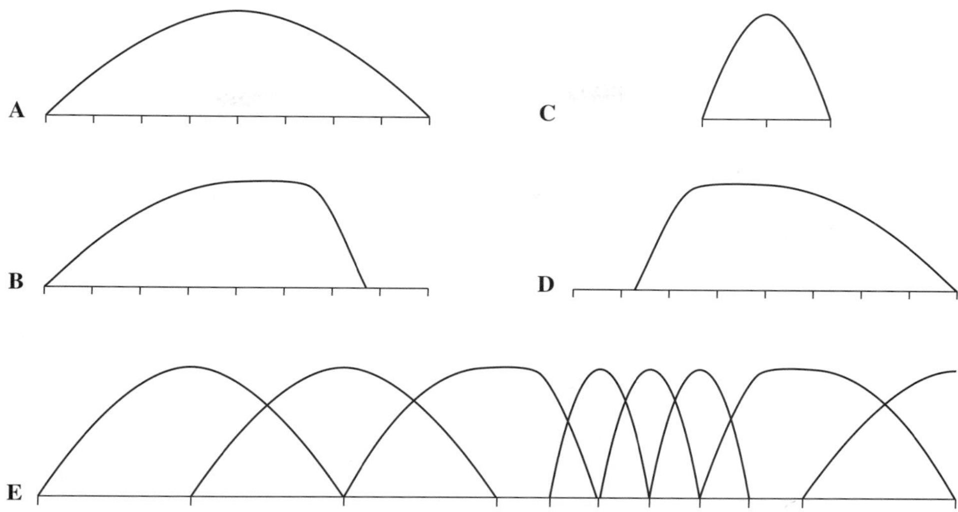

Figure 11.10 MPEG-1 Layer III allows adaptive window switching for the MDCT transform. Four window types are defined. A. Long (normal) window. B. Start window (long to short). C. Short window. D. Stop window (short to long). D. An example of a window sequence. (*Brandenburg and Stoll, 1994*)

to transform the spectrum of 576 values into 18 consecutive spectra of length 32. These spectra are converted into the time domain by executing a polyphase synthesis filter bank 18 times. The polyphase filter bank contains a frequency mapping operation (such as matrix multiplication) and an FIR filter with 512 coefficients.

MP3 files can be coded at a variety of bit rates. However, the format is not scalable with respect to variable decoding. In other words, the decoder cannot selectively choose subsets of the entire bitstream to reproduce different quality signals.

MP3 Bit Allocation and Huffman Coding

The allocation control algorithm suggested for the Layer III encoder uses dynamic quantization. A noise allocation iteration loop is used to calculate optimal quantization noise in each subband. This technique is referred to as noise allocation, as opposed to bit allocation. Rather than allocate bits directly from SNR values, in noise allocation the bit assignment is an inherent outcome of the strategy. For example, an analysis-by-synthesis method can be used to calculate a quantized spectrum that satisfies the noise requirements of the modeled masking threshold. Quantization of this spectrum is iteratively adjusted so the bit rate limits are observed. Two nested iteration loops are used to find two values that are used in the allocation: the global gain value determines quantization step size, and scale factors determine noise-shaping factors for each scale factor band. To form scale factors, most of the 576 spectral lines in long windows are grouped into 21 scale factor bands, and most of the 192 lines from short windows are grouped into 12 scale factor bands. The grouping approximates critical bands, and varies according to sampling frequency.

An inner iteration loop (called the rate loop) acts to decrease the coder rate until it is sufficiently low. The Huffman code assigns shorter codewords to smaller quantized values that occur more frequently. If the resulting bit rate is too high, the rate loop adjusts gain to yield larger quantization step sizes and hence small quantized values and smaller Huffman codewords and a lower bit rate. The process of quantizing spectral lines and determining the appropriate Huffman code can be time-consuming. The outer iteration loop (called the noise control loop) uses analysis-by-synthesis to evaluate quantization noise levels and hence the quality of the coded signal. The outer loop decreases the quantizer step size to shape the quantization noise that will appear in the reconstructed signal, aiming to maintain it below the masking threshold in each band. This is done by iteratively increasing scale factor values. The algorithm uses the iterative values to compute the resulting quantization noise. If the quantization noise level in a band exceeds the masking threshold, the scale factor is adjusted to decrease the step size and lower the noise floor. The algorithm then recalculates the quantization noise level. Ideally, the loops yield values such that the difference between the original spectral values and the quantized values results in noise below the masking threshold.

If the psychoacoustic model demands small step sizes and in contradiction the loops demand larger step sizes to meet a bit rate, the loops are terminated. To avoid this, the perceptual model can be modified and two loops tuned, to suit different bit rates; this tuning can require considerable development work. Nonuniform quantization is used such that step size varies with amplitude. Values are raised to the 3/4 power before quantizing to optimize the signal-to-noise ratio over a range of quantizer values (the decoder reciprocates by raising values to the 4/3 power).

Huffman and run-length entropy coding exploit the statistical properties of the audio signal to achieve lossless data compression. Most audio frames will yield larger spectral values at low frequencies and smaller (or zero) values at higher frequencies. To utilize this, the 576 spectral lines are considered as three groups and can be coded with different Huffman code tables. The sections from low to high frequency are BIG_VALUE, COUNT1, and RZERO, assigned according to pairs of absolute values ranging from 0 to 8191, quadruples of 0, –1, or +1 values, and the pairs of 0 values, respectively. The BIG_VALUE pairs can be coded using any of 32 Huffman tables, and the COUNT1 quadruples can be coded with either of two tables. The RZERO pairs are not coded with Huffman coding. A Huffman table is selected based on the dynamic range of the values. Huffman coding is used for both scale factors and coefficients.

The data rate from frame to frame can vary in Layer III; this can be used for variable bit-rate recording. The psychoacoustic model calculates how many bits are needed and sets the frame bit rate accordingly. In this way, for example, music passages that can be satisfactorily coded with fewer bits can yield frames with fewer bits. A variable bit rate is efficient for on-demand transmission. However, variable bit rate streams cannot be transmitted in real time using systems with a constant bit rate. When a constant rate is required, Layer III can use an optional bit reservoir to allow for more accurate coding of particularly difficult (large perceptual entropy) short window passages. In this way, the average transmitted data rate can be smaller than peak data rates. The number of bits per frame is variable, but has a constant long-term average. The mean bit rate is never allowed to exceed the fixed-channel capacity. In other words, there is reserve capacity in the reservoir. Unneeded bits (below the average) can be placed in the reservoir. When additional bits are needed (above the average), they are taken from the reservoir. Succeeding frames are coded with somewhat fewer bits than average to replenish the reservoir. Bits can only be borrowed from past frames; bits cannot be borrowed from future frames. The buffer memory adds throughput time to the codec. To achieve synchronization at the decoder, headers and side information are conveyed at the frame rate.

Frame size is variable; boundaries of main data blocks can vary whereas the frame headers are at fixed locations. Each frame has a synchronization pattern and subsequent side information discloses where a main data block began in the frame. In this way, main data blocks can be interrupted by frame headers.

In some codecs, the output file size is different from the input file size, and the signals are not time-aligned; the time duration of the codec's signal is usually longer. This is because of the block structure of the processing, coding delays, and the look-ahead strategies employed. Moreover, for example, an encoder might either discard a final frame that is not completely filled at the end of a file, or more typically pad the last frame with zeros. To maintain the original file size and time alignment, some codecs use ancillary data in the bitstream in a technique known as original file length (OFL). By specifying the number of samples to be stripped at the start of a file, and the length of the original file, the number of samples to be stripped at the end can be calculated. The OFL feature is available in the MP3PRO codec.

MP3 Stereo Coding

To take advantage of redundancies between stereo channels, and to exploit limitations in human spatial listening, Layer III allows a choice of stereo coding methods, with four basic modes: normal stereo mode with independent left and right channels; M/S stereo mode in which the entire spectrum is coded with M/S; intensity stereo mode in which the lower spectral range is coded as left/right and the upper spectral range is coded as intensity; and the intensity and M/S mode in which the lower spectral range is coded as M/S and the upper spectral range is coded as intensity. Each frame may have a different mode. The partition between upper and lower spectral modes can be changed dynamically in units of scale factor bands.

Layer III supports both M/S (middle/side) stereo coding and intensity stereo coding. In M/S coding, certain frequency ranges of the left and right channels are mixed as sum (middle) and difference (side) signals of the left and right channels before quantization. In this way, stereo unmasking can be avoided. In addition, when there is high correlation between the left and right channels, the difference signal is further reduced to conserve bits. In intensity stereo coding, the left and right channels of upper-frequency subbands are not coded individually. Instead, one summed signal is transmitted along with individual left- and right-channel scale factors indicating position in the stereo panorama. This method retains one spectral shape for both channels in upper subbands, but scales the magnitudes. This is effective for stationary signals, but less effective for transient signals because they may have different envelopes in different channels. Intensity coding may lead to artifacts such as changes in stereo imaging, particularly for transient signals. It is used primarily at low bit rates.

MP3 Decoder Optimization

MP3 files can be decoded with dedicated hardware chips or software programs. To optimize operation and decrease computation, some software decoders implement special features. Calculation of the hybrid synthesis filter bank is the most computationally complex aspect of the decoder. The process can be simplified by implementing a stereo downmix to monaural in the frequency domain, before the filter bank, so that only one filter operation must be performed. Downmixing can be accomplished with a simple weighted sum of the left and right channels. However, this is not optimal because, for example, an M/S-stereo or intensity-stereo signal already contains a sum signal. More efficiently, built in downmixing routines can calculate the sum signal only for those scale factor bands that are coded in left/right stereo. For M/S- and intensity-coded scale factor bands, only scaling operations are needed.

To further reduce computational complexity, the hybrid filter bank can be optimized. The filter bank consists of IMDCT and polyphase filter bank sections. As noted, the IMDCT is executed 32 times for 18 spectral values each to transform the spectrum of 576 values into 18 consecutive spectra of length 32. These spectra are converted into the time domain by executing a polyphase synthesis filter bank 18 times. The polyphase filter bank contains a frequency mapping operation (such as matrix multiplication) and a FIR filter with 512 coefficients. The FIR filter calculation can be simplified by reducing the number of coefficients, the filter coefficients can be truncated at the ends of the impulse response, and the impulse response can be modeled with fewer coefficients. Experiments have suggested that filter length can be reduced by 25% without yielding additional audible artifacts. More directly, computation can be reduced by limiting the output audio bandwidth. The high-frequency spectral values can be set to zero; an IMDCT with all input samples set to zero does not have to be calculated. If only the lower halves of the IMDCTs are calculated, the audio bandwidth is limited. The output can be downsampled by a factor of 2, so that computation for every second output value can be skipped, thus cutting the FIR calculation in half.

There are many nonstandard codecs that produce MP3-compliant bitstreams; they vary greatly in performance quality. LAME is an example of a fast, high-quality, royalty-free codec that produces a MP3-compliant bitstream. LAME is open-source, but using LAME may require a patent license in some countries. LAME is available at http://lame.sourceforge.net. MP3 Internet applications are discussed in Chap. 15.

MPEG-1 Psychoacoustic Model 1

The MPEG-1 standard suggests two psychoacoustic models that determine the minimum masking threshold for inaudibility. The models are only informative in the standard; their use is not mandated. The models are used only in the encoder. In both cases, the difference between the maximum signal level and the masking threshold is used by the bit allocator to set the quantization levels. Generally, model 1 is applied to Layers I and II and model 2 is applied to Layer III.

Psychoacoustic model 1 proposes a low-complexity method to analyze spectral data and output signal-to-mask ratios. Model 1 performs these nine steps:

1. *Perform FFT analysis:* A 512- or 1024-point fast Fourier transform, with a Hann window with adjacent overlapping of 32 or 64 samples, respectively, to reduce edge effects, is used to transform time-aligned time-domain data to the frequency domain. An appropriate delay is applied to time-align the psychoacoustic model's output. The signal is normalized to a maximum value of 96 dB SPL, calibrating the signal's minimum value to the absolute threshold of hearing.
2. *Determine the sound pressure level:* The maximum SPL is calculated for each subband by choosing the greater of the maximum amplitude spectral line in the subband or the maximum scale factor that accounts for low-level spectral lines in the subband.
3. *Consider the threshold in quiet:* An absolute hearing threshold in the absence of any signal is given; this forms the lower masking bound. An offset is applied depending on the bit rate.

4. *Finding tonal and nontonal components:* Tonal (sinusoidal) and nontonal (noise-like) components in the signal are identified. First, local maxima in the spectral components are identified relative to bandwidths of varying size. Components that are locally prominent in a critical band by +7 dB are labeled as tonal and their sound-pressure level is calculated. Intensities of the remaining components, assumed to be nontonal, within each critical band are summed and their SPL is calculated for each critical band. The nontonal maskers are centered in each critical band.
5. *Decimation of tonal and nontonal masking components:* The number of maskers is reduced to obtain only the relevant maskers. Relevant maskers are those with magnitude that exceeds the threshold in quiet, and those tonal components that are strongest within 1/2 Bark.
6. *Calculate individual masking thresholds:* The total number of masker frequency bins is reduced (for example, in Layer I at 48 kHz, 256 is reduced to 102) and maskers are relocated. Noise masking thresholds for each subband, accounting for tonal and nontonal components and their different downward shifts, are determined by applying a masking (spreading) function to the signal. Calculations use a masking index and masking function to describe masking effects on adjacent frequencies. The masking index is an attenuation factor based on critical-band rate. The piecewise masking function is an attenuation factor with different lower and upper slopes between –3 and +8 Bark that vary with respect to the distance to the masking component and the component's magnitude. When the subband is wide compared to the critical band, the spectral model can select a minimum threshold; when it is narrow, the model averages the thresholds covering the subband.
7. *Calculate the global masking threshold:* The powers corresponding to the upper and lower slopes of individual subband masking curves, as well as a given threshold of hearing (threshold in quiet), are summed to form a composite global masking contour. The final global masking threshold is thus a signal-dependent modification of the absolute threshold of hearing as affected by tonal and nontonal masking components across the basilar membrane.
8. *Determine the minimum masking threshold:* The minimum masking level is calculated for each subband.
9. *Calculate the signal-to-mask ratio:* Signal-to-mask ratios are determined for each subband, based on the global masking threshold. The difference between the maximum SPL levels and the minimum masking threshold values determines the SMR value in each subband; this value is supplied to the bit allocator.

The principal steps in the operation of model 1 can be illustrated with a test signal that contains a band of noise, as well as prominent tonal components. The model analyzes one block of the 16-bit test signal sampled at 44.1 kHz. Figure 11.11A shows the audio signal as output by the FFT; the model has identified the local maxima. The figure also shows the absolute threshold of hearing used in this particular example (offset by –12 dB). Figure 11.11B shows tonal components marked with a "+" and nontonal components marked with a "o." Figure 11.11C shows the masking functions assigned to tonal maskers after decimation. The peak SMR (about 14.5 dB) corresponds to that used for tonal maskers. Figure 11.11D shows the masking functions assigned to nontonal

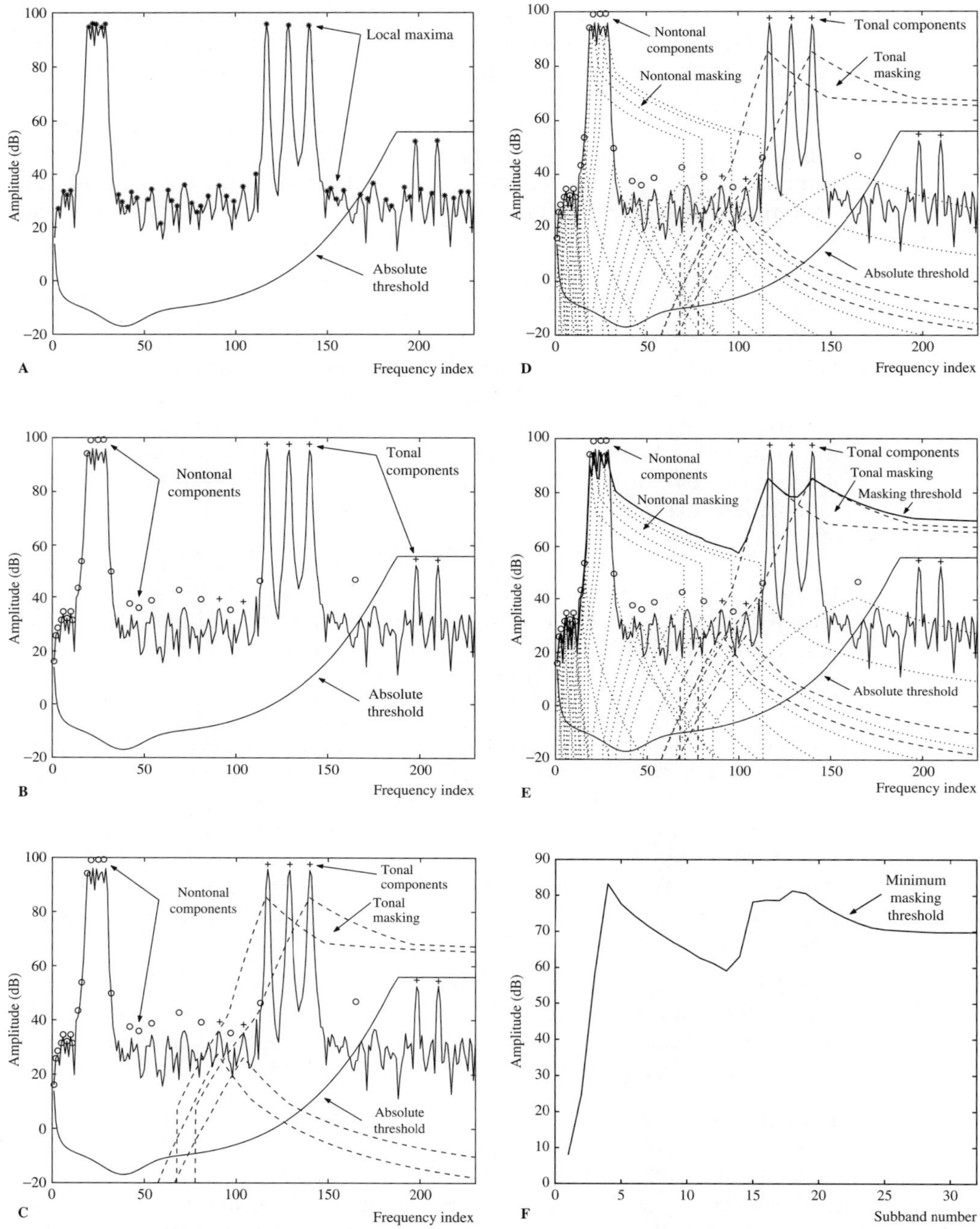

FIGURE 11.11 Operation of MPEG-1 model 1 is illustrated using a test signal. A. Local maxima and absolute threshold. B. Tonal and nontonal components. C. Tonal masking. D. Nontonal masking. E. Masking threshold. F. Minimum masking threshold.

maskers after decimation. The peak SMR (about 5 dB) corresponds to that used for nontonal maskers. Figure 11.11E shows the final global masking curve obtained by combining the individual masking thresholds. The higher of the global masking curve and the absolute threshold of hearing is used as the final global masking curve. Figure 11.11F shows the minimum masking threshold. From this, SMR values can be calculated in each subband.

To further explain the operation of model 1, additional comments are given here. The delay in the 512-point analysis filter bank is 256 samples and centering the data in the 512-point Hann window adds 64 samples. An offset of 320 samples (256 + (512 – 384)/2 = 320) is needed to time-align the model's 384 samples.

The spreading function used in model 1 is described in terms of piecewise slopes (in dB):

$$SF = 17(dz+1) - (0.4X[z(j)] + 6 \qquad \text{for } -3 \le dz \le -1 \text{ Bark}$$

$$SF = (0.4X[z(j)]) + 6)dz \qquad \text{for } -1 \le dz \le 0 \text{ Bark}$$

$$SF = -17dz \qquad \text{for } 0 \le dz \le 1 \text{ Bark}$$

$$SF = -(dz-1)(17 - 0.15X[z(j)]) - 17 \qquad \text{for } 1 \le dz \le 8 \text{ Bark}$$

where $dz = z(i) - z(j)$ is the distance in Bark between the maskee and masker frequency; i and j are index values of spectral lines of the maskee and masker, respectively. $X[z(j)]$ is the sound pressure level of the jth masking component in dB. Values outside –3 and +8 Bark are not considered in this model.

Model 1 uses this general approach to detect and characterize tonality in audio signals: An FFT is applied to 512 or 1024 samples, and the components of the spectrum analysis are considered. Local maxima in the spectrum are identified as having more energy than adjacent components. These components are decimated such that a tonal component closer than 1/2 Bark to a stronger tonal component is discarded. Tonal components below the threshold of hearing are discarded as well. The energies of groups of remaining components are summed to represent tonal components in the signal; other components are summed and marked as nontonal. A binary designation is given: tonal components are assigned 1, and nontonal components are assigned 0. This information is presented to the bit allocation algorithm. Specifically, in model 1, tonality is determined by detecting local maxima of 7 dB in the audio spectrum. To derive the masking threshold relative to the masker, a level shift Δ is applied; the nature of the shift depends on whether the masker is tonal or nontonal:

$$\Delta_T(z) = -6.025 - 0.275z \text{ dB}$$

$$\Delta_N(z) = -2.025 - 0.175z \text{ dB}$$

where z is the frequency of the masker in Bark.

Model 1 considers all the nontonal components in a critical band and represents them with one value at one frequency. This is appropriate at low frequencies where subbands and critical bands have good correspondence, but can be inefficient at high frequencies where there are many critical bands in each subband. A subband that is apart from the identified nontonal component in a critical band may not receive a correct nontonal evaluation.

MPEG-1 Psychoacoustic Model 2

Psychoacoustic model 2 performs a more detailed analysis than model 1, at the expense of greater computational complexity. It is designed for lower bit rates than model 1. As in model 1, model 2 outputs a signal-to-mask ratio for each subband; however, its approach is significantly different. It contours the noise floor of the signal represented by many spectral coefficients in a way that is more accurate than that allowed by coarse subband coding. Also, the model uses an unpredictability measure to examine the side-chain data for tonal or nontonal qualities. Model 2 performs these 14 steps:

1. *Reconstruct input samples:* A set of 1024 input samples is assembled.
2. *Calculate the complex spectrum:* The time-aligned input signal is windowed with a 1024-point Hann window; alternatively, a shorter window may be used. An FFT is computed and output represented in magnitude and phase.
3. *Calculate the predicted magnitude and phase:* The predicted magnitude and phase are determined by extrapolation from the two preceding threshold blocks.
4. *Calculate the unpredictability measure:* The unpredictability measure is computed using the Euclidian distance between the predicted and actual values in the magnitude/phase domain. To reduce complexity, the measure may be computed only for lower frequencies and assumed constant for higher frequencies.
5. *Calculate the energy and unpredictability in the partitions:* The energy magnitude and the weighted unpredictability measure in each threshold calculation partition are calculated. A partition has a resolution of one spectral line (at low frequencies) or 1/3 critical band (at high frequencies), whichever is wider.
6. *Convolve energy and unpredictability with the spreading function:* The energy and the unpredictability measure in threshold calculation partitions are each convolved with a cochlea spreading function. Values are renormalized.
7. *Derive tonality index:* The unpredictability measures are converted to tonality indices ranging from 0 (high unpredictability) to 1 (low unpredictability). This determines the relative tonality of the maskers in each threshold calculation partition.
8. *Calculate the required signal-to-noise ratio:* An SNR is calculated for each threshold calculation partition using tonality to interpolate an attenuation shift factor between noise-masking-tone (NMT) and tone-masking-noise (TMN). The interpolated shift ranges from 5.5 dB for NMT and upward. The final shift value is the higher of the interpolated value or a frequency-dependent minimum value.
9. *Calculate power ratio:* The power ratio of the SNR is calculated for each threshold calculation partition.
10. *Calculate energy threshold:* The actual energy threshold is calculated for each threshold calculation partition.
11. *Spread threshold energy:* The masking threshold energy is spread over FFT lines corresponding to threshold calculation partitions to represent the masking in the frequency domain.
12. *Calculate final energy threshold of audibility:* The spread threshold energy is compared to values in absolute threshold of quiet tables, and the higher value is used (not the sum) as the energy threshold of audibility. This is because it is wasteful to specify a noise threshold lower than the level that can be heard.

13. *Calculate pre-echo control:* A narrow-band pre-echo control used in the Layer III encoder is calculated, to prevent audibility of the error signal spread in time by the synthesis filter. The calculation lowers the masking threshold after a quiet signal. The calculation takes the minimum of the comparison of the current threshold with the scaled thresholds of two previous blocks.
14. *Calculate signal-to-mask ratios:* Threshold calculation partitions are converted to codec partitions (scale factor bands). The SMR (energy in each scale factor band divided by noise level in each scale factor band) is calculated for each partition and expressed in decibels. The SMR values are forwarded to the allocation algorithm.

The principal steps in the operation of model 2 can be illustrated with a test signal that contains three prominent tonal components. The model analyzes a set of 1024 input samples of the 16-bit test signal sampled at 44.1 kHz. Figure 11.12A shows the magnitude of the audio signal as output by the FFT; the phase is also computed. Following prediction of magnitude and phase, the unpredictability measure is computed, as shown in Fig. 11.12B, using the Euclidian distance between the predicted and actual values in the magnitude/phase domain. When the measure equals 0, the current value is completely predicted. Figure 11.12C shows the energy magnitude in each partition and the spreading functions that are applied. Figure 11.12D shows the tonality index derived from the unpredictability measure; the tonality index ranges from 0 (high unpredictability and noise-like) to 1 (low unpredictability and tonal). Figure 11.12E shows the spread masking threshold energy in the frequency domain and the absolute threshold of quiet; the higher value is used to find the energy threshold of inaudibility. Figure 11.12F shows signal-to-mask ratios (energy in each scale factor band divided by noise level in each scale factor band) in codec partitions.

To further explain the operation of model 2, additional comments are given here. The spreading function used in model 2 is:

$$10\log_{10} SF(dz) = 15.8111389 + 7.5(1.05dz + 0.474) - 17.5[1.0 + (1.05dz + 0.474)^2]^{1/2} + 8\,\mathrm{MIN}[(1.05dz - 0.5)^2 - 2(1.05dz - 0.5), 0]\ \mathrm{dB}$$

where dz is the distance in Bark between the maskee and masker frequency.

The spectral flatness measure (SFM), devised by James Johnston, measures the average or global tonality of the segment. SFM is the ratio of the geometric mean of the power spectrum to its arithmetic mean. The value is converted to decibels and referenced to –60 dB to provide a coefficient of tonality ranging continuously from 0 (nontonal) to 1 (tonal). This coefficient can be used to interpolate between TMN and NMT models. SFM leads to very conservative masking decisions for nontonal parts of a signal. More efficiently, specific tonal and nontonal regions within a segment can be identified. This local tonality can be measured as the normalized Euclidean distance between the actual and predicted values over two successive segments, for amplitude and phase. On the basis of this, tonality unpredictability can be computed for narrow frequency partitions and used to create tonality metrics that are used to interpolate between tone or noise models.

Specifically, in model 2, a tonality index is created, on the basis of the predictability of the audio signal's spectral components in a partition in two successive frames. Tonal

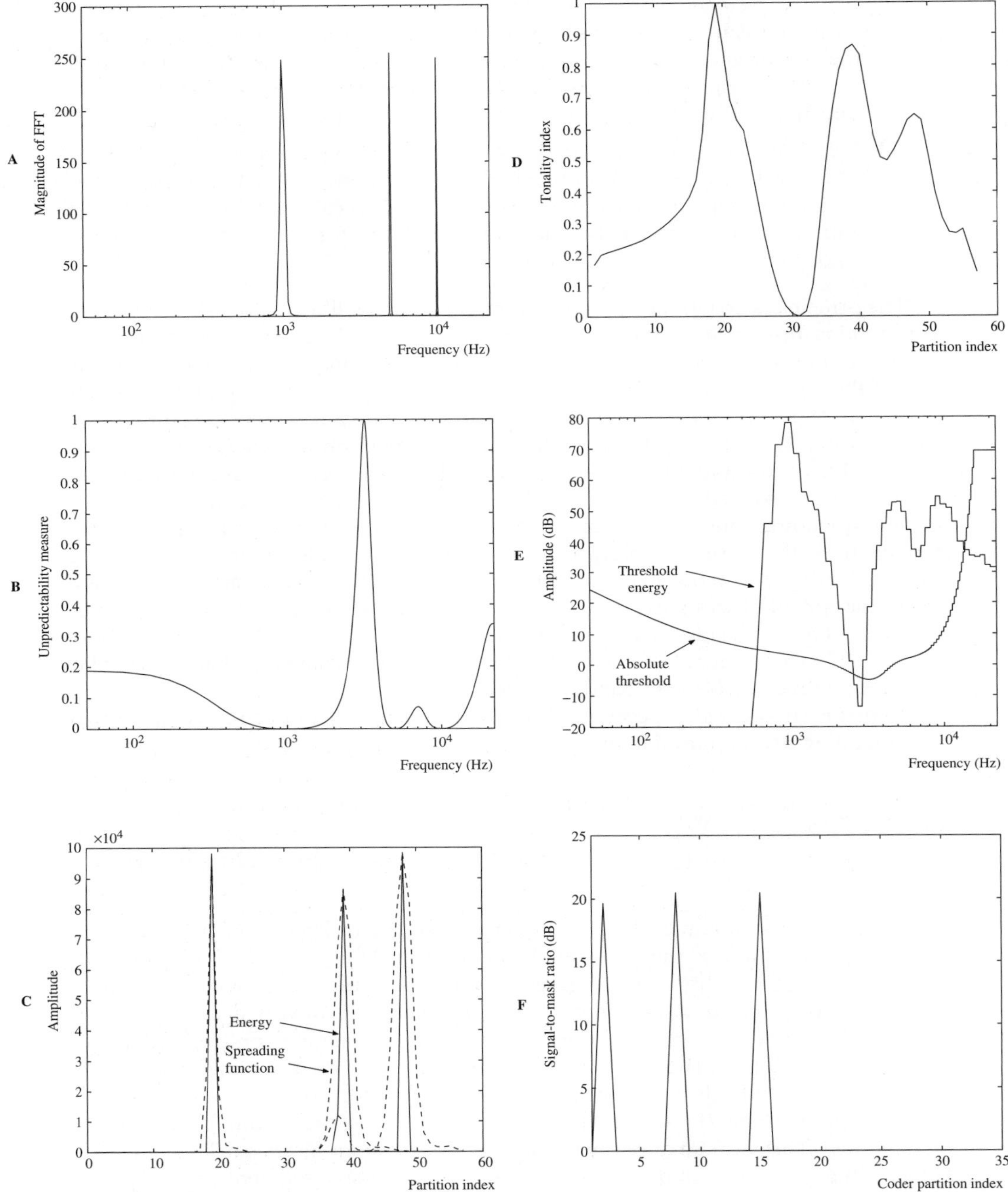

FIGURE 11.12 Operation of MPEG-1 model 2 is illustrated using a test signal. A. Magnitude of FFT. B. Unpredictability measure. C. Energy and spreading functions. D. Tonality index. E. Threshold energy and absolute threshold. F. Signal-to-mask ratios. (*Boley and Rao, 2004*)

components are more accurately predicted. Amplitude and phase are predicted to form an unpredictability measure C. When $C = 0$, the current value is completely predicted, and when $C = 1$, the predicted values differ from the actual values. This yields the tonality index T ranging from 0 (high unpredictability and noise-like) to 1 (low unpredictability and tonal). For example, the audio signal's strongly tonal and nontonal areas are evident in Fig. 11.12D. The tonality index is used to calculate a $\Delta(z)$ shift, for example, interpolating values from 6 dB (nontonal) to 29 dB (tonal).

When used in a Layer III encoder, model 2 is modified. The model is executed twice, once with a long block and once with a short 256-sample block. These values are used in the unpredictability measure calculation. A slightly different spreading function is used. The NMT shift is changed to 6.0 dB and a fixed TMN shift of 29.0 dB is used. As noted, a pre-echo control is calculated. Perceptual entropy is calculated as the logarithm of the geometric mean of the normalized spectral energy in a partition. This predicts the minimum number of bits needed for transparency. High values are used to identify transient attacks, and thus to determine block size in the encoder. In addition, model 2 accepts the minimum masking threshold at low frequencies where there is good correspondence between subbands and critical bands, and it uses the average of the thresholds at higher frequencies where subbands are narrow compared to critical bands.

Much research has been done since the informative model 2 was published in the MPEG-1 standard. Thus, most practical encoders use models that offer better performance, even if they are based on the informative model. An encoder that follows the informative documentation literally will not provide good results compared to more sophisticated implementations.

MPEG-2 Audio Standard

The MPEG-2 audio standard was designed for applications ranging from Internet downloading to high-definition digital television (HDTV) transmission. It provides a backward-compatible path to multichannel sound and a low sampling frequency provision, as well as a non-backward-compatible multichannel format known as Advanced Audio Coding (AAC). The MPEG-2 audio standard encompasses the MPEG-1 audio standard of Layers I, II, and III, using the same encoding and decoding principles as MPEG-1. In many cases, the same layer algorithms developed for MPEG-1 applications are used for MPEG-2 applications. Multichannel MPEG-2 audio is backward compatible with MPEG-1. An MPEG-2 decoder will accept an MPEG-1 bitstream and an MPEG-1 decoder can derive a stereo signal from an MPEG-2 bitstream. However, MPEG-2 also permits use of incompatible audio codecs.

One part of the MPEG-2 standard provides multichannel sound at sampling frequencies of 32, 44.1, and 48 kHz. Because it is backward compatible to MPEG-1, it is designated as BC (backward compatible), that is, MPEG-2 BC. Clearly, because there is more redundancy between six channels than between two, greater coding efficiency is achieved. Overall, 5.1 channels can be successfully coded at rates from 384 kbps to 640 kbps. MPEG-2 also supports monaural and stereo coding at sampling frequencies of 16, 22.05, and 24 kHz, using Layers I, II, and III. The MPEG-1 and -2 audio coding family is shown in Fig. 11.13. The MPEG-2 audio standard was approved by the MPEG committee in November 1994 and is specified in ISO/IEC 13818-3.

The multichannel MPEG-2 BC format uses a five-channel approach sometimes referred to as 3/2 + 1 stereo (3 front and 2 surround channels + subwoofer). The low-frequency

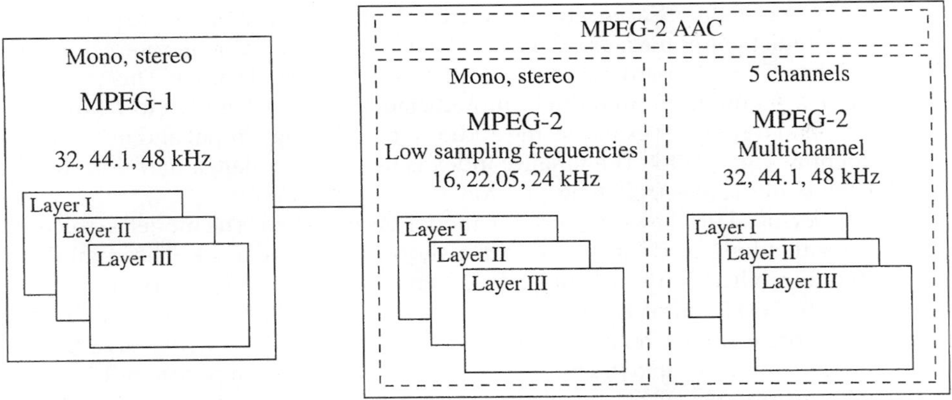

FIGURE 11.13 The MPEG-2 audio standard adds monaural/stereo coding at low sampling frequencies, multichannel coding, and AAC. The three MPEG-1 layers are supported.

effects (LFE) subwoofer channel is optional, providing an audio range up to 120 Hz. A hierarchy of formats is created in which 3/2 may be downmixed to 3/1, 3/0, 2/2, 2/1, 2/0, and 1/0. The multichannel MPEG-2 BC format uses an encoder matrix that allows a two-channel decoder to decode a compatible two-channel signal that is a subset of a multichannel bitstream. The multiple channels of MPEG-2 are matrixed to form compatible MPEG-1 left/right channels, as well as other MPEG-2 channels, as shown in Fig. 11.14. The MPEG-1 left and right channels are replaced by matrixed MPEG-2 left and right channels and these are encoded into backward-compatible MPEG frames with an MPEG-1 encoder. Additional multichannel data is placed in the expanded ancillary data field.

To efficiently code multiple channels, MPEG-2 BC uses techniques such as dynamic crosstalk reduction, adaptive interchannel prediction, and center channel phantom image coding. With dynamic crosstalk reduction, as with intensity coding, multichannel high-frequency information is combined and conveyed along with scale factors to direct levels

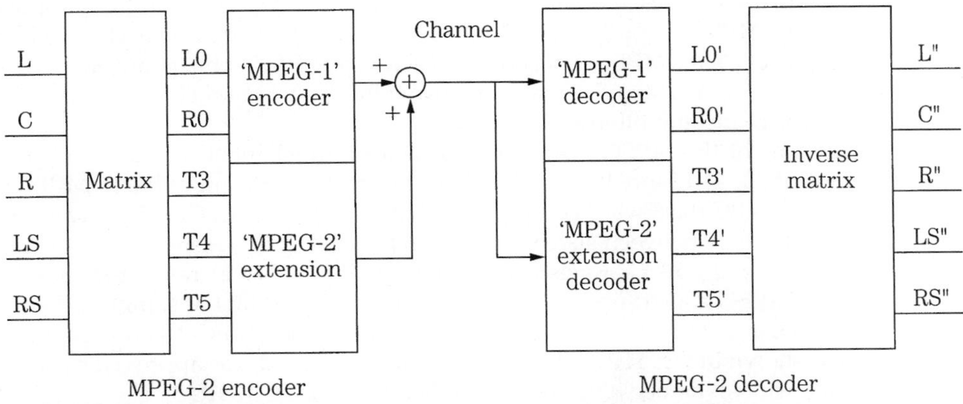

FIGURE 11.14 The MPEG-2 audio encoder and decoder showing how a 5.1-channel surround format can be achieved with backward compatibility with MPEG-1.

to different playback channels. In adaptive prediction, a prediction error signal is conveyed for the center and surround channels. The high-frequency information in the center channel can be conveyed through the front left and right channels as a phantom image.

MPEG-2 BC can achieve a combined bit rate of 384 kbps, using Layer II at a 48-kHz sampling frequency. MPEG-2 allows for audio bit rates up to 1066 kbps. To accommodate this, the MPEG- 2 frame is divided into two parts. The first part is an MPEG-1-compatible stereo section with Layer I data up to 448 kbps, Layer II data up to 384 kbps, or Layer III data up to 320 kbps. The MPEG-2 extension part contains all other surround data.

A standard two-channel MPEG-1 decoder ignores the ancillary information, and reproduces the front main channels. In some cases, the dematrixing procedure in the decoder can yield an artifact in which the sound in a channel is mainly phase canceled but the quantization noise is not, and thus becomes audible. This limitation of spatial unmasking in MPEG-2 BC is a direct result of the matrixing used to achieve backward compatibility with the original two-channel MPEG standard. In part, it can be addressed by increasing the bit rate of the coded signals.

MPEG-2 also specifies Layer I, II, and III at low sampling frequencies (LSF) of 16, 22.05, and 24 kHz. This extension is not backward compatible to MPEG-1 codecs. This portion of the standard is known as MPEG-2 LSF. At these low bit rates, Layer III generally shows the best performance. Only minor changes in the MPEG-1 bit rate and bit allocation tables are necessary to adapt this LSF format. The relative improvement in quality stems from the improved frequency resolution of the polyphase filter bank in low- and mid-frequency regions; this allows more efficient application of masking. Layers I and II fare better than Layer III in these applications because Layer III already has good frequency resolution. The bitstream is unchanged in the LSF mode and the same frame format is used. For 24-kHz sampling, the frame length is 16 ms for Layer I and 48 ms for Layer II. The frame length of Layer III is decreased relative to that of MPEG-1. In addition, the "MPEG-2.5" standard supports sampling frequencies of 8, 11.025, and 12 kHz with the corresponding decrease in audio bandwidth; implementations use Layer III as the codec. Many MP3 codecs support the original MPEG-1 Layer III codec as well as the MPEG-2 and MPEG-2.5 extensions for lower sampling frequencies.

The menu of data rates, fidelity, and layer compatibility provided by MPEG are useful in a wide variety of applications such as computer multimedia, CD-ROM, DVD-Video, computer disks, local area networks, studio recording and editing, multichannel disk recording, ISDN transmission, digital audio broadcasting, and multichannel digital television. Numerous C and C++ programs performing MPEG-1 and -2 audio coding and decoding can be downloaded from a number of Internet file sites, and executed on personal computers. The backward-compatible format, using Layer II coding, is used for the soundtracks of some DVD-Video discs. However, a matrix approach to surround sound does not preserve spatial fidelity as well as discrete channel coding.

MPEG-2 AAC

The MPEG-2 Advanced Audio Coding (AAC) format codes monaural, stereo, or multichannel playback for up to 48 channels, including 5.1-channel, at a variety of bit rates. AAC is known for its relatively high fidelity at low bit rates; for example, about 64 kbps per channel. It also provides high-quality 5.1-channel coding at an overall rate of 320 kbps or 384 kbps. AAC uses a reference model (RM) structure in which a set of tools (modules) has defined interfaces and can be combined variously in three different profiles.

Individual tools can be upgraded and used to replace older tools in the reference software. In addition, this modularity makes it easy to compare revisions against older versions. AAC also comprises the kernel of audio tools used in the MPEG-4 standard for coding high-quality audio. AAC also supports lossless coding. AAC is specified in Part 7 of the MPEG-2 standard (ISO/IEC 13818-7), which was finalized in April 1997.

MPEG-2 AAC coding is not backward compatible with MPEG-1 and was originally designated as NBC (non-backward compatible) coding. An AAC bitstream cannot be decoded by an MPEG-1-only decoder. By lifting the constraint of compatibility, better performance is achieved compared to MPEG-2 BC. MPEG-2 AAC supports standard sampling frequencies of 32, 44.1, and 48 kHz, as well as other rates from 8 kHz to 96 kHz, yielding maximum bit rates of 48 kbps and 576 kbps, respectively. Its input channel configurations are: 1/0 (monaural), 2/0 (two-channel stereo), different multichannel configurations up to 3/2 + 1, and provision for up to 48 channels. Matrixing is not used. Downmixing is supported. To improve error performance, the system is designed to maintain bitstream synchronization in the presence of bit errors, and error concealment is supported as well.

To allow flexibility in audio quality versus processing requirements, AAC coding modules are used to create three profiles: main profile, scalable sampling rate (SSR) profile, and low-complexity (LC) profile. The main profile employs the most sophisticated encoder using all the coding modules except preprocessing to yield the highest audio quality at any bit rate. A main profile decoder can also decode the low-complexity bitstream. The SSR profile uses a gain control tool to perform polyphase quadrature filtering (PQF), gain detection, and gain modification preprocessing; prediction is not used and temporal noise shaping (TNS) order is limited. SSR divides the audio signal into four equal frequency bands each with an independent bitstream and decoders can choose to decode one or more streams and thus vary the bandwidth of the output signal. SSR provides partial compatibility with the low-complexity profile; the decoded signal is bandlimited. The LC profile does not use preprocessing or prediction tools and the TNS order is limited. LC operates with low memory and processing requirements.

AAC Main Profile

A block diagram of a main profile AAC encoder and decoder is shown in Fig. 11.15. An MDCT with 50% overlap is used as the only input signal filter bank. It uses lengths of 1024 for stationary signals or 128 for transient signals, with a 2048-point window or a block of eight 256-point windows, respectively. To preserve interchannel block synchronization (phase), short block lengths are retained for eight-block durations. For multichannel coding, different filter bank resolutions can be used for different channels. At 48 kHz, the long-window frequency resolution is 23 Hz and time resolution is 21 ms; the short window yields 187 Hz and 2.6 ms. The MDCT employs time-domain aliasing cancellation (TDAC). Two alternate window shapes are selectable on a frame basis in the 2048-point mode; either sine or Kaiser–Bessel-derived (KBD) windows can be employed. The encoder can select the optimal window shape on the basis of signal characteristics. The sine window is used when perceptually important components are spaced closer than 140 Hz and narrow-band selectivity is more important than stopband attenuation. The KBD window is used when components are spaced more than 220 Hz apart and stopband attenuation is needed. Window switching is seamless, even with the overlap-add sequence. The shape of the left half of each window must match

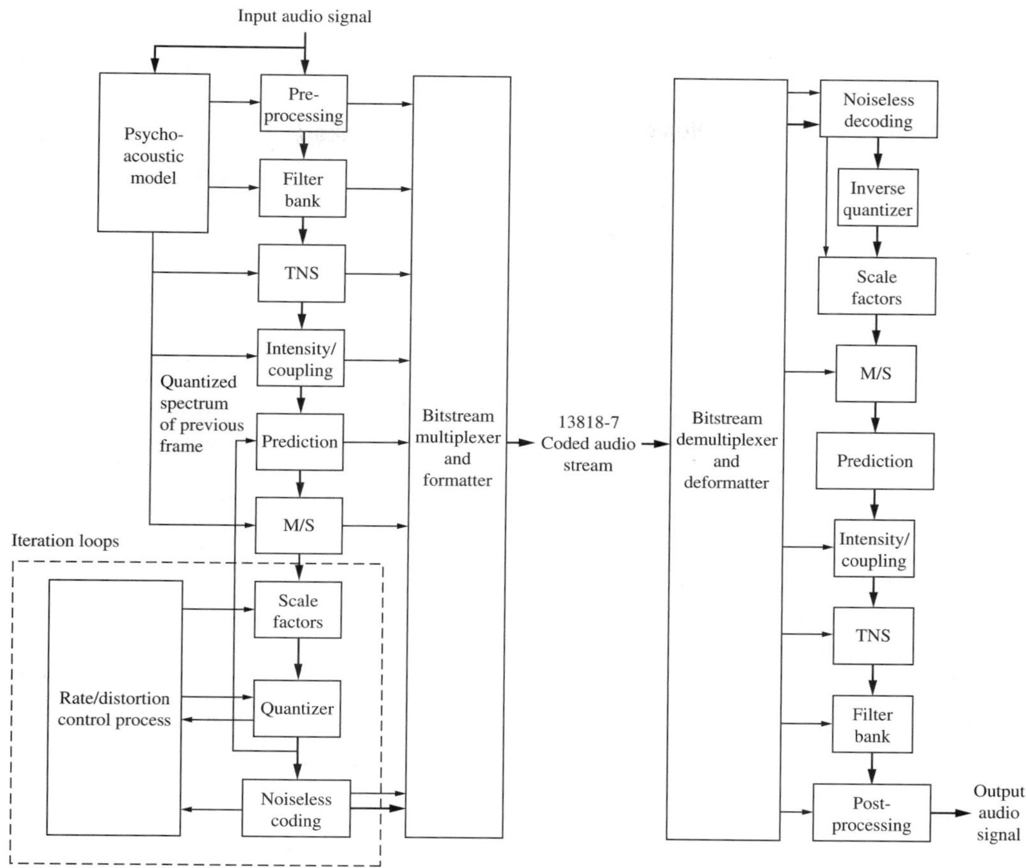

Figure 11.15 Block diagram of MPEG-2 AAC encoder and decoder. Heavy lines denote data paths, light lines denote control signals.

the shape of the right half of the preceding window; a new window shape is thus introduced as a new right half.

The suggested psychoacoustic model is based on the MPEG-1 model 2 and examines the perceptual entropy of the audio signal. It controls the quantizer step size, increasing step size to decrease buffer levels during stationary signals, and correspondingly decreasing step size to allow levels to rise during transient signals.

A second-order backward-adaptive predictor is applied to remove redundancy in stationary signals found in long windows; residues are calculated and used to replace frequency coefficients. Reconstructed coefficients in successive blocks are examined for frequencies below 16 kHz. Values from two previous blocks are used to form one predicted value for each current coefficient. The predicted value is subtracted from the actual target value to yield a prediction error (residue) which is quantized. Coefficient residues are grouped into scale factor bands that emulate critical bands. A prediction control algorithm determines if prediction should be activated in individual scale factor bands or in the frame at all, based on whether it improves coding gain.

AAC Allocation Loops

Two nested inner and outer loops iteratively perform nonuniform quantization and analysis-by-synthesis. The simplified nested algorithms are shown in Fig. 11.16. The inner loop (within the outer loop) begins with an initial quantization step size that is used to quantize the data and perform Huffman coding to determine the number of bits needed for coding. If necessary, the quantizer step size can be increased to reduce the number of bits needed. The outer loop uses scale factors to amplify scale factor bands to reduce audibility of quantization noise (inverse scale factors are applied in the decoder). Each scale factor band is assigned one multiplying scale factor. The scale factor is a gain value that changes the amplitude of the coefficients in the scale factor band; this shapes the quantization noise according to the masking threshold. The outer loop uses analysis-by-synthesis to determine the resulting distortion and this is compared to the distortion allowed by the psychoacoustic model; the best result so far is stored. If distortion is too high in a scale factor band, the band is amplified (this increases the bit rate) and the outer loop repeats. The two loops work in conjunction to optimally distribute quantization noise across the spectrum.

The width of the scale factor bands is limited to 32 coefficients, except in the last scale factor band. There are 49 scale factor bands for long blocks. Scale factor bands can

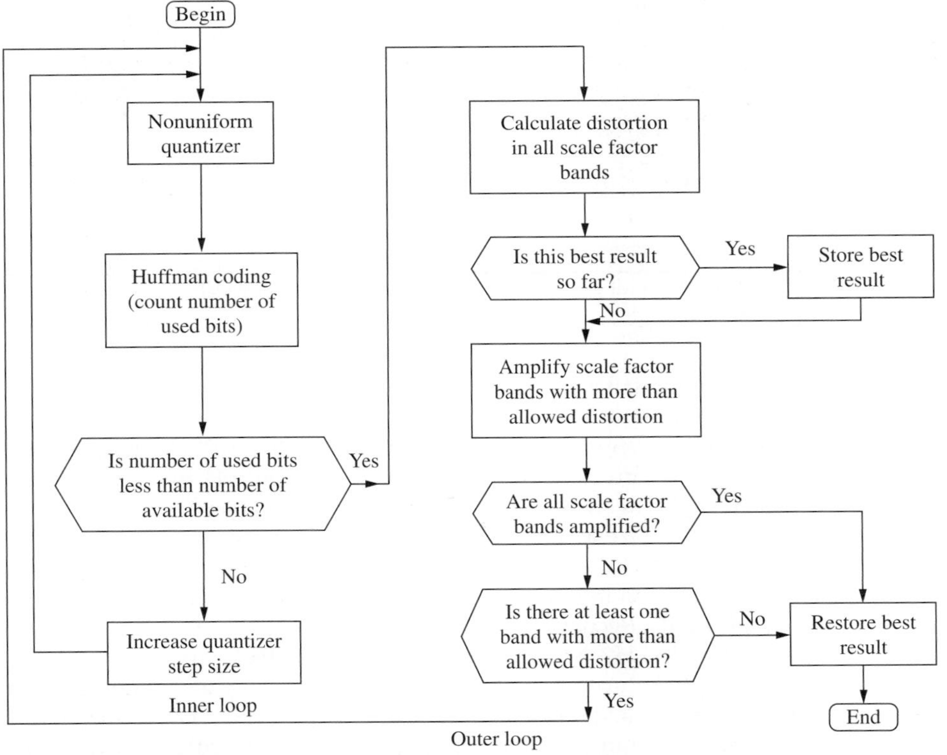

FIGURE 11.16 Two nested inner and outer allocation loops iteratively perform nonuniform quantization and analysis-by-synthesis.

be individually amplified in increments of 1.5 dB. Noise shaping results because amplified coefficients have larger values and will yield a higher SNR after quantization. Because inverse amplification must be applied at the decoder, scale factors are transmitted in the bitstream. Designers should note that scale factors are defined with opposite polarity in MPEG-2 AAC and MPEG-1/2 Layer III (larger scale factor values represent larger signals in AAC, whereas it is the opposite in Layer III).

Huffman coding is applied to the quantized spectrum, scale factors, and directional information. Twelve Huffman codebooks are available to code pairs or quadruples of quantized spectral values. Two codebooks are available for each maximum value, each representing a different probability function. A bit reservoir accommodates instantaneously variable bit rates, allowing bits to be distributed across consecutive blocks for more effective coding within the average bit-rate constraint. A frame output consists of spectral coefficients and control parameters. The bitstream syntax defines a lower layer for raw audio data, and a higher layer contains audio transport data. In the decoder, current spectral components are reconstructed by adding a prediction error to the predicted value. As in the encoder, the coefficients are calculated from preceding values; no additional information is required.

AAC Temporal Noise Shaping

The spectral predictability of signals dictates the optimal coding strategy. For example, consider a steady-state sine wave comprising a flat temporal envelope, and a single spectral line—an impulse which is maximally nonflat spectrally. This sine wave is most easily coded directly in the frequency domain or by using linear prediction in the time domain. Conversely, consider a transient pulse signal comprising an impulse in the time domain, and a flat power spectrum. This pulse would be difficult to code directly in the frequency domain and difficult to code with prediction in the time domain. However, the pulse could be optimally coded directly in the time domain, or by using linear prediction in the frequency domain.

In the AAC codec, predictive coding is used to examine coefficients in each block. Transient signals will yield a more uniform spectrum and allow transients to be identified and more efficiently coded as residues. When coding transients, by analyzing the spectral data from the MDCT, temporal noise shaping (TNS) can be used to control the temporal shape of the quantization noise within each window to achieve perceptual noise shaping. By using the duality between the time and frequency domains, TNS provides improved predictive coding. When a time-domain signal is coded with predictive coding, the power spectral density of the quantization noise in the output signal will be shaped by the power spectral density of the input signal. Conversely, when a frequency-domain signal is coded with predictive coding, the temporal shape of the quantization noise in the output signal will follow the temporal shape of the input signal.

In particular, TNS shapes the temporal envelope of the quantization noise to follow the transient's temporal envelope and thus conceals the noise under the transient. This can overcome problems such as pre-echo. As noted, this is accomplished with linear predictive coding of the spectral signal; for example, using open-loop differential pulse-code modulation (DPCM) encoding of spectral values. Corresponding DPCM decoding is performed in the decoder to create the output signal. During encoding, TNS replaces the target spectral coefficients with the forward-prediction residual (prediction error). In the AAC main profile, up to 20 successive coefficients in a block can be examined to predict the next coefficient and the prediction value is subtracted from the target coefficient

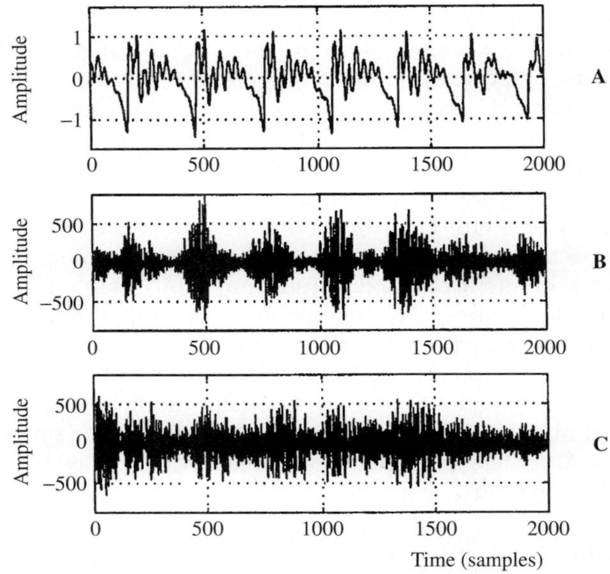

FIGURE 11.17 An example showing how TNS shapes quantization noise to conceal it under the transient envelope. A. The original speech signal. B. The quantization coding noise shaped with TNS. C. The quantization coding noise without TNS; masking is not utilized as well. (*Herre and Johnston, 1997*)

to yield a spectral residue, which is quantized and encoded. A filter order up to 12 is allowed in the LC and SSR profiles. During decoding, the inverse predictive TNS filtering is performed to replace the residual values with spectral coefficients.

It should be emphasized that TNS prediction is done over frequency, and not over time. Thus the prediction error is shaped in time as opposed to frequency. Time resolution is increased as opposed to frequency resolution; temporal spread of quantization noise is reduced in the output decoded signal. TNS thus allows the encoder to control temporal pre-echo quantization noise within a filter-bank window by shaping it according to the audio signal, so that the noise is masked by the temporal audio signal, as shown in Fig. 11.17. TNS allows better coding of both transient content and pitch-based signals such as speech. The impulses which comprise speech are not always effectively coded with traditional transform block switching and may demand instantaneous increases in bit rate. TNS minimizes unmasked pre-echo in pitch-based signals and reduces the peak bit demand. With TNS, the codec can also use the more efficient long-block mode more often without introducing artifacts, and can also perform better at low sampling frequencies. TNS effectively and dynamically adapts the codec between high-time resolution for transient signals and high-frequency resolution for stationary signals and is more efficient than other designs using switched windows. As explained by Juergen Herre, the prediction filter can be determined from the range of spectral coefficients corresponding to the target frequency range (for example, 4 kHz to 20 kHz) and by using DPCM predictive coding methods such as calculating the autocorrelation function of the coefficients and using the Levinson–Durban recursion algorithm.

A single TNS prediction filter can be applied to the entire spectrum or different TNS prediction filters can be uniquely applied to different parts of a spectrum, and TNS can be omitted for some frequency regions. Thus the temporal quantization noise control can be applied in a frequency-dependent manner.

AAC Techniques and Performance

The input audio signal can be applied to a four-band polyphase quadrature mirror filter (PQMF) bank to create four equal-width, critically sampled frequency bands. This is used for the scalable sampling rate (SSR) profile. An MDCT is used to produce 256 spectral coefficients from each of the four bands, for a total of 1024 coefficients. Positive or negative gain control can be applied independently to each of the four bands. With SSR, lower sampling rate signals (with lower bit rates) can be obtained at the decoder by ignoring the upper PQMF bands. For example, bandwidths of 18, 12, and 6 kHz can be obtained by ignoring one, two, or three bands. This allows scalability with low decoder complexity.

Two stereo coding techniques are used in AAC: intensity coding and M/S (middle/side) coding. Both methods can be combined and applied to selective parts of the signal's spectrum. M/S coding is applied between channel pairs that are symmetrically placed to the left and right of the listener; this helps avoid spatial unmasking. M/S coding can be selectively switched in time (block by block) and frequency (scale factor bands). M/S coding can control the imaging of coding noise that is separate from the imaging of the masking signal. High-frequency time-domain imaging must be preserved in transient signals. Intensity stereo coding considers that perception of high-frequency sounds is based on their energy-time envelopes. Thus, some signals can be conveyed with one set of spectral values, shared among channels. Envelope information is maintained by reconstructing each channel level. Intensity coding can be implemented between channel pairs, and among coupling channel elements. In the latter, channel spectra are shared between channel pairs. Also, coupling channels permit downmixing in which additional audio elements such as a voice-over can be added to a recording. Both of these techniques can be used on both stereo and 5.1 multichannel content.

In one listening test, multichannel MPEG-2 AAC at 320 kbps outperformed MPEG-2 Layer II BC at 640 kbps. MPEG-2 Layer II at 640 kbps did not outperform MPEG-2 AAC at 256 kbps. For five full-bandwidth channels, MPEG-2 AAC claims "indistinguishable quality" for bit rates as low as 256 kbps to 320 kbps. Stereo MPEG-2 AAC at 128 kbps is said to provide significantly better sound quality than MPEG-2 Layer II at 192 kbps or MPEG-2 Layer III at 128 kbps. MPEG-2 AAC at 96 kbps is comparable to MPEG-2 Layer II at 192 kbps or MPEG-2 Layer III at 128 kbps. Spectral band replication (SBR) can be applied to AAC codecs. This is sometimes known as High-Efficiency AAC (HE AAC) or aacPlus. With SBR, a bit rate of 24 kbps per channel, or 32 kbps to 40 kbps for stereo signals, can yield good results. The MPEG-4 and MPEG-7 standards are discussed in Chap. 15.

ATRAC Codec

The proprietary ATRAC (Adaptive TRansform Acoustic Coding) algorithm was developed to provide data reduction for the SDDS cinematic sound system and was subsequently employed in other applications such as the MiniDisc format. ATRAC uses a modified discrete cosine transform and psychoacoustic masking to achieve a 5:1 compression ratio; for example, data on a MiniDisc is stored at 292 kbps. ATRAC transform coding is based on nonuniform frequency and time splitting concepts, and assigns bits according to rules fixed by a bit allocation algorithm. The algorithm both observes the fixed threshold of hearing curve, and dynamically analyzes the audio program to take advantage of psychoacoustic effects such as masking. The original codec version is sometimes known as ATRAC1. ATRAC was developed by Sony Corporation.

An ATRAC encoder accepts a digital audio input and parses it into blocks. The audio signal is divided into three subbands, which are then transformed into the frequency domain using a variable block length. Transform coefficients are grouped into 52 subbands (called block floating units or BFUs) modeled on the ear's critical bands, with particular resolution given to lower frequencies. Data in these bands is quantized according to dynamic sensitivity and masking characteristics based on a psychoacoustic model. During decoding, the quantized spectra are reconstructed according to the bit allocation method, and synthesized into the output audio signal.

ATRAC differs from some other codecs in that psychoacoustic principles are applied to both the bit allocation and the time-frequency splitting. In that respect, both subband and transform coding techniques are used. In addition, the transform block length adapts to the audio signal's characteristics so that amplitude and time resolution can be varied between static and transient musical passages. Through this processing, the data rate is reduced by 4/5. The ATRAC encoding algorithm can be considered in three parts: time-frequency analysis, bit allocation, and quantization of spectral components. The analysis portion of the algorithm decomposes the signal into spectral coefficients grouped into BFUs that emulate critical bands. The bit allocation portion of the algorithm divides available bits between the BFUs, allocating more bits to perceptually sensitive units. The quantization portion of the algorithm quantizes each spectral coefficient to the specified word length.

The time-frequency analysis, shown in Fig. 11.18, uses subband and transform coding techniques. Two quadrature mirror filters (QMFs) divide the input signal into three subbands: low (0 Hz to 5.5125 kHz), medium (5.5125 kHz to 11.025 kHz), and high (11.025 kHz to 22.05 kHz). The QMF banks ensure that time-domain aliasing caused by the subband decomposition will be canceled during reconstruction. Following splitting, contents are examined to determine the length of block durations. Signals in each

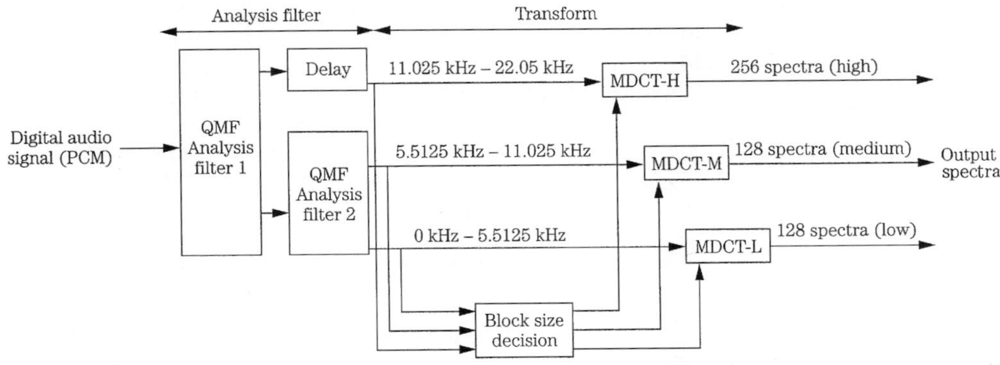

FIGURE 11.18 The ATRAC encoder time-frequency analysis block contains QMF filter banks and MDCT transforms to analyze the signal.

of these bands are then placed in the frequency domain with the MDCT algorithm. The MDCT allows up to a 50% overlap between adjacent time-domain windows; this maintains frequency resolution at critical sampling. A total of 512 coefficients are output, with 128 spectra in the low band, 128 spectra in the mid band, and 256 spectra in the high band.

Transform coders must balance frequency resolution with temporal resolution. A long block size achieves high frequency resolution and quantization noise is readily masked by simultaneous masking; this is appropriate for a steady-state signal. However, transient signals require temporal resolution, otherwise quantization noise will be spread in time over the block of samples; a pre-echo can be audible prior to the onset of the transient masker. Thus, instead of a fixed transform block length, the ATRAC algorithm adaptively performs nonuniform time splitting with blocks that vary according to the audio program content. Two modes are used: long mode (11.6 ms in the high-, medium-, and low-frequency bands) and short mode (1.45 ms in the high-frequency band, and 2.9 ms in the mid- and low-frequency bands). The long block mode yields a narrow frequency band, and the short block mode yields wider frequency bands, trading time and frequency resolution as required by the audio signal. Specifically, transient attacks prompt a decrease in block duration (to 1.45 ms or 2.9 ms), and a more slowly changing program promotes an increase in block duration (to 11.6 ms). Block duration is interactive with frequency bandwidth; longer block durations permit selection of narrower frequency bands and greater resolution. This time splitting is based on the effect of temporal pre-masking (backward masking) in which tones sounding close in time exhibit masking properties.

Normally, the long mode provides good frequency resolution. However, with transients, quantization noise is spread over the entire signal block and the initial quantization noise is not masked. Thus, when a transient is detected, the algorithm switches to the short mode. Because the noise is limited to a short duration before the onset of the transient, it is masked by pre-masking. Because of its greater extent, post-masking (forward masking) can be relied on to mask any signal decay in the long mode. The block size mode can be selected independently for each band. For example, a long block mode might be selected in the low-frequency band, and short modes in the mid- and high-frequency bands.

The MDCT frequency domain coefficients are then grouped into 52 BFUs; each contains a fixed number of coefficients. As noted, in the long mode, each unit conveys 11.6 ms of a narrow frequency band, and in the short mode each block conveys 1.45 ms or 2.9 ms of a wider frequency band. Fifty-two nonuniform BFUs are present across the frequency range; there are more BFUs at low frequencies, and fewer at high frequencies. This nonlinear division is based on the concept of critical bands. In the ATRAC model, for example, the band centered at 150 Hz is 100 Hz wide, the band at 1 kHz is 160 Hz wide, and the band at 10.5 kHz is 2500 Hz wide. These widths reflect the ear's decreasing sensitivity to high frequencies.

Each of the 512 spectral coefficients is quantized according to scale factor and word length. The scale factor defines the full-scale range of the quantization. It is selected from a list of possibilities and describes the magnitude of the spectral coefficients in each of the 52 BFUs. The word length defines the precision within each scale; it is calculated by the bit allocation algorithm as described below. All the coefficients in a given BFU are given the same scale factor and quantization word length because of the psychoacoustic similarity within each group. Thus the following information is coded for each frame of 512 values: MDCT block size mode (long or short), word length for each BFU, scale factor for each BFU, and quantized spectral coefficients.

The bit allocation algorithm considers the minimum threshold curve and simultaneous masking conditions applicable to the BFUs, operating to yield a reduced data rate. Available bits must be divided optimally between the block floating units. BFUs coded with many bits will have low quantization noise, but BFUs with few bits will have greater noise. ATRAC does not specify an arbitrary bit allocation algorithm; this allows improvement in future encoder versions. The decoder is completely independent of any allocation algorithm, also allowing future improvement. To some extent, because the time-frequency splitting relies on critical band and pre-masking considerations, the choice of the bit allocation algorithm is less critical. However, any algorithm must minimize perceptual error.

One example of a bit allocation model declares both fixed and variable bits, as shown in Fig. 11.19. Fixed bits are allocated mainly to low-frequency BFU regions,

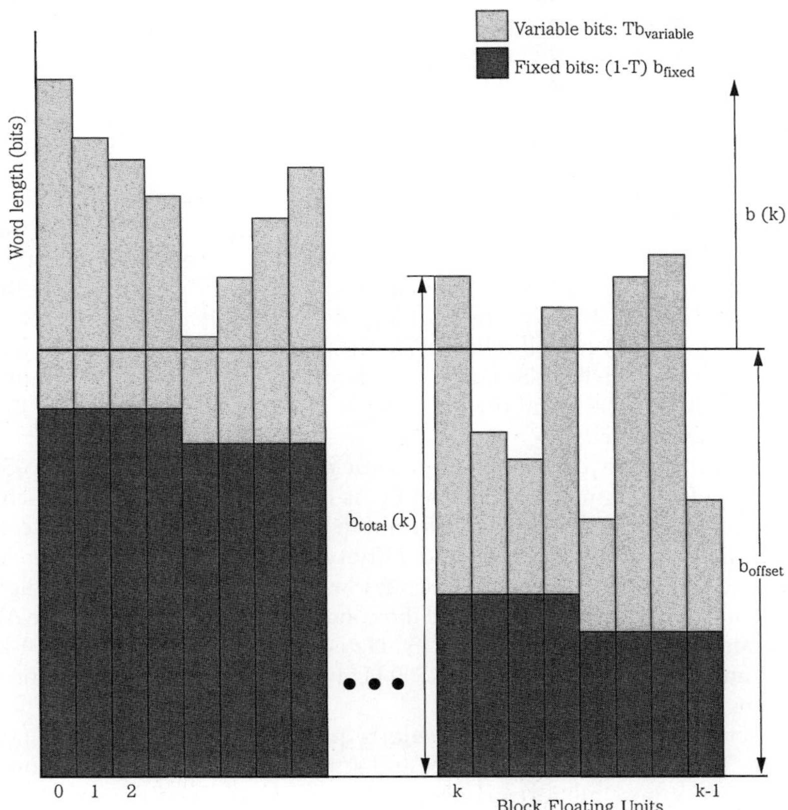

Figure 11.19 An example of a bit-allocation algorithm showing the bit assignment, using both fixed and variable bits. Fixed bits are weighted toward low-frequency BFU regions. Variable bits are assigned according to the logarithm of the spectral coefficients in each BFU. (*Tsutsui et al., 1996*)

emphasizing their perceptual importance. Variable bits are assigned according to the logarithm of the spectral coefficients in each BFU. The total bit allocation b_{total} for each BFU is the weighted sum of the fixed bits $b_{fixed}(k)$ and the variable bits $b_{variable}(k)$ in each BFU. Thus, for each BFU k:

$$b_{total}(k) = Tb_{variable} + (1 - T)b_{fixed}$$

The weight T describes the tonality of the signal, taking a value close to 0 for nontonal signals, and a value close to 1 for tonal signals. Thus the proportion of fixed bits to variable bits is itself variable. For example, for noise-like signals the allocation emphasizes fixed bits, thus decreasing the number of bits devoted to insensitive high frequencies. For pure tones, the allocation emphasizes variable bits, concentrating available bits to a few sensitive BFUs with tonal components.

However, the allocation method must observe the overall bit rate. The previous equation does not account for this and will generally allocate more bits than available. To maintain a fixed and limited bit rate, an offset b_{offset} is devised, and set equal for all BFUs. The offset is subtracted from $b_{total}(k)$ for each BFU, yielding the final bit allocation $b_{final}(k)$:

$$b_{final}(k) = \text{Integer}[b_{total}(k) - b_{offset}]$$

If the final value describes a negative word length, that BFU is given zero bits. Because low frequencies are given a greater number of fixed bits, they generally need fewer variable bits to achieve the offset threshold, and become coded (see Fig. 11.19). To meet the required output bit rate, the global bit allocation can be raised or lowered by correspondingly raising or lowering the threshold of masking. As noted, ATRAC does not specify this, or any other arbitrary allocation algorithm.

The ATRAC decoder essentially reverses the encoding process, performing spectral reconstruction and time-frequency synthesis. Time-frequency synthesis is shown in Fig. 11.20. The decoder first accepts the quantized spectral coefficients, and uses the word length and scale factor parameters to reconstruct the MDCT spectral coefficients. To reconstruct the audio signal, these coefficients are first transformed back into the time domain by the inverse MDCT (IMDCT), using either long or short mode blocks as specified by the received parameters. The three time-domain subband signals are synthesized into the output signal using QMF synthesis banks, obtaining a full spectrum,

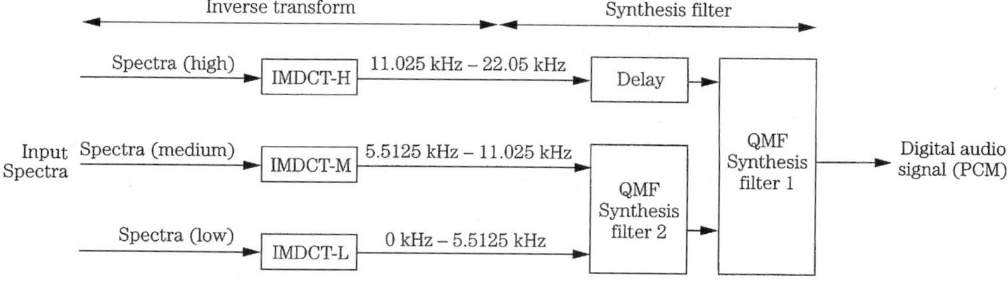

FIGURE 11.20 The ATRAC decoder time-frequency synthesis block contains QMF banks and MDCT transforms to synthesize and reconstruct the signal.

16-bit digital audio signal. Wideband quantization noise introduced during encoding (to achieve data reduction) is limited to critical bands, where it is masked by signal energy in each band.

Other versions of ATRAC were developed. ATRAC3 achieves twice the compression of ATRAC1 while providing similar sound quality operating at bit rates such as 128 kbps. The broadband audio signal is split into four subbands using a QMF bank; the bands are 0 Hz to 2.75625 kHz, 2.75625 kHz to 5.5125 kHz, 5.5125 kHz to 11.025 kHz, and 11.025 kHz to 22.05 kHz. Gain control is applied to each band to minimize pre-echo. When a transient occurs, the amplitude of the section preceding the attack is increased. Gain is correspondingly decreased during decoding, effectively attenuating pre-echo. The subbands are applied to fixed-length MDCT with 256 components. Tonal components are subtracted from the signal and analyzed and quantized separately. Entropy coding is applied. In addition, joint stereo coding can be used adaptively for each band.

The ATRAC3plus codec is designed to operate at generally lower bit rates; rates of 48, 64, 132, and 256 kbps are often used. The broadband audio signal is processed in 16 subbands; a window of up to 4096 samples (92 ms) can be used and bits can be allocated unequally over two channels.

The ATRAC Advanced Lossless (AAL) codec provides scalable lossless compression. It codes ATRAC3 or ATRAC3plus data as well as residual information that is otherwise lost. The ATRAC3 or ATRAC3plus data can be decoded alone for lossy reproduction or the residual can be added for lossless reproduction.

Perceptual Audio Coding (PAC) Codec

The Perceptual Audio Coding (PAC) codec was designed to provide audio coding with bit rates ranging from 6 kbps for a monophonic channel to 1024 kbps for a 5.1-channel format. It was particularly aimed at digital audio broadcast and Internet download applications, at a rate of 128 kbps for two-channel near-CD quality coding; however, 96 kbps may be used for FM quality. PAC employs coding methods that remove signal perceptual irrelevancy, as well as source coding to remove signal redundancy, to achieve a reduction ratio of about 11:1 while maintaining transparency. PAC is a third-generation codec with PXFM and ASPEC as its antecedents, the latter also providing the ancestral basis for MPEG-1 Layer III. PAC was developed by AT&T and Bell Laboratories of Lucent Technologies.

The architecture of a PAC encoder is similar to that of other perceptual codecs. Throughout the algorithm, data is placed in blocks of 1024 samples per channel. An MDCT filter bank converts time-domain audio signals to the frequency domain; a hybrid filter is not used. The MDCT uses an adaptive window size to control quantization noise spreading, where the spreading is greater in the time domain with a longer 2048-point window and greater in the frequency domain with a series of shorter 256-point windows. Specifically, a frequency resolution of 1024 uniformly spaced frequency bands (a window of 2048 points) is usually employed. When signal transient characteristics suggest that pre-echo artifacts may occur, the filter bank adaptively switches to a transform with 128 bands. In either case, the perceptual model calculates a frequency-domain masking threshold to determine the maximum quantization noise that can be added to each frequency band without an audible penalty. The perceptual model used in PAC to code monophonic signals is similar to the MPEG-1 psychoacoustic model 2.

The audio signal, represented as spectral coefficients, is requantized to one of 128 exponentially distributed quantization step sizes according to noise allocation determinations. The codec uses a variety of frequency band groupings. A fixed "threshold calculation partition" is a set of one-to-many adjacent filter bank outputs arranged to create a partition width that is about 1/3 of a critical band. Fixed "coder bands" consist of a multiple of four adjacent filter bank outputs, ranging from 4 to 32 outputs, yielding a bandwidth as close to 1/3 critical band as possible. There are 49 coder bands for the 1024-point mode and 14 coder bands for the 128-point filter mode. An iterative rate control loop is used to determine quantization relative to masking thresholds. Time buffering may be used to smooth the resulting bit rate. Coder bands are assigned one scale factor. "Sections" are data dependent groupings of adjacent coder bands using the same Huffman codeword. Coefficients in each coder band are encoded using one of 16 Huffman codebooks.

At the codec output, a formatter generates a packetized bitstream. One 1024-sample block (or eight 128-sample blocks) from each channel are placed in one packet, regardless of the number of channels. The size of a packet corresponding to each 1024 input samples is thus variable. Depending on the reliability of the transmission medium, additional header information is added to the first frame, or to every frame. A header may contain data such as synchronization, error correction, sample rate, number of channels, and transmission bit rate.

For joint-stereo coding, the codec employs a binary masking level difference (BMLD) using M (monaural, L+R), S (stereo, L-R) and independent L and R thresholds. M-S versus L-R coding decisions are made independently for each band. The multichannel MPAC codec (for example, coding 5.1 channels) computes individual masking thresholds for each channel, two pairs (front and surround) of M-S thresholds, as well as a global threshold based on all channels. The global threshold takes advantages of masking across all channels and is used when the bit pool is close to depletion.

PAC employs unequal error protection (UEP) to more carefully protect some portions of the data. For example, corrupted control information could lead to a catastrophic loss of synchronization. Moreover, some errors in audio data are more disruptive than others. For example, distortion in midrange frequencies is more apparent than a loss of stereo separation. Different versions of PAC are available for DAB and Internet applications; they are optimized for different transmission error conditions and error concealment. The error concealment algorithm mitigates the effect of bit errors and corrupted or lost packets; partial information is used along with heuristic interpolation. There is slight audible degradation with 5% random packet losses and the algorithm is effective with 10 to 15% packet losses.

As with most codecs, PAC has evolved. PAC version 1.A is optimized for unimpaired channel transmission of voice and music with up to 8-kHz bandwidth; bit rates range from 16 kbps to 32 kbps. PAC version 1.B uses a bandwidth of 6.5 kHz. PAC version 2 is designed for impaired channel broadcast applications, with bit rates of 16 kbps to 128 kbps for stereo signals. PAC version 3 is optimized for 64 kbps with a bandwidth of about 13 kHz. PAC version 4 is optimized for 5.1-channel sound. EPAC is an enhanced version of PAC optimized for low bit rates. Its filter switches between two different filter-bank designs depending on signal conditions. At 128 kbps, EPAC offers CD-transparent stereo sound and is compliant with RealNetwork's G2 streaming Internet player. In some applications, monaural MPAC codecs are used to code multichannel audio using a perceptual model with provisions for spatial coding conditions such as binaural unmasking effects and binary-level masking differences. Signal pairs are coded and masking thresholds are computed for each channel.

AC-3 (Dolby Digital) Codec

Many data reduction codecs are designed for a variety of applications. The AC-3 (Dolby Digital) codec in particular is widely used to convey multichannel audio in applications such as DTV, DBS, DVD-Video, and Blu-ray. The AC-3 codec was preceded by the AC-1 and AC-2 codecs.

The AC-1 (Audio Coding-1) stereo codec uses adaptive delta modulation, as described in Chap. 4, combined with analog companding; it is not a perceptual codec. An AC-1 codec can code a 20-kHz bandwidth stereo audio signal into a 512-kbps bitstream (approximately a 3:1 reduction). AC-1 was used in satellite relays of television and FM programming, as well as cable radio services.

The AC-2 codec is a family of four single-channel codecs used in two-channel or multichannel applications. It was designed for point-to-point transmission such as full-duplex ISBN applications. AC-2 is a perceptual codec using a low-complexity time-domain aliasing cancellation (TDAC) transform. It divides a wideband signal into multiple subbands using a 512-sample 50% overlapping FFT algorithm performing alternating modified discrete cosine and sine transform (MDCT/MDST) calculations; a 128-sample FFT can be used for low-delay coding. A window function based on the Kaiser–Bessel kernel is used in the window design. Coefficients are grouped into subbands containing from 1 to 15 coefficients to model critical bandwidths. The bit allocation process is backward-adaptive in which bit assignments are computed equivalently at both the encoder and decoder. The decoder uses a perceptual model to extract bit allocation information from the spectral envelope of the transmitted signal. This effectively reduces the bit rate, at the expense of decoder complexity. Subbands have preallocated bits, with the lower subbands receiving a greater share. Additional bits are adaptively drawn from a pool and assigned according to the logarithm of peak energy levels in subbands. Coefficients are quantized according to bit allocation calculations, and blocks are formed. Algorithm parameters vary according to sampling frequency. At sampling frequencies of 48, 44.1, and 32 kHz, the following apply: bytes/block: 168, 184, 190; total bits: 1344, 1472, 1520; subbands: 40, 43, 42; adaptive bits: 225, 239, 183.

The AC-2 codec provides high audio quality with a data rate of 256 kbps per channel. With 16-bit input, reduction ratios include 6.1:1, 5.6:1, and 5.4:1 for sample rates of 48, 44.1, and 32 kHz, respectively. AC-2 is also used at 128 kbps and 192 kbps per channel. AC-2 is a registered .wav type so that AC-2 files are interchangeable between computer platforms. The AC-2 .wav header contains an auxiliary data field at the end of each block, selectable from 0 to 32 bits. For example, peak levels can be stored to facilitate viewing and editing of .wav files. AC-2 codec applications include PC sound cards, studio/transmitter links, and ISDN linking of recording studios for long distance recording. The AC-2 bitstream is robust against errors. Depending on the implementation, AC-2 delay varies between 7 ms and 60 ms. AC-2A is a multirate, adaptive block codec, designed for higher reduction ratios; it uses a 512/128-point TDAC filter. AC-2 was introduced in 1989.

AC-3 Overview

The AC-3 coding system (popularly known as Dolby Digital) is an outgrowth of the AC-2 encoding format, as well as applications in commercial cinema. AC-3 was first introduced in 1992. AC-3 is a perceptual codec designed to process an ensemble of audio channels. It can code from 1 to 7 channels as 3/3, 3/2, 3/1, 3/0, 2/2, 2/1, 2/0, 1/0, as well as an optional low-frequency effects (LFE) channel. AC-3 is often used to provide a 5.1 multichannel surround format with left, center right, left-surround, right-surround, and an LFE channel. The frequency response of the main channels is 3 Hz to 20 kHz, and the frequency response of the LFE channel is 3 Hz to 120 Hz. These six channels (requiring 6×48 kHz $\times$ 18 bits = 5.184 Mbps in uncompressed PCM representation) can be coded at a nominal rate of 384 kbps, with a bandwidth reduction of about 13:1. However, the AC-3 standard also supports bit rates ranging from 32 kbps to 640 kbps. The AC-3 codec is backward compatible with matrix surround sound formats, two-channel stereo, and monaural reproduction; all of these can be decoded from the AC-3 data stream. AC-3 does not use 5.1 matrixing in its bitstream. This ensures that quantization noise is not directed to an incorrect channel, where it could be unmasked. AC-3 transmits a discrete multichannel coded bitstream, with digital downmixing in the decoder to create the appropriate number (monaural, stereo, matrix surround, or full multichannel) of reproduction channels.

AC-3 contains a dialogue normalization level control so that the reproduced level of dialogue (or any audio content) is uniform for different programs and channels. With dialogue normalization, a listener can select a playback volume and the decoder will automatically replay content at that average relative level regardless of how it was recorded. AC-3 also contains a dynamic range control feature. Control data can be placed in the bitstream so that a program's recorded dynamic range can be varied in the decoder over a ±24-dB range. Thus, the decoder can alter the dynamic range of a program to suit the listener's preference (for example, a reduced dynamic range "midnight mode"). AC-3 also provides a downmixing feature; a multichannel recording can be reduced to stereo or monaural. The mixing engineer can specify relative interchannel levels. Additional services can be embedded in the bitstream including verbal description for the visually impaired, dialogue with enhanced intelligibility for the hearing impaired, commentary, and a second stereo program. All services may be tagged to indicate language. AC-3 facilitates editing on a block level, and blocks can be rocked back and forth at the decoder, and read as forward and reverse audio. Complete encoding/decoding delay is typically 100 ms.

Because AC-3 eliminates redundancies between channels, greater coding efficiency is achieved relative to AC-2; a stereo version of AC-3 provides high quality with a data rate of 192 kbps. In one test, AC-3 at 192 kbps scored 4.5 on the ITU-R impairment scale. Differences between the original and coded files were perceptible to expert listeners, but not annoying. The AC-3 format also delivers data describing a program's original production format (monaural, stereo, matrix, and the like), can encode parameters for selectable dynamic range compression, can route low bass only to those speakers with subwoofers, and provide gain control of a program.

AC-3 uses hybrid backward/forward adaptive bit allocation in which an adaptive allocation routine operates in both the encoder and decoder. The model defines the spectral envelope, which is encoded in the bitstream. The encoder contains a core psychoacoustic model, but can employ a different model and compare results. If desired, the encoder can use the data syntax to code parameter variations in the core model, or convey explicit delta bit allocation information, to improve results. Block diagrams of an AC-3 encoder and decoder are shown in Fig. 11.21.

AC-3 achieves its data reduction by quantizing a frequency-domain representation of the audio signal. The encoder first uses an analysis filter bank to transform time-domain PCM samples into frequency-domain coefficients. Each coefficient is represented in binary exponential notation as a binary exponent and mantissa. Sets of exponents are

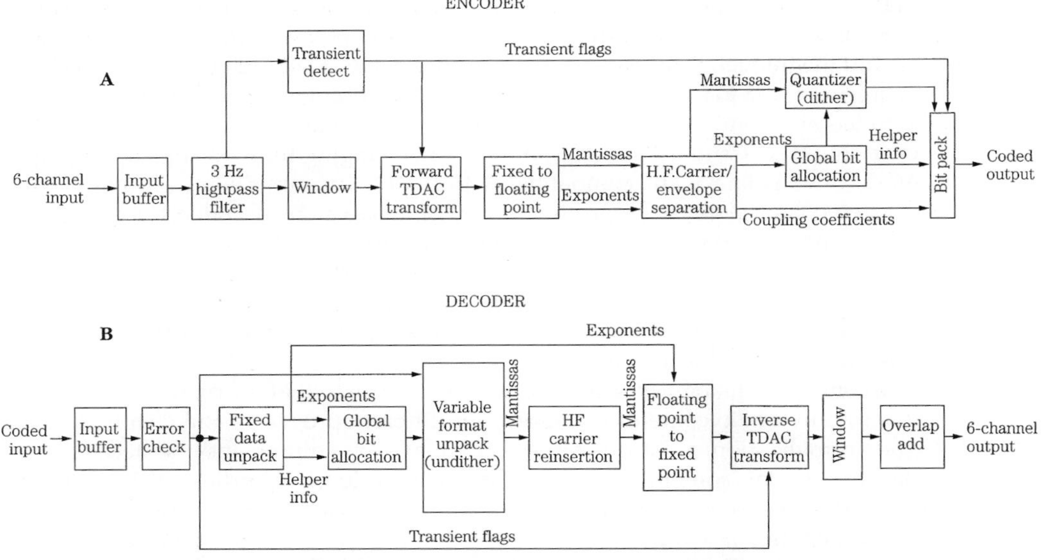

Figure 11.21 The AC-3 (Dolby Digital) adaptive transform encoder and decoder. This codec can provide 5.1-channel surround sound. A. AC-3 encoder. B. AC-3 decoder.

encoded into a coarse representation of the signal spectrum and referred to as the spectral envelope. This spectral envelope is used by the bit allocation routine to determine the number of bits needed to code each mantissa. The spectral envelope and quantized mantissas for six audio blocks (1536 audio samples) are formatted into a frame for transmission.

The decoding process is the inverse of the encoding process. The decoder synchronizes the received bitstream, checks for errors, and de-formats the data to recover the encoded spectral envelope and quantized mantissas. The bit allocation routine and the results are used to unpack and de-quantize the mantissas. The spectral envelope is decoded to yield the exponents. Finally, the exponents and mantissas are transformed back to the time domain to produce output PCM samples.

AC-3 Theory of Operation

Operation of the AC-3 encoder is complex, with much dynamic optimization performed. In the encoder, blocks of 512 samples are collected and highpass filtered at 3 Hz to eliminate dc offset and analyzed with a bandpass filter to detect transients. Blocks are windowed and processed with a signal-adaptive transform codec using a critically sampled filter bank with time-domain aliasing cancellation (TDAC) described by Princen and Bradley. An FFT is employed to implement an MDCT algorithm. Frequency resolution is 93.75 Hz at 48 kHz; each transform block represents 10.66 ms of audio, but transforms are computed every 5.33 ms so the audio block rate is 187.5 Hz. Because there is a 50% long-window overlap (an optimal window function based on the Kaiser–Bessel kernel is used in the window design), each PCM sample is represented in two sequential transform blocks; coefficients are decimated by a factor of two to yield

256 coefficients per block. Aliasing from sub-sampling is exactly canceled during reconstruction. The transformation allows the redundancy introduced in the blocking process to be removed. The input to the TDAC is 512 time-domain samples while the output is 256 frequency-domain coefficients. There are 50 bands between 0 Hz and 24 kHz; the bandwidths vary between 3.4 and 1.4 of critical bandwidth values.

Time-domain transients such as an impulsive sound might create audible quantization artifacts. A transient detector in the encoder, using a high-frequency bandpass filter, can trigger window switching to dynamically halve the transform length from 512 to 256 samples for a finer time resolution. The 512-sample transform is replaced by two 256-sample transforms, each producing 128 unique coefficients; time resolution is doubled, to help ensure that quantization noise is concealed by temporal masking. Audio blocks are 5.33 ms, and transforms are computed every 2.67 ms at 48 kHz. Short blocks use an asymmetric window that uses only one-half of a long window. This yields poor frequency selectivity and does not give a smooth crossfade between blocks. However, because short blocks are only used for transient signals, the signal's flat and wide spectrum does not require selectivity and the transient itself will mask artifacts. This block switching also simplifies processing, because groups of short blocks can be treated as groups of long blocks and no special handling is needed.

Coefficients are grouped into subbands that emulate critical bands. Each frequency coefficient is processed with floating-point representation with mantissa (0 to 16 bits) and exponent (5 bit) to maintain dynamic range. Coefficient precision is typically 16 to 18 bits but may reach 24 bits. The coded exponents act as scale factors for mantissas and represent the signal's spectrum; their representation is referred to as the spectral envelope. This spectral envelope coding permits variable resolution of time and frequency. Unlike some codecs, to reduce the number of exponents conveyed, AC-3 does not choose one exponent, based on the coefficient with the largest magnitude, to represent each band. In AC-3, fine-grained exponents are used to represent each coefficient, and efficiency is achieved by differential coding and sharing of exponents across frequency and time. The spectral envelope is coded as the difference between adjacent filters; because the filter response falls off at 12 dB/bin, maximum deltas of 2 (1 represents a 6-dB difference) are needed. The first dc term is coded as an absolute, and other exponents are coded as one of five changes (±2, ±1, 0) from the previous lower frequency exponent, allowing for ±12 dB/bin differences in exponents.

AC-3 Exponent Strategies and Bit Allocation

For improved bit efficiency, the differential exponents are combined into groups in the audio block. One, two, or four mantissas can use the same exponent. These groupings are known as D15, D25, or D45 modes, respectively, and are referred to as exponent strategies. The number of grouped differential exponents placed in the audio block for a channel depends on the exponent strategy and the frequency bandwidth of that channel. The number of exponents in each group depends only on the exponent strategy, which is based on the need to minimize audibility of quantization noise. In D15 exponent coding of the spectral envelope (2.33 bits per exponent), groups of three differentials are coded in a 7-bit word; D15 provides fine frequency resolution at the expense of temporal resolution. It is used when the audio signal envelope is relatively constant over many audio blocks. Because D15 is conveyed when the spectrum is stable, the estimate is coded only occasionally, for example, once every six blocks (32 ms) yielding 0.39 bits per audio sample.

The spectral estimate must be updated more frequently when transient signals are coded. The spectral envelope must follow the time variations in the signal; this estimate is coded with less frequency resolution. Two methods are used. D25 provides moderate frequency resolution and moderate temporal resolution. Coefficients are shared across frequency pairs. A delta is coded for every other frequency coefficient; the data rate is 1.17 bits per exponent. D25 is used when the spectrum is stable over two to three blocks, and then significantly changes. In the D45 coding, one delta is coded for every four coefficients; the data rate is 0.58 bits per exponent. D45 provides high temporal resolution and low frequency resolution. It is used when transients occur within single audio blocks. The encoder selects the exponent coding method (D15, D25, D45, or REUSE) for every audio block and places this in a 2-bit exponent strategy field. Because the exponent selection is coded in the bitstream, the decoder tracks the results of any encoder methodology. Examples of the D15, D25, and D45 modes are shown in Fig. 11.22.

Exponents can also be shared across time. Signals may be stationary for longer than a 512-sample block and have similar spectral content over many blocks. Thus, exponents can be reused for subsequent blocks. In most cases, D15 is coded in the first block in a frame, and reused for the next five blocks. This can reduce the exponent data rate by a factor of 6, to 0.10 bits per exponent.

Most encoders use a forward-adaptive psychoacoustics model and bit allocation that analyzes the signal spectrum and quantizes mantissa values and sends them to the decoder; all modeling and allocation is done in the encoder. In contrast, the AC-3 encoder contains a forward-backward adaptive psychoacoustics model that determines the masked threshold and quantization, and the decoder also contains the core backward-adaptive model. This reduces the amount of bit allocation information that must be conveyed. The encoder bases bit allocation on exponent values, and because the decoder receives the exponent values, it can backward-adaptively recompute the corresponding bit allocation. The approach allows overall lower bit rates, at the expense of increased decoder complexity. Also, this limits the ability to revise the psychoacoustic model in the encoder while retaining compatibility with existing decoders. This is addressed with a parametric model, and by providing for a forward-adaptive delta correction factor.

In the encoder, the forward-adaptive model uses an iterative rate control loop to determine the model's parameters defining offsets of maskers from a masking contour. These are used by the backward-adaptive model, along with quantized spectral envelope information, to estimate the masked threshold. The perceptual model's parameters and an optional forward-adaptive delta bit allocation that can adjust the masking thresholds are conveyed to the decoder.

The encoder's bit allocation delta parameter can be used to upgrade the encoder's functionality. Future encoders could employ two perceptual models, the original version and an improved version, and the encoder could convey a masking level adjustment.

The bit allocation routine analyzes the spectral envelope of the audio signal according to masking criteria to determine the number of bits to assign to each mantissa. Ideally, allocation is calculated so that the SNR for quantized mantissas is greater than or equal to the SMR. There are no preassigned mantissa or exponent bits; assignment is performed globally on all channels from a bit pool. The routine interacts with the parametric perceptual model to estimate audible and inaudible spectral components. The estimated noise level threshold is computed for 50 bands of nonuniform bandwidth (approximately 1.6 octaves). The bit allocation for each mantissa is determined by a

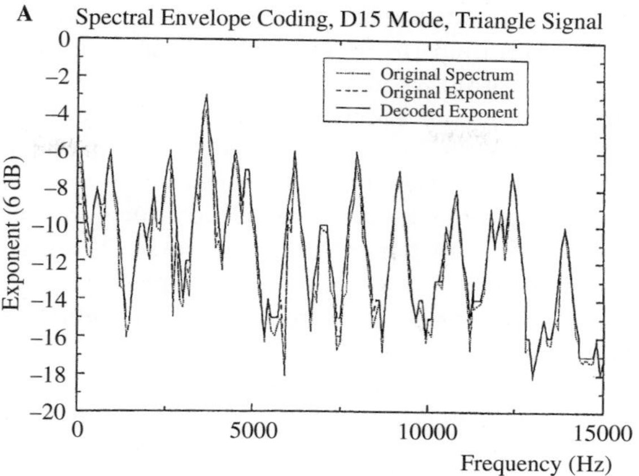

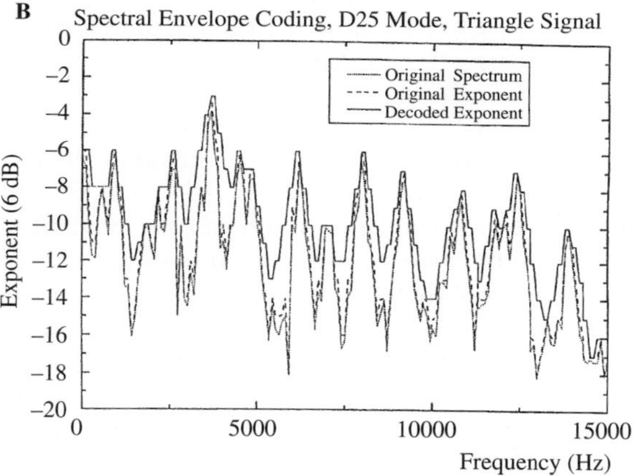

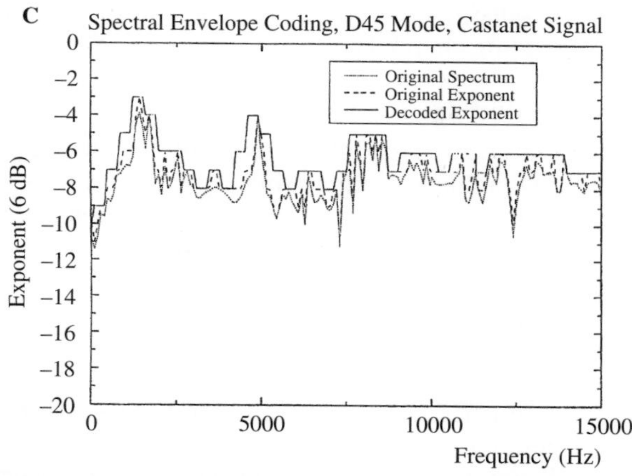

FIGURE 11.22 Three examples of spectral envelope (exponent) coding strategies used in the AC-3 codec. A. D15 mode coding for a triangle signal. B. D25 mode coding for a triangle signal. C. D45 mode coding for a castanet signal. (*Todd et al., 1994*)

lookup table based on the difference between the input signal power spectral density (psd) along a fine-grain uniform frequency scale, and estimated noise level threshold along a banded coarse-grain frequency scale. The routine considers the decoded spectral envelope to be the power spectral density of the signal. This signal is effectively convolved, using two IIR filters for the two-slope spreading function, with a simplified spreading function that represents the ear's masking response. The spreading function is approximated by two masking curves: a fast-decaying upward curve and slow-decaying upward curve, which is offset downward in level. The curves are referred to as a fast leak and slow leak. Convolution is performed starting at the lowest frequency psd. Each new psd is compared to current leak values and judged to be significant or not; this yields the predicted masking value for each band. The shape of the spreading function is conveyed to the decoder by four parameters.

This predicted curve is compared to a hearing threshold, and the larger of the values is used at each frequency point to yield a global masking curve. The resulting predicted masking curve is subtracted from the original unbanded psd to determine the SNR value for each transform coefficient. These are used to quantize each coefficient mantissa from 0 to 16 bits. Odd-symmetric quantization (the mantissa range is divided by an odd number of steps that are equal in width and symmetric about zero) is used for all low-precision quantizer levels (3-, 5-, 7-, 11-, 15-level) to avoid biases in coding, and even-symmetric quantization (the first step is 50% shorter and the last step is 50% longer than the rest) is used for all other quantization levels (from 32-level through 65,536-level).

Bits are taken iteratively from a common bit pool available to all channels. Mantissa quantization is adjusted to use the available bit rate. The effect of interchannel masking is relatively slight. To minimize audible quantization noise at low frequencies where frequency resolution is relatively low, the noise from coefficients is examined and bits are iteratively added to lower noise where necessary. Quantized mantissas are scaled and offset. Subtractive dither is optionally employed when zero bits are allocated to the mantissa. A pseudo-random number generator can be used. A mode bit indicates when dither is used and provides synchronization to the decoder's subtractive dither circuit. In addition, the carrier portion of high-frequency localization information is removed, and the envelope is coded instead; high-frequency multichannel carrier content may be combined into a coupling channel.

AC-3 Multichannel Coding

One of the hallmarks of AC-3 coding is its ability to efficiently code an ensemble of multiple channels to a single low bit-rate bitstream. To achieve greater bit efficiency, the encoder may use channel coupling and rematrixing on selective frequencies while preserving perceptual spatial accuracy. Channel coupling, based on intensity stereo coding, combines the high-frequency content of two or more channels into one coupling channel. The combined coupling channel, along with coupling coordinates, the quantized power spectral ratios between each channel's original signal and the coupled channel, are conveyed for decoding and reconstructing the original energy envelopes. Coupling may be performed at low bit rates, when signal conditions exceed the bit rate. It efficiently (but perhaps not transparently) codes a multichannel signal, taking into account high-frequency directionality limitations in human hearing, without reducing audio bandwidth. Above 3 kHz, the ear is unable to distinguish the fine temporal structure of high-frequency waveforms and instead relies on the envelope energy of the sounds. In other words, at high frequencies, we are not sensitive to phase, but only to amplitude. Directionality is determined by the interaural time delay of the envelope and by perceived frequency response based on head shadowing. As a result, the ear cannot independently detect the direction of two high-frequency sounds that are close in frequency.

Coupling can yield significant bit-rate reduction. As Steve Vernon points out, given the critical band resolution, high-frequency mantissa magnitudes do not need to be accurately coded for individual frequency bins, as long as the overall signal energy in each critical band is correct. Because 85% of transform coefficients are above 3 kHz, large amounts of error are acceptable in high-frequency bands.

The AC-3 codec can couple channels at high frequencies; care is taken to avoid phase cancellation of the common channels. The coupling strategy is determined wholly in the encoder. Coupling coordinates for each individual channel are used to code the ratio of original signal power in a band to the coupling channel power in each band. Coupling channels are encoded in the same way as individual channels with a spectral envelope comprising exponents and mantissas. Frequencies below a certain coupling frequency (a range of 3 kHz to 20 kHz is possible, with 10 kHz being typical) are encoded as individual channels, and encoded as coupling coordinates above the coupling frequency. With a 10-kHz coupling frequency, the data rate for exponents and mantissas is nearly halved. Half of the exponents and mantissas for each coupled channel are discarded. Only the exponents and mantissas for the coupling channel, and scale factors, are conveyed.

Figure 11.23A shows an example of channel coupling of three channels. Phase is adjusted to prevent cancellation when summing channels. The monaural coupled channel is created by summing the input coupled-channel spectral coefficients above the coupling frequency. The signal energy in each critical band of the input channels and coupled channel are computed and used to determine coupling coordinate scale factors. The energy in each band of each coupled channel is divided by the energy in each corresponding band of the coupling channel. The scale factors thus represent the amount that the decoder will scale the coupling-channel bands to re-create bands of the original energy. The dynamic range of the scaling factors may range from –132 to +18 dB with a resolution between 0.28 dB to 0.53 dB. One to 18 frequency bands may be used, with 14 bands being typical. During decoding, individual channel-coupling coordinates are multiplied by the coupling-channel coefficients to regenerate individual high-frequency coefficients. Figure 11.23B shows how three coupled channels are decoded.

The encoder may also use rematrixing coding, similar to M/S coding, to exploit the correlation between channel pairs. Rather than code the spectra of right and left (R, L) channels independently, the sum and difference (R′, L′) is coded. For example, when the R and L channels are identical, no bits are allocated to the R′ channel. Rematrixing is applied selectively to up to four frequency regions; the regions are based on coupling information. Rematrixing is not used in coupling channels but can be used simultaneously with coupling. If used with coupling, the rematrixing frequency region ends at the start of the coupling region. Rematrixing coding is only used in 2/0 channel mode, and is compatible with Dolby Pro Logic and other matrix systems.

If matrixing is not used and a decoded perceptually coded monaural signal is played through a matrix decoder, small differences in quantization noise between the two channels could cause the matrix system to derive a surround channel from the noise differences between channels, and thus direct unmasked quantization noise to the surround

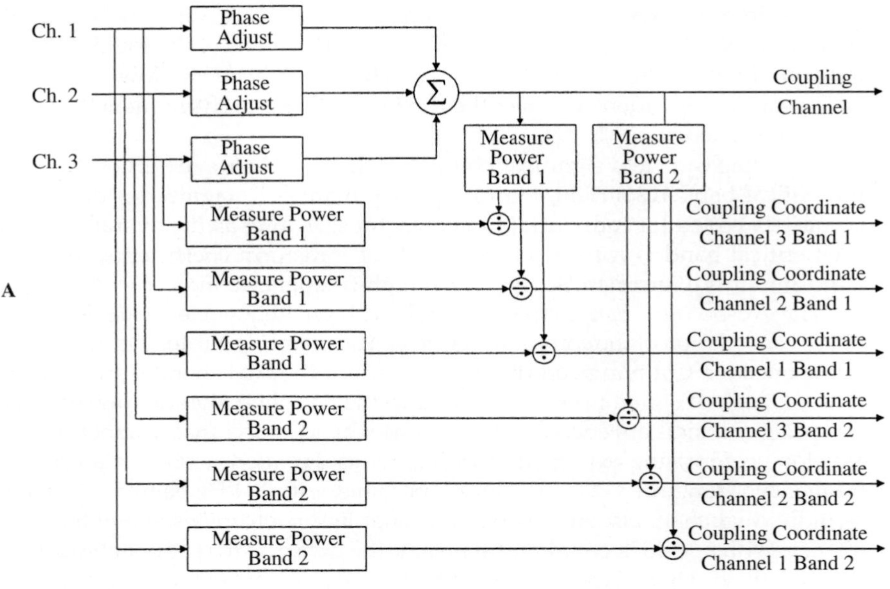

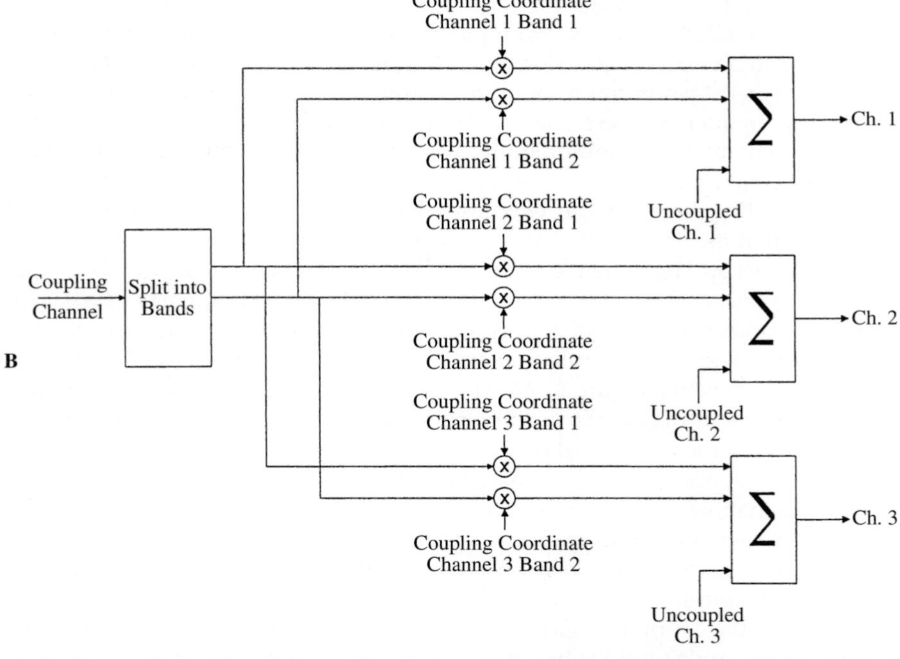

Figure 11.23 The AC-3 codec can couple audio channels at high frequencies. In this illustration, three channels are coupled to form one coupling channel. A. Encode coupling process. B. Decode coupling process. (*Fielder, 1996*)

channels. Moreover, the noise will be modulated by the center-channel signal, further increasing audibility. Rematrixing ensures that when the input signals are highly correlated, the quantization noise is also highly correlated. Thus, when a rematrixed two-channel AC-3 signal is played through a matrix system, quantization noise will not be unmasked.

AC-3 Bitstream and Decoder

Data in an AC-3 bitstream is contained in frames, as shown in Fig. 11.24. Each frame is an independently encoded entity. A frame contains a synchronization information (SI) header, bitstream information (BSI) header, 32 milliseconds of audio data as quantized frequency coefficients, auxiliary field, and CRCC error detection data. The frame period is 32 milliseconds at a 48-kHz sampling frequency. The SI field contains a 16-bit synchronization word, 2-bit sampling rate code, and 6-bit frame size code. The BSI field describes the audio data with information such as coding mode, timecode, copyright, normalized dialogue level, and language code. Audio blocks are variable length, but six transform blocks, with 256 samples per block (1536 in total), must fit in one frame. Audio blocks mainly comprise quantized mantissas, exponent strategy, differentially coded exponents, coupling data, rematrixing data, block switch flags, dither flags, and bit allocation parameters. Each block contains data for all coded channels. A frame holds a constant number of bits based on the bit rate. However, audio block boundaries are not fixed. This allows the encoder to globally allocate bitstream resources where needed.

One 16-bit CRCC word is contained at the end of each frame, and an additional 16-bit CRCC word may be optionally placed in the SI header. In each case, the generating polynomial is $x^{16} + x^{15} + x^2 + 1$. Error detection and the response to errors vary in different AC-3 applications. Error concealment can be used in which a previous audio segment is repeated to cover a burst error. The decoder does not decode erroneous data; this could lead to invalid bit allocation. The first block always contains a complete refresh of

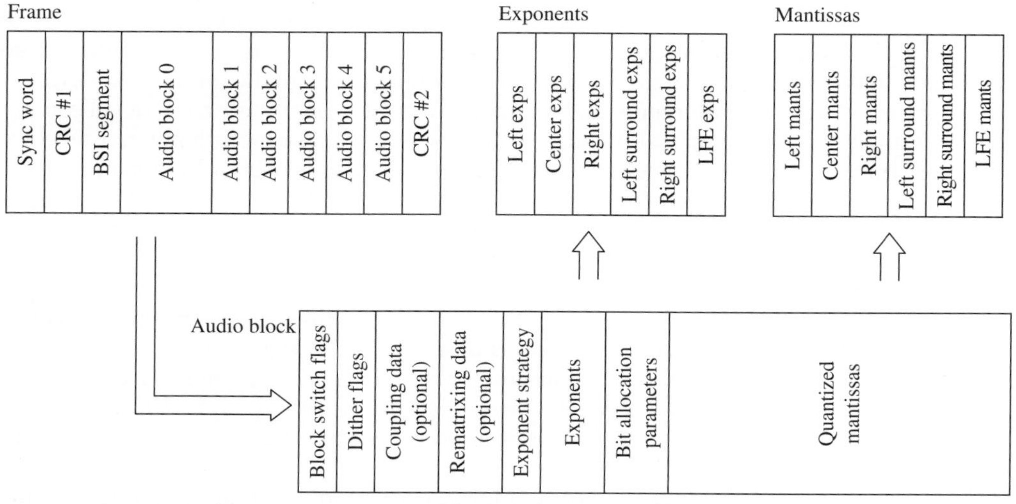

FIGURE 11.24 Structure of the AC-3 bitstream showing blocks in an audio frame. (*Vernon, 1999*)

all decoding information. Unused block areas may be used for auxiliary data. Audio data is placed in the bitstream so that the decoder can begin to output a signal before starting to decode the second block. This reduces the memory requirements in the decoder because it does not need to store parameters for more than one block at a time.

One or more AC-3 streams may be conveyed in an MPEG-2 transport stream as defined in the ISO/IEC 13818-1 standard. The stream is packetized as PES packets. It is not necessary to unambiguously indicate that a stream is AC-3. The MPEG-2 standard does not contain codes to indicate an AC-3 stream. One or more AC-3 streams may also be conveyed in an AES3 or IEC958 interface. Each AC-3 frame represents 1536 encoded audio samples (divided into 6 blocks of 256 samples each). AC-3 frame boundaries occur at a frequency of exactly once every 1536 IEC958 frames.

The decoder receives data frames; it can accept a continuous data stream at the nominal bit rate, or chunks of data burst at a high rate. The decoder synchronizes the data and performs error correction as necessary. The decoder contains the same backward-adaptive perceptual model as the encoder and uses the spectral envelope to perform backward-adaptive bit allocation. The quantized mantissas are denormalized by their exponents. However, using delta bit allocation values, the encoder can change model parameters and thus modify the threshold calculations in the decoder. As necessary, decoupling is performed by reconstructing the high-frequency section (exponents and mantissas) of each coupled channel. The common coupling-channel coefficients are multiplied by the coupling coordinates for the individual channel. As necessary, dynamic range compression and other processing is performed on blocks of audio data. Coefficients are returned to fixed-point representation, dither is subtracted, and carrier and envelope information is reconstructed. The inverse TDAC transform, window, and overlap operations produce data that is buffered prior to outputting continuous PCM data. Blocks of 256 coefficients are transformed to a block of 512 time samples; with overlap-and-add, 256 samples are output. LFE data is padded with zeros before the inverse transform so the output sampling frequency is compatible with the other channels. Multiple channels are restored and dynamic range parameters are applied.

When mantissa values are quantized to 0 bits, the decoder can substitute a dither value. Dither flags placed by the encoder determine when dither substitution is used. At very low data rates, the dither is a closer approximation to the original unquantized signal in that frequency range than entirely removing the signal would be. Also, dither is preferred over the potentially audible modulation in cases where a low-level component might appear on and off in successive blocks. Dither substitution is not used for short transform blocks (when a soft-signal block is followed by a loud-signal block) because the interleaved small and large values in the coefficient array and their exponents can cause the dither signal to be scaled to a higher, too-audible level in the short block.

The decoder can perform downmixing that is required when the number of output channels is less than the number of coded channels. In this way, the full multichannel program can be reproduced over fewer channels. The decoder decodes channels and routes them to output channels using downmix scale factors supplied by the encoder that set each channel's relative level. Downmixes can be sent to eight different output configurations. There are two two-channel modes, a stereo or a matrix surround mode; the LFE is not included in these downmixes.

As noted, the AC-3 codec can code original content in L/R matrix surround sound formats, two-channel stereo, or monaural form. Conversely, all of these formats can be downmixed at the decoder from a multichannel AC-3 bitstream, using information in the BSI header. A decoder may also use metadata (entered during encoding) for volume normalization. Program material from different sources and using different reference volume levels can be replayed at a consistent level. Because of the importance of dialogue, its level is used as the reference; program levels are adjusted so that their long-term averaged dialogue levels are the same during playback. A parameter called dialnorm can control relative volume level of dialogue (or other audio content) with values from –1 dBFS to –31 dBFS in 1-dB steps. The value does not measure attenuation directly but rather the LAeq value of the content. Program material with a dialogue level higher than a dialnorm setting is normalized downward to that setting. Metadata also controls the dynamic range of playback (dynrng). The ATSC DTV standard sets the normalized dialogue level at –31 dBFS LAeq below digital full scale to represent optimal dialogue normalization. The decoder can thus maintain intelligibility in a variety of playback content. For material other than speech, more subjective settings are used. Dynamic range control compresses the output at the decoder; this accounts for different listening conditions with background noise levels. Because this metadata may be included in every block, the control words may occur every 5.3 ms at a 48-kHz sampling rate. A smoothing effect is used between blocks to minimize audibility of gain changes. Not all decoders allow the user to manually access functions such as dynamic range control.

Algorithms such as AC-3 can be executed on general-purpose processors, or high performance dedicated processors. Both floating- and fixed-point processors are used; floating-point processors are generally more expensive, but can execute AC-3 in fewer cycles with good fidelity. However, the higher cost as well as relatively greater memory requirements may make fixed-point processors more competitive for this application. In some cases, a DSP plug-in system is used for real-time encoding on a host PC.

AC-3 Applications and Extensions

AC-3, widely known as Dolby Digital, is used in a variety of applications. Dolby Digital provides 5.1 audio channels when used for theatrical motion picture film coding. Data is optically printed between the sprocket holes, with a data rate of approximately 320 kbps. Existing analog soundtracks remain unaffected, providing compatibility. The Dolby Digital Surround EX system adds a center-surround channel. The new channel is matrixed within the left and right surround channels. A simple circuit is used to retrieve the center surround channel.

The Dolby Digital Plus (also known as Enhanced AC-3 or E-AC-3) is an extension format. It primarily adds spectral coding that is used at low bit rates and also supports two additional channels (up to 7.1) and higher bit rates. A core bitstream is coded with Dolby Digital and the two-channel extension is coded as Dolby Digital Plus. Legacy devices that decode only Dolby Digital will not decode Dolby Digital Plus. The Dolby Digital Plus bitstream can be downconverted to yield a Dolby Digital bitstream. However, if the original bitstream contains more than 5.1 channels, the additional channel content appears as a downmix to 5.1 channels.

The Dolby TrueHD is a lossless extension that uses Dolby Digital encoding as its core and Meridian Lossless Packing (MLP) as the lossless codec. Dolby TrueHD can accommodate up to 24-bit word lengths and a 192-kHz sampling frequency. Multichannel MLP bitstreams include lossless downmixes. For example, a MLP bitstream holding eight channels also contains two- and six-channel lossless downmixes for compatibility with playback systems with fewer channels.

In consumer applications, Dolby Digital is used to code 5.1 audio channels (or fewer) for cable and satellite distribution and in home theater products such as DVD-Video and Blu-ray discs. Dolby Digital is an optional coding format for the DVD-Audio standard. Dolby Digital is also used to code the audio portion of the digital television ATSC DTV standard. It was selected as the audio standard for DTV by the ATSC Committee in 1993, and codified as the ATSC A/52 1995 standard. DVD is discussed in Chap. 8 and DTV is discussed in Chap. 16. Dolby Digital is not suitable for many professional uses because it was not designed for cascaded operation. In addition, its fixed frame length of 32 ms at 48 kHz does not correspond with video frame boundaries (33.37 ms for NTSC and 40 ms for PAL). Video editing results in data errors and mutes, or loss of synchronization.

Dolby E is a codec used in professional audio applications. It codes up to eight audio channels, provides a 4:1 reduction at 384 kbps to 448 kbps rates, and allows eight to ten generations of encoding and decoding. Using Dolby E, eight audio channels along with metadata describing the contents can be conveyed along a single AES3 interface operating in data mode. A 5.1-channel mix can be easily conveyed and recorded on two-channel media. The data can be recorded, for example, on a VTR capable of recording digital audio; Dolby E's frame rate can be set to match all standard video frame rates to facilitate editing. The metadata carries AC-3 features such as dialogue normalization and dynamic range compression. These coding algorithms were developed by Dolby Laboratories.

DTS Codec

The DTS (Digital Theater Systems) codec (also known as Coherent Acoustics) is used to code multichannel audio in a variety of configurations. DTS can operate over a range of bit rates (32 kbps to 4.096 Mbps), a range of sampling frequencies (from 8 kHz to 192 kHz, based on multiples of 32, 44.1, and 48 kHz) and typically encodes five channels plus an LFE channel. For example, when used to code 48-kHz/16-bit, 5.1-channel soundtracks, bit rates of 768 kbps or 1509 kbps are often used. A sampling frequency of 48 kHz is nominal in DTS coding. Some of the front/surround channel combinations are: 1/0, 2/0, 3/0, 2/1, 2/2, 3/2, 3/3, all with optional LFE. A rear center-channel can be derived using DTS-ES Matrix 6.1 or DTS-ES Discrete 6.1. In addition, the DTS-ES Discrete 7.1 mode is available.

The DTS subband codec uses adaptive differential pulse-code modulation (ADPCM) coding of time-domain data. Input audio data is placed in frames. Five different time durations (256, 512, 1024, 2048, 4096 samples per channel) are available depending on sampling frequency and output bit rate. The frame size determines how many audio samples are placed in a frame. Generally, large frames are used when coding at low bit rates. Operating over one frame at a time, for sampling frequencies up to and including 48 kHz, a multirate filter bank splits the wideband input signal into 32 uniform subbands. A frame containing 1024 samples thus places 32 samples in each subband. Either of two polyphase filter banks can be selected. One aims for reconstruction precision while the other promotes coding gain; the latter filter bank has high stopband rejection ratios. In either case, a flag in the bitstream informs the decoder of the choice. Each subband is ADPCM coded, representing audio as a time-domain difference signal; this essentially removes redundancies from the signal. Fourth-order forward adaptive linear prediction is used.

With tonal signals, the difference signal can be more efficiently quantized than the original signal, but with noise-like signals the reverse might be true. Thus ADPCM can be selectively switched off in subbands and adaptive PCM coding used instead. In a side chain the audio signal is examined for psychoacoustic and transient information; this can be used to modify the ADPCM coding. For example, the position of a transient in an analysis window is marked and transient modes can be calculated for each subband. A global bit management system allocates bits over all the coded subbands in all the audio channels; its output adapts to signal and coding conditions to optimize coding. The algorithm calculates normalizing scale factors and bit-allocation indices and ultimately quantizes the ADPCM samples using from 0 to 24 bits. Because the difference signals are quantized, rather than the actual subband samples, the encoder does not always only use SMR to determine bit allocation. At low bit rates, quantization can be determined by SMR, SMR modified by subband prediction gain, or a combination of both. At high bit rates, quantization determination can combine SMR and differential minimum mean square error. Twenty-eight different mid-tread quantizers can be selected to code the differential signal. The statistical distribution of the differential codes is nonuniform, so for greater coding efficiency, codewords can be represented using variable-length entropy coding; multiple code tables are available.

The LFE channel is coded independently of the main channels by decimating a full-bandwidth input PCM bitstream to yield an LFE bandwidth; ADPCM coding is then applied. The decimation can use either 64- or 128-times decimation filters, yielding LFE bandwidths of 150 Hz or 80 Hz. The encoder can also create and embed downmixing data, dynamic range control, time stamp, and user-defined information. Sample-accurate synchronization of audio to video is possible. A joint-frequency mode can code the combined high-frequency subbands (excluding the bottom two subbands) from multiple audio channels; this can be applied for low bit-rate applications. Audio frequencies above 24 kHz can be coded as an extension bitstream added to the standard 48-kHz sampling-frequency codec, using side-chain ADPCM encoders. The decoder demultiplexes the data into individual subbands. This data is requantized into PCM data, and then inverse-filtered to reconstruct full bandwidth audio signals as well as the LFE channel. The DTS codec is often used to code the multichannel soundtracks for DVD-Video and Blu-ray titles.

The DTS codec supports XCH and X96 extensions. XCH (Channel Extension) is also known as DTS-ES. It adds a discrete monaural channel as a rear output. X96 (Sampling Frequency Extension) is also known as Core+96k or DTS-96/24. It extends the sampling frequency from 48 kHz to 96 kHz by secondarily encoding a residual signal following the baseband encoding. The residual is formed by decoding the encoded baseband and subtracting it from the original signal.

DTS-HD bitstreams contain a core of DTS legacy data (5.1-channel, 48-kHz, typically coded at 768 kbps or 1.509 Mbps) and use the extension option in the DTS bitstream structure. The DTS-HD codec supports XXCH, XBR, and XLL extensions. XXCH (Channel Extension) adds additional discrete channels. XBR (High Bit-Rate Extension) allows an increase in bit rate. XLL (Lossless Extension) is also known as DTS-HD Master Audio. DTS-HD Master Audio is a lossless codec extension to DTS. It uses a substream to accommodate lossless audio compression. Coding can accommodate up to 24-bit word lengths and a 192-kHz sampling frequency. Multichannel bitstreams include lossless downmixes. The DTS-HD High Resolution Audio codec is a lossy extension format to DTS. Its substream codes data supporting additional channels and higher sampling frequencies. A legacy decoder operates on the core bitstream while an HD-compatible decoder operates on both the core and extension.

The DTS codec used for some commercial motion picture soundtracks is different from that used for consumer applications. Some motion picture soundtracks employ the apt-X100 codec, which uses more conventional ADPCM coding. The apt-X100 system is an example of a subband coder providing 4:1 reduction; however, it does not explicitly reply on psychoacoustic modeling. It operates entirely in the time domain, and uses predictive coding and adaptive bit allocation with adaptive differential pulse code modulation. The audio signal is split into four subbands using QMF banks and analyzed in the time domain. Linear prediction ADPCM is used to quantize each band according to content. Backward adaptive quantization is used in which accuracy of the current sample is compared to the previous sample, and correction is applied with adaption multipliers taken from lookup tables in the encoder. This codes the difference in audio signal levels from one sample to the next; the added noise is white. A 4-bit word is output for every 16-bit input word. The decoder demultiplexes the signal, and applies ADPCM decoding and inverse filtering. A primary asset is low coding delay; for example, at a 32-kHz sampling frequency, the coding delay is a constant 3.8 ms. A range of sampling frequencies and reduction ratios can be used. The apt-X100 algorithm was developed by Audio Processing Technology. The DTS algorithms are licensed by Digital Theater Systems, Inc.

A diverse number of lossy codecs have been developed for various applications. Although many are based on standards such as MPEG, some are not. Ogg Vorbis is an example of a royalty-free, unpatented, public domain, open-source codec that provides good-quality lossy data compression. It can be used at bit rates ranging from 64 kbps to 400 kbps. Its applications include recording and playback of music files, Internet streaming, and game audio. Ogg Vorbis is free for use in commercial and noncommercial applications, and is available at www.vorbis.com. The latency of Ogg Vorbis encoding makes it unsuitable for speech telephony. Other open-source codecs such as Speex may be more suitable. Speex is available at www.speex.org.

Meridian Lossless Packing

Meridian Lossless Packing (MLP) is a proprietary audio coding algorithm used to achieve lossless data compression. It is used in applications such as the DVD-Audio and Blu-ray disc formats. MLP reduces average and peak audio data rates (bandwidth) and storage capacity requirements. Unlike lossy perceptual coding methods, MLP preserves bit-for-bit content of the audio signal. However, MLP offers relatively less compression than lossy methods, and the degree of compression varies according to the signal content. In addition, with MLP, the output bit rate continually varies according to audio signal conditions; however, a fixed data rate mode is provided.

MLP supports all standard sampling frequencies and quantization may be selected for 16 to 24 bits in single-bit steps. MLP can code both stereo and multichannel signals simultaneously. The degree of compression varies according to the nature of the music data itself. For example, without compression, 96-kHz/24-bit audio consumes 2.304 Mbps per channel, thus a 6-channel recording would consume 13.824 Mbps. This would exceed, for example, the DVD-Audio maximum bit rate of 9.6 Mbps. Even if the high bit rate could be recorded, it would allow only about 45 minutes of music on a single-layer DVD-Audio disc with a capacity of 4.7 Gbytes. In contrast, MLP would allow 6-channel 96-kHz/24-bit

	Data-Rate Reduction: Bits/Sample/ Channel	
Sampling Frequency (kHz)	Peak	Average
48	4	5–11
96	8	9–13
192	9	9–14

Table 11.2 MLP compression of a two-channel signal is relatively more efficient for higher sampling frequencies.

recordings. It may achieve 38 to 52% of bandwidth reduction, reducing bandwidth to 6.6 Mbps to 8.6 Mbps, allowing a playing time of 73 to 80 minutes on a DVD disc. In the highest quality DVD-Audio two-channel stereo mode of 192-kHz/24-bit, MLP would provide a playing time of about 117 minutes, versus a playing time of 74 minutes for PCM coding.

Generally, more compression is achieved with higher sampling rates such as 96 kHz and 192 kHz, and less compression is achieved for lower sampling rates such as 44.1 kHz. This is shown in Table 11.2. Data-rate reduction is measured in bit/sample/channel, both in peak and average rates, where peak rates reflect performance on signals that are difficult to encode. In this table, for example, a peak-data rate reduction of 9 bits/sample means that a 192-kHz 24-bit channel can be coded to fit into a 24 – 9 = 15-bit PCM channel. The peak-rate reduction is important when encoding for any format because of the format's bit-rate channel limit. The average-rate reduction indicates the average bit savings and is useful for estimating the storage capacity needed for archiving, mastering, and editing applications. For example, a 192 kHz 24-bit signal might be coded at an average bit rate of 24 – 11.5 = 12.5 bits per channel. This table shows examples of compression for a two-channel recording. Compression increases if the number of channels increases, if channels are correlated, if any of the channels have a low bandwidth, and if any of the channels are lightly used (as with some surround channels). With MLP, the output peak rate is always reduced relative to the input peak rate.

MLP supports up to 63 audio channels, with sampling frequencies ranging from 32 kHz to 192 kHz and word lengths from 14 to 24 bits. Any word length shorter than 24 bits can be efficiently transmitted and decoded as a 24-bit word by packing zeros in the unused LSBs, without increasing the bit rate. Moreover, this is done automatically without the need for special flagging of word length. In addition, mixed sampling frequencies and word lengths are also automatically supported. The MLP core accepts and delivers PCM audio data; other types of data such as low-bit, video, or text cannot be applied to the encoder.

MLP does not discard data during coding; instead, it "packs" the data more efficiently. In that respect, MLP is analogous to computer utilities such as Zip and Stuffit that reduce file sizes without altering the contents (algorithmically, MLP operates quite differently). Data reduction aside, MLP offers other specific enhancements over PCM. Whereas a PCM signal can be subtly altered by generation loss, transmission errors, and other causes as it passes through a production chain, MLP can ensure that the output signal is exactly the same as the input signal by checking the MLP-coded file and confirming its bit accuracy.

The MLP encoder inserts proprietary check data into the bitstream; the decoder uses this check data to verify bit-for-bit accuracy. The bitstream is designed to be robust

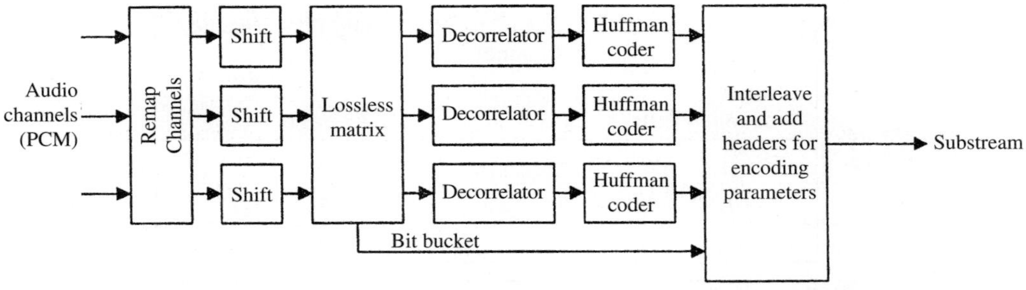

FIGURE 11.25 An example of an MLP encoder showing a lossless matrix, decorrelators, and Huffman coders. (*Stuart, 1998b*)

against transmission errors from disc defect or broadcast breakup. Full restart points in the bitstream occur at intervals of between 10 ms to 30 ms, when the content is easy to compress, and hard to compress, respectively. Errors cannot propagate beyond a restart point. Following a dropout, lossless operation will resume. MLP uses full CRCC checking and minor transmission errors are recovered in less than 2 ms. Recovery from a single bit error occurs within 1.6 ms. Full recovery from burst errors can occur within 10 ms to 30 ms; recovery time may be longer (36 ms to 200 ms) for a simple decoder. Interpolation may be used to prevent clicks or pops in the audio program. The restart points also allow fast recueing of the bitstream on disc. Startup time is about 30 ms and is always less than 200 ms (for complex passages that are difficult to compress). Similarly, fast-speed reviewing of audio material can be accommodated.

MLP applies several processes to the audio signal to reduce its overall data rate, and also to minimize the peak data rate in the variable rate of output data. A block diagram of the encoder is shown in Fig. 11.25. Data may be remapped to expedite the use of substreams. Data is shifted to account for unused capacity, for example, when less than 24-bit quantization is used or when the signal is not full scale. A lossless matrixing technique is used to optimize the data in each channel, reducing interchannel correlations. The signal in each channel is then decorrelated (inter-sample correlation is reduced) using a separate predictor for each channel. The encoder makes an optimum choice by designing a decorrelator in real time from a large variety of FIR or IIR prediction filters (this process takes into account the generally falling high-frequency response of music signals) to code the difference between the estimated and actual signals. A separate predictor is used for each encoded channel. Either IIR or FIR filters, up to 8th order, can be selected, depending on the type of audio signal being coded. MLP does not make assumptions about data content, nor search for patterns in the data. The decorrelated audio signal is further encoded with entropy coding to more efficiently code the most likely occurring successive values in the bitstream. The encoder can choose from several entropy coding methods including Huffman and Rice coding, and even PCM coding. Multiple data streams are interleaved.

The stream is packetized for fixed or variable data rate. (Data is encoded in blocks of 160 samples; blocks are assembled into packets and the length of packets can be adjusted, with a default of 10 blocks, or 1600 samples.) A first-in, first-out (FIFO) buffer of perhaps 75 ms is used in the encoder and decoder to smooth the variable data rate. To allow more rapid startup, the buffer is normally almost empty, and fills when the look-ahead encoder determines that a high entropy segment is imminent. To account for this, the decoder buffer empties. This helps to maintain the data rate below a preset limit, and to reduce the peak data rate.

Because data is matrixed into multiple substreams, each buffered separately, simple decoders can access a subset of the signal. MLP can use lossless matrixing of two substreams to encode both a multichannel mix and stereo downmix. Downmix instructions determine some coefficients for the matrices and the matrices perform a transformation so that two channels on one substream decode to provide a stereo mix, and also combine with another substream to provide a multichannel mix. Thus, the addition of the stereo mix adds minimal extra data, for example, one bit per sample.

The MLP bitstream contains a variety of data in addition to the compressed audio data. It also contains instructions to the decoder describing the sampling rate, word length, channel use, and so on. Auxiliary data from the content provider can describe whether speaker channels are from hierarchical or binaural sources, and can also carry copyright, ownership, watermarking data, and accuracy warranty information. There is also CRCC check data, and lossless testing information.

A fixed bit rate can be achieved in a single-pass process if the target rate is not too severe. When the desired rate is near the limit of MLP, a two-pass process is used in which packeting is performed in a second pass with data provided by the encoder. In any case, a fixed bit rate yields less compression than a variable bit rate.

A fixed rate gives predictable file sizes, allows the use of MLP with motion video, and ensures compatibility with existing interfaces. A specific bit rate may be manually requested by the operator. To achieve extra compression, for example, to fit a few more minutes onto a disc data layer, small adjustments may be made to the incoming signal. For example, word length could be reduced by one bit by requantizing (and redithering and perhaps noise shaping) a 24-bit signal to 23 bits. Alternatively, a channel could be lowpass-filtered, perhaps from 48 kHz to 40 kHz. MLP also allows mixed stream rates in which the option of a variable rate accompanies a fixed-stream rate.

An MLP transcoder can re-packetize a fixed-rate bitstream into a variable-rate stream, and vice versa; this does not require encoding or decoding. Editing can be performed on data while it is packed in the MLP format. The MLP decoder automatically recognizes MLP data and reverts to PCM when there is no MLP data. Different tracks can intermingle MLP and PCM coding on one disc. Finally, MLP is cascadable; files can be encoded and decoded repeatedly in cascade without affecting the original content.

MLP is optionally used on DVD-Audio and Blu-ray titles. The MLP bitstream can also be carried on AES3, S/PDIF, IEEE 1384, and other interconnections. MLP was largely developed by Robert Stuart, Michael Gerzon, Malcolm Law, and Peter Craven.

Other lossless codecs have been developed. These codecs include FLAC (Free Lossless Audio Codec), LRC (Lossless Real-time Coding), Monkey's Audio, WavPack, LPAC, Shorten, OptiFROG, and Direct Stream Digital (used in SACD). A lossless WMA codec is part of the Windows Media 9 specification. ALS (Audio Lossless Coding) is used in MPEG-4 ALS. The FLAC lossless codec is free, open-source, and can be used with no licensing fees. FLAC can support PCM samples with a word length from 4 to 32 bits per sample, a wide range of sampling frequencies, and 1 to 8 channels in a bitstream. FLAC is available at http://flac.sourceforge.net. FLAC is free for any commercial or noncommercial applications.

CHAPTER 12

Speech Coding for Transmission

The purpose of speech coding is to represent speech signals in a format that can be used efficiently for communication. Speech codecs must provide good speech intelligibility, while using signal compression to reduce the bit rate. Although speech codecs have goals that are similar to music codecs, their design is fundamentally different. In particular, unlike music codecs, most speech codecs employ source modeling that tailors the codec to the characteristics of the human voice. Speech codecs thus perform well with speech signals, and relatively poorly with nonspeech signals.

Because most transmission channels have a very limited bit-rate capability, speech coding presents many challenges. In addition to the need to balance speech intelligibility with bit-rate reduction, a codec must be robust over difficult signal conditions, must observe limitations in computational complexity, and must operate with a low delay for real-time communication. Also, wireless systems must overcome the severe error conditions that they are prone to.

Speech Coding Criteria and Overview

A successful speech codec must provide high intelligibility. This includes the ability of the listener to understand the literal spoken content as well as the ability to ascertain the speaker's identity, emotions, and vocal timbre. Beyond intelligible and natural-sounding speech, the speech sound should have a pleasant and listenable quality. Many speech codecs can employ perceptual enhancement algorithms to subjectively improve sound quality. For example, the perception of noise and coding artifacts can be reduced. Many perceptual enhancements are performed only in the decoder as post-processing. As described later, the performance of speech coding systems can be assessed through subjective listening tests.

A successful speech codec must provide robust performance. A codec should be able to convey speech in the presence of loud background noise and microphone distortion. A codec must provide good performance for different human talkers. For example, the speech sounds of adult males, adult females, and children have very different characteristics. Also, different languages present different coding demands. To a lesser extent, speech codecs should be able to convey nonspeech signals such as dual-tone multi-frequency (DTMF) signaling tones and music. Although fidelity is not expected, the output signal should provide nominal quality. A speech codec should at least be capable of conveying music without overly annoying artifacts.

A speech codec should allow low algorithmic complexity; this directly impacts power consumption and cost. Generally, analysis-by-synthesis techniques are more computationally intensive; thus care must be taken in their design. Other features such as echo cancellation and bandwidth enhancement also add to complexity. From an implementation standpoint, a speech codec should be realizable, and provide good performance, with low computational complexity and low memory size requirements.

A speech codec should introduce minimal delay. All codecs introduce a time delay, the time needed for a signal to enter the encoder and output the decoder. This occurs in the transmission path as algorithmic signal processing (and CPU processing) occurs during encoding and decoding. A codec operating in real time must minimize this coding throughput delay. A coding delay of 40 ms is considered acceptable for speech communications. Low-delay codecs may operate with a delay of less than 5 ms. However, the external communications network may add an additional delay to any signal transmission. Data buffering in the encoder is a primary contributor to delay. Any frame-based algorithm incurs finite delay. Speech codecs operate on frames of input samples; for example, 160 to 240 samples may be accumulated (20 ms to 30 ms) before encoding is performed. In most codec designs, greater processing delay achieves greater bit-rate reduction. Backward-adaptive codecs have an inherent delay based on the size of the excitation frame; delay may be between 1 ms and 5 ms. Forward-adaptive codecs have an inherent delay based on the size of the short-term predictor frame and interpolation; delay may be between 50 ms and 100 ms. In non-real-time applications such as voice mail, coding delay is not a great concern.

As much as possible, a codec must accommodate errors introduced along the transmission channel that may result in clicks or pops. Forward error-correction techniques can provide strong error protection, but may require significant redundancy and a correspondingly higher bit rate. Error-correction coding also introduces overhead data and additional delay.

In many mobile speech transmission systems, unequal error protection (UEP) is used to minimize the effects of transmission bit errors. More extensive error correction is given to high-importance data, and little or no correction is given to low-importance data. This approach is easy to implement in scalable systems (described later), because data hierarchy is already prioritized; core data is protected whereas enhancement-layer data may not be. Sometimes listening tests are used to determine the sensitivity of data types to errors.

A speech coding system is shown in Fig. 12.1. It contains the same elemental building blocks as the music coding systems described in earlier chapters. For example, the encoder functionally contains a lowpass filter, a sample-and-hold circuit, and an

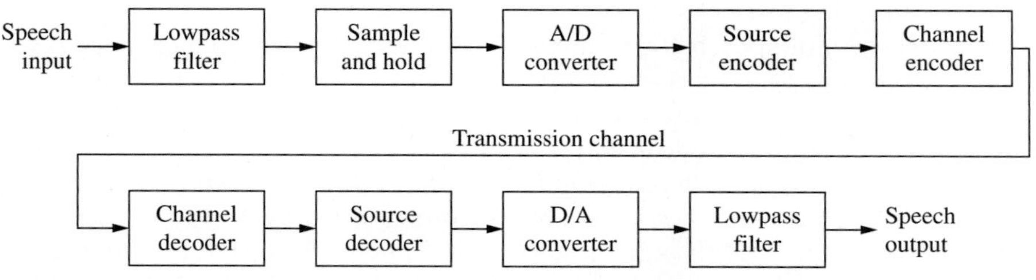

Figure 12.1 A speech coding system employs the same general processes as a music coding system. However, the signal coding methods, particularly the codecs employed, are different.

analog-to-digital converter. Similarly, the decoder contains a digital-to-analog converter and an output lowpass filter. Historically, telephone systems have provided a narrow audio bandwidth from approximately 300 Hz to 3400 Hz; a sampling frequency of 8 kHz is used. This can allow good intelligibility, while respecting the small signal transmission bandwidth available. As with music coding systems, the coded speech signal is channel-coded to optimize its travel through the transmission channel; for example, error-correction processing is applied.

Speech coding for transmission is highly constrained by bit-rate limitations; thus, the transmitted bit rate must be reduced. For example, with an 8-kHz sampling frequency, if 16-bit pulse-code modulation (PCM) words were used, the bit rate input to the source codec would be 128 kbps. However, the speech codec employs lossy coding methods that enable an output bit rate of, for example, 2 kbps to 15 kbps. The speech decoder may use this data-reduced signal to generate a speech signal at the original bit rate, for example, of 128 kbps.

Traditional telephone systems use bit rates such as 64 kbps (PCM) and 32 kbps adaptive differential pulse-code modulation (ADPCM). However, these rates are too high for mobile telephone systems. Simple linear-prediction vocoders may provide a low bit rate of 4 kbps, but speech quality is poor. The bit rate of modern parametric codecs can be quite low, for example, from 2 kbps to 4 kbps, while providing good intelligibility. Many coding systems are scalable; the conveyed bitstream contains embedded lower-rate bitstreams that can be extracted even if the full-rate bitstream is not available. In many cases, the embedded bitstreams can be used to improve the sound quality of the base layer bitstream.

As described later, many speech codecs employ a model of the human speech system. The speech system is considered as streams of air passing through the vocal tract. The stream of air from the lungs is seen as a white-noise excitation signal. The vocal tract is modeled as a time-varying filter. The output is thus the filter's response to the excitation signal. In particular, as output speech, the signal comprises voiced glottal pulses (modeled as a periodic signal), unvoiced turbulence (modeled as white noise), and a combination of these. At the decoder, the speech signal is reproduced from parameters that represent, from moment to moment, the excitation signal and the filter.

Waveform Coding and Source Coding

Both speech codecs and music codecs have a common goal: to convey an audio signal with the highest possible fidelity and at the lowest possible bit rate. However, their methods are very different. In particular, many speech codecs use a model of the human vocal system, whereas music codecs such as MP3 codecs use a model of the human hearing system.

Many music recording systems use PCM or perceptual coding algorithms such as MP3. These can be classified as waveform codecs because the shape of the music waveform is coded. Among other reasons, this is because the music waveform is highly unpredictable and can take on a wide range of characteristics. Thus, the codec must use a coding approach that can accommodate any possible waveform. Waveform codecs typically represent the waveform as quantized samples or quantized frequency coefficients. Waveform coding requires a relatively higher bit rate to achieve adequate signal quality; for example, a bit rate of 32 kbps may be needed.

ADPCM-algorithms use predictive coding and are used in landline telephone systems. ADPCM is a waveform coding technique and does not use source modeling; thus it can operate equally with voice and music signals. Compared to other speech codecs that use source modeling, ADPCM is relatively inefficient and uses a relatively high bit rate, for example, from 24 kbps to 32 kbps. Many speech codecs such as ADPCM, particularly in stationary telephone systems, use companding principles. For example, the μ-law and A-law dynamic compression algorithms use PCM coding and 8-bit word lengths to provide sufficient fidelity for telephone speech communication, as well as bandwidth reduction that is adequate for some applications. In addition, various forms of delta modulation (DM) were sometimes used for speech coding. Delta modulation is a form of predictive coding but does not use source modeling. ADPCM and DM are described in Chap. 4.

Other speech coding methods provide much greater bandwidth reduction. In particular, they are used in digital mobile telephone systems that require high channel capacity and very low bit rates. These speech coding systems model human speech with speech-specific parameter estimation and also use data compression methods to convey the speech signal as modeled parameters at a low bit rate. The method is somewhat akin to perceptual coding, because there is no attempt to convey a physically exact representation of the original waveform. But in perceptual coding, psychoacoustics is used to transmit a signal that is relevant to the human listener; this is destination coding. In speech coding, speech modeling is used to represent a signal that is based on the properties of the speech source; this is source coding. Source coding is used in mobile (cell) telephones and in Voice over IP (VoIP) applications.

Source encoding and decoding is a primary field of study that is unique to speech coding. Speech codecs are optimized for a particular kind of source, namely a speech signal. Generally speech signals are more predictable than music because the signal always only originates from the human vocal system, as opposed to diverse musical instruments. Thus, speech signals can be statistically characterized by a smaller number of quantized parameters.

Parametric source codecs represent the audio signal with parameters that coarsely describe the characteristics of an excitation signal. These parameters are taken from a model of the speech source. Codec performance is greatly determined by the sophistication of the model. During encoding, parameters are estimated from the input speech signal. A time-varying filter models the speech source, and linear prediction is used to generate filter coefficients. These parameters are conveyed to the decoder, which uses the parameters to synthesize a representative speech signal. Linear-prediction codecs are the most widely used type of parametric codecs. Because they code speech as a set of parameters extracted from the speech signal rather than as a waveform, speech codecs offer significant data reduction. The bit rate in parametric codecs can be quite low, for example, from 2 kbps to 4 kbps, and can yield good speech fidelity. Because of the nature of the coding technique, higher bit rates do not produce correspondingly higher fidelity.

Hybrid codecs combine the attributes of parametric source codecs and waveform codecs. A speech source model is used to produce parameters that accurately describe the source model's excitation signal. Along with additional parameters, these are used by the decoder to yield an output waveform that is similar to the original input waveform. In many cases, a perceptually weighted error signal is used to measure the difference. Code excited linear prediction (CELP)–type codecs are widely used examples of hybrid codecs. Hybrid codecs typically operate at medium bit rates.

Whereas single-mode codecs use a fixed encoding algorithm, multimode codecs pick different algorithms according to the dynamic characteristics of the audio signal, as well as varying network conditions. Different coding mechanisms are selected either by source control based on signal conditions or by network control based on transmission channel conditions. The selection information is passed along to the decoder for proper decoding. Although each coding mode may have a fixed bit rate, the overall operation yields a variable bit rate. This efficiency leads to a lower average bit rate.

Human Speech

As noted, speech coding differs from music coding because speech signals only originate from the human vocal system, rather than many kinds of musical instruments. Whereas a music codec must code many kinds of signals, a speech codec only codes one kind of signal. This similarity in speech signals encourages use of source coding in which a model of the human speech system is used to encode and decode the signals. Therefore, an understanding of speech codecs begins with the physiological characteristics and operation of the human vocal tract.

On one hand, speech sounds are easily described. As with all sounds, they are changes in air pressure. Air flow from the lungs is disturbed by constrictions in the body and expelled at the mouth and nostrils. In practice, the human speech system is complex. The formation of speech sounds requires precise control of air flow by many component parts that comprise a time-varying acoustic filter. Figure 12.2 shows a simplified view of the human vocal tract. The energy source of the system is the air flow from the diaphragm and lungs. Air flows through the trachea and across the vocal cords (also called vocal folds), a pair of elastic muscle bands and mucous membrane, contained in the larynx, at the "Adam's apple." The vibration of the vocal cords, as they rapidly open and close, forms a noise-like excitation signal. The glottis, the opening formed by the vocal cords, can impede air flow and thus cause quasi-periodic pulses, or

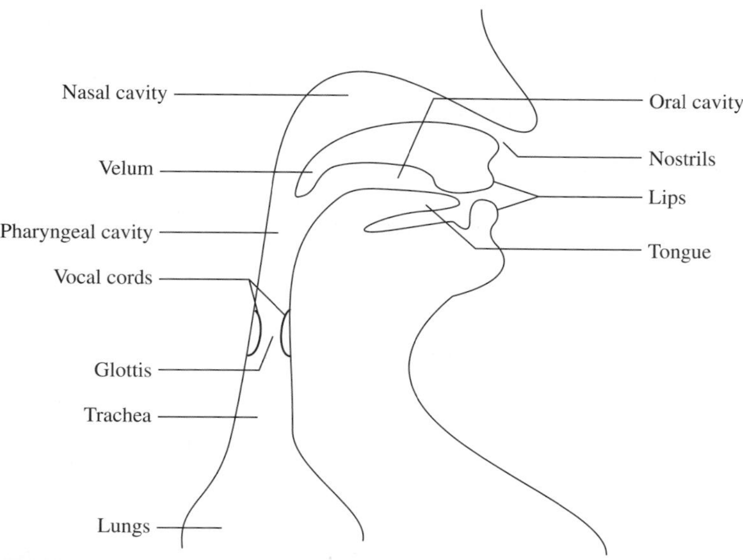

FIGURE 12.2 Representation of the human vocal tract. The system can be modeled in speech codecs as a time-varying acoustic filter.

it can allow air to flow freely. When pulses are created, the tension of the vocal cords determines the pitch of the vibration. The velum controls air flow into the nasal cavity. Air flows through the oral and nasal cavities where it is modulated by the tongue, lips, teeth, and jaw, causing localized constriction, and thus changes in air pressure as air passes through. In addition, the oral and nasal cavities act as resonators. The resonant frequencies change according to the position and shape of the vocal tract components. These frequencies are also called formant frequencies.

Speech sounds can be considered as voiced or unvoiced. Voiced sounds are formed by spectral shaping of glottal quasi-periodic pulses. In other words, the vibration of the vocal cords interrupts the air flow, creating a series of pulses. Voiced sounds are periodic and thus have a pitch. The fundamental frequency typically ranges from 50 Hz to 250 Hz for men and from 150 Hz to 500 Hz for women. Voiced sounds include English-language vowels and some consonants. Unvoiced sounds are not produced by the vocal cords. Rather, they are produced by a steady flow of air passing by stationary vocal cords. The passage through constrictions and other parts of the vocal tract creates aperiodic, noise-like turbulence. Unvoiced sounds include "s" and "f" and "p" sounds. These sounds are also known as fricatives. Sudden releases of air pressure are called stop consonants or plosives; they can be voiced or unvoiced. In addition, transitions from voiced to unvoiced and vice versa are difficult to classify.

Try this experiment: place your fingers on your throat. When making a sustained "aaaaaa" vowel sound, you will feel the vocal-cord vibrations of this voiced sound; moreover, the mouth is open. When making an "mmmmmm" sound, note that it is voiced, but there is no air flow from the closed mouth. Rather, the velum opens so air flows into the nasal cavity to generate a nasal consonant. When making an "ssssss" sound, note that the vocal cords are silent, and the unvoiced consonant sound comes from your upper mouth. Some sounds, such as a "zzzzzz" sound, are mixed sounds. The beginning of the sustained sound is unvoiced in the upper mouth, followed by voiced vocal-cord vibration. As noted, the place where the constriction occurs will influence the sound. As you say the words, "beet," "bird," and "boot," notice how the vowel sound is formed by the placement of the tongue constriction in the front, middle, and back of the throat, respectively. Similarly, although "p" and "t" and "k" are all plosives, they are created by different vocal tract constrictions: closed lips, tongue against the teeth, and tongue at the back of the mouth. Clearly, even a single spoken sentence requires complex manipulation of the human speech system.

Speech sounds can be classified as phonemes; these are the distinct sounds that form the basis of any language vocabulary. Phonemes are often written with a /*/ notation. Vowels (for example, /a/ in "father") are voiced sounds formed by quasi-periodic excitation of the vocal cords. They are differentiated by the amount of constriction, by the position of the tongue, for example, its location from the roof of the mouth, and by lip rounding. Nasal vowels are formed when the velum allows air to the nasal passage. Fricatives (for example, /s/ in "sap" and /z/ in "zip") are consonants formed by turbulence in air flow caused by constriction in the vocal tract position of the tongue and teeth and lips. They may be voiced or unvoiced. Nasal consonants (for example, /n/ in "no") are voiced consonants where the mouth is closed and the velum allows air flow through the nasal cavity and nostrils. Stop consonants, or plosives (for example, /t/ in "tap" and /d/ in "dip") are formed in two parts: a complete constriction at the lips, behind the teeth or at the rear roof of the mouth, followed by a sudden release of air. They are short-duration transients and affected by the sounds preceding and succeeding

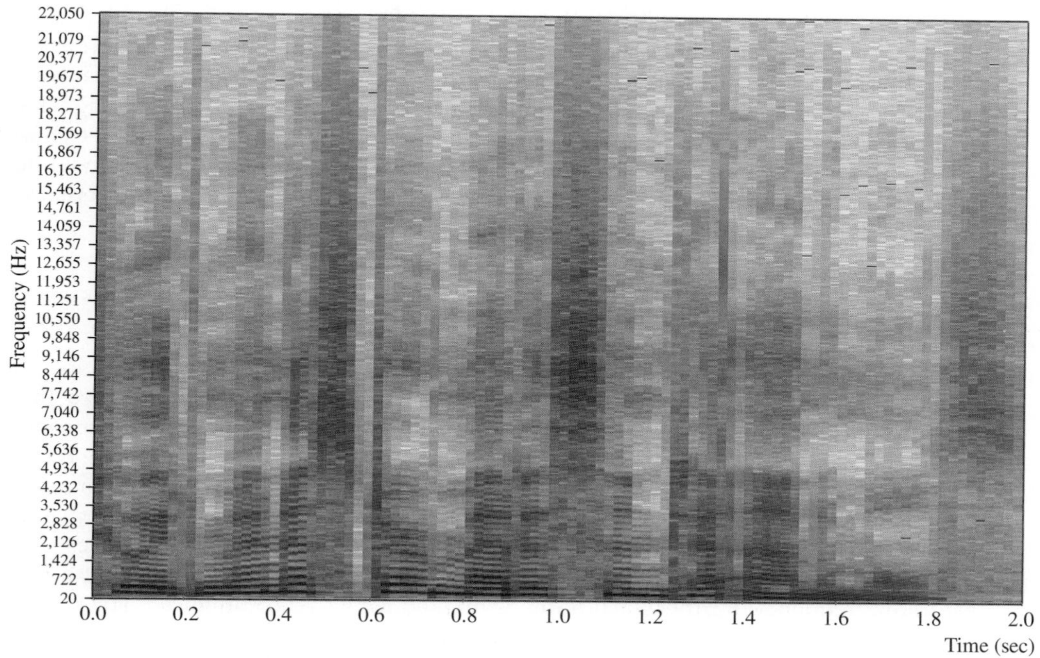

Figure 12.3 A spectrogram of the phrase "to administer medicine to animals" spoken by a female talker. Darker areas represent regions of higher signal energy.

them. They may be voiced or unvoiced. Affricates are a blending of a plosive followed by a fricative. Semivowels (for example, /r/ in "run") are dynamic-voiced consonants. They are greatly affected by the sounds preceding and succeeding them. Diphthongs (for example, /oI/ in "boy") are dynamic-voiced phonemes comprised of two vowel sounds. The classification of a sound as a vowel or a diphthong can vary according to regional accents or by individual speakers.

Many speech transmission systems convey audio frequencies ranging from 300 Hz to 3400 Hz, and although this is adequate for high intelligibility, human speech actually covers a much wider frequency range. Figure 12.3 shows a spectrogram of the phrase "to administer medicine to animals" spoken by a female talker. Darker areas represent regions of higher energy; dark and light banding indicates pitch harmonics. Although the energy in a speech signal is predominantly contained in lower frequencies, there is considerable high-frequency energy.

Source-Filter Model

The source-filter model used in many speech codecs provides an approximation of the human vocal tract, and thus can produce signals that approximate speech sounds. This design approach is why speech codecs using such a model perform relatively poorly on music signals. The source-filter model is based on the observation that the vocal tract functions as a time-varying filter that spectrally shapes the different sounds of speech. Different speech codecs may use source-filter model approaches that are similar, but the codecs differ in how the speech parameters are created and coded for transmission.

Vocal cords, a fundamentally important source of human speech, are modeled as an excitation signal of spectrally flat sound. The phonemes comprising speech sounds are the result of how this source excitation signal is spectrally shaped by the vocal-tract filter. The excitation signal for voiced sounds (such as vowels) is periodic. This signal is characterized by regularly spaced harmonics (impulses in the time domain). The excitation signal for fricatives (sounds such as "f," "s," and "sh") is similar to Gaussian white noise. The excitation signal for voice fricatives (sounds such as "v" and "z") contains both noise and harmonics.

Channel, Formant, and Sinusoidal Codecs

The channel vocoder, introduced in 1939, was one of the first speech codecs (speech codecs are sometimes referred to as vocoders or voice coders). It demonstrated that speech could be efficiently conveyed as frequency components. A channel vocoder analysis encoder is shown in Fig. 12.4A. The input speech signal, considered one frame at a time (for example, 10 ms to 30 ms), is split into multiple frequency bands using fixed bandpass filters. For example, a design may use 20 nonlinearly spaced frequency bands over a 0-Hz to 4-kHz range. Bands may be narrower at lower frequencies. The magnitude of the energy in each subband channel is estimated and output at a low sampling frequency. Further, each frame is marked as voiced or unvoiced. For frames that are voiced, the pitch period is also estimated. Thus, two types of signals are output: magnitude signals and excitation parameters. The magnitude signals can be quantized logarithmically, or the logarithmic difference between adjacent bands can be quantized, for example, using ADPCM coding. Excitation parameters of unvoiced frames may be represented by a signal bit denoting that status. Voiced frames are marked as such, and the pitch period is quantized, for example, with an 8-bit word.

A channel vocoder synthesis decoder is shown in Fig. 12.4B. The excitation parameters are used to determine the type of excitation signal. For example, unvoiced frames are generated from white noise. Voiced frames are generated from pitch-adjusted harmonic excitation. These sources are applied to each channel and their magnitudes are scaled in each of these subbands. The addition of these subbands produces the speech output. An example of a modern speech vocoder is the version developed by the Joint Speech Research Unit (JSRU). It uses 19 subbands, codes differences between adjacent subbands, and operates at 24 kbps.

Another type of speech vocoder, a formant speech codec, improves the efficiency of channel vocoders. Rather than code all of the spectral information across the speech frequency band, a formant vocoder only codes the most significant parts. The human vocal tract has several natural resonance frequencies for voiced sounds, called formants. These are seen as maxima in a spectral envelope; generally there are three significant peaks below 3 kHz. A formant vocoder analysis encoder is shown in Fig. 12.5A. The encoder identifies formants on a frame-by-frame basis. This can be accomplished with formant tracking, that is, by finding the peaks in the spectral envelope. The encoder then codes the formant frequencies and amplitudes. As with channel vocoders, voicing and pitch parameters are also conveyed.

A formant vocoder synthesis decoder is shown in Fig. 12.5B. The excitation signal is generated based on voicing and pitch parameters, and is used to scale formant amplitude

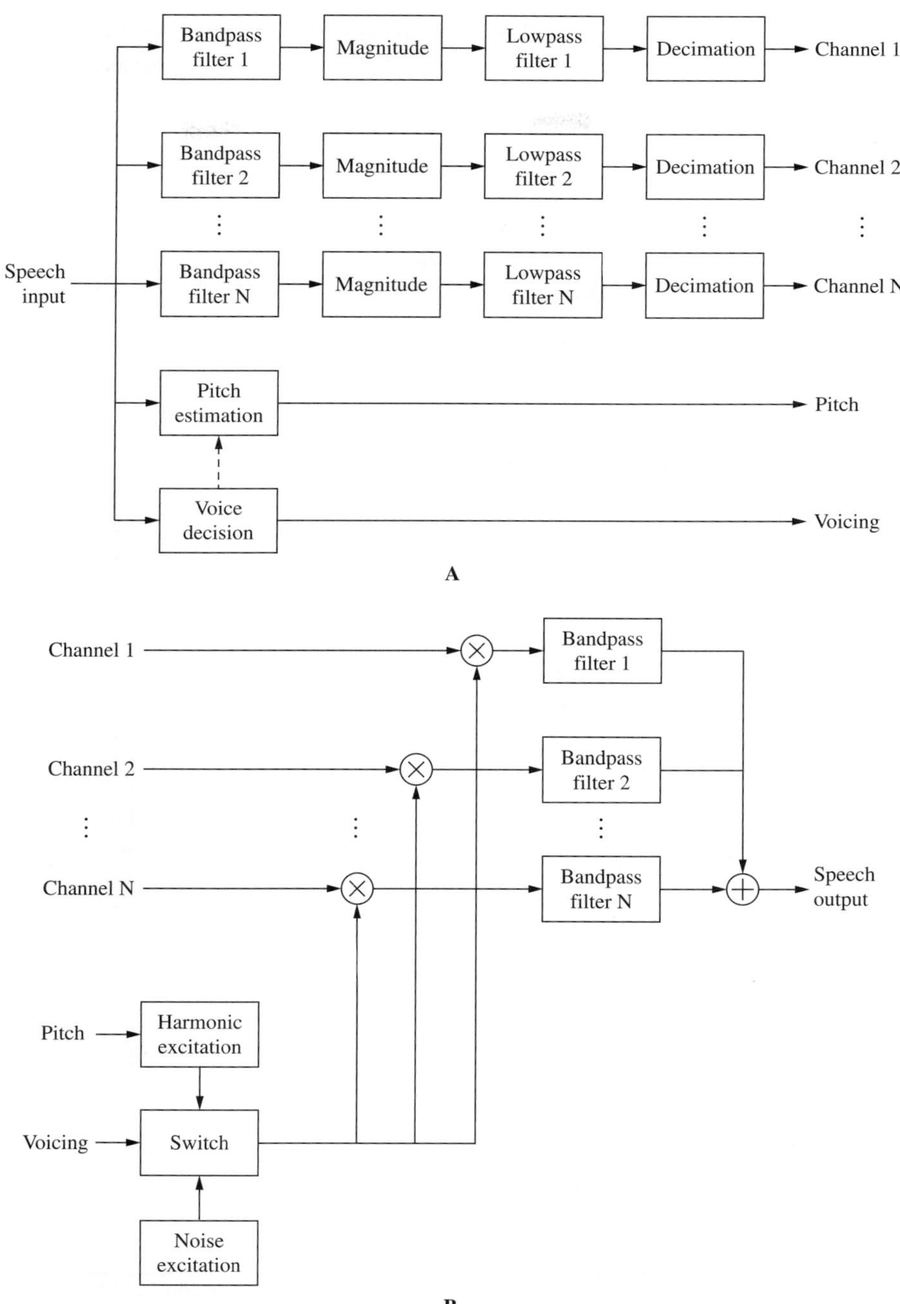

FIGURE 12.4 Channel vocoders convey speech signals as frequency components. A. Channel vocoder analysis encoder. B. Channel vocoder synthesis decoder.

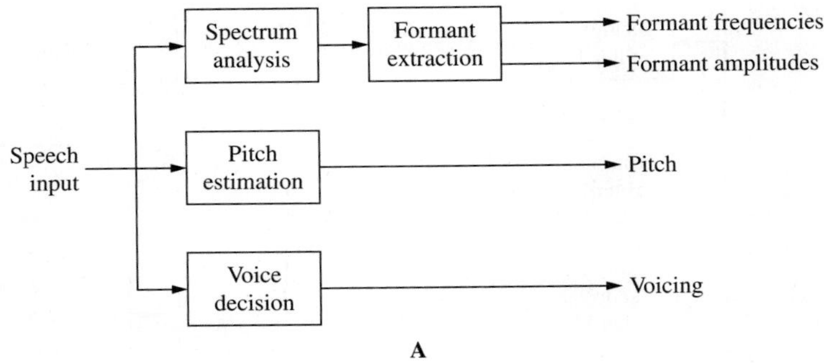

A

Formant frequencies

F_3 F_2 F_1

Formant amplitudes

A_1

A_2

A_3

Adjustable band-pass filter

Adjustable band-pass filter

Adjustable band-pass filter

Speech output

Pitch

Harmonic excitation

Voicing

Switch

Noise excitation

B

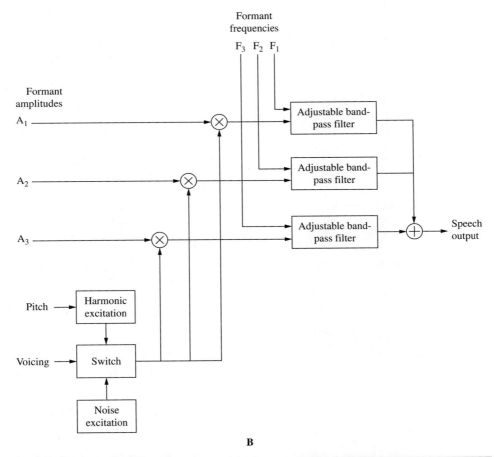

Figure 12.5 Formant speech codecs code only the most significant parts of the speech spectrum, natural resonant frequencies called formants. A. Formant vocoder analysis encoder. B. Formant vocoder synthesis decoder.

information. This is input to a bank of adjustable bandpass filters that are tuned with formant frequency information and their outputs are summed. Difficulties in formant tracking, such as when spectral peaks do not correlate well with formants, can cause poor performance. For that reason, formant vocoders are more widely used for speech synthesis where formants are known, rather than for speech coding.

Sinusoidal speech codecs use sinusoidal modeling to code speech. In this codec, the excitation signal is comprised of the sum of sinusoidal components where amplitude, frequency, and phase are varied to fit the source signal. For example, for unvoiced speech, various sine components with random phases are used. For voiced speech, in-phase, harmonically related sine components, placed at multiples of the pitch frequency, are used. The speech signal input to the encoder can be applied to a Fourier transform; its output can determine the signal peaks. Also, voicing and pitch information are obtained. Amplitudes are further transformed to the cepstral domain for encoding. At the decoder, the cepstral, voicing and pitch parameters are restored to sinusoidal parameters with an inverse transform, and their amplitudes, frequencies, and phases are used to synthesize the speech output.

Predictive Speech Coding

Speech waveforms are generally simple, being comprised of periodic fundamental frequencies interrupted by bursts of noise. They can be modeled as pulse trains, noise, or a combination of the two. In speech coding, the signal is assumed to be the output response to an input excitation signal. By transmitting parameters that characterize the excitation and vocal-tract effects, the speech signal can be synthesized at the decoder.

The statistics of a speech signal vary over time. In fact, clearly, it is the changes in the signal that convey its information. As one listens to a spoken sentence, the phonemes and frequency content continually change. However, over shorter periods of time, speech signals do not change significantly; thus speech signals are said to be quasi-stationary. Some speech signals, such as vowels, have a very high correlation from sample to sample. For these reasons, many speech codecs operate across short segments of time (perhaps 20 ms or less) and use predictive coding to remove redundancies. With predictive coding, a current input sample value is predicted using values of previously reconstructed samples. The difference between the actual current value and the predicted value is quantized; this is generally more bit-efficient than coding values themselves. The decoder subsequently uses the difference value to reconstruct the signal; this technique is used in ADPCM codecs.

Moreover, at low bit rates, rather than coding the difference signal directly, it is more efficient to code an excitation signal that the decoder can use to synthesize a signal that is close to the original speech signal. This is a characteristic of linear prediction. A short-term predictor describes the spectral envelope of the speech signal while a long-term predictor describes the spectral fine structure. In some designs (multipulse and regular-pulse excitation), the long-term predictor is omitted or the sequence of the predictors is reversed. An analysis-by-synthesis encoder using short- and long-term prediction is shown in Fig. 12.6.

Both the long- and short-term predictors are continuously adapted according to the speech signal to improve the efficiency of the prediction; that is, their filter coefficients are changed. The short-term predictor models the vocal tract. Because the vocal tract can only change its physical shape, and hence its output sound, relatively slowly, the

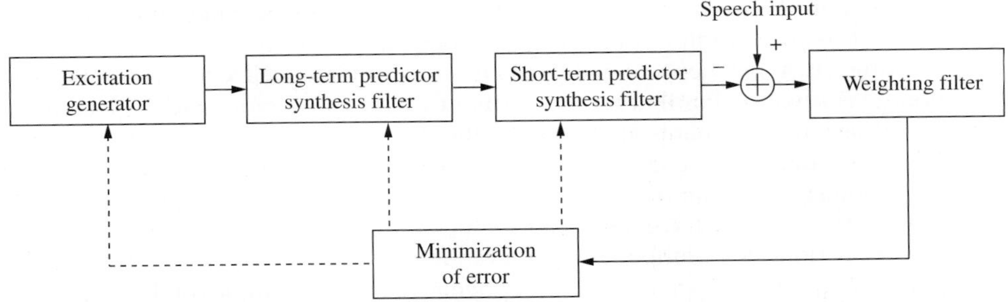

FIGURE 12.6 An analysis-by-synthesis encoder using short- and long-term prediction. Both predictors are continuously adapted according to signal conditions to improve the prediction accuracy.

parameters can be updated relatively slowly. The short-term predictor is often adapted at rates ranging from 30 to 500 times per second. Because an abrupt change in the filter coefficients may produce an audible artifact, values may be interpolated.

A long-term prediction filter is generally used to model the periodicity of the excitation signal in code-excited codecs; a small codebook cannot efficiently represent periodicity. A delay might range from 2.5 ms to 18 ms. This accounts for periodicity of pitch from 50 Hz to 400 Hz, which is the range for most human speakers. The coefficients are revised, for example, at rates of 50 to 200 times per second.

Predictor parameters can be determined from a segment (perhaps 10 ms to 30 ms) of the speech signal. In forward-adaptive prediction, the original input speech signal is used, as shown in Fig. 12.7A. A delay must be used to accumulate the required segment of signal and the parameters are conveyed as part of the output bitstream. In backward-adaptive prediction, the previously reconstructed signal is used to estimate the parameters, as shown in Fig. 12.7B. Because the reconstructed signal is used, it is not necessary

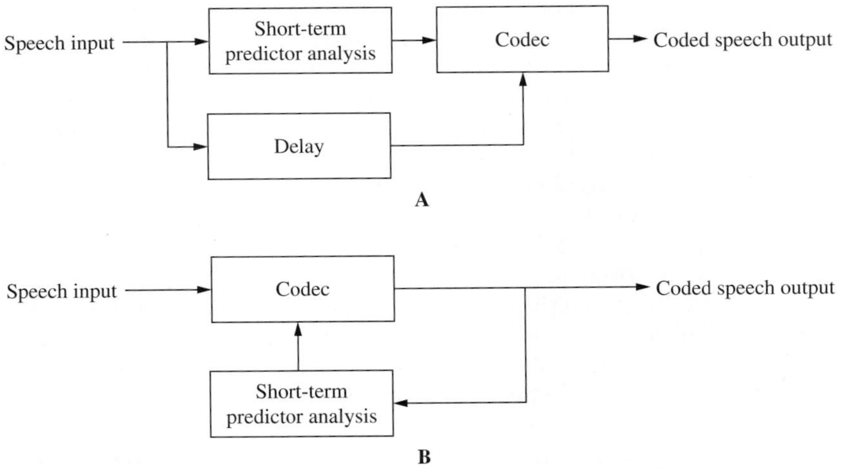

FIGURE 12.7 Either forward- or backward-adaptive methods can be used to determine predictor parameters. A. In forward-adaptive prediction, the original speech signal is used to estimate predictor parameters. B. In backward-adaptive prediction, the previously reconstructed signal is used.

to explicitly convey the parameters in the output bitstream. Either of these methods can be used to estimate the short-term predictor parameters.

The long-term predictor parameters can be estimated using an analysis-by-synthesis method; linear equations are solved for an optimal delay period. Backward-adaptive prediction is not used because the values are sensitive to errors in transmission. Some designs, in place of the long-term synthesis filter, use an adaptive codebook that contains versions of the previous excitation. In a code-excited codec, the excitation parameters can be determined from the excitation structure and by searching for the codebook sequence to yield the minimal error with the weighted speech signal.

The error between the original and the reconstructed signal can be minimized by using a criterion such the mean-squared error. Except at very low bit rates, this results in a quantization noise that has equal energy at all frequencies. Clearly, low bit rates lead to higher noise levels. The audibility of noise can be minimized by observing masking properties. For example, quantization noise can be spectrally shaped so that its distribution falls within formant peaks in the speech signal's spectral envelope as shown in Fig. 12.8. Noise shaping can be performed with a suitable error-weighting filter. The mean-squared error is increased by noise shaping, but audibility of noise is reduced.

In code-excited or vector-excited codecs, both the encoder and decoder contain a codebook of possible sequences. The encoder conveys an index (the one selected to yield the smallest weighted mean-square error between the original speech signal and the reconstructed signal) to a vector in the codebook. This is used to determine the excitation signal for each frame. The code-excited technique is very efficient; an index might occupy 0.2 to 2 bits per sample.

In some cases, a speech enhancement post-filter is used to emphasize the perceptually important formant peaks of the speech signal. This can improve intelligibility and reduce audibility of noise. The post-filter parameters can be derived from the short-term or long-term prediction parameters. Post-filters introduce distortion and can degrade nonspeech signals.

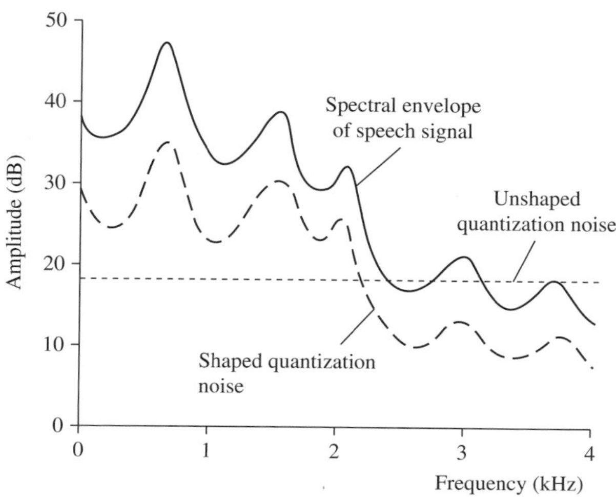

FIGURE 12.8 Noise shaping can be used in a perceptually weighted filter to reduce audibility of noise. The level of shaped noise is equal to that of unshaped noise, but it is less audible.

Linear Predictive Coding

Linear predictive coding (LPC), also referred to as linear prediction, is often used for speech coding. LPC provides an efficient technique for analyzing speech. It operates at a low bit rate and is computationally efficient. It generates estimates of speech parameters using a source-filter model to represent the vocal tract. Linear prediction uses the sum of weighted past time-domain speech signal samples to predict the current speech signal. This inherently reduces the redundancy in the short-term correlation between samples.

In a simple codec, linear prediction can be used in the encoder to analyze the speech signal in a frame, and thus estimate the coefficients of a time-varying filter. These filter coefficients are conveyed along with a scale factor. The decoder generates a white noise signal, multiplies it by the scale factor, and filters it with a filter using the conveyed filter coefficients. This process is updated and repeated for every frame, producing a continuously varying speech signal. Linear prediction is a spectrum estimation technique. Although the output and input signals may differ considerably when viewed in the time domain, the spectrum of the output signal is similar to the spectrum of the original signal. Because the phase differences are not perceived, the output signal sounds similar to the input. Because only the filter coefficients along with a scale factor are conveyed, the bit rate is greatly reduced from the original sampled signal. For example, a frame holding 256 sixteen-bit samples (4096 bits) might be conveyed with 40 bits for the coefficients and 5 bits for the scale factor. In practice, many refinements are applied to this simple codec example to improve its efficiency and performance.

In linear predictive speech codecs, the speech signal is decomposed into a spectral envelope and a residual signal. In some designs, the spectral envelope is represented by LP coefficients (or reflection coefficients). In other designs, the signal is represented by line spectral frequencies (LSFs). The residual signal (error or difference signal) of the LPC encoder can be represented in a variety of ways. LPC codecs use a pulse train and noise representation, ADPCM codecs use a waveform representation. CELP codecs use a codebook representation, and parametric codecs use a parametric representation. Parametric codecs represent the LPC residual signal with harmonic waveforms or sinusoidal waveforms. In particular, the residual signal is conveyed by parametric representation, for example, by sinusoidal harmonic spectral magnitudes. Phase information in the residual is not conveyed. The choice of codec depends on application and bit rate.

Essentially, LPC speech codecs model the human vocal system as a buzzer placed at one end of a tube. More specifically, the model convolves the glottal vibration with the response of the vocal tract. The vocal cords and the space between them, the glottis, produce the buzz which is characterized by changes in loudness and pitch. The throat and mouth of the vocal tract is modeled as a tube; it produces resonances, that is, formants. The tube model works well on vowel sounds but less well on nasal sounds. The model can be improved by adding a side branch representing the nasal cavity, but this increases computational complexity. In practice, nasal sounds can be accounted for by the residue, as described below.

In LPC the signal is decomposed into two components: a set of LP coefficients and a prediction error (residue) signal. Residual signals are modeled with a pulse train (voiced) or noise (unvoiced); either can be switched in depending on the voicing determination. The coefficients are used in an analysis filter to generate the prediction error signal. The analysis filter has the approximate inverse characteristic of the response of the vocal tract; the prediction error signal acts as the glottal vibration.

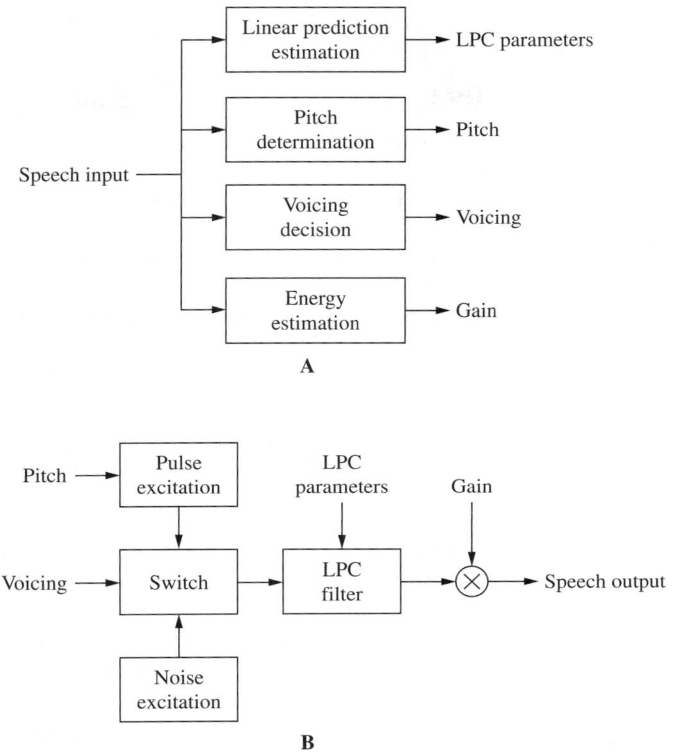

FIGURE 12.9 In linear-predictive coding (LPC) the speech signal is decomposed into a spectral envelope and a residual signal. A. Linear-predictive encoders use a source-filter model to analyze the speech signal. B. Linear-predictive decoders reverse the encoding process and synthesize a speech signal.

An LPC encoder is shown in Fig. 12.9A; four elements are used with a source-filter model to analyze the speech signal. All analysis is done on sampled data segmented into frames. Linear prediction estimates the formants in the spectral envelope and uses inverse filtering to remove their effect from the signal, leaving a residue signal. The encoder then estimates the pitch period and gain of the residue. The frame is classified as voiced or unvoiced using information such as the significance of the harmonic structure of the frame, the number of zero crossings of the time waveform (voiced speech generally has a lower number than unvoiced), or the energy in the frame (voiced speech has higher energy). The formants and residual signal and other information are represented by values that can be stored or transmitted.

An LPC decoder is shown in Fig. 12.9B; it reverses the encoding process and synthesizes a speech signal. The decoder uses the residue to form a source signal and then uses the formants to filter the source to create speech, thus operating like the original model of the human speech system. The voicing input dictates whether the frame is treated as voiced or unvoiced. When voiced, the excitation signal is periodic and pulses are generated at the proper pitch period. When unvoiced, a random noise excitation is used.

In either case, the excitation signal is shaped by the LPC filter and its gain is adjusted to scale the signal amplitude of the output speech.

In the process of estimating speech formants, each signal sample is expressed as a linear combination of previous signal samples. This process uses a difference equation known as a linear predictor (hence the name, linear predictive coding). The process produces prediction coefficients that characterize the formants. The encoder estimates the coefficients by minimizing the mean-square error between the actual input signal and the predicted signal. Methods such as autocorrelation, covariance, or recursive lattice formulation can be used to accomplish this. The process is performed in frames; the frame rate may be 30 to 50 frames per second. Overall bit rate might be 2.4 kbps.

The performance of the LPC technique depends in part on the speech signal itself. Vowel sounds (modeled as a buzz) can be represented as accurate predictor coefficients. The frequency and amplitude of this type of signal can be easily represented. Some speech sounds are produced with turbulent air flow in the human voice system; this hissy sound is found in fricatives and stop consonants. This chaotic sound is different from a periodic buzz. The encoder differentiates between buzz and hiss sounds. For example, for a buzz sound, it may estimate frequency and loudness; for a hiss sound, it may estimate only loudness. For example, the LPC-10e algorithm (found in Federal Standard 1015, described later) uses one value to represent the frequency of buzz, and a value of 0 to signify hiss.

Performance is degraded for some speech sounds that do not fall within the LPC model, for example, sounds that simultaneously contain buzzing and hissing sounds, some nasal sounds, consonants with some tongue positions, and tracheal resonances. Such sounds mean that inaccuracies occur in the estimation of the formants, so more information must be coded in the residue. Good quality can result when this additional information is coded in the residue, but residue coding may not provide adequate data reduction.

As noted, linear predictive encoders generate a residual signal by filtering the signal with an inverse lowpass filter. If this signal was conveyed to the decoder and used as the excitation signal, the speech output would be identical to the original windowed speech signal. However, for this application, the bit rate would be prohibitive. Thus, means are taken to convey the residual signal with a lower bit rate. The closer the coded residual signal is to the original residual, the better the performance of the codec. The performance of linear predictive coding can thus be improved by using sophisticated (and efficient) means to code the residual signal. In particular, as described later, CELP codecs use codebooks to provide better speech quality than LPC methods, while maintaining a low bit rate.

The Federal Standard 1015, LPC-10e, describes an early, low-complexity LPC codec; it has been superseded by the MELP codec. LPC-10e operates with a frame period of 22.5 ms; with 54 bits per frame, the overall bit rate is 2.4 kbps. It estimates pitch from 50 Hz to 400 Hz using an average magnitude difference function (AMDF); its minima occur at pitch periods for voiced signals. Pitch is coded with 6 bits. Voicing determinations using a linear discriminant classifier are made at the beginning and end of each frame. Factors such as the ratio of AMDF maxima to minima, low-frequency energy content, number of zero crossings, and reflection coefficients are considered in voicing. With voiced speech, low-frequency content is greater than high-frequency content; the first reflection coefficient is used to estimate this. Also in voiced speech, low-frequency peaks are more prominent; the second reflection coefficient estimates this. Tenth-order LPC analysis is used. LPC coefficients are given as log-area ratios for the first two coefficients and as reflection coefficients for the eight higher-order coefficients. Only the first four coefficients (4th order) are coded with unvoiced speech segments while all ten coefficients (10th order) are coded for voiced speech segments.

Code Excited Linear Prediction

The code excited linear prediction (CELP) codec and its variants are the most widely used types of speech codecs. The CELP algorithm was originally devised by Manfred Schroeder and Bishnu Atal in 1983 and was published in 1985. Originally a particular algorithm, the CELP term now refers to a broad class of speech codecs. Variants of CELP include ACELP, CS-ACELP, RCELP, QCELP, VSELP, FS-CELP, and LD-CELP. CELP-based codecs are used in many mobile phone standards including the Groupe Speciale Mobile (GSM) standard. CELP is used in Internet telephony, and it is the speech codec in software packages such as Windows Media Player. Part of CELP's popularity stems from its ability to perform well over a range of bit rates, for example, from 4.8 kbps (Department of Defense) to 16 kbps (G.728).

The CELP algorithm is based on linear prediction. It principally improves on LPC by keeping codebooks of different excitation signals in the encoder and decoder. Simply put, the encoder analyzes the input signal, finds a corresponding excitation signal in its codebooks, and outputs an indentifying index code. The decoder uses the index code to find the same excitation signal in its codebooks and uses that signal to excite a formant filter (hence the name, code excited linear prediction). As a complete system, CELP codecs combine a number of coding techniques in a novel architecture. As noted, the vocal tract is modeled with a source-filter model using linear prediction, and excitation signals contained in codebooks are used as input to the linear-prediction model. Moreover, the encoder uses a closed-loop analysis-by-synthesis search performed in a perceptually weighted domain, and vector quantization is employed to improve coding efficiency.

The heart of the CELP algorithm is shown in Fig. 12.10. A codebook contains hundreds of typical residual waveforms stored as code vectors. Codebooks are described in more detail later. Each residual signal corresponds to a frame, perhaps 5 ms to 10 ms, of the original speech signal and is represented by a codebook entry. An analysis-by-synthesis method is used in the encoder to select the codebook entry that most closely corresponds to the residual. A codebook outputs a residual signal that undergoes synthesis that is performed in a local decoder contained in the encoder. The selection is performed by matching a code vector-generated synthetic speech signal to the original speech signal; in particular, a perceptually weighted error is minimized.

Clearly, because a codebook is used, there are a finite number of possible residual signals. The decoder contains a gain stage to scale the residual signal, a long-term prediction filter, and a short-term prediction filter. The long-term prediction filter contains time-shifted versions of past excitations and acts as an adaptive codebook. The synthesized signal is subtracted from the original speech signal and the difference is applied to a perceptual weighting filter. This sequence is repeated until the algorithm identifies the codebook entry that generates the minimum residual signal. For example, every codebook entry can be sequentially tested. In practice, methods are used to make the codebook search more efficient. The entry yielding the best match is conveyed as the final output. At the receiver, an identical codebook and decoder are used to synthesize the output speech signal.

The final excitation signal is obtained by summing the pitch prediction with an innovation signal from a fixed codebook. The fixed codebook is a vector quantization dictionary that is hard-wired into the codec. The codebook can be stored explicitly (as

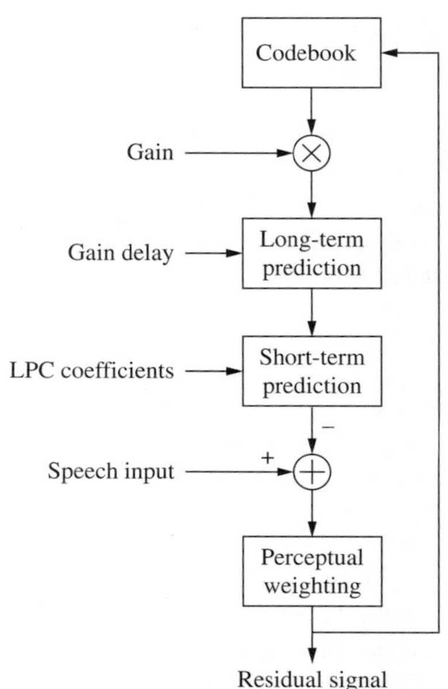

Figure 12.10 In the code excited linear prediction (CELP) algorithm, a closed-loop analysis-by-synthesis search is performed in a perceptually weighted domain. The output residual signal is minimized by searching for and selecting the optimal codebook entry.

in Speex) or it can be algebraic (as in ACELP). The innovation signal is the part of the signal that could not be obtained from linear prediction or pitch prediction, and its quantization accounts for most of the allocated bits in a coded signal.

CELP codecs also shape the error (noise) signal so that its energy predominantly lies in frequency regions that are less audible to the ear, thus perceptually minimizing the audibility of noise. For example, the mean square of the error signal in the encoder can be minimized in the perceptually weighted domain. By applying a perceptual noise-weighting filter to the error, the magnitude of the error is greater in spectral places where the magnitude of the speech signal is louder; thus the speech signal can mask the error. The weighting filter can be derived from the linear-prediction coefficients, for example, by using bandwidth expansion. Although far from optimal, the simple weighting filter does improve subjective performance.

In some cases, performance of speech codecs such as CELP is augmented by regenerating additional harmonic content that was lost during coding. In other cases, narrow-band coding is improved by estimating a wideband spectral shape from the narrow-band signal. During LPC coding, wideband LPC envelope descriptions are also entered to create a shadow codebook. During decoding, the wideband envelopes are combined with the narrow-band excitation.

CELP Encoder and Decoder

The CELP encoder models the vocal tract using linear prediction and an analysis-by-synthesis technique to encode the speech signal. It accomplishes this by perceptually optimizing the decoded (synthesis) signal in a closed loop. This produces the optimal parameterization of the excitation parameters that are ultimately coded and output by the encoder. When decoded, these yield synthesized speech that best matches the original input speech signal. This kind of perceptually optimized coding may imply that a human listener selects the "best sounding" decoded signal overall. Since this is not possible, the encoder parses the signal into small segments and performs sequential searches using a perceptual-weighting filter in a closed loop that contains a decoding algorithm. The loop compares an internally generated synthesized speech signal to the original input speech signal. The difference forms an error signal which is iteratively used to update parameters until the closest match is found and the error signal is minimized, thus yielding optimal parameters. The error signal is perceptually weighted by a filter that is approximately the inverse of the input speech signal spectrum. This reduces the effect of spectral peaks; large peaks would disproportionately influence the error signal and parameter estimation.

A simplified CELP encoder is shown in Fig. 12.11A. The analysis-by-synthesis process begins with an initial set of coding parameters generated by linear-prediction analysis to determine the vocal system's impulse response. The excitation codebook

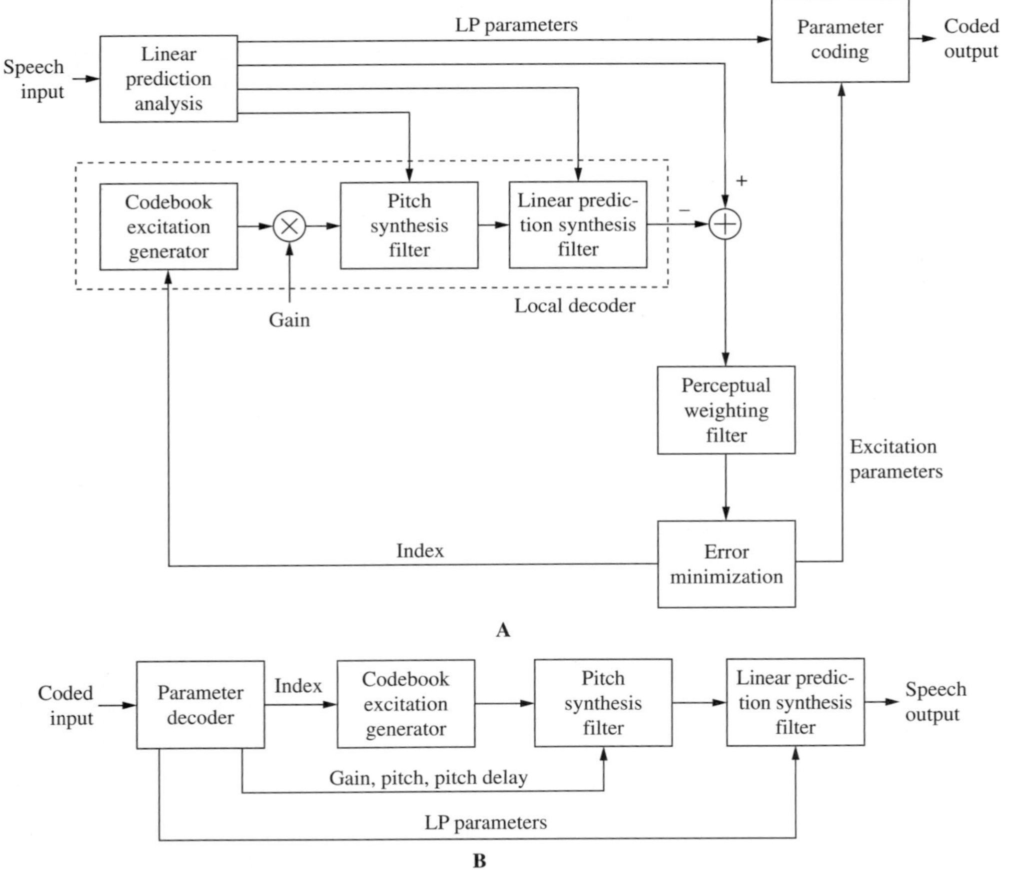

FIGURE 12.11 A CELP encoder and decoder. A. CELP encoders use linear prediction and a local analysis-by-synthesis decoder to minimize an error signal and output coding parameters. B. CELP decoders use received parameters to find the corresponding excitation residual and use that signal to excite the formant filter and output the speech signal with a reconstructed LPC filter.

generator and linear-prediction filter form a local decoder. They output a synthesized speech signal that is subtracted from the original speech input. The difference forms an error signal that is used to improve the parameters by minimizing the perceptually weighted error. The loop iteratively applies the minimized error to the local decoder until its output yields excitation parameters that optimally minimize the energy in the error signal. The parameters are output from the encoder. The LPC parameters are updated for every frame; excitation parameters are updated more often, in subframe intervals. In some cases, the perceptual-weighting computation is applied to the original input speech signal, rather than to the residual signal. In this way, the computation is performed once, rather than on every loop iteration.

The CELP decoder can be considered as a synthesizer; it uses a speech synthesis technique to output speech signals. A basic CELP decoder is shown in Fig. 12.11B. The decoder receives quantized excitation and LPC parameters. Gain, pitch, and pitch delay information is applied to the formant filter. LPC coefficients, representing the vocal tract, are used to reconstruct the linear-prediction filter. The codebook uses the conveyed index code to find the corresponding excitation residuals and uses those signals to excite the formant filter. The output of the formant filter is applied to the linear-prediction filter to generate the output synthesized speech signal. This is done by predicting the input signal using the linear combination of past samples. As with any prediction, there is a prediction error; linear prediction strives to provide the most accurate prediction coefficients and thus minimize the error signal. The Levinson–Durbin algorithm can be used to perform the computation. Other techniques are used to ensure that the filter is stable.

CELP Codebooks

Linear-predictive encoders such as CELP generate a residual signal. If this signal were used at the decoder as the excitation signal, the speech output would be identical to the original windowed speech signal. However, the bit rate would be prohibitively high. Thus, the residual signal is coded at a lower bit rate; the closer the coded residual signal is to the original residual, the better the performance of the codec. CELP codecs use codebooks to provide high speech quality, while maintaining a low bit rate.

CELP encoders contain codebooks with tables of pre-calculated excitations, each with an identifying index codeword. The encoding analyzer compares the residue to the entries in the codebook table and chooses the entry that is the closest match. It should provide the lowest quantization error and hence yield perceptually improved coding. The encoder then efficiently sends just the index code that represents that excitation entry. The system is efficient because only a codeword, as well as some auxiliary information, is conveyed.

Testing all of the entries in a large codebook would require considerable computation and time. Much work in CELP design is devoted to minimizing this computational load. For example, codebooks are structured to be more computationally efficient. Also, codewords are altered to reduce the complexity of computation; for example, codewords may be convolved with CELP filters. Overlapping codebooks are efficient in terms of computation and storage requirements. Rather than store codewords separately, overlapping codebooks can store all entries in one array with each codeword value overlapping the next.

Generally, single-stage codebooks may range in size from 256 (8-bit) to 4096 (12-bit) entries. At least theoretically, performance would be improved if the codebook were large enough to contain every possible residue signal. However, the larger the codebook, the more computation and time required to search it, and the longer the code needed to

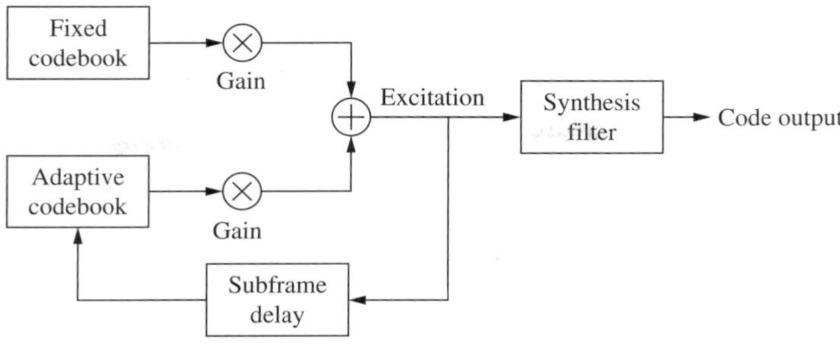

FIGURE 12.12 Both fixed and adaptive codebooks can be used in CELP decoders. Two smaller codebooks can be searched more efficiently than one larger codebook, and their different approaches improve coding performance.

identify the index. In particular, the codebook would have to contain an individual code for every voice pitch, making the codebook inefficiently large. Codebooks are generated (trained) using statistical means to improve searching. For example, the LBG clustering algorithm (named after Linde, Buzo, and Gray) is often used to generate codebooks. Beginning with a set of codebook vectors, the algorithm assigns training vectors and recomputes the codebook vectors on the basis of the assigned training vectors. This is repeated until quantization error is minimized.

CELP algorithms can avoid the problem of large codebooks by using two small codebooks instead of one larger one. In particular, fixed and adaptive codebooks can be used, as shown in Fig. 12.12. The fixed (innovation) codebook is set by the algorithm designers and contains the entries needed to represent one pitch period of residue. It generates signal components that cannot be derived from previous frames; these components are random and stochastic in nature. A vector quantization can be used. In addition to linear prediction, an adaptive (pitch) prediction codebook is used during segments of voiced sounds when the signal is periodic. The adaptive codebook is empty when coding begins, and during operation it is gradually filled with delayed, time-shifted versions of excitations signal coded in previous frames. This is useful for coding periodic signals.

Vector Quantization

Vector quantization (VQ) encodes entire groups of data, rather than individual data points. Analyzing the relationships among the group elements and coding the group is more efficient than coding specific elements. Moreover, instead of coding samples, VQ systems generally code parameters that define the samples. In the case of speech coding, vector quantization can be used to efficiently code line spectral frequencies that define a vocal tract. VQ is used in CELP and other codecs.

The basic structure of a VQ encoder and decoder is shown in Fig. 12.13. A speech signal is applied to the encoder. Parameter extraction is performed on a frame basis, for example, with linear-predictive coding to yield parameter vectors. These are compared to vector entries in a fixed, precalculated codebook. A distance metric is used to find the codebook entry that is the best match for the current analyzed vector. The encoder outputs an index codeword that represents the selected codebook entry for that frame.

The codeword is received by the decoder which contains the same codebook as the encoder. The codeword is used as an index to retrieve the corresponding vector from the codebook. This vector best matches the original vector extracted from the speech

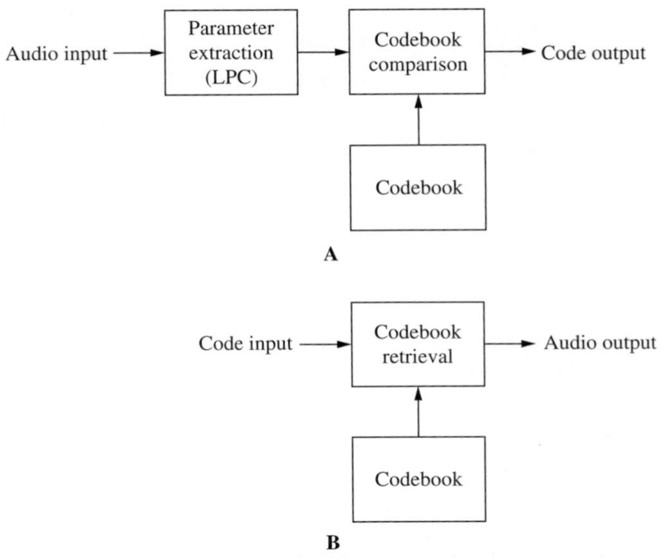

Figure 12.13 Vector quantization (VQ) analyzes the parameter relationships among the group elements; coding the group is more efficient than coding specific elements. VQ is used in CELP and other codecs. A. VQ encoder. B. VQ decoder.

signal. This vector is used as the basis for output speech synthesis. For example, voicing and pitch information is also applied.

Examples of CELP Codecs

As noted, many types of CELP codecs are in wide use. Federal Standard 1016 describes a CELP codec operating at 4.8 kbps. The frame rate is 30 ms with four subframes per frame. The adaptive codebook contains 256 entries; the fixed codebook contains 512 entries. The total delay is 105 ms. The ITU-T G.728 CELP codec operates at 16 kbps; it is designed for low delay (LD) of less than 2 ms. Frame length is 2.5 ms with four subframes per frame. To achieve low delay, only the excitation is conveyed. The fixed codebook coefficients are derived at the decoder using prior decoded speech data. The ITU G.723.1 CELP codec operates at 5.3 kbps and 6.3 kbps; it is used for VoIP and video conferencing. Frame rate is 30 ms with four subframes per frame. Two different codecs are used for the two bit rates, and the system can switch between them on a frame-by-frame basis. Total delay is 37.5 ms. The ETSI GSM Enhanced Full Rate Algebraic CELP codec operates at 12.2 kbps with an additional 10.6 kbps used for error correction; the total bit rate is 22.8 kbps. The codec is used for mobile telephone applications. The frame rate is 20 ms with five subframes per frame. Linear-prediction parameters are coded as line spectral pairs (LSPs).

Scalable Speech Coding

Some speech codecs generate embedded (nested) data to allow scalability or other functionality. In addition to core-layer data, the conveyed bitstream contains one or more embedded lower-rate bitstreams that can also be extracted and decoded. In many cases, embedded enhancement bitstreams, and their resulting higher bit rate, can be used to improve the sound quality of the core layer bitstream. For example, the embedded layers might reduce coding artifacts or engage a higher sampling frequency to allow a wider audio bandwidth.

A scalable encoder contains a core encoder nested inside enhancement encoders. It outputs a multi-layer data structure; for example, a three-layer bitstream may contain a core layer and two enhancement layers. The layers are hierarchical; higher layers can only be decoded if lower layers are also received. Using rate adaptation, higher layers may be omitted if dictated by transmission bandwidth or other limitations. The granularity of the coding is determined by the number of layers and increase in bit rate for each layer. The decoder will output a signal that depends on the number of layers transmitted and the capabilities of the decoder.

Scalability allows great diversity in transmission conditions and receiver capability. For example, a VoIP-based telecommunications system can choose the appropriate bitstream according to capacity limitations, while maintaining the best possible sound quality. The technique can also be used to introduce new coding types or improvements while maintaining backward core compatibility. For example, a standard narrow-band codec may comprise the core layer while bandwidth extension uses higher layers.

Scalability can be applied in many different ways. For example, a subband CELP codec may use a QMF filter to split the speech signal into low and high bands. Each band is separately coded with a codec optimized for that frequency range. For example, a CELP codec may be used for the low band and a parameter model codec used for the high band because of reduced tonality of higher-frequency signals.

The Internet is a packet-switched network; in contrast to a circuit-switched system, data is formatted and transmitted in discrete data packets. Speech information comprises the payload, and there is also header information. A packet may contain one or more speech frames. In the case of scalable (embedded) VoIP formats, the various layers must be distinguishable and separable. For example, a packet might contain a single layer; if layers were unneeded, the system could easily drop those packets. However, because of the relatively large amount of header data for each packet (40 to 60 bytes), this approach is inefficient; a speech frame may comprise 80 bytes. A more efficient approach places many speech frames in each packet; this is effective for streaming applications. However, this approach can be sensitive to packet loss, and may incur longer delays. For conversational applications, relatively fewer frames are placed in each packet, and for embedded formats, the various layers are placed in the same packet. For example, if the frame size is approximately 20 ms to 30 ms, one frame can be placed in each packet. The payload would identify the different embedded layers in a frame.

Clearly, the transport protocol must support the data hierarchy and features such as scalability. In addition, the system may prioritize payload data. For example, in the event of network congestion or receiver capability or customer preferences, a rate adaptation unit in the network can transmit core data while dropping enhancement data. Some data protocol versions such as IPv4 and IPv6 support Differentiated Services. In this case, only complete packets can be labeled, not specific data within. Thus, different layers would be placed in different packets. For example, a scalable MPEG-4 bitstream could be conveyed by multiplexing audio data in Low-overhead MPEG-4 Audio Transport Multiplex (LATM) format and each multiplexed layer is placed in different Real Time Protocol (RTP) packets. In this way, Differentiated Services can distinguish between layers.

G.729.1 and MPEG-4 Scalable Codecs

Scalability features are used in many speech codecs. The ITU-T G.729.1 Voice over IP codec is an example of a scalable codec; it is scalable in terms of bit rate, bandwidth, and complexity. It operates from 8 kbps to 32 kbps, supports 8-kHz and 16-kHz sampling frequency (input and output), and provides an option for lower coding delay. The 8-kbps core codec of G.729.1 provides backward compatibility with the bitstream format of the G.729 codec, which is widely used in VoIP applications. G.729.1 has two modes: a narrow-band output at 8 kbps and 12 kbps providing an audio bandwidth of 300 Hz to 3400 Hz and a wideband output from 14 kbps to 32 kbps providing an audio bandwidth of 50 Hz to 7000 Hz.

A G.729.1 encoder is shown in Fig. 12.14A. The input and output sampling frequency is 16 kHz. The encoder uses a QMF analysis filter bank to divide the audio signal into two subbands, split at 4 kHz. The lower band (0–4 kHz) is processed by a 50-Hz highpass filter and applied to a CELP encoder. The upper band (4–8 kHz) is processed by a 3-kHz lowpass filter and further applied to time-domain bandwidth extension (TDBWE) processing. The upper-band signal (prior to TDBWE) and the residual error of the lower-band signal are together applied to a time-domain aliasing cancellation (TDAC) encoder, a transform codec using the modified discrete-cosine transform. Further, frame erasure concealment (FEC) processing is used to generate concealment and recovery parameters used to assist decoding in the presence of frame erasures.

A G.729.1 decoder is shown in Fig. 12.14B; its operation depends on the bit rate of the received signal. At 8 kbps and 12 kbps, the lower-band signal (50 Hz to 4 kHz) is generated by the CELP decoder and a QMF synthesis filter bank upsamples the signal to a 16-kHz sampling frequency. At 14 kbps, the lower-band 12 kbps signal is combined with a reconstructed higher-band signal from TDBWE processing to form a signal with higher bandwidth (50 Hz to 7 kHz). For received bit rates from 16 kbps to 32 kbps, the upper-band signal and the lower-band residual signal are decoded with TDAC and pre- and post-echo processing reduces artifacts. This signal is combined with the CELP output signal; the TDBWE signal is not used. The decoder can also perform frame erasure concealment using transmitted parameters. As noted, the system can operate in two bit-rate modes: narrow band or wideband. To avoid audible artifacts when switching between modes, the decoder applies crossfading in the post-filter, and fades in the high band when switching to wideband mode.

The G.729.1 bitstream is hierarchical, enabling scalability. There are 12 layers. The first two layers comprise the narrow-band mode (8 kbps and 12 kbps), and the additional 10 layers comprise the wideband mode (14–32 kbps in steps of 2 kbps). Layer 1 is a core layer (160 bits) that is interoperable with the G.729 format. Layer 2 (12 kbps) is a narrow-band enhancement layer (38 CELP bits and 2 FEC bits) using cascaded CELP parameters and signal class information. Layer 3 (14 kbps) is a wideband extension layer (33 TDBWE bits and 7 FEC bits) with time-domain bandwidth extension parameters and phase information. Layers 4 to 12 (greater than 14 kbps) are wideband extension layers (TDAC1 with 35 TDAC bits and 5 FEC bits, thereafter TDAC2–TDAC9 with 40 TDAC bits) with time-domain aliasing cancellation parameters. Nominally, the maximum bit rate is 32 kbps; after encoding, the instantaneous bit rate can be set as low as 8 kbps. The frame rate is 20 ms.

The G.729.1 codec has a maximum delay of 48.9375 ms; of this, 40 ms is due to TDAC windowing. Delay is reduced in other modes. For example, a delay of 25 ms is possible. Computational complexity varies according to the bit rate. At 32 kbps, total encoder/decoder complexity is 35.8 weighted million operations per second.

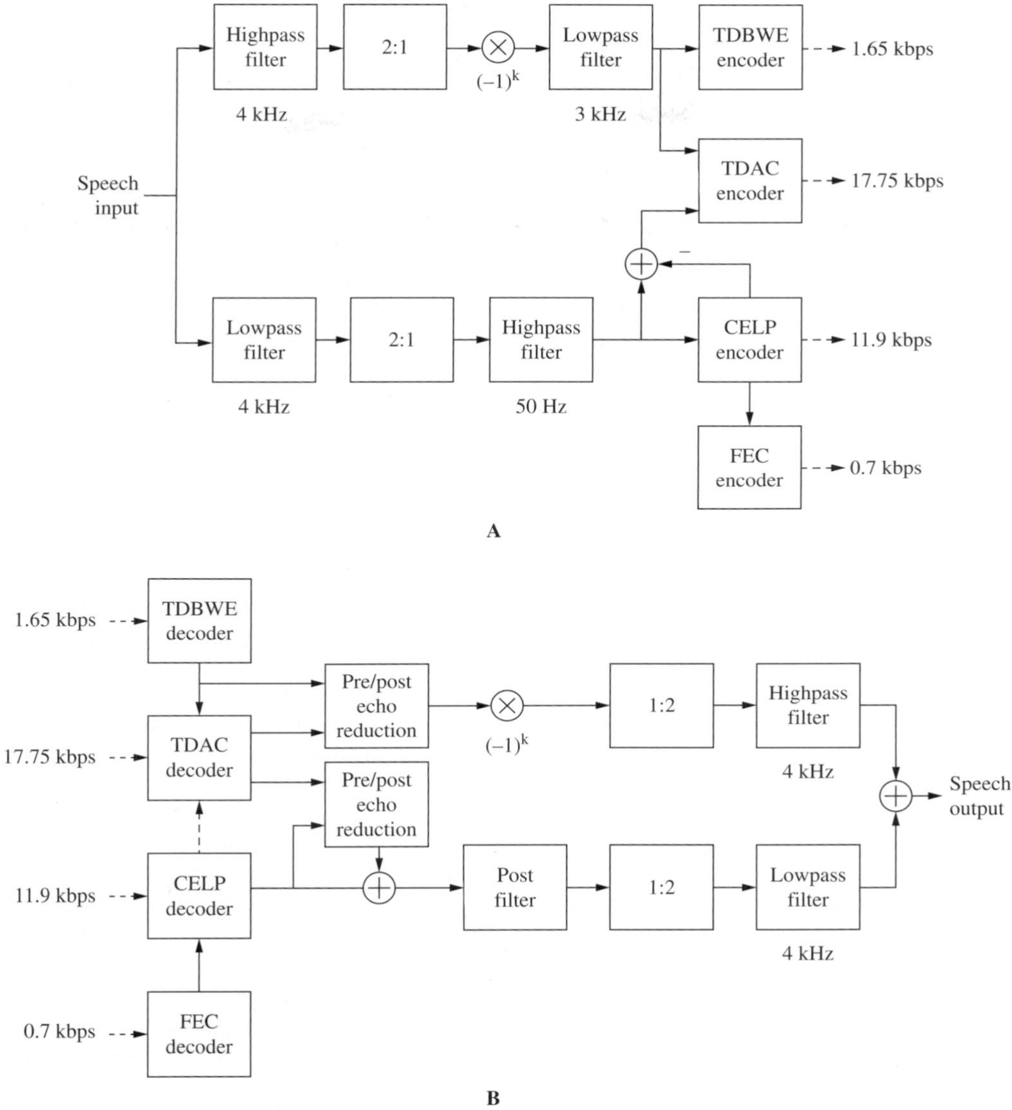

FIGURE 12.14 The ITU-T G.729.1 Voice over IP codec is an example of a scalable codec; it is scalable in terms of bit rate, bandwidth, and complexity. A. G.729.1 encoder. B. G.729.1 decoder.

Listening evaluations for narrow-band clean speech show that at 8 kbps, G.729.1 has a higher mean opinion score (MOS) than G.729 Annex A, but a lower score than G.729 Annex E. At 12 kbps, the G.729.1 score is more similar to G.729 Annex E. These evaluations are at a 0% frame error rate (FER). At a 6% FER, the 12 kbps G.729.1 scores much higher than Annex A or E. For wideband clean speech, G.729.1 scores higher than G.722.1 at 32 kbps. For wideband music, at 32 kbps, music quality of G.729.1 is scored similar to G.722.1 at 32 kbps. When comparing codecs, it is important to remember that relative quality varies for different frame-error rates.

Several scalable codecs have been standardized by the Moving Pictures Expert Group of the International Organization for Standardization (ISO/MPEG) and the International Telecommunication Union–Telecommunication Standardization Sector (ITU-T). The MPEG-4 standard (ISO/IEC 14496-3 2005) describes coding tools that can be used for scalable coding. The MPEG-4 CELP codec operates in narrow-band (8-kHz sampling frequency) or wideband mode (16-kHz sampling frequency). In the MPE mode, two types of enhancement layers can be added to the core. In narrow-band mode, the bandwidth is extended, and in wideband mode, the audio signal quality is improved. As another example, the MPEG-4 HVXC narrow-band codec provides a low bit rate of 2 kbps. An optional enhancement layer increases the bit rate to 4 kbps. MPEG-4 is discussed in more detail in Chap. 15.

Bandwidth Extension

Bandwidth extension (BWE) can be used to improve sound quality in speech codecs. In particular, it supplements the essential data in a narrow-band (NB) signal with additional data. The resulting wideband (WB) signal requires only a modest amount of additional data to greatly extend the frequency range of the audio signal, for example, from a range of 300 Hz to 3400 Hz to a range of 50 Hz to 7000 Hz. In some cases, the wideband signal is further augmented to produce a super-wideband signal with a high-frequency extension, for example, to 15 kHz. High-frequency extension can improve intelligibility, particularly the ability to distinguish between fricatives (such as "f" and "s" sounds). Such WB BWE systems can be backward compatible with existing NB systems. Bandwidth extension for speech codecs is similar to spectral band replication techniques used for music signals. One difference is that BWE systems can use signal source models and SBR systems cannot.

Bandwidth extension can be implemented in a number of ways. For example, processing can be placed in both the transmitter (encoder) and receiver (decoder) with transmitted side information (as scalable data) or without (as watermarked data). It can be placed only in the receiver as post-processing with no transmitted side information. It can be placed only in the network with a NB/WB transcoder and no transmitted side information.

In bandwidth extension, the parameters of the narrow-band signal can be used to estimate a wideband signal. In the case of speech BWE, the estimation can use a speech model similar to the source-filter model used in speech codecs. As noted, speech originates from noise-like signals from the vocal cords and is further processed by the vocal tract and tract cavities. This can be source-modeled with a signal generator and a synthesis filter to shape the spectral envelope of the signal. This model is also used for bandwidth extension; the source component and the filter component are treated separately. The bandwidth of the excitation signal is increased. For example, this can be accomplished with sampling-frequency interpolation, and the narrow-band signal is applied to analysis filters for resynthesis with lower and higher frequencies. Extended-frequency harmonics can be generated by distortion or filtering, or by synthetic means, or by mirroring, shifting, or frequency scaling. In addition, feature extraction performed on the narrow-band signal can be used to statistically estimate the spectral envelope. The estimation also incorporates a prior knowledge of both the features and the

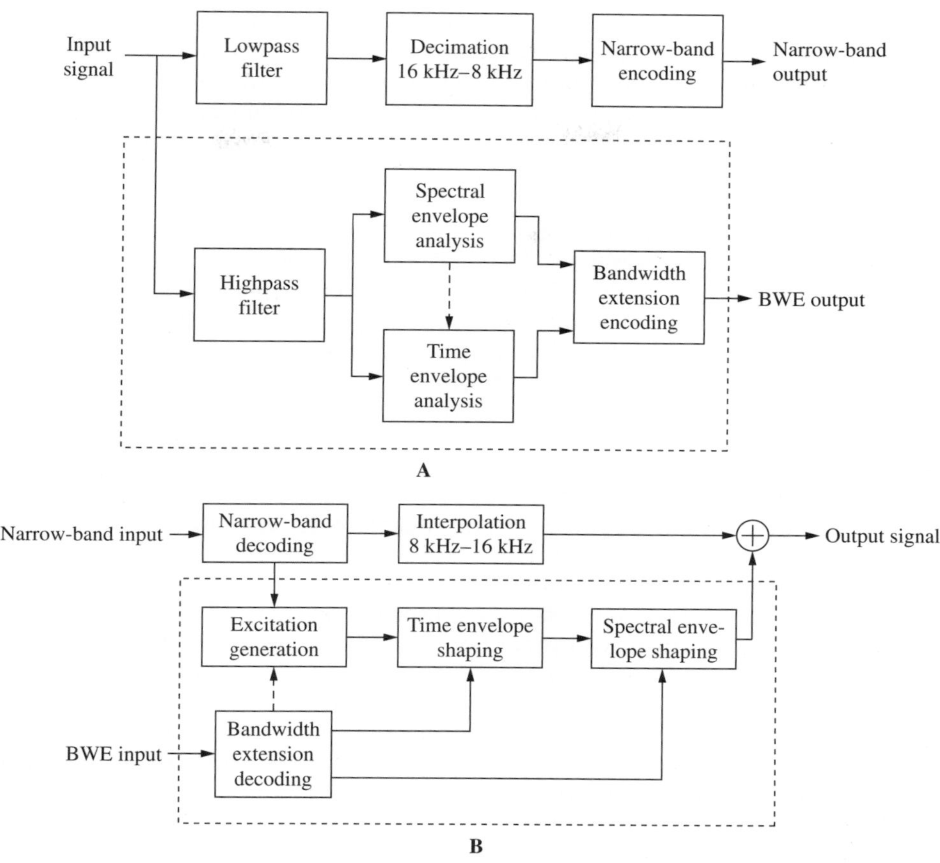

FIGURE 12.15 Bandwidth extension (BWE) supplements the essential data in a narrow-band (NB) signal with additional data to produce a wideband (WB) signal. A. Narrow-band encoder with bandwidth extension; BWE information is transmitted as side information. B. Narrow-band decoder with bandwidth extension; the lower subband narrow-band signal is decoded and interpolated to allow addition of the high subband extension.

estimated signal. The spectral envelope is critical in providing improved audio quality. For example, in some systems, only BWE envelope information is transmitted, and excitation extension is only done in the receiver.

A narrow-band encoder with bandwidth extension is shown in Fig. 12.15A. BWE information is transmitted as side information. The speech signal, with a sampling frequency of 16 kHz, is separated into low and high subbands. The lower subband forms the narrow-band signal. It is decimated to a sampling frequency of 8 kHz and applied to a narrow-band encoder. The higher subband forms the BWE signal. It is analyzed to determine the spectral and time envelopes (the spectral envelope information may be used in the time-envelope analysis), and quantized BWE parameters are generated. In some cases, a time-envelope prediction residual is transmitted.

A narrow-band decoder with bandwidth extension is shown in Fig. 12.15B. The lower-subband narrow-band signal is decoded and interpolated to 16 kHz to allow addition of the higher-subband extension. The higher-subband BWE parameters are decoded and an excitation signal is generated. This signal is noise-like and may be further refined from parameters from the narrow-band decoder. The subframe time envelope of the excitation signal is shaped by gain factors from the BWE linear-prediction decoder. The spectral envelope of the excitation signal is shaped by lowpass filter coefficients from the BWE decoder using an all-pole synthesis filter. The extension signal is added to the narrow-band signal.

In one bandwidth extension implementation that is part of the G.729.1 standard, a FIR filter-bank equalizer is used for spectral envelope shaping in place of a lowpass synthesis filter in the decoder. Filter coefficients are adapted by comparing a transmitted target spectral envelope to the spectral envelope of the synthesized excitation signal, which uses parameters from the narrow-band decoder.

Bandwidth extension techniques can also employ watermarking in which the BWE information is hidden in the narrow-band signal. This can promote backward compatibility with narrow-band speech telecommunication systems. As with any watermarking system, the watermark data must not audibly interfere with the host signal. In this case, the BWE data must not degrade intelligibility of the narrow-band signal in receivers that do not make use of the BWE data. Also, its implementation must make the watermark data robust so that it can be conveyed through the complete transmission channel and its associated transcoding, processing, and error conditions. For example, the watermark signal may pass through speech encoding and decoding stages. A system may have unequal error protection. The narrow-band data may have more robust error correction while enhancement layers may have much less. A BWE system must also operate within acceptable computational complexity and latency limits, and the watermark must provide sufficient capacity for useful BWE data.

A watermarking system must analyze the narrow-band host signal to determine how the added watermark can be placed in the signal so the watermark is inaudible. Perceptual analysis can be used to determine suitable masking properties of the host, or linear prediction can be used to control noise shaping. Techniques such as spread spectrum and dither modulation can be used. Fig. 12.16A shows a BWE watermark encoder. Operation follows the same steps as any narrow-band encoder. The input speech signal is lowpass filtered and decimated to reduce the sampling frequency to 8 kHz. In a side chain, the input signal is also applied to a BWE encoder, and this information is applied to watermark embedding processing that places the BWE information in the host signal. The embedding process employs analysis of the host signal to modulate and shape the watermarked BWE signal. Figure 12.16B shows a BWE watermark decoder. The host narrow-band signal is decoded and interpolated to restore the sampling frequency to 16 kHz. The decoded narrow-band signal is applied to a watermark detector. The BWE watermark signal is recovered and decoded, and the high-frequency signal components are added to the narrow-band signal. If the watermark signal is not available or properly recovered, the receiver can switch to a separate receiver-only BWE strategy or simply output the narrow-band signal without BWE.

Several WB speech codecs are in use. The G.722.1, introduced in 1999, provides good speech quality at 24 kbps and 32 kbps. The AMR-WB (Adaptive Multi-Rate Wideband) codec, standardized as G.722.2, uses bit rate modes ranging from 6.6 kbps to 23.85 kbps. The AMR-WB+ extension uses bit rate modes ranging from 6.6 kbps to 32 kbps and supports monaural and stereo signals, and audio bandwidths ranging from 7 kHz to 16 kHz.

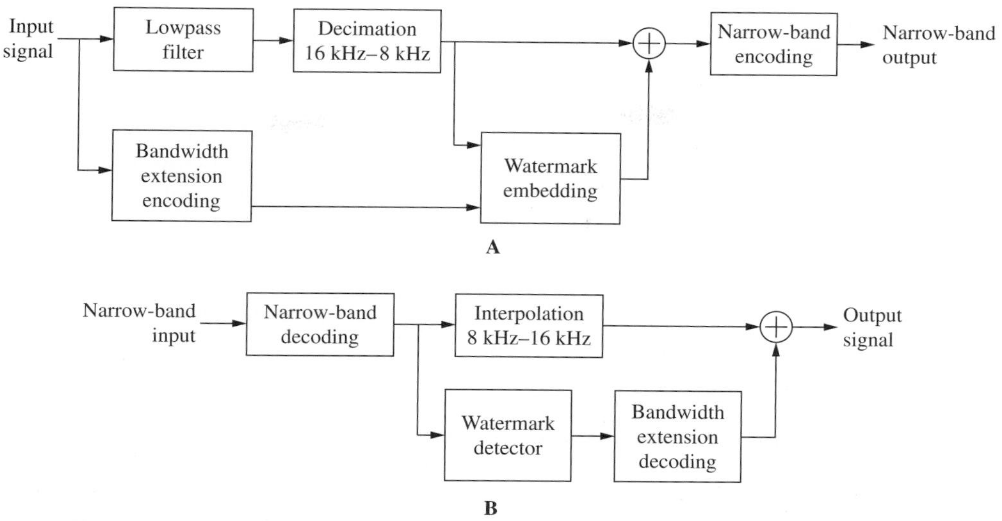

Figure 12.16 Bandwidth extension (BWE) techniques can be used in a watermarking system. The system analyzes the narrow-band host signal to determine how the added watermark can be placed in the signal so the watermark is inaudible. A. Speech encoder using watermarked bandwidth extension. B. Speech decoder using watermarked bandwidth extension.

Echo Cancellation

Codec performance can be improved by accounting for distortions such as acoustic echo and background noise. Echo cancellation addresses a potentially serious issue in speech communication. An acoustic echo is a separately audible delayed reflection from a surface. Acoustic echoes can be produced during hands-free operation, for example, by the acoustic coupling between the loudspeaker and the microphone. Echoes can also be produced electrically. For example, impedance mismatches from hybrid transformers in a long-distance transmission medium can yield line echoes. Both acoustic and electric echoes with long delays can be annoying and can greatly degrade speech intelligibility.

Various techniques have been devised to electrically minimize the effects of echoes. Voice-controlled switching is a simple technique. When a signal is detected at the far speaker side, the microphone signal can be attenuated. Conversely, during near-end activity, the loudspeaker can be attenuated. However, the result is a half-duplex communication and unnatural changes in noise. The latter can be mitigated by adding comfort noise. Echo suppressors are used to mitigate line echoes using voice-operated switches. When an echo is detected (as opposed to an original speech signal), a large loss is switched into the line or the line is momentarily switched open. Errors in operation can occur when the original speech signal's level is low or when the echo signal's level is high.

Adaptive echo cancellation is another approach. Adaptive echo cancelers do not interrupt or alter the communication path. Rather, a synthetic echo is generated from the incoming speech signal, and then later subtracted from the outgoing signal to cancel the echo. Operating in parallel with the actual echo signal path, an adaptive filter uses the known loudspeaker signal to generate a replica of the acoustic echo.

The replica echo signal is subtracted from the microphone signal to produce an output speech signal. However, it is difficult to track the original echo path and generate an accurate replica; echo attenuation is insufficient because of a residual echo. Many adaptive algorithms iteratively update filter coefficients using control mechanisms to optimize step size. Some techniques include the delay coefficients method, double-talk detection, remote single-talk detection, two-echo path model, and dynamic regularization. Small steps improve accuracy, but can fail to track changes in the acoustic echo path whereas large steps can cause the algorithm to diverge in noisy conditions. Some designs thus use time-varying adaptive step size or adaptive memory. Algorithms include NLMS (normalized least mean square), APA (affine projection), FDAF (frequency domain adaptive filter), and RLS (recursive least square). In some designs, voice-controlled switches are combined with echo cancelers. A frequency-selective adaptive post-filter can be used to minimize residual echo after echo cancellation, using the power spectral density of the residual echo. Operation is similar to that of a noise-suppression filter, and in fact noise suppression can be performed simultaneously. Estimation of the residual echo PSD signal is difficult in the context of background noise.

Regarding background noise, for example, a noise signal can be extracted during a pause in speech, and used as a power density spectrum for noise filtering. Alternatively, during speech, it can be assumed that only noise is present in at least some FFT bins, particularly the lower bins, and the noise power density spectrum can be devised for those bins. To make this determination, for example, it can be assumed that a speech spectrum varies faster than even that of nonstationary noise.

Voice Activity Detection

Voice Activity Detection (VAD) monitors the input signal to determine if the signal is speech or background noise. When no voice activity and only background noise is detected, the signal is coded with the lowest bit rate needed to convey background noise or a silence compression mode is engaged. During these inactive periods, the decoder may generate a random comfort noise. VAD can be automatically engaged when in variable-bit-rate mode, or engaged separately in constant-bit-rate mode. Similarly, when the background noise is stationary, discontinuous transmission (DTX) can be used to halt data transmission, thus further lowering the bit rate.

Variable Bit Rate

In some applications such as voice mail, a variable bit rate (VBR) can be used. This can be more efficient than a constant bit rate and can yield a lower average bit rate. With variable bit rate, a codec can vary its bit rate dynamically according to the nature of the audio signal. A more complex signal demands a higher bit rate, but a less complex signal can be coded with a lower bit rate. Thus, the overall bit rate can be reduced while maintaining an audio quality level. For example, transients and vowels may require a higher bit rate. Fricatives such as "s" and "f" sounds may be coded at a lower bit rate. However, VBR may result in an instantaneous maximum bit rate that is too high. For example, this can overload a real-time channel such as a VoIP channel. Also, when a certain quality level is specified in VBR, a particularly complex audio signal may result in a too-high average bit rate. In many codecs, this problem is addressed by an average-bit-rate (ABR) feature. A target average bit rate is set, and the algorithm monitors and adjusts the variable bit rate to obtain it. This is done in real time with small adjustments. The overall quality of the encoding will be less than would have been obtained with an exactly optimal VBR setting (resulting in the desired average bit rate) without ABR.

Speech Recognition

Many speech recognition systems identify short-term features of speech using cepstral analysis. Other systems may use statistical techniques. In any case, the performance of speech recognition systems can be degraded by poor acoustic conditions such as background noise and room reverberation. In a speech recognition system, the speech signal is applied to front-end processing where short-term spectral analysis is used for feature extraction. In particular, the signal spectrum is transformed to the cepstral domain to obtain the Mel frequency cepstral coefficients. For example, the signal can be preemphasized with a highpass filter and input to a Hamming window and FFT to provide a magnitude spectrum. This is applied to a Mel-scale filter bank to emulate critical bands; then the logarithm is taken and this is input to a discrete cosine transform. For example, 13 cepstral coefficients may be calculated including the zeroth coefficient which represents the spectrum mean value. In addition to these short-term static parameters, dynamic parameters may be obtained by examining static contours over several frames. These are sometimes referred to as delta (first derivative) and delta-delta (second derivative) coefficients. For example, a feature vector may contain 39 components; analysis may yield a feature vector every 10 ms. In some cases, the front end is augmented by a noise reduction stage and the cepstral output is applied to blind equalization to yield adaptive equalization.

The output of the cepstral analysis feature extraction stage is used in two modes, as shown in Fig. 12.17. In the training mode, feature vectors are used to estimate reference patterns that may represent a word or smaller part of speech. Over time, usually with many human talkers, a reference model is assembled and used as a speech pattern database representing the distribution of the characteristics of many human talkers. Hidden Markov models may be used as reference patterns, using sequences of multistate phonemes. In the recognition mode, the input feature vectors are compared to the reference features and the likelihood of correct representation is calculated using a Viterbi or other algorithm. The pattern sequence with the highest likelihood forms the output result.

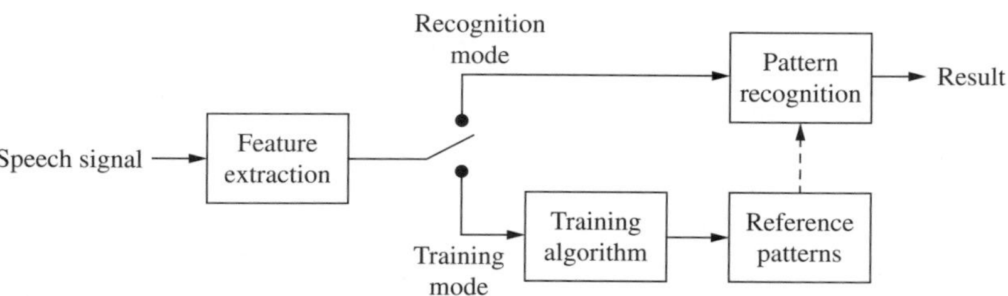

Figure 12.17 A speech recognition system showing a training mode, and recognition mode. Short-term spectral analysis can be used for feature extraction.

Speex Codec

The Speex speech codec is used in Voice over IP (VoIP), and streaming applications such as teleconferencing and video games, as well as file-based data reduction applications such as podcasts. It is not used for cellular telephony. Speex is an open-source codec that can be used freely. Most speech codecs are licensed under patent agreements. Speex is open-source software that does not have any patent or royalty restrictions and can be licensed under the BSD license. Speex shares the same philosophical basis as the Vorbis audio format. Speex was developed by the Xiph.org Foundation; the file name extension is .spx. Speex can be used with the Ogg container format (media type is audio/ogg) or transported directly over UDP/RTP (media type is audio/x-speex). Speex can be found in DirectShow, OpenACM (NetMeeting), Microsoft Windows, Linux (Ekiga), and libvorbis (Xiph.org's reference implementation).

Speex uses CELP as its coding basis. Among other advantages, this allows Speex to be robust to lost packets, an important consideration for VoIP usage; corrupted packets are blocked from arrival by the User Datagram Protocol (UDP). Speex can be used at different bit rates ranging from 2 kbps to 44 kbps, and at different sampling frequencies including 8 kHz (narrow band), 16 kHz (wideband), 32 kHz (ultra-wideband), and up to 48 kHz. Narrow band and wideband coding can be contained in the same bitstream. Speex also supports dynamic bit-rate switching, variable bit rate, voice activity detection, and intensity stereo encoding options. In the Speex codec, the delay equals the frame size plus time required to process a frame. This delay is about 30 ms at an 8-kHz sampling frequency and 34 ms at a 16-kHz sampling frequency.

Operation of Speex is controlled by a number of variables. A quality parameter ranges from 0 to 10; this parameter is in integer when in constant-bit-rate (CBR) mode and floating-point number when in variable-bit-rate (VBR) mode. The complexity allowed in the encoder can be varied. To do this, the way that the search is performed is controlled by an integer ranging from 1 to 10 (this is similar to the –1 to –9 options used in gzip compression). Encoder complexity determines CPU requirements; it is about five times higher for a complexity of 10 compared to a complexity of 1. Higher complexities are used for nonspeech signals such as DTMF tones. They are also used when encoding in non-real-time. Generally, a complexity of 2 to 4 is often used.

Quantifying Performance of Speech Codecs

The performance of speech coding systems can be assessed through subjective listening tests. For example, the mean opinion score (MOS) is often used. On this 5-point scale, a score of 4.0 or higher denotes high-quality or near-transparent performance. Scores between 3.5 and 4.0 are often permissible for speech communications systems. Scores below 3.0 denote low quality. For example, a signal may be highly intelligible, but the speech may sound synthetic or otherwise unnatural. A circuit-merit (CM) quality scale can be used to rate subjective performance. The average of CM scores obtained from listeners provides a mean opinion score. The ranking of CM scores is shown in Table 12.1.

The diagnostic rhyme test (DRT) is a standardized subjective method used to evaluate speech intelligibility. In this American National Standards Institute (ANSI) test, listeners are presented with word pairs. They must choose which word they perceive.

CM5	Excellent	Speech perfectly understandable
CM4	Good	Speech easily understandable, some noise
CM3	Fair	Speech understandable with a slight effort, occasional repetitions needed
CM2	Poor	Speech understandable only with considerable effort, frequent repetitions needed
CM1	Unsatisfactory	Speech not understandable

Table 12.1 Ranking of CM scores for subjective performance of speech codecs.

The words are different only in their leading consonants and pairs are selected so that six attributes of speech intelligibility are evaluated in their present or absent states. The diagnostic acceptability measure (DAM) test is also used to evaluate speech codec performance.

Historically, telephone systems provided a narrow bandwidth from approximately 300 Hz to 3400 Hz. It is generally agreed that the lower cutoff frequency, although it causes a "thin" sound because of missing lower frequencies and thus decreases sound quality, does not greatly affect speech intelligibility. It is said that the upper cutoff frequency of 3400 Hz allows 97% of speech sounds to be understood. Recognition of sentences can achieve 99% intelligibility because of assistance by context. Newer wideband telephone speech systems provide a response from 50 Hz to 7 kHz. Although the improvement in measured intelligibility is slight, these systems provide a more natural speech sound with clearer sound quality. Among other things, this suggests the need for more refined telephony sound-quality measures. The subjective evaluation of music signals is discussed in Chap. 10.

Speech Coding Standards

Speech codecs can be classified as wideband codecs, and narrow-band codecs. Examples of wideband codecs are AMR-WB (for WCDMA networks); VMR-WB (for CDMA2000 networks); and G.722, G.722.1 and Speex (for VoIP and video teleconferencing). Examples of narrow-band codecs are SMV (for CDMA networks); Full Rate, Half Rate, EFR, and AMR (for GSM networks); G.723.1, G.726, G.728, G.729, and iLBC (for VoIP and video teleconferencing); and FNBDT (for military applications).

The Telecommunications Industry Association (TIA) has developed speech coding standards for North America including TDMA and CDMA systems. TIA is part of the American National Standards Institute.

The G series of International Telecommunications Union (ITU) recommendations define speech coding standards including G.711 (μ law), G.728 (LD-CELP), and G.729A (CS-ACELP).

The European Telecommunications Standards Institute (ETSI) publishes standards for digital cellular applications in Europe. The Groupe Speciale Mobile (GSM) has developed standards such as GSM 06.60 and 06.90.

The International Standards Organization (ISO) contains the Moving Picture Experts Group (MPEG). The MPEG 4 standard describes the Harmonic Vector Excitation (HVXC) codec and narrow-band and wideband CELP codecs.

Standard	Application	Bit Rate (kbps)
ITU-T G.711 PCM	General purpose	64
FS 1015 LPC	Secure communication	2.4
ETSI GSM 6.10 RPE-LTP	Digital mobile radio	13
ITU-T G.726 ADPCM	General purpose	16, 24, 32, 40
TIA IS54 VSELP	North American TDMA cellular telephony	7.95
ETSI GSM 6.20 VSELP	GSM cellular telephony	5.6
RCR STD-27B VSELP	Japanese cellular telephony	6.7
FS 1016 CELP	Secure communication	4.8
ITU-T G.728 LD-CELP	General purpose	16
TIA IS96 VBR-CELP	North American CDMA cellular telephony	0.8, 2, 4, 8.5
ITU-T G.723.1 MP-MLQ/ ACELP	Multimedia communications, videophones	5.3, 6.3
ITU-T G.729 CS-ACELP	General purpose	8
ETSI GSM EFR ACELP	General purpose	12.2
TIA IS641 ACELP	North American TDMA cellular telephony	7.4
FS MELP	Secure communications	2.4
ETSI AMR-ACELP	General purpose	4.75, 5.15, 5.90, 6.70, 7.40, 7.95, 10.2, 12.2

Table 12.2 Summary of speech coding standards.

The U.S. Department of Defense Voice Processing Consortium (DDVPC) standardized the Federal Standard FS1015 LPC-10e and FS1016 CELP codecs, as well as the MELP and MELPe MIL-STD-3005 standard, adopted as the STANAG-4591 standard.

The Research and Development Center for Radio Systems of Japan (RCR) develops digital cellular standards for Japan.

A list of some speech coding standards is shown in Table 12.2.

CHAPTER 13

Audio Interconnection

Analog signals can be conveyed from one device to another with relative ease, but the transfer of audio signals in the digital domain is a good deal more complex. Sampling frequency, word length, control and synchronization words, and coding must all be precisely defined to permit successful interfacing. Above all, the data format itself takes precedence over the physical medium or interconnection it happens to currently occupy. Numerous data formats have been devised to connect digital audio devices, both between equipment of the same manufacturer, and between equipment of different manufacturers. Using appropriate interconnection protocols, data can be conveyed in real time over long and short distances, using proprietary or open channel formats. Fiber-optic communication provides very high bandwidth and is particularly effective when data is directed over long distances. In many applications, files are transmitted in non-real-time; these file formats are described in Chap. 14.

Audio Interfaces

Perhaps the most fundamental interconnection in a studio is the connection of one hardware device to another, so that digital audio data can be conveyed between them in real time. Clearly, a digital connection is preferred over an analog connection; the former can be transparent, but the latter imposes degradation from digital-to-analog and analog-to-digital conversion.

To convey digital data, there must be both a data communications channel and common clock synchronization. One hard-wired connection can provide both functions, but as the number of devices increases, a separate master clock signal is recommended. In addition, the interconnection requires an audio format recognized by both transmitting and receiving devices. Data flow is usually unidirectional, directed point to point (as opposed to a networked or bus distribution), and runs continuously without handshaking. Perhaps two, or many audio channels, as well as auxiliary data, can be conveyed, usually in serial fashion. The data rate is determined by the signal's sampling frequency, word length, number of channels, amount of auxiliary data, and modulation code. When the receiving device is a recorder, it can be placed in record mode, and in real time can copy the received data stream. Given correct operation, the received data will be a clone of the transmitted data.

One criterion for successful transmission of serial data over a coaxial cable is the cable's attenuation at one-half the clock frequency of the transmitted signal. Very generally, the maximum length of a cable can be gauged by the length at which the cable attenuates the frequency of half the clock frequency by 30 dB. Professional interfaces can permit cable runs of 100 to 300 meters, but consumer cables might be limited to less than 10 meters. Fiber cables are much less affected by length loss and permit much longer cable runs. Examples of digital audio interfaces are: SDIF-2, AES3 (AES/EBU), S/PDIF, and AES10 (MADI).

SDIF-2 Interconnection

The SDIF-2 (Sony Digital InterFace) protocol is a single-channel interconnection protocol used in some professional digital products. For example, it allows digital transfer from recorder to recorder. A stereo interface uses two unbalanced BNC coaxial cables, one for each audio channel. In addition, there is a separate coaxial cable for word clock synchronization, a symmetrical square wave at the sampling frequency that is common to both channels. The word clock period is 22.676 μs at a 44.1-kHz sampling frequency, the same period as one transmitted 32-bit word. Any sampling frequency can be used. The signal is structured as a 32-bit word, as shown in Fig. 13.1. The most significant bit (MSB) through bit 20 are used for digital audio data, with MSB transmitted first, with nonreturn to zero (NRZ) coding. The data rate is 1.21 Mbps at a 44.1-kHz sampling frequency, and 1.53 Mbps at a 48-kHz sampling frequency.

When 16-bit samples are used, the remaining four bits are packed with binary 0s. Bits 21 through 29 form a control (or user) word. Bits 21 through 25 are held for future expansion; bits 26 and 27 hold an emphasis ID determined at the point of A/D conversion; bit 28 is the dubbing prohibition bit; and bit 29 is a block flag bit that signifies the beginning of an SDIF-2 block. Bits 30 through 32 form a synchronization pattern. This field is uniquely divided into two equal parts of 1.5T (one and one-half bit cell) forming a block synchronization pattern. The first word of a block contains a high-to-low pulse and the remaining 255 words have a low-to-high pulse.

This word structure is reserved for the first 32-bit word of each 256-word block. The digital audio data and synchronization pattern in subsequent words in a block are structured identically. However, the control field is replaced by user bits, nominally set to 0. The block flag bit is set to 1 at the start of each 256-word block. Internally, data is processed in parallel; however, it is transmitted and received serially through digital input/output (DI/O) ports. For two-channel operation, SDIF-2 is carried on a single-ended 75-Ω coaxial cable, as a transistor-transistor logic (TTL) compatible signal. To ensure proper operation, all three coaxial cables should be the same length. Some multitrack recorders use a balanced/differential version of SDIF-2 with RS-422-compatible signals. A twisted-pair ribbon cable is used, with 50-pin D-sub type connectors, in addition to a BNC word clock cable. The SDIF-2 interface was introduced by Sony Corporation.

Both SDIF-3 and MAC-DSD are used to convey DSD (Direct Stream Digital) audio data, as used in the Super Audio CD (SACD) disc format. SDIF-3 is an interface designed to carry DSD data. It conveys one channel of DSD audio per cable, and employs 75-Ω unbalanced coaxial cable and represents data in phase-modulated form (as opposed to DSD-raw, NRZ unmodulated form). A word clock of 44.1 kHz can be used for synchronization; alternatively, a 2.8224-MHz clock can be used. MAC-DSD (Multi-channel Audio Connection for DSD) is a multichannel interface for DSD data for professional applications. It uses twisted-pair Ethernet (100Base-TX using CAT-5 cabling terminating in 8-way RJ45 jacks and using a PHY physical layer interface) interconnection but it is used for point-to-point transfer rather than as a network node. MAC-DSD can transfer 24 channels of DSD audio in both directions simultaneously

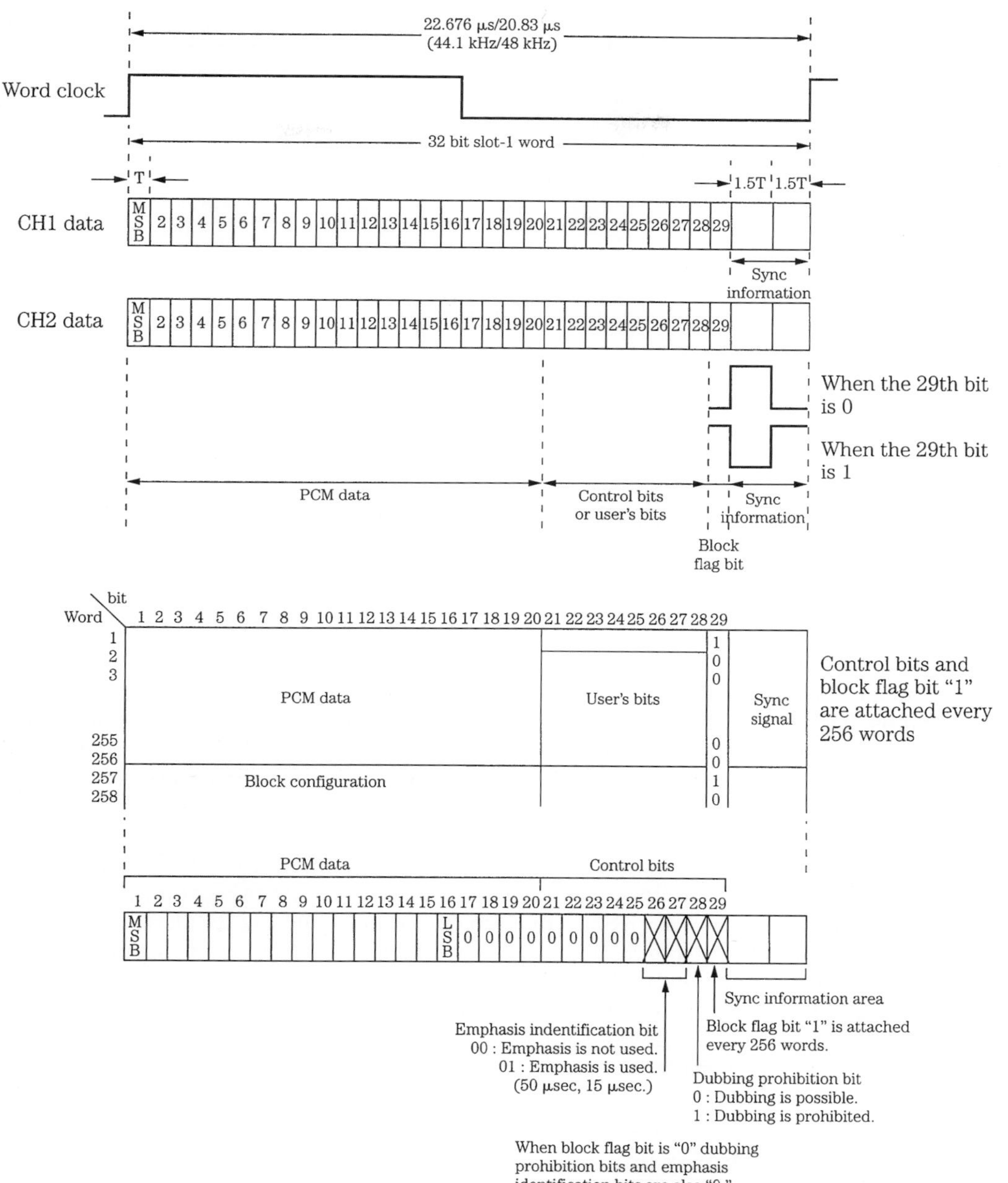

Figure 13.1 The SDIF-2 interface is used to interconnect digital audio devices. The control word conveys nonaudio data in the interface.

Word	B31-B28	B27-B24	B23-B20	B19-B16	B15-B12	B11-B8	B7-B4	B3-B0
0	5h	5h	5h	5h	5h	5h	5h	5h
1	Dh	5h	5h	5h	5h	5h	5h	5h
2	Reserved for destination MAC address							
3	Reserved for source MAC address				Reserved for destination MAC address			
4	Reserved for source MAC address							
5					Length - always 1438 bytes (0x59E)			
6	Reserved for networking headers							
7								
8								
9								
10								
11								
12	Reserved	Reserved	Reserved	Frame type				
13-364	352 interleaved 32-bit data blocks (352 DSD samples, 24 channels, plus 88 bytes aux data)							
365	CRCC							

FIGURE 13.2 The structure of a MAC-DSD audio data frame. Audio data, control data, and check bits are interleaved within 32-bit data blocks. Each data block corresponds to a DSD sampling period.

along with $64f_s$ (2.8224 MHz) DSD sample clocks in both directions. Point-to-point latency is less than 45 μs. The PHY device transmits data in frames with a bit rate of 100 Mbps. Each frame is up to 1536 bytes with a 64-bit preamble and 96 bit-period interframe gap. User data is 1528 bytes and maximum bit rate is 98.7 Mbps; 24 channels of DSD audio yields a bit rate of 67.7 Mbps. Other capacity is used for error correction, frame headers, and auxiliary data. Audio data, control data, and check bits are interleaved in 32-bit blocks, one per DSD sample period. The structure of a MAC-DSD audio data frame is shown in Fig. 13.2. If multiple MAC-DSD links are used (for more than 24 channels) differences in latency are overcome with a 44.1-kHz synchronization signal. Connections between a source/destination device and a hub use standard CAT-5 cable such that pin-outs on hub devices are reversed to connect inputs to outputs. In peer-to-peer interconnections between two source/destination devices, a crossover cable is used such that pin-outs at one end are reversed to connect inputs to outputs. Unlike typical Ethernet cables, in these crossover cables both two data pairs and clock signal connections are reversed.

AES3 (AES/EBU) Professional Interface

The Audio Engineering Society (AES) has established a standard interconnection generally known as the AES3 or AES/EBU digital interface. It is a serial transmission format for linearly represented digital audio data. It permits transmission of two-channel digital audio information, including both audio and nonaudio data, from one professional audio device to another. The specification provides flexibility within the defined standard for specialized applications; for example, it also supports multichannel audio and higher sampling frequencies. The format has been codified as the AES3-1992 standard;

this is a revised version of the original AES3-1985 standard. In addition, other standards organizations have published substantially similar interface specifications: The International Electrotechnical Commission (IEC) IEC-60958 professional or broadcast use (known as type I) format, the International Radio Consultative Committee (CCIR) Rec. 647 (1990), the Electronic Industries Association of Japan (EIAJ) EIAJ CP-340-type I format, the American National Standards Institute (ANSI) ANSI S4.40-1985 standard, and the European Broadcasting Union (EBU) Tech. 3250-E.

The AES3 standard establishes a format for nominally conveying two channels of periodically sampled and uniformly quantized audio signals on a single twisted-pair wire. The format is intended to convey data over distances of up to 100 meters without equalization. Longer distances are possible with equalization. Left and right audio channels are multiplexed, and the channel is self-clocking and self-synchronizing. Because it is independent of sampling frequency, the format can be used with any sampling frequency; the audio data rate varies with the sampling frequency. A sampling frequency of 48 kHz ±10 parts per million is often used but 32, 44.1, 48, and 96 kHz are all recognized as standard sampling frequencies by the AES for pulse-code modulation (PCM) applications in standards document AES5-1998. Moreover, AES3 has provisions for sampling frequencies of 22.05, 24, 88.2, 96, 176.4, and 192 kHz. Sixty-four bits are conveyed in one sampling period; the period is thus 22.7 μs with a 44.1-kHz sampling frequency. AES3 alleviates polarity shifts between channels, channel imbalances, absolute polarity inversion, gain shifts, as well as analog transmission problems such as hum and noise pickups, and high-frequency loss. Furthermore, an AES3 data stream can identify monaural/stereo, use of pre-emphasis, and the sampling frequency of the signal.

The biphase mark code, a self-clocking code, is the binary frequency modulation channel code used to convey data over the AES3 interconnection. There is always a transition (high to low, or low to high) at the beginning of a bit interval. A binary 1 places another transition in the center of the interval; a binary 0 has no transition in the center. A transition at the start of every bit ensures that the bit clock rate can be recovered by the receiver. The code also minimizes low-frequency content, and is polarity-free (information lies in the timing of transitions, not their direction). All information is contained in the code's transitions. Using the code, a properly encoded data stream will have no transition lengths greater than one data period (two cells), and no transition lengths shorter than one-half coding period (one cell). This kind of differential code can tolerate about twice as much noise as channels using threshold detection. However, its bandwidth is large, limiting channel rate; logical 1 message bits cause the channel frequency to equal the message bit rate. The overall bit rate is 64 times the sampling frequency; for example, it is 3.072 Mbps at a 48-kHz sampling frequency. Channel codes are discussed in Chap. 3.

AES3 Frame Structure

The AES3 specification defines a number of terms. An audio sample is a signal that has been periodically sampled, quantized, and digitally represented in a two's complement manner. A subframe is a set of audio sample data with other auxiliary information. Two subframes, one for each channel, are transmitted within the sampling period; the first subframe is labeled 1 and the second is labeled 2. A frame is a sequence of two subframes; the rate of transmission of frames corresponds exactly to the sampling rate of the source. With stereo transmissions, subframe 1 contains left A-channel data and subframe 2 contains right B-channel data, as shown in Fig. 13.3A. For monaural, the rate remains at the two-channel rate, and the audio data is placed in subframe 1.

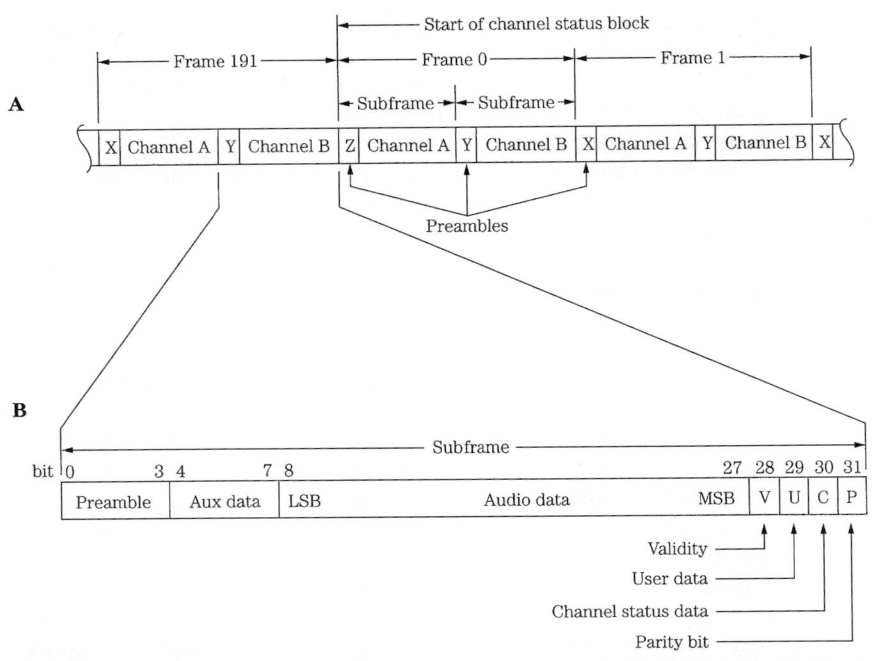

Figure 13.3 The professional AES3 serial interface is structured in frames and subframes as well as channel status blocks formed over 192 frames. A. There are two subframes per frame; each subframe is identified with a preamble. B. The interface uses a subframe of 32 bits.

A block is a group of channel status data bits and an optional group of user bits, one per subframe, collected over 192 source sample periods. A subframe preamble designates the starts of subframes and channel status blocks, and synchronizes and identifies audio channels. There are three types of preambles. Preamble Z identifies the start of subframe 1 and frame 0, which is also the start of a channel status block. Preamble X identifies the start of subframe 1 otherwise, and Preamble Y identifies the start of subframe 2. Preambles occupy four bits; they are formed by violating the biphase mark coding in specific ways. Preamble Z is 3UI/1UI/1UI/3UI where UI is a unit interval. Preamble X is 3UI/3UI/1UI/1UI. Preamble Y is 3UI/2UI/1UI/2UI.

The format specifies that a subframe has a length of 32 bits, with fields that are defined as shown in Fig. 13.3B. Audio data might occupy up to 24 bits. Data is linearly represented in two's complement form, with the least significant bit (LSB) transmitted first. If the audio data does not require 24 bits, then the first four bits can be used as an auxiliary data sample, as defined in the channel status data. For example, broadcasters might use the four auxiliary bits for a low bit-rate talkback feed. When devices use 16-bit words, the last 16 bits in the data field are used, with the others set to 0. Four bits conclude the subframe: Bit V—An audio sample validity bit is 0 if the transmitted audio sample is error-free, and 1 if the sample is defective and not suitable for conversion to an analog signal. Bit U—A user data bit can optionally be used to convey blocks of user data. A recommended format for user data is defined in the AES18-1992 standard, as described below. Bit C—A channel status bit is used to form blocks describing information about the interconnection channel and other system parameters, as described below. Bit P—A subframe parity bit provides even parity for the subframe; the bit can detect when an odd number of errors have occurred in the transmission.

AES3 Channel Status Block

The audio channel status bit is used to convey a block of data 192 bits in length. An overview of the block is shown in Fig. 13.4. Received blocks of channel status data are accumulated from each of the subframes to yield two independent channel status data blocks, one for each channel. At a sampling frequency of 48 kHz, the blocks repeat at 4-ms intervals. Each channel status data block consists of 192 bits of data in 24 bytes,

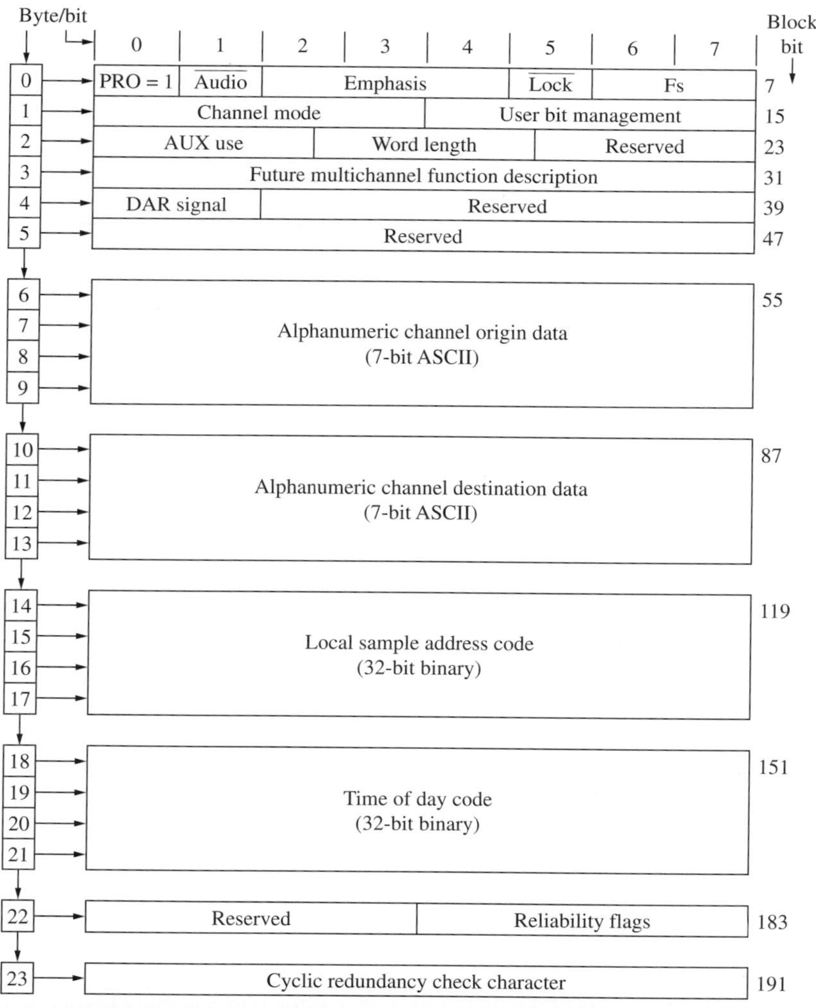

FIGURE 13.4 Specification of the 24-byte channel status block used in the AES3 serial interface.

transmitted as one bit in each subframe, and collected from 192 successive frames. The block rate is 250 Hz at a 48-kHz sampling frequency. The channel status block is synchronized by the alternate subframe preamble occurring every 192 blocks.

There are 24 bytes of channel status data. The first six bytes (outlined at the top of Fig. 13.4) are detailed in Fig. 13.5. Byte 0 of the channel status block contains information that identifies the data for professional use, as well as information on sampling frequency and use of pre-emphasis. With any AES3 communication, bit 0 in byte 0 must be set to 1 to signify professional use of the channel status block. Byte 1 specifies the signal mode such as stereo, monaural, or multichannel. Byte 2 specifies the maximum audio word length and number of bits used in the word; an auxiliary coordination signal can be specified. Byte 3 is reserved for multichannel functions. Byte 4 identifies multichannel modes, type of digital audio reference signal (Grade 1 or 2) and alternative sampling frequencies. Byte 5 is reserved. Bytes 6 through 9 contain alphanumeric

Byte 0		
	Bit 0	PRO = 1 (Professional)
	0	Consumer use of channel status block
	1	Professional use of channel status block
	Bit 1	$\overline{\text{Audio}}$
	0	Digital audio
	1	Nonaudio
	Bit 234	Emphasis
	000	Emphasis not indicated
	100	No emphasis
	110	50/15 μs emphasis
	111	CCITT J.17 emphasis
	Bit 5	$\overline{\text{Lock}}$
	0	Sampling frequency locked
	1	Sampling frequency unlocked
	Bit 67	Encoded sampling frequency
	00	Not indicated
	01	48 kHz
	10	44.1 kHz
	11	32 kHz

Byte 1		
	Bit 0123	Encoded channel mode
	0000	Not indicated
	0001	Two-channel
	0010	Single-channel
	0011	Primary/Secondary
	0100	Stereophonic
	0101	Reserved
	0110	Reserved
	0111	Single-channel double sampling frequency
	1000	Single-channel double sampling frequency
	1001	Single-channel double sampling frequency
	1111	Multichannel mode. Vector to byte 3 for channel identification
	Bit 4567	User bits management
	0000	Default
	0001	192-bit block
	0010	AES18
	0011	User-defined

FIGURE 13.5 Description of the data contained in bytes 0 to 5 in the 24-byte channel status block used in the AES3 serial interface.

Byte 2	Bit 012	Auxiliary sample bits	
	000	Maximum 20 bits	
	001	Maximum 24 bits	
	010	Maximum 20 bits (single coordination)	
	011	Reserved	
	Bit 345	Sample word length (24)	Sample word length (20)
	000	Default	Default
	001	23 bits	19 bits
	010	22 bits	18 bits
	011	21 bits	17 bits
	100	20 bits	16 bits
	101	24 bits	20 bits
	Bit 67	Reserved	

3	Bit 0-7	Vector from byte 1 (multichannel modes)
Byte 4	Bit 01	Digital audio reference signal (AES11)
	00	Default
	01	Grade 1
	10	Grade 2
	11	Reserved
	Bit 2	Reserved
	Bit 3456	Sampling frequency
	0000	Not indicated (default)
	1000	24 kHz
	0100	96 kHz
	1100	192 kHz
	0010	Reserved
	1010	Reserved
	0110	Reserved
	1110	Reserved
	0001	Reserved
	1001	22.05 kHz
	0101	88.2 kHz
	1101	176.4 kHz
	0011	Reserved
	1011	Reserved
	0111	Reserved
	1111	User defined
	Bit 7	Sampling frequency scaling flag
	0	No scaling
	1	Sampling frequency is 1/1.001 times
5	Bit 0-7	Reserved

FIGURE 13.5 *(Continued)*

channel origin code, and bytes 10 through 13 contain alphanumeric destination code; these can be used to route a data stream to a destination, then display its origin at the receiver. Bytes 14 through 17 specify a 32-bit sample address. Bytes 18 through 21 specify a 32-bit time-of-day timecode with 4-ms intervals at a 48-kHz sampling frequency; this timecode can be divided to obtain video frames. Byte 22 contains data reliability flags for the channel status block, and indicates when an incomplete block is transmitted. The final byte, byte 23, contains a cyclic redundancy check code (CRCC) codeword with the generation polynomial $x^8 + x^4 + x^3 + x^2 + 1$ across the channel status block for error detection.

Three levels of channel status implementation are defined in the AES3 standard: minimum, standard, and enhanced. These establish the nature of the data directed to the receiving units. With the minimum level, the first bit of the channel status block is set to 1 to indicate professional status, and all other channel status bits are set to 0. With standard implementation, all channel status bits in bytes 0, 1, and 2 (used for sampling frequency, pre-emphasis, monaural/stereo, audio resolution, and so on) and CRCC data in byte 23 must be transmitted; this level is the most commonly used. With enhanced implementation, all channel status bits are used.

As noted, audio data can occupy 24 bits per sample. When the audio data occupies 20 bits or less, the four remaining bits can be optionally used as an auxiliary speech-quality coordination channel, providing a path so that verbal communication can accompany the audio data signal. Such a channel could use a sampling frequency that is 1/3 that of the main data rate, and use 12-bit coding; one 4-bit nibble is transmitted in each subframe. Complete words would be collected over three frames, providing two independent speech channels. The resolution of the main audio data must be identified by information in byte 2 of the channel status block.

AES3 Implementation

In many ways, in its practical usage an AES3 signal can be treated similarly to a video signal. The electrical parameters of the format follow those for balanced-voltage digital circuits as defined by the International Telegraph and Telephone Consultative Committee (CCITT) of the International Telecommunication Union (ITU) in Recommendation V.11. Driver and receiver chips used for RS-422 communications, as defined by the Electronic Industries Association (EIA), are typically used; the EBU specification dictates the use of a transformer. The line driver has a balanced output with internal impedance of 110 Ω ± 20% from 100 kHz to 128× the maximum frame rate. Similarly, the interconnecting cable's characteristic impedance is 110 Ω ± 20% at frequencies from 100 kHz to 128× the maximum frame rate. The transmission circuit uses a symmetrical differential source and twisted-pair cable, typically shielded, with runs of 100 meters. Runs of 500 meters are possible when adaptive equalization is used. The waveform's amplitude (measured with a 110-Ω resistor across a disconnected line) should lie between 2 V and 7 V peak-to-peak. The signal conforms to RS-422 guidelines.

Jitter tolerance in AES3 can be specified with respect to unit intervals (UI), the shortest nominal time interval in the coding scheme; there are 128 UIs in a sample frame. Output jitter is the jitter intrinsic to the device as well as jitter passed through from the device's timing reference. Peak-to-peak output jitter from an AES3 transmitter should be less than 0.025 UI when measured with a jitter highpass-weighting filter. An AES3 receiver requires a jitter tolerance of 0.25 UI peak-to-peak at frequencies above 8 kHz, increasing with an inverse of frequency to 10 UI at frequencies below 200 Hz. Some manufacturers use an interface with an unbalanced 75-Ω coaxial cable (such as 5C2V type), signal level of 1 V peak-to-peak, and BNC connectors. This may be preferable in a video-based environment, where switchers are used to route and distribute digital audio signals, or where long cable runs (up to 1 kilometer) are required. This is described in the AES3-ID document.

The receiver should provide both common-mode interference and direct current rejection, using either transformers or capacitors. The receiver should present a nominal resistive impedance of 110 Ω ± 20% to the cable over a frequency range from 100 kHz to 128× the maximum frame rate. A low-capacitance (less than 15 pF/foot) cable is greatly preferred, especially over long cable runs. Shielding is not critical, thus an unshielded twisted-pair (UTP) cable is used. If shielding is needed, braid or foil shielding is preferred over server (spiral) shielding. More than one receiver on a line can cause transmission errors. Receiver circuitry must be designed with phase-lock loops to reduce jitter. The receiver must also synchronize the input data to an accurate clock reference with low jitter; these tolerances are further defined in the AES11 standard. Input (female) and output (male) connectors use an XLR-type connector with pin 1 carrying the ground signal, and pins 2 and 3 carrying the unpolarized signal.

A simple multichannel version of AES3 uses a number of two-channel interfaces. It is described in AES-2id-1996 and combines 16 channels using a 50-pin D-sub type connector. Byte 3 of the channel status block indicates multichannel modes and channel numbers. AES42 is based on AES3. It can be used to connect a digital microphone in which A/D conversion occurs at the microphone. It adds provision for control, powering, and synchronization of microphones. The power signal can be modulated to convey remote control information. Microphones using this interface are sometimes known as an AES3-MIC microphone.

Low bit-rate data can be conveyed via AES3. Because data rate is reduced, a number of channels can be conveyed in a nominally two-channel interface, packing data in the PCM data area. SMPTE 337M and the similar IEC 61937 standard describe a multichannel interface; SMPTE 338M and 339M describe data types. For example, Dolby E data could be conveyed for professional applications and Dolby Digital, DTS, or MPEG for consumer applications. Dolby E can carry up to eight channels of audio plus associated metadata via conventional two-channel interfaces. For example, it allows 5.1-channel programs to be conveyed along one AES3 digital pair at a typical bit rate of 1.92 Mbps.

AES10 (MADI) Multichannel Interface

The Multichannel Audio Digital Interface (MADI) extends the AES3 protocol to provide a standard means of interconnecting multichannel digital audio equipment. MADI, as specified in the AES10 standard, allows up to 56 channels of linearly represented, serial data to be conveyed along a single length of BNC-terminated cable for distances of up to 50 meters. Word lengths of up to 24 bits are permitted. In addition, MADI is transparent to the AES3 protocol. For example, the AES3 validity, user, channel status, and parity bits are all conveyed. An interconnection with the AES3 format requires two cables for every two audio channels (for send and return), but a MADI interconnection requires only two audio cables (plus a master synchronization signal) for up to 56 audio channels. The MADI protocol is documented as the AES10-1991 and ANSI S4.43-1991 standards.

To reduce bandwidth requirements, MADI does not use a biphase mark code. Instead, an NRZI code is used, with 4/5 channel coding based on the Fiber Distributed Data Interface (FDDI) protocol that yields low dc content; NRZI code is described in Chap. 3. Each 32-bit subframe is parsed into 4-bit words that are encoded to 5-bit channel words. The link transmission rate is fixed at 125 Mbps regardless of the sampling rate or number of active channels. One sampling period carries 56 channels, each with eight 5-bit channel symbols, that is, 32 data bits or 40 channel bits. Because of the 4/5 encoding scheme, the data transfer rate is thus 100 Mbps. Although AES3 is self-clocking, MADI is designed to run asynchronously. To operate asynchronously, a MADI receiver must extract timing information from the transmitted data so the receiver's clock can be synchronized. To ensure this, the MADI protocol stipulates that a synchronization symbol is transmitted at least once per frame. Moreover, a dedicated master synchronization signal (such as defined by AES11) must be applied to all interconnected MADI transmitters and receivers.

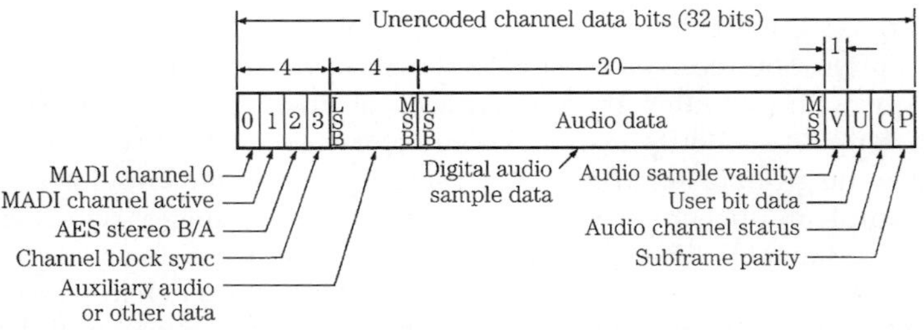

FIGURE 13.6 The AES10 (MADI) interface is used to connect multichannel digital audio equipment. The MADI channel format differs from the AES3 subframe format only in the first four bits.

The MADI channel format is based on the AES3 subframe format. A MADI channel differs from a subframe only in the first four bits, as shown in Fig. 13.6. Each channel therefore consists of 32 bits, with four mode identification bits, up to 24 audio bits, as well as the V, U, C, and P bits. The mode identification bits provide frame synchronization, identify channel active/inactive status, identify A and B subframes, and identify a block start. The 56 MADI channels are transmitted serially, starting with channel 0 and ending with channel 55, with all channels transmitted within one sampling period; the frame begins with bit 0 of channel 0. Because biphase coding is not used in the MADI format, preambles cannot be used to identify the start of each channel. Thus in MADI a 1 setting in bit 0 in channel 0 is used as a frame synchronization bit identifying channel 0, the first to be transmitted in a frame. Bit 0 is set to 0 in all other channels. Bit 1 indicates the active status of the channel. If the channel is active it is set to 1, and if inactive it is set to 0. Further, all inactive channels must have a higher channel number than the highest-numbered active channel. The bit is not dynamic, and remains fixed after power is applied. Bit 2 identifies whether a channel is A or B in a stereo signal; this also replaces the function of the preambles in AES3. Bit 3 is set when the user data and status data carried within a channel falls at the start of a 192-frame block. The remainder of the MADI channel is identical to an AES3 subframe. This is useful because MADI and AES3 are thus compatible, allowing free exchange of data.

MADI uses coaxial cables to support 100 Mbps. The interconnection is designed as a transmitter-to-receiver single-point to single-point link; for send and return links, two cables are required. Standard 75-Ω video coaxial cable with BNC connector terminations is specified; peak-to-peak transmitter output voltage should lie between 0.3 V and 0.6 V. Fiber-optic cable can also be used; for example, an FDDI interface could be used for distances of up to 2 kilometers. Alternatively, the Synchronous Optical NETwork (SONET) could be used. As noted, a distributed master synchronizing signal must be applied to all interconnected MADI transmitters and receivers. Because of the asynchronous operation, buffers are placed at both the transmitter and receiver, so that data can be reclocked from the buffers according to the master synchronization signal.

The audio data frequency can range from 32 kHz to 48 kHz, a ±12% variation is permitted. Higher sampling frequencies could be supported by transmitting at a lower rate, and using two consecutive MADI channels to achieve the desired sampling rate.

S/PDIF Consumer Interconnection

The S/PDIF (Sony/Philips Digital InterFace) interconnection protocol is designed for consumer applications. The IEC-60958 consumer protocol (known as type II) is a substantially identical standard. In some applications, the EIAJ CP-340 type II protocol is used. IEC-958 was named IEC-60958 in 1998. These consumer standards are very similar to the AES3 standard, and in some cases professional and consumer equipment can be directly connected. However, this is not recommended because important differences exist in the electrical specification, and in the channel status bits, so unpredictable results can occur when the protocols are mixed. Devices that are designed to read both AES3 and S/PDIF data must reinterpret channel status block information according to the professional (1) or consumer (0) status is the block's first bit.

The overall structure of the consumer channel status block is shown in Fig. 13.7. It differs from the professional channel status block (see Fig. 13.4). The serial bits are arranged as twenty-four 8-bit bytes; only the first four bytes are defined.

Figure 13.8 provides specific details on bytes 0 through 23. They differ from the professional AES3 channel status block (see Fig. 13.5). Byte 0, bit 0 is set to 0, indicating consumer use; bit 1 specifies whether the data is audio (0) or nonaudio (1); bit 2 is the copyright or C bit, and indicates copy-protected (0) or unprotected (1); bit 3 shows use of pre-emphasis (if bit 1 shows audio data and bits 4 and 5 show two-channel audio); bits 6 and 7 set the mode, that defines bytes 1 through 3. Presently, only mode 00 is specified. Byte 1, bits 0 through 6, define a category code that identifies the type of equipment transmitting the data stream; byte 1, bit 7 (the 15th bit in the block) is the generation or L bit, and indicates whether data is original or copied. If a recorder with an S/PDIF input receives an AES3 signal, it can read the professional pre-emphasis indicator as a copy-prohibit instruction, and thus refuse to record the data stream. Likewise, a professional recorder can correctly identify a consumer data stream by examining bit 0 (set to 0), but misinterpret a consumer copy-inhibit bit as a sign that emphasis is not indicated. In mode 00, the category code in byte 1 defines a variety of transmitting formats including CD, DAT, synthesizer, sample rate converter, and broadcast reception. Byte 2 specifies

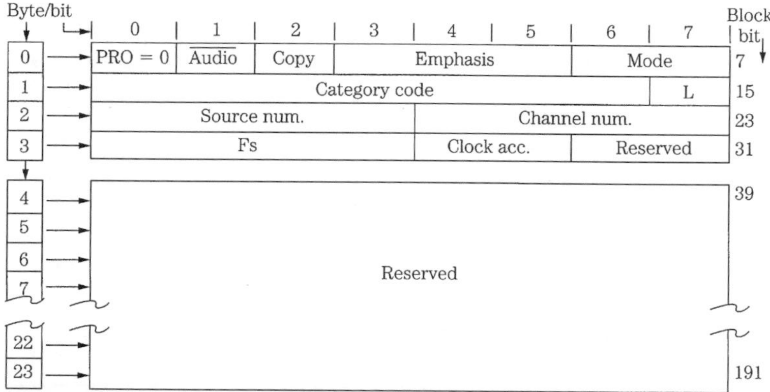

FIGURE 13.7 Specification of the 24-byte channel status block used in the consumer S/PDIF serial interface. (*Sanchez, 1994*)

BYTE 0	
bit 0	**PRO = 0 (consumer)**
0	Consumer use of channel status block
1	Professional use of channel status block
bit 1	**Audio**
0	Digital audio
1	Nonaudio
bit 2	**Copy/Copyright**
0	Copy inhibited/copyright asserted
1	Copy permitted/copyright not asserted
bits 3 4 5	**Pre-emphasis - if bit 1 is 0 (digital audio)**
0 0 0	None-2 channel audio
1 0 0	50/15 µs-2 channel audio
0 1 0	Reserved-2 channel audio
1 1 0	Reserved-2 channel audio
× × 1	Reserved-4 channel audio
bits 3 4 5	**If bit 1 is 1 (nonaudio)**
0 0 0	Digital data
× × ×	All other states of bits 3–5 are reserved
bits 6 7	**Mode**
0 0	Mode 0 (defines bytes 1–3)
× ×	All other states of bits 6–7 are reserved

BYTE 1		
bits 0 1 2 3 4 5 6		**Category Code**
0 0 0 0	0 0 0	General
	0 0 1	Experimental
	× × ×	Reserved
0 0 0 1	× × ×	Solid-state memory
0 0 1 ×	× × ×	Broadcast reception of digital audio
0 1 0 ×	× × ×	Digital/digital converters
0 1 1 0	0 × ×	A/D converters w/o copyright
	1 × ×	A/D converters w/ copyright (using copy and L bits)
0 1 1 1	× × ×	Broadcast reception of digital audio
1 0 0 ×	× × ×	Laser-Optical
1 0 1 ×	× × ×	Musical instruments, mics, etc.
1 1 0 ×	× × ×	Magnetic tape or disk
1 1 1 ×	× × ×	Reserved
bit 7		**L: Generation Status.**
		Only category codes: 001××××, 0111×××, 100××××
0		Original/Commercially prerecorded data
1		No indication or 1st generation or higher
		All other category codes
0		No indication or 1st generation or higher
1		Original/Commercially prerecorded data
The subgroups under the category code groups listed above are described in tables below. Those not listed are reserved.		
The copy and L bits form a copy protection scheme for original works. Further explanations can be found in the amendment (TC84) to IEC-958.		

BYTE 1 - Category Code 001	
bits 3 4 5 6	**Broadcast reception of digital audio**
0 0 0 0	Japan
0 0 1 1	United States
1 0 0 0	Europe
0 0 0 1	Electronic software delivery
× × × ×	All other states are reserved

BYTE 1 - Category Code 100	
bits 3 4 5 6	**Laser-Optical**
0 0 0 0	CD-compatible with IEC-908
1 0 0 0	CD-not compatible with IEC-908 (MO)
1 0 0 1	MD-MiniDisc
× × × ×	All other states are reserved

BYTE 1 - Category Code 101	
bits 3 4 5 6	**Musical Instruments, mics, etc.**
0 0 0 0	Synthesizer
1 0 0 0	Microphone
× × × ×	All other states are reserved

BYTE 1 - Category Code 010	
bits 3 4 5 6	**Digital/digital conversion and signal processing**
0 0 0 0	PCM encoder/decoder
0 0 1 0	Digital sound sampler
0 1 0 0	Digital signal mixer
1 1 0 0	Sample-rate converter
× × × ×	All other states are reserved

BYTE 1 - Category Code 110	
bits 3 4 5 6	**Magnetic tape or disk**
0 0 0 0	DAT
1 0 0 0	Digital audio sound VCR
× × × ×	All other states are reserved

BYTE 2	
bits 0 1 2 3	**Source Number**
0 0 0 0	Unspecified
1 0 0 0	1
0 1 0 0	2
1 1 0 0	3
0 0 1 0	4 to
0 1 1 1	14 (binary - 0 is LSB, 3 is MSB)
1 1 1 1	15
bit 4 5 6 7	**Channel number**
0 0 0 0	Unspecified
1 0 0 0	A (Left in 2 channel format)
0 1 0 0	B (Right in 2 channel format)
1 1 0 0	C to
0 1 1 1	N (binary - 4 is LSB, 7 is MSB)
1 1 1 1	O

BYTE 3	
bits 0 1 2 3	**Sampling Frequency**
0 0 0 0	44.1 kHz
0 1 0 0	48 kHz
1 1 0 0	32 kHz
1 1 0 0	Sample-rate converter
× × × ×	All other states are reserved
bits 4 5	**Clock Accuracy**
0 0	Level II, ±1000 ppm (default)
0 1	Level III, variable pitch
1 0	Level I, ±50 ppm-high accuracy
1 1	Reserved
bits 6 7	
× ×	Reserved

BYTE 4	
bit 0	**Word Length Status**
0	Maximum = 20 bits (default)
1	Maximum = 24 bits
bits 1 2 3	**(With 24-bit set in bit 0)**
0 0 0	Word length not indicated (default)
0 0 1	23 bits
0 1 0	22 bits
0 1 1	21 bits
1 0 0	20 bits
1 0 1	24 bits
× × ×	Reserved
bits 1 2 3	**(With 20-bit set in bit 0)**
0 0 0	Word length not indicated (default)
0 0 1	19 bits
0 1 0	18 bits
0 1 1	17 bits
1 0 0	16 bits
1 0 1	20 bits
× × ×	Reserved

BYTES 5–23
Reserved

FIGURE 13.8 Description of the data contained in the 24-byte channel status block used in the S/PDIF serial interface. (*Sanchez, 1994*)

source number and channel number, and byte 3 specifies sampling frequency and clock accuracy.

The category code, as noted, defines different types of digital equipment. This in turn defines the subframe structure, and how receiving equipment will interpret channel status information. For example, the category code for CD players (100) defines the subframe structure with 16 bits per sample, a sampling frequency of 44.1 kHz, control bits derived from the CD's Q subcode, and places CD subcode data in the user bits. Subcode is transmitted as it is derived from the disc, one subcode channel bit at a time, over 98 CD frames. The P subcode, used to identify different data areas on a disc, is not transmitted. The start of subcode data is designated by a minimum of sixteen 0s, followed by a high start bit. Seven subcode bits (Q–W) follow. Up to eight 0s can follow for timing purposes, or the next start bit and subcode field can follow immediately. The process repeats 98 times until the subcode is transmitted. Subcode blocks from a CD have a data rate of 75 Hz. There is one user bit per audio sample, but there are fewer subcode bits than audio samples ($12 \times 98 = 1176$) so the remaining bits are packed with 0s.

Unlike the professional standard, the consumer interface does not require a low-impedance balanced line. Instead, a single-ended 75-Ω coaxial cable is used, with 0.5 V peak-to-peak amplitude, over a maximum distance of 10 meters. To ensure adequate transmission bandwidth, video-type cables are recommended. Alternatively, some consumer equipment uses an optical Toslink connector and plastic fiber-optic cable over distances less than 15 meters. Glass fiber cables and appropriate code/decode circuits can be used for distances over 1 kilometer.

The IEC-61937 specification describes a multichannel interface. Low bit-rate data such as Dolby Digital, DTS, MPEG, or ATRAC can be substituted for the PCM data originally specified in the IEC-60958 protocol, as shown in Fig. 13.9. Because the data rate is reduced, a number of channels (such as 5.1 channels) can be conveyed

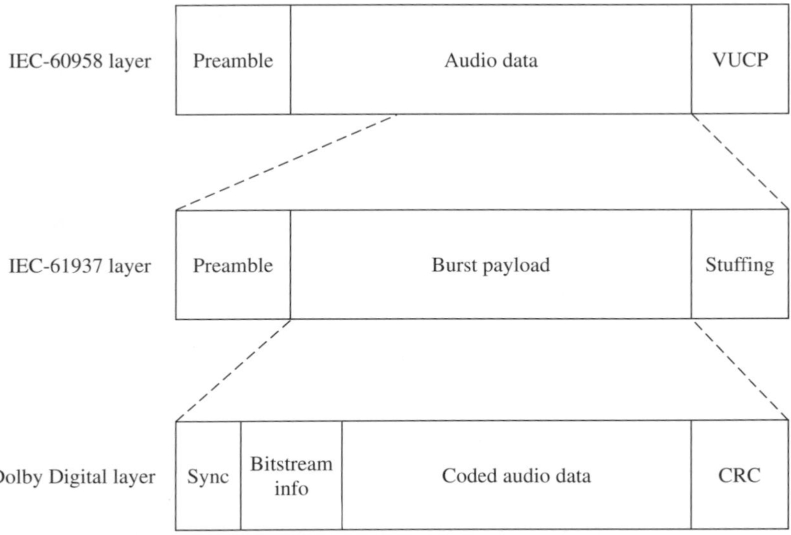

FIGURE 13.9 The protocol stack used by the IEC-61937 standard to convey multichannel data (such as Dolby Digital) via the IEC-60958 standard.

in a nominally two-channel optical or coaxial interface. Receiving equipment reads channel-status information to determine the type of bitstream (IEC-60958 or IEC-61937). If the latter, the bitstream is directed to the appropriate decoder (such as Dolby Digital or DTS).

Serial Copy Management System

The Serial Copy Management System (SCMS) is used on many consumer recorders to limit the number of copies that can be derived from a recording. A user can make digital copies of a prerecorded, copyrighted work, but the copy itself cannot be copied; first-generation copying is permissible, but not second-generation copying. For example, a user can digitally copy from a CD to a second media, but a copy-inhibit flag is set in the second media's subcode so that it is impossible to digitally copy from the second media. However, a SCMS-equipped recorder can record any number of digital copies from an original source. SCMS does not affect analog copying in any way. SCMS is a fair solution because it allows a user to make a digital copy of purchased software, for example, for compilation of favorite songs, but helps prevent a second party from copying music that was not paid for. On the other hand, SCMS might prohibit the recopying of original recordings, a legitimate use. Use of SCMS is mandated in the United States by the Audio Home Recording Act of 1992, as passed by Congress to protect copyrighted works.

The SCMS algorithm is found in consumer-grade recorders with S/PDIF (IEC-60958 type II) interfaces; it is not present in professional AES3 (IEC-60958 type I) interfaces. In particular, SCMS resides in the channel status bits as defined in IEC-60958 type II, Amendment No. 1 standard. This data is used to determine whether the data is copyrighted, and whether it is original or copied. The SCMS circuit first examines the channel status block (see Fig. 13.7) in the incoming digital data to determine whether it is a professional bitstream or a consumer bitstream. In particular, when byte 0, bit 0 is a 1, the bitstream is assumed to adhere to the AES3 standard; SCMS takes no action. SCMS signals do not appear on AES3 interfaces, and the AES3 standard does not recognize or carry SCMS information; thus, audio data is not copy-protected, and can be indefinitely copied. When bit 0 is set to 0, the SCMS identifies the data as consumer data. It examines byte 0, bit 2, the copyright or C bit; it is set to 0 when copyright is asserted, and set to 1 when copyright is not asserted. Byte 1, bit 7 (the 15th bit in the block) is the generation or L bit; it is used to indicate the generation status of the recording.

For most category codes, an L bit of 0 indicates that the transmitted signal is a copy (first-generation or higher) and a 1 means that the signal is original. However, the L bit may be interpreted differently by some product categories. For example, the meaning is reversed for laser optical products other than the CD: 0 indicates an original, and 1 indicates a copy. The L bit is thus interpreted by the category code contained in byte 1, bits 0 to 6 that indicate the type of transmitting device. In the case of the Compact Disc, because the L bit is not defined in the CD standard (IEC 908), the copy bit designates both the copyright and generation. Copyright is not asserted if the C bit is 1; the disc is copyrighted and original if the C bit is 0; if the C bit alternates between 0 and 1 at a 4-Hz to 10-Hz rate, the signal is first-generation or higher, and copyright has been asserted. Also, because the general category and A/D converter category without copyrighting cannot carry C or L information, these bits are ignored and the receiver sets C for copyright, and L to original.

Generally, the following recording scenario exists when bit 0 is set to 0, indicating a consumer bitstream: When bit C is 1, incoming audio data will be recorded no matter what is written in the category code or L bit, and the new copy can in turn be copied an unlimited number of times. When bit C is 0, the L bit is examined; if the incoming signal is a copy, no recording is permitted. If the incoming signal is original, it will be recorded, but the recording is marked as a copy by setting bits in the recording's subcode; it cannot be copied. When no defined category code is present, one generation of copying is permitted. When there is a defined category code but no copyright information, two generations are permitted. However, different types of equipment respond differently to SCMS. For example, equipment that does not store, decode, or interpret the transmitted data, is considered transparent and ignores SCMS flags. Digital mixers, filters, and optical disc recorders require different interpretations of SCMS; the general algorithm used to interpret SCMS code is thus rather complicated.

By law, the SCMS circuit must be present in consumer recorders with the S/PDIF or IEC-60958 type II interconnection. However, some professional recorders, essentially upgraded consumer models, also contain an SCMS circuit. If recordists use the S/PDIF interface, copy-inhibit flags are sometimes inadvertently set, leading to problems when subsequent copying is needed.

High-Definition Multimedia Interface (HDMI) and DisplayPort

The High-Definition Multimedia Interface (HDMI) provides a secure digital connection for television and computer video, and multichannel audio. HDMI conveys full-bandwidth video and audio and is often used to connect consumer devices such as Blu-ray players and television displays. HDMI is compliant with HDCP (High-bandwidth Digital Content Protection). HDMI Version 1.0 was introduced in 2003, and has been improved in subsequent versions. For example, Version 1.0 permitted a throughput of 4.9 Gbps, and Version 1.3 permitted 10.2 Gbps. Version 1.4 adds an optional Ethernet channel and an optional audio return channel between connected devices. Version 1.4 also defines specifications for conveying 3D movie and game formats in 1080p/24 and 720p/60 resolution; Version 1.4a supports 3D broadcast formats.

The DisplayPort interface conveys video and audio data via a single cable connection and is an open-standard alternative to HDMI. DisplayPort is compatible with HDMI 1.3 and adapters can be used to interconnect them. DisplayPort 1.1 provides a throughput of 10.8 Gbps and also supports HDCP.

Musical Instrument Digital Interface (MIDI)

The Musical Instrument Digital Interface (MIDI) is widely used to interconnect electronic music instruments and other audio production equipment, as well as music notation devices. MIDI is not a digital audio interface because audio signals do not pass through it. Instead, MIDI conveys control information as well as timing and synchronization information to control musical events and system parameters of music devices. For example, striking a key on a MIDI keyboard generates a note-on message, containing information such as the note's pitch and the velocity with which the key was struck. MIDI allows one instrument to control others, eliminating the requirement that each

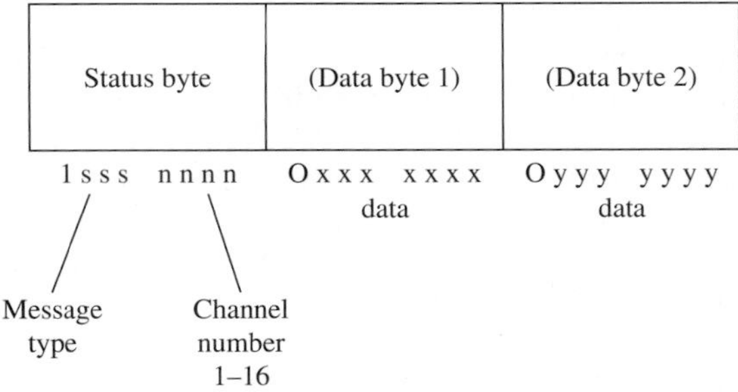

FIGURE 13.10 The general MIDI message protocol conveys data in sequences of up to three words, comprising one status word and two data words.

instrument have a dedicated controller. Since a MIDI file notates musical events, MIDI files are very small compared to WAV files and can be streamed with low overhead or downloaded very quickly. However, at the client end, the user must have a MIDI synthesizer installed on the local machine to render the data into music. Many sound cards contain MIDI input and output ports within a 15-pin joystick connector; an optional adapter is required to provide the 5-pin DIN jack comprising a MIDI port. A MIDI port provides 16 channels of communication, allowing many connection options.

MIDI is an asynchronous, unidirectional serial interface. It operates at 31,250 bits/second (31.25 baud). The data format uses 8-bit bytes; most messages are conveyed as a one-, two-, or three-word sequence, as shown in Fig. 13.10. The first word in a sequence is a status word describing the type of message and channel number (up to 16). Status words begin with a 1 and may convey a message such as "note on" or "note off." Subsequent words (if any) are data words. Data words begin with a 0 and contain message particulars. For example, if the message is "note on," two data bytes describe which note and its playback level.

MIDI can also convey system exclusive messages with data relating to specific types of equipment. These sequences can contain any number of bytes. MIDI can also convey synchronization and timing information. Status bytes can convey MIDI beats, song pointers can convey specific locations from the start of a song, and MIDI Time Code (MTC) can convey SMPTE/EBU timecode. MIDI Show Control (MSC) is used to control multimedia and theatrical devices such as lights and effects. General MIDI (GM) standardizes many functions such as instrument definitions; it is used in many game and Internet applications where conformity is needed. MIDI connections typically use a 5-pin DIN connector, but XLR connectors can also be used. A twisted-pair cable is used over a maximum length of 50 feet. As shown in Fig. 13.11, many MIDI devices have three MIDI ports for MIDI In, MIDI Out, and MIDI Thru, to receive data, output data and to duplicate and pass along received data, respectively.

The original MIDI specification did not describe what instrument sounds (called patches) should be included on a synthesizer. Thus, a song recorded with piano, bass, and drum sounds on one synthesizer could be played back with different instruments on another. The General MIDI specification set a new standard for governing the ordering and naming of sounds in a synthesizer's memory banks. This allows a song written on one manufacturer's instrument to play back with the correct sounds on another manufacturer's instrument. General MIDI provides 128 musical sounds and 47 percussive sounds.

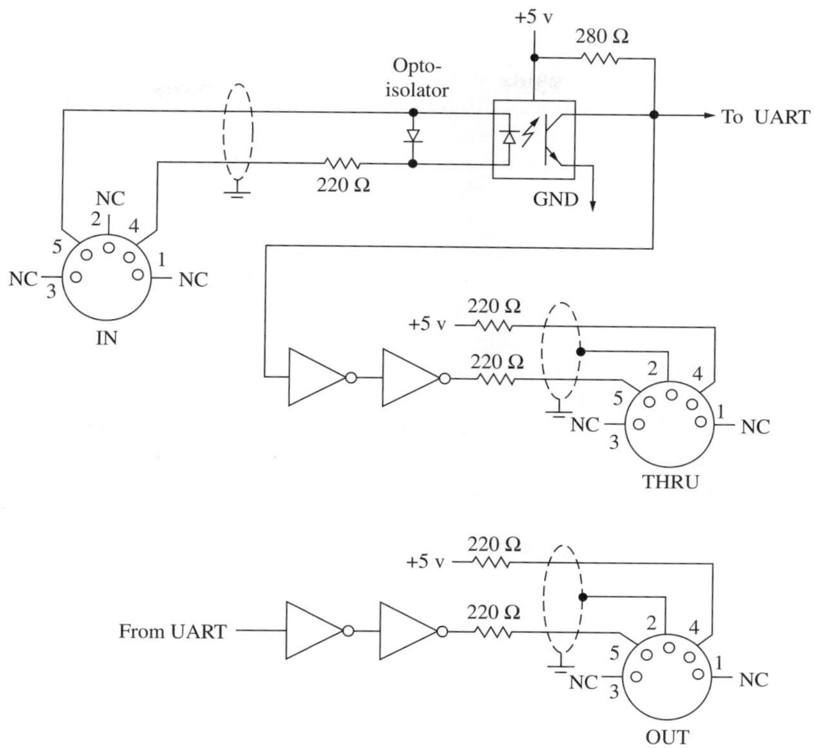

Figure 13.11 The MIDI electrical interface uses In, Thru, and Out ports, typically using 5-pin DIN connectors.

AES11 Digital Audio Reference Signal

The AES11-1997 standard specifies criteria for synchronization of digital audio equipment in studio operations. It is important for interconnected devices to share a common timing signal so that individual samples are processed simultaneously. Timing inaccuracies can lead to increased noise, and even clicks and pops in the audio signal. With a proper reference, transmitters, receivers, and D/A converters can all work in unison. Devices must be synchronized in both frequency and phase, and be SMPTE time synchronous as well. It is relatively easy to achieve frequency synchronization between two sources; they must follow a common clock, and the signals' bit periods must be equal. However, to achieve phase-synchronization, the bit edges in the different signals must begin simultaneously.

When connecting one digital audio device to another, the devices must operate at a common sampling frequency. Also, equally important, bits in the sending and received signals must begin simultaneously. These synchronization requirements are relatively easy to achieve. Most digital audio data streams are self-clocking; the receiving circuits read the incoming modulation code, and reference the signal to an internal clock to produce stable data. In some cases, an independent synchronization signal is transmitted.

In either case, in simple applications, the receiver can lock to the bitstream's sampling frequency.

However, when numerous devices are connected, it is difficult to obtain frequency and phase synchronization. Different types of devices use different timebases hence they exhibit noninteger relationships. For example, at 44.1 kHz, a digital audio bitstream will clock 1471.47 samples per NTSC video frame; sample edges align only once every 10 frames. Other data, such as the 192-sample channel status block, creates additional synchronization challenges; in this case, the audio sample clock, channel status, and video frame will align only once every 20 minutes.

To achieve synchronization, a common clock with good frequency stability should be distributed through a studio. In addition, external synchronizers are needed to read SMPTE timecode, and provide time synchronization between devices. Figure 13.12 shows an example of synchronization for an audio/video studio. Timecode is used to provide general time lock; a master oscillator (using AES11 or video sync) provides a stable clock to ensure frequency lock of primary devices (the analog multitrack recorder is locked via an external synchronizer and synthesizers are not locked). It is important that the timecode reference is different from the frequency lock reference. In addition, most timecode sources are not sufficiently accurate to provide frequency and phase-locked references through a studio.

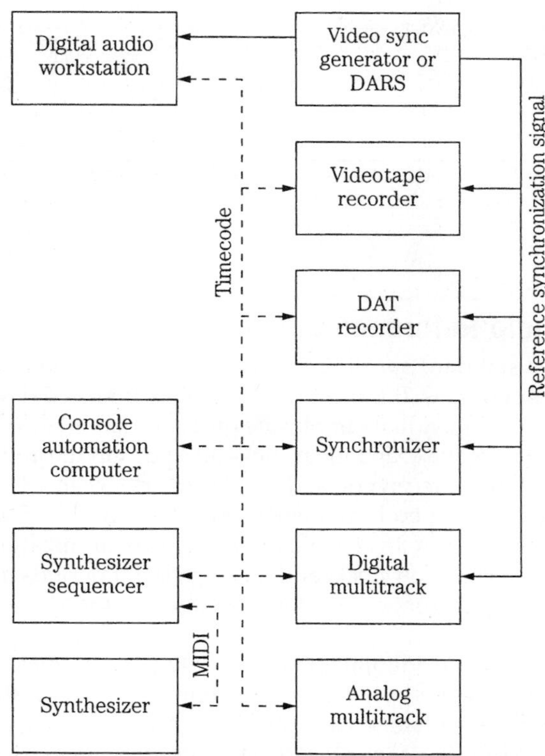

FIGURE 13.12 An example of synchronization in an audio/video studio showing a reference synchronization signal, timecode, and MIDI connections.

Although an AES3 line could be used to distribute a very stable clock reference, a dedicated word clock is preferred. Specifically, the AES11 Digital Audio Reference Signal (DARS) has been defined, providing a clocking signal with high frequency stability for jitter regulation. Using this reference signal, any sample can be time-aligned to any other sample, or with the addition of a timecode reference, aligned to a specific video frame edge. AES11 uses the same format and electrical configuration, and connectors as AES3; only the preamble is used as a reference clock. The AES11 signal is sometimes called "AES3 black" because when displayed it looks like an AES3 signal with no audio data present.

The AES11 standard defines Grade 1 long-term frequency accuracy (accurate to within ±1 ppm) and Grade 2 long-term accuracy (accurate to within ±10 ppm), where ppm is one part per million. Use of Grade 1 or 2 is identified in byte 4, bits 0 and 1 of the channel status block: Grade 1 (01), Grade 2 (10). With Grade 1, for example, a reference sampling clock of 44.1 kHz would require an accuracy of ±0.0441 Hz. A Grade 1 system would permit operation with 16- or 18-bit audio data; 20-bit resolution might require a more accurate reference clock, such as one derived from a video sync pulse generator (SPG). Timecode references lack sufficient stability. When synchronizing audio and video equipment, the DARS must be locked to the video synchronization reference. Frequency stability of the clocking in several audio and video interface signals is summarized in Table 13.1.

A separate word clock cable is run from the reference source to each piece of digital equipment through a star-type architecture, using a distribution amplifier as with a video sync pulse generator, to the reference clock inputs of all digital audio devices. Only a small buffer is needed to reclock data at the receiver. For example, a 5-ms buffer would accommodate 8 minutes of program that varies by ±10 ppm from the master reference. When A/D or D/A converters do not have internal clocks, and derive their clocks from the DARS, timing accuracy of the DARS must be increased; any timing errors are applied directly to the reconstructed audio signal. For 16-bit resolution at the converter, the AES11 standard recommends that peak sample clock modulation, sample to sample, be less than ±1 ns at all modulation frequencies above 40 Hz, and that random clock jitter, sample to sample, be less than ±0.1 ns per sample clock period, as shown in Fig. 13.13. Jitter is discussed in more detail in Chap. 4.

Digital Signal	Frequency Stability	Jitter Performance
AES3/IEC 958 Type 1	10^{-5}	Very good
S/PDIF/CP-340/IEC 958 Type II	10^{-3} to 10^{-5}	Varies
AES11 DARS	10^{-5} to 10^{-6}	Excellent
NTSC video	3×10^{-6}	Excellent
PAL video	2×10^{-7}	Excellent
Film sync	10^{-3}	Poor
SMPTE timecode	NA	Varies; usually poor
MIDI timecode	NA	Varies; usually poor
Analog tape recorder	Varies	Varies; usually poor

Table 13.1 Frequency stability of clocking in some audio and video interface signals.

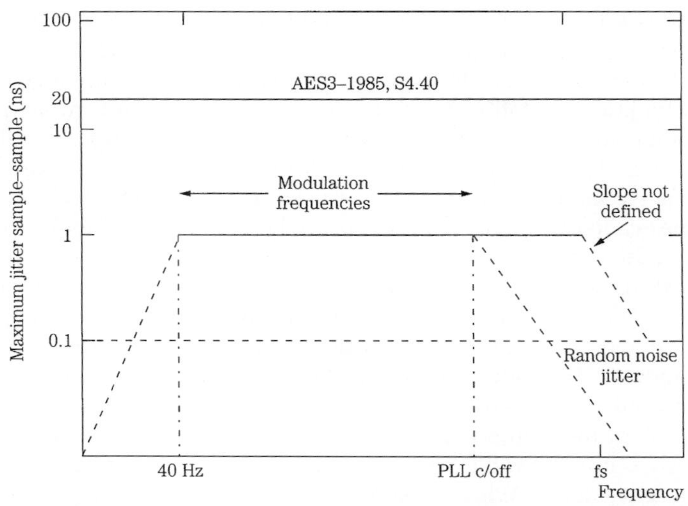

FIGURE 13.13 Timebase tolerances must be increased when DARS is used to clock sample converters. Recommended jitter limits for sample clocks in 16-bit A/D and D/A converters are depicted.

AES18 User Data Channels

The AES18-1996 standard describes a method for formatting the user data channels found within the AES3 interface. This format is derived from the packet-based High-Level Data Link Control (HDLC) communications protocol. It conveys text and other message data that might be related or unrelated to the audio data. The user data channel is a transparent carrier providing a constant data rate when the AES3 interface operates at an audio sampling frequency of 48 kHz ± 12.5%. A message is sent as one or more data packets, each with the address of its destination; a receiver reads only the messages addressed to it. Packets can be added or deleted as is appropriate as data is conveyed from one device to another. A packet comprises an address byte, control byte, address extension byte (optional), and an information field that is no more than 16 bytes. Multiple packets are placed in an HDLC frame; it contains a beginning flag field, packets, a CRCC field, and an ending flag. As described above, each AES3 subframe contains one user bit. User data is coded as an NRZ signal, LSB leading. Typical user bit applications can include messages such as scripts, subtitles, editing information, copyright, performer credits, switching instructions, and other annotations.

AES24 Control of Audio Devices

The AES24-1-1999 standard describes a method to control and monitor audio devices via digital data networks. It is a peer-to-peer protocol, so that any device may initiate or accept control and monitoring commands. The standard specifies the formats, rules, and meanings of commands, but does not define the physical manner in which commands (or audio signals) are transmitted; thus, it is applicable to a variety of communication networks. Using AES24, devices from different manufacturers can be controlled and monitored with a unified command set within a standard format. Each AES24 device is uniquely addressable by the transport network software, such as through a port number. Devices may be signal processors, system controllers, or other components, with or without user interfaces. Devices contain hierarchical objects that may include functions or controls such as gain controls, power switches, or pilot lamps. Object-to-object communication is provided because every object has a unique address. Object addresses have two forms: an object path (a pre-assigned text string) and an object address (a 48-bit unsigned integer).

Messages pass from one object to another; all messages share a common format and a set of exchange rules. In normal message exchange, an object creates and sends a message to another object, specifying a target method. Upon receiving the message, the required action is performed by the target method; if the original message requested a reply when the action has been completed, the target returns a reply stating the outcome. Because the standard is an abstraction, the action may be completed by any means, such as voltage-controlled amplifiers or digital signal processing software. AES24 subnetworks can be connected to form AES24 internetworks. Complex networks with bridges, hubs, and repeaters are possible. However, each device must be reachable by any other device. In one application, a PC-based controller may run a program with a graphical display of faders and meters (each one is an object). It communicates with external hardware processing devices (containing other objects) using the AES24 protocol. Other objects supervise initialization and configuration processes.

Sample Rate Converters

In a monolithic digital world, there would be one sampling frequency. In practice, the world is populated by many different sampling frequencies. Although 44.1 kHz and 48 kHz are the most common, 32 kHz is used in many broadcast applications. The AES5-1984 standard originally defined the use of these sampling frequencies. Sound cards use frequencies of 44.1, 22.05, and 11.025 kHz (among others), and 44.056 kHz is often used with video equipment. The DVD-Audio format defines several sampling frequencies including 88.2, 96, 176.4, and 192 kHz. The Blu-ray format can use 48-, 96-, and 192-kHz sampling frequencies. In addition, in many applications, vari-speed is used to bend pitch, producing radically diverse sampling rates.

Devices generally cannot be connected when their sampling rates differ. Even when sources are recorded at a common sampling frequency, their data streams can be asynchronous and thus differ by a few Hertz; they must be synchronized to an exact common frequency. In addition, a signal can be degraded by jitter. This changes the accuracy of the signal's sample rate. In some cases, sample rate can be changed with little effort. For example, a 44.1-kHz signal can be converted to 44.056 kHz by removing a sample approximately every 23 ms, or about every 1000 samples. More typically, dedicated converters are needed.

A synchronous sample rate converter converts one rate to another using an integer ratio. The output rate is fixed in relation to the input rate; this limits applications. An asynchronous sample rate converter (ASRC) can accept a dynamically changing input sampling frequency, and output a constant and uninterrupted sampling frequency, at the same or different frequency. The input and output rates can have an irrational ratio relationship. In other words, the input and output rates are completely decoupled. In addition, the converter will follow any slow rate variations. This solves many interfacing problems.

Conceptually, sample rate conversion works like this: a digital signal is passed through a D/A converter, the analog signal is lowpass-filtered, and then passed through an A/D converter operating at a different sampling frequency. In practice, these functions can be performed digitally, through interpolation and decimation. The input sampling frequency is increased to a very high oversampling rate by inserting zeros in the bitstream, and then is digitally lowpass-filtered. This interpolated signal, with a very high data rate, is then decimated by deleting output samples and digitally lowpass-filtered to decrease output sampling frequency to a rate lower than the oversampling rate. The resolution of the conversion is determined by the number of samples available to the decimation filter. For example, for 16-bit accuracy, the difference between adjacent interpolated samples must be less than 1 LSB at the 16-bit level. This in turn determines the interpolation ratio; an oversampling ratio of 65,536 is required for 16-bit accuracy.

This ratio could be realized with a time-varying finite impulse response (FIR) filter of length 64, in which only nonzero data values are considered, but the required master clock signal of 3.27 GHz is impractical. Another approach uses polyphase filters. A lowpass filter highly oversampling by factor N can be decomposed into N different filters, each filter using a different subset of the original set of coefficients. If the subfilter coefficients are relatively fixed in time, their outputs can be summed to yield the original filter. They act as a parallel filter bank differing in their linear-phase group delays. This can be used for sample rate conversion. If input samples are applied to the polyphase filter bank, samples can be generated at any point between the input samples by selecting the output of a particular polyphase filter. An output sample late in the input sampling period would require a short filter delay (a large offset in the coefficient set), but an early output sample would demand a long delay (a short offset). That is, the offset of the coefficient set is proportional to the timing of the input/output sample selection. As before, accurate conversion requires 216 polyphase filters; in practice, reduction methods reduce the number of coefficients.

To summarize, by adjusting the interpolation and decimation processes, arbitrary rate changes can be accommodated. These functions are effectively performed through polyphase filtering. Input data is applied to a highly oversampled digital lowpass filter. It has a passband of 0 Hz to 20 kHz, and many times the filter coefficients needed to provide this response (equivalent to thousands of polyphase filters). Depending on the instantaneous temporal relationship between the input/output frequency ratio, a selected set of these coefficients processes input samples and compute the amplitude of output samples, at the proper output frequency.

The computation of the ratio between the input and output samples is itself digitally filtered. Effectively, when the frequency of the jitter is higher than the cutoff frequency of the polyphase selection process, the jitter is attenuated; this reduces the effect of any jitter on the input clock. Short periods of sample-rate conversion can thus be used to synchronize signals. An internal first-in first-out (FIFO) buffer is used to absorb data during dynamically changing input sample rates. Input audio samples enter the buffer at the input sampling rate, and are output at the output rate. For example, a timing error of one sample period can exist at the input, but the sampling rate converter can correct this by distributing the content of 99 to 101 input samples over 100 output samples. In this way, the ASRC isolates the jittered clock recovered from the incoming signal, and synchronizes the signal with an accurate low-jitter clock. Devices such as this make sample rate conversion essentially transparent to the user, and overcome many interfacing problems such as jitter. Because rate converters mathematically alter data values, when sampling rate conversion is not specifically required, it is better to use dedicated reclocking devices to solve jitter problems.

Fiber-Optic Cable Interconnection

With electric cable, information is transmitted by means of electrons. With fiber optics, photons are the carrier. Signals are conveyed by sending pulses of light through an optically clear fiber. Bandwidth is fiber optic's forte; transmission rates of 1 Gbps are common. Either glass or plastic fiber can be used. Plastic fiber is limited to short distances (perhaps 150 feet) and is thicker in diameter than glass fiber; communications systems use glass fiber. The purity of a glass fiber is such that light can pass through 15 miles of it before the light's intensity is halved. In comparison, 1 inch of window glass will halve light intensity. Fibers are pure to within 1 part per billion, yielding absorption losses less than 0.2 dB/km. Fiber-optic communication is not affected by electromagnetic and radio-frequency interference, lightning strikes and other high voltage, and other conditions hostile to electric signals. For example, fiber-optic cable can be run in the same conduit as high-power cables, or along the third rail of an electric railroad. Moreover, because a fiber-optic cable does not generate a flux signal, it causes no interference of its own. Because fiber-optic cables are nonmetallic insulators, ground loops are prevented. A fiber is safe because it cannot cause electrical shock or spark. Fiber-optics also provide low propagation delay, low bit-error rates, small size, light weight, and ruggedness.

Although bandwidth is very high, the physical dimensions of the fiber-optic cables are small. Fiber size is measured in micrometers, with a typical fiber diameter ranging from 10 μm to 200 μm. Moreover, many individual fibers might be housed in a thin cable. For example, there may be 6 independent fibers in a tube, with 144 tubes within one cable.

Any fiber-optic system, whether linking stereos, cities, or continents, consists of three parts: an optical source acting as an optical modulator to convert an electrical signal to a light pulse; a transmission medium to convey the light; and an optical receiver to detect and demodulate the signal. The source can be a light-emitting diode (LED), laser diode, or other component. Fiber optics provides the transmission channel. Positive-intrinsic-negative (PIN) photodiodes or avalanche photodiodes (APD) can serve as receivers. As with any data transmission line, other components such as encoding and decoding circuits are required. In general, low-bandwidth systems use LEDs and PINs with TTL interfaces and multimode fiber. High-bandwidth systems use lasers and APDs with emitter-coupled logic (ECL) interfaces and single-mode fiber.

Laser sources are used for long-distance applications. Laser sources can be communication laser diodes, distributed feedback lasers, or lasers similar to those used in optical disc pickups. Although the low-power delivery available from LEDs limits their applications, they are easy to fabricate and useful for short-distance, low-bandwidth transmission, when coupled with PIN photodiodes. Over longer distances, LEDs can be used with single-mode fiber and avalanche photodiodes. Similarly, selection of the type of detector often depends on the application. Data rate, detectivity, crosstalk, wavelength, and available optical power are all factors.

Fiber-Optic Cable

Optical fiber operates as a light pipe that traps entering light. The glass or plastic rod, called the core, is surrounded by a reflective covering, called the cladding, that reflects light back toward the center of the fiber, and hence to the destination. A protective buffer

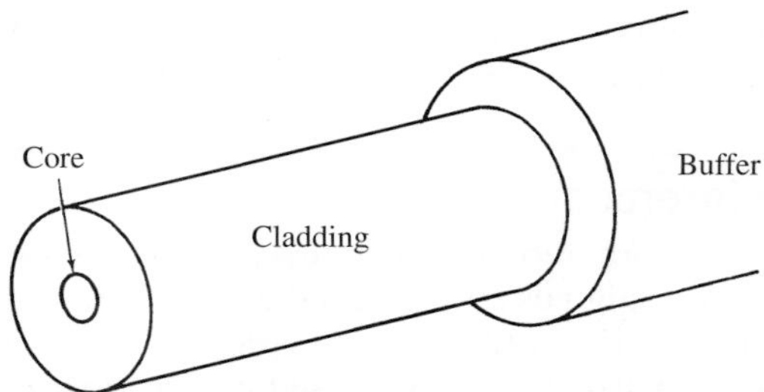

FIGURE 13.14 A fiber-optic cable is constructed with fiber core and cladding, surrounded by a buffer.

sleeve surrounds the cladding. A fiber-optic cable is shown in Fig. 13.14. The cladding comprises a glass or plastic material with an index of refraction lower than that of the core. This boundary creates a highly efficient reflector. When light traveling through the core reaches the cladding, the light is either partly or wholly reflected back into the core. If the angle of the ray with the boundary is less than the critical angle (determined from the refractive indexes of the core and the cladding), the ray is partly refracted into the cladding, and partly reflected into the core.

If the ray is incident on the boundary at an angle greater than the critical angle, the ray is totally reflected back into the core. This is known as total internal reflection (TIR). Thus, all rays at incident angles greater than the critical angle are guided by the core, affected only by absorption and connector losses. TIR is shown in Fig. 13.15 (see also Fig. 6.3). The principle of TIR is credited to British physicist John Tyndall. In 1870, using a candle and two beakers of water, he demonstrated that light could travel contained within a stream of flowing water.

In 1873, James Clerk Maxwell proved that the equations describing the behavior of electric waves apply equally to light. Moreover, he showed that light travels in modes—mathematically, eigenvalue solutions to his electromagnetic field equations that

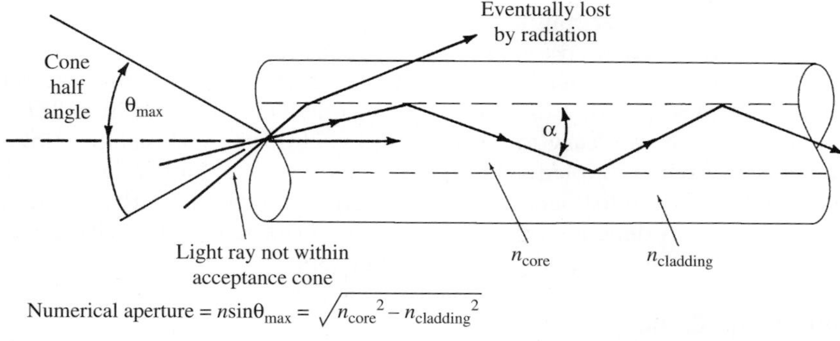

FIGURE 13.15 Total internal reflection (TIR) characterizes the propagation of light through the fiber. The numerical aperture of a stepped index optical fiber is a measure of its light acceptance angle and is determined by the refractive indexes of the core and the cladding.

characterize wave guides. In the case of optical fiber, this represents one or more paths along the light wave guide. Multimode fiber-optic cable has a core diameter (perhaps 50 μm to 500 μm) that is large compared to the wavelength of the light source; this allows multiple propagation modes. The result is multiple path lengths for different modes of the optical signal; simply put, most rays of light are not parallel to the fiber axis.

Multimode fiber is specified according to the reflective properties of the boundary: stepped index and graded index. In stepped index fiber, the boundary between the core and cladding is sharply defined, causing light to reflect angularly. Light with an angle of incidence less than the critical angle will pass into the cladding. With graded index fiber, the index of refraction decreases gradually from the central axis outward. This gradual interface results in smoother reflection characteristics. In either case, in a multimode fiber, most light travels within the core.

Performance of multimode fiber is degraded by pulse-broadening caused by intermodal and intramodal dispersion, both of which decrease the bandwidth of the fiber. Stepped index fiber is inferior to graded index in this respect. With intermodal dispersion (also called modal dispersion), some light reaches the end of a multimode fiber earlier than other light due to path length differences in the internal reflective angles. This results in multiple modes. In stepped index cable, there is delay between the lowest-order modes, those modes that travel parallel to the fiber axis, and the highest-order modes, those propagating at the critical angle. In other words, reflections at steeper angles follow a longer path length, and leave the cable after light traveling at shallow angles. A stepped index fiber can exhibit a delay of 60 ns/km. This modal dispersion significantly reduces the fiber's available bandwidth per kilometer, and is a limiting factor. This dispersion is shown in Fig. 13.16A.

Multimode graded index fiber has reduced intermodal dispersion. This is achieved by compensating for high-order mode delay to ensure that these modes travel through a lower refractive index material than the low-order modes, as shown in Fig. 13.16B. The high-order modes travel at a greater speed than the lower-order modes, compensating for their longer path lengths. Specifically, light travels faster near the cladding, away from the center axis where the index of refraction is higher. The velocity of higher mode light traveling farther from the core more nearly equals that of lower mode light

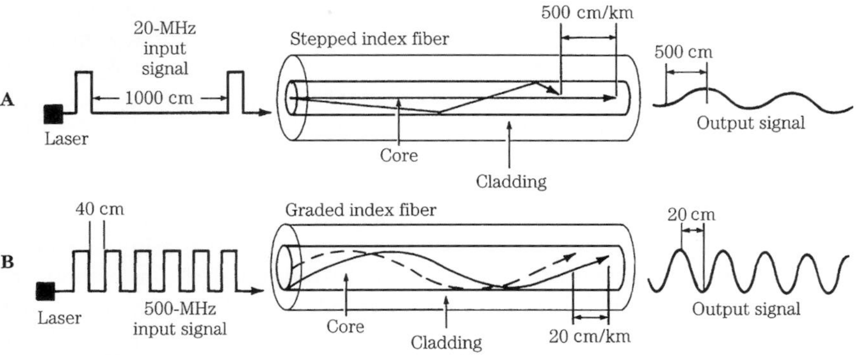

FIGURE 13.16 Fiber-optic cables are available as stepped index or graded index fibers. A. Stepped index fiber suffers from modal dispersion. B. Graded index fiber provides a higher transmission bandwidth.

in the optically dense center. Pulse-broadening is reduced, hence the data rate can be boosted. By selecting an optimal refractive index profile, this delay can be reduced to 0.5 ns/km.

Intramodal dispersion is caused by irregularities in the index of refraction of the core and cladding. These irregularities are wavelength-dependent, thus the delay varies according to the wavelength of the light source. Fibers are thus manufactured to operate at preferred light wavelengths.

In stepped and graded index multimode fibers, the degree of spreading is a function of cable length; bandwidth specification is proportional to distance. For example, a fiber can be specified at 500 kbps for 1 km. It could thus achieve a 500-kbps rate over 1 km or, for example, a 5-Mbps rate over 100 meters. In a multimode system, using either stepped or graded index fiber, wide core fibers carry several light waves simultaneously, often emitted from an LED source. However, dispersion and attenuation limit applications. Multimode systems are thus most useful in applications with short to medium distances and lower data rates.

Single-mode fiber was developed to eliminate modal dispersion. In single-mode systems, the diameter of the stepped index fiber core is small (perhaps 2 μm to 10 μm) and approaches the wavelength of the light source. Thus only one mode, the fundamental mode, exists through the fiber there is only one light path, so rays travel parallel to the fiber axis. For example, a wideband 9/125-μm single-mode fiber contains a 9-μm diameter light guide inside a 125-μm cladding. Because there is only one mode, modal dispersion is eliminated. In single-mode fibers, a significant portion of the light is carried in the cladding. Single-mode systems often use laser drivers; the narrow beam of light propagates with low dispersion and attenuation, providing higher data rates and longer transmission distances. For example, high-performance digital and radio frequency (RF) applications such as CATV (cable TV) would use single-mode fiber and laser sources.

The amount of optical power loss due to absorption and scattering is specified at a fixed wavelength over a length of cable, typically 1 km, and is expressed as decibels of optical power loss per km (dB/km). For example, a 50/125-μm multimode fiber has an attenuation of 2.5 dB/km at 1300 nm, and 4 dB/km at 850 nm. (Because light is measured as power, 3 dB represents a doubling or halving of power.) Generally, a premium glass cable can have an attenuation of 0.5 dB/km, and a plastic cable can exhibit 1000 dB/km.

Most fibers are best suited for operation in visible and near infrared wavelength regions. Fibers are optimized for operation in certain wavelength regions called windows where loss is minimized. Three commonly used wavelengths are approximately 850, 1300, and 1550 nm (353,000, 230,000, and 194,000 GHz, respectively). Generally, 1300 nm is used for long-distance communication; small fiber diameter (less than 10 μm) and a laser source must be used. Short distances, such as LANs (local-area networks), use 850 nm; LED sources can be used. The 1550-nm wavelength is often used with wavelength multiplexers so that a 1550-nm carrier can be piggybacked on a fiber operating at 850 nm or 1300 nm, running either in a reverse direction or as additional capacity.

Single-mode systems can operate in the 1310-nm and 1550-nm wavelength ranges. Multimode systems use fiber optimized in the 800-nm to 900-nm range. Generally, multimode plastic fibers operate optimally at 650 nm. In general, light with longer wavelengths passes through fiber with less attenuation. Most fibers exhibit medium losses (3 to 5 dB/km) in the 800-nm to 900-nm range, low loss (0.5 to 1.5 dB/km) at 1150-nm to 1350-nm region, and very low loss (less than 0.5 dB/km) at 1550 nm.

Fiber optics lends itself to time-division multiplexing, in which multiple independent signals can be transmitted simultaneously. One digital bitstream operating, for example, at 45 MHz, can be interleaved with others to achieve an overall rate of 1 GHz. This signal is transmitted along a fiber at an operating wavelength. With wavelength division multiplexing (WDM), multiple optical signals can be simultaneously conveyed on a fiber at different wavelengths. For example, transmission windows at 840, 1310, and 1550 nm could be used simultaneously. Independent laser sources are tuned to different wavelengths and multiplexed, and the optical signal consisting of the input wavelengths is transmitted over a single fiber. At the receiving end, the wavelengths are demultiplexed and directed to separate receivers or other fibers.

Connection and Installation

Fiber-optic interconnection provides a number of interesting challenges. Although the electrons in an electrical wire can pass through any secure mechanical splice, the light in a fiber-optic cable is more fickle through a transition. Fiber ends must be clean, planar, smooth, and touching—not an inconsiderable task considering that the fibers can be 10 μm in diameter. (In some cases, rounded fiber ends yield a more satisfactory transition.) Thus, the interfacing of fiber to connectors requires special consideration and tools. Fundamentally, fiber and connectors must be aligned and mechanically held together.

Fibers can be joined with a variety of mechanical splices. Generally, fiber ends are ground and polished and butted together using various devices. A V-groove strip holds fiber ends together, where they are secured with an adhesive and a metal clip. Ribbon splice and chip array splicers are used to join multifiber multimode fibers. A rotary splicer is used for single-mode fibers; fibers are placed in ferrules that allow the fibers to be independently positioned until they are aligned and held by adhesive. Many mechanical splices are considered temporary, and are later replaced by permanent fusion splicing; it is preferred because it reduces connector loss. With fusion splicing, fiber ends are heated to the fiber melting point, and fused together to form a continuous fiber. Specialized equipment is needed to perform the splice. Losses—as low as 0.01 dB—can be achieved.

Some fiber maintenance is simple; for example, to test continuity, a worker can shine a flashlight on one cable end, while another worker watches for light on the other end. Other measures are more sophisticated. An optical time-domain reflector (OTDR) is used to locate the position of poor connections or a break in a very long cable run. Light pulses are directed along a cable and the OTDR measures the delay and strength of the returning pulse. With short runs, a visual fault inspection can be best; a bright visible light source is used, and the cable and connectors are examined for light leaks. An optical power meter and a light source are used to measure cable attenuation and output levels. An inspection microscope is used to examine fiber ends for scratches and contamination. Although light disperses after leaving a fiber, workers should use eye protection near lasers and fiber ends. A fiber link does not require periodic maintenance. Perhaps the primary cause of failure is a "back hoe fade," which occurs when a cable is accidentally cut during digging.

Various connectors and couplers are used in fiber installations. The purpose of an optical connector is to mechanically align one fiber with another, or with a transmitting or receiving port. Simple connectors allow fibers to be connected and disconnected. They properly align fiber ends; however, connector loss is typically between

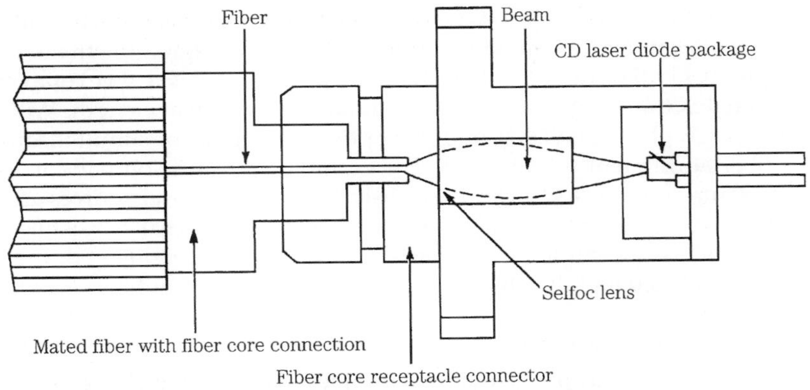

FIGURE 13.17 Laser-diode packages consolidate a laser diode, lens, and receptacle connector.

0.5 dB and 2.0 dB, using butt-type connectors. Directional couplers connect three or more ports to combine or separate signals. Star couplers are used at central distribution points to distribute signals evenly to several outputs. Transmitters, such as the one shown in Fig. 13.17, and receivers integrate several elements into modules, simplifying installation. Subsystem integration provides data links in which the complete path from one buffer memory to another is integrated.

Design Example

Generally, the principal design criteria of a fiber-optic cable installation are not complex. Consider a system with a 300-foot fiber run from a microphone panel to a recording console. In addition, midway in the cable, a split runs to a house mixing console. Design begins by specifying the signal-to-noise (S/N) ratio of the fiber (as opposed to the S/N ratio of the audio signal) to determine the bit-error rate (BER). For example, to achieve a BER of 10^{-10} (1 error for every 10^{10} transmitted bits), an S/N of 22 dB would be needed. A multimode graded index fiber would be appropriate in terms of bandwidth and attenuation. For example, the manufacturer's bandwidth length product (band factor) might be 400 MHz. For a fiber length of 300 feet (0.09 km), the cable's optical bandwidth is the band factor divided by its length, yielding 4.4 GHz. Dividing this by 1.41 yields the cable's electrical bandwidth of 3.1 GHz. The total system bandwidth is determined by:

$$1/BW_{\text{total}}^{\;2} = 1/BW_{\text{trans}}^{\;2} + 1/BW_{\text{rec}}^{\;2} + 1/BW_{\text{cable}}^{\;2}$$

where the transmitter, receiver, and cable bandwidths are all considered. If BW_{trans} is 40 MHz, BW_{rec} is 44 MHz, and BW_{cable} is 3.1 GHz, total bandwidth BW_{total} is 30 MHz.

If the cable's attenuation is specified as 5.5 dB/km, this 0.09-km length would have 0.5 dB of attenuation. The splitter has a 3-dB loss. Total coupling loss at the connection points is 5.5 dB, and temperature and aging loss is another 6 dB. Total loss is 15 dB. The system's efficiency can be determined: The adjusted output power is the difference between the transmitter output power and the detector sensitivity. For example, selected parts might yield figures of –8 dB and –26 dB, respectively, yielding an adjusted output power of 18 dB. The received signal level would yield an acceptable S/N ratio of 56 dB. The power margin is the difference between the adjusted power output and the total loss. In this case, 18 dB – 15 dB = 3 dB, a sufficient power margin, indicating a good choice of cable, transmit and receive parts with respect to system loss.

As with any technology involving data protocol, standards are required. In the case of fiber optics, Fiber Distributed Data Interface (FDDI) and FDDI II are two standards governing fiber-optic data communications networks. They interconnect processors and can form the basis for a high-speed network, for example, linking workstations and a file server. FDDI uses LED drivers transmitting at a nominal wavelength of 1310 nm over a multimode fiber with a 62.5-μm core and 125-μm cladding, with a numerical aperture of 0.275. Connections are made with a dual-fiber cable (separate send and receive) using a polarized duplex connector. FDDI offers data bandwidths of 100 MHz with up to 500 station connections. The FDDI II standard adds greater flexibility to fiber-optic networking. For example, it allows a time-division multiplexed mode providing individual routes with a variety of data rates.

The Synchronous Optical NETwork (SONET) provides wideband optical transmission for commercial users. The lowest bandwidth optical SONET protocol is OC-1 with a bandwidth of 51.84 Mbps. A 45-Mbps T-3 signal can be converted to an STS-1 (synchronous transport signal level 1) signal for transmission over an OC-1 cable. Other protocols include OC-3 (155.52 Mbps) for uncompressed NTSC digital video, and OC-12 (622.08 Mbps) for high-resolution television. SONET is defined in the ANSI T1.105-1991 standard. Telecommunications is discussed in more detail in Chap. 15.

CHAPTER 14

Personal Computer Audio

Early personal computers were little more than programmable calculators, with almost no audio capabilities. A PC typically contained a tiny speaker that was used to emit prompt beeps. The idea of making music was virtually improbable. In one celebrated demonstration, an early enthusiast programmed his computer so that when an AM radio was placed nearby, the interference from the meticulously programmed digital circuits produced rhythmic static in the radio that, in the broadest sense of the word, could be interpreted as the musical theme from Rossini's *William Tell* overture.

With the advent of more powerful hardware and software, personal computers became more adept at making music. In particular, audio chip sets became staples in PCs. Containing synthesis chips and A/D and D/A converters, these systems are used to play MIDI files, sound effects, and to record and play back WAV files. Simultaneously, CD-ROM and DVD-ROM drives became common, and are used to play audio CDs as well as DVD-Video movies. Multichannel speakers and subwoofers are common accessories in the PC market. The PC industry has adopted audio technologies such as the CD, DVD, Blu-ray, 3D audio, and sound synthesis, and contributed computer technologies such as the PCI bus, Win98, MMX, AC '97, IEEE 1394, and USB. These individually diverse improvements allow an integration of technologies, and bring a new level of fidelity and features to PC digital audio and multimedia in general. Moreover, the personal computer has become the nucleus of both professional and personal recording studios in a wide variety of applications.

PC Buses and Interfaces

Internal and external peripherals can be interfaced to a host computer with any of several interconnections. IBM-PC computers historically contained the ISA (Industry Standard Architecture) local bus; its 11-Mbyte/second transfer rate limits audio applications. A stereo CD data stream, at 1.4 Mbps, along with sound effects and several voices might consume 40% of the bus, leaving little capacity for the multitasking operating system. The EISA (Extended ISA) local bus offers a 32-Mbyte/second rate and other improvements, but is also limited for audio applications. The PCI (Peripheral Component Interconnect) is a high-performance local interconnection bus. PCI provides a maximum 132-Mbyte/second transfer rate, 32-bit pathways, good noise immunity, and efficiently integrates the computer's processor and memory with peripherals. The PCI bus is lightly taxed with even multiple audio streams. Moreover, the PCI bus outperforms the ISA and EISA bus in terms of latency. An audio output must be free of any interruptions in the flow of audio data. Because of its high performance, the PCI bus allows cooperative signal processing in which tasks are balanced between the host processor and an audio accelerator. For these and other reasons, the PCI bus replaced the ISA bus. Microsoft's PC98 specification decreed that new logos would no longer be issued to ISA audio products. The PCI bus is found on both Macintosh and IBM-PC computers.

The ATA (Advanced Technology Attachment) bus was designed as the expansion slot format for the PC providing a 16-bit path width and burst speeds up to 8.3 Mbyte/second; it is also known as the IDE (Integrated Drive Electronics) bus. Faster variants include EIDE (Enhanced Integrated Drive Electronics) or ATA-2 and Ultra ATA (ATA-3 or Ultra DMA). ATA-2 accommodates burst speeds up to 16 Mbytes/second, and ATA-3 accommodates speeds up to 33.3 Mbytes/second. The PCMCIA (Personal Computer Memory Card International Association) is used on notebook computer expansion ports.

Some computers use SCSI (Small Computer System Interface) connections. SCSI (pronounced "scuzzy") is a high-speed data transfer protocol that allows multiple devices to access information over a common parallel bus. Transmitting (smart) devices initiate SCSI commands; for example, to send or request information from a remote device. Likewise, receiving (dumb) devices accept SCSI commands; for example, a hard disk can only receive commands. Some devices (sometimes called logical devices) are classified as transmitter/receivers; computers are grouped in this category.

The SCSI protocol allows numerous devices to be daisy-chained together; each device is given a unique identification number; numbers can follow any order in the physical chain. A SCSI cable can extend to 12 meters. A device must have two SCSI ports to allow chaining; otherwise, the device must fall at the end of the chain. Generally, the first and last physical devices (determined physically, not by identification number) in a chain must be terminated; intermediate devices should not be terminated. Termination can be internal or external; devices that are externally terminated allow greater liberty of placement in the chain.

SCSI defines a number of variants, differing principally in data path width and speed, as well as physical connectors. The basic SCSI-1 width (number of bits in the parallel data path) is 8 bits (narrow). In addition, 16-bit (wide) and 32-bit (very wide) versions are used. The basic SCSI-1 data transmission rate is 12.8 Mbps, that is, 1.6 Mbyte/second; all data transfer is asynchronous. Other speeds include Fast, Ultra, Ultra 2, and Ultra 3. For example, Ultra 2 can accommodate transfers of up to 80 Mbyte/second. Narrow SCSI devices use 50-pin connectors, and wide devices use 68-pin connectors.

Alternatively, the Enhanced-IDE/ATAPI interface can be used to connect computers and peripherals at speeds of 13.3 Mbyte/second over short (18 inch) cables. Fast ADA-2 hard-drive interfaces can operate at 16.6 Mbyte/second. However, unlike SCSI, these are simple interfaces that do not support intelligent multitasking. To overcome many interconnection installation problems in PCs, the Plug and Play standard was devised. The operating system and system BIOS automatically configure jumper, IRQ, DMA address, SCSI IDs, and other parameters for the plug-in device. Many DVD-ROM drives use E-IDE/ATAPI or SCSI-2 interfaces. The SFF 8090i Mt. Fuji standard allows the ATAPI and SCSI interfaces to fully support DVD reading of regional codes, decryption authentication, CSS and CPRM data, BCA area, and physical format information.

IEEE 1394 (FireWire)

The Institute of Electrical and Electronics Engineers (IEEE) commissioned a technical working group to design a new data bus transmission protocol. Based on the FireWire protocol devised by Apple Computer, the IEEE 1394-1995 High Performance Serial Bus standard defines the physical layer and controller for both a backplane bus and a serial data bus that allows inexpensive, general purpose, high-speed data transfer; the latter is discussed here. The IEEE 1394 cable bus is a universal, platform-independent digital interface that can connect digital devices such as personal computers, audio and video products for multimedia, digital cable boxes and HDTV tuners, printer and scanner products, digital video cameras, displays, and other devices that require high-speed data transfer. For example, a digital camcorder can be connected to a personal computer for transferring video/audio data. Wireless FireWire technology has also been developed, using IEEE 802.15.3 technology.

The IEEE 1394 cable standard defines three data rates: 98.304, 196.608, and 393.216 Mbps. These rates are rounded to 100, 200, and 400 Mbps, and are referred to as S100, S200, and S400. The latter is also known as FireWire 400. These rates permit transport, for example, of 65, 130, and 260 channels of 24-bit audio, sampled at 48 kHz. The IEEE 1394b standard enables rates from 800 Mbps to 3.2 Gbps. More details of IEEE 1394b are given below. The 1394c standard merges FireWire and Ethernet technologies.

The IEEE 1394 connecting cable is thin (slightly thicker than a phone cable) and uses copper wires; there are two separately shielded twisted-pair transmission lines for signaling, two power conductors, and a shield. A four-conductor version of the standard cable with no power connectors is used in some consumer audio/video components. It is defined in IEEE 1394.1 and is sometimes known as i.Link. (IEEE 1394, FireWire and i.Link are all functionally equivalent and compatible.) The two twisted pairs are crossed in each cable assembly to form a two-way transmit-receive connection. The IEEE 1394 connector (the same on both cable ends) is small and rugged (and derived from Nintendo's GameBoy connector). It uses either a standard friction detent or a special side-locking tab restraint, as shown in Fig. 14.1. Since it can deliver electrical power (8 to 40 volts, up to 1.5 amperes), an IEEE 1394 cable can be used as the sole connecting cable to some devices. It is similar to a SCSI cable in that it can be used as a point-to-point connection between two devices, or devices can be connected with branches or daisy-chained along lengths of cable. However, unlike SCSI, no addressing (device ID) or termination is needed. In some cases, IEEE 1394 is used with other cabling such as Category 5 twisted pair and optical fiber. With 50-mm multimode fiber, for example, cable runs of hundreds of meters are permitted. However, more robust clocking methods are required.

A cable can run for 4.5 meters between two devices without a repeater box (this length is called a hop) and there may be up to 16 cable hops in a line, extending a total distance of up to 72 meters with standard cable (longer distances are possible with higher quality cable). Up to 63 devices can be directly connected into a local cluster

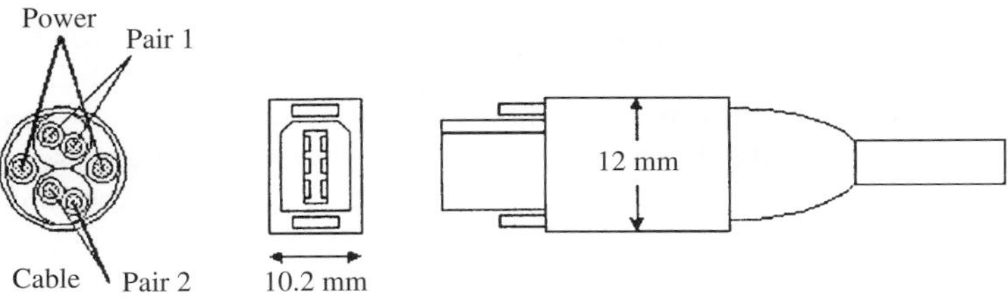

FIGURE 14.1 The IEEE 1394 connector (the same on both cable ends) forms a two-way transmit-receive connection carrying two twisted-pair signal cables, and power.

before a bus bridge is required, and up to 1023 clusters can be connected via data bridges. IEEE 1394 can be used equally well to connect two components or run throughout an entire home or office to interconnect electronic appliances. IEEE 1394 is "hot-pluggable" so that users can add or remove devices from a powered bus. It is also a scalable architecture, so users can mix multiple speed devices on one bus.

IEEE 1394 defines three layers: physical, link, and transaction. The physical layer defines the signals required by the IEEE 1394 bus. The link layer formats raw data (from the physical layer) into recognizable IEEE 1394 packets. The transaction layer presents packets (from the link layer) to the application.

The IEEE 1394 bus data rate is governed by the slowest active node. However, the bus can support multiple signaling speeds between individual node pairs. Considering that the rate of a stereo digital audio bitstream may be 1.4 Mbps, a compressed video stream may also be 1.4 Mbps, and an uncompressed broadcast-quality video stream may be 200 Mbps, an IEEE 1394 interface can accommodate multimedia data loads.

When a connection is made, the bus automatically reinitializes the entire bus, recognizing the new device and integrating it with other networked devices, in less than 125 μs. Similarly, upon disconnection, IEEE 1394 automatically reconfigures itself. Asynchronous transmission performs simple data transfers. As with all asynchronous transmission, a chunk of data is transferred after acknowledgement that a previously transmitted chunk has been received. Since this timing cannot be predicted because of other network demands, timing of delivery is random; this is problematic for real-time audio and video data.

Real-time, data-intensive applications such as uncompressed digital video can be provided bandwidth and low latency, using isochronous transmission. With isochronous transmission, synchronized data such as audio and video will be conveyed with sufficiently small discrepancy so that they can be synchronized at the output. Instead of a send/acknowledgement method, isochronous transmission guarantees bus bandwidth for a device. Nodes request a bandwidth allocation and the isochronous resource manager uses a Bandwidth Available register to monitor the bandwidth available to all isochronous nodes. All bus data is transmitted in 32-bit words called quadlets and bandwidth is measured in bandwidth allocation units. A unit is about 20 ns, the time required to send one data quadlet at 1600 Mbps. The isochronous resource manager uses its Channels Available register to assign a channel number (0 to 63) to a node requesting isochronous bandwidth. This channel number identifies all isochronous packets. When a node completes its isochronous transfer, it releases its bandwidth and channel number. A bus manager such as a PC is not needed; any "talker" device can act as the isochronous resource manager to create a single, fixed isochronous channel.

Using the common timebase of isochronous transmission, data packets can be encoded with a small equalizing time delay, so the output is exactly synchronized with the clock source—a critical feature for audio and video data words that must arrive in order and on time. This feature is particularly crucial when an IEEE 1394 cable is handling simultaneous transmissions between different devices. Isochronous transmissions are given priority status in the time-multiplexed data stream so that an audio and video transfer, for example, is not disrupted by a control command. Whereas some interfaces are expensive because large memory buffers are needed to temporarily store data at either cable end, IEEE 1394 does not need large buffers. Its just-in-time data delivery allows devices to exchange data or commands directly between their internal memories.

The IEEE 1394 specification includes an "Audio and Music Data Transmission Protocol" (known as the A/M protocol) that defines how real-time digital audio can be conveyed over IEEE 1394 using isochronous packets. Data types include IEC-958 and raw audio samples (in data fields up to 24-bits in length) as well as MIDI files (with up to 3 bytes per field); it is standardized as IEC 61883-1/FDIS. This protocol, and multichannel versions, provide sufficient bandwidth to convey DVD-Audio data streams of 9.6 Mbps. Copy-protection methods are used to guard against piracy of these high-quality DVD-Audio signals.

The mLAN specification can be used to send multiple sample-accurate AES3 signals, raw audio, MIDI, and other control information over an IEEE 1394 bus. The mLAN protocol uses an isochronous transfer mode that ensures on-time delivery and also prevents collisions and reduces latency and jitter. The specification reduces jitter to 20 ns, and when phase-locked loops are used at individual mLAN nodes, jitter can be further reduced to 1 ns. Up to 63 devices can be connected in any topology, ports are hot-pluggable and software patching is routinely used. A portion of mLAN was adopted by the 1394 Trade Association as a supplemental standard for handling audio and music control data over the IEEE 1394 bus. mLAN was developed by Yamaha Corporation.

The first implementations of the 1394b standard, with throughput of 800 Mbps and 1600 Mbps, are also known as S800 and S1600. IEEE 1394b allows daisy-chaining of multiple peripherals. It also allows cable lengths of up to 800 meters for networks using twisted pair CAT-5 and plastic optical fiber. The Bus Owner Supervisor/Selector (BOSS) protocol allows data packets to be transmitted more efficiently, using less network bandwidth. IEEE 1394b ports use a different physical configuration than 1394; adapter cables are needed for compatibility.

IEEE 1394 is a non-proprietary standard and many standards organizations and companies have endorsed the standard. The Digital VCR Conference selected IEEE 1394 as its standard digital interface. An EIA subcommittee selected IEEE 1394 as the point-to-point interface for digital TV as well as the multi-point interface for entertainment systems. Video Electronics Standards Association (VESA) adopted IEEE 1394 for home networking. The European Digital Video Broadcasters also endorsed IEEE 1394 as their digital television interface. Microsoft first supported IEEE 1394 in the Windows 98 operating system, and it is supported in newer operating systems. IEEE 1394 was first supported by the Macintosh operating system in 1997; Apple Computer supports IEEE 1394 on its computer motherboards.

IEEE 1394 may appear in PCs, satellite receivers, camcorders, stereos, VCRs, printers, hard drives, scanners, digital cameras, set-top boxes, music keyboards and synthesizers, cable modems, CD-ROM drives, DVD players, DTV decoders, and monitors. In some applications, such as when connecting to displays, Digital Transmission Content Protection (DTCP) technology is used to encrypt data, allowing secure, two-way transmission of digital content across an IEEE 1394 interface.

Digital Transmission Content Protection (DTCP)

The security of transmitted data is an important issue in many applications. The Digital Transmission Content Protection (DTCP) system was devised for secure (anti-piracy) transmission in the home environment over bi-directional digital lines such as the IEEE 1394 bus. Sometimes known as "5C," it was devised by a consortium of companies including Sony, Toshiba, Intel, Hitachi, and Matsushita, as well as the Motion Picture

Association of America. DTCP prevents unauthorized copying of digital content while allowing legitimate copying for purposes such as time-shifting. Connected devices trade keys and authentication data, the transmitting device encrypts the signal, and the receiving device decrypts it. Devices such as video displays identify themselves as playback-only devices and can receive all data. Recorders can only receive data marked as copy-permitted, and must update and pass along Copy Control Information (CCI). DTCP does not affect other copy protection methods that may be employed on DVD or Blu-ray discs, satellite broadcasts, and so on. DTCP uses encryption on each digital link. Each device on a link obeys embedded CCI that specifies permitted uses of content: Copy-Never (no copies allowed, display only), Copy-One-Generation (one copy allowed), Copy-No-More (prevents making copies of copies), and Copy-Freely (no copy restrictions). Two-way "challenge and response" communication provided by the Authentication and Key Exchange system enables source components in the home to confirm the authenticity of receiving components.

DTCP can revoke the privileges of rogue equipment attempting to defeat the system, obtain encryption keys, and so on. To do this, each piece of consumer equipment contains System Renewability Messages (SRMs), a list of serial numbers of individual pieces of equipment used in piracy. SRMs are updated through packaged software, transmissions, and new equipment, and are automatically passed along to other components. Source components re-encrypt data to be transmitted to receiving components; encryption keys are changed as often as every 30 seconds. The Digital Transmission Licensing Administrator (DTLA) was established to license the content-protection system and to generate and distribute cryptographic materials such as keys and certificates. DTCP is designed for use in HDTV receivers, set-top boxes, digital recorders, satellite receivers, and other consumer components. Encryption and watermarking are discussed in Chap. 15.

Universal Serial Bus (USB)

The Universal Serial Bus (USB) was designed to replace older computer serial (and parallel) I/O buses, to provide a faster, more user-friendly interconnection method, and to overcome the limitation of too-few free interrupts available for peripherals. Computer keyboards, mice, cable modems, telephones, ROM drives, flash memories, printers, scanners, digital cameras, multimedia game equipment, MIDI devices, and loudspeakers are all candidates for USB. Unlike IEEE 1394, which permits interconnection between any two enabled devices, USB requires a microprocessor-based controller, and hence it is used primarily for PC peripherals. A few USB ports can replace disparate back-panel connectors.

The original USB specification, known as USB 1.1, provides low-speed interconnection. The newer USB 2.0 specification (sometimes known as "Hi-Speed USB") provides data rates that are 40-times faster than USB 1.1. The USB 1.1 specification provides a transfer rate of 12 Mbps (about 1 Mbps of this is used for overhead). This transfer rate is sufficient for applications employing S/PDIF, AC-3, and MPEG-1, as well as some MPEG-2 applications. There is also a 1.5-Mbps subchannel available for low data-rate devices, such as a mouse. USB 2.0 offers transfer rates up to 480 Mbps. This allows connection to high-speed ROM drives and in particular allows rapid transfer of large video files. USB 2.0 is fully compatible with 1.1 devices; however, 1.1 devices cannot operate at the faster speed. USB 2.0 uses the same connectors and cables as USB 1.1.

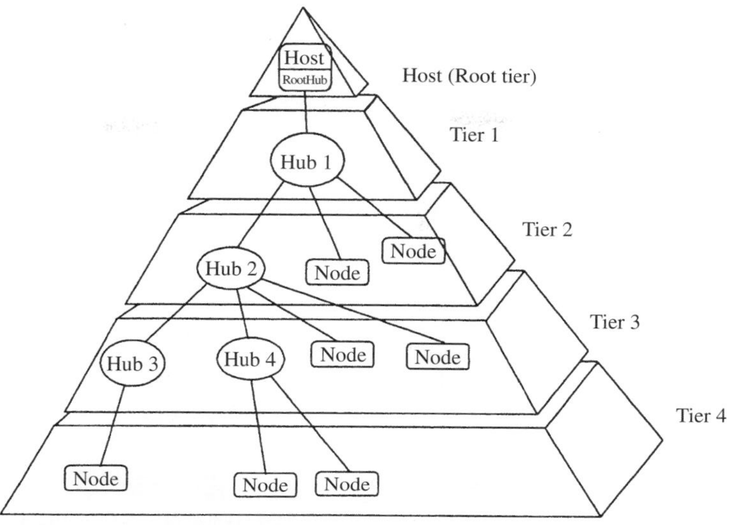

FIGURE 14.2 The USB interconnect uses a tiered-star topology, with a hub at the center of each star. Each cable forms a point-to-point connection between the host and a hub or function, or a hub connected to another hub or function.

USB is SCSI-like in its ability to support up to 127 devices per port/host in a plug-and-play fashion. Moreover, USB devices can be hot-swapped without powering down the system. USB detects when a device is added or withdrawn, and automatically reinitializes the system. USB uses a tiered star topology in which only one device, such as a monitor or DVD-ROM drive, must be plugged into the PC's host (root) connector, as shown in Fig. 14.2. There is only one host in any system. It becomes a hub and additional devices can be connected directly to that hub or to additional hubs, using cable runs of 5 meters (full-speed devices) and 3 meters (low-speed devices). The host polls connected devices and initiates all data transfers.

USB hubs may be embedded in peripheral devices or exist as stand-alone hubs. Hubs contain an upstream connection port (pointed toward the PC) as well as multiple downstream ports to connect peripheral devices. USB uses a four-wire connector; a single twisted pair carries bidirectional data (one direction per wire), and there are 5-V power and ground conductors to deliver electrical power to low-power (500 mA, or 100 mA for a bus-powered hub) peripheral devices. The typical detachable cable is known as an "A to B" cable. "A" plugs are always oriented upstream toward the host ("A" receptacles are downstream outputs from the host or hub). "B" plugs are always oriented downstream toward the USB device ("B" receptacles are upstream inputs to the device or hub). An "A" and "B" cable assembly is shown in Fig. 14.3. All detachable cables must be full-speed.

USB host controllers manage the driver software and bandwidth required by each peripheral connected to the bus, and allocate electrical power to USB devices. Both USB host controllers and USB hubs can detect attachments and detachments (for device identification and dynamic reconfiguration) of peripherals using biased termination at cable ends. Hubs are required for multiple connections to the host connector. All devices

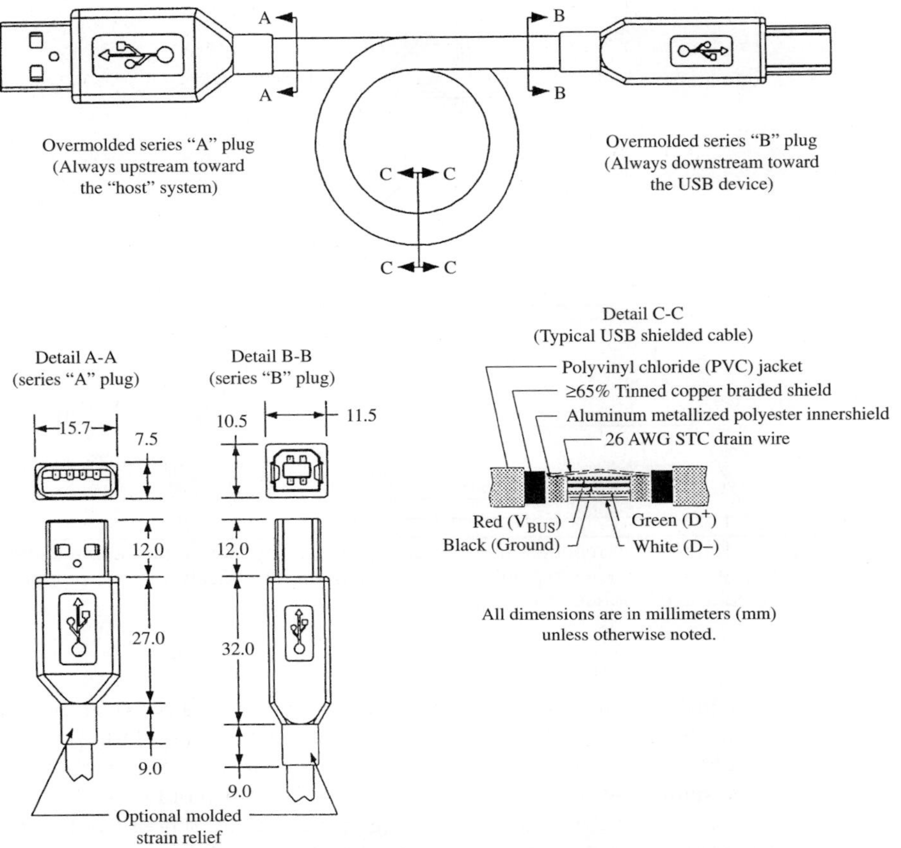

FIGURE 14.3 Physical specifications for the USB detachable cable. "A" plugs orient upstream toward the host and "B" plugs orient downstream. The cable carries one twisted-pair cable, and power.

have an upstream connection; hubs include downstream connections. Upstream and downstream connectors are polarized to prevent loop-back. Hubs can have up to seven connectors to nodes or other hubs, and may be self-powered or powered by the host. Two PCs can be connected to each other with a specialized USB peripheral known as a USB bridge (sometimes called a USB-to-USB adapter). A direct PC-to-PC connection using an illegal "A to A" cable could short the two PCs' power supplies together, creating a fire hazard.

The USB On-the-Go (USB OTG) supplement to the USB specification is used for portable devices such as cell phones and digital cameras. It allows limited hosting capabilities for direct point-to-point communication with selected peripherals. Devices can be either a host or peripheral and can dynamically switch roles. In addition, a smaller connector and low-power features are implemented. USB OTG is designed to supplant the many proprietary connections used in docking stations and slot connectors. USB OTG devices are compliant with the USB 2.0 specification.

USB is well-suited for transport of audio signals. Standardized audio transport mechanisms are used to minimize software driver complexity. A robust synchronization scheme for isochronous transfers is incorporated in the USB specification. In particular, USB provides asynchronous transfer, but isochronous transfer is used for relatively higher-bandwidth (audio and video) devices. Isochronous transfer yields low jitter but increased latency. The transfer rate for a 16-bit, 48-kHz stereo signal is 192 bytes/ms. To maintain a correct phase relationship between physical audio channels, an audio function is required to report its internal delay to every audio streaming interface. The delay is expressed in number of frames (ms) and is caused by the need to buffer frames to remove packet jitter and by some audio functions that introduce additional delay (in integer numbers of frames) as they interpret and process audio streams. Host software uses delay information to synchronize different audio streams by scheduling correct packet timings. Phase jitter is limited to ±1 audio sample.

USB has many practical audio applications. For example, USB allows designers to bypass potentially poor D/A converters on motherboards and sound cards, and instead use converters in the peripheral device. For example, USB loudspeakers have D/A converters and power amplifiers built-in, so the speaker can receive a digital signal from the computer. USB loudspeakers obviate the need for sound cards in many implementations, and simplify the connection of PCs to outboard converters, digital signal processors, Dolby Digital decoders, and other peripherals.

Sound Card and Motherboard Audio

Most computer motherboards and sound cards contain A/D and D/A converters and hardware- and software-based processing to permit recording and playback of stereo 8- or 16-bit audio at multiple sampling rates. They also allow playback via wavetable synthesis, sampled sound, or FM synthesis. They provide digital I/O; ROM drive interfaces; a software-controlled audio mixer; onboard power amplifiers; and may also provide analog line-in and line-out, S/PDIF input and output, microphone input, and a gamepad/joystick/MIDI connector. Sound cards plug into an expansion slot and are accompanied by the appropriate software that is bundled with the card. The most basic part of the software regimen is the device drivers needed to control the various audio components.

Sound synthesis capabilities are used to create sound effects when playing MIDI files or playing video games. A sampled sound synthesizer plays stored sound files—for example, SND or AIFF (Audio Interchange File Format) files. Most chip sets support sample-based wavetable synthesis; this allows synthesis and playback of both music and sound effects via software. With wavetable synthesis, a particular waveform is stored in ROM, and looped through to create a continuous sound. Some audio chip sets support physical model-based waveguide synthesis in which mathematical models are used to emulate musical instrument sounds. In terms of synthesis ability, a chip may support 128 wavetable instruments, with many variation sounds and multiple drum sets using onboard RAM. Moreover, 64 voices may be supported, with multi-timbral capability on 16 channels. Most chips have a MIDI interface for connection to an external MIDI hardware instrument such as a keyboard. A chip set may also contain built-in 3D stereo enhancement circuitry. These proprietary systems, in greater or lesser degrees, increase depth and breadth of the stereo soundstage, and broaden the "sweet spot" where channel separation is perceived by the listener. Some chip sets include a DSP chip that allows hardware data reduction during recording and playback, and others provide resident non-real-time software data reduction algorithms. In addition, some chips provide voice recognition capability. These topics are discussed in more detail below.

Music Synthesis

From their origins as simple MIDI synthesizers employing FM synthesis, computer sound systems have grown in complexity. Diverse synthesis methods are employed, and the quality of rendered music varies dramatically with the type of synthesizer hardware installed. Many synthesizers generate audio signals from a file consisting of a table of audio samples. Traditionally, these tables are filled with single cycles of simple basis waveforms such as sinusoids or triangle waves. Complex sounds are generated by dynamically mixing the simple signals using more sophisticated algorithms. This is known as traditional wavetable synthesis. During playback, a pointer loops through the table continuously reading samples and sending them to a D/A converter. Different pitches are obtained by changing the rate at which the table is read. For higher pitches, the processor skips samples; for example, to double the frequency, every other sample is read. Noninteger higher pitches are accomplished by skipping samples and interpolating new values. For lower pitches, the processor adds samples by interpolating values in-between those in the table. In both cases, the sample rate at the D/A converter remains constant. By dynamically mixing or crossfading different basis waveforms, complex sounds can be generated with low data storage overhead; a table may be only 512 samples in length.

Sample-based synthesis uses short recordings of musical instruments and other natural sounds for the basis waveforms. However, instead of perhaps 512 samples per table, these synthesizers may use thousands of samples per table. Because musical events may last several seconds or more, considerable memory would be required. However, most musical events consist of an attack transient followed by a steady-state sustain and decay. Therefore, only the transient portion and a small section of the steady-state portion need to be stored. Sound is created by reading out the transient, then setting up a wavetable-like loop through the steady-state section.

Because sample-based synthesis also uses wave lookup tables, the term "wavetable synthesis" became synonymous with both technologies and the two terms are used interchangeably. However, contemporary "wavetable synthesis" chips are really sample-based instruments. The quality of the synthesis depends on the quality of the initial recording, size of the table, and location of the loop points. Short, low-resolution tables produce poor quality tones. Although a sound card may claim "CD-quality," table resolution may only be 8, 12, or 14 bits, and not the 16-bit CD standard.

In physical modeling synthesis, the sound of a vibrating system (such as a musical instrument) is created using an analogous software model. For example, a plucked string vibrates as transverse waves propagating along the length of a string. The vibrations decay as the waves lose energy. A string model might consist of an impulse, traveling through a circular delay line (or memory buffer), with its output connected back to the input through an attenuator and filter. The filtering and attenuation cause the impulse to simultaneously decay in amplitude and become more sinusoidal in nature, emulating a vibrating string. The length of the buffer controls pitch, and the filter and attenuator provide the correct timbre and decay. Physical modeling is attractive because it is easily implemented in software. It can produce consistent results on different computers, and the coded algorithms are quite short; however, relatively fast processors are needed to synthesize sound in real time.

In pure software synthesis, sounds are created entirely on the host computer, rather than in a dedicated sound chip. Software synthesizers can be distributed on disc or as downloadable files. Some downloadable software synthesizers are available as browser plug-ins for network applications.

Surround Sound Processing

Stereo playback can convey a traditional concert soundstage in which the sound comes mainly from the front. However, stereo signals cannot convey ambient sounds that come from all around the listener. Stereo's lack of spatiality undermines sonic realism, for example, in a game where aircraft fly overhead from front to back. The lack of spatiality is exacerbated by the narrow separation between speakers in most PC playback systems. To provide a more convincing sound field, various algorithms can modify stereo signals so that sound played over two speakers can seem to come from around the listener. Alternatively, adequately equipped PCs can convey discrete 5.1-channel playback.

Stereo surround sound programs process stereo signals to enlarge the perceived ambient field. Other 3D positioning programs seek to place sounds in particular locations. Psychoacoustic cues such as interaural time delays and interaural intensity differences are used to replicate the way sources would sound if they were actually in a 360-degree space. This processing often uses a head-related transfer function (HRTF) to calculate the sound heard at the listener's ears relative to the spatial coordinates of the sound's origin. When these time, intensity, and timbral differences are applied to the stereo signal, the ear interprets them as real spatial cues, and localizes the sound outside the stereo panorama. These systems can process sound during real-time playback, without any prior encoding, to position sound statically or dynamically. Results differ, but are generally quite good if the listener is seated exactly between the speakers—the ergonomic norm for most PC users. In some cases, the surround process must be encoded in the source media itself. There are a wide variety of 3D audio plug-in and chip options from numerous companies. Surround programs written for DirectSound contain compatible positioning information required by the spatial positioning programs.

Home theater systems employ 5.1-channel playback with left and right front channels, center front channel, left and right rear channels, and a low-frequency effects channel. Dolby Digital (also known as AC-3) and DTS both employ 5.1-channel processing. Dolby Digital was selected as the audio coding method for DVD-Video, as well as DTV; Dolby Digital and DTS are used in the Blu-ray format. Dolby Digital and DTS are described in Chap. 11. Many PCs can play back 5.1-channel audio tracks on both movie and game titles. Similarly, with DTV tuners, PCs can play back 5.1-channel broadcasts; 5.1 playback is the norm for home media PCs.

Although 5.1-channel playback improves realism, it presents the practical problem of arraying six loudspeakers around a PC. Thus, a number of surround synthesis algorithms have been developed to specifically replay multichannel formats such as Dolby Digital over two speakers, creating "virtual speakers" to convey the correct spatial sense. This multichannel virtualization processing is similar to that developed for surround synthesis. Dolby Laboratories grants a Virtual Dolby certification for both the Dolby Digital and ProLogic processes. Although not as realistic as physical speakers, virtual speakers can provide good sound localization around the PC listener.

Audio Codec '97 (AC '97)

The Audio Codec '97 (AC '97) component specification (also known as MC '97 or Modem Codec '97) describes a two-chip partitioned architecture that provides high-quality PC audio features. It is used in motherboards, modems, and sound cards. High-speed PC buses and clocks, and digital grounds and the electromagnetic noise they radiate, are anathema to high-quality audio. With legacy systems, integrated hardware is placed on the ISA bus so that analog audio circuitry is consolidated with digitally intensive bus interfaces and digital synthesizer circuits, resulting in audio signal degradation. The Audio Codec '97 specification segregates the digital portion of the audio system from the analog portion. AC '97 calls for a digital chip (with control and processing such as equalization, reverberation, and mixing) on the bus itself, and an analog chip (for interfacing and conversion) off the bus and near the I/O connectors. AC '97 supports all Windows drivers and bus extensions. AC '97 is also backward-compatible with legacy ISA applications.

The AC '97 specification defines the baseline functionality of an analog I/O chip and the digital interface of a controller chip. The analog chip is purposefully small (48-pins) so that it can be placed near the audio input and output connectors, and away from digital buses. The larger (64-pin) digital controller chip can be located near the CPU or system bus, and is specifically dedicated to interfacing and digital processing. The two chips are connected via an AC-Link; it is a digital 5-wire, bidirectional, time-division-multiplexed (TDM) serial link that is impervious to PC electrical noise. The five wires carry the clock (12.288 MHz), a sync signal, a reset signal, and two data wires that carry sdata_out (containing the DC97 output) and sdata_in (containing the codec output). The fixed bitstream of 12.288 Mbps is divided into 256-bit frames (frame frequency is 48 kHz). Every frame is subdivided into 13 slots, from which slot 0 (16 bits) is used to specify which audio codec is communicating with the controller. The remaining 240 bits are divided into twelve 20-bit slots (slots 1–12), used as data slots. Each data slot (48 kHz, 20 bits/sample) is used to transmit a raw PCM audio signal (960 kbps). Several data slots in the same frame can be combined into a single high-quality signal (the maximum is 4 slots, obtaining a 192-kHz, 20 bit/sample, stereo signal).

The specification provides for four analog line-level stereo inputs, two analog line-level monaural inputs, 4- or 6-channel output, I^2S input port, S/PDIF output port, USB and IEEE 1394 ports, and a headphone jack. It allows digital audio to be looped through system memory where it can be processed and output to any internal or external destination. The specification uses a fixed nominal 48-kHz sampling frequency for compatibility with DVD-Video movies with surround soundtracks coded at 48 kHz; 16- and 20-bit resolution is supported. Recordings with a 44.1-kHz sampling frequency are automatically upsampled to 48 kHz. From an audio fidelity standpoint it is preferable to perform digital sample rate conversion and digital mixing at a common rate, rather than operate multiple A/D and D/A converters at different sampling rates, and perform analog mixing.

The AC '97 specification allows for the development of a wide range of chips, with many different functions, while retaining basic compatibility. For example, a baseline chip set might simply connect the computer to a basic analog input/output section. A more sophisticated chip set might perform digital mixing, filtering, compressing, expanding, reverberation, equalization, room analysis, synthesis, other DSP functions, and also provide 20-bit conversion, pseudo-balanced analog I/O, and digital interfacing to other protocols. AC '97 can be used for high-quality stereo playback, 3D audio, multiplayer gaming, and interactive music and video. AC '97-compliant PCs may contain ROM drives, DTV tuner cards, audio/video capture and playback cards, and Dolby Digital decoders. AC '97 calls for audio specifications such as a signal/noise ratio of 90 dB, frequency response from 20 Hz to 19.2 kHz (±0.25 dB), and distortion figure of 0.02%. The AC '97 specification is available via a royalty-free reciprocal license and may be downloaded from Intel's Web site.

The PC 99 specification is an Intel blueprint for PC designers. It removes audio from the ISA bus and charts a convergence path with its Entertainment PC 99 system requirements. This specification recommends USB-compliance, three IEEE 1394 ports for positional 3D audio and external D/A conversion, and an audio accelerator, and it also endorses a large-screen monitor, as well as support for DVD-Video, DBS, and DTV.

High Definition Audio (HD Audio)

The High Definition Audio (HD Audio) specification, among other improvements, specifies hardware that can play back more audio channels, and at a higher quality, than AC '97. The HD Audio specification supersedes AC '97 and is not backward-compatible with it. Link protocol and operation between the two specifications is not possible. For example, AC '97 or HD Audio codecs cannot be mixed with the same controller or on the same link. Unlike AC '97, HD Audio provides a uniform programming interface and also provides extended features. HD Audio is sometimes referred to as Azalia, its code name during development. The HD Audio specification was released by Intel Corporation in 2004.

HD Audio can support 15 input and 15 output streams simultaneously. There can be up to 16 channels per stream. The inbound link transfer rate is 24 Mbps per SDI (serial data input) signal, and the outbound rate is 48 Mbps per SDO (serial data output) signal. Sampling frequencies can range from 6 kHz to 192 kHz and sample resolution can be 8-, 16-, 20-, 24-, and 32- bits. HD Audio allows simultaneous playback of two different audio streams directed to two locations in the PC. Microphone array inputs are supported, to allow improved voice capture, for example, with noise cancellation or beam-forming. A jack retasking feature allows a computer to sense when a device is plugged into an audio jack, determine its type, and change the jack function if necessary. For example, if a microphone is plugged into a speaker jack, the computer will change the jack to function as a microphone input. The specification also supports all Dolby audio technologies.

As with AC '97, HD Audio defines the architecture, programming interfaces, and a link-frame format that are used by a host controller and a codec linked on the PCI bus. (A "codec" here refers to any device connected to the controller via the link, such as A/D and D/A converters, and does not refer to signal-processing algorithms such as an MP3 codec.) The controller is a bus-mastering I/O peripheral that is attached to system memory via a PCI or other interface. The controller implements the memory mapped registers that comprise the programming interface. The controller contains one or more DMA engines, each of which can transfer an audio stream from the codec or from the memory to the codec. A stream is a logical or virtual input or output connection that contains channels of data. For example, a simple stereo output stream contains left and right audio channels, each directed to a separate D/A converter. Each active stream must be connected through a DMA engine in the controller.

The codec extracts one or more audio streams from the link and converts them to an analog output signal through one or more converters. Likewise, a codec can accept an

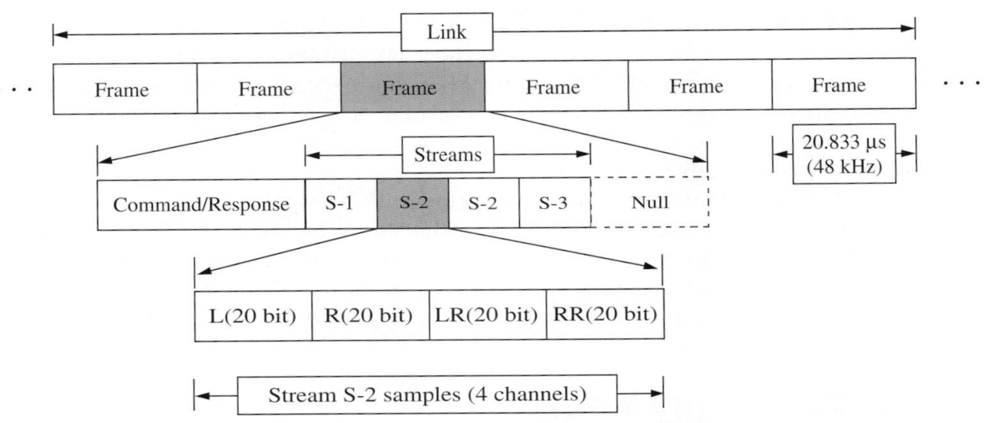

Figure 14.4 The data frame composition used in the HD Audio specification, defining how streams and channels are transferred on a link.

analog input signal, convert it to digital and transfer it as an audio stream. A codec can also deliver modem signals, or deliver unmultiplexed digital audio signals such as S/PDIF. Up to 15 codecs can be connected to the controller.

The link physically connects the controller and the codecs and conveys serial data between them. A time-multiplexed link protocol supports various sampling frequencies and bit resolutions using isochronous (no flow control) transport with a fixed data transfer rate. Generally, all channels in a stream must have the same sampling frequency and bit resolution. Signals on the link are transmitted as a series of data packets called frames. Each next frame occurs at 20.833 µs, corresponding to a 48-kHz sampling frequency, as shown in Fig. 14.4.

Each frame contains command information and as many stream sample blocks as needed. The total number of streams that can be supported is limited by the content of all the streams; unused capacity is filled with null data. Frames occur at a fixed rate of 48 kHz so if a stream has a sampling frequency less than or more than 48 kHz, there is less than or more than one sample block in each frame (multiple blocks are transmitted at one time in the packet). For example (see Fig. 14.4), some frames contain two sample blocks (S2) and some frames contain none. A single stream sample block can contain one sample for each of the multiple channels in the stream. For example (see Fig. 14.4), the illustrated S2 stream contains four channels (L, R, LR, and RR) and each channel has 20-bit samples; the stream thus uses 80 bits per sample block. This stream has a 96-kHz sampling frequency since two sample blocks are conveyed per 20.833 µs frame.

Samples are packed in containers that are 8, 16, or 32 bits wide; the smallest container size which will fit the sample is used. For example, 24-bit samples are placed in 32-bit containers; samples are padded with zeros at the LSB to left-justify the sample in the container. A block contains sets of samples to be played at a point in time. A block size equals the container size multiplied by the number of channels; for example, a 24-bit, 3-channel, 96-kHz stream has a block size of 12 bytes. The same stream has a packet size of 24 bytes.

The specification allows considerable flexibility in system design and application. For example, a stereo analog signal might be input through an A/D converter for internal recording at 44.1 kHz and 16 bits; a 7.1-channel signal is output from a DVD disc at 96 kHz and 24 bits; a stereo 22-kHz, 16-bit signal is mixed into the front two channels of the 7.1-channel playback; a connected headset plays a communications stream from a net meeting—all functions occurring simultaneously. Also, the sequential order of processing can be specified; for example, a compressor/limiter can be placed before equalization, which will yield a different result from the reverse order.

Implementations of controllers and codecs are available from numerous companies. HD Audio is supported by a Universal Audio Architecture (UAA) class driver in Microsoft Windows XP SP3 and in Windows Vista; the AppleHDA driver is included in Mac OS X; Linux and other open operating systems support HD Audio. The HD Audio specification is available for download from the Intel Web site.

Windows DirectX API

The DOS programming environment provided simple and low-level access to functions, allowing full implementation of audio and video features. The Windows operating system added considerable complexity. The Windows Multimedia API allowed access to sound-card functionality. However, developers could not access peripherals directly—they were limited to whatever access functions Windows provided. For example, the Multimedia API provides no direct means to mix audio files.

Microsoft's DirectX API suite was designed overcome these kinds of limitations, and promote high-performance multimedia application development in Windows. DirectX APIs effectively provide real-time, low-level access to peripherals specifically used in intensive audio/video applications. DirectX APIs divide multimedia tasks into components including DirectSound, DirectSound3D, DirectMusic, DirectShow, DirectDraw, Direct-Play, DirectInput, and DirectSetup.

DirectSound provides device-independent access to audio accelerator hardware. It provides functions for mixing audio files and controlling each file's volume, balance, and playback rate within the mix. This allows real-time mixing of audio streams as well as control over effects like panning. DirectSound also provides low-latency playback so that sounds can be synchronized with other multimedia events.

The DirectSound3D API is an extension to DirectSound. It is a set of functions that allow application programmers to add 3D audio effects to the audio content, imparting 3D sound over two speakers or headphones. The programmer can establish the 3D coordinates (x, y, z) of both the listener and sound source. It does not assume that the listener's central axis is in the center of the screen. The DirectSound3D API allows processing to be done natively on the local CPU, or on an expansion card's hardware DSP chip. In this way, the maximum number of applications can employ 3D audio, with appropriate degrees of processing overhead depending on the resources available.

DirectMusic is an extension to DirectX that provides wavetable synthesis with support for Downloadable Sounds (DLS), interactive music composition, and DLS and other authoring tools. DLS is an extension to the MIDI specification that defines a file format, device architecture, and API. DLS lets synthesizer developers add custom wavetables to the General MIDI sounds already stored in a sound card's ROM. Using system memory, DLS-compatible devices can automatically download sounds from discs, the Internet, or other sources. DirectMusic also provides a music composition engine that lets developers specify the style and characteristics of a musical accompaniment, and also change these parameters in terms of tempo or voicing.

DirectShow (Version 5.2 and later) supports DVD decoders and DVD applications. It demultiplexes the MPEG-2 bitstream from the disc so that audio, video, sub-picture and other decoding can be performed in real time with dedicated hardware or software means; in either case, the interface is the same. DirectShow supports aspects of DVD playback such as navigation, regional management, and exchange of CSS encrypted data.

Vendors provide DirectX drivers in addition to their standard Windows drivers. For example, a SoundBlaster DirectX driver provides access to fast SRAM on a SoundBlaster card, accelerating audio functionality. If a vendor does not supply a DirectX driver, DirectX provides an emulated driver. Applications may use this driver when audio acceleration hardware is not available. Although the emulated driver is slower, the developer retains access to the enhanced functionality. DirectX thus gives developers access to low-level hardware functions.

MMX

Companies have developed single-chip solutions to relieve the central processor of audio computation burdens. However, simultaneously, processors have become more adept at performing multimedia calculations. For example, the Intel Multimedia Extensions (MMX) instruction set contained in Pentium processors is expressly designed to accelerate graphics, video, and audio signal processing. Among other attributes, these 57 instructions allow a Pentium processor to simultaneously move and process eight bytes—seven more than previous Pentiums. In particular, this capability is called Single Instruction, Multiple Data (SIMD) and is useful for processing complex and multiple audio streams. MMX instructions also double the onboard L1 memory cache to 32 kbytes and provide other speed advantages. Using Intel's benchmarks, media software written for MMX will run 40 to 66% faster for some tasks. This efficiency allows faster execution and frees other system resources for still more sophisticated processing. Some Intel MMX processors will play DVD-Video movies and decode their Dolby Digital soundtracks. However, software-based processing on the host CPU has its limitations. If a 500-MHz processor devotes half its power to processing surround sound, wavetable synthesis, and video decoding, it effectively becomes a 250-MHz processor for other simultaneous applications.

File Formats

Interfaces such as AES3 convey digital audio data in real time. In other applications, transfer is not in real time (it can be faster or slower). Defined file formats are needed to transfer essence (content data such as audio) along with metadata (nonaudio data such as edit lists). In this way, for example, one creator or many collaborators can author projects with an efficient workflow. Moreover, essence can be transferred from one platform to another. In still other applications, file formats are specifically designed to allow streaming. In a multimedia environment, audio, video, and other data is intermingled.

Media content data such as audio, video, still pictures, graphics, and text is sometimes known as essence. Other related data can be considered as data describing data, and is called metadata. Metadata can hold parameters (such as sampling frequency, downmixing, and number of channels) that describe how to decode essence, can be used to search for essence, and can contain intellectual property information such as copyright and ownership needed to access essence. Metadata can also describe how to assemble different elements (this metadata is sometimes called a composition), and provides information on synchronization.

In some cases, audio data is stored as a raw data, headerless sound file that contains only amplitude samples. However, in most cases, dedicated file formats are used to provide compatibility between computer platforms so that essence can be stored, then transmitted or otherwise moved to other systems, and be compatibly processed or replayed. In addition to audio (or video) data, many file formats contain an introductory header with metadata such as the file's sampling frequency, bit resolution, number of channels, and type of compression (if any), title, copyright, and other information. Some file formats also contain other metadata. For example, a file can contain an edit decision list with timecode and crossfade information, as well as equalization data. Macintosh files use a two-part structure with a data fork and resource fork; audio can be stored in either mode. Many software programs can read raw or coded files and convert them into other formats. Some popular file formats include WAV, AIFF, SDII, QuickTime, JPEG, MPEG, and OMFI.

WAV and BWF

The Waveform Audio (WAV) file format (.wav extension) was introduced in Windows 3.1 as the format for multimedia sound. The WAV file interchange format is described in the Microsoft/IBM Multimedia Programming Interface and Data Specifications document. The WAV format is the most common type of file adhering to the Resource Interchange File Format (RIFF) specification; it is sometimes called RIFF WAV. WAV is widely used for uncompressed 8-, 12-, and 16-bit audio files, both monaural and multichannel, at a variety of sampling frequencies. RIFF files organize blocks of data into sections called chunks. The RIFF chunk at the beginning of a file identifies the file as a WAV file and describes its length. The Format chunk describes sampling frequency, word length, and other parameters. Applications might use information contained in chunks, or ignore it entirely. The Data chunk contains the amplitude sample values. PCM or non-PCM audio formats can be stored in a WAV file. For example, a format specific field can hold parameters used to specify a data-compressed file such as Dolby Digital or MPEG data. The format chunk is extended to include the additional content, and a "cbSize" descriptor is included along with data describing the extended format. Eight-bit bytes are represented in unsigned integer format and 16-bit bytes are represented in signed two's complement format.

Figure 14.5 shows one cycle of a monaural 1-kHz square wave recorded at 44.1 kHz and 16 bits and stored as a WAV file. The left-hand field represents the file in hexadecimal form and the right-hand side in ASCII form. The first 12 bytes (each pair of hex numbers is a byte) (52 49 46 46) represent the ASCII characters for RIFF, the file type. The next four bytes (CE 00 00 00) represent the number of bytes of data in the remainder of the file (excluding the first eight header bytes). This field is expressed in "little-endian" form (the term taken from *Gulliver's Travels* and the question of which end of a soft-boiled egg should be cracked first). In this case, the least significant bytes are listed first. (Big-endian represents most significant bytes first.) The last four bytes of this chunk (57 41 56 45) identify this RIFF file as a WAV file. The next 24 bytes is the Format chunk holding information describing the file. For example, 44 AC 00 00 identifies the sampling frequency as 44,100 Hz. The Data chunk (following the ASCII word "data") describes the chunk length, and contains the square-wave data itself, stored in little-endian format. This file concludes with an Info chunk with various text information.

```
 0: 52 49 46 46 CE 00 00 00  57 41 56 45 66 6D 74 20  RIFFÎ...WAVEfmt
10: 10 00 00 00 01 00 01 00  44 AC 00 00 88 58 01 00  .......D¬..^X..
20: 02 00 10 00 64 61 74 61  5A 00 00 00 FF 7F FF 7F  ....dataZ...ÿ▒ÿ▒
30: FF 7F FF 7F FF 7F FF 7F  FF 7F FF 7F FF 7F FF 7F  ÿ▒ÿ▒ÿ▒ÿ▒ÿ▒ÿ▒ÿ▒ÿ▒
40: FF 7F FF 7F FF 7F FF 7F  FF 7F FF 7F FF 7F FF 7F  ÿ▒ÿ▒ÿ▒ÿ▒ÿ▒ÿ▒ÿ▒ÿ▒
50: FF 7F FF 7F FF 7F FF 7F  01 80 01 80 01 80 01 80  ÿ▒ÿ▒ÿ▒ÿ▒. . . .
60: 01 80 01 80 01 80 01 80  01 80 01 80 01 80 01 80  . . . . . . . .
70: 01 80 01 80 01 80 01 80  01 80 01 80 01 80 01 80  . . . . . . . .
80: 01 80 01 80 FF 7F 4C 49  53 54 48 00 00 00 49 4E  . . ÿ▒LISTH...IN
90: 46 4F 49 43 52 44 0B 00  00 00 32 30 30 34 2D 30  FOICRD....2004-0
A0: 36 2D 31 37 00 00 49 45  4E 47 0F 00 00 00 4B 65  6-17..IENG....Ke
B0: 6E 20 43 20 50 6F 68 6C  6D 61 6E 6E 00 00 49 53  n C Pohlmann..IS
C0: 46 54 10 00 00 00 53 6F  75 6E 64 20 46 6F 72 67  FT....Sound Forg
D0: 65 20 34 2E 30 00                                  e 4.0.
```

FIGURE 14.5 One cycle of a 1-kHz square wave recorded at 44.1 kHz and saved as a WAV file. The hexadecimal (left-hand field) and ASCII (right-hand field) representations are shown.

The Broadcast Wave Format (BWF) audio file format is an open-source format based on the WAV format. It was developed by the European Broadcasting Union (EBU) and is described in the EBU Tech. 3285 document. BWF files may be considered as WAV files with additional restrictions and additions. BWF uses an additional "broadcast audio extension" header chunk to define the audio data's format, and contains information on the sound sequence, originator/producer, creation date, a timecode reference, and other data, as shown in Table 14.1. Each file is time-stamped using a 64-bit value. Thus, one

Data	Size (bytes)	Description
ckID	4	Chunk ID = bext
ckSize	4	Size of chunk
Description	256	Description of the sound clip
Originator	32	Name of the originator
OriginatorReference	32	Unique identifier of originator (issued by EBU)
OriginationDate	10	yyyy-mm-dd
OriginationTime	8	hh-mm-ss
TimeReferenceLow	4	Low byte of first sample count since midnight
TimeReferenceHigh	4	High byte of first sample count since midnight
Version	2	BWF version number, e.g., &0001 is Version 1
UMID	64	UMID according to SMPTE 330M. If only 32-byte UMID, then second half should be padded with zeros
Reserved	190	Reserved for extensions. Set to zero in Version 1
CodingHistory	Unrestricted	Series of ASCII strings, each terminated by CR/LF (carriage return, line feed) describing each stage of audio coding history, according to EBU R-98

TABLE 14.1 BWF extension chunk format.

advantage of BWF over WAV is that BWF files can be time-stamped with sample accuracy. Applications can read the time stamp and places files in a specific order or at a specific location, without a need for an edit decision list. The BWF specification calls for a 48-kHz sampling frequency and at least a 16-bit PCM word length. Multichannel MPEG-2 data is supported. It defines parameters such as surround format, downmix coefficients, and channel ordering; multichannel data is written as multiple monaural channels, and not interleaved. BWF files use the same .wav extension as WAV files, and BWF files can be played by any system capable of playing a WAV file (they ignore the additional chunks). However, a BWF file can contain either PCM or MPEG-2 data. Some players will not play MPEG files with a .wav extension. The AES46 standard describes a radio traffic audio delivery extension to the BWF file format. It defines a CART chunk that describes cartridge labels for automated playback systems; information such as timing, cuing, and level reference can be placed in the label.

MP3, AIFF, QuickTime, and Other File Formats

As noted, MPEG audio data can be placed in AIFF-C, WAV, and BWF file formats using appropriate headers. However, in many cases, MP3 files are used in raw form with the .mp3 extension. In this case, data is represented as a sequence of MPEG audio frames; each frame is preceded by a 32-bit header starting with a unique synchronization pattern of eleven "1s." These files can be deciphered and playback can initiate from any point in the file by locating the start of the next frame.

The ID3 tag is present in many MP3 files; it can include data such as title, artist, and genre information. This tag is usually located in the last 128 bits of the file. Since it does not begin with an audio-frame synchronization pattern, audio decoders do not play this nonaudio data. An example of an ID3 tag data is shown in Table 14.2. The ID3v2.2 tag is a more complex tag structure.

The AU (.au) file format was developed for the Unix platform, but is also used on other systems. It supports a variety of linear audio types as well as compressed files with ADPCM or μ law.

Sign	Length (bytes)	Description
A	3	Tag identification; normally ASCII TAG
B	30	Title
C	30	Artist
D	30	Album
E	4	Year
F	30	Comment string (only 28 bytes followed by "\0" in some versions)
G	1	This may represent track number or may be part of comment string
H	1	Genre

TABLE 14.2 Example of data found in an ID3 tag structure.

The Voice VOC (.voc) file format is used in some sound cards. VOC defines eight block types that can vary in length; sampling frequency and word length are specified. Sound quality up to 16-bit stereo is supported, along with compressed formats. VOC files can contain markers for looping, synchronization markers for multimedia applications, and silence markers.

The AIFF (.aiff or .aif) Audio Interchange File Format is native to Macintosh computers, and is also used in PC and other computer types. AIFF is based on the EA IFF 85 standard. AIFF supports many types of uncompressed data with a variety of channels, sampling frequencies, and word lengths. The format contains information on the number of interleaved channels, sample size, and sampling frequency, as well as the raw audio data. As in WAV files, AIFF files store data in chunks. The FORM chunk identifies the file format, COMM contains the format parameter information, and SSND contains audio data. A big-endian format is used. Markers can be placed anywhere in a file, for example, to mark edit points. However, markers can be used for any purpose, as defined by the application software.

The AIFF format is used for some Macintosh sound files and is recognized by numerous software editing systems. Because sound files are stored together with other parameters, it is difficult to add data, for example, as in a multitrack overdub, without writing a new file. The AIFF-C (compressed) (or AIFC) file format is an AIFF version that allows for compressed audio data. Several types of compression are used in Macintosh applications including Macintosh Audio Compression/Expansion (MACE), IMA/ADPCM, and μ law. AIFF is defined in the context of the EA IFF 85 standard.

The Sound Designer II (SDII or SD2) file format (.sd2) was developed by Digi-Design. It is the successor to the Sound Designer I file format, originally developed for their Macintosh-based recording and editing systems. Audio data is stored separately from file parameters. SDII stores monaural or stereo data. For the latter, tracks are interleaved as left/right samples; samples are stored as signed values. In some applications, left and right channels are stored in separate files; .L and .R suffixes are used. The format contains information on the file's sampling frequency, word length and data sizes. Parameters are stored in three STR fields. SDII is used to store and transfer files used in editing applications, and to move data between Macintosh and PC platforms.

QuickTime is a file format (.mov) and multimedia extension to the Macintosh operating system. It is cross-platform and can be used to play videos on most computers. More generally, time-based files, including audio, animation, and MIDI can be stored, synchronized and controlled, and replayed. Because of the timebase inherent in a video program, the video itself can be used to control preset actions. QuickTime movies can have multiple audio tracks. For example, different language soundtracks can accompany a video. In addition, audio-only QuickTime movies may be authored. Videos can be played at 15 fps or 30 fps. However, frame rate, along with picture size and resolution, may be limited by hard-disk data transfer rates. QuickTime itself does not define a video compression method; for example, a Sorenson Video codec might be used. Audio codecs can include those in MPEG-1, MPEG-2 AAC, or Apple Lossless Encoder. Using the QuickTime file format, MPEG-4 audio and video codecs can be combined with other QuickTime-compatible technologies such as Macromedia Flash. Applications that use the Export component of QuickTime can create MPEG-4 compatible files. Any .mp4 file containing compliant MPEG-4 video and AAC audio should be compatible with QuickTime.

Hardware and software tools allow users to record video clips to a hard disk, trim extraneous material, compress video, edit video, add audio tracks, then play the result as a QuickTime movie. In some cases, the software can be used as an off-line video editor, and used to create an edit decision list with timecode, or the finished product can be output directly. Audio files with 16-bit, 44.1-kHz quality can be inserted in QuickTime movies. QuickTime also accepts MIDI data for playback. QuickTime can also be used to stream media files or live events in real time. QuickTime also supports iTunes and other applications. The 3GPP and 3GPP2 standards are used to create, deliver, and play back bandwidth-intensive multimedia over 3G wireless mobile cellular networks such as GSM. The standards are based on the QuickTime file format and contain MPEG-4 and H.263 video, AAC and AMR audio, and 3G text; 3GPP2 can also use QCELP audio. QuickTime software, or other software that uses QuickTime exporters, can be used to create and play back 3GPP and 3GPP2 content.

As noted, QuickTime is also available for Windows so that presentations developed on a Macintosh can be played on a PC. The Audio Video Interleaved (AVI) format is similar to QuickTime, but is only used on Windows computers.

The RealAudio (.ra or .ram) file format is designed to play music in real time over the Internet. It was introduced by RealNetworks in 1995 (then known as Progressive-Networks). Both 8- and 16-bit audio is supported at a variety of bit rates; the compression algorithm is optimized for different modem speeds. RA files can be created from WAV, AU, or other files or generated in real time. Streaming technology is described in Chap. 15.

The JPEG (Joint Photographic Experts Group) lossy video compression format is used primarily to reduce the size of still image files. Compression ratios of 20:1 to 30:1 can be achieved with little loss of quality, and much higher ratios are possible. Motion JPEG (MJPEG) can be used to store a series of data-reduced frames comprising motion video. This is often used in video editors where individual frame quality is needed. Many proprietary JPEG formats are in use. The MPEG (Moving Picture Experts Group) lossy video compression methods are used primarily for motion video with accompanying audio. Some frames are stored with great resolution, then intervening frames are stored as differences between frames; video compression ratios of 200:1 are possible. MPEG also defines a number of compressed audio formats such as MP3, MPEG AAC, and others; MPEG audio is discussed in more detail in Chap. 11. MPEG video is discussed in Chap. 16.

Open Media Framework Interchange (OMFI)

In a perfect world, audio and video projects produced on workstations from different manufacturers could be interchanged between platforms with complete compatibility. That kind of common cross-platform interchange language was the goal of the Open Media Framework Interchange (OMFI). OMFI is a set of file format standards for audio, text, still graphics, images, animation, and video files. In addition, it defines editing, mixing, and processing notation so that both content and description of edited audio and video programs can be interchanged. The format also contains information identifying the sources of the media as well as sampling and timecode information, and accommodates both compressed and uncompressed files. Files can be created in one format, interchanged to another platform for editing and signal processing, and then returned to the original format without loss of information. In other words, an OMFI file contains all the information needed to create, edit, and play digital media presentations. In most cases, files in a native format are converted to the OMFI format, interchanged via direct transmission or removable physical media, and then converted to the new native format. However, to help streamline operation, OMFI is structured to facilitate playback directly from an interchanged file when the playback platform has similar characteristics as the source platform. To operate efficiently with large files, OMFI is able to identify and extract specific objects of information such as media source information, without reading the entire file. In addition, a file can be incrementally changed without requiring a complete recalculation and rewriting of the entire file.

OMFI uses two basic types of information. "Compositions" are descriptions of all the data required to play or edit a presentation. Compositions do not contain media data, but point to them and provide coordinated operation using methods such as timecode-based edit decision lists, source/destination labels, and crossfade times. "Physical sources" contain the actual media data such as audio and video, as well as identification of the sources used in the composition. Data structures called media objects (or mobs) are used to identify compositions and sources. An OMFI file contains objects—information that other data can reference. For example, a composition mob is an object that contains information describing the composition; an object's data is called its values. An applications programming interface (API) is used to access object values and translate proprietary file formats into OMFI-compatible files.

OMFI allows file transfer such as via disc or transmission on a network. Common file formats included in the OMFI format are TIFF (including RGB, JPEG, and YCC) for video and graphics, and AIFC and WAV for audio. Clearly, a common transmission method must be used to link stations; for example, Ethernet, FDDI, ATM, or TCP/IP could be used. OMFI can also be used in a client-server system that allows multi-user real-time access from the server to the client.

OMFI was migrated to Microsoft's Structured Storage container format to form the core of the Advanced Authoring Format (AAF). AAF also employs Microsoft's Component Object Model (COM), an inter-application communication protocol supported by most popular programming languages including C++ and Java. The Bento format, developed by Apple, links elements of a project within an overall container; it is used with OMFI. Both OMFI Version 1 and OMFI Version 2 are used. They differ to the extent that they can be incompatible. OMFI was developed by Avid Corporation.

Advanced Authoring Format (AAF)

The Advanced Authoring Format (AAF) is a successor to the OMFI specification. Introduced in 1998, AAF is an open-source interchange protocol for professional multimedia post-production and authoring applications (not delivery). Essence and metadata can be exchanged between multiple users and facilities, diverse platforms, systems, and applications. It defines the relationship between content elements, maps elements to a timeline, synchronizes content streams, describes processing, tracks the history of the file, and can reference external essence not in the file. For example, it can convey complex project structures of audio, video, graphics, and animation that enable sample-accurate editing from multiple sources, along with the compositional information needed to render the materials into finished content. For example, an AAF file might contain 60 minutes of video, 100 still images, 20 minutes of audio, and references to external content, along with instructions to process this essence in a certain way and combine them into a 5-minute presentation. AAF also uses variable key length encoding; chunks can be appended to a file and be read only by those with the proper key.

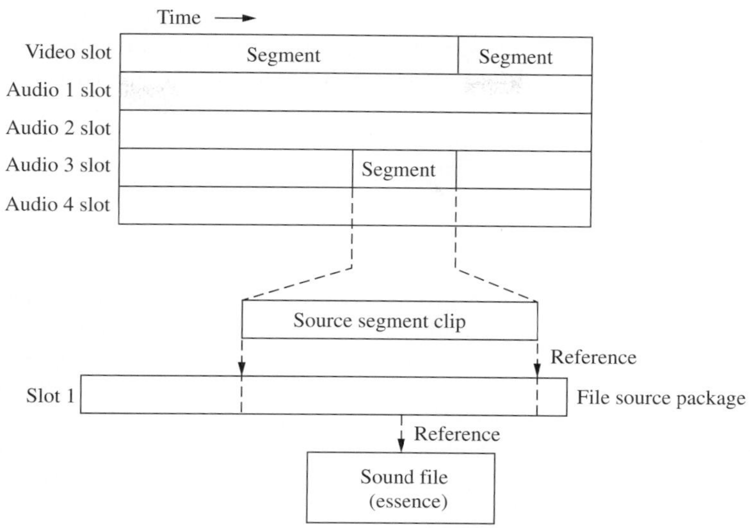

FIGURE 14.6 An example of how audio post-production metadata might appear in an AAF file.

Those without the key skip the chunk and read the rest of the file. In this way, different manufacturers can tailor the file for specific needs while still adhering to the format, and new metadata can be added to the standard as well.

Considerable flexibility is possible. For example, a composition package has slots that may be considered as independent tracks, as shown in Fig. 14.6. Each slot can describe one kind of essence. Slots can be time-independent, can use a timecode reference, or be event-based. Within a slot, a segment might hold a source clip. That clip might refer to a section of another slot in another essence package, which in turn refers to an external sound file.

AAF uses a Structured Storage format implemented in a C++ environment. As an AAF file is edited, its version control feature can keep a record of the revisions. The AAF Association was founded by Avid and Microsoft.

Material eXchange Format (MXF)

The Material eXchange Format (MXF) is a media format for the exchange of program material. It is used in professional post-production and authoring applications. It provides a simplified standard container format for essence and metadata, and is platform-independent. It is closely related to AAF. Files comprise a sequence of frames where each frame holds audio, video, and data content along with frame-based metadata. MXF files can contain any particular media format such as MEG, WAV, AVI, or DPX; and MPX files may be associated with AAF projects. MXF and AAF use the same internal data structure, and both use the SMPTE Metadata Dictionary to define data and workflow information. However, MXF promotes a simpler metadata that is compatible with AAF. MXF files can be read by an AAF workstation, for example, and be integrated into a project. An AAF file might comprise only metadata, and simply reference the essence in MXF files.

MXF's data structure offers a partial retrieve feature so that users can retrieve only sections of a file that are pertinent to them without copying the entire file. Because essence is placed in a temporal streaming format, data can be delivered in real time, or conveyed with conventional file-transfer operations. Because data is sequentially arranged, MXF is often used for finished projects and for writing to media. MXF was developed by the Pro-MPEG Forum, EBU, and SMPTE.

AES31

The AES31 specification is a group of non-proprietary specifications used to interchange audio data and project information between devices such as workstations and recorders. It allows simple exchange of one audio file or exchanges of complex files with editing information from different devices. There are four independent stages with interchange options. AES31-1 defines the physical data transport describing how files can be moved via removable media or high-speed network. It is compatible with the Microsoft FAT32 disk filing structure. AES31-2 defines an audio file format, describing how BWF data chunks should be placed on a storage media or packaged for network transfer. AES31-3 defines a simple project structure using a sample-accurate audio decision list (ADL) that can contain information such as levels and anti-click and creative crossfades, and allows files to be played back in synchronization; files use the .adl extension. It uses a URL to identify a file source, whether residing locally or on a network. It also provides interchange between NTSC and PAL formats. For user convenience, edit lists are conveyed with human-readable ASCII characters; the Edit Decision Markup Language (EDML) is used. AES31-4 defines an object-oriented project structure capable of describing a variety of characteristics. In AES31, a universal resource locator can access files on different media or platforms; the locater specifies the file, host, disk volume, directory, subdirectories, and the file name with a .wav extension.

Digital Audio Extraction

In digital audio extraction (DAE), music data is copied from a CD with direct digital means, without conversion to an analog signal, using a CD- or DVD-ROM drive to create a file (such as a WAV file) on the host computer. The copy is ostensibly an exact bit-for-bit copy of the original data—a clone. Performing DAE is not as easy as copying a file from a ROM disc. Red Book data is stored as tracks, whereas Yellow Book and DVD data is stored as files. CD-ROM (and other computer data) is formatted in sectors with unique addresses that inherently provide incremental reading and copying. Red Book does not provide this. Its track format supposes that the data spiral will be mainly read continuously, as when normally listening to music, thus there is no addressing provision. With DAE, the system reads this spiral discontinuously, and must piece together a continuous file. Moreover, when writing a CD-R or CD-RW disc, the data sent to the recorder must be absolutely continuous. If there is an interruption at the writing laser, the recording is unusable. One application of DAE is the illegal copying of music. DAE is often known as "ripping," which also generally refers to the copying of music into files.

CD-ROM (and DVD-ROM) drives are capable of digitally transferring information from a CD-ROM disc to a host computer. The transfer is made via an adapter or interface connected to the drive. CD-ROM drives can also play back CD-Audio discs by reading the data through audio converters and analog output circuits, as in regular CD-Audio players. Users may listen to the audio CD through the computer sound system or a CD-ROM headphone jack. With DAE software, a CD-ROM drive may also deliver this digital audio data to a computer. In this case, the computer requests audio data from the CD-ROM drive as it would request computer data. Because of the differences between the CD-ROM and CD-Audio file formats, DAE requires special logic, either in hardware, firmware (a programmable chip in the data path) or a software program on the host. This logic is needed to correct synchronization errors that would result when a CD-Audio disc is read like a CD-ROM disc.

As described in Chap. 7, CD-Audio data is organized in frames, each with 24 bytes of audio information. There is no unique identifier (address) for individual frames. Each frame also contains eight subcode bits. These are collected over 98 frames to form eight channels, each 98 bits in length. Although Red Book CD players do not do this, computers can collect audio bytes over 98 frames, yielding a block of 2352 audio bytes. In real playing time, this 98-frame block comprises 1/75th of a second. In computer parlance, this block is sometimes called a sector or raw sector. For CD-Audio, there is no distinction between a sector and a raw sector because all 2352 bytes of data are "actual" audio data. The subcode contains the time in minutes:seconds:frames (called MSF time) of the subcode block. This MSF time is displayed by CD-Audio players as track time.

In the CD-Audio player, audio data and subcode are separated and follow different processing paths. When a computer requests and receives audio data, it is not accompanied by any time or address data. Moreover, if the computer were to separately request subcode data, it would not be known which group of 98 frames the time/address refers to, and in any case the time/address cannot be placed in the 2352-byte sector, which is occupied by audio data. In contrast, CD-ROM data is organized in blocks, each with 2352 bytes; 2048 of these bytes contain the user data (CD-ROM Mode 1). The remaining bytes are used for error correction and to uniquely identify that block of data with an address; this is called a sector number. A 2352-byte block is called a raw sector. The 2048 bytes of data in the raw sector are called a sector. Importantly, the block's data and its address are intrinsically combined. When a computer requests and receives a block of CD-ROM data, both the data and its disc address are present and can be placed in a 2352-byte sector. In this way, the computer can identify the contents and position of data read from a CD-ROM disc.

In the CD-ROM format the unique identifier or address is carried as part of the block itself. The CD-ROM drive must be able to locate specific blocks of data because of the way operating systems implement their File I/O Services. (File I/O Services are also used to store data on hard disks.) As files are written and deleted, the disk becomes fragmented. The File I/O Services may distribute one large file across the disc to fill in these gaps. The file format used in CD-ROM discs was designed to mimic the way hard-disk drives store data—in sectors. The storage medium must therefore provide a way to randomly access any sector. By including a unique address in each block, CD-ROM drives can provide this random access capability.

CD-Audio discs do not need addresses because the format was designed to deliver continuous streams of music data. A 4-minute song comprises about 42 million bytes of data. Attaching unique addresses to every 24-byte frame would needlessly waste space. On the other hand, some type of random access is required, so the user can skip to a different song or fast-forward through a song. However, these accesses do not need to be completely precise. This is where the MSF subcode time is used. A CD-Audio player finds a song, or a given time on the disc, by reading the subcode and placing the laser pickup close to the accompanying block that contains the target MSF in its subcode. However, because each MSF time is only accurate to 1/75th of a second, the resolution is only 98 frames. A CD-Audio player can only locate data within a block 1/75th of a second prior to the target MSF time.

Because CD-ROM drives evolved from CD-Audio players, they incur the same problem. If asked to read a certain sector on a CD-ROM disc, they can only place the pickup just before the target sector, not right on it. However, since the CD-ROM blocks are addressable (uniquely identifiable) the CD-ROM drive can read through the partial sector where it starts, and wait until the correct sector appears. When the correct sector starts, it will begin streaming data into its buffer. This method is reliable because each CD-ROM block is intrinsically addressed; the address of each data block is known when the block is received.

CD-ROM drives can also mimic sector reading from a CD-Audio disc. CD-ROM drives respond to two types of read commands: Read Sector and Read Raw Sector. Read Sector commands will transfer only the 2048 actual data bytes from each raw sector read. In normal CD-ROM operations, the Read Sector command is typically used. If needed, an application may issue a Read Raw Sector command, which will transfer all 2352 bytes from each raw sector read. The application might be concerned with the CD-ROM addresses or error correction.

By reading 98 CD-Audio frames to form a 2352-byte block, the computer can issue a Read Raw Sector command to a CD-Audio disc in a CD-ROM drive. This works, but not perfectly. The flow, as documented by Will Pirkle, works like this:

1. The host program sends a Read Raw Sector command to read a sector (e.g., #1234) of an audio disc in a CD-ROM drive.
2. The CD-ROM drive converts the sector number into an approximate MSF time.
3. The CD-ROM searches the subcode until it locates the target MSF, and it moves the pickup to within 1/75th of a second before the correct accompanying block (sector) and somewhere inside the previous block (sector #1233).
4. The CD-ROM begins reading data. It loads the data into its buffer for output to the host computer. It does not know the address of each received block, so it cannot locate the next sector properly. The CD-ROM drive reads out exactly 2352 bytes and then stops, somewhere inside sector #1234.

In this example, the program read one block's worth of data (2352 bytes); however, it did not get the correct (or one complete) block. It read part of the previous block and part of the correct block. This is the crux of the DAE problem.

In DAE, large data files are moved from the ROM drive to either a CD/DVD recorder or a hard disk. Large-scale data movement cannot happen continuously in the PC environment. The flow of data is intermittent and intermediate buffers (memory storage locations) are filled and emptied to simulate continuous transfer. Thus, the ROM drive will read sectors in short bursts, with gaps of time between the data transfers. This exacerbates the central DAE problem because the ROM drive must start and stop its reading mechanism millions of times during the transfer of one 4-minute song. The DAE software must sort the partial blocks, which will occur at the beginning and end of each burst of data, and piece the data together, not repeating redundant data, or skipping data. Any device performing DAE, be it the ROM drive, the adapter, the driver, or host software, faces the same problem of dealing with the overlapping of incorrect data.

The erroneous data read from the beginning and end of the data buffers is a result of synchronization error. If left uncorrected, these synchronization errors manifest themselves as periodic clicks, pops, or noise bursts in the extracted music. Clearly, DAE errors must be corrected by matching known good data to questionable data. This matching procedure is essential in any DAE system. ROM drives may also utilize high-accuracy motors to try to place the laser pickup as close as possible to a target CD-Audio sector. Even highly accurate motors cannot guarantee perfect sector reads. Some type of overlap and compare method must be employed.

This overlap technique has no other function except to perform DAE. A ROM drive, adapter, or host software that implements DAE must apply several engineering methods to properly correct the synchronization errors that occur when a CD-Audio disc is accessed like a CD-ROM disc. Only then may these accesses transfer correct digital audio data from a given CD-Audio source. The ATAPI specification defines drive interfacing. The command set in newer revisions of the specification provides better head-positioning and supports error-correction and subcode functions and thus simplifies software requirements and expedites extraction.

Flash Memory

Portable flash memory offers a small and robust way to store data to nonvolatile solid-state memory (an electrically erasable programmable read-only memory or EEPROM), via an onboard controller. A variety of flash memory formats have been developed, each with a different form factor. These formats include the Compact Flash, Memory Stick, SD, and others. Flash memories on cards are designed to directly plug into a receiving socket. Other flash memory devices interface via a USB or other means. Flash memory technology allows high storage capacity of many gigabytes using semiconductor traces of 130-nm thickness, or as thin as 80 nm. Data transfer rates of 20 Mbps and higher are available. Some cards also incorporate WiFi or Bluetooth for wireless data exchange and communications. Flash memory operates with low power consumption and does not generate heat.

The Compact Flash format uses Type I and Type II form factors. Type II is larger than Type I and requires a wider slot; Type I cards can use Type II slots. The SD format contains a unique, encrypted key, using the Content Protection for Recordable Media (CPRM) standard. The SDIO (SD Input/Output) specification uses the standard SD socket to also allow portable devices to connect to peripherals and accessories. The MultiMediaCard format is a similar precursor to SD, and can be read in many SD devices. Memory Stick media is available in unencrypted Standard and encrypted MagicGate forms. Memory Stick PRO provides encryption and data loss protection features. Specifications for some flash memory formats are given in Table 14.3.

	Compact Flash	SD Memory	Smart-Media	Multi-Media	Memory Stick
Size (mm)	42.8 × 36.4 × 3.3	24 × 32 × 2.1	37 × 45 × 0.76	24 × 32 × 1.4	21.5 × 50 × 2.8
Weight (g)	11.4	2	2	1.5	4
Number of pins	50	9	22	7	10
Copyright protection	ID	SDMI	ID	ID	SDMI (MagicGate)
Developer	SanDisk (1994)	Matsushita, Toshiba, SanDisk (2000)	Toshiba (1995)	Siemens, SanDisk (1997)	Sony (1998)

Table 14.3 Specifications for removable flash memory formats.

Hard-Disk Drives

Both personal computers and dedicated audio workstations are widely used for audio production. From a hardware standpoint, the key to both systems is the ability to store audio data on magnetic hard-disk drives. Hard-disk drives offer fast and random access, fast data transfer, high storage capacity, and low cost. A hard-disk-based audio system consolidates storage and editing features and allows an efficient approach to production needs. Optical disc storage has many advantages, particularly for static data storage. But magnetic hard-disk drives are superior for the dynamic workflow of audio production.

Magnetic Recording

Magnetic media is composed of a substrate coated with a thin layer of magnetic material, such as gamma ferric oxide (Fe_2O_3). This material is composed of particles that are acicular (cigar-shaped). Each particle exhibits a permanent magnetic pole structure that produces a constant magnetic field. The orientation of the magnetic field can be switched back and forth. When a media is unrecorded, the magnetic fields of the particles have no net orientation. To record information, an external magnetic field orients the particles' magnetic fields according to the alignment of the applied field. The coercivity of the particles describes the strength of the external field that is needed to affect their orientation. Further, the coercivity of the particles exhibits a Gaussian distribution in which a few particles are oriented by a weak applied field, and the number increases as the field is increased, until the media saturates and an increase in the external field will no longer change net magnetization.

Saturation magnetic recording is used when storing binary data. The force of the external field is increased so that the magnetic fields in virtually all the particles are oriented. When a bipolar waveform is applied, a saturated media thus has two states of equal magnitude but opposite polarity. The write signal is a current that changes polarity at the transitions in the channel bitstream. Signals from the write head cause entire regions of particles to be oriented either positively or negatively. These transitions in magnetic polarity can represent transitions between 0 or 1 binary values.

During playback, the magnetic medium with its different pole-oriented regions passes before a read head, which detects the changes in orientation. Each transition in recorded polarity causes the flux field in the read head to reverse, generating an output signal that reconstructs the write waveform. The strength of the net magnetic changes recorded on the medium determines the medium's robustness. A strongly recorded signal is desired because it can be read with less chance of error. Saturation recording ensures the greatest possible net variation in orientation of domains; hence, it is robust.

Hard-Disk Design

Hard-disk drives offer reliable storage, fast access time, fast data transfer, large storage capacity, and random access. Many gigabytes of data can be stored in a sealed environment at an extremely low cost and in a relatively small size.

In most systems, the hard disk media is nonremovable; this lowers manufacturing cost, simplifies the medium's design, and allows increased capacity. The media usually comprises a series of disks, usually made of rigid aluminum alloy, stacked on a common spindle. The disks are coated on the top and bottom with a magnetic material such as ferric oxide, with an aluminum-oxide undercoat. Alternatively, metallic disks can be electroplated with a magnetic recording layer. These magnetic thin-film disks allow closer spacing of data tracks, providing greater data density and faster track access. Thin-film disks are more durable than conventional oxide disks because the data surface is harder; this helps them to resist head crashes. Construction of a hard-disk drive is shown in Fig. 14.7.

Hard disks rotate whenever the unit is powered. This is because the mass of the system might require several seconds to reach proper rotational speed. A series of read/write heads, one for each magnetic surface, are mounted on an arm called a head actuator. The actuator moves the heads across the disk surfaces in unison to seek data. In most designs, only one head is used at a time (some drives used for digital video are an exception); thus, read/write circuitry can be shared among all the heads. Hard-disk heads float over the magnetic surfaces on a thin cushion of air, typically 20 μm or less.

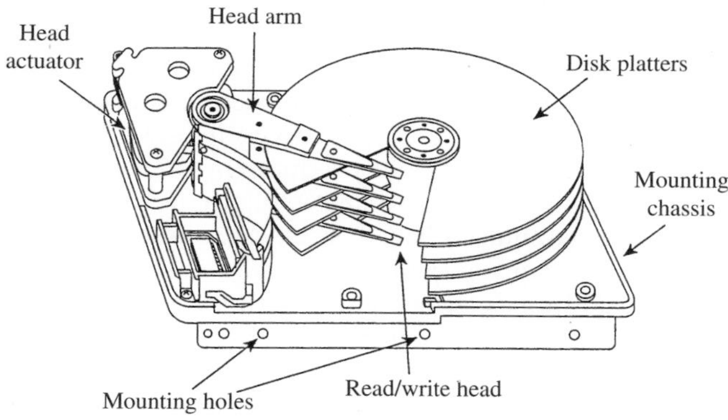

FIGURE 14.7 Construction of a hard-disk drive showing disk platters, head actuator, head arm, and read/write heads. (*Mueller, 1999*)

The head must be aerodynamically designed to provide proper flying height, yet negotiate disk surface warping that could cause azimuth errors, and also fly above disk contaminants. However, the flying height limits data density due to spacing loss. A special, lubricated portion of the disc is used as a parking strip. In the event of a head crash, the head touches the surface, causing it to burn (literally, crash and burn). This usually catastrophically damages both the head and disks, necessitating, at best, a data-recovery procedure and drive replacement.

Some disk drives use ferrite heads with metal-in-gap and double metal-in-gap technology. The former uses metal sputtered in the trailing edge of the recording gap to provide a well-defined record pulse and higher density; the latter adds additional magnetic material to further improve head response. Some heads use thin-film technology to achieve very small gap areas, which allows higher track density. Some drives use magneto-resistive heads (MRH) that use a nano-sized magnetic material in the read gap with a resistance that varies with magnetic flux. Typically, only one head is used at a time. The same head is used for both reading and writing; precompensation equalization is used during writing. Erasing is performed by overwriting. Several types of head actuator designs are used; for example, a moving coil assembly can be used. The moving coil acts against a spring to position the head actuator on the disk surface. Alternatively, an electric motor and carriage arrangement could be used in the actuator. Small PCMCIA drives use a head-to-media contact recording architecture, thin-film heads, and vertical recording for high data density.

To maintain correct head-to-track tolerances, some drives calibrate their mechanical systems according to changes in temperature. With automatic thermal recalibration, the drive interrupts data flow to perform this function; this is not a hardship with most data applications, but can interrupt an audio or video signal. Some drives use smart controllers that do not perform thermal recalibration when in use; these drives (sometimes called AV drives) are recommended for critical audio and video applications.

Data on the disk surface is configured in concentric data tracks. Each track comprises one disk circumference for a given head position. The total tracks provided by all the heads at a given radius position is known as a cylinder—a strictly imaginary construction. Most drives segment data tracks into arcs known as sectors. A particular physical address within a sector, known as a block, is identified by a cylinder (positioner address), head (surface address), and sector (rotational angle address). Modified frequency modulation (MFM) coding as well as other forms of coding such as 2/3 and 2/7 run-length limited codes are used for high storage density.

Hard-disk drives were developed for computer applications where any error is considered fatal. Drives are assembled in clean rooms. The atmosphere inside the drive housing is evacuated, and the unit is hermetically sealed to protect the media from contamination. Media errors are greatly reduced by the sealed disk environment. However, an error correction encoding scheme is still needed in most applications. Manufactured disk defects, resulting in bad data blocks, are logged at the factory, and their locations are mapped in directory firmware so the drive controller will never write data to those defective addresses. Some hard-disk drives use heat sinks to prevent thermal buildup from the internal motors. In some cases, the enclosure is charged with helium to facilitate heat dissipation and reduce disk drag.

Data can be output in either serial or parallel; the latter provides faster data transfer rates. For faster access times, disk-based systems can be designed to write data in a logically organized fashion. A method known as spiraling can be used to minimize interruptions in data transfer by reducing sector seek times at a track boundary. Overall, hard disks should provide a sustained transfer rate of 5 to 20 Mbyte/s and access time (accounting for seek time, rotational latency, and command overhead) of 5 ms. Some disk drives specify burst data transfer rates; a sustained rate is a more useful specification. A rotational speed of 7,200 rpm to 15,000 rpm is recommended for most audio applications. In any case, RAM buffers are used to provide a continuous flow of output data.

Hard-disk drives are connected to the host computer via SCSI, IDE (ATA), Firewire, USB, EIDE (ATA-2), and Ultra ATA (ATA-3 or Ultra DMA) connections. ATA drives offer satisfactory performance at a low price. SCSI-2 and SCSI-3 drives offer faster and more robust performance compared to EIDE; up to 15 drives can be placed on a wide SCSI bus. In practice, the transfer rate of a hard disk is faster than that required for a digital audio channel. During playback, the drive delivers bursts of data to the output buffer which in turn steadily delivers output data. The drive is free to access data randomly distributed on different platters. Similarly, given sufficient drive transfer rate, it is possible to record and play back multiple channels of audio. High-performance systems use a Redundant Array of Independent Disks (RAID) controller; RAID level 0 is often used in audio or video applications. A level 0 configuration uses disk mirroring, writing blocks of each file to multiple drives, to achieve fast throughput.

It is sometimes helpful to periodically defragment or optimize a drive; a bundled or separate utility program places data into continuous sections for faster access. Most disk editing is done through an edit decision list in which in/out and other edit points are saved as data addresses. Music plays from one address, and as the edit point approaches, the system accesses the next music address from another disk location, joining them in real time through a crossfade. This allows nondestructive editing; the original data files are not altered. Original data can be backed up to another medium, or a finished recording can be output using the edit list.

Highly miniaturized hard-disk drives are packaged to form removable cards that can be connected via USB or plugged directly into a receiving slot. This drive technology is sometimes known as "IBM Microdrive." The magnetic platter is nominally "1 inch" in diameter; the card package measures 42.8 × 36.4 × 5 mm overall. Other devices use platters with 17-mm or 18-mm diameter; outer card dimensions are 36.4 × 21.4 × 5 mm. Both flash memory and Microdrive storage are used for portable audio and video applications. Microdrives offer low cost per megabyte stored.

Digital Audio Workstations

Digital audio workstations are computer-based systems that provide extensive audio recording, storage, editing, and interfacing capabilities. Workstations can perform many of the functions of a traditional recording studio and are widely used for music production. The low cost of personal computers and audio software applications has encouraged their wide use by professional music engineers, musicians, and home recordists. The increase in productivity as well as creative possibilities, are immense.

Digital audio workstations provide random access storage, and multitrack recording and playback. System functions may include nondestructive editing; digital signal processing for mixing, equalization, compression, and reverberation; subframe synchronization to timecode and other timebase references; data backup; networking; media removability; external machine control; sound-cue assembly; edit decision list I/O; and analog and digital data I/O. In most cases, this is accomplished with a personal computer, and dedicated audio electronics that are interfaced to the computer, or software plug-in programs that add specific functionality. Workstations provide multitrack operation. Time-division multiplexing is used to overcome limitations of a hard-wired bus structure. In this way, the number of tracks does not equal the number of audio outputs. In theory, a system could have any number of virtual tracks, flexibly directed to inputs and outputs. In practice, as data is distributed over a disk surface, access time limits the number of channels that can be output from a drive. Additional disk drives can increase the number of virtual tracks available; however, physical input/output connections ultimately impose a constraint. For example, a system might feature 256 virtual tracks with 24 I/O channels.

Digital audio workstations use a graphical interface, with most human action taking place with a mouse and keyboard. Some systems provide a dedicated hardware controller. Although software packages differ, most systems provide standard "tape recorder" transport controls along with the means to name autolocation points, and punch-in and punch-out indicators. Time-scale indicators permit material to be measured in minutes, seconds, bars and beats, SMPTE timecode, or feet and frames. Grabber tools allow regions and tracks to be moved; moves can be precisely aligned and synchronized to events. Zoomer tools allow audio waveforms to be viewed at any resolution, down to the individual sample, for precise editing. Other features include fading, crossfading, gain change, normalization, tempo change, pitch change, time stretch and shrink, and morphing.

Digital audio workstations can provide mixing console emulation including virtual mixing capabilities. An image of a console surface appears on the display, with both audio and MIDI tracks. Audio tracks can be manipulated with faders, pan pots, and equalization modules, and MIDI tracks addressed with volume and pan messages. Other console controls include mute, record, solo, output channel assignment, and automation. Volume Unit (VU) meters indicate signal level and clipping status. Nondestructive bouncing allows tracks to be combined during production, prior to mix down. Digital signal processing can be used to perform editing, mixing, filtering, equalization, reverberation, and other functions. Digital filters may provide a number of filter types, and filters can provide smoothing function between previous and current filter settings.

Although the majority of workstations use off-the-shelf Apple or PC computers, some are dedicated, stand-alone systems. For example, a rack-mount unit can contain multiple hard-disk drives, providing recording capability for perhaps 96 channels. A remote-control unit provides a control surface with faders, keyboard, track ball, and other controllers.

Audio Software Applications

Audio workstation software packages permit audio recording, editing, processing, and analysis. Software packages permit operation in different time modes including number of samples, absolute frames, measures and beats, and different SMPTE timecodes. They can perform audio file conversion, supporting AVI, WAV, AIFF, RealAudio, and other file types. For example, AVI files allow editing of audio tracks in a video program. In most cases, software can create samples and interface with MIDI instruments.

The user can highlight portions of a file, ranging from a single sample to the entire piece, and perform signal processing. For example, reverberation could be added to a dry recording, using presets emulating different acoustical spaces and effects. Time compression and expansion allow manipulation of the timebase. For example, a 70-second piece could be converted to 60 seconds, without changing the pitch. Alternatively, music could be shifted up or down a standard musical interval, yielding a key change. A flat vocal part can be raised in pitch. A late entrance can be moved forward by trimming the preceding rest. A note that is held too long can be trimmed back. A part that is rushed can be slowed down while maintaining its pitch. Moreover, all those tools can be wielded against musical phrases, individual notes, or parts of notes.

In many cases, software provides analysis functions. A spectrum analysis plug-in can analyze the properties of sound files. This software performs a fast Fourier transform (FFT), so a time-based signal can be examined in the frequency domain. A spectrum graph could be displayed either along one frequency axis, or along multiple frequency axes over time, in a "waterfall" display. Alternatively, signals could be plotted as a spectrogram, showing spectral and amplitude variations over time. Advanced users can select the FFT sample size and overlap, and apply different smoothing windows.

Noise reduction plug-in software can analyze and remove noise such as tape hiss, electrical hum, and machinery rumble from sound files by distinguishing the noise from the desired signal. It analyzes a part of the recording where there is noise, but no signal, and then creates a noiseprint by performing an FFT on the noise. Using the noiseprint as a guide, the algorithm can remove noise with minimal impact on the desired signal. For the best results, the user manually adjusts the amount of noise attenuation, attack and release of attenuation, and perhaps changes the frequency envelope of the noiseprint. Plug-ins can also remove clicks and pops from a vinyl recording. Clicks can be removed with an automatic feature that detects and removes all clicks, or individual defects can be removed manually. The algorithm allows a number of approaches. For example, a click could be replaced by a signal surrounding the click, with a signal from the opposite channel, or a pencil tool could be used to draw a replacement waveform. Conversely, other plug-ins add noises to a new recording, to make it sound vintage.

One of the most basic and important functions of a workstation is its use as an audio editor. With random access storage, instantaneous auditioning, level adjustment, marking, and crossfading, nondestructive edit operations are efficiently performed. Many editing errors can be corrected with an "undo" command.

Using an edit cursor, clipboard, cut and paste, and other tools, sample-accurate cutting, copying, pasting, and splicing are easily accomplished. Edit points are located in ways analogous to analog tape recorders; sound is "scrubbed" back and forth until the edit point is found. In some cases, an edit point is assigned by entering a timecode number. Crossfade times can be selected automatically or manually. Edit splices often contain four parameters: duration, mark point, crossfade contour, and level. Duration sets the time of the fade. The mark point identifies the edit position, and can be set to various points within the fade. Crossfade contour sets the gain-versus-time relationship across the edit. Although a linear contour sets the midpoint gain of each segment at –6 dB, other midpoint gains might be more suitable. The level sets the gain of any segments edited together to help match them.

Most editing tasks can be broken down into cut, copy, replace, align, and loop operations. Cut and copy functions are used for most editing tasks. A cut edit moves marked audio to a selected location, and removes it from the previous location. However, cut edits can be broken down into four types. A basic cut combines two different segments. Two editing points are identified, and the segments joined. A cut/insert operation moves a marked segment to a marked destination point. Three edit points are thus required. A delete/cut edit removes a segment marked by two edit points, shortening overall duration. A fourth cut operation, a wipe, is used to edit silence before or after a segment.

A copy edit places an identical duplicate of a marked audio section in another section. It thus leaves the original segment unchanged; the duration of the destination is changed, but not that of the source. A basic copy operation combines two segments. A copy/insert operation copies a marked segment to a destination marked in another segment. Three edit points are required.

A "replace" command exchanges a marked source section with a marked destination section, using four edit points. Three types of replace operations are performed. An exact replace copies the source segment and inserts it in place of the marked destination segment. Both segment durations remain the same. Because the duration of the destination segment is not changed, any three of the four edit points define the operation. The fourth point could be automatically calculated. A relative replace edit operation permits the destination segment to be replaced with a source segment of a different duration; the location of one edit point is simply altered. A replace-with-silence operation writes silence over a segment. Both duration and timecode alignment are unchanged.

An "align" edit command slips sections relative to timecode, slaving them to a reference, or specifying an offset. Several types of align edits are used. A synchronization align edit is used to slave a segment to a reference timecode address. One edit point defines the synchronization reference alignment point in the timecode, and the other marks the segment timecode address to be aligned. A trim alignment is used to slip a segment relative to timecode; care must be taken not to overlap consecutive segments. An offset alignment changes the alignment between an external timecode and an internal timecode when slaving the workstation.

A "loop" command creates the equivalent of a tape loop in which a segment is seamlessly repeated. In effect, the segment is sequentially copied. The loop section is marked with the duration and destination; the destination can be an unused track, or an existing segment.

When any edit is executed, the relevant parameters are stored in an edit list and recalled when that edit is performed. The audio source material is never altered. Edits can be easily revised; moreover, memory is conserved. For example, when a sound is copied, there is no need to rewrite the data. An extensive editing session would result in a short database of edit decisions and their parameters, as well as system configuration, assembled into an edit list. Note that in the above descriptions of editing operations, the various moves and copies are virtual, not physical.

Professional Applications

Much professional work is done on PC-based workstations, as well as dedicated workstations. A range of production and post-production applications are accommodated: music scoring, recording, video sweetening, sound design, sound effects, edit-to-picture, Foley, ADR (automatic dialogue replacement), and mixing. To achieve this, a workstation can combine elements of a multitrack recorder, sequencer, drum machine, synthesizer, sampler, digital effects processor, and mixing board, with MIDI, SMPTE, and clock interfaces to audio and video equipment. Some workstations specialize in more specific areas of application.

Soundtrack production for film and video benefits from the inherent nature of synchronization and random access in a workstation. Instantaneous lockup and the ability to lock to vari-speed timecode facilitates production, as does the ability to slide individual tracks or cues back and forth, and the ability to insert or delete musical passages while maintaining lock. Similarly, a workstation can fit sound effects to a picture in slow motion while preserving synchronization.

In general, to achieve proper artistic balance, audio post-production work is divided into dialogue editing, music, effects, Foley, atmosphere, and mixing. In many cases, the audio elements in a feature film are largely re-created. A workstation is ideal for this application because of its ability to deal with disparate elements independently, locate, overlay, and manipulate sound quickly, synthesize and process sound, adjust timing and duration of sounds, and nondestructively audition and edit sounds. In addition, there is no loss of audio quality in transferring from one digital medium to another.

In Foley, footsteps and other natural sound effects are created to fit a picture's requirements. For example, a stage with sand, gravel, concrete, and water can be used to record footstep sounds in synchronization with the picture. With a workstation, the sounds can be recorded to disk or memory, and easily edited and fitted. Alternatively, a library of sounds can be accessed in real time, eliminating the need for a Foley stage. Hundreds of footsteps or other sounds can be sequenced, providing a wide variety of effects. Similarly, film and video requires ambient sound, or atmosphere, such as traffic sounds or cocktail party chatter. With a workstation, library sounds can be sequenced, then overlaid with other sounds and looped to create a complex atmosphere, to be triggered by an edit list, or crossfaded with other atmospheres. Disk recording also expedites dialogue replacement. Master takes can be assembled from multiple passes by setting cue points, and then fitted back to picture at locations logged from the original synchronization master. Room ambience can be taken from location tapes, then looped and overlaid on re-recorded dialogue.

A workstation can be used as a master MIDI controller and sequencer. The user can remap MIDI outputs, modify or remove messages such as aftertouch, and transmit patch changes and volume commands as well as song position pointers. It can be advantageous to transfer MIDI sequences to the workstation because of its superior timing resolution. For example, a delay problem could be solved by sliding tracks in fractions of milliseconds.

Workstations are designed to integrate with SMPTE timecode. SMPTE in- and out-points can be placed in a sequence to create a hit list. Offset information can be assigned flexibly for one or many events. Blank frame spaces can be inserted or deleted to shift events. Track times can be slid independently from times of other tracks, and sounds can be interchanged without otherwise altering the edit list.

Commercial production can be expedited. For example, an announcer can be recorded to memory, assigning each line to its own track. In this way, tags and inserts can be accommodated by shifting individual lines backward or forward; transfers and back-timing are eliminated. Likewise "doughnuts" can be produced by switching sounds, muting and soloing tracks, changing keys without changing tempos, cutting and pasting, and manipulating tracks with fade-ins and fade-outs. For broadcast, segments can be assigned to different tracks and triggered via timecode. In music production, for example, a vocal fix is easily accomplished by sampling the vocal to memory, bending the pitch, and then flying it back to the master—an easy job when the workstation records timecode while sampling.

Audio for Video Workstations

Some digital audio workstations provide digital video displays for video editing, and for synchronizing audio to picture. Using video capture tools, the movie can be displayed in a small insert window on the same screen as the audio tools; however, a full-screen display on a second monitor is preferable. Users can prepare audio tracks for QuickTime movies, with random access to many takes. Using authoring tools, audio and video materials can be combined into a final presentation. Although the definitions have blurred, it is correct to classify video edit systems as linear, nonlinear, off-line, and on-line.

Nonlinear systems are disk-based, and linear systems are videotape-based. Off-line systems are used to edit audio and video programs, then the generated edit decision list (EDL) can be transferred to a higher-quality on-line video system for final assembly. In many cases, an off-line system is used for editing, and its EDL is transferred to a more sophisticated on-line system for creation of final materials. In other cases, and increasingly, on-line, nonlinear workstations provide all the tools, and production quality, for complete editing and post-production.

Audio for video workstations offer a selection of frame rates and sampling frequencies, as well as pre-roll, post-roll, and other video prerequisites. Some workstations also provide direct control over external videotape recorders. Video signals are recorded using data reduction algorithms to reduce throughput and storage demands. Depending on the compression ratio, requirements can vary from 8 Mbytes/min to over 50 Mbytes/min. In many cases, video data on a hard disk should be defragmented for more optimal reading and writing.

Depending on the application, either the audio or video program can be designated as the master. When a picture is slaved to audio, for example, the movie will respond to audio transport commands such as play, rewind, and fast forward. A picture can be scrubbed with frame accuracy, and audio dropped into place. Alternatively, a user can park the movie at one location while listening to audio at another location. In any case, there is no waiting while audio or videotapes shuttle while assembling the program. Using machine control, the user can audition the audio program while watching the master videotape, then lay back the final synchronized audio program to a digital audio or video recorder.

During a video session, the user can capture video clips from an outside analog source. Similarly, audio clips can be captured from analog sources. Depending on the onboard audio, or sound card used, audio can be captured in monaural or stereo, at a variety of sampling rates, and word length resolutions. Similarly, source elements that are already digitized can be imported directly into the application. For example, sound files in the AIFF or WAV formats can be imported, or tracks can be imported from CD-Audio or CD-ROM.

Once clips are placed in the project window, authoring software is used to audition and edit clips, and combine them into a movie with linked tracks. For example, the waveforms comprising audio tracks can be displayed, and specific nuances or errors processed with software tools. Audio and video can be linked and edited together, or unlinked and edited separately. In- and out-points and markers can be set, identifying timed transition events such as a fade. Editing is nondestructive; the recorded file is not manipulated, only the directions for replaying it; edit points can be revised ad infinitum. Edited video and audio clips can be placed relative to other clips using software tools, and audio levels can be varied. The finished project can be compiled into a QuickTime movie, for example, or with appropriate hardware can be recorded to an external videotape or optical disc recorder.

CHAPTER 15

Telecommunications and Internet Audio

The telephone was invented in 1876, either by Alexander Graham Bell or Elisha Gray. Bell's attorney registered his telephone patent at the U.S. Patent Office a few hours after Gray's attorney had registered a caveat, an announcement of a telephone invention he intended to file a patent for. The men fought for ownership of the invention in litigation that lasted for years. Finally, Bell was given credit for inventing the telephone. It was later determined that Gray's proposed apparatus would have worked, but Bell's apparatus, as represented, would not.

Whoever invented it, telephone technology evolved tremendously, and progressed from a wired analog system to one that is wired and wireless, and digital. The system's sophistication is impressive; by entering a string of digits, a user's voice is quickly routed over networks to another user and conveyed with good sound quality. The telephone system became the basis for the Internet, a technology that revolutionized the dissemination of information. It would be a tame prediction to say that soon every bit of human knowledge and information will be accessible on the Internet, and that includes music. Whether as MP3 or MPEG-4 files, or proprietary protocols, the Internet can be used to efficiently download recorded music, and to stream live music as it is performed. Whether the connection is between two computers or millions of them, across a room or around the world, via dedicated systems or ordinary telephone lines, digital audio comprises ever more important telecommunications content.

Telephone Services

The digital transmission systems used to convey voice communications synonymously carry high bandwidth digital audio transmission. Both analog and digital signals attenuate over long distances; whereas analog signals are thus subject to increased noise, digital signals can be regenerated with great precision. Telephone companies almost exclusively use electronic switching and fiber-optic cable between telephone exchanges. As far as the telephone company is concerned, any phone call is treated simply as data; a long-distance call has the same sound quality as a local call. However, in most cases, the connection from the consumer's landline phone to the exchange is still analog.

Some information regarding the plain old telephone system (POTS): the telephone lines provided by the phone company are known as subscriber loops or central office lines. They connect users to central offices via analog lines. Trunk lines connect users to a private branch exchange (PBX) where switching routing takes place. Circuit switching is used, in which a continuous connection is temporarily established between users, exchanging data at the same rate. The audio frequency response for typical subscribers is 300 Hz to 3400 Hz. Speech is coded with 8-bit PCM, at a sampling frequency of 8 kHz to yield a bit rate of 64 kbps; bandwidth is sharply limited to 3.4 kHz. Amplitude companding and other means are used to improve dynamic range and fidelity.

Before the 1960s, telephone companies used copper wires to convey analog communications between switching offices, one pair of wires per phone conversation. Bundles of wires, each with thousands of pairs, ran through underground conduits. To increase capacity, Bell Laboratories turned to digital communications, and devised the T-carrier network. Illinois Bell installed the first T-carrier digital transmission system in 1962. Today, most long-distance telephone carriers use a variety of T-carrier lines, using copper, optical fiber, microwave radio, and coaxial cables to convey digital audio, video, and data.

T-1 serial digital communication circuits provide a dedicated point-to-point bandwidth of 1.544 Mbps; the signal they convey is called DS-1. Each end of a T-1 line terminates in a customer service unit (CSU), as shown in Fig. 15.1. The CSU interfaces data from the customer premises equipment (CPE) or a data service unit (DSU) and encodes it for transmission along the T-1; a multiplexer or local area network (LAN) bridge could be used. For example, long-haul T-1 lines can connect a long-distance provider's office in Miami with an office in Seattle; each office is known as a point of presence (POP). Each POP then communicates with local access networks, where the T-1 line terminates with a CSU. The CSU then communicates via a DSX-1 signal that interfaces to data signals such as RS-449 or V.35. A T-1 user sees two conventional copper wire pairs, one for send, and one for receive, and an RJ-48C plug.

A T-1 circuit is capable of carrying Compact Disc data (1.41 Mbps) without data reduction, 24 voice channels, or a single video program with data reduction. A T-1 circuit comprises 24 subchannels (called DSOs or slots) each carrying 64 kbps (8 bits with 8-kHz sampling). Communications such as data or normal voice traffic can employ these subchannels. Data bytes are applied to frames, with one frame holding 24 DSOs plus one framing bit. Thus a frame holds 193 bits ($24 \times 8 + 1$); the frame rate of 8 kHz yields a 1.544-Mbps overall rate.

Many different applications, such as voice, data, or video, can share one T-1 line, with individual channels assigned to one or multiplexed DSOs, as needed. T-1 is full duplex (bi-directional) and the assignments need not be identical. A line might convey a

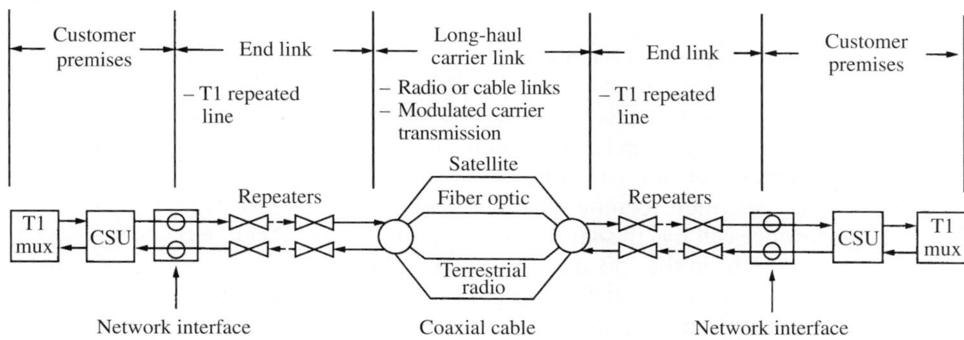

FIGURE 15.1 A T-1 circuit provides long-haul digital communication at a data rate of 1.544 Mbps; a CSU terminates each end of the line.

few high-quality audio channels in one direction, with many voice channels in the other. In some cases a Fractional T-1 line can be used; only a few DSO slots are assigned, as needed. T-1 lines are very reliable, with a bit-error rate (BER) of 10^{-9}. Individual DSOs can be sent to different destinations with the digital access cross-connect system (DACS).

Other services include the T-0 (or DS-0) service. Operating at 64 kbps, a T-0 line can deliver one 56-kbps digitized voice channel, sampled at 4 kHz, with 7 bits; that is, one standard voice telephone call. Some video teleconferencing systems use six DS-0 slots (384 kbps). The T-1C service operates at 3.152 Mbps. The T-2 service operates at 6.312 Mbps; it is not offered commercially. The T-3 (or DS-3) service operates at 44.736 Mbps; it can deliver 28 T-1 channels or 672 voice channels, or one compressed television channel. The T-4 service operating at 274.760 Mbps delivers 168 T-1 channels or 4032 voice channels. The overall data rates have numerical differences from multiples of the basic 64 kbps; this is due to framing information that must be added. All DS levels use alternate mark inversion (AMI) channel coding; the signal has three levels: positive, negative, and ground reference. The T-1 through T-4 rates in Europe are 2.048, 8.448, 34.368, and 565.000 Mbps, respectively.

ISDN

Integrated Services Digital Network (ISDN) is a dial-up telephone service with full duplex operation. ISDN provides a digital connection between the user and the telephone exchange, and ultimately to long-haul digital transmission systems. An overall rate of about 144 kbps is supported. Although ISDN uses existing telephone lines, specialized equipment is needed at the send and receive ends. Basic rate ISDN (sometimes called 2B+D) is intended for home use. It uses standard copper pair wire to provide two 64-kbps circuits (B or bearer channels) to send and receive audio or other information, and one 16-kbps (D or data channel) circuit for dialing and other signaling functions. Audio, video, fax, telephone, or other data can be transmitted. Primary rate ISDN is offered to business customers. It exists as either coaxial cable or fiber cable. It can provide 23 (or more) 64-kbps channels, along with a 64-kbps D channel (23B+D), totaling 1.544 Mbps of bandwidth (equivalent to a T-1 line).

With regular switched telephone lines, communication can be briefly interrupted by concurrent signaling and call-directing data (for example, tones in the same bandwidth as the communication data); this is called in-band signaling. With ISDN, the bearer channels are independent of the signaling channel; this is called out-of-band signaling. The advantage is that the bearer channels are specifically designated for the user and are uninterrupted by any signaling data. Timecode information can be transmitted over ISDN, using, for example, an RS-232 data channel. In addition, numerical data and compressed video signals can be sent over ISDN, with speeds exceeding that of dial-up modems that must convert data for analog transmission through the subscriber loop.

Although simple telephones could be used in a basic voice-only ISDN hookup, to exchange audio data the user supplies an A/D and D/A converter, data reduction encoder/decoder (codec), and ISDN hookup as shown in Fig. 15.2. Data between the user and the local telephone exchange is sent over copper wires; lines are terminated at a dedicated terminal adapter (TA) interface. The copper wire pair connecting to the user is called the user (U) interface and a network termination (NT-1) converts this to a four-wire ST interface. The outgoing data is converted to a telecommunications format where it is directed over a carrier to the receiving party, which must have corresponding ISDN

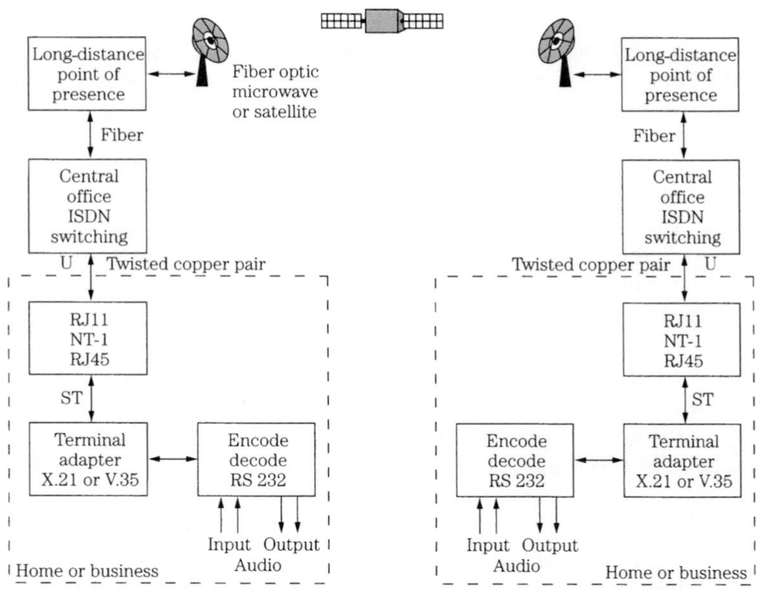

Figure 15.2 An example of an ISDN connection between two users. Each user needs appropriate interfacing equipment. The long-distance connection can be via fiber-optic cable, microwave, or other means.

service and equipment. Using an inverse multiplexer (I-MUX), multiple digital lines can be combined to create a higher data-rate service.

For example, a 384-kbps signal could be transmitted over six 64-kbps channels. The I-MUX resynchronizes the different lines (they might go through different long-distance routing) to produce a coherent stream at the receiver. I-MUX devices are placed at the terminal adapter or the codec. Depending on the number of channels and fidelity required, numerous data reduction methods can be used. The coding delay in an ISDN system must be accounted for. For example, when recording an overdub from a remote location, the user might send the mix over ISDN and listen to a delayed mix over the monitors. The received (delayed) overdub part is then remixed and recorded. Later, the overdubbed part can be slipped back into synchronization with the mix.

Similar to ISDN, the Switched-56 dial-up service provides a bi-directional 56-kbps data channel. The 56k Digital Data Service (DDS) is a dedicated 56-kbps service, with long-distance calls billed by the minute, at prices similar to those of regular long distance calls. Channel service units (CSU) are used to interface the user's application to a Switched-56 line; they can be purchased or leased. Units are required at both the send and receive ends of the line.

Various commercial systems provide multiple channels of full duplex digital audio connection over reserved T-1 telephone lines. For example, a connection might use Dolby AC-2 data reduction at 256 kbps for dual monaural, with timecode occupying a third channel. Alternatively, a monaural or stereo voice-over service can be provided over Switched-56 and ISDN lines using MPEG data reduction. In addition, video services are available. Timecode can be conveyed through the system, permitting synchronized overdubs.

Asymmetric Digital Subscriber Line (ADSL)

Companies invest billions of dollars in fiber optics, switchers, and routers to develop high-speed Internet access. But the most stubborn problem is "the last mile"—the actual hookup to the consumer's home. Several methods are used to cost-effectively establish that final link. Existing telephone modems are too slow. Attention has turned to cable modems connected to CATV (cable TV) lines to take advantage of cable's high bandwidth. Other groups are looking at electric power service lines to move data to and from homes. In addition, small dish satellites, as well as terrestrial wireless delivery, are employed.

Asymmetric Digital Subscriber Line (ADSL) provides high-capacity data channels, along with regular telephone service, on a single pair of copper wires. Like ISDN, ADSL needs two modems (one in the home, and the other at the telephone switching office) but unlike ISDN, ADSL uses existing wires and existing phone numbers. However, ADSL requires the installation of DSLAMs (Digital Subscriber Line Access Multipliers) in telephone switching offices. ADSL was originally developed by telephone companies in 1989 to provide interactive television, and thus bandwidth is its forte. Moreover, unlike dial-up modems that require dialing and connecting waits, ADSL provides a continuous always-on connection. However, as the name "asymmetric" implies, the data rate is not evenly balanced. The rate out of the home is much lower than the incoming rate. Since most applications require much higher input than output, asymmetry is acceptable for most users.

ADSL uses DMT (discrete multi-tone) technology as the coding method. DMT divides the telephone bandwidth into discrete frequency subchannels and simultaneously transmits data over each band. DMT also monitors and analyzes each subchannel, continuously and variably distributing data to each subchannel for optimal data rate. DMT is part of the ADSL standard, formally known in North America as T1.413. Many systems require a signal splitter, a separate line to the computer, as well as a modem and software. Nonstandard proprietary equipment may be used.

The Universal ADSL Working Group (UAWG) was established to simplify ADSL and lower costs. Its goal was to replace competing incompatible ADSL standards with a simplified consumer ADSL, known as Universal ADSL. Sometimes called ADSL "lite," it is compatible with standard ADSL but eliminates a splitter and separate wires. The modem is supplied by the telephone company or purchased in retail stores and installed by users as a plug-and-play feature. PCs and other devices may have ADSL lite modems built-in. The lite version may offer relatively low data rates, making high-quality live video problematic, but it is acceptable for other applications.

Whichever method is employed, ADSL competes with cable modems. Both have pros and cons. Both systems are inherently asymmetric. CATV was originally designed to move data in one direction—into homes. Cable modems also suffer from the party-line syndrome in which many homes share a common line back to the head-end. The more users who log on, the slower the traffic. ADSL's growth has been slowed by competing implementations, but cable modems have had a unified standard, called Multimedia Cable Network System (MCNS), for years. However, although the standard is in place, there are still practical incompatibility issues that remain unsettled. An initiative known as OpenCable may resolve this. Although cable passes by 90% of all U.S. homes, it is only connected to about 60% of them, and not all of these homes have access to cable-modem quality cable.

ADSL is generally not as fast as cable modem hookups. But ADSL users enjoy a private line to the central switching computer. However, ADSL's speed depends on the distance away from the switching computer; top speeds are possible only for relatively short connecting distances. Rates are slower for longer distances. Beyond that, service may not be possible. Perhaps 30% of all U.S. homes are either too far from a switching office, or use lines that cannot accommodate ADSL. For all ADSL users, the return path is always relatively slow, and that limits applications. For example, a high-quality videoconference between PC users would require high bandwidth in both directions. Cable companies, telephone companies, and satellite companies (and others) will continue to compete as data access providers to American homes.

Cellular Telecommunications

A number of technologies have been developed for cellular telecommunications. Three representative examples are summarized here. The Group Special Mobile (GSM) system is a Time Division Multiple Access (TDMA) technology that was first developed in Europe and has been adopted worldwide. Frequency Division Duplexing (FDD) is used for both the uplink and downlink. The aggregate data rate is 270.8 kbps and the per-user data rate is 33.85 kbps with eight users per channel. The supported user data rates for full-rate mode are 12.2 kbps and 13 kbps. Gaussian Minimum Shift Keying (GMSK) modulation is used.

The Edge system is an improvement on GSM technology. GSM was primarily designed for CS (circuit-switched) communication such as voice and fax. Edge is designed for greater efficiency with PS (packet-switched) communication such as multimedia and other Internet traffic. Edge provides higher capacity and greater data throughput data rate. PS data flow is bursty and application of dynamic resource allocation to user needs is key to good performance; bandwidth on demand is required. Although Edge retains the channel baud rate and time-slot structure of GSM, it uses 8-PSK modulation and other means to accommodate varying throughout demands. Nine coding schemes are available. The maximum data rate using 8-PSK modulation is 473.6 kbps (59.2 kbps × 8). Data rate can change by changing the modulation method and the number of parity bits used. A technique called incremental redundancy is used to maximize the data rate. To determine link quality, a packet with minimal overhead is sent and this high data rate is used if the communication is successful. If not, coding overhead is increased until communication is successful.

Wideband Code Division Multiple Access (CDMA), also known as WCDMA, is a 3G technology. It uses FDD for both uplink and downlink communication. The 3G UTRA and GSM standards are specified by the Third-Generation Partnership Project (3GPP).

Networks and File Transfers

Arguably, the first digital network was installed by Samuel Morse, linking Washington, DC to Baltimore with 35 miles of copper cable. His first transmitted telegraphic message on May 24, 1844, was: "What hath God wrought?" Today, thousands of networks interconnect millions of communications devices worldwide.

Whereas most dedicated audio interfaces operate point-to-point, with continuous data flow, and in real time, computer networks are typically asynchronous, transmit data in discrete packets, and can interconnect many disparate devices. Successful communication across a network calls for arbitration to avoid usage conflicts. Although a dedicated audio interface uses a dedicated audio format, a network is not concerned with the type of data being transmitted, and uses a common file structure for all data types such as e-mail, graphics, audio, and video. In addition, unlike dedicated audio interfaces, data delivery over a network is usually not continuous, and often not in real time. For example, delivery depends on network data rate as well as on current network traffic; transfer can occur at speeds well below, or well above real time. In some applications, a portion of the network bandwidth can be reserved to allow continuous real-time multimedia exchanges, such as video conferencing. Networked data transfer is increasingly supplanting the traditional practice of hand-carrying physical media, using the "sneaker net."

Whereas a dedicated digital audio link provides bandwidth that is sufficient to transmit a program in real time, a network is designed to interconnect multiple devices as networked nodes. For example, personal computers equipped with network interface cards, each with a unique address, can form nodes on a common network. Data is sent when a path is available, at the speed determined by the network interface. File exchange as well as random-access functions among computers are both permitted. The task of conveying high-bandwidth data (one channel of 44.1 kHz, 16-bit samples requires 705.6 kbps) in real time (and over an extended time) without interruption requires special consideration in the network design. In addition, one node may transmit multiple audio channels or a video channel; bandwidth of 30 Mbps to 40 Mbps might be needed. It might be necessary to lock out lower priority transfers, while a high priority time-sensitive transfer is in progress. A typical office LAN will not suffice. On the other hand, a multimedia network allows one major concession—data reduction. Low bit-rate coding of audio and video data can effectively multiply the bandwidth of the network.

Networks can be configured in a variety of ways. Most commonly, a bus configuration places nodes along a serial bus, a ring configuration places nodes along a closed circuit, and a star configuration gives each node direct access to the central controller. A star configuration is preferred because a disabled node does not affect other node performance. A central concentrator unit is needed to monitor and direct bus traffic in a star configuration. When the maximum number of nodes are placed on a network, or the system's longest length is reached, it can be extended with a repeater, a device that receives signals, then resynchronizes and retransmits them. A bridge isolates network segments so that data is only transmitted across the bridge when its destination is another segment. A router is a computer with two network interfaces. It can pass data between different types of networks; moreover, it can optimize the routing for faster communication. A hybrid configuration is sometimes recommended. For example, in addition to a ring joining workstation nodes and a central storage area, nodes might be star-configured to central storage using SCSI. An example of local networks bridged to other local networks, also showing different network configurations, is shown in Fig. 15.3.

Generally, it is the job of a network to break a message into data packets of uniform size and code them with a destination address and a header that describes where each packet fits within the message. The packets are transmitted, and received where they are assembled into the message. A packet that is corrupted with errors can be quickly retransmitted. Packet-switching is extremely efficient at conveying bursty communication through a distributed computer environment. However, because the arrival speed of packets cannot be guaranteed, real-time transfer is not always possible. All networks

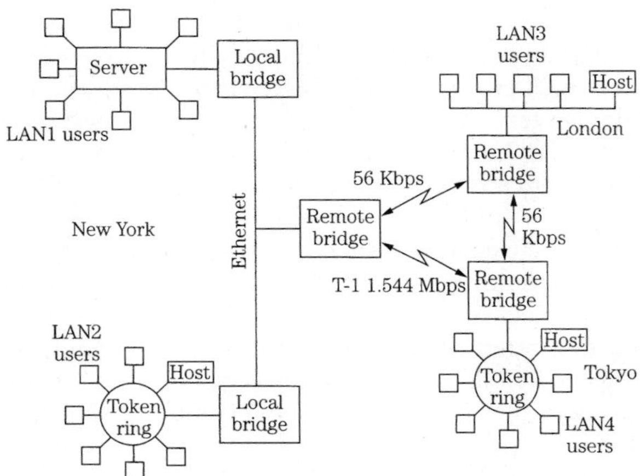

Figure 15.3 An example of star, ring, and bus network configurations, connected through long-distance bridging circuits.

must define rules for access to physical storage and terminal devices; this limits bandwidth of the network. Two common control methods are token passing and collision detection. With token passing, when a node finishes a transmission, it sends a token (a bit pattern) to the next node; a node can transmit only when it has the token. Each token ring node has both an input and output, so that connections are passed from one node to another, forming a ring. The ARCnet, operating at 2.5 Mbps uses token passing. A Fiber Distributed Data Interface (FDDI) network configured in a ring and using token ring can achieve 100 Mbps; FDDI is defined in the ISO 9314 standard. With Carrier Sense Multiple Access with Collision Detection (CSMA/CD), transmission occurs on command; if a collision occurs, priority is assigned, and data is retransmitted. Utilization (measuring successful transmission) of CSMA/CD networks decreases considerably with heavy use, as more collisions occur. A CSMA/CD protocol can achieve an overall efficiency of approximately 40%.

In client-server networks, most applications, data, and the network operating system, are centrally placed on a common file server. In peer-to-peer networks, data is kept locally under the control of individual computers. For example, a user could access files, ROM drives, and a printer—all in different physical locations; any computer can act as the server. Because of the size of audio and multimedia files, emphasis on distributed storage is often most efficient, in which nodes act as peers, each serving as both client and server, each with local storage. A distributed network with modest bandwidth of 10 Mbps to 15 Mbps might be adequate for many applications. For example, this would allow transfer of audio data at a rate several times faster than real time. However, this cannot always be achieved when multiple users are on the network, particularly when relatively slow SCSI hard-drive transfer rates limit data transfer speed. In many applications, data reduction algorithms are used to reduce the size of files, speeding up communications.

A wide area network (WAN) connects multiple stations beyond a single building, reaching up to global distances. For example, a WAN network can use an ISDN interconnection. A local area network (LAN) distributes data through an office, building, or campus. A LAN can use FDDI, CDDI, or ATM protocols.

Ethernet

Ethernet is a kind of computer network used to connect personal computers, printers, disk drives, and other equipment over coaxial cable, twisted pair, and optical fiber. Ethernet uses asynchronous transmission and collision detection. The following bit rates are often used: Ethernet at 10 Mbps, Fast Ethernet at 100 Mbps, Gigabit Ethernet ("GigE") at 1000 Mbps and 10 Gigabit Ethernet (10GE) at 10,000 Mbps. Gigabit Ethernet is based on IEEE 802.3 Ethernet and ANSI X3T11 Fibre-Channel. Generally, because of technical overhead, Ethernet is not used as a facility backbone to carry audio/video files. In many cases, a Storage Area Network (SAN) using Host Bus Adapter (HBA) cards and 1000BaseT or Fibre-Channel fiber-optic cable, is a better choice for moving large, multiple files among many storage devices.

An FDDI installation uses duplex (two-conductor) fiber-optic cables to permit 2-kilometer runs between nodes, and up to 200 kilometers on a single ring. It uses a token ring arbitration method. The FDDI standard specifies bandwidth of 100 Mbps. FDDI uses a Token Ring protocol. Copper Distributed Data Interface (CDDI) is a copper implementation of FDDI, offering 100 Mbps performance on Class 5 twisted-pair cable. It is more economical than FDDI, but does not provide the immunity to radio frequency (RF) and magnetic interference afforded by optical fiber, and it is limited to 50- to 100-meter runs between nodes. CDDI and FDDI can be freely mixed using converters. Fiber-optic technology is discussed in Chap. 13.

Generally, unshielded, twisted-pair cable (similar to that used in telephony) is classified in five levels: Level 1 (voice), Level 2 (RS-232 low-speed data), Level 3 (10-Mbps Ethernet), Level 4 (16-Mbps Token Ring), and Level 5 (100-Mbps high speed). For example, a 10Base-T network implementation requires Level 3 unshielded twisted-pair cable. It is used in a star configuration with a multi-port hub; it can be terminated with RJ-45 modular plugs. Daisy-chained Ethernet cabling is specified as 10Base2 (thin coaxial), or RG58 coaxial cable; it is limited to 185-meter segments. It is fitted with 50-Ω BNC plugs and attached to each node with a BNC T adapter. 10Base5 (thicker coaxial), also known as RG8 or RG11 cable, is used in Ethernet backbones; it is limited to 500-meter segments. Category 5 (Cat5e) cable is satisfactory for short runs of Gigabit Ethernet; Category 6 cable is needed for more sophisticated installations.

Asynchronous Transfer Mode (ATM)

Asynchronous Transfer Mode (ATM) is a high-bandwidth network standard using low-delay switching, variable bandwidth, and multiplexing to provide flexible and efficient communications. Audio, video, and other data can be simultaneously delivered at a rate of 1 Gbps or more. ATM is used primarily in wide area networking and local area network backbones with many users. Traditional networks pool many channels of information, with each receiver picking its information from the stream; considerable routing data is needed, and this increases overhead. The ATM architecture uses switching and multiplexing to form a temporary dedicated, virtual channel within the transmission path bandwidth. The virtual channel is allocated a sufficient data rate, providing bandwidth on demand. Data packets, called cells, carry the name of the channel with much less routing data required. In addition, many ATM users can share a channel through multiplexing, yet maintain a fixed-time relationship between data cells. In this way, slower moving data can be efficiently combined with fast data, with every time slot of the channel packed with data. In particular, this helps ensure continuous data flow, an important criterion for real-time audio and video transfers. ATM is sometimes referred to as Broadband ISDN.

Some implemented ATM channels provide transmission rates of 155 Mbps, 620 Mbps, or 2.4 Gbps along copper or fiber cables. However, data is transmitted at variable data rates, using only the rate needed at that moment. Billing is based, for example, on the number of cells transmitted. Moreover the data flow can be asymmetric; for example, a remote user can query a database with little data, but receive considerable data in return.

Specifically, ATM transmits data in fixed-length data fields. It uses 53-byte (8-bit) cells each composed of 48 bytes of user (payload) data, a 5-byte routing and control header, and an 8-bit error correction header. ATM switchers can thus route cells efficiently to proper destinations where they are assembled into useful information. In addition, packets are sequentially delivered along the virtual channel. Corrupted data cells are detected by the receiver, which requests a retransmission of cells. The ATM data protocol is based on the Open Systems Interconnection (OSI) model that specifies the physical medium and links.

Each ATM header is composed of six data fields. The Generic Flow Control (GFC) controls usage between several terminals with the same access connection. The Virtual Path Identifier (VPI) designates the path, or bundle of individual channels connecting two points in an ATM network. The virtual channel identifier (VCI) is a number allocated to every connection on the ATM during the call. The Payload Type (PT) defines maintenance cells, congestion conditions, and other types of cells. The Cell Loss Priority (CLP) assigns a priority rating to a cell, designating which cells can be discarded if necessary. The Header Error Check (HEC) is a checksum used to detect errors in the 5-byte header.

The AES47 standard describes the transmission of audio over ATM channels. PCM samples can be inserted into ATM cells in a number of ways including time order, channel order, or in multichannel groups. Data in the AES3 format can be conveyed, and there is a provision to carry up to 60 audio channels.

Bluetooth

The Bluetooth specification (BTSIG99) is an open standard describing a protocol used for wireless short-range audio and data communication. The short-range proximity network uses a peer-to-peer method; two devices can establish a link when they come within range of each other. This forms a wireless personal area network (WPAN). Radio transmission is used to transmit and receive signals. Class 1 devices have a maximum permitted power of 100 mW and a range of 100 meters; Class 2 devices have a maximum power of 2.5 mW and a range of 10 meters; Class 3 devices have a maximum power of 1 mW and a range of 1 meter. Because of relatively low power consumption, Bluetooth is often used in small, battery-powered devices. Bluetooth is a worldwide standard and is not constrained by otherwise incompatible telecommunications standards. Moreover, Bluetooth operates in a spectrum that can be used without license worldwide, with certain restrictions. Bluetooth can replace wired connections and is used in mobile phones, MP3 players, car audio receivers, hands-free headsets, printers, keyboards, video game controllers, and other mobile and fixed devices.

When a Bluetooth connection is established, one device is a master and the other is a slave. The device initiating the communication is usually the master, but either device may assume either status. The master device sets up the frequency hopping spread spectrum (FHSS) synchronization and pattern, and controls which devices can transmit. When conveying data, a slave can only transmit in response to a master device.

When conveying voice, a slave periodically transmits in its designated time slot. Each device has a 48-bit address. A device can be a master for one link and a slave for another. A master may directly communicate with up to 7 active slaves and up to 255 parked slaves which the master can bring into activity. All slaves linked to a single master form a piconet; all devices are synchronized. A device may be linked to more than one piconet, establishing synchronization with more than one master; overlapping piconets are called scatternets. In a scatternet, each piconet retains one master and has its own synchronization pattern. A device can act as a bridge, operating as a master in one piconet and a slave in another.

When in active mode, a slave device can continuously receive transmissions from the master device and remain synchronized; response time is fast but power consumption is relatively high. In sniff mode, a slave device is considered active but is only active periodically and the master only transmits packets at certain times. Response time is somewhat slower, but power consumption is also somewhat lower. In hold mode, a slave device is considered active but is inactive for a certain time interval. Response time is slow, but power consumption is low. When in parked mode, a slave device is considered inactive on the piconet, but it is synchronized with a master using a beaconing method. Response time is slow, but power consumption can be very low. By placing devices in and out of parked mode, a master can communicate with many devices, while observing the limit of seven active devices. An adaptive transmission power feature lets each slave use a received signal strength indictor to tell a master that its power is not appropriate and that it must cut or boost its transmission power to that slave.

Bluetooth uses frequency division spread spectrum communications, also known as frequency hopping spread spectrum (FHSS). Hopping occurs at a nominal rate of 1600 times per second. Bluetooth operates in the ISM (Industrial, Scientific, Medical) frequency band at 2.4 GHz to 2.4835 GHz. The spectrum is divided into 79 channels; bandwidth is limited to 1 MHz per channel. Each channel is divided in time (TDMA) into time slots; each slot corresponds to an RF hop frequency. In basic mode, Gaussian frequency shift keying modulation is used. Version 1.2 enables a gross data rate of 1 Mbps (maximum unidirectional asynchronous bit rate is 723.2 kbps), and Version 2.0+EDR (Enhanced Data Rate) enables 3 Mbps (maximum unidirectional asynchronous bit rate is 2.1 Mbps). Bluetooth data can be conveyed in asynchronous connectionless (ACL) mode and synchronous connection-oriented (SCO) mode. The ACL mode uses packet-switching; headers contain destination addresses. Up to three SCO links are allowed in a Bluetooth channel, each link operating at up to 64 kbps. An SCO link between two devices may be used to convey real-time audio. Both synchronous and asynchronous audio channels may operate simultaneously.

Audio in conventional (narrow-band) Bluetooth is conveyed with a sampling frequency of 8 kHz yielding 64 kbps with companding for voice communication. The Advanced Audio Distribution Profile (A2DP) can also be used to convey high-quality audio via Bluetooth ACL mode in either monaural or stereo. For example, music can be streamed from a phone to a car audio receiver. The maximum bit rate is 320 kbps for monaural and 512 kbps for stereo. Use of a SBC subband codec is mandatory. SBC was specifically designed for use in Bluetooth and features four or eight subbands, adaptive bit allocation, simple adaptive block PCM quantizers, and low complexity.

Block diagrams of an SBC encoder and decoder are shown in Fig. 15.4. In the SBC encoder, the input PCM signal is split by a polyphase analysis filter bank into subband signals. A scale factor is calculated for each subband, and scale factors are used to calculate

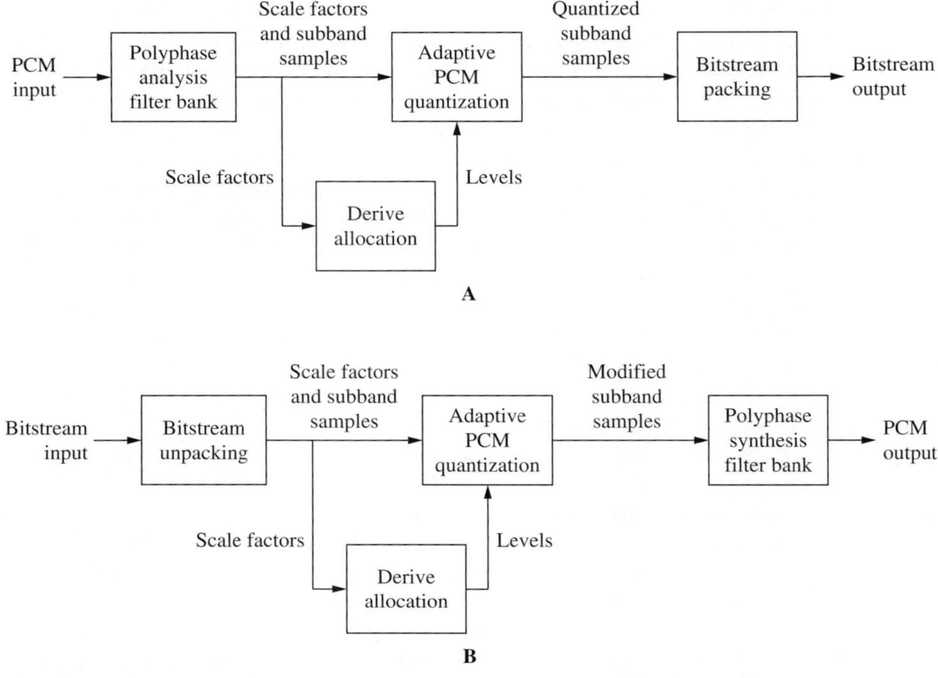

FIGURE 15.4 An SBC subband codec is used in the Bluetooth A2DP specification. It can be used to convey high-quality audio via Bluetooth. A. SBC subband encoder. B. SBC subband decoder.

bit allocation and levels for each subband. Subband samples are scaled and quantized and a bitstream is output. In the SBC decoder, scale factors are used to calculate bit allocation. The number of quantization levels is derived for each subband, subband samples are calculated, and a polyphase synthesis filter bank is used to output the PCM signal.

Other codecs including MPEG-1/2 Audio, MPEG-2, 4 AAC, and ATRAC can be used optionally when both the source and receiver support the optional codec. When an optional codec is not fully supported, audio data is transcoded into SBC.

The decoder in an SBC receiver must mandatorily support sampling frequencies of 44.1 kHz and 48 kHz. The encoder in the SBC source must mandatorily support at least one of the sampling frequencies of 44.1 kHz or 48 kHz. Lower sampling frequencies are optional. The decoder in the receiver must support monaural, dual channel, stereo, and joint stereo modes. The encoder in the source must support monaural and at least one of the other modes. Both the encoder and decoder must support block lengths of 4, 8, 12, and 16. The decoder must support four and eight subbands, and the encoder must support at least eight subbands. Similar requirements are imposed for other codecs. For example, both the encoder and decoder must support at least one of the MPEG-1/2 Layer I, II, or III codecs. For MPEG-2, both the encoder and decoder must support AAC LC; other object types including those of MPEG-4 are optional. The Audio/Video Control Transport Protocol (AVCTP) can be used to control devices, and the AVCTP and Audio/Video Distribution Transport Protocol (AVDTP) can disseminate audio.

Bluetooth was conceived by Ericsson in 1994, and Version 1.0 was published in July 1999. The specification is publicly available and is royalty-free. The Bluetooth trademark is owned by Ericsson and licensed to adopters of the Bluetooth Special Interest Group; the founding companies of the SIG are Ericsson, Intel, IBM, Nokia, and Toshiba. The Bluetooth specification is described in the IEEE 802.15.1 standard. The specifications for a number of networking technologies are shown in Table 15.1.

As noted, Bluetooth uses frequency hopping spread spectrum (FHSS) transmission. This helps overcome interference from other broadcast channels as well as microwave ovens and other interfering devices in the unlicensed 2.4-GHz to 5-GHz band. The spectrum is divided into different channels, each at a different frequency. A message packet is transmitted on a channel, then a new channel is selected for the next packet, and the process repeats; the message is thus spread across the entire spectrum. The receiver

Standard	Description	Media	Maximum Data Rate	Maximum Distance
IEEE 1394a	Firewire	Standard Twisted Pair (STP)	400 Mbps	4.5 meters/hop to 16 hops
IEEE 1394b Gigabit	Firewire	Fiber, CAT-5	3.2 Gbps	100 meters/hop to 16 hops
Ethernet 10BASE-T	Ethernet	STP, CAT-5, Fiber, Coax	10 Mbps	500 meters (CAT-5), 2 km (fiber)
Ethernet 100BASE-T	Fast Ethernet	STP, CAT-5, Fiber	100 Mbps	500 meters (CAT-5), 2 km (fiber)
Gigabit Ethernet	1G Ethernet	CAT-5, Fiber	1 Gbps	5 km (fiber)
10 Gigabit Ethernet	10G Ethernet (WAN)	Fiber	10 Gbps	40 km
IEEE 802.11a	High speed wireless Ethernet	Wireless	54 Mbps	100 meters
IEEE 802.11b	Wireless Ethernet	Wireless	11 Mbps	100 meters
IEEE 802.11g	High speed wireless Ethernet	Wireless	54 Mbps	100 meters
IEEE 802.11n	Very high speed Ethernet	Wireless	600 Mbps	100 meters
Home RF	Ethernet	Wireless	10 Mbps	100 meters
Bluetooth	Personal area network	Wireless	3 Mbps	100 meters
Home PNA	Ethernet via telephone lines	Telephone lines (CAT-5)	10 Mbps	150 meters
Home Plug	Ethernet via power lines	Power lines	10 Mbps	1 km

Table 15.1 Specifications for several networking technologies. (*Moses, 2002*)

uses the same channel pattern to receive each packet. FHSS reduces interference from another device operating at one channel frequency, and collisions rarely occur with other FHSS devices in the same band. A single packet loss can be retransmitted at a new frequency. Adaptive hopping can be used to avoid an interfering frequency, and more power can be delivered to any particular broadcast frequency. The broadcast signal appears as noise-like impulsive interference to a narrow-band receiver.

The invention of frequency hopping is often attributed to actress Hedy Lamarr and musician George Antheil. Their 1942 patent (2,292,387) proposed a "secret communication system" to guide torpedoes by radio. Lamarr proposed that the signal be switched from one frequency to another, and Antheil used his expertise with player pianos to devise a synchronization method. FHSS also provides some security against eavesdropping; only the receiver knows the hopping pattern.

Direct sequence spread spectrum (DSSS) techniques are used to code data into patterns of bits, and portions of redundant data are delivered to different carrier frequencies in a wideband. A spreading ratio as defined by the level of redundancy is a factor that determines the system's robustness.

IEEE 802.11 Wireless LAN (Wi-Fi)

Increasingly, data that was once confined to copper wires or optical fiber is being conveyed wirelessly via radio. Wi-Fi is used for wireless LAN (WLAN) applications. Wi-Fi technology is described in the IEEE 802.11 standard. Four standards are widely used: 802.11a, b, g, and n. A Wi-Fi system establishes a local area in which devices such as cell phones, computers, and printers can wirelessly communicate with each other. Wi-Fi is used to convey audio data between computers and peripherals, as well as audio playback devices. Wi-Fi is sometimes known as wireless Ethernet because it shares some parts of its specification with Ethernet (IEEE 802.3). For example, Wi-Fi uses carrier detection and collision avoidance, as well as authentication and encryption.

Some terminology: a WPAN is a wireless personal area network, usually operating over personal distances. A WLAN is a wireless local area network operating over longer distances. A hot spot, for example, can encompass a home, office, or school. A WMAN is a wireless metropolitan area network operating over a large area. A WWAN is a wireless wide area network, potentially providing a global reach, for example, a cellular telephone system such as the General Packet Radio Service (GPRS).

Specifications for the four Wi-Fi standards, IEEE 802.11a, b, g, and n, are summarized in Table 15.2. They may differ in terms of operating frequency band and data transfer rate. In addition, they provide different levels of performance and various

Wi-Fi Version	Frequency Band (GHz)	Maximum Data Rate (Mbps)	Modulation
802.11a	5, 3.7	54	OFDM
802.11b	2.4	11	DSSS
802.11g	2.4	54	OFDM, DSSS
802.11n	2.4, 5 (selectable or concurrent)	600	OFDM

Table 15.2 Wi-Fi standards and specifications.

features and security enhancements. For example, an 802.11b system might provide a theoretical maximum bit rate of 11 Mbps over a 300-meter range, while an 802.11a system might provide a bit rate of 54 Mbps over 50 meters. Actual user data throughput is much less. Lower bit rate 802.11a systems use FHSS and DSSS techniques, and higher bit rate 802.11b systems use only DSSS. Higher transmission frequencies allow higher bit rates, but that generally dictates a shorter operating range.

The 802.11a standard operates in the 5.15–5.35-GHz and 5.725–5.825-GHz bands. It uses orthogonal frequency division multiplexing (OFDM) with 52 subcarriers (48 main plus 4 pilot) using Binary Phase Shift Keying/Quadrature Phase Shift Keying (BPSK/QPSK), and 16-Quadrature Amplitude Modulation (16-QAM) for higher bit rates, and Forward Error Correction (FEC) convolutional coding. A variety of data rates are supported: 6, 9, 12, 18, 24, 36, 48, and 54 Mbps; rates of 6, 12, and 24 Mbps are mandatory.

Different data rates are supported in the 802.11b standard. The basic rates of 1 Mbps and 2 Mbps use a DSSS system; extensions allow rates of 5.5 Mbps and 11 Mbps in which complementary code keying (CCK) modulation is used. Some worldwide frequency bands are 2.4–2.4835 GHz (United States, Canada, Europe), 2.471–2.497 GHz (Japan), 2.4465–2.4835 GHz (France), and 2.445–2.475 GHz (Spain).

The 802.11g standard provides data rate extensions in the 2.4-GHz frequency band. Supported data rates are 1, 2, 5.5, 6, 9, 11, 12, 18, 24, 36, 48, and 54 Mbps; rates of 1, 2, 5.5, 6, 9, 11, 12, and 24 Mbps are mandatory. DSSS/CCK and OFDM are supported. 802.11g hardware is backward-compatible with 802.11b hardware.

The 802.11n standard uses four spatial streams with a channel width of 40 MHz to provide a significant increase in overall data rate. A multiple-input multiple-output feature employs multiple antennas and Spatial Division Multiplexing (SDM). Wi-Fi standards are developed, and Wi-Fi products are certified, by the Wi-Fi Alliance trade group.

MediaNet

The MediaNet is an example of a dedicated high-speed multimedia LAN network. It lets multiple users simultaneously access materials on central disk drives. MediaNet is implemented on CDDI and FDDI protocols, and supports the Apple Filing Protocol, Networked File System, and ATM. Using CDDI, the network can simultaneously handle multiple channels of compressed audio and video with throughputs of 24 Mbps from node to node. With ATM, transfer rates of 120 Mbps can be achieved. Computers can be interfaced either as servers or clients; both have SCSI controllers and ports for connection to hard-disk drives. Disk drives are local in node Apple computers; however, any node can access data on any other node hard drive in the same way that one would access a printer or other network device. Token ring is used for data traffic control. Using such networks, multimedia data can be directly manipulated remotely without copying files from place to place. MediaNet was developed by Sonic Solutions.

Internet Audio

The Internet is a global collection of interconnected networks that permits transmission of diverse data to one or many users. A network is a collection of computers sharing resources between them; the Internet is a network of networks. The advantage of internetworking is clear—the more systems online, the greater the resources available to any single user.

The Internet was born in September 1969 when packet switching was demonstrated at the University of California, Los Angeles. In December 1969, the Advanced Research Projects Agency (ARPA) of the Department of Defense officially debuted ARPANET by linking four computers in California and Utah with packet-switching lines. In 1984, the MILnet was partitioned off for military use. In 1987, seeking a way to link five supercomputers as well as regional networks, the National Science Foundation (NSF) founded the NSFNET. In 1990, the NSF decommissioned the ARPANET and greatly expanded its own network; the NSFNET became the high-speed backbone of the U.S. portion of the Internet.

The Internet operates on protocols defined by the Department of Defense. They are provided in Request for Comments (RFC) documents published by the Defense Data Network Information Center. Some of the basic specifications and documents that define implementation of the Internet are shown in Table 15.3. The communication and message routing standard that forges the links that form the Internet is a set of documents called the Transmission Control Protocol/Internet Protocol (TCP/IP). Using these protocols, networks can share information resources, thus forming the Internet. For example, the NSFNET ties together regional domestic networks such as WESTnet, SURAnet, and NEARnet, as well as wide area networks such as BITNET, FIDOnet, and USENET that provide links to foreign networks. The result is a complex global map of computer systems.

<table>
<tr><td colspan="3">File Transfer</td><td colspan="3">Electronic Mail</td><td colspan="3">Terminal Emulation</td><td colspan="3">Network Management</td></tr>
<tr><td colspan="3">File Transfer
Protocol
(FTP)
MIL-STD-1780
RFC 959</td><td colspan="3">Simple Mail
Transfer Protocol
(SMTP)
MIL-STD-1781
RFC 821</td><td colspan="3">TELNET
Protocol
MIL-STD-1782
RFC 854</td><td colspan="3">Simple Network
Management Protocol
(SNMP)
RFC 1098</td></tr>
<tr><td colspan="6">Transmission Control Protocol (TCP)
MIL-STD-1778
RFC 793</td><td colspan="6">User Datagram Protocol (UDP)
RFC 768</td></tr>
<tr><td colspan="4">Address Resolution
ARP RFC 826
RARP RFC 903</td><td colspan="4">Internet Protocol (IP)
MIL-STD-1777
RFC 791</td><td colspan="4">Internet Control Message
Protocol
(ICMP)
RFC 792</td></tr>
<tr><td colspan="12">Network Interface Cards:
Ethernet, StarLAN, Token ring, ARCNET
RFC 894, RFC 1042, RFC 1051</td></tr>
<tr><td colspan="12">Transmission Media:
Twisted pair, coaxial, fiber-optic cable</td></tr>
</table>

Table 15.3 The Department of Defense defined the protocols used in the Internet; these are defined in military specifications and RFC documents.

The Internet is a packet-switched network. A user sends information to a local network, which is controlled by a central server computer. At the server, the Transmission Control Protocol (TCP) parses the message, placing it in packets according to the Internet Protocol (IP), with the proper address on each packet. The network sends the packets to a router computer that reads the address and sends the packets over data lines to other routers, each determining the best path to the address. Packets may travel along different routes. This helps spread loads across the network and reduces average travel time. However, real-time transmission can be difficult because packets can be delivered out of order, delivered multiple times, or dropped altogether. When the packets arrive at the destination address, the information is assembled and acted upon.

To illustrate the structure used to route data, the IP header is shown in Table 15.4. This header is contained in an IP datagram, contained in every information field conveyed along the Internet, and is based on a 32-bit word. The Version specifies the current IP software used; the Internet Header Length (IHL) specifies the length of the header; Type of Service flags indicate reliability and other parameters; Total Length gives length of datagram; Identifier identifies the datagram; Flags specify if fragmentation is permitted; Fragment Offset specifies where a fragment is placed; Time to Live measures gateway hops; Protocol identifies the next protocol such as TCP following the IP; the Checksum can be used for error detection; Source Address identifies the originating host; Destination Address identifies the destination host; Options can contain a route specification and padding completes the 32-bit byte; the Data field contains the TCP header and user data, with a maximum total of 65,535 8-bit octets in the datagram.

<table>
<tr><td>Version
4</td><td>IHL
4</td><td>Type of Service
8</td><td colspan="2">Total Length
15</td></tr>
<tr><td colspan="3">Identifier
16</td><td>Flags
3</td><td>Fragment Offset
13</td></tr>
<tr><td colspan="2">Time to Live
8</td><td>Protocol
8</td><td colspan="2">Header Checksum
16</td></tr>
<tr><td colspan="5">Source Address
32</td></tr>
<tr><td colspan="5">Destination Address
32</td></tr>
<tr><td colspan="5">Options and Padding
32</td></tr>
<tr><td colspan="5">User Data
Multiple of 8, less than
65,535 octets</td></tr>
</table>

Table 15.4 The internet Protocol (IP) header field contains 32-bit bytes to identify and route IP datagrams over the Internet.

Packet networks such as the Internet operate on a first-come, first-serve basis thus throughput rate can be unpredictable. There is no bandwidth reservation; the Internet cannot guarantee a percentage of the network throughput to the sender. The number of packets delivered per second is continuously variable. Buffers at the receiver can smooth discontinuities due to bursty delivery, but add delay time to the throughput. Finally, packet networks such as the Internet operate point to point. A message addressed to multiple receivers must be sent multiple times; this greatly increases bandwidth requirements for multicasting. To overcome this, new transmission protocols have been developed for multicasting as described below.

In the same way that the Post Office does not need a special truck to send a letter from one address to another, but rather routes it through an existing infrastructure, the Internet sends information over its infrastructure according to standardized addresses. For example, a user at the University of California (UC), Berkeley, sending a message to Dartmouth would log onto the UC Berkeley campus network, and a router may direct it to BARRNet, the Bay Area Regional Research Network, which may route it to the NSFnet backbone, which may route it to NEARnet, the New England Academic and Research Network, which may route it to the Dartmouth campus network, and to the individual user account.

Each Internet address is governed by the Domain System Structure, a method that uniquely identifies host computers and individual users. When the Internet was first created, six high-level domains were created to distinguish between types of users: com (commercial), edu (education), gov (government), mil (military), org (other), and net (network). Many other domains have subsequently been created.

Access to the Internet requires a computer, a communications link such as a LAN connection or a modem, and a gateway. Historically, universities, corporations, and governments have provided Internet access to their students and employees via in-house computer systems. Subsequently, the Internet was made available through networks such as America Online, as well as communications companies such as AT&T. Broadband connections have replaced modem connections.

Internet bit rates are often variable and the average rate is not always sufficient to convey high-quality music in real time. The system is also susceptible to time delays from encoding and decoding latencies, routing and switching latencies, and other transmission limitations. Because the Internet is a packet-based system, it is inherently difficult to convey a continuous, time-sensitive music signal. One method to overcome these limitations is to convey music instructions instead of music waveforms. For example, with a maximum bit rate of 32.5 kbps, MIDI can convey 16 multiplexed channels of information. Although latencies from 1 ms to 20 ms are not uncommon, real-time, interactive networked performances among MIDI instruments and live performers have demonstrated the system's utility.

Voice over Internet Protocol (VoIP)

Voice over Internet Protocol (VoIP) is a transmission technology mainly used to deliver speech and voice messaging over IP networks such as the Internet using packet-switched methods. In other words, VoIP is an "Internet telephone call." In VoIP, an audio signal is converted from analog to digital form, compressed and translated into IP packets, and transmitted over the Internet; the process is reversed at the receiver. VoIP is separate from the public switched telephone network (PSTN). VoIP technology is also referred to as Internet telephony, IP telephony, broadband telephony, and voice over broadband (VoBB). VoIP promotes bandwidth efficiency and low calling cost, and users range from individuals to large corporations. In some VoIP systems, the voice data is encrypted. In transport mode, only the payload audio data is encrypted and the packet header data is not. In tunnel mode, the entire packet is contained in a new protected packet.

VoIP can be implemented using either open-source or proprietary standards. Some examples include H.323, IMS, SIP, RTP, and the Skype network. A VoIP service provider may be needed. Several different methods are used to connect to these services. An analog telephone adapter can be placed between a telephone jack and an IP network broadband connection; this provides VoIP to all connected telephones. Alternatively, dedicated VoIP phones connect directly to the IP network through Wi-Fi or Ethernet; they can place VoIP calls without a computer. Alternatively, Internet phone "softphone" software can be used on a computer to place VoIP calls; no other hardware is needed. The IP Multimedia Subsystem (IMS) is used to coordinate mobile telephony with Internet telecommunications. This unifies voice and data applications such as telephone calls, voice mail, faxes using the T.38 protocol, email, and Web conferencing.

VoIP offers the advantage of lower cost. For example, Internet access is billed by data (Mbyte) which is cheaper than telephone billing by time, and multiple simultaneous calls may be placed on one line. However, the IP network is less reliable than the traditional circuit-switched telephone system. IP data packets are more prone to congestion; for example, lost or delayed packets may result in a momentary audio dropout. The IEEE 802.11e amendment to the IEEE 802.11 standard provides modifications to the Media Access Control (MAC) layer, which mitigates data-delay problems in VoIP.

Digital Rights Management

File formats such as MP3, WMA, and AAC are widely used to disseminate music over the Internet via downloading. The compressed files permit efficient storage of music on distribution servers, as well as efficient transmission over IP networks. Unfortunately, they can also be used to freely share copyrighted music. Digital Rights Management (DRM) systems can be used to encode music for Internet delivery, for storage in portable players, or for recording on discs. A DRM system establishes usage rules; for example, it might limit playback to one or a few devices, limit the number of playbacks, establish a time limit, or impose no restrictions at all.

DRM systems control the use of intellectual property content, restrict its copying, and identify illegal copies in e-commerce systems. In short, DRM provides secure exchange of intellectual property in digital form. Although no single standard has been universally accepted, various systems have been devised so that copyright-protected music can be distributed via the Internet. They define how content can move from various formats to various devices. Ideally, a DRM system should be platform-independent, have minimal object code file size, require minimal computation, and be revocable and renewable. A DRM system should also be reliable, flexible, unobtrusive, and easy to use. DRM is more than copy-protection; it establishes copy limitations and defines how a copyright holder can be paid for the copy.

DRM systems rely on numerous technologies. Cryptography can be used to allow secure delivery between authorized parties. Content is transmitted and stored in an encrypted form to prevent unauthorized copying, playback or transmission, and recordings cannot be downloaded or played by others. Authentication is used so that properly encoded data can be read only by compliant devices and media. Cryptography is discussed below. Watermarks can be placed in files to prevent unauthorized copying, and to identify copies made illegally in spite of the DRM protection. Watermarks are discussed below. Audio fingerprints can be used to identify content, and to limit playback. A content-based identification system extracts unique signature characteristics of the content and stores them in a database. Audio fingerprints are discussed below.

Other e-commerce considerations include methods for electronic transaction. For example, anti-copy protection can be placed in every audio file, so that only a legally designated user can play it. Users must first obtain a "passport" for authorization to download music. The passport resides on the hard disk and it stamps each downloaded file so that only that user can play it and so it can be traced if necessary. If the content owner permits it, the user may make a one-time copy of the music for playback on any player.

A Rights Expression Language (REL) is used to establish and communicate usage rules. In some case, XML (eXtensible Markup Language) is used as a format. Both open-source and royalty-bearing RELs are used. XrML (eXtensible rights Markup Language) is a rights expression language that describes aspects such as digital property rights language and self-protecting documents. This protocol does not define implementation aspects such as encryption or watermarking. XrML uses a license and grant method; a user is granted a right, for example, to download a product resource. This right is contained in a license issued to the user (or principal). Multiple rights may be granted relating to different content resources; for example, such as different rights to play or copy different material. A grant may contain conditions that must be fulfilled before the right is authorized. A user may need an encryption key to unlock content resource. XrML is managed by Content-Guard.

The MPEG-21 standard is an example of a framework that describes e-business mechanisms including rights management and transaction metadata using a rights data dictionary and a royalty-bearing REL based on XrML. However, any XML-based REL can be used within MPEG-21. MPEG-21 defines the description of content as well as methods for searching and storing content and observing its copyright. MPEG-21 is formally titled "Multimedia Framework" and is described in the ISO/IEC 21000-N standard.

Extensible Media Commerce Language (XMCL) is a language used to specify rights in digital media. It can be used in conjunction with ODRL. Open Digital Rights Language (ODRL) is an XML scheme for expressing digital rights. A signed license authorizes use of a resource; each copy is cryptographically unique to the licensee. The Light Weight Digital Rights Management (LWDRM) system marks content with the user's digital signature. If a user's copy of protected content is released for general distribution, copies can be traced back to the original user. Files are protected by Advanced Encryption Standard (AES) encryption as well as proprietary watermarking. Local media format (LMF) files are unique to the computer they are created on and cannot be played elsewhere. Signed Media Format (SMF) files may be fair-use copies and have different levels of playback restrictions. XMCL was developed by RealNetworks.

A number of open source DRM systems have been developed. Open IPMP (Intellectual Property Management and Protection) describes DRM tools for user and content identification and management such as cryptography, digital signatures, and secure storage. Symmetric encryption is used for content, and asymmetric encryption is used for licenses. ODRL is used within Open IPMP. Open IPMP conforms to the Internet Streaming Media Alliance, ISMA 1.0. Media.

The consumer's convenience and fair use must be respected. DRM systems must balance their effectiveness versus usability. Generally, they discourage casual copiers, but fall short of complete protection against professional pirates and determined hackers. Legal recourse is relied upon to limit the activities of the latter two groups. One pitfall is that attack software developed by a few pirates can be disseminated to casual copiers as easily as the music files themselves.

Audio Encryption

In many applications it is important to protect copyrighted music content against both dedicated and casual piracy. Encryption may be used to prevent copying, watermarking may identify properties of the content, and fingerprinting can identify the content itself. To be useful, these abilities must be robust even as content moves across many platforms and media in various incarnations. Ideally, a content-protection system would offer these capabilities: rights management, usage control, authentication, piracy deterrence, and tracking. However, these technologies must also be supported by legislation.

Cryptography provides one method to protect the content of files. With cryptography, data is encrypted prior to storage or transmission, then decrypted prior to use. One early advocate was Julius Caesar; he encrypted his messages to his generals by replacing every A with a D, every B with an E, and so on, using rotational encoding. Only a legitimate recipient (someone who knew the "shift by 3" rule) could decipher his messages. This provided good security for the Roman Empire, circa 50 B.C. Today, encryption codes are more complex. The encryption algorithm is a set of mathematical rules for rendering information unintelligible. It may consist of a mathematically difficult problem such as prime number factorization, or discrete logarithms. The original message is called the plaintext and the encrypted message is the ciphertext. To decrypt a file, a "key" is used to decipher the data. For example, the Caesar "shift by n" code can use different values of n, where n is the key.

A modern key may be a binary number, perhaps from 40 to 128 bits in length. In many applications, there is a public key that many users can employ, as well as private keys reserved for privileged access. In many audio file formats, audio data is placed into frames; the frame may also include a frame header with auxiliary information about the frame to assist decoding. A key could be placed in the header, informing the decoder that the file may be decoded if the user is authorized to do so. Illegitimate files, even copied from legitimate sources, would lack the key. Likewise, unauthorized decoders could not play back legitimate files. However, if the audio data is separated from the key, or if the key is hacked, the audio file could be played or copied without restriction. Encryption may also require a severe computation overhead, or a small increase in file size. Of course, encryption does not offer protection after a file has been decrypted.

Generally, publicly known encryption codes are preferred. Although the codes can be attacked literally using textbook means, the codes have been well-studied by experts and the codes' weaknesses are well understood. In that respect, many people know how the lock works, but the secret is in the keys—not the lock. A critical aspect to any cryptographic system is key management. If keys are not guarded, then even strong algorithms can be unlocked. Any system thus must consider that a weak algorithm/strong key management is preferable over strong algorithm/weak key management. In other words, the security of a strong system lies in the secrecy of the key rather than with the supposed secrecy of the algorithm.

In symmetric encryption, the original information is encrypted using a key, and the ciphertext is decrypted using the same key. Because the sender and receiver use the same key, it is vital, and sometimes difficult, to keep the key a secret. Data Encryption Standard (DES) is an example of symmetric encryption; it is a block cipher with 64-bit block size and 56-bit keys. DES is identical to the ANSI standard Data Encryption Algorithm (DEA) defined in ANSI X3.92-1981. Triple DES (DES3) uses an encrypt/decrypt/encrypt sequence with 64-bit blocks and two or three different and unrelated 56-bit keys. Blowfish uses symmetric encryption; it is a block cipher using 64-bit block size and key lengths from 32 to 448 bits. International Data Encryption Algorithm (IDEA) is an example of symmetric encryption; it is a block cipher using 128-bit keys.

In asymmetric encryption, sometimes called public-key encryption, public keys and private keys are used. The public keys are freely given to users and shared openly. However, each user also has a private key that is kept secret. The public and private keys are related by a complex mathematical transform. To illustrate: a sender has a public and private key, as does the receiver. The sender and receiver exchange their public keys, so each has three keys. The sender encrypts messages using the receiver's public key. The receiver decrypts it with a private key; only the receiver can decrypt a message that was decrypted with the receiver's public key. Not even the sender can decrypt the message. Asymmetric encryption is often used to make online financial transactions such as credit-card purchases on the Internet. Asymmetric encryption algorithms are often used for key management applications, that is, for sending secret keys needed by symmetric encryption algorithms. The first public-key encryption algorithm was devised by RSA Laboratories in 1977.

In a "brute force" attack, an opponent tries every possible key until the plaintext is revealed. Simply put, if $f(x) = y$ and the attacker knows y and can compute f, he or she can find x by trying every possible x. Depending on the hardware resources employed, a brute force attack on a code might discover the key to a 40-bit code in 2 ms to 2 seconds; a 56-bit key might require 2 minutes to 35 hours; a 128-bit key might require 10^{16} to 10^{19} years. Cryptography codes are classified as munitions by the International Traffic in Arms Regulations for the purpose of export. Sophisticated codes cannot be placed in products for export. The encryption systems used in the DVD and Blu-ray formats are discussed in Chapters 8 and 9, respectively.

Audio Watermarking

Digital audio watermarking offers another security mechanism, one that is intrinsic with the audio data and hidden in it. Watermarking ties ownership to content and can verify the content's authenticity. Specifically, the aim of watermarking is to embed an inaudible digital code (or tag) into an audio signal so that the identifying code can be extracted to provide an electronic audit trail to determine ownership, trace piracy, and so on. For example, if pirate recordings appeared, an embedded watermark could be recovered from them to trace the source of the original recording. Or, a player could check for watermarks before starting playback; if the watermark is missing or corrupted (as in an illegal copy), the player would refuse to play the copy. Source watermarks can be attached to a specific media (such as DVD or Blu-ray) to identify and protect the content. Transactional watermarks are independent of the media and track usage; for example, a watermark could monitor the number of Internet downloads. Watermarking is also known as steganography, the process of concealing information within data. In some cases, the watermark itself may be encrypted.

Because watermarks alter the audio data, they must ideally be transparently inaudible to the end user, and not affect the sound quality of the audio signal. A watermark should also be robust so that it can be recovered from the audio signal, and it should have survivability so that even after other processing, such as perceptual coding, the watermark is intact. For example, a watermark should be able to persist after D/A or A/D conversion, or transmission through an AM or FM radio broadcast path. In an extreme case, it should be possible to play a watermarked analog audio signal through a loudspeaker, and re-record the signal through a microphone, with an intact watermark. A watermark must also survive file format changes, for example, conversion from PCM to ADPCM to MP3. A watermark must also be secure so that it cannot be removed, altered, or defeated by an unauthorized user. If the watermark is attacked, it must result in a clearly degraded audio signal or produce other evidence that tampering has occurred. For example, it must overcome attempts to overwrite it with a counterfeit watermark. A watermark must also guard against false positives and false negatives. A receiver must not mistakenly identify watermark data where none exists, or misinterpret the watermark data. For example, the receiver should not refuse to play data that has a legitimate watermark, or interpret a counterfeit watermark as authentic. Finally, a watermark must not burden the file with excessive overhead data.

Simple watermarks can be realized by a few bits and this makes them easy to inaudibly embed and also makes them potentially more survivable. However, low bit-rate watermarks are more vulnerable. For example, the 16 permutations of a 4-bit least-significant bit (LSB) watermark could be easily tested and hacked. Thus, designers may consider security measures such as cryptographic keys. Watermarks can be applied to different file types such as PCM, WAV, and AIFF. In addition, watermarking can be applied to monaural or multichannel recordings at various sampling frequencies and word lengths.

Numerous audio (and video) watermarking methods have been developed. Systems may manipulate signal phase, place data in an undetectable frequency range, or employ spread-spectrum encoding. For example, with phase coding, watermark data is represented by a smooth (inaudible) phase change in certain frequency components. However, the integrity of phase coding may be compromised by low bit-rate coding algorithms. With spread-spectrum coding, watermark data is distributed over a relatively broad frequency range, thus hiding it. The watermark is coded with necessary error correction, then modulated and multiplied by a pseudo-random noise sequence. The resulting watermark signal occupies a certain spectrum and is attenuated and combined with the audio signal as noise. At the receiver, the watermark is extracted by multiplying the signal with the same pseudo-random sequence that is synchronized with the received signal; the watermark is thus de-spread. This technique can yield reliable results.

Many watermarking systems combine two or more methods to achieve a more robust solution. For example, spread-spectrum coding could use amplitude masking to shape the watermark and ensure its inaudibility. However, watermarks that use amplitude masking to prevent their audibility might risk being removed by perceptual codecs that eliminate masked or redundant signals. In another approach, watermarking data is pseudo-randomized and placed in the least-significant bits of certain words; the watermarking data adds benign noise to the audio signal. However, the watermark is relatively easy to attack. A buried data channel technique is described in Chap. 18. Watermarking techniques optimized for digital audio signals cannot be used on executable software code, ASCII code, or other code that has zero-tolerance for errors. Other approaches must be employed for these files.

There are a number of different watermarking techniques, each with a different intent. A copy-control watermark is designed to prevent casual unauthorized copying. This watermark is embedded in the recording, and is detected by a subsequent recorder. The watermark could be coded to allow no copies, one copy, a limited number of copies, or unlimited copying. For example, in one application, a rights holder may permit a user to make an unlimited number of copies, if a licensing fee is paid. A compliant recorder would detect the watermark and its copy code, and follow its dictates. If it made a subsequent copy, it would decrement the copy number in the watermark in the new copy.

A forensic watermark might be designed to deter professional piracy. This type of watermark holds a significant amount of data so the file can be authenticated and any tampering detected, using a verification key. The file may be designed to be intentionally fragile, so that if the file is subjected to processing, the original watermark is damaged or lost, and the tampering can be thus detected. For example, a player would detect tampering in the corrupted watermark in a pirated disc, and refuse to play that disc. An accompanying and more robust copy control watermark could identify the source.

Some MP3 ripping software automatically embeds song and artist information into the file using ID3 tags. A Track Unique IDentifier (TUID) tag can be added to a song to link it to the source album. Technologies such as these can be used by gated P2P "music download" services to track the content sold to authenticated customers.

Embedded watermarking technology has been selected for use in audio applications such as DVD-Audio, Super Audio CD (SACD) and Blu-ray disc. Similar watermarking methods can also be used for the CD.

Audio Fingerprinting

Audio fingerprinting, also known as content-based audio identification, analyzes an audio file to extract a unique signature and then uses that signature to search for a similar fingerprint in a database and thus identify the audio file. Fingerprints are primarily used for licensing and copyright enforcement. For example, a broadcast or webcast can be monitored to verify that paid advertisements are aired and song royalties are collected, or a network administrator could prohibit transmission of copyrighted material. Alternatively, fingerprints can be used to identify musical qualities, and recommend other similar music to a listener.

An audio fingerprint can be viewed as a summary that describes the characteristics of an audio file. It can provide an efficient means to search for and identify a file among many others. In many systems, features are extracted from known content and stored in a database during a training phase. Unknown content can be identified by extracting its features and comparing them to those in a large database containing perhaps a million fingerprints. The output of the system is metadata that identifies the unknown audio file, and a confidence measure. A system is shown in Fig. 15.5. The method is efficient because of the compact size of the fingerprints; it is faster to search and compare them rather than the entirety of the waveform files themselves. However, the system may be compromised when the unknown signal is distorted or fragmented. Although audio fingerprinting techniques have some similarities to audio watermarking, they are considered to be different applications.

Unlike a simple file name that can easily be changed to obscure a file's identity, the extracted information inherently identifies the content itself. Ideally, the reference

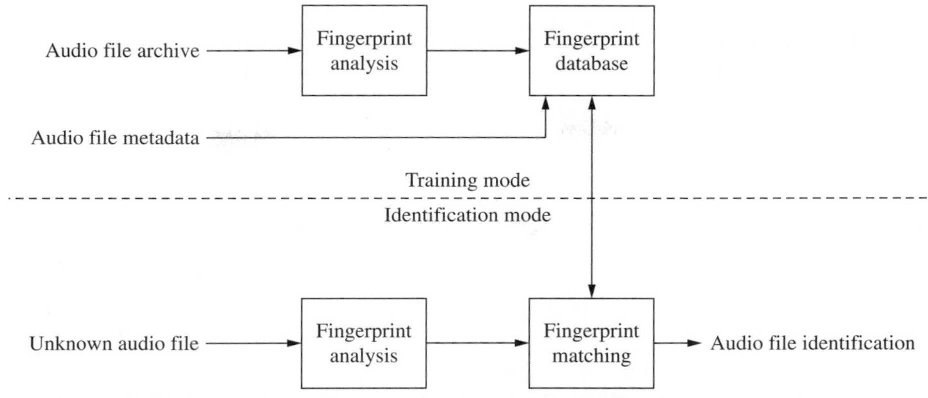

FIGURE 15.5 An audio fingerprint system operates in two modes. A training method is used to accumulate a database of fingerprints. An identification mode is used to identify an unknown audio file by generating and matching its fingerprint to one stored in the database.

characteristics are independent of variables such as audio format, sampling frequency, equalization, and signal distortion. A signature extraction should be possible even after a file has been processed or altered. Moreover, a signature should resist attempts to forge a signature or reuse it on other content. To do this, the features used in the fingerprint must remain relatively unaffected by changes in the signal. Similarly, a system must be reliable and avoid errors in identification. Large databases are more prone to false matches, and thus have lower reliability. A system should also be computationally efficient with compact fingerprint, low algorithmic complexity, and fast searching and matching techniques. Most systems are scalable and able to accommodate increasingly larger databases. Fingerprints themselves should be compact to provide efficiency, yet include sufficient parameters to completely characterize the audio file and allow accurate matching.

Fingerprint identification generally comprises two steps: extraction and modeling. Prior to extraction, the audio signal is placed into frames, windowed, and overlapped. The signal is transformed, for example, with fast Fourier transform (FFT) or discrete cosine transform (DCT). Feature extraction can employ many types of techniques. In many cases, critical-band spectrum analysis is used to generate perceptual parameters similar to those meaningful to the human hearing system. For example, a spectral flatness measure may be used to estimate the tonal or noise-like properties of the signal, or the energy in Bark bands may be calculated. Feature vectors are presented frame-by-frame to a fingerprint modeling algorithm. It consolidates the features from the entire file into a compact fingerprint. Various techniques can be used. For example, among other features, one model includes an average of the spectrum, average rate of zero crossing, and beats per minute. Another model uses sequences of indexes to sound classes that describe the audio signal over time.

The output stage of a fingerprint system searches the index of a database of known fingerprints in an attempt to identify the fingerprint from an unknown audio file. Metrics, such as Euclidean or Hamming distance, or correlation can be used. Various methods are used to optimize the indexing and searching of the database. When a match is found, its viability is measured against a threshold to determine the likelihood of correct identification. The MPEG-7 standard specifies numerous ways to perform audio fingerprinting and other related functions; it is described later.

Streaming Audio

When the size of the file is greatly reduced, so that it can be received as fast as it can be played, the file can be streamed. In this application, the music begins to play as soon as a buffer memory receiving the signal is filled. Data reduction algorithms can successfully reduce file size to permit downloading of files in a reasonable time. However, it is more difficult to stream audio files continuously, in real time.

Not only must the file size be small, it must also cope with the packet-switching transmission method of the Internet. Packets usually arrive sequentially, but not always, and some packets may be missing, and must be retransmitted, which incurs a delay. A buffer is needed because otherwise interruptions in the flow of data would cause interruptions in the playback signal. In addition, many different kinds of computers are used to play streaming files, with different processing power and different speeds of Internet connections. Finally, data speeds across an Internet path differ according to the path itself and traffic conditions. Thus streaming audio presents challenges, challenges that are multiplied when accompanied by streaming video, particularly if it is to be synchronized with audio.

The RealAudio format is used to stream prerecorded and live audio on the Internet. RealAudio files can be coded at a variety of bit rates, and the format supports scalability of the bitstream. A decoder can decode all the bits to reproduce a high-quality signal, or a subset of the bitstream for a lower-quality signal. This feature helps to sustain transmission when bit rate is variable or when available processor power varies. Files can be streamed from any HTTP computer; RealServer software permits better performance, more connections, and better security features. RealPlayer supports both streaming audio and video. Author, title, copyright, and other information can be included in files. Stored files can be downloaded and copied unless stored in a directory not accessible to users. This is easily accomplished by linking the Web page to a small metafile that links to the actual media files. As with any streaming file protocol, it is important to preprocess audio files prior to streaming. Typically, a midrange (2.5 kHz) boost is added, the dynamic range of the file is compressed, and the file is normalized to 95% of maximum. SureStream-encoded files can be used for multicast streaming. The stream contains scaled files; the server identifies the client's connection speed and transmits at the appropriate bit rate. RealAudio was developed by RealNetworks.

The Internet TCP protocol is efficient for packet transmission, providing robust error correction and verification. However, the required overhead slows processors and transmission throughput speeds in general. In addition, TCP allocates per-connection bandwidth proportional to available bandwidth, without considering how much bandwidth is needed. Thus, highly compressed files may be inefficiently allocated with too much bandwidth. An alternative to TCP is the User Datagram Protocol (UDP); it is a simpler packet protocol without error correction or verification. Clients must request a missing packet, rather than receiving it automatically. This promotes better throughput. RealSystem servers use UDP. RealNetworks has also developed Real-Time Streaming Protocol (RTSP), an open standard for transmitting time-based multimedia content, described below.

Among its diverse capabilities, Apple's QuickTime software offers several compression options for music and speech applications, for both downloading and streaming. QuickTime itself is not application software, but rather an extension of the computer operating system. Using the Sound Manager, media elements are stored in the movie

MOV file format, allowing synchronization among diverse elements. Over 35 different file formats are supported. The Netcasting feature allows QuickTime players to play streaming broadcasts of any QuickTime media type. Data fields can store user information and audio files can include video, text, and MIDI content. The QuickTime file format is the basis for the MPEG-4 file specification, discussed later. QuickTime is discussed in Chap. 14.

Shockwave is a streaming audio package based on MPEG Layer III coding; it allows compression at bit rates ranging from 8 kbps to 128 kbps. As noted, some streaming systems employ UDP to improve reliability of transmission, but this requires special server software, and streaming may be inhibited by a firewall. In contrast, Shockwave operates with any HTTP server; dropouts are mitigated by longer buffering of audio data. Shockwave audio files are coded as SWA files. When the browser loads the HTML page, it encounters an Embed tag that identifies the Shockwave file; the browser then loads the plug-in and downloads the file. When buffering is completed, the file begins to play. Shockwave integrates with Macromedia's Director program, which is used for multimedia authoring, in which a series of frames make up a "movie." Director movies are converted to Shockwave files following data compression. Shockwave was developed by Macromedia.

The Microsoft Windows Media (WMA) Technologies platform is used to provide streaming audio and video and other digital media. Media Technologies uses a data-reduction algorithm, sometimes known as MSAudio, to provide FM-quality audio at a bit rate of 28.8 kbps. In one mode, the algorithm streams 44.1-kHz sampled stereo audio at a bit rate of 20 kbps; it claims CD transparency at 160 kbps. Windows Media Encoder (WME) is used to create content; for example, WAV files can be compressed to WMA files. An ASX text metafile points to the WMA file. The system supports multi-bit-rate encoding to create multiple data-rate streams in a single media file. Users receive the optimal stream for their connection; content is coded in the Advanced Streaming Format (ASF) native file format. Using the open architecture of the DirectShow API, the Media Player can support files types including MP3, WAV, AVI, MIDI, QuickTime, and MPEG including MPEG-4 v.3 for video streaming. In addition, the Media Player automatically checks incoming streams for codec versions and then downloads new versions of the codec as needed. Content from 3 kbps of audio to 6 Mbps of audio/video is supported. Both on-demand ASF files and live-encoded streams are supported. The server can automatically select a stream from a range of bandwidths. If the connection conditions change, the server can switch to a higher or lower bit rate.

Windows Media Services is the server component of the system. A single server can support thousands of simultaneous user connections; both unicast and multicast modes are supported. Transmission protocols UDP, TCP, and HTTP are supported. The Windows Media Rights Manager system reduces piracy and enables digital distribution. Customers use an authorization mechanism to play content; content can be encrypted so that it is licensed for use only as the publisher intends.

G2 Music Codec for Streaming

Music codecs designed for Internet streaming must provide fast encoding and decoding, tolerance to lost data, and scalable audio quality. Fast encoding speed is important for live real-time streaming, particularly at multiple simultaneous data rates. Similarly, decoding must optimize subjective audio quality versus computing resources. Simple algorithms may yield poor audio results, while high computational complexity may tax the processor and slow the throughput rate.

The RealAudio G2 music codec is an example of a streaming codec. It is a transform codec that uses a combination of fast algorithms and approximations to achieve fast operation with high audio quality. For example, lookup tables or simpler approximations may be used, and prediction (with subsequent correction) may be used instead of iterative procedures. In this way, simultaneous encodings may be performed with reasonable computing power.

Whereas non-real-time data such as text or static images may be delivered via a network such that missing or corrupted data is retransmitted, with real-time audio (and video) it may not be possible for the server to resend lost data when using "best effort" protocols such as UDP. This is particularly problematic for coders that use predictive algorithms. They rely on past data thus when packets are lost, current and future audio quality is degraded. To help overcome this, interdependent data may be bundled into each packet. However, in that case a packet might comprise 200 ms or more of audio, and a lost packet would result in a long mute. Recovery techniques such as repeating or interpolating may not be sufficient.

The RealAudio G2 music codec limits algorithmic dependencies on prior data. It also encodes data in relatively small, independently decodable units such that a lost frame will not affect surrounding frames. In addition, compressed frames can be interleaved with several seconds of neighboring frames before they are grouped into packets for transmission. In this way, large network packets can be used. This promotes efficiency, yet large gaps in decoded audio are avoided because a lost packet produces many small gaps over several seconds rather than a single large gap. Interpolation then uses past and future data to estimate missing content, and thus cover small audio gaps. Overall, the G2 music codec can handle packet loss up to 10 to 15% with minimal degradation of audio quality.

Ideally, codecs provide both a wide frequency response and high fidelity of the coded waveform. But in many cases although a wide frequency response is obtained, the waveform is coded relatively poorly. Alternatively, codecs may support only a narrow frequency response while trying to optimize accuracy in that range; speech codecs are examples of the latter approach. To balance these demands on the bit rate, many codecs aim at a specific bit rate range, and deliver optimal performance at that rate. However, when used at a lower rate, the audio quality may be less than that of another codec targeted for that low rate. Conversely, a low bit-rate codec may not be optimal at higher bit rates or its computational complexity may make it inefficient at high rates.

The RealAudio G2 music codec was optimized for Internet bit rates of 16 kbps to 32 kbps to provide good fidelity over a relatively broad frequency response. The algorithm was also optimized for bit rates outside the original design to accommodate rates from 6 kbps to 96 kbps and average frequency responses ranging from 3 Hz to 22 kHz. These frequency responses are average because the codec can dynamically change response according to current data requirements. For example, if a musical selection is particularly difficult to code, the codec dynamically decreases the frequency response so that additional data is available to more accurately reproduce the selected frequency range. With this scalability, the codec can deliver the maximum quality possible for a given connection. RealNetworks G2 players use G2 music encoding. A version of Dolby's AC-3 format is employed if files are encoded for backward-compatibility with older RealPlayers. For speech coding, a CELP-based (Code Excited Linear Prediction) coder is employed.

The SureStream protocol is used by RealNetworks to deliver streaming RealAudio and RealVideo files over the Internet. It specifically addresses the issues of constrained and changing bandwidth on the Internet, as well as differences in connection rate. SureStream creates one file for all connection rate environments, streams to different connection rates in a heterogeneous environment, seamlessly switches to different streams based on changing network conditions, prioritizes key frames and audio data over partial frame data, thins video key frames for the lowest possible data rate throughput, and streams backward-compatible files to older RealPlayers.

Internet connectivity can range from cell phones to high-speed corporate networks, with a corresponding range of data transfer rates. A high-bandwidth stream would satisfy users with high-bandwidth connections, but would incur stream stoppages and continual re-buffering for users with low-speed connectivity. To remedy this, scalable streams can use stream-thinning in which the server reduces the bit rate to alleviate client re-buffering. However, when a stream has been optimized for one rate, and then it is thinned to a lower rate, content quality may suffer disproportionately. Moreover, this approach is more appropriate for streaming video, than it is to audio.

To serve different connection rates, multiple audio files can be created, and the server transmits at the most applicable bit rate. If this bandwidth negotiation is not dynamic, the server cannot adjust to changes in a user's throughput (due to congestion or packet loss). With SureStream, multiple streams at different bit rates may be simultaneously encoded and combined into a single file. In addition, the server can detect changes in bandwidth and make changes in combinations of different streams. This Adaptive Stream Management (ASM) in the RealSystem G2 protocol can thus deliver data packets efficiently. File format and broadcast plug-ins define the ASM rules that assign predefined properties such as "priority" and "average bandwidth" to groups of packets. If packets must be sent again, for example, the packet priorities identify which packets are most important to redeliver. With ASM, plug-ins use condition expressions to monitor and modify the delivery of packets based on changing network conditions. For example, an expression might define client bandwidth from 5 kbps to 15 kbps and packet loss as less than 2.5%. The client subscribes to the rule if that condition describes the client's current network connection. If network conditions change, the client can subscribe to a different rule.

The encoder can create multiple embedded copies at varying discrete bit rates. The server seamlessly switches streams based on the clients' current bandwidth or packet-loss conditions. The system must monitor bandwidth and losses quickly. If a bandwidth constriction is not detected quickly, for example, the router connecting to the user will drop packets, causing signal degradation. However, if bandwidth changes are measured too quickly, there may be unnecessary shifting of the stream's bit rate.

Audio Webcasting

The Internet is efficient at sending data such as email from one point to another; this is called unicasting. Likewise, direct music transactions can be accommodated. For some applications, point-to-point communication is not efficient. For example, it would be inefficient to use such methods to send the same message to millions of users because a separate transmission would be required for each one. IP multicasting is designed to overcome this. Using terrestrial broadcast, satellite downlink, or cable, multicasting can send one copy to many simultaneous users. Much like radio, information can be continually streamed using special routers. Instead of relying on servers to send multiple transmissions, individual routers are assigned the responsibility of replicating and delivering data streams. In addition, multicasting provides timing information to synchronize audio/video packets. This is advantageous even if a relatively few key routers are given multicasting capability. The Real-time Transport Protocol (RTP) delivers real-time synchronized data. The Real-time Transport Control Protocol (RTCP) works with RTP to provide quality-of-service information about the transmission path. The Real-time Transport Streaming Protocol (RTSP) is specifically designed for streaming applications. It minimizes overhead of multimedia delivery and uses high-efficiency methods such as IP multicast. The Resource Reservation Protocol (RSVP) works with RTP or RTSP to reserve network resources to ensure a specific end-to-end quality of service. In this way, music distribution systems can efficiently distribute music to consumers.

The RealSystem G2 supports IP multicast with two types of multicast: back-channel and scalable. Back-channel multicast uses the RealNetwork's Real Delivery Transport (RDT) data packet transport protocol. Data is multicast over UDP that does not guarantee delivery, but RDT includes a back-channel or resend channel. An RTSP or RealNetwork's PNA control channel is thus maintained between the server and the client; this communicates client information to the server. A packet resend option can be employed when multicast UDP packets are lost. In addition, back-channel content can be authenticated, and client connections are logged in the server. G2 back-channel multicast serves all bit rates defined in the SureStream protocol. However, the bandwidth is static and will not dynamically vary according to network conditions. Back-channel multicasting is preferred when content quality is important and audience size is limited. It is not suitable for large audiences because of the system resources required for each client connection.

Scalable multicast uses a data channel only (no TCP control channel between server and clients) that is multicast with RTP/RTCP as the data packet transport and session reporting protocol. Two RTP ports are used; one provides data transmission while the other reports on quality of data delivery. Data is delivered over UDP. Client connection statistics are not available during G2-scalable multicasting; however, the server logs all users. Scalable multicasting supports SureStream by multicasting each stream on a unique multicast address/port. The client chooses a preferred stream type, and the bit rate does not dynamically vary as with unicasting. Minimal server resources are required for any size audience; thus, it is preferred for large audience live events. However, there is no recovery mechanism for packet loss. Because G2-scalable multicasting is implemented with standard protocols such as RTP, RTCP, SDP, and SAP, it can interoperate with other applications that use these protocols.

MPEG-4 Audio Standard

The MPEG-4 standard, as with MPEG-1 and MPEG-2, defines ways to represent audio-visual content. It is also designed for multimedia communication and entertainment applications. Moreover, MPEG-4 retains backward-compatibility with its predecessors. However, MPEG-4 is a departure from the MPEG-1 and MPEG-2 standards because it offers considerably greater diversity of application and considerable interoperability with other technologies including other content delivery formats. MPEG-4 is a family of tools (coding modules) available for a wide variety of applications including interactive coding of audiovisual objects whose properties are defined with a scene description language. Objects can comprise music, speech, movies, text, graphics, animated faces, a

Web page, or a virtual world. A scene description supplies the temporal and spatial relationships between the objects in a scene. MPEG-4 provides an emphasis on very low bit rates and scalability of the coded bitstream that allows operation over the Internet and other networks. MPEG-4 supports high-quality audio and video, wired, wireless, streaming, and digital broadcasting applications. MPEG-4 promotes content-based multimedia interactivity. It specifies how to represent both natural and synthetic (computer-generated) audio and video material as objects, and defines how those objects are transmitted or stored and then composed to form complete scenes. An example of MPEG-4 architecture is shown in Fig. 15.6. For example, a scene might consist of a still image comprising a fixed background, a video movie of a person, the person's voice, a graphics insert, and running text. With MPEG-4, those five independent objects can be transmitted as multiplexed data streams along with a scene description.

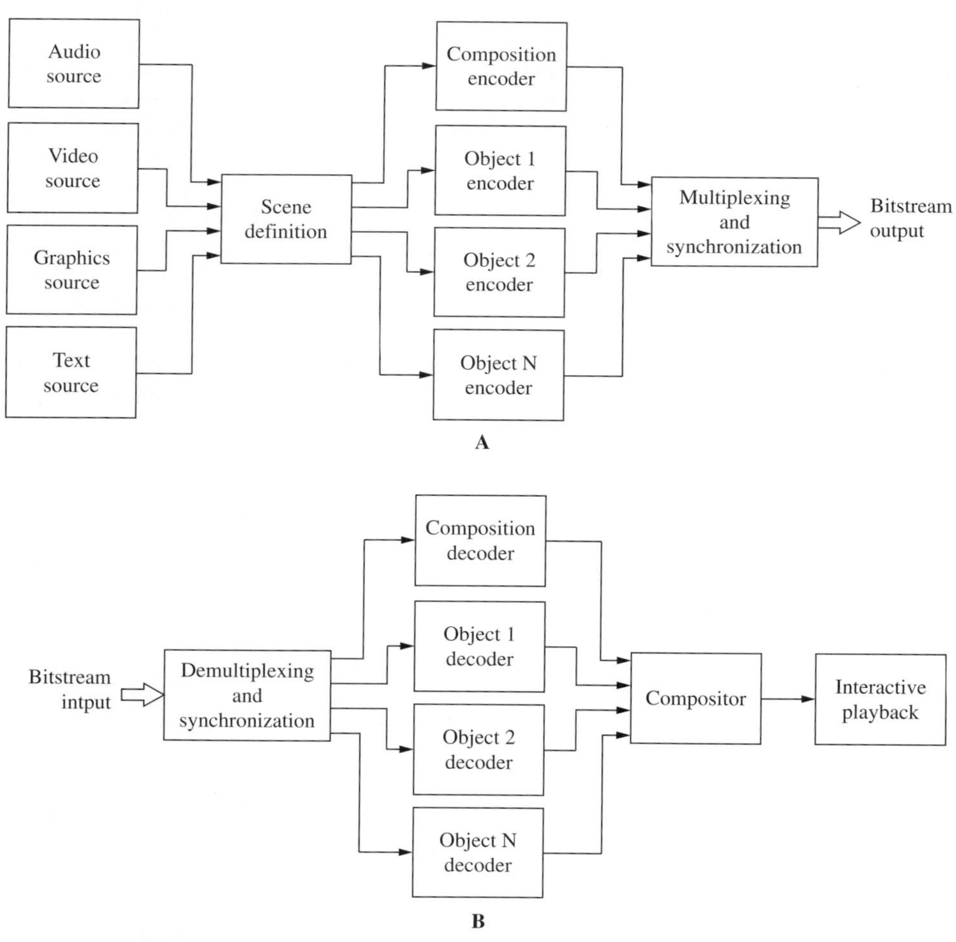

FIGURE 15.6 The MPEG-4 standard provides an object-based representation architecture. A. Encoder. B. Decoder.

Data streams are demultiplexed at the receiver, the objects are decompressed, composed according to the scene description, and presented to the user. Moreover, because the objects are individually represented, the user can manipulate each object separately. The scene descriptions that accompany the objects describe their spatial-temporal synchronization and behavior (how the objects are composed in the scene) during the presentation. MPEG-4 also supports management of intellectual property as well as controlled access to it. As with many other standards, MPEG-4 provides normative specifications to allow interoperability, as well as many non-normative (informative) details. Version 1 of the MPEG-4 standard (ISO/IEC-14496), titled "Coding of Audio-Visual Objects," was adopted in October 1998.

MPEG-4 Interactivity

MPEG-4 uses a language called BInary Format for Scenes (BIFS) to describe and dynamically change scenes. BIFS uses aspects of the Virtual Reality Modeling Language (VRML'97), but whereas VRML describes objects in text, BIFS describes them in binary code. Unlike VRML, BIFS allows real-time streaming of scenes. BIFS allows a great degree of interactivity between objects in a scene and the user. Objects may be placed anywhere in a given coordinate system, transforms may be applied to change the geometrical or acoustical aspects of an object, primitive objects can be grouped to form compound objects, streamed data can be applied to objects to modify their attributes, and the user's viewing and listening points can be placed anywhere in the scene. Up to 1024 objects may comprise a scene.

The user may interact with the presentation, either by using local processing or by sending information back to the sender. Alternatively, data may be used locally, for example, stored on a hard disk. MPEG-4 employs an object-oriented approach to code multimedia information for both mobile and stationary users. The standard uses a syntax language called MSDL (MPEG syntax description language) to be flexible and extensible.

Whereas MPEG-1 and MPEG-2 describe ways to data-reduce, transmit, and store frame-based video and audio content with minimal interactivity, MPEG-4 provides for control over individual data objects and the way they relate to each other, and also integrates many diverse kinds of data. However, MPEG-4 does not specify a transport mechanism. This allows for diverse methods (such as the MPEG-2 transport stream, asynchronous transfer mode, and real-time transport protocol on the Internet) to access data over networks and other means. Some examples of applications for MPEG-4 are Internet multimedia streaming (the MPEG-4 player operating as a plug-in for a Web browser), interactive video games; interpersonal communications such as videoconferencing and videophone, interactive storage media such as optical discs, multimedia mailing, networked database services, co-broadcast on HDTV, remote emergency systems, remote video surveillance, wireless multimedia, and broadcasting applications.

Users see scenes as they are authored; however, the author can allow interaction with the scene. For example, a user could navigate through a scene and change the viewing and listening point; a user could drag objects in the scene to different positions; a user could trigger a cascade of events by clicking on a specific object; and a user could hear a virtual phone ring and answer the phone to establish two-way communication. MPEG-4 uses Delivery Multimedia Integration Framework (DMIF) protocol to manage multimedia streaming. It is similar to FTP; however, whereas FTP returns data, DMIF returns pointers directed to streamed data.

There are many creative applications of MPEG-4 audio coding. For example, a string quartet might be transmitted as five audio objects. The listener could listen to any or all of the instruments, perhaps deleting the cello, so he or she could play along on his or her own instrument. In another example, the audio portion of a movie scene might contain four types of objects: dialogue, a passing airplane sound effect, a public-address announcement, and background music. The dialogue could be encoded in multiple languages, using different objects; the sound of the passing airplane could be processed with pitch-shifting and panning, and could be omitted in a low bit-rate stream; the public-address announcement could be presented in different languages and also reverberated at the decoder; and the background music could be coded with MPEG-2 AAC.

MPEG-4 Audio Coding

MPEG-4 audio consolidates high-quality music coding, speech coding, synthesized speech, and computer music in a common framework. It spans the range from low-complexity mobile-access applications to high-quality sound systems, building on the existing MPEG audio codecs. It supports high-quality monaural, stereo and multichannel signals using both MPEG-2 AAC and MPEG-4 coding. In particular, MPEG-4 supports very low bit rates; it codes natural audio at bit rates from 200 bps to 64 kbps/channel. When variable rate coding is employed, coding at less than 2 kbps (perhaps with an average bit rate of 1.2 kbps) is also supported. MPEG coding tools such as AAC are employed for bit rates above 16 kbps/channel and the MPEG-4 standard defines the bitstream syntax and the decoding processes as a set of tools. MPEG-4 also supports intermediate-quality audio coding, such as the MPEG-2 LSF (Low Sampling Frequencies) mode. The audio signals in this region typically have sampling frequencies starting at 8 kHz. Generally, the MPEG-4 audio tools can be grouped according to functionality: speech, general audio, scalability, synthesis, composition, streaming, and error protection. Eight profiles are supported: Main, Scalable, Speech, Synthesis, Natural, High Quality, Low Delay, and Mobile Audio Internetworking (MAUI).

MPEG-4 supports wideband speech coding, narrow-band speech coding, intelligible speech coding, synthetic speech, and synthetic audio. Four audio profiles are defined, each comprising a set of tools. The Synthesis Profile provides score-driven synthesis using SAOL (described below), wavetable synthesis, and a Text-to-Speech (TTS) interface to generate speech at very low bit rates. The Speech Profile provides a Harmonic Vector eXcitation Coding (HVXC) very-low bit-rate parametric speech coder, and a CELP narrow-band/wideband speech coder. The Scalable Profile, a superset of the speech and synthesis profiles, provides scalable coding of speech and music for networks, such as Internet and Narrow-band Audio DIgital Broadcasting (NADIB). The bit rates range from 6 kbps to 24 kbps, with bandwidths between 3.5 kHz and 9 kHz. The Main Profile is a superset of the other three profiles, providing tools for natural and synthesized audio coding. The general structure of the MPEG-4 audio encoder is shown in Fig. 15.7. Three general types of coding can be used. Para-

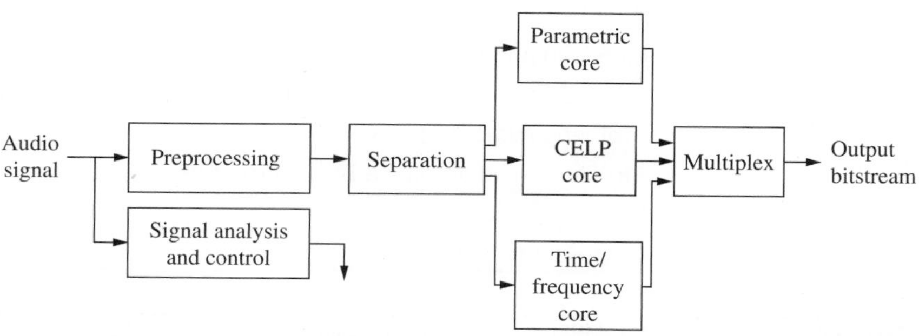

Figure 15.7 The MPEG-4 audio encoder can use three general classes of coding algorithms.

metric models are useful for low bit-rate speech signals and low bit-rate music signals with single instruments that are rich in harmonics. CELP coding uses a source vocal tract model for speech coding, as well as a quantization noise spectral envelope that tracks the spectral envelope of the audio signal. Time/frequency coding, such as in MPEG-1 and MPEG-2 AAC can also be used.

Speech coding at bit rates between 2 kbps and 24 kbps is supported by parametric Harmonic Vector eXcitation Coding (HVXC) for a recommended operating bit rate of 2 kbps to 4 kbps and CELP coding for an operating bit rate of 4 kbps to 24 kbps. In variable bit-rate mode, an average of about 1.2 kbps is possible. The HVXC codec allows user variation in speed and pitch; the CELP codec supports variable speed using an effects tool. In HVXC coding, voiced segments are represented by harmonic spectral magnitudes of linear-predictive coding (LPC) residual signals, and unvoiced segments are represented by a vector excitation coding (VXC) algorithm. In CELP coding, two sampling rates, 8 kHz and 16 kHz, are used to support narrow-band and wideband speech, respectively. Two different excitation modes are used: multipulse excitation (MPE) and regular pulse excitation (RPE). Both a narrow-band mode (8-kHz sampling frequency) and wideband mode (16-kHz sampling frequency) are supported. The synthetic coding is known as synthetic-natural-hybrid-coding (SNHC). Some of the MPEG-4 audio coding tools are shown in Table 15.5.

Type	Tools
Speech	Code Excited Linear Prediction (CELP)
	Harmonic Vector eXcitation Coding (HVXC)
	Text-to-Speech (TTS) Interface
	Variable Bit Rate HVXC
	Silence Compression
General Audio	MPEG-2 AAC Main
	MPEG-2 AAC Low Complexity (LC)
	MPEG-2 AAC Scalable Sampling Rate (SSR)
	MPEG-4 AAC Low Delay (LD)
	Perceptual Noise Substitution (PNS)
	Long-Term Prediction (LTP)
	Harmonic and Individual Lines plus Noise (HILN)
	Transform-domain Weighted INterleaved Vector Quantization (TwinVQ)
Scalability	Bit-Sliced Arithmetic Coding (BSAC)
Synthesis	Tools for Large Step Scalability (TLSS)
	Synthetic Audio (SA) Tools
	Structured Audio Sample Bank Format (SASBF)
	MIDI
Error Protection	Error Robustness Tools

TABLE 15.5 A partial list of MPEG-4 audio coding tools.

MPEG-4 provides scalable audio coding without degrading coding efficiency. With scalability, the transmission and decoding of the bitstream can be dynamically adapted to diverse conditions; this is known as large-step scalability. Speech and music can be coded with a scalable method in which only a part of the bitstream is sufficient for decoding at a lower bit rate and correspondingly lower quality level. This permits, for example, adaptation to changing transmission channel capacity and playback with decoders of different complexity. Bitstream scalability also permits operations with selected content. For example, part of a bitstream representing a certain frequency spectrum can be discarded. In addition, one or several audio channels can be singled out by the user for reproduction in combination with other channels. Scalability in MPEG-4 is primarily accomplished with hierarchical embedding coding such that the composite bitstream comprises multiple subset bitstreams that can be decoded independently or combined.

As shown in Fig. 15.8, the composite bitstream contains a base-layer coding of the audio signal and one or more enhancement layers. The enhancement layer bitstreams are generated by subtracting a locally decoded signal from a previous signal to produce

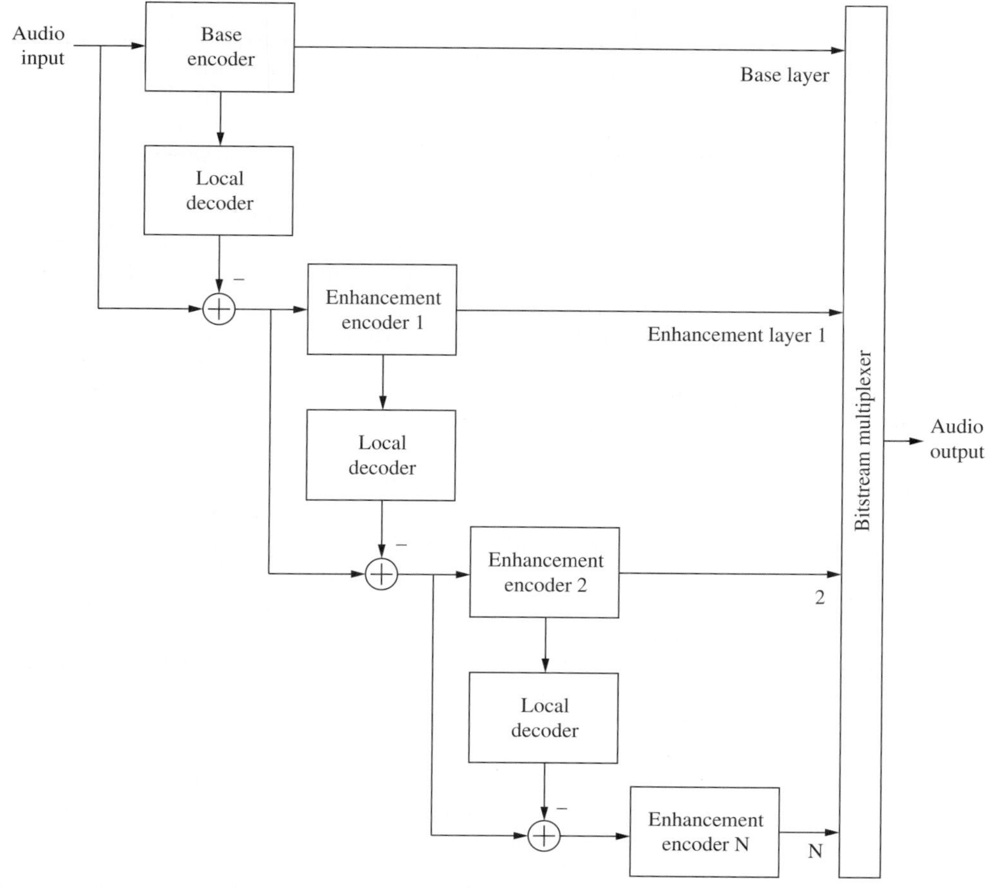

FIGURE 15.8 The MPEG-4 standard permits scalability of the audio bitstream. A base layer is successively recoded to generate multiple enhancement layers.

a residual or coding error signal. This residual signal is coded by an enhancement encoder to produce the next partial bitstream. This is repeated to produce all of the enhanced layers. A decoder can use the base layer for basic audio quality or add subsequent enhancement layers to improve audio quality. Scalability can be used for signal-to-noise or bandwidth improvements, as well as monaural to stereo capability. Although MPEG-4 imposes some restrictions on scalability, many coding configurations are possible. For example, either the Transform-domain Weighted INterleaved Vector Quantization (TwinVQ) or AAC codec could be used for both base and enhancement coding, or a TwinVQ or CELP codec could be used for the base layer and AAC used for enhancement layers. To permit scalability, the bitstream syntax of MPEG-4 differs somewhat from that of MPEG-2. The coding allows pitch changes and time changes, and is designed for error resilience.

MPEG-4 defines structured audio formats and an algorithmic sound language in which sound models are used to code signals at very low bit rates. Sound scene specifications are created and transmitted over a channel and executed at the receiver; synthetic music and sound effects can be conveyed at bit rates of 0.01 kbps to 10 kbps. The description of postproduction processing such as mixing multiple streams and adding effects to audio scenes can be conveyed as well.

As with other content, audio in MPEG-4 is defined on an object-oriented basis. An audio object can be defined as any audible semantic entity, for example, voices of one or more people, or one or more musical instruments, and so on. It might be a monaural or multichannel recording, and audio objects can be grouped or mixed together. An audio object can be more than one audio signal; for example, one multichannel MPEG-2 audio bitstream can be coded as one object.

MPEG-4 also defines audio composition functions. It can describe aspects of the audio objects such as synchronization and routing, mixing, tone control, sample rate conversion, reverberation, spatialization, flanging, filtering, compression, limiting, dynamic range control, and other characteristics.

MPEG-4's Structured Audio tools are used to decode input data and produce output sounds. Decoding uses a synthesis language called Structured Audio Orchestra Language (SAOL, pronounced "sail"). It is a signal-processing language used for music synthesis and effects post-production using MPEG-4. SAOL is similar to the Music V music-synthesis languages; it is based on the interaction of oscillator models. SAOL defines an "orchestra" comprising "instruments" that are downloaded in the bitstream, not fixed in the decoder. An instrument is a hardware or software network of signal-processing primitives that might emulate specific sounds such as those of a natural acoustic instrument. "Scores" or "scripts" in the downloading bitstream are used to control the synthesis. A score is a time-sequenced set of commands that invokes various instruments at specific times to yield a music performance or sound effects.

The Synthetic Audio Sample Bank Format (SASBF) describes a bank of wavetables storing specific instruments that can be reproduced with a very low data rate. The score description is downloaded in a language called Structured Audio Score Language (SASL). It can create new sounds and include additional control information to modify existing sounds. Alternatively, MIDI may be used to control the orchestra. A "wavetable bank format" is also standardized so that sound samples for use in wavetable synthesis may be downloaded, as well as simple processing such as filters, reverberation, and chorus effects.

MPEG-4 defines a signal-processing language for describing synthesis methods but does not standardize a synthesis method. Thus, any synthesis method may be contained in the bitstream. However, because the language is standardized, all MPEG-4 compliant synthesized music sounds the same on every MPEG-4 decoder. MPEG-4 also defines a text-to-speech (TTS) conversion capability. Text is converted into a string of phonetic symbols and the corresponding synthetic units are retrieved from a database and then concatenated to synthesize output speech. The system can also synthesize speech with the original prosody (such as pitch contour and phoneme duration) from the original speech, synthesize speech with facial animation, synchronize speech to moving pictures using text and lip shape information, and alter the speed, tone, and volume, as well as the speaker's sex and age. In addition, TTS supports different languages. TTS data rates can range from 200 bps to 1.2 kbps. Some possible applications for TTS include artificial storytellers, speech synthesizers for avatars in virtual-reality applications, speaking newspapers, and voice-based Internet.

Scalability options mean, for example, that a single audio bitstream could be decoded at a variety of bandwidths depending on the speed of the listener's Internet connection. An audio bitstream can comprise multiple streams of different bit rates. The signal is encoded at the lowest bit rate, and then the difference between the coded and original signal is coded.

MPEG-4 Versions

Version 1 of MPEG-4 was adopted in October 1998. Version 2 was finalized in December 1999 and adds tools to the MPEG-4 standard, but the Version 1 standard remains unchanged. In other words, Version 2 is a backward-compatible extension of Version 1. In December 1999, Versions 1 and 2 were merged to form a second edition of the MPEG-4 standard. A block diagram of the MPEG-4 general audio encoder is shown in Fig. 15.9. Long-term prediction (LTP) and perceptual noise substitution (PNS) tools are included along with a choice of AAC, BSAC, or TwinVQ tools; these differentiate this general audio encoder from the AAC encoder.

Version 2 extends the audiovisual capabilities of the standard. For example, Version 2 provides audio environmental spatialization. The acoustical properties of a scene such as a 3D model of a concert hall can be characterized with the BIFS scene description tools. Properties include room reverberation time, speed of sound, boundary material properties such as reflection and transmission, and sound source directivity. These scene description parameters allow advanced audiovisual rendering, detailed room acoustical modeling, and enhanced 3D sound presentation.

Version 2 also improves the error robustness of the audio algorithms; this is useful, for example, in wireless channels. Some aspects address the error resilience of particular codecs, and others provide general error protection. Tools reduce the audibility of artifacts and distortion caused by bit errors. Huffman coding for AAC is made more robust with a Huffman codeword reordering tool. It places priority codewords at regular positions in the bitstream; this allows synchronization that can overcome bit-error propagation. Scale-factor bands of spectral coefficients with large absolute values are vulnerable to bit errors so virtual codebooks are used to limit the value of the scale-factor bands. Thus large coefficients produced by bit errors can be detected and their effects concealed. A reversible variable-length coding tool improves the error robustness of Huffman-coded DPCM scale-factor bands with symmetrical codewords. Scale-factor data can be decoded forward or backward. Unequal error protection can also be applied by reordering the bitstream payload according to error sensitivity to allow adapted channel-coding techniques. These tools allow flexibility in different error-correction overheads and capabilities, thus accommodating diverse channel conditions.

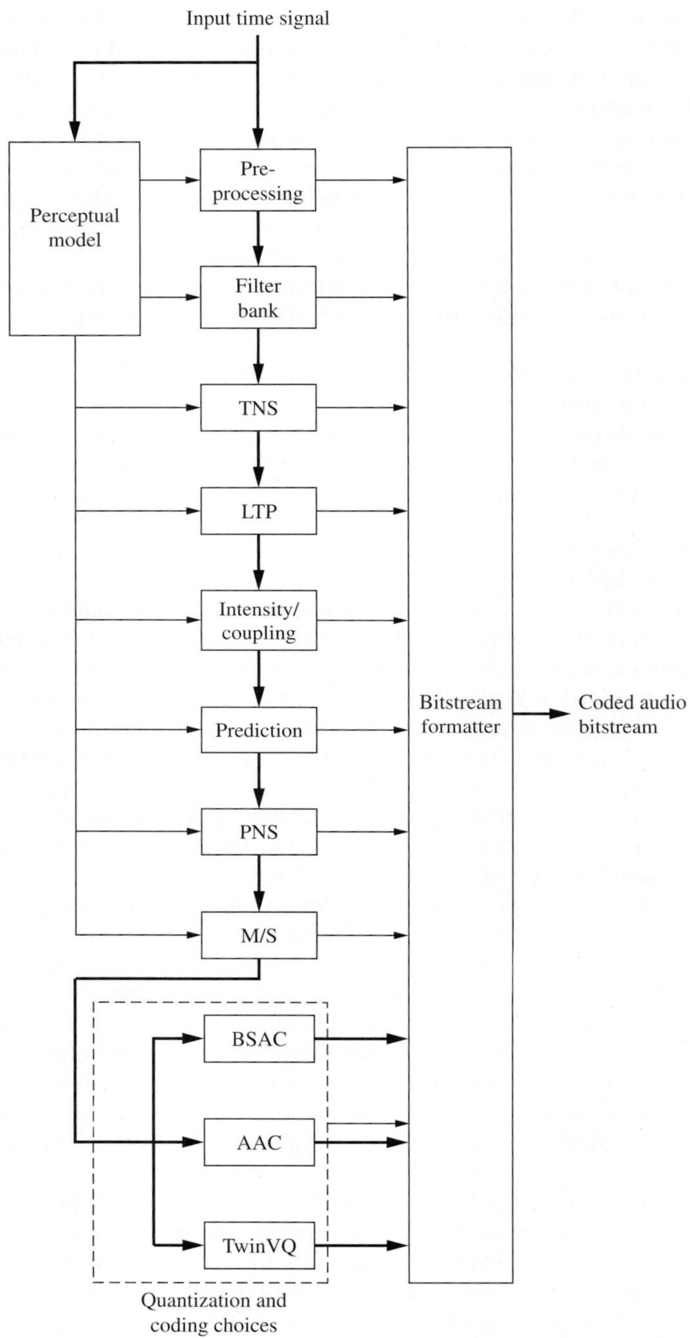

FIGURE 15.9 Block diagram of the MPEG-4 general audio encoder with a choice of three quantization and coding table methods. Heavy lines denote data paths, light lines denote control signals.

Version 2 also provides music and speech audio coding with low-delay coding (AAC-LD). LD is useful in real-time two-way communication where low latency is critical; the total delay at 48 kHz is reduced from 129.4 ms to 20 ms. However, coding efficiency is moderately diminished, very generally, by about 8 kbps/channel compared to the AAC main profile. Viewed another way, the performance of LD coding is comparable to that of MPEG-1/2 Layer III at 64 kbps/channel. In LD, the frame length is reduced from 1024 to 512 samples, or from 960 to 480 samples, and filter-bank window size is halved. Window-switching is not permitted; a look-ahead buffer otherwise used for block-switching is eliminated. To minimize pre-echo artifacts in transient signals, however, TNS is permitted along with window-shape adaptation. A low-overlap window is used for transient signals in place of the sine window used for nontransient signals. The use of the bit reservoir is constrained or eliminated entirely. Sampling frequencies up to 48 kHz are permitted.

Version 2 defines a Harmonic and Individual Lines plus Noise (HILN) coding tool. HILN is a parametric coding technique in which source models are used to code audio signals. An encoder is shown in Fig. 15.10. Frames of audio samples are windowed and overlapped; frame length is typically 32 ms at a 16-kHz sampling frequency. The signal is decomposed into individual sinusoids and output as sinusoidal parameters and a residual signal. The model parameters for the component's source models are estimated; the sinusoidal parameters are separated into harmonic and sinusoidal components, and the residual signal is estimated as a noise component. For coding efficiency, sinusoids with the same fundamental frequency are grouped into a single harmonic tone while other sinusoids are treated individually. Spectral modeling, for example, using the frequency response of an all-pole LPC synthesis filter, is used to represent the spectral envelope of the harmonic tone and noise components. The components are perceptually quantized; to decrease bit rate, only the most perceptually significant components are coded. HILN is used for very low bit rates of typically 4 kbps to 16 kbps with bit-rate scalability. With HILN coding, playback speed or pitch can be altered at the decoder without requiring a special-effects processor. HILN can be used in conjunction with the HVXC speech codec.

Version 2 allows substitution of Bit-Sliced Arithmetic Coding (BSAC), sometimes known as bit-plane coding, in AAC. BSAC is optimally used at high bit rates of 48 kbps to

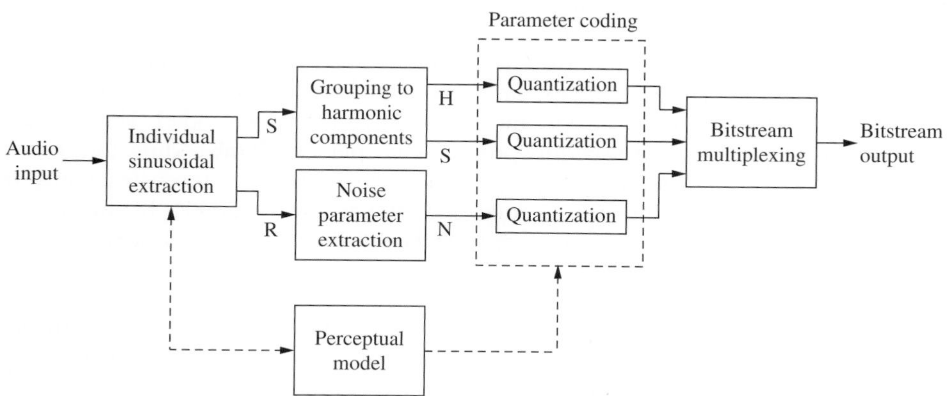

FIGURE 15.10 The MPEG-4 standard, Version 2 defines an HILN parametric audio encoder in which source models are used to code audio signals.

64 kbps/channel, and is less bit-efficient at low bit rates. BSAC replaces Huffman coding of quantized spectral coefficients with arithmetic coding to provide small-step scalability. A sign-magnitude format is used, and the magnitudes are expressed as binary integers up to 13 bits in length. The bits are spectrally processed in bit-plane slices from MSB to LSB significance. Bit slices are entropy-encoded using one of several BSAC models. A base layer contains bit slices of spectral data within an audio bandwidth defined by bit rate as well as scale factors and side information. Enhancement layers add LSB bit-slice information as well as higher frequency bands providing greater audio bandwidth. Data packets of consecutive frames are grouped to reduce the overhead of the packet payload. Using the bitstream layers, the decoder can reproduce a multiple resolution signal scalable in 1 kbps/channel increments.

Version 2 also allows substitution of Transform-domain Weighted INterleaved Vector Quantization (TwinVQ) coding in AAC. TwinVQ uses spectral flattening and vector quantization and can be used at low bit rates (for example, 6 kbps/channel). Version 2 can also improve audio coding efficiency in applications using object-based coding by allowing selection and/or switching between different coding techniques. Because of its underlying architecture and fixed-length codewords, TwinVQ provides inherently good error robustness.

The High-Efficiency AAC (HE AAC) extension uses spectral bandwidth replication (SBR) to improve AAC coding efficiency by 30%. HE AAC is backward-compatible to AAC and is sometimes used at bit rates of 24 kbps or 48 kbps/channel. The SBR tool was approved in March 2003. Other extensions include the use of parametric coding for high-quality audio, and lossless audio coding. HE AAC is also known as aacPlus.

The High-Definition AAC (HD AAC) extension provides scalable lossless coding. It provides a scalable range of coding resolutions from perceptually coded AAC to a fully lossless high-definition representation with intermediate near-lossless representations. The input signal is first coded by an AAC core-layer encoder, and then the scalable lossless coding algorithm uses the output to produce an enhancement layer to provide lossless coding. The two layers are multiplexed into one bitstream. Decoders can decode the perceptually coded AAC part only or use the additional scalable information to enhance performance to lossless or near-lossless sound-quality levels. In the scalable encoder, the audio signal is converted into a spectral representation using the integer-modified discrete cosine transform (IntMDCT). AAC coding tools such as middle/side (M/S) and temporal noise shaping (TNS) can be applied to the coefficients in an invertible integer manner. An error-mapping process maintains a link between the perceptual core and the scalable lossless layer. The map removes the information that has already been coded in the perceptual path so that only the IntMDCT residuals are coded. The residual can be coded using two bit-plane coding processes: bit-plane Golomb coding and context-based arithmetic coding, and a low-energy mode encoder. Lossless reconstruction is obtained when all of the bit planes are decoded, and lossy reconstruction is obtained (but superior to the AAC core decoding only) if only parts of the bit planes are decoded.

MPEG-4 Coding Tools

The MPEG-4 standard encompasses previous MPEG audio standards. In particular, within MPEG-4, MPEG-2 AAC is used for high-quality audio coding. As noted, MPEG-4 adds several specialized tools to MPEG-2 AAC to achieve more efficient coding. A perceptual noise substitution (PNS) tool can be used with AAC at bit rates below 48 kbps. Using the approach that "all noise sounds alike," PNS can efficiently code noise-like signal components with a parametric representation. This helps overcome the inefficiency in coding noise because of its lack of redundancy. PNS exploits the fact that noise-like signals are not perceived as waveforms, but by their temporal and spectral characteristics. When PNS is employed, instead of conveying quantized spectral coefficients for a scale-factor band, a noise substitution flag and designation of the power of the coefficients are conveyed. The decoder inserts pseudo-random values scaled by the proper noise power level. PNS can be used for stereo coding because the pseudo-random noise values are different in each channel, thus avoiding monaural localization. However, PNS cannot be used for M/S-stereo coding.

MPEG-4 also provides a long-term prediction (LTP) tool to be used with AAC. It is a low-complexity prediction method similar to the technique used in speech codecs. It is particularly efficient for stationary harmonic signals and to a lesser extent for nonharmonic tonal signals. In this tool, forward-adaptive long-term prediction is applied to preceding frames of spectral values to yield a prediction of the audio signal. The spectral values are mapped back to the time domain, and the reconstructed signal is analyzed in terms of the actual input signal. Parameters for amplitude scaling and delay comprise a prediction signal. The prediction and input signals are mapped to the frequency domain and subtracted to form a residual (difference) signal. Depending on which provides the most efficient coding on a scale-factor band basis, either the input signal or the residual can be coded and conveyed, along with side information.

As noted, MPEG-4 also provides TwinVQ as an alternate coding kernel in MPEG-2 AAC. It is used for bit rates ranging from 6 kbps to 16 kbps/channel for audio including music; it is often used in scalable codec designs. TwinVQ first normalizes the spectral coefficients from the AAC filter bank to a target range, then codes blocks of coefficients using vector quantization (sometimes referred to as block quantization). The normalization process applies LPC estimation to the signal's spectral envelope to normalize the amplitude of the coefficients; the LPC parameters are coded as line spectral pairs. Fundamental signal frequency is estimated and harmonic peaks are extracted; pitch and gain are coded. Spectral envelope values using Bark-related AAC scale-factor bands are quantized with vector quantization; this further flattens the coefficients. Coefficient coding is performed. Coefficients are interleaved in frequency and divided into subvectors; this allows a constant bit allocation per subvector. Perceptual shaping of quantization noise may be applied. Finally, subvectors are vector quantized. Quantization distortion is minimized by optimizing the choice of codebook indices. The TNS and LTP tools may be used with TwinVQ.

The MPEG-4 audio lossless coding (ALS) tool allows compression and bit-accurate reconstruction of the audio signal with resolutions up to 24 bits and sampling frequencies up to 192 kHz. Forward-adaptive linear predictive coding is employed using, for example, the Levinson–Durbin autocorrelation algorithm. The encoder estimates and quantizes adaptive predictor coefficients and uses the quantized predictor coefficients to calculate prediction residues. The residues are entropy-coded with one of several different Rice codes or other coding variations. The coded residues, as well as side information, such as code indices, predictor coefficients, and CRCC checksum, are combined to form the compressed bitstream. The decoder decodes the entropy-coded residue and uses the predictor coefficients to calculate the lossless reconstruction signal.

The Internet Streaming Media Alliance (ISMA) has specified a transport protocol for streaming MPEG-4 content. This protocol is based on RTSP and SDP for client-server hand shaking and RTP packets for data transmission. MP4 and XMT files are used for interoperability between authoring tools and MPEG-4 servers. MPEG-4 content can be conveyed in various ways such as via Internet RTP/UDP, MPEG-2 transport streams, ATM, and DAB. FlexMux, using elements defined in recommendation H.223, is a multiplexing tool defined by MPEG-4.

MPEG-7 Standard

The MPEG-7 standard is entitled "Multimedia Content Description Interface." Its aim is very different from the MPEG standards that describe compression and transmission methods. MPEG-7 provides ways to characterize multimedia content with standardized analysis and descriptions. For example, it allows more efficient identification, comparison, and searching of multimedia content in large amounts of data. MPEG-7 is described in the ISO/IEC 15938 standard; Version 1 was finalized in 2001.

Many search engines are designed to use database management techniques to find text and pictures on the Web. However, this content is mainly identified by annotative, text-based metadata that describes outward aspects of works such as keywords, title, author or composer, and year of creation. Creation of this metadata is often done manually, and a text-based description may not adequately describe all aspects of the file.

MPEG-7 describes more intrinsic characteristics in multimedia and allows content-based retrieval. In the case of audio, MPEG-7 provides ways to analyze audio waveforms and manually and automatically extract content-based information that can be used to describe and classify the signal. These descriptors are sometimes called feature vectors or fingerprints. The process of obtaining these descriptors from an audio file is called audio feature extraction or audio fingerprinting. A variety of descriptors may be extracted from one audio file. Descriptors might include key signature, instrumentation, melody, and other parameters that may be derived from the content itself.

Audio signals are described by low-level descriptors. These include basic descriptors (audio waveform, power), basic spectral descriptors (spectrum envelope, spectrum spread, spectrum flatness, spectrum centroid), basic signal parameters (fundamental frequency, harmonicity), temporal timbral descriptors (log attack time, temporal centroid), spectral timbral descriptors (harmonic spectral centroid, harmonic spectral deviation, harmonic spectral spread, harmonic spectral variation, spectral centroid), spectral basis representations (spectrum basis, spectrum projection), and silence descriptors. Other descriptors include lyrics, key, meter, and starting note. Musical instrument sounds can be classified in four ways as harmonic, sustained, coherent; percussive, nonsustained; nonharmonic, sustained, coherent; and noncoherent, sustained. In addition to music, sound effects and spoken content could also be described.

Descriptors also describe information such as artist and title, copyright pointers, usage schedule, user preferences, browsing information, storage format, and encoding. The standard also provides description schemes written in XML. They specify which low-level descriptors can be used in any description as well as relationships between descriptors or between other schemes. For example, description schemes can be used to characterize timbre, melody, tempo, and nonperceptual audio signal quality; these high-level tools can be used to compare and query music.

The MPEG-7 standard also provides a description definition language based on the XML Schema Language; it defines the syntax for using description tools. The language allows users to write their own descriptors and schemes. Binary-coded representations of descriptors are also provided. The standard does not specify how the waveform is coded; for example, the content may be either a digital or analog waveform. As with many other standards, although MPEG-7 includes standardized tools, informative instructions are provided for much of the implementation. This allows future improvement of the technology while retaining compatibility. Some parts of the standard such as syntax and semantics of descriptors are normative. MPEG-7 allows searches using audiovisual descriptions. For example, one could textually describe a visual scene as "the witches in Macbeth" to find a video of the Shakespeare tragedy, sketch a picture of the Eiffel Tower to find a photograph of the structure, or whistle a few notes to obtain a listing of recordings of Beethoven's Ninth Symphony. In addition to a search and retrieval function, MPEG-7 could be applied to content-based applications such as real-time monitoring and filtering of broadcasts, recognition, semi-automated editing, and playlist generation. There is no MPEG-3, MPEG-5, or MPEG-6 standard.

CHAPTER 16

Digital Radio and Television Broadcasting

Although the Internet is eroding their dominance, broadcast radio and television still play an important role in our lives. In some cases, they retain their supremacy. For example, adults spend about 1.5 hours in their cars every day, and of the time spent listening to audio content, 74% is broadcast radio and 6% is satellite radio.

Radio and television have historically been broadcast from terrestrial towers, transmitting on assigned frequencies to local markets. In addition, both services can be relayed nationwide by satellite. In addition, satellites are the workhorses of the global telecommunications industry. This chapter surveys audio broadcasting technology and some of its applications, with attention to digital audio radio (DAR) and digital television (DTV) broadcasting.

Satellite Communication

Outer space is only 62 miles away. If you could drive your car straight up, you could get there in about an hour. However, since today's cars cannot do that, we need rockets instead, and it is enormously expensive to move things into space—perhaps $10,000 per pound. Despite the cost, we routinely launch space vehicles, and the most commercially valuable ones are communication satellites. With satellite transmission, information is conveyed thousands of miles, to one receiver or to millions of receivers, using telecommunications satellites as unmanned orbiting relay stations.

Geostationary satellites use a unique orbit, rotating from west to east over the equator, moving synchronously with the earth's rotation. From the earth, they appear to be fixed in the sky; this is a geostationary orbit. Objects orbiting close to the earth rotate faster than the earth, and objects farther away rotate more slowly. The International Space Station (173 to 286 miles high) orbits the earth in 90 minutes. The moon (221,000 to 253,000 miles away) orbits in 27.3 days. At 22,236 miles above the earth, geostationary orbit (one orbit per day) is achieved; this is where geostationary satellites are parked. International law dictates a separation of 2° between vehicles. The unique properties of geostationary satellites make these positions quite valuable.

The conveyed signal has a line-of-sight characteristic similar to that of visible light; thus, it is highly directional. From their high altitude, geostationary communications satellites have a direct line of sight to almost half the earth's surface; three satellites would encompass the entire globe except for small Polar regions. A satellite's footprint describes the area over which its receiving and transmitting antennas are focused.

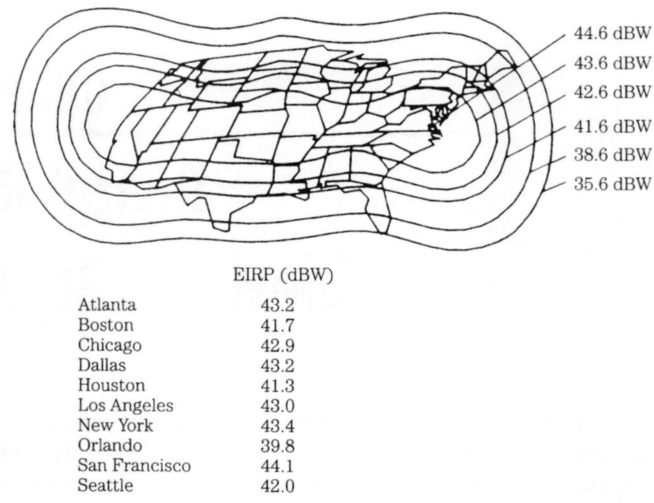

	EIRP (dBW)
Atlanta	43.2
Boston	41.7
Chicago	42.9
Dallas	43.2
Houston	41.3
Los Angeles	43.0
New York	43.4
Orlando	39.8
San Francisco	44.1
Seattle	42.0

Figure 16.1 Satellite downlink footprints are contoured to cover a specific geographic location, and for example, minimize interference with neighboring countries.

The footprint can cover an entire hemisphere, or a smaller region, with a gradual reduction in sensitivity away from the footprint's center area, as shown in Fig. 16.1. In this example, the footprint is characterized as effective isotropic radiated power (EIRP). Both earth stations (uplink and downlink) must lie within the satellite's footprint. Generally, C-band footprints cover larger geographical areas than Ku-band footprints. Because of the high orbit, a communications delay of 270 ms is incurred.

Satellite communications operate at microwave frequencies. Specifically, they occupy the super-high-frequency (SHF) band extending from 3 GHz to 30 GHz; the broadcast spectrum is shown in Fig. 16.2. Two Fixed Satellite Services (FSS) bands are in common domestic use: the C-band (3.4 GHz to 7.075 GHz) and the Ku-band (10.7 GHz to 18.1 GHz). In either case, several higher-frequency subbands (C-band: 5.725 GHz to 7.075 GHz; Ku-band: 12.7 GHz to 18.1 GHz) are used for uplink signals, and several lower-frequency subbands (C-band: 3.4 GHz to 4.8 GHz; Ku-band: 10.7 GHz to 12.7 GHz) are used for downlink signals. Many geostationary satellites share the same spectral space, and ground stations must rely on physical satellite spacing and antenna directionality to differentiate between satellites.

Most C-band transponders use a 36-MHz bandwidth placed on 40-MHz centers, although in some cases 72-MHz transponders are used. Ku-band transponder bandwidths are either 36-MHz or 72-MHz wide. The C-band affords superior propagation characteristics. However, Ku-band satellites can offset this with greater transponder antenna gain. Some satellites operate in both bands. Because the C-band must share its spectral space with other terrestrial applications, it suffers from the possibility of terrestrial microwave interference such as terrestrial microwave links. This necessitates a lower transmitting power and larger antenna diameter. C-band dishes are typically 2 meters in diameter or larger, and downlink stations must properly shield their antennas.

The shorter Ku-band wavelengths are more easily absorbed by moisture, thus the signal can be degraded by snow, rain, and fog. In particular, heavy rainfall might significantly degrade Ku-band signals. However, because the Ku-band is not shared with

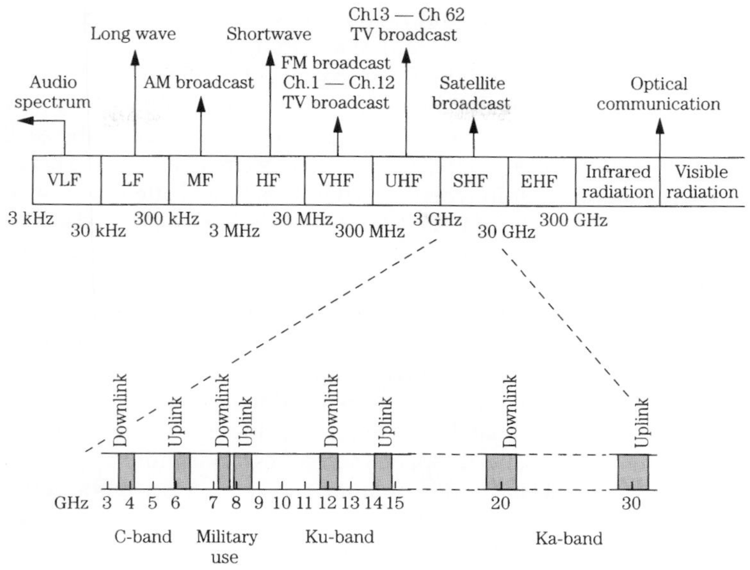

FIGURE 16.2 Satellite communications occupy the super-high-frequency band, with specific bands used for uplink and downlink transmissions. Only a few of the uplink and downlink bands are shown.

other terrestrial applications, it does not suffer from microwave interference, thus higher power can be applied. In addition, for a given size, Ku-band dishes provide higher gain than C-band dishes. Although Ku-band dishes are typically 1.8 meters in diameter, much smaller dishes are used in direct broadcast satellite applications. In some cases, a combination of bands is used. The Ku-band can be accessed via portable uplinks. The downlinked signal is converted to the C-band at a ground station, and uplinked via the C-band for distribution.

Audio content can be transmitted as voice-grade 7.5-kHz audio, 15-kHz audio, or other formats. The voice-grade format is coded with continuously variable slope delta (CVSD) modulation to attain a data rate of 32 kbps. The 7.5-kHz format is sampled at 16 kHz, and the 15-kHz format is sampled at 32 kHz; both use 15-bit quantization followed by μ-law companding to yield 11 bits plus a parity bit. Multiple channels are multiplexed into a T-1 (1.544 Mbps) stream and sent to the uplink station. The individual audio channels are multiplexed into a 7.68-MHz bitstream, modulated to a 70-MHz intermediate frequency (IF) carrier, upconverted and uplinked for satellite distribution. A single 15-kHz PCM channel requires a bit rate of 512 kbps; companding decreases this to 384 kbps; data reduction decreases this to 128 kbps.

A satellite's transponders receive the ground station's uplink signal and retransmit it back to earth where a downlink receives the signal. A communications satellite might have 48 or more transponders each capable of receiving multiple (8 to 12) data channels from an uplink, or transmitting those channels to a receiving downlink. Horizontally, vertically, and circularly polarized signals are broadcast to increase capacity in a frequency band. Depending on the transmitting power of the satellite, the signal can be received by equipment of greater or lesser sophistication. For example, a

20-W satellite transmitter would require a receiving dish several meters in diameter, and a 200-W transmitter would require a dish diameter of less than a meter. The transponder reliability rate exceeds 99% over years of service.

Satellites use solar cells to derive power from solar energy. So that correct attitude stabilization is maintained (the antennas must stay pointed at the earth) geostationary satellites must rotate once every 24 hours as they circle the earth. This action is provided by thrusters that create spin when the satellite is first orbited. In addition, in the case of cube-shaped, body-stabilized 3D satellites, three internal gyroscopes control position about three axes, providing correction when necessary. In this design, solar cells are mounted on large solar sails, and motors move the sails to face the sun. Cylindrically shaped satellites, called spinners, achieve stabilization by spinning the entire satellite body about an axis. In this design, solar cells are mounted directly on the satellite's body; antennas must be despun. Hydrazine fuel thrusters are used to maintain absolute position within a 40-mile square in the geostationary orbit, compensating for the pull of the sun and moon. Most satellite failures are due to fuel depletion, and the resulting drifting of the vehicle due to space mechanics. A satellite might measure 20 feet in height and weigh 15,000 pounds at launch. A satellite's weight determines its launch cost. Because of limited satellite life span of 15 years or less, any satellite system must budget for periodic renewal of its spacecraft.

Interestingly, twice each year all geostationary satellite downlink terminals undergo solar outages (for 5 minutes or so) when the sun, the relaying satellite, and the earth station are all in a straight line. The outage occurs when the shadow of the antenna's feed element is in the center of the dish; solar interference (noise power from the sun) degrades reception. Solar transit outages occur in the spring and fall. Beginning in late February, outages occur at the U.S.–Canada border and slowly move southward at 3° latitude per day, and beginning in early October outages begin at the U.S.–Mexico border and move northward at the same rate. In addition, eclipses of geostationary satellites occur about 90 evenings a year in the spring and fall when the earth blocks the sun's light to the satellite. Onboard batteries provide continuous power while the solar sails go dead for 70 minutes.

Instead of using a relatively few geostationary satellites, low earth orbit (LEO) satellite systems use a constellation of many satellites to permit low-cost access. Rather than sitting in fixed points, LEOs ride in low and fast orbits, moving freely overhead in the sky. With LEOs, a terrestrially transmitted signal is picked up by a satellite passing overhead, and it relays a signal as directed. Because of their close proximity, LEOs minimize communication latency (perhaps 100 ms or less). Because they are small, they can be launched more cheaply, and are more easily replaced. Because the system is distributed, any single satellite failure will have a minimal impact on overall operation. Some LEOs operate at a frequency of less than 1 GHz; they have relatively small bandwidths and are used for low bit-rate applications such as paging. Other LEOs operate anywhere from 1 GHz to 30 GHz.

Medium earth orbit (MEO) satellites use special orbital mechanics for more efficient landmass coverage. Because of its higher altitude, an MEO satellite's coverage footprint is much larger than that of a LEO; this reduces the number of deployed satellites. Instead of using circular orbits, MEO satellites might use elliptical orbits; for example, with apogees of 4000 miles and perigees of 300 miles. The apogees are near the northern extremity of the orbits, thus the satellites spend more time over domestically populated areas. Moreover, the orbits can be configured to be sun-synchronous; their orbital plane remains fixed relative to the sun throughout the year. In this way, the satellite's greatest orbital coverage can be optimized to the time of day with peak usage.

Direct Broadcast Satellites

In many applications, satellites are used to convey programs from one point to a few others. For example, a television channel provider can beam programming to local cable companies across the country, which in turn convey the programs to subscribers via coaxial cable. Direct broadcast satellite (DBS) is a point-to-multipoint system in which individual households equipped with a small parabolic antenna and tuner receive broadcasts directly from a geostationary satellite. The satellite receives digital audio and video transmissions from ground stations and relays them directly to individuals. The receiving system comprises an offset parabolic antenna that collects the microwave signals sent by the satellite, and a converter mounted at the antenna's focal point that converts the microwave signal to a lower frequency signal. Because of the high sensitivity of these devices and relatively high satellite transmitting power, the parabolic antenna can be 0.5 meter in diameter. The dishes are mounted outside the home with a southern exposure and are manually aligned with a diagnostic display showing received signal strength. Inside the home, a phase-locked loop tuner demodulates the signal from the converter into video and audio signals suitable for a home television or stereo. For areas not centrally located in the satellite's footprint, larger antennas of a meter or more in diameter can be used for favorable reception. Direct broadcast satellite systems transmit in the Ku-band, a higher frequency region than the very high frequency (VHF) and ultra high frequency (UHF) channels used for terrestrial television broadcasting. Bandwidth is 27 MHz per channel.

The DirecTV system is an example of a direct broadcast satellite system providing digital audio and video programming to consumers. Subscribers use a 0.5-meter-diameter satellite dish and receiver to receive over 200 channels of programming. Three co-located, body-stabilized HS 601 satellites orbit at 101° west longitude, each providing 16 high-power (120 W) transponders in the Ku-band (uplink: 17.2 GHz to 17.7 GHz; downlink: 12.2 GHz to 12.7 GHz). They beam their high-power signals over the continental United States, lower Canada, and upper Mexico. Signals originate from a broadcast center and are digitally delivered over the satellite link, then converted into conventional analog signals in the home, providing audio and video output. MPEG-2 coding is used to reduce a channel's nominal bit rate from 270 Mbps to a rate of 3.75 Mbps to 7.5 Mbps. The compressed data is time-division multiplexed. The bit rate of individual channels can be continuously varied according to content or channel format. The signal chain also includes Reed–Solomon and convolutional error correction coding, and quadrature phase-shift keying (QPSK) modulation. The advent of low-cost satellite-based distribution technology has changed the way that radio and television signals are received.

Digital Audio Radio

Analog broadcasting harkens back to the early days of audio technology. However, in August 1986, WGBH-FM in Boston simulcast its programming over sister station WGBX-TV using a pseudo-video PCM (F1) processor, coding the stereo digital audio signal as a television signal, thus experimentally delivering the first digital audio broadcast. Following this experiment, the broadcasting industry has developed digital audio radio (DAR) technologies, also known as digital audio broadcasting (DAB). Instead of using analog modulation methods such as AM or FM, DAR transmits audio signals digitally. DAR is designed to replace analog AM and FM broadcasting, providing a signal that is more robust against reception problems such as multipath interference. In addition to audio data, a DAR system supports auxiliary data transmission; for example, text, graphics, or still video images ("radio with pictures") can be conveyed.

The evolution of a DAR standard is complicated because broadcasting is regulated by governments and swayed by corporate concerns. Two principal DAR technologies have been developed: Eureka 147 DAB and in-band on-channel (IBOC) broadcasting known as HD Radio. The way to DAR has been labyrinthine, with each country choosing one method or another; there are no worldwide standards.

Digital Audio Transmission

Radio signals are broadcast as an analog or digital baseband signal that modulates a high-frequency carrier signal to convey information. For example, an AM radio station might broadcast a 980-kHz carrier frequency, which is amplitude modulated by baseband audio information. The receiver is tuned to the carrier frequency and demodulates it to output the original baseband signal; IF modulation techniques are used. Digital transmissions use a digital baseband signal, often in a data reduction format. The digital data modulates a carrier signal (a high-frequency sinusoid) by digitally manipulating a property of the carrier (such as amplitude, frequency, or phase); the modulated carrier signal is then transmitted. In addition, prior to modulation, multiple baseband signals can be multiplexed to form a digital composite baseband signal. An example of a transmit/receive signal chain is shown in Fig. 16.3.

In any digital audio broadcasting system, it is important to distinguish between the source coder and the channel coder. The source coder performs data reduction coding so the wideband signal can be efficiently carried over reduced spectral space. The channel coder prepares the rate-reduced signal for modulation onto radio frequency (RF) carriers, the actual broadcasting medium; this is needed for efficient, robust transmission. One important consideration of channel coding is its diversity to reduce multipath interference that causes flat or frequency-selective interference, called a fade, in the received signal. Channel coders can use frequency diversity in which the source coder data is encoded on several carrier frequencies spread across a spectral band; a fade will not affect all of the received carriers.

Using adaptive equalization, the receiver might use a training sequence placed at the head of each transmitted data block to recognize multipath interference and adjust its receiver sensitivity across the channel spectrum to minimize interference. In addition, because multipath interference can change over time (particularly in a mobile receiver), time diversity transmits redundant data over a time interval to help ensure proper reception; a cancellation at one moment might not exist a moment later. With space diversity, two or more antennas are used at the receiver (for example, on the windshield and rear bumper of a car), so the receiver can choose the stronger received signal. A fade at one spectral point at one antenna might not be present at the other antenna. Finally, in some systems, space diversity can be used in the transmission chain; multiple transmission antennas are used and the receiver selects the stronger signal.

Digital audio radio can be broadcast in a variety of ways. Like analog radio, DAR can be transmitted from transmission towers, but DAR is much more efficient. An analog

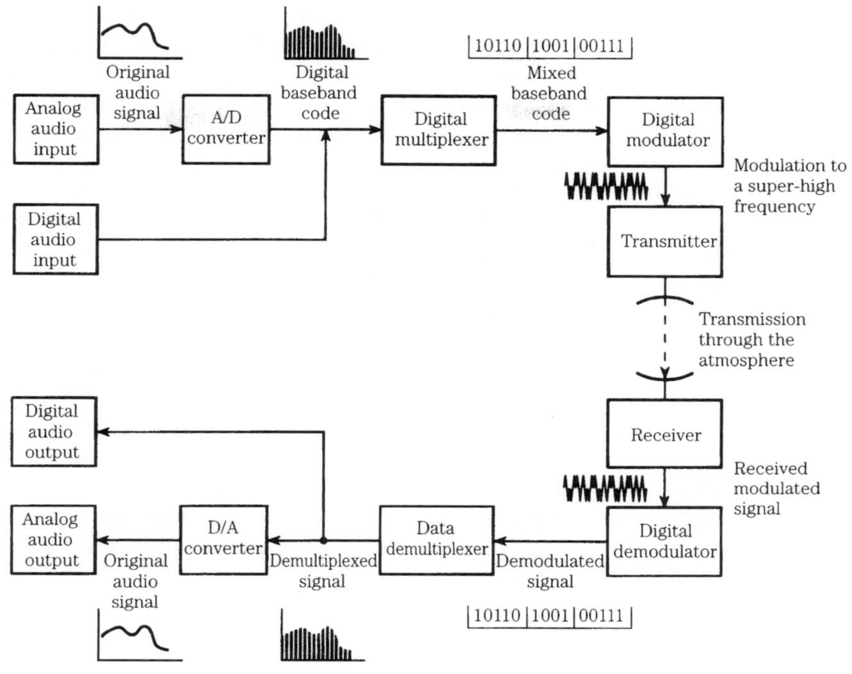

Figure 16.3 An example of signal processing in a transmit/receive signal path.

radio station might broadcast with 100,000 W of power. However, a DAR station might require only 1000 W, providing a significant savings in energy costs. Terrestrial transmission continues the tradition of locally originated stations in which independent stations provide local programming. To their advantage, terrestrial DAR systems can be implemented rather quickly, at a low overall cost.

DAR can also be broadcast directly from satellites, using a system in which programs are uplinked to satellites then downlinked directly to consumers equipped with digital radios. Receivers use low-gain, nondirectional antennas, for example, allowing automotive installation as small flush-mounted modules on the car roof. The resulting national radio broadcasting networks particularly benefit rural areas; they are ideal for long-distance motorists, and a satellite system could extend the effective range of terrestrial stations.

In some proposed satellite radio systems, the transmitting satellite would have multiple (perhaps 28) spot beams, each aimed at a major metropolitan area, as well as a national beam. In this way, both regional and national programming could be accommodated. For example, each beam could convey 16 separate stereo audio programs; listeners in a metropolitan area would receive 32 channels (16 local and 16 national). The use of spot beams was pioneered by the National Aeronautics and Space Administration (NASA); the TDRS (Tracking Data and Relay System) geostationary satellites use space-to-earth spot beam transponders. Although TDRS is used to track low-orbit spacecraft (replacing the ground-tracking stations previously used), it approximates operation of a direct radio broadcast system. NASA and the Voice of America demonstrated a direct broadcast system in the S-band (at 2050 MHz) using

MPEG-1 Layer II coding to deliver a 20-kHz stereo signal at a rate of 256 kbps. The experimental receiver used a short whip antenna, with reasonable indoor performance. The geostationary satellite used a 7-W transmitter.

Implementation of any commercial satellite system requires great capital investment in the satellite infrastructure. In addition, to ensure good signal strength in difficult reception areas such as urban canyons and tunnels, local supplemental transmitters known as gap fillers are needed. Using a system known as single-frequency networking, these transmitters broadcast the same information and operate on the same frequency with contiguous coverage zones. The receiver automatically selects the stronger signal without interference from overlapping zones.

Alternatively, digital audio programs can be broadcast over home cable systems. For example, digital audio programming originating from a broadcast center can be delivered via satellite to local cable providers. Time-division multiplexing is used to efficiently combine many digital audio channels into one wideband signal. At the cable head-end, the channels are demultiplexed, encrypted, and remodulated for distribution to cable subscribers over an unused television channel. At the consumer's home, a tuner is used to select a channel, the channel is decoded, decrypted, and converted to analog form for playback. In practice, a combination of all three systems, terrestrial, satellite, and cable, is used to convey digital audio (and video) signals. Figure 16.4 shows an example of transmission routing originating from an event such as a football game conveyed over various paths to consumers.

Spectral Space

A significant complication for any DAR broadcast system is where to locate the DAR band (perhaps 100 MHz wide) in the electromagnetic spectrum. Spectral space is a limited resource that has been substantially allocated. Furthermore, the frequency of the

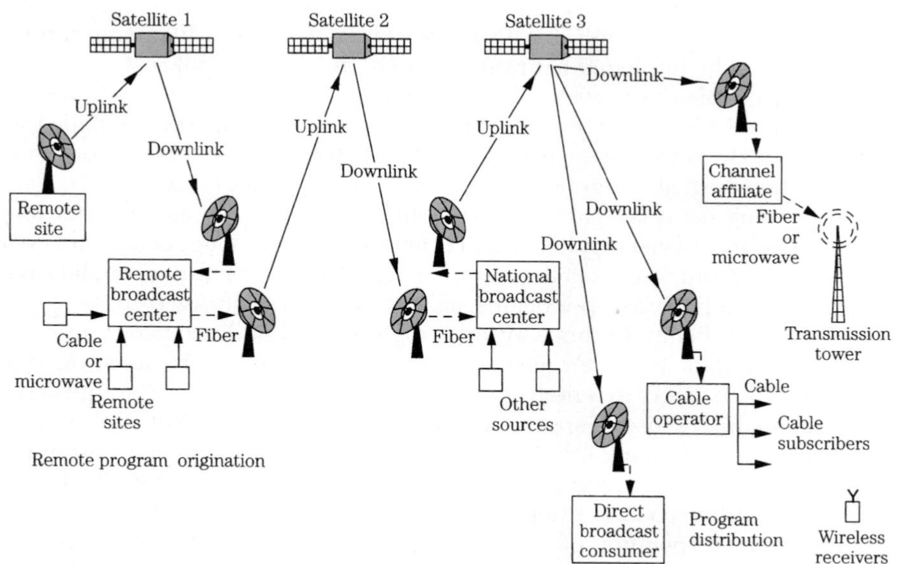

Figure 16.4 An example of transmission routing used to convey audio and video signals.

DAR transmission band will impact the technology's quality, cost, and worldwide compatibility. Any band from 100 MHz to 1700 MHz could be used for terrestrial DAR, but the spectrum is already crowded with applications. In general, lower bands are preferable (because RF attenuation increases with frequency) but are hard to obtain. The S-band (2310 MHz to 2360 MHz) is not suitable for terrestrial DAR because it is prone to interference. However, the S-band is suitable for satellite delivery. Portions of the VHF and UHF bands are allocated to DTV applications.

A worldwide allocation would assist manufacturers, and would ultimately lower cost, but such a consensus is impossible to obtain. The World Administrative Radio Conference (WARC) allocated 40 MHz at 1500 MHz (L-band) for digital audio broadcasting via satellite, but ultimately deferred selection for regional solution. Similarly, the International Radio Consultative Committee (CCIR) proposed a worldwide 60-MHz band at 1500 MHz for both terrestrial and satellite DAR. However, terrestrial broadcasting at 1500 MHz is prone to absorption and obstruction, and satellite broadcasting requires repeaters.

There is no realistic possibility of a worldwide satellite standard. In the United States, in 1995, the Federal Communications Commission (FCC) allocated the S-band (2310 MHz to 2360 MHz) spectrum to establish satellite-delivered digital audio broadcasting services. Canada and Mexico have allocated space at 1500 MHz. In Europe, both 1500-MHz and 2600-MHz regions have been developed. Ideally, whether using adjacent or separated bands, DAR would permit compatibility between terrestrial and satellite channels. In practice, there is not a mutually ideal band space, and any allocation will involve compromises.

Alternatively, new DAR systems can cohabit spectral space with existing applications. Specifically, the DAR system uses a shared-spectrum technique to locate the digital signal in the FM and AM bands. By using an in-band approach, power multiplexing can provide compatibility with analog transmissions, with the digital broadcast signal coexisting with the analog carriers. Because of its greater efficiency, the DAR signal transmits at lower power relative to the analog station. An analog receiver rejects the weaker digital signal as noise, but DAR receivers can receive both DAR and analog broadcasts. No matter how DAR is implemented, the eventual disposition of AM and FM broadcasting is a concern. A transition period will be required, lasting until analog AM and FM broadcasts gradually disappear. HD Radio is an example of an in-band system; it is described later.

Data Reduction

Digital audio signals cannot be practically transmitted in a PCM format because the bandwidth requirements would be extreme. A stereo DAB signal might occupy 2 MHz of bandwidth, compared to the approximately 240 kHz required by an analog FM broadcast. Thus, DAR must use data reduction to reduce the spectral requirement. For example, instead of a digital signal transmitted at a 2-Mbps rate, a data-reduced signal might be transmitted at 256 kbps. There are numerous perceptual coding methods suitable for broadcasting. For example, the MPEG algorithms use subband and transform coding with numerous data rates such as 256, 128, 96, 64, and 32 kbps. Although data reduction is used successfully in many applications, the bandwidth limitations of commercial radio broadcasting make it a particularly challenging application. In addition, an audio signal passing through a broadcast chain may undergo multiple data reduction encoding/decoding stages; this increases distortion and artifacts. Data reduction via perceptual coding is discussed in Chaps. 10 and 11.

Technical Considerations

The performance of a digital audio broadcasting system can be evaluated with a number of criteria including: delivered sound quality, coverage range for reliable reception, interference between analog and digital signals at the same or adjacent frequencies, signal loss in mountains or tunnels, deep "stoplight" fades, signal "flutter" produced by passing aircraft, data errors in the presence of man-made and atmospheric noise, interference from power lines and overhead signs, attenuation by buildings, multipath distortion during fixed and mobile reception, receiver complexity, and capacity for auxiliary data services. In addition, ideally, the same receiver can be used for both terrestrial and satellite reception.

Designers of DAR systems must balance many variables to produce a system with low error rate, moderate transmitted power levels, and sufficient data rate, all within the smallest possible bandwidth. As with any digital data system, a broadcasting system must minimize errors. The bit-error rate (BER) must be reduced through error-correction data accompanying the audio data, and is monitored for a given carrier-to-noise ratio (C/N) of the received signal. Transmitted digital signals are received successfully with low C/N, but analog signals are not. Generally, a BER of 10^{-4} at the receiver might be nominal, but rates of 10^{-3} and 10^{-2} can be expected to occur, in addition to burst errors.

Receiver performance also can be gauged by measuring the ratio between the energy per bit received to the power spectral density of the input noise in a 1-Hz bandwidth; this is notated as E_b/N_o. Designers strive to achieve a low BER for a given C/N or E_b/N_o. Digital transmission tends to have brick-wall coverage; the system operates well with a low BER within a certain range, then BER increases dramatically (yielding total system failure) when there is an additional small decrease in signal strength.

Most digital communications systems use pulse-shaping techniques prior to modulation to limit bandwidth requirements. Pulse shaping performs lowpass filtering to reduce high-frequency content of the data signal. Because this spreads the bit width, resulting in intersymbol interference, raised cosine filters are used so that the interference from each bit is nulled at the center of other bit intervals, eliminating interference.

Multipath interference occurs when a direct signal and one or more strongly reflected and delayed signals, for example, signals reflected from a building, destructively combine at the receiver. The delay might be on the order of 5 μs. In addition, other weak reflected signals might persist for up to 20 μs. The result at the receiver is a comb filter with 10-dB to 50-dB dips in signal strength, as shown in Fig. 16.5. This type of RF multipath is a frequency-selective problem, and short wavelengths, for example, in FM broadcasting, are more vulnerable. When the receiver is moving, multipath interference results in the amplitude modulation "picket fence" effect familiar to mobile analog FM listeners. Even worse, in a stoplight fade, when the receiver is stopped in a signal null, the signal is continuously degraded; a single, strong specular reflection completely cancels the transmitted signal's bandwidth. FM signals can become noisy, but because digital signals operate with a small C/N ratio, they can be lost altogether. Increasing power is not a remedy because both the direct and reflected signal will increase proportionally, preserving the interference nulls.

Another effect of multipath interference, caused by short delays, occurs in the demodulated bitstream. This is delay spread in which multiple reflections arrive at the receiver over a time interval of perhaps 15 μs. The result is intersymbol interference in the received data; bits arrive at multiple times. This can be overcome with

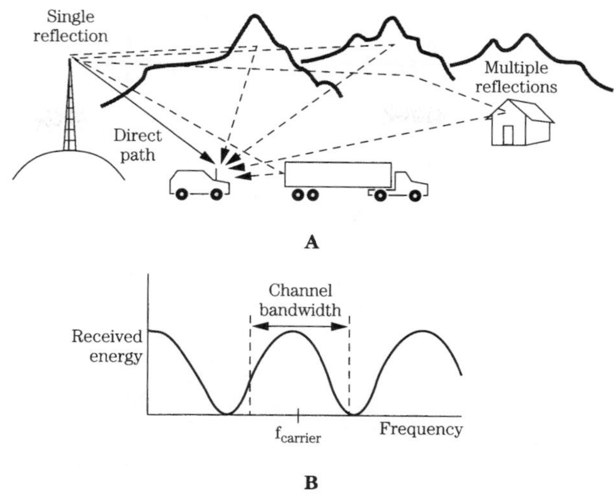

Figure 16.5 Multipath interference degrades signal quality at the receiver. A. A radio receives the direct transmission path, as well as delayed single and multiple reflection paths. B. The combined path lengths produce nulls in signal strength in the received channel bandwidth.

bit periods longer than the spread time. However, with conventional modulation, this would limit the bit rate to less than 100 kbps; thus, data reduction must also be used. Frequency diversity techniques are very good at combating multipath interference. By placing the data signal on multiple carriers, interference on one carrier frequency can be overcome.

Two types of multiplexing are used. The most common method is time-division multiplexing (TDM), in which multiple channels share a single carrier by time interleaving their data streams on a bit or word basis; different bit rates can be time-multiplexed. Frequency-division multiplexing (FDM) divides a band into subbands, and individual channels modulate individual carriers within the available bandwidth. A single channel can be frequency-multiplexed; this lowers the bit rate on each carrier, and lowers bit errors as well. Because different carriers are used, multipath interference is reduced because only one carrier frequency is affected. On the other hand, more spectral space is needed.

Phase-shift keying (PSK) modulation methods are commonly used because they yield the lowest BER for a given signal strength. In binary phase-shift keying (BPSK), two phase shifts represent two binary states. For example, a binary 0 places the carrier in phase, and a binary 1 places it 180° out of phase, as shown in Fig. 16.6A. This phase change codes the binary signal, as shown in Fig. 16.6B. The symbol rate equals the data rate. In quadrature phase-shift keying (QPSK), four phase shifts are used. Thus, two bits per symbol are represented. For example, 11 places the carrier at 0°, 10 at 90°, 00 at 180°, and 01 at 270°, as shown in Fig. 16.6C. The symbol rate is twice the transmission rate. QPSK is the most widely used method, especially for data rates above 100 Mbps. Higher-order PSK can be used (for example, 8-PSK, 16-PSK) but as the number of phases increases, higher E_b/N_o is required to achieve satisfactory BER. Other modulation methods include amplitude shift-keying (ASK) in which different carrier powers represent binary values, frequency shift-keying (FSK) in which the carrier frequency is varied

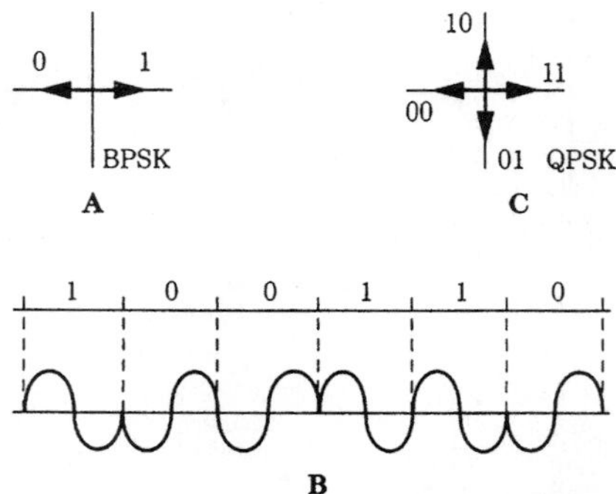

FIGURE 16.6 Phase-shift keying (PSK) is used to modulate a carrier, improving efficiency. A. Phasor diagram of binary phase-shift keying (BPSK). B. An example of a BPSK waveform. C. Phasor diagram of quadrature phase-shift keying (QPSK).

(FSK is used in modems), and quadrature amplitude modulation (QAM) in which both the amplitude and phase are varied.

The bandwidth (*BW*) for an M-PSK signal is given by:

$$\frac{D}{\log_2 M} < BW < \frac{2D}{\log_2 M}$$

where D is the data rate in bits per second. For example, a QPSK signal transmitting a 400-kbps signal would require a bandwidth of between 200 kHz to 400 kHz. A 16-PSK signal could transmit the same data rate in half the bandwidth, but would require 8 dB more power (E_b/N_o) for a satisfactory BER. Given the inherently high bandwidth of digital audio signals, data reduction is mandatory to conserve spectral space, and provide low BER for a reasonable transmission power level. As Kenneth Springer has noted, a 4,000,000-level PSK modulation would be needed to make a transmitted signal's bandwidth equal its original analog baseband; the required power would be prohibitive. But with a 4:1 data reduction, 256-PSK provides the same baseband. In practice, an error-corrected, data-reduced signal, with QPSK modulation, can be transmitted with lower power than an analog signal.

One of the great strengths of a digital system is its transmission power efficiency. This can be seen by relating coverage area to the C/N ratio at the receiver. A digital system might need a C/N of only 6 dB, but an FM receiver needs a C/N of 30 dB, a difference of 24 dB, to provide the same coverage area. The field strength for a DAR system can be estimated from:

$$E = V_i + NF + C/N - \frac{96.5}{F_{\text{MHZ}}}$$

where E = minimum acceptable field strength at the receiver in dBu and V_i = thermal noise of receiver into 300 Ω in dBu where

$$V_i = 20\log\left(\frac{kTRB}{10^{-6}}\right)^{1/2}$$

where $k = 1.38 \times 10^{-23}$ W/kHz
T = temperature in Kelvin (290 K at room temperature)
R = input impedance of the receiver
B = bandwidth of the digital signal
NF = noise figure of the receiver
C/N = carrier-to-noise ratio for a given BER
F_{MHz} = transmission frequency

For example, if a DAR signal is broadcast at 100 MHz, with 200-kHz bandwidth, into a receiver with 6-dB noise figure, with C/N of 6 dB, then $E = 5.5$ dBu. In contrast, an FM receiver might require a field strength of 60 dBu for good reception, and about 30 dBu for noisy reception.

Eureka 147 Wideband Digital Radio

The Eureka 147 digital audio broadcasting (DAB) system was selected as the European standard in 1995 for broadcasting to mobile, portable, and fixed receivers. The Eureka 147 technology is suitable for use in terrestrial, satellite hybrid (satellite and terrestrial), and cable applications. Canada, Australia, and parts of Asia and Africa have also adopted the Eureka 147 system for the broadcast of DAB signals. Eureka is a research and development consortium of European governments, corporations, and universities, established in 1985 to develop new technologies, through hundreds of projects ranging from biotechnology to transportation. Project number 147, begun in 1986, aimed to develop a wideband digital audio broadcasting system (formally known as DAB, as opposed to DAR). A prototype Eureka 147/DAB system was first demonstrated in a moving vehicle in Geneva in September 1988 and many improvements followed. System specifications were finalized at the end of 1994.

In traditional radio broadcasting, a single carrier frequency is used to transmit a monaural or stereo audio program, with one carrier per radio station. This method allows complete independence of stations, but poses a number of problems. For example, reception conditions at the receiver might produce multipath interference at the desired carrier frequency, in part because the station's bandwidth is narrow (e.g., approximately 240 kHz for analog FM radio). In addition, wide guard bands must be placed around each carrier to prevent adjacent interference. In short, independent carrier transmission methods are not particularly robust, and are relatively inefficient from a spectral standpoint. Eureka 147 employs a different method of transmission coding which overcomes many problems incurred by traditional broadcast methods.

Eureka 147 digitally combines multiple audio channels, and the combined signal is interleaved in both frequency and time across a wide broadcast band. A receiver does not tune to a single carrier frequency. Rather, it performs partial fast Fourier transforms (FFTs) on a broadcast band and decodes the appropriate channel data from among many carriers. This innovative approach provides spectrum- and power-efficient transmission and reliable reception even over a multipath fading channel. A block diagram of a Eureka 147 transmitter is shown in Fig. 16.7A. Audio data as well as other data is individually encoded with channel coders and interleaved. A multiplexer combines many different services to create a main service channel (MSC). The multiplexer output is frequency interleaved and synchronization symbols are added. Channel coding is applied: coded orthogonal frequency-division multiplexing (COFDM) with QPSK

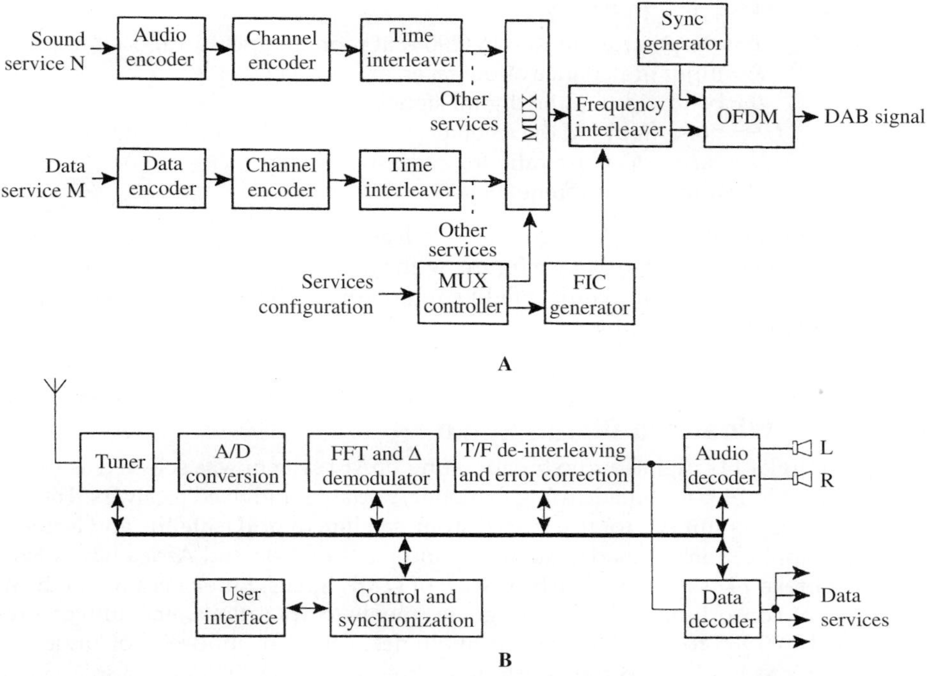

Figure 16.7 Block diagrams of the Eureka 147/DAB system. A. DAB transmitter. B. DAB receiver.

modulation is employed for each carrier to create an ensemble DAB signal. In orthogonal coding, carriers are placed at 90°-phase angles such that the carriers are mutually orthogonal and the demodulator for one carrier is not affected by the modulation of other carriers.

COFDM processing divides the transmitted information into many bitstreams, each with a low bit rate, which modulate individual orthogonal carriers so that the symbol duration is longer than the delay spread of the transmission channels. The carriers in this ensemble signal may be generated by an FFT. Depending on the transmission mode employed, there may be 1536, 384, or 192 carriers. In the presence of multipath interference, some of the carriers undergo destructive interference while others undergo constructive interference. Because of frequency interleaving among the carriers, successive samples from the same service are not affected by a selective fade, even in a fixed receiver. Time interleaving further assists reception, particularly in mobile receivers.

Carrier frequency centers are separated by the inverse of the time interval between bits, and separately modulated within their fractional spectral space, with a portion of the overall signal. This reduces the data rate on any one carrier, which promotes long bit periods. This frequency diversity yields immunity to intersymbol interference and multipath interference. To increase robustness over that provided by frequency diversity, and further minimize the effect of intersymbol and inter-carrier multipath interference, a guard interval is used; each modulation symbol is transmitted for a period that is longer than an actively modulated symbol period. During the interval, the phase of the carrier is unmodulated. This reduces the capacity of the channel but protects against reflection delays less than the duration of the guard interval. For example, a guard interval of 1/4 the active period protects against delays of 200 μs. The guard interval also decreases the burden placed on error correction. Convolutional coding adds redundancy, using, for example, a code with a constraint length of 7.

A collection of error-correction profiles is used, optimized for the error characteristics of MPEG Layer II encoded data. The coders aim to provide graceful degradation as opposed to brick-wall failure. Thus, stronger protection is given to data for which an error would yield catastrophic muting, and weaker protection where errors would be less audible. Specifically, three levels of protection are used within an MPEG frame: the frame header and bit allocation data are given the strongest protection, followed by the scale factors, and subband audio samples, respectively. For example, errors in scale-factor data may lead to improperly raised subband levels, whereas errors in a subband audio sample will be confined to a sample in one subband, and will likely be masked.

A block diagram of a Eureka 147 receiver is shown in Fig. 16.7B. The DAB receiver uses an analog tuner to select the desired DAB ensemble; it also performs down-conversion and filtering. The signal is quadrature-demodulated and converted into digital form. FFT and differential demodulation is performed, followed by time and frequency de-interleaving and error correction. Final audio decoding completes the signal chain. Interestingly, a receiver may be designed to simultaneously recover more than one service component from an ensemble signal.

The DAB standard defines three transmission mode options, allowing a range of transmitting frequencies up to 3 GHz. Mode I with a frame duration of 96 ms, 1536 carriers, and a nominal frequency range of less than 375 MHz, is suited for a terrestrial VHF network because it allows the greatest transmitter separations. Mode II with a frame duration of 24 ms, 384 carriers, and a nominal frequency range of less than 1.5 GHz, is suited for UHF and local radio applications. Mode III with a frame duration of 24 ms (as in Mode II), 192 carriers, and a nominal frequency range of less than 3 GHz, is suited for cable, satellite, and hybrid (terrestrial gap filler) applications. In all modes, the transmitted signal uses a frame structure with a fixed sequence of symbols. The gross capacity of the main service channel is about 2.3 Mbps within a 1.54-MHz bandwidth DAB signal. The net bit rate ranges from approximately 0.6 Mbps to 1.7 Mbps depending on the error correction redundancy used.

The Eureka 147 system's frequency diversity provides spectral efficiency that exceeds that of analog FM broadcasting. In addition, time interleaving combats fading experienced in mobile reception. The transmission power efficiency, as with many digital radio systems, is impressive; it can be 10 to 100 times more power-efficient than FM broadcasting; a Eureka 147 station could cover a broadcast market with a transmitter power of less than 1000 W. A principal feature of Eureka 147 is its ability to support both terrestrial and satellite delivery on the same frequency; the same receiver can be used to receive a program from either source. Eureka 147 uses MPEG-1 Layer II bit rate reduction in its source coding to minimize the spectrum requirements. Bit rate may range from 32 kbps to 384 kbps in 14 steps; nominally, a rate of 128 kbps/channel is used. Stereo or surround-sound signals could be conveyed. Nominally, a sampling frequency of 48 kHz is used; however, a 24-kHz sampling rate is optional. MPEG is discussed in Chap. 11.

One prototype Eureka 147 system, using L-band transmission, was evaluated in Canada. The COFDM used the following parameters: 7-MHz RF bandwidth, 448-quadrature phase-shift keying, subcarriers with 15.625-kHz spacing, 80-μs symbol length with a 16-μs guard interval, capacity to transmit in multiplex 33 monophonic channels (129 kbps) or 16 stereo pairs, and 1 data channel. A transmitter with power of 150 W (1.5 kW ERP) total, or 9.4 W (94 W ERP) per stereo channel, produced a coverage range of 50 km, where ERP is effective radiated power. The propagation and reception were similar to that of FM and UHF broadcasting; a local FM station required 40 kW ERP for 70-km coverage. Multipath errors were generally absent.

In another Canadian test, fixed and mobile receivers performed signal strength measurements using a 50-W transmitter and 16-dB antenna to broadcast nine CD-quality channels (with the power of 200 W/channel) with reliable coverage up to distances of 45 km from the transmitter. In addition, Canadian tests verified the system's ability to provide a single-frequency network in which multiple transmitters can operate on a single frequency, without interference in overlapping areas. The Canadian system also proposed a mixed mode of broadcasting in which a single receiver could receive either satellite or terrestrial digital audio broadcasts. In a test in London, a Eureka 147 transmitter with 100 W of power provided coverage over 95% of the London area, with antennas similar to those used in table radios.

Eureka 147 is inherently a wideband system, and it can operate at any frequency range up to 3 GHz for mobile reception and higher frequencies for fixed reception. Practical implementation requires a spectrum allocation outside the existing commercial broadcast bands. The narrowest Eureka 147 configuration uses 1.5 MHz to transmit six stereo channels. In practice, a much wider band would be required for most applications. A fully implemented Eureka 147 might occupy an entire radio band. Because spectral space is scarce, this poses a problem. In general, Eureka 147 can operate in a number of bands, ranging from 30 MHz to 3 GHz; however, a 100-MHz to 1700-MHz range is preferred. Domestic proponents argued for allocation of the L-band (1500 MHz) for Eureka 147, but the U.S. government, for example, was unwilling to commit that space; in particular, the U.S. military uses that spectral space for aircraft telemetry. Another proposal called for operation in the S-band (2300 MHz). Although the FCC authorized the use of the S-band for digital satellite radio, lack of suitable spectral space posed an insurmountable obstacle in the development of a Eureka 147 system in the United States.

Other drawbacks exist. In particular, the need to combine stations leads to practical problems in some implementations. Eureka 147's designers, taking a European community bias, envisioned a satellite delivery system that would blanket Europe with a single footprint. Terrestrial transmitters operating on the same frequencies would be used mainly as gap fillers, operating on the same frequency. This monolithic approach is opposed in the United States where independent local programming is preferred. With satellite delivery of Eureka 147, the concept of local markets becomes more difficult to implement, while national stations become easier to implement. This would redefine the existing broadcast industry. To address this issue, some researchers unsuccessfully advocated the use of Eureka 147 in the FM and AM bands as an in-band system.

Eureka 147 can be used with terrestrial transmission in which local towers supplement satellite delivery, with local stations coexisting with national channels. In January 1991, the National Association of Broadcasters (NAB) endorsed such a system, and proposed that the L-band be allocated. Existing AM and FM licensees would be given DAB space in the L-band, before phasing out existing frequencies. The plan called for the creation of "pods," in which each existing broadcaster would be given a digital channel; four stations would multiplex their signals over a 1.5-MHz-wide band. The power levels and location of pods would duplicate the coverage areas of existing stations. The NAB estimated that no more than 130 MHz of spectrum would be needed to accommodate all existing broadcasters in the new system.

However, broadcasters did not accept the multiplexing arrangement and the potential for new stations it allowed, and many argued for an in-band DAB system that would allow existing stations to phase in DAB, yet still provide AM and FM transmission. In March 1991, the Department of Defense indicated that the L-band was not available. In the face of these actions, in January 1992, the NAB reversed its position and instead proposed development of an in-band digital radio system that would operate in the FM and AM bands, coexisting with analog FM and AM stations. The NAB expressed concern over satellite delivery methods because they could negatively impact the infrastructure of terrestrial stations. Meanwhile, not bothered by America's political and commercial questions, other countries have argued that Eureka 147, practical problems aside, remains the best technical system available. The Eureka 147 system has been standardized by the European Telecommunications Standards Institute as ETS 300 401.

In-Band Digital Radio

The in-band digital radio system in the United States broadcasts digital audio radio signals in existing FM (88 MHz to 108 MHz) and AM (510 kHz to 1710 kHz) bands along with analog radio signals. Such systems are hybrids because the analog and digital signals can be broadcast simultaneously so that analog radios can continue to receive analog signals, while digital radios can receive digital signals. At the end of a transition period, broadcasters would turn off their analog transmitters and continue to broadcast in an all-digital mode.

Such in-band systems offer commercial advantages over a wideband system because broadcasters can retain their existing listener base during a transition period, much of their current equipment can be reused, existing spectral allocation can be used, and no new spectral space is needed. However, in-band systems are incompatible with wideband Eureka-type systems. An in-band system must equal the performance demonstrated by wideband systems, and provide, for example, robust rejection of multipath interference. Finally, an in-band system must surmount the inherently difficult task of simultaneously broadcasting digital audio signals in the same radio band as existing analog broadcasts.

In-band systems permit broadcasters to simultaneously transmit analog and digital programs. Digital signals are inherently immune to interference; thus, a digital receiver is able to reject the analog signals. However, it is more difficult for an analog receiver to reject the digital signal's interference. Coexistence can be achieved if the digital signal is broadcast at much lower power. Because of the broadcast efficiency of DAR, a low-power signal can maintain existing coverage areas for digital receivers, and allow analog receivers to reject the interfering signal.

With an in-band on-channel (IBOC) system, DAR signals are superimposed on current FM and AM transmission frequencies (in some other systems, DAR signals are placed on adjacent frequencies). In the United States, FM radio stations have a nominal bandwidth of 480 kHz with approximately a 240-kHz signal spectrum. FM radio stations are spaced 200 kHz apart, and there is a guard band of 400 kHz between

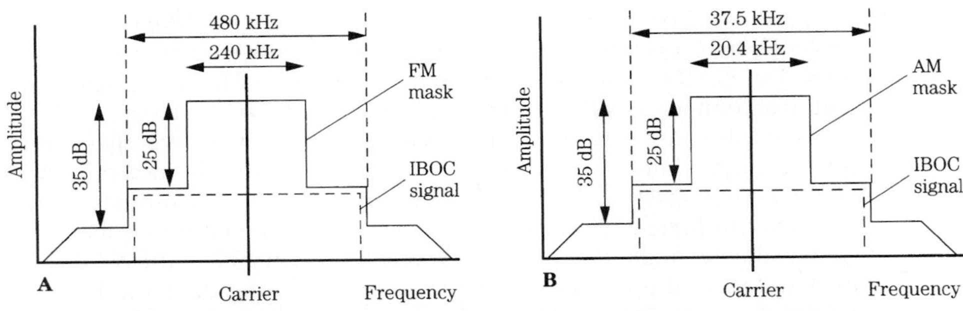

FIGURE 16.8 The FCC strictly defines the RF spectrum masks allowed to broadcasters. Any IBOC system must stay within the spectral mask. A. The FM spectrum mask. B. The AM spectrum mask.

co-located stations to minimize interference. The FM emissions mask specifies that the middle 240-kHz (±120 kHz from the center frequency) region has a 25-dB stronger power than the sidebands on either side of the middle band, that extend into two neighboring stations. AM stations nominally occupy a 37.5-kHz bandwidth, with stations spaced at 10 kHz intervals. This results in interference from overlapping sidebands. The AM emissions mask specifies that the middle 20.4 kHz (±10.2 kHz from the center frequency) region has a 25-dB stronger power than the sidebands on either side of the middle band. Some stations may be interference-limited by signals from a distant station via ground-wave propagation as well as reflected signals from the ionosphere at night.

In-band systems fit within the same bandwidth constraints as analog broadcasts, and furthermore, efficiently use the FCC-regulated RF mask in which the channel's spectrum widens as power decreases. For example, if a DAR signal is 25 dB below the FM signal, it could occupy a 480-kHz bandwidth, as shown in Fig. 16.8A. In the case of AM, if the DAR signal is 25 dB below the AM signal, the band can be 37.5 kHz wide, as shown in Fig. 16.8B. Because the digital signal's power can be lower, it can thus efficiently use the entire frequency mask area while optimizing data throughput and signal robustness. Clearly, because of the wider FM bandwidth, an FM in-band system is much easier to implement; rates of 256 kbps can be accommodated. The narrow AM channels can limit DAR data rates to perhaps 128 kbps or 96 kbps. In addition, existing AM radios are not as amenable to DAR signals as FM receivers. On the other hand, AM broadcast is not hampered by multipath problems, but multipath immunity is more difficult to achieve in a narrow-band in-band FM system, compared to a wideband DAR system. Any DAR system must rely on perceptual coding to reduce the channel data rate to 128 kbps or so, to allow the high-fidelity signal (along with nonaudio data) to be transmitted in the narrow bands available. Given the finite throughput of a broadcast channel, a higher bit rate provides better sound quality while lower bit rates allow greater error correction and hence more robust coverage in interference conditions.

The IBOC method is attractive because it fits within much of the existing regulatory statutes and commercial interests. No modifications of existing analog AM and FM receivers are required, and DAR receivers can receive both analog and digital signals. Moreover, because digital signals are simulcast over existing equipment, start-up costs are low. An in-band system provides improved frequency response, and lower noise and distortion within existing coverage areas. Receivers can be designed so that if the digital signal fails, the radio automatically switches to the analog signal.

Alternatively, in-band interstitial (IBI) systems transmit low power DAR signals on guard band frequencies adjacent to existing carriers. This helps reduce the problem of differentiating between the types of signals. In a single-channel IBI system, the DAR signal is placed in one adjacent channel (upper or lower). Alternatively, both adjacent channels can be used. This would reduce the number of available stations, but frequency hopping, switching from carrier to carrier, can be used to reduce multipath interference. In a single-channel multiplexed IBI system, various stations in a market would multiplex their DAR signals, and broadcast in adjacent channels across the band, providing greater frequency diversity, and protection against multipath interference.

The differentiation of the analog and digital signals presents technological challenges, particularly to an IBOC system. Specifically, the DAR signal must not interfere with the analog signal in existing receivers, and DAR receivers must use encryption methods to extract the DAR signal while ignoring the much stronger analog signal. FM receivers are good at rejecting amplitude noise; for example, their limiters reject a DAR signal using ASK modulation. Existing FM receivers see the weaker (30 dB or so) digital signal as noise, and reject it. With PSK modulation, the DAR signal might have to be 45 dB to 50 dB below the FM signal level. It is more difficult to extract the digital information from the analog signal. For example, an adaptive transversal filter could provide interference cancellation to eliminate the analog AM or FM signal so that on-channel digital information can be processed. Thanks to the military-industrial complex, signal extraction technology has been well developed. For example, the problem of retrieving signals in the presence of jamming signals has been carefully studied. In the case of IBOC, the problem is further simplified because the nature of the jamming signal (the analog signal) is known at the broadcast site, and can be determined at the receiver.

At a meeting of the CCIR, virtually every country supported the adoption of Eureka 147 as a worldwide standard, except the United States. That opposition, supported by prototype demonstrations of in-band systems, stalled any decision on the part of the CCIR. Critics argued that in-band DAR would be a minor improvement over analog AM and FM broadcasting because of interference and crosstalk problems, especially in mobile environments. Instead, they argued that L-band DAR should entirely replace AM and FM broadcasting, because it would be more effective. They argued that if marketplace and political realities are used as the primary constraint, technological effectiveness would be compromised. However, the NAB formally endorsed an in-band, on-channel system for the United States and the FCC authorized transmission of IBOC signals. To allow implementation of an IBOC system, the FCC determined that IBOC is an appropriate means of digital audio broadcast, established interference criteria to ensure compatibility of analog and digital signals, established a transition plan, determined that a commission-adopted transmission standard was necessary, established testing criteria and a time table to evaluate systems, and selected an IBOC system and transmission standard.

From 1994 to 1996, the National Radio Systems Committee (NRSC) developed field-testing and system evaluation guidelines, and supervised the field testing and evaluated test results from competing developers of first-generation digital radio systems:

AT&T/Lucent Technologies (IBOC in FM band), AT&T/Lucent Technologies/Amati Communications (IBOC in FM band), Thomson Consumer Electronics (Eureka 147 COFDM at 1.5 GHz), USA Digital Radio (IBOC in FM and AM band), and Voice of America/Jet Propulsion Laboratory (DBS at 2.3 GHz). The IBOC systems all provided good audio fidelity in unimpaired environments, but none of the systems proved to be viable for commercial deployment. Among other problems, the digital signal degraded the host analog signal. Second-generation IBOC systems sought to reduce spectral bandwidth while maintaining audio quality. In October 2002, the Federal Communications Commission approved broadcast of an IBOC technology which enables digital broadcasting in the AM and FM bands. iBiquity Digital Corporation is the developer and licenser of the IBOC system used in the United States, known as HD Radio. The NRSC is sponsored by the National Association of Broadcasters and the Electronics Industries Association.

HD Radio

HD Radio is an IBOC broadcast system authorized by the FCC for transmission in the United States. It is used by commercial broadcasters to transmit digital data over the existing FM and AM bands. Many radio stations currently use this technology to simultaneously broadcast analog and digital signals on the same frequencies in a hybrid mode. In the future, the technology can be switched to an all-digital mode. Audio signals, as well as data services, can be transmitted from terrestrial transmitters in the existing VHF radio band and received by mobile, portable, and fixed IBOC radios. HD Radio uses the proprietary High-Definition Coding (HDC) codec for data reduction. HDC is based on the MPEG-4 High-Efficiency AAC codec but is not compatible with it. HDC employs spectral band replication (SBR) to improve high-frequency response while maximizing coding accuracy at lower frequencies. HDC was jointly developed by iBiquity, and Coding Technologies which developed SBR and High-Efficiency AAC (also known as HE ACC and aacPlus) which provides improvements to the AAC codec at low bit rates. SBR and AAC are discussed in Chaps. 10 and 11.

In both hybrid and digital modes, the FM-IBOC signal uses orthogonal frequency-division multiplexing (OFDM). OFDM creates digital carriers that are frequency-division multiplexed in an orthogonal manner such that each carrier does not interfere with each adjacent subcarrier. With OFDM, the power of each subcarrier can be adjusted independently of other subcarriers. In this way, the subcarriers can remain within the analog FCC emissions mask and avoid interference. Instead of a single-carrier system where data is transmitted serially and each symbol occupies the entire channel bandwidth, OFDM is a parallel modulation system. The data stream simultaneously modulates a large number of orthogonal narrow-band subcarriers across the channel bandwidth. Instead of a single carrier with a high data rate, many subcarriers can each operate at low bit rates. The frequency diversity and longer symbol times promote a robust signal that resists multipath interference and fading. OFDM also allows for the use of on-channel digital repeaters to fill gaps in the digital coverage area. Power, directionality, and distance from the primary transmitter must be considered to avoid intersymbol interference.

HD Radio FM-IBOC

The FM-IBOC subcarriers are grouped into frequency partitions. Each partition has 18 data subcarriers carrying program content and one reference subcarrier carrying control information. In total, the subcarriers are numbered from 0 to ±546 at either end of the channel frequency allocation. Up to five additional subcarriers can be inserted, depending on the service mode, across the channel allocation at –546, –279, 0, 279, and 546. The subcarrier spacing is $\Delta f = 363.373$ Hz.

The FM-IBOC signal can be configured in a variety of ways by varying the robustness, latency, and throughput of the audio and data program content. Several digital program services are supported including main program service (MPS), personal data service (PDS), station identification service (SIS), and auxiliary application service (AAS). MPS delivers existing programming in digital form along with data that correlates to the programming. PDS lets listeners select on-demand data services. SIS provides control information that lets listeners search for particular stations. AAS allows custom radio applications. A layered stack protocol allows all of these services to be supported simultaneously. The protocol is based on the International Organization for Standardization Open Systems Interconnection (ISO OSI) model. There are five layers: Layer 5 (Application) accepts user content; Layer 4 (encoding) performs compression and formatting; Layer 3 (Transport) applies specific protocols; Layer 2 (Service Mux) formats data into frames; Layer 1 (Physical) provides modulation, error correction, and framing prior to transmission.

In the hybrid waveform, FM-IBOC places low-level primary main (PM) digital carriers in the lower and upper sidebands of the emissions spectrum as shown in Fig. 16.9. To help avoid digital-to-analog interference, no digital signal is placed directly at the analog carrier frequency, and the separation of the sidebands provides frequency diversity. The primary lower and upper carriers are modulated with redundant information; this allows adequate reception even when one sideband is impaired. Each PM sideband contains 10 frequency partitions (as described below), with subcarriers –356 through –545,

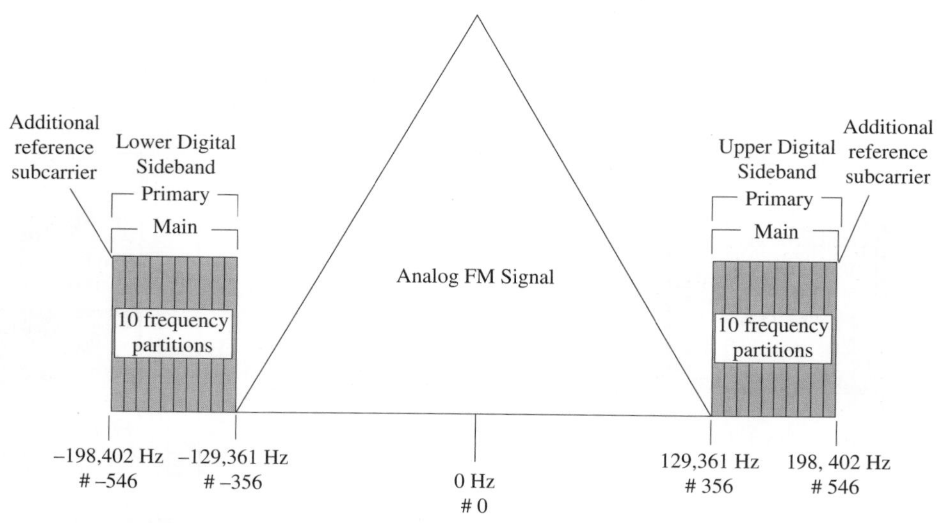

FIGURE 16.9 The HD Radio hybrid FM-IBOC waveform spectrum contains primary main (PM) digital carriers in regions between approximately ±129 kHz and ±198 kHz, around the center frequency. (*Peyla*)

and 356 through 545. Subcarriers –546 and 546 are reference subcarriers. Each subcarrier sideband is approximately 69-kHz wide. Specifically, subcarrier frequencies range from –198,402 Hz to –129,361 Hz and 198,402 Hz to 129,361 Hz relative to the 0-Hz center frequency. This placement minimizes interference to the host analog signal and adjacent channels and remains within the emissions mask. The power spectral density of each subcarrier is –45.8 dB relative to the analog signal. (A power of 0 dB would equal the total power in an unmodulated analog FM carrier.) The total average power in a PM sideband is thus 23 dB below the total power of the unmodulated FM carrier. Each subcarrier sideband operates independently of the other and reception can continue even if one sideband is lost. A digital audio bit rate of 96 kbps is possible, along with 3 kbps to 4 kbps of auxiliary data.

An extended hybrid waveform adds subcarriers to the inner edges of the primary main sidebands, as shown in Fig. 16.10. The additional spectrum area is called the primary extended (PX) sideband. One, two, or four frequency partitions can be added to each PM sideband, depending on the service mode. When four partitions are used, the PX subcarrier frequencies range from –128,997 to –101,744 and 128,997 to 101,774. The power spectral density of each PX subcarrier is –45.8 dB. The level of the subcarriers in the PX sidebands thus equals that of the subcarriers in the PM sidebands.

In the all-digital FM-IBOC mode, the analog signal is disabled, and the primary main region is fully expanded (at higher power) and lower-power secondary sidebands are placed in the center region of the emissions mask, as shown in Fig. 16.11. The primary sidebands have a total of 14 frequency partitions (10+4) and each secondary sideband contains 10 secondary main (SM) frequency partitions and four secondary

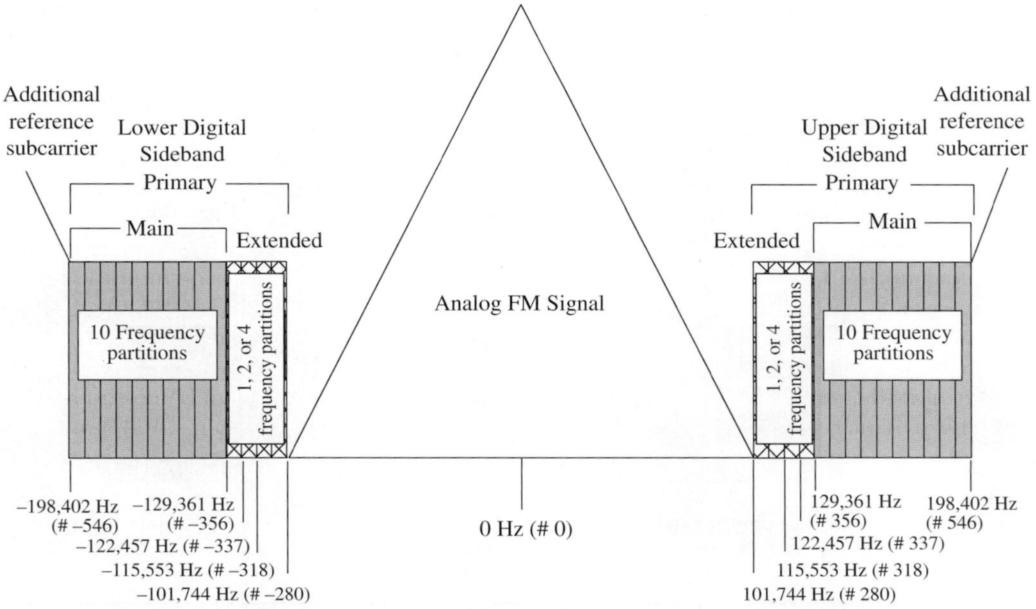

Figure 16.10 The HD Radio extended hybrid FM-IBOC waveform spectrum adds primary extended (PX) subcarriers to the inner edges of the primary main sidebands. (*Peyla*)

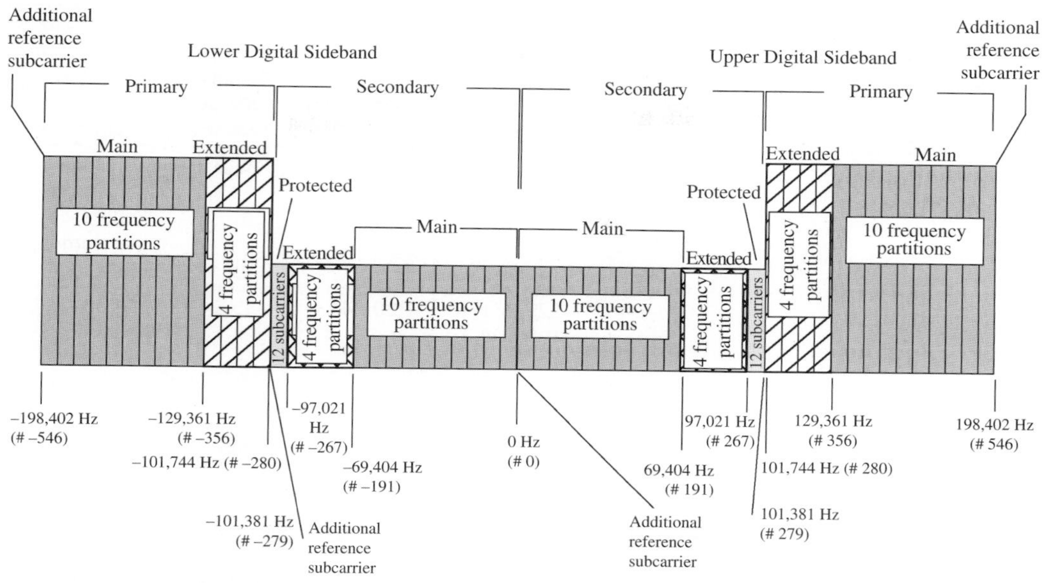

Figure 16.11 In the HD Radio all-digital FM-IBOC waveform spectrum, the analog signal is disabled, the primary main region is fully expanded at higher power, and lower-power secondary sidebands are placed in the center region of the emissions mask. (*Peyla*)

extended (SX) frequency partitions. In addition, each secondary sideband has one secondary protected (SP) region with 12 subcarriers and a reference subcarrier; these subcarriers do not contain frequency partitions. The SP sidebands are located in an area of the spectrum that is least likely to experience analog or digital interference. A reference subcarrier is placed at the 0-Hz position. The power spectral density of each subcarrier and other parameters is given in Table 16.1. The total average power in a primary digital subcarrier is at least 10 dB above the total power in a hybrid subcarrier. The secondary sidebands may use any of four power levels, using one power level for all the secondary sidebands. This selection yields a power spectral density of the secondary subcarriers from 5 dB to 20 dB below that of the primary subcarriers. The total frequency span of the all-digital FM-IBOC signal is 396,803 Hz. As the transition to digital broadcasting is completed, all stations will employ the all-digital system.

To promote time diversity, the audio program is simulcast on both the digital and analog portions of the hybrid FM-IBOC signal. The analog version is delayed by up to 5 seconds. If the received digital signal is lost during a transitory multipath fade, the receiver uses blending to replace impaired digital segments with the unimpaired analog signal. The backup analog signal also alleviates cliff-effect failure in which the audio signal is muted at the edge of the coverage area. Moreover, the blend feature provides a means of quickly acquiring the signal upon tuning or reacquisition.

During encoding, following perceptual coding to reduce the audio bit rate, data is scrambled and applied to a forward error-correction (FEC) algorithm. This coding is optimized for the nonuniform interference of the broadcast environment. Unequal error protection is used (as in cellular technology) in which bits are classified according to

Sideband	Number of Frequency Partitions	Frequency Partition Ordering	Subcarrier Range	Subcarrier Frequencies (Hz from channel center)	Frequency Span (Hz)	Power Spectral Density (dBc per subcarrier)	Comments
Upper primary main	10	A	365 to 546	129,361 to 198,402	69,041	–35.8	Includes additional reference subcarrier 546
Lower primary main	10	B	–356 to –546	–129,361 to –198,402	69,041	–35.8	Includes additional reference subcarrier –546
Upper primary extended	4	A	280 to 355	101,744 to 128,997	27,253	–35.8	None
Lower primary extended	4	B	–280 to –355	–101,744 to –128,997	27,253	–35.8	None
Upper secondary main	10	B	0 to 190	0 to 69,041	69,041	–40.8, –45.8, –50.8, –55.8	Includes additional reference subcarrier 0
Lower secondary main	10	A	–1 to –190	–363 to –69,041	68,678	–40.8, –45.8, –50.8, –55.8	None
Upper secondary extended	4	B	191 to 266	69,404 to 96,657	27,253	–40.8, –45.8, –50.8, –55.8	None
Lower secondary extended	4	A	–191 to –266	–69,404 to –96,657	27,253	–40.8, –45.8, –50.8, –55.8	None
Upper secondary protected	N/A	N/A	267 to 279	97,021 to 101,381	4,360	–40.8, –45.8, –50.8, –55.8	Includes additional reference subcarrier 279
Lower secondary protected	N/A	N/A	–267 to –279	–97,021 to –101,381	4,360	–40.8, –45.8, –50.8, –55.8	Includes additional reference subcarrier –279

TABLE 16.1 Spectral summary of information carried in the HD Radio all-digital FM-IBOC waveform. (*Peyla*)

importance; more important data is given more robust error-correction coding. Channel coding uses convolutional methods to add redundancy and thus improve reliability of reception. The signal is interleaved both in time and frequency in a way that is suitable for a Rayleigh fading model, to reduce the effect of burst errors. The output of the interleaver is structured in a matrix form and data is assigned to OFDM subcarriers.

FM transmitters can employ any of three methods to transmit FM-IBOC signals. With the high-level combining or separate amplification method, the output of the existing transmitter is combined with the output of the digital transmitter, and the hybrid signal is applied to the existing antenna. However, some power loss occurs due to power differences in the combined signal. In the low-level combining or common amplification methods, the output of the FM exciter is combined with the output of the IBOC exciter. The signal is applied to a common broadband linear amplifier. Overall power consumption and space requirements are reduced. Alternately, a separate antenna may be used for the IBOC signal; a minimum of 40 dB of isolation is required between the analog and digital antennas.

An FM-IBOC receiver receives both analog and digital broadcast signals. As in existing analog radios, the signal is passed through an RF front-end and mixed to an intermediate frequency. The signal is digitized at the IF section and down-converted to produce baseband components. The analog and digital components are separated. The digital signal is processed by a first-adjacent cancellation (FAC) circuit to reduce FM analog signal interference in the digital sidebands. The signal is OFDM demodulated, error-corrected, de-compressed, and passed to the output blend section.

HD Radio AM-IBOC

The AM-IBOC system has the same goals as FM-IBOC, to broadcast digital signals within an FCC-compliant emissions mask. However, the AM mask provides much less bandwidth, and the AM band is subject to degradation. For example, grounded conductive structures (such as power lines, bridges, and overpasses) disrupt the phase and magnitude of AM waveforms. For these and other reasons, AM-IBOC is presented with difficult issues. As in FM-IBOC, both hybrid and digital waveforms have been developed to allow an orderly transition to all-digital broadcasting.

The AM-IBOC hybrid mode places pairs of primary and secondary subcarriers in the lower and upper sidebands of the emissions spectrum, and tertiary sidebands within the main AM mask, as shown in Fig. 16.12. Each sideband is 5-kHz wide. The two sidebands in each pair are independent so reception continues even when one sideband is lost. Both secondary and tertiary sidebands are needed for stereo reproduction. Otherwise the system switches to monaural reproduction using the primary sideband. Each primary subcarrier has a power spectral density of –30 dB relative to the analog signal, as shown in Table 16.2. The power level of the primary subcarriers is fixed but the levels of the secondary, tertiary, and IBOC Data System (IDS) subcarriers are selectable. Higher power makes the digital signal more robust but can degrade analog reception in some radios. The primary subcarriers are placed at 10,356.1 Hz to 14,716.6 Hz and –10,356.1 Hz to –14,716.6 Hz relative to the center frequency. To avoid interference from the carriers of the first adjacent channel stations, digital subcarriers near ±10 kHz (at –54 to –56 and 54 to 56) are not transmitted. Also, the digital carrier at the center of the channel is not transmitted. A digital audio bit rate of 36 kbps is possible, along with 0.4 kbps

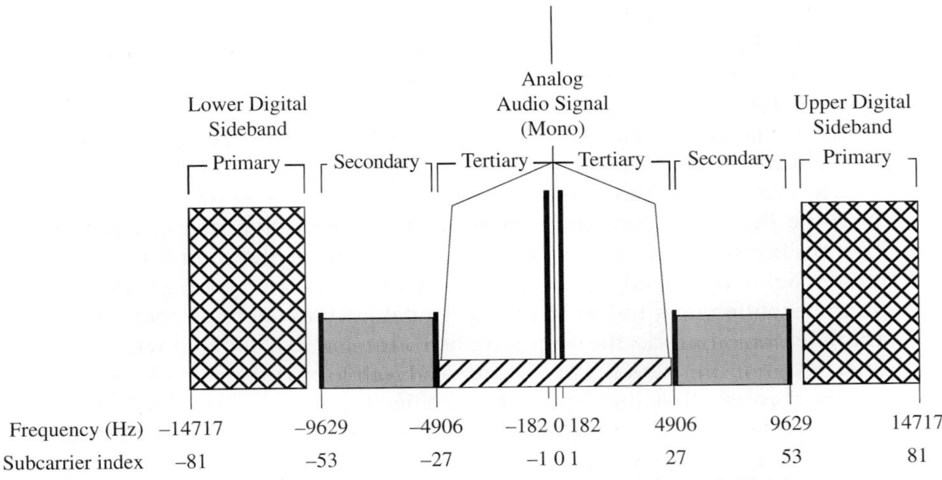

Figure 16.12 The HD Radio hybrid AM-IBOC waveform spectrum contains pairs of primary and secondary subcarriers in the lower and upper sidebands of the emissions spectrum, and tertiary sidebands within the main AM mask. (*Johnson*)

Sideband	Subcarrier Range	Subcarrier Frequencies (Hz from channel center)	Frequency Span (Hz)	Power Spectral Density (dBc per subcarrier)	Modulation Type
Primary upper	57 to 81	10356.1 to 14716.6	4360.5	–30	64-QAM
Primary lower	–57 to –81	–10356.1 to –14716.6	4360.5	–30	64-QAM
Secondary upper	28 to 52	5087.2 to 9447.7	4360.5	–43 or –37	16-QAM
Secondary lower	–28 to –52	–5087.2 to –9447.7	4360.5	–43 or –37	16-QAM
Tertiary upper	2 to 26	363.4 to 4723.8	4360.4	To be announced	QPSK
Tertiary lower	–2 to –26	–363.4 to –4723.8	4360.4	To be announced	QPSK
Reference upper	1	181.7	181.7	–26	BPSK
Reference lower	–1	–181.7	181.7	–26	BPSK
Upper IDS1	27	4905.5	181.7	–43 or –37	16-QAM
Upper IDS2	53	9629.4	181.7	–43 or –37	16-QAM
Lower IDS1	–27	–4905.5	181.7	–43 or –37	16-QAM
Lower IDS2	–53	–9629.4	181.7	–43 or –37	16-QAM

Table 16.2 Spectral summary of information carried in the HD Radio hybrid AM-IBOC waveform. (*Johnson*)

of auxiliary data. Alternatively, broadcasters may decrease the digital audio bit rate to 20 kbps and increase the auxiliary bit rate, or increase the digital audio bit rate to 56 kbps and reduce error-correction overhead.

The digital carriers maintain a quadrature (90°) phase relationship to the AM carrier; this minimizes interference in an analog receiver's envelope detector. In this way, low-powered carriers can be placed underneath the central analog carrier mask, in quadrature to the analog signal, in tertiary sidebands. The complementary modulation also expedites demodulation of the tertiary subcarriers in the presence of a strong analog carrier. However, the quadrature relationship dictates that the subcarriers can only convey one-half the data that could be conveyed on non-quadrature subcarriers. Control information is transmitted on reference subcarriers on each side of the AM carrier. There are also two additional subcarriers for IBOC Data System (IDS): primary and secondary, and secondary and tertiary sidebands at each side of the main carrier. Station identification service (SIS) and Radio Broadcast Data System (RBDS) data can be conveyed via these subcarriers.

The analog signal must be monophonic; AM stereo broadcasts cannot coexist with AM-IBOC. AM-IBOC reduces the total analog bandwidth to 10 kHz to reduce interference with the digital subcarriers (placed in the 5-kHz bands on either side of the analog channel). This has minimal impact on existing AM receivers, which limit audio bandwidth to 5 kHz. The analog bandwidth can be set at 5 kHz or 8 kHz. However, the latter decreases the robustness of the digital signal.

In the all-digital AM-IBOC mode, higher-power primary signals are broadcast in the center of the emissions mask, as shown in Fig. 16.13, along with secondary and tertiary digital sidebands. The unmodulated analog AM carrier is broadcast, and reference subcarriers with control information are placed at either side. Compared to the hybrid waveform, the secondary and tertiary sidebands now have one-half the number of

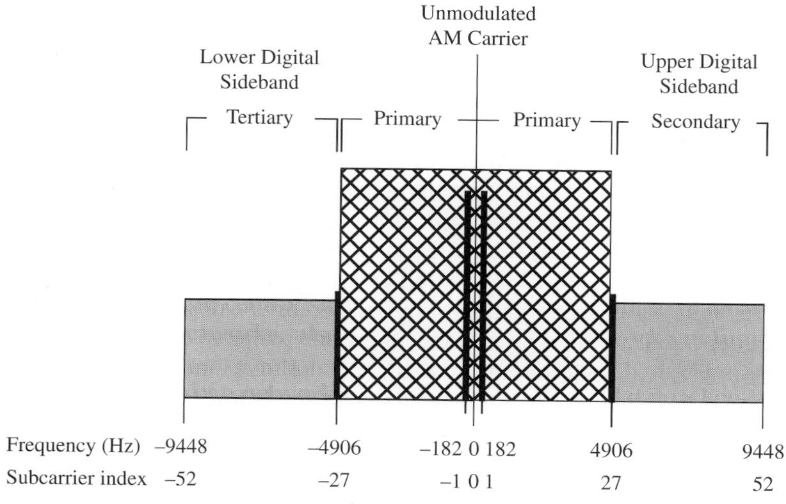

FIGURE 16.13 The HD Radio all-digital AM-IBOC waveform spectrum contains high-power primary signals broadcast in the center of the emissions mask, along with secondary and tertiary digital sidebands. (*Johnson*)

Sideband	Subcarrier Range	Subcarrier Frequencies (Hz from channel center)	Frequency Span (Hz)	Power Spectral Density (dBc per subcarrier)	Modulation Type
Primary upper	2 to 26	363.4 to 4723.8	4360.5	–15	64-QAM
Primary lower	–2 to –26	–363.4 to –4723.8	4360.5	–15	64-QAM
Secondary	28 to 52	5087.2 to 9447.7	4360.5	–30	64-QAM
Tertiary	–28 to –52	–5087.2 to –9447.7	4360.5	–30	64-QAM
Reference upper	1	181.7	181.7	–15	BPSK
Reference lower	–1	–181.7	181.7	–15	BPSK
IDS1	27	4905.5	181.7	–30	16-QAM
IDS2	–27	–4905.5	181.7	–30	16-QAM

Table 16.3 Spectral summary of information carried in the HD Radio all-digital AM-IBOC waveform. (*Johnson*)

subcarriers; this is because quadrature is not needed (the AM carrier is unmodulated). With higher power, the bandwidth is reduced compared to the hybrid waveform; this reduces adjacent channel interference. Parameters of the all-digital AM-IBOC waveform are given in Table 16.3.

As in FM-IBOC, the AM-IBOC specification allows for main program service (MPS), personal data service (PDS), station identification service (SIS), and auxiliary application service (AAS) features. This is supported via a layered stack protocol. The AM-IBOC also uses OFDM to modulate digital data onto RF subcarriers. Because of the narrow-band AM mask, QAM is used on each ODFM subcarrier. To help resist noise, the duration of QAM symbol pulses is designed to be longer than noise pulses; symbol durations are 5.8 ms and subcarrier spacing is 181.7 Hz.

As in the FM-IBOC system, the hybrid AM-IBOC system employs time diversity by delaying the analog signal in relation to the digital signal. The receiver can blend in the analog signal to replace corrupted digital signals. Because of the time diversity between the two signals, the AM signal can be free from the interference momentarily corrupting the primary digital signal. As in FM-IBOC, AM-IBOC encoding follows the steps of data reduction, scrambling, error-correction encoding, interleaving in time and frequency, structuring in matrix format, assignment to OFDM subcarriers, creation of baseband waveform, and transmission. Error correction and interleaving are optimized for conditions in the AM interference environment.

To pass the digital signal, AM transmitters must have sufficient bandwidth and low phase distortion. The transmitter must not vary the correct phase relationship between the analog and digital broadcast signals. The center carrier is used as a reference signal, so a low group delay is required. Tube transmitters do not have sufficient linearity to pass the IBOC signal. Multiphase pulse duration modulation and digitally modulated solid-state transmitters require only minor modifications at the input. A single transmitter for both the analog and digital signals is recommended.

USA Digital Radio and Lucent Technologies were two (of many) early developers of IBOC systems for both FM and AM digital radio transmission. USA Digital Radio publicly demonstrated their first IBOC system (Project Acorn) in April 1991. Subsequently, they developed narrow-band, split-channel, spread-spectrum IBOC systems. USA Digital Radio and Lucent pooled resources and iBiquity Digital Corporation is now the principal developer of HD Radio technology for the commercial market.

Direct Satellite Radio

To cost-effectively reach nationwide radio audiences, two companies, Sirius Satellite Radio and XM Satellite Radio, developed systems to broadcast digital audio channels using satellites as the relay/distribution method. This is sometimes known as satellite digital audio radio service (SDARS). This system creates a national radio service featuring transmissions to any subscriber with a suitable receiver. To facilitate this, miniature satellite antennas have been developed. These antennas can be mounted on a car's roofline, and receive a downlinked signal, and convey it to the vehicle's receiver and playback system. Alternatively, signals may be received by stationary or handheld receivers. The FCC mandated that a common standard be developed so that SDARS receivers can receive programming from both companies. These services were granted licenses by the FCC in the Wireless Communications Service (WCS) band, within the S-band: XM Satellite Radio (2332.5 MHz to 2345.0 MHz) and Sirius Satellite Radio (2320.0 MHz to 2332.5 MHz). In other words, both services employ a 12.5-MHz bandwidth.

Sirius XM Radio

Sirius Satellite Radio and XM Satellite Radio began broadcasting in 2001, operating as competitors. Following FCC approval, the companies merged in July 2008; both services are currently operated by Sirius XM Radio. The two services employ incompatible compression and access technologies; thus receivers are incompatible. Current receivers continue to receive either shared Sirius or XM content. Newer interoperable receivers receive both services. The services are also licensed to operate in Canada, and the signal footprints can be received in other parts of North America.

XM content originates from broadcast studios in Washington, D.C., where it is stored in the MPEG-1 Layer II format at 384 kbps. The signal is uplinked to two high-powered satellites in geostationary orbit at 115° and 85° west longitude. The earth view of these and other satellites is available at www.fourmilab.ch/earthview/satellite.html. Two older satellites provide in-orbit backup. Signals are downlinked from the two satellites to earth. Each satellite broadcasts the full-channel service. Receivers can de-interleave the bitstreams from both satellites. Beam-shaping is used to concentrate relatively higher power to the east and west coasts. Because of the relatively low elevation angle (look angle) of the geostationary orbit, a large number of terrestrial repeater transmitters (gap fillers) are needed to convey the signal to receivers that would otherwise be blocked from the direct line-of-sight signal. Specifically, there are approximately 800 repeaters in 70 U.S. cities; a large city may have 20 repeaters. The repeaters receive their downlink signals from the primary geostationary satellites and rebroadcast the signal on different frequencies. The signal output from the repeater, and one of the satellite signals, is delayed with respect to the other satellite signal (by 4 seconds). In this way, if the received signal is momentarily lost and then retrieved, the delayed signal can be resynchronized with data in the receiver's buffer. Thus, audible interruptions are minimized. Signals from repeaters often allow reception inside buildings and homes.

XM Radio uses a modified High-Efficiency AAC (also known as HE AAC and aacPlus) codec for data reduction of music signals. This algorithm was developed by the Fraunhofer Institute and Coding Technologies. Spectral band replication is used to provide bandwidth extension. This increases the bandwidth of the audio signal output by the receiver, preserving bits for coding lower frequencies. In addition, proprietary signal processing is used to pre-process the audio signal for improved fidelity following AAC decoding. Aspects such as stereo imaging and spatiality are specifically addressed. Content coded with matrix surround sound (such as Dolby Pro Logic) can be decoded at the system output. The pre-processing algorithms were devised by Neural Audio. AAC and SBR are discussed in more detail in Chaps. 10 and 11. Some voice channels are coded with a proprietary Advanced Multi-Band Excitation (AMBE) codec, developed by Digital Voice Systems.

The system payload throughput is 4.096 Mbps. About 25% of the broadcast bandwidth is devoted to error correction and concealment. The sampling frequency is 44.1 kHz for music channels and 32 kHz for voice channels. Six carriers are used in the 12.5-MHz bandwidth. Two carriers convey the entire programming content, and four carriers provide signal diversity and redundancy. Two-carrier groups convey one hundred 8-kbps streams that are processed to form a variable number of channels at bit rates ranging from 4 kbps to 64 kbps. Different music channels may be assigned different bit rates. Once determined, bit rates for individual channels are constant. Speech channels, using vocoder coding, are allotted fewer bits than music channels. A substream of 128 kbps is separately provided for the OnStar service. A local traffic service is provided with Traffic Message Channel (TMC) technology. Signals are encrypted so that only subscribers can listen.

From originating studios in New York City, the Sirius signal is uplinked from New Jersey to three FS-1300 satellites constructed by Space Systems/Loral. These satellites are geosynchronous but not geostationary. They circle the earth with an elliptical 24-hour orbit inclined relative to the equator. Their orbits are modeled after patterns developed by the Department of Defense, and move the satellites over North America in an elliptical "figure 8" pattern. With these patterns, the apogees (highest point of the orbit) are over North America and the ground tracks of the satellites loop over the continent. Two of the three satellites are always broadcasting over the United States, with at least one at a look angle between 60° and 90°. The third satellite ceases to broadcast as it moves out of the line of sight. A fourth satellite is held as a spare. Approximately 100 terrestrial repeaters are needed, mainly in dense urban areas. Repeater signals and some satellite signals are delayed to promote buffering in the receiver, minimizing audible dropouts in reception when the signal is interrupted. The repeaters receive their downlink signals from a separate geostationary telecommunications satellite.

Sirius uses a type of PAC4 algorithm, developed by Lucent and iBiquity, for data reduction of the audio signals. The bit rate of the audio data and associated data conveyed by the system is about 4.4 Mbps. Speech channels use fewer bits than music channels. A statistical multiplexing technique is used in bit allocation. Audio channels are grouped into bit pools, and bit rates are dynamically varied (several times a second) according to the quantity of data needed to code signals in the pool at any time. Signals are encrypted so that only subscribers can listen.

Digital Television (DTV)

The technology underlying the analog television NTSC (National Television Standards Committee) standard was introduced in 1939. Originally designed for viewing screens measuring 5 to 10 inches across, NTSC persisted as a broadcast standard for over 60 years. It accepted an upgrade to color in 1954 and stereo sound in 1984. However, except for low-power and specialized transmitters, analog broadcasting has ended in the United States. High-power, over-the-air NTSC broadcasting ceased in June 2009 and was replaced by broadcasts adhering to the ATSC (Advanced Television Systems Committee) standard. Using the ATSC standard, Digital Television (DTV) offers improved picture quality, surround sound, and widescreen presentation. In addition, it expedites convergence of the television industry and the computer industry. Television program distribution has also changed since 1939. Today, terrestrial broadcasters are only part of the digital mosaic of cable, satellite, and the Internet.

DTV and ATSC Overview

The path toward DTV began in 1986 when the FCC announced that it intended to allocate unused broadcast spectral space to mobile communications companies. Broadcasters argued that they needed that spectral space to broadcast future high definition television signals, and in 1987 they showed Congress prototypes of analog HD receivers. They were awarded the spectral space, along with the obligation to develop the new technology. It was widely believed that analog technology was the most rational course. Then, in 1990, a digital system was proposed by General Instrument Corporation. It inspired development of a new digital television standard. In 1992, four digital systems and two analog systems were proposed. In 1993, an FCC advisory committee determined that a digital system should be selected, but none of the proposed systems was clearly superior to the other. The FCC called for the developers to form a Grand Alliance, pooling the best aspects of the four systems, to develop a "best of the best" HDTV standard. In 1995, the Advanced Television Systems Committee (ATSC) recommended the DTV format, and in December 1996, the FCC approved the ATSC standard and mandated it for terrestrial broadcasting. Channel assignment and service rules were adopted in April 1997. As noted, nationwide ATSC broadcasts began in June 2009.

A bandwidth challenge confronted engineers when they began designing the DTV system. The spectral space provided by the FCC called for a series of spectral slots, each channel occupying a 6-MHz band. Using signal modulation techniques, each channel could accommodate 19.39 Mbps of audio/video data. This is a relatively high bit rate (the CD, for example, delivers 1.4 Mbps) but inadequate for high-resolution display. Specifically, a high-definition picture may require a bit rate as high as 1.5 Gbps

(1500 Mbps). Clearly, data compression is required. Using data compression, the high bit rate needed for HDTV could be placed within the spectral slot.

When devising the ATSC standard, DTV designers selected two industry-standard compression methods: MPEG-2 for video compression and Dolby Digital for audio compression. A video decoder that conforms to the MPEG-2 Main Profile at High Level (MPEG-2 MP@HL) standard can decode DTV bitstreams. Because a DTV video program may have resolution that is five times that of a conventional NTSC program, a bit rate reduction ratio of over 50:1 is needed to convey the signal in the 19.39-Mbps bandwidth. MPEG-2 source compression achieves this (at this relatively high bit rate). An MPEG-2 encoder analyzes each frame of the video signal, as well as series of frames, and discards redundancies and relatively less visible detail, coding only the most vital visual information. This data may be placed into digital files called GOPs (Groups of Pictures) that ultimately convey the series of video frames. Video compression is described below.

Beginning in July 2008, the ATSC standard also supports the H.264/MPEG-4 AVC video codec. This codec was developed by the ITU-T Video Coding Experts Group and the ISO/IEC Moving Picture Experts Group; the partnership is sometimes known as the Joint Video Team (JVT). A/72 Part 1 describes video characteristics of AVC and A/72 Part 2 describes the AVC video transport substream. H.264 is also used in the Blu-ray format.

The video formats within the ATSC standard can employ Dolby Digital audio coding. Signals can be decoded with the same Dolby Digital-equipped A/V receiver used in many home theaters. However, not all DTV programs are broadcast in surround sound. Conceptually similar to video bit-rate reduction, Dolby Digital analyzes the audio signal for content that is inaudible and either discards the signal or codes it so that introduced noise is not perceived. In particular, it relies on psychoacoustics to model the strengths and weaknesses of human hearing. Dolby Digital is described in Chap. 11.

Video Data Reduction

As we observed in Chap. 10, lossy and lossless coding principles may be applied to audio signals to decrease the bit rate needed to store or transmit the signal. Similarly, these techniques may be applied to video signals. Indeed, the high bandwidth of video dictates data reduction for most applications. Although the specific techniques are quite different, the goal remains the same—to decrease bit rate. In fact, proportionally, a video signal may undergo much greater reduction before artifacts intrude. The MPEG video coding standards for video data reduction are good examples of how such algorithms operate.

In normal daylight, the optical receptors of the retina collect visual information at a rate of about 800 Mbps. However, the neural network connecting the eye to the visual cortex reduces this by a factor of 100; most of the information entering the eye is not needed by the brain. By exploiting the limitations of the eye, video reduction algorithms can achieve significant data reduction using spatial and temporal coding. Psychovisual studies verify that the eye is more sensitive to variations in brightness than in color. For example, the retina has about 125 million photoreceptors; 94% of these photoreceptors are rod cells that perceive intensity even in dim light but have low resolution, while 6% are cone cells that yield color perception and high resolution but only in fairly bright light. Our sensitivity to variations in brightness and color has a dynamic range, and is not linear. For example, our visual sensitivity is good for shades of green, but poor for red and blue colors. On the other hand, we are sensitive to very gradual variations in brightness and color.

Audio engineers are accustomed to the idea of frequency, the number of waveform periods over time. Spatial frequency describes the rate of change in visual information, without reference to time. (The way a picture changes over time is referred to as temporal frequency.) The eye's contrast sensitivity peaks at spatial frequencies of about 5 cycles/degree, and falls to zero at about 100 cycles/degree, where detail is too small to be perceived. The former corresponds to viewing objects 2 mm in size at a distance of 1 meter; the latter corresponds to objects 0.1 mm in size. (Cycles/degree is used instead of cycles/meter because spatial resolution depends on how far away an object is from the eye.) The contrast in a picture can be described in horizontal and vertical spatial frequency. Moreover, because the eye is a lowpass filter, high spatial frequencies can be successfully quantized to achieve data reduction.

The eye can perceive about 1000 levels of gray; this requires 10 bits to quantize. In practice, eight bits is usually sufficient. Brightness (luminance) uses one set of receptors in the retina, and color (chrominance) uses three. Peak sensitivity of chrominance change occurs at 1 cycle/degree and falls to zero at about 12 cycles/degree. Clearly, because the eye's color resolution is less, chrominance can be coded with fewer bits.

Video data reduction exploits spatial and temporal correlation between pixels that are adjacent in space and time. Highly random pictures with high entropy are more difficult to reduce. Most pictures contain redundancy that can be accurately predicted. Video coding exploits redundancy in individual frames as well as redundancy in sequences of frames. In either case, psychovisual principles are applied to take advantage of limitations in the human visual system; in other words, visual perceptual coding is used. For example, the spatial coding aspect of MPEG video coding is similar to and based on JPEG (Joint Photographic Experts Group) coding of still images. This type of coding uses a transform to convert a still image into its component spatial frequencies. This expedites elimination of irrelevant detail and quantization of remaining values. Temporal coding reduces a series of images to key frames and then describes only the differences between other frames and the key frames. For example, a video of a "talking head" newscaster reporting the news would have small differences over many frames, while an action-packed car chase might have large differences. Even then, from one frame to the next, some parts of each frame merely move to a new location in the next frame. Video codecs take advantage of all of these video characteristics to remove redundancy while preserving the vital information.

An enormous amount of data is required to code a video program. A legacy analog composite video NTSC signal has a bandwidth of 4.2 MHz. To digitize this, the Nyquist theorem demands a sampling frequency of twice the bandwidth, or 8.4 MHz. At 8 bits/sample, this yields 67.2 Mbps. Because a color image comprises red, green, and blue components, this rate is multiplied by three, yielding 201.6 Mbps. A 650-Mbyte CD could store about 26 seconds of this digital video program. Data requirements can be reduced by dropping the frame rate, the size of the image can be reduced to a quarter-screen or smaller, and the number of bits used to code colors can be reduced. More subtly, file size can be reduced by examining the file for irrelevancy within a frame and over a series of frames, and redundant data can be subjected to data compression. These techniques can be particularly efficient at preserving good picture quality, with a low bit rate.

MPEG-1 and MPEG-2 Video Coding

The MPEG-1 and MPEG-2 video compression algorithms are fundamentally similar; this general description considers MPEG-1 and MPEG-2 together. Both use a combination of lossy and lossless techniques to reduce bit rate. They can perform over a wide range of bit rates and picture resolution levels. As with the audio portions of the standard, the MPEG video standard only specifies the syntax and semantics of the bitstream and the decoding process. The encoding algorithms are not fixed, thus allowing optimization for particular visual phenomena, and overall improvements in encoding techniques. Rather than rely exclusively on algorithmic data reduction, MPEG also uses substantial pre-processing of the video signal to reduce pixel count, number of coded frames, method of signal representation and other parameters. Both MPEG-1 and MPEG-2 analyze a video program for both interframe (spatial) and intraframe (temporal) redundancy. Taking into account the eye's limitations in these respects, the video signal's bit rate can be reduced. Figure 16.14 shows an MPEG video encoder that employs spatial and temporal processing for data reduction.

A color video signal can be represented as individually coded red, green, and blue signals, known as RGB, that represent the intensity of each color. Video reduction begins by converting RGB triplets into component video YCrCb triplets (Y is luminance, Cr is red chrominance difference, and Cb is blue chrominance difference) that separates brightness and color. This allows more efficient coding because the Y, Cb, and Cr signals have less correlation than the original, highly correlated RGB components; other coding approaches can be used instead of YCbCr coding. Because the eye is less sensitive to color, the color components can undergo a greater reduction. In particular, Cr and Cb values are sub-sampled to reduce color bandwidth. The Y luminance component represents brightness and ranges from black to white. The image is divided into blocks each corresponding to an 8 × 8 pixel area. To remove spatial redundancy, blocks are transformed, coefficients quantized, run-length coded,

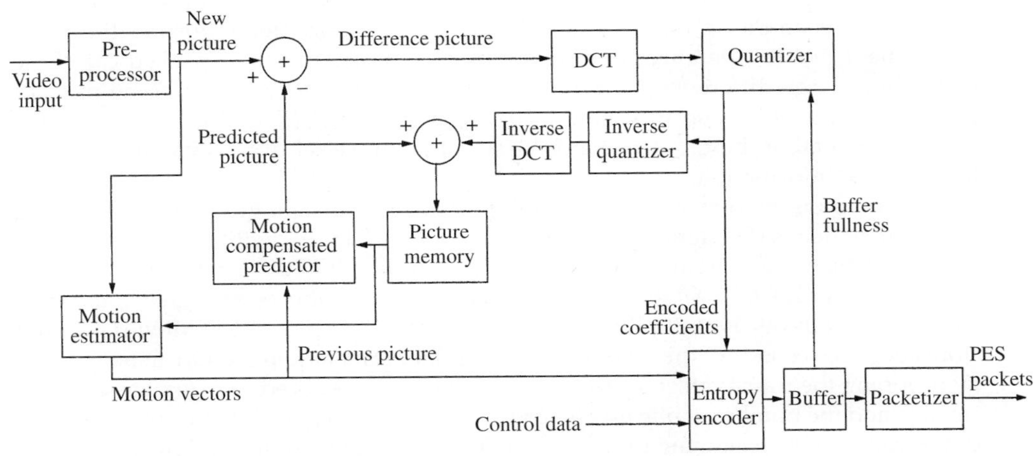

FIGURE 16.14 Block diagram of an MPEG video encoder using spatial and temporal processing for data reduction. An MPEG packetized elementary stream (PES) bitstream is output. (*Hopkins, 1994*)

and entropy coded. To remove temporal redundancy, blocks of 8×8 pixels are grouped into macroblocks of four luminance blocks (yielding a 16×16 pixel array) and a number of Cr and Cb chrominance blocks. A 4:2:0 format uses one each of Cr and Cb, a 4:2:2 format uses two each, and 4:4:4 uses four each. Each new picture is compared to a previous picture in memory and movement of macroblocks across the screen is analyzed. A predicted picture is generated using that analysis. The new picture is compared to the predicted picture to produce a transmitted difference signal.

The static image contained in one video frame or field can be represented as a two-dimensional distribution of amplitude values, or a set of two-dimensional frequency coefficients. The MPEG video algorithm uses the discrete cosine transform (DCT) to convert the former to the latter. The DCT is a type of discrete Fourier transform, using only cosine coefficients. Using the DCT, the spatial information comprising a picture is transformed into its spatial frequencies. Specifically, the prediction difference values (in P and B frames) as well as the undifferenced values (in I frames) are grouped into 8×8 blocks and a spatial DCT transform is applied to the blocks of values. The luminance and chrominance components are transformed separately.

The DCT transform requires the summation of $i \times j$ multiplicative blocks (where $i \times j$ is the pixel amplitude at a row i and column j position) to generate the value of the transform coefficient at each point. Following the DCT, as represented in Fig. 16.15, the spatial frequency coefficients are arranged with the dc value in the upper-left corner, and ac values elsewhere, with increasing horizontal frequency from left to right and increasing vertical frequency from top to bottom. The output values are no longer pixels; they are coefficients representing the level of energy in the frequency components. For example, an image of finely spaced stripes would produce a large value at the pattern repetition frequency, with other components at zero. Figure 16.16 shows the amplitudes of pixels in a complicated image, and its transform coefficients; the latter are highly ordered. The transform does not provide reduction. However, after it is transformed, the relative importance of signal content can be analyzed. For most images,

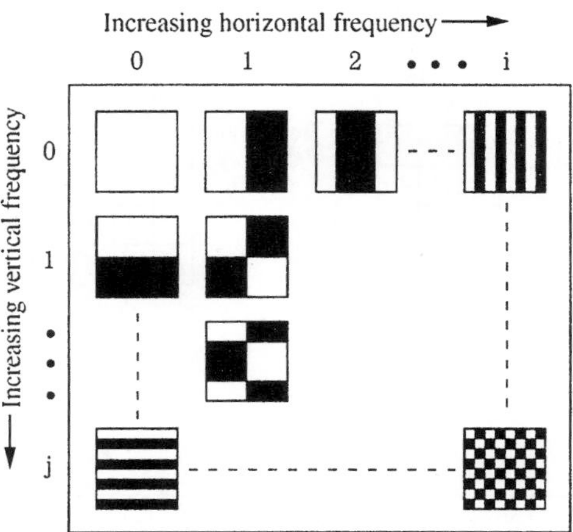

FIGURE 16.15 DCT transform coefficients can be represented as a block showing increasing horizontal and vertical spatial frequency.

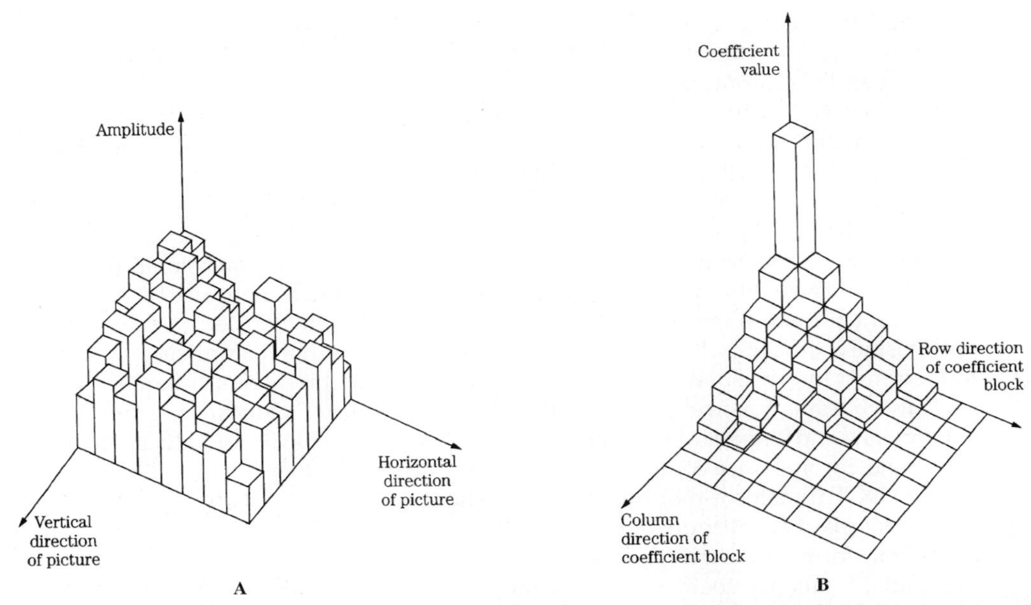

FIGURE 16.16 A complicated visual image can be applied to a two-dimensional DCT transform, resulting in an ordered representation in spatial frequency. A. Original sample block of image. B. Coefficient block following DCT transform.

the important information is concentrated in a small number of coefficients (usually the dc and low spatial frequency coefficients). When these DCT coefficients are quantized, the least perceptible detail is lost, but data reduction is achieved.

Figure 16.17 illustrates spatial coding of an image (for simplicity, this example will consider only luminance coding). The screen is divided into 48 blocks. In practice,

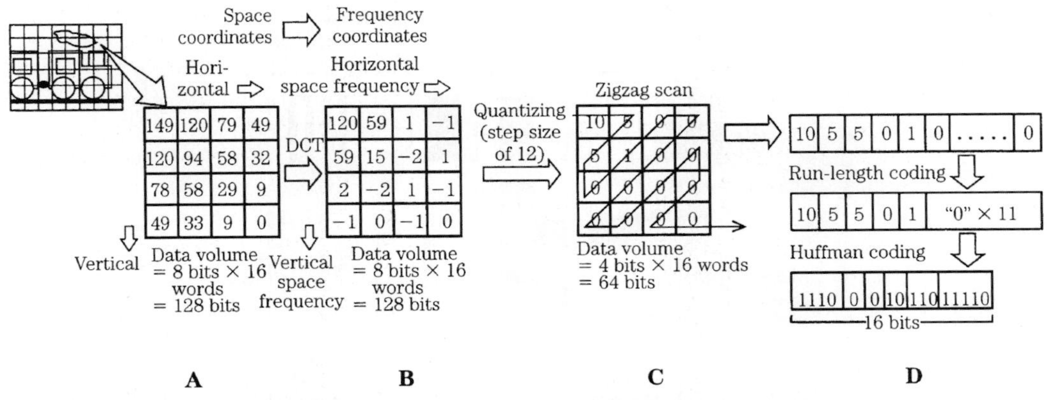

FIGURE 16.17 Example of intraframe encoding of an image using the MPEG video algorithm. A. An 8 × 8 block of pixels represents luminance values. B. A DCT transform presents the frequency coefficients of the block. C. The coefficients are quantized to achieve reduction and scanned with a zigzag pattern. D. The data sequence is applied to run-length and Huffman coding for data compression.

for example, a video frame of 720 × 480 pixels (such as DVD-Video) would yield 90 blocks (of 8 × 8 pixels) across the screen width and 60 blocks across its height. In this example, each block is represented by 16 pixels in 4 × 4 arrays. Each pixel has an 8-bit value from 0 (black) to 255 (white) representing brightness. Observe, for example, that the lower pixels in the detailed block (from the top of the locomotive's smokestack) have lower values, representing a darker picture area. Following the DCT, the spatial frequency coefficients are presented. The dc value takes the proportional average value of all pixels in the block and the ac coefficients represent changes in brightness, that is, detail in the block. For example if all the pixels were the same brightness of 100, the coefficient array would have a dc value of 100, and all the other blocks would be zero. In this example (Fig. 16.17), the proportional average brightness is 120, and the block does not contain significant high spatial frequencies.

Reduction is achieved by quantizing the coefficients and discarding low energy coefficients by truncating them to zero. In this example, a step size of 12 is applied; it is selected according to the average brightness value. In practice, the important low-frequency values are often coded with high precision (for example, a value of 8) and higher frequencies in the matrix are quantized more coarsely (for example, a value of 80). This effectively shapes the quantization noise into visually less important areas. The dc coefficients are specially coded to take advantage of their high spatial correlation. For example, Main Profile MPEG-2 allows up to 10-bit precision of the dc coefficient. Quantization can be varied from one block to another.

To achieve different step sizes, quantizer matrices can be used to determine optimal perceptual weighting of each picture; the DCT coefficients are weighted prior to uniform quantization. The weighting algorithm exploits the visual characteristics and limitations of human vision to place coding impairments in frequencies and regions where they are perceptually minimal. For example, a loss of high-frequency detail in dark areas of the picture is not as apparent as in bright areas. Also, high-frequency error is less visible in textured areas than in flat areas. With this quantization, data may be reduced by a factor of 2.

In video compression, because there are usually few high-frequency details, many of the transform coefficients are quantized to zero. To take advantage of this, the two-dimensional array is reduced to a one-dimensional sequence. Variable-length coding (VLC) is applied to the transform coefficients. They are scanned with a zigzag pattern (or alternate zigzag pattern when the picture is interlaced) and sequenced. The most important (non-zero) coefficients are grouped at the beginning of the sequence and the truncated high-frequency detail appears as a series of zeros at the end. Run-length coding is applied. It efficiently consolidates the long strings of zero coefficients, identifying the number (run) of consecutive zero coefficients. In addition, an end-of-block (EOB) marker signifies that all remaining values in the sequence are zero.

These strings are further coded with Huffman coding tables; the most frequent value is assigned the shortest codeword. Overall, a further reduction of four may be achieved. MPEG-1 requires a fixed output bit rate, whereas MPEG-2 allows a variable bit rate. In MPEG-2 encoders, prior to output, data is passed through a channel buffer to ensure that the peak data rate is not exceeded. For example, the DTV standard specifies a channel buffer size of 8 Mbits. In addition, if the bit rate decreases significantly, a feedback mechanism may vary upstream processing. For example, it may request finer quantization. In any case, the output bitstream is packetized as a stream of MPEG-compatible PES packets. The transmitted picture is also applied to an inverse DCT, summed with the predicted picture and placed in picture memory. This data is used in the motion compensated prediction loop as described below.

This description of the encoder's operation (focusing on spatial reduction) assumes that the video picture used to predict the new picture is unchanging. The result is similar to the JPEG coding of still images. That is, temporal changes in pictures, as in video, are not accounted for. It is advantageous, in video programs, to predict a new picture from a future picture, or from both a past and future picture. The majority of reduction achieved by MPEG video coding comes from temporal redundancy in sequential video frames. Because video sequences are highly correlated in time, it is only necessary to code interframe differences. Specifically, motion-compensated interframe coding, accounting for past and future correlation in video frames, is used to further reduce the bit rate.

In a motion-compensated prediction loop, the macroblocks in the current new frame are searched for and compared to regions in the previous frame. If a match is made within an error criteria, a displacement vector is coded describing how many pixels (direction and distance) the current macroblock has moved from the previous frame. Because frame-to-frame correlation is often high, predictive strategies are used to estimate the macroblock motion from past frames. A predicted picture is generated from the macroblock analysis. The new current picture is compared to this predicted picture and used to produce a transmitted difference signal.

Data in the current frame not found in other frames is considered as a residual error in the prediction, as shown in Fig. 16.18. Given the previous frame in memory, the displacement vectors are applied to create a motion-compensated frame. This predicted intermediate frame is subtracted from the actual current frame being coded. The additional detail is the residual prediction error that is coded with DCT as described above. If the estimation is good, the residue is small and little new information would be needed (this would be the case, for example, in slowly changing images). Conversely, fast motion would require more new information. The prediction is based on previous

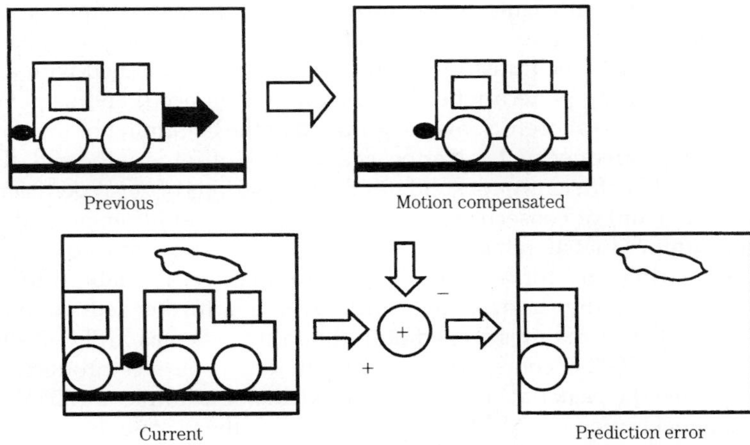

FIGURE 16.18 Interframe coding offers opportunity for efficient data reduction. Displacement vectors are applied to a previous frame to create a motion-compensated frame. Other residual data is coded as a prediction error.

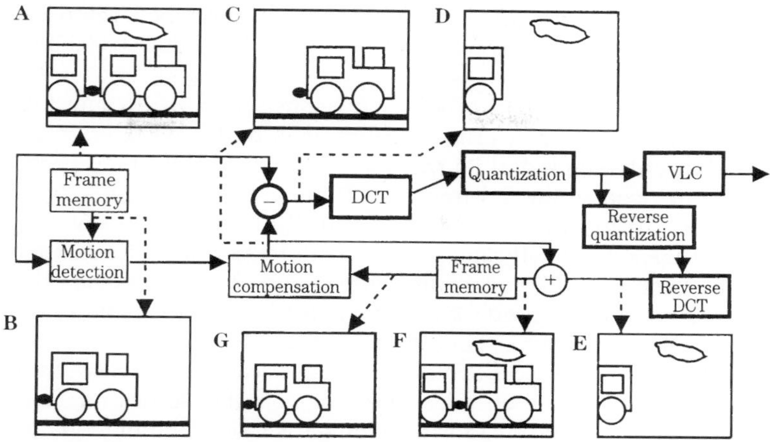

FIGURE 16.19 Summary of interframe coding showing how motion-compensation frames are created and data is coded with DCT and VLC. A. Current frame to be coded. B. Previous frame. C. Intermediate motion-compensated frame. D. Prediction-error frame. E., F., G. Frames within the coding loop.

information available in the loop, derived from previous pictures. During decoding, this error signal is added to the motion-compensated detail in the previous frame to obtain the current frame. Predicted information can be transmitted and used to reconstruct the picture because the same information used to make the prediction in the encoder is also available at the receiving decoder.

Figure 16.19 summarizes motion compensation and coding. The current frame (A) is compared to the previous frame (B) to create an intermediate motion-compensated frame (C) and displacement vectors. The prediction error (D) between the intermediate frame and the current frame is input to the DCT for coding. Feed-forward and feedback control paths use the frame error (produced by applying the reverse quantized data and reverse DCT) as input to reduce quantization error. Macroblocks that are new to the current frame also are coded with DCT and combined with the motion vectors. Generally, distortion over one or two frames is not visible at normal speed. For example, after a scene change, the first frame or two can be greatly distorted, without obvious perceptible distortion.

MPEG codes three frame types; they allow key reference frames so that artifacts do not accumulate, yet provide bit-rate savings. For example, accurate frames must appear regularly in the bitstream to serve as references for other motion-compensated frames. Frame types are intra, predicted, and bidirectional (I, P, and B) frames. Intra (I) frames are self-contained; they do not refer to other frames, are used as reference frames (no prediction used), and are moderately compressed with DCT coding without motion compensation. An I-frame is transmitted as a new picture, not as a difference picture. Predicted (P) frames refer to the most recently decoded I- or P-frame; they are more highly compressed using motion compensation. Bidirectional (B) frames are coded with interpolation prediction using motion vectors from both past and future I- or P-frames; they are very highly compressed. Very generally, P-frames require one-half as many bits as I-frames, and B-frames about one-fifth as many. Exact numbers depend on the picture itself, but at the MPEG-2 MP@ML level (described below) and a bit rate of 4 Mbps, an I-frame might contain 400,000 bits, a P-frame 200,000 bits, and a B-frame 80,000 bits.

An MPEG Group of Pictures (GOP) consists of an I-frame followed by some number of P- and B-frames that is determined by the encoder design. For example, 1 second of video coded at 30 frames per second might be represented as a GOP of 30 frames containing 2 I-frames, 8 P-frames (repeating every third frame), and 20 B-frames: IBBPBBPBBPBBPBBIBBPBBPBBPBBPBB. Every GOP must begin with an I-frame. A GOP can range from 10 to 15 frames for fast motion, and 30 to 60 for slower motion. Picture quality suffers with GOPs of less than 10 frames. I-frames use more bytes, but are used for random access, still frames, and so on. In many cases, the I-B-P sequence is altered. For example, I-frames might be added (in a process known as "I-frame forcing") on scene cuts, so each new scene starts with a fresh I-frame. Moreover, in variable bit-rate encoders, visually complex scenes might receive a greater number of I-frames. The creation of B frames is more computationally intensive; some low-cost video codecs only use I and P frames.

The MPEG video decoder must invert the processing performed by the encoder. Figure 16.20 shows a video decoder. Data bits are taken from a channel buffer and de-packetized and the run length is decoded. Coefficients and motion vectors are held in memory until they are used to decode the next picture. Data is placed into 8 × 8 arrays of quantized DCT coefficients and dequantized with step-size information input from auxiliary data. Data is transformed by the inverse discrete cosine transform (IDCT) to obtain pixel values or prediction errors to reconstruct the picture. A predicted picture is assembled from the received motion vectors, using them to move macroblocks of the previous decoded picture. Finally, pixel values are added to the predicted picture to produce a new picture.

MPEG-1 Video Standard

The MPEG-1 video standard, ISO/IEC 11172-3, describes a video compression method optimized for relatively low bit rates of approximately 1.4 Mbps. Moreover, MPEG-1 is a frame-based system, with a fixed data rate. One goal of MPEG-1 is the reduction of audio/video data rates to within the 1.41-Mbps data transfer rate of the Compact Disc, allowing storage of 72 minutes of video program on a CD as well as real-time playback. MPEG-1 also allows transmission over computer networks and other applications.

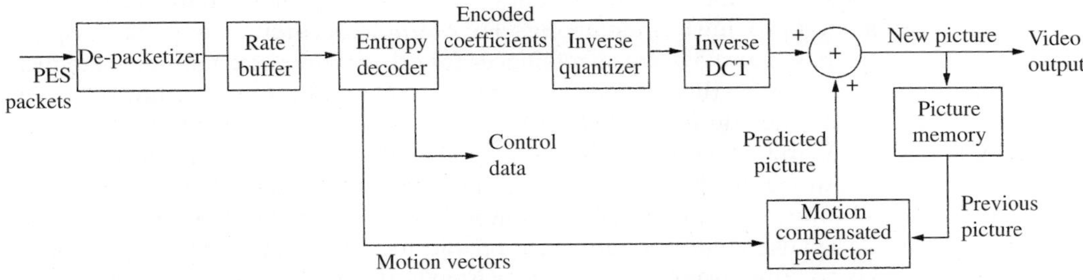

FIGURE 16.20 Block diagram of an MPEG video decoder using inverse DCT and motion vectors to reconstruct the image from input PES packets. (*Hopkins, 1994*)

Using the MPEG-1 video coding algorithm, for example, a video program coded at the professional rate of 165 Mbps can be reduced to approximately 1.15 Mbps, an overall reduction ratio of 140:1. Using an audio codec, audio data at 1.41 Mbps can be reduced to 0.224 Mbps, a overall ratio of 7:1. The video and audio data are combined into a single data stream with a total rate of 1.41 Mbps. The large amount of video reduction is partly achieved by aggressive pre-processing of the signal. For example, vertical resolution can be halved by discarding half of the video fields and horizontal resolution can be halved by sub-sampling. NTSC resolution might be 352 pixels by 240 lines. MPEG-1 supports only progressive-scan coding.

In MPEG-1, a rectangular spatial format is typically used, with a maximum picture area of noninterlaced 352 pixels by 240 lines (NTSC) or 352 pixels by 288 lines (PAL/SECAM), but the picture area can be used flexibly. For example, a low and wide area of 768 pixels by 132 lines, or a high and narrow area of 176 pixels by 576 lines could be coded. During playback, the entire decoded picture or other defined parts of it can be displayed, with the window's size and shape under program control. Audio and video synchronization is ensured by the MPEG-1 multiplexed bitstream. The video quality of MPEG-1 coding is similar to that of the VHS format. The Video CD format uses MPEG-1 coding. MPEG-1 is partly based on the H.261 standard.

MPEG-2 Video Standard

The MPEG-2 video standard, ISO/IEC 13818-2, is optimized for higher bit rates than MPEG-1. It also uses a variable data rate for greater coding efficiency and higher picture quality. MPEG-2 defines a toolbox of data-reduction processes. It specifies five profiles and four levels. A profile defines how a bitstream is produced. A level defines parameters such as picture size, resolution, and bit rate. MPEG-2 defines several hierarchical combinations, notated as a Profile@Level. The hierarchical profiles are: Simple, Main, SNR, Spatial, and High. The hierarchical levels are: Low, Main, High-1440, and High. Eleven different combinations are defined, as shown in Table 16.4. A compliant MPEG-2 decoder can decode data at its profile and level and should also be compatible with lower-mode MPEG bitstreams. Furthermore, because MPEG-2 is a superset of MPEG-1, every MPEG-2 decoder can decode an MPEG-1 bitstream. MPEG-2 supports both interleaved and progressive-scan coding. The MPEG standard defines a syntax for video compression, but implementation is the designer's obligation, and picture quality varies according to the integrity of any particular encoding algorithm.

MPEG-2 video compression is used in DBS direct satellite broadcasting, the DVD-Video and Blu-ray disc formats, and the DTV format, at different quality levels. Main Profile at Main Level (MP@ML) is known as the "standard" version of MPEG-2. It supports interlaced video, random picture access, and 4:2:0 YUV representation (luminance is full bandwidth, and chrominance is sub-sampled horizontally and vertically to yield a quarter of the luminance resolution). Main Level has a resolution of 720 samples per line with 576 lines per frame and 30 frames per second. Its video quality is similar to analog NTSC or PAL.

MP@ML is used in DVD-Video. MP@ML and MP@HL/H1440L can be used in Blu-ray. The Main Profile at High Level (MP@HL) mode is used in the ATSC DTV standard. At higher bit rates, MP@ML achieves video quality that rivals professional broadcast standards. MPEG-1 essentially occupies the MP@LL category. Generally, the hierarchical MPEG modes were designed to provide distribution to the end user. They were not designed for repeated coding and decoding as might occur in internal production studio

		Profiles				
		Simple I,P 4:2:0 Non-scalable	Main I,P,B 4:2:0 Non-scalable	SNR I,P,B 4:2:0 SNR Scalable	Spatial I,P,B 4:2:0 Spatially Scalable	High I,P,B 4:2:0 or 4:2:2 SNR and Spatially Scalable
Levels	High < 1920 × 1152/1080 60 fps	—	MP@HL < 80 Mbps 1080i 720p	—	—	HP@HL < 100 Mbps
	High-1440 < 1440 × 1152/1080 60 fps	—	MP@H-1440 < 60 Mbps	—	SSP@ H-1440 < 60 Mbps	HP@H-1440 < 80 Mbps
	Main < 720 × 576/480 30 fps	SP@ML < 15 Mbps	MP@ML < 15 Mbps	SNR@ML < 15 Mbps	—	HP@ML < 20 Mbps
	Low < 352 × 288/240 30 fps	—	MP@LL < 4 Mbps	SNR@LL < 4 Mbps	—	—

TABLE 16.4 MPEG-2 defines five profiles and four levels, yielding 11 different combinations.

applications. One exception is the 4:2:2 Profile at Main Level mode; this is used in some Betacam systems. Although MPEG-1 and MPEG-2 also define audio compression, other methods such as Dolby Digital can be used for audio compression.

ATSC Digital Television

The Advanced Television Systems Committee (ATSC) standard defines the transmission and reception of digital television, but it does not describe a single format. The lengthy development time of the technology, the sometimes contradictory interests of the television and computer industries, the desire for a highly flexible system and the inherent complexity of the technology itself, all led to an umbrella system with a range of picture resolution and features. Thus DTV can appear as SDTV (Standard Definition Television) or HDTV (High Definition Television) with numerous protocols within each system. Very generally, SDTV delivers about 300,000 pixels per frame and its resolution (for example, 480P) provides a picture quality similar to that of DVD-Video. HDTV can provide about 2 million pixels per frame; its resolution (for example, 1080i or 720p) is superior to any analog consumer video system.

The rationale for dual high-definition and standard-definition formats lies in the economics of the bitstream. HDTV requires more bandwidth than SDTV so that one HDTV channel occupies the entire broadcast bandwidth. Alternatively, within one broadcast slot, multiple simultaneous SDTV channels (four to six of them) can be multicast in the place of one HDTV channel. A local affiliate might broadcast several SDTV channels during the day, and then switch to one HDTV channel during prime time. Another station might broadcast HDTV all the time, or never. The FCC only requires that a broadcaster provide one free DTV channel. Thus stations may broadcast SDTV channels, provide one free SDTV channel, and charge fees for the others. Because one HDTV channel consumes a station's entire bandwidth, a (free) HDTV channel must rely on its advertising stream; this is an economic disincentive for broadcasting HDTV. In addition, DTV allows datacasting in which auxiliary information accompanies the program signal. Some examples of datacasting include electronic programming guides, program transcripts, stock quotes, statistics or historical information about a program, and commercial information. Interactive applications such as interactive Web links, email, online ordering, chatting, polling, and gaming are also possible.

In addition to its picture quality, and provisions for widescreen displays and surround sound, a DTV signal is more reliable, with less noise than analog TV. The picture quality is consistent over the broadcast coverage area. However, reception does not gracefully degrade outside the coverage area; reception is nominal or unusable. Severe multipath interference can also lead to unusable signals. Most DTV channels are broadcast in the spectrum encompassing VHF channels 2 to 13 (144-216 MHz) and UHF channels 14 to 51 (470-698 MHz) excluding channel 37 (608-614 MHz), which is used for radio astronomy. Generally, if consumers live in the coverage area of a DTV broadcaster, and get good analog reception, they can continue to use existing indoor or outdoor 75-Ω antennas to receive DTV broadcasts.

ATSC Display Formats and Specification

Within the SDTV and HDTV umbrella, the ATSC standard defines 18 basic display formats (totaling 36 with different frame rates). Table 16.5 shows these formats and some of the parameters that differentiate them. HDTV signals provide higher resolution than SDTV, as exemplified by the number of pixels (in both the horizontal and vertical dimensions) that comprise the displayed picture. For example, a 1080 × 1920 picture is the highest HD resolution, and 480 × 640 is the lowest SD resolution. The display can also assume a conventional 4:3 aspect ratio or a widescreen 16:9 aspect ratio.

The formats are also differentiated by progressive (P) and interleaved (I) scanning. Conventional TV displays use interleaving in which a field of odd-numbered scan lines are displayed, followed by a field of even-numbered lines, to display one video frame 30 times per second. This was originally devised to yield a flicker-free picture. However, computer displays use progressive scanning in which all lines are displayed in sequence. The question of whether DTV should use interleaved or progressive scanning inspired debate between the traditional broadcast industry (which favored interleaving), and the computer industry (which favored progressive scanning). Each felt that its more familiar technology would lead to a competitive advantage. In the end, both kinds of display technology were included. At the highest resolution of 1080p, a display must scan at about 66 kHz. At a resolution of 720p, the scanning frequency is about 45 kHz. Low-cost displays might scan at 33.75 kHz, which is capable of displaying a 1080i signal. Generally, broadcasters have embraced either the 1080i or 720p formats.

	Vertical Size (pixels)	Horizontal Size (pixels)	Aspect Ratio		Scan Pattern		Frame Rate per Second[1]		
			16:9	4:3	Progressive	Interlaced	24 or 23.976[2]	30 or 29.97[2]	60 or 59.94[2]
HDTV	1080[3]	1920	✔		✔		✔	✔	[5]
	720	1280	✔			✔		✔	
			✔		✔		✔	✔	✔
SDTV	480	704[4]	✔		✔		✔	✔	✔
			✔			✔		✔	
				✔	✔		✔	✔	✔
				✔		✔		✔	
		640		✔	✔		✔	✔	✔
				✔		✔		✔	

[1]With interlaced pictures, the field rate is twice the frame rate.

[2]Both scan rates are specified to ensure backward-compatibility with older formats like NTSC.

[3]1088 lines are coded to satisfy an MPEG-2 requirement that coded vertical size be a multiple of 16 (progressive scanning) or 32 (interlaced scanning).

[4]Non-square pixels.

[5]The 1080 format at 60 frames per second is defined by the ATSC, but it will not fit within the broadcast bandwidth currently allocated for DTV.

Table 16.5 ATSC digital television display formats.

The ATSC DTV standard describes methods for video transmission, audio transmission, data transmission, and broadcast protocols. Very generally, the system can be considered as three subsections of source coding and compression, service multiplex and transport, and RF/Transmission, as shown in Fig. 16.21. The source coding and compression subsystem encompasses bit-rate reduction methods. Video compression is based on MPEG-2 and audio compression is based on Dolby Digital (AC-3). In the service multiplex and transport subsystem, the data streams are divided into packets and each packet type is uniquely identified. Moreover, the different types of packets (video, audio, auxiliary data) are multiplexed into a single data stream. DTV employs the MPEG-2 transport stream syntax for the packetization and multiplexing of video, audio, and data signals for digital broadcasting. Channel coding and modulation is performed in the RF/Transmission subsystem. Additional information needed by the receiver's decoder such as error-correction parity is added. The modulation (or physical layer) creates the transmitted signal in either a terrestrial broadcast mode or high data rate mode.

The DTV standard specifies a bit rate of 384 kbps for main audio service, 192 kbps for two-channel associated dialogue service, and 128 kbps for single-channel associated service. The main service may contain from 1 to 5.1 audio channels; audio in multiple languages may be provided by supplying multiple main services. Examples

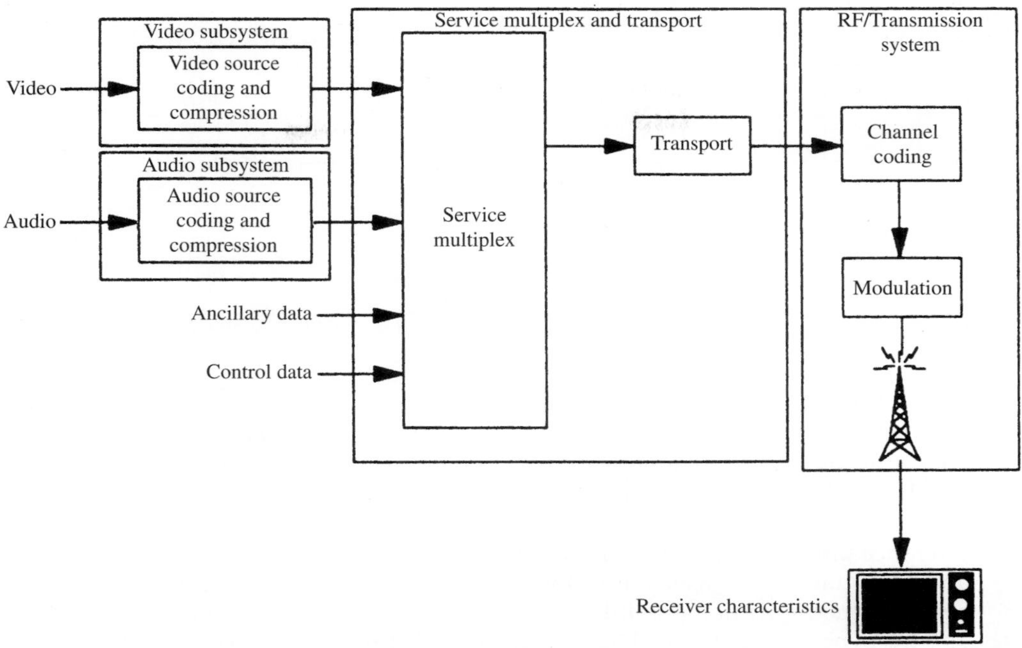

Figure 16.21 The ATSC DTV terrestrial television broadcasting model can be considered as source coding, multiplex and transport, and transmission subsections.

of associated services are data for the visually impaired, commentary, emergency, and voice-over announcements. The combined bit rate of main and associated service is 512 kbps. All audio is conveyed at a sampling frequency of 48 kHz. The main channels have an approximate high-frequency response of 20 kHz, and the low-frequency effects (LFE) channel is limited to 120 Hz. Either analog or digital inputs may be applied to a Dolby Digital encoder. When AES3 or other two-channel interfaces are used, the ATSC recommends that pair 1 carries left and right channels, pair 2 carries center and LFE, and pair 3 carries left surround and right surround. The implementation of Dolby Digital for DTV is specified in ATSC Document A/52. The Dolby Digital Plus (Enhanced AC-3) format may be used in some applications; it is backward-compatible with Dolby Digital. Dolby Digital is described in Chap. 11.

While MPEG-2 and Dolby Digital form the coding basis for DTV, they are only two algorithms within a much larger system. The DTV standard also defines the way in which the bits are formatted and how the digital signal is wirelessly broadcast, or conveyed over wired cable. The input to the transmission subsystem from the transport subsystem is a 19.39-Mbps serial data stream comprising 188-byte MPEG-compatible data packets. This stream is defined in the ISO/IEC 13818-1 standard. Each packet is identified by a header that describes the application of the elementary bitstream. These applications include video, audio, data, program, system control, and so on. Combinations of data types are combined to create programs that are ubiquitously conveyed by the transport protocol.

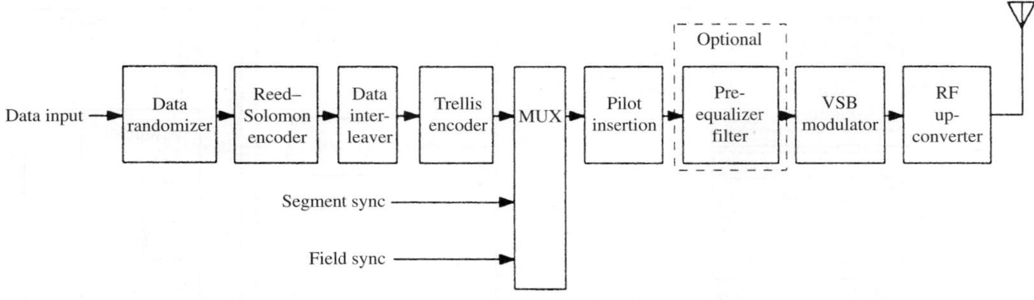

FIGURE 16.22 Block diagram of a VSB transmitter used to broadcast ATSC DTV signals.

Each fixed-length packet contains 188 bytes: a sync byte and 187 data bytes. This adheres to the MPEG-2 transport syntax. Packet contents are identified by the packet header, containing both a 4-byte link header and a variable-length adaptation header. The link header provides synchronization, packet identification, error detection, and conditional access. The adaptation header contains synchronization reference timing information, information for random entry into compressed data, and information for insertion of local programs. This MPEG-2 transport layer is also designed for interoperability with the Asynchronous Transport Mode (ATM) and the Synchronous Optical NETwork (SONET) protocols, as well as Direct Broadcast Satellite (DBS) systems. The DVD-Video and Blu-ray disc standards also use this transport layer.

The DTV standard defines two broadcast modes. The terrestrial broadcast mode employs 8 VSB (vestigial sideband modulation with eight discrete amplitude levels), as shown in Fig. 16.22. This mode delivers a payload data rate of 19.28 Mbps (from a net rate of 19.39 Mbps) in a 6-MHz channel; this accommodates one HDTV channel. The terrestrial broadcast mode can operate in an S/N environment of 14.9 dB. The high data rate mode is used for cable applications. It sacrifices transmission robustness for a higher payload data rate of 38.57 Mbps (from a net rate of 38.78 Mbps) in a 6-MHz channel; two HDTV channels can be transmitted in this mode. This mode is similar to the broadcast mode. However, principally, 16 VSB is employed, increasing the number of transmitted levels. The high data rate mode can operate in an S/N environment of 28.3 dB.

Both modes use the same symbol rate, data frame structure, interleaving, Reed–Solomon coding, and synchronization. Input data is randomized and processed for forward error-correction (FEC) with (207,187) Reed–Solomon coding (20 RS bytes are added to each packet). Packets are formed into data frames for transmission, with each data frame holding two interleaved packets, each packet containing 313 data segments, as shown in Fig. 16.23. A time stamp is used to synchronize video and audio signals.

The television receiver must receive the DTV signal, perform IF and other tuning functions, perform A/D conversion of the baseband signal, perform de-interleaving and error correction and concealment, decode compressed MPEG-2, Dolby Digital and other data, and process the signals into high-resolution picture and sound. In most cases, receivers can process each of the 18 ATSC formats, as well as the audio data services. Finally, only carefully engineered displays can provide the picture quality represented by the DTV signal.

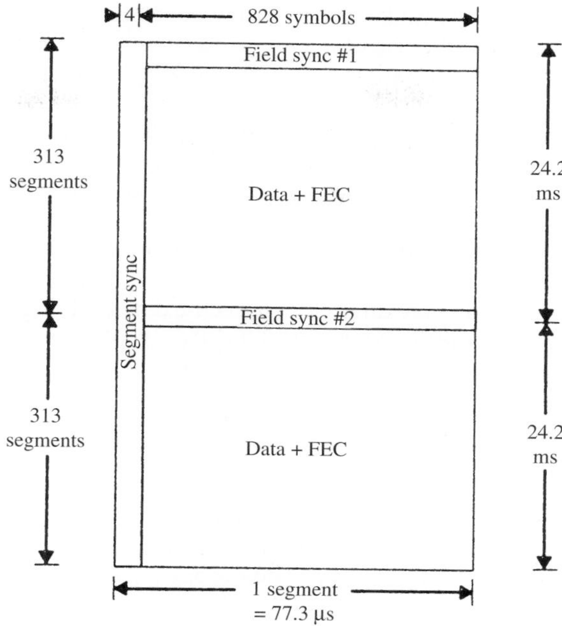

FIGURE 16.23 A VSB data frame holds two interleaved packets with data and error correction.

DTV Implementation

Commercial broadcast networks have implemented DTV in different ways. CBS, NBC, and PBS broadcast in 1080i, and ABC and Fox broadcast in 720p. Local network affiliate stations are the link between the networks and consumers. There are perhaps 1500 different affiliates and each must acquire the equipment needed to pass through the DTV signal from the network, as well as the equipment needed to originate their own DTV programs. Some local affiliates might economically downconvert the HDTV feed from the network, and rebroadcast it as SDTV.

DTV receivers are designed to receive all the DTV formats. Thus, whether it is a 1080i or 720p transmission, the receiver can input the data and create a display. However, although a model will receive HDTV signals, it may not necessarily display an HDTV signal. Instead, it may downconvert the signal to a lower-quality SDTV display. The best DTVs can receive all 18 DTV formats (as well as Dolby Digital) and display the HDTV 1080i and 720p formats in a 16:9 aspect ratio. Other DTV receivers may be digital, but they are not genuinely HDTVs.

To enable the local broadcast of DTV, the FCC provided most affiliates with access to a 6-MHz channel for digital broadcasting within a core digital TV spectrum of TV channels 2 to 51. Because of the limited availability of spectrum and the need to accommodate all existing facilities with minimal interference between stations, during the transition some broadcasters were provided DTV channels (52 to 69) outside of this core spectrum. These broadcasters will move their DTV operations to a channel in the core spectrum when one becomes available.

A number of security tools have been developed for DTV applications. Over-the-air broadcasters advocate the use of a copy-protection broadcast flag. The Redistribution Control Descriptor (RCD) can be placed in the MPEG-2 bitstream prior to modulation so legitimate receiving devices will ensure that the bitstream is only conveyed via secure digital interfaces. Other tools provide a secure path between devices. The content provider places a Copy Control Information (CCI) flag in the MPEG-2 bitstream. The High-bandwidth Digital Content Protection (HDCP) protocol can be used for viewing DTV signals via the Digital Visual Interface (DVI) and High-Definition Multimedia Interface (HDMI) while prohibiting digital copying of video data. With HDCP, the source device and display device must authenticate each other before the source will transmit encrypted data. The display device decrypts the data and displays it, while authentication is renewed periodically. Moreover, HDCP encrypts video data in a way that it cannot be recorded in native unencrypted form. The Digital Transmission Content Protection (DTCP) protocol can be used to limit the use of video bitstreams conveyed over an IEEE 1394 interface; copying can be permitted. Macrovision can be used to prevent analog copying.

In addition to the United States, the governments of Canada, Mexico, South Korea, and Taiwan have adopted the ATSC standard for digital terrestrial television. Japan, Brazil, and other Latin American countries employ the ISDB-T standard. China has adopted a DMB-T/H dual standard. The European DVB standard has been adopted by many countries, including many with PAL-based television systems.

CHAPTER 17

Digital Signal Processing

In many ways, digital signal processing (DSP) returns us to the elemental beginning of our discussion of digital audio. Although conversion, error correction, data reduction, and other concerns can be critical to a digitization system, it is the software-driven signal processing of digital audio data that is germane to the venture. Without the ability to manipulate the numbers that comprise digital audio data, its digitization would not be useful for many applications. Moreover, a discussion of digital signal processing returns us to the roots of digital audio in that the technology is based on the same elemental mathematics that first occupied us. On the other hand, digital signal processing is a science far removed from simple logic circuits, with sophisticated algorithms required to achieve its aim of efficient signal manipulation. Moreover, digital signal processing may demand very specialized hardware devices for successful operation.

Fundamentals of Digital Signal Processing

Digital signal processing is used to generate, analyze, alter, or otherwise manipulate signals in the digital domain. It is based on sampling and quantization, the same principles that make digital recording possible. However, instead of providing a storage or transmission means, it is a processing method. DSP is similar to the technology used in computers and microprocessor systems. However, whereas a regular computer processes data, a DSP system processes signals. In particular, a digital audio signal is a time-based sequence in which the ordering of values is critical. A digital audio signal only makes sense, and can be processed properly, if the sequence is properly preserved. DSP is thus a special application of general data processing. Simply stated, DSP uses a mathematical formula or algorithm to change the numerical values in a bitstream signal.

A signal can be any natural or artificial phenomenon that varies as a function of an independent variable. For example, when the variable is time, then changes in barometric pressure, temperature, oil pressure, current, or voltage are all signals that can be recorded, transmitted, or manipulated either directly or indirectly. Their representation can be either analog or digital in nature, and both offer advantages and disadvantages.

Digital processing of acquired waveforms offers several advantages over processing of continuous-time signals. Fundamentally, the use of unambiguous discrete samples promotes: the use of components with lower tolerances; predetermined accuracy; identically reproducible circuits; a theoretically unlimited number of successive operations on a sample; and reduced sensitivity to external effects such as noise, temperature, and aging. The programmable nature of discrete-time signals permits changes in function without changes in hardware. Digital integrated circuits are small, highly reliable, low in cost, and capable of complex processing. Some operations implemented with digital processing are difficult or impossible with analog means. Examples include filters with linear phase, long-term uncorrupted memory, adaptive systems, image processing, error correction, data reduction, data compression, and signal transformations. The latter includes time domain to frequency-domain transformation with the discrete Fourier transform (DFT) and special mathematical processing such as the fast Fourier transform (FFT).

On the other hand, DSP has disadvantages. For example, the technology always requires power; there is no passive form of DSP circuitry. DSP cannot presently be used for very high frequency signals. Digital signal representation of a signal may require a larger bandwidth than the corresponding analog signal. DSP technology is expensive to develop. Circuits capable of performing fast computation are required. Finally, when used for analog applications, analog-to-digital (A/D) and digital-to-analog (D/A) conversion are required. In addition, the processing of very weak signals such as antenna signals or very strong signals such as those driving a loudspeaker, presents difficulties; digital signal processing thus requires appropriate amplification treatment of the signal.

DSP Applications

In the 1960s, signal processing relied on analog methods; electronic and mechanical devices processed signals in the continuous-time domain. Digital computers generally lacked the computational capabilities needed for digital signal processing. In 1965, the invention of the fast Fourier transform to implement the discrete Fourier transform, and the advent of more powerful computers, inspired the development of theoretical discrete-time mathematics, and modern DSP.

Some of the earliest uses of digital signal processing included soil analysis in oil and gas exploration, and radio and radar astronomy, using mainframe computers. With the advent of specialized hardware, extensive applications in telecommunications were implemented including modems, data transfer between computers, and vocoders and transmultiplexers in telephony. Medical science uses digital signal processing in processing of X-ray and NMR (nuclear magnetic resonance) images. Image processing is used to enhance photographs received from orbiting satellites and deep-space vehicles. Television studios use digital techniques to manipulate video signals. The movie industry relies on computer-generated graphics and 3D image processing. Analytical instruments use digital signal transforms such as FFT for spectral and other analysis. The chemical industry uses digital signal processing for industrial process control. Digital signal processing has revolutionized professional audio in effects processing, interfacing, user control, and computer control. The consumer sees many applications of digital signal processing in the guise of personal computers, cell phones, gaming consoles, MP3 players, DVD and Blu-ray players, digital radio receivers, HDTV receivers and displays, and many other devices.

DSP presents rich possibilities for audio applications. Error correction, multiplexing, sample rate conversion, speech and music analysis and synthesis, data reduction and data compression, filtering, adaptive equalization, dynamic compression and expansion, reverberation, ambience processing, time alignment, acoustical noise cancellation, mixing and editing, encryption and watermarking, and acoustical analysis can all be performed with digital signal processing.

Discrete Systems

Digital audio signal processing is concerned with the manipulation of audio samples. Because those samples are represented as numbers, digital audio signal processing is thus a science of calculation. Hence, a fundamental understanding of audio DSP must begin with its mathematical essence.

When the independent variable, such as time, is continuously variable, the signal is defined at every real value of time (t). The signal is thus a continuous time-based signal. For example, atmospheric temperature changes continuously throughout the day. When the signal is only defined at discrete values of time (nT), the signal is a discrete time signal. A record of temperature readings throughout the day is a discrete time signal. As we observed in Chap. 2, using the sampling theorem, any bandlimited continuous time function can be represented without theoretical loss as a discrete time signal. Although general discrete time signals and digital signals both consist of samples, a general discrete time signal can take any real value but a digital signal can only take a finite number of values. In digital audio, this requires an approximation using quantization.

Linearity and Time-Invariance

A discrete system is any system that accepts one or more discrete input signals $x(n)$ and produces one or more discrete output signals $y(n)$ in accordance with a set of operating rules. The input and output discrete time signals are represented by a sequence of numbers. If an analog signal $x(t)$ is sampled every T seconds, the discrete time signal is $x(nT)$, where n is an integer. Time can be normalized so that the signal is written as $x(n)$.

Two important criteria for discrete systems are linearity and time-invariance. A linear system exhibits the property of superposition: the response of a linear system to a sum of signals is the sum of the responses to each individual input. That is, the input $x_1(n) + x_2(n)$ yields the output $y_1(n) + y_2(n)$. A linear system exhibits the property of homogeneity: the amplitude of the output of a linear system is proportional to that of the input. That is, an input $ax(n)$ yields the output $ay(n)$. Combining these properties, a linear discrete system with the input signal $ax_1(n) + bx_2(n)$ produces an output signal $ay_1(n) + by_2(n)$ where a and b are constants. The input signals are treated independently, the output amplitude is proportional to that of the input, and no new signal components are introduced. As described in the following paragraphs, all z-transforms and Fourier transforms are linear.

A discrete time system is time-invariant if the input signal $x(n - k)$ produces an output signal $y(n - k)$ where k is an integer. In other words, a linear time-invariant discrete (LTD) system behaves the same way at all times; for example, an input delayed by k samples generates an output delayed by k samples.

A discrete system is causal if at any instant the output signal corresponding to any input signal is independent of the values of the input signal after that instant. In other words, there are no output values before there has been an input signal. The output does not depend on future inputs. As some theorists put it, a causal system doesn't laugh until after it has been tickled.

Impulse Response and Convolution

The impulse response is an important concept in many areas, including digital signal processing. The impulse response $h(n)$ gives a full description of a linear time-invariant discrete system in the time domain. An LTD system, like any discrete system, converts

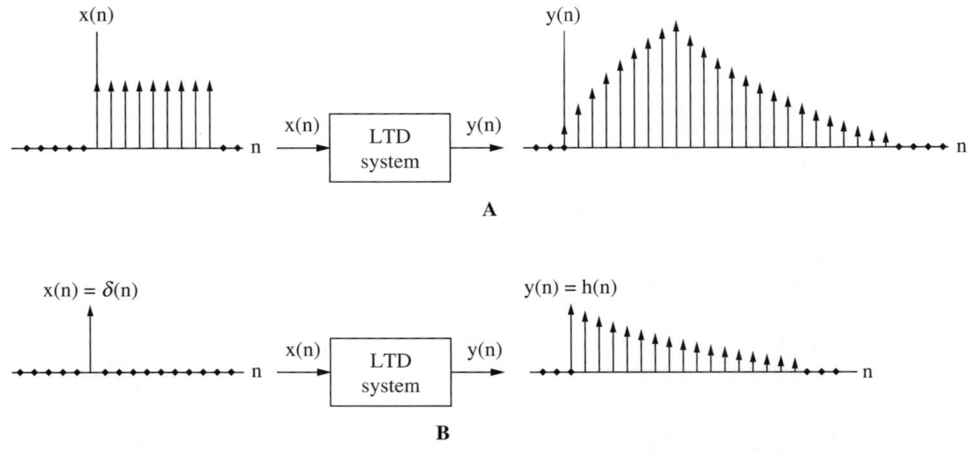

FIGURE 17.1 Two properties of linear time-invariant discrete (LTD) systems. A. LTD systems produce an output signal based on the input. B. LTD systems can be characterized by their impulse response, the output from a single pulse input.

an input signal into an output signal, as shown in Fig. 17.1A. However, an LTD system has a special property such that when an impulse (a delta function) is applied to an LTD system, the output is the system's impulse response, as shown in Fig. 17.1B. The impulse response describes the system in the time domain, and can be used to reveal the frequency response of the system in the frequency domain. Practically speaking, most digital filters are LTD systems, and yield this property. A system is stable if any input signal of finite amplitude produces an output signal of finite amplitude. In other words, the sum of the absolute value of every input and the impulse response must yield a finite number. Useful discrete systems are stable.

Furthermore, the sampled impulse response can be used to filter a signal. Audio samples themselves are impulses, represented as numbers. The signal could be filtered, for example, by using the samples as scaling values; all of the values of a filter's impulse response are multiplied by each signal value. This yields a series of filter impulse responses scaled to each signal sample. To obtain the result, each scaled filter impulse response is substituted for its multiplying signal sample. The filter response can extend over many samples; thus, several scaled values might overlap. When these are added together, the series of sums forms the new filtered signal values.

This is the process of convolution. The output of a linear system is the convolution of the input and the system's impulse response. Convolution is a time-domain process that is equivalent to the multiplication of the frequency responses of two networks. Convolution in the time domain is equivalent to multiplication in the frequency domain. Furthermore, the duality exists such that multiplication in the time domain is equivalent to convolution in the frequency domain.

Fundamentally, in convolution, samples (representing the signal at different sample times) are multiplied by weighting factors. These products are continually summed together to produce an output. A finite impulse response (FIR) oversampling filter (as described in Chap. 4) provides a good example. A series of samples are multiplied by the coefficients that represent the impulse response of the filter, and these products are summed. The input time function has been convolved with the filter's impulse in the

		0.5	0.5	0.5			
		×	×	×			
2	3	4					= 2.0
	2	3 +	4				= 3.5
		2 +	3 +	4			= 4.5
			2 +	3	4		= 2.5
				2	3	4	= 1.0

FIGURE 17.2 Convolution can be performed by folding, shifting, multiplying, and adding number sequences to generate an ordered weighted product.

time domain. For example, the frequency response of an ideal lowpass filter can be achieved by using coefficients representing a time-domain $\sin(x)/x$ impulse response. The convolution of the input signal with coefficients results in a filtered output signal.

Recapitulating, the response of a linear and time-invariant system (such as a digital filter) over all time to an impulse is the system's impulse response; its response to an amplitude scaled input sample is a scaled impulse response; its response to a delayed impulse is a delayed impulse response. The input samples are composed of a sequence of impulses of varying amplitude, each with a unique delay. Each input sample results in a scaled, time-delayed impulse response. By convolution, the system's output at any sample time is the sum of the partial impulse responses produced by the scaled and shifted inputs for that instant in time.

Because convolution is not an intuitive phenomenon, some examples might be useful. Mathematically, convolution expresses the amount of overlap of one function as it is shifted over another function. Suppose that we want to convolve the number sequence 0.5,0.5,0.5 (representing an audio signal) with 4,3,2 (representing an impulse response). We reverse the order of the second sequence to 2,3,4 and shift the sequence through the first sequence, multiplying common pairs of numbers and adding the totals, as shown in Fig. 17.2. The resulting values are the convolution sum and define the output signal at sample times.

To illustrate this using discrete signals, consider a network that produces an output $h(n)$ when a single waveform sample is input (refer to Fig. 17.1B). The output $h(n)$ defines the network; from this impulse response we can find the network's response to any input. The network's complete response to the waveform can be found by adding its response to all of the individual sample responses. The response to the first input sample is scaled by the amplitude of the sample and is output time-invariantly with it. Similarly, the inputs that follow produce outputs that are scaled and delayed by the delay of the input. The sum of the individual responses is the full response to the input waveform:

$$y(n) = \sum_{k=-\infty}^{+\infty} x(k)h(n-k)$$

$$= \sum_{k=-\infty}^{+\infty} x(n-k)h(k)$$

This is convolution, mathematically expressed as:

$$y(n) = x(n) * h(n) = h(n) * x(n)$$

where $*$ denotes the convolution sum.

The output signal is the convolution of the input signal with the system's impulse response. A convolution sum can be graphically evaluated by a process of folding, translating, multiplying, and shifting. The signal $x(n)$ is the input to a linear shift invariant system characterized by the impulse response $h(n)$, as shown in Fig. 17.3A. We can convolve $x(n)$ with $h(n)$ to find the output signal $y(n)$. Using convolution graphically, we first fold $h(n)$ to time-reverse it as shown in Fig. 17.3B. Folding is necessary to yield the correct time-domain response as we move the impulse response from left to right through the input signal. Also, $h(n)$ is translated to the right, to a starting time. To view convolution in action, Fig. 17.3C shows $h(n)$ shifting to the right, through $x(n)$, one sample at a time. The values of samples coincident in time are multiplied, and overlapping time values are summed to determine the instantaneous output value. The entire sequence is obtained by moving the reversed impulse response until it has passed through the duration of the samples of interest, be it finite or infinite in length.

More generally, when two waveforms are multiplied together, their spectra are convolved, and if two spectra are multiplied, their determining waveforms are multiplied. The response to any input waveform can be determined from the impulse response of the network, and its response to any part of the input waveform. As noted, the convolution of two signals in the time domain corresponds to multiplication of their Fourier transforms in the frequency domain (as well as the dual correspondence). The bottom line is that any output signal can be considered to be a sum of impulses.

Complex Numbers

Analog and digital networks share a common mathematical basis. Fundamentally, whether the discussion is one of resistors, capacitors, and inductors, or scaling, delay, and addition (all linear, time-invariant elements), processors can be understood through complex numbers. A complex number z is any number that can be written in the form $z = x + jy$ where x and y are real numbers, and where x is the real part, and jy is the imaginary part of the complex number. An imaginary number is any real number multiplied by j, where j is the square root of -1. There is no number that when multiplied by itself gives a negative number, but mathematicians cleverly invented the concept of an imaginary number. (Mathematicians refer to it as i, but engineers use j, because i

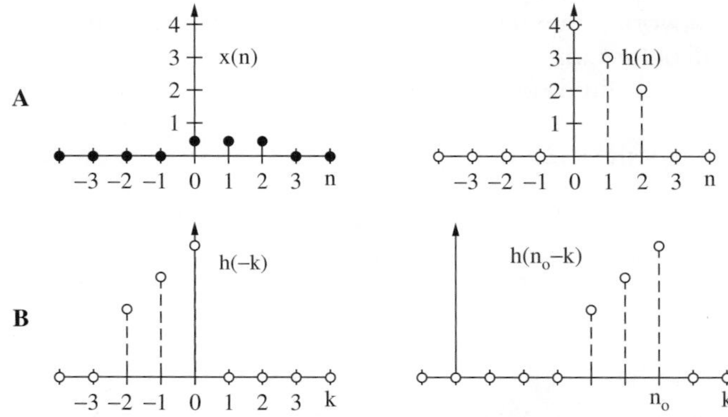

FIGURE 17.3 A graphical representation of convolution, showing signal $x(n)$ convolved with $h(n)$. A. An input signal $x(n)$ and the impulse response $h(n)$ of a linear time invariant system. B. The impulse response is folded and translated.

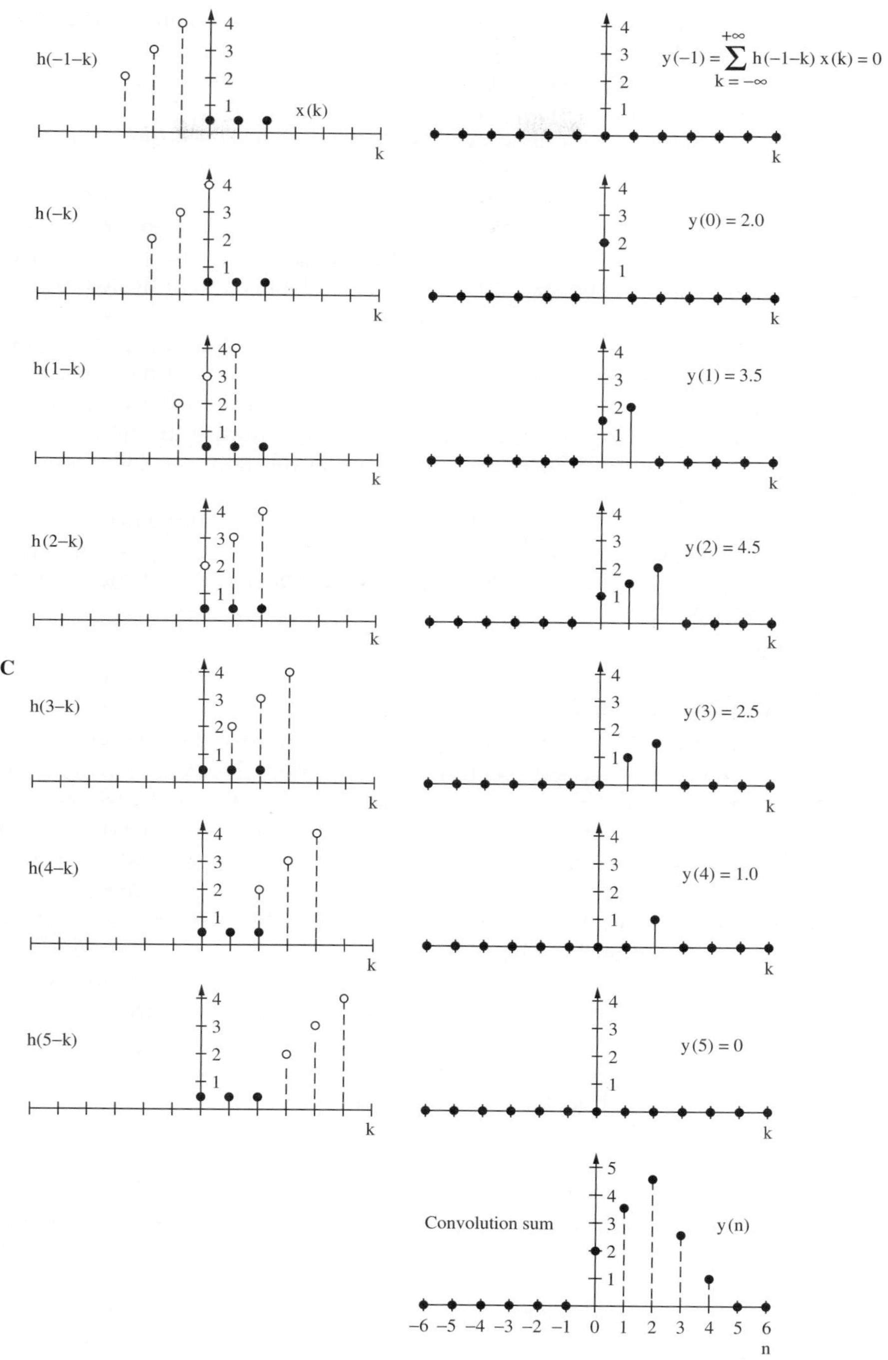

FIGURE 17.3 C. The convolution sum yields the output signal $y(n)$.

denotes current.) The form $x + jy$ is the rectangular form of a complex number, and represents the two-dimensional aspects of numbers. For example, the real part can denote distance, and the imaginary part can denote direction. A vector can be constructed, showing the indicated location.

A waveform can be described by a complex number. This is often expressed in polar form, with two parameters: r and θ. The form $re^{j\theta}$ also can be used. If a dot is placed on a circle and rotated, perhaps representing a waveform changing over time, the dot's location can be expressed by a complex number. A location of 45° would be expressed as $0.707 + 0.707j$. A location of 90° would be $0 + 1j$, 135° would be $-0.707 + 0.707j$, and 180° would be $-1 + 0j$. The size of the circle could be used to indicate the magnitude of the number.

The j operator can be used to convert between imaginary and real numbers. A real number multiplied by an imaginary number becomes complex, and an imaginary number multiplied by an imaginary number becomes real. Multiplication by a complex number is analogous to phase-shifting; for example, multiplication by j represents a 90° phase shift, and multiplication by $0.707 + 0.707j$ represents a 45° phase shift. In the digital domain, phase shift is performed by time delay. A digital network composed of delays can be analyzed by changing each delay to a phase shift. For example, a delay of 10° corresponds to the complex number $0.984 - 0.174j$. If the input signal is multiplied by this complex number, the output result would be a signal of the same magnitude, but delayed by 10°.

Mathematical Transforms

Signal processing, either analog or digital, can be considered in either of two domains. Together, they offer two perspectives on a unified theory. For analog signals, the domains are time and frequency. For sampled signals, they are discrete time and discrete frequency. A transform is a mathematical tool used to move between the time and frequency domains. Continuous transforms are used with signals continuous in time and frequency; series transforms are applied to continuous time and discrete frequency signals; and discrete transforms are applied to discrete time and frequency signals.

The analog relationships between a continuous signal, its Fourier transform, and Laplace transform are shown in Fig. 17.4A. The discrete-time relationships between a discrete signal, its discrete Fourier transform, and z-transform are shown in Fig. 17.4B. The Laplace transform is used to analyze continuous time and frequency signals. It maps a time-domain function $x(t)$ into a frequency domain, complex frequency function $X(s)$. The Laplace transform takes the form:

$$X(s) = \int_{-\infty}^{+\infty} x(t)e^{-st}\,dt$$

The inverse Laplace transform performs the reverse mapping. Laplace transforms are useful for analog design.

The Fourier transform is a special kind of Laplace transform. It maps a time-domain function $x(t)$ into a frequency-domain function $X(j\omega)$, where $X(j\omega)$ describes the spectrum (frequency response) of the signal $x(t)$. The Fourier transform takes the form:

$$X(jw) = \int_{-\infty}^{+\infty} x(t)e^{-j\omega t}\,dt$$

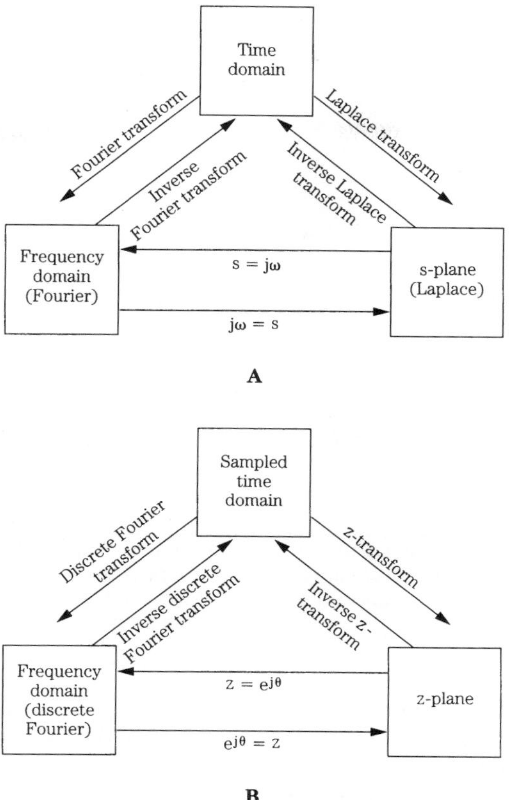

FIGURE 17.4 Transforms are used to mathematically convert a signal from one domain to another. A. Analog signals can be expressed in the time, frequency and *s*-plane domains. B. Discrete signals can be expressed in the sampled-time, frequency, and *z*-plane domains.

This equation (and the inverse Fourier transform), are identical to the Laplace transforms when $s = j\omega$; the Laplace transform equals the Fourier transform when the real part of s is zero. The Fourier series is a special case of the Fourier transform and results when a signal contains only discrete frequencies, and the signal is periodic in the time domain.

Figure 17.5 shows how transforms are used. Specifically, two methods can be used to compute an output signal: convolution in the time domain, and multiplication in the frequency domain. Although convolution is conceptually concise, in practice, the second method using transforms and multiplication in the frequency domain is usually preferable. Transforms also are invaluable in analyzing a signal, to determine its spectral characteristics. In either case, the effect of filtering a discrete signal can be predictably known.

The Fourier transform for discrete signals generates a continuous spectrum but is difficult to compute. Thus, a sampled spectrum for discrete time signals of finite duration is implemented as the discrete Fourier transform (DFT). Just as the Fourier transform generates the spectrum of a continuous signal, the DFT generates the spectrum of a discrete signal, expressed as a set of harmonically related sinusoids with unique

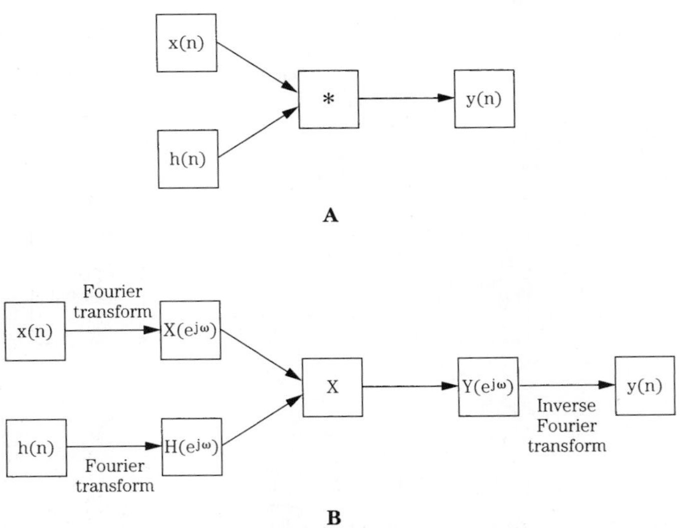

FIGURE 17.5 Given an input signal $x(n)$ and impulse response $h(n)$, the output signal $y(n)$ can be calculated through direct convolution or through Fourier transformation, multiplication, and inverse Fourier transformation. In practice, the latter method is often an easier calculation. A. Direct convolution. B. Fourier transformation, multiplication, and inverse Fourier transformation.

amplitude and phase. The DFT takes samples of a waveform and operates on them as if they were an infinitely long waveform composed of sinusoids, harmonically related to a fundamental frequency corresponding to the original sample period. An inverse DFT can recover the original sampled signal. The DFT can also be viewed as sampling the Fourier transform of a signal at N evenly spaced frequency points.

The DFT is the Fourier transform of a sampled signal. When a finite number of samples (N) are considered, the N-point DFT transform is expressed as:

$$X(m) = \sum_{n=0}^{N-1} x(n)e^{-j(2\pi/N)mn}$$

The $X(m)$ term is often called bin m, and describes the amplitude of the frequencies in signal $x(n)$, computed at N equally spaced frequencies. The $m = 0$, or bin 0 term describes the dc content of the signal, and all other frequencies are harmonically related to the fundamental frequency corresponding to $m = 1$, or bin 1. Bin numbers thus specify the harmonics that comprise the signal, and the amplitude in each bin describes the power spectrum (square of the amplitude). The DFT thus describes all the frequencies contained in signal $x(n)$. There are identical positive and negative frequencies; usually only the positive half is shown, and multiplied by 2 to obtain the actual amplitudes.

An example of DFT operations is shown in Fig. 17.6. The input signal to be analyzed is a simple periodic function $x(n) = \cos(2\pi n/6)$. The function is periodic over six samples because $x(n) = x(n+6)$. Three N-point DFTs are used, with $N = 6$, 12, and 16. In the first two cases, N is equal to 6 or is an integer multiple of 6; a larger N yields greater spectral resolution. In the third case, $N = 16$, the discrete spectrum positions cannot exactly represent the input signal; spectral leakage occurs in all bins. In all cases, the spectrum is symmetrical.

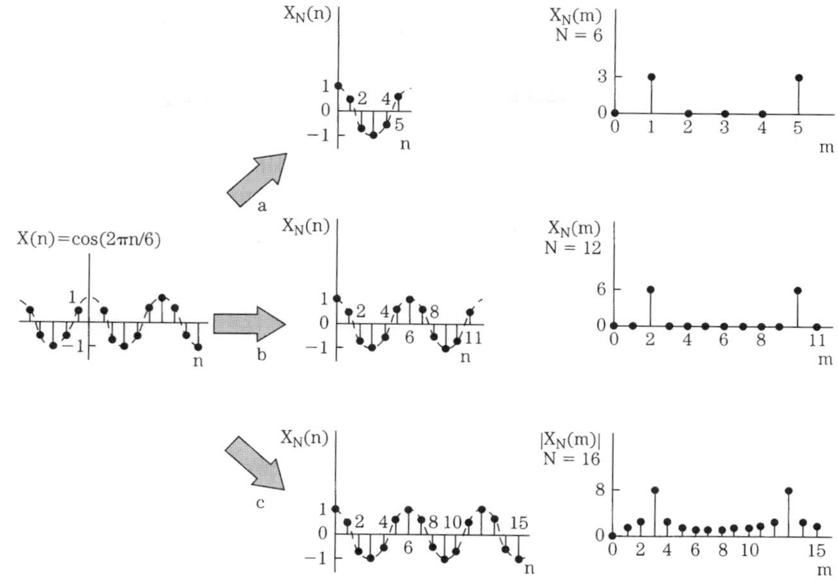

FIGURE 17.6 Example of a periodic signal applied to an *N*-point DFT for three different values of *N*. Greater spectral resolution is obtained as *N* is increased. When *N* is not equal to an integral number of waveform periods, spectral leakage occurs. (*Van den Enden and Verhoeckx, 1985*)

The DFT is computation-intensive, requiring N^2 complex multiplications and $N(N-1)$ complex additions. The DFT is often generated with the fast Fourier transform (FFT), a collection of fast and efficient algorithms for spectral computation that takes advantage of computational symmetries and redundancies in the DFT. The FFT requires $N\log_2 N$ computations, 100 times fewer than DFT. The FFT can be used when N is an integral power of 2; zero samples can be padded to satisfy this requirement. The FFT can also be applied to a sequence length that is a product of smaller integer factors. The FFT is not another type of transformation, but rather an efficient method of calculating the DFT. The FFT recursively decomposes an N-point DFT into smaller DFTs. This number of short-length DFTs are calculated, then the results are combined. The FFT can be applied to various calculation methods and strategies, including the analysis of signals and filter design.

The FFT will transform a time series, such as the impulse response of a network, into the real and imaginary parts of the impulse response in the frequency domain. In this way, the magnitude and phase of the network's transfer function can be obtained. An inverse FFT can produce a time-domain signal. FFT filtering is accomplished through multiplication of spectra. The impulse response of the filter is transformed to the frequency domain. Real and imaginary arrays, obtained by FFT transformation of overlapping segments of the signal, are multiplied by filter arrays, and an inverse FFT produces a filtered signal. Because the FFT can be efficiently computed, it can be used as an alternative to time-domain convolution if the overall number of multiplications is fewer.

The z-transform operates on discrete signals in the same way that the Laplace transform operates on continuous signals. In the same way that the Laplace transform is a

Property	Time Domain	z-Domain
Linearity	$ax(n) + by(n)$	$aX(z) + bY(z)$
Shift	$x(n - i)$	$z^{-i}X(z)$
Convolution	$x(n) * y(n)$	$X(z)Y(z)$

TABLE 17.1 Equivalent properties of discrete signals in the time domain and z-domain.

generalization of the Fourier transform, the z-transform is a generalization of the DFT. Whereas the Fourier transform operates on a particular complex value $e^{-j\omega}$, the z-transform operates with any complex value. When $z = e^{j\omega}$, the z-transform is identical to the Fourier transform. The DFT is thus a special case of the z-transform. The z-transform of a sequence $x(n)$ is defined as:

$$X(z) = \sum_{n=-\infty}^{+\infty} x(n)z^{-n}$$

where z is a complex variable and z^{-1} represents a unit delay element. The z-transform has an inverse transform, often obtained through a partial fraction expansion.

Whereas the DFT is used for literal operations, the z-transform is a mathematical tool used in digital signal processing theory. Several basic properties govern the z-domain. A linear combination of signals in the time domain is equivalent to a linear combination in the z-domain. Convolution in the time domain is equivalent to multiplication in the z-domain. For example, we could take the z-transform of the convolution equation, such that the z-transform of an input multiplied by the z-transform of a filter's impulse response is equal to the z-transform of the filter's output. In other words, the ratio of the filter output transform to the filter input transform (that is, the transfer function $H(z)$) is the z-transform of the impulse response. Furthermore, this ratio, the transfer function $H(z)$, is a fixed function determined by the filter. In the z-domain, given an impulse input, the transfer function equals the output. Furthermore, a shift in the time domain is equivalent to multiplication by z raised to a power of the length (in samples) of the shift. These properties are summarized in Table 17.1. For example, the z-transforms of $x(n)$ and $y(n)$ are $X(z)$ and $Y(z)$, respectively.

Unit Circle and Region of Convergence

The Fourier transform of a discrete signal corresponds to the z-transform on the unit circle in the z-plane. The equation $z = e^{j\omega}$ defines the unit circle in the complex plane. The evaluation of the z-transform along the unit circle yields the function's frequency response.

The variable z is complex, and $X(z)$ is the function of the complex variable. The set of z in the complex plane for which the magnitude of $X(z)$ is finite is said to be in the region of convergence. The set of z in the complex plane for which the magnitude of

$X(z)$ is infinite is said to diverge, and is outside the region of convergence. The function $X(z)$ is defined over the entire z-plane but is only valid in the region of convergence. The complex variable s is used to describe complex frequency; this is a function of the Laplace transform. S variables lie on the complex s-plane. The s-plane can be mapped to the z-plane; vertical lines on the s-plane map as circles in the z-plane.

Because there is a finite number of samples, practical systems must be designed within the region of convergence. The unit circle is the smallest region in the z-plane that falls within the region of convergence for all finite stable sequences. Poles must be placed inside the unit circle on the z-plane for proper stability. Improper placement of the poles constitutes an instability.

Mapping from the s-plane to the z-plane is an important process. Theoretically, this function allows the designer to choose an analog transfer function and find the z-transform of that function. Unfortunately, the s-plane generally does not map into the unit circle of the z-plane. Stable analog filters, for example, do not always map into stable digital filters. This is avoided by multiplying by a transform constant, used to match analog and digital frequency response. There is also a nonlinear relationship between analog and digital break frequencies that must be accounted for. The nonlinearities are known as warping effects and the use of the constant is known as pre-warping the transfer function.

Often, a digital implementation can be derived from an existing analog representation. For example, a stable analog filter can be described by the system function $H(s)$. Its frequency response is found by evaluating $H(s)$ at points on the imaginary axis of the s-plane. In the function $H(s)$, s can be replaced by a rational function of z, which will map the imaginary axis of the s-plane onto the unit circle of the z-plane. The resulting system function $H(z)$ is evaluated along the unit circle and will take on the same values of $H(s)$ evaluated along its imaginary axis.

Poles and Zeros

Summarizing, the transfer function $H(z)$ of a linear, time-invariant discrete-time filter is defined to be the z-transform of the impulse response $h(n)$. The spectrum of a function is equal to the z-transform evaluated on the unit circle. The transfer function of a digital filter can be written in terms of its z-transform; this permits analysis in terms of the filter's poles and zeros. The zeros are the roots of the numerator's polynomial of the transfer function of the filter, and the poles are the denominator's roots. Mathematically, zeros make $H(z) = 0$, and poles make $H(z)$ nonanalytic. When the magnitude of $H(z)$ is plotted as a function of z, poles appear at a distance above the z-plane and zeros touch the z-plane. One might imagine the flat z-plane and above it a flexible contour, the magnitude transfer function, passing through the poles and zeros, with peaks on top of poles, and valleys centered on zeros. Tracing the rising and falling of the contour around the unit circle yields the frequency response. For example, the gain of a filter at any frequency can be measured by the magnitude of the contour. The phase shift at any frequency is the angle of the complex number that represents the system's response at that frequency.

If we plot $|z| = 1$ on the complex plane, we get the unit circle; $|z| > 1$ specifies all points on the complex plane that lie outside the unit circle; and $|z| < 1$ specifies all points inside it. The z-transform of a sequence can be represented by plotting the locations of the poles and zeros on the complex plane.

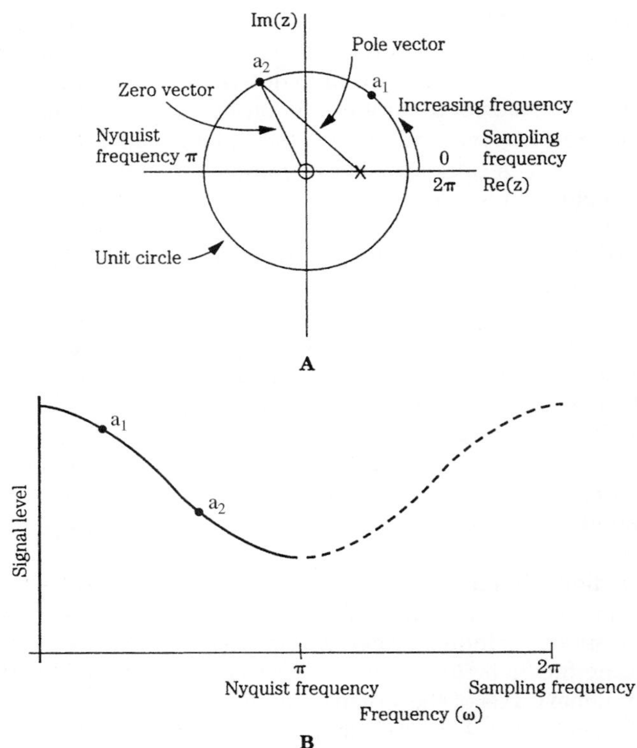

FIGURE 17.7 The frequency response of a filter can be obtained from the pole/zero plot. The amplitude of the frequency response (for example, at points a_1 and a_2) is obtained by dividing the magnitude of the zero vector by that of the pole vector at those points on the unit circle. A. An example of a *z*-plane plot of a lowpass filter, showing the pole and zero locations. B. Analysis of the *z*-plane plot reveals the filter's lowpass frequency response.

Figure 17.7A shows an example of a *z*-plane plot. Among other approaches, the response can be analyzed by examining the relationships between the pole and zero vectors. In the *z*-plane, angular frequency is represented as an angle, with a rotation of 360° corresponding to the sampling frequency. The Nyquist frequency is thus located at π in the figure. The example shows a single pole (×) and zero (o). The corresponding frequency response from 0 to the Nyquist frequency is seen to be that of a lowpass filter, as shown in Fig. 17.7B. The amplitude of the frequency response can be determined by dividing the magnitude of the zero vector by that of the pole vector. For example, points a_1 and a_2 are plotted on the unit circle, and on the frequency response graph. Similarly, the phase response is the difference between the pole vector's angle (from $\omega = 0$ radians) and the zero vector's angle. As the positions of the pole and zero are varied, the response of the filter changes. For example, if the pole is moved along the negative real axis, the filter's frequency response changes to that of a highpass filter.

Some general observations: zeros are created by summing input samples, and poles are created by feedback. A filter's order equals the number of poles or zeros it exhibits, whichever is greater. A filter is stable only if all its poles are inside the unit circle of the z-plane. Zeros can lie anywhere. When all zeros lie inside the unit circle, the system is called a minimum-phase network. If all poles are inside the unit circle and all zeros are outside, and if poles and zeros are always reflections of one another in the unit circle, the system is a constant-amplitude, or all-pass network. If a system has zeros only, except for the origin, and they are reflected in pairs in the unit circle, the system is phase-linear. No real function can have more zeros than poles. When the coefficients are real, poles and zeros occur in complex conjugate pairs; their plot is symmetrical across the real z-axis. The closer its location to the unit circle, the greater the effect of each pole and zero on frequency response.

DSP Elements

Successful DSP applications require sophisticated hardware and software. However, all DSP processing can be considered in three simple processing operations: summing, multiplication, and time delay, as shown in Fig. 17.8. With summing, multiple digital values are added to produce a single result. With multiplication, a gain change is accomplished by multiplying the sample value by a coefficient. With time delay $(n - 1)$, a digital value is stored for one sample period. The delay element (realized with shift registers or memory locations) is alternatively notated as z^{-1} because a delay of one sampling period in the time domain corresponds to multiplication by z^{-1} in the z-domain; thus $z^{-1}x(n) = x(n - 1)$. Delays can be cascaded, for example, a z^{-2} term describes a two-sample $(n - 2)$ delay. Although it is usually most convenient to operate with sample numbers, the time of a delay can be obtained by taking nT, where T is the sampling interval. Figure 17.9 shows two examples of simple networks and their impulse responses (see also Fig. 17.1B). LTD systems such as these are completely described by the impulse response.

In practice, these elemental operations are performed many times for each sample, in specific configurations depending on the desired result. In this way, algorithms can be devised to perform operations useful to audio processing such as reverberation, equalization, data reduction, limiting, and noise removal. Of course, for real-time operation, all processing for each sample must be completed within one sampling period.

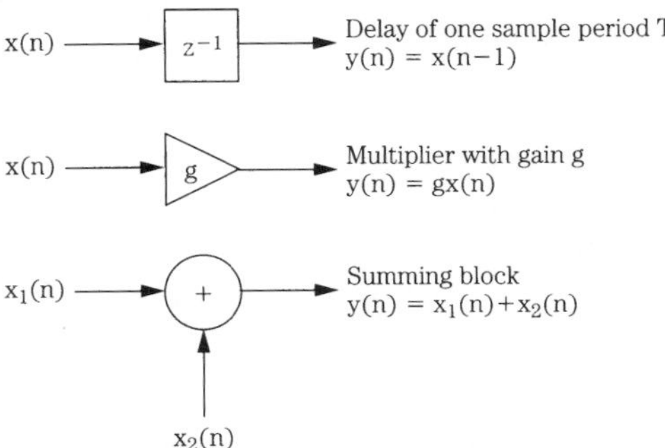

FIGURE 17.8 The three basic elements in any DSP system are delay, multiplication, and summation. They are combined to accomplish useful processing.

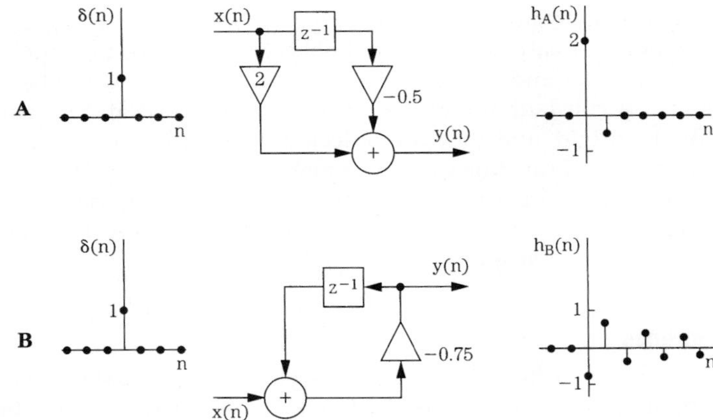

FIGURE 17.9 LTD systems can be characterized by their impulse responses. A. A simple nonrecursive system and its impulse response. B. A simple recursive system and its impulse response. (*Van den Enden and Verhoeckx, 1985*)

Digital Filters

Filtering (or equalization) is important in many audio applications. Analog filters using both passive and active designs shape the signal's frequency response and phase, as described by linear time-invariant differential equations. They describe the system's performance in the time domain. With digital filters, each sample is processed through a transfer function to affect a change in frequency response or phase. Operation is generally described in linear shift-invariant difference equations; they define how the discrete time signal behaves from moment to moment in the time domain. At an infinitely high sampling rate, these equations would be identical to those used to describe analog filters. Digital filters can be designed from analog filters; such impulse-invariant design is useful for lowpass filters with a cutoff frequency far below the sampling rate. Other filter designs make use of transformations to convert characteristics of an analog filter to a digital filter. These transformations map the frequency range of the analog domain into the digital range, from 0 Hz to the Nyquist frequency.

A digital filter can be represented by a general difference equation:

$$y(n)+b_1y(n-1)+b_2y(n-2)+\cdots+b_Ny(n-N)$$
$$=a_0x(n)+a_1x(n-1)+a_2x(n-2)+\cdots+a_Mx(n-M)$$

More efficiently, the equation can be written as:

$$y(n)=\sum_{i=0}^{M}a_ix(n-i)-\sum_{i=1}^{N}b_iy(n-i)$$

where x is the input signal, y is the output signal, the constants a_i and b_i are the filter coefficients, and n represents the current sample time, the variable in the filter's equation. A difference equation is used to represent $y(n)$ as a function of the current input, previous inputs, and previous outputs. The filter's order is specified by the maximum time duration (in samples) used to generate the output. For example, the equation:

$$y(n) = x(n) - y(n-2) + 2x(n-2) + x(n-3)$$

is a third-order filter.

To implement a digital filter, the z-transform is applied to the difference equation so that it becomes:

$$Y(z) = \sum_{i=0}^{M} a_i z^{-i} X(z) - \sum_{i=1}^{N} b_i z^{-i} Y(z)$$

where z^{-i} is a unit of delay i in the time domain. Rewriting the equation, the transfer function $H(z)$ can be determined by:

$$H(z) = \frac{Y(z)}{X(z)} = \frac{\sum_{i=0}^{M} a_i z^{-i}}{\left(1 + \sum_{i=1}^{N} b_i z^{-i}\right)}$$

As noted, the transfer function can be used to identify the filter's poles and zeros. Specifically, the roots (values that make the expression zero) of the numerator identify zeros, and roots of the denominator identify poles. Zeros constitute feed-forward paths and poles constitute feedback paths. By tracing the contour along the unit circle, the frequency response of the filter can be determined.

A filter is canonical if it contains the minimum number of delay elements needed to achieve its output. If the values of the coefficients are changed, the filter's response is altered. A filter is stable if its impulse response approaches zero as n goes to infinity. Convolution provides the means for implementing a filter directly from the impulse response; convolving the input signal with the filter impulse response gives the filtered output. In other words, convolution acts as the difference equation, and the impulse response acts in place of the difference equation coefficients in representing the filter. The choice of using a difference equation or convolution in designing a filter depends on the filter's architecture, as well as the application.

FIR Filters

As noted, the general difference equation can be written as:

$$\begin{aligned} y(n) + b_1 y(n-1) + b_2 y(n-2) + \cdots + b_N y(n-N) \\ = a_0 x(n) + a_1 x(n-1) + a_2 x(n-2) + \cdots + a_M x(n-M) \end{aligned}$$

Consider the general difference equation without b_i terms:

$$y(n) = \sum_{i=0}^{M} a_i x(n-i)$$

and its transfer function in the z domain:

$$H(z)=\sum_{i=0}^{M} a_i z^{-i}$$

There are no poles in this equation; hence there are no feedback elements. The result is a nonrecursive filter. Such a filter would take the form:

$$y(n)=ax(n)+bx(n-1)+cx(n-2)+dx(n-3)+\cdots$$

Any filter operating on a finite number of samples is known as a finite impulse response (FIR) filter. As the name FIR implies, the impulse response has finite duration. Furthermore, an FIR filter can have only zeros outside the origin, it can have a linear phase (symmetrical impulse response). In addition, it responds to an impulse once, and it is always stable. Because it does not use feedback, it is called a nonrecursive filter. A nonrecursive structure is always an FIR filter; however, an FIR filter does not always use a nonrecursive structure.

Consider this introduction to the workings of FIR filters: we know that large differences between samples are indicative of high frequencies and small differences are indicative of low frequencies. A filter changes the differences between consecutive samples. The digital filter described by $y(n) = 0.5[x(n) + x(n - 1)]$ makes the current output equal to half the current input plus half the previous input. Suppose this sequence is input: 1, 8, 6, 4, 1, 5, 3, 7; the difference between consecutive samples ranges from 2 to 7. The first two numbers enter the filter and are added and multiplied: (1 + 8)(0.5) = 4.5. The next computation is (8 + 6)(0.5) = 7.0. After the entire sequence has passed through the filter the sequence is: 4.5, 7, 5, 2.5, 3, 4, 5. The new intersample difference ranges from 0.5 to 2.5; this filter averages the current sample with the previous sample. This averaging smoothes the output signal, thus attenuating high frequencies. In other words, the circuit is a lowpass filter.

More rigorously, the filter's difference equation is:

$$y(n)=0.5[x(n)+x(n-1)]$$

Transformation to the z-domain yields:

$$Y(z)=0.5[X(z)+z^{-1}X(z)]$$

The transfer function can be written as:

$$H(z)=\frac{Y(z)}{X(z)}=\frac{(1+z^{-1})}{2}=\frac{(z+1)}{2z}$$

This indicates a zero at $z=-1$ and a pole at $z=0$, as shown in Fig. 17.10A. Tracing the unit circle, the filter's frequency response is shown in Fig. 17.10B; it is a lowpass filter. Finally, the difference equation can be realized with the algorithm shown in Fig. 17.10C.

Another example of a filter is one in which the output is formed by subtracting the past input from the present, and dividing by 2. In this way, small differences between samples (low-frequency components) are attenuated and large differences (high-frequency components) are accentuated. The equation for this filter is only slightly different from the previous example:

$$y(n)=0.5[x(n)-x(n-1)]$$

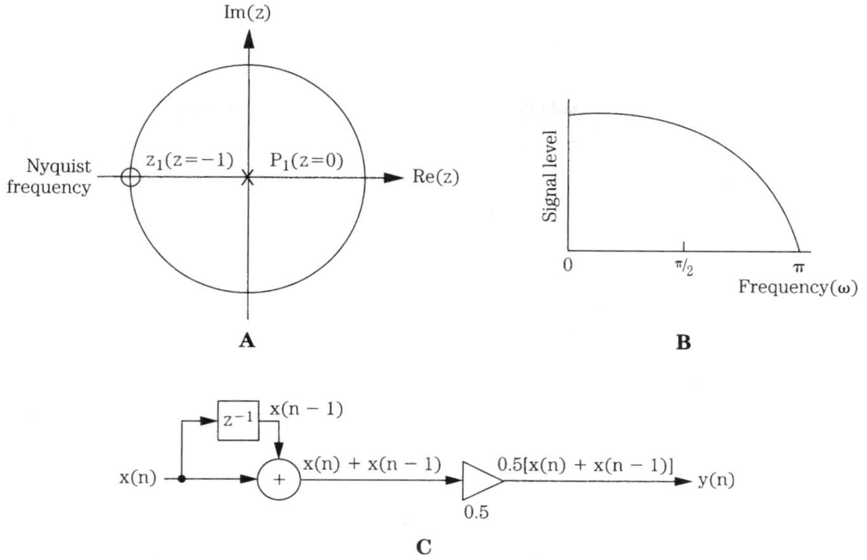

FIGURE 17.10 An example showing the response and structure of a digital lowpass FIR filter. A. The pole and zero locations of the filter in the *z*-plane. B. The frequency response of the filter. C. Structure of the lowpass filter.

Transformation to the *z*-plane yields:

$$Y(z) = 0.5[X(z) - z^{-1}X(z)]$$

The transfer function can be written as:

$$H(z) = \frac{Y(z)}{X(z)} = \frac{(1 - z^{-1})}{2} = \frac{(z - 1)}{2z}$$

This indicates a zero at $z = 1$ and a pole at $z = 0$, as shown in Fig. 17.11A. Tracing the unit circle, the filter's frequency response is shown in Fig. 17.11B; it is a highpass filter. The difference equation can be realized with the algorithm shown in Fig. 17.11C. This highpass filter's realization differs from that of the previous lowpass filter's realization only in the −1 multiplier. In both of these examples, the filter must store only one previous sample value. However, a filter could be designed to store a large (but finite) number of samples for use in calculating the response.

An FIR filter can be constructed as a multi-tapped digital filter, functioning as a building block for more sophisticated designs. The direct-form structure for realizing an FIR filter is shown in Fig. 17.12. This structure is an implementation of the convolution sum. To achieve a given frequency response, the impulse response coefficients of an FIR filter must be calculated. Simply truncating the extreme ends of the impulse response to obtain coefficients would result in an aperture effect and Gibbs phenomenon. The response will peak just below the cutoff frequency and ripples will appear in the passband and stopband. All digital filters have a finite bandwidth; in other words, in practice the impulse response must be truncated. Although the Fourier transform of

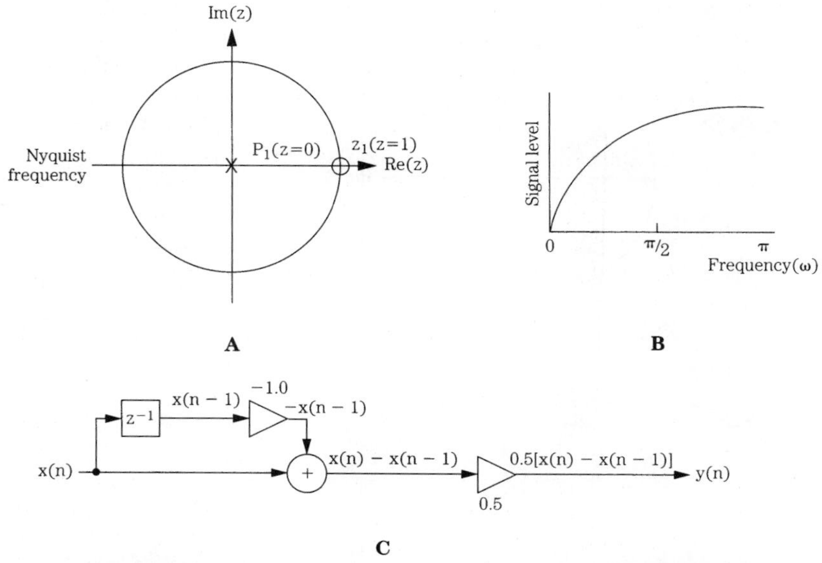

FIGURE 17.11 An example showing the response and structure of a digital highpass FIR filter. A. The pole and zero locations of the filter in the *z*-plane. B. The frequency response of the filter. C. Structure of the highpass filter.

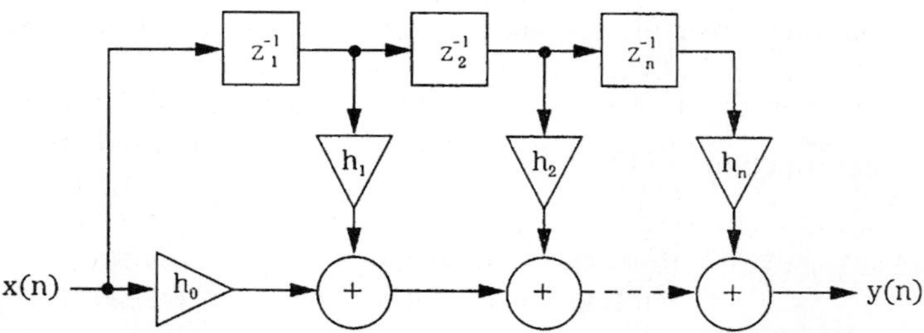

FIGURE 17.12 The direct-form structure for realizing an FIR filter. This multi-tapped structure implements the convolution sum.

an infinite (ideal) impulse response creates a rectangular pulse, a finite (real-world) impulse response creates a function exhibiting Gibbs phenomenon. This is not ringing as in analog systems, but the mark of a finite bandwidth system.

Choice of coefficients determines the phase linearity of the resulting filter. For many audio applications, linear phase is important; the filter's constant delay versus frequency linearizes the phase response and results in a symmetrical output response. For example, the steady-state response of a phase linear system to a square-wave input displays center and axial symmetry. When a system's phase response is nonlinear, the step response does not display symmetry.

The length of the impulse to be considered depends on the frequency response and filter ripple. It is important to provide a smooth transition between samples that are relevant and those that are not. In many applications, the filter coefficients are multiplied by a window function. A window can be viewed as a finite weighting sequence used to modify the infinite series of Fourier coefficients that define a given frequency response. The shape of a window affects the frequency response of the signal correspondingly by the frequency response of the window itself. As noted, multiplication in the time domain is equivalent to convolution in the frequency domain. Multiplying a time-domain signal by a time-domain window is the same as convolving the signals in the frequency domain. The effect of a window on a signal's frequency response can be determined by examining the DFT of the window.

Many DSP applications involve operation on a finite set of samples, truncated from a larger data record; this can cause side effects. For example, as noted, the difference between the ideal and actual filter lengths yields Gibbs phenomenon overshoot at transitions in the transfer function in the frequency domain. This can be reduced by multiplying the coefficients by a window function, but this also can change the transition bands of the transfer function.

For example, a rectangular window function can be used to effectively gate the signal. The window length can only take on integer values, and the window length must be an integer multiple of the input period. The input signal must repeat itself over this integer number of samples. The method works well because the spacing of the nulls in the window transform is exactly the same as the spacing of the harmonics of the input signal. However, if the integer relationship is broken, and there is not an exact number of periods in the window, spectrum nulls do not correspond to harmonic frequencies and there is spectral leakage.

Some window functions are used to overcome spectral leakage. They are smoothly tapered to gradually reduce the amplitude of the input signal at the endpoints of the data record. They attenuate spectral leakage according to the energy of spectral content outside their main lobe.

Alternatively, the desired response can be sampled, and the discrete Fourier transform coefficients computed. These are then related to the desired impulse response coefficients. The frequency response can be approximated, and the impulse response calculated from the inverse discrete Fourier transform. Still another approach is to derive a set of conditions for which the solution is optimal, using an algorithm providing an approximation, with minimal error, to the desired frequency response.

IIR Filters

The general difference equation contains $y(n)$ components that contribute to the output value. These are feedback elements that are delayed by a unit of time i, and describe a recursive filter. The feedback elements are described in the denominator of the transfer function. Because the roots cause $H(z)$ to be undefined, certain feedback could cause the filter to be unstable. The poles contribute an exponential sequence to each pole's impulse response. When the output is fed back to the input, the output in theory will never reach zero; this allows the impulse to be infinite in duration. This type of filter is known as an infinite impulse response (IIR) filter.

Feedback provides a powerful method for recalling past samples. For example, an exponential time-average filter adds the current input sample to the last output (as opposed to the previous sample) and divides the result by 2. The equation describing its operation is: $y(n) = x(n) + 0.5y(n - 1)$. This yields an exponentially decaying response in which each next output sample is one-half the previous sample value. The filter is called an infinite impulse response filter because of its infinite memory. In theory, the impulse response of an IIR filter lasts for an infinite time; its response never decays to zero. This type of filter is equivalent to an infinitely long FIR filter where:

$$y(n) = x(n) + 0.5x(n-1) + 0.25x(n-2) + 0.125x(n-3) + \cdots + (0.5)^M x(n-M)$$

In other words, the filter adds one-half the current sample, one-fourth the previous sample, and so on. The impulse response of a practical FIR filter decays exponentially, but it has a finite length, thus cannot decay to zero.

In general, an IIR filter can be described as:

$$y(n) = ax(n) + by(n-1)$$

When the value of b is increased relative to a, the lowpass filtering is augmented; that is, the cutoff frequency is lowered. The value of b must always be less than unity or the filter will become unstable; the signal level will increase and overflow will result.

An IIR filter can have both poles and zeros, can introduce phase shift, and can be unstable if one or more poles lie on or outside the unit circle. Generally, an all-pole filter is an IIR filter. IIR filters cannot achieve linear phase (the impulse response is asymmetrical) except in the case when all poles in the transfer function lie on the unit circle. This is realized when the filter consists of a number of cascaded first-order sections. Because the output of an IIR is fed back as an input (with a scaling element), it is called a recursive filter. An IIR filter always has a recursive structure, but filters with a recursive structure are not always IIR filters. Any feedback loop must contain a delay element. Otherwise, the value of a sample would have to be known before it is calculated—an impossibility.

Consider the IIR filter described by the equation:

$$y(n) = x(n) - x(n-2) - 0.25y(n-2)$$

It can be rewritten as:

$$y(n) + 0.25y(n-2) = x(n) - x(n-2)$$

Transformation to the z-plane yields:

$$Y(z) + 0.25z^{-2}Y(z) = X(z) - z^{-2}X(z)$$

The transfer function is:

$$\begin{aligned} H(z) &= \frac{Y(z)}{X(z)} \\ &= \frac{(1 - z^{-2})}{(1 + 0.25z^{-2})} \\ &= \frac{(z^2 - 1)}{(z^2 + 0.25)} \\ &= \frac{(z+1)(z-1)}{(z+0.5j)(z-0.5j)} \end{aligned}$$

There are zeros at $z = \pm 1$ and conjugate poles on the imaginary axis at $z = \pm 0.5j$, as shown in Fig. 17.13A. The frequency response can be graphically determined by dividing

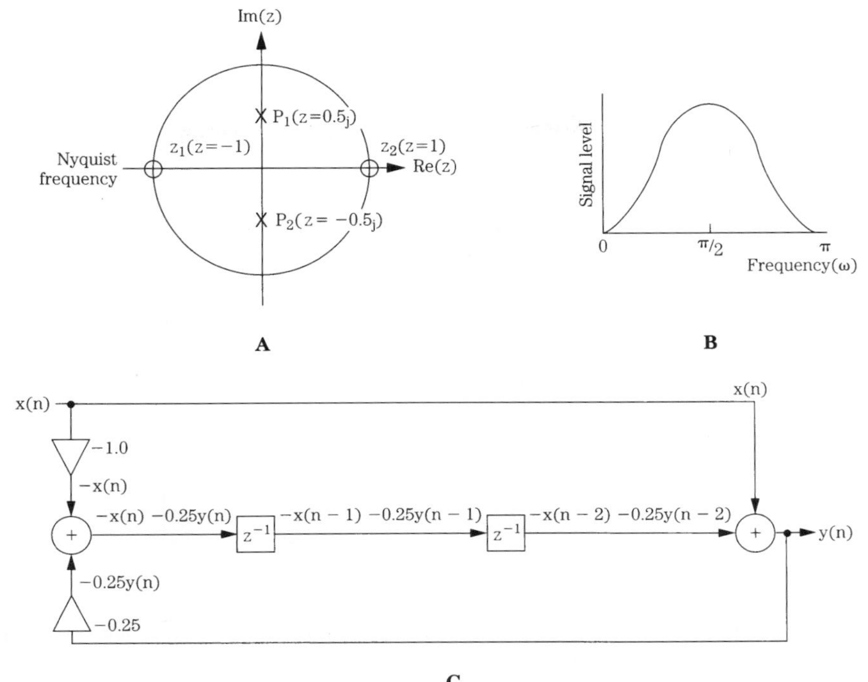

FIGURE 17.13 An example showing the response and structure of a digital bandpass IIR filter. A. The pole and zero locations of the filter in the *z*-plane. B. The frequency response of the filter. C. Structure of the bandpass filter.

the product of zero-vector magnitudes by the product of the pole-vector magnitudes; (phase response is the difference between the sum of their angles). The resulting bandpass frequency response is shown in Fig. 17.13B. A realization of the filter is shown in Fig. 17.13C. Delay elements have been combined to simplify the design. Similarly, the difference equation $y(n) = x(n) + x(n-2) + 0.25y(n-2)$ is also an IIR filter. The development of its pole/zero plot, frequency response, and realization are left to the ambition of the reader.

In general, it is easier to design FIR filters with linear phase and stable operation than IIR filters with the same characteristics. However, IIR filters can achieve a steeper roll-off than an FIR filter for a given number of coefficients. FIR filters require more stages, and hence greater computation, to achieve the same result. As with any digital processing circuit, noise must be considered. A filter's type, topology, arithmetic, and coefficient values all determine whether meaningful error will be introduced. For example, the exponential time-average filter described above will generate considerable error if the value of b is set close to unity, for a low cutoff frequency.

Filter Applications

An example of second-order analog filter is shown in Fig. 17.14A, and an IIR filter is shown in Fig. 17.14B; this is a biquadratic filter section. Coefficients determine the filter's response; in this example, with appropriate selection of the five multiplication coefficients, highpass, lowpass, bandpass, and shelving filters can be obtained. A digital

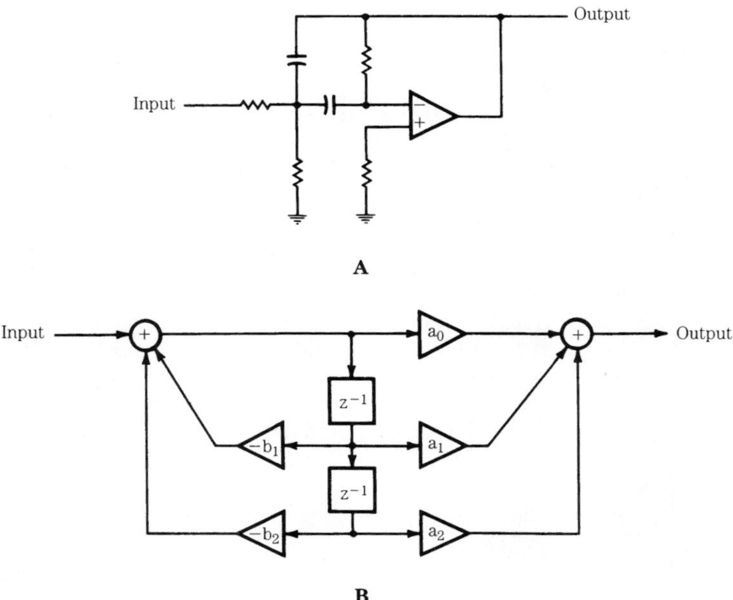

FIGURE 17.14 A comparison of second-order analog and digital filters. A. A second-order analog filter. B. An IIR biquadratic second-order filter section. (*Tydelski, 1988*)

audio processor might have several of these sections at its disposal. By providing a number of presets, users can easily select frequency response, bandwidth, and phase response of a filter. In this respect, a digital filter is more flexible than an analog filter that has relatively limited operating parameters. However, a digital filter requires considerable computation, particularly in the case of swept equalization. As the center frequency is moved, new coefficients must be calculated—not a trivial task. To avoid quantization effects (sometimes called zipper noise), filter coefficients and amplitude scaling coefficients must be updated at a theoretical rate equal to the sampling rate; in practice, an update rate equal to one-half or one-fourth the sampling rate is sufficient. To accomplish even this, coefficients are often obtained through linear interpolation. The range must be limited to ensure that filter poles do not momentarily pass outside the unit circle, causing transient instability.

Adaptive filters automatically adjust their parameters according to optimization criteria. They do not have fixed coefficients; instead, values are calculated during operation. Adaptive filters thus consist of a filter section and a control unit used to calculate coefficients. Often, the algorithm used to compute coefficients attempts to minimize the difference between the output signal and a reference signal. In general, any filter type can be used, but in practice, adaptive filters often use a transversal structure as well as lattice and ladder structures. Adaptive filters are used for applications such as echo and noise cancelers, adaptive line equalizers, and prediction.

A transversal filter is an FIR filter in which the output value depends on both the input value and a number of previous input values held in memory. Inputs are multiplied by coefficients and summed by an adder at the output. Only the input values are stored in delay elements; there are no feedback networks; hence it is an example of a nonrecursive filter. As described in Chap. 4, this architecture is used extensively to implement lowpass filtering with oversampling.

In practice, digital oversampling filters often use a cascade of FIR filters, designed so the sampling rate of each filter is a power of 2 higher than the previous filter. The number of delay blocks (tap length) in the FIR filter determines the passband flatness, transition band slope, and stopband rejection; there are $M + 1$ taps in a filter with M delay blocks. Many digital filters are dedicated chips. However, general purpose DSP chips can be used to run custom filter programs.

The block diagram of a dedicated digital filter (oversampling) chip is shown in Fig. 17.15. It demonstrates the practical implementation of DSP techniques. A central processor performs computation while peripheral circuits accomplish input/output and other functions. The filter's characteristic is determined by the coefficients stored in ROM; the multiplier/accumulator performs the essential arithmetic operations; the shifter manages data during multiplication; the RAM stores intermediate computation results; a microprogram stored in ROM controls the filter's operation. The coefficient word length determines filter accuracy and stopband attenuation. A filter can have, for example, 293 taps and a 22-bit coefficient; this would yield a passband that is flat to within ±0.00001 dB, with a stopband suppression greater than 120 dB. Word length of the audio data increases during multiplication; truncation would result in quantization error thus the data must be rounded or dithered. Noise shaping can be applied at the accumulator, using an IIR filter to redistribute the noise power, primarily placing it outside the audio band. Noise shaping is discussed in Chap. 18.

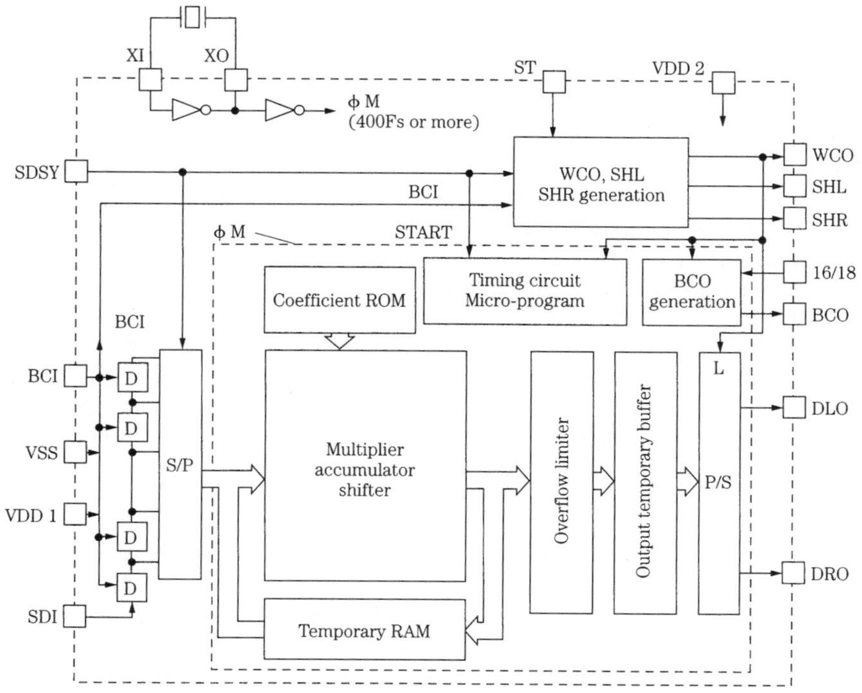

FIGURE 17.15 An example showing the functional elements of a dedicated digital filter (oversampling) chip.

Sources of Errors and Digital Dither

Unless precautions are taken, the DSP computation required to process an audio signal can add noise and distortion to the signal. In general, errors in digital processors can be classified as coefficient, limit cycle, overflow, truncation, and round-off errors. Coefficient errors occur when a coefficient is not specified with sufficient accuracy. For example, a resolution of 24 bits or more is required for computations on 16-bit audio samples. Limit cycle error might occur when a signal is removed from a filter, leaving a decaying sum. This decay might become zero or might oscillate at a constant amplitude, a condition known as limit cycle oscillation. This effect can be eliminated, for example, by offsetting the filter's output so that truncation always produces a zero output.

Overflow occurs when a register length is exceeded, resulting in a computational error. In the case of wraparound, when a 1 is added to the maximum value positive two's complement number, the result is the maximum value negative number. In short, the information has overflowed into a nonexistent bit. The drastic change in the amplitude of the output waveform would yield a loud pop if applied to a loudspeaker. To prevent this, saturating arithmetic can be used so that when the addition of two positive numbers would result in a negative number, the maximum positive sum is substituted instead. This results in clipping—a more satisfactory, or at least more benign, alternative. Alternatively, designers must provide sufficient digital headroom.

Truncation and round-off errors occur whenever the word length of a computed result is limited. Errors accumulate both inside the processor during calculation, and when word length is reduced for output through a D/A converter. However, A/D conversion always results in quantization error, and computation error can appear in different guises. For example, when two n-bit numbers are multiplied, the number of output bits will be $2n - 1$. Thus, multiplication almost doubles the number of bits required to represent the output. Although many hardware multipliers can perform double-precision computation, a finite word length must be maintained following multiplication, thus limiting precision. Discarded data results in an error analogous to that of A/D quantization. To be properly modeled, multiplication must be followed by quantization; multiplication does not introduce error, but inability to keep the extra bits does. It is important to note that unnecessary cumulative dithering should be avoided in interim calculations.

Rather than truncate a word, for example, following multiplication, the value can be rounded; that is, the word is taken to the nearest available value. This results in a peak error of 1/2 LSB, and an rms value of $1/(12)^{1/2}$, or 0.288 LSB. This round-off error will accumulate over successive calculations. In general, the number of calculations must be large for significant error to occur. However, in addition, dither information can be lost during computation. For example, when a properly dithered 16-bit word is input to a 32-bit processor, even though computation is of high precision, the output signal might be truncated to 16 bits for conversion through the output D/A converter. For example, a 16-bit signal that is delayed and scaled by a 12-dB attenuation would result in a 12-bit undithered signal. To overcome this, digital dithering to the resolution of the next processing (or recording) stage should be used in a computation.

To apply digital dithering, a pseudo-random number of the appropriate magnitude is added to each sample and then the new LSB is rounded up or down to the nearest quantization interval according to value of the portion to be discarded. In other words, a logical 1 is added to the new LSB and the lower portion is discarded, or the lower portion is simply discarded. The lower bit information thus modulates the new LSB and provides the linearizing benefit of dithering. A different pseudo-random number is used

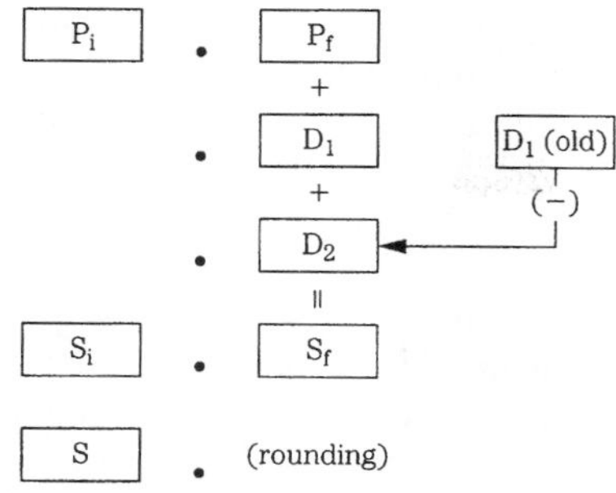

FIGURE 17.16 An example of digital redithering used during a truncation or rounding computation. (*Vanderkooy and Lipshitz, 1989*)

for each audio channel. To duplicate the effects of analog dither, two's complement digital dither values should be used to provide a bipolar signal with an average zero value. In some cases, noise shaping is also applied at this stage. Bipolar pseudo-random noise and a D/A converter can be used to generate analog dither.

As an example of digital dithering, consider a 20-bit resolution signal that must be dithered to 16 bits. IIR and noise-shaping filters with digital feedback can exhibit limit cycle oscillations with low-level signals if gain reduction or certain equalization processing (resulting in gain changes) is performed; digital dither can be used to randomize these cycles.

John Vanderkooy and Stanley Lipshitz have shown that truncated or rounded digital words can be redithered with rectangular pdf or triangular pdf fractional numbers, as shown in Fig. 17.16. A rectangular pdf dither word D_1 is added to a digital audio word with integer part P_i and fractional part P_f. The carry bit dithers the rounding process in the same way that 1 LSB analog rectangular pdf dither affects an A/D converter. When a statistically independent rectangular pdf dither D_2 is added, triangular pdf dither results. This triangular pdf dither noise power is $Q^2/6$ and rounding noise power is $Q^2/12$ so the total noise power is $Q^2/4$. The final sum has integer part S_i and fractional part S_f, which become S upon rounding.

In cases of gain fading, triangular pdf appears to be a better choice than rectangular pdf because it eliminates noise modulation as well as distortion, at the expense of a slightly higher noise floor. To minimize audibility of this noise penalty, highpass triangular pdf dither can be most appropriate in rounding. The triangular pdf statistics are not changed. However, dither samples are correlated. Average dither noise power is $Q^2/6$, with no noise at 0 Hz and double the average value at the Nyquist frequency. Hence the term: highpass dither. This shaping becomes more pronounced with noise increasingly shifted outside the audio band, as the oversampling ratio is increased, for example, by eight times. The audible effect of a noise penalty is lessened when oversampling is used because in-band noise is relatively decreased proportional to the oversampling rate. Further reduction of the $Q^2/12$ requantization noise power can be achieved through noise-shaping circuits, as described in Chap. 18.

As noted, the problem of error sources is applicable to both audio samples as well as the computations used to determine other system operators, such as filter coefficients. For example, improperly computed filter coefficients could shift the locations of poles and zeros, thus altering the characteristics of the response. In some cases, an insufficiently defined coefficient can cause a stable IIR filter to become unstable. On the other hand, the quantization of filter coefficients will not affect the linear operation of a circuit, or introduce artifacts that are affected by the input signal or that vary with time. The effect of coefficient errors in a filter is generally determined by the number of coefficients determining the location of each pole and zero. The fewer the coefficients, the lower the sensitivity. However, when poles and zeros are placed in locations where they are few, the effect of errors is greater.

DSP Integrated Circuits

A DSP chip is a specialized hardware device that performs digital signal processing under the control of software algorithms. DSP chips are stand-alone processors, often independent of host CPUs (central processing units), and are specially designed for operations used in spectral and numerical applications. For example, large numbers of multiplications are possible, as well as special addressing modes such as bit-reverse and circular addressing. When memory and input/output circuits are added, the result is an integrated digital signal processor. Such a general purpose DSP chip is software programmable, and thus can be used for a variety of signal-processing applications. Alternatively, a custom signal processor can be designed to accomplish a specific task.

Digital audio applications require long word lengths and high operating speeds. To prevent distortion from rounding error, the internal word length must be 8 to 16 bits longer than the external word. In other words, for high-quality applications, internal processing of 24 to 32 bits or more is required. For example, a 24-bit DSP chip might require a 56-bit accumulator to prevent overflow when computing long convolution sums.

Processor Architecture

DSP chips often use a pipelining architecture so that several instructions can be paralleled. For example, with pipelining, a fetch (fetch instruction from memory and update program counter), decode (decode instruction and generate operand address), read (read operand from memory), and execute (perform necessary operations) can be effectively executed in one clock cycle. A pipeline manager, aided by proficient user programming, helps ensure efficient processing.

DSP chips, like all computers, are composed of input and output devices, an arithmetic logic unit, a control unit, and memory, interconnected by buses. All computers originally used a single sequential bus (von Neumann architecture), shared by data, memory addresses, and instructions. However, in a DSP chip a particularly large number of operations must be performed quickly for real-time operation. Thus, parallel bus structures are used (such as the Harvard architecture) that store data and instructions in separate memories and transfer them via separate buses. For example, a chip can have separate buses for program, data, and direct memory access (DMA), providing parallel program fetches, data reads, as well as DMA operations with slower peripherals.

A block diagram of a general purpose DSP chip is shown in Fig. 17.17; many DSP chips follow a similar architecture. The chip has seven components: multiply-accumulate unit, data address generator, data RAM, coefficient RAM, coefficient address generator, program control unit, and program ROM. Three buses interconnect these components: data bus, coefficient bus, and control bus. In this example, the multiplier is asymmetrical; it multiplies 24-bit sample words by 12-bit coefficient words. The result of multiplication is carried out to 36 bits. For dynamic compression, the 12-bit words containing control information are derived from the signal itself; two words could be taken together to provide double precision when necessary. A 40-bit adder is used in this example. This adds the results of multiplications to other results stored in the 40-bit accumulator. Following addition in the arithmetic logic unit (ALU), words

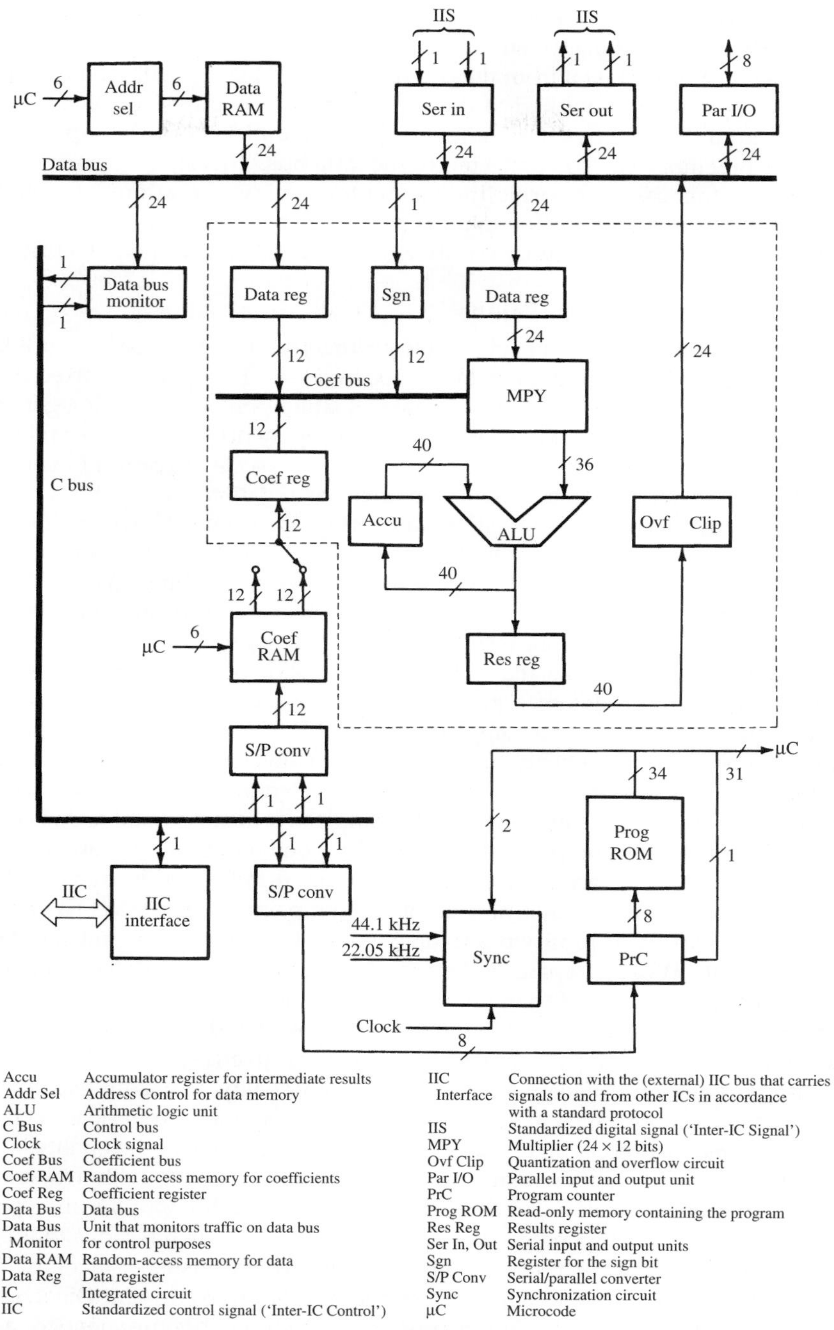

FIGURE 17.17 Block diagram of a digital signal processor chip. The section surrounded by a dashed line is the arithmetic unit where computation occurs. Independent buses are used for data, coefficients, and control.

must be quantized to 24 bits before being placed on the data bus. Different methods of quantization can be applied.

Coefficients are taken from the coefficient RAM, loaded with values appropriate to the task at hand, and applied to the coefficient bus. Twenty-four-bit data samples are moved from the data bus to two data registers. Parallel and serial inputs and outputs, and data memory can be connected to the data bus. This short memory (64 words by 24 bits, in this case) performs the elementary delay operation. As noted, to speed multiple multiplications, pipelining is provided. Necessary subsequent data is fed to intermediate registers simultaneously with operations performed on current data.

Fixed Point and Floating Point

DSP chips are designed according to two arithmetic types, fixed point and floating point, which define the format of the data they operate on. Fixed-point (or fixed-integer) chips use two's complement, binary integer data. Floating-point chips use integer and floating-point numbers (represented as a mantissa and exponent). The dynamic range of a fixed-point chip is based on its word length. Data must be scaled to prevent overflow; this can increase programming complexity. The scientific notation used in a floating-point chip allows a larger dynamic range, without overflow problems. However, the resolution of floating point representation is limited by the word length of the exponent.

Many factors are considered by programmers when deciding which DSP chip to use in an application. These include cost, programmability, ease of implementation, support, availability of useful software, availability of internal legacy software, time to market, marketing impact (for example, the customer's perception of 24-bit versus 32-bit), and sound quality. As Brent Karley notes, programming style and techniques are both critical to the sound quality of a given audio implementation. Moreover, the hardware platform can also affect sound quality. Neither can be ignored when developing high-quality DSP programs. The Motorola DSP56xxx series of processors is an example of a fixed-point architecture, and the Analog Devices SHARC series is an example of a floating-point architecture. Conventional personal computer processors can perform both kinds of calculations.

Three principal processor types are widely used in DSP-based audio products: 24-bit fixed point, 32-bit fixed point, and 32-bit floating point (24 mantissa bits with 8 exponent bits). Each variant of precision and architecture has advantages and disadvantages for sound quality. When comparing 32-bit fixed and 24-bit fixed, with the same program running on two DSP chips of the same architecture, the higher precision DSP will obviously provide higher quality audio. However, DSP chips often do not use the same architecture and do not support the same implementation of any given algorithm. It is thus possible for a 24-bit fixed-point DSP chip to exceed the sound quality of a 32-bit fixed-point chip. For example, some DSP chips have hardware accelerators that can off-load some processing chores to free up processing resources that can be allocated to audio quality improvements or extra features. Programming methods such as double-precision arithmetic and truncation-reduction methods may be implemented to improve quality. Optimization programming methods can be performed on any of the available DSP architectures, but are often not implemented (or only in limited numbers) because of processing or memory limitations.

For higher-quality audio implementations, a 24-bit fixed-point DSP chip can implement double-precision filtering or processing in strategic components of a given algorithm to improve the sound quality beyond that of a 32-bit fixed-point DSP chip. The combination of two 24-bit numbers into one 48-bit integer allows very high-fidelity processing. This is possible in some cases due to the higher processing power available in some 24-bit DSP chips over less efficient 32-bit DSP chips. Some manufacturers carefully hand-optimize their implementations to exceed the sound quality of competitors using 32-bit fixed-point DSP chips. Chip manufacturers often provide different versions of audio decoders and audio algorithms with and without qualitative optimizations to allow the customer to evaluate cost versus quality. In some double-precision 24-bit programming, some high-order bits are used for dynamic headroom, low-order "guard bits" store fractional values, and the balance represent the audio sample; a division of 8/32/8 might be used. In some cases, products with 24-bit chips using double-precision arithmetic are advertised as having "48-bit DSP." As discussed below, performance is also affected by the nature of the instruction set used in the DSP chip.

The comparison of fixed-point and floating-point processors is not straightforward. Generally, in fixed-point processors, there is a trade-off between round-off error and dynamic range. In floating-point processors, computation error is equivalent to the error in the multiplying coefficients. Floating-point DSP chips offer ease of programmability (they are more easily programmed in a higher-level language such as C++) and extended dynamic range. For example, in a floating-point chip, digital filters can be realized without concern for scaling in state variables. However, code that is generated in a higher-level language is sometimes not efficiently implemented due to the difference in hand assembly versus C++ compilers. This wastes processing resources that could be used for qualitative improvements. The dynamic range of a 24-bit fixed-point DSP chip is 144 dB and a 32-bit fixed-point chip provides 196 dB, as shown in Fig. 17.18. In contrast, a 32-bit floating-point DSP chip provides a theoretical dynamic range of 1536 dB. A 1536-dB dynamic range is not a direct advantage for audio applications, but the noise floor is maintained at a level correspondingly below the audio signal. If resources permit, improvements can be implemented in 32-bit floating-point DSP chips to raise the audio quality. The Motorola DSP563xx is an example of a 24-bit fixed-point chip, and the Texas Instruments TMS320C67xx is an example of a 32-bit floating-point chip.

Both platform and programming are critical to the audio quality capabilities of DSP software. Bit length, while an important component, is not the only variable determining audio-quality capabilities. With proper programming, both fixed-point and floating-point processors can provide outstanding sound quality. If needed, optimization programming methods such as double-precision filtering can improve this further.

The IEEE 752 floating-point standard allows either 32-bit single precision (24-bit mantissa and 8-bit exponent) or double precision (64-bit with a 52-bit mantissa and 11-bit exponent). The latter allows very high-fidelity processing, but is not available on all DSP chips.

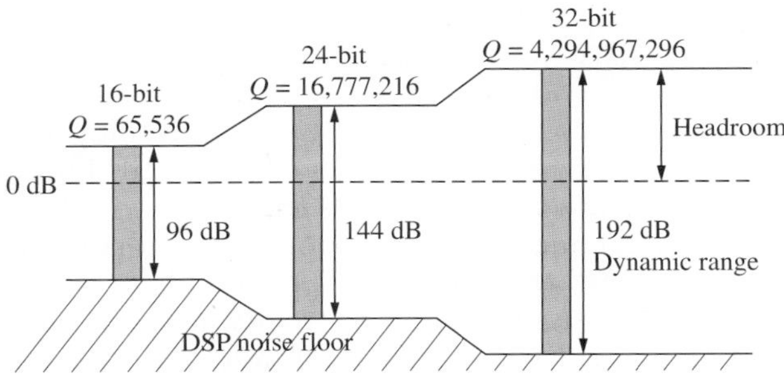

Figure 17.18 The dynamic range of fixed-point DSP chips is directly based on their word length. In some applications, double precision increases the effective dynamic range.

DSP Programming

As with all software programming, DSP programming allows the creation of applications with tremendous power and flexibility. The execution of software instructions is accomplished by the DSP chip using audio samples as the input signal. Although the architecture of a DSP chip determines its theoretical processing capabilities, performance is also affected by the structure and diversity of its instruction set.

DSP chips use assembly language programming, which is highly efficient. Higher-level languages (such as C++) are easy to use, document, debug, and are independent of hardware. However, they are less efficient and thus slower to execute, and do not take advantage of specialized chip hardware. Assembly languages are more efficient. They execute faster, require less memory, and take advantage of special hardware functions. However, they are more difficult to program, read, debug, are hardware-dependent, and are labor-intensive.

An assembly language program for a specific DSP chip type is written using the commands in the chip's instruction set. These instructions take advantage of the chip's specialized signal processing functions. Generally, there are four types of instructions: arithmetic, program control, memory access and transfer, and input/output. Instructions are programmed with mnemonics that identify the instruction, and reference a data field to be operated upon. Because many DSP chips operate within a host environment, user front-end programs are written in higher-level languages, reserving assembly language programming for the critical audio processing tasks.

An operation that is central to DSP programming is the multiply-accumulate command, often called MAC. Many algorithms call for two numbers to be multiplied, and the result summed with a previous operation. DSP chips feature a multiply-accumulate command that moves two numbers into position, multiplies them together, and accumulates the result, with rounding, all in one operation. To perform this efficiently, the arithmetic unit contains a multiplier–accumulator combination (see Fig. 17.17). This can calculate one subproduct and simultaneously add a previous subproduct to a sum, in one machine cycle. To accommodate this, DSP chips rely on a hardware-multiplier to execute the multiply-accumulate operation. Operations such as sum of products and iterative operations can be performed efficiently.

In most cases, the success of a DSP application depends on the talents of the human programmer. Software development costs are usually much greater than the hardware itself. In many cases, manufacturers supply libraries of functions such as FFTs, filters, and windows that are common to many DSP applications. To assist the development process, various software development tools are employed. A text editor is used to write and modify assembly language source code. An assembler program translates assembly language source code into object code. A compiler translates programs written in higher-level languages into object code. A linker combines assembled and compiled object programs with library routines and creates executable object code. The assembler and compiler use the special processing features of the DSP chip wherever possible. A simulator is used to emulate code on a host computer before running it on DSP hardware. A debugger is used to perform low-level debugging of DSP programs. In addition, DSP operating systems are available, providing programming environments similar to those used by general purpose processors.

Filter Programming

DSP chips can perform many signal processing tasks, but digital filtering is central to many applications. As an example of assembly language programming, consider the FIR transversal filter shown in Fig. 17.19A. It can be described with this difference equation:

$$y(n) = x(n-4)h(4) + x(n-3)h(3) + x(n-2)h(2) + x(n-1)h(1) + x(n)h(0)$$

The five input audio sample values, five coefficients describing the impulse response at five sample times, and the output audio sample value will be stored in memory locations; data value $x(n)$ will be stored in memory location XN; $x(n-1)$ in XNM1; $y(n)$ in YN; $h(0)$ in H0, and so on. The running total will be stored in the accumulator, as shown in Fig. 17.19B. In this case, intermediate calculations are held in the T register.

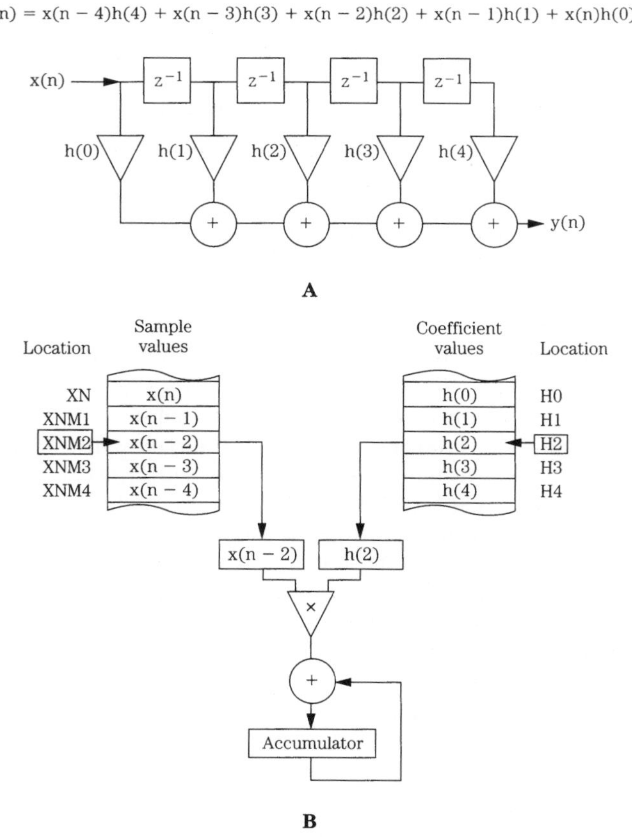

Figure 17.19 A five-coefficient FIR filter can be realized with assembly language code authored for a commercial DSP chip. A. Structure of the FIR filter. B. Audio sample and filter coefficient values stored in memory are multiplied and accumulated to form the output samples.

Texas Instruments Code

The code to execute this five-coefficient FIR filter on a Texas Instruments TMS320xxx chip is shown below:

```
1.  NXTPT IN XN,PA2   * get the new input value XN from port
2.        ZAC         * zero the accumulator
3.        LT XNM4     * x(n – 4)h(4)
4.        MPY H4      * multiply
5.        LTD XNM3    * x(n – 4)h(4) + x(n – 3)h(3)
6.        MPY H3      * multiply
7.        LTD XNM2    * similar to previous steps
8.        MPY H2
9.        LTD XNM1
10.       MPY H1
11.       LTD XN
12.       MPY H0
13.       APAC        * add the result of the last multiply
                      * to the accumulator
14.       SACH YN,1   * store the result in YN
15.       OUT YN,PA2  * output the response to port
16.       B NXTPT     * get the next point
```

Of particular interest is the LTD instruction that loads a value into a register, adds the result of the last multiply to the accumulator, and shifts the value to the next higher memory address.

Line 1 moves an input data word from an I/O port to memory location XN. Line 2 zeros the accumulator. Line 3 loads the past value XNM4 into the T register. Line 4 multiplies the contents of the T register with the coefficient in H4. Lines 5 and 6 multiply XNM3 and H3, and the LTD instruction adds the result of the line 4 multiplication to the accumulator. Similarly, the other filter taps are multiplied and accumulated in lines 7 through 12. Line 13 adds the final multiplication to the accumulator, so that the accumulator contains the filtered output value corresponding to the input value from line 1. Line 14 transfers this output to YN. Line 15 transfers location YN to the I/O port. Line 16 returns the program to line 1 where the next input value is received.

When the new input is received, past values are shifted one memory location; XN moves to XNM1, XNM1 to XNM2, XNM2 to XNM3, XNM3 to XNM4, and XNM4 is dropped. Although these shifts could be performed with additional instructions, as noted, the LTD instruction (an example of the power of parallel processing) shifts each value to the next higher address after it is transferred to the T register. After line 11, the data values are ready for the next pass. Other program sections not shown would set up I/O ports, synchronize I/O with the sampling rate of external hardware, and store filter coefficients. This particular filter example uses straight-line code with separate steps for each filter tap. It is very fast for short filters, but longer filters would be more efficiently performed with looped algorithms. The decision of which is more efficient for a given filter is not always obvious.

Motorola Code

The Motorola DSP56xxx family of processors are widely used to perform DSP operations such as filtering. Efficiency stems from the DSP chip's ability to perform multiply-accumulate operations while simultaneously loading registers for the next multiply-accumulate operation. Further, hardware do-loops and repeat instructions allow repetition of the multiply-accumulate operation.

To illustrate programming approaches, the difference equation for the five-coefficient FIR filter in the previous example (see Fig. 17.19) will be realized twice, using code written by Thomas Zudock. First, as in the previous example, the filter is implemented with the restriction that each operation must be executed independently in a straight-line fashion. Second, the filter is implemented using parallel move and looping capabilities to maximize efficiency as well as minimize code size.

First, some background on the DSP56xxx family syntax and architecture. The DSP56xxx has three parallel memories, x, y, and p. Typically, the x and y memories are used to hold data, and the p memory holds the program that performs the processing. The symbols r0 and r5 are address registers. They can be used as pointers to read or write values into any of the three memories. For example, the instruction "move x:(r0)+,x0" reads the value stored at the memory location pointed to by r0 in x memory into the x0 data register, then increments the r0 address register pointer to the data to be accessed. The x0 and y0 symbols are data registers, typically used to hold the next data that is ready for a mathematical operation. The accumulator is denoted by a; it is the destination for mathematical operations. The instruction "mac x0,y0,a" is an example of the multiply-accumulate operation; the data registers x0 and y0 are multiplied and then summed with the value held in accumulator a.

In a straight-line coded algorithm implementing the FIR example, each step is executed independently. At the start, x0 is already loaded with a new sample, r0 points to the location in x memory to save that sample, and r5 points to the location in memory where h(0) is located. The FIR straight-line code is:

```
1. move x0,x:(r0)+     ;save x(n) in memory, increment r0
2. move y:(r5)+,y0     ;get h(0) from memory, increment r5
3. mpy x0,y0,a         ;a =x0y0 = x(n)h(0)
4. move x:(r0)+,x0     ;get x(n – 1) from x memory, increment r0
5. move y:(r5)+,y0     ;get h(1) from y memory, increment r5
6. mac x0,y0,a         ;a = a + x0y0 = x(n)h(0) + x(n – 1)h(1)
7. move x:(r0)+,x0     ;get x(n – 2) from x memory, increment r0
8. move y:(r5)+,y0     ;get h(2) from y memory, increment r5
9. mac x0,y0,a         ;a = a + x0y0 = x(n)h(0) + x(n – 1)h(1) + · · ·
                       ; · · · + x(n – 2)h(2)
```

```
10. move x:(r0)+,x0     ;get x(n – 3) from x memory, increment r0
11. move y:(r5)+,y0     ;get h(1) from y memory, increment r5
12. mac x0,y0,a         ;a = a + x0y0 = x(n)h(0) + x(n – 1)h(1) + · · ·
                        ;· · · + x(n – 2)h(2) + x(n – 3)h(3)
13. move x:(r0)+,x0     ;get x(n – 4) from x memory, increment r0
14. move y:(r5)+,y0     ;get h(1) from y memory, increment r5
15. macr x0,y0,a        ;a = a + x0y0 = x(n)h(0) + x(n – 1)h(1) + · · ·
                        ;· · · + x(n – 2)h(2) + x(n – 3)h(3) + · · ·
                        ;· · · + x(n – 4)h(4), round to 24 bits
16. lua (r0)–,r0        ;decrement r0
```

In line 1, the just-acquired sample x(n), located in x0, is moved into the x memory location pointed to by r0, and r0 is advanced to point to the next memory location, the previous sample. In line 2, the filter coefficient in the y memory pointed to by r5, h(0), is read into data register y0, and r5 is incremented to point to the next filter coefficient, h(1). In line 3, x(n) and h(0) are multiplied and the result is stored in accumulator a. The initial conditions of the algorithm, and the status of the algorithm after line 3 can be described as:

Initial Status: $x0 = x(n)$ the sample just acquired

	x memory		y memory
r0 →	x(n – 5)	r5 →	h(0)
	x(n – 1)		h(1)
	x(n – 2)		h(2)
	x(n – 3)		h(3)
	x(n – 4)		h(4)

After line 3: $a = x0y0 = x(n)h(0)$

	x memory		y memory
	x(n)		h(0)
r0 →	x(n – 1)	r5 →	h(1)
	x(n – 2)		h(2)
	x(n – 3)		h(3)
	x(n – 4)		h(4)

The remainder of the program continues similarly, with the next data value loaded into x0 and the next filter coefficient loaded into y0 to be multiplied and added to accumulator a. After line 16, the difference equation has been fully evaluated and the filtered output sample is in accumulator a. This status can be described as:

After line 15:

$$a = x0y0 = x(n)h(0) + x(n-1)h(1) + x(n-2)h(2) + x(n-3)h(3) + x(n-4)h(4)$$

	x memory	y memory
	x(n)	r5 → h(0)
	x(n – 1)	h(1)
	x(n – 2)	h(2)
	x(n – 3)	h(3)
r0 →	x(n – 4)	h(4)

After the next input sample is acquired, the same process will be executed to generate the next output sample. Address registers r0 and r5 have wrapped around to point to the needed data and filter coefficients. The registers are set up for modulo addressing with a modulus of five (because there are five filter taps). Instead of continuing to higher memory, they wrap around in a circular fashion when incremented or decremented to repeat the process again.

Although this code evaluates the difference equation, the DSP architecture has not been used optimally; this filter can be written much more efficiently. Values stored in x and y memory can be accessed simultaneously with execution of one instruction. While values are being multiplied and accumulated, the next data value and filter coefficient can be simultaneously loaded in anticipation of the next calculation. Additionally, a hardware loop counter allows repetition of this operation, easing implementation of long difference equations. The operations needed to realize the same filter can be implemented with fewer instructions and in less time, as shown below. The FIR filter coded using parallel moves and looping is:

```
1. clr a x0,x:(r0)+ y:(r5)+,y0            ;initialize accumulator a to zero
                                          ;save x(n) in x memory, increment r0
                                          ;get h(0) from y memory, increment r5
2. rep #4                                 ;repeat the next instruction four times
3. mac x0,y0,a x:(r0)+,x0 y:(r5)+,y0;     a = a + x(n – D)h(D) : D = 0,1,2,3
                                          ;get x(n – D + 1) from x memory, increment r0
                                          ;get h(D + 1) from memory, increment r5
4. macr x0,y0,a (r0)–                     ;a = a + x(n – 4)h(4), round to 24 bits
                                          ;decrement r0
```

The parallel move and looping implementation is clearly more compact. The number of comments needed to describe each line of code is indicative of the power in each line. However, the code can be confusing because of its consolidation. As programmers become more experienced, use of parallel moves becomes familiar. Many programmers view the DSP56xxx code as having three execution flows: the instruction execution, x memory accesses, and y memory accesses.

The Motorola DSP563xx core uses code that is compatible with the DSP56xxx family of processors. For example, the DSP56367 is a 24-bit fixed-point, 150 MIPS DSP chip. It is used to process multichannel decoding standards such as Dolby Digital and DTS. It can simultaneously perform other audio processing such as bass management, sound-field effects, equalization, and ProLogic II.

Analog Devices SHARC Code

Digital audio processing is computation-intensive, requiring processing and movement of large quantities of data. Digital signal processors are specially designed to efficiently perform the audio and communication chores. They have a high degree of parallelism to perform simultaneous tasks (such as processing multichannel audio files), and balance performance with I/O requirements to avoid bottlenecks. In short, such processors are optimized for audio processing tasks.

The architecture used in the ADSP-2116x SHARC family of DSP chips is an example of how audio computation can be optimized in hardware. As noted, intermediate DSP calculations require higher precision than the audio signal resolution; for example, 16-bit processing is certainly inadequate for a 16-bit audio signal, unless double-precision processing is used. Errors from rounding and truncation, overflow, coefficient quantization, and limit cycles would all degrade signal quality at the D/A converter. Several aspects of the ADSP-2116x SHARC architecture are designed to preserve a large signal dynamic range. In particular, it aims to provide processing precision that will not introduce artifacts into a 20- or 24-bit audio signal.

This architecture performs native 32-bit fixed-point, 32-bit IEEE 754-1985 floating-point, and 40-bit extended precision floating-point computation, with 32-bit filter coefficients. Algorithms can be switched between fixed- and floating-point modes in a single instruction cycle. This alleviates the need for double-precision mathematics and the associated computational overhead. The chip has two DSP processing elements onboard, designated PEX and PEY, each containing an arithmetic logic unit, multiplier, barrel shifter, and register file. The chip also has circular buffer support for up to 32 concurrent delay lines. To help prevent overflow in intermediate calculations, the chip has four 80-bit fixed-point accumulators; two 32-bit values can be multiplied with a 16-bit guard band. Also, multiple link ports allow data to move between multiple processors.

Dual processing units allow single-instruction multiple-data (SIMD) parallel operation in which the same instruction can be executed concurrently in each processing element with different data in each element. Code is optimized because each instruction does double duty. For example, in stereo processing, the left and right channels could be handled separately in each processing element. Alternatively, a complex block calculation such as an FFT can be evenly divided between both processing elements. The code example shown below computes the simultaneous computation of two.FIR filters:

```
/* initialize pointers */
      b0 = DELAYLINE;
      I0 = @ DELAYLINE;
      b8 = COEFADR;
      I8 = @ COEFADR;
      m1= –2; m0=2; m8=2;
```

```
/* Circular Buffer Enable, SIMD MODE Enable */
        bit set MODE1 CBUFEN I PEYEN;
        nop;
        f0 = dm(samples);
        dm(i0,m0)=f0, f4=pm(i8,m8);
        f8=f0*f4, f0=dm(i0,m0), f4=pm(i8,m8);
        f12=f0*f4, f0=dm(i0,m0), f4=pm(i8,m8);
        Icntr=(TAPS–3), do macs until Ice;
/*FIR loop*/
macs:   f12=f0*f4, f8=f8+f12, f0=dm(i0,m0), f4=pm(i8,m8);
        f12=f0*f4, f8=f8+f12, f0=dm(i0,m1);
        f8=f8+f12;
```

To execute two filters simultaneously in this example, audio samples are interleaved in 64-bit data memory and coefficients are interleaved in 64-bit program memory, with 32 bits for each of the processing elements in upper and lower memory. A single-cycle, multiply-accumulate, dual-memory fetch instruction is used. It executes a single tap of each filter computing the running average while fetching the next sample and next coefficient.

Specialized DSP Applications

Beyond digital filtering, DSP is used in a wide variety of audio applications. For example, building on the basic operations of multiplication and delay, applications such as chorusing and phasing can be developed. Reverberation perhaps epitomizes the degree of time manipulation possible in the digital domain. It is possible to synthesize reverberation to both simulate natural acoustical environments, and create acoustical environments that could not physically exist. Digital mixing consoles embody many types of DSP processing in one consolidated control set. The power of digital signal processing also is apparent in the algorithms used to remove noise from audio signals.

Digital Delay Effects

A delay block is a simple storage unit, such as a memory location. A sample is placed in memory, stored, then recalled some time later and output. A delay unit can be described by the equation: $y(n) = x(n - m)$ where m is the delay in samples. Generally, when the delay is small, the frequency response of the signal is altered; when the delay is longer, an echo results. Just as in filtering, a simple delay can be used to create sophisticated effects. For example, Fig. 17.20A shows an echo circuit using a delay block. Delay mT is of duration m samples, and samples are multiplied by a gain coefficient (a scaling factor) less than unity. If the delay time is set between 10 ms and 50 ms, an echo results; with shorter fixed delays, a comb filter response results, as shown in Fig. 17.20B. Peaks and dips are equally spaced through the frequency response, from 0 Hz to the Nyquist frequency. The number of peaks depends on the delay time; the longer the delay, the greater the number of peaks.

If the delay time of the circuit in Fig. 17.20A is slowly varied between 0 ms and 10 ms, the time-varying comb filter creates a flanging effect. If the time delay is varied between

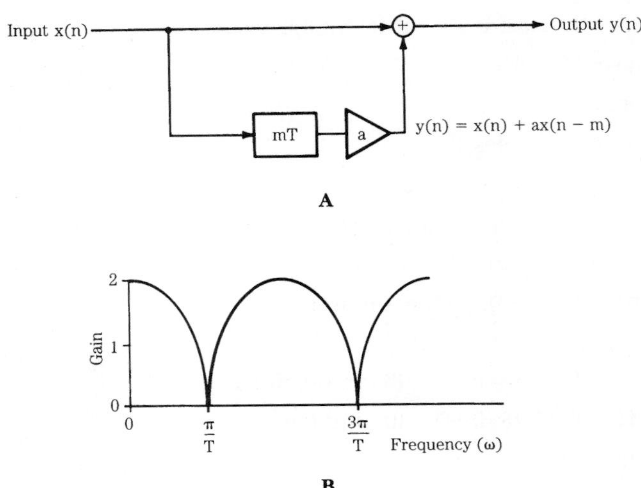

FIGURE 17.20 A delay block can be used to create an echo circuit. A. The circuit contains an *mT* delay and gain stage. B. With shorter delay times, a comb filter response will result. (*Bloom, 1985; Berkhout and Eggermont, 1985*)

10 ms and 25 ms, a doubling effect is achieved, giving the impression of an accompanying voice. A chorus effect is provided when the signal is directed through several such blocks, with different delay variations.

A comb filter can be either recursive or nonrecursive. It cascades a series of delay elements, creating a new response. Mathematically, we see that a nonrecursive comb filter, such as the one described above, can be designed by adding the input sample to the same sample delayed: $y(n) = x(n) + ax(n - m)$ where m is the delay time in samples. A recursive comb filter creates a delay with feedback. The delayed signal is attenuated and fed back into the delay: $y(n) = ax(n) + by(n - m)$. This yields a response as shown in Fig. 17.21. The number of peaks depends on the duration of the delay; the longer the delay, the greater the number of peaks.

An all-pass filter is one that has a flat frequency response from 0 Hz to the Nyquist frequency. However, its phase response causes different frequencies to be delayed by different amounts. An all-pass filter can be described as: $y(n) = -ax(n) + x(n - 1) + by(n - 1)$.

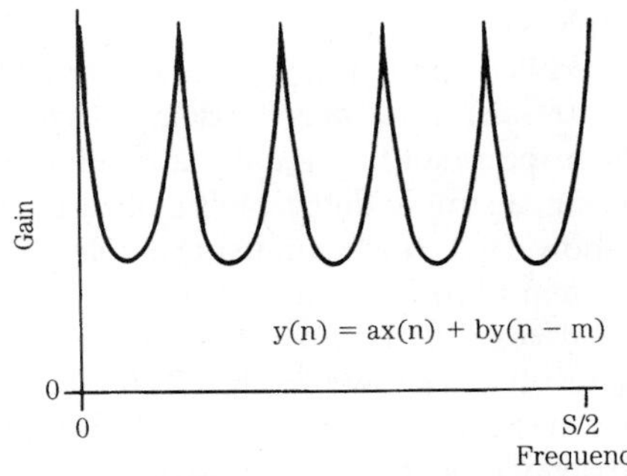

FIGURE 17.21 A recursive comb filter creates a delay with feedback, yielding a toothed frequency response.

If the delay in the above circuits is replaced by a digital all-pass filter or a cascade of all-pass filters, a phasing effect is achieved: $y(n) = -ax(n) + x(n - m) + by(n - m)$. The effect becomes more pronounced as the delay increases. The system exhibits nonuniformly spaced notches in its frequency response, varying independently in time.

Digital Reverberation

Reflections are characterized by delay, relative loudness, and frequency response. For example, in room acoustics, delay is determined by the size of the room, and loudness and frequency response are determined by the reflectivity of the boundary surfaces and surface construction. Digital reverberation is ideally suited to manipulate these parameters of an audio signal, to create acoustical environments. Because reverberation is composed of a large number of physical sound paths, a digital reverberation circuit must similarly process many data elements.

Both reverberation and delay lines are fundamentally short-memory devices. A pure delay accepts an input signal and reproduces it later, possibly a number of times at various intervals. With reverberation, the signal is mixed repetitively with itself during storage at continually shorter intervals and decreasing amplitudes. In a delay line, the processor counts through RAM addresses sequentially, returning to the first address after the last has been reached. A write instruction is issued at each address, and the sampled input signal is routed to RAM. In this way, audio information is continually stored in RAM for a period of time until it is displaced by new information. During the time between write operations, multiple read instructions can be issued sequentially with different addresses. By adjusting the numerical differences, the delay times for the different signals can be determined.

In digital reverberation, the stored information must be read out a number of times, and multiplied by factors less than unity. The result is added together to produce the effect of superposition of reflections with decreasing intensity. The reverberation process can be represented as a feedback system with a delay unit, multiplier, and summer. The processing program in a reverberation unit corresponds to a series and parallel combination of many such feedback systems (for example, 20 or more). Recursive configurations are often used. The single-zero unit reverberator shown in Fig. 17.22A generates an exponentially decreasing impulse response when the gain block is less than unity. It functions as a comb filter with peaks spaced by the reciprocal of the delay time. The echoes are spaced by the delay time, and the reverberation time can be given as: $RT = 3T/\log_{10}(a)$ where T is the time delay and a is the coefficient. Sections can be cascaded to yield more densely spaced echoes; however, resonances and attenuations can result. Overall, it is not satisfactory as a reverberation device.

Alternatively, a reverberation section can be designed from an all-pass filter. The filter shown in Fig. 17.22B yields an all-pass response by adding part of the unreverberated signal to the output. Total gain through the section is unity. In this section, frequencies that are not near a resonance and thus not strongly reverberated are not attenuated. Thus, they can be reverberated by a following section. Other types of configurations are possible; for example, Fig. 17.22C shows an all-pole section of a reverberation circuit which can be described as: $y(n) = -ax(n) + x(n - m) + ay(n - m)$.

The spacing between the equally spaced peaks in the frequency response is determined by the delay time, and their amplitude is set by the scaling coefficients. One reverberation algorithm devised by Manfred Schroeder uses four reverberant comb filters in parallel followed by two reverberant all-pass filters in series, as shown

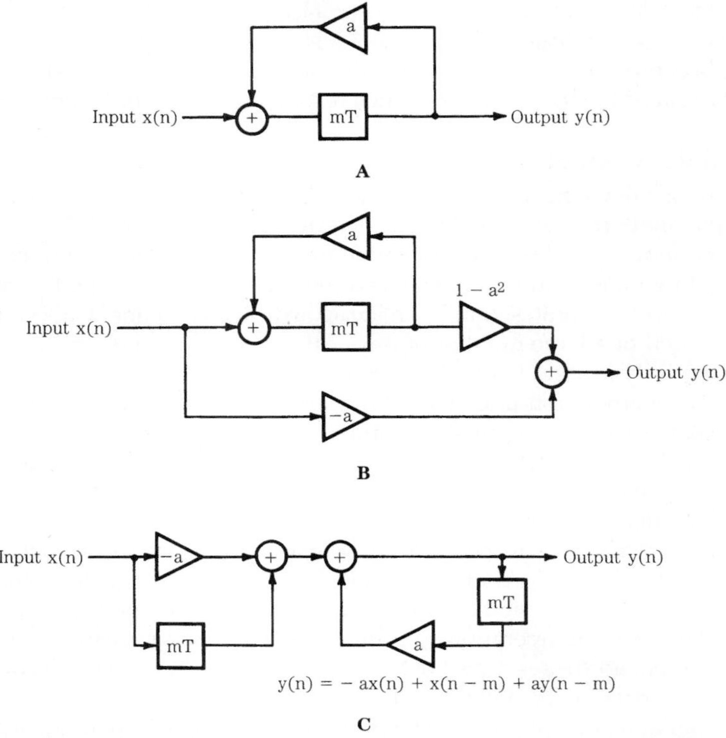

FIGURE 17.22 Reverberation algorithms can be constructed from recursive configurations. A. A simple reverberation circuit derived from a comb filter. B. A reverberation circuit derived from an all-pass filter. C. A reverberation circuit constructed from an all-pole section. (*Blesser et al., 1975*)

in Fig. 17.23. The four comb filters, with time delays ranging from 30 ms to 45 ms, provide sustained reverberation, and the all-pass networks with shorter delay times contribute to reverberation density. The combination of section types provides a relatively smooth frequency response, and permits adjustable reverberation times. Commercial implementations are often based on similar methods, with the exact algorithm held in secrecy.

Digital Mixing Consoles

The design of a digital mixing console is quite complex. The device is, in effect, a vast digital signal processor, requiring perhaps 10^9 instructions per second. Although many of the obstacles present in analog technology are absent in a digital implementation, seemingly trivial exercises in signal processing might require sophisticated digital circuits. In addition, the task of real-time computation of digital signal processing presents problems unique to digital architecture. The digital manifestation of audio components requires a rethinking of analog architectures and originality in the use of digital techniques.

A digital console avoids numerous problems encountered by its analog counterpart, but its simple task of providing the three basic functions of attenuating, mixing, and signal selection and routing entails greater design complexity. The problem is augmented by the requirement that all operations must be interfaced to the user's analog world. A virtual

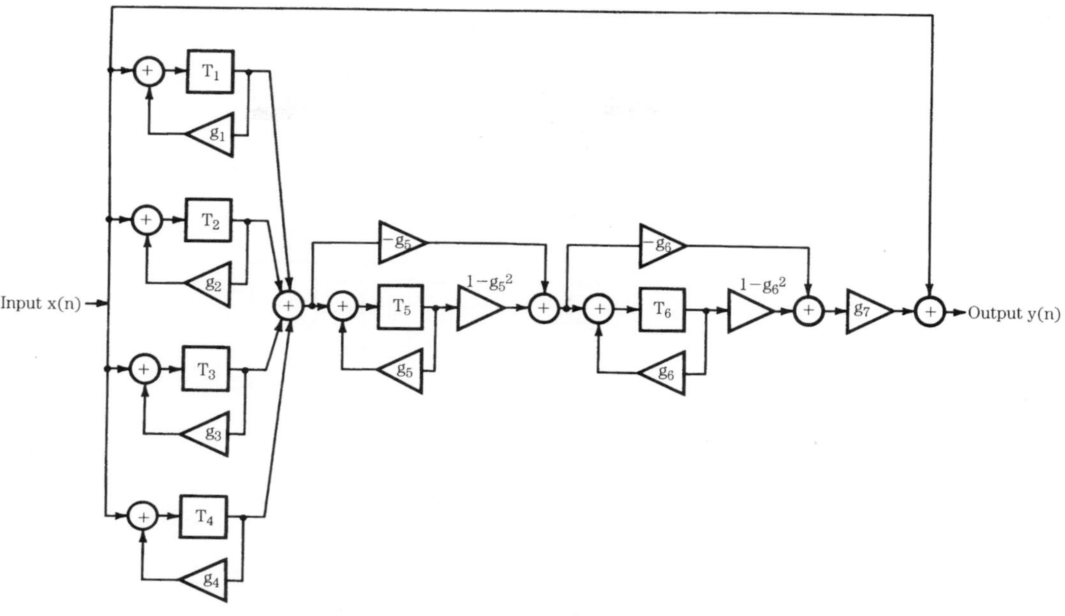

FIGURE 17.23 Practical reverberation algorithms yielding natural sound contain a combination of reverberation sections. (*Schroeder, 1962*)

mixer is much easier to design than one with potentiometers and other physical comforts. In either case, a digital console's software-defined processing promotes flexibility.

The processing tasks of a digital console follow those of any DSP system. The first fundamental task of a mixing console is gain control. In analog, this is realized with only a potentiometer. However, a digital preamplifier requires a potentiometer, an A/D converter to convert the analog position information of the variable resistor into digital form, and a multiplier to adjust the value of the digital audio data, as shown in Fig. 17.24A. The mixing console's second task, mixing, can be accomplished in the analog domain with several resistors and an operational amplifier. In a digital realization, this task requires a multiplexer and accumulator, as shown in Fig. 17.24B. The third task, signal selection and routing, is accomplished in analog with a multipole switch, but digital technology requires a demultiplexer and encoding circuit to read the desired selection, as shown in Fig. 17.24C. All processing steps must be accomplished on each audio sample in each channel in real time, that is, within the span of one sample period (for example, 10.4 μs at a 96-kHz sampling frequency). For real-time applications such as broadcasting and live sound, throughput latency must be kept to a minimum.

Any console must provide equalization faculties such as lowpass, highpass, and shelving filters. These can be designed using a variety of techniques, such as a cascade of biquadratic sections, a type of digital filter architecture. The fast processing speeds and necessity for long word length place great demands on the system. On the other hand, filter coefficients are readily stored in ROM, and no more than a few thousand characteristics are typically required for most applications.

Because the physical controls of a digital console are simply remote controls for the processing circuits, no audio signal passes through the console surface. It is thus possible

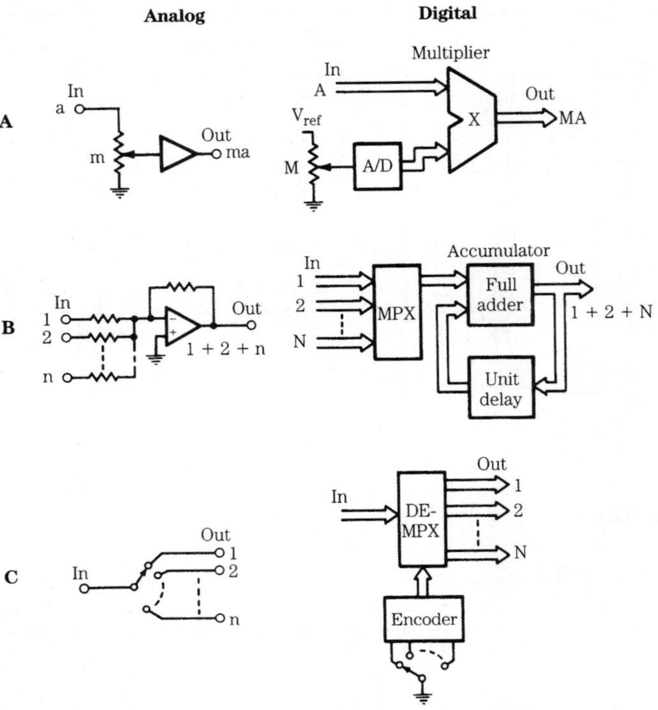

FIGURE 17.24 The three primary functions in signal processing can be realized in both analog and digital forms. A. Attenuating. B. Mixing. C. Selecting. (*Sakamoto et al., 1982*)

to offer assignable functions. Although the surface might emulate a conventional analog console with parallel controls, a set of fewer assignable controls can be used as a basic control set to view and control the mix elements. For example, after physical controls are used to record and equalize basic tracks, they can be assigned to virtual strips that are not assigned to physical controls, then individually reassigned back to physical controls as needed. In other words, the relationship between the number of physical channel strips and the number of console inputs and outputs is flexible. The console can provide a central control section that provides access to all of the assignable functions currently requested for a particular signal path.

Particular processing functions can be assigned to a channel, inserted in the signal path, and controlled from that channel's strip or elsewhere. Physical controls on a channel strip can provide different functions, with displays that reflect the value of the selected function. A set of controls might provide equalization, or auxiliary sends. In another example, one channel strip might be assigned to a monaural, or stereo channel. As a session progresses from tracking to mix-down, the console can gradually be reconfigured, for example, adding more sends, then later creating subgroups. There is no need to limit controls to the equivalent fixed analog circuit parameters. Thus, rotary knobs, for example, might have no end stops and can be programmed for various functions or resolution. For example, equalization might be accomplished in fractions of decibels, with selected values displayed.

Console setup can be accomplished via an alphanumeric keyboard, and particular configurations can be saved in memory, or preprogrammed in presets for rapid change-over in a real-time application. Flexibility would be more of a curse than a blessing if the operator were required to build a software console from scratch each time. Therefore, using the concept of a "soft" signal path, multiple console preset configurations may be available: track laying, mix-down, return to previous setting, and a fail-safe minimum mode.

A video display might show graphical representations of console signal blocks as well as their order in the signal path. This is useful when default settings of delay, filter, equalization, dynamics, and fader are varied. In addition, each path can be named, and its fader assignment shown, as well as its mute group, solo status, phase reverse, phantom power on/off, time delay, and other parameters.

Loudspeaker Correction

Loudspeakers are far from perfect. Their nonuniform frequency response, limited dynamic range, frequency-dependent directivity, and phase nonlinearity all degrade the reproduced audio signal. In addition, a listening room reinforces and cancels selected frequencies in different room locations, and contributes surface reflections, superimposing its own sonic signature on that of the audio signal. Using DSP, correction signals can be applied to the audio signal. Because the corrections are opposite to the errors in the acoustically reproduced signal, they theoretically cancel them.

Loudspeakers can be measured as they leave the assembly line, and deviations such as nonuniform frequency response and phase nonlinearity can be corrected by DSP processing. Because small variations exist from one loudspeaker to the next, the DSP program's coefficients can be optimized for each loudspeaker. Moreover, certain loudspeaker/room problems can be addressed. For example, floor-standing loudspeakers can have predetermined relationships between the drivers, the cabinet, and the reflecting floor surface. The path-length difference between the direct and reflected sound creates a comb-filtered frequency response that can be corrected using DSP processing.

A dynamic approach can be used in an adaptive loudspeaker/room correction system, in which the loudspeakers generate audio signals to correct the unwanted signals reflected from the room. Using DSP chips and embedded software programs, low-frequency phase delay, amplitude and phase errors in the audio band, and floor reflections can all be compensated for. Room acoustical correction starts with room analysis, performed on-site. Using a test signal, the loudspeaker/room characteristics are collected by an instrumentation microphone placed at the listening position, and processed by a program that generates room-specific coefficients. The result is a smart loudspeaker that compensates for its own deficiencies, as well as anomalies introduced according to its placement in a particular room.

Noise Removal

Digital signal processing can be used to improve the quality of previously recorded material or restore a signal to a previous state, for example, by removing noise. With DSP it is possible to characterize a noise signal and minimize its effect without affecting a music signal. For example, DSP can be used to reduce or remove noises such as clicks, pops, hum, tape hiss, and surface noise from an old recording. In addition, using methods borrowed from the field of artificial intelligence, signal lost due to tape dropouts can be synthesized with great accuracy. Typically, noise removal is divided into two tasks, detection and elimination of impulsive noise such as clicks, and background noise reduction.

Interactive graphical displays can be used to locate impulsive disturbances. A de-clicking program analyzes frequency and amplitude information around the click, and synthesizes a signal over the area of the click. Importantly, the exact duration of the performance is preserved. For example, Fig. 17.25 shows a music segment with a click and the success of various de-clicking methods. A basic sample-and-hold function

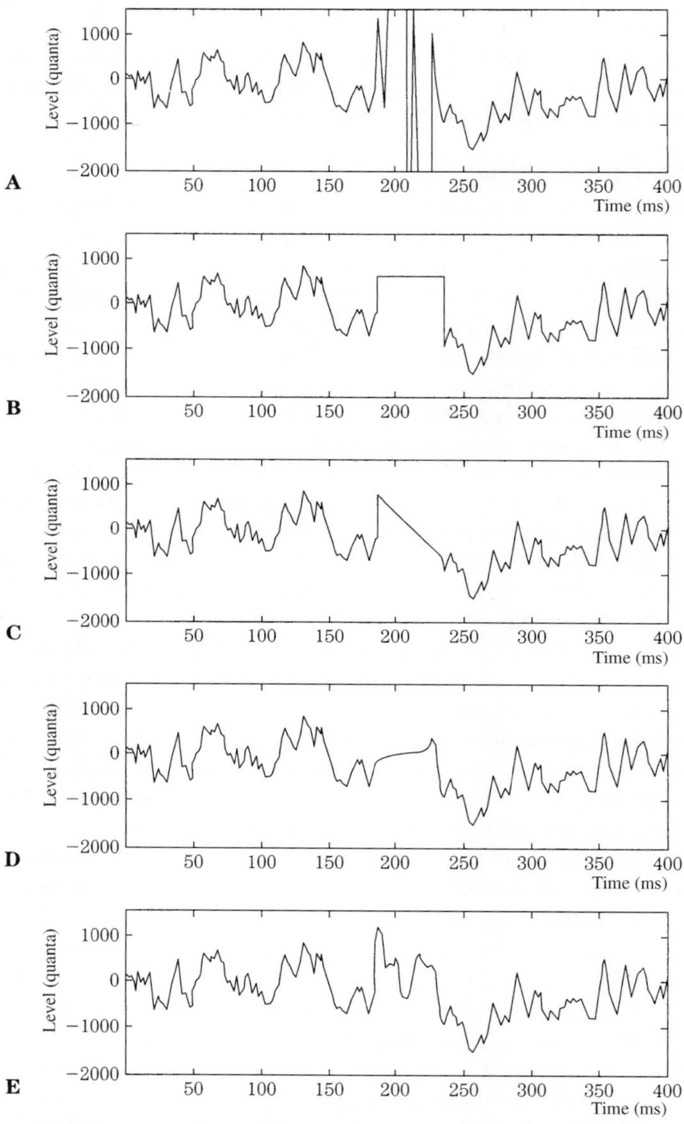

FIGURE 17.25 Examples showing how interpolation can be used to overcome an impulsive click. Higher-order interpolation can synthesize an essentially inaudible bridge. A. Original sample with click. B. Sample and hold with highpass click detection. C. Straight-line interpolation. D. Low-order interpolation. E. High-order interpolation.

produces a plateau that is preferable to a loud click. With linear interpolation the click is replaced by a straight line connecting valid samples. Interpolation can be improved by using a signal model to measure local signal characteristics (a few samples). Interpolation can be further improved by increasing the number of samples used in the model (hundreds of samples). Automatic de-clicking algorithms similarly perform interpolation over defects, but first must differentiate between unwanted clicks, and transient music information. Isolated clicks will exhibit short rise and decay times between amplitude extremes but a series of interrelated clicks can be more difficult to identify. When more than one copy of a recording is available (or when the same signal is recorded on two sides of a groove), uncorrelated errors are more easily identified. However, different recordings must be precisely time-aligned.

A background noise-removal process can begin by dividing the audio spectrum into frequency bands, then performing an analysis to determine content. The spectral composition of surface noise and hiss is determined and this fingerprint is used to develop an inverse function to remove the noise. Ideally, the fingerprint data is taken from a silent portion of the recording, perhaps during a pause. Because there is no music signal, the noise can be more accurately analyzed. Alternatively, samples can be taken from an area of low-amplitude music, and a noise template applied.

For removal of steady-state artifacts, the audio signal is considered in small sections. The energy in each frequency band is compared to that in the noise fingerprints and the system determines what action to take in each frequency band. For example, at a particular point in the program, music can dominate in a lower spectrum and the system can pass the signal in that band unprocessed, but hiss dominating at a higher frequency might trigger processing in that band. In many cases, some original noise is retained, because the result is generally more natural sounding. In some algorithms, a clean audio signal is estimated and then analyzed to create a masking curve. Only noise that is not masked by the curve is subjected to noise removal; this minimizes introduction of artifacts or damage to the audio signal.

Hiss from a noisy analog recording, wind noise from an outdoor recording, hum from an air conditioner, even the noise obscuring speech in a downed aircraft's flight recorder can all be reduced in a recording. In short, a processed recording can sound better than the original master recording. On the other hand, as with most technology, noise removal processing requires expertise in its application, and improper use degrades a recording. One question is the degree to which processing is applied. A 10 to 15% reduction in noise level can make the difference between an acceptable recording, and an unpleasantly noisy one. Additional noise removal can be problematic, even if it is only psychoacoustical tricks that make us think it is the audio signal that is being affected.

Thomas Stockham's experiments on recordings made by Enrico Caruso in 1906 resulted in releases of digitally restored material in which those acoustical recordings were analyzed and processed with a blind deconvolution method, so-called because the nature of the signals operated upon cannot be precisely defined. In this case, the effect of an unknown filter on an unknown signal was removed by observing the effect of the filtering. By averaging a large number of signals, constant signals could be identified relative to dynamic signals. In this way, the strong resonances of the horn used to make these acoustical recordings, and other mechanical noise, were estimated and removed from the recording by applying an inverse function, leaving a restored audio signal.

In essence, the acoustical recording process was viewed as a filter. The music and the impulse response of the recording system were convolved. From this analysis of the playback signal, the original performance's spectrum was multiplied by that of the recording apparatus, and then inverse Fourier transformed to yield the signal actually present on the historic recordings. Correspondingly, the undesirable effects of the recording process were removed from the true spectrum of the performance. The computation of a correction filter using FFT methods was not a trivial task. A long-term average spectrum was needed for both the recording to be restored, and for a model, higher-fidelity recording. In Stockham's work, a 1-minute selection of music required about 2500 FFTs for complete analysis. The difference between the averaged spectra of the old and new recordings was computed, and became the spectrum of the restoration filter. Restoration required multiplying segments of the recording's spectrum by the restoration spectrum and performing inverse FFTs, or by direct convolution using a filter impulse response obtained through inverse FFT.

Blind deconvolution uses techniques devised by Alan Oppenheim, known as generalized linearity or homomorphic signal processing. This class of signal processing is nonlinear, but is based upon a generalization of linear techniques. Convolved or multiplied signals are nonlinearly processed to yield additive representations, which in turn are processed with linear filters and returned via an inverse nonlinear operation to their original domain. These and other more advanced digital-signal processing topics are discussed in detail in other texts.

CHAPTER 18

Sigma-Delta Conversion and Noise Shaping

Design limitations and the relative cost of PCM converter architectures encouraged the development of sigma-delta converters. These converters are characterized by short word lengths, very high oversampling rates, and noise shaping. They demonstrate that conversion can be performed either with a high-resolution quantizer at a low sampling rate (as in traditional converters), or with a low-resolution quantizer at a high sampling rate (as in sigma-delta converters). Sigma-delta analog-to-digital (A/D) and digital-to-analog (D/A) converters both use conversion methods such as sigma-delta modulation with noise shaping, and process high sampling-rate signals with oversampling and decimation filters. These converters share the goal of translating nonideal converter errors into uncorrelated, benign noise. Sigma-delta converters are sometimes referred to as one-bit or multi-bit converters, depending on the specific architecture employed. In addition, nonoversampling noise shaping is critical when reducing word length during a data transfer, for example, when transferring a 24-bit master recording to a 16- or 20-bit format.

Sigma-Delta Conversion

Conversion of PCM audio data words, using a resistor ladder, is a traditional approach. A PCM system represents the analog waveform as an amplitude signal, storing information that measures the amplitude sample by sample. However, the method is flawed when quantization introduces differential nonlinearity errors in the amplitude representation. Moreover, because a multiplicity of bits are used to form the representation, and because each bit has an error unequal to the others, the overall error varies with each sample, and is thus difficult to correct. In practice, calibration procedures during manufacture, and sophisticated circuit design are required to achieve high performance. Understandably, manufacturers sought to develop alternative conversion methods, including sigma-delta converters. Sigma-delta methods are particularly desirable for A/D conversion because they obviate the need for an analog brick-wall anti-aliasing filter. But the method also offers many advantages in D/A conversion, particularly when designing a mixed-signal integrated circuit that combines A/D, D/A, and DSP functions.

It is not easy to see how one bit (or a few) can replace 16 or more bits. Consider this analogy: traditional ladder converters are like a row of light bulbs, each connected to a switch. Sixteen bulbs, for example, each with a different brightness, can be lighted in various combinations to achieve 2^{16} or 65,536 different brightness levels.

However, relative differences in individual bulb intensities will introduce error into the output brightness level. Any particular switch combination may not exactly produce the desired brightness. Similarly, ladder converters introduce error as they attempt to reproduce the audio signal.

Sigma-delta technology uses a wholly different approach. Instead of many bulbs and switches, only one bulb and one switch are used. Brightness is varied by simply switching the bulb on and off. For example, if the bulb is dynamically switched on and off equally, the output is at half-brightness. If the bulb's on-time is increased, brightness will increase. Similarly, sigma-delta converters may ideally use one bit to represent audio amplitude, with very fast switching and very accurate timing. Sigma-delta technology is an inherently precise way to represent an audio waveform.

PCM conversion divides the signal in multiple amplitude steps. However, sigma-delta conversion divides the signal in time, keeping amplitude changes constant. Non-intuitively, a high- or low-level pulse signal can represent an audio signal. For example, pulse-density modulation (PDM) can be used. Fig. 18.1 shows how a single constant-width pulse, with either a high or low level, can reconstruct a waveform. Alternatively, a pulse-width modulation (PWM) signal can be used to reconstruct the output signal (variations are pulse-edge and pulse-length modulation).

Sigma-delta converters are called one-bit converters when one quantized audio bit is output, and called multi-bit converters when the audio signal is quantized to several bits (perhaps four). True one-bit sigma-delta converters output full-scale positive or negative pulses that ensure perfect linearity. However, their inherently coarse quantization yields a higher noise floor. Thus, higher orders of noise shaping are required for good in-band dynamic range performance. In addition, one-bit converters are susceptible to phase jitter on the modulator clock. Relatively small jitter levels can yield a full-scale error in the one-bit signal. Multi-bit sigma-delta converters use multiple quantization levels and this yields a relatively lower noise floor both in-band and out of band. This allows the use of a relatively lower oversampling ratio, as well as lower-order noise shaping. In addition, the smaller quantization levels of multi-bit converters

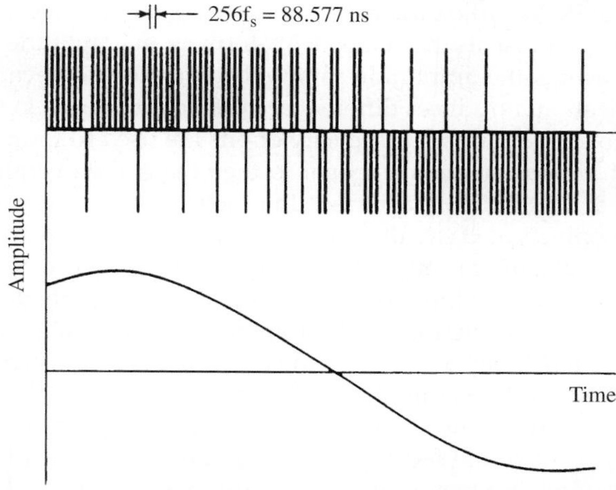

FIGURE 18.1 Pulse-density modulation can be used to code and reconstruct an analog waveform.

make them more tolerant of phase jitter. However, care must be taken to achieve good linearity. Generally, multi-bit sigma-delta converters are employed in higher-quality audio applications.

One-bit sigma-delta converters are inherently linear; any level mismatch results in an offset. However, multi-bit sigma-delta converters can have element mismatches that result in noise and distortion. In this architecture, a capacitor or resistor element is assigned to a fixed code; small variations yield a mismatch. For example, assuming 32 elements, a 1% mismatch yields about 0.1% THD+N distortion. The distortion is also signal-dependent. To address this mismatch, codes can be used to randomly assign different elements in each cycle, but this increases noise. Assuming 32 elements and an oversampling ratio of 128, a 1% mismatch would yield a dynamic range of only 80 dB. Dynamic element matching (DEM) is a common solution used to reduce the effect of mismatch errors in the conversion elements of multi-bit converters. With DEM, analog capacitor elements are swapped so that the mismatch error is averaged close to zero. DEM can use open-loop noise shaping after the modulator to convert the error into low-level benign noise. For example, elements can be selected cyclically, depending on the weighting of the input data. This averages the mismatch, and the noise from mismatching is first-order noise shaped. In other designs, the mismatch shaping function can be placed inside the sigma-delta feedback loop. The randomized errors are shaped by the modulator, further reducing the in-band level. Assuming 32 elements and an oversampling ratio of 128, a 1% mismatch would yield a dynamic range of 120 dB.

Sigma-delta conversion methods are widely used in digital audio products. These techniques are extremely competitive in both A/D and D/A applications because they obviate the need for analog brick-wall filters. Although sigma-delta methods use familiar techniques such as oversampling, highly sophisticated processing is required to implement noise shaping and thus decrease the high in-band noise levels otherwise present in sigma-delta conversion. A variety of sigma-delta A/D and D/A architectures have been devised, with different algorithms and orders of noise shaping.

Delta Modulation

In PCM a signal is sampled and quantized into discrete steps; the maximum signal amplitude determines the maximum quantizer range. Traditional PCM A/D and D/A converters are shown in Fig. 18.2A. Quantization error is uniformly present across the Nyquist frequency band from 0 to $f_s/2$ Hz and cannot be removed from the signal; f_s is the sampling frequency. If quantization is performed at a higher sampling frequency $R \times f_s$ Hz where R is the oversampling rate, the error is spread across the band to $R \times f_s/2$ Hz, hence the noise in the audio band is reduced by 3 dB for every factor of 2 oversampling.

With a maximum-amplitude sinusoidal input signal, oversampling will increase the signal-to-error ratio as follows:

$$S/E = 6.02(N + 0.5L) + 1.76 \text{ dB}$$

where N = the number of quantization bits
L = the number of octaves of oversampling

For example, an oversampling A/D converter can perform as well as a longer word length A/D converter, yielding a benefit of 0.5 bit/oversampling octave. However, the benefit is limited. For example, a 10-bit improvement would require an L of 20 octaves,

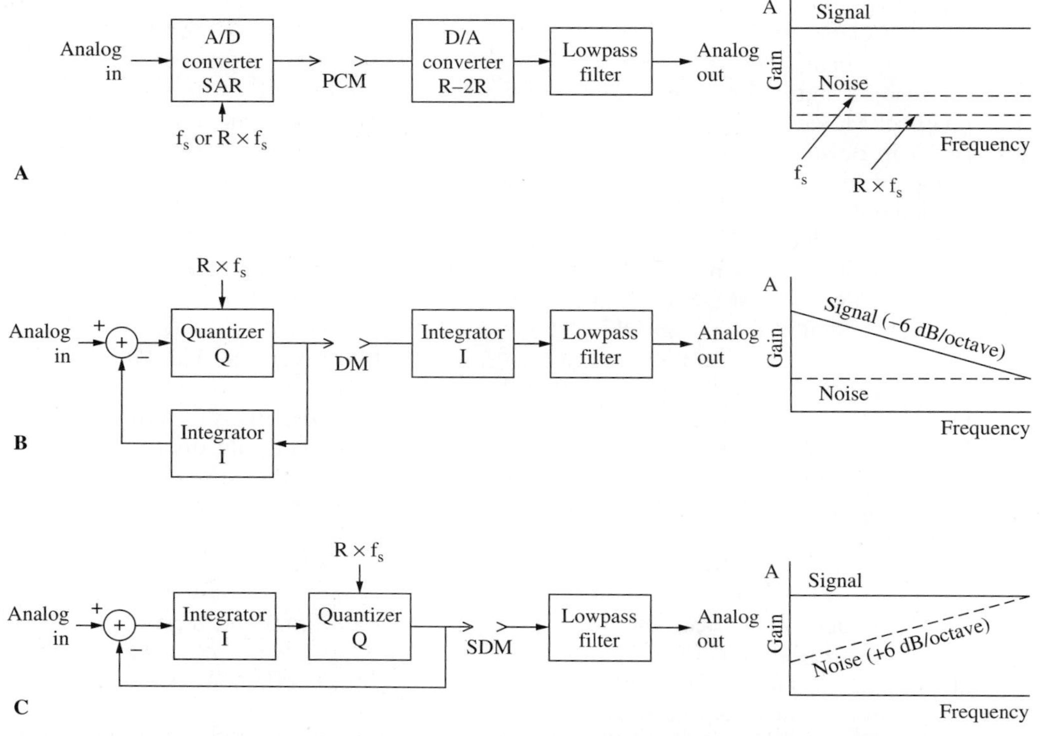

FIGURE 18.2 A comparison of modulation methods used in A/D and D/A conversion. A. Pulse-code modulation (PCM) yields a flat signal and noise floor. B. Delta modulation (DM) yields a signal that falls with frequency. C. Sigma-delta modulation (SDM) yields a noise floor that rises with frequency.

equivalent to an impossible oversampling factor of 1 million. It is the aim of noise shaping to introduce a highpass function in the noise spectrum and thus improve oversampling performance.

Delta modulation and sigma-delta modulation (also called delta-sigma) were developed in the 1940s and 1960s, respectively, for voice telephony applications. Limitations prohibited their use in high-quality music applications until the emergence of high-speed digital signal processing techniques in the 1980s, that improved audio performance. Differential pulse-code modulation (DPCM) is a technique in which the derivative of the signal is quantized. When signal changes between sample periods are small, the quantizer's word length can be reduced. With very high oversampling rates, the changes between sample periods are made very small, thus the quantizer can be reduced to one bit. A one-bit DPCM coder is known as a delta modulator (DM). In other words, DM codes the differences in the signal amplitude, its slope, instead of the signal amplitude itself. In contrast, a sigma-DPCM technique places an integrator at the input to the quantizer. Instead of coding the slope of the signal, sigma-DPCM, like PCM, codes its amplitude. However, with sigma-DPCM, the signal can be quantizing to only one or a few bits; both implementations are generally known as sigma-delta modulation. As with delta modulation, sigma-delta modulation requires high oversampling rates.

A delta-modulation encoder and decoder are shown in Fig. 18.2B; this is known as a single integration modulator. The analog input signal is compared to the integrated output pulses and the delta (difference) signal is applied to the quantizer. The quantizer generates a positive pulse when the difference signal is negative, and a negative pulse when the difference signal is positive. This difference error signal moves the integrator step by step closer to the present value input, tracking the derivative of the analog input signal. The integrator's output is a past approximation to the input, thus the coder operates similarly to other integrating feedback loops such as phase-locked loops.

A delta-modulation decoder consists of an integrator and a lowpass filter. When the one-bit pulses are integrated using a time constant that is long compared to the sample period, a step waveform is produced. An output analog waveform is produced by lowpass filtering this step waveform. Requantization error at the output of the integrator is white. As in PCM, oversampling decreases the error level by 3 dB for every factor of two oversampling. The dynamic range can be improved, that is, the requantization error made smaller, by making the delta (difference or step size) value smaller. The limit to which the delta can be reduced is given by the maximum derivative of the signal. The coded signal amplitude decreases at 6 dB/octave; thus, S/N decreases as the signal frequency increases.

The maximum derivative occurs at maximum signal frequency and maximum signal amplitude. Exceeding this limit causes slope overload distortion. For music signals, the delta value must be high, resulting in a high quantization error level. When the signal has reduced high-frequency content, as in speech, delta can be reduced. In any case, the success of DM hinges on assumptions about the nature of the encoded signal. This dependence of the dynamic range on the signal spectrum, and good performance only when the signal has lowpass characteristics, as well as correlated patterns at low signal levels, limits single-integration DM applications.

To convert a maximum amplitude 16-bit word, a one-bit modulator would have to perform 2^{16} toggles per conversion period. With a sampling frequency of 44.1 kHz, this would demand an unrealistic toggle rate of approximately 2.9 GHz. As the rate is slowed to accommodate hardware limitations, noise levels increase to an intolerable level. Looked at in another way, bit reduction at a high sampling frequency is required to output a one-bit signal from a high-bit source; this greatly degrades the signal's dynamic range.

Sigma-Delta Modulation

Sigma-delta modulation (SDM) was developed to overcome the limitations of delta modulation. Sigma-delta systems quantize the delta (difference) between the current signal and the sigma (sum) of the previous difference. A first-order (single integration) sigma-delta modulation encoder and decoder are shown in Fig. 18.2C. An integrator is placed at the input to the quantizer; signal amplitude is constant with increasing frequency. The input to the midrise quantizer is the integral of the difference between the input and the quantized output as applied via negative feedback. If the input signal in one sample period is greater than the value accumulated in the feedback loop over previous samples, the converter outputs a "1." Otherwise, if it is lower, the output is a "0." In other words, the difference between the input signal and the accumulated error is quantized. When the error is sufficiently large, the quantizer changes state to reduce the error. Thus, the difference between the input signal and the output signal approaches zero; the average value of the output approximates the input. There is little dc error in the output signal. However, the frequency spectrum of the quantizing error rises with increasing frequency (6 dB/octave).

The integrator forms a lowpass filter on the difference signal thus providing low-frequency feedback around the quantizer. This feedback results in a reduction of quantization noise at low (in-band) frequencies. Unlike PCM and DM, the noise floor is not flat, but shaped by a first-order highpass characteristic as analyzed below. In practice, the in-band noise floor level is not satisfactory with first-order sigma-delta modulation. In addition, quantization noise is highly correlated in a first-order modulator. Further noise shaping must be achieved with higher-order (multiple integration) sigma-delta modulation coders. Michael Gerzon and Peter Craven have shown that shaped noise in multi-bit noise shapers will occupy equal areas above and below the original noise level. Since this relationship occurs on a logarithmic vertical axis and linear frequency axis, shaping increases the total noise power. The benefit, of course, is lower in-band noise.

Like PCM, sigma-delta modulation quantizes the signal amplitude directly, and not its derivative as in DM. Thus the maximum quantizer range is determined by the maximum signal amplitude and is not dependent on signal spectrum. As with delta modulation, to achieve high resolution, high oversampling rates are required. For example, with an audio band of 24 kHz and 64 times oversampling, the internal sampling frequency rises to 3.072 MHz, thus quantization noise is spread from dc to 1.536 MHz. However, because the quantizer has only two states, the quantization error, and the resulting noise level, is high. To overcome this, sigma-delta modulators add noise shaping to move the noise power to higher, out-of-band frequencies. The noise is shaped by the inverse of the loop transfer function; when a lowpass filter is placed in the loop, the noise spectrum increases with frequency. In many designs, a multi-bit quantizer is used within the loop and a low-bit PCM signal is coded; the dynamic range increases with 6 dB for every quantization bit. As in one-bit designs, the noise floor increases by 6 dB/octave in multi-bit designs, and noise shaping is required.

A one-bit sigma-delta modulation decoder theoretically requires only a lowpass filter to decode the signal, to remove high-frequency (out-of-band) components. In other words, it averages the output signal to produce an analog waveform. An integrator is not needed in the decoder (as in DM) because the signal's amplitude is coded, not its slope. In multi-bit designs, a D/A converter is needed in the decoder to decode the low-bit PCM signal.

Analysis of a First-Order Sigma-Delta Modulator

The signal and noise transfer functions for a first-order sigma-delta modulator are analyzed in Fig. 18.3. $X(z)$ is the z-transform of the input sequence and $Y(z)$ represents the output sequence. The quantization error is represented as white noise $N(z)$. For a zero noise source, the transfer function shows a lowpass characteristic. For a zero signal, the transfer function shows a highpass characteristic. In other words, as the loop integrates the difference between the input signal and the sampled signal, it lowpass filters the signal and highpass filters the noise. If the system is designed so the signal's frequency content is less than the filter's cutoff frequency, the signal will not be affected. Given a first-order noise-shaping loop, with a maximum amplitude sinusoidal input, the maximum signal-to-error ratio will be:

$$S/E = 6.02(N + 1.5L) - 3.41 \text{ dB}$$

where N = the number of quantization bits
L = the number of octaves of oversampling

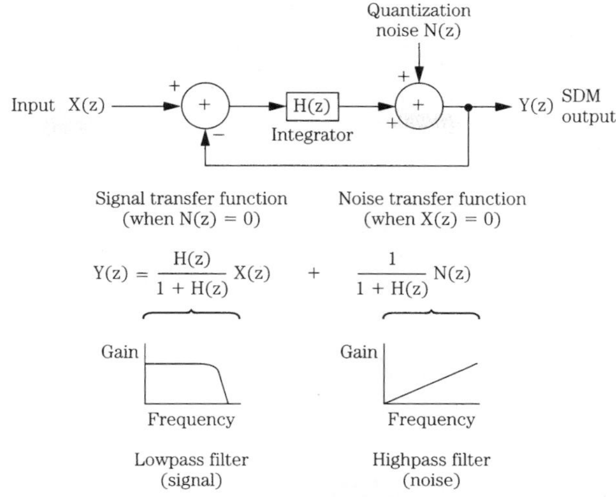

FIGURE 18.3 A *z*-transform analysis of a sigma-delta modulator. (*Hauser, 1991*)

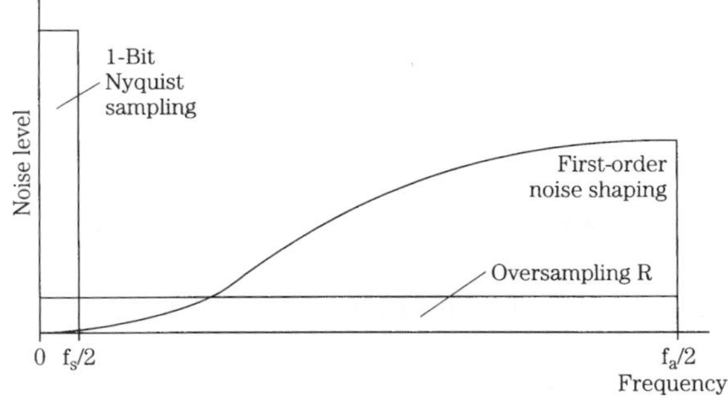

FIGURE 18.4 With one-bit conversion, quantization noise is quite high. In-band noise is reduced with oversampling. With noise shaping, quantization noise is shifted away from the audio band, further reducing in-band noise.

For example, an A/D converter with sigma-delta first-order noise shaping provides a benefit of 1.5 bits/octave compared to an oversampling converter without a noise-shaping loop.

The performance of a sigma-delta converter relies on both oversampling, and its noise-shaping characteristic. The quantization noise floor falls from 0 to $f_s/2$ Hz and is quite high, as shown in Fig. 18.4. Quantization noise is reduced with an R oversampling ratio because noise is spread over a 0-Hz to $f_a/2$-Hz spectrum where $f_a = R \times f_s$ Hz. With sigma-delta noise shaping, in-band noise (0 to $f_s/2$) is further decreased and out-of-band noise is increased.

Figure 18.5 summarizes the mathematical basis of first-order sigma-delta noise shaping. Quantization noise is assumed to be random, and the quantizer is modeled as an additive noise source. Note that the $(1 - z^{-1})$ factor doubles quantized noise power; however, the same factor also shifts the noise to higher frequencies. This sigma-delta modulator forms the basis for many A/D and D/A one-bit and multi-bit converters.

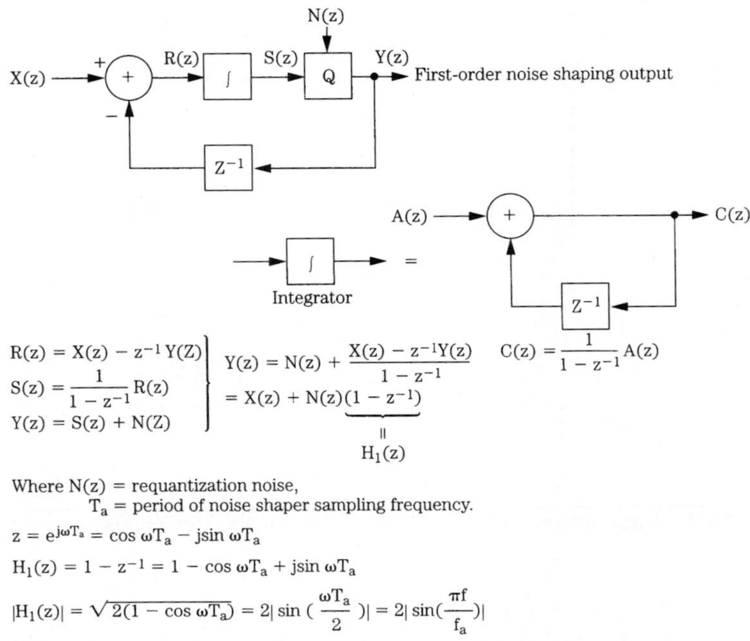

FIGURE 18.5 Analysis of a first-order sigma-delta noise shaper.

Higher-Order Noise Shaping

As noted, first-order (single integration) sigma-delta modulation is not satisfactory for high-fidelity audio performance. Higher-order loops further decrease in-band quantization noise, with the penalty of increased total noise power. For example, a second-order loop would yield:

$$S/E = 6.02(N + 2.5L) - 11.14 \text{ dB}$$

This provides a benefit of 2.5 bits/octave, with a fixed-noise penalty approximately equal to two equivalent bits.

The input/output characteristic of a basic sigma-delta noise shaper of nth order is:

$$Y(z) = X(z) + (1 - z^{-1})^n N(z)$$

where $Y(z)$ = the noise-shaped output
$X(z)$ = the input signal
n = the order of the differentiation
$N(z)$ = the quantization noise (assumed to be white)

This characteristic can be theoretically composed of n cascaded digital differentiators. As n increases, the slope in frequency of the shaping function increases, thus it is more effective in suppressing low-frequency noise. However, the out-of-band noise could overly burden subsequent analog filters. A successful noise-shaping circuit thus seeks to balance a high oversampling rate with noise-shaping order to reduce in-band noise and shift it away from the audible range. Higher-order noise-shaping loops can

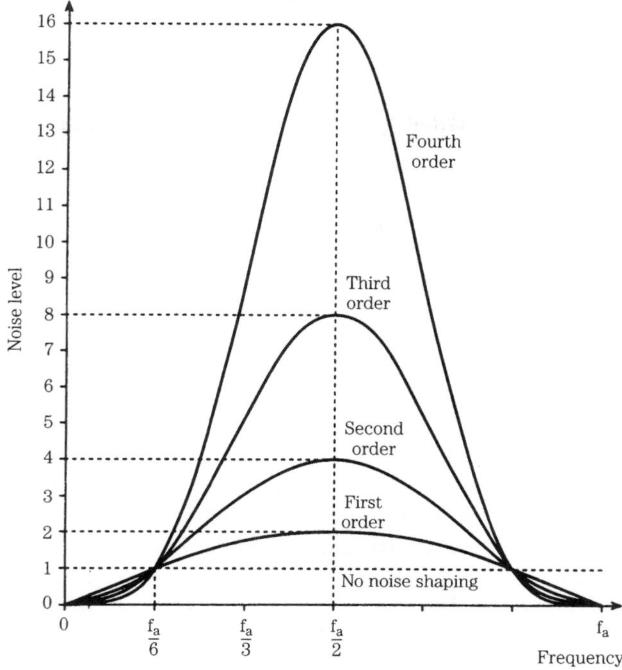

FIGURE 18.6 Higher orders of noise shaping result in more pronounced shifts in requantization noise.

remove even more in-band noise overall, but relatively more noise is present near the Nyquist frequency. Hence these algorithms are more effective at high oversampling rates so there is more spectral space between the highest audio frequency and the Nyquist frequency; this allows the use of very simple analog lowpass filters.

Using first-order and higher-order noise-shaping algorithms, a series of noise-shaping curves can be generated, as shown in Fig. 18.6. As higher orders of noise shaping are used, the in-band noise level is decreased. The frequency response curves described by sigma-delta noise-shaping equal a unity value at $f_a/6$ Hz where f_a is the noise-shaping oversampling frequency. Noise is reduced only for $f < f_a/6$ Hz, and increased for $f_a/6 < f < f_a/2$ Hz. The noise level reaches a maximum at $f_a/2$ Hz. As the oversampling rate is increased, the portion of the noise curve in the audio band is relatively reduced. Although the shape of the noise curve remains the same, high oversampling rates relatively decrease in-band noise.

In a traditional noise-shaper design, the poles of the loop filter are at 0 Hz, as in an ideal integrator; this results in zeros in the audio band. In some noise-shaper designs, a technique called zero-shifting is used to modify the rising noise spectrum by shifting one or more zeros to the edge of the audio band (for example, 18 kHz). For example, when two zeros are shifted in a third-order noise filter, noise in the range from 13 kHz to 20 kHz can be reduced, but increased below 13 kHz. Overall, the noise measurement is enhanced. However, suppression of idle patterns and thresholding effects can be diminished; thus, the zero-shifting technique must be used with care.

Idle Tones and Limit Cycles

The low-level linearity of low-order (particularly first-order) sigma-delta noise-shaping algorithms can be degraded by idle tones. The quantization noise is not always random, and instead can be correlated to the input signal. This is an idle tone, a high-frequency oscillation of the modulator output. Given a zero input signal, a noise shaper can output an alternating 1010 pattern. A very low-level input might result in a similar 1010 pattern, but disturbed by double 1s and 0s. If the period of the repetition of such patterns is long enough, they yield energy that might fall in the audio baseband, being audible as a deterministic or oscillatory tone, rather than as noise. Because they occur when the channel is idling, these nonlinear patterns are called idle tones, or idling patterns, and result in idle channel noise. The double codes will be generated, or not, depending on the duration of the input signal. The phenomenon is especially characteristic of low-amplitude, high-frequency sine waves. Because the phenomenon has a frequency-dependent threshold level, below which the signal is not coded, the effect is sometimes called thresholding.

First-order sigma-delta noise shapers particularly exhibit these effects because of their stable 1010 patterns. Higher-order noise shapers are much less prone to the problem because their output patterns are less stable. Thus, for example, virtually all sigma-delta converters use at least a second-order modulator loop. However, in many multistage designs, the effect can occur in each of the cascaded low-order stages. Thus, it is important to add a dither signal in the first stage to disturb any fixed patterns and remove the correlation. Dither is applied most effectively in multi-bit converters, and less effectively in true one-bit converters. Generally, multi-bit converters are completely linearized with triangular pdf dither, which eliminates noise modulation but slightly increases the level of the noise floor. It is felt that true one-bit converters cannot be fully dithered because this would overload the modulator. In true one-bit converters, dither has both advantages and disadvantages and the optimum dithering technique has not yet been found; other methods are sometimes used to linearize the converter.

A limit cycle is a repeating output sequence. It will yield spurious spectral lines and potentially audible distortion on the output. Limit cycles may occur even in simple noise shapers for some input signals such as a dc input. Their occurrence also depends on the initial state of the noise-shaper filter. Limit cycles can generally be avoided by adding sufficient dither. For example, high-order noise shapers can use small dither levels, such as a rectangular pdf dither with a peak-to-peak amplitude of 0.01 LSB.

Higher-order sigma-delta modulation loops offer wider dynamic range and overall better performance, but loops greater than second-order can be unstable. The quantizer is nonlinear because its effective "gain" inversely varies with the input level. A large input signal can overload the loop, reducing the gain of the quantizer, thus causing instability that will persist even after the signal is withdrawn. For example, a powering-up transient could cause a converter to oscillate. Converters thus sense instability by counting the number of consecutive ones or zeroes in the bitstream, and reset when necessary.

One-Bit D/A Conversion with Second-Order Noise Shaping

A sigma-delta D/A converter comprises a digital interpolation filter, sigma-delta modulator, one-bit or multi-bit converter, and a low-order lowpass output filter. The input digital word, with long word length, is applied to the interpolation filter that lowpass filters the signal and increases its sampling rate with oversampling. The digital modulator noise-shapes the signal and reduces the word length to one or a few bits. For example, its transfer function can be implemented with an IIR filter. The signal is effectively lowpass-filtered and the quantization noise is highpass-filtered. The analog output filter removes the high-frequency shaped quantization noise as well as high-frequency images.

One implementation of a true one-bit D/A conversion method comprises an oversampling filter, second-order sigma-delta noise shaping, and pulse-density modulation (PDM) output. The sampling frequency is increased from 44.1 kHz to 11.2896 MHz, an increase of 256 times. At the same time, the 16-bit signal is converted to a one-bit signal that reconstructs the audio waveform. The requantization error of the output signal is corrected by feedback. Instead of outputting a signal with conventional quantization error, the error undergoes sigma-delta processing to attenuate its in-band level.

The output bit, operating at a frequency of 11.2896 MHz, is converted to an analog signal using a simple switched capacitor network. Specifically, a capacitor is charged and discharged according to the 1 or 0 value of the data. The result is an analog waveform that reflects the encoded waveform through time-averaging of the output bit. The network's operation is accurate, and hence the error of the signal is low. There are only positive and negative full-scale reference points. Errors in the reference values will generate a gain offset error, but not a linearity error. The offset error can easily be removed. In practice, nonlinearities could result from idle patterns in the noise-shaping circuitry.

Figure 18.7 represents the operation of a one-bit PDM converter. It performs noise shaping through feedback loops and generates a one-bit signal for conversion to analog. The noise shaper consists of two integration (filter) loops to reduce in-band requantization noise. The output (H) of the quantizer is +1 if its input is positive (MSB = 0), and −1 if its input is negative (MSB = 1); the one-bit code output from the quantizer is simply a sign bit. Following a limiting operation designed to prevent overflow, the remainder of each sample is fed back as a quantization error. The error signal (I) is fed back into the double integration loops. Values inside the loop exceed the unit value; in other words, wider data buses are required. In this case, a 21-bit data bus is used within the loop; signals are processed in two's complement form.

Also, if large values are input to the circuit, the limiter would be needed to prevent overloading of the loops. Ideally, with no input signal, the coder should output only a tone at $R \times f_s/2$ Hz where R is the oversampling rate. However, idle tones can also occur at additional frequencies. To overcome this, dither can be added to the input data so the circuit always operates with a changing signal even when the audio signal is zero or dc.

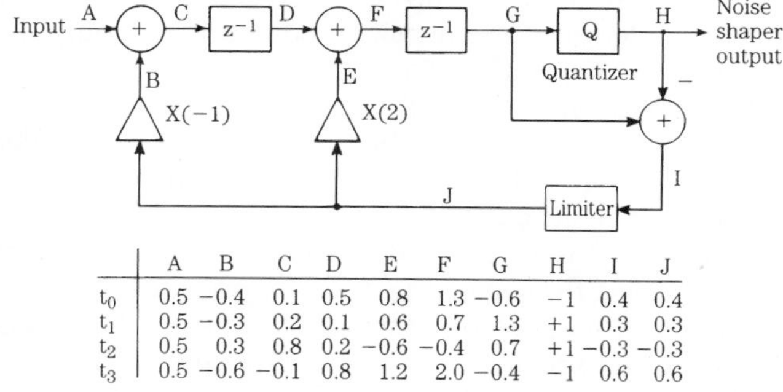

	A	B	C	D	E	F	G	H	I	J
t_0	0.5	−0.4	0.1	0.5	0.8	1.3	−0.6	−1	0.4	0.4
t_1	0.5	−0.3	0.2	0.1	0.6	0.7	1.3	+1	0.3	0.3
t_2	0.5	0.3	0.8	0.2	−0.6	−0.4	0.7	+1	−0.3	−0.3
t_3	0.5	−0.6	−0.1	0.8	1.2	2.0	−0.4	−1	0.6	0.6

FIGURE 18.7 Operation of a second-order noise-shaping circuit.

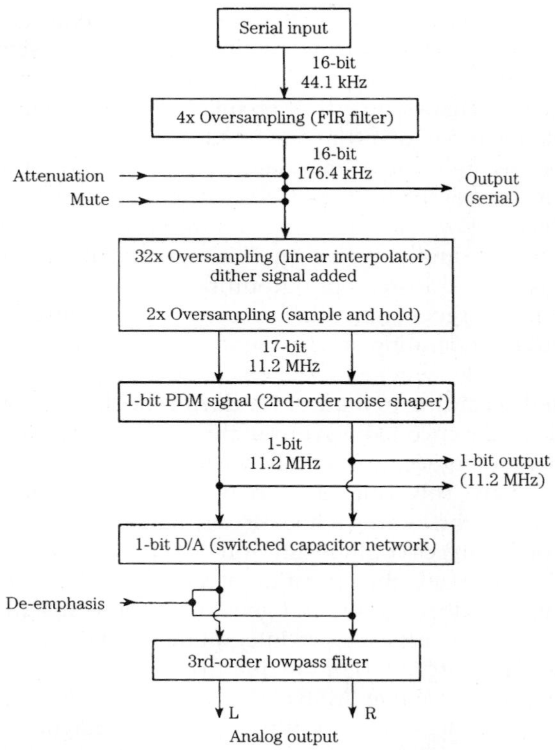

Figure 18.8 Processing elements in a second-order, pulse-density modulation D/A conversion system.

Figure 18.8 shows the complete system, including the one-bit PDM noise-shaper modeled above. The first of the three oversampling stages performs four-times oversampling to attenuate image spectra; in addition, first-order noise shaping is performed in the filter. The second stage performs 32-times oversampling. A dither signal (–20 dB at 352 kHz) is added to prevent idle tones from causing nonlinearity. Two-times oversampling is performed in the third stage. This 17-bit signal (dither adds one bit to the original 16-bit signal) undergoes second-order noise shaping as described above, and a single bit is output from the quantizer. Finally, D/A conversion is accomplished at a one-bit D/A converter via pulse-density modulation that outputs two-valued (±) data at 256-times oversampling, or 11.2896 MHz. A third-order analog lowpass filter removes out-of-band high-frequency components.

The output signal conveys the audio waveform through the density of pulses above and below zero (see Fig. 18.1); this pulse-density modulation signal is converted to an analog signal using a switched dual-capacitor network. Two control signals representing the data stream's logic 0 and logic 1 values control the switching of the capacitors, subject to a clock pulse. During the negative half of the clock, the first capacitor discharges while the second capacitor charges. During the positive half, if the data is 1, the first capacitor is charged by taking a fixed amount of charge from the summing node of an operational amplifier. If the data is 0, a fixed charge is transferred into the summing node from the second capacitor. In this way, there are only positive and negative full-scale reference points, and intermediate points are determined by time averaging. There is no MSB change around zero, for example, because zero is represented by an equal number of positive and negative full-scale pulses. Zero-cross distortion is thus eliminated.

In this one-bit converter, the quantization noise introduced by the word length reduction is spectrally shaped by a lowpass feedback loop around the quantizer (see Fig. 18.7). Second-order noise shaping is performed:

$$Y(z) = X(z) + (1 - z^{-1})^2 N(z)$$

where $Y(z)$ = the noise-shaped output
$X(z)$ = the input signal
$N(z)$ = the quantization noise

The requantization noise of the output signal is corrected by feedback, and noise shaping attenuates its in-band level. As a result, the spectrum of the noise in the output signal is shifted away from the audio band. Figure 18.9 summarizes the mathematical basis of second-order noise shaping.

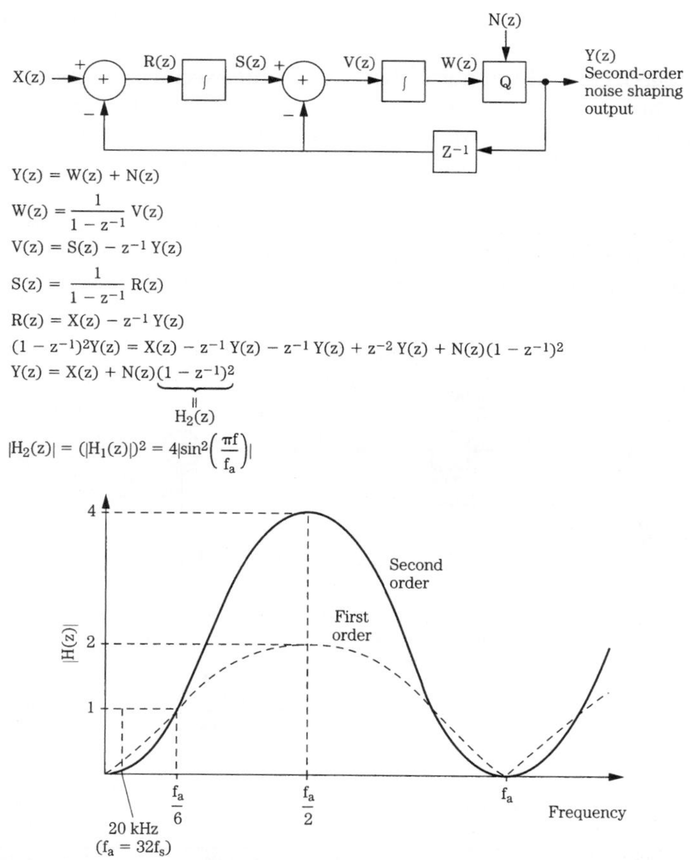

FIGURE 18.9 Analysis of a second-order noise shaper.

Multi-Bit D/A Conversion with Third-Order Noise Shaping

When the noise-shaping circuits in sigma-delta modulators exceed second order, the noise-shaping feedback loops can pose overload or oscillation problems. For example, an overload would effectively reduce the magnitude of loop gain, lowering the crossover frequency, where phase shift is too large for stability. When a loop filter $H(z)$ has three or more integrations, its phase shift can be 180° at the frequency where the loop gain magnitude reaches unity at the crossover frequency.

Higher-order noise shapers overcome this problem with a more complex loop architecture that often uses multistage noise shaping. The loop filter $H(z)$ provides a high-order lowpass response at low baseband frequencies, but gain drops to a first- or second-order lowpass response nearer the crossover frequency. This approach provides good in-band noise shaping, with sufficient conditional stability.

The MASH system is a multistage third-order noise-shaping method. It is an example of a practical multi-bit converter. One implementation of this design accepts 16-bit words at a nominal sampling frequency, and a digital filter performs eight-times oversampling and outputs 24-bit words. Noise-shaping circuits output 11-valued data, at a 32-times oversampling rate. D/A conversion is accomplished via pulse-width modulation (PWM), outputting the data at a 768-times oversampling rate.

As noted, generally, if noise-shaping circuits exceed second-order they can be prone to oscillation from instability. To avoid such errors, this third-order implementation uses a multistage configuration. A simplified schematic for the complete noise shaper is shown in Fig. 18.10; it contains a first-order noise shaper in parallel with a second-order noise shaper. The input signal is applied to quantizer Q_1 after the error signal through the delay block is subtracted from the input. The signal output from the first loop is also applied to the second loop. The output of quantizer Q_2 is differentiated and added to the output of the first loop to form the final output signal. Thus, the requantization error

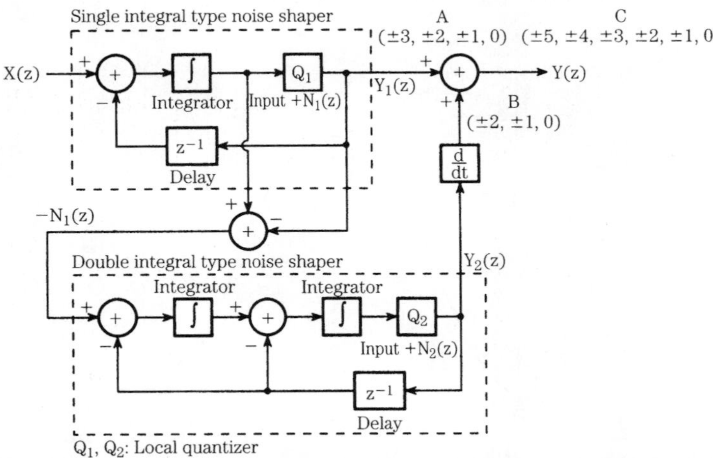

FIGURE 18.10 A multistage third-order noise-shaping circuit with output of 11 data values before PWM reconstruction.

of the first loop is requantized by the second, and canceled by adding the requantized noise to the first loop's signal. The outputs of each stage can be characterized as:

$$Y_1(z) = X(z) + (1 - z^{-1})N_1(z)$$
$$Y_2(z) = -N_1(z) + (1 - z^{-1})^2 N_2(z)$$

where $Y_1(z)$ and $Y_2(z)$ = the outputs of stages 1 and 2, respectively
$X(z)$ = the input signal
$N_1(z)$ and $N_2(z)$ = quantization noise of the local quantizers Q_1 and Q_2, respectively

When both sides of the second equation are multiplied by $(1 - z^{-1})$ and this is added to the first equation, observe that the quantization error of the first stage can be canceled. By passing the output of the second stage through a differentiator and adding it to the output of the first stage, the overall circuit output is:

$$Y(z) = X(z) + (1 - z^{-1})^3 N_2(z)$$

In other words, the quantization error $N_2(z)$ is output with a third-order differential characteristic (18 dB/octave), achieving reduced in-band noise compared to first- and second-order characteristics.

Input data is linearly requantized into a multi-bit output digital signal with seven values (±3, ±2, ±1, 0) at the main loop, and at the sub-loop the requantization error is requantized into five values (±2, ±1, 0). When these output values are added together, the digital signal output from the circuit represents 11 values (±5, ±4, ±3, ±2, ±1, 0). These different data values are shown graphically in Fig. 18.11. Using a vertical scale to represent amplitude of the values, it can be seen that the main loop outputs seven

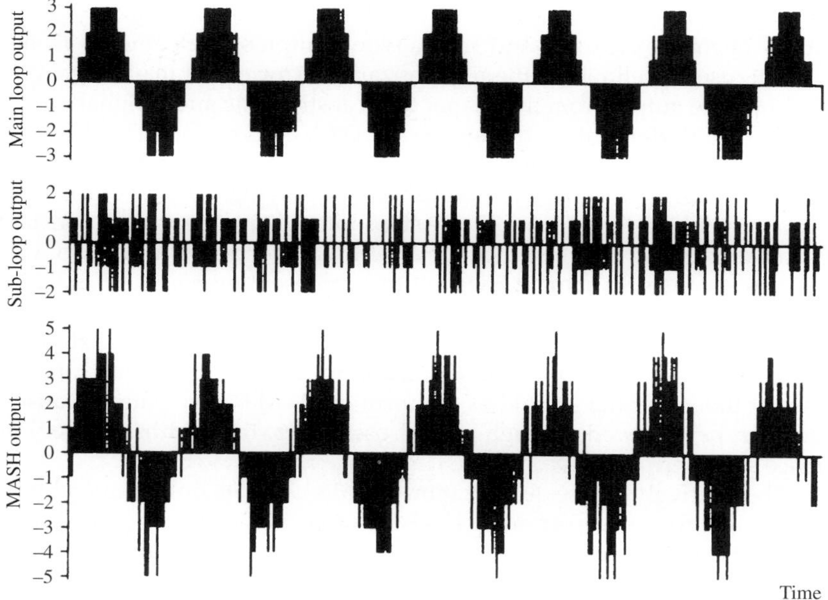

FIGURE 18.11 A graphical representation of the data values in a multistage noise shaper.

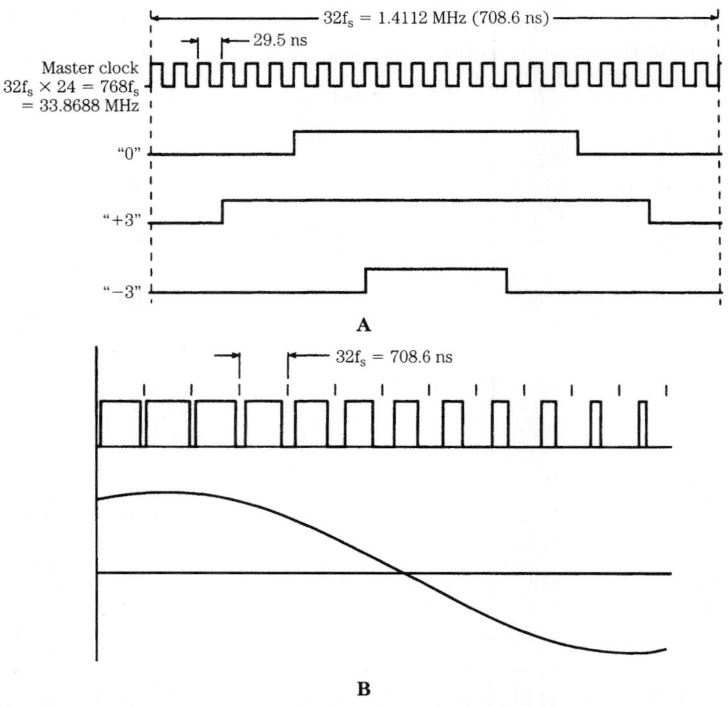

FIGURE 18.12 Pulse-width modulation data is output from a MASH converter. A. Examples of pulse-width modulation data. B. Reconstruction of the analog waveform.

values of rough accuracy, and the sub-loop outputs five values with high-frequency content, used to eliminate the requantization error of the main loop. When summed, 11 values are output from the shaper to reconstruct the audio signal.

The final element in the system is D/A conversion. The 11-valued signal is converted into pulses, each with a width corresponding to one value, as shown in Fig. 18.12A. This can be accomplished by applying the 4-bit output of the shaper to a lookup table to map 11 amplitude steps into 22 time steps with constant amplitude. For example, the figure shows the PWM waveforms resulting from the 0, +3, and −3 output values. In actuality, waveforms representing ±5, ±4, ±3, ±2, ±1, and 0 are all output. The widest pulses translate into a large positive output, and the narrowest pulses translate into a large negative output, as shown in Fig. 18.12B. The width of the pulses carries the vital information; the amplitude of this signal can only be high or low. At this point the signal has the form of PWM binary data. Because timing accuracy can be achieved through crystal oscillators, the widths are very accurate, and hence signal error is low.

The relatively coarse quantization permits accurate pulse timing by synchronizing pulse edges to the oversampling clock. Positive- and negative-going pulses are output, to cancel common noise. This 33.8688-MHz ($768 \times f_s$) data forms a PWM representation of the waveform; Fig. 18.13 shows the spectrum of a 20-kHz input signal. Proof of performance can be evaluated by measuring the in-band noise of the system; it is below −100 dB. Figure 18.14 summarizes the mathematical basis of third-order noise shaping.

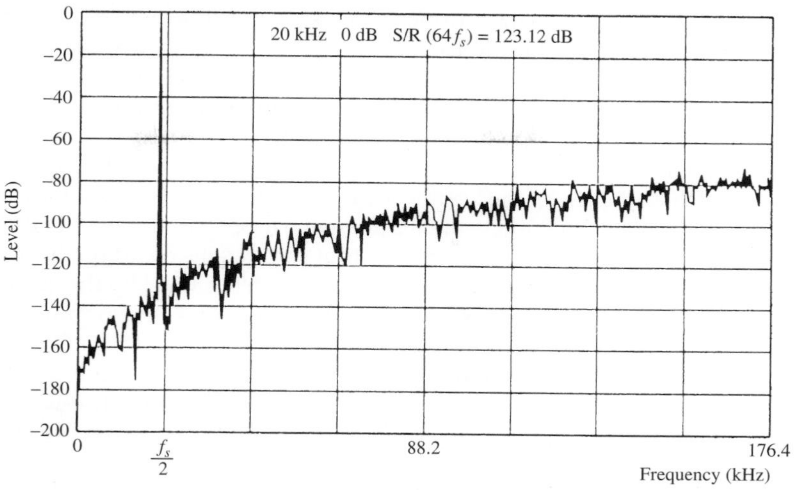

FIGURE 18.13 Reproduction of a 20-kHz waveform showing the effect of third-order noise shaping.

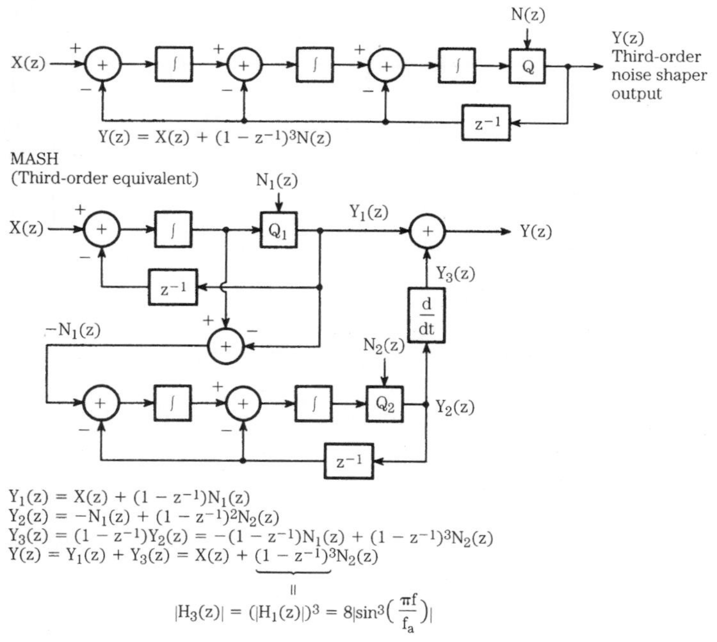

FIGURE 18.14 Analysis of a third-order noise shaper.

Multi-Bit D/A Conversion with Quasi Fourth-Order Noise Shaping

Victor Advanced Noise Shaping (VANS) is an example of a multi-bit D/A converter architecture using eight-times oversampling, a quasi fourth-order noise shaper, and pulse-edge modulation conversion. The noise shaper uses four loop filters in a configuration that yields in-band performance equivalent to fourth-order noise shaping.

The VANS circuit is designed to operate like a fourth-order noise shaper at audible frequencies, gradually shifting toward second-order noise shaping at higher frequencies. This provides stability, yet improves performance in the audio band.

Thirty-two times oversampling is performed, and the output clock frequency is 16.9344 MHz ($384 \times f_s$). With pulse-edge modulation, input data is converted into a binary pulse train with 15 discrete values (±7, ±6, ±5, ±4, ±3, ±2, ±1, 0). A differential configuration is used in which the rise of the leading edge of a pulse, and the fall of the trailing edge of a pulse, are output by two independent pulse-edge modulation converters. This determines the width of the pulse. Two converters output pulse trains based on the input signal, and an analog subtractor generates a composite signal determined by the leading and trailing edges of the pulse trains. This signal can take either a positive or a negative value. For example, data representing a –1 value would generate a short negative-going pulse, but data representing a +5 value would generate a longer positive-going pulse. When time averaged, these values create the analog waveform, as in pulse-width modulation. Figure 18.15 summarizes the mathematical basis of this quasi fourth-order noise shaping.

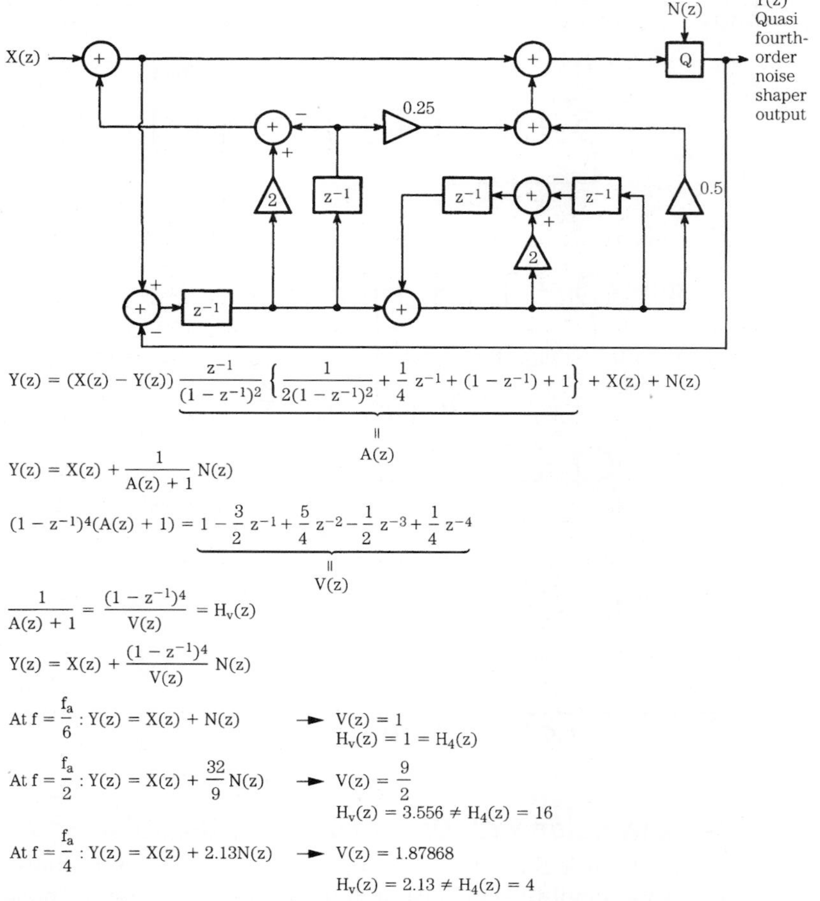

FIGURE 18.15 Analysis of a quasi fourth-order noise shaper.

Generally, even higher orders of noise shaping can be successfully employed, yielding very low in-band noise floors. For example, a fifth-order sigma-delta modulator is used in some Direct Stream Digital (DSD) encoders used in the Super Audio CD (SACD) format.

Sigma-Delta A/D Conversion

Traditional successive approximation A/D converters compare the unknown input with accurately known fractions of a reference voltage. Starting with the largest fraction and rejecting any fraction that causes the sum to be larger than the unknown input, k iterations are required for a k-bit word conversion. The input oversampling rates (and conversely, the order of the input filters) are limited by the relatively low speed at which these A/D converters can operate. Hence, analog brick-wall filters are used. Such A/D converters, either directly or through associated circuitry such as brick-wall filters, can contribute substantial distortion to the signal.

One way to improve the linearity of conversion is to increase word length. Longer word-length ladder A/D converters were introduced, and these converters improve performance, but resolution is generally constrained to 18 or 20 bits. Thus oversampling A/D converters, using sigma-delta architectures, were introduced to remedy the ills of traditional A/D converters and also provide lower cost. First- and second-order A/D converters provide limited quality, and idle tones can produce audible tones in the noise floor. Attention turned to higher-order (fifth- and sixth-order) A/D converters that have reduced idle tones, and reduced sensitivity to clock jitter. Care must be taken to prevent oscillation from modulation overload.

In theory, oversampling A/D conversion is simple: the input signal is first passed through a low-order analog anti-aliasing filter, and then sampled at a very high rate to extend the Nyquist frequency. After quantization, the signal passes through a digital filter to prevent aliasing and reduce the sampling frequency to a standard frequency (such as 48 kHz) for storage or processing using normal methods.

In practice, other factors play a role. Only coarse quantization is possible at the highly oversampled rate; this results in a high noise floor. Although noise is spread over a large oversampled spectrum, it is unsatisfactorily high. Noise shaping must be used to reduce in-band noise. In addition, a conventional digital filter with satisfactory passband response and stopband attenuation cannot operate at this highly oversampled rate. Rather, a digital decimation filter, operating as a lowpass filter, is used; its computation requirements are far easier. When a sigma-delta quantizer is used, in conjunction with noise shaping, the decimation filter must remove out-of-band quantization noise; this effectively increases the resolution of the digital output.

An analog lowpass filter is required at the converter's input to remove the frequency components that cannot be removed by the digital filter. However, because the preliminary sampling rate is high, the analog lowpass filter is low order. The filter must remove any frequency components outside the audio band to prevent aliasing at the resulting lower sampling rate. This would occur when the output of the digital filter is resampled (undersampled) at the lower downstream sampling rate.

Oversampling A/D converters are unusual in that the basic A/D elements of anti-alias filtering, sampling, and quantization are merged throughout the subsections of the converter. For example, anti-alias filtering occurs in both the input analog filter, and in the digital decimation filter. Although traditional A/D converters only perform quantization, oversampling A/D converters are complete signal acquisition interfaces.

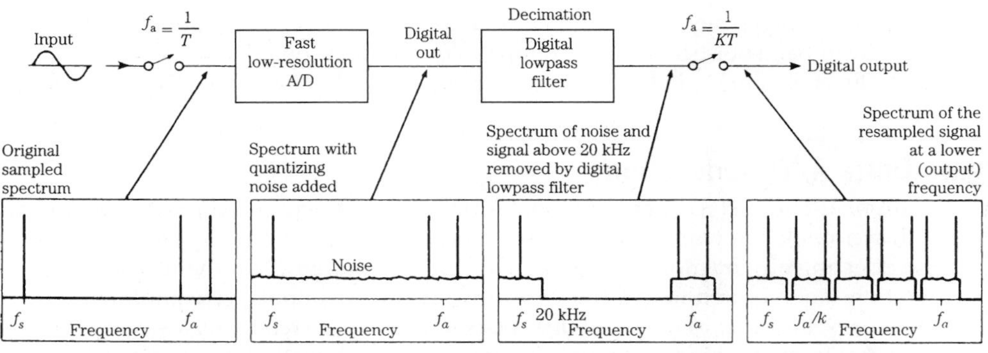

FIGURE 18.16 Diagram showing the theory of oversampling A/D conversion. (*Adams, 1986*)

A diagram illustrating oversampling A/D conversion is shown in Fig. 18.16. The input signal is first passed through an analog anti-aliasing filter, and the input signal is sampled at a very fast rate (for example, $f_a = 64 \times f_s$) to extend the Nyquist frequency. The signal is applied to a coarse quantizer such as a sigma-delta converter, which adds (shaped) noise to the signal. The digital data is lowpass-filtered with a cutoff at the Nyquist frequency; this removes out-of-band noise components to prevent anti-aliasing. Finally, the signal is resampled at a lower rate (such as 48 kHz) for storage or processing using normal methods. A decimation lowpass filter bandlimits the wideband signal (to 20 kHz in Fig. 18.16) so that aliasing will not occur when the signal is subsampled at the lower output frequency. A sample-and-hold circuit is not needed because an input sample can be taken during every internal clock cycle. In successive approximation converters, the sampled analog value must be held for the number of clock cycles equal to the number of bits being converted.

Sigma-Delta A/D Modulator

A sigma-delta modulator can be used to create true one-bit coding from the lowpass-filtered input analog signal. A first-order sigma-delta A/D modulator is shown in Fig. 18.17.

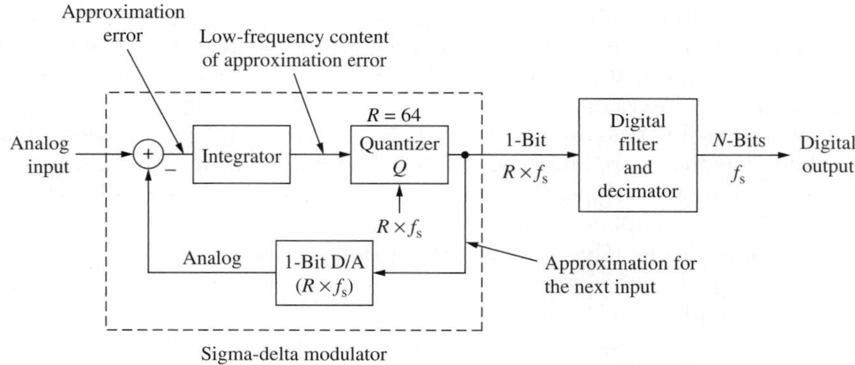

FIGURE 18.17 A first-order sigma-delta modulation circuit showing a one-bit D/A converter in the feedback loop.

In this converter design, the sigma-delta modulator is followed by a digital filter and decimation stage. Because the input sampling rate is high, a simple one-pole RC anti-alias filter suffices. The modulator accepts a sampled analog signal, performs quantizing, and outputs a one-bit signal at a rate determined by the sampling clock. A low-resolution (one-bit quantizer) D/A converter operating at a high sampling rate is placed in a feedback loop. The input to the loop filter is the difference between the input signal and the quantized output converted back to an analog signal; this difference is theoretically equal to the quantization error. The average value of the D/A output (and the modulator output) must approach that of the input signal.

Because a coarse (one-bit A/D) quantizer is used, quantization error at sampling time is large. The coarse output signal is subsequently averaged by the decimation filter, interpolating over several samples (64 or so) to achieve a precise result. High resolution (manifested as dynamic range) is achieved through noise shaping. The integrator can be viewed in the frequency domain as an analog loop filter $H(z)$. The noise-shaping characteristic in this sigma-delta modulator is the inverse of the transfer function of the filter. A filter with higher gain at low frequencies is thus desired to attenuate audio band noise. This transfer function is essentially a lowpass filter to the signal and a highpass filter to quantization noise; thus, the noise is shifted to a higher frequency. The higher the oversampling rate and order of noise shaping, the higher the resolution of the converter. For example, with an oversampling rate of 64, an ideal second-order modulator yields a signal-to-noise ratio of about 80 dB, equivalent to a 13-bit A/D resolution.

Instead of a one-bit code, converters might produce a multi-bit word of three or four bits using, for example, a sigma-delta modulator modified to contain a multi-bit quantizer. Several quantizer output bits are applied to an internal D/A converter and its analog output is subtracted from the analog input signal, thus producing a quantization error signal. This error signal is applied to the loop filter, and quantized to minimize error and thus yield an output that approximates the input. In this architecture, the dynamic range increases in proportion to the resolution of the quantizer; however, this must be balanced against operating speed. Dynamic range can also be increased by using higher-order filters. Because the output is proportional to the signal's amplitude rather than slope, it is like a PCM converter. Unlike a traditional PCM converter, the noise floor rises with increasing frequency, at 6 dB/octave. Alternatively, a differential pulse code modulator differs from a delta modulator only in that the error signal is quantized to more than one bit. However, such an architecture is still slew-rate limited.

Numerous sigma-delta methods have been applied to A/D conversion, all using a high input sampling rate, and noise shaping. These methods include: single and dual integrator loops, cascaded first-order sigma-delta loops, and multi-bit quantizers with loop filters. The first two methods use true one-bit coders with inherent linearity. The third method uses several bits, and noise is reduced in proportion to the number of quantizer levels used. However, the converter's linearity depends on the linearity of the quantizer. In any case, noise performance hinges on the oversampling rate and order of noise shaping used. Some converter architectures use several first- or second-order sigma-delta coders in combination to achieve higher order, stable noise shaping.

Given a second-order sigma-delta modulator, Charles Thompson has demonstrated that M-bit resolution requires an oversampling rate:

$$R = \left[\left(\frac{\pi^2}{5^{1/2}}\right) \times 10^{6M/20}\right]^{2/5}$$

where R is the oversampling rate defined by:

$$R = f_a / f_s$$

where f_a = the oversampling frequency
f_s = the output sampling frequency

Thus, 16-bit resolution would require an oversampling rate of 150. A 100-kHz output sampling frequency would necessitate a filter sampling frequency of 15 MHz; this is difficult to achieve. If the order of noise shaping is raised to third-order, the required oversampling rate is described by:

$$R = \left[\left(\frac{\pi^3}{7^{1/2}} \right) \times 10^{6M/20} \right]^{2/7}$$

Thus, the required oversampling ratio is 48; this is well within practical design limits.

Depending on its order and design, a sigma-delta feedback loop generally performs the following operations: subtraction of output from input to find the approximation error, filtering to extract the low-frequency content of the approximation error, sigma-delta D/A conversion of the output code into a signal to subtract it from the input analog signal, and quantization to output an approximation for the next input sample. In practice, a third-order loop can be used to shape the noise toward higher frequencies, where it is removed by the subsequent decimation (undersampling) filter. As with any noise-shaping loop, the signal must be properly dithered to overcome idle tones and other artifacts. In some cases, a dither signal can be applied so that its fundamental and harmonics can be removed by the decimation filter.

Digital Filtering and Decimation

As Robert Adams has pointed out, oversampling converters provide high resolution not by decreasing the error between the analog input and the digital output, but by making the error occur more often. In this way the error spectrum moves beyond the audio passband and although the total noise power is high, the in-band noise power is low. The high bit rate is reduced to more manageable rates through decimation in which a discrete time signal is sampled at a rate lower than the original rate. Decimation provides both an averaging (lowpass) filter and rate reduction. It removes the high-frequency shaped noise, and provides an anti-aliasing function for the final sampling rate. Looked at in another way, decimation removes the redundant information created by oversampling.

Decimation can be described through a simple example. Sixteen one-bit values could be reduced through a 16:1 decimation to a single multi-bit value; for example, values 1,0,1,0,0,1,0,1,1,0,1,1,1,1,0,0 would be decimated to 9/16, or 0.5625. Because there is only one (multi-bit) output value for every 16 input values, the decimator has decreased the sampling rate by 16:1. As Sangil Park has shown, it is also important to note that decimation has increased resolution; in this example, the input signal is only one bit, but the decimation (averaging) process yields 4-bit resolution ($2^4 = 16$) while reducing the sampling rate. Thus, oversampling followed by decimation demonstrates how speed can be exchanged for resolution. The meaning of the word decimation, incidentally, originally referred to a form of harsh discipline administered by the Roman army to punish cowardice. Soldiers selected for decimation were placed in groups of

10 and drew lots. The soldier on whom the lot fell was executed by his nine comrades, usually by clubbing or stoning.

The decimation process lowpass filters the signal and noise in the one-bit code, band-limiting the code prior to sample-rate reduction to remove alias components. Decimation also replaces the one-bit coding with 16-bit coding, for example, at a lower sampling rate. However, the computation rate of the filter is not trivial; output samples cannot be discarded (providing decimation) until the filtering computation is complete.

Ideally, the decimation filter would provide a sharp lowpass cutoff at half the output sampling frequency, thus upholding the Nyquist sampling theorem. However, as Robert Adams has shown, this is not always efficient. For example, an FIR filter would require many coefficients because of the high ratio of the input sampling rate to the output sampling rate. Still, when an FIR filter is used, filter outputs are only computed at the lower output sampling rate. An FIR filter is well-suited for decimation. If an IIR filter is used, the feedback loop dictates that an output value must be computed for every input. The decimation function cannot be combined as part of an IIR filter. A practical approach uses two or more stages of decimation, operating at intermediate sampling frequencies. For example, the first stage might use an FIR filter for decimation and a second stage might use an IIR filter for digital filtering. Alternatively, two-stage FIR filters, or two-stage IIR filters can be used, with both stages performing some decimation.

If the first stage resamples at an intermediate frequency f_i it would appear that all frequencies above $f_i/2$ must be rejected to prevent subsequent aliasing. However, only certain portions of the spectrum will alias in the audio band, thus the decimation filter need only attenuate those frequency bands. In particular, these alias bands can be identified:

$$f_{\text{alias}} = I \times f_i \pm BW \text{ Hz}$$

where I = any integer
f_i = the decimation filter's intermediate resampling frequency
BW = the audio bandwidth (for example, 20 kHz)

For example, if $f_i = 96$ kHz, the bands of interest will lie at 96, 2×96, 3×96 kHz, and so on, each occupying a width of 40 kHz. The decimation filter can be designed so that its frequencies of maximum attenuation will coincide with these potentially aliasing frequency bands. A filter with pockets of attenuation, rather than attenuation across the entire stopband, is much easier to implement. As the sampling rate is decreased from one stage to the next, the pockets become proportionally wider and filter complexity increases, but intense computation is performed at the slower rate. In this way, each filter must only reject the signals that would be aliased by the immediate next decimation. Subsequent filters will reject signals that would alias with later decimation. A comb filter is an expedient choice because its design does not require a multiplier (all coefficients are unity).

However, as Sangil Park points out, comb filters cannot wholly remove out-of-band quantization noise so they are followed by additional filter stages of other design. These additional stages can also be needed to compensate for high-frequency drooping caused by the comb filter. A final filter, operating at the slowest sampling rate, could provide a true lowpass characteristic, and correct any frequency-response deviations. A comb filter of length R is an FIR filter with coefficients equal to unity; its transfer function is:

$$H(z) = \sum_{n=0}^{R-1} z^{-n}$$

In other words, this expression shows a moving average. For example, if $R = 4$:

$$y(n) = x(n) + x(n-1) + x(n-2) + x(n-3)$$

In recursive form, the transfer function can be written as:

$$H(z) = \frac{(1 - z^{-R})}{(1 - z^{-1})}$$

This can be expressed in terms of integration followed by differentiation:

$$Y(z) = (1 - z^{-1})^{-1}(1 - z^{-R})X(z)$$

This single-stage comb filter decimator can be easily realized, as shown in Fig. 18.18A. Not only is no storage required for the filter coefficients, but the burden

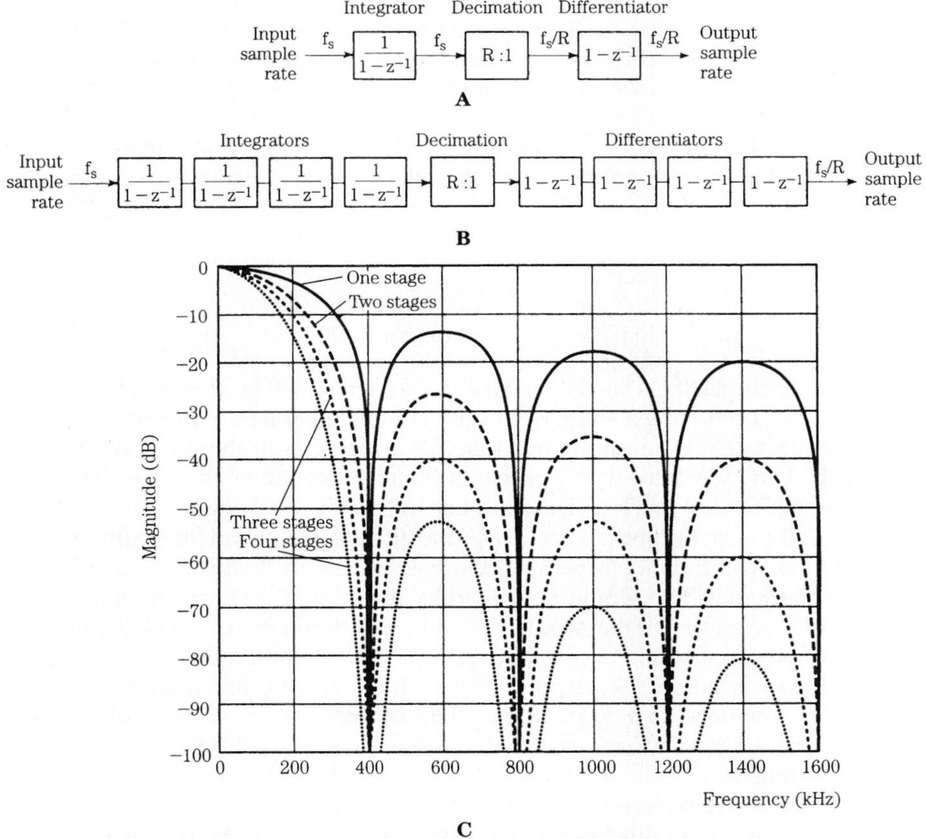

FIGURE 18.18 Comb filters can be used in decimation. A. Block diagram of a one-stage comb filter. B. Block diagram of a cascaded four-stage comb filter. C. Spectrum showing the response of one-, two-, three-, and four-stage cascaded comb filter sections. (*Park, 1990b*)

of intermediate computations is decreased owing to the low sampling rate at the differentiator. In addition, the same topology can be used for higher orders of rate change. As noted, in practice, a single comb filter stage does not provide sufficient stopband attenuation to prevent aliasing, thus cascaded stages are often used, as shown in Fig. 18.18B. In this example, four sections are cascaded, requiring eight data registers and $4(R + 1)$ additions per input sample. As noted, the comb filter is designed for maximum attenuation at higher frequency components that would alias after rate decimation. Figure 18.18C shows the spectrum with one-, two-, three-, and four-stage cascaded comb filter sections.

In some decimator designs, the cascaded comb filter is followed by an FIR filter. The intermediate-rate output from the comb filter is further decimated and the FIR section provides sharp filtering when the sampling frequency is reduced to nominal values (for example, 48 kHz). The decimation factor is typically lower in the FIR section as compared to that in the comb filter section. However, the FIR filter must provide extreme stopband attenuation. In addition, the FIR section can provide compensation for audio band droop caused by the comb filter. FIR computation also provides a linear phase response.

Consider an example in which coding takes place at 64×48 kHz = 3.072 MHz. The decimation filter can have two stages. With a $64 \times f_s$ Hz input bitstream, the first stage can generate a multi-bit output sample at a sampling frequency of $2 \times f_s$ Hz. The second stage of the decimation filter can use a multi-bit multiplier with convolution performed at the output sampling frequency of f_s Hz. In all, the decimation filter provides a stopband from 20 kHz to the half-sampling frequency of 1.536 MHz. The analog filter at the system's input is modest, perhaps first- or second-order, ensuring phase linearity in the audio band.

The use of one-bit coding as the intermediate phase of A/D conversion simplifies the filter design. For example, a new output sample is not required for every input bit. Because the decimation factor is 64 (in this example), an output is required only for every 64 input bits. In practice, the decimation filtering might be carried out in two stages. An FIR filter would commonly be used for downsampling, because its nonrecursive operation would simplify computation to one sample every $1/f_s$ second. Following decimation, the result can be rounded to 16 bits, and output at a 48-kHz sampling frequency. Figure 18.19 summarizes the operation of a sigma-delta A/D converter in the frequency domain.

Digital audio equipment containing A/D (and D/A) converters must have a stable sampling clock that in turn is phase-locked to a distributed master clock. The individual clocks must have very low jitter levels to prevent generated sidebands from rising to audibility. For example, a 16-bit A/D converter might require jitter of less than 20 ps. Jitter is proportionally greater per period for a sigma-delta A/D converter than a ladder converter. Amplitude errors attributable to jitter increase as the input signal frequency increases. However, because the slew rate of the input signal is equal in either type of converter, the amplitude error resulting from sinusoidal jitter is also equal in both cases.

In the case of noise-induced jitter, added noise is distributed over the sigma-delta converter's increased Nyquist frequency range and lowpass-filtered by the decimation circuit. Hence overall in-band jitter-induced noise is less than in some traditional converters. Thus analysis would show that oversampling sigma-delta A/D converters are generally no more sensitive to sinusoidal jitter than a traditional converter and are less susceptible to random noise clock jitter. However, actual performance depends on a converter's specific design. For example, true one-bit converters are generally more susceptible to jitter than multi-bit converters. Timebase correction is discussed in Chap. 4.

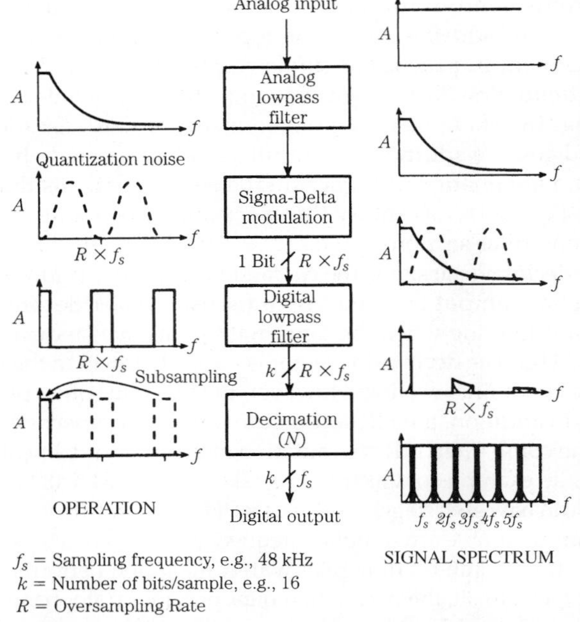

FIGURE 18.19 Summary of spectral characteristics of a one-bit A/D converter.

Sigma-Delta A/D Converter Chip

The block diagram of a sigma-delta A/D converter chip is shown in Fig. 18.20. It is a linear 16-bit converter, using 64-times oversampling, providing output sampling frequencies up to 100 kHz, operating at up to 6.4 MHz. As with other sigma-delta A/D converters,

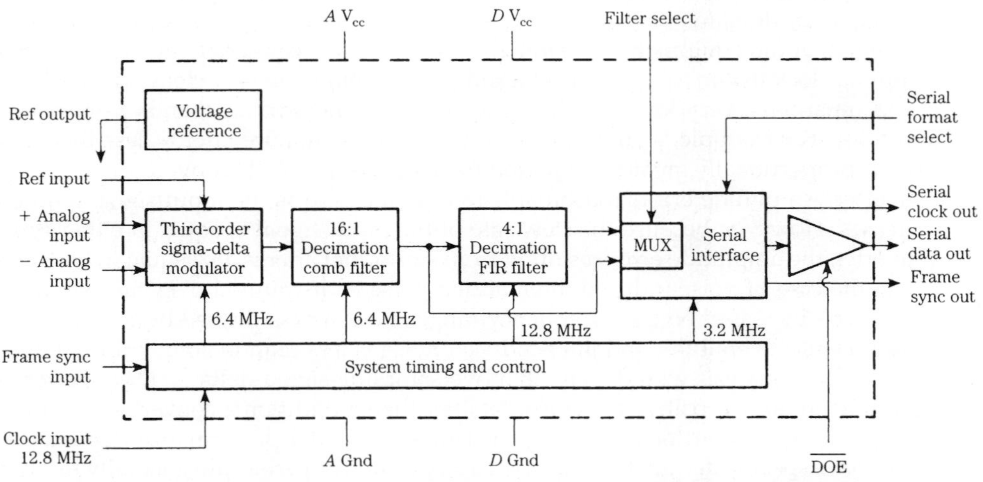

FIGURE 18.20 Internal block diagram of a DSP56ADC16 sigma-delta A/D converter. (*Kloker et al., 1989*)

the input signal is oversampled to extend the noise spectrum well beyond the audio band. Noise shaping reduces noise in the audio band, and lowpass-filtering removes out-of-band quantization noise. Finally, the signal is decimated to reduce the sample rate commensurate with the audio band and to increase resolution.

The converter is designed around four major blocks: third-order sigma-delta modulator and noise shaper, 16:1 decimation comb filter, 4:1 decimation FIR filter, and serial interface. The third-order noise shaper places an 18 dB/octave characteristic on the quantization noise. The analog front end to the converter consists of three differential, switched-capacitor, linear integrators. Filtering and decimation are performed in two steps to reduce the complexity of the digital filter. For example, to achieve the desired stopband attenuation and filter steepness, a single-stage FIR with over 2800 taps would be required. Use of a multirate decimation filter system also allows a dual mode application.

The output of the modulator is filtered by a fourth-order comb filter and decimated; the sampling rate is decreased by a factor of 16:1. A comb filter is used because it contains only adders and delay, without need for multiplication. The first stage comb filter accomplishes initial filtering as well as decimation of the input sampling rate by a factor of 16:1. Its z-domain transfer function can be expressed as:

$$H(z) = \frac{(1 - z^{-16})^4}{(1 - z^{-1})^4}$$

The equivalent frequency domain transfer function is:

$$H(f) = \left[\frac{\sin(16\pi f/f_s)}{16\sin(\pi f/f_s)}\right]^4$$

where f_s = the filter's sampling frequency.

An FIR filter is used to decimate the signal by a 4:1 factor with a lowpass response. Overall, a 64:1 decimation ratio is achieved. In other words, 63 of every 64 output samples are discarded. A stopband attenuation of –96 dB is achieved. To compensate for the response (passband droop) of the fourth-order comb filter, the FIR uses an inverse equalization response to achieve an overall flat response. FIR images occur at multiples of the comb filter output sampling rate; these are also zeros in the fourth-order comb response. The FIR stopband attenuates the comb response, leaving a negligible alias component at the overlap of the two responses. In all, this digital filter section is the equivalent of a 30th-order analog Bessel filter. The output sampling frequency is 100 kHz, with 16-bit resolution and S/N ratio of 90 dB.

Because the cutoff frequencies of the comb and FIR filters are scaled by the input sampling rate, the converter can be used with any arbitrary sampling rate without changing component values. For further flexibility, this A/D converter chip is designed so the 16:1 comb filter can be connected directly to a serial output. This permits operation at faster speed (output sampling frequency of 400 kHz) at the expense of lower resolution (12 bit, and S/N of 72 dB). This is useful for ultrasonic applications and where lower resolution is tolerable. A general application for this chip using its full resolution is shown in Fig. 18.21; the A/D converter is connected to a DSP processor.

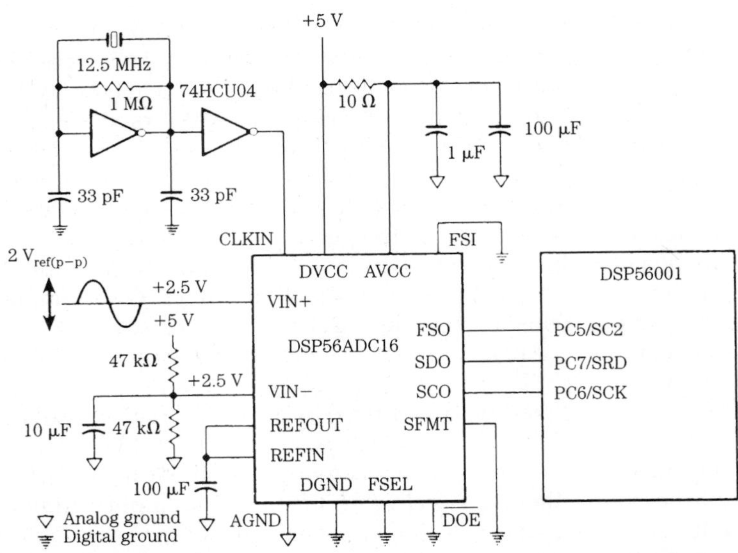

FIGURE 18.21 Application circuit showing an interconnection of a sigma-delta A/D converter (single-ended mode) and DSP processor.

Sigma-Delta D/A Converter Chip

A typical sigma-delta D/A converter comprises a digital interpolation filter, sigma-delta modulator, and switched-capacitor filter. The interpolation filter raises the input sampling frequency to the modulation rate. The modulator reduces the word length to one or a few bits and reduces in-band noise. The switched-capacitor elements filter out-of-band noise and perform signal D/A reconstruction.

One example of a multi-bit sigma-delta D/A converter uses a second-order mismatch shaping function inside the feedback loop of a high-order modulator. This feature moves element mismatch noise to higher frequencies where it is removed along with other sigma-delta noise by lowpass-filtering. This feature is used in lieu of dynamic element matching (DEM) after the modulator. PCM or DSD data at sample rates up to 200 kHz is input via a serial port and passes through an interpolator and volume control, as shown in Fig. 18.22. DSD data is volume-adjusted and upsampled by a factor of 2. Data is applied to a sixth-order sigma-delta modulator with integrated second-order mismatch noise shaping. To ensure stability, a fallback second-order sigma-delta modulator can be used. The mismatch noise shaping is not changed when in the fallback mode. When processing SACD data, the modulator also uses a fifth-order Butterworth lowpass filter with a corner frequency of 50 kHz.

The mismatch shaper effectively provides 16 second-order loops with the first and second integrals using 16 elements. The main quantizer outputs the number of elements that the mismatch shaper should turn on. The shaper can override this value to optimize noise shaping. The number of elements actually turned on is used in the main feedback loop. Even with an element mismatch of 5%, a signal-to-noise ratio of 129 dB is still achieved. Mismatch shaping can continue for full-scale signals. Some DEM

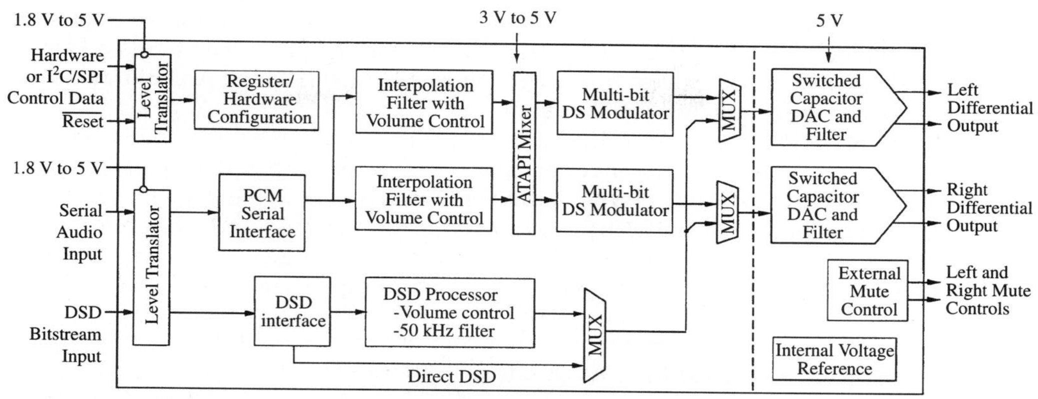

FIGURE 18.22 System architecture of a multi-bit sigma-delta D/A converter with mismatch shaping in the feedback loop. (*Deuwer et al., 2003*)

designs can introduce a data-dependent noise floor when given a high-level signal and all elements are turned on. The analog output stage comprises a 16-element switched-capacitor D/A converter operating at 6 MHz.

In this design, the noise shaper is inside the main loop; a balance is struck between quantization error and element mismatch error, determined by the number of elements the mismatch shaper turns on. When the quantizer's output is not followed, quantization error increases. The noise contribution from the main loop quantization error, assuming no mismatch, is set equal to the noise from mismatch shaping error, assuming worst case element mismatch.

As with other multi-bit converters, this converter has relatively low quantization noise, low sensitivity to clock jitter, and fewer idle tones compared to many one-bit converters. This design outputs a bitstream compatible with SACD without a decimation filter following the multi-bit conversion. A dynamic range of 120 dB (A-weighted) and distortion level of –105 dB THD+N can be achieved. Converters such as this are used for CD/SACD/DVD/Blu-ray playback.

Sigma-Delta A/D-D/A Converter Chip

Because of the high degree of integration permitted by sigma-delta conversion methods, it is possible to place a linear, 16-bit sigma-delta analog-to-digital and digital-to-analog converter on a single chip. One such chip permits input-output sampling frequencies up to 50 kHz with 16-bit resolution, and frequencies of 100 kHz with 12-bit resolution. Third-order noise shaping is used on the A/D side, and fourth-order noise shaping is used on the D/A side. The A/D section uses 64-times oversampling and 64-times decimation. A digital compensation circuit is used to equalize the response to within ±0.025-dB ripple in the passband, with phase linearity.

The D/A section uses two digital anti-imaging interpolation filters, along with an FIR compensation filter for flat passband response. The D/A section provides the output signal. An analog sixth-order Bessel lowpass filter is provided on-chip, as is a temperature-compensated voltage reference for stable coding and clocking. This reference can operate in a master–slave configuration to ensure gain matching and tracking between multiple devices. Likewise, sampling coherency can be preserved between multiple converter chips to ensure interchannel phase accuracy. Digital data can be shifted into and out of the converters with either MSB or LSB first. An SSI bus can be implemented in several different modes.

The DSP56ADA16 provides a dynamic range of 96 dB and signal-to-noise ratio of 90 dB. As with all sigma-delta converters, this converter pair is based on digital filtering techniques, thus approximately 90% of the chip is given to digital circuitry. This promotes compatibility, reliability, increased functionality, and reduced chip cost. Two of these chips form a complete conversion circuit for a stereo signal, and together with a DSP56xxx chip form a complete digital signal processing system.

Noise Shaping of Nonoversampling Quantization Error

As noted, noise shaping is prerequisite in any sigma-delta system to preserve dynamic range when a signal is represented with a reduced number of bits. For example, the noise-shaping characteristic of sigma-delta converters allows one-bit quantization. However, noise shaping can be applied in a variety of ways. For example, a noise-shaping feedback loop can be placed around a quantizer, as shown in Fig. 18.23. This noise-shaping loop uses the known characteristics of the error generated by the word length reduction (requantization) to alter the spectrum of the requantization noise error.

Recursion places the error information back into the signal, much like negative feedback is used to reduce distortion in analog amplifiers. The quantizer's output error is fed back through a filter and subtracted from the quantizer's input. Because only the difference between the input and output of the quantizer is fed back, the input signal is not affected. The configuration alters the frequency response of the error signal, but not that of the audio signal. It has the effect of passing the noise through the filter, not the signal.

However, with proper dither, the error is white, and the $H(z)$ filter in the feedback loop spectrally shapes the output error by $1 - H(z)$. That is, the output error e becomes: $[1 - H(z)]e$. The noise is shaped by the inverse of the loop transfer function; when a lowpass filter is placed in the loop, the noise spectrum rises with frequency. A filter with high gain at low frequencies yields improved baseband attenuation of noise. Higher-order functions perform a higher-order difference operation on quantizer error, with

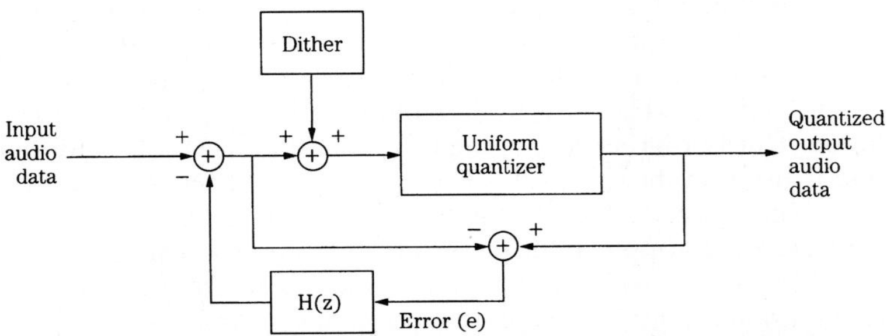

FIGURE 18.23 A requantization topology showing dithering and noise shaping. This processing reduces quantization distortion artifacts and can be used to reduce the noise floor in perceptually critical frequency regions.

greater attenuation of baseband noise. The frequency response of the requantization noise can be creatively manipulated by the filter in the feedback loop. For example, the filter's parameters could be dynamically adapted so that the error noise is always optimally masked by the audio signal. The feedback loop must incorporate at least a one-sample z^{-1} delay; the error cannot be processed until after it has been created by quantization. Theory also dictates that $1 - H(z)$ must be minimum phase (all poles and zeros within the z-plane unit circle) to preserve the capacity of the channel.

Referring again to Fig. 18.23, John Vanderkooy and Stanley Lipshitz have pointed out that $H(z)$ represents a loop error that is subtracted from the input at each next sample. This corrects for any such errors on average and gives a highpass shape to both quantization and dither signals present inside the loop. A digital dither signal applied as shown (inside the shaping loop) is identical to a highpass-filtered dither signal applied at a point outside the loop prior to the quantizer. Figure 18.24A shows the spectrum of the quantized output of an undithered noise shaper when a 937.5-Hz signal of 1-LSB peak amplitude (approximately –90.3 dBFS) is passed though an undithered requantizer. The spectrum shows many correlated errors with this low-level input signal.

When triangular pdf digital dither is applied, a highly uncorrelated spectrum results, as shown in Fig. 18.24B. The quantizer and the dither signal noise are both shaped by the loop. A rectangular pdf dither signal could be applied, but could result in noise modulation and limit cycle oscillation. The latter is a repeating output sequence that will produce spectral lines that can yield audible distortion. Alternatively, a high-pass triangular pdf dither could be applied; requantization noise is shaped as before,

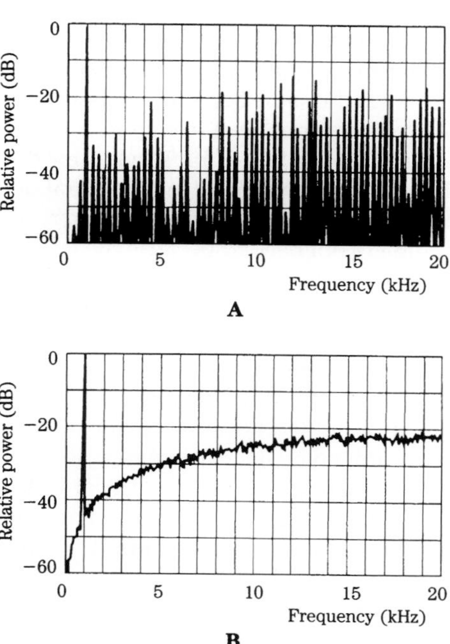

FIGURE 18.24 Dither profoundly affects the spectrum of the signal output from a noise-shaping circuit. A. Spectrum of a signal with an undithered noise shaper. B. Spectrum of the signal with a triangular pdf-dithered noise shaper. (*Vanderkooy and Lipshitz, 1989*)

but the higher frequency dither signal is shaped to even higher frequencies. However, correlation can result in higher overall noise. In this example, triangular pdf dither with a white spectrum appears to yield the best results.

Psychoacoustically Optimized Noise Shaping

The goal of noise-shaping systems is to dither the audio signal, then shape quantization noise to yield a less audible noise floor. These systems consider the fact that total noise power does not fully describe audibility of noise; perceived loudness also depends on spectral characteristics. Oversampling noise shapers reduce audio-band quantization noise and increase noise beyond the audio band, where it is inaudible. Nonoversampling noise shapers only redistribute noise energy within the audio band itself. For example, the difference in quantization noise between a 20-bit input signal and a 16-bit output signal can be reshaped to minimize its audibility. In particular, psychoacoustically optimized noise-shaping systems use a feedback filter designed to shape the noise according to an equal-loudness contour or other perceptual weighting function. In addition, such systems can use masking properties to conceal requantization noise.

Sixteen-bit master recordings are not adequate for subsequent music distribution on 16-bit media; for example, for replication of 16-bit CDs. When using a digital console or hard-disk recorder to add equalization, change levels, or perform other digital signal processing, error accumulates in the 16th bit due to computation. It is desirable to use a longer word length, such as 20 bits, that allows processing prior to 16-bit storage. Furthermore, with proper transfer, much information contained in the four LSBs can be conveyed in the upper 16 bits. However, the problem of transferring 20 bits to 16 bits is not trivial. Simple truncation of the four least-significant bits greatly increases distortion. If the 16th bit is rounded, the improvement is only modest.

It is thus important to redither the signal during the requantization that occurs in the transfer. This provides the same benefits as dithering during the original recording. If the most significant bit has not been exercised in the recording, it is possible to bit-shift the entire program upward, thus preserving more of the dynamic range. This is accomplished with a simple gain change in the digital domain. It can be argued that in some cases, for example, when transferring from an analog master tape, a 20-bit interface and noise shaping are not needed because the tape's noise floor makes it self-dithering. However, even then it is important to preserve the analog noise floor which contains useful audio information.

Nonoversampling noise-shaping systems are often used when converting a professional master recording to a consumer format such as a CD. With linear conversion and dither, a 16-bit recording can provide a distortion floor below –110 dBFS. Noise shaping cannot decrease total unweighted noise, but given a 20-bit master recording, subjective performance can be improved by decreasing noise in the critical 1-kHz to 5-kHz region, at the expense of increasing noise in the non-critical 15-kHz region, and increasing total unweighted noise power as well. Because noise shaping removes requantization noise in the most critical region, this noise cannot mask audible details, thus improving subjective resolution. However, the benefit is realized only when output D/A converters exhibit sufficient low-level linearity, and high S/N ratio is available. Indeed, any subsequent requantization must preserve the most critical noise floor improvements, and not introduce other noise that would negate the advantage of a shaped noise floor. For example, 19-bit resolution in D/A converters may be required to fully preserve noise-shaping improvements in a 16-bit recording.

When reducing word length, the audio signal must be redithered for a level appropriate for the receiving medium, for example, 16 bits for CD storage; white triangular pdf dither can be used. A nonoversampling noise-shaping loop redistributes the spectrum of the requantization noise. As noted earlier in this chapter, sigma-delta noise shapers used in highly oversampled converters yield a contour with a gradually increasing spectral characteristic. This characteristic will not specifically reduce noise in the 1-kHz to 5-kHz region. To take advantage of psychoacoustics, higher-order shapers are used in nonoversampling shapers to form more complicated weighting functions. In this way, the perceptually weighted output noise power is minimized. A digital filter $H(z)$ in a feedback loop (see Fig. 18.23) accomplishes this, in which the filter coefficients determine a response so that the output noise is weighted by $1 - H(z)$, the inverse of the desired psychoacoustic weighting function. The resulting weighted spectrum ideally produces a noise floor that is equally audible at all frequencies.

As Robert Wannamaker suggests, a suitable filter design begins with the selection of a weighting function. This design curve is inverted, and normalized to yield a zero average spectral power density that represents the squared magnitude of the frequency response of the minimum-phase noise shaper. The desired response is specified, and an inverse Fourier transform is applied to produce an impulse response. The response is windowed to produce a number of filter coefficients corresponding to $1 - H(z)$, then $H(z)$ is derived from this, yielding an FIR filter.

Theory shows that as very high-order filters $H(z)$ are used to approximate the optimal filter weighting function, the unweighted noise power increases, tending toward infinity with an infinite filter order. For example, although an optimal approximation might yield a 27-dB decrease in audible weighted noise (using a particular weighting curve that reflects the ear's high-frequency roll-off), other weighting functions must be devised, with more modest performance. For example, using a nine-coefficient FIR shaping filter, perceived noise can be decreased by 17 dB compared to unshaped requantization noise. Total unweighted noise power is increased by a reasonable 18 dB compared to an unshaped spectrum. In other words, the output is subjectively as quiet as an unshaped truncated signal with an additional three bits. In this way, audio data with resolution of 19 bits can be successfully transferred to a 16-bit CD. Similar techniques, of course, are applicable to DVD and Blu-ray authoring, when 16-, 20-, or 24-bit words may be used.

Methods that decrease audible noise while increasing total noise (at higher inaudible frequencies) perform a delicate balance. For example, a very high total noise power might damage tweeters, and some listeners suggest that aggressively boosted high-frequency noise produces artifacts, or perhaps masks otherwise audible information. In practice, depending on the design, the weighting function often approximates a proprietary contour. For example, Fig. 18.25 shows a proprietary noise-shaping contour, plotted with linear frequency for clarity. In some cases, this curve is fixed; in other cases, the curve is adaptively varied according to signal conditions. Similarly, in some designs, an adaptive dither signal is correlated to the audio signal so the audio signal masks the added dither noise. For example, the audio signal can be spectrally analyzed so that dither frequencies slightly higher in frequency can be generated.

Figure 18.26 shows a 1-kHz sine wave with –90-dBFS amplitude. Measurements are made with a 16-kHz lowpass filter, to approximate the ear's averaging response. A 20-bit recording is quite accurate; when truncated to 16 bits, quantization is clearly evident; when dithered (±1 LSB triangular pdf) to 16 bits, quantization noise is alleviated, but

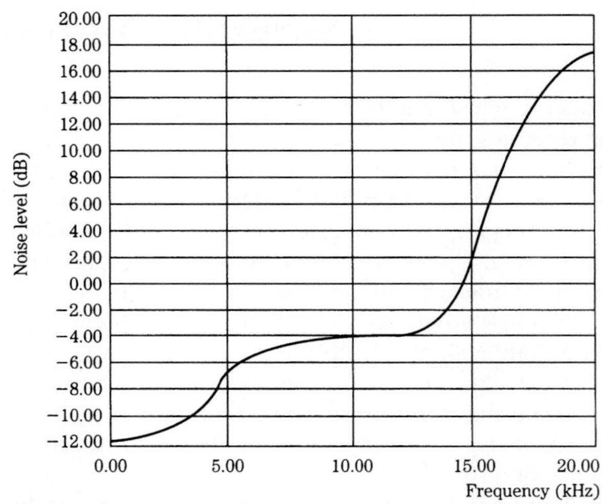

Figure 18.25 An equal-loudness noise-shaping curve. This frequency response plot uses a linear scale to better illustrate the high-frequency contour. (*Akune et al., 1992*)

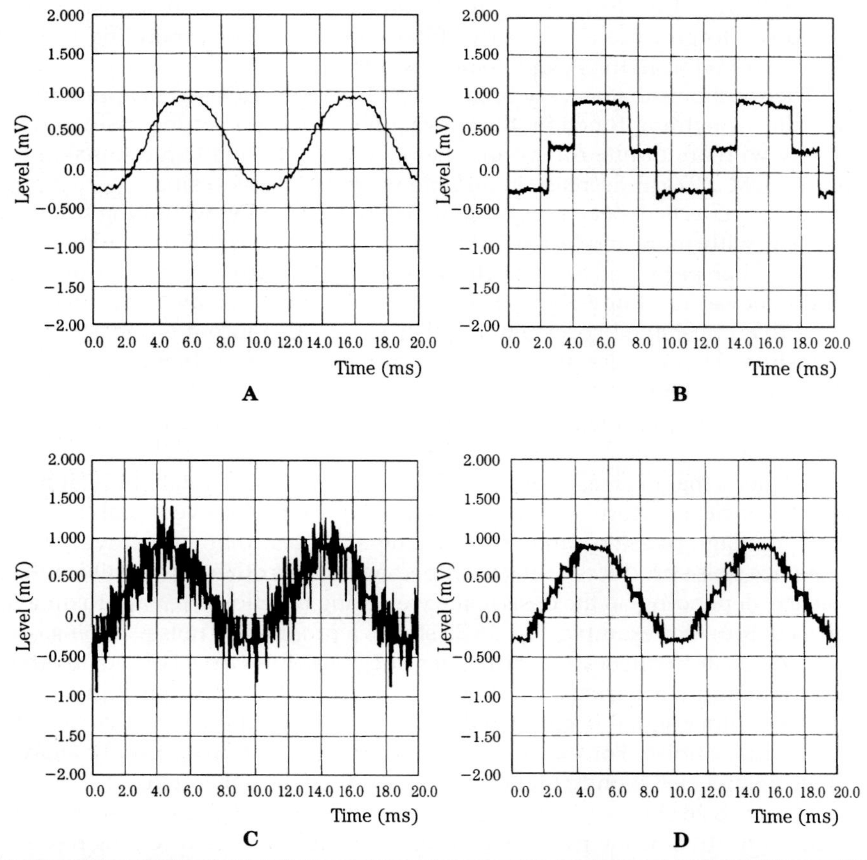

Figure 18.26 An example of noise shaping showing a 1-kHz sine wave with −90-dBFS amplitude. Measurements are made with a 16-kHz lowpass filter. A. Original 20-bit recording. B. Truncated 16-bit signal. C. Dithered 16-bit signal. D. Noise shaping preserves information in the lower 4 bits.

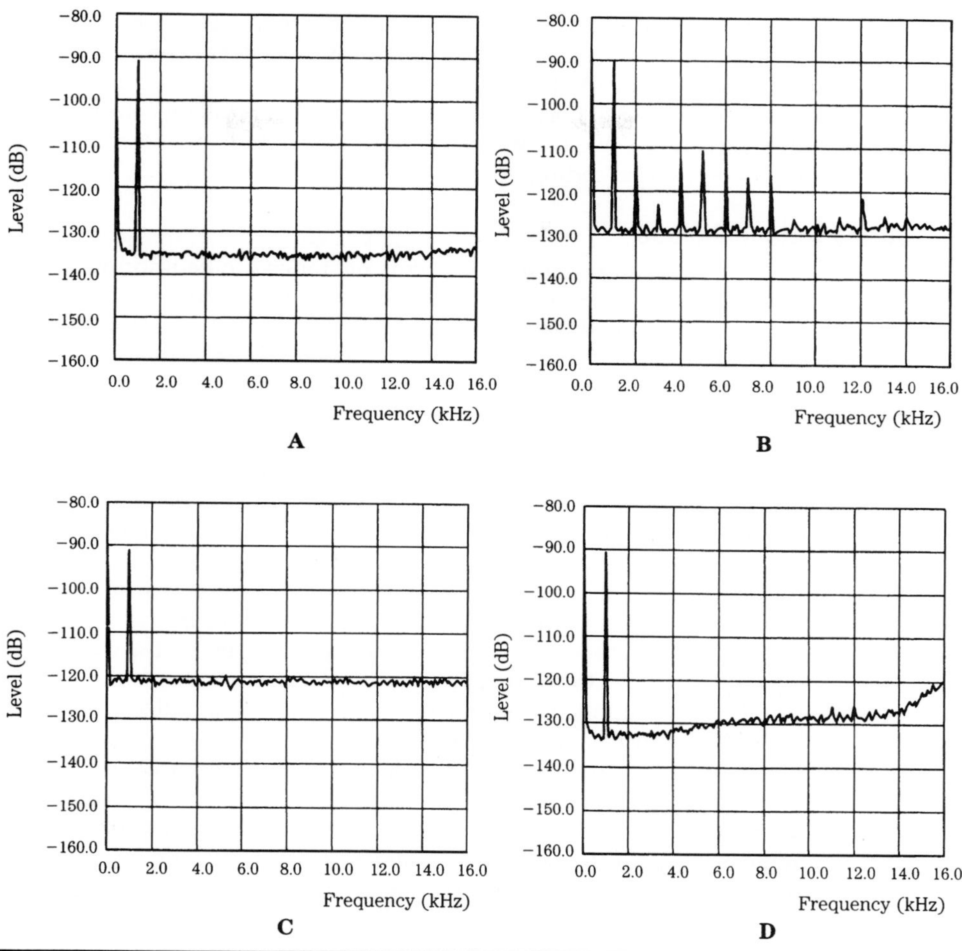

FIGURE 18.27 An example of noise shaping showing the spectrum of a 1 kHz, –90-dBFS sine wave (from Fig. 18.26). A. Original 20-bit recording. B. Truncated 16-bit signal. C. Dithered 16-bit signal. D. Noise shaping reduces low- and mid-frequency noise, with an increase at higher frequencies.

noise is increased; when noise shaping is applied, the noise in this lowpass-filtered measurement is reduced. This 16-bit representation is quite similar to the original 20-bit representation. Figure 18.27 shows the spectrum of the same –90-dBFS sine wave, with the four representations. The 20-bit recording has low error and noise; truncation creates severe quantization error; dithering removes the error but increases noise; noise shaping reduces low- and mid-frequency noise, with an increase at higher frequencies.

In one implementation of a psychoacoustic noise shaper, adaptive error-feedback filters are used to optimize the requantization noise spectrum according to equal-loudness contours as well as masking analysis of the input signal. An algorithm analyzes the signal's masking properties to calculate simultaneous masking curves. These are adaptively combined with equal-loudness curves to calculate the noise-shaping filter's coefficients, to yield the desired contour. This balance is dynamically and continuously varied according to the power of the input signal. For example, when power is low, masking is minimal, so the equal-loudness contour is used. Conversely, when power is high, masking is prevalent so the masking contour is more prominently used.

The input signal is converted into critical bands, convolved with critical-band masking curves, and converted to linear frequency to form the masking contour and hence the noise-shaping contour. In other words, masking analysis follows the same processing steps as used in perceptual coding.

Buried Data Technique

With proper dithering and noise shaping, dynamic range can be improved. However, processing can also be applied to use this dynamic range for purposes other than conventional audio headroom. Michael Gerzon and Peter Craven have demonstrated how data can be "buried" in a bitstream. The data is coded with psychoacoustic considerations so the data is inaudible under the masking curve of the audio program; the added data signal is randomized to act as shaped noise. For example, the method could be used to place new information on conventional audio CDs, without significantly degrading the quality of the audio program. In particular, this coding technique replaces several of the least-significant bits of the 16-bit format with independent data. Clearly, if unrelated data simply displaced audio data, and the disc was played in a conventional CD player, the result would be unlistenable. For example, nonstandard data in the four least-significant bits would add about 27 dB of noise to the music, as well as distortion caused by truncating the 16-bit audio signal. The buried data method makes buried data discs compatible with conventional CD players. However, a separate decoder is needed to utilize the buried data.

An example of the subtractively dithered, noise-shaped quantizer used to encode buried data is shown in Fig. 18.28. For example, a 16-bit signal is quantized with an M-bit step size (rounding the signal to the nearest integer multiple of M) to yield a (16–M)-bit signal. The buried data is coded to be pseudo-random, to make it noise-like with a uniform probability density function. This signal is used as subtractive M-bit dither to remove the artifacts caused by quantization. Specifically, the data dither is subtracted prior to quantization, and then added after quantization, replacing the

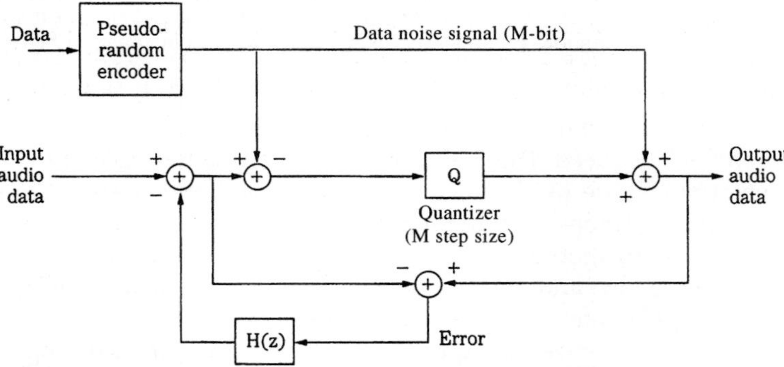

FIGURE 18.28 A buried data channel encoder converts added data to a pseudo-random noise signal, which is used as a dither signal. This is subtracted from the audio signal prior to quantization and added to the signal after quantization. Noise shaping is performed around the quantizer. (*Gerzon and Craven, 1995*)

M least-significant bits of the signal. The quantizing error signal is statistically independent of the input audio signal. To reduce the audibility of the resulting increase in the noise floor, a noise-shaping filter is applied in a loop around the quantizer so that the shaped noise is subtracted from the input signal. The transfer function $H(z)$ is selected so that $1 - H(z)$ yields a noise floor that ideally lies below the threshold of audibility. Through noise shaping, the noise created by four bits of buried data per channel (conveying 352.8 kbps with stereo channels) can be reduced to yield an overall S/N ratio of about 91 dB, a level that is similar to conventional CDs. Two bits of buried data provide a buried channel rate of 176.4 kbps, while maintaining an S/N ratio of 103 dB.

The average bit rate of the buried data could be increased by variably "stealing" bits from the original program only when their absence will be psychoacoustically masked by the music signal. By using an adaptive noise-shaping filter and a variable quantizer step size, the noise-shaping characteristic is varied according to the analyzed masking properties of the signal and the noise can be maintained below the masking threshold. The overall buried data rate could exceed 500 kbps, with 800 kbps possible during loud passages, depending on the music program. Combining methods, for example, buried data might consist of two 2-bit fixed channels, and a variable rate channel; side information would indicate the variable data rate. A buried data CD could be played in a regular CD player; the fidelity of music with limited dynamic range might not be affected at all.

More significantly, a CD player with appropriate decoding (or a player outputting buried data to an external decoder) could play the original music signal, and process buried data as well. The possibilities for buried data are numerous; many audio improvements can be more useful than the lost dynamic range. For example, buried 4-bit data could be used to convey multiple (5.1 channel) audio channels for surround-sound playback; the main left/rights channels are conventionally coded, the buried data carries four additional channels. A hybrid disc would compatibly deliver stereo reproduction with a conventional CD player, and surround sound with a 5.1-channel CD player.

Alternatively, one or two bits of buried data could carry dynamic range compression or expansion information. Depending on the playback circumstances, the dynamic range of the music could be adjusted for the most desirable characteristics. Because the range algorithms are calculated prior to playback, they are much more effective than conventional real-time dynamic processing. Buried data could convey additional high-frequency information above the Nyquist frequency, and provide a gentle bandlimiting roll-off rate. Any of these applications could be combined, within the limits of the buried data's rate. For example, two ambience channels and dynamic range control data could be delivered simultaneously. Techniques such as these demonstrate the utility of noise shaping and further underscore the power of digital signal processing in digital audio applications.

APPENDIX

The Sampling Theorem

Discrete time sampling is founded on the concept of a rectangular impulse of infinitesimal width. In practice, the width of a rectangular pulse is considered to be of finite width τ. Although data signals are often characterized by their time-domain properties, the transmission channel is usually best described by its frequency-domain properties. Specifically, it is important to know the bandwidth required for transmission of a sampled signal. The Fourier transform describes a time-domain function in the frequency domain. Given a single rectangular pulse of duration τ, the transformation of the pulse yields a $\sin(x)/x$ function.

This $\sin(x)/x$ function is composed of a fundamental cosine wave and its harmonics, its maximum value occurs at $x = 0$, and it approaches zero as x approaches $\pm\infty$. The width of the center lobe is exactly at $2/\tau$, and the frequency response passes through zero at multiples of $1/\tau$. Importantly, it demonstrates the fundamental nature of sampling as a modulation process; the frequency pattern of the function shows that the rectangular time pulse modulates the amplitude of a carrier frequency. The center frequency can be shifted without altering the shape of the envelope. Clearly, this spectrum extends to infinity; thus, ideal transmission of the pulse would require a system with infinite bandwidth. However, only the central lobe is required; thus, a finite bandwidth will suffice.

Given an understanding of the properties of a single pulse, it is useful to examine a series of such pulses with a periodic repetition of T. This leads to the creation of a practical sampling signal as a periodic series of pulses of fixed amplitude and finite width. The frequency spectrum of this function is defined at discrete values of n; that is, as equally spaced spectral lines with amplitudes corresponding to the discrete frequency components. Spectral lines are spaced according to the period T. They fall within a $|\sin(x)/x|$ envelope that is determined by the pulse duration τ, duty ratio τ/T, and pulse amplitude with zero-crossings at frequencies that are multiples of $1/\tau$.

The spectral response of a series of sampling pulses thus creates spectral lines with amplitudes that follow the same contour as that of a single pulse. The spectrum bandwidth is not affected by the pulse repetition frequency; rather, the bandwidth is determined by the pulse width τ. The shorter the duration of the pulse, the greater the frequency spread of the bandwidth. It is the case that transmission of narrow pulses requires a channel with a higher bandwidth. From a frequency-domain standpoint, wider pulses might appear advantageous; however, as viewed in the time domain, narrow pulses permit a greater repetition rate and, for example, permit time multiplexing of channels.

In any case, it is not a higher repetition rate that necessitates a higher bandwidth, but the narrow width of the pulses. Similarly, aperture error can be minimized by decreasing the duration of the pulse width. In the case of ideal sampling with a pulse of infinitesimal width and infinite bandwidth, its spectral lines are placed at multiples of the sampling frequency, as in natural sampling; however, the amplitudes of the lines remain constant across the spectrum.

Given a sampling signal, it is possible to define the sampled signal as the multiplication of the sampling signal, and the message signal. Moreover, we can obtain an expression for the frequency spectrum of the sampled signal. The multiplication of these two time-domain functions can be represented as the convolution of their spectra. The spectrum of the sampled signal contains both positive and negative sidebands centered at the impulses defined by the sampling function, and placed at multiples of the sampling frequency. The spectra are strictly bandlimited. In addition, their amplitude again follows the $\sin(x)/x$ contour predicted by the Fourier transform of the sampling signal.

The spectrum of the original message signal is repeated at multiples of the sampling frequency within the envelope of the sampled signal. When proper signal bandlimiting is provided, the spectrum repeats itself without overlap; if the signal's bandwidth is less than the Nyquist (half-sampling) frequency, the image sidebands are separated by a guard band. Note that complete information of the message signal is held in each sideband. The complete signal can be retrieved by removing higher frequency sideband spectra, leaving only the first sideband. As noted, complete information is contained in each sideband, thus, for example, a negative sideband could theoretically be used. The $\sin(x)/x$ function occurs repeatedly in sampling theory and in fact is often called the sampling function.

Given that any practical channel is bandlimited, it is important to know the maximum transmission rate afforded by a channel. Nyquist demonstrated that a message of S Hz can be completely characterized by samples taken at a frequency of $2S$ Hz. Moreover, the $\sin(x)/x$ function can be used as an interpolation function to reconstruct the original signal from the sample values. Each sample is multiplied by its interpolation function, and added to the functions of all other samples to obtain the signal waveform. Importantly, the $\sin(x)/x$ function represents the response of an ideal lowpass filter of bandwidth S Hz. In other words, the original signal can be reconstructed exactly by passing the representing samples through a lowpass filter with a bandwidth of S Hz. Thus, as Nyquist stated, a bandlimited signal can be completely reconstructed from samples. This is the key component that permits the transformation of analog signals and digital sequences. The following paragraphs summarize the sampling theorem.

The sampling process defines the values $f(nT)$ of $f(t)$ at regular time intervals. This is equivalent to a multiplication of $f(t)$ and $\delta_S(t)$, as shown in Fig. A.1. Therefore:

$$f_S(t) = \sum_{n=-\infty}^{+\infty} Tf(nT)\delta(t-nT)$$

where $\delta(t)$ is a delta function. The Fourier transform $F_S(\omega)$ of $f_S(t)$ is:

$$F_S(\omega) = \int_{-\infty}^{+\infty} \sum_{n=-\infty}^{+\infty} Tf(nT)\delta(t-nT)e^{-j\omega t}dt$$

$$= \sum_{n=-\infty}^{+\infty} F(\omega + n\omega_S)$$

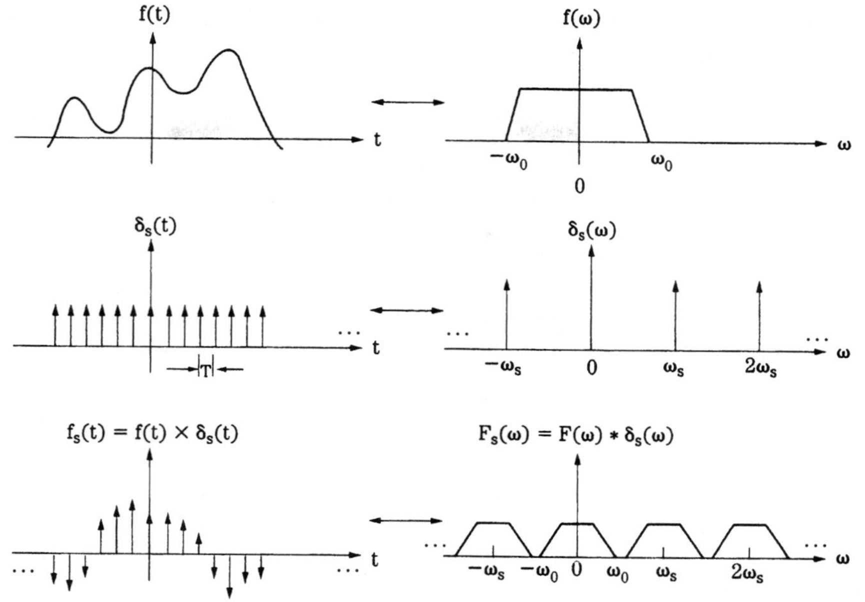

FIGURE A.1 The sampling process.

where $\omega_S = 2\pi/T$. When we multiply two functions in time, we are convolving their transforms in the frequency domain. For this reason, we see the spectrum $F(\omega)$ repeated at multiples of the sampling frequency.

We can recover $f(t)$ from $F_S(\omega)$ by first multiplying it with a gating function $G(\omega)$, as shown in Fig. A.2. We have:

$$F(\omega) = F_S(\omega)G(\omega)$$

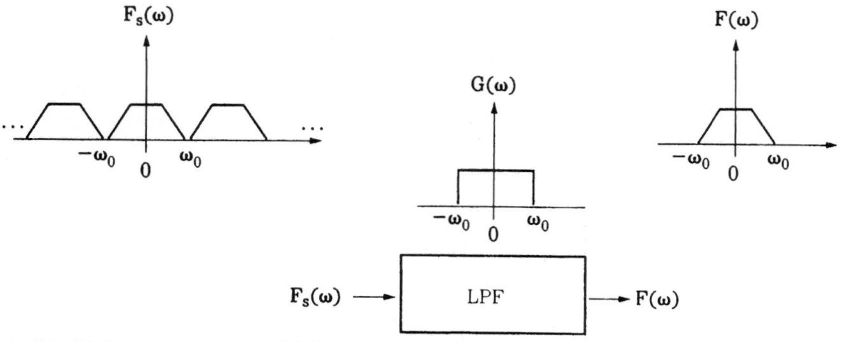

FIGURE A.2 Recovery of $f(t)$ through lowpass filtering.

That is, we are lowpass filtering $F_S(\omega)$. Using the fact that the inverse transform of $G(\omega)$ is:

$$\frac{\sin \omega_0 t}{\pi t}$$

and if $F_S(\omega) \leftrightarrow f_S(t)$ and $G(\omega) \leftrightarrow g(t)$, then:

$$F_S(\omega)G(\omega) \leftrightarrow f_S(t) * g(t)$$

where $*$ denotes convolution, and we obtain:

$$f(t) = \sum_{n=-\infty}^{+\infty} Tf(nT)\delta(t-nT) * \frac{\sin \omega_0 t}{T_0 \omega_0 t/2}$$

$$= \sum_{n=-\infty}^{+\infty} Tf(nT) \frac{\sin \omega_0 (t-nT)}{T_0 \omega_0 (t-nT)/2}$$

or

$$f(t) = \sum_{n=-\infty}^{+\infty} f(nT) \frac{\sin \omega_0 (t-nT)}{\omega_0 (t-nT)}$$

where

$$\frac{T}{T_0/2} = 1$$

This result, known as the sampling theorem, relates the samples $f(nT)$ taken at regular intervals to the function $f(t)$. The same result can also be interpreted in the following way: $f(t)$ is represented by the Fourier series, which has the sampling function:

$$\frac{\sin(x)}{x}$$

as its basis, and each coordinate is weighted by $f(nT)$. Note that we can only reconstruct the original waveform $f(t)$ from the sampled values if it is bandlimited. In practice, this condition is satisfied by passing $f(t)$ through an anti-aliasing filter.

The quantity:

$$\frac{1}{T} = 2\left(\frac{1}{T_0}\right)$$

is called the Nyquist rate. This can also be written as $f_S = 2f_0$. Hence, the sampling frequency should be twice as high as the maximum input frequency anticipated.

Bibliography

1 Sound and Numbers

Backus, J., *The Acoustical Foundations of Music,* W.W. Norton, 1969.

Bartee, T. C. (ed.), *Data Communications, Networks, and Systems,* Howard W. Sams, 1985.

Bartee, T. C., *Digital Computer Fundamentals,* McGraw-Hill, 1981.

Blesser, B. A., "Elementary and Basic Aspects of Digital Audio," *AES Digital Audio Collected Papers,* Rye, 1983.

Booth, T. L., *Digital Networks and Computer Systems,* John Wiley & Sons, 1971.

Eargle, J., *Handbook of Recording Engineering,* 3rd ed., Van Nostrand Reinhold, 1996.

Everest, F. A., and K. C. Pohlmann, *Master Handbook of Acoustics*, 5th ed., McGraw-Hill, 2009.

Roth, Jr., C. H., *Fundamentals of Logic Design,* West, 1979.

Woram, J. W., *Sound Recording Handbook,* Howard W. Sams, 1989.

2 Fundamentals of Digital Audio

Adams, R. W., "Nonuniform Sampling of Audio Signals," *JAES*, vol. 40, no. 11, November 1992.

Blesser, B. A., "Advanced Analog-to-Digital Conversion and Filtering: Data Conversion," *AES Digital Audio Collected Papers,* Rye, 1983.

Blesser, B. A., "Digitization of Audio: A Comprehensive Examination of Theory, Implementation, and Current Practice," *JAES,* vol. 26, no. 10, October 1978.

Cauchy, A. L., "Memoire sur Diverses Formules d'Analyse," *Comptes Rendus Acad. Sci.* Paris, vol. 12, 1841.

Clarke, A. B., and R. L. Disney, *Probability and Random Processes for Engineers and Scientists,* John Wiley & Sons, 1970.

Jerri, A. J., "The Shannon Sampling Theorem—Its Various Extensions and Applications: A Tutorial Review," *Proc. IEEE,* vol. 65, no. 11, November 1977.

Karl, J. H., *An Introduction to Digital Signal Processing*, Academic Press, 1989.

Kotelnikov, V. A., "On the Transmission Capacity of "Ether" and Wire in Electro-communications," *Izd. Red. Upr. Svyazi RKKA,* Moscow, 1933.

Lipshitz, S. P., R. A. Wannamaker, and J. Vanderkooy, "Quantization and Dither: A Theoretical Survey," *JAES,* vol. 40, no. 5, May 1992.

Maher, R. C., "On the Nature of Granulation Noise in Uniform Quantization Systems," *JAES*, vol. 40, no. 1/2, January/February 1992.

Mathews, M., *The Technology of Computer Music*, MIT Press, 1969.

Nyquist, H., "Certain Factors Affecting Telegraph Speed," *Bell Sys. Tech. J.*, vol. 3, no. 2, April 1924.

Nyquist, H., "Certain Topics in Telegraph Transmission Theory," *Trans. AIEE*, vol. 47, no. 2, April 1928.

Ogura, K., "On a Certain Transcendental Integral Function in the Theory of Interpolation," *Tohoku Math. J.*, vol. 17, 1920.

Schuchman, L. "Dither Signals and Their Effect on Quantization Noise," *IEEE Trans. Commun. Tech.*, vol. COM-12, December 1964.

Shannon, C. E., "Communication in the Presence of Noise," *Proc. Inst. Radio Eng.*, vol. 37, 1949.

Shannon, C. E., "A Mathematical Theory of Communication," *Bell Sys. Tech. J.*, vol. 27, no. 3, July 1948.

Stockham, Jr., T. G., "The Promise of Digital Audio," *AES Digital Audio Collected Papers*, Rye, 1983.

Stuart, J. R., "Coding for High-Resolution Audio Systems," *JAES*, vol. 52, no. 3, March 2004.

Talambiras, R., "Some Considerations in the Design of Wide-Dynamic-Range Audio Digitizing Systems," *AES 57th Conv.*, preprint 1226, Los Angeles, May 1977.

Tanabe, H., and T. Wakuri, "On the Quality of Some Digital Audio Equipment Measured by the High Accuracy Dynamic Distortion Measuring System," *AES 72nd Conv.*, preprint 1909, Anaheim, October 1982.

Vanderkooy, J., and S. P. Lipshitz, "Dither in Digital Audio," *JAES*, vol. 35, no. 12, December 1987.

Vanderkooy, J., and S. P. Lipshitz, "Resolution Below the Least Significant Bit in Digital Audio Systems with Dither," *JAES*, vol. 32, no. 3, March 1984; Erratum, *JAES*, vol. 32, no. 11, November 1984.

Wannamaker, R. A., "Efficient Generation of Multichannel Dither Signals," *JAES*, vol. 52, no. 6, June 2004.

Wannamaker, R. A., "Subtractive and Nonsubstractive Dithering: A Mathematical Comparison," *JAES*, vol. 52, no. 12, December 2004.

Whittaker, E. T., "On the Functions Which Are Represented by the Expansions of the Interpolation-Theory," *Proc. R. Soc. Edinburgh*, vol. 35, 1915.

Yoshikawa, S., et al., "Does High Sampling Frequency Improve Perceptual Time-Axis Resolution of a Digital Audio Signal?" *AES 103rd Conv.*, preprint 4562, New York, September 1997.

Zayed, A. I., *Advances in Shannon's Sampling Theorem*, CRC Press, 1993.

3 Digital Audio Recording

Black, H. S., *Modulation Theory*, Van Nostrand Reinhold, 1953.

Blesser, B. A., and F. F. Lee, "An Audio Delay System Using Digital Technology," *JAES*, vol. 19, no. 5, May 1971.

Blesser, B. A., and B. N. Locanthi, "The Application of Narrow-Band Dither Operating at the Nyquist Frequency in Digital Systems to Provide Improved Signal-to-Noise Ratio over Conventional Dithering," *JAES*, vol. 35, no. 6, June 1987.

Cabot, R. C., "Fundamentals of Modern Audio Measurement," *Proc. AES 12th UK Conf.*, London, April 1997.

Cabot, R. C., "Noise Modulation in Digital Audio Equipment," *AES 90th Conv.*, preprint 3021, Paris, February 1991.

Craven, P. G., "Antialias Filters and System Transient Response at High Sample Rates," *JAES*, vol. 52, no. 3, March 2004.

Davidson, M. et al., "High Density Magnetic Recording Using Digital Block Codes of Low Disparity," *IEEE Trans. Magn.*, vol. MAG-12, no. 5, September 1976.

Doi, T. T., "Channel Codings for Digital Audio Recordings," *JAES*, vol. 31, no. 4, April 1983.

Doi, T. T., "Recent Progress in Digital Audio Technology," *AES Digital Audio Collected Papers*, Rye, 1983.

Finger, R. A., "On the Use of Computer Generated Dithered Test Signals," *JAES*, vol. 35, no. 6, June 1987.

Gersho, A., "Quantization," *IEEE Commun. Soc. Mag.*, vol. 15, September 1977.

Gilchrist, N. H. C., "The Subjective Effect and Measurement of ADC–DAC Transfer Characteristic Discontinuity," *JAES*, vol. 36, no. 9, September 1988.

Hamill, D. C., "Transient Response of Audio Filters," *Wireless World*, August 1981.

Harris, S., "Techniques to Measure and Maximize the Performance of a 120-dB, 24-Bit, 96-kHz A/D Converter Integrated Circuit," *AES 103rd Conv.*, preprint 4530, New York, September 1997.

Haykin, S., *Communication Systems*, John Wiley & Sons, 1983.

Hiroshi, I. et al., "Pulse-Code Modulation Recording System," *JAES*, vol. 21, no. 7, September 1973.

Hnatek, E. R., *A User's Handbook of D/A and A/D Converters*, John Wiley & Sons, 1976.

Hoeschele, D., *Analog-to-Digital, Digital-to-Analog Conversion Techniques*, John Wiley & Sons, 1968.

Horiguchi, T., and K. Morita, "An Optimization of Modulation Codes in Digital Recording," *IEEE Trans. Magn.*, vol. MAG-12, no. 6, November 1976.

Immink, K. A. S., *Coding Techniques for Digital Recorders*, Prentice Hall, 1991.

Iwamura, H., et al., "Pulse-Code-Modulation Recording System," *JAES*, vol. 21, no. 9, September 1973.

Jayant, N., and L. Rabiner, "The Application of Dither to the Quantization of Speech Signals," *Bell Sys. Tech. J.*, vol. 51, 1972.

Korn, I., *Digital Communications*, Van Nostrand Reinhold, 1985.

Krause, M., and H. Petersen, "How Can the Headroom of Digital Recordings Be Used Optimally?" *JAES*, vol. 38, no. 11, November 1990.

Lagadec, R., and T. G. Stockham, Jr., "Dispersive Models for A-to-D and D-to-A Conversion Systems," *AES 75th Conv.*, preprint 2097, Paris, March 1984.

Lindholm, D. A., "Power Spectra of Channel Codes for Digital Magnetic Recording," *IEEE Trans. Magn.*, vol. MAG-14, no. 5, September 1978.

Lindsey, W. C., and M. K. Simon, *Telecommunications Systems Engineering*, Prentice Hall, 1973.

MacArthur, J., "20/20 Vision," *Studio Sound*, January 1993.

Marchant, A. B., *Optical Recording: A Technical Overview*, Addison-Wesley, 1990.

Matick, R. E., *Computer Storage Systems and Technology*, John Wiley & Sons, 1977.

Meyer, J., "Time Correction of Anti-Aliasing Filters Used in Digital Audio Systems," *JAES*, vol. 32, no. 3, March 1984.

Miller, J. W., "DC-Free Encoding for Data Transmission System," U.S. Patent 4,027,335, May 31, 1977.

Morris, D. J., *Pulse Code Formats for Fiber Optical Data Communication*, Marcel Dekker, 1983.

Muraoka, T., Y. Yamada, and M. Yamazaki, "Sampling Frequency Considerations in Digital Audio Standards," *JAES*, vol. 26, no. 4, April 1978; Erratum, *JAES*, vol. 26, no. 7/8, July/August 1978.

Myers, J. P., and A. Feinburg, "High-Quality Professional Recording Using New Digital Techniques," *JAES*, vol. 20, no. 10, October 1972.

Nakajima, H., T. T. Doi, J. Fukuda, and A. Iga, *Digital Audio Technology*, TAB Books, 1983.

Owen, F. F. E., *PCM and Digital Transmission Systems*, McGraw-Hill, 1982.

Patel, A. M., "Zero-Modulation Encoding in Magnetic Recording," *IBM J. Res. Dev.*, July 1975.

Peled, A., and B. Liu, *Digital Signal Processing: Theory, Design, and Implementation*, John Wiley & Sons, 1976.

Picot, J. P., *Introduction a l'Audio Numerique*, Editions Frequencies, 1984.

Preis, D., and P. J. Bloom, "Perception of Phase Distortion in Anti-Alias Filters," *JAES*, vol. 32, no. 11, November 1984.

Putzeys, B., "Design Techniques for High-Performance Discrete A/D Converters," *AES 114th Conv.*, preprint 5823, Amsterdam, March 2003.

Rathmell, J., J. Scott, and A. Parker, "TDFD-Based Measurement of Analog-to-Digital Converter Nonlinearity," *JAES*, vol. 45, no. 10, October 1997.

Reefman, D., et al., "A New Digital-to-Analogue Converter Design Technique for HiFi Applications," *AES 114th Conv.*, preprint 5846, Amsterdam, March 2003.

Salman, W. P., and M. S. Solotareff, *Le Filtrage Numerique*, Eyrolles, 1978.

Sato, N. "PCM Recorder—A New Type of Audio Magnetic Tape Recorder," *JAES*, vol. 21, no. 9, September 1973.

Tanaka, S., "Method and Apparatus for Encoding Binary Data," U.S. Patent 4,728,929, March 1, 1985.

Takahashi, Y., et al., "Study and Evaluation of New Method of ADPCM Encoding," *AES 86th Conv.*, preprint 2813, Hamburg, March 1989.

Takashi, T., et al., "A Coding Method in Digital Magnetic Recording," *IEEE Trans. Magn.*, vol. MAG-8, no. 3, September 1972.

Van der Kam, J. J., "A Digital 'Decimating' Filter for Analog-to-Digital Conversion of Hi-Fi Audio Signals," *Philips Tech. Rev.*, vol. 42, no. 6/7, 1986.

Widrow, B., "A Study of Rough Amplitude Quantization by Means of Nyquist Sampling Theory," *IRE Trans. Circuit Theory*, vol. CT-3, December 1956.

Wong, P. W., "Quantization Noise, Fixed-Point Multiplicative Roundoff Noise, and Dithering," *IEEE Trans. ASSP*, vol. 38, no. 2, February 1990.

4 Digital Audio Reproduction

Adams, R. W., "Clock Jitter, D/A Converters, and Sample-Rate Conversion," *The Audio Critic*, no. 21, Spring 1994.

Adams, R. W., "Design and Implementation of an Audio 18-Bit Analog-to-Digital Converter Using Oversampling Techniques," *JAES*, vol. 34, no. 3, March 1986.

Angus, J., "A New Method for Analyzing the Effects of Jitter in Digital Audio Systems," *AES 103rd Conv.*, preprint 4607, New York, September 1997.

Baldwin, G. L., and S. K. Tewksbury, "Linear Delta Modulator Integrated Circuit with 17-Mbit/s Sampling Rate," *IEEE Trans. Commun.*, vol. COM-22, no. 7, July 1974.

Benjamin, E., and G. Benjamin, "Theoretical and Audible Effects of Jitter on Digital Audio Quality," *AES 105th Conv.*, preprint 4826, San Francisco, September 1998.

Bristow-Johnson, R., "Effect of DAC Deglitching on Frequency Response," *JAES*, vol. 36, no. 11, November 1988.

Cabot, R. C., "Fundamentals of Modern Audio Measurement," *JAES*, vol. 47, no. 9, September 1999.

Charbonnier, A., and J. Petit, "Sub-Band ADPCM Coding for High Quality Audio Signals," *IEEE ICASSP*, 1988.

Couch, L., *Digital and Analog Communication Systems*, Macmillan, 1983.

Darling, T. F., and M. O. J. Hawksford, "Oversampled Analog-to-Digital Conversion for Digital Audio Systems," *JAES*, vol. 38, no. 12, December 1990.

Dijkmans, E. C., and P. J. A. Naus, "The Next Step Towards Ideal A/D and D/A Converters," *Proc. AES 7th Int. Conf.*, Toronto, May 1989.

Dunn, J., "Digital-to-Analog Converter Measurements," *Audio Precision Technote TN-25*, 2001.

Dunn, J., "Jitter Theory—Part One, Interface Jitter," *Audio Precision Tech Note*, vol. 14, no. 1, December 1999.

Fielder, L. D., "Evaluation of the Audible Noise and Distortion Produced by Digital Audio Converters," *JAES*, vol. 35, no. 7/8, July/August 1987.

Finger, R. A., "Review of Frequencies and Levels for Digital Audio Performance Measurements," *JAES*, vol. 34, no. 1/2, January/February 1986.

Gundry, K. J., D. P. Robinson, and C. C. Todd, "Recent Developments in Digital Audio Techniques," *AES 73rd Conv.*, preprint 1956, Eindhoven, March 1983.

Halbert, J. M., and M. A. Shill, "An 18-Bit Digital-to-Analog Converter for High Performance Digital Audio Applications," *JAES*, vol. 36, no. 6, June 1988.

Harris, S., "The Effects of Sampling Clock Jitter on Nyquist Sampling Analog-to-Digital Converters, and on Oversampling Delta-Sigma ADCs," *JAES*, vol. 38, no. 7/8, July/August 1990.

Hawksford, M. O. J., "Jitter Simulation in High Resolution Digital Audio," *AES 121st Conv.*, preprint 6864, San Francisco, October 2006.

Hawksford, M. O. J., "Ultrahigh-Resolution Audio Formats for Mastering and Archival Applications," *JAES*, vol. 55, no. 10, October 2007.

Kakiuchi, S., et al., "Application of Oversampling A/D and D/A Conversion Techniques to RDAT," *AES 83rd Conv.*, preprint 2520, New York, October 1987.

Lipshitz, S. P., and J. Vanderkooy, "Are Digital-to-Analog Converters Getting Worse?" *AES 84th Conv.*, preprint 2586, Paris, March 1988.

Lipshitz, S. P., and J. Vanderkooy, "Pulse-Code Modulation—An Overview," *JAES*, vol. 52, no. 3, March 2004.

Nagata, A., et al., "Over-Sampling Filter for Digital Audio Use," *AES 79th Conv.*, preprint 2289, New York, October 1985.

Naylor, J. R., "A Dual Monolithic 18-Bit Analog-to-Digital Converter for Digital Audio Applications," *Proc. AES 7th Int. Conf.*, Toronto, May 1989.

Nishimura, A. and N. Koizumi, "Measurement of Sampling Jitter in Analog-to-Digital and Digital-to-Analog Converters Using Analytic Signals," *AES 112th Conv.*, preprint 5586, Munich, May 2002.

Nishiguchi, N., K. Akagiri, and T. Suzuki, "A New Audio Bit Rate Reduction System for the CD-I Format," *AES 81st Conv.*, preprint 2375, Los Angeles, November 1986.

Pohlmann, K. C., "Multibit Conversion," *Advanced Digital Audio*, K. C. Pohlmann (ed.), Howard W. Sams, 1991.

Pohlmann, K. C., "Pulse Modulation and Sampling Systems," *Advanced Digital Audio*, K. C. Pohlmann (ed.), Howard W. Sams, 1991.

Proakis, J., *Digital Communications*, McGraw-Hill, 1983.

Roden, M. S., *Analog and Digital Communication Systems*, Prentice Hall, 1985.

Schindler, H. R., "Delta Modulation," *IEEE Spectr.*, October 1970.

Schott, W., "Philips Oversampling System for Compact Disc Decoding," *Audio*, vol. 68, no. 4, April 1984.

Shelton, W. T., "Progress Towards a System of Synchronisation in Digital Audio," *AES 82nd Conv.*, preprint 2484, London, March 1987.

Steele, R., *Delta Modulation Systems*, Halsted Press, 1975.

Stockham, Jr., T. G., "A/D and D/A Converters: Their Effect on Digital Audio Fidelity," *Digital Signal Processing*, L. R. Rabiner and C. M. Rader (eds.), IEEE Press, 1972.

Sun, M. T., and L. Wu, "Efficient Design of the Oversampling Filter for Digital Audio Applications," *AES 81st Conv.*, preprint 2378, Los Angeles, November 1986.

Takasaki, Y., *Digital Transmission Design and Jitter Analysis*, Artech House, 1991.

Trischitta, P. R., and E. L. Varma, *Jitter in Digital Transmission Systems*, Artech House, 1989.

Van de Plassche, R. J., and E. C. Dijkmans, "A Monolithic 16-Bit D/A Conversion System for Digital Audio," *AES Digital Audio Collected Papers*, Rye, 1983.

Wong, W. K., "Optimization Techniques for High Order Phase-Locked Loop Type Jitter Reduction Circuit for Digital Audio," *IEEE Trans. Consum. Electron.*, vol. 42, no. 1, February 1996.

Yamada, M., and K. Odaka, "A New Audio Digital Filter with Compensation of Phase for A/D and D/A Conversion," *AES 83rd Conv.*, preprint 2528, New York, October 1987.

5 Error Correction

Berlekamp, E. R., *Algebraic Coding Theory*, McGraw-Hill, 1968.

Berlekamp, E. R., "Error Correcting Code for Digital Audio," *AES Digital Audio Collected Papers*, Rye, 1983.

Buddine, L., and E. Young, *The Brady Guide to CD-ROM*, Prentice Hall, 1987.

Cabot, R. C., "Measuring AES–EBU Digital Audio Interfaces," *JAES*, vol. 38, no. 6, June 1990.

Clark, Jr., G. C., and J. B. Cain, *Error-Correction Coding for Digital Communications*, Plenum Press, 1981.

Doi, T. T., "Error Correction for Digital Audio Recordings," *AES Digital Audio Collected Papers*, Rye, 1983.

Feher, K., *Telecommunications Measurements, Analysis and Instrumentation*, Prentice Hall, 1987.

Gallager, R. G., *Information Theory and Reliable Communication*, John Wiley & Sons, 1968.

Hagelbarger, D. W., "Recurrent Codes: Easily Mechanized, Burst-Correcting, Binary Codes," *Bell Sys. Tech. J.*, vol. 38, no. 4, July 1959.

Hamming, R. W., *Coding and Information Theory*, Prentice Hall, 1980.

Hamming, R. W., "Error Detecting and Error Correcting Codes," *Bell Sys. Tech. J.*, vol. 29, no. 1, January 1950.

Hoeve, H., J. Timmermans, and L. B. Vries, "Error Correction and Concealment in the Compact Disc System," *Philips Tech. Rev.*, vol. 40, no. 6, 1982.

Huffman, W. C., and V. Pless, *Fundamentals of Error-Correcting Codes*, Cambridge University Press, 2003.

Imai, H. (ed.), *Essentials of Error-Control Coding Techniques*, Academic Press, 1990.

Lin, S., *An Introduction to Error-Correcting Codes*, Prentice Hall, 1970.

Maher, R. C., "A Method for Extrapolation of Missing Digital Audio Data," *JAES*, vol. 42, no. 5, May 1994.

McEliece, R. J., *The Theory of Information and Coding*, Addison-Wesley, 1977.

Morelos-Zaragoza, R. H., *The Art of Error Correcting Coding*, John Wiley & Sons, 2002.

Odaka, K., et al., "Error Correctable Data Transmission Method," U.S. Patent 4,413,340, November 1, 1983.

Peterson, W. W., "Error-Correcting Codes," *Scientific American*, vol. 206, no. 2, February 1962.

Pless, V., *Introduction to the Theory of Error-Correcting Codes*, 3rd ed., John Wiley & Sons, 1998.

Reed, I. S., and G. Solomon, "Polynomial Codes over Certain Finite Fields," *J. SIAM*, vol. 8, 1960.

Shenton, D., E. DeBenedictis, and B. N. Locanthi, "Improved Reed–Solomon Decoding Using Multiple Pass Decoding," *JAES*, vol. 33, no. 11, November 1985.

Viterbi, A. J., "Coding and Interleaving for Correcting Burst and Random Errors in Recording Media," *AES Digital Audio Collected Papers*, Rye, 1983.

Viterbi, A. J., *Coherent Communication*, McGraw-Hill, 1966.

Vries, L. B., and K. Odaka, "CIRC—The Error Correcting Code for the Compact Disc Digital Audio System," *AES Digital Audio Collected Papers*, Rye, 1983.

Watkinson, J., "Inside CD, Part 4," *Hi-Fi News & Record Review*, February 1987.

6 Optical Disc Media

Blake, L., "Digital Sound in the Cinema," *Mix*, vol. 19, no. 10, October 1995.

Bouwhuis, G., et al., *Principles of Optical Disc Systems*, Adam Hilger, 1985.

Bradley, A., *Optical Storage for Computers: Technology and Application*, Ellis Horwood, 1989.

Doi, T. T., T. Itoh, and H. Ogawa, "A Long Play Digital Audio Disc System," *JAES*, vol. 27, no. 12, December 1979.

Freese, R. P., "Optical Disks Become Erasable," *IEEE Spectr.*, February 1988.

Hartmann, M., B. Jacobs, and J. Braat, "Erasable Magneto-Optical Recording," *Philips Tech. Rev.*, vol. 42, no. 2, August 1985.

Immink, K. A. S., and J. Braat, "Experiments Toward an Erasable Compact Disc Digital Audio System." *JAES*, vol. 32, no. 7/8, July/August 1984.

JAES Staff, "High-Density Optical-Disk Formats," *JAES*, vol. 53, no. 7/8, July/August 2005.

Koechner, W., *Solid-State Laser Engineering, Springer Series in Optical Sciences*, Springer-Verlag, 1988.

Kramer, P., "Reflective Optical Record Carrier," U.S. Patent 5,068,846, November 26, 1991.

Kurahashi, A., et al., "Development of an Erasable Magneto-Optical Digital Audio Recorder," *AES 79th Conv.*, preprint 2296, New York, October 1985.

Marshall, G. F., *Laser Beam Scanning, Opto-Mechanical Devices, Systems, and Data Storage Optics*, Marcel Dekker, 1985.

Mecca, C. M. J., et al., "Interference of Converging Spherical Waves with Application to the Design of Compact Disks," *Opt. Commun.*, vol. 182, August 2000.

Meyer-Arendt, J. R., *Introduction to Classical and Modern Optics*, Prentice Hall, 1972.

Morgan, J., *Introduction to University Physics*, vol. 2, Allyn and Bacon, 1964.

Murata, S., et al., "Multimedia Type Digital Audio Disc System," *IEEE Trans. Consum. Electron.*, vol. 35, no. 3, August 1989.

Nöldeke, C., "Compact Disc Diffraction," *The Physics Teacher*, October 1990.

Pohlmann, K. C. "Optical Storage Technologies," *Mix*, vol. 16, no. 7, July 1992.

Thomas, G. E., "Future Trends in Optical Recording," *Philips Tech. Rev.*, vol. 44, no. 2, 1988.

Uhlig, R. E., "Feasibility of Digital Sound on Motion Picture Film," *Proc. AES 7th Int. Conf.*, Toronto, May 1989.

Varela, A., "Film Sound Goes Digital," *Post*, September 1990.

Verdeyen, J. T., *Laser Electronics*, Prentice Hall, 1981.

Webb, R. H., *Elementary Wave Optics*, Academic Press, 1969.

Yariv, A., *Optical Electronics*, 3rd ed., CBS College Publishing, 1985.

7 Compact Disc

AES17-1998, "AES Standard Method for Digital Audio Engineering—Measurement of Digital Audio Equipment," *JAES*, vol. 46, no. 5, May 1998.

AES28-1997, "AES Standard for Audio Preservation and Restoration—Method for Estimating Life Expectancy of Compact Discs (CD-ROM) Based on Effects of Temperature and Relative Humidity," *JAES*, vol. 45, no. 5, May 1997.

Baert, L., L. Theunissen, and G. Vergult, *Digital Audio and Compact Disc Technology*, 2nd ed., Butterworth-Heinemann Ltd, 1992.

Bouwhuis, G., et al., *Principles of Optical Disc Systems*, Adam Hilger Ltd., 1985.

Carasso, M. G., J. B. H. Peck, and J. P. Sinjou, "The Compact Disc Digital Audio System," *Philips Tech. Rev.*, vol. 40, no. 6, 1982.

Carasso, M. G., et al., "Disc-Shaped Optically Readable Record Carrier Used as a Data Storage Medium," U.S. Patent 4,238,843, December 9, 1980.

Funasaka, E., and H. Suzuki, "DVD-Audio Format," *AES 103rd Conv.*, preprint 4566, New York, September 1997.

Goedhart, D., R. J. van de Plassche, and E. F. Stikvoort, "Digital-to-Analog Conversion in Playing a Compact Disc," *Philips Tech. Rev.*, vol. 40, no. 6, 1982.

Heemskerk, J. P. J., and K. A. S. Immink, "Compact Disc System Aspects and Modulation," *Philips Tech. Rev.*, vol. 40, no. 6, 1982.

IEC 908: *Compact Disc Digital Audio System*, 1987.

Immink, K. A. S., "The Compact Disc Story," *JAES*, vol. 46, no. 5, May 1998.

Immink, K. A. S., et al., "Method of Coding Binary Data," U.S. Patent 4,501,000, February 19, 1985.

Isailovic, J., *Videodisc and Optical Memory Systems*, Prentice Hall, 1985.

ISO 9660: 1988 (E), Information Processing—Volume and File Structure of CD-ROM for Information Interchange, 1988.

ISO/IEC 10149: 1989 (E), Information Technology—Data Interchange on Read-Only 120-mm Optical Data Disks (CD-ROM), 1989.

Lambert, S., and S. Ropiequet, *CD-ROM, The New Papyrus*, Microsoft Press, 1986.

Matull, J., "ICs for Compact Disc Decoders," *Electrical Components and Applications*, vol. 4, no. 3, May 1982.
Microsoft MS-DOS CD-ROM Extensions Documentation, Version 2.20; Microsoft Corporation, 1990.
Miyaoka, S., "Manufacturing Technology of the Compact Disc," *AES Digital Audio Collected Papers*, Rye, 1983.
Nakajima, H., and H. Ogawa, *Compact Disc Technology*, Ohmsha Ltd., 1992.
Nishio, A., et al., "Direct Stream Digital Audio System," *AES 100th Conv.*, preprint 4163, Copenhagen, May 1996.
Ogawa, H., and K. A. S. Immink, "EFM—The Modulation Method for the Compact Disc Digital Audio System," *AES Digital Audio Collected Papers*, Rye, 1983.
Ogawa, H., K. Odaka, and M. Yamamoto, "Digital Disk Recording and Reproduction," *Audio Engineering Handbook*, K. B. Benson (ed.), McGraw-Hill, 1988.
Oppenheim, C. (ed.), *CD-ROM: Fundamentals to Applications*, Butterworths, 1988.
Pahwa, A., *The CD-Recordable Bible*, Eight Bit Books, 1994.
Philips International, Inc., *Compact Disc-Interactive: A Designer's Overview*, McGraw-Hill, 1988.
Pohlmann, K. C., "The Compact Disc Formats: Technology and Applications," *JAES*, vol. 36, no. 4, April 1988.
Pohlmann, K. C., *The Compact Disc Handbook*, 2nd ed., A-R Editions, 1992.
Roth, J. P., *Essential Guide to CD-ROM*, Meckler Publishing, 1986.
Sekiguchi, K., Y. Maruyama, and M. Tsubaki, "An Extension of the CD Mastering System Format for CD-ROM Mastering," *AES 83rd Conv.*, preprint 2557, New York, October 1987.
Sherman, C. (ed.), *The CD-ROM Handbook*, 2nd ed., McGraw-Hill, 1988.
Sony Corporation, *Super Audio CD*, August 1999.
Stuart, J. R., "Coding Methods for High-Resolution Recording Systems," *AES 103rd Conv.*, preprint 4639, New York, September 1997.
Ten Kate, R., "Disc Technology for Super-Quality Audio Applications," *AES 103rd Conv.*, preprint 4565, New York, September 1997.
Van der Meer, J., "The Full Motion System for CD-I," *IEEE Trans. Consum. Electron.*, vol. 38, no. 4, November 1992.
Verbakel, J., et al., "Super Audio CD Format," *AES 104th Conv.*, preprint 4705, Amsterdam, May 1998.
Verkaik, W., "Compact Disc (CD) Manufacturing—An Industrial Process," *AES Digital Audio Collected Papers*, Rye, 1983.
Vries, L. B. "The Error Control System of Philips Compact Disc," *AES 64th Conv.*, preprint 1548, New York, 1979.
Vries, L. B., et al., "The Compact Disc Digital Audio System: Modulation and Error Correction," *AES 67th Conv.*, preprint 1674, New York, October/November 1980.

8 DVD

Bennett, H., "In DVD's Own Image," *EMedia Professional*, vol. 11, no. 7, July 1998.
Carriere, J., et al., "Principles of Optical Disk Data Storage," *Progress in Optics*, vol. 41, E. Wolf (ed.), Elsevier, 2000.
DeLancie, P., "Digital Versatile Disc," *Mix*, vol. 20, no. 10, October 1996.
DVD Forum, www.dvdforum.org.
Electronic Industries Alliance, www.eia.org.

Fuchigami, N., T. Kuroiwa, and B. H. Suzuki, "DVD-Audio Specifications," *JAES*, vol. 48, no. 12, December 2000.

Gonzales, R. C., and R. E. Woods, *Digital Image Processing*, Addison Wesley, 1992.

Guenette, D. R., and D. J. Parker, "CD, CD-ROM, CD-R, CD-RW, DVD, DVD-R, DVD-RAM: The Family Album," *EMedia Professional*, vol. 10, no. 4, April 1997.

Hawksford, M. O., "Multi-Channel High-Definition Digital Audio Systems for High-Density Compact Disk," *AES 101st Conv.*, preprint 4362, Los Angeles, November 1996.

Hayashi, H., et al., "DVD Players Using a Viterbi Decoding Circuit," *IEEE Trans. Consum. Electron.*, vol. 44, no. 2, May 1998.

Held, G., *Data Compression: Techniques and Applications, Hardware and Software Considerations*, 3rd. ed., John Wiley & Sons, 1991.

Immink, K. A. S., "The Digital Versatile Disc (DVD): System Requirements and Channel Coding," *SMPTE J.*, vol. 105, no. 8, August 1996.

Immink, K. A. S., "EFM Coding: Squeezing the Last Bits," *IEEE Trans. Consum. Electron.*, vol. 43, no. 3, August 1997.

Immink, K. A. S., "EFMPlus: The Coding Format of the Multimedia Compact Disc," *IEEE Trans. Consum. Electron.*, vol. 41, no. 3, August 1995.

Lin, J.-Y., "Sub-Picture Decoder Architecture for DVD," *IEEE Trans. Consum. Electron.*, vol. 44, no. 2, May 1998.

Lubell, P. D., "The Gathering Storm in High Density Compact Disks," *IEEE Spectr.*, vol. 32, no. 8, August 1995.

McMurdie, M., and R. Griffith, "Packet Writing and UDF," *EMedia Professional*, vol. 10, no. 5, May 1997.

Moving Pictures Expert Group, DVD Resources, www.mpeg.org/MPEG/ dvd.html.

Nakamura, M., et al., "A Study on a High Density Digital Video Disc Player," *IEEE Trans. Consum. Electron.*, vol. 41, no. 3, August 1995.

Parker, D. J., "The Many Faces of High-Density Rewritable Optical," *EMedia Professional*, vol. 11, no. 1, January 1998.

Parker, D. J., "Writable DVD," *EMedia Professional*, vol. 12, no. 1, January 1999.

Pushic, D., "Mastering Machines, Mastering Choices," *EMedia Professional*, vol. 10, no. 8, August 1997.

Shinoda, M., et al., "Optical Pick-Up for DVD," *IEEE Trans. Consum. Electron.*, vol. 42, no. 3, August 1996.

Starrett, R. A., "Packet Writing," *Emedia Professional*, vol. 10, no. 5, May 1997.

Stuart, J. R., "High-Quality Audio Application of DVD," www.meridian-audio.com/ara/, 1996.

Taylor, J., *DVD Demystified*, 3rd ed., McGraw-Hill, 2005.

Yamada, M., et al., "DVD/CD/CD-R Compatible Pick-Up with Two-Wavelength Two-Beam Laser," *IEEE Trans. Consum.* Electron., vol. 44, no. 3, August 1998.

Yasuda, M., et al., "MPEG2 Video Decoder and AC-3 Audio Decoder LSIs for DVD Player," *IEEE Trans. Consum. Electron.*, vol. 43, no. 3, August 1997.

9 Blu-Ray

Blu-ray Disc Association, "Audio Visual Application Format Specifications for AVCREC," March 2008.

Blu-ray Disc Association, "Blu-ray Disc Format: 1.A Physical Format Specifications for BD-RE," 2nd ed., February 2006.

Blu-ray Disc Association, "Blu-ray Disc: 1.C Physical Format Specifications for BD-ROM," 5th ed., March 2007.

Blu-ray Disc Association, "Blu-ray Disc Format: 2.B Audio Visual Application Format Specifications for BD-ROM," March 2005.

Blu-ray Disc Association, "Blu-ray Disc Format: 2.B Audio Visual Application Format Specifications for BD-ROM Version 2.4," May 2010.

Blu-ray Disc Association, "Blu-ray Disc Recordable Format: Part 1 Physical Specifications," February 2006.

Blu-ray Disc Association, "Blu-ray Disc Rewritable Format: Audio Visual Application Format Specifications for BD-RE," Version 2.1, March 2008.

Blu-ray Disc Association, "Blu-ray Disc Rewritable Format: Audio Visual Application Format Specifications for BD-RE," Version 3.0, March 2008.

Blu-ray Disc Founders, "Blu-ray Disc Format: General," August 2004.

Blu-ray Disc Founders, "Blu-ray Disc Format: 3. File System Specifications for BD-RE, R, ROM," August 2004.

Blu-ray Disc Founders, "Blu-ray Disc Format: 4. Key Technologies," August 2004.

Dell Inc., "Blu-ray Disc Next-Generation Optical Storage: Protecting Content on the BD-ROM," October, 2006.

ISO/IEC 14496-10:2005 "Information Technology—Coding of Audio-Visual Objects—Part 10: Advanced Video Coding." *International Organization for Standardization*, Geneva, 2005.

Johnson, M., C. Crawford, and C. Armbrust, *High-Definition DVD Handbook: Producing for HD DVD and Blu-ray Disc*, McGraw-Hill, 2007.

O'Hara, G., "Understanding the Blu-ray Format," *Mix*, vol. 33, no. 12, December 2009.

SMPTE, "421M-2006 Television—VC-1 Compressed Video Bitstream Format and Decoding Process," *Society of Motion Picture and Television Engineers*, 2006.

Taylor, J., et al., *Blu-ray Disc Demystified*, McGraw-Hill, 2009.

Zink, M., P. Starner, and B. Foote, *Programming HD-DVD and Blu-ray Disc: The HD Cookbook*, McGraw-Hill, 2008.

10 Low Bit-Rate Coding: Theory and Evaluation

Aarts, R. M., et al., "A Unified Approach to Low- and High-Frequency Bandwidth Extension," *AES 115th Conv.*, preprint 5921, New York, October 2003.

Agerkvist, F. T., "A Time-Frequency Auditory Model Using Wavelet Packets," *JAES*, vol. 44, no. 1/2, January/February 1996.

Ah med, N., T. Natarajan, and K. R. Rao, "Discrete Cosine Transform," *IEEE Trans. Comput.*, vol. C-23, no. 1, January 1974.

Allen, J., "Cochlear Modeling," *IEEE ASSP Mag.*, vol. 2, no. 1, January 1985.

Angus, J., et al., "Krasner's Audio Coder Revisited," *AES 126th Conv.*, preprint 7715, Munich, May 2009.

Annadana, R., et al., "New Enhancements to the Audio Bandwidth Extension Toolkit (ABET)," *AES 124th Conv.*, preprint 7488, Amsterdam, May 2008.

Annadana, R., et al., "New Results in Low Bit Rate Speech Coding and Bandwidth Extension," *AES 121st Conv.*, preprint 6876, San Francisco, October 2006.

Annadana, R., et al., "A Novel Audio Post-Processing Toolkit for the Enhancement of Audio Signals Coded at Low Bit Rates," *AES 123rd Conv.*, preprint 7221, New York, October 2007.

Audio Precision, "Testing Reduced Bit-Rate Codecs Using the System One DSP Program CODEC.DSP," *Tech Notes TN-14*, 1994.

Barbedo, J. G. A., and A. Lopes, "A New Cognitive Model of Objective Assessment of Audio Quality," *JAES*, vol. 53, no. 1/2, January/February 2005.

Baumgarte, F., "Improved Audio Coding Using a Psychoacoustic Model Based on a Cochlear Filter Bank," *IEEE Trans. Speech Audio Proc.*, vol. 10, no. 7, October 2002.

Beaton, R. J., et al., "Objective Perceptual Measurement of Audio Quality," *AES Collected Papers on Digital Audio Bit-Rate Reduction*, N. Gilchrist and C. Grewin (eds.), Audio Engineering Society, 1996.

Bech, S., and N. Zacharov, *Perceptual Audio Evaluation: Theory, Method and Application*, John Wiley & Sons, 2006.

Beerends, J. G., et al., "Degradation Decomposition of the Perceived Quality of Speech Signals on the Basis of a Perceptual Modeling Approach," *JAES*, vol. 55, no. 12, December 2007.

Beerends, J. G., and J. A. Stemerdink, "A Perceptual Audio Quality Measure Based on a Psychoacoustic Sound Representation," *JAES*, vol. 40, no. 12, December 1992.

Benjamin, E., "Evaluating Digital Audio Artifacts with PEAQ," *AES 113th Conv.*, preprint 5711, Los Angeles, October 2002.

Berger, T., *Rate Distortion Theory*, Prentice Hall, 1971.

Boley, J., and M. Lester, "Statistical Analysis of ABX Results Using Signal Detection Theory," *AES 127th Conv.*, preprint 7826, New York, October 2009.

Brandenburg, K., "Evaluation of Quality for Audio Encoding at Low Bit Rates," *AES 82nd Conv.*, preprint 2433, London, March 1987.

Brandenburg, K., "High Quality Sound Coding at 2.5 Bits/sample," *AES 84th Conv.*, preprint 2582, Paris, March 1988.

Brandenburg, K., "Perceptual Models for the Prediction of Sound Quality of Low Bit-Rate Codecs," *Proc. AES 12th Int. Conf.*, Copenhagen, June 1993.

Brandenburg, K., and D. Seitzer, "Low Bit Rate Coding of High-Quality Digital Audio: Algorithms and Evaluation of Quality," *Proc. AES 7th Int. Conf.*, Toronto, May 1989.

Burstein, H., "Approximation Formulas for Error Risk and Sample Size in ABX Testing," *JAES*, vol. 36, no. 11, November 1988.

Burstein, H., "Transformed Binomial Confidence Limits for Listening Tests," *JAES*, vol. 37, no. 5, May 1989.

Cabot, R. C., "Performance Assessment of Reduced Bit Rate Codecs," *AES UK Conf.*, London, May 1994.

Cambridge, P., and M. Todd, "Audio Data Compression Techniques," *AES 94th Conv.*, preprint 3584, Berlin, March 1993.

Campbell M., and C. Greated, *The Musician's Guide to Acoustics*, Schirmer Books, 1987.

Capellini, V., *Data Compression and Error Control Techniques with Applications*, Academic Press, 1985.

Cellier, C., P. Chenes, and M. Rossi, "Lossless Audio Bit Rate Reduction," *Proceedings of Managing the Bit Budget, AES UK Conference*, paper MBB-11, May 1994.

Cellier, C., P. Chenes, and M. Rossi, "Lossless Audio Data Compression for Real Time Applications," *AES 95th Conv.*, preprint 3780, New York, October 1993.

Choi, I., et al., "Objective Measurement of Perceived Auditory Quality in Multichannel Audio Compression Coding Systems," *JAES*, vol. 56, no. 1/2, January/February 2008.

Colomes. C., et al., "A Perceptual Model Applied to Audio Bit-Rate Reduction," *JAES*, vol. 43, no. 4, April 1995.

Colomes, C., et al., "Perceptual Quality Assessment for Digital Audio: PEAQ—The New ITU Standard for Objective Measurement of the Perceived Audio Quality," *Proc. AES 17th Int. Conf.*, Florence, September 1999.

Cook, P. R. (ed.), *Music, Cognition, and Computerized Sound: An Introduction to Psychoacoustics*, MIT Press, 1999.

Cox, R. V., "The Design of Uniformly and Nonuniformly Spaced Pseudoquadrature Mirror Filters," *IEEE Trans. ASSP*, vol. ASSP-34, no. 5, October 1986.

Craven, P. G., and M. Gerzon, "Lossless Coding for Audio Discs," *JAES*, vol. 44, no. 9, September 1996.

Craven, P. G., and J. R. Stuart, "Cascadable Lossy Data Compression Using a Lossless Kernel," *AES 102nd Conv.*, preprint 4416, Munich, March 1997.

Craven, P. G., M. J. Law, and J. R. Stuart, "Lossless Compression Using IIR Prediction Filters," *AES 102nd Conv.*, preprint 4415, Munich, March 1997.

Crochiere, R. E., "On the Design of Sub-Band Coders for Low-Bit-Rate Speech Communication," *Bell Sys. Tech. J.*, vol. 56, no. 5, May/June 1977.

Crochiere, R. E., "Sub-Band Coding," *Bell Sys. Tech. J.*, vol. 60, no. 7, September 1981.

Davisson, L. D., and R. M. Gray (eds.), *Data Compression*, Dowden, Hutchinson & Ross, 1976.

Deutsch, D., "Auditory Illusions, Handedness, and the Spatial Environment," *JAES*, vol. 31, no. 9, September 1983.

Egan, J. P., and H. W. Hake, "On the Masking Pattern of a Simple Auditory Stimulus," *JASA*, vol. 22, no. 5, September 1950.

Eng, K. H. C., et al., "A New Bit Allocation Method for Low Delay Audio Coding at Low Bit Rates," *AES 112th Conv.*, preprint 5573, Munich, May 2002.

Erne, M., "Perceptual Audio Coders: What to Listen For," *AES 111th Conv.*, preprint 5489, New York, September 2001.

Esteban, D., and C. Galand, "Application of Quadrature Mirror Filters to Split Band Voice Coding Schemes," *Proc. ICASSP*, Hartford, CT, 1977.

Feiten, B., "Spectral Properties of Audio Signals and Masking with Aspect to Bit Rate Reduction," *AES 86th Conv.*, preprint 2795, Hamburg, March 1989.

Ferreira, A. J. S., "Accurate Spectral Replacement," *AES 118th Conv.*, preprint 6383, Barcelona, May 2005.

Ferreira, A. J. S., "The Perceptual Audio Coding Concept: From Speech to High-Quality Audio Coding," Proc. *AES 17th Int. Conf.*, Florence, September 1999.

Ferreira, A. J. S., "Tonality Detection in Perceptual Coding of Audio," *AES 98th Conv.*, preprint 3947, Paris, February 1995.

Fielder, L. D., "Evaluation of the Audible Distortion and Noise Produced by Digital Audio Converters," *JAES*, vol. 35, no. 7/8, July/August 1987.

Fielder, L. D., "Human Auditory Capabilities and Their Consequences in Digital-Audio Converter Design," *Proc. AES 7th Int. Conf.*, Toronto, May 1989.

Fletcher, N, "Auditory Patterns," *Rev. Modern Physics*, vol. 12, pp. 47–65, 1940.

Forshay, S. E., "Audio Data Compression," *Broadcast Engineering*, September 1994.

Frindle, P., "Are We Measuring the Right Things? Artifact Audibility Versus Measurement," *Proc. AES 12th UK Conf.*, London, April 1997.

Furmann, A., et al., "On the Correlation between the Subjective Evaluation of Sound and the Objective Evaluation of Acoustic Parameters for a Selected Source," *JAES*, vol. 38, no. 11, November 1990.

Galand, C. R., and H. J. Nussbaumer, "New Quadrature Mirror Filter Structures," *IEEE Trans. ASSP*, vol. ASSP-32, no. 3, June 1984.

Geddes, E. R., and L. W. Lee, "Auditory Perception of Nonlinear Distortion—Theory," *AES 115th Conv.*, preprint 5890, New York, October 2003.

Gerzon, M., "The Gentle Art of Digital Squashing," *Studio Sound*, May 1990.

Gerzon, M., "Problems of Error Masking in Audio Data Compression Systems," *AES 90th Conv.*, preprint 3013, Paris, February 1991.

Gilchrist, N., "ATLANTIC Audio: Preserving Technical Quality During Low Bit-Rate Coding and Decoding," *AES 104th Conv.*, preprint 4694, Amsterdam, May 1998.

Greenwood, D. D., "Critical Bandwidth and Frequency Coordinates of the Basilar Membrane," *J. Acoust. Soc. Amer.*, no. 33, pp. 1344–1356, 1961.

Grewin, C., "Methods for Quality Assessment of Low Bit-Rate Audio Codecs," *Proc. AES 12th Int. Conf.*, Copenhagen, June 1993.

Groschel, A., et al., "Enhancing Audio Coding Efficiency of MPEG Layer-2 with Spectral Band Replication (SBR) for Digital Radio (DAB) in a Backwards Compatible Way," *AES 114th Conv.*, preprint 5850, Amsterdam, March 2003.

Grusec, T. L., "Subjective Evaluation of High Quality Audio Systems," *Can. Acous.*, vol. 22, September 1994.

Grusec, T., L. Thibault, and R. J. Beaton, "Sensitive Methodologies for the Subjective Evaluation of High Quality Audio Coding Systems," *Proc. AES UK DSP Conf.*, London, September 1992.

Grusec, T., L. Thibault, and G. Soulodre, "EIA/NRSC DAR Systems Subjective Tests, Part I: Audio Codec Quality," *IEEE Trans. Broadcast.*, vol. 43, September 1997.

Handel, S., *Listening*, MIT Press, 1993.

Hans, M., and R. W. Schafer, "An MPEG Audio Layered Transcoder," *AES 105th Conv.*, preprint 4812, San Francisco, September 1998.

Hellman, R. P., "Asymmetry of Masking between Noise and Tone," *Perception and Psychophysics*, vol. 11, pp. 241–246, 1972.

Heo, I., and K.-M. Sung, "An Improved Weighting Curve Based on Equal-Loudness Contour," *Proc. AES 34th Int. Conf.*, Jeju Island, Korea, August 2008.

Herre, J., "Temporal Noise Shaping, Quantization and Coding Methods in Perceptual Audio Coding: A Tutorial Introduction," *Proc. AES 17th Int. Conf.*, Florence, September 1999.

Herre, J., and J. D. Johnston, "Enhancing the Performance of Perceptual Audio Coders by Using Temporal Noise Shaping," *AES 101st Conv.*, preprint 4384, Los Angeles, November 1996.

Herre, J., and J. D. Johnston, "Exploiting Both Time and Frequency Structure in a System That Uses an Analysis /Synthesis Filter Bank with High-Frequency Resolution," *AES 103rd Conv.*, preprint 4519, New York, September 1997.

Herre, J., and M. Schug, "Analysis of Decompressed Audio—The Inverse Decoder," *AES 109th Conv.*, preprint 5256, Los Angeles, September 2000.

Herre, J., et al., "Analysis Tool for Real Time Measurements Using Perceptual Criteria," *Proc. AES 11th Int. Conf.*, Portland, May 1992.

Hollier, M. P., M. O. Hawksford, and D. R. Guard, "Algorithms for Assessing the Subjectivity of Perceptually Weighted Audible Errors," *JAES*, vol. 43, no. 12, December 1995.

Humes, L. E., "Psychoacoustic Foundations of Clinical Audiology," *Handbook of Clinical Audiology*, 3rd ed., J. Katz (ed.), Williams and Wilkins, 1985.

Isherwood, D., et al., "Augmentation, Application and Verification of the Generalized Listener Selection Procedure," *AES 115th AES Conv.*, preprint 5984, New York, October 2003.

ITU-R Rec. BS.1116, "Methods for the Subjective Assessment of Small Impairments in Audio Systems Including Multichannel Sound Systems," International Telecommunication Union, Geneva, 1994.

Iwadare, M., et al., "A 128 kb/s Hi-Fi Audio CODEC Based on Adaptive Transform Coding with Adaptive Block Size," *IEEE J. Sel. Areas in Commun.*, vol. 10, no. 1, January 1992.

Jang, D., et al., "A Multichannel Audio Codec System for Multichannel Audio Authoring," *AES 106th Conv.*, preprint 4914, Munich, May 1999.

Jayant, N. S., and P. Noll, *Digital Coding of Waveforms: Principles and Applications to Speech and Video*, Prentice Hall, 1984.

Jeong, H., and I. Jeong-Guon, "Implementation of a New Algorithm Using the STFT with Variable Frequency Resolution for the Time-Frequency Auditory Model," *JAES*, vol. 47, no. 4, April 1999.

Jerger, J. (ed.), *Modern Developments in Audiology*, 2nd ed., Academic Press, 1973.

Jiao, Y., et al., "Optimal Hierarchical Bandwidth Limitation of Surround Sound," *AES 124th Conv.*, preprint 7475, Amsterdam, May 2008.

Johnston, J. D., "Estimation of Perceptual Entropy Using Noise Masking Criteria," *Proc. ICASSP*, pp. 2524–2527, May 1988.

Johnston, J. D., "Perceptual Transform Coding of Wideband Stereo Signals," *ICASSP-89 Record*, pp. 1993–1996, May 1989.

Johnston, J. D., "Rate Loop Processor for Perceptual Encoder/Decoder," U.S. Patent 5,627,938, May 6, 1997.

Johnston, J. D., "Transform Coding of Audio Signals Using Perceptual Noise Criteria," *IEEE J. Sel. Areas Comm.*, vol. 6, no. 2, pp. 314–323, February 1988.

Johnston, J. D., and R. E. Crochiere, "An All-Digital 'Commentary Grade' Subband Coder," *J. Audio Eng. Soc.*, vol. 27, no. 11, pp. 855–865, 1979.

Kate, W. R., "Maintaining Audio Quality in Cascaded Psychoacoustic Coding," *AES 101st Conv.*, preprint 4387, Los Angeles, November 1996.

Katz, J., "Clinical Audiology," *Handbook of Clinical Audiology*, 3rd ed., J. Katz (ed.), Williams and Wilkins, 1985.

Keyhl, M., J. Herre, and C. Schmidmer, "NMR Measurements of Consumer Recording Devices Which Use Low Bit-Rate Audio Coding," *AES 94th Conv.*, preprint 3616, Berlin, March 1993.

Knapen, E., et al., "Lossless Compression of One-Bit Audio," *JAES*, vol. 52, no. 3, March 2004.

Koller, J., et al., "Robust Coding of High Quality Audio Signals," *AES 103rd Conv.*, preprint 4621, New York, September 1997.

Komly, A., "Assessing the Performance of Low-Bit-Rate Audio Codecs: An Account of the Work of ITU-R Task Group 10/2," *AES Collected Papers on Digital Audio Bit-Rate Reduction*, N. Gilchrist and C. Grewin (eds.), Audio Engineering Society, 1996.

Krämer, U., et al., "Ultra Low Delay Audio Coding with Constant Bit Rate," *AES 117th Conv.*, preprint 6197, San Francisco, October 2004.

Krasner, M. A., "The Critical Band Coder–Digital Encoding of Speech Signals Based on the Perceptual Requirements of the Auditory System," *Proc. ICASSP*, pp. 327–331, April 1980.

Krasner, M. A., "Digital Encoding of Speech and Audio Signals Based on the Perceptual Requirements of the Auditory System," Technical Report 535, MIT Lincoln Laboratory, June 1979.

Kulesza, M., and A. Czyzewski, "Audio Codec Employing Frequency-Derived Tonality Measure," *AES 127th Conv.*, preprint 7877, New York, October 2009.

Kulesza, M., and A. Czyzewski, "Tonality Estimation and Frequency Tracking of Modulated Tonal Components," *JAES*, vol. 57, no. 4, April 2009.

Kurniawati, E., S. Ng, and S. George, "A Study of MPEG Surround Configurations and Its Performance Evaluation," *AES 126th Conv.*, preprint 7667, Munich, May 2009.

Kurniawati, E., et al., "Decoder-Based Approach to Enhance Low Bit-Rate Audio," *AES 116th Conv.*, preprint 6177, Berlin, May 2004.

Kurth, F., "An Audio Codec for Multiple Generations Compression without Loss of Perceptual Quality," *AES 17th Int. Conf.*, Florence, September 1999.

Kurth, F., and V. Hassenrik, "A Dynamic Embedding Codec for Multiple Generations Compression," *AES 109th Conv.*, preprint 5257, Los Angeles, September 2000.

Larsen, E., et al., "Efficient High-frequency Bandwidth Extension of Music and Speech," *AES 112th Conv.*, preprint 5627, Munich, May 2002.

Lauber, P., and R. Sperschneider, "Error Concealment for Compressed Digital Audio," *AES 111st Conv.*, preprint 5460, New York, September 2001.

Lee, L. W., and E. R. Geddes, "Auditory Perception of Nonlinear Distortion," *AES 115th Conv.*, preprint 5891, New York, October 2003.

Liebchen, T., "Lossless Audio Coding Using Adaptive Multichannel Prediction," *AES 113th Conv.*, preprint 5680, Los Angeles, October 2002.

Liebetrau, J., D. Beer, and M. Lubkowitz, "Psychoacoustical Bandwidth Extension of Lower Frequencies," *AES 127th Conv.*, preprint 7890, New York, October 2009.

Lin, C.-S., and C. Kyriakakis, "High-Definition Audio Spectrum Separation via Frequency Response Masking Filter Banks," *JAES*, vol. 57, no. 10, October 2009.

Liu, C.-M., W.-C. Lee, and T.-W. Chang, "The Efficient Temporal Noise Shaping Method," *AES 116th Conv.*, preprint 6051, Berlin, May 2004.

Liu, C.-M., et al., "Compression Artifacts in Perceptual Audio Coding," *AES 121st Conv.*, preprint 6872, San Francisco, October 2006.

Liu, C. M., et al., "High Frequency Reconstruction by Linear Extrapolation," *AES 115th Conv.*, preprint 5968, New York, October 2003.

Lookabaugh, T., and M. Perkins, "Application of the Princen–Bradley Filter Bank to Speech and Image Compression," *IEEE Trans. ASSP*, vol. 38, no. 11, November 1990.

Lyman, S., "An Introduction to Audio Subjective Testing," *CBC Engineering Review*, vol. 33, 1993–94.

Madisetti, V. K., and D. B. Williams (eds.), *The Digital Signal Processing Handbook*, CRC Press, 1998.

Maher, R. C., "Lossless Audio Coding," *Lossless Compression Handbook*, K. Sayood (ed.), Academic Press, 2003.

Malvar, H. S., "Lapped Transforms for Efficient Transform/Subband Coding," *IEEE Trans. ASSP*, vol. 38, no. 6, June 1990.

Marins, P., F. Rumsey, and S. Zielinski, "The Relationship between Basic Audio Quality and Selected Artifacts in Perceptual Audio Codecs—Part II: Validation Experiment," *AES 122nd Conv.*, preprint 7079, Vienna, May 2007.

Marins, P., F. Rumsey, and S. Zielinsky, "The Relationship between Selected Artifacts and Basic Audio Quality in Perceptual Audio Codecs," *AES 120th Conv.*, preprint 6745, Paris, May 2006.

Marins, P., F. Rumsey, and S. Zielinski, "Unraveling the Relationship between Basic Audio Quality and Fidelity Attributes in Low Bit-Rate Multichannel Audio Codecs," *AES 124th Conv.*, preprint 7335, Amsterdam, May 2008.

Marston, D., F. Kozamernik, G. Stoll, and G. Spikofski, "Further EBU Tests of Multichannel Audio Codecs," *AES 126th Conv.*, preprint 7730, Munich, May 2009.

Mason, A. J., et al., "The ATLANTIC Audio Demonstration Equipment," *JAES*, vol. 48, no. 1/2, January/February 2000.

Meyer, E. B., and D. R. Moran, "Audibility of a CD-Standard A/D/A Loop Inserted into High-Resolution Audio Playback," *JAES*, vol. 55, no. 9, September 2007.

Moehrs, S., et al., "Analysing Decompressed Audio with the "Inverse Decoder"—Towards an Operative Algorithm," *AES 112th Conv.*, preprint 5576, Munich, May 2002.

Moore, B. C. J. (ed.), *Handbook of Perception and Cognition*, vol. 6, Hearing, Academic Press, 1995.

Moore, B. C. J., *An Introduction to the Psychology of Hearing*, Academic Press, 1982.

Moore, B. C. J., "Masking in the Human Auditory System," *AES Collected Papers on Digital Audio Bit-Rate Reduction*, N. Gilchrist and C. Grewin (eds.), Audio Engineering Society, 1996.

Najaf-Zadeh, H., et al., "Incorporation of Inharmonicity Effects into Auditory Masking Models," *AES 113th Conv.*, preprint 5685, Los Angeles, October 2002.

Nguyen, T. Q., and P. P. Vaidyanathan, "Two-Channel Perfect-Reconstruction FIR QMF Structures Which Yield Linear-Phase Analysis and Synthesis Filters," *IEEE Trans. ASSP.*, vol. 37, no. 5, May 1989.

Paillard, B., et al., "PERCEVAL: Perceptual Evaluation of the Quality of Audio Signals," *JAES*, vol. 40, no. 1/2, January/February 1992.

Petrovsky, A., E. Azarov, and A. Petrovsky, "Harmonic Representation and Auditory Model-Based Parametric Matching and its Application in Speech/Audio Analysis," *AES 126th Conv.*, preprint 7705, Munich, May 2009.

Painter, T., and A. Spanias, "Perceptual Coding of Digital Audio," *Proc. IEEE*, vol. 88, no. 4, April 2000.

Prakash, V., et al., "Removal of Birdie Artifact in Perceptual Audio Coders," *AES 116th Conv.*, preprint 6064, Berlin, May 2004.

Precoda, K., and T. H. Meng, "Listener Differences in Audio Compression Evaluations," *JAES*, vol. 45, no. 9, September 1997.

Princen, J., and A. Bradley, "Analysis /Synthesis Filter Band Design Based on Time-Domain Aliasing Cancellation," *IEEE Trans. ASSP*, vol. ASSP-34, no. 5, October 1986.

Princen, J., A. Johnson, and A. Bradley, "Subband/Transform Coding Using Filter Band Designs Based on Time Domain Aliasing Cancellation," *Proc. IEEE Int. Conf. ASSP*, pp. 2161–2164, Dallas, 1987.

Purat, M., et al., "Lossless Transform Coding of Audio Signals," *AES 102nd Conv.*, preprint 4414, Munich, March 1997.

Puri, A., and T. Chen (eds.), *Advances in Multimedia: Standards, Systems, and Networks*, Marcel Dekker, 1999.

Purnhagen, H., et al., "Object-Based Analysis/Synthesis Audio Coder for Very Low Bit Rates," *AES 104th Conv.*, preprint 4747, Amsterdam, May 1998.

Radocy, R. E., and J. D. Boyle, *Psychological Foundations of Musical Behavior*, 3rd ed., Charles C Thomas, 1996.

Recommendation ITU-R BS.1387-1, "Method for Objective Measurements of Perceived Audio Quality," 1998–2001.

Robinson, D. W., and R. S. Dadson, "A Re-Determination of the Equal-Loudness Relations for Pure Tones," *British J. of Applied Physics*, vol. 7, pp. 166–181, 1956.

Rödén, J., et al., "A Study of the MPEG Surround Quality versus Bit-Rate Curve," *AES 123rd Conv.*, preprint 7219, New York, October 2007.

Rothweiler, J. H., "Polyphase Quadrature Filters—A New Subband Coding Technique," *Proc. ICASSP*, pp. 1280–1283, April 1983.

Rumsey, F., "Putting Low-Bit-Rate Audio to Work," *AES Collected Papers on Digital Audio Bit-Rate Reduction*, N. Gilchrist and C. Grewin (eds.), Audio Engineering Society, 1996.

Ryden, T., "Using Listening Tests to Assess Audio Codecs," *AES Collected Papers on Digital Audio Bit-Rate Reduction*, N. Gilchrist and C. Grewin (eds.), Audio Engineering Society, 1996.

Seefeldt, A., M. Vinton, and C. Robinson, New Techniques in Spatial Audio Coding," *AES 119th Conv.*, preprint 6587, New York, October 2005.

Schroeder, E. F., and W. Voessing, "High Quality Digital Audio Encoding with 3.0 Bits/Sample Using Adaptive Transform Coding," *AES 80th Conv.*, preprint 2321, Montreaux, March 1986.

Schroeder, M. R., B. S. Atal, and J. L. Hall, "Optimizing Digital Speech Coders by Exploiting Masking Properties of the Human Ear," *JASA*, vol. 66, no. 6., pp. 1647–1652, December 1979.

Schulz, D. "Improving Audio Codecs by Noise Substitution," *JAES*, vol. 44, no. 7/8, July/August 1996.

Seitzer, D., K. Brandenburg, R. Kapust, E. Eberlein, H. Gerhaeuser, H. Popp, and H. Schott, "Real-Time Implementation of Low Complexity Adaptive Transform Coder," *AES 84th Conv.*, preprint 2581, Paris, March 1988.

Sinha, D., and A. Ferreira, "A New Broadcast Quality Low Bit Rate Audio Coding Scheme Utilizing Novel Bandwidth Extension Tools," *AES 119th Conv.*, preprint 6588, New York, October 2005.

Sinha, D., et al., A Novel Integrated Audio Bandwidth Extension Toolkit (ABET)," *AES 120th Conv.*, preprint 6788, Paris, May 2006.

Smyth, N., and D. Trainor, "Reducing the Complexity of Sub-band ADPCM Coding to Enable High-Quality Audio Streaming from Mobile Devices," *AES 127th Conv.*, preprint 7891, New York, October 2009.

Soulodre, G. A., et al., "Subjective Evaluation of State-of-the-Art Two-Channel Audio Codecs," *JAES*, vol. 46, no. 3, March 1998.

Sporer, T., and K. Brandenburg, "Constraints of Filter Banks Used for Perceptual Measurement," *JAES*, vol. 43, no. 3, March 1995.

Sporer, T., J. Liebetrau, and S. Schneider, "Statistics of MUSHRA Revisited," *AES 127th Conv.*, preprint 7825, New York, October 2009.

Sporer, T., U. Gbur, J. Herre, and R. Kapust, "Evaluating a Measurement System," *JAES*, vol. 43, no. 5, May 1995.

Stautner, J. P., "Scalable Audio Compression for Mixed Computing Environments," *AES 93rd Conv.*, preprint 3357, San Francisco, October 1992.

Stautner, J. P., and D. M. Horowitz, "Efficient Data Reduction for Digital Audio Using a Digital Filter Array," *Proc. ICASSP*, pp. 13–16, 1986.

Storer, J. A., and M. Cohn (eds.), *Proceedings of the Data Compression Conference*, IEEE Computer Society Press, Snowbird, UT, March 1992.

Stuart, J. R., "Lossless Coding for DVD-Audio," Ad Hoc Committee 5 of WG-4, Version 0.5, June 30, 1998.

Tobias, J. V. (ed.), *Foundations of Modern Auditory Theory*, vol. 1, Academic Press, New York, 1970.

Trahiotis, C., "Progress and Pitfalls Associated with Scientific Measures of Auditory Acuity," *AES Digital Audio Collected Papers*, Rye, 1983.

Tribolet, J. M., and R. E. Crochiere, "Frequency Domain Coding of Speech," *IEEE Trans. Acoust. Speech Signal Process.*, ASSP-27, pp. 512–530, 1979.

Tseng, H.-Y., H.-W. Hsu, and C.-M. Liu, "High Quality, Low Power QMF Bank Design for SBR, Parametric Coding, and MPEG Surround Decoders," *AES 122nd Conv.*, preprint 7000, Vienna, May 2007.

Tsoukalas, D. E., J. Mourjopoulos, and G. Kokkinakis, "Perceptual Filters for Audio Signal Enhancement," *JAES*, vol. 45, no. 1/2, January/February 1997.

Vaidyanathan, P. P., "Multirate Digital Filters, Filter Banks, Polyphase Networks, and Applications: A Tutorial," *Proc. IEEE*, vol. 78, no. 1, January 1990.

Vaidyanathan, P. P., Multirate Systems and Filter Banks, Prentice Hall, 1993.

Vaidyanathan, P. P., "Quadrature Mirror Filter Banks, M-Band Extensions and Perfect-Reconstruction Techniques," *IEEE ASSP Mag.*, vol. 4, no. 3, July 1987.

Van de Par, S., J. Koppens, A. Kohlrausch, and W. Oomen, "A New Perceptual Model for Audio Coding Based on Spectro-Temporal Masking," *AES 124th Conv.*, preprint 7336, Amsterdam, May 2008.

Van Schijndel, N. H., et al., "Adaptive RD Optimized Hybrid Sound Coding," *JAES*, vol. 56, no. 10, October 2008.

Veldhuis, R., M. Breeuwer, and R. van der Waal, "Subband Coding of Digital Audio Signals without Loss of Quality," *Proc. IEEE Int. Conf. ASSP*, Glasgow, 1989.

Viemeister, N. F., "An Overview of Psychoacoustics and Auditory Perception," *Proc. AES 8th Int. Conf.*, May 1990.

Weiss, J., and D. Schremp, "Putting Data on a Diet," *IEEE Spectr.*, vol. 30, August 1993.

Werner, M., and G. Schuller, "An Enhanced SBR Tool for Low-Delay Applications," *AES 127th Conv.*, preprint 7876, New York, October 2009.

Wever, E. G., and M. Lawrence, *Physiological Acoustics*, Princeton University Press, 1954.

Witten, J. H., R. M. Neal, and J. G. Cleary, "Arithmetic Coding for Data Compression," *Communication of the ACM*, June 1987.

Yen Pan, D., "Digital Audio Compression," *Digit. Tech. J.*, vol. 5, no. 2, Spring 1993.

Zernicki, T., and M. Bartkowiak, "Audio Bandwidth Extension by Frequency Scaling of Sinusoidal Partials," *AES 125th Conv.*, preprint 7622, San Francisco, October 2008.

Zielinski, S. K., "Comparison of Basic Audio Quality and Timbral and Spatial Fidelity Changes Caused by Limitation of Bandwidth and by Down-Mix Algorithms in 5.1 Surround Audio Systems," *JAES*, vol. 53, no. 3, March 2005.

Zielinski, S. K., "Development and Initial Validation of a Multichannel Audio Quality Expert System," *JAES*, vol. 53, no. 1/2, January/February 2005.

Zielinski, S. K., P. Hardisty, C. Hummersone, and F. Rumsey, "Potential Biases in MUSHRA Listening Tests," *AES 123rd Conv.*, preprint 7179, New York, October 2007.

Zwicker, E., "Subdivision of the Audible Frequency Range into Critical Bands," *JASA*, vol. 33, no. 2, February 1961.

Zwicker, E., and H. Fastl, *Psychoacoustics*, 2nd ed., Springer, 1999.

Zwicker, E., and U. T. Zwicker, "Audio Engineering and Psychoacoustics: Matching Signals to the Final Receiver, The Human Auditory System," *JAES*, vol. 39, no. 3, March 1991.

11 Low Bit-Rate Coding: Codec Design

AT&T Bell Labs, "ASPEC," Doc. No. 89/ 205, ISO-IEC/JTC1/SC2/WG8 MPEG-AUDIO, October 18, 1989.

Baron, S., and W. R. Wilson, "MPEG Overview," *SMPTE J.*, June 1994.

Bergher, L., et al., "Dolby AC-3 and MPEG-2 Audio Decoder IC with 6-Channels Output," *IEEE Trans. Consum. Electron.*, vol. 43, no. 3, August 1997.

Boley, J., and V. Rao, University of Miami, Frost School of Music, research project, 2004.

Bosi, M., and R. E. Goldberg, *Introduction to Digital Audio Coding and Standards*, Kluwer Academic, 2003.

Bosi, M., et al., "ISO/IEC MPEG-2 Advanced Audio Coding," *JAES*, vol. 45, no. 10, October 1997.

Brandenburg, K., "MP3 and AAC Explained," *Proc. AES 17th Int. Conf.*, Florence, September 1999.

Brandenburg, K., "OCF—A New Coding Algorithm for High Quality Sound Signals," *Proc. ICASSP*, pp. 141–144, 1987.

Brandenburg, K., and M. Bosi, "Overview of MPEG Audio: Current and Future Standards for Low-Bit-Rate Audio Coding," *JAES*, vol. 45, no. 1/2, January/February 1997.

Brandenburg, K., and J. D. Johnston, "Perceptual Coding of Audio Signals," U.S. Patent 5,040,217, August 13, 1991; Re. 36,714, May 23, 2000.

Brandenburg K., and J. D. Johnston, "Second Generation Perceptual Audio Coding: The Hybrid Coder," *AES 88th Conv.*, preprint 2937, Montreux, March 1990.

Brandenburg, K., and D. Seitzer, "OCF: Coding High Quality Audio with Data Rates of 64 kbit/sec," *AES 85th Conv.*, preprint 2723, Los Angeles, November 1988.

Brandenburg, K., and T. Sporer, "NMR and Masking Flag: Evaluation of Quality Using Perceptual Criteria," *Proc. AES 11th Int. Conf.*, Portland, May 1992.

Brandenburg, K., and G. Stoll, "ISO-MPEG-1 Audio: A Generic Standard for Coding of High-Quality Digital Audio," *JAES*, vol. 42, no. 10, October 1994.

Brandenburg, K., et al., "ASPEC: Adaptive Spectral Entropy Coding of High Quality Music Signals," *AES 90th Conv.*, preprint 3011, Paris, February 1991.

Brandenburg, K., et al., "Low Bit Rate Codecs for Audio Signals—Implementation in Real Time," *AES 85th Conv.*, preprint 2707, Los Angeles, November 1988.

Breebaart, J. et al., "Background, Concept, and Architecture for the Recent MPEG Surround Standard on Multichannel Audio Compression," *JAES*, vol. 55, no. 5, May 2007.

Burgel, C., et al., "Beyond CD-Quality: Advanced Audio Coding (AAC) for High Resolution Audio with 24 Bit Resolution and 96 kHz Sampling Frequency," *AES 111th Conv.*, preprint 5476, New York, December 2001.

Chen, J., and H.-M. Tai, "Real-Time Implementation of the MPEG-2 Audio CODEC on a DSP," *IEEE Trans. Consum. Electron.*, vol. 44, no. 3, August 1998.

Chen, Y.-C., "Fast Time-Frequency Transform Algorithms and Their Applications to Real-Time Software Implementation of AC-3 Audio Codec," *IEEE Trans. Consum. Electron.*, vol. 44, no. 2, May 1998.

Datta, J., "Study of Window Performance with Respect to the Auditory Masking Curve for Low Bit-Rate Coding Applications," *AES 93rd Conv.*, preprint 3370, San Francisco, October 1992.

Davidson, G., L. D. Fielder, and M. Antill, "High-Quality Audio Transform Coding at 128 kbits/s," *Proc. IEEE Int. Conf. ASSP*, Albuquerque, 1990.

Davidson, G., L. D. Fielder, and M. Antill, "Low Complexity Transform Coder for Satellite Link Applications," *AES 89th Conv.*, preprint 2966, Los Angeles, September 1990.

Davis, M., "The AC-3 Multichannel Coder," *AES 95th Conv.*, preprint 3774, New York, October 1993.

DeLancie, P., "Meridian Lossless Packing," *Mix*, vol. 22, no. 12, December 1998.

Eberlein, E., et al., "Layer III: A Flexible Coding Standard," *AES 94th Conv.*, preprint 3493, Berlin, March 1993.

Elder, A. G., and S. G. Turner, "A Real-Time PC-Based Implementation of AC-2 Digital Audio Compression," *AES 95th Conv.*, preprint 3773, New York, October 1983.

Fejzo, Z., "DTS-HD: Technical Overview of Lossless Mode of Operation," *AES 118th Conv.*, preprint 6445, Barcelona, May 2005.

Fielder, L. D., and G. A. Davidson, "AC-2: A Family of Low Complexity Transform Based Music Coders," *Proc. AES 10th Int. Conf.*, London, September 1991.

Fielder, L. D., et al., "AC-2 and AC-3: Low-Complexity Transform-Based Audio Coding," *AES Collected Papers on Digital Audio Bit-Rate Reduction*, N. Gilchrist and C. Grewin (eds.), Audio Engineering Society, 1996.

Fielder, L. D., et al., "Introduction to Dolby Digital Plus, an Enhancement to the Dolby Digital Coding System," *AES 117th Conv.*, preprint 6196, San Francisco, October 2004.

Finger, R. A., "DCC Format" *Matsushita Technical Lecture*, Osaka, October 15, 1991.

Fletcher, J., "ISO/MPEG Layer 2—Optimum Re-Encoding of Decoded Audio Using a MOLE Signal," *AES 104th Conv.*, preprint 4706, Amsterdam, May 1998.

Gaston, L., and R. Sanders, "Evaluation of HE-ACC, AC-3, and E-AC-3 Codecs," *JAES*, vol. 56, no. 3, March 2008.

Gerzon, M. A., et al., "The MLP Lossless Compression System for PCM Audio," *JAES*, vol. 52, no. 3, March 2004.

Grill, B., et al., "Closing the Gap between the Multichannel and the Stereo Audio World: Recent MP3 Surround Extensions," *AES 120th Conv.*, preprint 6754, Paris, May 2006.

Herre, J., et al., "MPEG Surround—The ISO/MPEG Standard for Efficient and Compatible Multichannel Audio Coding," *JAES*, vol. 56, no. 11, November 2008.

Herre, J., et al., "MP3 Surround: Efficient and Compatible Coding of Multichannel Audio," *AES 116th Conv.*, preprint 6049, Berlin, May 2004.

Herre, J., et al., "The Reference Model Architecture for MPEG Spatial Audio Coding," *AES 118th Conv.*, preprint 6447, Barcelona, May 2005.

Hilpert, J., et al., "Implementing ISO/MPEG-2 Advanced Audio Coding in Realtime on a Fixed Point DSP," *AES 105th Conv.*, preprint 4822, San Francisco, September 1998.

ISO/IEC 11172-3:1993, "Information Technology—Coding of Moving Pictures and Associated Audio for Digital Storage Media at up to about 1.5 Mbit/s, Part 3: Audio," *International Organization for Standardization*, Geneva, 1993.

ISO/IEC 13818-3: 1997, "Information Technology—Generic Coding of Moving Pictures and Associated Audio Information, Part 3: Audio," *International Organization for Standardization*, Geneva, 1997.

ISO/IEC 13818-3: 1997, "Information Technology—Generic Coding of Moving Pictures and Associated Audio Information, Part 7: Advanced Audio Coding (AAC)," *International Organization for Standardization*, Geneva, 1997.

ISO/IEC JTC1/SC29/WG11 N1420, "Overview of the Report on the Formal Subjective Listening Tests of the MPEG-2 AAC Multichannel Audio Coding," Maceio, Brazil, November 1996.

ITU-R Rec. BS.1115, "Low Bit Rate Audio Coding," International Telecommunication Union, Geneva, 1994.

Johnston, J. D., et al., "AT&T Perceptual Audio Coding (PAC)," *AES Collected Papers on Digital Audio Bit-Rate Reduction*, N. Gilchrist and C. Grewin (eds.), June 1996.

Jhung, Y., and S. Park, "Architecture of Dual Mode Audio Filter for AC-3 and MPEG," *IEEE Trans. Consum. Electron.*, vol. 43, no. 3, August 1997.

Kate, W. R., et al., "5-Channel MUSICAM Codec," *AES 94th Conv.*, preprint 3671, Berlin, March 1993.

Kawakami, D., "The Sony MiniDisc (MD)," *Broadcast Engineering*, February 1993.

Kim, J. H., E. Oh, and J. Robilliard, "Enhanced Stereo Coding with Phase Parameters for MPEG Unified Speech and Audio Coding," *AES 127th Conv.*, preprint 7875, New York, October 2009.

Kim, S.-Y., et al., "A Real-Time Implementation of the MPEG-2 Audio Encoder," *IEEE Trans. Consum. Electron.*, vol. 43, no. 3, August 1997.

Ko, W.-S., et al., "A VLSI Implementation of Dual AC-3 and MPEG-2 Audio Decoder," *IEEE Trans. Consum. Electron.*, vol. 44, no. 3, August 1998.

Lau, W., and A. Chwu, "A Common Transform Engine for MPEG & AC3 Audio Decoder," *IEEE Trans. Consum. Electron.*, vol. 43, no. 3, August 1997.

Liu, C.-M., and W.-C. Lee, "The Design of a Hybrid Filter Bank for the Psychoacoustic Model in ISO/MPEG Phases 1,2 Audio Encoder," *IEEE Trans. Consum. Electron.*, vol. 43, no. 3, August 1997.

Liu, C.-M., et al., Efficient Bit Reservoir Design for MP3 and AAC," *AES 117th Conv.*, preprint 6200, San Francisco, October 2004.

Lokhoff, G. C. P., "DCC—Digital Compact Cassette," *IEEE Trans. Consum. Electron.*, vol. 37, no. 3, August 1991.

Lokhoff, G. C. P., "The Digital Compact Cassette," *AES Collected Papers on Digital Audio Bit-Rate Reduction*, N. Gilchrist and C. Grewin (eds.), Audio Engineering Society, 1996.

Lokhoff, G. C. P., "Precision Adaptive Subband Coding (PASC) for the Digital Compact Cassette (DCC)," *IEEE Trans. Consum. Electron.*, vol. 38, no. 4, November 1992.

Maeda, Y., "MiniDisc System," *J. Acoust. Soc. Jpn.*, vol. 49, no. 4, April 1993.

MPEG Audio Web Page, http://www.tnt.uni-hannover.de/project/mpeg/audio/.

Najaf-Zadeh, H., et al., "Use of Auditory Temporal Masking in the MPEG Psychoacoustic Model 2," *AES 114th Conv.*, preprint 5840, Amsterdam, 2003.

Nishida, S., et al., "New Developments for the Mini Disc System," *IEEE Trans. Consum. Electron.*, vol. 40, no. 3, August 1994.

Nithin, S., et al., "Low Complexity Bit Allocation Algorithms for MP3/AAC Encoding," *AES 124th Conv.*, preprint 7339, Amsterdam, May 2008.

Noll, P., "MPEG Digital Audio Coding," *IEEE Sig. Proc. Mag.*, September 1997.

Pan, D., "A Tutorial on MPEG/Audio Compression," *IEEE Multimedia*, Summer 1995.

Petitcolas, F. A. P., "MPEG for Matlab," www.petitcolas.net/fabien/software/mpeg/.

Pohlmann, K. C., "MiniDisc Technology, Parts 1–5," *Mix*, vol. 16, no. 11, November 1992—vol. 17, no. 3, March 1993.

Pohlmann, K. C., "The PASC Algorithm, Part 1," *Mix*, vol. 16, no. 3, March 1992.

Pras, A., R. Zimmerman, D. Levitin, and C. Guastavino, "Subjective Evaluation of MP3 Compression for Different Musical Genres," *AES 127th Conv.*, preprint 7879, New York, October 2009.

Rault, J. B., et al., "MUSICAM (ISO/MPEG Audio) Very Low Bit-Rate Coding at Reduced Sampling Frequency," *AES 95th Conv.*, preprint 3741, New York, October 1993.

Ryu, S.-U., and K. Rose, "A Frame Loss Concealment Technique for MPEG-AAC," *AES 120th Conv.*, preprint 6662, Paris, May 2006.

Schneider, A., K. Krauss, and A. Ehret, "Evaluation of Real-Time Transport Protocol Configurations Using aacPlus," *AES 120th Conv.*, preprint 6789, Paris, May 2006.

Schroeder, E. F., and J. Boehm, "Original File Length (OFL) for mp3, mp3PRO and other Audio Codecs," *AES 114th Conv.*, preprint 5830, Amsterdam, March 2003.

Shilen, S., "Guide to MPEG-1 Audio Standard," *IEEE Trans. Broadcast.*, vol. 40, December 1994.

Smyth, S. M. F., et al., "DTS Coherent Acoustics," *AES 100th Conv.*, preprint 4293, Copenhagen, May 1996.

Smythe, M., and S. Smythe, "Apt-X100: A Low-Delay, Low Bit-Rate, Sub-Band ADPCM Audio Coder for Broadcasting," *Proc. AES 10th Int. Conf.*, London, September 1991.

Sony Corporation, "Overview to the Technology behind MiniDisc," Audio Development Group, 1992.

Sony Corporation, "Recordable MiniDisc Technical Information," MiniDisc Division, Recording Media Group, 1992.

Soulodre, G., and M. Lavoie, "Subjective Evaluation of MPEG Layer II with Spectral Band Replication," *AES 117th Conv.*, preprint 6185, San Francisco, October 2004.

Sreenivas, T. V., and M. Dietz, "Improved AAC Performance at <64 kb/s using VQ," *AES 104th Conv.*, preprint 4750, Amsterdam, May 1998.

Stoll, G., "ISO-MPEG-2 Audio: A Generic Standard for the Coding of Two-Channel and Multichannel Sound," *AES Collected Papers on Digital Audio Bit-Rate Reduction*, N. Gilchrist and C. Grewin (eds.), Audio Engineering Society, 1996.

Stoll, G., "A Perceptual Coding Technique Offering the Best Compromise between Quality, Bit-Rate, and Complexity for DSB," *AES 94th Conv.*, preprint 3458, Berlin, March 1993.

Stoll, G., M. Link, and G. Thiele, "Masking-Pattern Adapted Subband Coding: Use of the Dynamic Bit-Rate Margin," *AES 84th Conv.*, preprint 2585, Paris, March 1988.

Stoll, G., et al., "Generic Architecture of the ISO/MPEG Audio Layer I and II: Compatible Developments to Improve the Quality and Addition of New Features," *AES 95th Conv.*, preprint 3697, New York, October 1993.

Stuart, J. R., "Further Evaluation of MLP for DVD-Audio," Ad Hoc Committee 5 of WG-4, July 12, 1998.

Sugiyama, A., et al., "Adaptive Transform Coding with an Adaptive Block Size (ATCABS)," *Proc. IEEE Int. Conf. ASSP*, Albuquerque, NM, 1990.

Ten Kate, W. R., et al., "Scalability in MPEG Audio Compression: From Stereo via 5.1-Channel Surround Sound to 7.1-Channel Augmented Sound Fields," *AES 100th Conv.*, preprint 4196, Copenhagen, May 1996.

Thiele, G., M. Link, and G. Stoll, "Low Bit Rate Coding of High Quality Audio Signals," *AES 82nd Conv.*, preprint 2432, London, March 1987.

Thiele, G., G. Stoll, and M. Link, "Low Bit-Rate Coding of High Quality Audio Signals: An Introduction to the MASCAM System," *EBU Review*, UHF Satellite Sound Broadcasting, no. 230, August 1988.

Todd, C. C., et al., "AC-3: Flexible Perceptual Coding for Audio Transmission and Storage," *AES 96th Conv.*, preprint 3796, Amsterdam, February 1994.

Tsurushima, K., et al. "MiniDisc: Disc-Based Digital Recording for Portable Audio Applications," *AES Collected Papers on Digital Audio Bit-Rate Reduction*, N. Gilchrist and C. Grewin (eds.), 1996.

Tsutsui, K., et al., "ATRAC: Adaptive Transform Acoustic Coding for MiniDisc," *AES Collected Papers on Digital Audio Bit-Rate Reduction*, N. Gilchrist and C. Grewin (eds.), 1996.

United States Advanced Television Systems Committee, "Digital Audio Compression (AC-3)," *ATSC Draft Standard*, Doc. T3/ S7-016, July 25, 1994.

Van de Kerkhof, L., "Compatible 5.1 Channel Extension to the MPEG Layer II Audio Coding Standard," *Proc. Tirrenia Int. Workshop Digit. Commun.*, Pisa, September 1993.

Vernon, S., "Design and Implementation of AC-3 Coders," *IEEE Trans. Consum. Electron.*, vol. 41, no. 3, August 1995.

Vernon, S., "Dolby Digital: Audio Coding for Digital Television and Storage Applications," *Proc. AES 17th Int. Conf.*, Florence, September 1999.

Watkinson, J., *The MPEG Handbook*, 2nd ed., Focal Press, 2004.

Wirtz, G. C., "Digital Compact Cassette: Audio Coding Technique," *AES 91st Conv.*, preprint 3216, New York, October 1991.

Wirtz, G. C., "Digital Compact Cassette: Background and System Description," *AES 91st Conv.*, preprint 3215, New York, October 1991.

Witte, F. O., and W. Sinnhofer, "Single-Chip Implementation of an ISO/MPEG Layer III Decoder," *AES 96th Conv.*, preprint 3805, Amsterdam, February/March 1994.

Wylie, F. "aptX100: Low-Delay, Low-Bit-Rate Subband ADPCM Digital Audio Coding," *AES Collected Papers on Digital Audio Bit-Rate Reduction*, N. Gilchrist and C. Grewin (eds.), Audio Engineering Society, 1996.

Wylie, F., "Predictive or Perceptual Coding—apt-X and apt-Q," *AES 100th Conv.*, preprint 4200, Copenhagen, May 1996.

Ziegler, T., et al., "Enhancing mp3 with SBR: Features and Capabilities of the new mp3PRO Algorithm," *AES 112th Conv.*, preprint 5560, Munich, May 2002.

12 Speech Coding for Transmission

Aarts, R. M., E. Larsen, and O. Ouweltjes, "A Unified Approach to Low- and High-Frequency Bandwidth Extension," *AES 115th Conv.*, preprint 5921, New York, October 2003.

Ahn, C. Y., "Preprocessing Method For Enhancing Digital Audio Quality In Speech Communication System," *Proc. AES 29th Int. Conf.*, Seoul, September 2006.

Aoki, N., "A Band Extension Technique for G.711 Speech Based on Full Wave Rectification and Steganography," *Proc. AES 29th Int. Conf.*, Seoul, September 2006.

Atal, B. S., "High-Quality Speech at Low Bit Rates: Multi-Pulse and Stochastically Excited Linear Predictive Coders," *Proc. ICASSP*, 1986.

Atal, B. S., "Predictive Coding of Speech at Low Bit Rates," *IEEE Trans. Commun.*, COM-30, 1982.

Atal, B. S., and M. R. Schroeder, "Predictive Coding of Speech and Subjective Error Criteria," *IEEE Trans. ASSP*, no. ASSP-27, 1979.

Azzali, A., et al., "Comparison of Different Listening Systems for Speech Intelligibility Tests," *AES 118th Conv.*, preprint 6356, Barcelona, May 2005.

Backman, J., "Mobile Phone Audio: The Shape of Things to Come," *Proc. AES 34th Int. Conf.*, Jeju Island, Korea, August 2008.

Beerends, J. G., R. van Buuren, J. van Vugt, and J. Verhave, "Objective Speech Intelligibility Measurement on the Basis of Natural Speech in Combination with Perceptual Modeling," *JAES*, vol. 57, no. 5, May 2009.

Biswas, A., and A. C. den Brinker, "Perceptually Biased Linear Prediction," *JAES*, vol. 54, no. 12, December 2006.

Campbell, J. P., T. E. Tremain, and V. C. Welch, "The Proposed Federal Standard 1016 4800 bps Voice Coder: CELP," *Speech Technology*, May 1990.

Chen, S., and H. Leung, "Artificial Bandwidth Extension of Telephony Speech by Data Hiding," *Proc. Int. Sym. Circuits and Systems (ISCAS)*, vol. 4, 2005.

Cheng, Q., and J. Sorensen, "Spread Spectrum Signaling for Speech Watermarking," *Proc. ICASSP*, Salt Lake City, vol. 5, May 2001.

Chu, W. C., *Speech Coding Algorithms*, John Wiley & Sons, 2003.

Cox, R. V., W. B. Kleijn, and P. Kroon, "Robust CELP Coders for Noisy Backgrounds and Noisy Channels," *Proc. ICASSP*, vol. 2, Glasgow, May 1989.

Dietz, M., et al., "Spectral Band Replication: A Novel Approach in Audio Coding," *AES 112th Conv.*, preprint 5553, Munich, May 2002.

Ding, H., "Wideband Audio over Narrowband Low-Resolution Media," *Proc. ICASSP*, vol. 1, Montreal, May 2004.

Dunn, C., "Aspects of Scalable Audio Coding," *AES 122nd Conv.*, preprint 7081, Vienna, May 2007.

Ehara, H., T. Morii, and K. Yoshida, "Predictive Vector Quantization of Wideband LSF Using Narrowband LSF for Bandwidth Scalable Coders," *Speech Comm.*, vol. 49, no. 6, 2007.

Erskine, D. C., "Real-Time CELP Speech Coding in a Voice Response Environment," *AES 91st Conv.*, preprint 3186, New York, 1991.

Feiten, B., et al., "Audio Adaptation According to Usage Environment and Perceptual Quality Metrics," *IEEE Trans. Multimedia*, vol. 7, no. 3, 2005.

Furui, S., and M. M. Sondhi (eds.), *Advances in Speech Signal Processing*, Marcel Dekker, 1991.

Geiser, B., and P. Vary, "Backwards Compatible Wideband Telephony in Mobile Networks: CELP Watermarking and Bandwidth Extension," *Proc. ICASSP*, Honolulu, vol. 4, 2007.

Geiser, B., et al., "Bandwidth Extension for Hierarchical Speech and Audio Coding in ITU-T Rec. G.729.1," *IEEE Trans. Audio, Speech, and Language Processing*, vol. 15, no. 8, 2007.

Geissner, E., et al. "Subjective Evaluation of Speech Quality in a Conversational Context," *AES 124th Conv.*, preprint 7405, Amsterdam, May 2008.

Goldberg, R., and L. Riek, *A Practical Handbook of Speech Coders*, CRC Press, 2000.

Hermansky, H., "Perceptual Linear Prediction (PLP) Analysis of Speech," *JASA*, vol. 87, no. 4, 1990.

Hermansky, H., and N. Morgan, "RASTA Processing of Speech," *IEEE Trans. on Speech and Audio Processing*, vol. 2, no. 4, 1994.

Hersent, O., J.-P. Petit, and D. Gurle, *IP Telephony: Deploying Voice-over-IP Protocols*, John Wiley & Sons, 2005.

Hirvonen, T., J. Ahonen, and V. Pulkki, "Perceptual Compression Methods for Metadata in Directional Audio Coding Applied to Audiovisual Teleconference," *AES 126th Conv.*, preprint 7706, Munich, May 2009.

Hiwasaki, Y., et al., "A G.711 Embedded Wideband Speech Coding for VoIP Conferences," *IEICE Trans. Information and Systems*, vol. E89-D, no. 9, 2006.

Hiwasaki, Y., et al., "Scalable Speech Coding Technology for High-Quality Ubiquitous Communications," *NTT Tech. Rev.*, vol. 2, no. 3, 2004.

Hossain, Z., R. Islam, and K. M. Morshed, "Analysis of CLEP Coding with Different Bit Rate and Maximization of Speech Quality at Low Bit Rate," *National Conference on Communication and Information Security, (NCCIS)*, Dhaka, Bangladesh, November 2007.

Hoyle, R. D., and D. D. Falconer, "A Comparison of Digital Speech Coding Methods for Mobile Radio Systems," *IEEE J. Selected Areas Commun.*, June 1987.

ITU-T Rec. G.729.1, An 8–32 kbit/s Scalable Wideband Coder Bitstream Interoperable with G.729, International Telecommunication Union (ITU), 2006.

ITU-T Rec. G.764, Voice Packetization–Packetized Voice Protocols, International Telecommunication Union (ITU), 1990.

ITU-T Rec. P800, Methods for Subjective Determination of Transmission Quality, International Telecommunication Union (ITU), 1996.

Iyenger, V., and P. Kabal, "A Low Delay 16 kbits/sec Speech Coder," *Proc. ICASSP*, April 1988.

Jayant, N. (ed.), *Signal Compression: Coding of Speech, Audio, Text, Image and Video*, World Scientific Publishing, 1997.

Jax, P., and P. Vary, "Bandwidth Extension of Speech Signals: A Catalyst for the Introduction of Wideband Speech Coding?" *IEEE Comm. Mag.*, vol. 44, no. 5, 2006.

Jax, P., and P. Vary, "An Upper Bound on the Quality of Artificial Bandwidth Extension of Narrowband Speech Signals." *Proc. ICASSP*, Orlando, vol. 1, 2002.

Kamaruzzaman, M., and H. Taddei, "Embedded Speech Codec Based on Speex," *AES 116th Conv.*, preprint 6173, Berlin, May 2004.

Kemp, D. P., R. A. Sueda, and T. E. Tremain, "An Evaluation of 4800 bps Voice Coders," *Proc. ICASSP*, vol. 1, May 1989.

Kim, D. Y., et al., "Bandwidth Extension for Scalable Audio Coding," *Proc. AES 29th Int. Conf.*, Seoul, September 2006.

Kitawaki, N., and H. Nagabuchi, "Quality Assessment of Speech Coding and Speech Synthesis Systems," *IEEE Comm. Mag.*, October 1988.

Kleijn, W. B., D. J. Krasinski, and R. H. Ketchum, "Fast Methods for the CELP Speech Coding Algorithm," *IEEE Trans. ASSP*, ASSP-38, 1990.

Koishada, K., V. Cuperman, and A. Gersho, "A 16 kbit/s Bandwidth Scalable Audio Coder Based on the G.729 Standard," *Proc. ICASSP*, Instanbul, 2000.

Kovesi, B., D. Massaloux, and A. Sollaud, "A Scalable Speech and Audio Coding Scheme with Continuous Bitrate Flexibility," *Proc. ICASSP*, Montreal, 2004.

Kroon, P., and B. S. Atal, "Strategies for Improving the Performance of CELP Coders at Low Bit Rates," *Proc. ICASSP*, New York, 1988.

Kulesza, M., and A. Czyzewski, "Speech Codec Enhancements Utilizing Time Compression and Perceptual Coding," *AES 122nd Conv.*, preprint 7004, Vienna, May 2007.

Kulesza, M., G. Szwoch, and A. Czyzewski, "A Hybrid Speech Codec Employing Parametric and Perceptual Coding Techniques," *AES 121st Conv.*, preprint 6956, San Francisco, October 2006.

Kurittu, A., "Validation of ITU-T P.563 Single-Ended Objective Speech Quality Measurement," *JAES*, vol. 54, no. 11, November 2006.

Kurittu, A., et al., "Application and Verification of the Objective Quality Assessment Method According to ITU Recommendation Series ITU-T P.862," *JAES*, vol. 54, no. 12, December 2006.

Larsen, E., and R.M. Aarts (eds.), *Audio Bandwidth Extension*, John Wiley & Sons, 2004.

Le Dinh, K., C. T. Le Dinh, and R. Lefebvre, "A New Bandwidth Extension for Audio Signals without Using Side-Information," *AES 126th Conv.*, preprint 7654, Munich, May 2009.

Lee, W. C. Y., *Wireless & Cellular Telecommunications*, 3rd ed., McGraw-Hill, 2005.

Maher, R. C., "Wavetable Synthesis Strategies for Mobile Devices," *JAES*, vol. 53, no. 3, March 2005.

Makhoul, J. I., "Linear Prediction: A Tutorial Review," *Proc. IEEE*, vol. 63, April 1975.

Makhoul, J. I., and M. Berouti, "Adaptive Noise Spectral Shaping and Entropy Coding in Predictive Coding of Speech," *IEEE Trans. ASSP*, ASSP-27, 1979.

Markel, J. D., and A. H. Gray, *Linear Prediction of Speech*, Springer-Verlag, 1976.

Martin, R., U. Heute, and C. Antweiler (eds)., *Advances in Digital Speech Transmission*, John Wiley & Sons, 2008.

Mattila, V.-V., and A. Kurittu, "Practical Issues in Objective Speech Quality Assessment with ITU-T P.862," *AES 117th Conv.*, preprint 6209, San Francisco, October 2004.

Meana, H. P. (ed.), *Advances in Audio and Speech Signal Processing*, IGI Global, 2007.

Mermelstein, P., "A New CCITT Coding Standard for Digital Transmission of Wideband Audio Signals," *IEEE Comm. Mag.*, January 1988.

Neuendorf, M., et al., "A Novel Scheme for Low Bit Rate Unified Speech and Audio Coding—MPEG RM0," *AES 126th Conv.*, preprint 7713, Munich, May 2009.

Nomura, T., et al., "A Bitrate and Bandwidth Scalable CELP Coder," *Proc. ICASSP*, vol. 1, Seattle, 1998.

Park, S.-H., Y. B. Kim, and Y. S. Seo, "Multi-Layer Bit-Sliced Bit-Rate Scalable Audio Coding," *AES 103rd Conv.*, preprint 4520, New York, September 1997.

Philippe, P., D. Virette, and B. Kövesi, "Time-Varying Transform for High Quality Audio Communication Codecs," *AES 124th Conv.*, preprint 7333, Amsterdam, May 2008.

Raake, A., M. Wältermann, and S. Spors, "Which Wideband Speech Codec? Quality Impact Due to Room-Acoustics at Send Side and Presentation Method," *AES 127th Conv.*, preprint 7823, New York, October 2009.

Ragot, S., et al., "A 8-32 kbit/s Scalable Wideband Speech and Audio Coding Candidate for ITU-T G.729EV Standardization," *Proc. ICASSP*, Toulouse, 2006.

Ragot, S., et al., "ITU-T G.729.1: An 8-32 kbit/s Scalable Coder Interoperable with G.729 for Wideband Telephony and Voice over IP," *Proc. ICASSP*, Honolulu, 2007.

Sagi, A., and D. Malah, "Bandwidth Extension of Telephone Speech Aided by Data Embedding," *EURASIP J. Advances in Signal Processing*, vol. 2007, no. 1, 2007.

Sagi, A., and D. Malah, "Data Embedding in Speech Signals Using Perceptual Masking," *Proc. EUSIPCO*, Vienna, 2004.

Schroeder, M. R., and B. S. Atal, "Code-Excited Linear Prediction (CELP): High-quality Speech at Very Low Bit Rates," *Proc. ICASSP*, Tampa, 1985.

Schroeder, M. R., B. S. Atal, and J. L. Hall, "Optimizing Digital Speech Coders by Exploiting Masking Properties of the Human Ear," *JASA*, vol. 66, no. 6, 1979.

Shao, C., and M. Bouchard, A Perceptual Post Filter for Wideband Speech and Audio ACELP Codecs," *AES 118th Conv.*, preprint 6379, Barcelona, May 2005.

Soong, F. K., R. V. Cox, and N. S. Jayant, "Subband Coding of Speech Using Backward Adaptive Prediction and Bit Allocation," *Proc. ICASSP*, vol. 4, April 1985.

Spanias, A., T. Painter, and V. Atti, *Audio Signal Processing and Coding*, Wiley-Interscience, 2007.

Taddei, H., D. Massaloux, and A. Le Guyader, "A Scalable Three Bitrate (8, 14.2, and 24 kbit/s) Audio Coder," *AES 107th Conv.*, preprint 5034, New York, 1999.

Taddei, H., et al., "Mode Adaptive Unequal Error Protection for Transform Predictive Speech and Audio Coders," *Proc. ICASSP*, Orlando, 2002.

Taka, M., and X. Maitre, "CCITT Standardizing Activities on Speech Coding," *Proc. ICASSP*, April 1986.

Taniguchi, T, S. Unagami, and R. M. Gray, "Multimode Coding: Application to CELP," *Proc. ICASSP*, vol. 1, May 1989.

Taniguchi, T. S., et al., "Improved CELP Speech Coding at 4 kb/s and Below," *Proc. Int. Conf. on Spoken Language Processing*, November 1992.

Tapia, J., et al., "Introduction to the OpenCORE Audio Components used in the Android Platform," *Proc. AES 34th Int. Conf.*, Jeju Island, Korea, August 2008.

Turnbull, R., P. Hughes, and S. Hoare, "Audio Enhancement for Portable Device-Based Speech Applications," *AES 124th Conv.*, preprint 7350, Amsterdam, May 2008.

Valin, J.-M., and C. Montgomery, "Improved Noise Weighting in CELP Coding of Speech—Applying the Vorbis Psychoacoustic Model to Speex," *AES 120th Conv.*, preprint 6746, Paris, May 2006.

Vary, P., and R. Martin, *Digital Speech Transmission: Enhancement, Coding and Error Concealment*, John Wiley & Sons, 2006.

13 Audio Interconnection

Adams, R., and T. Kwan, "Theory and VLSI Architectures for Asynchronous Sample-Rate Converters," *JAES*, vol. 41, no. 7/8, July/August 1993.

Adams, R., et al., "An Integrated AES/EBU Receiver/Sample-Rate Converter," *AES 103rd Conv.*, preprint 4536, New York, September 1997.

AES3-1992, "AES Recommended Practice for Digital Audio Engineering—Serial Transmission Format for Two-Channel Linearly Represented Digital Audio Data," *JAES*, vol. 40, no. 3, March 1992.

AES3-1992 Amendment 1-1997, Amendment 2-1998, Amendment 3-1999, *JAES*, vol. 47, no. 3, March 1999.

AES-2id-1996, "AES Information Document for Digital Audio Engineering—Guidelines for the Use of the AES3 Interface," *JAES*, vol. 44, no. 10, October 1996.

AES-3id-1995, "AES Information Document for Digital Audio Engineering—Transmission of AES3 Formatted Data by Unbalanced Coaxial Cable," *JAES*, vol. 43, no. 10, October 1995.

AES5-1998, "AES Recommended Practice for Professional Digital Audio—Preferred Sampling Frequencies for Applications Employing Pulse-Code Modulation," *JAES*, vol. 46, no. 10, October 1998.

AES10-1991, "AES Recommended Practice for Digital Audio Engineering—Serial Multichannel Audio Digital Interface (MADI)," *JAES*, vol. 39, no. 5, May 1991.

AES11-1997, "AES Recommended Practice for Digital Audio Engineering—Synchronization of Digital Audio Equipment in Studio Operations," *JAES*, vol. 45, no. 4, April 1997.

AES17-1991, "AES Standard Method for Digital Audio Engineering—Measurement of Digital Audio Equipment," *JAES*, vol. 39, no. 12, December 1991.

AES18-1996, "AES Recommended Practice for Digital Audio Engineering—Format for the User Data Channel of the AES Digital Audio Interface," *JAES*, vol. 44, no. 11, November 1996.

AES24-1-1999, "AES Standard for Sound System Control—Application Protocol for Controlling and Monitoring Audio Devices via Digital Data Networks—Part 1: Principles, Formats and Basic Procedures," *JAES*, vol. 47, no. 4, April 1999.

Ajemian, R. G., "Audio Rides the Light of Fiber Optics: A Tutorial," *JAES*, vol. 57, no. 6, June 2009.

Ajemian, R. G., "Fiber-Optic Connector Considerations for Professional Audio," *JAES*, vol. 40, no. 6, June 1992.

Ajemian, R. G., and A. B. Grundy, "Fiber-Optics—The New Medium for Audio: A Tutorial," *JAES*, vol. 38, no. 3, March 1990.

Allard, F. C., *Fiber Optics Handbook for Engineers and Scientists*, McGraw-Hill, 1990.

Angelici, M., et al., "New Architecture for an AES-EBU Digital Audio Receiver," *IEEE Trans. Consum. Electron.*, vol. 43, no. 3, August 1997.

Cabot, R. C., "Measuring AES-EBU Digital Audio Interfaces," *JAES*, vol. 38, no. 6, June 1990.

Cabot, R. C., "Testing Digital Audio Devices in the Digital Domain," *AES 86th Conv.*, preprint 2800, Hamburg, March 1989.

Caine, R., "Timing in Digital Audio Systems—The Importance of AES11," *AES 100th Conv.*, preprint 4221, Copenhagen, May 1996.

Chan, C., "Monitoring Digital Audio/Video Signals," *Broadcast Engineering*, November 1994.

Cheo, P. K., *Fiber Optics, Devices and Systems*, Prentice Hall, 1985.

Cherin, A. H., *An Introduction to Optical Fibers*, McGraw-Hill, 1983.

Commerce Clearing House, "Technical Reference Document for Digital Audio Recorders," Copyright Law Reports, No. 16,080, 1992.

Dick, B., "Building Fiber-Optic Transmission Systems—Parts 1, 2, 3" *Broadcast Engineering*, November/December 1991, January 1992.

Dunn, C., and M. O. J. Hawksford, "Is the AES/EBU/SPDIF Digital Audio Interface Flawed?" *AES 93rd Conv.*, preprint 3360, San Francisco, October 1992.

EBU (European Broadcasting Union), Specification of the Digital Audio Interface, *EBU Doc. Tech.*, 3250, 1992.

Finger, R. A., "The Revised Two-Channel Digital Audio Interface," *JAES*, vol. 40, no. 3, March 1992.

Frankel, D., "ISDN Reaches the Market," *IEEE Spectr.*, vol. 32, no. 6, June 1995.

Geckeler, S., *Optical Transmission Systems*, Artech House, 1987.

Harris, S., et al., "A Monolithic 24-Bit, 96-kHz Sample-Rate Converter with AES3 Receiver and AES3 Transmitter," *AES 106th Conv.*, preprint 4965, Munich, May 1999.

Haynes, D., "Tying It All Together: Synchronization Issues for All-Digital Studios," *Mix*, vol. 16, no. 7, July 1992.

Izawa, T., and S. Sudo, *Optical Fibers: Materials and Fabrication*, KTK Scientific Publishers, 1987.

JAES Staff, "Moving Digital Audio, Part 1," *JAES*, December 2002.

JAES Staff, "Moving Digital Audio, Part 2," *JAES*, March 2003.

Jones, Jr., W. B., *Introduction to Optical Fiber Communication Systems*, Holt, Reinhart and Winston, 1988.

Karley, B. A., "Fiber Optics," *Advanced Digital Audio*, K. C. Pohlmann (ed.), Howard W. Sams, 1991.

Kirby, D. G., "Twisted-Pair Cables for AES/EBU Digital Audio Signals," *JAES*, vol. 43, no. 3, March 1995.

Koch, A., R. Lagadec, and D. Pelloni, "Method and Apparatus for Converting an Input Scanning Sequence into an Output Scanning Sequence," U.S. Patent 4,825,398, April 25, 1989.

Lagadec, R., "Digital Sampling Frequency Conversion," *AES Digital Audio Collected Papers*, Rye, 1983.

Lagadec, R., and H. Kunz, "Process and Apparatus for Translating the Sampling Rate of a Sampling Sequence," U.S. Patent 4,748,578, May 31, 1988.

Lambert, M., "Digital Audio Interfaces," *JAES*, vol. 38, no. 9, September 1990.

Lin, C., *Optoelectronic Technology and Lightwave Communications Systems*, Van Nostrand Reinhold, 1989.

Mahlke, G., and P. Gossing, *Fiber Optic Cables: Fundamentals, Cable Technology, Installation Practice*, John Wiley & Sons, 1987.

Miller, C. M., *Optical Fiber Splices and Connectors, Theory and Method,* Marcel Dekker, 1986.

Murata, H., *Handbook of Optical Fibers and Cables*, Marcel Dekker, 1988.

Page, M., et al., "Multi-Channel Audio Connection for Direct Stream Digital," *AES 113th Conv.*, preprint 5691, Los Angeles, October 2002.

Paulson, C. R., "Fiber Optic Transmission Systems," *National Association of Broadcasters Engineering Handbook*, 8th ed., NAB, 1992.

Personick, S. D., *Fiber Optics Technology and Applications*, Plenum Press, 1985.

Rabiner, L. R., "Digital Techniques for Changing the Sampling Rate of a Signal," *AES Digital Audio Collected Papers*, Rye, 1983.

Robjohns, H., "Digital Interconnections—Parts 1–2," *Audio Media,* August/September 1994.

Rumsey, F., "Audio Programme Interchange: Networks, File Formats, and Real-Time Transfer," *AES 95th Conv.*, preprint 3737, New York, October 1993.

Rumsey, F., *Digital Audio Operations*, Focal Press, 1991.

Sakura, S., et al., "Fiber Optics Links for Digital Audio Interface," *IEEE Trans. Consum. Electron.*, vol. CE-34, no. 3, August 1988.

Sanchez, C. W., "An Understanding and Implementation of the SCMS Serial Copy Management System for Digital Audio Transmission," *JAES*, vol. 42, no. 3, March 1994.

Shelton, W. T., "Synchronization of Digital Audio," *Proc. AES 7th Int. Conf.*, Toronto, May 1989.

Talbot, D., "Fiber-Optic Transmission and Professional Audio," *JAES*, vol. 42, no. 5, May 1994.

14 Personal Computer Audio

Anderson, D., *USB System Architecture (USB 2.0)*, Addison-Wesley Developer's Press, 2001.

Audio and Music Data Transmission Protocol (Version 1.0, May 1997), 1394 Trade Association, www.1394TA.org.

Bagnaschi, C. L., "A Magnetic Storage Disk-Based Digital Audio Recording, Editing, and Processing System," *AES 83rd Conv.*, preprint 2505, New York, October 1987.

Barish, J., "A Survey of New Sound Technology for PCs," *AES 105th Conv.*, preprint 4844, San Francisco, September 1998.

Chan, C., "Hard Drives," *Broadcast Engineering*, September 1994.

De Lancie, P., "Audio in Mac Authoring," *Mix*, vol. 18, no. 7, July 1994.

Dunn, J., "Sample Clock Jitter and Real-Time Audio over the IEEE 1394 High Performance Serial Bus," *AES 106th Conv.*, preprint 4920, Munich, May 1999.

Elen, R. G., "The Integration of Large-Scale Studio Systems," *Recording Engineer/Producer*, February 1988.

Fujimori, J., and R. Foss, "mLAN: Current Status and Future Directions," *AES 113th Conv.*, preprint 5699, Los Angeles, October 2002.

Fujimori, J., and S. Kakiuchi, "Digital Audio Transmission Over IEEE 1394: Protocol, Design and Implementation," *AES 103rd Conv.*, preprint 4547, New York, September 1997.

Fujimori, J., and Y. Osakabe, "Digital Audio and Performance Data Transmission Protocol Over IEEE 1394," *AES 101st Conv.*, preprint 4346, Los Angeles, November 1996.

IEC 61883-1 (1998-02), Consumer Audio/Video Equipment—Digital Interface—Part 1: General.

Ingebretsen, R. B., and T. G. Stockham, Jr., "Random Access Editing of Digital Audio," *JAES*, vol. 32, no. 3, March 1984.

Intel Corporation, Audio Codec '97 (AC '97) Component Specification, http://www.intel.com/pcsupp/platform/ac97.

Jorgensen, F., *The Complete Handbook of Magnetic Recording*, 3rd ed., TAB Books, 1988.

Kientzle, T., *A Programmer's Guide to Sound*, Addison-Wesley, 1997.

Kunzman, A. J., and A. T. Wetzel, "1394 High Performance Serial Bus: The Digital Interface for ATV," *IEEE Trans. Consum. Electron.*, vol. 41, no. 3, August 1995.

Lambert, M., "Defining a Digital Audio Workstation," *Mix*, vol. 18, no. 7, July 1994.

Leider, C., *Digital Audio Workstation*, McGraw-Hill, New York, 2004.

Lidbetter, P. S., "Digital Tape Transfer Console," *AES 79th Conv.*, preprint 2276, New York, October 1985.

Lowman, C., *Magnetic Recording*, McGraw-Hill, 1972.

McNally, G. W., "Fast Edit Point Location and Cueing in Disc-Based Digital Audio System," *AES 78th Conv.*, preprint 2232, Anaheim, May 1985.

McNally, G. W., P. S. Gaskell, and A. J. Stirling, "Digital Audio Editing," *AES 77th Conv.*, preprint 2214, Hamburg, March 1985.

Moorer, J. A., "The Lucasfilm Audio Signal Processor," *Comput. Music J.*, Fall 1982.

Moorer, J. A., and J. Borish, "An Optical Disk Recording, Archiving, and Editing Device for Digital Audio Signal Processing," *AES 81st Conv.*, preprint 2376, Los Angeles, November 1986.

Moorer, J. A., et al., "The Digital Audio Processing Station: A New Concept in Audio Postproduction," *JAES*, vol. 34, no. 6, June 1986.

Moses, B., "Home Networking Using IEEE 1394 in Combination with other Networking Technologies," *AES 17th UK Conf.*, London, April 2002.

Moses, B., "Implementing Digital Audio Devices for the IEEE 1394 High Performance Serial Bus," *AES 105th Conv.*, preprint 4761, San Francisco, September 1998.

Moses, B, and G. Bartlett, "Audio Distribution and Control Using the IEEE 1394 Serial Bus," *AES 103rd Conv.*, preprint 4548, New York, September 1997.

Mueller, S., *Upgrading and Repairing PCs*, 11th ed., Que, 1999.

OMF Interchange Specification, Version 1.0, Avid Technology, May 1993.

Pirkle, W., and K. Pohlmann, "PC Sound: Better All Around," *PC*, vol. 18, no. 3, February 9, 1999.

Pohlmann, K., and W. Pirkle, "The Shifting Soundscape," *PC*, vol. 17, no. 1, January 6, 1998.

Roach, D., S. Janus, and W. Jones, *High Definition Audio for the Digital Home*, Intel Press, 2006.

Schwartz, R., "File Interchange Formats," *Mix*, vol. 18, no. 7, July 1994.

Snell, J. M., "Professional Real-Time Signal Processor for Synthesis, Sampling, Mixing, and Recording," *AES 83rd Conv.*, preprint 2508, New York, October 1987.

USB Implementers Forum, "Universal Serial Bus Device Class Definition for Audio Data Formats," Release 1.0, March 1998.

USB Implementers Forum, "Universal Serial Bus Device Class Definition for Audio Devices," Release 1.0, March 1998.

Universal Serial Bus Specification, Revision 1.1, September 1998.

Whitaker, J., "Hard Disk Recording Technology," *Recording Engineer/Producer*, March 1988.

Wickelgren, I. J., "The Facts About FireWire," *IEEE Spectr.*, vol. 34, no. 4, April 1997.

15 Telecommunications and Internet Audio

AES47-2002, "AES Standard for Digital Audio—Digital Input-Output Interfacing—Transmission of Digital Audio over Asynchronous Transfer Mode (ATM) Networks," *JAES*, vol. 51, no. 7/8, July/August 2003.

AES-R4-2002, "Guidelines for AES Standard for Digital Audio—Digital Input-Output Interfacing—Transmission of Digital Audio over Asynchronous Transfer Mode (ATM) Networks, AES47," *JAES*, vol. 51, no. 7/8, July/August 2003.

Aikawa, M., et al., "A Lightweight Encryption Method Suitable for Copyright Protection," *IEEE Trans. Consum. Electron.*, vol. 44, no. 3, August 1998.

Allamanche, E., et al., "MPEG-4 Low-Delay Audio Coding Based on the AAC Codec," *AES 106th Conv.*, preprint 4929, Munich, May 1999.

Bai, M. R., and M.-C. Chen, "Intelligent Preprocessing and Classification of Audio Signals," *JAES*, vol. 55, no. 5, May 2007.

Barbedo, J. G. A., and A. Lopes, "Automatic Musical Genre Classification Using a Flexible Approach," *JAES*, vol. 56, no. 7/8, July/August 2008.

Bargar, R., et al., "Networking Audio and Music Using Internet2 and Next-Generation Internet Capabilities," *JAES*, vol. 47, no. 4, April 1999.

Bluetooth Audio Video Working Group, "Advanced Audio Distribution Profile Specification," Adopted Version 1.0, May 2003.

Boccuzzi, J., *Signal Processing for Wireless Communications*, McGraw-Hill, 2008.

Bonime, A., and K. Pohlmann, *Writing for New Media*, John Wiley & Sons, 1998.

Bouillot, N., et al., "AES White Paper: Best Practices in Network Audio," *JAES*, vol. 57, no. 9, September 2009.

Bouillot, N., et al., Performance Metrics for Network Audio Systems: Methodology and Comparison," *AES 127th Conv.*, preprint 7940, New York, October 2009.

Brandenburg, K., and B. Grill, "First Ideas on Scalable Audio Coding," *AES 97th Conv.*, preprint 3924, San Francisco, November 1994.

Burred, J. J., and A. Lerch, "Hierarchical Automatic Audio Signal Classification," *JAES*, vol. 52, no. 7/8, July/August 2004.

Campbell, Jr., J. R., "Speaker Recognition: A Tutorial," *Proc. IEEE*, vol. 85, no. 9, 1997.

Campbell, K., "Deploying Large Scale Audio IP Networks," *AES 126th Conv.*, preprint 7652, Munich, May 2009.

Chigwamba, N., and R. Foss, "Enhancing End-User Capabilities in High Speed Audio Networks," *AES 123rd Conv.*, preprint 7208, New York, October 2007.

Chon, S. B., et al., "A Bit Reduction Algorithm for Spectral Band Replication using the Masking Effect," *Proc. AES 34th Int. Conf.*, Jeju Island, Korea, August 2008.

D'Aguanno, A., and G. Haus, "Audio Fingerprint and its Applications to Peer-to-Peer Systems," *AES 124th Conv.*, preprint 7321, Amsterdam, May 2008.

Dahlman, E., S. Parkvall, J. Skold, and P. Beming, *3G Evolution,* 2nd ed., Academic Press, 2007.

Den Brinker, A. C., et al., "Parametric Coding for High-Quality Audio," *AES 112th Conv.*, preprint 5554, Munich, May 2002.

Denckla, B., "Subtractive Dither for Internet Audio," *JAES*, vol. 46, no. 7/8, July/August 1998.

Dietz, M., et al., "Audio Compression for Network Transmission," *JAES*, vol. 44, no. 1/2, January/February 1996.

Dietz, M., et al., "Bridging the Gap: Extending MPEG Audio Down to 8 kbit/s," *AES 102nd Conv.*, preprint 4508, Munich, March 1997.

Feiten, B., et al., "Dynamically Scalable Internet Audio Transmission," *AES 104th Conv.*, preprint 4686, Amsterdam, May 1998.

Floros, A., and T. Karoubalis, "Delivering High-Quality Audio over WLANs," *AES 116th Conv.*, preprint 5996, Berlin, May 2004.

Geiger, R., et al., "ISO/IEC MPEG-4 High-Definition Scalable Advanced Audio Coding," *JAES*, vol. 55, no. 1/2, January/February 2007.

Gish, H., M.-H. Siu, and R. Rohlicek, "Segregation of Speaker for Speech Recognition and Speaker Identification," *Proc. ICASSP*, Toronto, May 1991.

Grill, B., "A Bit-Rate Scalable Perceptual Coder for MPEG-4 Audio," *AES 103rd Conv.*, preprint 4620, New York, September 1997.

Heber, K. E., "Mobile Internet Audio: A Report on the State of Technology," *Proc. AES 36th Int. Conf.*, Dearborn, June 2009.

Herre, J., and D. Schulz, "Extending the MPEG-4 AAC Codec by Perceptual Noise Substitution," *AES 104th Conv.*, preprint 4720, Amsterdam, May 1998.

Hsu, H.-W., C.-M. Liu, and W.-C. Lee, "Fast Complex Quadrature Mirror Filterbanks for MPEG-4 HE-AAC," *AES 121st Conv.*, preprint 6871, San Francisco, October 2006.

ISO/IEC JTC1/SC29/WG11 MPEG94/443, "Requirements for Low Bitrate Audio Coding/MPEG-4 Audio," 1994.

ISO/IEC JTC1/SC29/WG11 MPEG, IS13818, "Generic Coding of Moving Pictures and Associated Audio, Part 3: Audio," 1994.

Jonsson, L., and M. Coinchon, "EBU Tech. doc. 3326 for Interoperability between Audio over IP Units," *AES 124th Conv.*, preprint 7322, Amsterdam, May 2008.

Keiser, B. E., and E. Strange, *Digital Telephony and Network Integration*, 2nd ed., Van Nostrand Reinhold, 1995.

Kemp, T., M. Schmidt, M. Westphal, and A. Waibel, "Strategies for Automatic Segmentation of Audio Data," *Proc. ICASSP 2000*, Instanbul, 2000.

Kim, H.-G., N. Moreau, and T. Sikora (eds.), *MPEG-7 Audio and Beyond: Audio Content Indexing and Retrieval*, John Wiley & Sons, 2005.

Kinoshita, S., "The RealPush Network: A New Push-Type Content Delivery System Using Reliable Multicasting," *IEEE Trans. Consum. Electron.*, vol. 44, no. 4, November 1998.

Koenen, R., "MPEG-4: Multimedia for Our Time," *IEEE Spectr.*, vol. 36, no. 2, February 1999.

Kraemer, U., et al., "Network Music Performance with Ultra-Low-Delay Audio Coding under Unreliable Network Conditions," *AES 123rd Conv.*, preprint 7214, New York, October 2007.

Lee, G. W., J. S. Lee, Y. C. Park, and D. H. Youn, "Quality Improvement of Very Low Bit Rate HE-AAC Using Linear Prediction Module," *AES 125th Conv.*, preprint 7624, San Francisco, October 2008.

Levy, M., and M. Sandler, "Application of Segmentation and Thumbnailing to Music Browsing and Searching," *AES 120th Conv.*, preprint 6642, Paris, May 2006.

Liebchen, T., "MPEG-4 Lossless Coding for High-Definition Audio," *AES 115th Conv.*, preprint 5872, New York, October 2003.

Liebchen, T., et al., "MPEG-4 Audio Lossless Coding," *AES 116th Conv.*, preprint 6047, Berlin, May 2004.

Liebchen, T., et al., The MPEG-4 Audio Lossless Coding (ALS) Standard—Technology and Applications," *AES 119th Conv.*, preprint 6589, New York, October 2005.

Lindsay, A. T., and J. Herre, "MPEG-7 and MPEG-7 Audio—An Overview," *JAES*, vol. 49, no. 7/8, July/August 2001.

Liu, C.-M., et al., "Bit Reservoir Design for HE-AAC," *AES 118th Conv.*, preprint 6382, Barcelona, May 2005.

Liu., C.-M., et al., "Design of MPEG-4 AAC Encoder," *AES 117th Conv.*, preprint 6201, San Francisco, October 2004.

Machine Listening Group, MPEG-4 Structured Audio (MP4 Structured Audio), MIT Media Laboratory, http: //sound.media.mit.edu/eds/mpeg4/.

Manjunath, M., P. Salembier, and T. Sikora, *MPEG-7 Multimedia Content Description Interface*, John Wiley & Sons, 2001.

Maufer, T. A., *Deploying IPMulticast in the Enterprise*, Prentice Hall, 1998.

Miller, B. A., and C. Bisdikian, *Bluetooth Revealed: The Insider's Guide to an Open Specification for Global Wireless Communications*, 2nd ed., Prentice Hall, 2002.

Miller, M. A., *Internetworking: A Guide to Network Communications*, M&T Books, 1991.

Moskowitz, S. A., "Steganographic Method and Device," U.S. Patent 5,687,236, November 11, 1997.

Neubauer, C., and J. Herre, "Audio Watermarking of MPEG-2 AAC Bit Streams," *AES 108th Conv.*, preprint 5101, Paris, February 2000.

Neubauer, C., and J. Herre, "Digital Watermarking and Its Influence on Audio Quality," *AES 105th Conv.*, preprint 4823, San Francisco, September 1998.

Neubauer, C., et al., "Technical Aspects of Digital Rights Management Systems," *AES 113th Conv.*, preprint 5688, Los Angeles, October 2002.

O'Donnell, M., Audio-Over-IP Acceptance Test Strategy," *AES 127th Conv.*, preprint 7944, New York, October 2009.

Patterson, J., and R. Melcher, *Audio on the Web: The Official IUMA Guide*, Peachpit Press, 1998.

Pennycook, B., "Audio and MIDI Markup Tools for the World Wide Web," *JAES*, vol. 44, no. 4, April 1996.

Pereira, F., and T. Ebrahimi (eds.), "The MPEG-4 Book," Prentice Hall, 2002.

Peterson, D. M., *TCP/IP Networking*, McGraw-Hill, 1995.

Purat, M., and T. Ritter, "Comparison of Receiver-Based Concealment and Multiple Description Coding in an 802.11-Based Wireless Multicast Audio Distribution Network," *AES 127th Conv.*, preprint 7943, New York, October 2009.

RealNetworks, "RealAudio G2 Music Codec—A Musical Stream Apart," www.real.com/devzone/library/whitepapers/music.html.

RealNetworks, "SureStream—Delivering Superior Quality and Reliability," www.real.com/devzone/library/whitepapers/surestrm.html.

Rumsey, F., "Audio On Data Networks," *JAES*, vol. 46, no, 5, May 1998.

Rumsey, F., "Audio Networking for the Pros," *JAES*, vol. 57, no. 4, April 2009.

Rumsey, F., "Searching, Analyzing, and Recommending Audio Content," *JAES*, vol. 57, no. 3, March 2009.

Ryu, S.-U., and K. Rose, "Enhanced Accuracy of the Tonality Measure and Control Parameter Extraction Modules in MPEG-4 HE-AAC," *AES 119th Conv.*, preprint 6586, New York, October 2005.

Sandford, II, M. T., "Data Embedding," U.S. Patent 5,659,726, August 19, 1997.

Scheirer, E. D., and L. Ray, "Algorithmic and Wavetable Synthesis in the MPEG-4 Multimedia Standard," *AES 105th Conv.*, preprint 4811, San Francisco, September 1998.

Schildbach, W., K. Krauss, and J. Rödén, Transcoding of Dynamic Range Control Coefficients and Other Metadata into MPEG-4 HE AAC," *AES 123rd Conv.*, preprint 7217, New York, October 2007.

Schmidt, G. R., and M. K. Belmonte, "Scalable, Content-Based Audio Identification by Multiple Independent Psychoacoustic Matching," *JAES*, vol. 52, no. 4, April 2004.

Schmidt, J., and E. F. Schröder, "New and Advanced Features for Audio Rendering in the MPEG-4 Standard," *AES 116th Conv.*, preprint 6058, Berlin, May 2004.

Schneier, B., *Applied Cryptography: Protocols, Algorithms, and Source Code in C*, 2nd ed., John Wiley & Sons, 1995.

Schnell, M., et al., "Enhanced MPEG-4 Low Delay AAC—Low Bit-Rate High-Quality Communication," *AES 122nd Conv.*, preprint 6998, Vienna, May 2007.

Schnell, M., et al., "MPEG-4 Enhanced Low Delay AAC—A New Standard for High Quality Communication," *AES 125th Conv.*, preprint 7503, San Francisco, October 2008.

Schuijers, E., et al., "Advances in Parametric Coding for High-Quality Audio," *AES 114th Conv.*, preprint 5852, Amsterdam, March 2003.

SDMI Portable Device Specification—Part 1, Version 1.0, www.sdmi.org, 1999.

Shimada, O., et al., "A Low Power SBR Algorithm for the MPEG-4 Audio Standard and Its DSP Implementation," *AES 116th Conv.*, preprint 6048, Berlin, May 2004.

Short, K., Garcia, R., and M. Daniels, Scalability in KOZ Audio Compression Technology," *AES 119th Conv.*, preprint 6598, New York, October 2005.

Sincaglia, N., "Content Management Using Native XML and XML-Enabled Database Systems in Conjunction with XML Metadata Exchange Standards," *AES 123rd Conv.*, preprint 7237, New York, October 2007.

Su, A., and Y.-S. Shao, "Real-Time Internet MPEG-4 SA Player and the Streaming Engine," *AES 116th Conv.*, preprint 6025, Berlin, May 2004.

Szwoch, G., et al., "A Double-Talk Detector Using Audio Watermarking," *JAES*, vol. 57, no. 11, November 2009.

Wolters, M., et al., "A Closer Look into MPEG-4 High Efficiency AAC," *AES 115th Conv.*, preprint 5871, New York, October 2003.

Wong, K. K., et al., "Adaptive Water Marking," *IEEE Trans. Consum. Electron.*, vol. 43 no. 4, November 1997.

Wray, S., "Low Bit Rate Audio Coding for Digital Wireless Systems," *AES 124th Conv.*, preprint 7485, Amsterdam, May 2008.

Xu, C., et al., "Applications of Digital Watermarking Technology in Audio Signals," *JAES*, vol. 47, no. 10, October 1999.

Yang, C.-H., et al., "Design of HE-AAC Version 2 Encoder," *AES 121st Conv.*, preprint 6873, San Francisco, October 2006.

Yu, R., et al., "MPEG-4 Scalable to Lossless Audio Coding," *AES 117th Conv.*, preprint 6183, San Francisco, October 2004.

16 Digital Radio and Television Broadcasting

Advanced Television Systems Committee, "ATSC Digital Television Standard," Doc. A/53, September 1995.

Advanced Television Systems Committee, "ATSC Standard Digital Audio Compression (AC-3)," Doc. A/52, December 1995.

Advanced Television Systems Committee, "Guide to the Use of the ATSC Digital Television Standard," Doc. A/54, October 1995.

Advanced Television Systems Committee, www.atsc.org.

Bech, S., "Calibration of Relative Level Differences of a Domestic Multichannel Sound Reproduction System," *JAES*, vol. 46, no. 4, April 1998.

Bell, T. E., "The HDTV 'Test Kitchens'," *IEEE Spectr.*, vol. 32, no. 4, April 1995.

Benoit, H., *Digital Television: MPEG-1, MPEG-2 and Principles of the DVB System*, John Wiley & Sons, 1997.

Bhatt, B., et al., "Digital Television: Making it Work," *IEEE Spectr.*, vol. 34, no. 10, October 1997.

Blonstein, L., *Communication Satellites*, Halsted Press, 1988.

Booth, S. A.,"Digital TV in the United States," *IEEE Spectr.*, vol. 36, no. 3, March 1999.

Chen, Z., J. Wang, and K. Feher, "Effect of HPA Non-Linearities on Crosstalk and Performance of Digital Radio Systems," *IEEE Trans. Broadcast.*, vol. 34, no. 3, September 1988.

Conway, F., and W. Kwong, "CBS Engineering Experimentation with DAB," *CBC Engineering Review*, vol. 31, 1991.

Conway, F., and J. C. Lee, "Digital Radio Broadcasting Service Considerations in Canada and the Requirement for an International System and Frequency Band Standard," *CBC Engineering Review*, vol. 33, 1993–1994.

Conway, F., and B. Sawyer, "Canadian Broadcasting Corporation Experiments with L-Band for Terrestrial Digital Radio Broadcasting," *CBC Engineering Review*, vol. 32, 1992–1993.

Cook, Jr., J. H., G. Springer, and J. B. Vespoli, "Satellite Earth Stations," *National Association of Broadcasters Engineering Handbook*, 8th ed., NAB, 1992.

Cupo, R. L., et al., "An OFDM All-Digital In-Band On-Channel (IBOC) AM and FM Radio Solution Using the PAC Encoder," *IEEE Trans. Broadcast.*, vol. 44, no. 1, March 1998.

Detweiler, J. R., "Conversion Requirements for AM & FM IBOC Transmission," iBiquity Digital Corporation.

Dick, B., "Satellite Transmission: C-Band vs. Ku-Band," *Sound & Video Contractor*, June 20, 1988.

Douglas, R. L., *Satellite Communications Technology*, Prentice Hall, 1988.

Elbert, B. R., *Introduction to Satellite Communication*, Artech House, 1987.

Elliott, C., "High-Quality Multimedia Conferencing through a Long-Haul Packet Network," *ACM Multimedia*, 1993.

FCC Advisory Committee on Advanced Television Service, "Federal Communications Commission Advanced Television System Recommendation," *IEEE Trans. Broadcast.*, vol. 39, 1993.

Feher, K., *Advanced Digital Communications Systems and Signal Processing Techniques*, Prentice Hall, 1987.

Feher, K., *Digital Communication Satellite/Earth Station Engineering*, Prentice Hall, 1983.

Finger, R. A., "Video CD: A Coding Challenge," *Audio*, vol. 78, no. 12, December 1994.

Forrest, J. R., "Commercial Broadcasting for Europe," *IEEE Trans. Broadcast.*, vol. 34, no. 4, December 1988.

Foster, E. J., "Understanding MPEG-2," *Audio*, vol. 81, no. 9, September 1997.

Fujimoto, M., et al., "Small and Light Weight DBS and FSS Converters," *IEEE Trans. Consum. Electron.*, vol. 36, no. 3, August 1990.

Grand Alliance, "The U.S. HDTV Standard," *IEEE Spectr.*, vol. 32, no. 4, April 1995.

Grusec, T., et al., "EIN /NRSC DAR System Subjective Tests, Part I: Audio Codec Quality," *IEEE Trans. Broadcast.*, vol. 43, no. 3, September 1997.

Grusec, T., et al., "EIN /NRSC DAR System Subjective Tests, Part II: Transmission Impairments," *IEEE Trans. Broadcast.*, vol. 43, no. 4, December 1997.

Haskell, B. G., et al., *Digital Video: An Introduction to MPEG-2 (Digital Multimedia Standards Series)*, Chapman & Hall, 1996.

Heymann, R., et al., "A Multipurpose Four IC Satellite Concept," *IEEE Trans. Consum. Electron.*, vol. 36, no. 3, August 1990.

Hoeg, W., "Dynamic Range Control (DRC) for Multichannel Audio Systems," *JAES*, vol. 46, no. 4, April 1998.

Hopkins, R. "Digital Terrestrial HDTV for North America: The Grand Alliance HDTV System," *EBU Technical Review*, Summer, 1994.

Hopkins, R., "Digital Terrestrial HDTV for North America: The Grand Alliance HDTV System," *IEEE Trans. Consum. Electron.*, vol. 40, no. 3, August 1994.

ISO/IEC 11172-3:1993, "Information Technology—Coding of Moving Pictures and Associated Audio for Digital Storage Media at up to about 1.5 Mbit/s, Part 1: Systems," International Organization for Standardization, Geneva, 1993.

ISO/IEC 11172-3:1993, "Information Technology—Coding of Moving Pictures and Associated Audio for Digital Storage Media at up to about 1.5 Mbit/s, Part 2: Video," International Organization for Standardization, Geneva, 1993.

ISO/IEC 13818-3: 1997, "Information Technology—Generic Coding of Moving Pictures and Associated Audio Information, Part 2: Video," International Organization for Standardization, Geneva, 1997.

ITU-R Rec. BS.1196, "Audio Coding for Digital Terrestrial Television Broadcasting," International Telecommunication Union, Geneva, 1995.

Johnson, S. A., "The Structure and Generation of Robust Waveforms for AM In Band On Channel Digital Broadcasting," iBiquity Digital Corporation.

Jurgen, R. K., "Broadcasting with Digital Radio," *IEEE Spectr.*, vol. 33, no. 3, March 1996.

Killen, H. B., *Digital Communications with Fiber Optics and Satellite Applications*, Prentice Hall, 1988.

Klank, O., and D. Rottman, "DSR-Receiver for Digital Sound Broadcasting via the European Satellites TV-SAT/DF," *IEEE Trans. Consum. Electron.*, vol. 35, no. 3, August 1989.

Konishi, Y., "Special Issue on Satellite Broadcasting," *IEEE Trans. Broadcast.*, vol. 34, no. 4, December 1988.

Konishi, Y., and Y. Fukuoka, "Satellite Receiver Technologies," *IEEE Trans. Broadcast.*, vol. 34, no. 4, December 1988.

Kroeger, B., and D. Cammarata, "Robust Modem and Coding Techniques for FM Hybrid IBOC DAB," *IEEE Trans. Broadcast.*, vol. 43, no. 4, December 1997.

Kroeger, B., and P. J. Peyla, "Compatibility of FM Hybrid In-Band On-Channel (IBOC) System for Digital Audio Broadcasting," *IEEE Trans. Broadcast.*, vol. 43, no. 4, December 1997.

Langhans, R., and M. Shumila, "Digital Audio Transmission System Using Satellite Distribution," *AES 74th Conv.*, preprint 2018, New York, October 1983.

Le Floch, B., R. Halbert-Lassalle, and D. Castelain, "Digital Sound Broadcasting to Mobile Receivers," *IEEE Trans. Consum. Electron.*, vol. 35, no. 3, August 1989.

LeGall, D., "MPEG—A Video Compression Standard for Multimedia Applications," *Comm. ACM*, vol. 34, no. 4, April 1991.

Mailhot, J. N., and H. Derovanessian, "The Grand Alliance HDTV Video Encoder," *IEEE Trans. Consum. Electron.*, vol. 41, no. 4, November 1995.

Matsushita, M., and S. Yokoyama, "Experience on Operating a DBS System (DB-2) in Japan," *IEEE Trans. Broadcast.*, vol. 34, no. 4, December 1988.

McNally, G. W., "Digital Audio in Broadcasting," *IEEE ASSP Mag.*, vol. 2, no. 4, October 1985.

Meares, D. J., and G. Theile, "Matrixed Surround Sound in an MPEG Digital World," *JAES*, vol. 46, no. 4, April 1998.

Miller, J. E., "Application of Coding and Diversity to UHF Satellite Sound Broadcasting Systems," *IEEE Trans. Broadcast.*, vol. 34, no. 4, December 1988.

Mitchell, J. L., and W. B. Pennebaker (eds.), *MPEG Video: Compression Standard (Digital Multimedia Standards Series)*, Chapman & Hall, 1996.

Muller-Romer, F., "Directions in Digital Audio Broadcasting," *JAES*, vol. 41, no. 3, March 1993.

National Association of Broadcasters, "Digital Audio Broadcasting: Status Report and Outlook," NAB, Washington, D.C., 1990.

National Association of Broadcasters, *NAB Guide to Advanced Television Systems*, 2nd ed., NAB, 1991.

National Association of Broadcasters, "Understanding DAB: A Guide for Broadcast Managers and Engineers," NAB, Washington, D.C., 1992.

Noah, J. O., "A Rational Approach to Testing MPEG-2," *IEEE Spectr.*, vol. 34, no. 5, May 1997.

Perry, T. S., "HDTV and the New Digital Television," *IEEE Spectr.*, vol. 32, no. 4, April 1995.

Peyla, P. J., "The Structure and Generation of Robust Waveforms for FM In Band On Channel Digital Broadcasting," iBiquity Digital Corporation.

Pizzi, S., "Digital Audio Applications in Radio Broadcasting," *Proc. AES 7th Int. Conf.*, Toronto, May 1989.

Pizzi, S., *Digital Radio Basics*, Telephony Intertec, 1992.

Pohlmann, K. C., "The DAB Debate," *Mix*, vol. 15, no. 9, September 1991.

Pohlmann, K. C., "Eureka 147," *Mix*, vol. 17, no. 5, May 1993.

Pohlmann, K. C., "In-Band Digital Radio," *Mix*, vol. 17, no. 6, June 1993.

Sawaguchi, M. M., et al., "HDTV Drama Multi-channel Sound Production with 3-2, 3-1, 2 CH," *AES 100th Conv.*, preprint 4223, Copenhagen, May 1996.

Schroder, E. F., and P. A. Ratliff, "Digital Audio Broadcasting (DAB)," *AES Collected Papers on Digital Audio Bit-Rate Reduction*, N. Gilchrist and C. Grewin (eds.), Audio Engineering Society, 1996.

Sheffield, E. G., and J. Kean, "Participant Testing of AM Broadcast Transmission Bandwidth and Audio Performance Measurements of Broadcast AM Receivers," *JAES*, vol. 57, no. 5, May 2009.

Sinha, D., and C.-E.W. Sundberg, "Unequal Error Protection (UEP) for Perceptual Audio Coders," *AES 104th Conv.*, preprint 4754, Amsterdam, May 1998.

Sinha, D., et al., "The Perceptual Audio Coder (PAC)," *The Digital Signal Processing Handbook*, V. K. Madisetti and D. B. Williams (eds.), CRC/IEEE Press, 1998.

Sukow, R., et al., "Radio's Digital Evolution," *Broadcasting*, October 17, 1988.

Sun, H., "Hierarchical Decoder for MPEG Compressed Video Data," *IEEE Trans. Consum. Electron.*, vol. 39, no. 3, August 1993.

Sundberg, C.-E.W. et al., "Technology Advances Enabling 'In-Band On-Channel' DSB Systems," *1998 Int. Conf. Broadcast Asia, Conf. Rec.*, Singapore, 1998.

Sweet, W., "Chiariglione and the Birth of MPEG," *IEEE Spectr.*, vol. 34, no. 9, September 1997.

Taura, K., et al., "A Digital Audio Broadcasting (DAB) Receiver," *IEEE Trans. Consum. Electron.*, vol. 42, no. 3, August 1996.

Ten Kate, W. R., "Compatibility Matrixing of Multichannel Bit-Rate-Reduced Audio Signals," *JAES*, vol. 44, no. 12, December 1996.

Tsunashima, K., et al., "An Integrated DTV Receiver for ATSC Digital Television Standard," *IEEE Trans. Consum. Electron.*, vol. 44, no. 3, August 1998.

Watson, A. B. (ed.), *Digital Images and Human Vision*, Bradford Books, 1993.

Weissleder, H., et al., "DAB in CATV Networks," *IEEE Trans. Consum. Electron.*, vol. 44, no. 3, August 1998.

Wilson, D., "Industry Evaluation of In-Band On-Channel Digital Audio Broadcast Systems," *JAES*, vol. 51, no. 5, May 2003.

17 Digital Signal Processing

Adams, R. W., "A New Windowing Technique for Digital Harmonic-Distortion Measurement," *JAES*, vol. 36, no. 5, May 1988.

Andreas, D. C., J. Dattorro, and J. W. Mauchly, "Digital Signal Processing for Audio Applications," U.S. Patent 5,517,436, May 14, 1996.

Baudot, M. D., "Hardware Design of a Digital Mixer for Musical Applications," *AES 83rd Conv.*, preprint 2506, New York, October 1987.

Berkhout, P. J., and L. D. J. Eggermont, "Digital Audio Systems," *IEEE ASSP Mag.*, vol. 2, no. 4, October 1985.

Berkovitz, R., "Digital Equalization of Audio Systems," *AES Digital Audio Collected Papers*, Rye, 1983.

Betts, D., and G. Reid, "DSP and Audio Restoration," *Studio Sound*, March 1993.

Blesser, B. A., K. Baeder, and R. Zaorski, "A Real-Time Digital Computer for Simulating Audio Systems," *JAES*, vol. 23, no. 9, November 1975.

Bloom, P. J., "High-Quality Digital Audio in the Entertainment Industry: An Overview of Achievements and Challenges," *IEEE ASSP Mag.*, vol. 2, no. 4, October 1985.

Booty, M., "Digital Signal Processing: Programming and Interfacing," *Advanced Digital Audio*, K. C. Pohlmann (ed.), Howard W. Sams, 1991.

Chen, T., et al., "Architecting a Versatile Multi-Channel Multi-Decoder System on DSP," *AES 106th Conv.*, preprint 4960, Munich, May 1999.

Chen, W., "Performance of Cascade and Parallel IIR Filters," *JAES*, vol. 44, no. 3, March 1996.

Cabot, R., "Practical Performance of Digital Systems," *Recording Engineer/Producer*, March 1988.

Datta, J., "Digital Signal Processing: Theory," *Advanced Digital Audio*, K. C. Pohlmann (ed.), Howard W. Sams, 1991.

Dattorro, J., "Effect Design Part 1: Reverberator and Other Filters," *JAES*, vol. 45, no. 9, September 1997.

Dattorro, J., "Effect Design Part 2: Delay-Line Modulation and Chorus," *JAES*, vol. 45, no. 10, October 1997.

Dattorro, J., "The Implementation of Recursive Digital Filters for High Fidelity Audio," *JAES*, vol. 36, no. 11, November 1988.

Dattorro, J., "Using Digital Signal Processor Chips in a Stereo Audio Time Compressor/Expander," *AES 83rd Conv.*, preprint 2500, New York, October 1987.

Elliot, S. J., and P. A. Nelson, "Multiple-Point Equalization in a Room Using Adaptive Digital Filters," *JAES*, vol. 37, no. 11, November 1989.

El-Sharkawy, M., *Digital Signal Processing Applications with Motorola's DSP56002 Processor*, Prentice Hall, 1996.

Georgopoulos, V. C., and D. Preis, "A Comparison of Computational Methods for Instantaneous Frequency and Group Delay of Discrete-Time Signals," *JAES*, vol. 46, no. 3, March 1998.

Griesinger, D., "Practical Processors and Programs for Digital Reverberation," *Proc. AES 7th Int. Conf.*, Toronto, May 1989.

Griesinger, D., "Theory and Design of a Digital Audio Signal Processor for Home Use," *JAES*, vol. 37, no. 1/2, January/February 1989.

Hamada, O., N. Kitazato, and T. Nakagami, "Digital Signal Processing LSIs Suitable for Digital Audio Equipment," *AES 79th Conv.*, preprint 2269, New York, October 1985.

Hamming, R. W., *Digital Filters*, Prentice Hall, 1977.

Hutchings, H. J., "Digital Filters Explained," *Electronics & Wireless World*, December 1985.

IEEE, "Programs for Digital Signal Processing," *IEEE Trans. ASSP DSP Comm.*, IEEE Press, 1979.

Jackson, L. B., *Digital Filters and Signal Processing: With Matlab Exercises*, 3rd ed., Kluwer Academic, 1996.

Kahrs, M., and K. Brandenburg (eds.), *Applications of Digital Signal Processing to Audio and Acoustics*, Kluwer Academic, 1998.

Kalliris, G., et al., "Z Language, A New DSP Dedicated Programming Environment," *AES 94th Conv.*, preprint 3514, Berlin, March 1993.

Karjalainen, M., et al., "Comparison of Loudspeaker Equalization Methods Based on DSP Techniques," *JAES*, vol. 47, no. 1/2, January/February 1999.

Kuo, S. M., and W. S. Gan, *Digital Signal Processors: Architectures, Implementations, and Applications*, Pearson Prentice Hall, 2005.

Lagadec, R., "Measuring the Phase Linearity of Digital Audio Systems," *AES 74th Conv.*, preprint 2040, New York, October 1983.

Lee, D. H., et al., "Design of a Digital Artificial Reverberation System Using a Dual All-Pass Filter," *JAES*, vol. 57, no. 3, March 2009.

Lindquist, C. S., *Adaptive and Digital Signal Processing*, vol. 2, Steward & Sons, 1989.

Menesi, B., and F. Takacs, "Processing of Audio Signals with Extended Precision by TMS 32010," *AES 82nd Conv.*, preprint 2475, London, March 1987.

Moorer, J. A., "The Audio Signal Processor: The Next Step in Digital Audio," *AES Digital Audio Collected Papers*, Rye, 1983.

Moorer, J. A., "48-Bit Integer Processing Beats 32-Bit Floating Point for Professional Audio Applications," *AES 107th Conv.*, preprint 5038, New York, September 1999.

Morris, R., *Digital Signal Processing Software*, Carleton University, 1983.

Motorola, "DSP56001 and DSP56000 Data Sheets and Applications Notes," Motorola Microprocessor Products Group, 1990.

Mourjopoulos, J., et al., "Noisy Audio Signal Enhancement Using Subjective Spectra," *AES 92nd Conv.*, preprint 3240, Vienna, March 1992.

Oppenheim, A. V. (ed.), *Applications of Digital Signal Processing*, Prentice Hall, 1978.

Oppenheim, A. V., and R. W. Schafer, *Digital Signal Processing*, Prentice Hall, 1975.

Park, S. "170 MIPS Real-Time Adaptive Digital Filter Board," *AES 91st Conv.*, preprint 3107, New York, October 1991.

Parks, T. W., and C. S. Burrus, *Digital Filter Design*, John Wiley & Sons, 1987.

Peek, J. B. H., "Digital Signal Processing—Growth of a Technology," *Philips Tech. Rev.*, vol. 42, no. 4, 1985.

Peled, A., and B. Liu, *Digital Signal Processing: Theory, Design and Applications*, Robert E. Krieger Publishing, 1985.

Persoon, E. H. J., and C. J. B. Vandenbulcke, "Digital Audio: Examples of the Application of the ASP Integrated Signal Processor," *Philips Tech. Rev.*, vol. 42, no. 6/7, 1986.

Pohlmann, K. C., "Multiple-Accumulate Commands," *Mix*, vol. 16, no. 10, October 1992.

Potchinkov, A., "Frequency-Domain Equalization of Audio Systems Using Digital Filters Part 1: Basics of Filter Design," *JAES*, vol. 46, no. 11, November 1998.

Potchinkov, A., "Frequency-Domain Equalization of Audio Systems Using Digital Filters Part 2: Examples of Equalization," *JAES*, vol. 46, no. 12, December 1998.

Rabiner, L. R., and B. Gold, *Theory and Application of Digital Signal Processing*, Prentice Hall, 1975.

Ramarapu, P. K., and R. C. Maher, "Methods for Reducing Audible Artifacts in a Wavelet-Based Broad-Band Denoising System," *JAES*, vol. 46, no. 3, March 1998.

Reviriego, P, J. Parera, and R. Garcia, "Linear-Phase Crossover Design Using Digital IIR Filters," *JAES*, vol. 46, no. 5, May 1998.

Rich, D. A., "The Present State of CD Player Technology," *The Audio Critic*, vol. 15, Spring/Winter 1990–1.

Robinson, E. A., and M. T. Silvia, *Digital Signal Processing and Time Series Analysis*, Holden Day, 1978.

Sakamoto, N., S. Yamaguchi, and A. Kurahashi, "A Professional Digital Audio Mixer," *JAES*, vol. 30, no. 1/2, January/February 1982.

Schroeder, M. R., "Improved Quasi-Stereophony and 'Colorless' Artificial Reverberation," *JASA*, vol. 33, August 1961.

Schroeder, M. R., "Natural Sounding Artificial Reverberation," *JAES*, vol. 10, no. 7, July 1962.

Schroeder, M. R., and B. S. Atal, "Computer Simulation of Sound Transmissions in Rooms," *IEEE Int. Conv. Rec.*, Pt. 7, 1963.

Schuck, P. L., "Digital FIR Filters for Loudspeaker Crossover Networks II: Implementation Example," *Proc. AES 7th Int. Conf.*, Toronto, May 1989.

Snell, J. M., "Professional Real-Time Signal Processor for Synthesis, Sampling, Mixing, and Recording," *AES 83rd Conv.*, preprint 2508, New York, October 1987.

Steiglitz, K., *A Digital Signal Processing Primer*, Addison-Wesley, 1996.

Stockham, Jr., T. G., T. M. Cannon, and R. B. Ingebretsen, "Blind Deconvolution through Digital Signal Processing," *Proc. IEEE*, vol. 63, no. 4, April 1975.

Strawn, J. (ed.), *Digital Audio Signal Processing*, A-R Editions, 1985.

Texas Instruments, "Digital Signal Processing Applications with the TMS320 Family: Theory, Algorithms, and Implementations," Texas Instruments, Inc., 1990.

Texas Instruments, Second Generation TMS320 User's Guide, Texas Instruments, Inc., 1989.

Tomarakos, J., and C. Duggan, "32-Bit SIMD SHARC Architecture Digital Audio Signal Processing Applications," *JAES*, vol. 48, no. 3, March 2000.

Tydelski, P., "Equalizer Trends: Digital Filtering," *Recording Engineer/Producer*, February 1988.

Van den Enden, A. W. M., and N. A. M. Verhoeckx, "Digital Signal Processing: Theoretical Background," *Philips Tech. Rev.*, vol. 42, no. 4, December 1985.

Van Meerbergen, J. L., "Developments in the Integrated Digital Signal Processors, and the PCB 5010," *Philips Tech. Rev.*, vol. 44, no. 1, 1988.

Vanderkooy, J., and S. P. Lipshitz, "Digital Dither: Processing with Resolution Far Below the Least Significant Bit," *Proc. AES 7th Int. Conf.*, Toronto, May 1989.

Varga, I., "Adaptive Filtering for Noise Reduction in Audio Signals," *AES 92nd Conv.*, preprint 3247, Vienna, March 1992.

Vaseghi, S. V., and R. Frayling-Cork, "Restoration of Old Gramophone Recordings," *JAES*, vol. 40, no. 10, October 1992.

Wilson, R., G. Adams, and J. Scott, "Application of Digital Filters to Loudspeaker Crossover Networks," *JAES*, vol. 37, no. 6, June 1989.

Zucker, I., "Reproducing Architectural Acoustical Effects Using Digital Soundfield Processing," *Proc. AES 7th Int. Conf.*, Toronto, May 1989.

18 Sigma-Delta Conversion and Noise Shaping

Adams, R. W., "Companded Predictive Delta Modulation: A Low-Cost Conversion Technique for Digital Recording," *JAES*, vol. 32, no. 9, September 1984.

Adams, R. W., "Design and Implementation of an Audio 18-Bit Analog-to-Digital Converter Using Oversampling Techniques," *JAES*, vol. 34, no. 3, March 1986.

Adams, R. W., "An IC Chip Set for 20-Bit A/D Conversion," *JAES*, vol. 38, no. 6, June 1990.

Adams, R. W., "Unusual Applications of Noise-Shaping Principles," *AES 101st Conv.*, preprint 4356, Los Angeles, November 1996.

Adams, R. W., et al., "A 116-dB SNR Multi-Bit Noise-Shaping DAC with 192 kHz Sample Rate," *AES 106th Conv.*, preprint 4963, Munich, May 1999.

Adams, R. W., et al., "Theory and Practical Implementation of a Fifth-Order Sigma-Delta A/D Converter," *JAES*, vol. 39, no. 7/8, July/August 1991.

Akune, M., R. Heddle, and K. Akagiri, "Super Bit Mapping: Psychologically Optimized Digital Recording," *AES 93rd Conv.*, preprint 3371, San Francisco, October 1992.

Analog Devices, "Sigma-Delta ADCs and DACs," Analog Devices Application Note AN-283.

Angus, J., "A Comparison of the 'Pruned Tree' versus 'Stack' Algorithms for Look-Ahead Sigma-Delta Modulators," *JAES*, vol. 54, no. 6, June 2006.

Angus, J., "A New Method of Applying High Levels of Dither to Delta-Sigma Modulators," *AES 117th Conv.*, preprint 6296, San Francisco, October 2004.

Angus, J., "The Practical Performance Limits of Multi-Bit Sigma-Delta Modulation," *AES 110th Conv.*, preprint 5394, Amsterdam, May 2001.

Ardalan, S. H., "Analysis of Delta-Sigma Modulators with Bandlimited Gaussian Inputs," *Proc. of IEEE Conf. ASSP*, vol. III, 1988.

Ardalan, S. H., and J. J. Paulos, "An Analysis of Nonlinear Behavior in Delta-Sigma Modulators," *IEEE Trans. Circuits and Systems*, vol. CAS-34, no. 6, June 1987.

Boser, B. E., and B. A. Wooley, "The Design of Sigma-Delta Modulation Analog-to-Digital Converters," *IEEE Journal of Solid State Circuits*, vol. 23, no. 6, December 1988.

Bourdopoulos, G., et al., *Delta-Sigma Modulators, Modeling, Design and Applications*, Imperial College Press, London, 2003.

Bruekers F., et al., "Improved Lossless Coding of One-Bit Audio Signals," *AES 103rd Conv.*, preprint 4563, New York, September 1997.

Carley, L. R., "An Oversampling Analog-to-Digital Converter Topology for High-Resolution Signal Acquisition Systems," *IEEE Trans. Circuits and Systems*, vol. CAS-34, no. 1, January 1987.

Craven, P., "Toward the 24-Bit DAC: Novel Noise-Shaping Topologies Incorporating Correction for the Nonlinearity in a PWM Output Stage," *JAES*, vol. 41, no. 5, May 1993.

Crochiere, R. E., and L. R. Rabiner, "Interpolation and Decimation of Digital Signals—A Tutorial Review," *Proc. IEEE*, vol. 69, no. 3, March 1981.

Darling, T. F., and M. O. J. Hawksford, "Oversampled Analog-to-Digital Conversion for Digital Audio Systems," *JAES*, vol. 38, no. 12, December 1990.

Duewer, B., et al., "A Multi-bit Delta-Sigma DAC with Mismatch Shaping in the Feedback Loop," *AES 115th Conv.*, preprint 5919, New York, October 2003.

Dunn, C., and M. Sandler, "A Comparison of Dithered and Chaotic Sigma-Delta Modulators," *JAES*, vol. 44, no. 4, April 1996.

Dunn, C., and M. Sandler, "Psychoacoustically Optimal Sigma-Delta Modulation," *JAES*, vol. 45, no. 4, April 1997.

Eastty, P., "An Inaudible Buried Data Channel in Digital Audio: A Development Using Advanced Software Tools," *AES 101st Conv.*, preprint 4369, Los Angeles, November 1996.

Gerzon, M. A., and P. G. Craven, "A High-Rate Buried-Data Channel for Audio CD," *JAES*, vol. 43, no. 1/2, January/February 1995.

Gerzon, M. A., et al., "Psychoacoustic Noise Shaped Improvements in CD and Other Linear Digital Media," *AES 94th Conv.*, preprint 3501, Berlin, March 1993.

Goldberg, J. M., and M. B. Sandler, "Noise Shaping and Pulse-Width Modulation for an All-Digital Audio Power Amplifier," *JAES*, vol. 39, no. 6, June 1991.

Gray, R. M., "Oversampled Sigma-Delta Modulation," *IEEE. Trans. Commun.*, vol. COM-35, 1987.

Green, S., "A New Perspective on Decimation and Interpolation Filters," Cirrus Logic, July 2004.

Greenfield, R. G., and M. O. J. Hawksford, "On the Dither Performance of High-Order Digital Equalization of Loudspeaker Systems," *JAES*, vol. 43, no. 11, November 1995.

Harris, S., "How to Achieve Optimum Performance from Delta-Sigma A/D and D/A Converters," *JAES*, vol. 41, no. 10, October 1993.

Hauser, M. W., "Principles of Oversampling A/D Conversion," *JAES*, vol. 39, no. 1/2, January/February 1991.

Hauser, M. W., and R. W. Brodersen, "Monolithic Decimation Filtering for Custom Delta-Sigma A/D Converters," *Proc. IEEE Conf. ASSP*, vol. III, 1988.

Hawksford, M. O. J., "Chaos, Oversampling, and Noise Shaping in Digital-to-Analog Conversion," *JAES*, vol. 37, no. 12, December 1989.

Hawksford, M. O. J., "Dynamic Model-Based Linearization of Quantized Pulse-Width Modulation for Applications in Digital-to-Analog Conversion and Digital Power Amplifier Systems." *JAES*, vol. 40, no. 4, April 1992.

Hawksford, M. O. J., "Parallel Look-Ahead Digital SDM with Energy-Balance Binary Comparator," *JAES*, vol. 56, no. 12, December 2008.

Hawksford, M. O. J., "SDM Versus LPCM: The Debate Continues," *AES 110th Conv.*, preprint 5397, Amsterdam, May 2001.

Hawksford, M. O. J., "Time-Quantized Frequency Modulation, Time-Domain Dither, Dispersive Codes and Parametrically Controlled Noise Shaping in SDM," *JAES*, vol. 52, no. 6, June 2004.

Hawksford, M. O. J., and W. Wingerter, "Oversampling Filter Design in Noise-Shaping Digital-to-Analog Conversion," *JAES*, vol. 38, no. 11, November 1990.

Hoare, S., and J. Angus, "The Performance of Look-Ahead Delta-Sigma Modulators with Unstable Noise Shaping Filters," *AES 121st Conv.*, preprint 6865, San Francisco, October 2006.

Kloker, K. L., B. L. Lindsley, and C. D. Thompson, "VLSI Architectures for Digital Audio Signal Processing," *Proc. AES 7th Int. Conf.*, Toronto, May 1989.

Komamura, M., "Wide-Band and Wide-Dynamic-Range Recording and Reproduction of Digital Audio," *JAES*, vol. 34, no. 1/2, January/February 1995.

Kuroda, N., JVC Corporation, personal correspondence, November 1989.

Lipshitz, S. P., and J. Vanderkooy, "Towards a Better Understanding of 1-Bit Sigma-Delta Modulators," *AES 110th Conv.*, preprint 5398, Amsterdam, May 2001.

Lipshitz, S. P., and J. Vanderkooy, "Towards a Better Understanding of 1-Bit Sigma-Delta Modulators, Part 2," *AES 111th Conv.*, preprint 5477, New York, September 2001.

Lipshitz, S. P., and J. Vanderkooy, "Towards a Better Understanding of 1-Bit Sigma-Delta Modulators, Part 3," *AES 112th Conv.*, preprint 5620, Munich, May 2002.

Lipshitz, S. P., J. Vanderkooy, and R. A. Wannamaker, "Minimally Audible Noise Shaping," *JAES*, vol. 39, no. 11, November 1991.

Lipshitz, S. P., R. A. Wannamaker, and J. Vanderkooy, "Dithered Noise Shapers and Recursive Digital Filters," *JAES*, vol. 52, no. 11, November 2004.

Løkken, I., A. Vinje, and T. Sæther, "Noise Power Modulation in Dithered and Undithered High-Order Sigma-Delta Modulators," *JAES*, vol. 54, no. 9, September 2006. Corrections, vol. 54, no. 10, October 2006.

Matsuya, Y., et al., "A 16-Bit Oversampling A/D Conversion Technology Using Triple-Integration Noise Shaping," *IEEE Journal Solid-State Circuits*, vol. SC-22, no. 6, December 1987.

Mladenov, V., J. D. Reiss, and G. Tsenov, "A Comparison of Theoretical, Simulated, and Experimental Results Concerning the Stability of Sigma Delta Modulators," *AES 124th Conv.*, preprint 7440, Amsterdam, May 2008.

Motorola, "DSP56ADC16 16-Bit Sigma-Delta Analog-to-Digital Converter," Motorola Semiconductor Technical Data, Ref. DSP56ADC16/D, 1989.

Naus, P. J. A., et al., "Low Signal-Level Distortion in Sigma-Delta Modulators," *AES 84th Conv.*, preprint 2584, Paris, March 1988.

Nguyen, K., et al., "A 113 dB SNR Oversampling Sigma-Delta DAC for CD/DVD Application," *IEEE Trans. Consum. Electron.*, vol. 44, no. 3, August 1998.

Ning, H., A. Buzo, and F. Kuhlmann, "Multi-Loop Sigma-Delta Quantization: Spectral Analysis," *Proc. IEEE Conf. ASSP,* vol. III, 1988.

Norsworthy, S. R., et al. (eds.), *Delta-Sigma Data Converters, Theory, Design and Simulation*, IEEE Press, New York, 1997.

Oomen, A. W. J., et al., "A Variable-Bit-Rate Buried-Data Channel for Compact Disc," *JAES*, vol. 43, no. 1/2, January/February 1995.

Park, S., "Principles of Sigma-Delta Modulation for Analog-to-Digital Converters," Motorola Applications Note APR8/D, 1990.

Reefman, D., and E. Janssen, "One-Bit Audio: An Overview," *JAES*, vol. 52, no. 3, March 2004.

Reefman, D., and P. Nuijten, "Why Direct Stream Digital is the Best Choice as a Digital Audio Format," *AES 110th Conv.*, preprint 5396, Amsterdam, May 2001.

Reiss, J. D., "Understanding Sigma–Delta Modulation: The Solved and Unsolved Issues," *JAES*, vol. 56, no. 1/2, January/February 2008.

Richards, M., "Improvements in Oversampling Analogue to Digital Converters," *AES 84th Conv.*, preprint 2588, Paris, March 1988.

Seitzer, D., et al., "Low Bit Rate Codecs for Audio Signals Implementation in Real Time," *AES 85th Conv.*, preprint 2707, New York, November 1988.

Spinnler, W., et al., "VLSI Implementation Aspects of Low Bit Rate Codecs for High Quality Audio Channels," *AES 86th Conv.*, preprint 2750, Hamburg, March 1989.

Stikvoort, E. F., "Higher Order One Bit Coder for Audio Applications," *AES 84th Conv.*, preprint 2583, Paris, March 1988.

Stikvoort, E. F., "Some Remarks on the Stability and Performance of the Noise Shaper or Sigma-Delta Modulator," *IEEE Trans. Commun.*, vol. 36, no. 10, October 1988.

Story, M., "Audio Analog-to-Digital Converters," *JAES*, vol. 52, no. 3, March 2004.

Stuart, J. R., and R. J. Wilson, "Dynamic Range Enhancement Using Noise-Shaped Dither at 44.1, 48, and 96 kHz," *AES 100th Conv.*, preprint 4236, Copenhagen, May 1996.

Thompson, C. D., "A VLSI Sigma-Delta A/D Converter for Audio and Signal Processing Applications," *Proc. IEEE Conference*, ASSP, May 1989.

Uchimura, K., et al., "Oversampling A-to-D and D-to-A Converters with Multistage Noise Shaping Modulators," *IEEE Trans. ASSP*, vol. 36, no. 12, December 1988.

Vanderkooy, J., and S. P. Lipshitz, "Digital Dither: Signal Processing with Resolution Far Below the Least Significant Bit," *Proc. AES 7th Int. Conf.*, Toronto, May 1989.

Wannamaker, R. A., "Psychoacoustically Optimal Noise Shaping," *JAES*, vol. 40, no. 7/8, July/August 1992.

Welland, D. R., et al., "A Stereo 16-Bit Delta-Sigma A/D Converter for Digital Audio," *JAES*, vol. 37, no. 6, June 1989.

Wong, P. W., and R. M. Gray, "FIR Filters with Sigma-Delta Modulation Encoding," *IEEE Trans. ASSP*, vol. 38, no. 6, June 1990.

Index

B

C

D

E

F

J

K

L

P

Q

R

S